CW01460027

Materials Chemistry

Bradley D. Fahlman

Materials Chemistry

Fourth Edition

Bradley D. Fahlman
Department of Chemistry and Biochemistry
Central Michigan University
Mount Pleasant, MI, USA

ISBN 978-3-031-18783-4 ISBN 978-3-031-18784-1 (eBook)
https://doi.org/10.1007/978-3-031-18784-1

This Springer imprint is published by the registered company Springer Nature Switzerland AG
The registered company address is: Gewerbestrasse 11, 6330 Cham, Switzerland

Preface

As we surge ahead with global electrification of the transportation industry, advanced materials will continue to play a critical role—not only for the battery packs—but also to expand the electrical grid needed to enable the impending surge in power use. However, beyond the transportation sector, breakthroughs in materials design will also be essential to help us address the effects of climate change and energy shortages, homeland security threats, and to fight next-generation diseases. An increasing focus on sustainability within the framework of a circular economy will also be needed to ensure that nonrenewable resources remain available for future generations. It is truly exciting to think about the potential breakthroughs coming in the next few years as artificial intelligence (AI) and machine learning (ML) have greatly expanded our capacity in the design of novel materials. The next steps in our technological journey are governed only by our creativity and steadfast ability to apply fundamental scientific knowledge. This latter quality represents the overarching goal of this textbook—to provide a suitable breadth and depth coverage of various classes of materials to help enable the development of next-generation devices and applications.

As with previous editions, *Materials Chemistry* focuses on the 'structure vs. property' assessment of both inorganic and organic materials, from their bulk to nanostructure regime. The content presented herein is most appropriate for junior or senior undergraduate students, as well as first-year graduate students in chemistry, physics, and engineering disciplines. In addition, this textbook has also been widely used by researchers in industry as a primer to learn about various materials classes or characterization techniques. Each chapter includes a comprehensive reference list, which is essential for more detailed research.

For such a broad and rapidly evolving field as materials chemistry, it is a daunting task to remain contemporary. However, we have attempted to do so through extensive updates and expansions to provide an innovative and practical perspective throughout. After providing a historical perspective in the first chapter, this edition provides an expanded discussion of sustainability—a topic that is often overlooked in our zeal to fabricate faster and better devices—but absolutely essential to ensure the survival of future generations. In addition to reorganization and rewriting for clarity throughout, this edition includes the following new sections:

- Materials design via combinatorial and machine learning approaches (Chap. 1)
- Kroger-Vink notation for defects (Chap. 2)
- Additive manufacturing (AM) for metallurgy (Chap. 3)
- Negative capacitance field effect transistors (NCFETs; Chap. 4)
- 3-D printing of polymers (Chap. 5)
- New toxicity studies of nanomaterials (Chap. 6)
- Core-shell quantum dots (Chap. 6)
- Heat-assisted magnetic recording (Chap. 6)
- Dynamic light scattering (DLS) techniques (Chap. 7).

In addition, the section related to nanotoxicity was extensively updated to reflect the latest discoveries. This book also continues to be the first to provide detailed processing information for the latest technology node of integrated circuit design, based on FinFETs. Over 100 additional references were added throughout this edition, and the end-of-chapter questions and bibliographies were extensively revised to reflect the current state of the art.

As with previous editions, Appendices are also provided that contain an updated and expanded timeline of major materials development and the complete transcript of Nobel Laureate Feynman's speech "There's Plenty of Room at the Bottom". A collection of materials-related laboratory experiments is also included, which are suitable for both undergraduate or graduate curricula.

I wish to thank my wife, Diyonn, for her love and continued support and encouragement. Our frequent travels across an expansive globe continually remind me how fortunate I am to have found my true soulmate. Our two kitties, Cash and Sissy, also played a role in the 4th edition by reminding me when to take cuddle breaks!

I also am extremely appreciative for the input provided by students and instructors who have either adopted, or have considered the adoption, of *Materials Chemistry*. I continue to offer thanks to every reader of this book and solicit your comments to my email (fahlmanb@gmail.com). Please let me know what you think of this edition; I promise to incorporate your suggestions to strengthen future editions.

Mount Pleasant, MI, USA
January 2023

Bradley D. Fahlman, Ph.D.

Contents

Chapter 1
What is "Materials Chemistry"?

Life in the twenty-first century is ever-dependent on an unlimited variety of advanced materials. In our consumptive world, it is easy to take for granted the macro-, micro-, and nanoscale building blocks that comprise any item ever produced. We are spoiled by the technology that adds convenience to our lives such as microwave ovens, laptop computers, cell phones and tablets, and improved modes of transportation. However, we rarely take time to think about and appreciate the materials that constitute these modern engineering feats.

The term *material* may be broadly defined as any solid-state substance or device that may be used to address a current or future societal need.[1] For instance, simple building materials such as glass, wood, aluminum, etc. address our need of shelter. Other more intangible materials such as nanoscale components are also considered as materials, even if a particular nanomaterial may not yet be widely proven for applications.[2]

A general classification of various types of materials is illustrated in Fig. 1.1. Although this indicates sharp distinctions between various classes, there is often ambiguity regarding the proper taxonomy for a specific material. For example, a *thin film* is defined as having a thickness of less than 1 μm; however, if the thickness drops to below 100 nm, the film may be more accurately classified as a 2-D nanomaterial.[3] Likewise, liquid crystals are best described as having properties intermediate between amorphous and crystalline phases, and composites typically involve both inorganic and organic components.

The broadly defined discipline of *materials chemistry* is focused on understanding the relationships between the arrangement of atoms, ions, or molecules comprising a material, and its physical and chemical properties. By this designation, common disciplines such as polymer, solid-state, and surface chemistry would all be placed within the scope of materials chemistry. This broad field consists of studying the structure and properties of existing materials, synthesizing and characterizing new materials, and using advanced computational techniques to understand the properties of existing materials or those that have not yet been fabricated.

B. D. Fahlman, *Materials Chemistry*,
https://doi.org/10.1007/978-3-031-18784-1_1

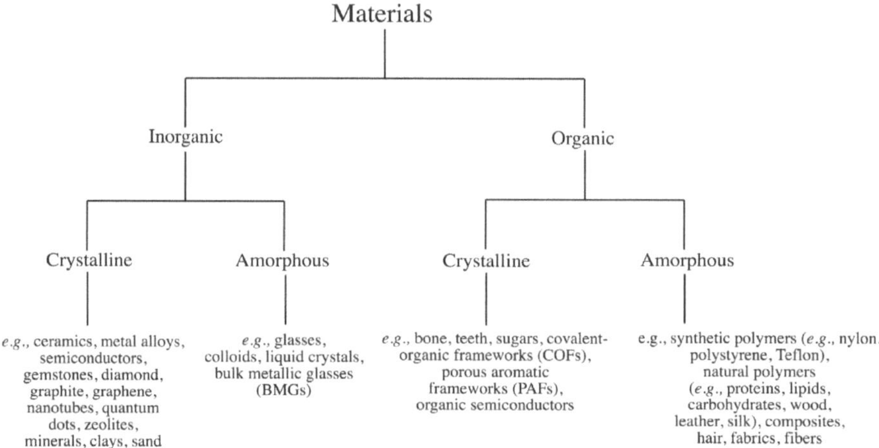

Fig. 1.1 Classification scheme for the various types of materials

1.1 Historical Perspectives

Although the study of materials chemistry is a relatively new entry in both undergraduate and graduate curricula, it has always been an important part of chemistry. An interesting timeline of materials developments from Prehistoric times to the present may be found in Fig. 1.2. By most accounts, Neolithic man (10,000–300 BC) was the first to realize that certain materials such as limestone, wood, shells, and clay were most easily shaped into materials used as utensils, tools, and weaponry. Applications for metallic materials date back to the Neolithic Age (4000–2500 BC), where copper was used for a variety of ornamental, functional, and protective applications. This civilization was the first to realize fundamental properties of metals, such as malleability and thermal conductivity. More importantly, Neolithic man was the first to practice *top-down* materials synthesis, with the development of techniques to extract copper from oxide ores such as malachite for use in various applications.

Metal alloys were first used in the Bronze Age (2500 BC–800 BC), brought on by the serendipitous discovery that the physical properties of copper could be drastically altered by doping with other elements. Artifacts from the Middle East dating back to 3000 BC are found to consist of arsenic-doped copper, due to the wide availability of lautite and domeykite ores, which are rich in both arsenic and copper. However, due to arsenic-related casualties, these alloys were quickly replaced with tin–copper alloys (bronze) that were widely used due to a lower melting point, higher hardness, and lower brittleness relative to their arsenic forerunner.

The Iron Age (800 BC–43 AD) first brought about applications for iron-based materials. Since Earth's crust contains significantly more iron than copper (Table 1.1), it is not surprising that bronze was largely replaced by iron for materials applications. An iron silicate material, known today as *wrought iron*, was accidentally discovered as a by-product from copper processing. However, this material was softer than

Fig. 1.2 Timeline of major developmental efforts related to materials science

bronze, so it was not used extensively until the discovery of steel by the Hittites in 1400 BC. The incorporation of steelmaking techniques throughout other parts of the world was likely an artifact of the war-related emigration of the Hittites from the Middle East in 1200 BC. The Chinese built upon the existing iron-making technology, by introducing methods to create iron alloys that enabled the molding of iron into desired shapes (i.e., cast iron production). Many other empirical developments were practiced at that time through other parts of the world; however, it must be stated that it was only in the eighteenth and nineteenth century AD that scientists began to understand why these diverse procedures were effective.

The materials applications shown in Fig. 1.2 are still of current interest, including the design of improved ceramics and glasses, originally discovered by the earliest civilizations. Although building and structural materials such as ceramics, glasses, and asphalt have not dramatically changed since their invention, the rapid evolution

Table 1.1 Natural abundance of elements in Earth's crust[4]

Element	Natural Abundance (%)
Oxygen	46.1
Silicon	28.2
Aluminum	8.2
Iron	5.6
Calcium	4.2
Sodium	2.4
Magnesium	2.3
Potassium	2.1
Titanium	0.57
Hydrogen	0.14
Copper	0.005
Total	99.8

of electronic devices and modes of transportation have required advanced nanoscale materials. Many new architectures for advanced materials design are surely yet undiscovered, as scientists are now attempting to mimic the profound structural order existing in living creatures and plant life, which is evident as one delves into their microscopic and nanoscale regimes.

As society moves onto newer technologies, existing materials become obsolete, or their concepts are converted to new applications. A prime example of this is related to phonographs that were commonplace in the early to mid-1900s. However, with the invention of magnetic tape by Marvin Camras in 1947, there was a sharp drop in record usage due to the preferred tape format. The invention of compact disk technology in 1982 drove the last nail in the coffin of vinyl record sales. Until only recently, records were found only in antique shops and garage sales.[5] The needles that were essential to play records no longer have wide marketability for this application, but have inspired another use at the micro-and nanoscale regime: *atomic force microscopy*, more generally referred to as *scanning probe microscopy* (SPM, see Sect. 7.5). This characterization technique uses a tip, analogous to the record needle that was once used in phonographs, to create images of the surface topology of a sample—even including the controlled placement of individual atoms (Fig. 1.3)! Hence, even though the needs and desires of society are constantly changing, the antiquated materials that are being replaced may still be of benefit for the design of new materials and technology.

The early world of materials discovery consisted solely of empirical observations, without an understanding of the relationship between material structure and properties. Each civilization had specific needs (e.g., materials for shelter, clothing, warfare), and adapted whatever materials were available at the time to address these necessities. Although this suitably addressed whatever issues were of societal concern at the time, such a trial-and-error manner of materials design resulted in slow growth.

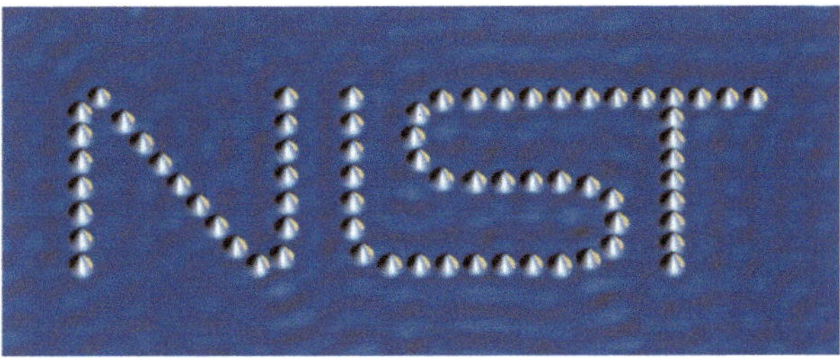

Fig. 1.3 A 40-nm wide logo for NIST (National Institute of Standards and Technology), made by the manipulation of Co atoms on a Cu(111) surface. The ripples in the background are due to electrons in the fluid-like layer at the copper surface, bouncing off the cobalt atoms—much like the patterns produced when pebbles are dropped in a pond. Image provided courtesy of J. A. Stroscio and R. J. Celotta (NIST, Gaithersburg, MD)

Interestingly, until the nineteenth century, the practice of chemistry was viewed as a religion, derived from alchemy that focused on a spiritual quest to make sense of the universe.[6] The alchemists searched for numerous intriguing discoveries including the keys to immortality, a "philosopher's stone" to transform base metals such as lead into noble metals such as silver and gold, and magic potions to cure diseases. However noble these pursuits were, they remained unaccomplished due to the lack of an underlying chemical theory to guide their experimentation. In addition, their trial-and-error methodology involved only qualitative characterization, and it was extremely difficult to control the reaction conditions, making it virtually impossible to repeat the exact procedure numerous times.

As a result, from 1000 BC to 1700 AD, only a few new substances were discovered which later turned out to be elements such as copper, iron, and mercury. Although this foundation resulted in the development of many experimental techniques of modern chemistry, it is not hard to see that true progress toward new material design may only be accomplished through foresight, based on an intimate understanding of specific relationships between the structure and property of a material. However, as you will see throughout this text, even with such knowledge, many important materials discoveries have been made by accident—the result of an unplanned occurrence during a carefully designed synthesis of an unrelated substance!

1.2 "Bottom-Up" Materials Synthesis

The development of new materials is governed by societal needs and the availability of resources. However, the adoption of a material depends primarily on its cost. When a new technology or material is introduced, there is almost always a high price associated with its adoption. For example, consider the cost of computers and plasma televisions when they first became available—worth tens of thousands of dollars!

The market price of a device is governed by the costs of its subunits. Shortly after the invention of germanium-based transistors in the late 1940s, the price of an individual transistor was approximately US $8–10. However, as germanium was substituted with silicon and fabrication techniques were improved, the price of these materials has exponentially decreased to its current price of one-millionth of a penny! This has allowed for an unprecedented growth in computational expediency, without a concomitant increase in overall price.

There are two rationales for the synthesis of materials, "top-down" and "bottom-up"; Fig. 1.4 illustrates examples of materials synthesized from both approaches. Whereas the transformation of complex natural products into desirable materials occurs primarily via a top-down approach (e.g., gemstones from naturally occurring mineral deposits, etching features on silicon wafers for computer chip production), the majority of synthetic materials are produced using the bottom-up approach. This route is even practiced by children who assemble individual LEGO™ building blocks into more complex architectures. Indeed, the relatively new field of nanotechnology has drastically changed the conception of bottom-up processes, from the historical approach of combining/molding bulk precursor compounds, to the self-assembly of individual atoms and molecules. This capability of being able to manipulate the design of materials from the atomic level provides an unprecedented control over resultant properties. This will allow possibilities for an unlimited number of future applications, including faster electronic devices, efficient drug-delivery agents, and "green" energy alternatives such as hydrogen-based fuel cells and high-capacity batteries.

The recent discovery of self-repairing/autonomic healing materials is an example of the next generation of "smart materials", which offer bottom-up structural control of their properties. Analogous to the way our bodies are created to heal themselves, these materials are designed to undergo spontaneous physical change, with little or no human intervention. Imagine a world where cracks in buildings repair themselves, or automobile bodies transform to showroom condition shortly following an accident. Within the next few decades, these materials could be applied to eliminate defective parts on an assembly line, and could even find use in structures that are at present impractical or impossible to repair, such as integrated circuits or implanted medical devices. An exciting world lies ahead of us; as we learn more about how to reproducibly design materials with specific properties from simple atomic/molecular subunits, the applications will only be limited by our imaginations!

1.3 Materials Discovery

From the medical doctor that must properly diagnose an illness, to the lawyer that must properly follow logic to defend his/her client, critical thinking skills are necessary to be successful in any career path. These skills are also very applicable for the design of new materials—the topic of this textbook. Figure 1.5 illustrates one example of a critical-thinking flowchart that could be applied to the design of a new

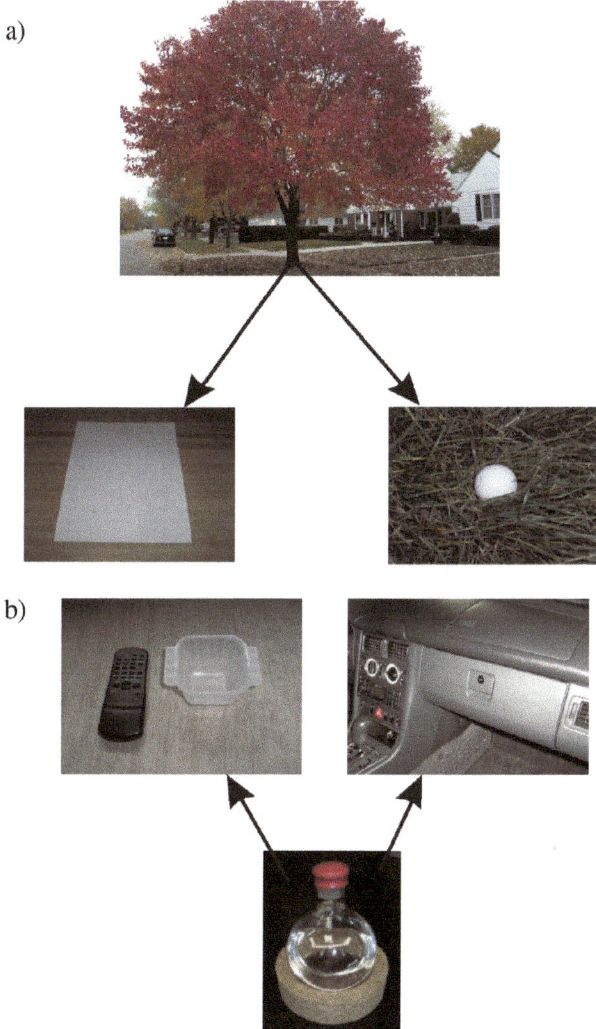

Fig. 1.4 Illustrations for the "top-down" and "bottom-up" approach to materials synthesis. **a** The top-down route is often used to transform naturally occurring products into useful materials. Representations shown above include the conversion of wood into paper products, as well as certain golf ball covers.[7] **b** The bottom-up route of materials synthesis is most prevalent. The representation shown above is the fabrication of plastics and vinyl found in common household products and automotive interiors, through polymerization processes starting from simple monomeric compounds (see Chap. 5)

material. Although there are many possibilities for such development, the following are essential components of any new development:

(i) Define the societal need and what type of material is being sought. That is, determine the desired properties of the new material.
(ii) Perform a comprehensive literature survey to determine what materials are currently being used. This must be done for the new product to successfully

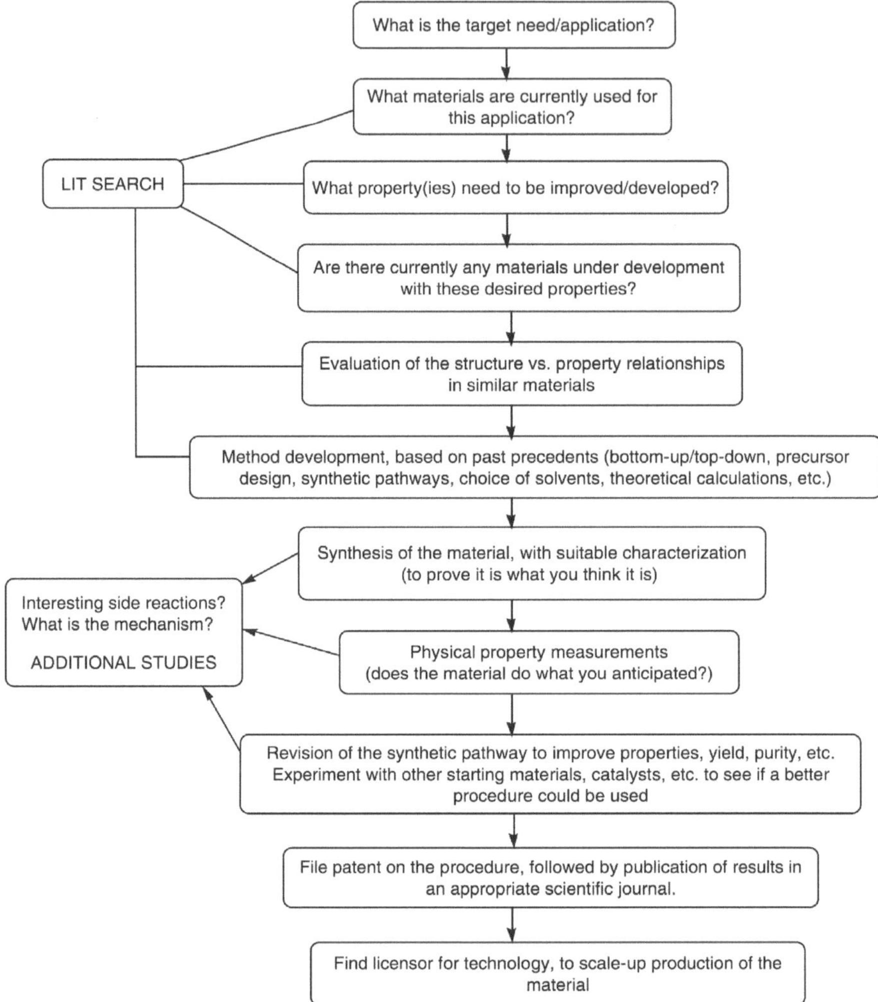

Fig. 1.5 An example of a critical thinking scheme for the design of a new material

compete in the consumer/industrial market. It is essential to search both scientific (e.g., http://www.pubs.acs.org—for all journals published by the American Chemical Society) and patent literature (e.g., http://www.uspto.gov), so that extensive research efforts are not wasted by reinventing something that already exists.

(iii) It should be noted that any exercise in critical thinking will result in more questions than originally anticipated. This is illustrated in the flowchart above, where one will look for interesting products/reactions and begin to think about the mechanism of the process. Such a "first-principle" understanding of the process is essential to increase yields of the material, and scale-up the technology for industrial applications.

(iv) After the new technology is protected by filing patents, publication in scientific literature is also important to foster continual investigations and fabricate new/improved materials. Top journals such as *Nature, Science, The Journal of the American Chemical Society, The Chemistry of Materials, Advanced Materials,* and *Nano Letters* publish articles every week related to new developments in the most active areas of science. In recent years, the number of materials-related papers has increased exponentially. The continual compounding of knowledge fosters further development related to the synthesis, characterization, and modeling of materials. However, this may only take place as active researchers share their results with their worldwide colleagues. The rise in popularity of open-access journals has helped expand the reach of literature precedents to those who do not subscribe to traditional journals[8].

However, these traditional methods of materials discovery take time and resources—both of which are in short supply as our hunger for advanced applications continues to increase. To fuel advancements related to energy, transportation, medicine, and everything in between, new materials must be developed more quickly and efficiently. The Materials Genome Initiative (MGI) was established in 2011 to help expedite the discovery, design, development, and fabrication of new materials (Fig. 1.6).[9] To date, there have been a number of advancements, including the establishment of the Materials Innovation Infrastructure (MII), which consists of computational and experimental tools and data. Since the property of a material is governed not only by its composition, but by its processing parameters, it is essential that materials research and development be unified among industry and academic institutions, with seamless sharing of their data. In addition, powerful computational techniques must be employed to scour the literature and materials databases to predict the optimized composition and properties of new materials based on the synthesis and processing parameters of similar materials. The use of computational tools alongside synthetic methods is thought to lower the overall cost and rate of materials design relative to synthetic methods alone, which tend to be based on trial-and-error from a smaller pool of literature precedents.

Similar to drug discovery methodologies, it is possible to design new materials through a combinatorial approach, allowing for high-throughput materials discovery. However, these efforts are not new. The first precedent for combinatorial materials

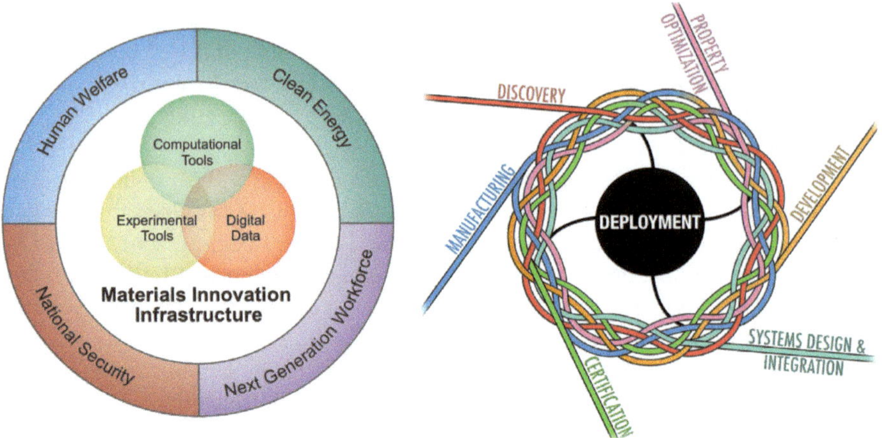

Fig. 1.6 An illustration of the Materials Genome Initiative goals (left) and operational paradigm (right), which aims to integrate the discovery, development, property optimization, manufacturing, systems integration, and certification stages of new materials. Reproduced with permission from "Materials Genome Initiative Strategic Plan" by the National Science and Technology Council, Nov. 2021

synthesis dates to the work of Boettcher and coworkers in 1955 with the fabrication of Ag-Sn–Pb thin films.[10] The design of varying compositions of cuprate superconductors by Schulz et al. sparked a renaissance in combinatorial approaches to develop new materials for energy storage, catalysis, electronic materials, and polymers.[11] These efforts have become more popular in recent years with advancements in supercomputers, machine learning, artificial intelligence, and robotics. Using powerful computational algorithms, it is now possible to guide the synthesis of new materials from a dataset of desired compositions and properties.[12] However, to accelerate materials discovery, researchers must have access to computational databases as well as synthesis and characterization (e.g., electron microscopy imaging, X-ray diffraction, etc.) libraries. As illustrated in Fig. 1.7, there are a variety of machine learning (ML) applications currently under investigation, which include[13]:

(i) Use of ML algorithms as a surrogate model to describe materials properties that guides the empirical and/or experimental synthesis.
(ii) Use of active learning to continuously improve the quality of ML algorithms through a feedback loop to select the most promising candidates that are fed into the next iteration of calculations or experiments. This is often carried out using *Bayesian optimization* algorithms.[14]
(iii) The use of generative algorithms to expand the exploration space of possible candidates during materials discovery.
(iv) The use of ML and autonomous robotic synthesis with an empirical-experimental feedback loop to optimize the discovery and synthesis of new materials.
(v) Use of quantum mechanical calculations (e.g., density functional theory, DFT) to identify promising new candidates for new materials. This is typically performed in tandem with experimentation and characterization to improve the quality and predictive accuracy for latter empirical iterations.

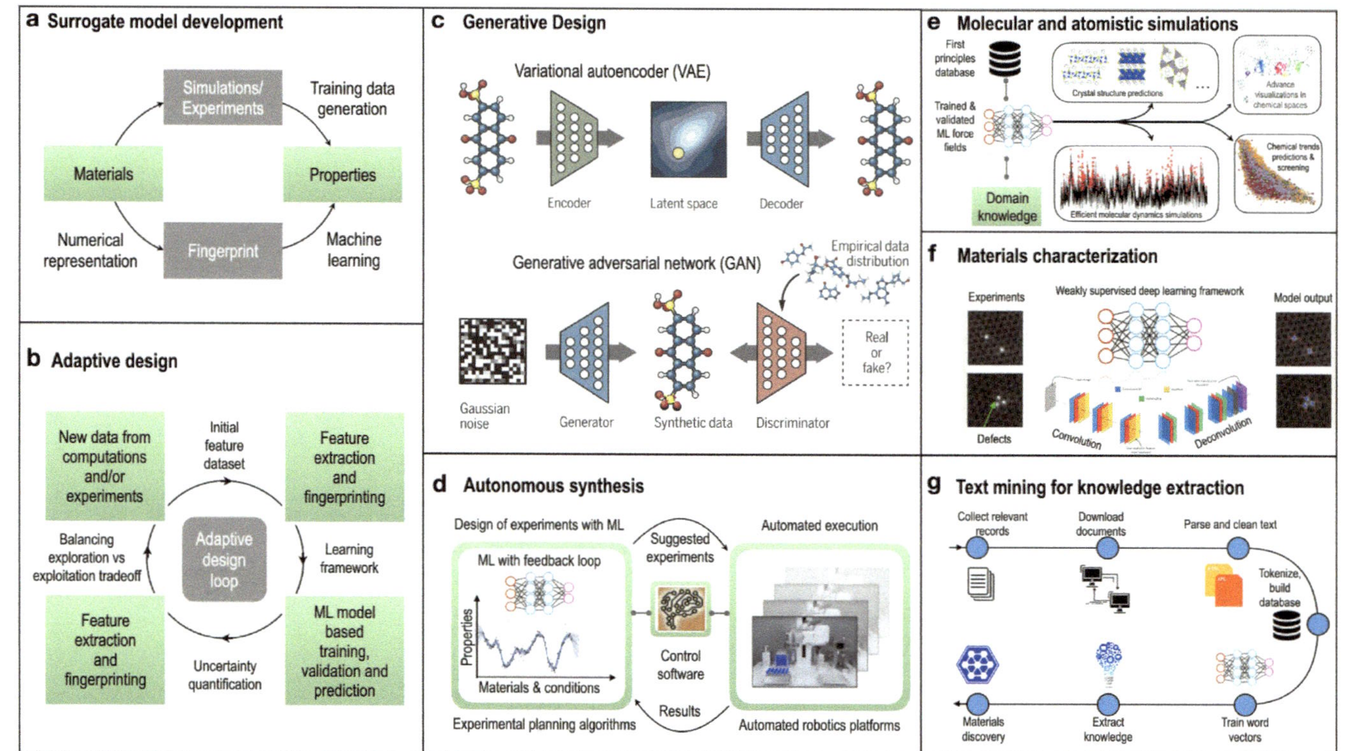

Fig. 1.7 Comparison of different methods for materials discovery using machine learning. Reproduced with permission from https://doi.org/10.1016/j.commatsci.2021.110360 © Elsevier

(vi) Use of high-resolution imaging techniques to provide structure–property data for ML algorithms. Feedback loops between empirical models and characterization data are used to expedite materials discovery.
(vii) Obtaining data from existing published data through automated text extraction, which is fed into ML algorithms for materials discovery using any of the above strategies.

Due to the expanse and complexity of the composition-structure–property (CSP) relationship of materials, the databases used for ML must also include a variety of synthetic conditions. Mirkin and coworkers have successfully applied a ML strategy for the synthesis and screening of *megalibraries* of novel nanoparticles with tunable morphologies, sizes, and compositions.[15] Through the use of a closed-loop empirical process, it was discovered that algorithms exhibited a high degree of predictive accuracy for nanoparticle synthesis.[16] The use of a closed-loop autonomous system for materials exploration and optimization (CAMEO) has been recently reported, which may truly revolutionize the design and synthesis of new materials.[17] As shown in Fig. 1.8, this strategy utilizes a human–machine partnership to predict and synthesize new materials through an active learning approach. With continuing advancements in robotics and augmented/virtual reality, it will be exciting to see how machine-learning materials discovery continues to develop.

1.4 Materials Sustainability

From the plastics that protect our electronic devices and encase our foods, to the metals that provide strength to buildings and various modes of transportation, we rely on the use of modern materials without giving much thought to their production. However, to ensure that our planet remains inhabitable for future generations, it is essential that we also understand the environmental impacts of materials production, use, and disposal. As a recent example, consider the environmental impacts resulting from the COVID-19 pandemic (Fig. 1.9). Although it is hard to think of any benefits, the global economic slowdown resulted in decreased automotive greenhouse gas emissions as fewer people traveled for leisure and more workers opted to work from home. However, it is likely that these gains were minimal due to increased emissions from the burgeoning varieties of ship-to-home services.

The long-term planning of resources is known as *sustainable development*, which entails balancing local and global efforts to meet the needs of the present without compromising the ability of future generations to meet their own needs.[18] Accordingly, the selection of a material, processing technique, or end-use for a product should coincide with the *three pillars of sustainability* (Fig. 1.10):

- Environmental—pollution prevention, natural resource use
- Social—better quality of life for all members of society
- Economic—fair distribution and efficient allocation of resources.

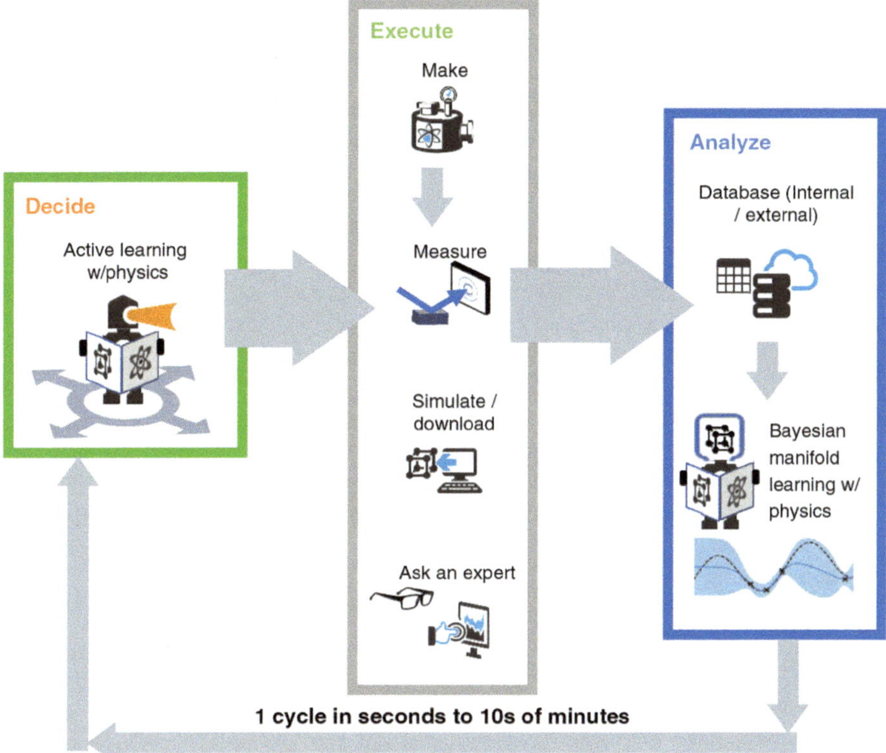

Fig. 1.8 Illustration of closed-loop autonomous materials exploration and optimization (CAMEO). © Springer Nature

Assessing the overall sustainability of a device or material is never an easy task, especially for the complex devices we encounter in today's society. For instance, consider the overall sustainability of a solar panel. You might be tempted to surmise that solar power is clearly more sustainable than burning fossil fuels. However, this comparison is not as straightforward as you might think. Clearly, the replacement of fossil-fuel based electrical power stations with photovoltaic cells would result in lower greenhouse gas (GHG) emissions during their operation. However, how many GHGs were released during the energy-intensive processing steps required to fabricate ultra-high purity silicon needed for solar panels? What about the energy and resources used during the transformation from ultra-high purity silicon into a multilayered solar panel? What are the energy consumption and environmental impacts of recycling solar panels after reaching their lifetime of use? It is important to realize that every device or material, even those that are more environmentally friendly during their use phase, has an environmental impact.

Keeping with our example of solar panels, how would their use be related to the social pillar? Consider Burma, located in southeast Asia surrounded by Bangladesh,

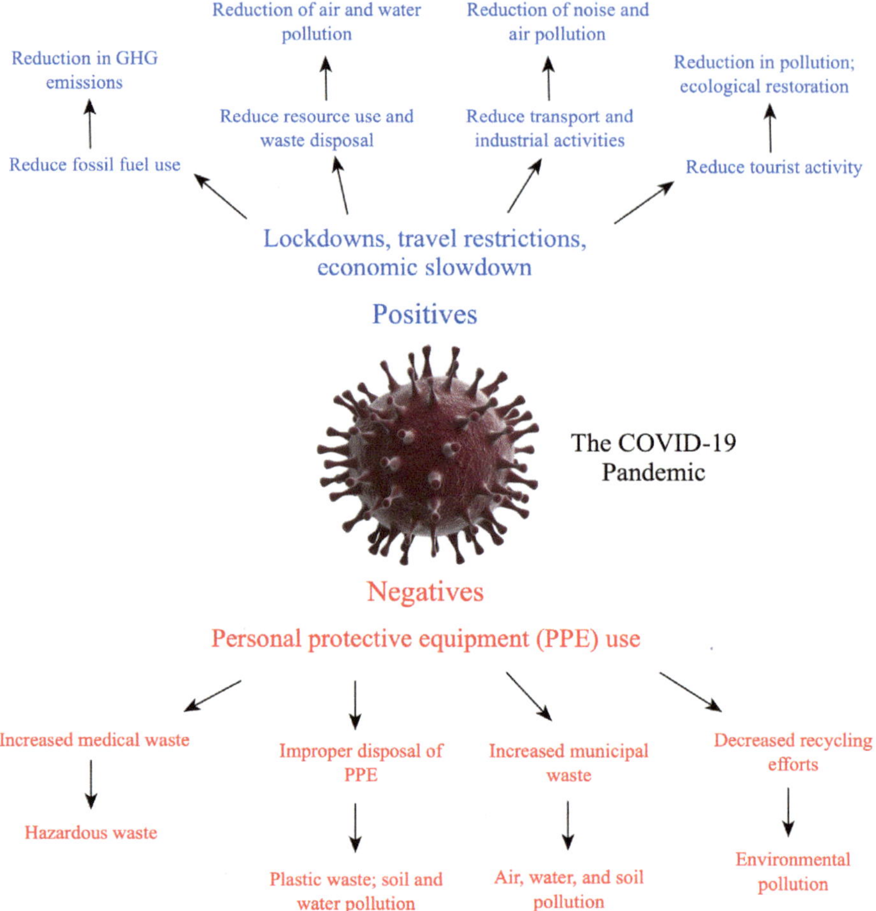

Fig. 1.9 Some environmental impacts of the COVID-19 pandemic

India, China, and Thailand. Of the total population of 52 million, it is estimated that 70% live in rural areas—often without access to electricity. With copious amounts of sunlight, solar energy is an attractive option for this isolated region, which is now being used for some remote water pumping stations. Hence, although solar energy was once reserved exclusively for industrialized nations, it is becoming closer to fulfilling the social pillar of sustainability as this technology flows to developing countries.

The environmental impact of a product is often described by a simple relationship (1.1):

$$I = P \times C \times T \tag{1.1}$$

where: I = environmental impact
P = population
A = consumption
T = technology.

As the world population approaches 8 billion, such a rapidly growing global economy will impose energy constraints, with the effects of climate change and environmental pollution threatening the availability of resources worldwide. Industrialized countries such as the United States, Europe, and Canada are societies of consumers, with access to tremendous technological advancements. However, this improved quality of life also leads to significant environmental impact as natural resources become depleted or polluted due to the need to satisfy growing consumption levels. For instance, the production of a single vehicle is estimated to require an average of 600,000 gallons of water, and 682 lb of mine pollution is generated from a single Pb-acid battery. Although the population of underdeveloped countries may outnumber those in the western world, technological advancements are yet to be equitably distributed to enable high consumption levels. However, as the quality of life improves for all nations, what will this mean for the limited supplies of global natural

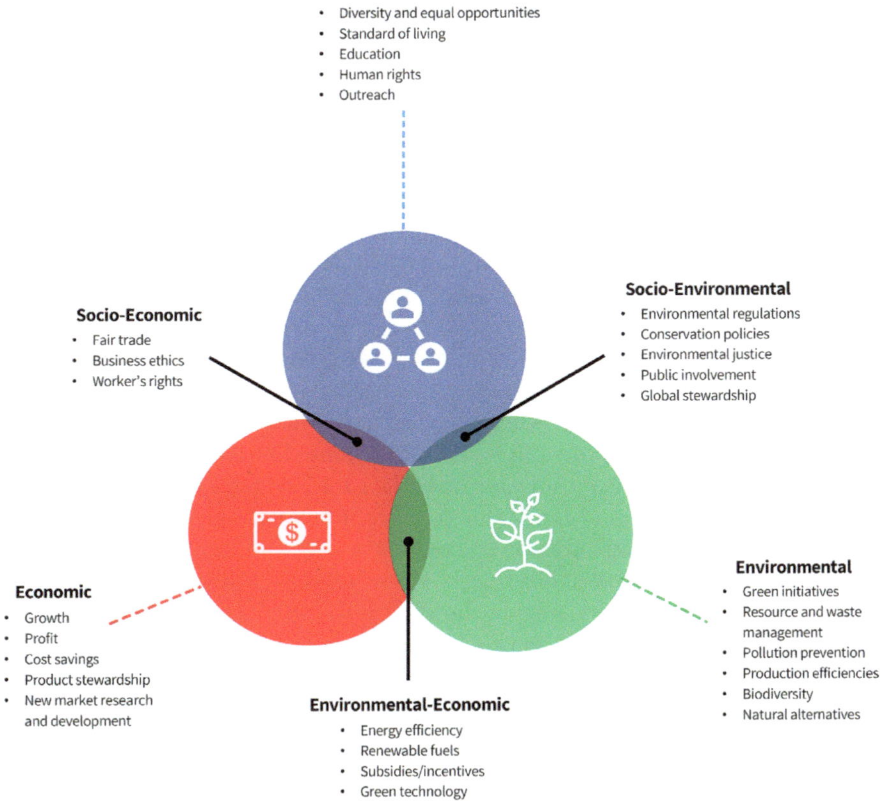

Fig. 1.10 An illustration of the three pillars of sustainability. Adapted from Mark Fedkin. Adopted from the University of Michigan Sustainability Assessment [Rodriguez et al., 2002]

resources? Enter the realm of politics, ethics, and human justice. As an example, if underdeveloped countries can only afford to use coal for electricity, it is unfair for industrialized countries to demand they reduce their GHG emissions by shifting to natural gas or solar. The economic pillar of sustainability must therefore consider the environmental health of the planet within a realistic economic framework for the incorporation of certain technologies in less affluent countries.

The sustainability of a product must take into account its production, use, and disposal/re-use. The common expression of **cradle-to-grave** is used to analyze the life cycle of an item, starting with the raw materials from which it came and ending with its ultimate disposal. Take a few minutes and examine your surroundings. The computers, cell phones, and clothing that we take for granted originated from somewhere, but where? What will happen to these materials after you are finished with them? Sustainable development attempts to transition from the historical cradle-to-grave scenario to that of *cradle-to-cradle*, where the end of the life cycle of one material dovetails with the beginning of another.

As a simple illustration, consider the aluminum can that holds your favorite beverage (Fig. 1.11). The raw material for aluminum is bauxite ore, an aluminum oxide mineral commonly found in Australia, China, Brazil, or India. After bauxite is mined, it must be refined to yield pure aluminum metal—processes that will be detailed in Chap. 4. In general, these processes involve the leaching of bauxite ore with a hot solution of sodium hydroxide, NaOH at elevated temperatures and pressures. If iron-rich bauxite is mined, the resultant red mud waste residue must be safely stored. It is estimated that over 70 million tons of red mud is produced annually from bauxite leaching, with the majority produced in Australia and China.[19] Most of the red mud is placed in landfills or settling ponds. However, the possible failure of these storage dams poses considerable risk to all three pillars of sustainability (Fig. 1.12).

Fortunately, aluminum represents a material that is largely recycled. Through recycling efforts, the cans are cleaned, re-melted, and re-cast into sheets, cans, or other shapes for use in another application. Other metals may also be acquired from closed-loop electronics recycling efforts, known as *urban mining*. It is estimated that one ton of cell phones contains 80 times more gold than a mine. With portable electronic devices now designed for easier disassembly, and robots programmed for efficient material reclamation, it is much less expensive and energy intensive to obtain gold from urban mining relative to traditional mining. Indeed, existing electronics are a rich source of the coinage metals; out of the *ca.* 1.7 billion cell phones sold each year, one could acquire up to 128,000 lb of gold, 56 M lb of copper, and 1.3 M lb of silver! However, recycling efforts must continue to improve; currently only 17% of the *ca.* 54 M tons of discarded electronics are recycled each year.

The four primary phases that occur during the life cycle of a material are listed below. Transportation-related emissions are also generated as components are shipped between the mine site or source-material fabrication plant, device production facility, consumer market, and recycling facility.

- *Material extraction and processing*—mining, refining, smelting, which use metal ores, sand, limestone, and other minerals. As these steps occur at high temperatures, significant amounts of fossil fuels are burned, resulting in the release of

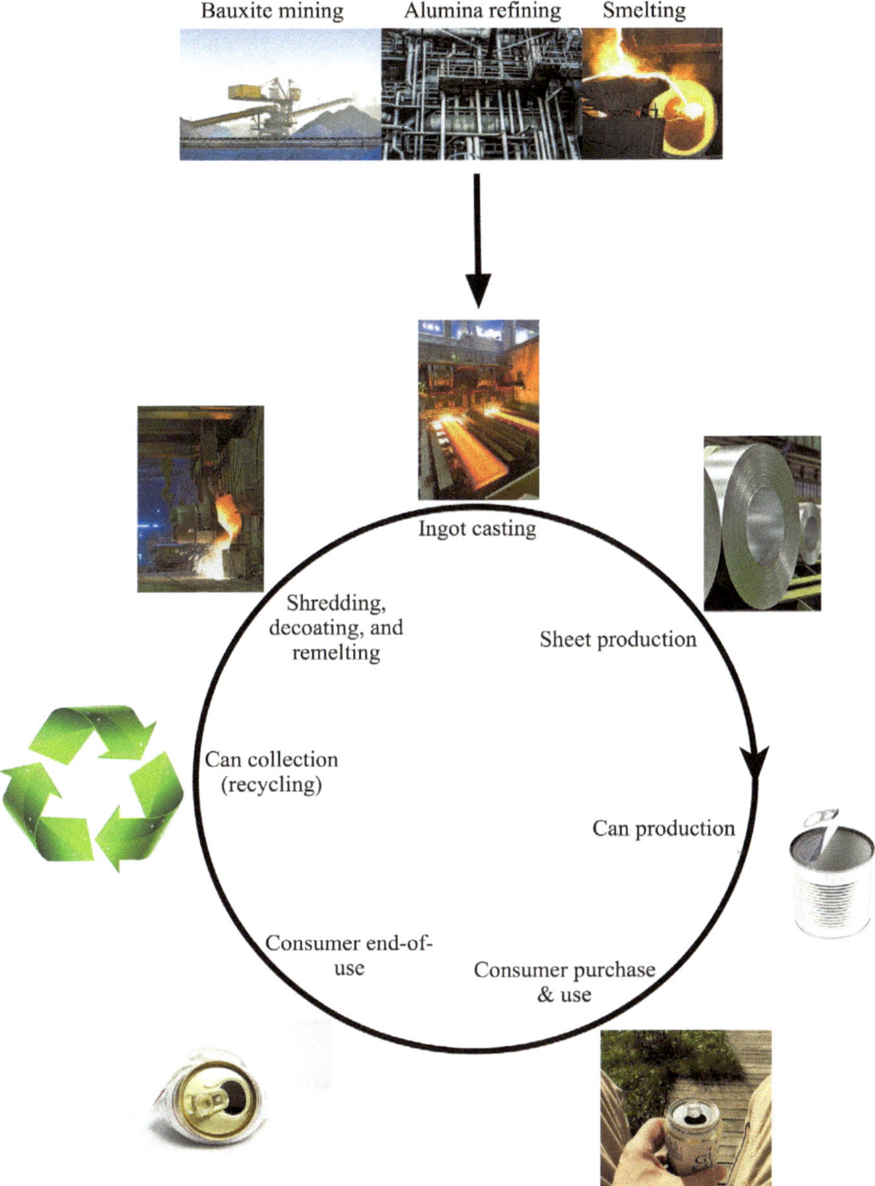

Fig. 1.11 Illustration of the steps involved in a cradle-to-cradle life cycle for an aluminum can

Fig. 1.12 Aerial image of a waste pond failure in Henan, China in August 2016. © Caixin Online

NO_x, SO_x, and CO_2 emissions. Mining waste in the form of slag or tailings are also generated during these processing steps, which often pollutes surface water and groundwater. For source materials such as polymers or ceramics, these are often synthesized by using a combination of gaseous, liquid, and solid chemicals and organic solvents. These processes often require high temperatures and sophisticated reaction conditions such as inert ambient environments and characterization techniques that employ high-field magnets (NMR) or light sources (UV–Vis, IR spectroscopies).

- *Product manufacturing*—depending on the composition of the final product, a variety of energy-intensive steps utilize metals, polymers (plastics), ceramics, and/or glasses that were isolated/refined or synthesized in the previous step. A variety of gases and chemicals are used for product fabrication, with fossil fuels often used as the source of electricity, which release CO_2, NO_x and SO_x emissions. A variety of gaseous, liquid, and solid waste is also generated.
- *Use by consumers*—fossil-fuel based transportation is used to transport products from their fabrication plants to stores across the world. Packaging materials such as plastics and paper are disposed, which often end up in landfills or as urban debris.
- *Recycling and/or disposal*—electricity and high temperatures are needed for disassembly, cleaning, and repurposing recycled materials. Discarded waste must be transported to landfills; the action of microorganisms releases GHGs (CH_4 and CO_2) as organic matter is decomposed over time. Some materials such as polystyrene plastics that are used for packing peanuts or bottle caps, are not

accepted in most plastic recycling bins. As a result, this type of plastic is a major component of landfills, urban litter, and marine debris.

As you would expect, the life cycle of modern devices such as portable electronics, vehicles, or televisions, which are assemblages of many different types of materials from varying parts of the world, would be much more complex than that of aluminum metal discussed above. The production of advanced devices from raw materials accounts for more than 90% of the energy consumed over their lifetime! This is not the case with low-tech products such as light bulbs that consume much more energy over their lifetimes than was spent for their fabrication. Automobiles used to be in the same 'low-tech' category, controlled by analog devices; however, microprocessors now monitor and control every aspect of modern vehicles from the fuel injection system to tailpipe emissions. The increased energy consumption during the production of high-tech devices is primarily due to:

- More diverse materials are needed, which requires greater costs for mining/purification, as well as ceramics and plastics manufacturing.
- Microprocessors (computer chips) originate from the energy-intensive conversion of sand into ultra-high purity silicon, and must go through hundreds of complex steps required to fabricate the integrated circuits (detailed in Chap. 5).
- Complex devices require many hours of design with teams of people using multiple high-speed computers that run 24/7.

While it is quite easy to determine how much energy an electronic device consumes during its operation, it is very difficult to calculate the energy used in its fabrication. For instance, a new cell phone begins its production many years before it is released, in the hands of engineers who plan its features and design the complex architecture and computer chips that it will employ. It is hard to estimate how much energy this initiative consumes, since it involves the electricity to power the buildings and laboratories used for research and development, administrators, and the sales team who not only use electricity in their offices, but also consume fossil fuels during their extensive travel. Overlooking these pre-manufacturing activities simplifies the situation somewhat, but we still have the problem of globalization. That is, the silicon employed for the computer chips may be purified in Michigan, the circuit board built in California, the lithium for the battery mined and purified in Chile, and the plastics synthesized in China. Some variability in these locations depends on the company's supply chain, which will vary dramatically among companies. The amount of energy required to mine lithium metal in South America would be very different than what is required in Canada. Hence, this makes a general life-cycle analysis very difficult to predict with any level of accuracy without knowing more information about the manufacturing practices of each materials supplier. As an example of how complex the situation is for a single company, as of 2021, Apple reported the use of 800 global production facilities and over 200 suppliers of the raw materials and components needed for their product lines!

More companies are being transparent about the environmental footprint of their products, but there is no way to accurately include information about the energy consumption of the supply chain companies. For instance, Apple reported that the iPad Pro (12.9") is responsible for 138 $kgCO_2e$ over its lifetime, with 80% of those emissions from manufacturing, 10% from transport, 9% from consumer use, and <1% from end-of-life processing. In contrast, the iPhone 13 Pro Max (128 GB) with its smaller energy footprint is reported to release 74 $kgCO_2e$, with 80% generated from production, 15% from consumer use, 4% from transport, and <1% from end-of-life processing.[20]

Based on the environmental emissions data above, an iPhone consumes a total of 172 kWh of electricity over its lifetime, which corresponds to 619 MJ of energy (495 MJ from production alone).[21] To put this in perspective, a gallon of gasoline contains 131 MJ of energy.[22] In other words, the energy contained in 4 gallons of gasoline (& the emissions that were released from its combustion) was needed to fabricate a single iPhone. While this may not seem too significant, bear in mind that there are currently over 7 billion cell phones in use on the planet, with approximately 2 billion new phones sold every year. Further, there are many more tablets, laptops, and other electronic devices that each require more than energy to produce than cell phones. In fact, a 27" iMac computer (with retina 5 K display) requires approximately 3,200 MJ of electrical energy (or 379 $kgCO_2e$) to fabricate—6.5 times more energy than an iPhone. Folding in consumer use, transportation and recycling, the iMac will consume a total of 6900 MJ of electrical energy (824 $kgCO_2e$) over its lifetime!

The environmental impact numbers we have discussed thus far only deal with the direct fabrication, use, and recycling of electronic devices. However, the full life cycle of a device also includes many other energy-intensive activities that are needed to extract, refine, and transport the raw materials from various parts of the world to the central fabrication facility. How much energy and environmental emissions does it take to extract lithium metal from an ore in Chile? It depends on how difficult the ore is to reach, and what specific techniques the company uses to break apart the ore, extract the metal, and then refine/purify the metal once it is removed. The same may be said about other components of the phone such as the outer screen. Whereas Samsung doesn't expend much energy in attaching the glass to the case in its final assembly plant, how much energy did the glass manufacturer consume to convert sand into a high-strength glass, and then ship large crates of the material to China for final assembly? The answers to these questions are not easily obtainable and illustrate just how complicated it is to determine the full life cycle analysis (LCA) of a high-tech device in our globalized society.

In order to simplify the environmental assessment of a material, full LCAs are often replaced with "eco-audits," which focus on certain impacts that occur during processing, use, and recycling. As an interesting case study, consider the comparative lifecycle GHG emissions of electric vehicles relative to conventional gasoline-powered automobiles. It is widely believed that electric cars are significantly more sustainable since they feature no tailpipe greenhouse gas emissions during their operation. Furthermore, gasoline vehicles can only use *ca.* 14–30% of the energy supplied by fuel, whereas electric cars are able to utilize between 74 and 94% of stored energy from the battery.[23] However, how much energy was used to produce electric vehicles? Also, how much electricity does it require to charge their battery packs? These are all important questions that must be addressed to assess their comparative carbon footprint.

From a materials standpoint, the primary difference between conventional and electric vehicles (EVs) is the battery pack. A Li-ion rechargeable battery consists of an iron or aluminum case, which envelopes two electrodes—the cathode and anode (Fig. 1.13). A separator is used to isolate the two electrodes, thus preventing a short-circuit and dangerous heat build-up. An electrolyte composed of an inorganic salt dissolved in organic liquids is used as an ionic conductor to maintain charge neutrality during battery cycling. Whereas the anode features a carbon coating on a copper current collector, the cathode is significantly more complex. The active cathode material is $LiMO_2$ (where M = Co, Ni, or Mn), which may be considered the primary component of Li-ion batteries with respect to its overall mass. Table 1.2 lists the weight of battery components for various sized electric vehicles. The electrodes and electrolyte comprising the battery cells contribute about 60% of the total weight of the battery.

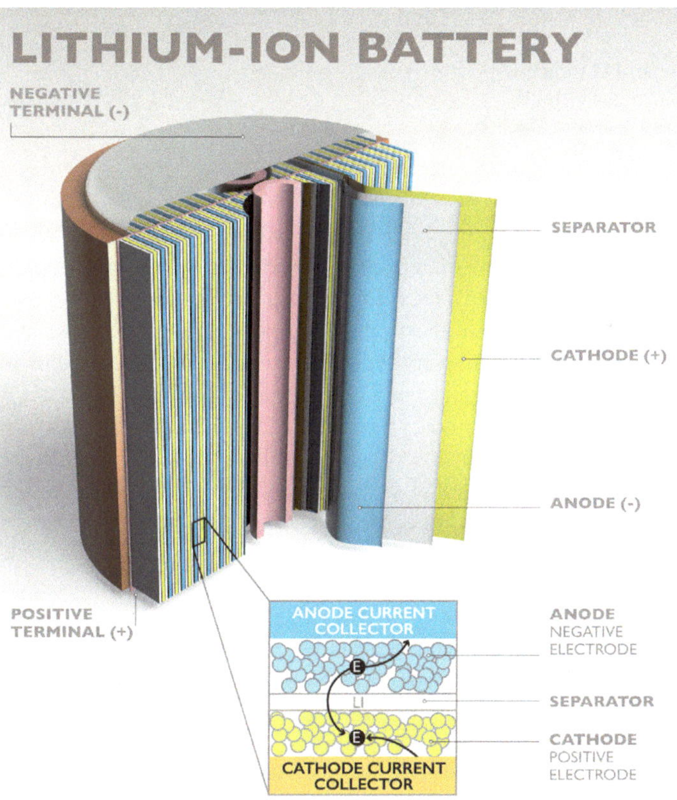

Fig. 1.13 Schematic of a lithium-ion battery. © Nablys/Shutterstock.com

Table 1.2 Weight of battery components and representative electric vehicles

Component	16.7 kWh Battery—Mini car[a] (kg)	39 kWh Battery—Medium car[b] (kg)	90 kWh Battery—Large car[c] (kg)	118 kWh Battery—Luxury car[d] (kg)
Battery cells	102	152	290	445
Module packaging	32	48	90	140
Battery management system	9.1	9.4	12	14
Cooling system	8.1	10.5	19	30
Battery packaging	26	33	60	90
Battery weight	177	253	471	719
Typical vehicle weight	923	1247	1700	2000
EV weight	1100	1580	2218	2300

[a] Example: Smart EQ fortwo
[b] Example: Kia Nissan LEAF.
[c] Example: Ford Mustang Mach-E
[d] Example: Lucid Air Dream Edition

The Li-ion battery (LiB) supply chain is illustrated in Fig. 1.14, which includes mining and refining of the raw materials, electrode fabrication, and cell/battery pack manufacturing. Current battery recycling efforts are focused on the cathode since these materials represent the highest percentage of total battery mass and cost. It is estimated that the LiB recycling market will exceed $2B by 2023 but will depend on the specific cathode chemistry used in battery packs. For instance, it is currently more expensive to recover Mn and Li from $LiFePO_4$ and $LiMn_2O_4$ than mining.[24] The globalization of electrode metals used in LiBs is illustrated in Fig. 1.15[25]; key lithium compounds of Li_2O, LiOH, and Li_2CO_3 flow primarily from the Americas to Asia where most battery manufacturing occurs. In contrast, Cobalt originates primarily in the Republic of Congo and is produced as a byproduct/co-product of nickel, copper, and platinum group metals (PGMs). Therefore, cobalt availability depends on that of other metals.

It is estimated that battery applications now account for 52% of worldwide demand for cobalt and 39% for lithium. Approximately 30 pounds of cobalt ore are needed to produce the refined metal used for each battery pack. In addition, some rare-earth (RE) metals such as neodymium, dysprosium, and terbium are also employed in the high-powered magnets within the electric motor. Since 97% of the current RE supply is currently acquired from a single nation (China), future geopolitical conflicts could severely alter the use of advanced technologies such as wind turbines, EVs, and consumer electronics. In order to circumvent these potential issues, recycling

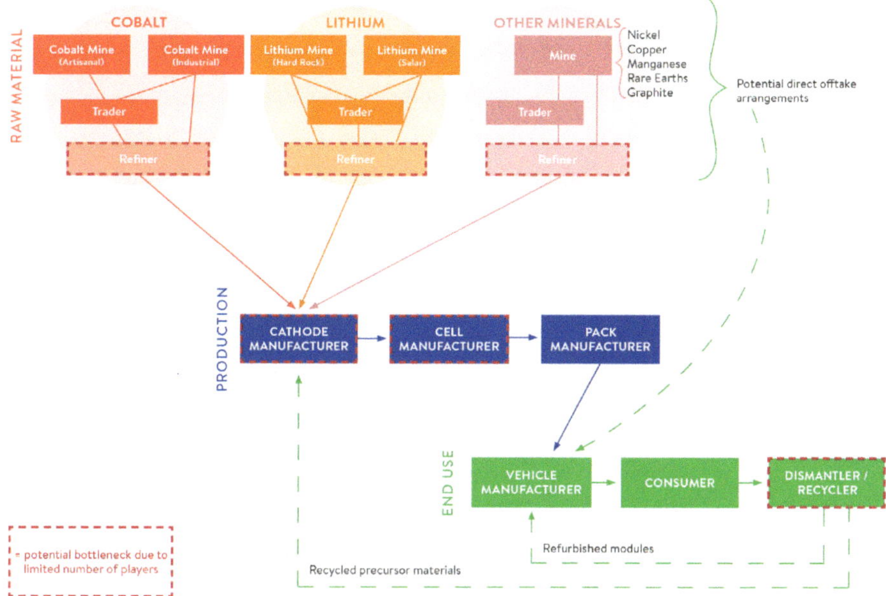

Fig. 1.14 A schematic of the Li-ion battery supply chain. Dashed lines represent recycling pathways. Reproduced with permission from the Center for Law, Energy & the Environment, UC Berkeley School of Law

efforts have been intensified across many high-technology sectors, including the battery industry. The use of recycled metals in the production of new battery cells is reported to result in a 51% natural resource savings, which is due to decreased dependence on mineral ores, reduced fossil fuel use, and nuclear energy demand.[26] However, improvements in recycling efficiencies are needed; for instance, 20 tons of spent Li-ion batteries yield only one ton of lithium metal.

For both conventional (fossil-fuel powered) and electric vehicles, the energy consumed during its average lifetime is assumed to cover 12 years and 180,000 km (108,000 mi). Based on the work of Ellingsen et al.,[27] greenhouse emissions from the production, use, and EOL segments for conventional vehicles are projected to be initially less than their EV counterparts (Fig. 1.16). This is a consequence of the high-energy inputs that are needed to produce EVs, thus requiring a significant driving distance to recover these production-based emissions. For instance, small vehicles could require 70,000 km (42,000 mi) before emissions begin to favor EVs. In contrast, large/luxury cars may require only 40,000–44,000 km (24,000–26,000 mi), since the increased fossil fuel emissions outweigh those required for battery production. However, although EVs require more energy-intensive processes for production and EOL recycling, their significantly lower use-phase impact is projected to result in a 20–27% lower lifecycle impact than conventional vehicles at their end of life.

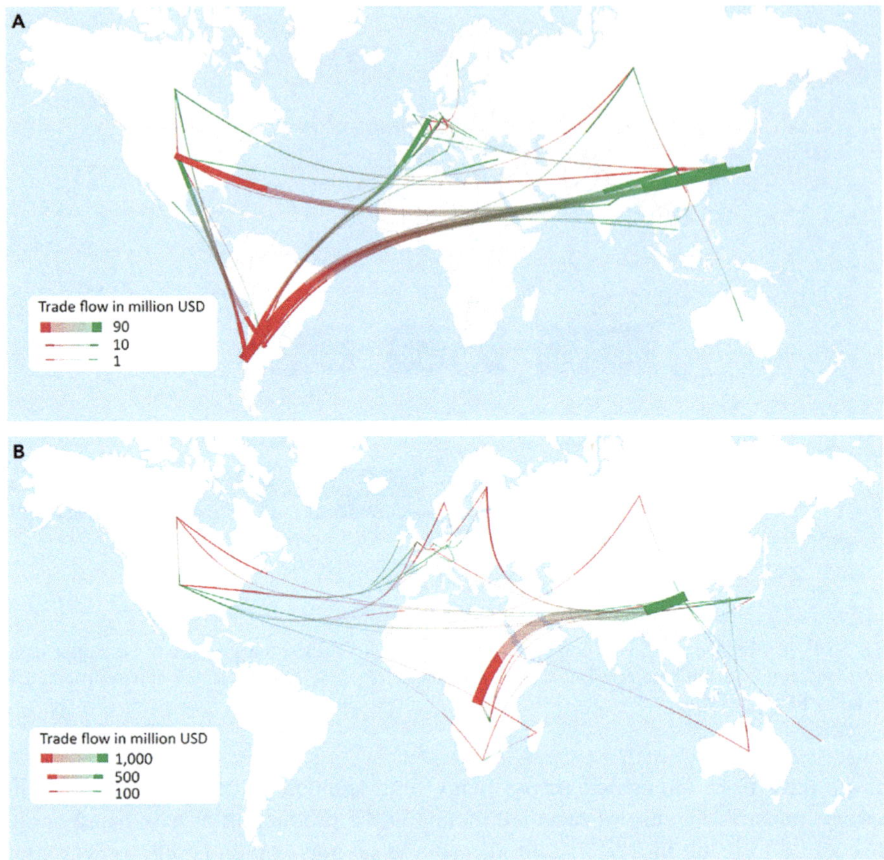

Fig. 1.15 Global trade flows for **a** lithium and **b** cobalt ores. The width of flows are proportional to trade value in U.S. dollars; importers are shown in green and exporters in red. Reproduced with permission from Olivetti, E. A.; Ceder, G.; Gaustad, G. G.; Fu, X. *Joule* **2017**, *1*, 229. © 2017 Elsevier

So, electric vehicles are truly more environmentally friendly than conventional cars, correct? Not so fast! The above calculation assumed that EVs were charged with the *European electricity mix*, which generates 26% from coal, 27% from nuclear energy, 27% from renewables, 17% from natural and derived gas, 2% from oil, and 1% from other fuels. As you can see, 54% of this mix is acquired from renewable and nuclear sources, which results in fewer GHG emissions. However, many parts of the world, including the U.S. and Canada, utilize coal for electricity generation. Figure 1.17 shows how the comparative GHG emissions would change with the makeup of electricity generation used to recharge an EV. If coal were used as the source of electricity, conventional vehicles would release *fewer* GHG emissions than their EV counterparts over the vehicle lifetime. If natural gas were used, a shift in favor EVs would occur as the mileage approached its mid-life of use. However,

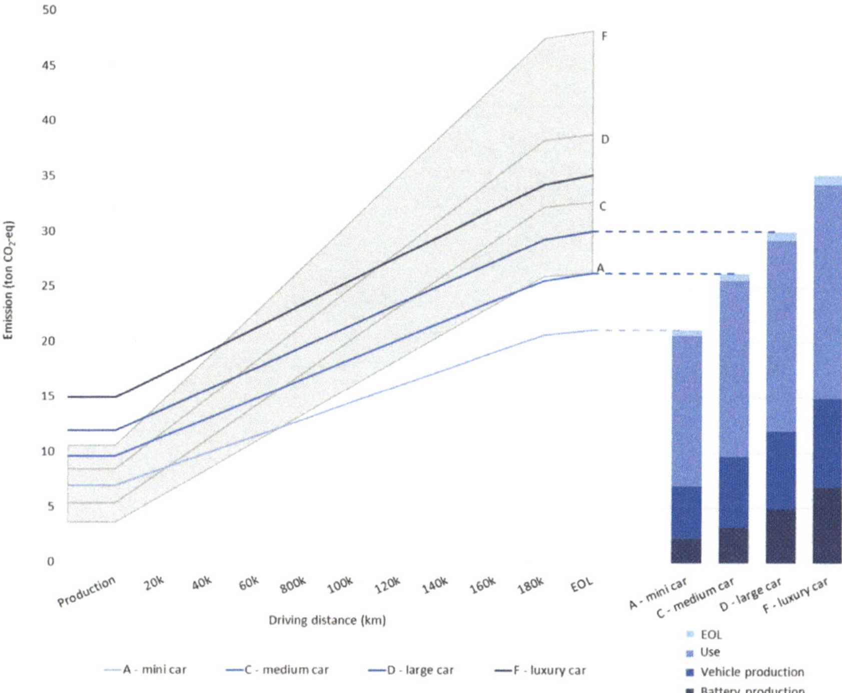

Fig. 1.16 Comparative lifecycle impacts for conventional and electric vehicles. The chart on the left displays cumulative emissions for production, use, and end-of-life (EOL) phases for conventional mini (A), medium (B), and large-sized (C) vehicles, as well as luxury-brand (D) vehicles. The right bar charts illustrate the emissions for battery production, as well as those during production, use, and EOL of EVs for analogous vehicle types (A–D). © IOP Open Access Publishing, 2016. Reproduced with permission from *Environ. Res. Lett.* **2016**, *11*, 054,010

the use of wind-generated electricity for battery charging would result in the most dramatic decrease in emissions (ca. 70%) throughout the lifetime of a vehicle. The use of wind power for all phases of production and EOL for EVs would result in ca. 85% fewer GHG emissions than conventional vehicles at their end of life. Hence, the source of electricity is an important consideration in determining whether an electric vehicle is more environmentally friendly than gasoline or diesel vehicles. Although it will take time to transition the power grid from fossil fuels to renewable sources, it is estimated that global GHG emissions will still be lowered in most world regions by switching to EVs even if power-sector decarbonization occurs more slowly than electrification efforts.[28]

This case study has revealed the complexities in assessing the sustainability of a technology or process. Not only must materials be chosen that are naturally abundant and easy to acquire, refine, and recycle, but one must also consider the source of electricity that is used during production and recycling efforts. As we will see throughout this book, the production of modern materials requires a plethora of specialized

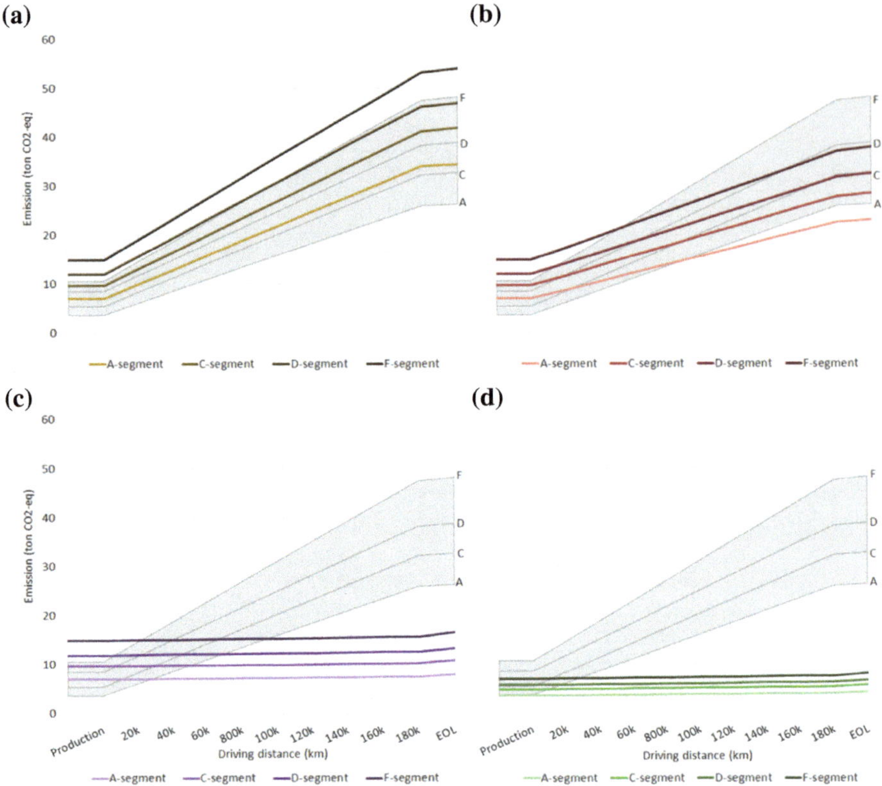

Fig. 1.17 Electric vehicle lifecycle emissions based on electricity sources used for battery charging. Shown are: a) coal, b) natural gas, c) wind, d) wind power used for all lifecycle phases; i.e., battery/EV production, use, and recycling. Reproduced with permission from *Environ. Res. Lett.* **2016**, *11*, 054,010. © IOP Open Access Publishing, 2016

techniques, all involving inputs of energy and generation of waste that may pollute atmosphere, water, or land resources. To lower our global carbon footprint and offset the effects of global climate change, we need to incorporate sustainable materials, use sources of energy with fewer GHG emissions, and develop recycling efforts to protect our inventory of renewable resources and reduce environmental waste.

Topics for Further Discussion

1. What are some differences between "Materials Chemistry" and "Solid-State Chemistry"?
2. What is meant by "top-down" and "bottom-up" synthetic approaches? Provide applications of each (both man-made or natural materials).
3. Are complex liquids such as crude oil or detergents considered 'materials'? Explain your reasoning.

4. After reviewing Appendix A, is there a relationship between the major materials-related discoveries and societal foci/needs? Elaborate.
5. Current problems or needs in our world are related to drinking water availability, "green" renewable energy sources, increasing computational speeds, and many others. What are some current areas of research focus that are attempting to address these issues? One should search Chemical Abstracts, as well as books and the Internet for this information.
6. Science is rapidly becoming interdisciplinary. Describe the origin of chemistry disciplines (inorganic, organic, physical) to their current status, including multidisciplinary programs.
7. When a new technology is introduced, the consumer price is astronomical. What are the factors that govern when and how much this price will be lowered? Cite specific examples.
8. It is more sustainable to use an electric or gas stove for cooking? Explain your reasoning and assumptions.
9. How does the annual worldwide supply of lithium metal compare with its demand? What about the worldwide supply vs. demand for rare-earth metals? Considering the production estimates of EVs and hybrids within the next 5 years, will we face a shortage of these metals? Elaborate.
10. Perform a LCA for an everyday household item of your choice. Ensure that you consider the environmental impacts during the isolation and refining of its base materials, overall production, and disposal/recycling. Transportation impacts that occur during resource mining, production, and consumer use of the finished product should also be included.
11. Considering the three pillars of sustainability shown in Fig. 1.10, is solar power always more sustainable than coal-fired power plants? Explain.
12. Evaluate the overall sustainability of the alternative energies listed below, considering the environmental impacts during their fabrication, installation and use, as well as end-of-use scenarios.

    (a) wind farms
    (b) solar parks
    (c) hydroelectric plants.

13. How would the use of solar-powered electricity affect the trends shown in Fig. 1.17? Describe your rationale and assumptions.
14. Describe some current recycling efforts that are being used for lithium-ion batteries (LiBs). How does the recycling rate for lead-acid batteries compare to LiBs? What are the largest technical and economic barriers that are currently preventing the widescale recycling of LiBs?

Reference and Notes

1. It should be noted that liquid substances such as crude oil, or gaseous compounds such as propane, should not be classified as materials; these compounds are more properly defined as *precursors* for materials.

2. For instance, the existence of the C_{60} nanomaterial was first theorized by Osawa in 1970 and was finally discovered empirically by Nobel Laureates Smalley, Curl & Kroto in 1985. However, other than some niche biomedical investigations initiated in 2003, there are currently no widespread applications for this nanomaterial. There remains ongoing research devoted to studying the feasibility of fullerenes for a myriad of possible applications.
3. There is often a "grey area" concerning the best definition for small particulate matter. Most structures are commonly referred to as "nanoscale materials," fueled by the popularity of the nanotechnology revolution. However, the "nano" prefix (e.g., nanoparticles) should only be used to describe materials with architectural dimensions (e.g., diameters, thicknesses, etc.) of less than 100 nm; intermediate dimensions between 100 and 1000 nm should instead be referred to as "sub-micron."
4. CRC Handbook of Chemistry and Physics, 103rd edition; CRC Press: New York, 2022.
5. A recent revival of vinyl record sales has occurred among audiophiles, who claim that sound quality from this medium is superior to CD, MP3, or tape formats in terms of crispness and clarity.
6. Balata golf ball covers may be fabricated from sap extracts of balata/bully trees in South America to produce a thin, resilient material. It is worth mentioning that balata-based materials are now produced artificially.
7. For a review of alchemy, see: http://www.levity.com/alchemy/index.html.
8. Open-access journals charge authors a fee to publish their peer-reviewed articles, which may then be accessed for free. In contrast, traditional journals do not charge authors, but the peer-reviewed papers are only available to institutions who pay an annual subscription fee.
9. A strategic plan for the Materials Genome Initiative (MGI) was published in Nov., 2021. More details about the MGI and the strategic plan may be found at https://www.mgi.gov.
10. https://pubs.rsc.org/en/content/articlelanding/2012/CS/C1CS15178A.
11. https://www.sciencedirect.com/science/article/pii/S1369702105711214?via%3Dihub.
12. https://www.nature.com/articles/s41524-019-0205-0.
13. https://doi.org/10.1016/j.commatsci.2021.110360.
14. For an example of using Bayesian optimization for nanoporous materials, see: https://chemrxiv.org/engage/api-gateway/chemrxiv/assets/orp/resource/item/60e8e7e4551cb02a05ae0c32/original/bayesian-optimization-of-nanoporous-materials.pdf.
15. https://doi.org/10.1073/pnas.1815358116.
16. https://www.science.org/doi/10.1126/sciadv.abj5505.
17. https://doi.org/10.1038/s41467-020-19597-w.
18. http://www.un-documents.net/wced-ocf.htm.
19. https://www.intechopen.com/chapters/62120.

20. The unit 'kgCO_2e' (kg of equivalent CO_2) refers to the relative emissions of greenhouse gases (carbon dioxide, methane (CH_4), and/or nitrous oxide (N_2O)) per unit of fuel that is consumed.
21. The conversion from kg CO_2e to kWh is based on electricity generation with the EPAs eGRID emission factors from 2018 data published in 2020, using the U.S. average electricity source emissions of 0.429 kg CO_2e per kWh.
22. The units MJ refer to megajoules, which is 1×10^6 J. A joule is the standard unit of energy and corresponds to the energy required to lift a small apple (with a mass of 100 g) vertically through one meter of air. A megajoule (MJ) corresponds to the kinetic energy of a one-tonne (1000 kg) vehicle moving at 100 mph (160 km/h).
23. For gasoline vehicles, engine losses (radiator, exhaust heat, combustion, friction), power loss to wheels (wind resistance, rolling resistance, braking), drivetrain losses, and parasitic losses (water pump, alternator, etc.) account for 70–86% of the energy supplied by the gasoline. In comparison, electric vehicles only suffer from electric drive system losses, power to wheels, and energy lost in charging the battery. The use of regenerative braking and start/stop technologies reduces the power-to-wheel losses by up to 17%. See http://fueleconomy.gov for more details.
24. For a recent report of potential recycling routes for Li, Co, Ni, and Mn from LiBs, see: Chan, K. H.; Anawati, J.; Malik, M.; Azimi, G. *ACS Sustainable Chem. Eng.* **2021**, *9*, 4398.
25. For other descriptions of the metal supply chains related to LiBs, see: (a) Sun, X.; Liu, Z.; Zhao, F.; Hao, H. *Environ. Sci. Technol.* **2021**, *55*, 12,180. (b) van den Brink, S.; Kleijn, R.; Sprecher, B.; Tukker, A. Resources, Convervation and Recycling 2020, 156, 104,743. (c) https://www.nrel.gov/docs/fy19osti/73374.pdf (d) https://www.usitc.gov/publications/332/working_papers/no_id_069_gvc_lithium-ion_batteries_electric_vehicles_final_compliant.pdf.
26. Dewulf, J.; Van der Vorst, G.; Denturck, K.; Van Langenhove, H.; Ghyoot, W.; Tytgat, J.; Vandeputte, K. *Resources, Conservation and Recycling* **2010**, *54*, 229.
27. Ellingsen, L. A-W.; Singh, B.; Stromman, A. H. *Environ. Res. Lett.* **2016**, *11*, 054,010.
28. Knobloch, F.; Hanssen, S. V.; Lam A.; Pollitt, H.; Salas, P.; Chewpreecha, U.; Huijbregts, M. A. J.; Mercure, J. -F. *Nature Sustainability* **2020**, *3*, 437.

Further Reading

1. D.B. Lowe, *The Chemistry Book: From Gunpowder to Graphene, 250 Milestones in the History of Chemistry* (Sterling, New York, 2016)
2. W.H. Brock, *The History of Chemistry: A Very Short Introduction* (Oxford University Press, New York, 2016)

3. H. Aldersey-Williams, *Periodic Tales: A Cultural History of the Elements, From Arsenic to Zinc* (Harper-Collins, New York, 2011)
4. W.H. Brock, *The Norton History of Chemistry* (W. W. Norton, New York, 1993)
5. S.G. Brush, *The History of Modern Science: A Guide to the Second Scientific Revolution, 1800–1950* (Iowa State University Press, Ames, IA, 1988)
6. J. L. Heilbron, *Oxford Companion to the History of Modern Science* (Oxford University Press, New York, 2003)
7. E. Mount, *Milestones in Science and Technology: The Ready Reference Guide to Discoveries, Inventions, and Facts*, 2nd edn. (Oryx Press, Phoenix, 1994)
8. P. De Luna, *Accelerated Materials Discovery: How to Use Artificial Intelligence to Speed Up Development* (De Gruyter, Boston, 2022)
9. A.K. Bierman, R.N. Assali, *The Critical Thinking Handbook* (Prentice Hall, New Jersey, 1996)
10. S. West, *Critical Thinking Skills: Practical Strategies for Better Decision Making, Problem-Solving and Goal Setting* (Amazon Digital Services, 2017)
11. M.N. Browne, S.M. Keeley, *Asking the Right Questions: A Guide to Critical Thinking*, 5th edn. (Prentice Hall, New Jersey, 1998)
12. R. Paul, *Critical Thinking: What Every Person Needs to Survive in a Rapidly Changing World* (Foundation for Critical Thinking, 1993)
13. M.F. Ashby, *Materials and Sustainable Development, 2nd edn.* (Elsevier, New York, 2022)
14. M.F. Ashby, *Materials and the Environment, 3rd edn.* (Elsevier, New York, 2021)
15. M. Irimia-Vladu, E.D. Glowacki, N.S. Sariciftci, S. Bauer, *Green Materials for Electronics* (Wiley, New York, 2017)
16. R. Thompson, *Sustainable Materials, Processes and Production* (Thames & Hudson, London, 2013)
17. R.T. Wright, D.F. Boorse, *Environmental Science: Toward a Sustainable Future*, 13th edn. (Pearson, New York, 2016)
18. J.M. Allwood, J.M. Cullen, *Sustainable Materials Without the Hot Air: Making Buildings, Vehicles and Products Efficiently and with Less New Material*, 2nd edn. (UIT Cambridge, Cambridge, UK, 2015)

Chapter 2
Solid-State Chemistry

Among the states of matter, solids possess the most structural diversity. Whereas gases and liquids consist of discrete molecules that are randomly distributed due to thermal motion, solids consist of molecules, atoms, or ions that are positioned in specific 3-D arrays. To fully understand the properties of solid materials, one must have a thorough knowledge of the structural interactions between its constituents. This chapter will outline the various types of solids, including structural classifications and nomenclature for both crystalline and amorphous solids. The material in this key chapter will set the groundwork for the rest of this textbook, which describes a variety of materials classes.

2.1 Structural Order

A solid is a material with a defined shape and volume. If a solid is composed of long-range regularly repeating units, it is classified as a *crystalline* material. Crystalline solids are only produced when the constituent atoms, ions, or molecules have an opportunity to organize themselves into regular arrangements, or *lattices*. For example, crystalline minerals found in nature have been formed through many years of extreme temperature and pressure, or slow evaporation processes. Most naturally occurring crystalline solids are an aggregate of individual microcrystalline units; single crystals without significant defects are extremely rare in nature and require special growth techniques (see Sect. 2.4.1).

If there is no long-range structural order throughout the solid, the material is best described as *amorphous*. These materials possess short-range order over distances of only a few Angstroms; this lack of long-range translational order (periodicity) separates this class of materials from their crystalline counterparts. Due to the popularity of minerals, salts, and gemstones, there is a common misconception that most solids are crystalline in nature. In fact, many solids generated from chemical reactions will be amorphous by default, unless special procedures are used to facilitate

B. D. Fahlman, *Materials Chemistry*,
https://doi.org/10.1007/978-3-031-18784-1_2

Table 2.1 Glass transition temperatures of various materials

Material	Intermolecular bonding	T_g(°C)
SiO_2	Covalent	1430
Borosilicate glass	Covalent	550
$Pd_{0.4}Ni_{0.4}P_{0.2}$	Metallic	580
BeF_2	Ionic	570
As_2S_3	Covalent	470
Polystyrene	Van der Waal	370
Se_{∞}	Covalent	310
Poly(vinyl chloride)	Van der Waal	81
Polyethylene	Van der Waal	−30

molecular ordering (i.e., crystal formation). Although the crystalline state is more thermodynamically-favorable than the disordered state, the formation of amorphous materials is favored in kinetically bound processes (e.g., chemical vapor deposition, sol–gel, solid precipitation, etc.).[1]

Some materials featuring extended networks of molecules such as glasses may never exist in the crystalline state. In these solids, the molecules are so entangled or structurally complex that crystallization may not occur as the temperature is slowly decreased. Due to the rigidity of the solid, but proclivity to remain in the amorphous state, these compounds have been incorrectly referred to as supercooled liquids. It was even thought that a slow flow of glass over hundreds of years has caused nineteenth century stained glass windows to have a proportionately thicker base.[2] However, it is now well understood that the glass structure remains intact unless its threshold transition temperature is exceeded. This parameter is known as the *glass transition temperature*, T_g, and corresponds to the temperature below which molecules have very little mobility.

Other amorphous solids such as polymers—being rigid and brittle below T_g, and elastic above it—also exhibit this behavior. Table 2.1 lists the glass transition temperatures of common solid materials. It should also be noted that whereas crystalline solids exhibit a discrete melting point, amorphous solids undergo a solid–liquid phase transition over a range of temperatures. Although most solid-state textbooks deal almost exclusively with crystalline materials, this text will attempt to address both the crystalline and amorphous states, describing the structure/property relationships of major amorphous classes such as polymers and glasses.

2.2 Bonding in Solids I: Intermolecular Forces

Every amorphous and crystalline solid possesses certain types of inter- and intramolecular interactions between its subunits that govern its overall properties. Depending on the nature and strength of these interactions, a variety of physical, optical, and electronic properties may be observed. For example, intramolecular forces (i.e., atomic separations/interatomic bonding energies) directly influence the conductivity, thermal expansion, and elasticity of a material; in contrast, intermolecular forces will govern the phase transitions, solubility, and vapor pressure of a material. As expected, these associations not only govern the behavior of a material in the solid state, but also for the less-ordered liquid phase. For example, the hydrogen bonding interactions between neighboring water molecules within an ice lattice are also important in the liquid phase, resulting in high surface tension and finite viscosity. For the gaseous state, intermolecular forces are largely broken, resulting in minimal impact on its properties.[3]

2.2.1 *Ionic Solids*

These solids are characterized by cationic and anionic species that are associated through electrostatic interactions. All predominantly ionic salts possess crystalline structures, as exhibited by common Group 1–17 or 2–17 binary salts such as NaCl and $CaCl_2$ (Fig. 2.1). The melting points of these solids are extremely high, as very strong electrostatic attractions between counterions must be overcome. Although oppositely charged ions have attractive interactions, like charges repel one another. In the determination of the lattice energy, *U*, the sizes and charges of the ions are most important (2.1, the Born-Lande equation). That is, the lattice energy for MgO would be much greater than BaO, since the ionic bonding is much stronger for the magnesium salt due to its high charge/small size (large charge density). By contrast, the salt MgN does not exist, even though Mg^{3+} and N^{3-} would be very strongly attracted through electrostatic interactions. The ionization energy required to produce the trivalent magnesium ion is too prohibitive.

$$U = \frac{N \times M \times Z_{cation} \times Z_{anion}}{r_o}\left(\frac{e^2}{4\pi\varepsilon_o}\right)\left(1 - \frac{1}{\rho}\right) \qquad (2.1)$$

where: N = Avogadro's number (6.02×10^{23} molecules mol^{-1})

$Z_{cation,\ anion}$ = magnitude of ionic charges

r_o = average ionic bond length

e = electronic charge (1.602×10^{-19} C)

$4\pi\varepsilon_o$= permittivity of a vacuum ($1.11 \times 10^{-10}\ C^2\ J^{-1}\ m^{-1}$)

M = the Madelung constant (see text)

ρ = the Born exponent[4]; related to the corresponding closed-shell electronic configurations of the cations and anions (e.g., [He] = 5; [Ne] = 7; [Ar] or $[3d^{10}]$[Ar]= 9; [Kr] or $[4d^{10}]$[Kr] = 10; [Xe] or $[5d^{10}]$[Xe] = 12)

The Madelung constant appearing in (2.1) is related to the specific arrangement of ions in the crystal lattice. The Madelung constant may be considered as a decreasing series, which takes into account the repulsions among ions of similar charge, as

Fig. 2.1 Ionic model for sodium chloride. This is a face-centered arrangement of chloride ions (white), with sodium ions occupying the octahedral interstitial sites (red). The attractive electrostatic forces, *a*, between adjacent Na^+ and Cl^- ions, and repulsive forces, *r*, between Na^+ ions are indicated

well as attractions among oppositely charged ions. For example, in the NaCl lattice illustrated in Fig. 2.1, each sodium or chloride ion is surrounded by six ions of opposite charge, which corresponds to a large attractive force. However, farther away there are 12 ions of the same charge that results in a weaker repulsive interaction. As one considers all ions throughout the infinite crystal lattice, the number of possible interactions will increase exponentially, but the magnitudes of these forces diminish to zero.

It is noteworthy that the calculated lattice energy is quite often smaller than the empirical value. Whereas the ions in ionic compounds may be treated as hard spheres in the calculation, there is often a degree of covalency in the bonding motif. In particular, Fajans' rules describe the degree of covalency as being related to the charge density of the cation and the polarizability of the anion. In general, polarizability increases down a periodic group due to lower electronegativities and valence electrons being housed in more diffuse orbitals, thus experiencing a much less effective nuclear charge, Z_{eff}. For instance, a compound such as LiI would exhibit a significant degree of covalent bonding due to the strong polarizing potential of the very small cation, and high polarizability of iodide. This is reflected in its lower melting point (459 °C) relative to a more purely ionic analogue, LiF (m.p. = 848 °C). The % ionic character of a bond is given by (2.2); there is always a degree of covalency in ionic bonding, even for classically defined ionic compounds such as NaCl or LiF (Fig. 2.2).

$$\%I = \left(1 - e^{-0.25(X_A - X_B)^2}\right) \times 100\% \tag{2.2}$$

where: X_i = electronegativity of element i.

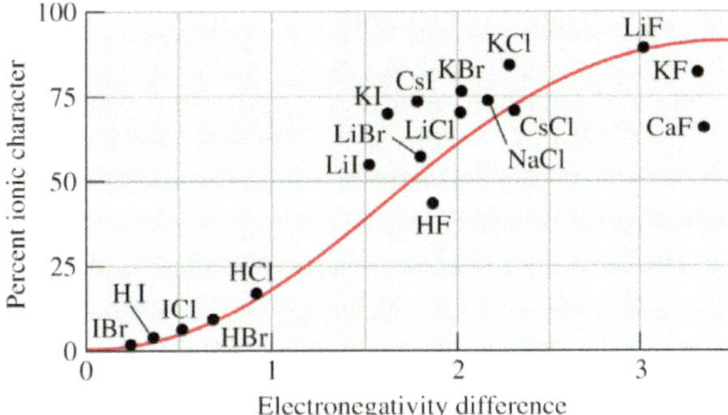

Fig. 2.2 Trends in the ionic character for various binary compounds

Ionic solids are soluble in polar solvents due to ion–dipole interactions between component ions and solvent molecules. Since the lattice energy of the crystal must be overcome in this process, the solvation of the ions (e.g., presence of $[(H_2O)_mNa]^+$*(aq)* and $[(H_2O)_nCl]^-$*(aq)* ions during the dissolution of NaCl) represents a significant exothermic process that acts as the driving force for solution formation.

2.2.2 *Metallic Solids*

Metallic solids are characterized by physical properties such as high thermal and electrical conductivities, malleability, and ductility (i.e., able to be drawn into a thin wire). Chemically, metals tend to have low ionization energies that often result in metals being easily oxidized by the surrounding environmental conditions. This explains why metals are found in nature as complex geological formations of oxides, sulfates, silicates, aluminates, etc. It should be noted that metals or alloys may also exist as liquids; mercury represents the only example of a pure metal that exists as a liquid at STP. The liquid state of Hg is a consequence of the electronic configurations of its individual atoms. The 6s valence electrons are shielded from the nuclear charge by a filled shell of 4f electrons. This shielding causes the effective nuclear charge (Z_{eff}) to be higher for these electrons, resulting in less sharing/delocalization of valence electrons relative to other metals. Further, relativistic contraction of the 6 s orbital causes these electrons to be situated closer to the nucleus, making them less available to share with neighboring Hg atoms.[5] In fact, mercury is the only metal that does not form diatomic molecules in the gas phase. Energetically, the individual atoms do not pack into a solid lattice since the lattice energy does not compensate for the energy required to remove electrons from the valence shell.

Metallic bonding

The most simplistic bonding description for metallic solids is known as the *free-electron model.* This considers the solid as a close-packed array of atoms, with valence electrons completely delocalized throughout the extended structure. Since the delocalization of electrons occurs more readily for valence electrons farther from the nucleus (experiencing a lesser Z_{eff}), metallic character increases going down a group of the periodic table. Perhaps the best example of this phenomenon is observed for the Group 14 congeners. As you move from carbon to lead, the elemental properties vary from insulating to metallic, through a transitional range of semiconducting behavior for Si and Ge.

The close chemical association among neighboring atoms in the solid gives rise to physical properties such as high melting points and malleability. The nondirectional bonding in metals allows for two modes of deformation to occur when a metal is bent. Either the atomic spacing between neighboring metal atoms in the crystal lattice may change (*elastic deformation*), or planes of metal atoms may slide past one another (*plastic deformation*). Whereas elastic deformation results in a material with "positional memory" (e.g., springs), plastic deformation results in a material that stays malformed. We will consider the bonding modes of metals and nonmetals in more detail later in this chapter. Although metals are primarily characterized by crystalline structures, amorphous alloys may also be produced, known as *bulk metallic glasses (BMGs)*, which will be described in more detail in Sect. 3.2.4.

2.2.3 Covalent Network Solids

These solids are characterized by very strong, directional covalent bonds between their constituent atoms. This bonding array generally leads to high melting points and bulk hardness. Due to the arrangement of the atoms comprising these solids, a variety of physical properties may be observed, as evidenced by the very different properties exhibited by the three *allotropes* (i.e., discrete structural forms) of carbon. For instance, diamond is an extremely hard, insulating material that is transparent to light, whereas graphite is a soft, black solid that is electrically conductive along the graphitic layers of the extended solid. Buckminsterfullerene (C_{60}) is very different from either of these carbon forms, being soluble in aromatic solvents, and thereby capable of undergoing chemical reactions. Other examples of covalent network solids are quartz $(SiO_2)_x$, $(BN)_x$, $(ZnS)_x$, $(HgS)_x$, and the two allotropes of selenium—grey (Se_∞) and red $(Se_8)_x$. It should be noted that although the discrete units of the extended solid are covalently bound, there may also be layers that are held together by weaker intermolecular forces such as van der Waal interactions (Fig. 2.3).

2.2.4 Molecular Solids

This class of solids features discrete molecules that are held together by rather weak intermolecular forces such as dipole–dipole, London dispersion, or hydrogen bonding. Since these forces are much weaker than ionic or metallic bonding interactions, molecular solids are usually characterized by low melting points. Examples

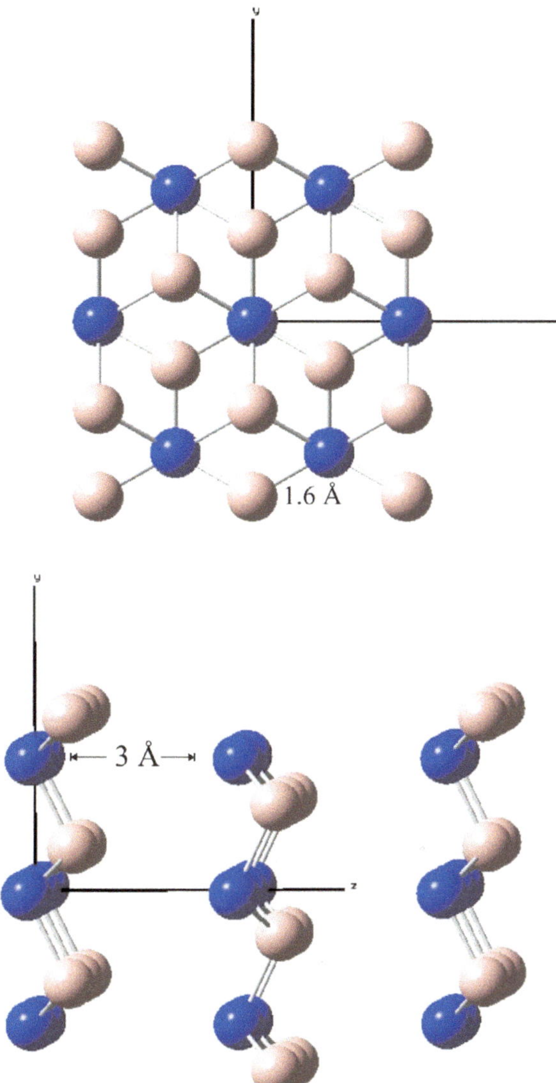

Fig. 2.3 Top and side views of hexagonal boron nitride $(BN)_x$, which exhibits the graphitic structure. Shown are the relatively weak van der Waal interactions that hold together adjacent layers/sheets of covalently-bound B-N units

include dry ice (CO_2), ice (H_2O), solid methane (CH_4), sugar (comprising various arrangements/conformations of $C_6H_{12}O_6$ molecules), and polymers. For polymeric materials, the melting points vary significantly depending on the nature of interactions among the polymer subunits. As Chap. 5 will delineate, intermolecular forces also greatly affect many other physical properties of these materials.

Molecular solids may exhibit either crystalline or amorphous structures, depending on the complexity of the individual molecules comprising the bulk material. As with all solids, the more complex the subunits are, the harder it is for them to organize themselves in a repeatable fashion, resulting in an amorphous structure. Unlike purely ionic solids, molecular compounds may be soluble in either nonpolar or polar solvents, as long as the solvent polarity between solute and solvent is matched ("like dissolves like").

Both dipole–dipole and London dispersion forces are subclasses of van der Waal interactions. When two polar molecules approach one another, a natural attraction known as dipole–dipole forces is created between oppositely charged ends. The relative intensity of dipole–dipole forces may be represented by (2.3):

$$-\frac{2}{3}\left(\frac{\mu_1^2\mu_2^2}{4\pi\varepsilon_o}\right)\frac{1}{k\mathrm{Tr}^6} \tag{2.3}$$

where μ_1 and μ_2 are the molecular dipole moments (Debyes), r is the average distance of separation (Å), T is the temperature (K), and k is the Boltzmann constant (1.38065×10^{-23} J K^{-1}).

In addition to the mutual attraction between polar molecules, there may also be an interaction between solute and liquid/gaseous solvent molecules. In highly polar solvents such as water or alcohols, a dense shell of solvent molecules will surround polar molecules. Although this solute/solvent interaction assists in the dissolution of molecules in the solvent, the dipole–dipole interactions between individual solute molecules are suppressed.

In contrast to dipole–dipole forces, London dispersion interactions are much weaker in nature since they involve nonpolar molecules that do not possess permanent dipole moments. The only modes for molecular attraction are through *polarization* of electrons, which leads to the creation of small dipole–dipole interactions and mutual attractive forces. Since electron polarization occurs much more readily for electrons farther from the nucleus, this effect is more pronounced for molecules that are larger with a greater number of electrons, especially positioned on atoms with a high atomic number, consisting of more diffuse orbitals. These "induced dipole" forces are responsible for the liquefaction of gases such as He and Ar at low temperatures and pressures. The relative strength of London dispersion forces is described by (2.4):

$$-\frac{3}{2}\left(\frac{\mathrm{I}_1\mathrm{I}_2}{\mathrm{I}_1+\mathrm{I}_2}\right)\frac{\alpha_1\alpha_2}{\mathrm{r}^6} \tag{2.4}$$

where I_1 and I_2 are the ionization potentials of the molecules, and α_1 and α_1 are the polarizabilities of the molecules.

If both polar and nonpolar molecules are present, a dipole-induced dipole interaction may occur. For this situation, the strength of association may be represented by (2.5), which is dependent on both the dipole moment of the polar molecule, and the polarizability of the nonpolar component. Once again, this relation does not include the interactions between polar solute molecules and solvent molecules.

$$-2\left(\frac{\mu_1^2}{4\pi\varepsilon_o}\right)\frac{\alpha_2}{r^6} \tag{2.5}$$

Hydrogen bonding may be considered a special case of dipole–dipole forces, where there exist relatively strong interactions between extremely polar molecules. This interaction is often designated by A–H–B, where a hydrogen bond is formed between a Lewis basic group (B) and the hydrogen covalently bonded to an electronegative group (A). In general, the magnitudes of these interactions (*ca.* 12–30 kJ mol^{-1}) are much less than a covalent bond. However, the linear $[F–H–F]^-$ anion present in concentrated hydrofluoric acid has a bond energy of *ca.* 50 kJ mol^{-1}, representing the strongest hydrogen bond ever discovered. The degree of hydrogen bonding has an influence on many physical properties of a compound such as melting and boiling points, dielectric constants, vapor pressure, thermal conductivity, index of refraction, viscosity, and solubility behavior.

The potential energy between pairs of non-bonded neutral atoms or molecules as a function of internuclear/intermolecular separation may be described as a combination of attraction and repulsion terms, referred to as the *Lennard–Jones potential* (2.6):

$$V(r) = 4\varepsilon\left[\left(\frac{\sigma}{r}\right)^{12} - \left(\frac{\sigma}{r}\right)^{6}\right] \tag{2.6}$$

where V(r) is the potential energy as a function of atomic separation, r; σ is the Lennard–Jones size parameter, the intermolecular separation for which the energy is zero ($\sigma = 2^{-1/6}r_o$, where r_o is the intermolecular separation at minimum energy), and ε is the Lennard–Jones energy constant, the minimum energy of the potential energy well.

At farther atomic separations, electron-nuclei attractive forces will dominate; however, as the atoms closely approach one another, there will be increasing mutual repulsion among negatively charged electrons and positively charged nuclei, resulting in an exponential increase in the total potential energy (Fig. 2.4). At an intermediate atomic separation distance, a potential energy well will be generated, corresponding to bond formation between the two atoms. The atomic separation, r_o, at which the force is zero, is referred to as the *equilibrium bond length.* As one would expect, the value of r_o will increase concomitantly with temperature, as atomic motions become greater with increasing thermal energy. The value of the potential energy at the equilibrium bond length, $V(r_o)$, is termed the *binding energy*. For two polar molecules, the long-range electrostatic interactions between molecular dipoles must be accounted for. Hence, another term referred to as the Stockmayer potential must be added to (2.6). The δ term in (2.7) is the polarity correction term, based on the magnitude and directions of the polar dipoles.

$$V(r) = 4\varepsilon\left[\left(\frac{\sigma}{r}\right)^{12} - \left(\frac{\sigma}{r}\right)^{6} + \delta\left(\frac{\sigma}{r}\right)^{3}\right] \tag{2.7}$$

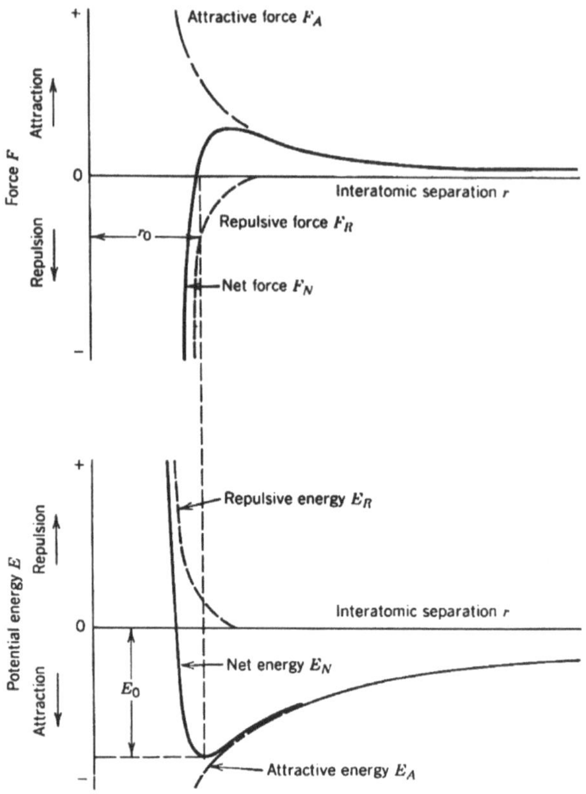

Fig. 2.4 Force and potential energy diagrams for a diatomic molecule, with respect to the interatomic separation. The equilibrium bond distance corresponds to the minimum of the potential energy well and the maximum attractive force between the two atoms. As the atoms are brought even closer together, the interatomic bonding becomes less stable due to exponential increases in repulsive forces and potential energy

2.3 Bonding in Solids II: Band Theory

Band theory is used to describe the electronic and optical properties of solids based on the quantum mechanical wave function of an electron within a large periodic lattice of atoms or molecules. Chemists are familiar with traditional molecular orbital diagrams that are comprised of linear combinations of atomic orbitals (L.C.A.O.–M.O. theory). For diatomic molecules, the overlap of two energetically similar s-orbitals results in σ (bonding) and σ* (antibonding) molecular orbitals. The overlap of p-orbitals may result in σ/σ* (via overlap of p_z atomic orbitals), as well as π/π* orbitals via π_x and π_y interactions. The overlap for metals containing d-orbitals is more complex, which gives rise to σ, π and δ bonding/antibonding molecular orbitals (Fig. 2.5).

A key concept in LCAO–MO bonding theory is the formation of the same number of molecular orbitals as the number of atomic orbitals that are combined (e.g., there are 12 M.O.s formed when two 4 s and two 3d atomic orbitals combine in the Ti_2 molecule, see Fig. 2.6a). As the number of atoms increases to infinity within a crystal lattice, the $\Delta E \approx 0$ between energy levels within bonding and antibonding regions (Fig. 2.6b). This is an application of the *Pauli exclusion principle*, which states for N electrons, there must be N/2 available states to house the electron density.[6]

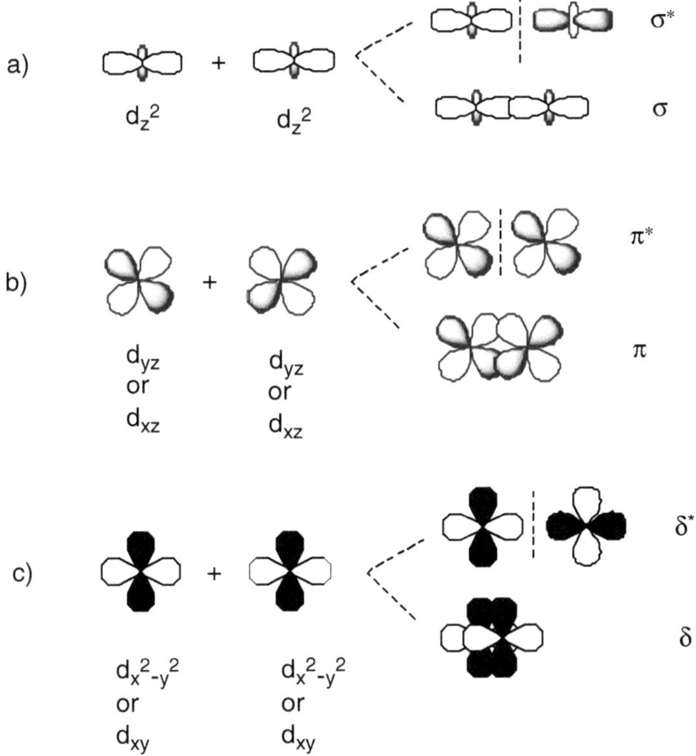

Fig. 2.5 Illustration of d-orbital overlap between adjacent metal atoms in an extended metallic network

The electron-occupied band is known as the *valence band*, whereas the unfilled band is referred to as the *conduction band*. The energy gap, if present, between these levels is known as the *band gap*, E_g (Fig. 2.7a). For metals with a partially filled valence band (e.g., Li, Na, Fe, etc.), or overlapping filled valence and empty conduction bands (e.g., Mg, Ca, Zn, etc.—Fig. 2.7b), there is no bandgap. Since electrical conductivity corresponds to promotion of electrons from valence to conduction bands, this corresponds to metals being conductive even at absolute zero. In contrast, the band gaps for semiconductors and insulators are on the order of 190 kJ mol^{-1} and > 290 kJ mol^{-1}, respectively. Whereas semiconductors become conductive at elevated temperatures due to thermal promotion of electrons between valence and conduction bands, insulators remain non-conductive due to the overwhelming energy gap that must be overcome by valence electrons.

It should be noted that at extremely high temperatures, electrons may be removed entirely from the crystalline lattice, referred to as *thermionic emission*. The minimum energy required to remove an electron from the Fermi level to the vacuum level (a position outside of the solid, Fig. 2.7a) is referred to as its *work function*,[7] and varies depending on the composition and crystallographic orientation of the solid.

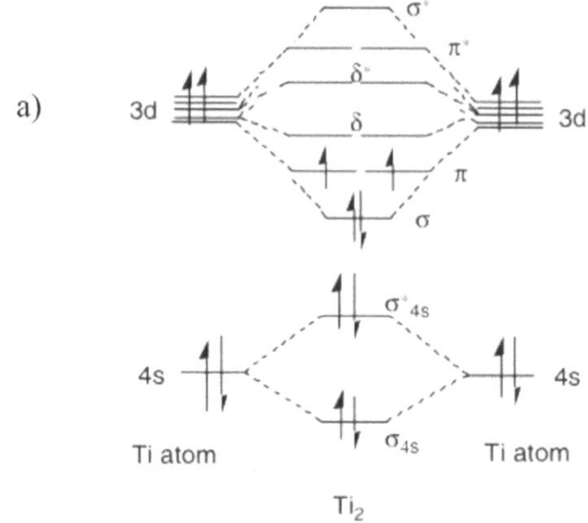

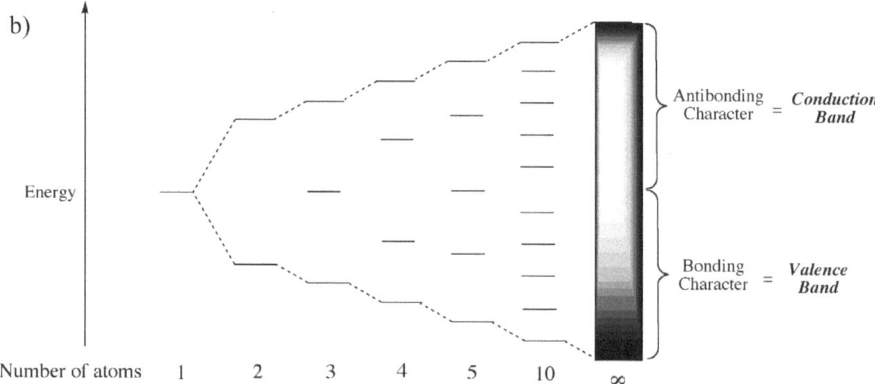

Fig. 2.6 Molecular orbital (MO) diagram for the homonuclear diatomic molecule Ti_2 showing a discrete bandgap between filled and empty MOs. Also shown is the band diagram for a metallic solid, illustrating the continuum between valence and conduction bands (*i.e.*, no bandgap) for increasingly larger numbers of atoms

For instance, some representative work functions for metals and semiconductors are Li: 2.93 eV, Na: 2.36 eV, Al: 4.1 eV, Ag(110): 4.64 eV, Ag(111): 4.74 eV, W: 4.35 eV, Si: 4.7 eV, Ge: 5 eV.

At absolute zero, the highest occupied energy level is referred to as the *Fermi level* (known as the Fermi surface in 3-D), derived from Fermi–Dirac statistics.[8] The Fermi–Dirac distribution function, *f(E)*, describes the probability that a given available energy state will be occupied at a given temperature (2.8):

$$f(E) = \frac{1}{e^{(E - E_F)/RT} + 1} \tag{2.8}$$

where k is the Boltzmann's constant (1.38×10^{-23} J K^{-1}), E is an available energy state, and E_F is the Fermi level.

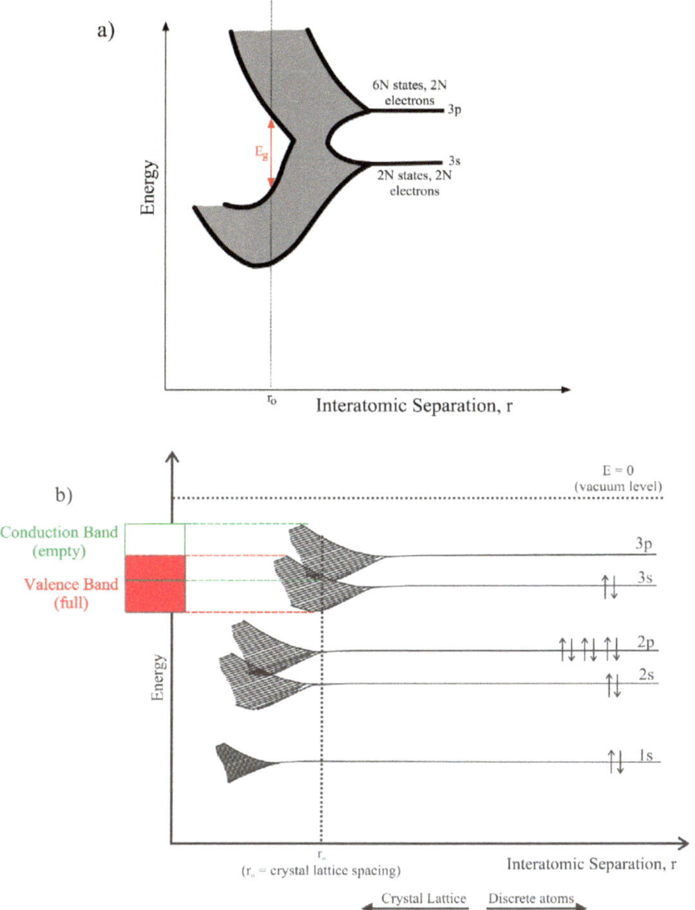

Fig. 2.7 **a** The splitting of 3 s and 3p atomic orbitals of Si_x into bands, showing the presence of a bandgap at the equilibrium bond distance, r_o. **b** The band diagram for Mg metal, showing the formation of valence/conduction bands and zero bandgap through overlap of 3 s/3p atomic orbitals on neighboring Mg atoms

At absolute zero, electrons will fill up all available energy states below the Fermi level, E_F. At low temperatures, all energy states below the Fermi level will have an electron occupation probability of 1, and those above E_F will essentially be zero. However, at elevated temperatures, the probability of having electron density in energy levels above E_F increases (Fig. 2.8).

In metals, the position of the Fermi level provides information regarding the thermal motion of conduction electrons, known as the *electron velocity, v*, through the extended crystal lattice (2.9). For instance, the Fermi energies for copper (7 eV) and gold (5.5 eV) correspond to velocities of 1.6×10^6 m s^{-1} and 1.4×10^6 m s^{-1}, respectively. However, it should be noted that the average speed of electron flow,

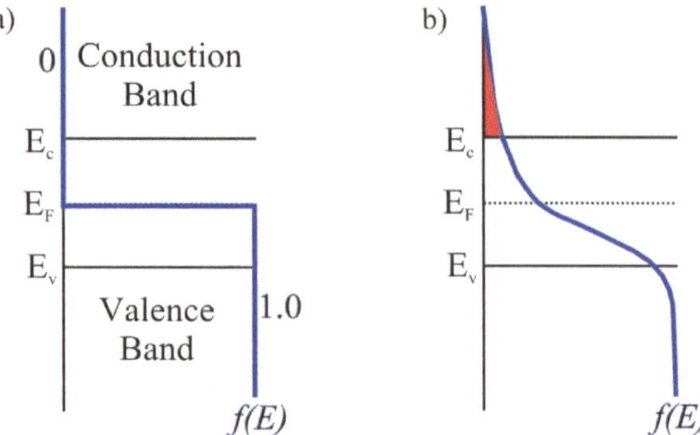

Fig. 2.8 Schematic of the Fermi–Dirac probability function at 0 K (left), and at T > 0 K (right) showing the promotion of electron density from the valence to conduction bands

known as the *drift velocity,* within electrical wires is much less, typically on the order of 100 μm s^{-1} (i.e., 6 mm min^{-1}) for DC voltage—much slower than one would think![9]

$$\nu = \sqrt{\frac{2E_F}{m_e}} \tag{2.9}$$

where m_e is the mass of an electron mass, 9.1066 × 10^{-28} g

As one might imagine, although the Fermi function, *f(E)*, may predict a finite probability for electrons to populate the conduction band, there may not be available empty energy levels to accommodate the electrons. Hence, one must also consider the *density of states* (DOS), or the number of available energy states per unit volume in an energy interval (Fig. 2.9). To determine the conduction electron population of a solid, one would simply multiply *f(E)* by the density of states, *g(E)*. For a metal, the DOS starts at the bottom of the valence band and fills to the Fermi level; since the valence and conduction bands overlap, the Fermi level lies within the conduction band and there is electrical conductivity at 0 K (Fig. 2.10a). In contrast, the DOS for conduction electrons in semiconductors begins at the top of the bandgap (Fig. 2.10b), resulting in appreciable electrical conductivity only at elevated temperatures. It should be noted that although the Fermi level of metals is on the order of 2–11 eV, thermal energy (kT) is only 0.026 eV at 300 K. Hence, only a tiny fraction of electrons (< 0.5%) that are positioned at energy levels within kT of the Fermi level may participate in electrical or thermal conductivity of the solid.

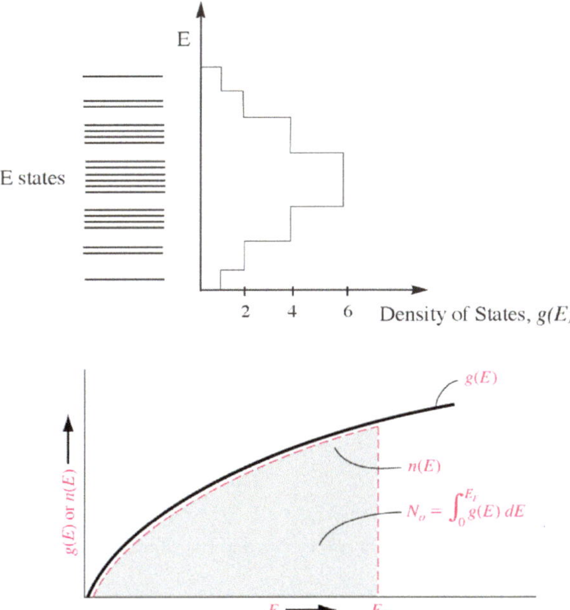

Fig. 2.9 Top: Schematic of the density of states (DOS)—the number of available energy states per unit volume in an energy interval. Bottom: Density of states and electrons in a continuous energy system at T = 0 K. Reproduced with permission from Neamen, D. A. *Semiconductor Physics and Devices*, 3rd ed., McGraw-Hill: New York, 2003. © 2003 McGraw-Hill

2.4 The Crystalline State

Single crystals are composed of an infinite array of ions, atoms, or molecules, known as a *crystal lattice*. The strength of the interactions between these components is known as the *lattice energy* and is dependent on the nature and degree of interactions between adjacent species. For example, the extremely high melting points of salts are directly associated with the strength of the ionic bonds between adjacent ions. For molecular species, it is the degree of intermolecular interactions such as van der Waal and hydrogen bonding forces that controls the lattice energy. Ionic and covalent crystals have similar lattice energies (*ca.* 700–900 kJ mol^{-1}), followed by metallic crystals (*ca.* 400–500 kJ mol^{-1}). By contrast, molecular crystals such as solid carbon dioxide are much more readily broken apart (*ca.* 5–20 kJ mol^{-1}); a consequence of the weak van der Waal interactions consisting between the discrete molecules that comprise the lattice.

The ions, molecules, or atoms pack in an arrangement that minimizes the total free energy of the crystal lattice. For ionic crystals, there is an overall balance of charge among all ions throughout the lattice. Non-ionic crystals exhibit a greater variety of packing interactions between constituent molecules. One of the most influential forces found in these lattices is hydrogen bonding. The molecules will pack in such a

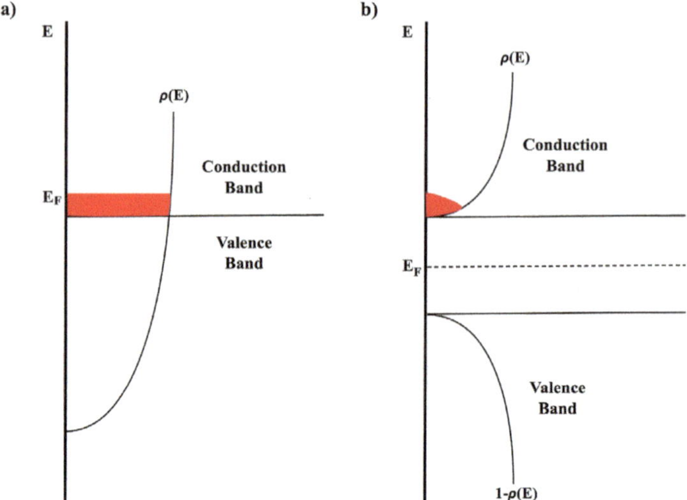

Fig. 2.10 Bands and Fermi level, E_F, for **a** a metal, and **b** a semiconductor

manner to balance the number of hydrogen bond donor and acceptor groups. Often, a residual polar solvent, capable of participating in hydrogen bonding, will play an important role in the observed packing arrangement. Depending on the polarity of the encapsulated solvent, a variety of arrangements of molecules will be observed in the crystal lattice, with hydrophobic and hydrophilic groups being preferentially aligned with respect to each other and the solvent.

Depending on how strongly a solvent is contained within the crystal lattice, sometimes the encapsulated solvent is lost, an occurrence referred to as *efflorescence*. By contrast, if the solid contains ions with a high charge density (high charge/size ratio) and is soluble in water, the crystals will readily adsorb water from the atmosphere and may even be transformed to a solution. An example of such a *deliquescent* crystal is calcium chloride, which is employed as a dehydrating agent for removal of moisture from a flow of inert gases.

The overall shape or form of a crystal is known as the *morphology*. Often, there is more than one crystalline form of the same substance. Each form is known as a *polymorph*, differing in both the arrangement of constituents as well as unit cell dimensions.[10] Although polymorphs differ in both the shape and size of the unit cell, most compounds may exhibit this behavior under appropriate experimental conditions. Common reasons for a varying crystal structure are similar ionic ratios for anions and cations in ionic crystals, or variations in temperature or pressure during crystal growth. These latter effects alter the amount of disorder within the crystal lattice, allowing for the migration of atoms/ions/molecules into lattice positions that are thermodynamically disfavored at lower temperatures and/or pressures.[11]

Most often, the energy for the interconversion between polymorphs is small, resulting in phase changes that occur after only moderate changes in temperature or pressure. In general, exposing a crystal to an applied pressure forces neighboring atoms closer together, causing a decrease in the volume of the unit cell, and an increase in the coordination number of individual atoms. For instance, silicon is transformed from a 4-coordinate polymorph at ambient pressure to a variety of higher-coordinate phases at pressures at elevated pressures.[12]

If the solid is an element, polymorphs are known as *allotropes*. One of the best-known examples for elemental polymorphism (allotropy) is observed for Group 16 elements (chalcogens). For instance, consider the allotropism exhibited by sulfur. The most stable form under ambient conditions is cyclooctasulfur, a yellow powder consisting of arrangements of 8-membered rings (Fig. 2.11a). At elevated temperatures, the S_8 rings begin to open and cross-link with one another, resulting in a highly viscous reddish solution (Fig. 2.11b). If this solution is quickly quenched by pouring into cold water, a hard, bright yellow solid will be formed, composed of infinite chains of disordered sulfur atoms (Fig. 2.11c). This latter form is known as *catenasulfur*, or "plastic sulfur"; however, since it is not thermodynamically stable at room temperature, it will slowly convert back to the powdery S_8 form. The heavier Group 16 congeners also possess this structural diversity, with the relative thermodynamic stability of a particular allotrope being governed by the structure with the lowest overall free energy.

If the chemical contents of a polymorph are different than other forms, it is designated as a *pseudopolymorph.*[13] This often occurs due to the presence of differing amounts of solvent (e.g., clathrates, host–guest, or inclusion compounds), which will alter physical properties of the crystals such as melting points and solubilities. Polymorphism and pseudopolymorphism may be observed when different experimental conditions are used for synthesis. For example, if crystals are grown by sublimation, changing the temperature will often yield different crystal structures, possibly even metastable phases that are kinetically favored.

When different compounds yield almost identical crystals, the forms are referred to as *isomorphs*. The word "almost" is indicated here, as isomorphs are not exactly the same; although the arrangement of atoms/ions in the lattices are identical, one or more of the atoms in the lattice have been replaced with another component. For example, alums of the general formula $(M)_2(SO_4) \cdot (M)_2(SO_4)_3 \cdot 24\ H_2O$ may crystallize as isomorphs where one of the monovalent or trivalent metals is substituted with another.

The conversion between polymorphs may be observed by differential scanning calorimetry (DSC), which shows peaks corresponding to endothermic (melting) or exothermic ((re)crystallization) events associated with structural changes. Figure 2.12 shows the differential scanning calorimetry (DSC) curves of two polymorphs (designated as $\mathbf{I}_R$ and $\mathbf{II}_o$) of picryltoluidine, obtained via crystallization from different solvents. Whereas form $\mathbf{I}_R$ exhibits an endothermic melting peak at 166 °C, form $\mathbf{II}_o$ melts at 163 °C. Further, the enthalpies of fusion for $\mathbf{I}_R$ and $\mathbf{II}_o$ forms are

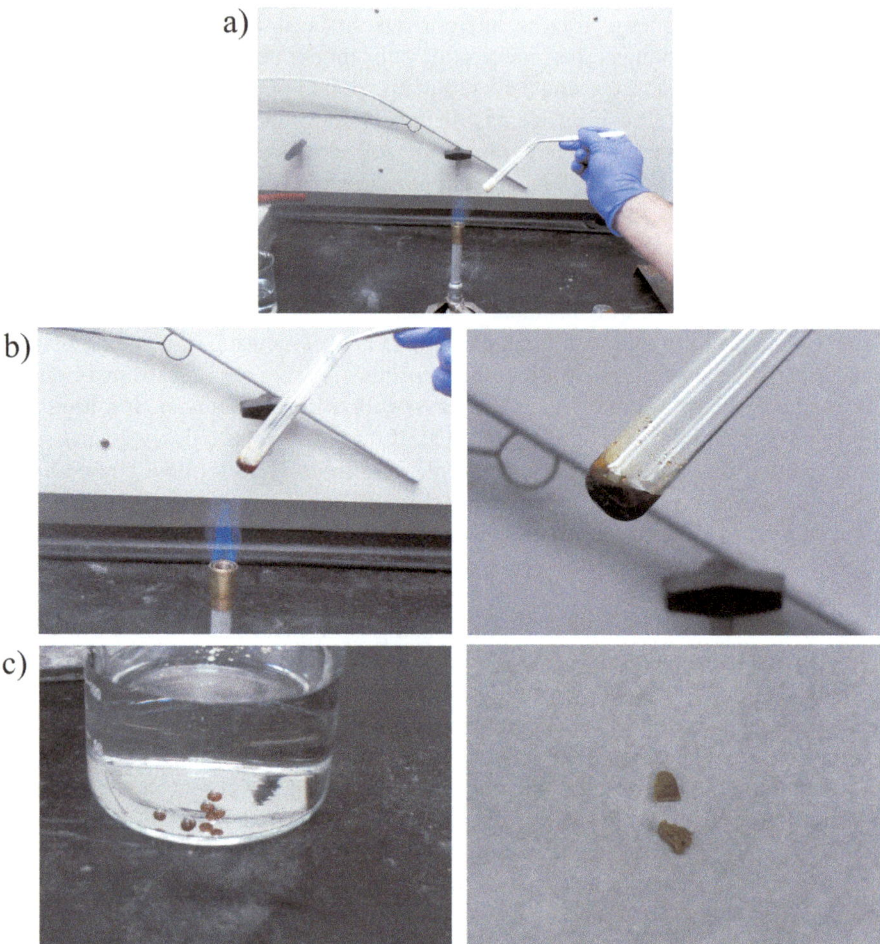

Fig. 2.11 Illustration of allotropic transformations exhibited by elemental sulfur. Shown are: **a** cyclooctasulfur (S_8) at room temperature/pressure, **b** breaking apart of discrete S_8 rings at elevated temperature to form a viscous liquid, and **c** formation of $S\infty$ (catenasulfur or "plastic sulfur") via quenching in cold water. S_8 allotrope

31.3 kJ mol^{-1} and 28.6 kJ mol^{-1}, respectively. The *heat of fusion rule* states that if the higher melting form has the lower enthalpy of fusion, the two forms are *enantiotropic* (i.e., two forms have differing stabilities at specific temperature ranges). On the other hand, if the lower melting form has the lower enthalpy of fusion (as exhibited here), the two forms are *monotropic* (i.e., one form is more stable at all temperatures).[14]

The rate of a polymorphic phase transition depends on nucleation and growth processes, which are related to the mobility of atoms/molecules in the solid state.

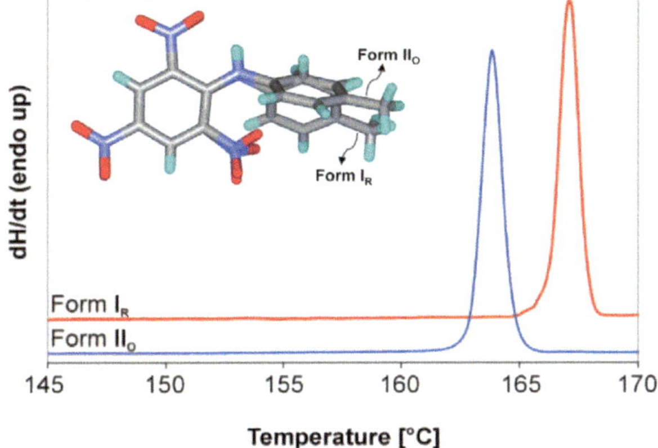

Fig. 2.12 Differential scanning calorimetry (DSC) curves of two polymorphs of picryltoluidine. Crystals of form $\mathbf{I}_{\mathrm{R}}$ were obtained from methanol, whereas form $\mathbf{II}_{\mathrm{o}}$ was obtained from an acetone–water solution. Whereas form $\mathbf{I}_{\mathrm{R}}$ exhibits an endothermic melting peak at 166 °C, form $\mathbf{II}_{\mathrm{o}}$ melts at 163 °C. Further, the enthalpies of fusion for $\mathbf{I}_{\mathrm{R}}$ and $\mathbf{II}_{\mathrm{o}}$ forms are determined as 31.3 kJ mol^{-1} and 28.6 kJ mol^{-1}, respectively. The *heat of fusion rule* states that if the higher melting form has the lower enthalpy of fusion, the two forms are *enantiotropic* (*i.e.*, two forms have differing stabilities at specific temperature ranges). On the other hand, if the lower melting form has the lower enthalpy of fusion (as exhibited here), the two forms are *monotropic* (*i.e.*, one form is more stable at all temperatures). Reproduced with permission from *Cryst. Growth Des.* **2008**, *8*, 1977. © 2008 American Chemical Society

The Avrami equation (2.10) may be applied to describe the degree of transformation, X, as a function of time, t:

$$X(t) = 1 - e^{kt^n} \tag{2.10}$$

where n and k are constants related to the relative importance of nucleation and growth, respectively.

External pressure may also be used to convert one form into another. When this medium is used, a polymorph with higher density will typically result due to the local confinement of lattice species through the externally applied pressure.[15]

2.4.1 Crystal Growth Techniques

Crystal growth involves a phase change from liquid or gas to a solid, such as the precipitation of a solute from solution or the formation of a solid from sublimation of a gas. This occurs through two processes, nucleation and growth, being favored by using supersaturated solutions and/or temperature gradients. When several molecules in the gas phase or in solution approach each other in appropriate orientations, they form a submicroscopic nucleus upon which additional molecules may adsorb en route toward an ordered extended crystal structure. The probability that a crystal will form depends on the nature and concentration of the solute (i.e., the distance between solute molecules), as well as solvent conditions such as temperature, pH, ionic strength, viscosity, polarity, etc. To grow single crystals suitable for X-ray diffraction analysis, relatively few nuclei should be formed rather than multiple sites of nucleation that will yield microcrystalline solids.

High quality crystals may only be obtained when the rate of deposition onto a nucleation site is kept at a rate sufficiently low to allow oriented growth. A high growth rate may lead to defects in the crystal, forming multi-branched or dendritic crystallites through rapid growth in too many directions. As molecules in the gas-phase or solvent interact with the surface of the growing crystal, they may or may not be preferentially adsorbed. That is, a nucleation site that contains steps, ledges, or surface depressions provides more efficient crystal growth due to the prolonged interaction of suspended molecules with a greater surface area.

Experimentally, the successful growth of single crystals on the order of 0.01–0.1 mm^2 is not trivial and has long been considered as a "black art"! Fig. 2.13 illustrates common techniques that may be applied for crystal growth via sublimation or from solution. Perhaps the most important starting point is with a solution that is filtered to remove most suspended nuclei. For air-sensitive solutions, this requires careful manipulation using filtering cannulas and Schlenk techniques. Most of the solvent is then removed to create a nearly supersaturated solution, and then left undisturbed. Another method that is used to grow single crystals from saturated solutions consists of layering a "nonsolvent" onto the top of the saturated solution. Since the compound of interest is not soluble in the layered nonsolvent, crystal formation may begin at the interfacial region. If the nonsolvent is volatile, vapor diffusion may provide another route for the growth of crystals.

Depending on the nature of the suspended molecules, crystal formation may begin immediately, or may even take months to occur. Many organometallic chemists have been surprised to find large crystals at the bottom of flasks placed in the back of the freezer, after months of observation and concluding that no crystals would ever be realized. Sometimes, fortuitous crystal growth may also be realized from unexpected sources. Crystals may be formed from the incorporation of impurities such as dust or vacuum grease, or from surface scratches on the inside walls of the flask. Surprisingly, NMR tubes are notorious for the formation of large crystals, discovered only as the tubes are about to be cleaned! Quite often, chemists set these tubes aside for weeks after the analysis, creating an undisturbed environment for crystal growth. NMR tubes are long and narrow, suppressing convection currents, and solvents very slowly evaporate through the low-permeable cap. Hence, the overall take-home message for crystal growth is to exercise patience; in the process of impatiently checking for crystal growth, additional nucleation sites are often introduced, resulting in the formation of small crystals. Fortunately, many institutions now possess CCD X-ray diffractometers that allow for enough data to be obtained from even microcrystalline solids, in a fraction of the time required for older 4-circle instruments.

Although much crystallization from a solution is performed at low temperatures, crystal growth may also be performed under high-temperature conditions, such as that described for traditional ceramics processing (Sect. 2.5). The Czochralski (CZ) method for purification of silicon uses a seed crystal on the surface of a molten solid. As the crystal is slowly pulled from the melt, there is extension of the crystal due

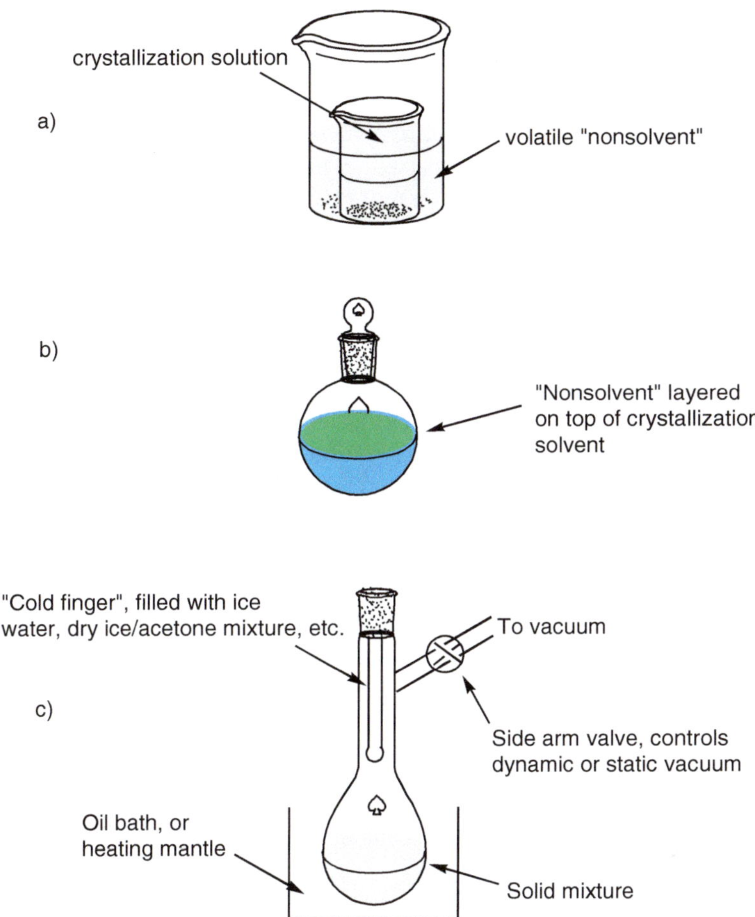

Fig. 2.13 Schematic of three common methods used to grow single crystals. Shown is **a** *diffusion* where vapors from a volatile "nonsolvent" meet the crystallization solvent; **b** *interfacial* where the non-solvent is layered on top of the crystallization solvent; **c** *sublimation* where the solid mixture is heated and the vapors form crystallites on the surface of a cold finger. For this latter technique, the crystallization flask may be opened to vacuum throughout the sublimation process (dynamic vacuum), or closed after maintaining initial vacuum to allow slower crystal growth (static vacuum)

to preferred alignment of other silicon atoms present in the melt (see Chap. 4). The empirical use of seed crystals is frequently used for crystal growth at all temperature regimes. Scratching the inside of the flask, or even using boiling stones, has been successful at inducing crystal formation, through the introduction of nucleation sites. It must be noted that pH is often not an important factor for inorganic or organic crystals; however, for protein crystallization, which is beyond the scope of this book, many such atypical conditions must also be considered.

2.4.2 Crystal Structures

A crystal is composed of an infinite 3-D lattice of repeating units, of which the smallest building block is known as the *asymmetric unit.* When acted upon by crystal symmetry operations such as rotation axes or mirror planes, the asymmetric unit is duplicated to produce the contents of a *unit cell* (Fig. 2.14). For any crystal lattice, it is possible to define an infinite number of unit cells (Fig. 2.15). However, by convention, it is chosen to be a repeatable unit that possesses the same symmetry elements of the bulk crystal, and will generate the entire extended crystal lattice via translations along the unit cell axes. The structure that exhibits these properties while having the smallest possible volume is known as the *primitive* unit cell.

Figure 2.16 provides a schematic of the defining vectors and angles for a unit cell. It is convenient to describe these units as having three vectors (**a**, **b**, and **c**) that may or may not be aligned along the Cartesian axes, based on the values of unit cell angles. Depending on the geometry and volume of the unit cell, there are seven crystal systems that may be generated (Table 2.2).

For simplicity, fractional coordinates are used to describe the lattice positions in terms of crystallographic axes, **a**, **b**, and **c** (Table 2.3). For instance, the fractional coordinates are (1/2, 1/2, 1/2) for an object perfectly in the middle of a unit cell, midway between all three crystallographic axes. To characterize crystallographic planes, integers known as *Miller indices* are used. These numbers are in the format (*hkl*) and correspond to the interception of unit cell vectors at (*a/h*, *b/k*, *c/l*). Figure 2.17 illustrates examples of the (001), (011) and (221) planes; since (*hkl*) intercepts the unit cell at {(*a*, *b*, *c*): (1/*h*, 1/*k*, 1/*l*)}, a zero indicates that the plane is parallel to the particular axis, with no interception along $\pm\infty$.[16] A Miller index with capped integers such as $(01\overline{1})$ indicates that the crystallographic axis is intercepted

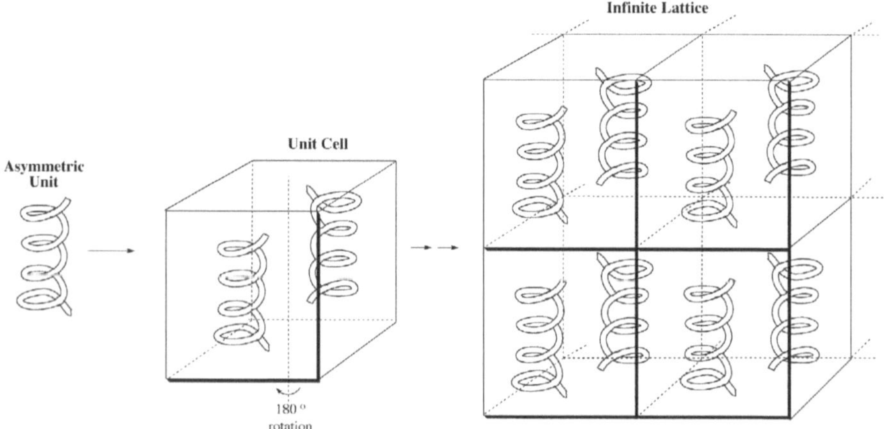

Fig. 2.14 Illustration of the asymmetric unit of a unit cell

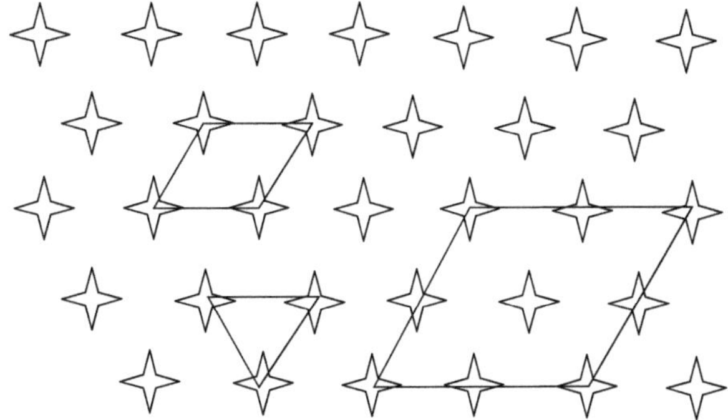

Fig. 2.15 Representation of some unit cell selections from a two-dimensional array. The triangle and smaller rhombus selections are known as "primitive" unit cells, as each contains one object per unit cell (3 × 1/3 for the triangle, 4 × 1/4 for the rhombus). By contrast, there are four objects per unit cell for the larger rhombus (4 × 1/4 for corner objects; 4 × 1/2 for those on the edges; one in the center). When each of these selections is translated along the two-dimensional axes, they reproduce the positions of all objects in the lattice

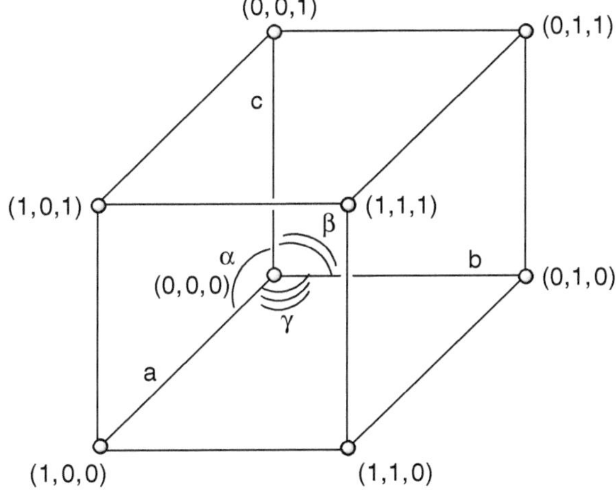

Fig. 2.16 Illustration of definitive axes and angles for a unit cell. The angles and side lengths shown above are not representative of all unit cells; seven types of cell dimensions are possible (see text)

Table 2.2 Unit cell definitions for the seven crystal systems

Crystal system	Unit cell vector lengths	Unit cell vector angles
Cubic (isometric)	$\|\mathbf{a}\| = \|\mathbf{b}\| = \|\mathbf{c}\|$	$\alpha = \beta = \gamma = 90°$
Tetragonal	$\|\mathbf{a}\| = \|\mathbf{b}\| \neq \|\mathbf{c}\|$	$\alpha = \beta = \gamma = 90°$
Orthorhombic	$\|\mathbf{a}\| \neq \|\mathbf{b}\| \neq \|\mathbf{c}\|$	$\alpha = \beta = \gamma = 90°$
Trigonal (rhombohedral)	$\|\mathbf{a}\| = \|\mathbf{b}\| = \|\mathbf{c}\|$	$\alpha = \beta = \gamma \neq 90°, \gamma < 120°$
Hexagonal	$\|\mathbf{a}\| = \|\mathbf{b}\| \neq \|\mathbf{c}\|$	$\alpha = \beta = 90°, \gamma = 120°$
Monoclinic	$\|\mathbf{a}\| \neq \|\mathbf{b}\| \neq \|\mathbf{c}\|$	$\alpha = \gamma = 90°, \beta \neq 90°$
Triclinic	$\|\mathbf{a}\| \neq \|\mathbf{b}\| \neq \|\mathbf{c}\|$	$\alpha \neq 90°, \beta \neq 90°, \gamma \neq 90°$

Table 2.3 Fractional unit cell coordinates of interstitial sites

Crystal system	Octahedral Interstitials	Tetrahedral Interstitials
FCC	(1/2, 0, 0), (0, 1/2, 0), (1, 1/2, 0), (1/2, 1, 0), (0, 0, 1/2), (0, 1, 1/2), (1, 1, 1/2), (1, 0, 1/2), (1/2, 0, 1), (0, 1/2, 1), (1, 1/2, 1), (1/2, 1, 1), (1/2, 1/2, 1/2)	(1/4, 1/4, 1/4), (1/4, 3/4, 1/4), (3/4, 3/4, 1/4), (3/4, 1/4, 1/4), (1/4, 1/4, 3/4), (1/4, 3/4, 3/4), (3/4, 3/4, 3/4), (3/4, 1/4, 3/4)
BCC	(1/2, 0, 0), (1, 1/2, 0), (1/2, 1/2, 0), (1/2, 1, 0), (0, 1/2, 0), (1, 0, 1/2), (1/2, 0, 1/2), (0, 0, 1/2), (1, 1, 1/2), (0, 1, 1/2), (1, 1/2, 1/2), (1/2, 1, 1/2), (0, 1/2, 1/2), (1/2, 0, 1), (1, 1/2, 1), (1/2, 1, 1), (0, 1/2, 1), (1/2, 1/2, 1)	(1/2, 1/4, 0), (1/2, 3/4, 0), (1/4, 1/2, 0), (3/4, 1/2, 0), (1/2, 1/4, 1), (1/2, 3/4, 1), (1/4, 1/2, 1), (3/4, 1/2, 1), (3/4, 0, 1/4), (3/4, 0, 3/4), (1/4, 0, 1/4), (1/4, 0, 3/4), (1, 1/4, 1/4), (1, 3/4, 1/4), (1, 1/4, 3/4), (1, 3/4, 3/4), (0, 1/4, 1/4), (0, 3/4, 1/4), (0, 1/4, 3/4), (0, 3/4, 3/4), (1/4, 1, 1/4), (1/4, 1, 3/4), (3/4, 1, 1/4), (3/4, 1, 3/4)
HCP	(1/3, 2/3, 1/4), (1/3, 2/3, 3/4)	(0, 0, 3/8), (1, 0, 3/8), (0, 1, 3/8), (1, 1, 3/8), (0, 0, 5/8), (1, 0, 5/8), (0, 1, 5/8), (1, 1, 5/8), (2/3, 1/3, 1/8), (2/3, 1/3, 7/8)

in the negative region (interception of the **c** axis at − 1, in this case). Symmetry-equivalent planes are indicated by $\{hkl\}$; for example, $\{100\}$ for a cubic unit cell would have six equivalent planes: (100), $(\bar{1}00)$, , (010), $(00\bar{1})$, (001), and $(00\bar{1})$.

Crystallographic directions correspond to vectors between two lattice points. To properly index directions, the vector should first be (re)positioned so its starting point is at (0, 0, 0) of the unit cell. The vector is then projected onto each of the three crystallographic axes, multiplying/dividing by a common factor to remove fractional terms. The three terms, expressed as integers, are enclosed within square brackets of the form [*abc*] (Fig. 2.16e). For instance, the [211] direction would correspond to the vector $\mathbf{a} + (1/2)\mathbf{b} + (1/2)\mathbf{c}$. Families of equivalent directions are indicated by <abc >. For example, [100], [100], [010], [010], [001], and [001] directions in a cubic crystal are designated as <100> .

For hexagonal crystal planes, a slightly different indexing nomenclature is used relative to cubic crystals. To index a plane in the hexagonal system, four axes are used, called Miller–Bravais indices. In addition to both **a** and **b** axes, another axis is used, that bisects the γ-angle on the hexagonal *basal plane*—designated as $\mathbf{a}_3$ in Fig. 2.18.[17] The use of three axes on the basal plane is required due to the threefold rotation axis present in the lattice. Examples of hexagonal plane indices are shown in Fig. 2.17, of the form (*hkil*), where $i = -h-k$. Lattice directions are indexed analogously to cubic crystals, using only **a**, **b**, and **c** crystallographic axes.

For crystals comprised of metal atoms, the lattice may be simplified as a regular packing of hard spheres. In three dimensions, there are only two ways that spheres may pack: either simple cubic or close-packed. For example, think about dropping marbles into a large beaker partially filled with water. As the marbles settle to the bottom of the container, they will seek their most stable resting position. These positions will be mostly within the voids formed between three spheres (close-packed),

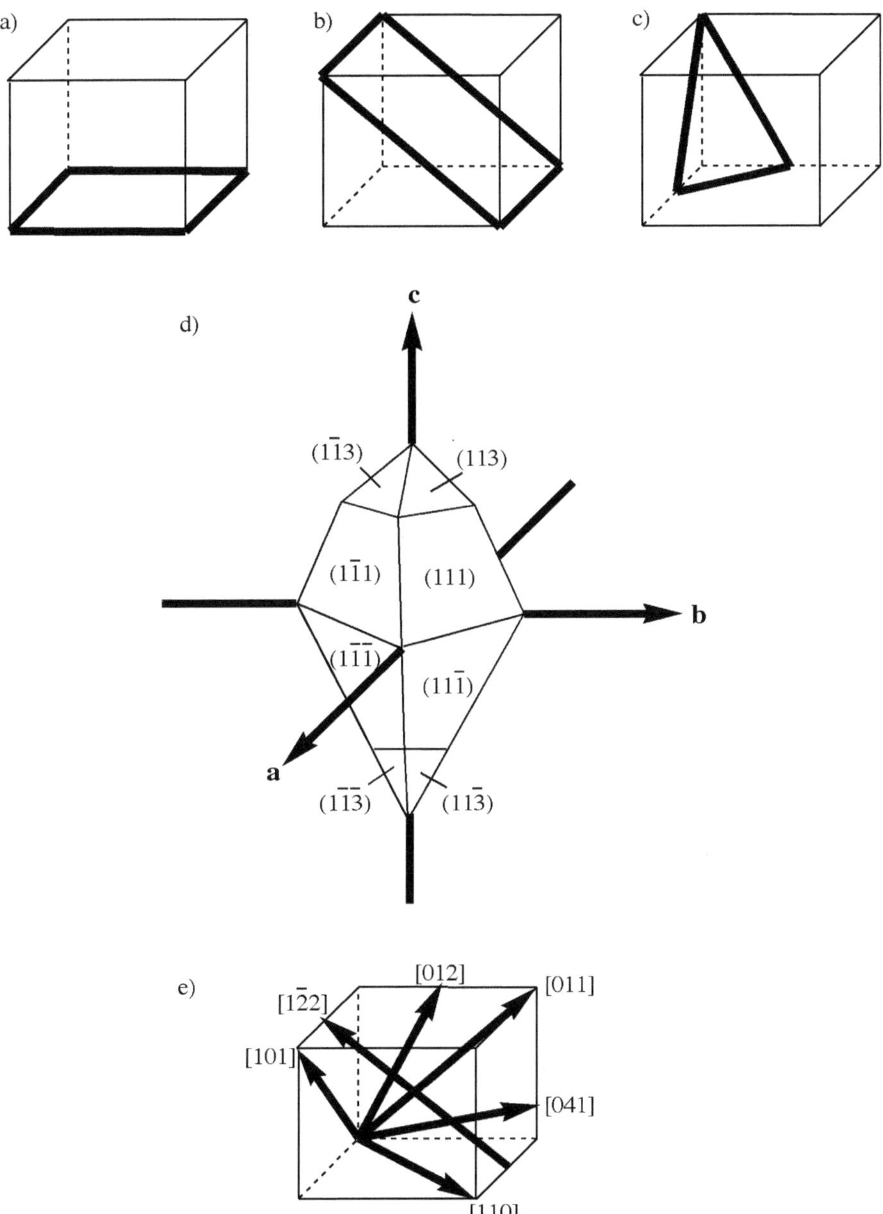

Fig. 2.17 Representations of unit cell planes. Shown are: **a** (100), **b** (011), **c** (221), **d** {111} and **e** {113} families of planes. Also illustrated are examples of unit cell directions

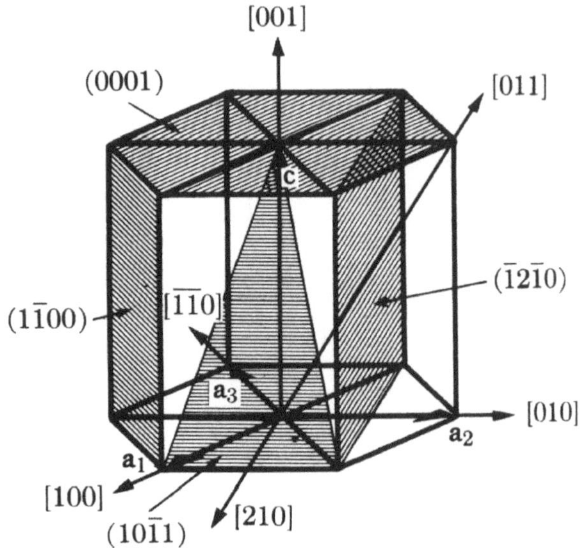

Fig. 2.18 Examples of planes (in parentheses) and directions [in brackets] of a hexagonal unit cell. Planes are designated by the $(h\ k\ i\ l)$ designation, where $i = -h - k$. Reproduced with permission from Cullity, B. D. *Elements of X-Ray Diffraction*, 2nd ed., Addison-Wesley: Reading, MA. © 1978 Prentice-Hall

rather than on top of the crests of individual spheres (simple cubic packing). Whereas the coordination number (i.e., number of nearest neighbors) of individual species for simple cubic packing is 6, the coordination number for close-packing is 12.

Since a simple cubic arrangement would have a significant amount of excess void-space, this presents a much less efficient manner to pack individual spheres. This larger volume of empty space in simple packing would result in a lower density for the solid relative to its close-packing analog. It should also be noted that the sliding of one layer over another would be more preferred for simple cubic relative to close-packing. This corresponds to a greater degree of crystal defects that are possible for these solids.

Although marble migration is only mentioned as an example, this process is not far removed from the crystallization process. In the formation of single crystals, individual ions/atoms/molecules slowly interact with one another and nucleate from thermodynamically favored positions. Sometimes metastable phases may be obtained if one does not allow such preferential migration to occur, through rapid cooling events, for instance.

Figure 2.19 serves to aid in the visualization of sphere packing sequences in three-dimensions. After a few spheres (in A sites) are arranged in a close-packed arrangement, the next layer packs into B sites, formed as voids above three A spheres. For the third layer of spheres, there are two possibilities that may occur. The atoms/ions in the third layer may sit directly above the spheres in the A layer, or may be placed in C sites. If the packing is described by the former case (ABABAB…), the crystal lattice is described as a hexagonal close-packed (hcp) arrangement. By contrast, if the packing scheme occupies all possible sites (ABCABCABC…, or ACBACBACB…), then the arrangement is termed cubic close-packed, or face-centered cubic (fcc). The

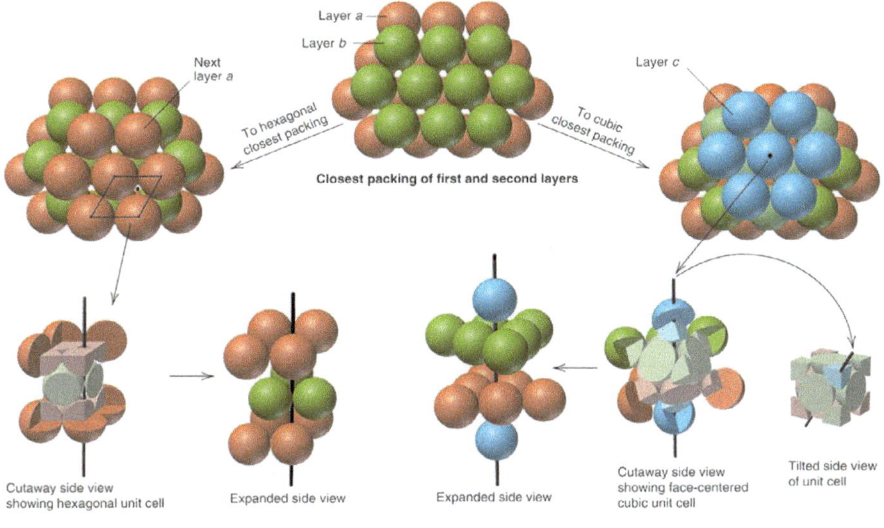

Fig. 2.19 Representations of hexagonal close-packing (hcp) and face-centered cubic (fcc) packing. Reproduced with permission from Silberberg, M. S. *Chemistry: The Molecular Nature of Matter*, 2nd ed., McGraw-Hill: New York, 2000

periodic table shown in Fig. 2.20 demonstrates that the close-packing motifs are the most prevalent for natural forms of the elements, which represents the most efficient means to pack multiple layers of hard spheres.

Since hcp crystals have vacant C sites, there are uniaxial channels through the crystalline solid that influences their physical properties. For example, Be, Mg, and Ti alloys that crystallize in hcp arrangements possess extremely low densities; other hcp metals (e.g., Co) have the possibility for anisotropic magnetic properties.

Although it is easy to visualize a fcc unit cell, with atoms/ions on each corner and on each face of a cube, the ABCABC… packing scheme is more difficult to unravel. Once the unit cell is viewed along the cell corners, perpendicular to the (110) plane, the packing layers are readily observed (Fig. 2.21a). When viewed along a vector perpendicular to the packing planes, the atoms/ions occupying the B and C sites are observed as inverted triangles (Fig. 2.21b). Common examples of this packing scheme are diamond, ZnS (zinc blende form), HgS, CuI, AlSb, BeSe, etc.

Hexagonal close-packing is illustrated in Fig. 2.22, showing a unit cell, with translation to form a hexagonal arrangement with a sixfold rotation axis. Using simple geometry, it may be proven that the coordinates of the B sites are $\{(a, b, c) = (2/3, 1/3, 1/2)$ or $(1/3, 2/3, 1/2)\}$. There are two possibilities due to lattice equivalency; however, once one of these is designated for B sites, the other positions are designated as C sites and remain vacant in hcp crystal lattices. Although the packing sequence is different between hcp and fcc lattices, there is a similarity between these close-packed structures. The atoms in the (111) planes of B and C layers shown in Fig. 2.20

H hcp																	He hcp
Li bcc	Be hcp											B tetrag	C hcp	N hcp	O cubic	F	Ne fcc
Na bcc	Mg hcp											Al fcc	Si diam	P orthor	S orthor	Cl tetrag	Ar fcc
K bcc	Ca fcc	Sc hcp	Ti hcp	V bcc	Cr bcc	Mn bcc	Fe bcc	Co hcp	Ni fcc	Cu fcc	Zn hcp	Ga orthor	Ge diam	As rhomb	Se hcp	Br orthor	Kr fcc
Rb bcc	Sr fcc	Y hcp	Zr hcp	Nb bcc	Mo bcc	Tc hcp	Ru hcp	Rh fcc	Pd fcc	Ag fcc	Cd hcp	In tetrag	Sn tetrag	Sb rhomb	Te hcp	I orthor	Xe fcc
Cs bcc	Ba bcc	La hcp	Hf hcp	Ta bcc	W bcc	Re hcp	Os hcp	Ir fcc	Pt fcc	Au fcc	Hg rhomb	Tl hcp	Pb fcc	Bi rhomb	Po monoc	At	Rn fcc
Fr bcc	Ra bcc	Ac fcc	Rf	Db	Sg	Bh	Hs	Mt									

Ce fcc	Pr hcp	Nd hcp	Pm	Sm rhomb	Eu bcc	Gd hcp	Tb hcp	Dy hcp	Ho hcp	Er hcp	Tm hcp	Yb fcc	Lu hcp
Th fcc	Pa tetrag	U orthor	Np orthor	Pu monoc	Am hcp	Cm	Bk	Cf	Es	Fm	Md	No	Lr

Fig. 2.20 *Preferred* crystal structures of the elements. Shown are the ordinary forms of each element in its solid state (e.g., graphite, α-iron, etc.). The following abbreviations are used: fcc, bcc (body-centered cubic), hcp, cubic (simple cubic), diam (diamond or zinc blende), monoc (monoclinic), rhomb (rhombohedral), tetrag (tetragonal), and orthor (orthorhombic)

are arranged in a hexagonal pattern, analogous to the stacking planes for hcp shown in Fig. 2.21.

Although ccp and hcp arrangements have been shown in detail, there are many other possibilities for the packing of species in a crystal lattice. In 1849, Bravais discovered that there are only 14 possible arrangements of points in three-dimensional space, assuming that each point has an identical environment. Hence, regardless of the identity of the species comprising an infinite crystal lattice, the unit cell must be described by one of these Bravais lattices (Fig. 2.23). Other designations that are not listed are not oversights, but may be further simplified. For instance, a "base-centered tetragonal" unit cell does not appear as one of the 14 Bravais lattices. As shown in Fig. 2.24, a primitive unit cell may be defined from a base-centered tetragonal array—the former is preferentially chosen due to its smaller volume.

A subclass of polymorphism known as *polytypism* is found for one-dimensional close-packed and layered structures such as SiC, CdI_2, ZnS, GaSe, and micas/clay minerals such as kaolins. These crystal structures exhibit differing three-dimensional lattices that vary depending on the stacking order of the two-dimensional sheets that comprise the crystal, known as modular layers. Figure 2.25 illustrates the crystal structures for two common polytypes of SiC, an important material used in ceramic brakes, abrasive machining, and high-temperature electronic applications. Polytypes are defined by the Ramsdell notation < number> < letter> , where the number indicates how many planes are stacked in a discrete unit cell, and the letter indicates the crystal lattice type (C = cubic, H = hexagonal, R = rhombohedral). Amazingly,

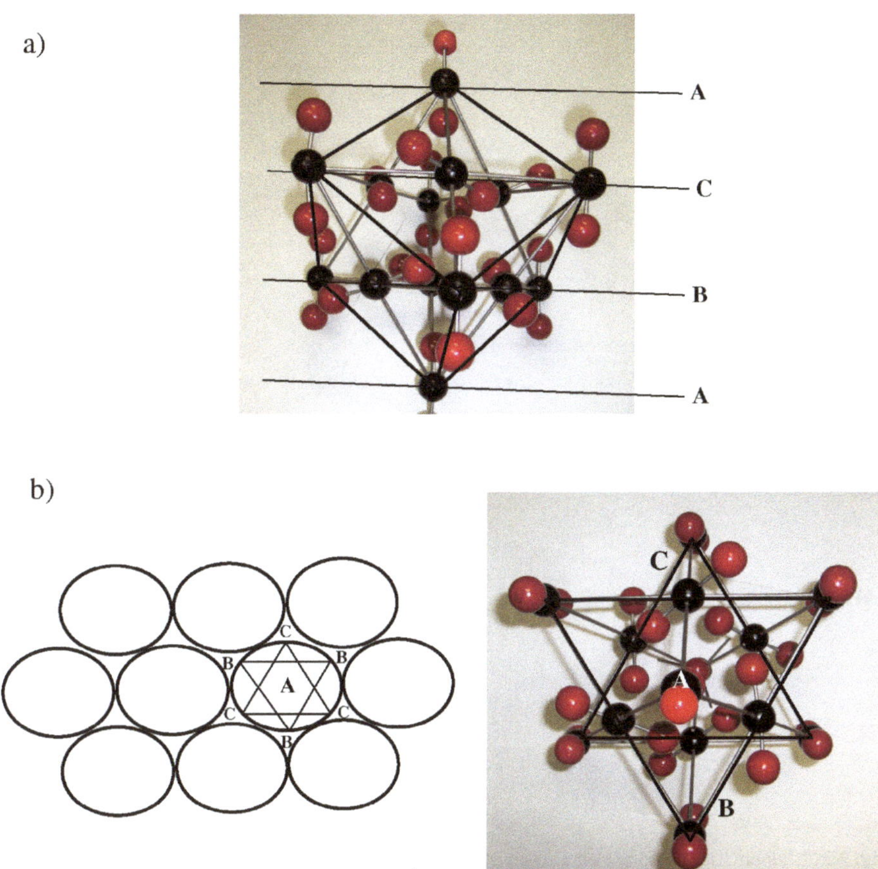

Fig. 2.21 Packing model and illustration for cubic close-packed (face-centered cubic, fcc). Shown is solid carbon dioxide (dry ice), that consists of individual CO_2 molecules that pack in an fcc lattice. The black carbon atoms shown in **a** illustrate a traditional representation of the lattice, consisting of a cube with atoms on each of the faces. The packing layers are shown from two different views; **a** parallel to the ABC... layers (i.e., (111) planes) and **b** perpendicular to these layers

there are over 200 known polytypes of SiC; the most common include 3C (i.e., fcc with ABC... packing), 2H, 4H, 6H, 8H, 9R, 10H, 14H, 15R, 19R, 20H, 21H, and 24R.[18]

2.4.3 Crystal Symmetry and Space Groups

Crystallography employs two terms to describe the symmetry of the crystal lattice: *point groups* and *space groups*. Chemists should be quite familiar with the notion of point groups, as this designation is used to predict molecular reactivities and

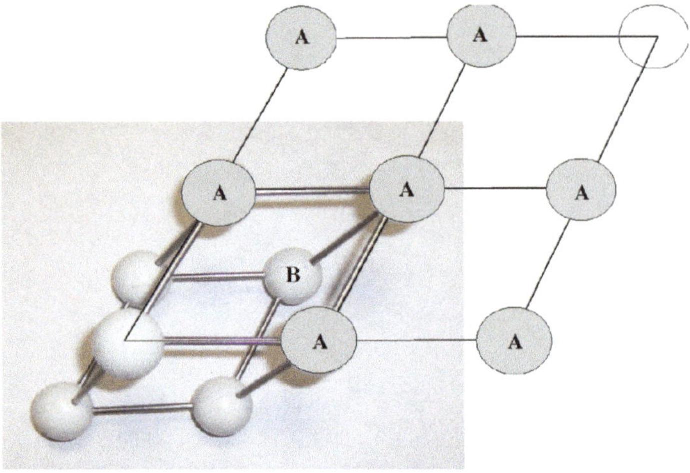

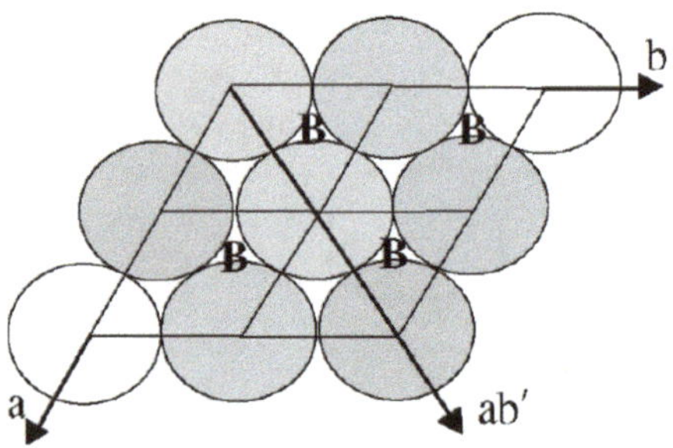

Fig. 2.22 Schematic of the packing in an hcp unit cell, showing equivalent close-packed B sites. The hexagon of A atoms is referred to as the basal plane, as this plane marks the top and bottom of a unit cell. The fractional coordinates of each set of B sites are {(*a*, *b*, *c*): (1/3, 2/3, 1/2) shown, or (2/3, 1/3, 1/2)}

IR/Raman absorption bands. A flowchart for the assignment of a molecular point group is shown in Fig. 2.26. Examples of this notation, given by Schoenflies symbols, are C_{3v} for ammonia and $Os_3(CO)_9(C_6H_6)$, O_h for $[CoF_6]^{3-}$, and D_{3h} for $(CH_3)_8Si_5O_6$ molecules (Fig. 2.27).

By definition, a symmetry operation is an event that results in the transposing of one item into another that is indistinguishable from the original. This operation may take place about a point, line, or plane of symmetry. When a set of symmetry operations is applied to components of the unit cell of a crystal lattice, the resulting symmetry is designated as a *crystallographic point group*. Whereas a molecular point group operation will reproduce an individual molecule, a crystallographic point group operation must leave the entire crystal unchanged.

Illustrative examples of crystallographic symmetry operations are shown in Fig. 2.28. An integer label, n, indicates the regeneration of an equivalent lattice

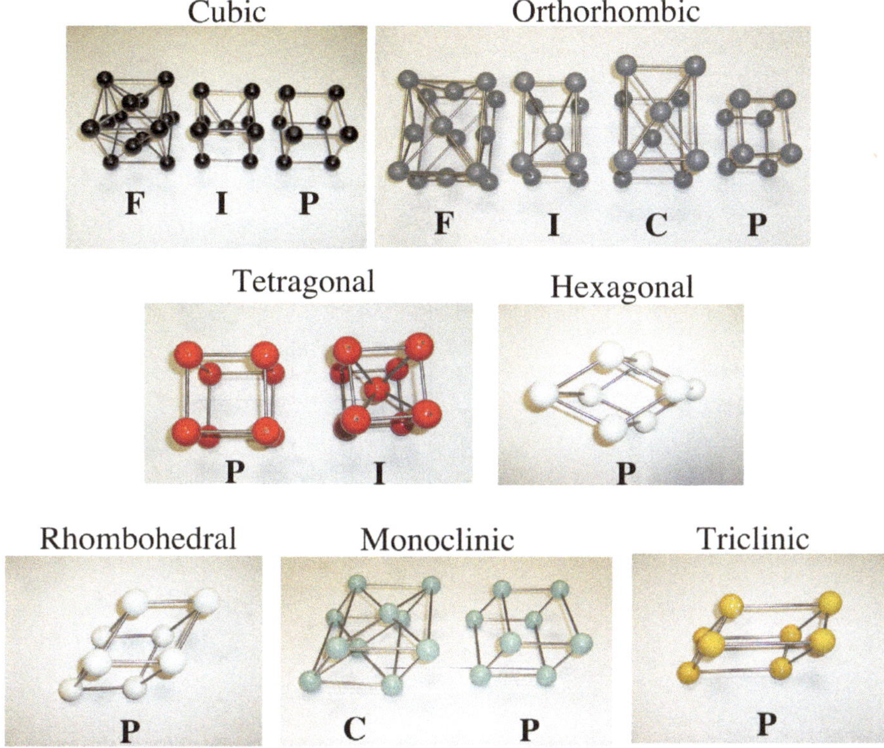

Fig. 2.23 Models of the 14 Bravais lattices. The various types of Bravais centering are given the symbols *P* (primitive/simple), *F* (face-centered), *I* (body-centered), and *C* (base-centered). The primitive rhombohedral Bravais lattice is often given its own symbol, *R*, and corresponds to a primitive unit cell possessing trigonal symmetry

point when an object in the crystal lattice is rotated 360°/n about an axis. A rotation-inversion axis is designated by $\overline{n}$, featuring rotation about an axis (360°/n), followed by reflection in a plane ⊥ to the rotation axis. If individual lattice points are simply reflected through a plane of symmetry, the operation is symbolized by m, denoting the presence of a mirror plane. Although individual atomic and ionic lattice points are not affected by $\overline{n}$ and m operations, molecular lattice points exhibit a change in handedness following these operations (see Fig. 2.41b–c). Point groups that include these operations, known as *improper symmetry operations*, must exclude all chiral molecules, as they would then be superimposible on their mirror images. Two important restrictions apply to crystallographic point group symmetry operations:

1. The symmetry operations must be compatible with infinite translational repeats in a crystal lattice;
2. A symmetry operation cannot induce a higher symmetry than the unit cell possesses.

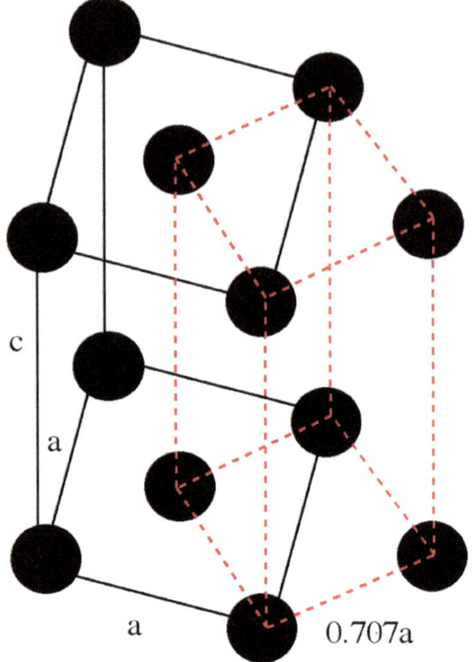

Fig. 2.24 Equivalency of a base-centered tetragonal and a primitive tetragonal unit cell

The point group symmetry describes the non-translational symmetry of the crystal; however, the infinite crystal lattice is generated by translational symmetry (see below). Only 2, 3, 4, and six-fold rotation axes are compatible with translational symmetry, so point groups containing other types of rotation axes (i.e., 5 or $\overline{5}$) are not relevant (Fig. 2.29).[19] Furthermore, since a "linear 3-D crystal" does not exist, the linear point groups are also irrelevant. The remaining point group symmetry operations yield a total of 32 crystallographic point groups, designated by Hermann-Mauguin symbols (Table 2.4). Since these point groups can be deduced from the macroscopic crystal symmetry, they are also referred to as the *32 crystal classes*. For the same reason, the symmetry elements that give rise to the crystal classes are sometimes referred to as external symmetry elements.

One way to visualize the 32 crystallographic point groups is to use stereographic projections (Fig. 2.30). The symbolism used to illustrate rotation axes are as follows:

Two-fold (diad): ⬮
Three-fold (triad): ▲
Four-fold (tetrad): ◆
Six-fold (hexad): ⬢

Mirror planes are indicated as solid lines and may be positioned perpendicular or parallel to the plane of the paper.

As illustrated by the stereographic projections in Fig. 2.29, the 6 mm crystallographic point group is observed to have a primary sixfold axis of rotation, with two

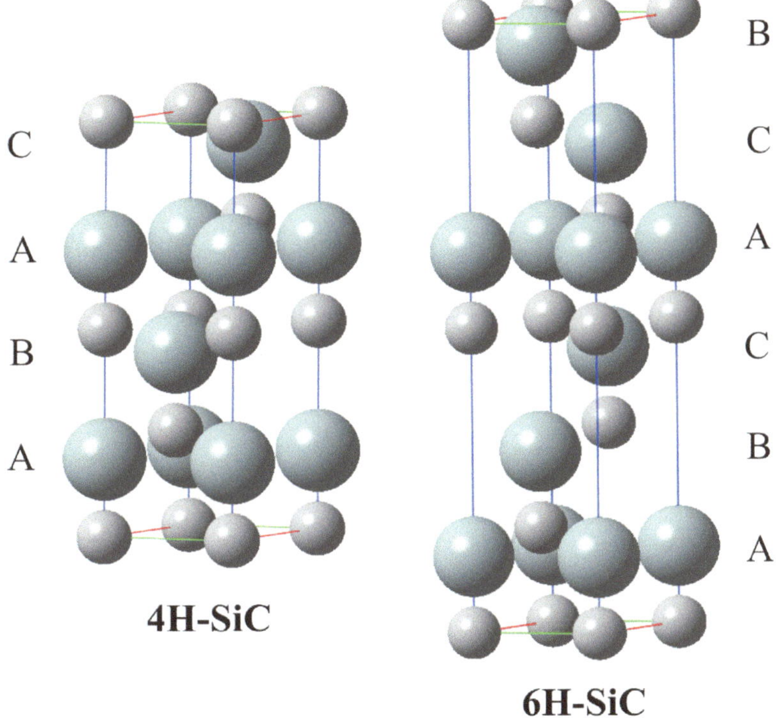

Fig. 2.25 Unit cells for the 4H and 6H polytypes of silicon carbide, as viewed along the $(11\bar{2}0)$ plane. Note that each carbon atom is situated in the center of a Si_4 tetrahedron

mirror planes that are parallel to the rotation axis. In contrast, the 6/mmm point group has a mirror plane perpendicular to the rotation axis (in the plane of the paper), as well as other mirror planes that lie parallel to the primary rotation axis. Carrying out the first two symmetry operations automatically generates the third term of the point group symbol, if present. For instance, for the mm2 point group, the twofold rotation axis is generated by the presence of two mutually perpendicular mirror planes.

The 32 crystallographic point groups are useful to define the contents of a discrete unit cell; however, there must be additional symmetry elements that take into account the translational symmetry of an extended periodic lattice. Accordingly, both *glide planes* and *screw axes* feature translation in addition to mirror and rotation operations, respectively (Fig. 2.31). For glide planes, the translation is designated as *a*, *b*, or *c* if movement is halfway along the **a**, **b**, or **c** unit cell axes, respectively. If the translation is along the diagonals ½ (**a** + **b**), ½ (**a** + **c**), or ½ (**b** + **c**), the glide plane is designated as *n*. Lastly, if the translation is along ¼ (**a** + **b**), ¼ (**a** + **c**), or ¼ (**b** + **c**), the glide plane is given the symbol *d*.[20] For screw axes, the nomenclature is of the form n_x, indicating a 360°/n rotation, followed by a *x/n* translation along one of the unit cell

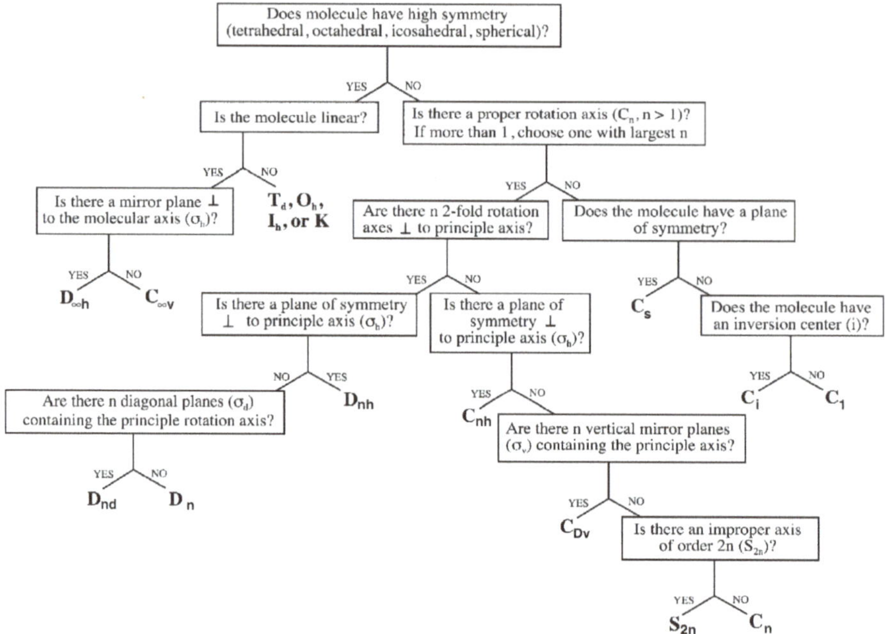

Fig. 2.26 Scheme for determining the point group symmetry of a molecule

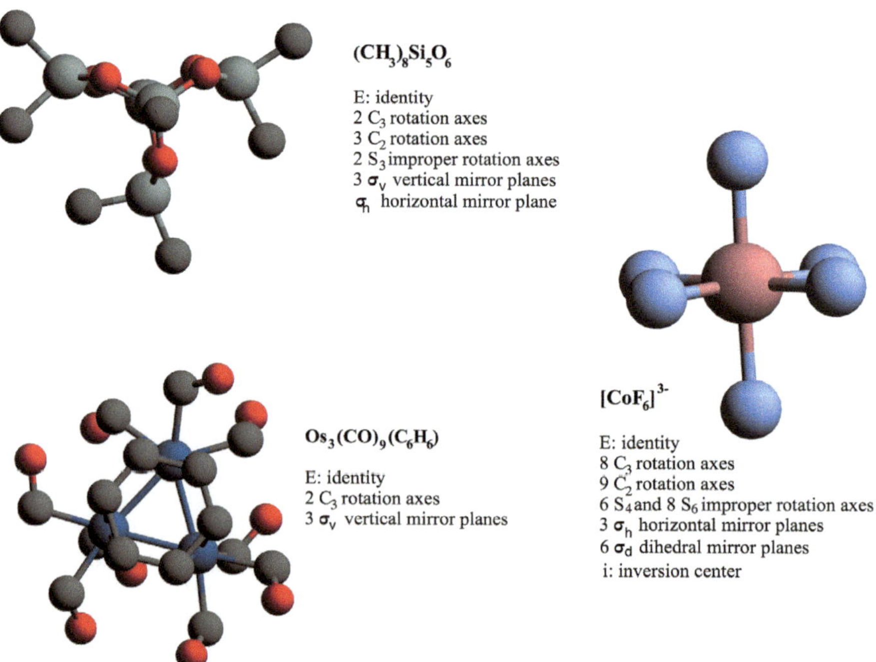

Fig. 2.27 Examples of symmetry group elements and molecular point group assignments

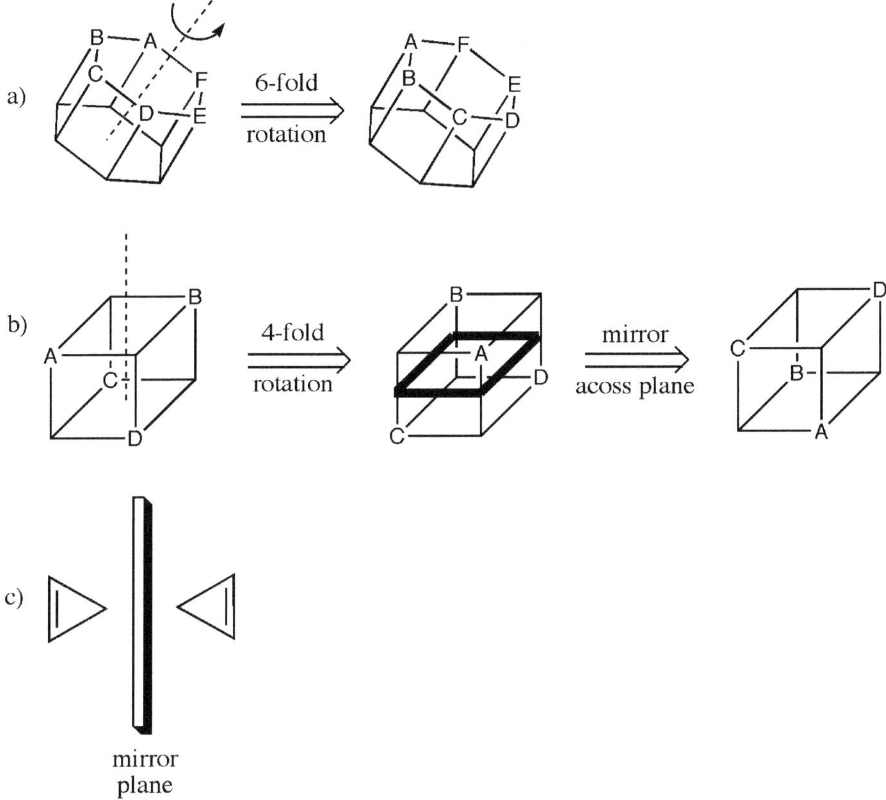

Fig. 2.28 Illustrations of: **a** 6, **b** $\bar{4}$, , and **c** m crystallographic symmetry operations

vectors, **a**, **b**, or **c**. For example, the 6_1 and 6_3 screw axes would imply six-fold axes of rotation followed by 1/6 and ½ translations, respectively.

It is noteworthy to point out that two sequential screw-axis or glide-plane operations will yield the original object that has been translated along one of the unit cell vectors. For example, a 6_3 axis yields an identical orientation of the molecule only after 6 repeated applications—3 unit cells away (i.e., $6 \times 1/2 = 3$). Since glide planes feature a mirror plane prior to translation, the first operation will cause a change in handedness of the molecule. By contrast, screw axis operations do not alter the stereoisomerism of the molecule.

Both glide and screw axes are not point group operations because they involve translations. That is, one cannot distinguish between analogous rotation and screw axes, or between glide and mirror planes, by simply looking at the crystal faces. You may notice that of the symmetry elements discussed, both glide planes and screw axes are absent from the list of point group symbols listed in Table 2.5. For the purposes of determining the crystallographic point group, screw axes are treated as rotation axes (e.g., $6_3 \equiv 6$), and glide planes treated as mirror planes (e.g., $b \equiv m$).

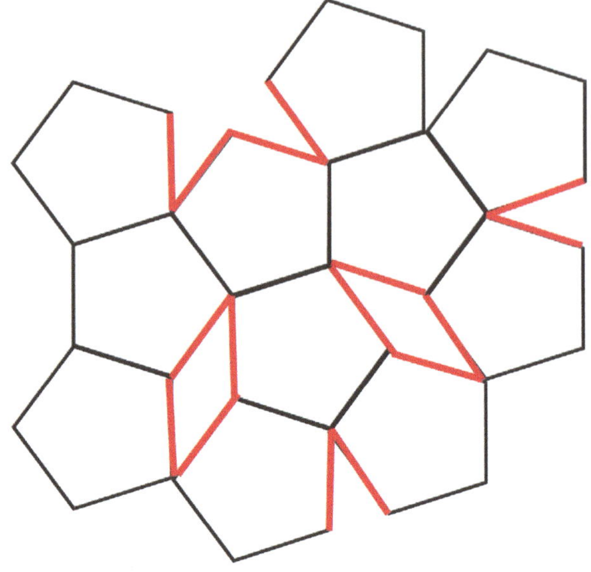

Fig. 2.29 The translational incompatibility of a five-fold rotation axis, which leaves voidspaces among the lattice objects

Table 2.4 The 32 crystallographic point groups

Crystal system (Bravais lattices) {Defining Symmetry Elements}	Crystallographic point groups[a] (molecular point groups[b])
Cubic (*P*, *I*, *F*) {Four ⊥ three-fold rotation axes}	23, *m*3, 432, $\bar{4}3m$, *m*3*m* (T, T_h, O, T_d, O_h)
Tetragonal (*P*, *I*) {One four-fold rotation axis}	4, $\bar{4}3$ *m*, 4/*m*, 422, 4 *mm*, $\bar{4}2$ m, 4/*mmm* (C_4, S_4, C_{4h}, D_4, C_{4v}, D_{2d}, D_{4h})
Orthorhombic (*P*, *C*, *I*, *F*) {Three ⊥ two-fold rotation axes or three ⊥ m's}	222, *mm*2, *mmm* (D_2, C_{2v}, D_{2h})
Trigonal/Rhombohedral (*P*) {One three-fold rotation axis}	3, $\bar{3}$, 32, 3 *m*, $\bar{3}m$(C_3, C_{3i}, D_3, C_{3v}, D_{3d})
Hexagonal (*P*) {One six-fold rotation axis}	6, $\bar{6}$, 6/*m*, 622, 6 *mm*, $\bar{6}m2$, 6/*mmm* (C_6, C_{3h}, C_{6h}, D_6, C_{6v}, D_{3h}, D_{6h})
Monoclinic (*P*, *C*) {One two-fold rotation axis}	2, *m*, 2/*m* (C_2, C_s, C_{2h})
Triclinic (*P*) {N/A}	1, $\bar{1}$ (C_1, C_i)

[a] The H–M symbolism derived from crystal symmetry operations

[b] The analogous Schoenflies symbolism derived from molecular symmetry operations

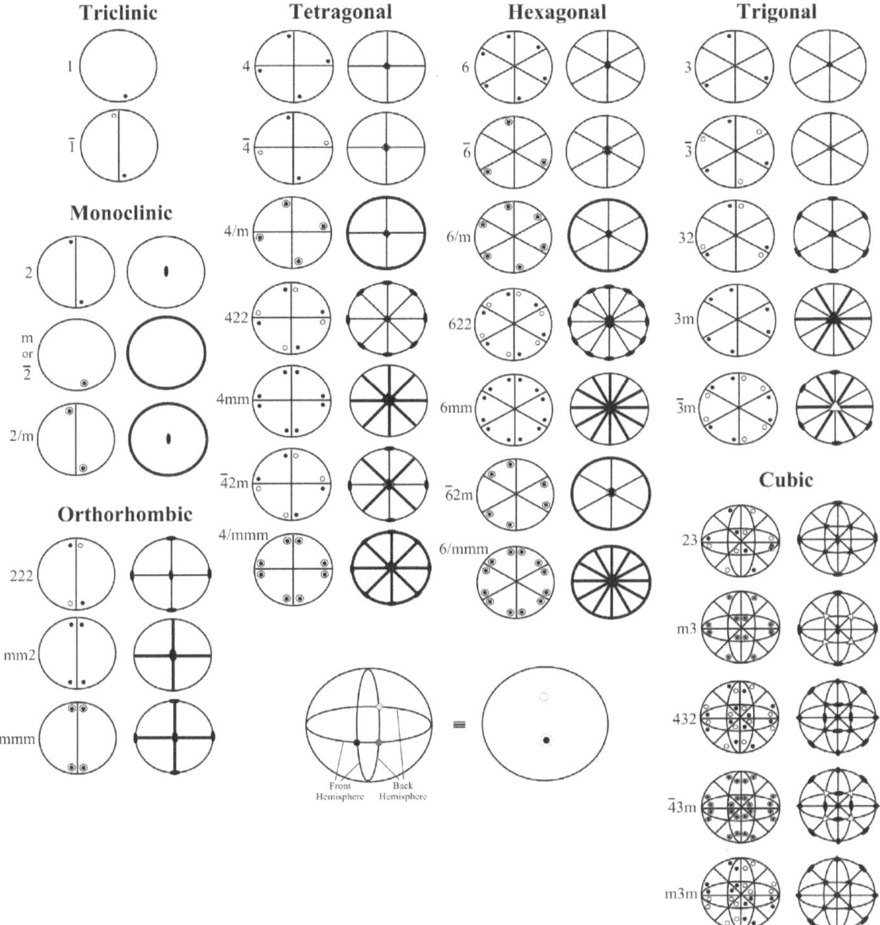

Fig. 2.30 Stereographic projections for the 32 crystallographic point groups

When the symmetry elements are applied to species arranged periodically on a crystal lattice, the result is a *space group*. The combination of the 32 crystallographic point groups with the 14 Bravais lattices yields a total of 230 possible space groups for crystals, designated by the Hermann-Mauguin (H-M) space group symbol. The 73 different space groups that can be generated from point groups only, without using glide planes and/or screw axes, are called *symmorphic* space groups.

The first letter of the H-M symbol is a single letter that refers to the Bravais centering, L. The letters used are P (primitive), A ((100) face-centered), B ((010) face-centered), C ((001) face-centered), F (face-centered), and I (body-centered). The remaining three letters refer to the crystal system as well as symmetry elements contained in the lattice. Table 2.5 lists the symmetry elements corresponding to each of the primary, secondary and tertiary terms of the space group symbol, L(1°)(2°)(3°).

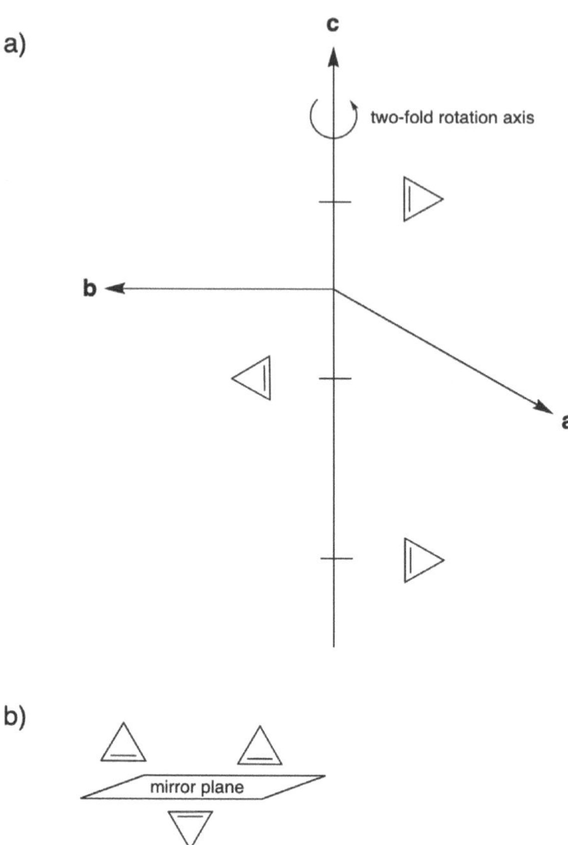

Fig. 2.31 Illustrations of: **a** 2_1 screw axis (two-fold rotation axis and 1/2 translation), listing the orientations and (x, y, z) coordinates for the original/translated object, **b** glide plane

Table 2.5 Space group symmetry element symbolism

Crystal System	Symmetry Direction[a] (symbol: L 1° 2° 3°)[b]		
	1°	2°	3°
Triclinic	N/A	N/A	N/A
Monoclinic	[010] (b-unique)	N/A	N/A
Orthorhombic	[100]	[010]	[001]
Tetragonal	[001]	[100]/[010]	[110]
Hexagonal/Trigonal	[001]	[100]/[010]	[120]/[110]
Cubic	[100]/[010]/[001]	[111]	[110]

[a] Mirror and glide planes will be perpendicular to the indicated directions, whereas rotation and screw axes will be aligned parallel to the directions
[b] L refers to the Bravais lattice centering (i.e., P = primitive, I = body-centered, F = face-centered, C = c-centered)

Both rotation and screw axes are parallel, whereas mirror/glide planes are perpendicular to the directions listed in Table 2.5. Note that the only space groups possible for a triclinic crystal are P1 and P$\overline{1}$. For monoclinic crystals, there are three possible crystallographic point groups (2, m, and 2/m), which are combined with varying combinations of lattice centering and glide planes/screw axes to yield 13 possible space groups (P2, P2_1, C2, Pm, Pc, Cm, Cc, P2/m, P2_1/m, C2/m (or B2/m), P2/c, P2_1/c, and C2/c (or B2/b)). More explicitly, the P2 space group would represent a primitive monoclinic unit cell, with a twofold rotation axis parallel to **b** ([010] direction). In comparison, the Cm space group indicates a C-centered unit cell with a mirror plane perpendicular to **b**. In addition to above triclinic and monoclinic space groups, there are 59 orthorhombic, 68 tetragonal, 25 trigonal, 27 hexagonal, and 36 cubic space groups for a grand total of 230.

To extract the relevant symmetry elements from a space group symbol, the following procedure should be followed. First, the centering is readily determined from the first term of the symbol. The relevant crystal system may be determined by comparing the remaining terms with the 32 crystallographic point groups (Table 2.4), treating screw axes and glide planes as rotation axes and mirror planes, respectively. For instance, the Pmmm space group implies that it is a primitive unit cell; the mmm is found in Table 2.4 for an orthorhombic unit cell. Hence, this would translate to a primitive orthorhombic unit cell. Next, using Table 2.5 for an orthorhombic system, the Pmmm space group would be defined by mirror planes perpendicular to each of the **a**, **b**, and **c** axes. As another example, let's consider a very common space group, P2_1/c. This can be simplified to the 2/m crystallographic point group, which is in the monoclinic crystal system. Hence, this space group consists of a primitive monoclinic unit cell, with a two-fold screw axis parallel to **b** (rotation of 180° about **b**, followed by translation along **b** of 1/2 the unit cell distance). In addition, a c-glide plane is perpendicular to the screw axis (**ab** mirror plane, with 1/2 translation along **c**). For a crystal classified within the F432 space group (e.g., sodium phosphate, Na_3PO_4), we can determine this is comprised of a fcc unit cell, with a four-fold rotation axis parallel to **a**, **b**, and **c**, a three-fold rotation axis parallel to the cube diagonal [111], and a two-fold rotation axis parallel to the **ab** plane [110].

Table 2.6 lists the various symbols that are used to graphically describe space groups. As an example, consider the representation below, which describes general positions of atoms/ions/molecules within a unit cell, projected onto the **ab** plane.

Points A–D, E–H, I/J, and K/L are equivalent positions in the unit cell, as they are duplicated by translations of one full unit along **a** or **b**. The relationship between points A and E looks like a simple twofold rotation axis; however, these points differ by 1/2 a unit cell length in the **c**-direction. Next, let's consider the relationship between points E and I. The apostrophe symbol indicates these points are mirror images of each other, suggesting that a mirror plane passes between them. However, these points also vary in their position along **c**, which implies the presence of a c-glide plane. Lastly, let's consider the relationship between points A and I (or E and L), which are located at the same position along **c**. To generate position I, one could mirror point A across the **a**-axis, followed by translation of 1/2 along **a**, an a-glide plane. Since we have a two-fold rotation axis and two glide planes, the crystal system

Table 2.6 Common symbols used for space group representations

Symmetry operation	Symbols
Mirror plane (m) ⊥ plane of projection	————————
Glide plane (a, b, c) ⊥ plane of projection	- - - - - - - - - - - - - - -
Glide plane (n) ⊥ plane of projection	– · – · – · – · – · – · –
Mirror plane (m) // plane of projection	
Glide plane (a, b, c) // plane of projection (translation of 1/2 in the direction of the arrow)	
Glide plane (n) // plane of projection (translation of 1/2 in the direction of the arrow)	
Rotation axes ⊥ plane of projection	2 3 4 6
Screw axes ⊥ plane of projection	2_1 3_1 3_2 4_1 4_2 4_3 6_1 6_2 6_3 6_4 6_5
Rotation axes // plane of projection	← 2 → 4
Screw axes // plane of projection	← 2_1 → 4_1
Center of symmetry (inversion center: i)	○
Rotation axis ⊥ plane of projection with center of symmetry	2/m 3/m 4/m 6/m

is likely orthorhombic (Table 2.5; defined by three two-fold rotation axes or mirror planes). To summarize, we would have a primitive orthorhombic unit cell, with a c-glide ⊥ **a**, an a-glide ⊥ **b**, and a 2_1 screw axis // **c**. Hence, the space group symbol is $Pca2_1$, and would be illustrated by the following symmetry operations.

There are four general positions (E, C, I, and K) that lie within the unit cell, which are defined as: $0 \leq x \leq 1; 0 \leq y \leq 1; 0 \leq z \leq 1$. The asymmetric unit for this unit cell is defined as: $0 \leq x \leq 1/4; 0 \leq y \leq 1; 0 \leq z \leq 1$. The coordinates of the four general positions are as follows: (x, y, z), (−x, −y, z + 1/2), (x + 1/2, −y, z), and (−x + 1/2, y, z + 1/2). A position lying exactly on a glide plane or the screw axis is called a special point, which always decreases the overall multiplicity (number of equivalent positions generated from the symmetry operation). For instance, if we had a molecule located at (1/2, 1/2, z), directly on the 2_1 screw axis, the ensuing symmetry operation would not generate another equivalent molecule; hence, a multiplicity of 1 rather than 2.

The effect of Bravais centering is illustrated in Fig. 2.32. As the degree of centering increases, so will the number of general positions within the unit cell. For instance, a primitive orthorhombic cell of the Pmm2 space group contains 4 atoms/ions/molecules per unit cell. By adding either an A-centered, C-centered, or body-centered units, there are now 8 species per unit cell. That is, for a C-centered unit cell, there are 4 general positions for each of the (0, 0, 0) and (1/2, 1/2, 0) sets. For a face-centered unit cell, there are 4 times the number of general positions since, by definition, a fcc array contains 4 species per unit cell relative to primitive (1/u.c.) and A, C, I cells (2/u.c.).

2.4.4 Archetypical Interstitial Crystal Lattices

For crystal lattices of ionic solids, the unit cell is best described as one of the 14 Bravais lattices for the larger ion (typically anion), with the smaller ion (cation) occupying vacant sites within the lattice, known as *interstitial sites*. Based on the number of nearest neighbors immediately surrounding these positions, interstitial sites may exist with coordination numbers of 3 (trigonal), 4 (tetrahedral), 6 (octahedral), and 8 (cubic). An example of a cubic site is a species that is housed within the (1/2, 1/2, 1/2) position of a unit cell, surrounded by eight nearest neighbors.

Figure 2.33 provides representations of the interstitial sites within fcc, bcc, and hcp lattices as projected onto the (a, b) plane, with fractional heights along the c-axis indicated in parentheses. Table 2.8 lists the fractional coordinates of the interstitials for each type of lattice. A fcc unit cell (Fig. 2.32a) has a total of 8 tetrahedral interstitial sites per unit cell. Each of the tetrahedral interstitial sites has a radius equal to 22.5% of the host atom/ion radius. The four octahedral sites in fcc are at the edges of the unit cell (*i.e.,* (1/2, 0, 0) and 11 others, each 1/4 inside the unit cell), plus a position in the center (1/2, 1/2, 1/2). The octahedral sites have a radius equal to 41.4% of the host atom/ion radius.

For a bcc unit cell (Fig. 2.32b), there are four tetrahedral interstitials on each of the six cell faces (each 1/2 inside the unit cell), giving rise to 12 tetrahedral sites per unit cell. There is one octahedral site on each of the six cubic faces (each 1/2 inside the unit cell), as well as one on each of the 12 cell edges (each 1/4 inside the unit cell), totaling 6 octahedral sites per unit cell. The radii of the tetrahedral and octahedral sites are 29% and 15.5% of the host atom/ion radius, respectively. It should be noted that a simple cubic crystal has a single cubic interstitial site at (1/2, 1/2, 1/2); this has a radius equal to 73% of the size of the host atoms/ions in the unit cell.

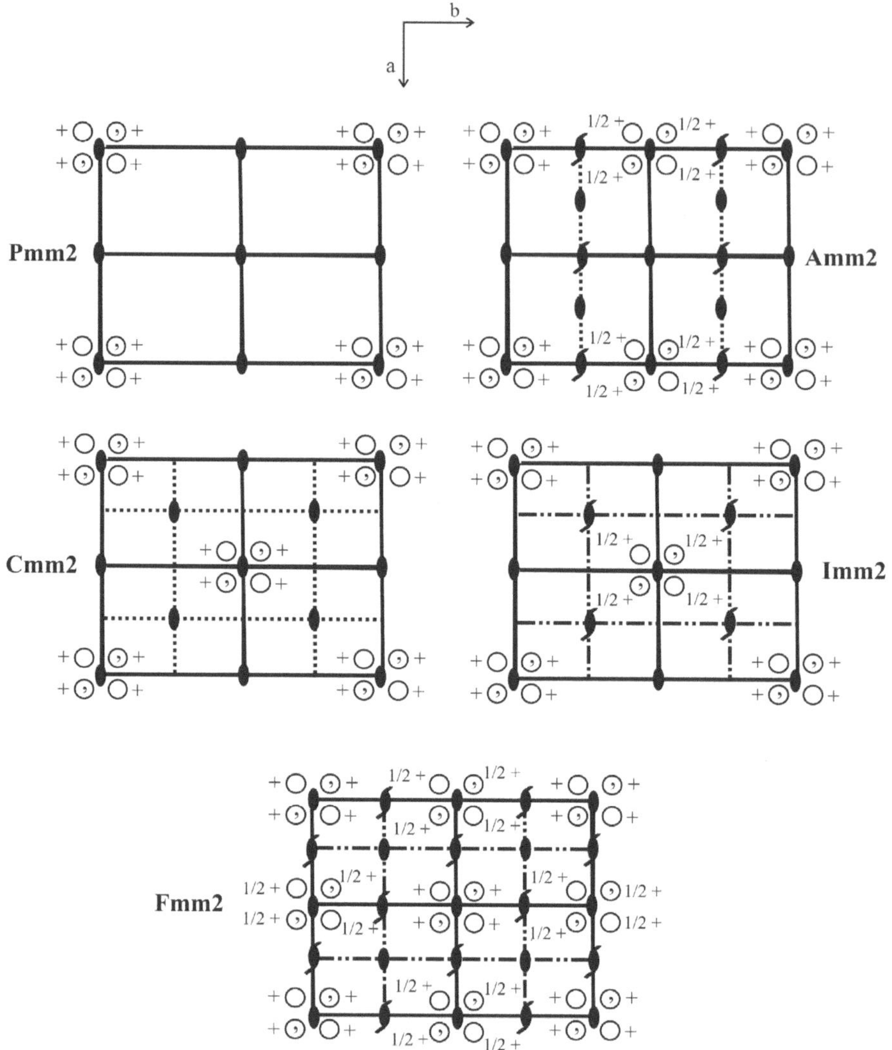

Fig. 2.32 2-D representations of the Pmm2, Amm2, Cmm2, Imm2 and Fmm2 space groups, showing general positions and symmetry elements

Figure 2.32c illustrates a hcp unit cell defined by lattice species at (0, 0, 0) and (2/3, 1/3, 1/2). There are four tetrahedral sites and two octahedral sites per unit cell. The sizes of tetrahedral or octahedral holes within a hcp and fcc array are equivalent; respectively accommodating a sphere with dimensions of 0.225 or 0.414 times (or slightly larger) the size of a close-packed lattice atom/ion.

The nickel arsenide (NiAs) structure (Fig. 2.34; $P6_3/mmc$) is an important hcp example; in this case, the cations form the backbone lattice, and the larger anions

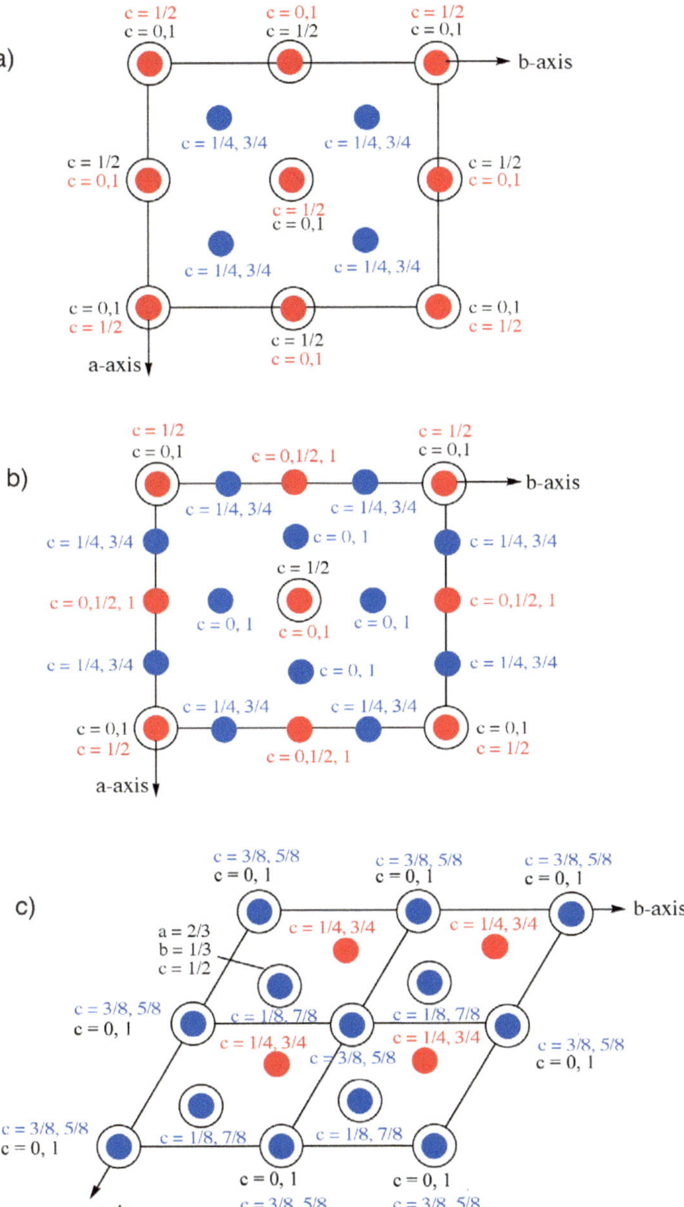

Fig. 2.33 Illustrations of the locations of interstitial sites within **a** fcc, **b** bcc, and **c** hcp unit cells. The positions of black spheres are the cubic close-packed lattice positions, whereas red and blue indicate octahedral and tetrahedral interstitial positions, respectively

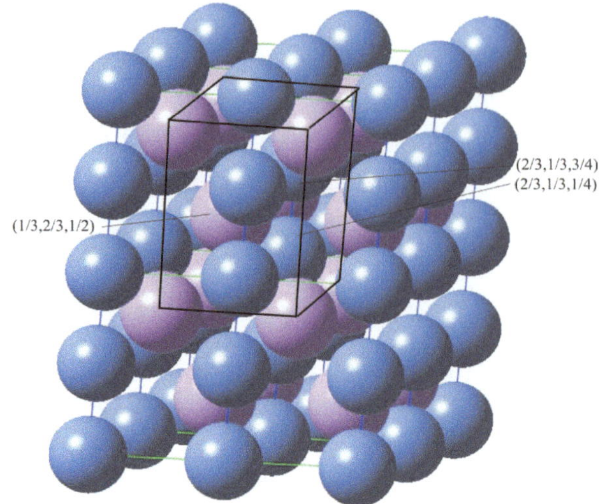

Fig. 2.34 Schematic of the NiAs structure. This consists of a hcp array of As^{3-} (purple) ions, with Ni^{3+} (blue) ions occupying all the available octahedral interstitial sites

occupy both octahedral sites. This structure, also associated with metal chalocogenides such as CoSe, NiTe, $Co_xNi_{1-x}As$, FeS, NiSe, $PtSb_xBi_{1-x}$ and $Pd_xNi_{1-x}Sb$, are only adopted for weakly ionic compounds. Since the octahedral sites are extremely close to one another, purely ionic compounds would be much too unstable due to strong anion-anion repulsions.

The crystal structure for α-alumina (*corundum*) is quite similar to NiAs; in this case, 2/3 of the available octahedral interstitial sites are occupied with Al^{3+}, within a hcp framework array of O^{2-} ions (Fig. 2.35; R $\overline{3}$ c). Alumina has a diverse range of uses including heterogeneous catalysis supports, aluminum metal production, and as an abrasive refractory ceramic material for grinding/cutting tooling and protective coating applications. Alumina is also used as a polishing agent within CD/DVD repair kits and toothpaste formulations.

Another common hcp-based variety, cadmium iodide (CdI_2), exhibits partially covalent bonding. This is a very common crystal structure not just for metal(II) halides (e.g., MgI_2, TiI_2, VI_2, MnI_2, FeI_2, CoI_2, PdI_2, $TiCl_2$, VCl_2, $MgBr_2$, $TiBr_2$, $FeBr_2$, $CoBr_2$), but for metal(II) hydroxides (e.g., $Mg(OH)_2$, $Ni(OH)_2$, $Ca(OH)_2$), metal(IV) dichalcogenides (i.e., ME_2, M = Grps 4, 5, 9, 10, 14; E = S, Se, Te), and intermetallics (e.g., Cd_2Ce, Cd_2La). This structure is based on a hcp array of the anionic species, with the cation occupying the octahedral sites in alternate layers (Fig. 2.36; P $\overline{3}$ m1). Interestingly, if the I^- anions in the CdI_2 structure are replaced with smaller Cl^- ions, a *cubic* structure (R $\overline{3}$ 2/m) may result; [21] for instance, $CdCl_2$, $MgCl_2$, $CoCl_2$, $NiCl_2$, $MnCl_2$, NbS_2, TaS_2, or NiI_2. This structure consists of a fcc array of anions, with the cations occupying octahedral holes in alternate layers.

For ionic crystals, the preference of a cation to occupy a certain interstitial site is primarily governed by the ionic radius ratio of the cation/anion (r_+/r_-).[22] Since anions are most often larger than cations, this ratio is usually less than 1 (Table 2.7). An exception may be found for unusually large cations with small anions such as

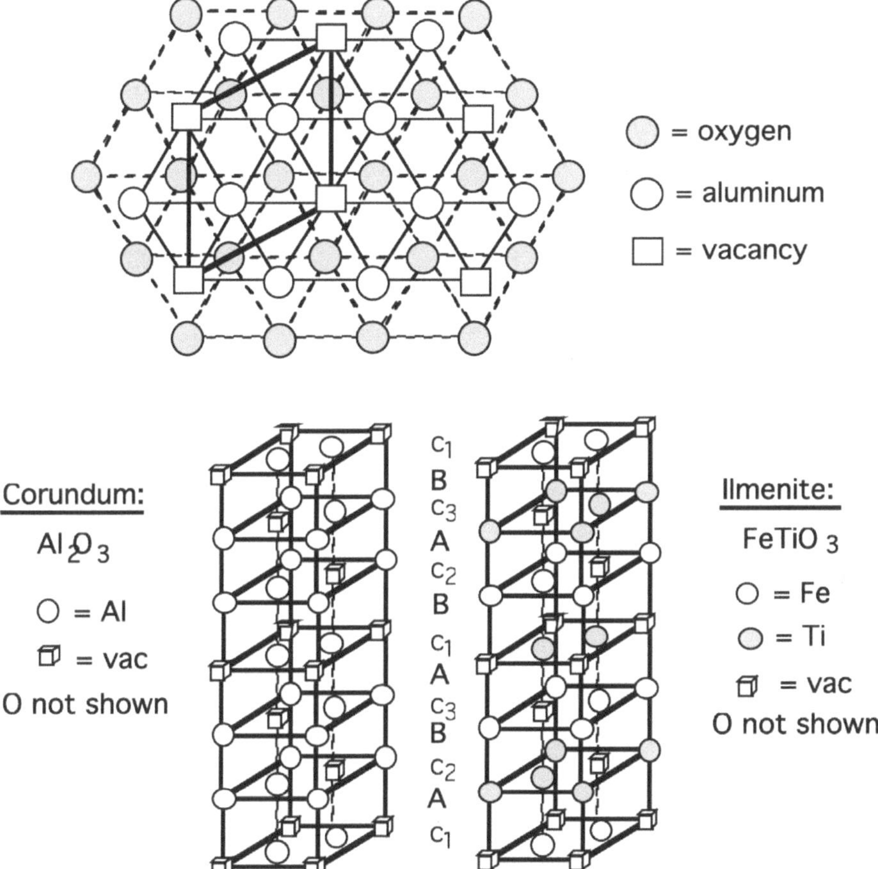

Fig. 2.35 Schematic of the a-Al_2O_3 (aluminum oxide, corundum) crystal structure. Shown is the hcp array of O^{2-} ions with Al^{3+} ions and vacancies in the octahedral c-sites above the A plane of the oxide ions. The comparative stacking sequences of Al_2O_3 and $FeTiO_3$ (ilmenite) are also illustrated

CsCl ($r_+/r_- = 1.08$). As the value of this ratio decreases, the size of the anions become significantly larger than the cations, and the cation will prefer to occupy a smaller interstitial site. To rationalize this preference, consider a simple cubic arrangement of bowling balls (cf. large anions) with a small golf ball (cf. small cation) in the middle. Due to their large size differences, this arrangement would not be stable, as the golf ball would rattle around the cubic "cage" formed by the bowling balls. Rather, a smaller close-packed interstitial site such as trigonal or tetrahedral would best contain the smaller golf ball.

For compound unit cells, it is important to point out the occupation of the atoms/ions occupying the Bravais framework and the other species in interstitial sites. For example, the *CsCl structure* (Pm $\overline{3}$ m) is best described as a simple cubic

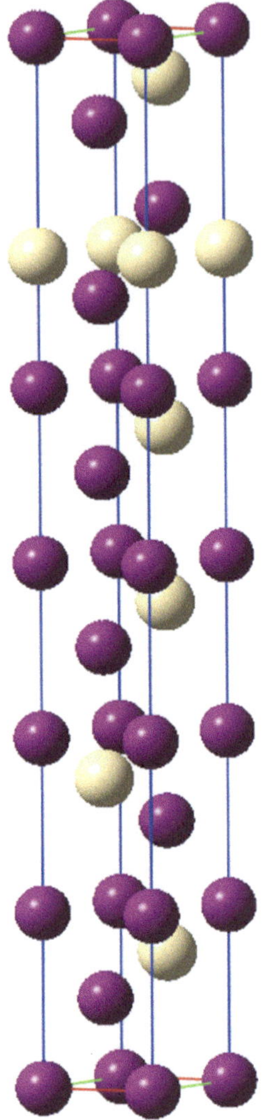

Fig. 2.36 Schematic of one common polymorph of the CdI_2 structure. This consists of a hcp array of I^- ions (purple), with Cd^{2+} ions (yellow) occupying octahedral interstitial sites in alternating layers

arrangement of Cl^- ions, and a Cs^+ ion in a cubic interstitial site. Alternatively, one could also designate this structure as having Cs^+ ions at the corners of a cube, and Cl^- in the interstitial cubic site. Even though the overall arrangement of ions in this structure is bcc, such a compound unit cell cannot be assigned to this arrangement since the ions are not equivalent. Many other compounds crystallize in the CsCl structure, such as CsBr, CsI, RbCl (at high temperature/pressure), and intermetallics such as β-AgCd, AgCe, AlFe, AlNd, β-AlNi, AlSc, β-AuCd, AuMg, BaCd, BeCu, BeCo, BePd, CaTi, CdCe, CeMg, CoFe, CoSc, CuEu, β-CuPd, β-CuZn, β-GaNi, GaRh, HgMg, HgSr, HoIn, HoTl, InLn, LiTi, MgPr, MgTi, β-MnPd, β-MnRh, OsTi, RhY, α-RuSi, RuTi, SbTi, ScRh, and SrTi.

Table 2.7 Ionic radii ratios corresponding to interstitial sites

Radii ratio range (r_+/r_-)	Geometry of interstitial site (coordination number)
< 0.225	Trigonal (3)
< 0.414	Tetrahedral (4)
< 0.732	Octahedral (6)
> 0.732	Cubic (8)

It is not immediately apparent why many different values for individual ionic radii appear in reference books. The value for an ionic radius is dependent on its lattice arrangement, as determined from electron density maps and empirical X-ray diffraction data. Some general trends for cationic radii are:

1. For a given species and charge, the radius increases as the coordination number increases.
2. For a given charge, the radius decreases with increasing effective nuclear charge, Z_{eff}.[23]
3. For a given species, the radius decreases with increasing ionic charge.
4. For a given species and charge, the radius is larger for high-spin (weak field) ions than for low-spin (strong field) ions.

Most inorganic chemistry texts list cut-off values for the r_+/r_- ratios corresponding to the various geometries of interstitial sites (Table 2.7). For instance, the halite or *rocksalt* structure (Fm $\overline{3}$ m) exhibited by MX (M = Grp I, Mg, Pb, Ag; X = F, Cl, Br, I) are predicted to have occupation of octahedral interstitial sites. Indeed, these structures are described as a fcc array of the halide ion (except for very small F^- ions), with the cation occupying all the octahedral interstitial sites (i.e., 4 MX units per unit cell).

However, it should also be pointed out that deviations in these predictions are found for many crystals due to covalent bonding character. In fact, the bonding character for compounds is rarely 100% covalent or ionic in nature, especially for inorganic species. For instance, consider the zinc sulfide (ZnS) crystal structure. The ionic radius ratio for this structure is 0.52, which indicates that the cations should occupy octahedral interstitial sites. However, due to partial covalent bonding character, the anions are closer together than would occur from purely electrostatic attractions. This results in an "effective radius ratio" that is decreased, and a cation preference for tetrahedral sites rather than octahedral. The resulting *wurtzite* (α-ZnS; $P6_3mc$) crystal structure illustrated in Fig. 2.37 is best described as a hcp lattice of sulfide ions with zinc ions occupying 1/2 of the 8 available tetrahedral interstitial sites. The wurtzite crystal structure is also adopted for other compounds such as β-AgI, ZnO, α-CdS, CdSe, α-SiC, GaN, AlN, ω-BN, and BeO. As you might expect, a hybrid of ionic/covalent bonding will greatly affect the physical properties of the solid; for example, the hardness of ZnS is significantly greater than what would be expected for a purely ionic solid.

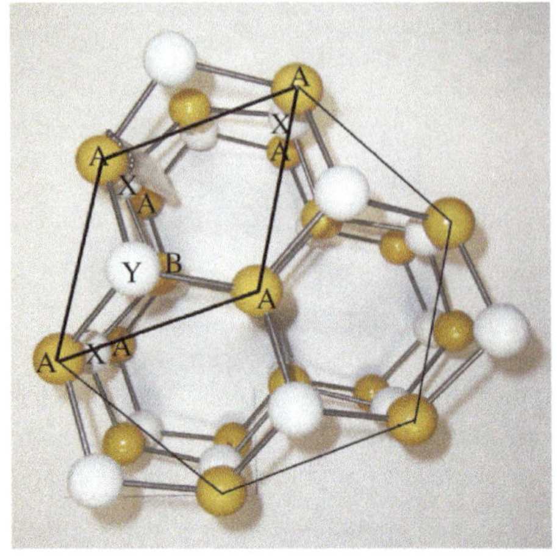

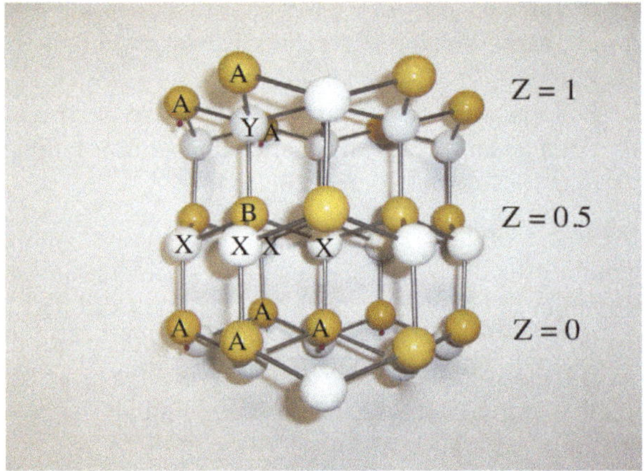

Fig. 2.37 Model of the wurtzite (ZnS) crystal structure. The framework is based on an hcp lattice of S^{2-} anions (yellow; the unit cell consists of A and B ions) with zinc ions occupying tetrahedral interstitial sites (white, labeled as X and Y ions)

Interestingly, zinc sulfide (β-ZnS) may also crystallize in a cubic lattice, which consists of a fcc array of S^{2-}, with Zn^{2+} occupying 1/2 of the available tetrahedral sites (Fig. 2.38; F $\overline{4}$ 3 m). This structure is known as *sphalerite* or *zincblende*, which is shared with other compounds such as α-AgI, β-BN, CuBr, and β-CdS. Important semiconductors such as GaAs, β-SiC, and InSb also adopt the sphalerite crystal structure. It should be noted that when the same atom occupies *both* the fcc and tetrahedral interstitials of the sphalerite structure, it is described as the *diamond lattice*, which is also adopted by elemental forms (referred to as *allotropes*) of silicon, germanium, and tin.

If the cation in the crystal lattice exhibits a cubic environment (coordination number of 8), the *fluorite* structure is commonly observed (Fig. 2.39; Fm $\overline{3}$ m). Lattices of this variety consist of a fcc arrangement of cations, with all 8 tetrahedral interstitial sites (e.g., (1/4, 1/4, 1/4), etc.) occupied by the anionic species. Of course, this will only be prevalent when the size of the anion is much smaller than the cation, such as CaF_2. Other examples of fluorite lattices include intermetallics (e.g., $PtGa_2$, $SnMg_2$, LiMgP, HoOF, $GeLi_5P_3$, $AuIn_2$), oxides (e.g., ZrO_2 (cubic zirconia), CeO_2,

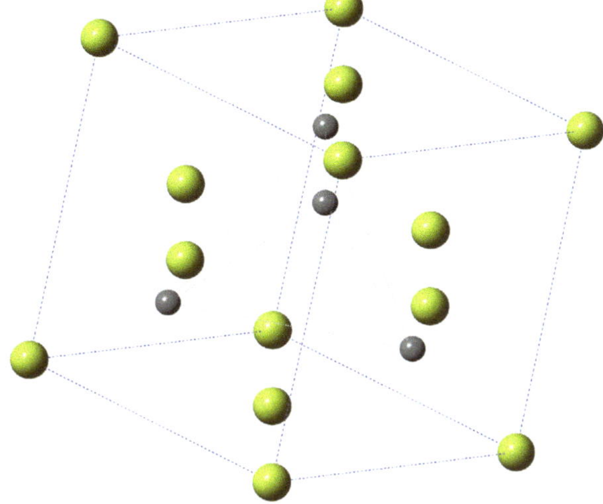

Fig. 2.38 Illustration of the sphalerite (ZnS) crystal structure, which consists of a diamond lattice. The Zn^{2+} cations (grey) form a fcc array with S^{2-} anions (yellow) occupying ½ of the available tetrahedral interstitial sites

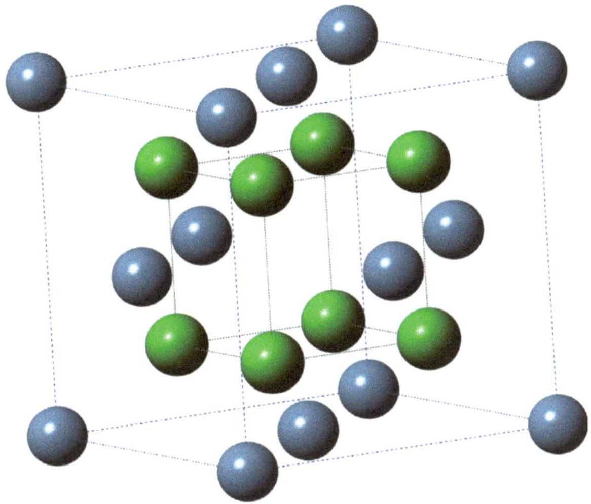

Fig. 2.39 Unit cell representation for the fluorite structure of CaF_2 (blue = Ca^{2+}; green = F^-)

UO_2), hydrides (e.g., CeH_2, NbH_2), and nitrides (e.g., UN_2). For structures with relatively smaller cations, the anions will form the fcc lattice, with cations situated within the interstitials. Since the relative positions of cations and anions are reversed in the latter case, the *anti* prefix is used, designating the structure as *antifluorite* (*e.g.*, alkali metal oxides, Li_2O). For intermetallic compounds of stoichiometry AB_2, which contain metals differing significantly in electronegativity, either the *pyrite* (FeS_2; Pa $\overline{3}$) structure (Fig. 2.40; e.g., $AuSb_2$, $PdAs_2$, $PdBi_2$, $PdSb_2$, $PtAs_2$, $PtBi_2$, $PtSb_2$, and $RuSn_2$), or *calcium carbide* (CaC_2; I4/mmm) structure (Fig. 2.41; e.g., Ag_2Er, Ag_2Ho, Ag_2Yb, $AlCr_2$, $AuEr_2$, $AuHo_2$, Au_2Yb, Hg_2MgSi_2W, $MoSi_2$, and $ReSi_2$) is favored.

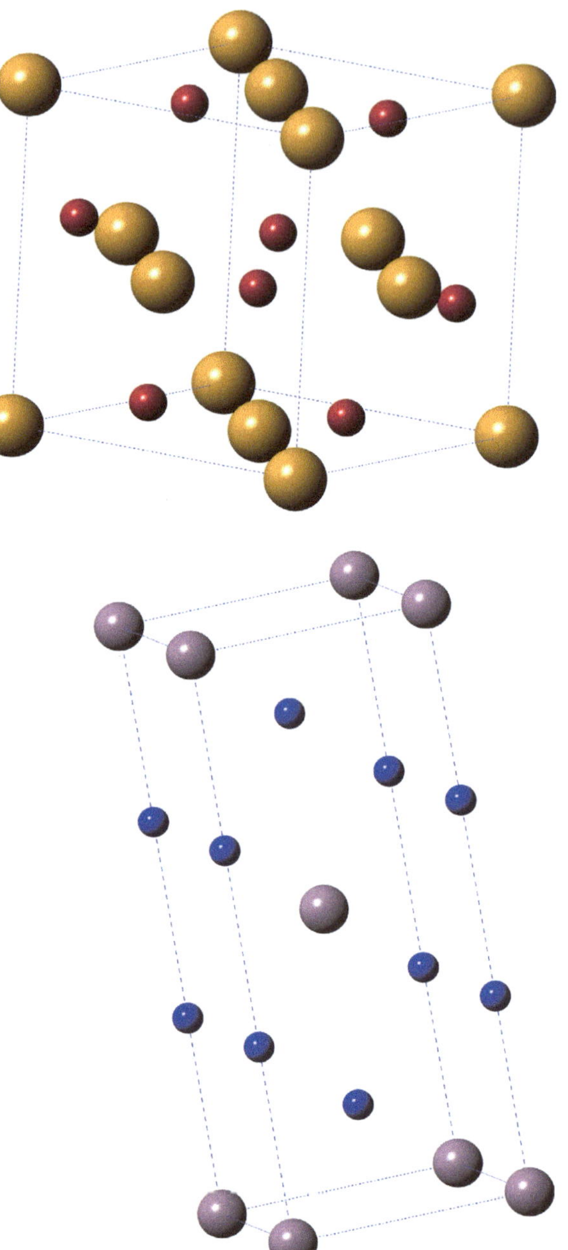

Fig. 2.40 Illustration of the pyrite structure of $AuSb_2$. This consists of a fcc array of one metal (Au, gold) with eight other metal atoms (Sb, red) positioned at the following interstitial sites: (0.385, 0.385, 0.385), (0.385 + ½, 0.385, 0.115), (0.115, 0.385 + ½, 0.385), (0.115 + ½, 0.385 + ½, 0.115), (0.115 + ½, 0.115 + ½, 0.115 + ½), (0.115, 0.115 + ½, 0.385 + ½), (0.385 + ½, 0.115, 0.115 + ½), (0.385, 0.115, 0.385 + ½)

Fig. 2.41 Illustration of the calcium carbide structure of $MoSi_2$. This consists of a body-centered tetragonal array of one species (Mo, purple) with the other species (Si, blue) positioned at the following interstitial sites: (0, 0, 1/3), (1, 0, 1/3), (0, 1, 1/3), (1, 1, 1/3), (0, 0, 2/3), (1, 0, 2/3), (0, 1, 2/3), (1, 1, 2/3), (1/4, 1/4, 1/4), (1/4, 1/4, 3/4)

Metal oxide lattices

The vast majority of catalysts used in heterogeneous catalytic processes are based on metal oxides, either as the catalytically active species (e.g., TiO_2) or as a high surface area support material (e.g., MgO). There is ongoing interest in the preparation of these catalysts with specific reproducible properties—a challenge that has been possible through increasing knowledge regarding the structure/property relationships of these materials.

In this section, we will describe a number of important crystals that are composed of a close-packed array of oxide anions, with cations situated in vacant interstitial sites. Often, there are two or more different types of cations that occupy the vacancies. One example is the *normal spinel* structure consisting of a fcc array of oxide ions (as well as S^{2-} (e.g., $FeCr_2S_4$, $CuCr_2S_4$, Fe_3S_4) or Se^{2-} (e.g., $ZnCr_2Se_4$)), with 1/8 of the tetrahedral holes occupied by M^{2+} ions, and 1/2 of the octahedral holes occupied with M^{3+} ions. The *inverse spinel* structure features the divalent cations switching places with half of the trivalent ions (i.e., M^{3+} positioned within tetrahedral sites and M^{2+} within octahedral sites).

The complicated unit cell for normal spinel is shown in Fig. 2.42 (Fd3m), which is comprised of a large fcc array of tetrahedrally-coordinated cations, and 8 octant sub-units that contain O^{2-} and M^{2+}/M^{3+} cations. The ionic count per unit cell (u.c.) is as follows:

M^{2+}: fcc array (4 ions/u.c.) + one ion in the center of 4/8 octant sub-units = 8/u.c.

M^{3+}: 4 ions at alternating corners of 4/8 octant sub-units = 16/u.c.

O^{2-}: 4 ions at alternating corners in all octant sub-units = 32/u.c.

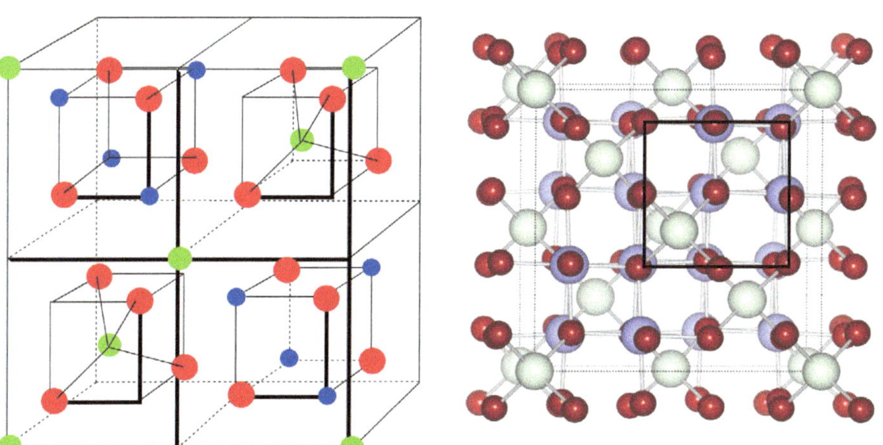

Fig. 2.42 Illustrations of the AB_2O_4 (binary: A = B; ternary: A ≠ B) normal spinel lattice. For clarity, the representation on the left shows the front half of the unit cell. The A ions (green), generally in the + 2 oxidation state, occupy 1/8 of the available tetrahedral sites; the B ions (blue), generally in the + 3 oxidation state, occupy 1/2 of the available octahedral sites within a fcc oxygen (red) sublattice. Reproduced with permission from *Phys. Rev. B* **2007**, *76*, 165,119. © 2007 The American Physical Society

Hence, the normal spinel structure may also be described as $[M^{2+}]_8(M^{3+})_{16}O_{32}$ or $[M^{2+}{}_{8/3}M^{3+}{}_{16/3}](M^{2+}{}_{16/3}M^{3+}{}_{32/3})O_{32}$, where brackets and parentheses indicate ions in tetrahedral and octahedral sites, respectively. More simply, one can describe the AB_2O_4 stoichiometry of the spinel structure as a fcc of O^{2-} ions, with A^{2+} occupying 1/8 of the tetrahedral sites and B^{3+} occupying 1/2 of the octahedral sites (Fig. 2.41).

Magnetic tape has long employed the inverse spinel magnetite, Fe_3O_4, which contains iron cations in mixed oxidation states $[Fe^{3+}](Fe^{2+}/Fe^{3+})O_4$. Another magnetic iron oxide used for these applications is known as maghemite (γ -Fe_2O_3), which has the structure $[Fe^{3+}]_8(Fe^{3+}{}_{40/3}\square_{8/3})O_{32}$. The $\square$ symbol indicates a vacancy; hence, this structure is considered Fe^{2+}-deficient magnetite, a defect inverse spinel structure. Since these magnetic ferrites (and others such as Co_3O_4) contain only one metal in two oxidation states of the general form M_3O_4, they are referred to as binary spinels.

In addition to the above ferrites that contain M^{2+}/M^{3+} cations, there are other common types of ternary spinels, such as $M^{+}{}_2M^{6+}O_4$ (*e.g.*, Na_2MoO_4, Ag_2MoO_4) and $M^{4+}M^{2+}{}_2O_4$ (*e.g.*, $TiZn_2O_4$, $SnCo_2O_4$). As we will detail at the end of this chapter, one interesting spinel of the form $LiMn_2O_4$ is used as a cathode (reduction site) material for lithium-ion batteries.[24] Based on the chemical formula, Mn ions within MnO_2 sub-units exhibit an average oxidation state of 3.5 (experimentally: 3.55, due to lithium oxides on the surface[25]); the variation between stable Mn^{3+}/Mn^{4+} oxidation states allows for electron transfer to occur in the solid state.[26] It should be noted that structures such as $BaFe_2O_4$, as well as oxygen-deficient analogues such as $BaFe_{12}O_{19}$ or $Ba_2Mn_2Fe_{12}O_{22}$, used in magnetic stripe cards are not spinel lattices. Rather, these structures consist of a hcp array of oxide anions, with some of the oxides replaced with Ba^{2+}.

The equilibrium distribution of cations in tetrahedral and octahedral sites within the spinel lattice depends on the size of the ionic radii, electrostatic energies, and polarization effects. As one would expect, the degree of cation disorder will significantly influence the magnetic properties of magnetic ferrite spinels.[27] One may intentionally prepare spinels containing nonequilibrium cation distributions through rapid quenching of sintered powders. For instance, when ground mixtures of MnO and Al_2O_3 powders are sintered and quenched the resulting $(Mn_xAl_{1-x})_3O_4$ spinel has an oxygen-deficient metal:oxygen ratio of 3:3.7.[28] Another mixed ternary system, $Mn_{1.5-0.5x}Co_{1+0.5x}Ni_{0.5}O_4$ ($0 \leq x \leq 1.0$), is used in negative temperature coefficient (NTC) thermistors[29] for applications in a variety of electrical products, communication and industrial equipment, and automobiles.

The Boltzmann distribution (2.11) relates the dependence of temperature on the amount of cation disorder in spinels, where E_i is the activation energy required to exchange interstitial sites of M^{2+} and M^{3+} cations (i.e., between tetrahedral and octahedral sites, respectively). Energy values for these interchanges generally fall in the range 0.11–0.14 eV.

$$\frac{i(1+i)}{(1-i)^2} = e^{\frac{-E_i}{kT}}, \tag{2.11}$$

where: $i = 0$ (normal spinel) or $i = 1$ (inverse spinel)

Another important oxide lattice is the *rutile* structure, common for both oxides and fluorides with the general formula MO_2 (M = Ti, Cr, Ge, Ir, Mo, Nb, Os, Pb, Ru, Sn, Te) and $M'F_2$ (M' = Co, Zn, Fe, Mg, Mn, Ni, Pd).[30] Rutile is the most stable polymorph of TiO_2, but may be converted to other phases such as anatase (see below) at elevated temperatures. Finely divided powders of rutile are used as a white pigment in paint formulations, plastics, and paper, and is also found in some sunscreens to protect against harmful UV exposure. The rutile structure (Fig. 2.43a; $P4_2/mnm$) consists of a primitive tetragonal unit cell with each cation surrounded by six O^{2-} ions. Figure 2.43b shows a space-filling view of the lattice, which illustrates the hcp arrangement of oxide ions and metal ions residing in one-half of the available octahedral holes. The MO_6 units form chains of edge-sharing octahedra along the c-axis, linked by vertices (Fig. 2.44).

TiO_2 also adopts the *anatase* structure, composed of a body-centered tetragonal array of cations with oxygen ions occupying octahedral holes (Fig. 2.45; $I4_1/amd$). In contrast to rutile, which contains two MO_2 units per unit cell, anatase contains four TiO_2 units. This corresponds to a larger unit cell volume (136.25 $Å^3$) relative to rutile (62.1 $Å^3$). Like rutile, the TiO_6 units of anatase form polyhedron chains in its extended structure (Fig. 2.46), but exhibit a much greater density of edge-sharing octahedra than rutile. As will be described in more detail in Chap. 4, anatase outperforms rutile as an important photocatalyst for water splitting (primarily for fuel cell applications), self-cleaning coatings, degradation of organic dyes in solution, and air/water purification. However, at all temperatures and pressures, the rutile phase is the thermodynamically favored phase; anatase and other polymorphs such as brookite[31] are considered metastable phases that may be kinetically stabilized at lower temperatures. The interconversion between anatase and rutile phases is quite complex and is influenced by a variety of factors such as the presence of dopants and synthetic/post-annealing conditions.[32] In particular, it is widely known that oxygen defects facilitate the anatase-to-rutile transformation, which are generated by atmospheric conditions or the presence of other impurities.[33]

The largest, and most widely studied, category of metal oxide crystals is the *perovskites*. These oxides have the general formula ABO_3, where A is a cation of larger size than B. These lattices consist of a ccp arrangement of both oxide anions and the larger cation (Fig. 2.47; $Pm\bar{3}m$). The smaller cation occupies the octahedral hole at the position (1/2, 1/2, 1/2). It should be noted that most perovskite lattices are comprised of distorted cubic unit cells. The degree of structural distortion, d, may be predicted by (2.12):

$$d = \frac{r_A + r_o}{\sqrt{2}(r_B + r_o)}, \tag{2.12}$$

where: r_A, r_B, and r_o are the ionic radii of A and B cations, and O anions, respectively.

As the value of d approaches 1.0, the perovskite will become more perfectly cubic; however, if $d \leq 0.81$, the ionic radius of the A site will be smaller than ideal, resulting in BO_6 octahedra becoming tilted to fill the available volume. Stable perovskites are predicted to have values in the range $0.78 \leq d \leq 1.05$. However, it should be noted

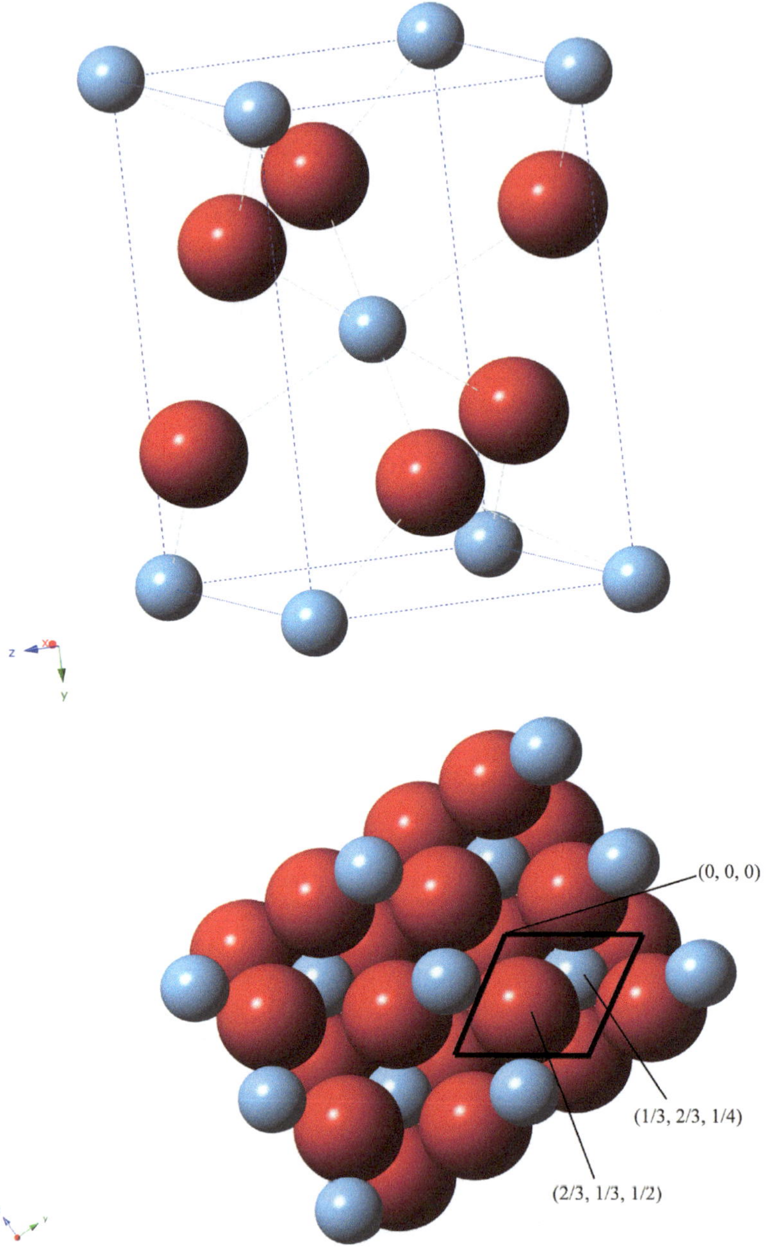

Fig. 2.43 Schematic of the rutile structure. This is defined as a primitive tetragonal unit cell with M^{4+} ions at (0, 0, 0) and (1/2, 1/2, 1/2), and O^{2-} ions at (0.3, 0.3, 0), (0.7, 0.7, 0), (0.8, 0.2, 1/2), and (0.2, 0.8, 1/2). The space-filling representation shows the hcp array of oxide ions, with Ti^{4+} located in an octahedral interstitial site

Fig. 2.44 Connectivity of MO_6 octahedra in rutile, composed of chains of edge-sharing octahedra along the c-axis, linked by vertices

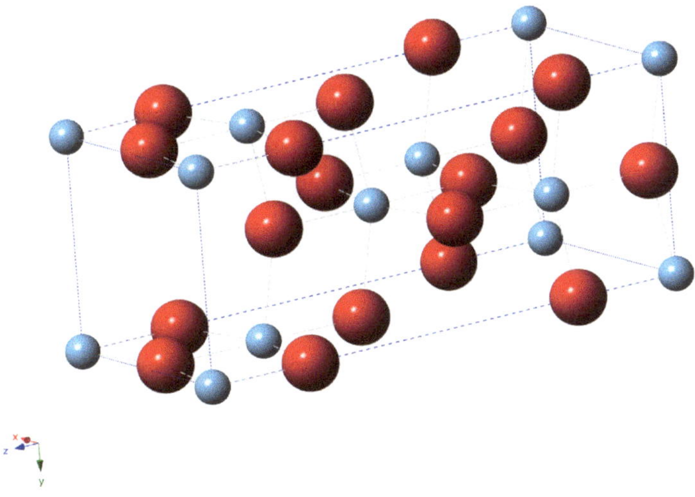

Fig. 2.45 Schematic of the anatase (TiO_2) crystal structure, which consists of four TiO_2 compounds per unit cell. For a listing of the relative coordinates for Ti^{4+} and O^{2-} ions, see the.cif files located online (http://rruff.geo.arizona.edu/AMS/minerals/Anatase)

Fig. 2.46 Connectivity of TiO_6 octahedra in anatase, consisting of chains of edge-sharing octahedra along the c-axis, linked by vertices and edges

that values outside of this range may also yield stable structures, as the above formula does not consider the influence of bond covalency within the perovskite lattice.

Approximately 90% of the metallic elements of the periodic table are known to form stable oxides with the perovskite structure. Further, it is possible to partially substitute A and B cations to yield a perovskite of the formula $A_{1-x}A'_xB_{1-x}B'_xO_3$. It should be noted that the perovskite structure is not only obtained for oxides, but also for some nitrides (e.g., Ca_3GeN), halides (e.g., $KMgF_3$), hydrides (e.g., $BaLiH_3$), and carbides (e.g., $MgCNi_3$). Recently, oxynitride perovskites (e.g., $SrMO_2N$; M = Nb, Ta) have received considerable attention due to their potential applications for nontoxic inorganic pigments and photocatalysts.[34] The reduced electronegativity of the nitride ion, relative to the oxide anion, increases the covalency of the

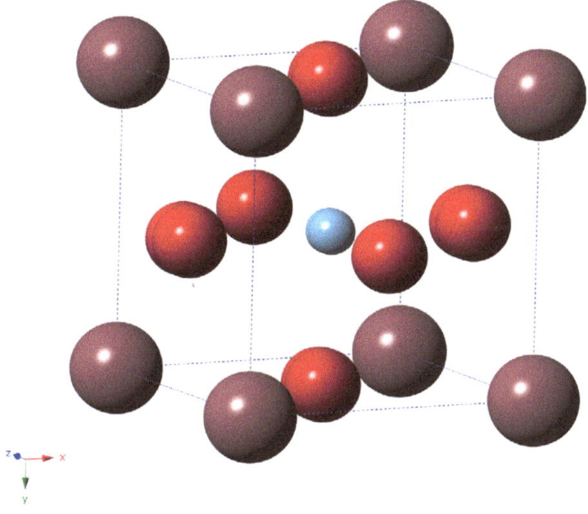

Fig. 2.47 Unit cell illustration of the perovskite $BaTiO_3$, illustrating the ABO_3 stoichiometry (blue = Ti^{4+}; purple = Ba^{2+}, red = O^{2-})

cation–anion bonds thus affecting its overall structure and physical/optical properties. The ability of inducing structural distortions by cationic substitution results in a diverse range of applications for perovskites. In addition to numerous applications in catalysis, sensors, fuel cells, ferroelectric/piezoelectric ceramics, and electronics,[35] the perovskite backbone is a key component in modern *high-temperature superconductive* (HTS) materials, which we will now delve into in more detail.

2.4.5 *Superconductivity of Perovskites: Toward a Room-Temperature Superconductor*

By definition, a superconductor exhibits no resistance to electrical conductivity. When a superconductor is placed in a weak external magnetic field, **H**, and cooled below its transition temperature, the magnetic field is repelled. This phenomenon is referred to as the *Meissner effect* and is the most intriguing property of superconductors—the ability to levitate on top of a magnetic surface (Fig. 2.48a). It should be noted that the Miessner effect does not completely repel **H**; that is, the field is able to penetrate the surface of the superconductor to a depth known as the London penetration depth, λ. For most superconductors, λ is on the order of 50 nm decaying exponentially beyond this region toward the bulk of the superconductor structure.

The Meissner effect should be distinguished from the diamagnetism exhibited by a perfect electrical conductor (Fig. 2.48b). According to Lenz's Law, when a magnetic field is applied to a conductor, it will induce an electrical current in the conductor that creates an opposing magnetic field. In a perfect conductor, a large current can be induced; however, the resulting magnetic field will exactly cancel the applied field.

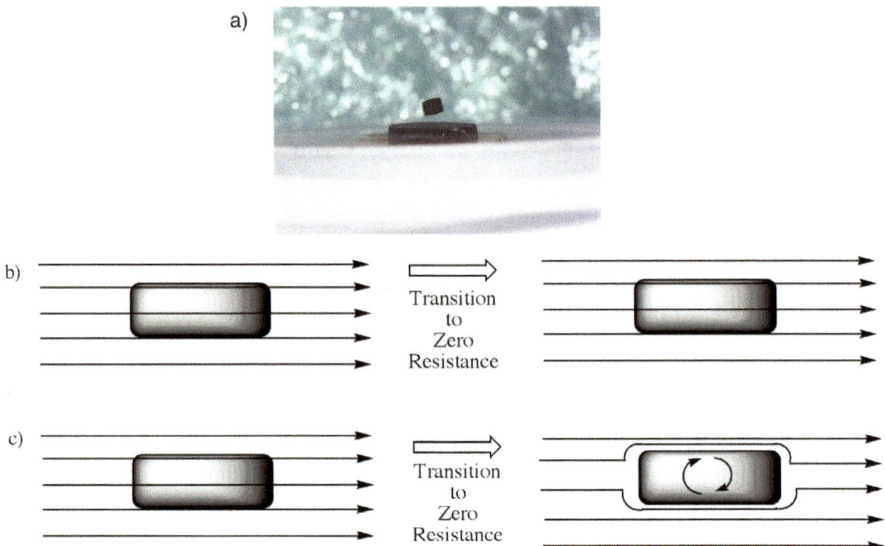

Fig. 2.48 **a** Photograph of the Meissner effect for a rare-earth magnet above a sample of YBCO immersed in liquid nitrogen. The onset of strong diamagnetism ("superdiamagnetism," as observed by the repulsion of an external magnetic field) is the most reliable method to determine superconductive behavior. The schematic illustrates the different behavior toward an applied external magnetic field for **b** a perfect conductor, and **c** a superconductor

In contrast, the Meissner effect is the spontaneous repulsion of the applied magnetic field that occurs only once the transition to superconductivity has been achieved.

Many pure transition metals (*e.g.*, Ti, Zr, Hf, Mo, W, Ru, Os, Ir, Zn, Cd, Hg) and main group metals (*e.g.*, Al, Ga, In, Sn, Pb) exhibit superconductivity, many only when exposed to high-pressure conditions. These materials are referred to as Type I or *soft superconductors*.

Binary and ternary alloys and oxides of these elements, as well as pure V, Nb, Gd, and T_c are referred to as Type II or *high-field superconductors*. In contrast to Type I, these materials exhibit conductive characteristics varying from normal metallic to superconductive, depending on the magnitude of the external magnetic field. It is noteworthy to point out that metals with the highest electrical conductivity (e.g., Cu, Au) do not naturally possess superconductivity.

Although superconductivity was first discovered in 1911 for supercooled liquid mercury, it was not until 1957 that a theory was developed for this phenomenon.[36] Proposed by Bardeen, Cooper and Schrieffer, the *BCS theory* purports that at a certain critical temperature, T_c, the electrons within thermal energy (kT_c) of the Fermi level are able to correlate their motion in pairs, referred to as *Cooper pairs*. It is not readily apparent why this should occur, since electrons are mutually repulsive due to their like negative charges. The formation of Cooper pairs is thought to result from electron–phonon (i.e., lattice vibration) coupling. That is, an electron moving through the lattice attracts the positively charged nuclei of the lattice atoms, causing them to be distorted from their original position. This creates a small attractive force toward another electron of opposite spin, whose motion becomes correlated with that of the original electron (Fig. 2.49a). The primary experimental evidence that supports the concept of phonon-facilitated Cooper pair formation is known as the isotope effect (Fig. 2.49b). That is, the linear inverse relationship of critical temperature and mass of lattice atoms suggests that electron/lattice coupling interactions are a key component to superconductivity.

Whereas individual electrons are fermions (1/2 spin) and must obey the Pauli exclusion principle, Cooper pairs exhibit boson-like properties and are hence able to condense into the same energy level. At absolute zero, the condensed Cooper pairs form a single energy state that lies kT_c below the Fermi level (Fig. 2.50). As a consequence of condensation, a number of forbidden energy levels appear within the conduction band, centered about the Fermi level. The energy of this bandgap (*ca.* 7/2 kT_c) corresponds to the minimum energy required to break up a Cooper

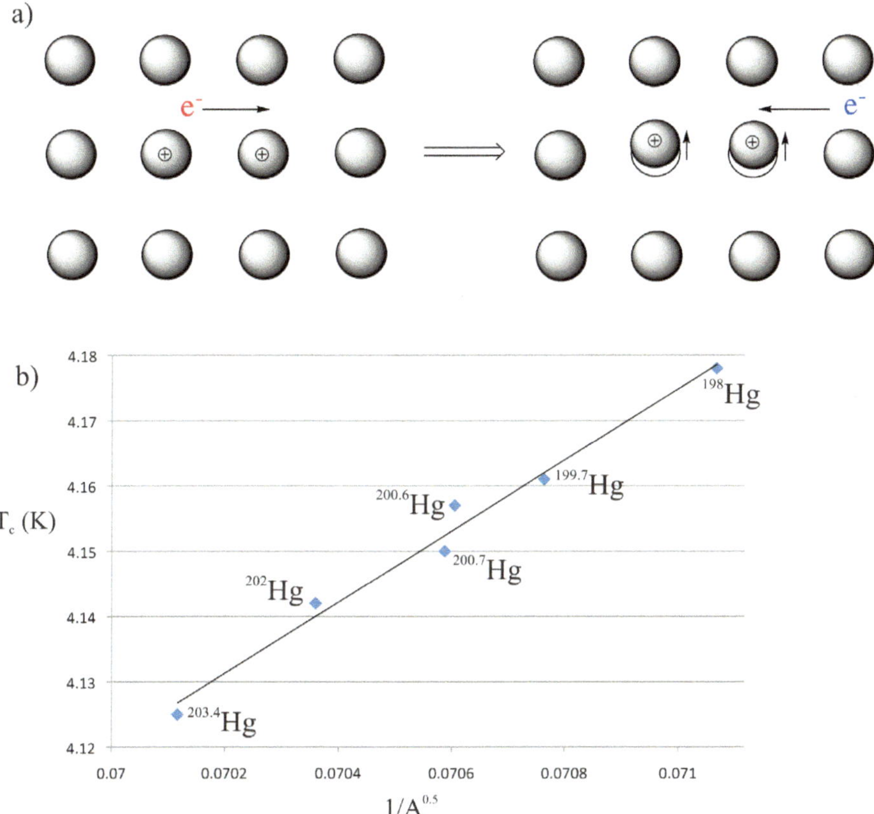

Fig. 2.49 Illustration of Cooper pair formation by electron–phonon coupling, and experimental evidence for their formation ("the isotope effect"). Data taken from Maxwell, E. *Phys. Rev.* **1950**, *78*, 477, and Reynolds, C. A. et al. *Phys. Rev.* **1950**, *78*, 487

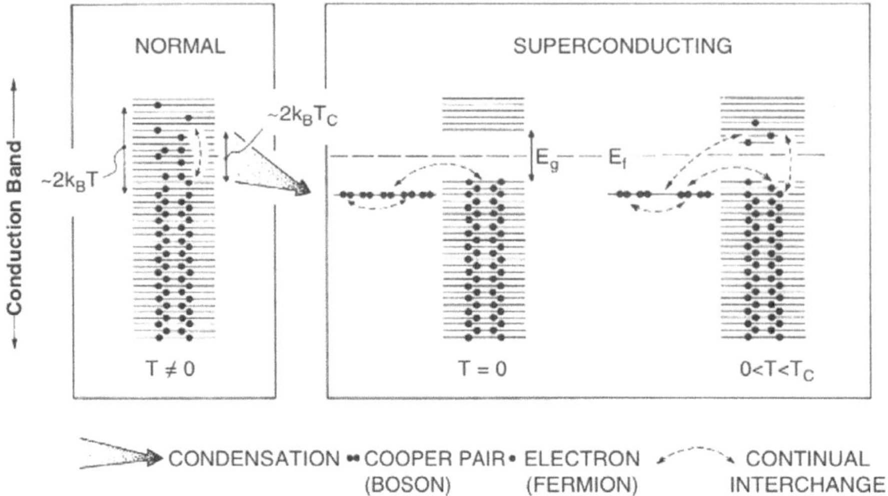

Fig. 2.50 Schematic of Cooper pair formation for a metallic superconductor. At a certain critical temperature, T_c, the electrons within thermal energy (kT_c) of the Fermi level are able to correlate their motion in pairs (Cooper pairs). This process is referred to as condensation. At absolute zero, the condensed Cooper pairs form a single energy state that lies kT_c below the Fermi level. Due to condensation, a number of forbidden energy levels (bandgap, E_g) appear within the conduction band, centered about the Fermi level. The energy of this bandgap corresponds to the minimum energy required to break up a Cooper pair and release the electrons into the vacant quantum levels. At $0 < T < T_c$, the superconductor metal is in an excited state, wherein some electrons, primarily from broken Cooper pairs, have been promoted across the bandgap into vacant energy states. Reproduced with permission from Hurd, C. M. *Electrons in Metals*, Wiley: New York, 1975. © John Wiley & Sons Limited

pair and release the electrons into the vacant quantum levels. The energy gap may be measured by microwave absorption spectroscopy and represents another key experimental finding that supports the BCS theory. As the critical temperature is approached, the energy gap decreases; at $0 < T < T_c$, the superconductor metal is in an excited state, wherein a number of electrons, primarily from broken Cooper pairs, have been promoted across the bandgap into vacant energy states. This indicates that the binding energy of the Cooper pairs is decreased as the temperature increases, caused by greater phonon vibrations that interrupt electron correlation. At $T > T_c$, the binding energy of the Cooper pairs has been exceeded and the electrons behave as discrete carriers, resulting in bulk resistivity due to e^-/e^- collisions. According to this theory, a superconductor's electrical resistance will be zero because the Cooper pair condensate moves as a coherent quantum mechanical entity, which lattice vibrations and impurities cannot disrupt by scattering in the same manner as individual conduction electrons.

In order to exhibit superconductive behavior, early Type I and II materials needed to be cooled below a critical temperature (T_c) ranging from 0.015 K (for W) to 23 K (for Nb_3Ge). An intriguing goal of current research is to design *high-temperature superconductors*, HTS, with the ultimate goal of increasing the T_c to room temperature, which would trivialize resistance-free applications such as power grid lines and levitated trains. In 1986, Muller and Bednorz at IBM made an important discovery toward this goal—the first high-temperature superconductor, $La_{2-x}Sr_xCuO_4$ (LSCO), with a critical temperature of 35 K. A year later, the first material with a critical point above the boiling point of nitrogen (77 K) was discovered, known as $YBa_2Cu_3O_{7-\delta}$ (YBCO), with a critical point of 92 K. In more recent years, the highest-temperature cuprate-based superconductors have been synthesized with a general formula $M_uN_wCa_xCu_yO_z$ (where M = Y, Bi, Tl, or Hg; N = Ba or Sr; $u = 1$ or 2; $w = 2$ or 4; $x = 0$, 1, or 2; $y = 1$, 2, or 3; $z = 3$, 4, 6, 7, 9, 10, or 15). For instance, $TlBa_2Ca_2Cu_3O_9$ ($T_c = 133$ K), $Hg_{0.8}Tl_{0.2}Ba_2Ca_2Cu_3O_{8.33}$ ($T_c =$ 138 K), and $(Hg_{0.75}Pb_{0.15}Tl_{0.1})Ba_2Ca_2Cu_3O_{8+}$ ($T_c = 142$ K).[37] To date, the highest-temperature superconductor is hydrogen sulfide (H_2S), with a T_c of 203 K at 90 kPa pressure (Fig. 2.51).[38]

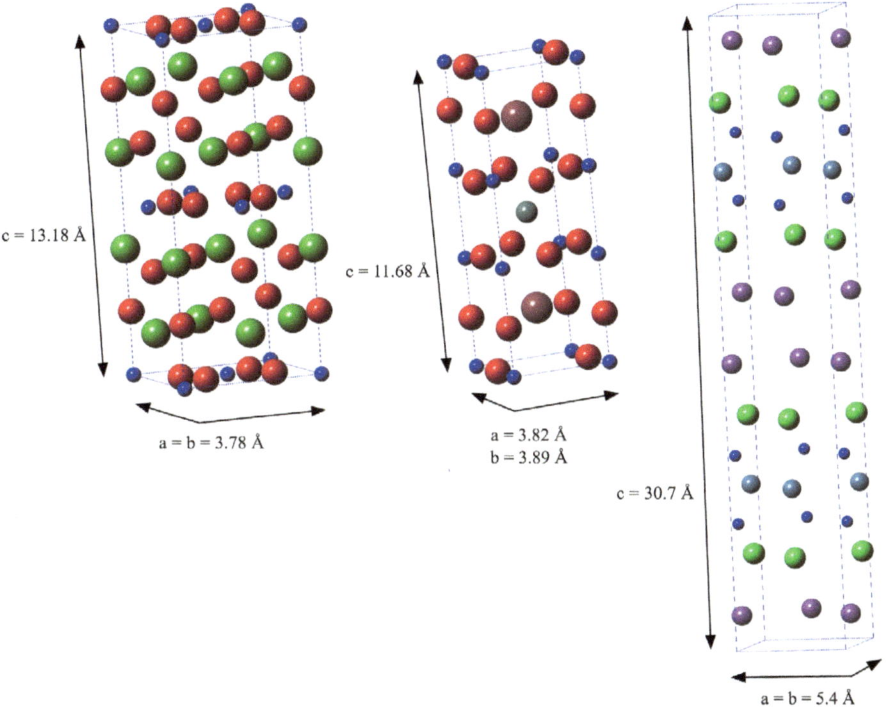

Fig. 2.51 Comparative unit cells for common cuprate-based superconductors. Shown are (left–right): LSCO (shown without Sr^{2+}; green = La^{3+}, blue = Cu^{2+}, red = O^{2-}), YBCO (blue-green = Y^{3+}, purple = Ba^{2+}, blue = Cu^{2+}, red = O^{2-}), and BSCCO (shown without O^{2-} ions for clarity; purple = Bi^{3+}, green = Sr^{2+}, blue-green = Ca^{2+}, blue = Cu^{2+})

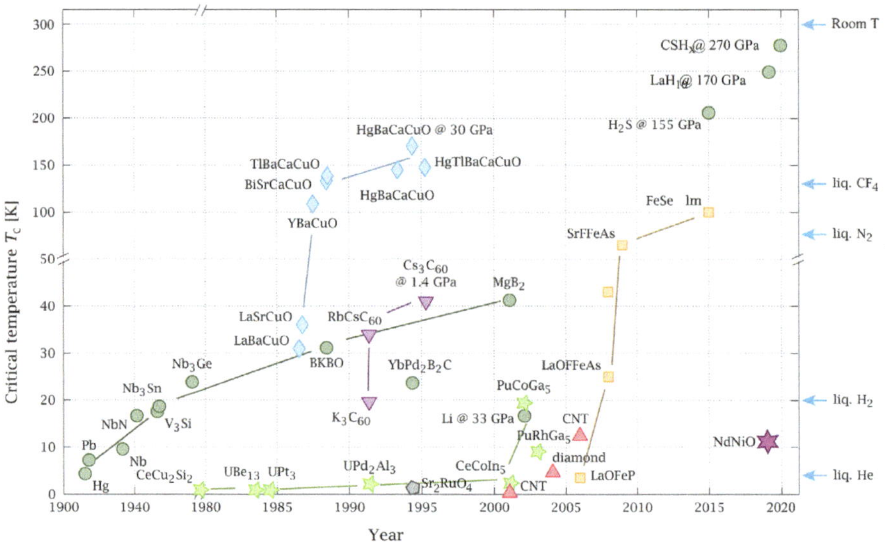

Fig. 2.52 Timeline of superconducting materials. Colors represent different classes of materials. Image obtained under CC BY-SA 4.0, courtesy of PJRay

Figure 2.52 shows a comparison between the major stacked cuprates that played an important role in the early development of room-temperature superconductors. The presence of three or four metals in the crystal structure is referred to as a ternary or quaternary metal oxide, respectively. Unlike the BCS theory for type-I superconductors, there is no prevailing theory to explain high-temperature superconductivity. The lattices for these compounds are distorted perovskites, where "CuO_2" layers are thought to be responsible for the superconductive behavior, and the other metal and metal oxide layers act as charge reservoirs, inducing redox behavior in the copper oxide layers. The Cu^{2+} ion is square planar with an electronic configuration of [Ar]$3d^9$ (one unpaired electron, $S = 1/2$).

For $YBa_2Cu_3O_{7-\delta}$ at an oxygen concentration corresponding to $\delta \geq 0.7$, the solid is insulating. This is due to effective antiferromagnetic ordering of the unpaired electron spins among neighboring Cu^{2+} ions in the CuO_2 layer. However, as additional O is added to the lattice, YBCO undergoes the *Mott transition* of insulating to metallic conductivity—thought to result from a change in the crystal symmetry from tetragonal to orthorhombic. The additional electronegative O atoms in the lattice serve to inject excess "holes" via oxidation of some Cu^{2+} centers to Cu^{3+}; hence, at $T < T_c$, the system is *perfectly diamagnetic* as the unpaired electrons condense into Cooper pairs and Cu^{3+} ions are formally d^8 (low-spin, diamagnetic). In these structures, the holes are also found to segregate themselves into stripes that alternate with antiferromagnetic regions in the material, widely referred to as a "stripe phase", thought to be important in the mechanism for high-T_c superconductivity.[39] Others have shown that lattice vibrations (phonons) play an unconventional role in superconductivity—in particular, electron–phonon coupling interactions.[40] More recently, scientists have employed the Fermi-Hubbard model, which represents electrons as Fermi–Dirac particles ("fermions") that jump between lattice positions and only interact with one another when occupying the same lattice site.[41] As you can see, there is no unifying theory that has yet been adopted to explain the superconductivity of high-T_c materials.

In terms of formal charges on the ions, p-type (or hole-doped) YBCO may be written as: $Y^{3+}(Ba^{2+})_2(Cu^{2+})_2Cu^{3+}(O^{2-})_{7-\delta}$. YBCO becomes superconductive at $\delta \leq 0.4$, with its most pronounced superconductivity at $\delta = 0.05$.[42] It should be noted that there are other examples of p-type superconductors that involve metal doping rather than varying oxygen content, such as $La_{2-x}Sr_xCuO_4$ ($T_c = 34$ K at x = 0.15).[43] Similarly, electron-doped (n-type) superconductors may be synthesized such as $Nd_{2-x}Ce_xCuO_4$ ($T_c = 20$ K), written formally as $Nd^{3+}{}_{2-x}Ce^{4+}{}_x(e^-)_xCu^{2+}O_4$.

A recent class of superconductors that contain iron has been synthesized only within the last few years.[44] The compound LaFePO was discovered in 2006, with a critical temperature of 4 K; fluorine doping to yield $LaFe[O_{1-x}F_x]$ increases the T_c to 26 K. Since 2008, analogous compounds of general formula (Ln)FeAs(O, F) (Ln = Ce, Sm, Nd, Pr) have been prepared that exhibit superconductivity at temperatures up to 56 K. Other compounds such as $(Ba, K)Fe_2As_2$ have T_c values up to 38 K, and MFeAs (M = Li, Na) have a T_c around 20 K.[45] What is most intriguing about iron-based superconductors is that *ferromagnetism* (see Chap. 3) directly competes against Cooper pair formation. Interestingly, these structures exhibit tetragonal-orthorhombic transitions, analogous to cuprate superconductors; however, there appears to be participation from all five 3d orbitals in the Fermi level, relative to just $d_{x^2-y^2}$ orbitals in the cuprates. Not only will further discoveries in this field be important in developing a unifying theory for HTS, but the physical properties of alternative HTS materials may be more conducive for applications. That is, cuprates suffer from a high degree of brittleness that precludes the facile production of superconductive power lines.[46]

There are already commercial applications that employ superconductive materials; for example, MAGLEV trains have been operable for many years in Japan, China, and England. However, reports of deleterious effects of radio waves may slow the widespread use of this technology. Beyond transportation applications, HTS cables (Fig. 2.53) are being developed to deliver power to homes and businesses through the utility power grid. In 2008, HTS cable spanning nearly half a mile was used in the Long Island Power Authority (LIPA) grid network.[47] This project represented the world's first installation of a superconducting cable in a live grid at transmission voltages, which served the equivalent of 300,000 homes. The AmpaCity HTS cable in Essen, Germany, is currently the world's longest, with a capacity of 40 MW. However, there are current plans for HTS installations in Russia, Amsterdam, Chicago, and Japan that will span much longer distances.[48] In order to develop a widespread resilient and ultra-efficient electric grid, many government

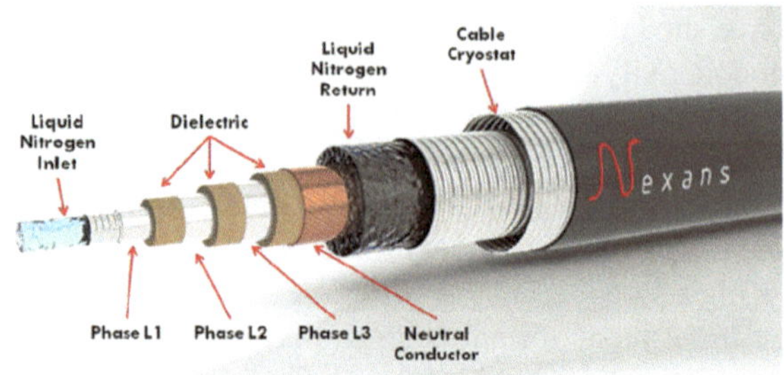

Fig. 2.53 Photograph of a high-temperature superconductor (HTS) wire used for electrical power applications. LN refers to the flow direction of the liquid nitrogen cryogenic cooling fluid. Photograph courtesy of Nexans

agencies have established priorities to develop HTS wires and novel cryogenic dielectric materials, as well as associated electrical applications such as cables, fault current limiters, and transformers.[49]

2.4.6 Crystal Imperfections

All crystals will possess a variety of defects in isolated or more extensive areas of their extended lattice. Surprisingly, even in solids with a purity of 99.9999%, there are on the order of 6×10^{16} impurities per cm^3! However, impurities are not always a disadvantage. Often, these impurities are added deliberately to a solid to improve its electrical, magnetic, or optical properties.

There are four main classifications of imperfections that exist in crystalline solids:

(a) *Point defects* —interstitial/substitutional dopants, Schottky/Frenkel defects, F- or H-centers, voids (vacancies)
(b) *Linear defects* —edge and screw dislocations
(c) *Surface defects* —grain boundaries, twin boundaries, surfaces, interfaces
(d) *Volume defects*—pores, cracks, twins.

Of these four types, point/linear defects may only be observed at the atomic level, requiring sophisticated electron microscopy. In contrast, planar defects are often visible using a light microscope, and bulk defects may sometimes be observed by the naked eye.

Although a solution is typically envisioned as a solid solute dissolved in a liquid solvent (*e.g.*, sugar dissolved in water), *solid solutions* are formed upon the placement of foreign atoms/molecules within a host crystal lattice. If the regular crystal lattice is composed of metal atoms, then this solution is referred to as an *alloy*. Solutions that contain two or more species in their crystal lattice may either be *substitutional* or *interstitial* in nature (Fig. 2.54), corresponding to shared occupancy of regular lattice sites or vacancies between lattice sites, respectively.

Substitutional solid solutions feature the actual replacement of solvent atoms/ions that comprise the regular lattice with solute species, known as *dopants*. The dopant species is typically arranged in a random fashion among the various unit cells of

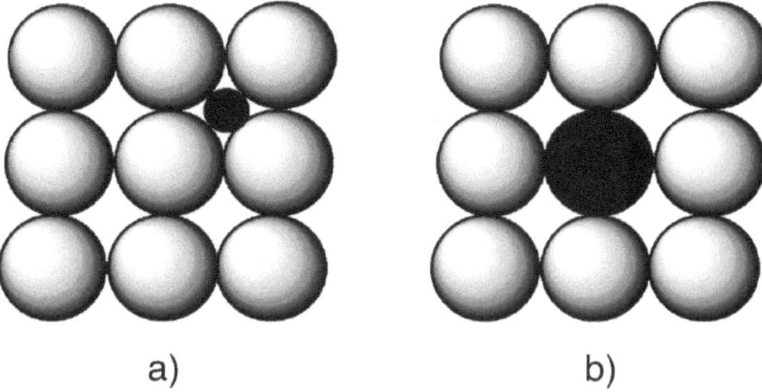

Fig. 2.54 Illustration of the difference between **a** interstitial and **b** substitutional defects in crystalline solids

the extended lattice. Examples of these types of lattices are illustrated by metal-doped aluminum oxide, constituting gemstones such as emeralds and rubies (both due to Cr-doping). For these solids, small numbers of formal Al^{3+} lattice sites are replaced with solute metal ions. We will describe later how this relates to the color of crystalline gemstones.

Formation and thermodynamics of defect formation

In order to form a stable substitutional solid solution of appreciable solubility, the following *Hume-Rothery* rules, must be satisfied. Although these requirements are often cited for metallic alloys, they apply to any solid solution, including ceramic lattices.

1. The percentage difference between solute and solvent atomic radii should be less than 15% (2.13). If there were a larger mismatch between the dopant and solvent atomic radii, the rate of diffusion would either be too slow (for large r_{solute}), or the dopant would position itself in an interstitial site rather than replacing a solvent atom (for small r_{solute}). For instance, although most gemstones feature the replacement of formally Al^{3+} ions in an aluminum oxide (alumina) based crystal with M^{3+} ions (M = Cr, Ti, Fe), the substitution of aluminum ions with Li^+ or W^{3+} ions would cause too drastic a perturbation of the crystal structure. Whereas small lithium atoms/ions would have an opportunity to diffuse into the lattice forming interstitial solutions, the large tungsten ions would only adsorb to the surface of individual crystals.

$$\left|\frac{r_{\text{solute}}}{r_{\text{solvent}}}\right| \times 100\% \leq 15\% \qquad (2.13)$$

2. The crystal structures of the dopant and solvent atoms must be matched. That is, the density of the host solvent unit cell must be voluminous enough to accommodate the solute atoms. As we will see more in Chap. 7, the resolution of electron microscopes is now suitable for the easy visualization of small atomic cluster

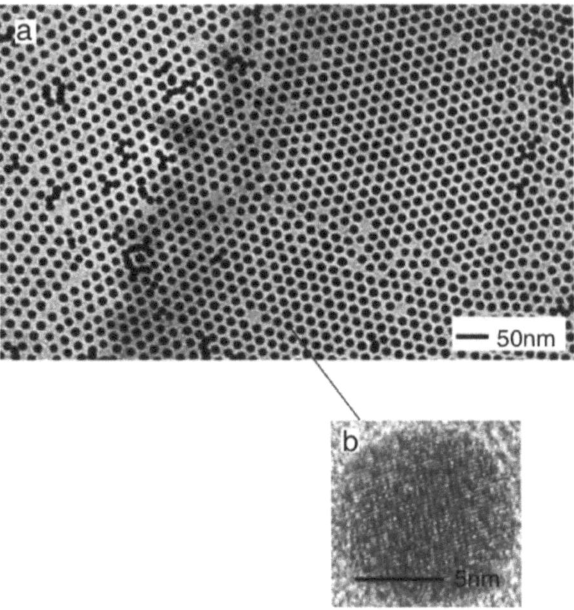

Fig. 2.55 Transmission electron microscope image showing a 2D hexagonal superlattice of Fe–Pd alloy nanoparticles. Image **b** confirms that each nanoparticle consists of only one type of crystal lattice. Reproduced with permission from *Chem. Mater.* **2004**, *16*, 5149. © 2004 American Chemical Society

arrays. Figure 2.55 illustrates a well-ordered array of Fe–Pd alloy nanoparticles. Interestingly, even though Fe (bcc) and Pd (fcc) do not share the same crystal structure, each nanoparticle crystallite comprises only one lattice, indicating that the Fe and Pd metals form a solid solution. Interestingly, it is common for the bcc lattice of iron to change to fcc when alloyed with metals such as Pt, Pd, Cu, or Ni.[50]

3. In order for dopant atoms to be stabilized within a host lattice, both solvent/solute species must have similar electronegativities. If this prerequisite were not met, electron density would transfer to the more electronegative atoms, forming a compound with an entirely new lattice structure and distinct properties. For instance, the reaction of metallic aluminum and nickel results in nickel aluminide, Ni_3Al, a compound with both ceramic and metallic properties. Such *transformational alloys* are in contrast to interstitial and substitutional alloys, in which the original solvent lattice framework is not significantly altered.
4. The solute and solvent atoms should have similar valences in order for maximum solubility, rather than compound formation. In general, a greater solubility will result from the dissolution of a higher-valence solute species in a lower-valence solvent lattice, than vice versa. For instance, the solubility limit of Zn in Cu is 38.4 at% Zn,but only 2.3 at% Cu for Cu in Zn. Solubilities also decrease with an increase in periodic separation; for example, the solubility maximum is 38.4% Zn in Cu, 19.9% Ga in Cu, 11.8% Ge in Cu, and only 6.9% As in Cu.

As the temperature of the crystal is increased, the atoms in the crystal vibrate about their equilibrium positions generating vacancies or voids in the lattice. As shown in

Fig. 2.56, a vacancy may be formed when an energetic surface atom breaks bonds with neighboring atoms and forms a new bond with other atoms on the surface. Once this happens, atoms in the bulk diffuse upward to fill the vacancy. As this atomic diffusion continues toward the surface, the vacancy moves in the opposite direction, toward the bulk of the crystal.

The Arrhenius equation (2.14) is used to calculate the equilibrium number of vacancies or voids in the crystal lattice at a specific temperature. Since the activation energy of defects is often significantly greater than the thermally induced kinetic energy of lattice atoms, the most pronounced atomic migration occurs along dislocations and voids in the crystal, since fewer atoms are involved in the atomic displacement (i.e., E_a is much lower).

$$\frac{N_V}{N_T} = e^{-(E_a/kT)} \tag{2.14}$$

where N_V is the number of vacancies, N_T is the total number of atoms in the crystal lattice, E_a is the activation energy for the diffusion process, k is the Boltzmann constant (1.38×10^{-23} J atom^{-1} K^{-1}), and T is the temperature (K).

For a typical solid, there is one vacancy per 10^{15} lattice atoms at room temperature; however, at a temperature just below its melting point, there will be one vacancy per *ca.* 10,000 lattice atoms.

The number of point defects at thermal equilibrium is related to the balance between the enthalpy required to produce the defect and the increase in entropy from crystal disorder.[51] Equation 2.15 represents the total Gibbs free energy for the formation of n vacancies.

$$\Delta G = \Delta H - T\Delta S \tag{2.15}$$

where T is the absolute temperature of the reaction (in K); ΔG, ΔH, and ΔS are Gibbs free energy, enthalpy, and entropy, respectively.

A process will proceed spontaneously when $\Delta G < 0$ and will continue to proceed until the Gibbs free energy is at its minimum. At this point, the $\Delta G = 0$ and the system is at equilibrium. Hence, the equilibrium concentration of defects will be reached when the Gibbs free energy of the system is at its minimum (Fig. 2.57).

Since bonds are broken during the formation of point defects, the enthalpy (ΔH) of a solid is directly proportional to the number of defects added; hence, defect formation is entropically driven (−TΔS), but limited in number based on the magnitude of ΔH. There are two entropic contributions to defect formation: configurational and vibrational—i.e., disorder caused by the movement of atoms around defect sites. The configurational entropy (S), or entropy of mixing, is described by (2.16).

$$S = k_B \ln(W) \tag{2.16}$$

where k_B = Boltzmann constant (1.38×10^{-23} J/K) and $W = \frac{N!}{n!(N-n)!}$ N = number of lattice sites and n = number of defects. The thermodynamic probability, W, represents the number of distinguishable ways in which n defects may be distributed over ($N + n$) lattice sites.

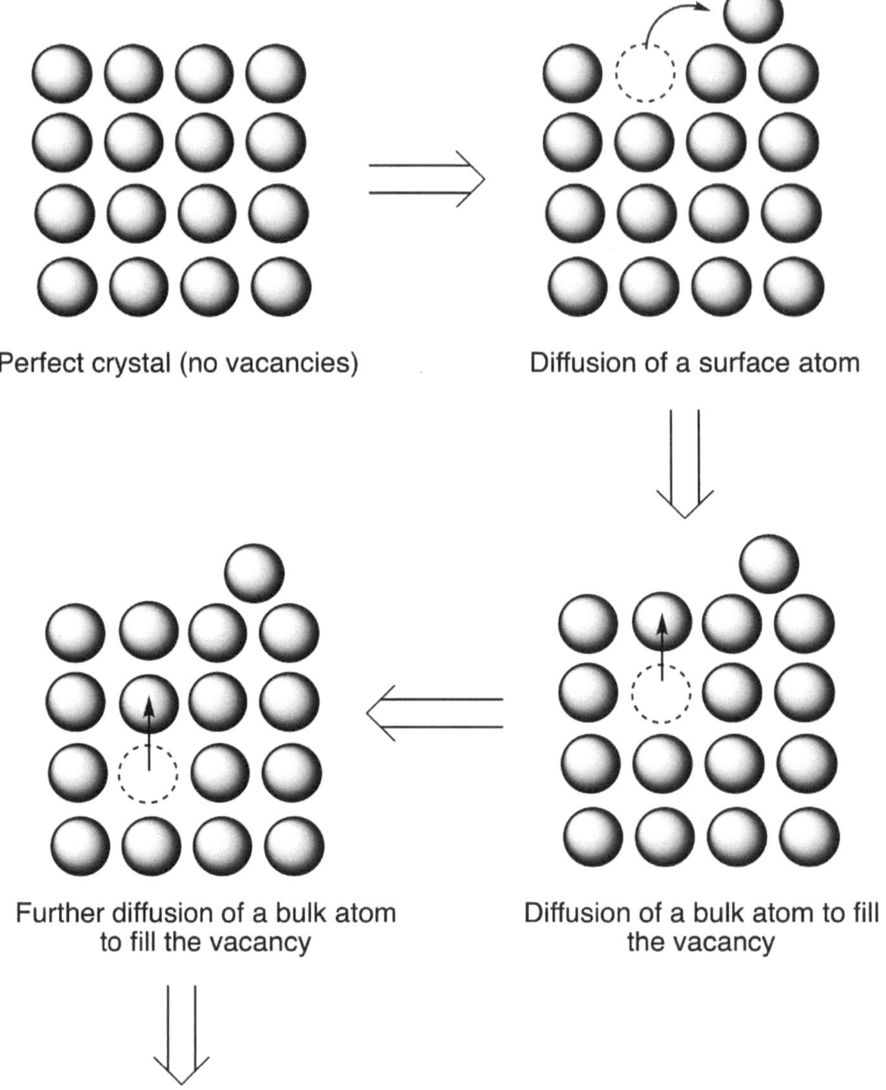

Fig. 2.56 Illustration of vacancy formation and migration in a crystalline solid

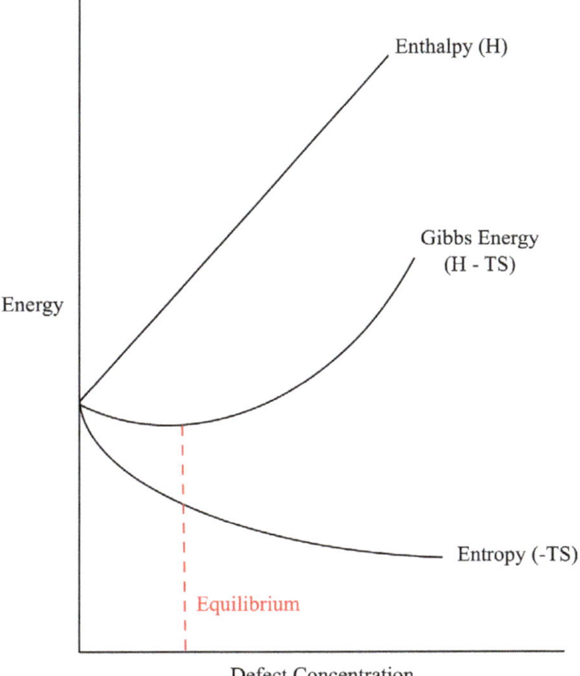

Fig. 2.57 Thermodynamics of defect formation in a solid, showing that defect formation is entropically favored

Therefore, at T > 0 K, point defects may be positioned at any site in the lattice, resulting in a random distribution of impurities or dopants. However, as the temperature is lowered, each lattice position may no longer be equivalent (Fig. 2.58), and ordered arrays known as *superlattices* may be formed. Examples of superlattice behavior are found for Au–Cu alloys used in jewelry, gold fillings, and other applications (Fig. 2.59). For an 18-karat Au/Cu dental alloy, a superlattice will be present at temperatures below 350 °C; at higher temperatures, a random substitutional alloy will be formed. The AuCu I superlattice (Fig. 2.59a) consists of alternating planes of copper and gold atoms, forming a tetragonal unit cell that has been elongated along both **a** and **b** axes. In fact, the hardening of gold alloys is thought to arise from superlattice ordering and precipitation-hardening mechanisms.[52] A more complex superlattice may also be observed in Au/Cu alloys (Fig. 2.59b), consisting of a periodic array of multiple unit cells with Cu and Au atoms exchanging positions between corners and faces. In this case, hardening is thought to occur from the existence of relatively high-energy antiphase boundaries (APBs) between adjacent arrays.[53]

In contrast to substitutional solid solutions, there must be a significant size difference between solute and solvent species for appreciable interstitial solubility (2.17). Accordingly, the most common interstitial solutes are hydrogen, carbon, nitrogen, and oxygen. If the dopant species is identical to the lattice atoms, the occupancy is

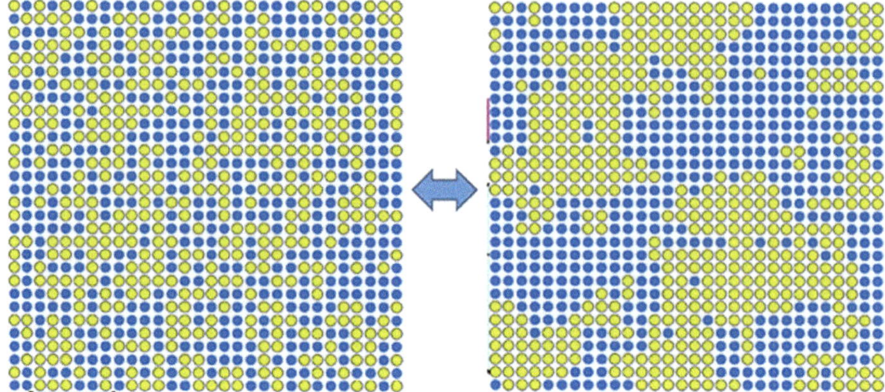

Fig. 2.58 Illustration of an order/disorder phase transition for a binary alloy. At higher temperatures, the alloy components are randomly positioned (left) due to thermal population of degenerate configurational entropy states. However, as the temperature is reduced, the attractive interactions among atoms within each alloy component removes the configurational degeneracy of lattice positions, causing separation of the components (right). Reproduced with permission from https://doi.org/10.3389/fchem.2020.00757

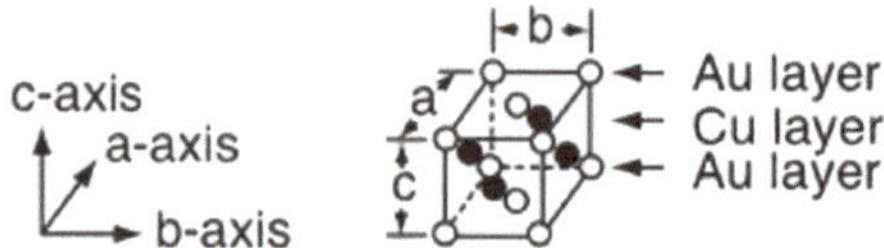

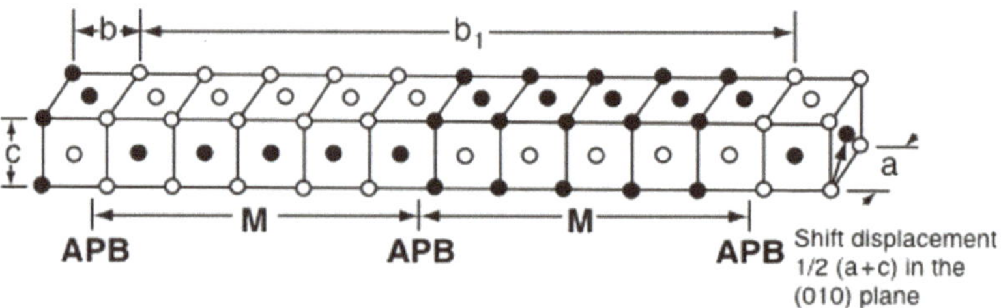

Fig. 2.59 Unit cell representations of two varieties of AuCu superlattices. For the AuCu II superlattice, M refers to the length of repeat unit, and APB indicates the antiphase boundaries between adjacent periodic arrays. Republished with the permission of the International and American Associations for Dental Research, from "Determination of the AuCu Superlattice Formation Region in Gold–Copper–Silver Ternary System", Uzuka, T.; Kanzawa, Y.; Yasuda, K. *J. Dent. Res.* **1981**, *60*, 883; permission conveyed through Copyright Clearance Center, Inc.

referred to as *self-interstitial*. This will result in a large local distortion of the lattice since the lattice atom is significantly larger than intersitial sites. Consequently, the energy of self-interstitial formation is *ca.* 3 times greater than that required to form vacancies, resulting in a very low concentration (i.e., < 1 cm^{-3} at room temperature).

$$\text{Interstitial solubility} = \frac{r_{solute}}{r_{solvent}} \leq 0.59 \tag{2.17}$$

As one would expect, smaller atoms diffuse more readily than larger ones. For instance, the interdiffusion of a carbon impurity atom within an $\alpha-$Fe lattice at 500 °C is $2.4 \times 10^{-12}\ m^2\ s^{-1}$, relative to $3.0 \times 10^{-21}\ m^2\ s^{-1}$ for self-diffusion of Fe atoms. Whereas carbon may migrate via *interstitial diffusion* requiring minimal lattice distortion, Fe diffusion occurs via *vacancy diffusion* (Fig. 2.60), which necessitates a much greater perturbation of the lattice since strong Fe–Fe metallic bonds must first be broken. Figure 2.61 compares the diffusion coefficients for various interstitial and substitutional impurities. Not surprisingly, the diffusion coefficient is shown to increase exponentially with temperature. An increasing slope for the diffusion coefficient vs. temperature line indicates a higher the activation energy that is required for defect diffusion, with substitutional impurities displaying much larger slopes than interstitial defects. The activation energy for self-diffusion increases as the melting point of the solid increases, which is related to the strength of lattice bonds in the crystal. The slope of the diffusion coefficient will also be much greater in bulk regions of the crystal than on the surface or near grain boundaries where diffusion may occur more readily.

Dopant diffusion follows Fick's first and second laws,[54] which rely on the temperature, diffusivity of the dopant, and concentration profiles of the crystal, which can be quite complicated for some solids such as alloys. However, there are some general trends that are observed for most crystals. For instance, dopant diffusion will generally be faster in open, less-dense crystal structures, lower-melting solids, crystals with weaker bonding (intermolecular forces), and solids involving smaller diffusing species such as cations.

The migration of a lattice atom/ion into an available interstitial site will leave behind a vacancy (Fig. 2.62); the formation of such an interstitial/vacancy pair is known as a *Frenkel defect*. In contrast, *Schottky defects* are formed through the migration of a cation–anion pair from the crystal lattice framework, leaving behind two vacant lattice sites. It requires considerably less energy to form a Frenkel defect (*ca.* 130 kJ mol^{-1}), relative to a Schottky defect (300 kJ mol^{-1}) since fewer lattice species are involved. Whereas "soft" metals (e.g., Ag) with a higher degree of covalency prefer Frenkel defects, "hard" metals such as Na prefer Schottky defects. It should be noted that cationic Frenkel defects are more common than anionic since the former are smaller and more readily fit into interstitial sites. However, most crystals do not generally contain large numbers of Schottky or Frenkel defects. At room temperature, only 1 in 10^{15} lattice sites are vacant in NaCl, which translates to *ca.* 10^4 Schottky defects in 1 mg of table salt. Even at 1000 K, there are typically only 1–2 defects per 100,000,000 lattice sites!

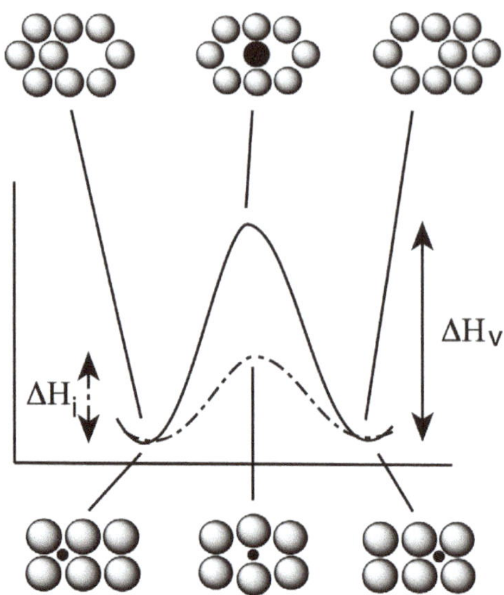

Fig. 2.60 Comparative illustration of the energetics involved for the atomic diffusion of interstitial and substitutional impurities

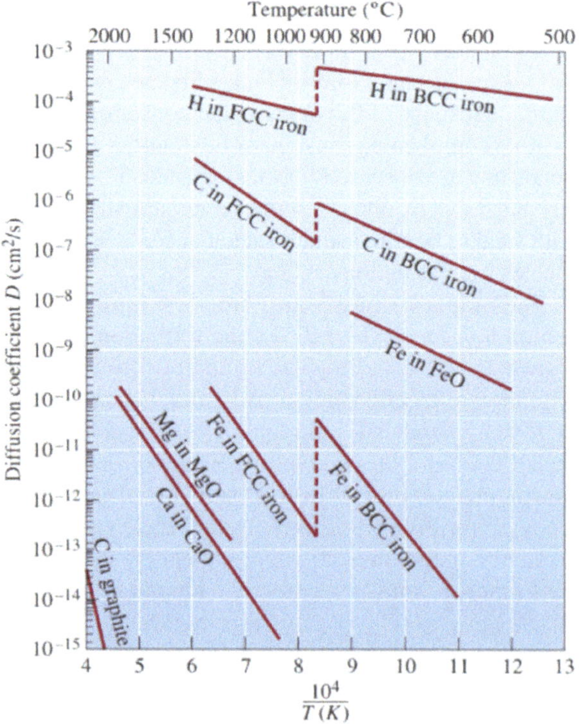

Fig. 2.61 Comparative diffusion coefficients for various substitutional and interstitial dopants

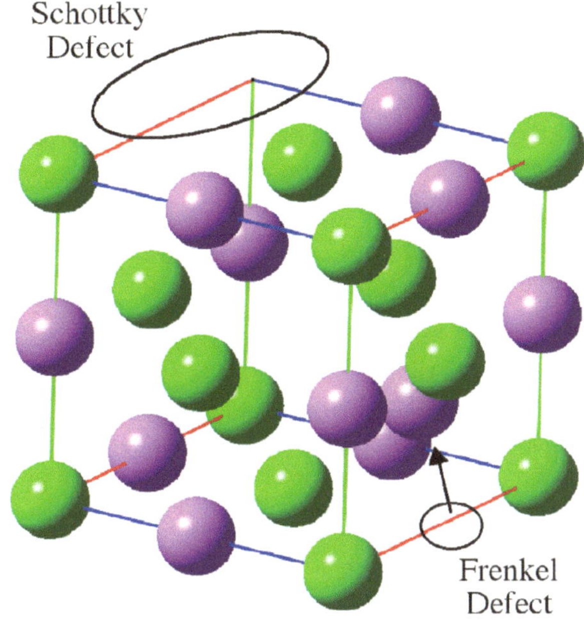

Fig. 2.62 Illustration of a unit cell of an ionic crystal with Frenkel and Schottky defects

The Kroger-Vink notation for defects

Point defects are denoted by a symbol that indicates its charge and placement in the crystal lattice. The Kroger-Vink notation consists of the following features:

- An elemental symbol or V for a vacancy
- A superscript of · indicates a net charge of +1; a superscript of ‘ indicates a net charge of −1. A superscript of x indicates there is no net charge
- A subscript indicates the site preference in the crystal lattice:
  - s = surface site
  - i = interstitial site
  - elemental symbol = normal lattice site

  To illustrate this notation, consider the following examples:

- V'_{Na} = a sodium ion vacancy with a net charge of −1
- $V^{\bullet}_{Cl}$ = chloride ion vacancy with a charge of +1
- Fe^{x}_{Fe} = an iron atom in its normal lattice site with no net charge
- $Ag^{\bullet}_{i}$ = a silver atom in an interstitial site with a net charge of +1.

For reactions that involve defects, the following rules must be met:

- The compounds should remain electrically neutral
- The ratio(s) of the number of regular cationic/anionic sites must remain constant
- The number and type of atoms involved in the defect reaction must be conserved (mass balanced); vacancies and electronic defects (electrons and holes) do not count toward mass balancing.

Equation 2.18 shows the Kroger-Vink notation for the formation of Schottky disorder in a MO salt (where M = metal).

$$M_M^x + O_O^x \rightleftharpoons V''_M + V_O^{\cdot\cdot} + M_M^x + O_O^x \tag{2.18}$$

This notation indicates the movement of both a metal and oxygen ion from their regular lattice sites. Since a M^{2+} ion has been removed from the lattice, the formed vacancy will have a charge of −2 (V''_M); in contrast, the vacancy formed from the displacement of a O^{2-} ion will have a +2 charge ($V^{\cdot\cdot}_O$). In order to compensate for the vacancies, two additional atoms are shown on the reactant side, which are needed at the surface to maintain an overall mass balance.

Equations 2.19 and 2.20 denote the Kroger-Vink notation for cation and anion Frenkel ("anti-Frenkel") disorder, respectively.

$$M_M^x \rightleftharpoons V''_M + M_i^{\cdot\cdot} \tag{2.19}$$

$$O_O^x \rightleftharpoons V_o^{\cdot\cdot} + O''_i \tag{2.20}$$

The notation for Frenkel defect formation shows the migration of a divalent metal ion from a regular lattice position into an interstitial site ($M^{\cdot\cdot}_i$). This leaves behind a vacancy, which has a -2 charge (V''_M). For an anti-Frenkel disorder, an oxygen ion moves into an interstitial site, leaving behind a positively charged vacancy ($V^{\cdot\cdot}_O$). Anion Frenkel defects are common for crystals with relatively small anions, such as CaF_2 (fluorite), UO_2, or TiO_2.

For ionic crystals, the overall charge of the crystal must be charge-balanced. Oftentimes, a dopant of the same oxidation state is introduced to a crystal, which will simply replace the lattice ion. However, *aliovalent substitution* occurs when the doping and lattice species differ in oxidation state, which induces the formation of vacancies in a crystal lattice. For instance, during the aliovalent substitution of ZrO_2 (i.e., Zr^{4+}) with Y_2O_3 (i.e., Y^{3+}) there must be a concomitant formation of anion vacancies to balance the crystal charge (2.21):

$$Y_2O_3 \rightleftharpoons 2Y'Z_r + V_O^{\cdot\cdot} + 3O^{xo} \tag{2.21}$$

In contrast, cation vacancies will be produced if higher-valent cations are used as dopants. For instance, consider the doping of KCl with $CaCl_2$ (2.22):

$$CaCl_2 \rightleftharpoons Ca_K^{\cdot} + V_K' + 2Cl_{Cl}^{x} \tag{2.22}$$

In this reaction, the 1:1 ratio of K:Cl sites must be maintained. Two Cl^- anions occupy the existing Cl sites; hence, two cation sites must be created—one occupied by Ca^{2+} ($Ca^{\bullet}_K$) and one left vacant (V'_K). This aliovalent substitution results in a net + 1 charge; hence, a vacancy must be introduced to offset this charge.

When the composition of a crystal is defined by a distinct chemical formula (e.g., SiO_2), it is known as a *stoichiometric compound*. If the composition of the crystal is altered upon doping or thermal treatment, the resulting solid may deviate from the original chemical formula, forming a *nonstoichiometric* solid. For instance, the compound $K_{1-2x}Ca_xCl$ will be formed from the aliovalent doping shown in (2.22). Nonstoichiometry and the existence of point defects in a solid are often closely related, and are prevalent for transition metal (e.g., Ti, Zr, Hf, V, W, Zn, Fe) and main group (e.g., Si, Al) oxides, sulfides, and hydrides. For example, the formation of x anion vacancies per each quartz (SiO_2) unit cell will result in the nonstoichiometric compound SiO_{2-x}. As seen earlier, nonstoichiometric cuprates are used for high-temperature superconductor applications; other applications for this class of solids include photocatalysis and energy storage.

Bulk Defects

Bulk defects are produced through the propagation of microscopic flaws in the lattice. For crystals with a planar defect such as polycrystalline solids, the *grain boundary* marks the interface between two misaligned portions of the bulk crystal (Fig. 2.63). The size of the individual microcrystals (or grains) that comprise a larger aggregate greatly affects many properties of the bulk crystal. Both optical microscopy and X-ray diffraction are used to determine the grain sizes; most commercial metals and alloys consist of individual crystallites with diameters ranging from 10 to 100μm, each corresponding to millions of individual metal atoms. Since energy is required to form a surface, grains tend to grow in size at the expense of smaller grains to minimize energy. This growth process occurs by diffusion, which is accelerated at elevated temperatures.

A decrease in the size of these microscopic grains or crystallites results in an increase in both strength and hardness of the bulk material, due to closer packing among neighboring grains. The density of atoms at a solid surface, or in the region surrounding a grain boundary is always smaller than the bulk value. That is, the atoms in these regions are coordinatively unsaturated and contain dangling bonds (Fig. 2.64). Hence, surfaces and interfaces are very reactive, often resulting in the concentration of impurities in these regions.

A special type of grain boundary, known as *twinning*, occurs when two crystals of the same type intergrow, so that only a slight misorientation exists between

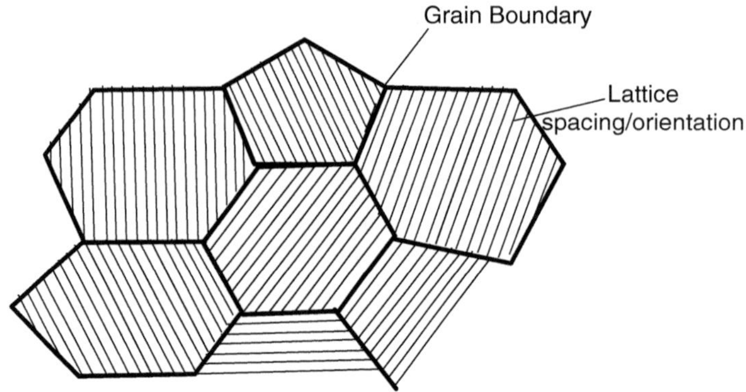

Fig. 2.63 Illustration of grain boundaries between individual crystalline domains

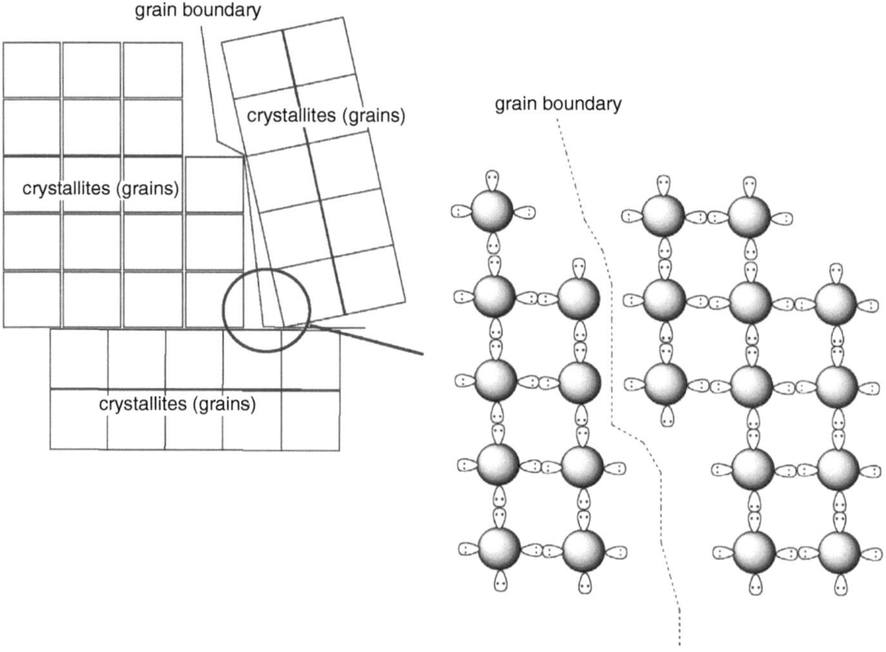

Fig. 2.64 Schematic of a polycrystalline solid, with grain boundaries formed from dangling bonds between neighboring metal atoms

them. Twinned crystals may form by inducing alterations in the lattice during nucleation/growth (e.g., impurity incorporation during slow cooling), or from the application of an external force (*stress*). The twin boundary is a highly symmetrical interface, often with the crystal pairs related to one another by a mirror plane or rotation axis. Accordingly, twinning poses a problem in determining the correct crystal structure via

X-ray diffraction due to the complexity created by overlapping reciprocal lattices.[55] Due to the symmetric equivalence of the polycrystals, twin boundaries represent a much lower-energy interface than typical grain boundaries that are formed when crystals of arbitrary orientation grow together.

A stress exerted on a material results in a structural deformation referred to as *strain*, whose magnitude is related to the bonding interactions among the atoms comprising the solid. For example, a rubbery material will exhibit a greater strain than a covalently bound solid such as diamond. Since steels contain similar atoms, most will behave similarly as a result of an applied stress. There are four modes of applying a load, referred to as tension, compression, shear, and torsional stresses (Fig. 2.65). Both tension and compression stresses are applied parallel to the long axis of the material, resulting in elongation or contraction of the material along the direction of the stress, respectively. In contrast, shear stress is applied at some angle with respect to the long axis and will cause the material to bend. The resultant flex is referred to as *shear strain*.

For small stresses, a material will generally deform elastically, involving no permanent displacement of atoms and reversal of the deformation upon removal of the shear stress. The linear relationship between stress and strain in these systems is governed by Hooke's law (2.23). The stiffer the material, the greater will be its *Young's modulus*, or slope of the stress *vs.* strain curve. It should be noted that some materials such as concrete do not exhibit a linear stress/strain relationship during elastic deformation. In these cases, the modulus is determined by taking the slope of a tangential line drawn at a specific level of stress.

$$\sigma = E \times \varepsilon \tag{2.23}$$

where σ is tensile stress in units of force per unit area (S.I. unit: 1 Pa = 1 N/m^2), E is the Young's modulus, or modulus of elasticity (e.g., 3 GPa for Nylon, 69 GPa for aluminum, and 407 GPa for tungsten), and ε is the strain, defined as the geometrical change in shape of an object in response to an applied stress.

Poisson's ratio is used to describe the lateral distortion that is generated in response to a tensile strain. The values for elastomeric polymers are *ca.* 0.5, metals 0.25–0.35, polymeric foams 0.1–0.4, and cork is near zero. Interestingly, *auxetic* materials exhibit a negative Poisson's ratio, becoming thicker under tension (Fig. 2.66). Although this phenomenon was first discovered for foam-like structures more than 40 years ago,[56] these materials also exist in nature. For instance, biomaterials such as cowhide and human tendons exhibit auxetic properties. There are now many classes of materials such as fcc metals, silicates (e.g., $Na_2Al_2Si_3O_{10}$), and polymers that display this property.[57] Such materials exhibit interesting mechanical properties such as high-energy shock absorption and fracture resistance, which may prove useful for applications such as packing material, personal protective gear, tennis shoes, and body armor. The popular waterproof and breathable fabric Gore-Tex® is an auxetic material, composed of a fluorinated polymer.

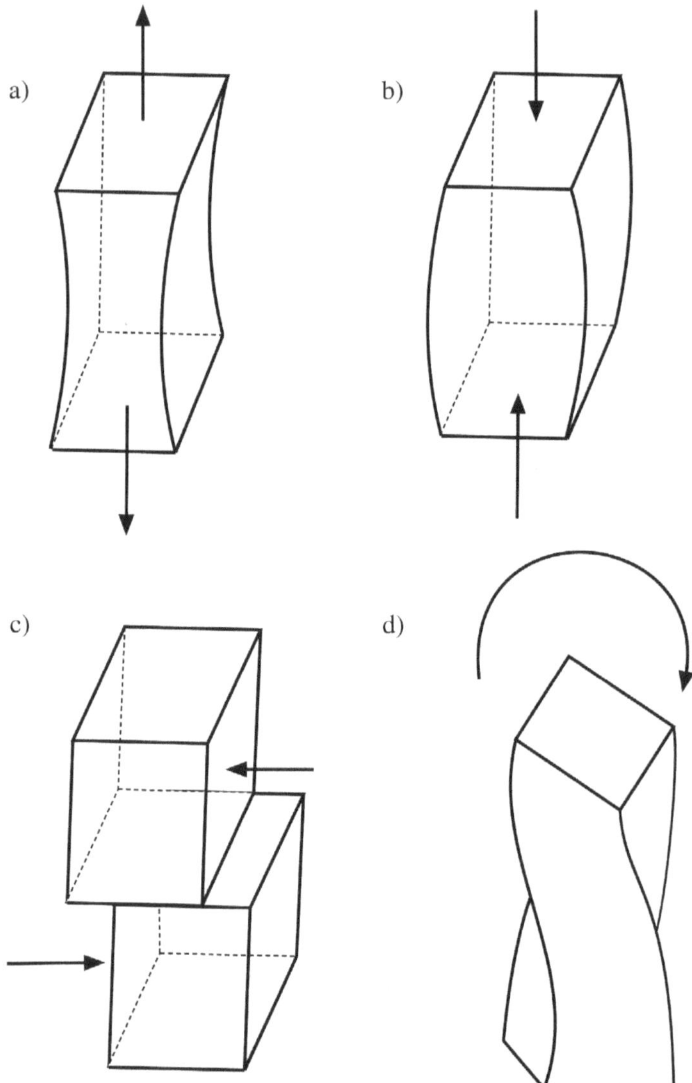

Fig. 2.65 Illustration of various types of loads (stresses), which result in material strain. Shown are **a** tensile stress, **b** compressive stress, **c** shear stress, and **d** tortional stress

For large stresses, a material will deform plastically, involving the permanent displacement of atoms. The onset of plastic deformation is referred to as the *yield point* (or *yield strength*) of the material. For most metals, there is a gradual transition from elastic to plastic deformation; however, some steels exhibit very sharp transitions. After the yield point is reached, plastic deformation continues until the material reaches its fracture point. Accordingly, the *tensile strength* represents the

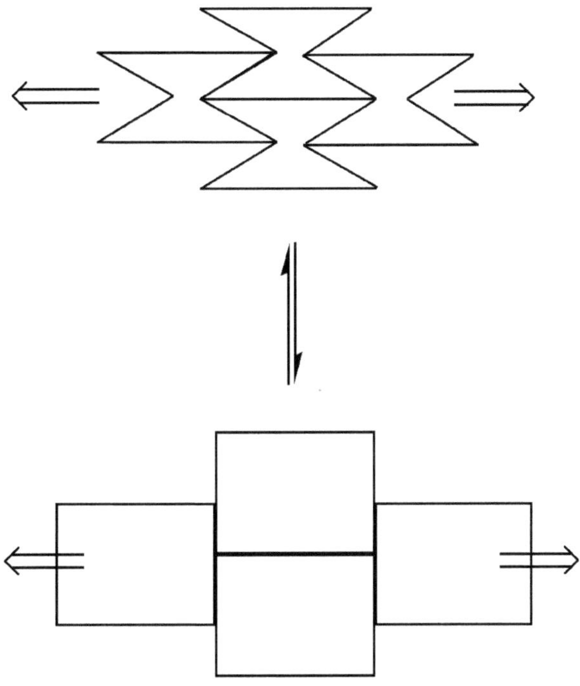

Fig. 2.66 Illustration of the deformation modes exhibited by an auxetic material. These materials possess hinge-like structures that flex upon elongation

maximum strain in the stress *vs.* strain curve (Fig. 2.67); this property with respect to its weight is referred to as the *specific strength* (2.24). In Chap. 3, we will discuss some strategies used to increase the yield and tensile strengths of metallic alloys—of extreme importance for structural engineering applications.

$$\text{specific strength} = \frac{\text{tensile strength}}{\text{specific gravity}} \tag{2.24}$$

The plastic deformation of crystalline solids is referred to as *slip* and involves the formation and movement of *dislocations*. Edge and screw dislocations are abrupt changes in the regular ordering of atoms along an axis in the crystal, resulting from breaking/reforming large numbers of atomic bonds (Fig. 2.68). Dislocations may be created by shear force acting along a line in the crystal lattice. As one would expect, the stress required to induce such dislocations is extremely high—of the same magnitude as the strength of the crystal. Accordingly, it is more likely that dislocations arise from irregularities (e.g., steps, ledges) at grain boundaries or crystal surfaces for poly- and single crystals, respectively, which may then propagate throughout the crystal lattice.

During plastic deformation, existing dislocations serve as nucleation sites for new dislocations to form; hence, the dislocation density of the material increases significantly. Whereas the dislocation density (in units of dislocation distance per unit volume; mm/mm^3 or mm^{-2}) of pure metallic crystals is on the order of 10^3 mm^{-2},

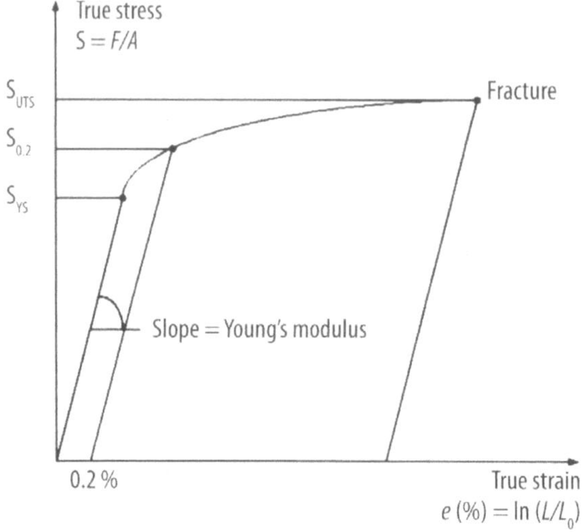

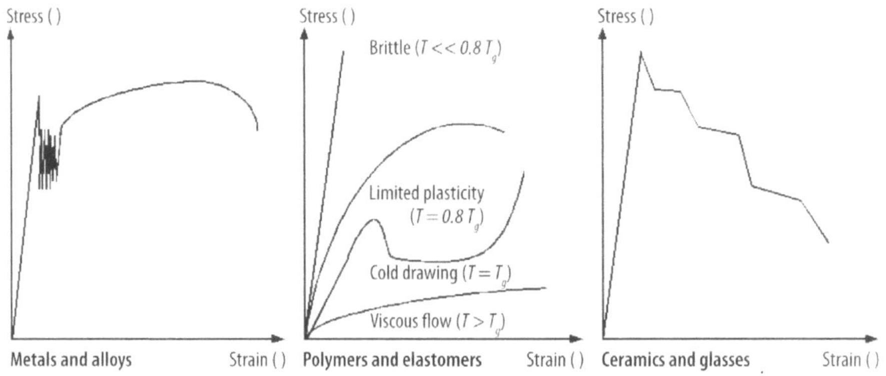

Fig. 2.67 Illustration of a true stress *vs.* strain curve and comparison of stress–strain curves for various materials. UTS = ultimate strength, and YS = yield strength. The tensile strength is the point of rupture, and the offset strain is typically 0.2%—used to determine the yield strength for metals without a well-defined yield point.[58] Reproduced with permission from Cardarelli, F. *Materials Handbook*, 2nd ed., Springer: New York, 2008. © 2008 Springer Science & Business Media

the density may reach 10^{10} mm^{-2} in heavily deformed metals. It should be noted that line defects may not always be detrimental. As we will see in Chap. 3, the interactions among neighboring dislocations are responsible for work hardening of metals.

In single crystals, there are preferred planes where dislocations may propagate, referred to as *slip planes*. For a particular crystal system, the planes with the greatest atomic density will exhibit the most pronounced slip. For example, slip planes for bcc and fcc crystals are {110} and {111}, respectively; other planes, along with those present in hcp crystals, are listed in Table 2.8. Metals with bcc or fcc lattices have significantly larger numbers of slip systems (planes/directions) relative to hcp. For

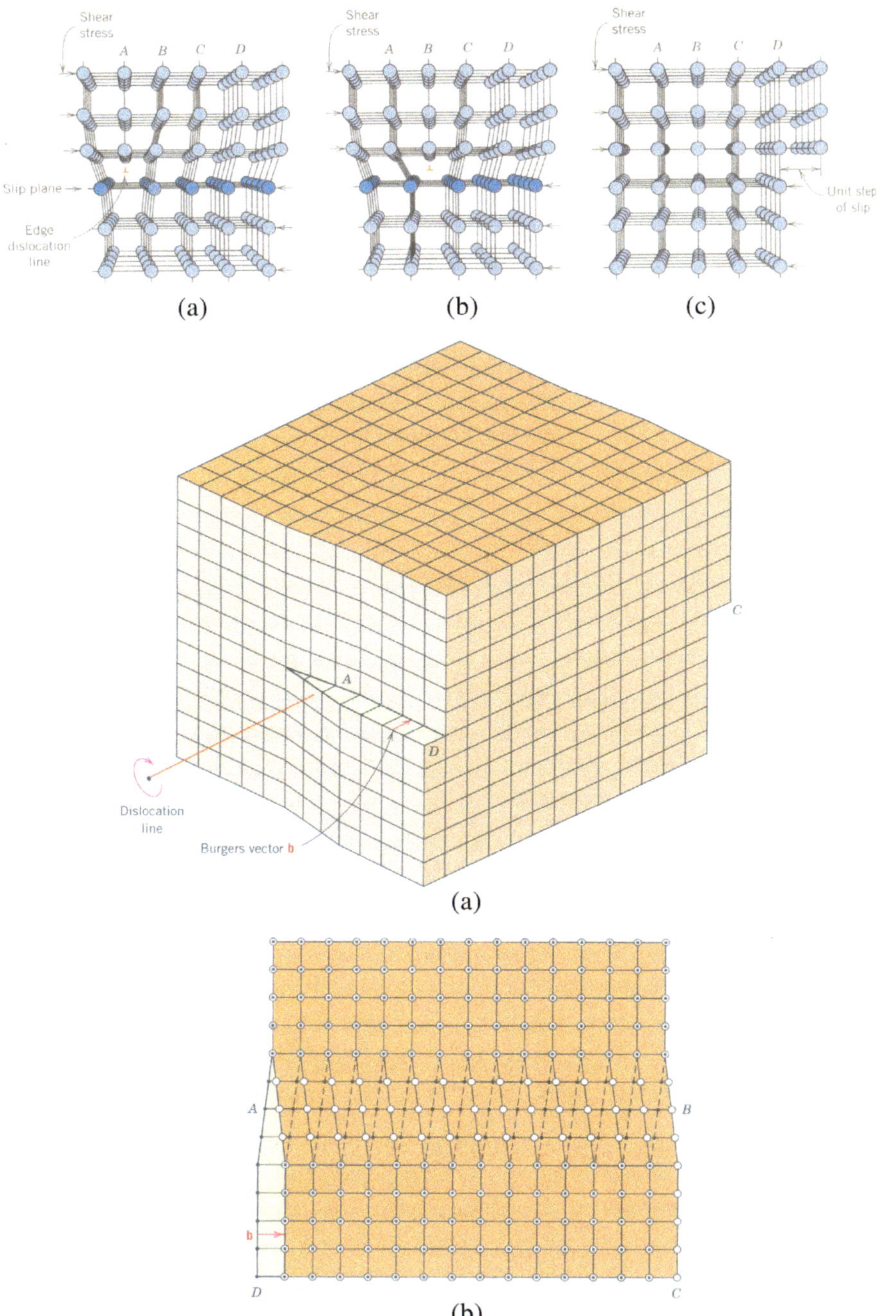

Fig. 2.68 Illustration of dislocations. Shown are (top) edge dislocations and (bottom) screw dislocations. Reproduced with permission from Callister, W. D. *Materials Science and Engineering: An Introduction*, 7th ed., Wiley: New York, 2007. © 2007 John Wiley & Sons, Inc

Table 2.8 Slip systems for BCC, HCP and FCC crystals

Crystal	Slip Plane	Slip Direction
Body-centered cubic (BCC)	{110} {211} {321}	<111> <111> <111>
Hexagonal close-packed (HCP)	{0001} {1010} {1011}	<1120> <1120> <1120>
Face-centered cubic (FCC)	{111}	<110>

example, fcc metals have 12 slip systems: four unique {111} planes, each containing three <110> slip directions. In contrast, hcp metals only have 3–6 slip systems (Fig. 2.69). Consequently, fcc metals are ductile due to more numerous routes for plastic deformation; in contrast, hcp metals are relatively brittle.

2.4.7 Physical Properties of Crystals

Hardness

Thus far, we have examined the 3-D arrangements of atoms, ions, or molecules comprising a crystal lattice alongside various types of defects (Fig. 2.70). The macroscopic physical properties of crystalline materials are directly related to these arrangements. For instance, the overall hardness of a crystal depends on the nature of the interactions among the discrete components of the crystal lattice. Those crystals possessing covalent interactions (e.g., diamond) will have a high hardness, whereas those containing only van der Waals forces will be soft (e.g., talc). A variety of scales (Table 2.9) may be used to assign the relative hardness of a material. The Mohs scale is generated by a qualitative assessment of how easily a surface is scratched by harder materials, with the hardest material (diamond) given a value of 10. The hardness of a material is directly proportional to its tensile strength; depending on which method is used, proportionality factors may be calculated. For instance, if the Brinell hardness value is known, the tensile strength is simply 500 times that value.

Tests such as Vickers, Knoop, and Brinell use an indentation technique that impinges a hard tip (e.g., diamond) into the sample with a known load (Fig. 2.71a). After a designated period of time, the load is removed, and the indentation area is measured. The hardness, H, is defined as the maximum load, L, divided by the residual indentation area, A_r (2.25). The coefficient, F, varies depending on which indentation method is used. This value (14.229 for Knoop and 1.854 for Vickers) is related to the geometry of the pyramidal probe, which will affect the penetration depth under the same load. Since a spherical probe is used for the Brinell test, a more complex formula is used to calculate the hardness (2.26), where D is the diameter of the spherical indenter and D_i is the diameter of the indenter impression (both in mm):

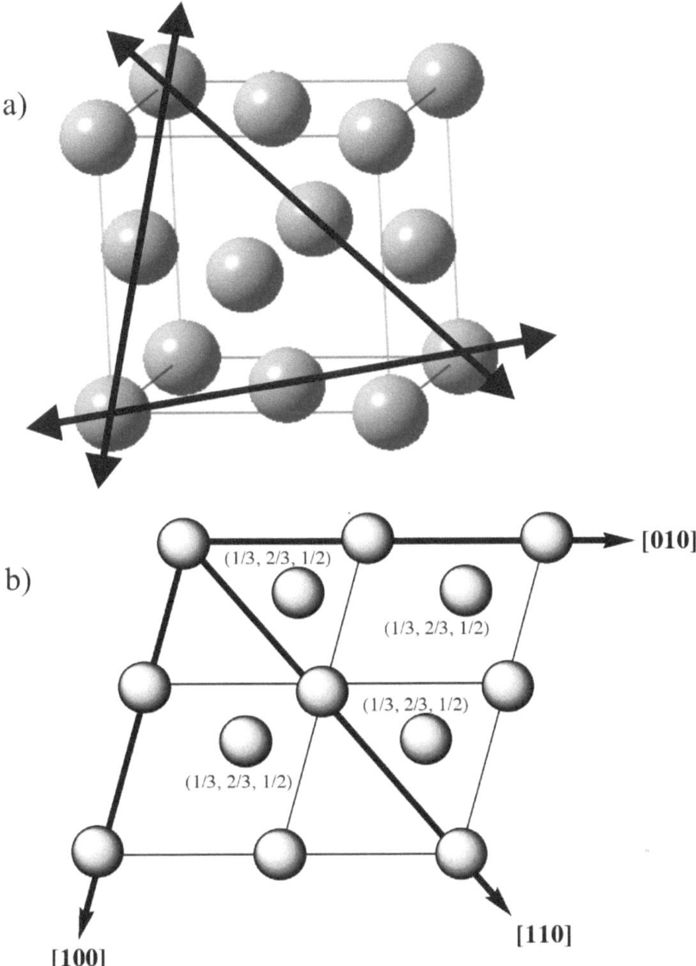

Fig. 2.69 **a** The {111} <110 > slip system for a face-centered cubic crystal. Note that there are 3 unique {111} planes, giving rise to 12 total slip systems for fcc. **b** The {001} <1 00> slip system for a hexagonal close-packed crystal. Shown is a 2 × 2 array of unit cells projected onto the (001) plane. Bold arrows indicate the three slip directions lying in each of the planes

$$H = F\left(\frac{\mathrm{L}}{\mathrm{A_r}}\right) \tag{2.25}$$

$$H = \frac{\mathrm{L}}{\frac{\rho}{2}\mathrm{D}\left(D - \sqrt{\mathrm{D}^2} - \mathrm{D}_i^2\right)} \tag{2.26}$$

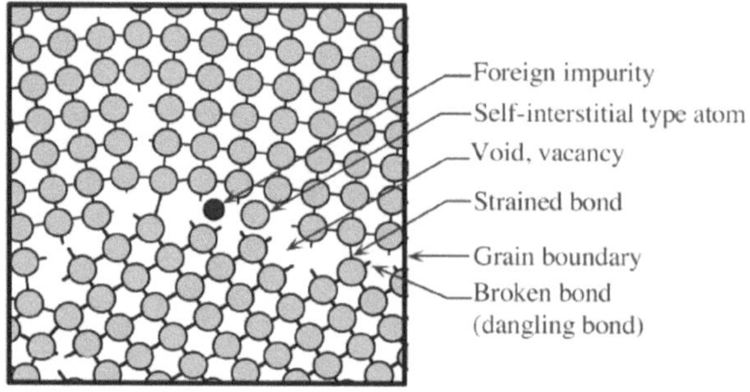

Fig. 2.70 Summary of crystal defects

Table 2.9 Hardness scales

Solid	Mohs	Vickers	Knoop
Talc	1	27	N/A
Graphite	1.5	37	N/A
Gypsum	2	61	N/A
Fingernail	2.5	102	117
Calcite	3	157	169
Fluorite	4	315	327
Apatite	5	535	564
Knife blade	5.5	669	705
Feldspar	6	817	839
Pyrex glass	6.5	982	929
Quartz	7	1,161	N/A
Topaz/Porcelain	8	1,567	N/A
Sapphire/Corundum	9	2,035	N/A
Diamond	10	N/A	N/A

N/A indicates the hardness value is above/below the acceptable range of the particular hardness scale. Values were obtained from the conversion site: http://www.efunda.com/units/hardness/convert_hardness.cfm?HD=HM&Cat=Steel#ConvInto

Quite often, an indentation is so small that it is difficult to resolve with a normal light microscope. To circumvent these problems, software is now capable of monitoring the load and displacement of the probe during the measurement and relating this to the contact area. Such an analysis without the need for visual confirmation is necessary for *nanoindentation* techniques used for nanomaterials and thin films. As its name implies, the hardness of a material is evaluated by the depth and symmetry of the cavity created from controlled perforation of a surface with a nanosized tip (Fig. 2.71b). It should be noted that although we focused on crystalline solids, hardness measurements are also easily performed on amorphous solids such as glasses and polymers.

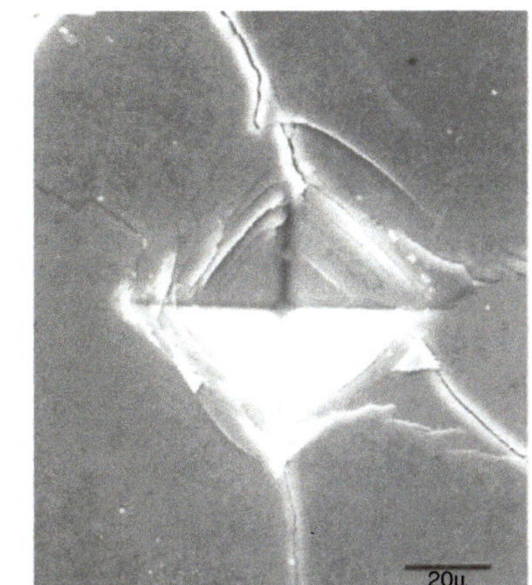

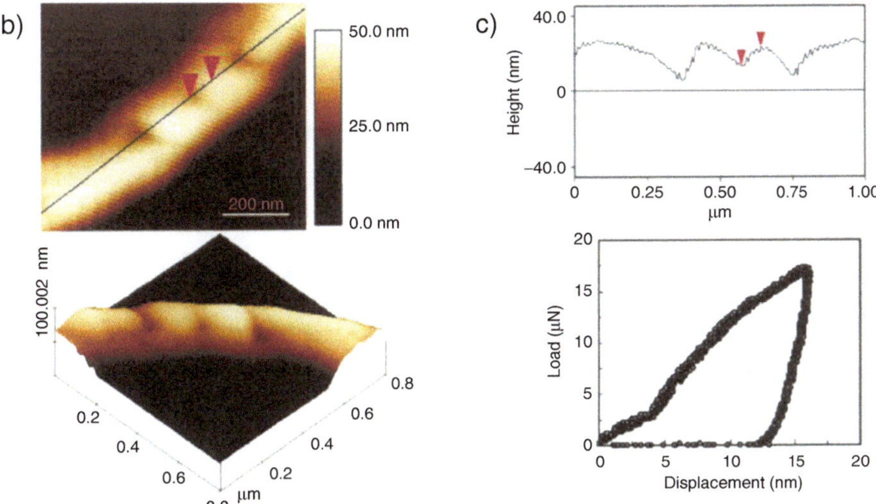

Fig. 2.71 Examples of indentation processes to determine surface hardness. Shown are: **a** Vickers indentation on a SiC–BN composite, **b** atomic force microscope images of the nanoindentation of a silver nanowire, and **c** height profile and load–displacement curve for an indent on the nanowire. Reproduced with permission from *Nano Lett.* **2003**, *3*, 1495. © 2003 American Chemical Society

Cleavage and fracturing

The intermolecular forces in a crystal lattice are often not uniform in all directions. If a solid consists of strong interactions among neighbors in specific layers, and weak interactions among molecules in neighboring layers (e.g., graphite, Fig. 2.72), a *cleavage plane* (*c.f.* slip planes, discussed earlier) is created where little force is needed to separate the crystal into two units. Since the cleavage planes are parallel to

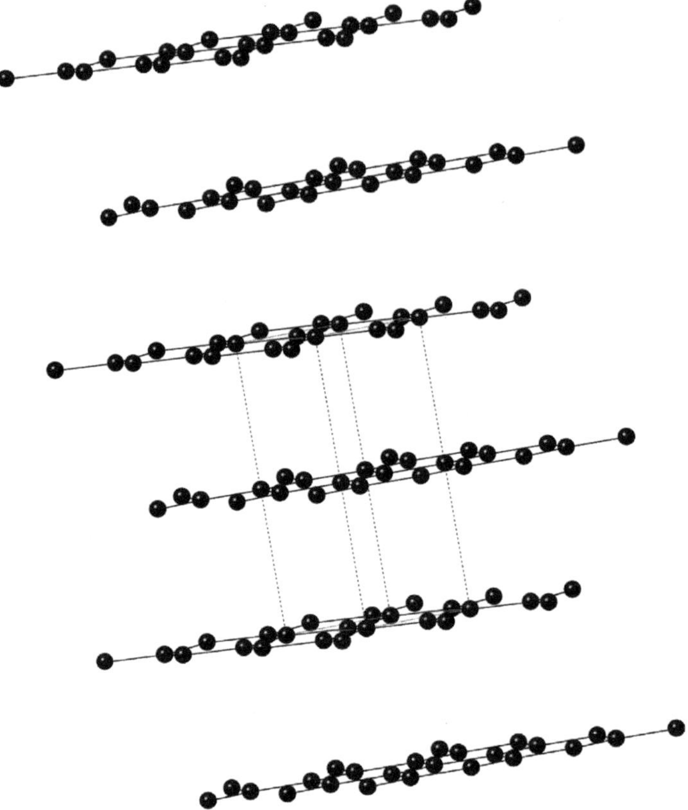

Fig. 2.72 Illustration of the graphite crystal structure, showing a relatively large separation (0.34 nm) between adjacent layers of sp^2-hybridized hexagons of carbon atoms

crystal faces, the fragments formed upon cleavage will retain the symmetry exhibited by the bulk crystal. Whereas cleavage describes the formation of a smooth piece of the original crystal when subjected to an external stress, a *fracture* refers to chipping a crystal into rough, jagged pieces. Although preferential cracking will occur along the cleavage plane, smaller fractured pieces will also be formed. In general, as one increases both the magnitude and obliqueness of the applied stress, the amount of fracturing will increase, relative to cleavage.

On occasion, the cleavage plane may be easily observed due to a fibrous network lattice. One example of such a crystal is ulexite, sodium calcium borate of the chemical formula $NaCa(B_5O_6)(OH)_6 \cdot 5H_2O$. Upon visual inspection, it is obvious that an external stress will cause a preferential cleavage in directions parallel to the crystallite fibers (Fig. 2.73a). However, more intriguing is the interesting optical properties exhibited by this crystal, commonly designated as the "T.V. rock" (Fig. 2.73b, c). If the crystal is surrounded by a medium of a lower refractive index (e.g., air), light is propagated through the individual fibers by internal reflection. This is analogous to fiber optic cables that will be described a bit later in this chapter. Figure 2.73b shows that light passing through the crystal will exhibit concentric circles. This is due to the difference in the effective path lengths of light passing either directly through individual fibers or crossing grain boundaries *en route* through the crystal via adjacent fibers.[59]

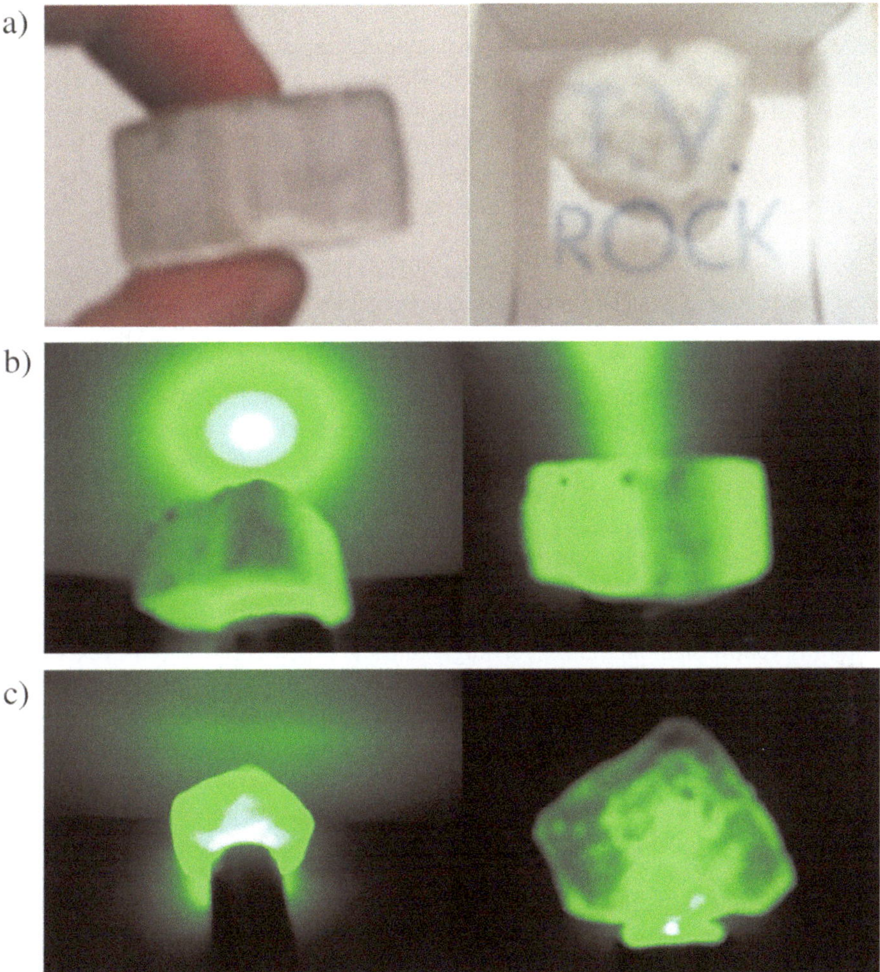

Fig. 2.73 Optical properties of ulexite, the "T.V. rock." Shown are **a** the fibrous morphology of the crystal, and projection of the words "T.V." on the surface of the rock from transmission through parallel fibers; **b** passage, and **c** blockage of a green laser beam when impinged on the crystal at angles parallel and perpendicular to the fibers, respectively

Color

In a previous section, we discussed substitutional impurities within crystal lattices without mentioning the changes in physical properties that this creates. Perhaps the most obvious outcome of a crystal impurity is the resultant color. In this section, we will answer the common question: "Why are some crystals colorless, while others are colored?"

Many crystals such as diamonds, quartz and corundum are normally colorless. In these crystals, the constituent atoms form a rigid, regular framework of covalent or ionic interactions. Since visible light (350–700 nm) is not energetically sufficient to cause bond rupturing and/or electronic transitions of the constituent metal atoms/ions, this energy is not absorbed by pure crystals, giving rise to a colorless state. However, when an impurity is added to the lattice, visible radiation may be suitably energetic to cause lattice alterations and/or electronic transitions, yielding an observable color change.

For example, the substitutional doping of B or N atoms in diamond yields a blue or yellow color. As you will recall from this chapter, a crystal lattice is composed of valence and conduction bands. For a pristine diamond, the bandgap is larger than the visible region of the electromagnetic (EM) spectrum. Hence, no visible light is absorbed and the diamond appears colorless (Fig. 2.74a). However, as nitrogen atoms replace carbon in the diamond lattice, mid-gap states in the bandgap become populated (Fig. 2.74b). Visible light is sufficient to promote electrons from the mid-gap states into the conduction band. Since this energy gap is in the blue region of the EM spectrum, yellow light is observed. The doping of a N atoms per 1,000,000 carbon atoms gives a yellow color; however, as the number of N dopants increases to a few per 1,000 carbon atoms, a green color is observed. Eventually, as the concentration of N dopants increases, all visible light may be absorbed resulting in a black diamond. Naturally yellow diamonds are exceedingly rare and expensive, with prices exceeding $10,000/carat. However, it is straightforward to fabricate yellow diamonds in a lab; synthetic yellow diamonds are much less expensive, *ca.* > $2,000/carat.

Everyone is familiar with the coloration phenomenon of gemstones such as ruby. In these crystals, the brilliant colors are due to the presence of transition metal dopants. Table 2.10 lists some common gemstones, and the respective host crystal and dopants that give rise to their characteristic colors. Whereas crystals of pure corundum (α-alumina) are colorless, a small amount (<1%) of chromium doping yields the familiar reddish/pink color. This color change is only possible if the periodic framework of the crystal is altered, through the incorporation of additional dopant atoms/ions or vacancies in the lattice. For ruby, a transition metal ion, Cr^{3+}, replaces Al^{3+} yielding electronic d–d transitions that were unattainable for the original main-group ion.

In a pure crystal of Al_2O_3 (as well as Fe_2O_3 and Cr_2O_3 that share the *corundum* structure) the oxide ions form an hcp array with the metal ions filling in 2/3 of the available octahedral intersitial sites (Fig. 2.75). The formal electronic configuration of Al^{3+} ions is [Ne], indicating that all electrons are paired. Since irradiation of

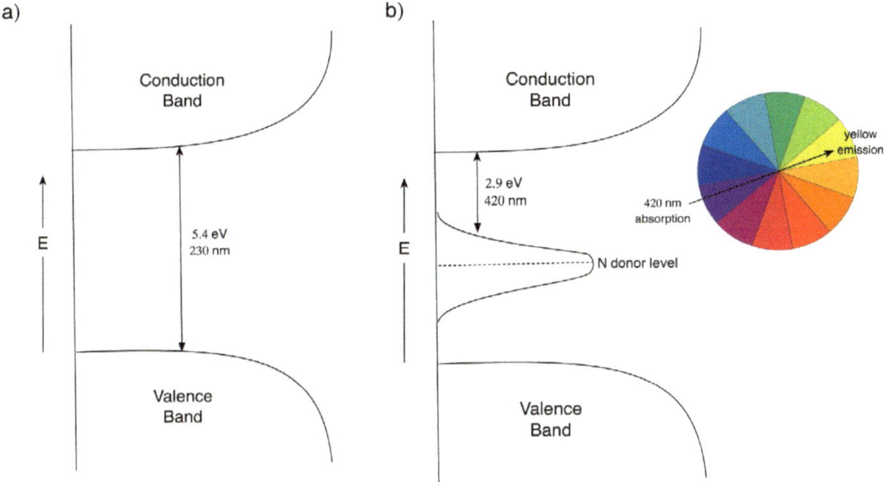

Fig. 2.74 Illustration of the color origin for yellow diamonds

Table 2.10 Active dopants in gemstone crystals

Gemstone	Color	Host crystal	Impurity ion(s)
Ruby	Red	Aluminum oxide	Cr^{3+}
Sapphire	Blue	Aluminum oxide	Fe^{2+}, Ti^{4+}
Emerald	Green	Beryllium aluminosilicate	Cr^{3+}
Aquamarine	Blue-green	Beryllium aluminosilicate	Fe^{2+}
Garnet	Red	Calcium aluminosilicate	Fe^{3+}
Topaz	Yellow	Aluminum fluorosilicate	Fe^{3+}
Tourmaline	Pink/red	Calcium lithium boroaluminosilicate	Mn^{2+}
Turquoise	Blue-green	Copper phosphor aluminate	Cu^{2+}

the crystal with visible light is not energetic enough to cause promotion of electrons into empty excited-state orbitals, the crystal appears colorless. However, if Al^{3+} ions are replaced with Cr^{3+} at a concentration of only 0.05 wt% (i.e., 1.58×10^{19} Cr^{3+} ions/cm^3), the crystal will appear brilliantly red. In these *ruby* crystals, each of the Cr^{3+} ions has a configuration of [Ar]3d^3. Although general chemistry tends to simplify the d-orbitals as being a set of five degenerate orbitals, transition metal complexes exhibit splitting of the d-orbital energy levels. This results in facile electronic transitions upon exposure to visible light, explaining the bright colors exhibited by many transition metal compounds.

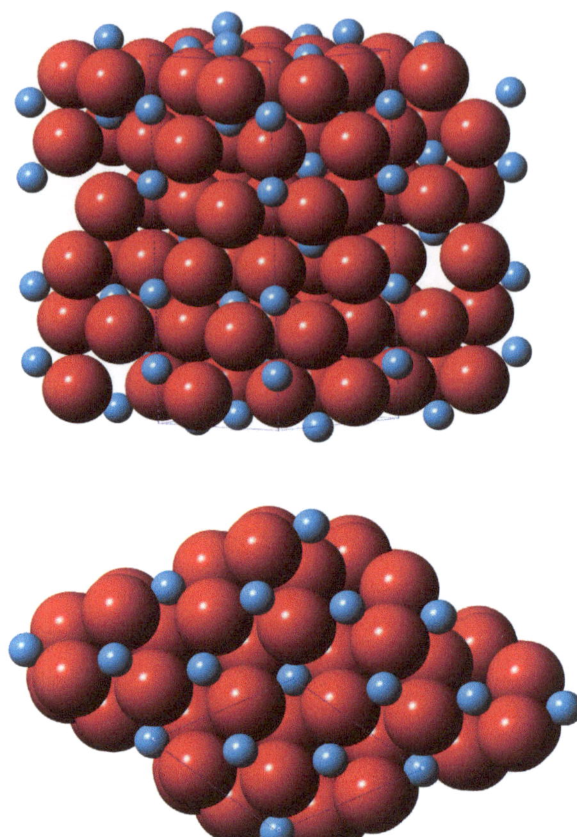

Fig. 2.75 Representations of α-alumina (corundum), as viewed along the [111] (top) and [001] (bottom) directions. Al^{3+} ions (blue spheres) are shown to occupy octahedral sites between the hcp layers of O^{2-} ions (red spheres)

A simple theory, referred to as *crystal field theory*,[60] is often used to account for the colors and magnetic properties of transition metal complexes. This theory is based on the electrostatic repulsions that occur between electrons in d-orbitals of a transition metal, and electrons contained in ligand orbitals. Figure 2.76 shows the splitting of the d-orbitals resulting from the electrostatic repulsions between the metal and ligand electrons for an octahedral Cr^{3+} complex. Since the d_{z^2} and $d_{x^2-y^2}$ orbitals are located directly along the internuclear bond axes, a greater electrostatic repulsion will occur resulting in an increase in energy. The energy gap between the two sets of d-orbitals is designated as 10 Dq or Δ_o (o = octahedral complex; Δ_t refers to a tetrahedral complex, etc.). Visible light is capable of being absorbed by the complex, causing the excitation of electrons into empty d_{z^2} or $d_{x^2-y^2}$ orbitals. As you are aware, the color we observe will be the reflected, or complementary, color of that being absorbed. For instance, absorbed wavelengths in the 490–560 nm regime (green) will appear red, whereas absorption of 560–580 nm (yellow) radiation will appear blue/violet, and so on.

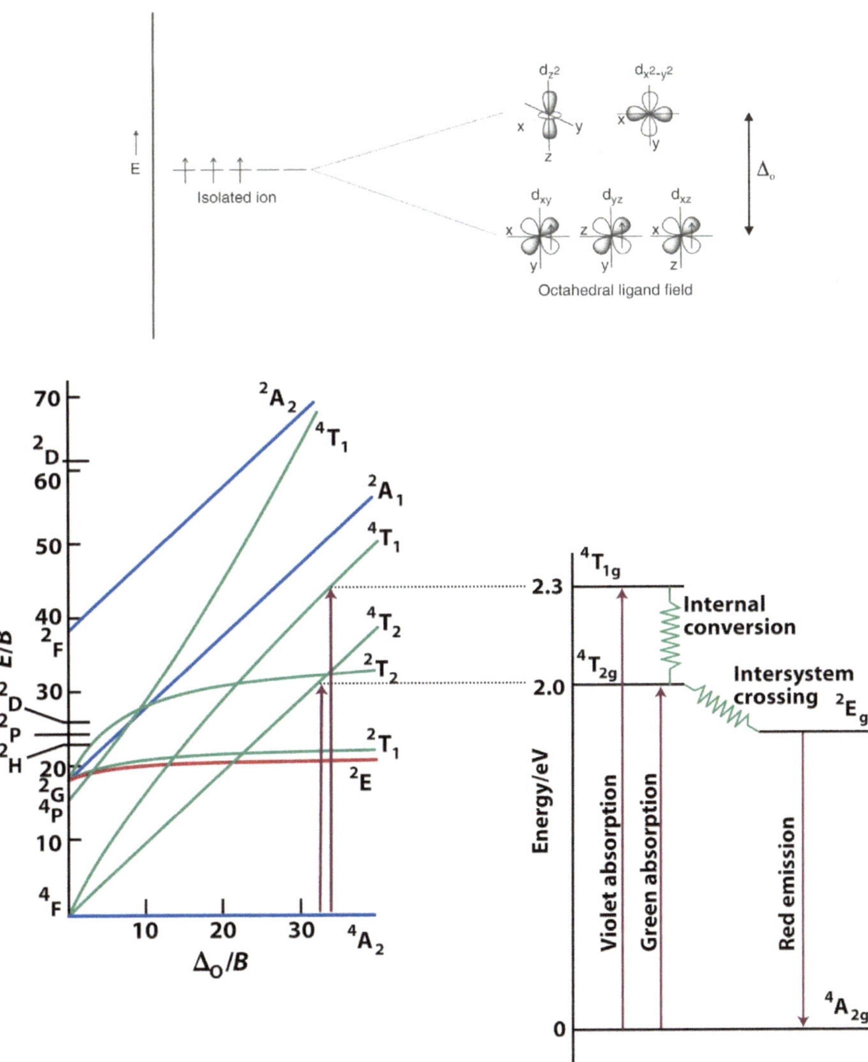

Fig. 2.76 Top: Energy splitting diagram for an octahedral Cr^{3+} ion in the ruby lattice. Bottom: The Tanabe-Saguno diagram for a d^3 transition metal (the parity subscript g has been omitted from the term symbols for clarity) along with an illustration of the transitions responsible for the absorption and luminescence of Cr^{3+} in ruby. Reproduced from Atkins, P. et al. *Inorganic Chemistry*, 4th ed., W. H. Freeman: New York, 2006

Figure 2.76 also illustrates the Tanabe-Saguno diagram for the d^3 Cr^{3+} ion of ruby, showing the ground-state molecular term symbol as $^4A_{2g}$ (g = gerade, since an octahedral ligand field has a center of symmetry), with two spin-allowed transitions to $^4T_{2g}$ (green, 550 nm) and $^4T_{1g}$ (blue, 420 nm); the transition from $^4A_{2g} \rightarrow {}^2E_g$ is spin-forbidden.[61] It should be noted that the Laporte selection rule disfavors electronic transitions between the ground $^4A_{2g}$ and excited ^{4} T states, since they both exhibit even parity. However, the absorption of energy and electronic excitation occurs because Cr^{3+} doping distorts the perfect octahedral environment of the corundum host, mixing in states of odd parity. Rather than simple relaxation back to the ground state and accompanying fluorescent emission, there is a fast (10^7 s^{-1}) *intersystem crossing* (ISC) into the metastable doublet state, ^2E_g. Even though this non-radiative decay process[62] is spin-forbidden, it is driven by *spin–orbit coupling*, which becomes more pronounced with increasing nuclear charge (i.e., Z^4). Since the transition from the ^2E_g intermediate state to the ground state is also spin-forbidden, the electrons experience a finite lifetime in the doublet intermediate state before relaxing to the ground state, with emission of red light (λ = 694 nm). The relatively long lifetime of an excited state (*ca.* 3 ms for ruby) is characteristic of *phosphorescence*, relative to *fluorescence* in which electrons exhibit fast relaxation (*ca.* 5 ps–20 ns) from excited to ground states.

Interestingly, if Cr^{3+} is substituted for Al^{3+} in the *beryl* ($Be_2Al_2Si_6O_{18}$, Fig. 2.77) base lattice of *emerald* gemstones, the crystal appears green rather than red. Since the coordination spheres about the Cr^{3+} centers for both ruby and emerald are distorted octahedra, the shift in the absorption wavelength must result from the lattice structure. In the beryl lattice, the Be^{2+} ions pull electron density away from the oxygen ions, which will cause less electron–electron repulsions between the Cr^{3+} d-orbitals and lone pairs of the oxygen ligands. This will correspond to a decrease in the Δ_o value, the absorption of lower-energy wavelengths, and a shift of the reflected color from red to green. It should be noted that red phosphorescence is also present in emerald; however, this is outweighed by the strong yellow/red absorption (from the $^2E_g \rightarrow {}^4A_{2g}$ relaxation) that yields the familiar green color.

Not only does the observed color depend on the nature of the transition metal impurity, but on the oxidation state of the dopant. For instance, upon doping beryl with Fe^{2+} (aquamarine) or Fe^{3+} (heliodor), the observed color is blue or yellow, respectively. For Mn^{2+} (morganite) and Mn^{3+} (red beryl) impurities, the observed colors are pink and red, respectively. As a general rule, as the oxidation state of the transition metal ion increases, the ligand ions are drawn in closer to the metal center. This will result in more electron–electron repulsions between the metal and ligand, and a larger Δ_o. The increase in the energy gap between d-orbitals causes the absorption of higher energy wavelengths, and a corresponding red shift for the observed/transmitted color. As you might expect, it should be possible to change the color of such a crystal through heating in an oxidizing or reducing environment. This is precisely the operating principle of "mood rings" that respond to differences in body temperature. The color change resulting from a temperature fluctuation is referred to as *thermochromism*. When the application of an external pressure causes a color change, the term *piezochromism* is used.

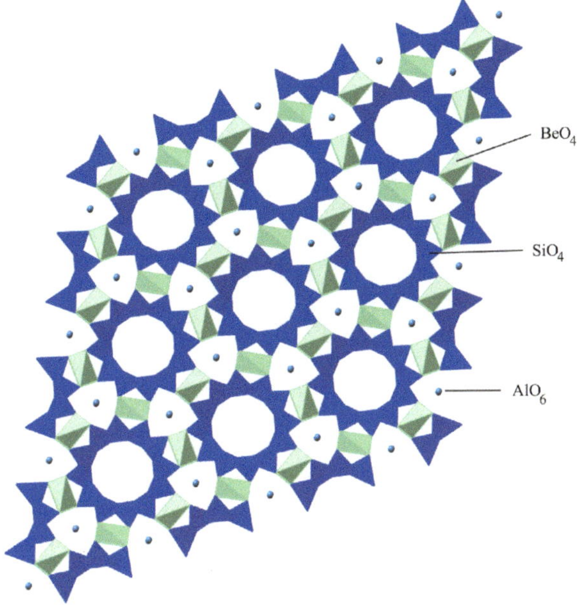

Fig. 2.77 Extended polyhedral representation of the beryl (emerald) crystal lattice (as viewed along [001]), showing locations of AlO_6, BeO_4, and SiO_4 polyhedra

Another factor that must be mentioned relative to our discussion of color is the wavelength of light used to irradiate the crystal. Alexandrite, Cr^{3+}-doped chysoberyl ($BeAl_2O_4$), has two equivalent transmission windows at the red and blue-green regions. As illustrated in Fig. 2.78, alexandrite appears brilliantly red upon exposure to incandescent lamplight and bright green upon exposure to sunlight or fluorescent light. Exposure to white light will also yield a purple color (blue + red), depending on the viewing angle and illumination source. The dramatic color intensity is due to the internal reflection of the incident light from the multifaceted crystal faces. Interestingly, in addition to its use as an expensive decorative gemstone, alexandrite may also be employed as a tunable solid-state laser, with $\lambda_{laser} = 750 \pm 50$nm.[63]

Thus far, we have only considered the colors responsible for doping a crystal with one type of metal ion. However, crystals such as blue sapphire contain two metal dopants that yield the desirable deep blue color due to charge-transfer effects. If two adjacent Al^{3+} sites in alumina are replaced with Fe^{2+} and Ti^{4+}, an internal redox reaction may occur, whereby the iron is oxidized and the titanium is reduced (2.27):

$$Fe^{2+} + Ti^{4+} \rightarrow Fe^{3+} + Ti^{3+} \quad (2.27)$$

As this process requires energy corresponding to orange/yellow wavelengths of visible light, the complementary color of blue is reflected. Although blue sapphire is an example of heteronuclear charge transfer involving two different transition metal ions, there are also precedents for homonuclear charge transfers involving two oxidation states of the same metal (e.g., Fe^{2+} and Fe^{3+} in magnetite, Fe_3O_4). Lastly, it should be noted that charge transfer may also occur between the metal and

Fig. 2.78 Photographs of chrysoberyl (alexandrite) taken under daylight (left) and incandescent light (right). ©NickKnight/Shutterstock.com

ligands. For instance, the blue color of the gemstone lapis lazuli (i.e., $(Na, Ca)_8(Al, Si)_{12}O_{24}(S, SO_4))$, is due to ligand–metal charge transfer (LMCT) initiating from the sulfur atoms in the ligand.[64]

Colored crystals need not be gemstones; in fact, colorless crystals of common salts such as KCl or NaCl may be suitably altered to exhibit color. When solid KCl is heated to 500 °C in the presence of potassium vapor, the crystal becomes a violet color. This occurs due to the ionization of gaseous potassium atoms that abstract a Cl^- anion from the crystal lattice. The electron formed in the oxidation process becomes trapped in the anion vacancy, forming a F-center defect[65] to rebalance the overall charge of the crystal (2.28). In fact, this trapped electron represents a real-world example of a particle-in-a-box! It should be noted that X_2^- species may also be generated in response to a photoinduced loss of M^+ from a lattice (2.28), denoted as an H-center defect.

$$[(KCl)_n(KCl)](s) + K(g) \rightarrow [(KCl)_n(K)(e^-)](s) + KCl(s) \tag{2.28}$$

$$MX_n \rightarrow [(M+)_{n-x}(X^-)_{n-2x}(X_2^-)_x] \tag{2.29}$$

Anion vacancies may also be generated by irradiation of a crystal with ionizing radiation such as X-rays. This high-energy radiation will cause the removal of a halide ion from the lattice and will promote some of the lattice electrons from valence to conduction bands. At this point, the electrons are free to diffuse through the crystal, where they remain mobile until they find an anion vacancy site. At low temperatures (e.g., in liquid nitrogen), electrons may even become localized by polarizing their surroundings; that is, displacing the surrounding ions, to give self-trapped electrons. For each type of electron trap, there is a characteristic activation energy that must be overcome for the release of the electron. As an irradiated crystal is heated, electrons are released from their traps by thermal activation, leading to a change in the observed color. The free electrons may migrate once again through the crystal until they recombine with an anion hole. This phenomenon has been studied in detail for aptly named "chameleon diamonds", which undergo color changes from greyish-green to yellow when they are heated/cooled (*thermochromic* behavior) or kept in the dark (*photochromic* behavior). In these diamonds, the color change is thought to arise from electron traps created by the complexation of H, N, and Ni impurities.[66]

Properties resulting from crystal anisotropy

Crystals are classified as being either *isotropic* or *anisotropic* depending on the equivalency of their crystallographic axes. All crystals that do not belong to the cubic crystal system possess anisotropic symmetry. Since electromagnetic radiation partially comprises an electrical component, the velocity of light passing through a material is partially dependent upon the electrical conductivity of the material. The relative speed at which electrical signals travel through a material varies with the type of signal and its interaction with the electronic structure of the solid, referred to as its *dielectric constant* (ε or κ).

Anisotropic crystals are composed of a complex crystal lattice orientation that has varying electrical properties depending upon the direction of the impinging light ray. As a result, the refractive index will vary with direction when light passes through an anisotropic crystal, giving rise to direction-specific trajectories and velocities. This effect is most easily observed in crystals when there are large differences in the refractive indices of the crystallographic axes. This phenomenon is referred to as *birefringence* and is illustrated by double refraction exhibited by optical calcite shown in Fig. 2.79. Examples of the two distinct refractive indices for representative crystals are calcite (1.6584 and 1.4864), tourmaline (1.669 and 1.638), and rutile (2.616 and 2.903). The birefringence exhibited by a crystal is dependant on the difference in the refractive indices experienced by the extraordinary and ordinary rays as they propagate through a crystal.

If birefringence occurs in a colored crystal, the observed color is often dependent on the viewing angle, as already discussed above for alexandrite. This phenomenon is known as *pleochroism* and is caused by the incident light beam following

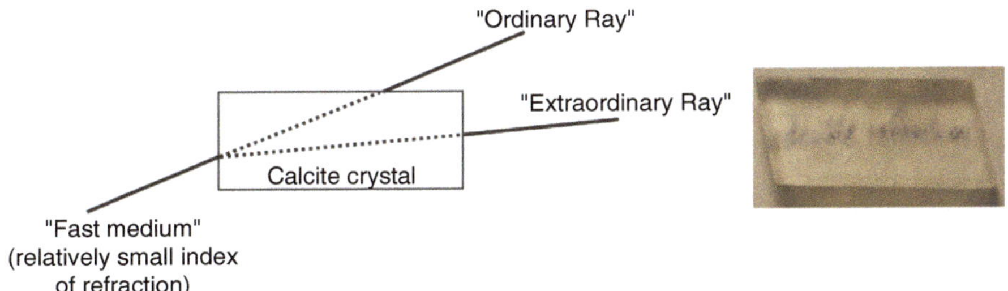

Fig. 2.79 Illustration of the double refraction phenomenon exhibited by calcite crystals

Fig. 2.80 Photographs of **a** dichroism exhibited by tourmaline, and **b** trichroism exhibited by iolite

different paths within the crystal, with each path absorbing different colors of light. Whereas tetragonal, trigonal, and hexagonal crystals often show two colors (*dichroic*, Fig. 2.80a), orthorhombic, monoclinic, and triclinic crystals may display three colors (*trichroic*, Fig. 2.80b). It should be noted that cubic crystals do not exhibit pleochroism, since all unit cell axes are equivalent.

Crystals that do not possess a center of symmetry (i.e., noncentrosymmetric) exhibit interesting properties when exposed to pressure—a phenomenon known as *piezoelectricity*, from the Greek word *piezein* (to squeeze). As pressure is applied, the crystal changes shape slightly by the movement of ions. The ionic migration causes some of the positive and negative ions to move in opposite directions, causing a polarization of charge. Conversely, if a piezoelectric crystal is placed in an electric field, the ions move toward opposite electrodes, thereby changing the shape of the crystal. With the exception of the point group 432, all noncentrosymmetric crystallographic point groups will exhibit piezoelectricity.

At temperatures greater than the Curie temperature, T_c, the lattice atoms can migrate and cancel the effects of an external stress. However, at $T < T_c$, the cubic perovskite crystal becomes tetragonal; therefore, the central cation (e.g., Ti^{4+} in $BaTiO_3$) becomes displaced, resulting in a net dipole moment and disruption of the charge symmetry of the crystal. The magnitude of the piezoelectric effect is not trivial; a 1 cm^3 quartz crystal exposed to a 2 kN (450 lb-force) external force will generate 12 kV!

Piezoelectricity is the operating principle of quartz watches. In these devices, a tiny crystal of quartz oscillates at a frequency of 32 kHz in response to an electrical charge generated from the battery. In general, the overall size and composition of a piezoelectric crystal will affect its oscillation frequency. Since quartz loses very little energy upon vibration, the integrated circuit (IC, see Chap. 4) within a watch is used to reduce the repeatable oscillations into electric pulses, which are displayed as hours, minutes, and seconds on the watch face. The loud pop one hears when the ignition button is pressed on a gas grill is the sound generated by a small spring-loaded hammer hitting a piezoelectric crystal, generating thousands of volts across the faces of the crystal—comparable to the voltage generated by an automotive spark plug! Room humidifiers also operate via the induction of *ca.* 2,000,000 vibrations per second of a piezoelectric crystal, which is strong enough to cause atomization of water molecules.

As one speaks into a microphone, small changes in air pressure surrounding the piezoelectric crystal cause tiny structural distortions that generate a very small voltage. Upon amplification, these voltage changes are used to transmit sounds; this principle has also been exploited for SONAR applications ("SOund Navigation And Ranging"). The military has also investigated piezoelectric crystals mounted in the boots of soldiers, whereby simple walking/running would provide current to power electronic devices embedded in warfighter uniforms such as sensors, IR shields, artificial muscles, etc.

Natural crystals that exhibit piezoelectricity include quartz (point group: 32), Rochelle salt (potassium sodium tartrate; orthorhombic space group, 222), berlinite ($AlPO_4$), cane sugar, topaz ($Al_2SiO_4(F, OH)_2$), tourmaline group minerals (Ca, K, Na)(Al, Fe, Li, Mn)$_3$(Al, Cr, Fe, V)$_6$(BO_3)$_3$(Si, Al, B)$_6O_{18}$(OH, F)$_4$), and even bone! Synthetic examples include perovskites (*e.g.*, $BaTiO_3$, $PbTiO_3$, $Pb(Zr_xTi_{1-x})O_3$ (lead zirconate titanate, PZT), $KNbO_3$, $LiTaO_3$, $BiFeO_3$) and polyvinylidene fluoride (PVDF). The latter structure is not crystalline, but is comprised of a polymeric array of intertwined chains that attract/repel one another when an electrical field is applied. This results in a much greater piezoelectric effect than quartz.

Other noncentrosymmetric crystals that alter their shape in response to changes in temperature are referred to as *pyroelectric*. These crystals are used in infrared detectors; as an intruder passes the detector, the body warmth raises the temperature of the crystal, resulting in a voltage that actuates the alarm. Even for such miniscule temperature changes of a thousandth of a degree, a voltage on the order of 15 mV may result, which is readily measured by electronic components. Crystals that exhibit this effect are $BaTiO_3$, $PbTi_{1-x}Zr_xO_3$, and PVDF. Of the 32 crystallographic point groups discussed earlier, 20 are piezoelectric and ten are pyroelectric (Table 2.11).

The pyroelectric crystal classes are denoted as *polar*, each possessing a spontaneous polarization that gives rise to a permanent dipole moment in their unit cells. If this dipole can be reversed by the application of an external electric field (generating a hysteresis loop, Fig. 2.81), the crystal is referred to as *ferroelectric*.[67] These crystals will exhibit a permanent polarization even in the absence of an applied electrical field. In comparison, an antiferroelectric material will exhibit a double-hysteresis loop, resulting from the spontaneous anti-parallel alignment of dipoles and zero net polarization.

As a ferroelectric example, let us consider $BaTiO_3$. At temperatures below the Curie temperature (T_c) of 130 °C, a tetragonal unit cell is preferred rather than cubic. Since the central Ti^{4+} ion is not centered with respect to the surrounding negatively

Table 2.11 Piezoelectric and Pyroelectric Crystal Systems

Crystal System	Crystallographic point group		
	Centrosymmetric	Non-centrosymmetric Piezoelectric	Non-centrosymmetric Pyroelectric
Triclinic	$\bar{1}$	1	1
Monoclinic	2/m	2, m	2, m
Orthorhombic	mmm	222, mm2	mm2
Tetragonal	4/m, 4/mmm	4, $\bar{4}$, 422, 4 mm, 42 m	4, 4 mm
Trigonal	$\bar{3}$, $\bar{3}$ m	3, 32, 3 m	3, 3 m
Hexagonal	6/m, 6/mmm	6, $\bar{6}$, 622, 6 mm, 6m2	6, 6 mm
Cubic	M $\bar{3}$, m $\bar{3}$ m	23, $\bar{4}$ 3 m	N/A

charged O^{2-} ions, a finite polarization vector is established. Thermal motion tends to destroy the ferroelectric ordering of unit cells, giving rise to *paraelectric* behavior above T_c, in which there is no net polarization. It should be noted that all ferroelectric crystals are *both* piezoelectric and pyroelectric, but the reverse is not necessarily true. The analogous concept of *ferromagnetism* will be discussed in more detail in Chap. 3; *multiferroic* materials, which exhibit more than one symmetry-related ordering property such as ferroelectricity, ferromagnetism, or ferroelasticity, are the focus of significant current research interest.[68]

2.5 Ceramics

Ceramics refer to a broad category of inorganic materials that possess a high hardness (close to diamond on the Moh's scale) and brittleness, pronounced resistance to heat and corrosion, and are electrically/thermally insulating. Amorphous glasses are sometimes classified as a type of ceramic; however, ceramics tend to include only inorganic materials with at least some degree of crystallinity. Most often, these materials possess both amorphous and polycrystalline regions, with the latter exhibiting abrupt changes in crystal orientation across individual grain boundaries in the extended lattice.

Ceramics may be classified based on their applications. Traditional varieties are used primarily for low-tech applications such as whiteware, cement, and abrasives, while advanced ceramics are employed for a broader range of applications (Fig. 2.82). Compositionally, there are four general categories of ceramics: *oxides* (e.g., alumina, zirconia, magnesia, aluminum titanate (Al_2O_3/TiO_2), lead zirconate titanate ($Pb[Zr_xTi_{1-x}]O_3$), *silicates* (e.g., porcelain, magnesium silicates (MgO/SiO_2), mullite (Al_2O_3/SiO_2)), *nitrides* (e.g., SiN, Si/Al oxynitrides, AlN), and *carbides* (e.g., BC, SiC, WC).

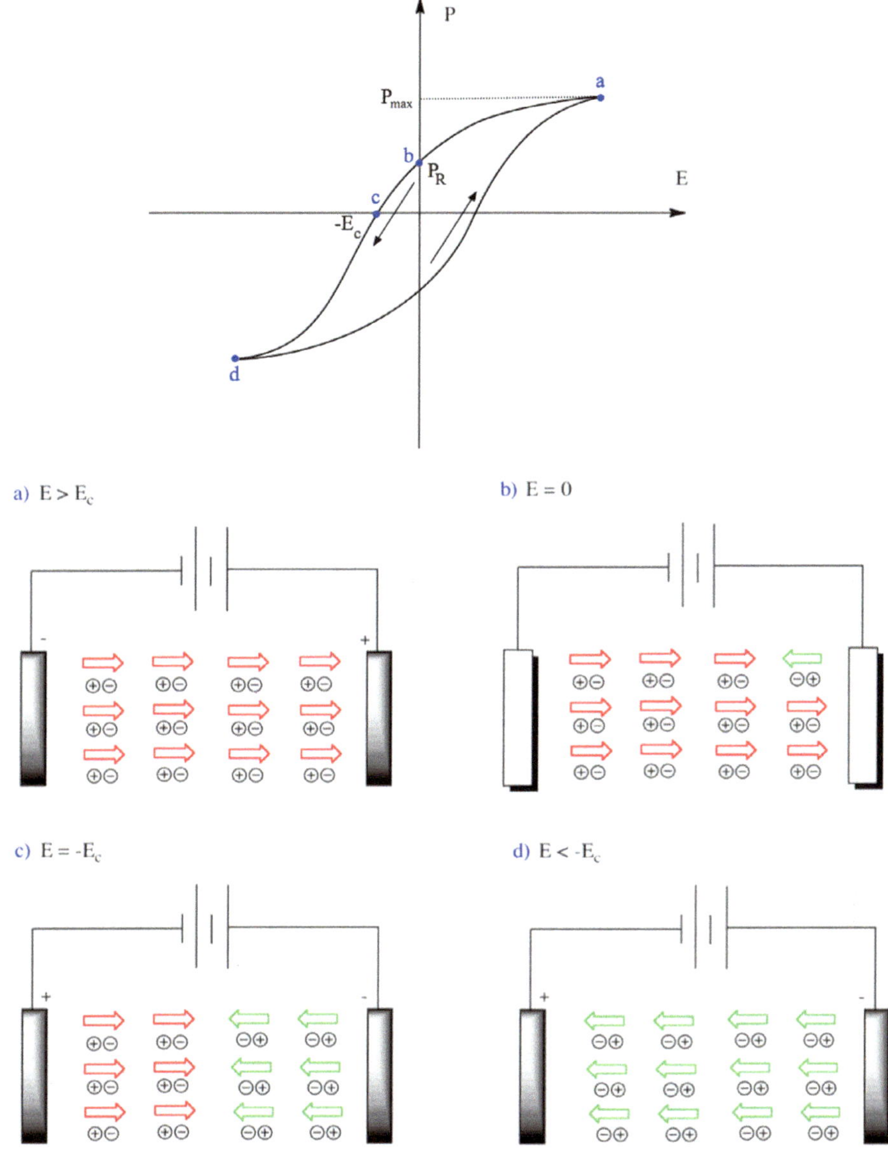

Fig. 2.81 A typical polarization hysteresis curve exhibited by a ferroelectric material. Initially, the dipole moment within the ferroelectric material is oriented in the direction of a positive applied electric field, leading to saturation polarization (position a). At zero applied field (position b), ferroelectric materials retain a remnant polarization, P_R). This may be compared to typical dielectric materials such as polymers, which cannot sustain a polarization without an applied voltage. At a critical negative electric field, known as the coercive field, E_c, the dipole moments point in opposite directions and thus cancel out, leading to no net polarization (position c). At higher negative potentials, the dipoles are polarized in the opposite direction (position d)

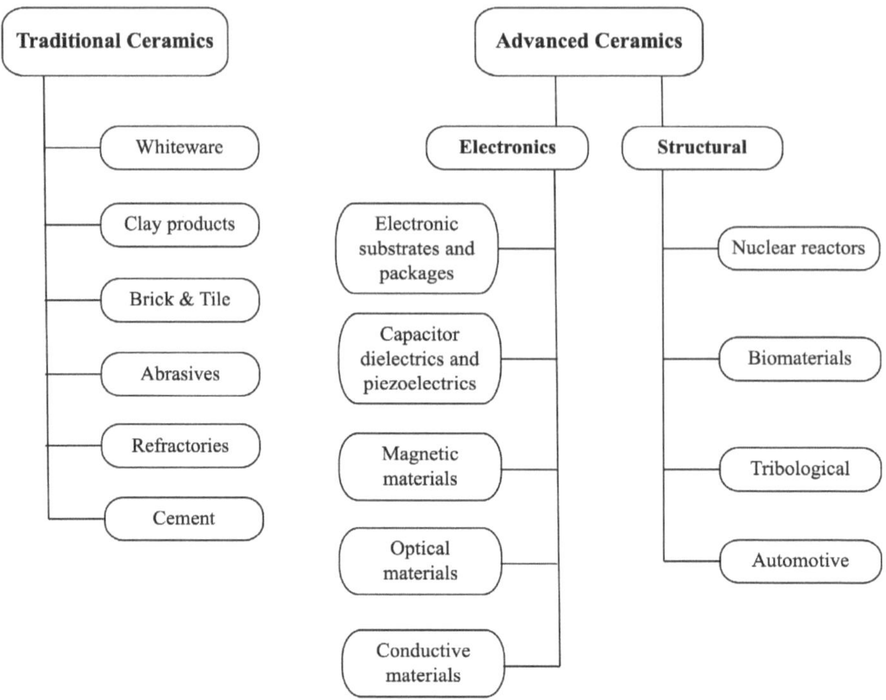

Fig. 2.82 Classification of ceramics based on their applications

Perhaps the least sophisticated, but most widely used, means for ceramics fabrication involves solid-state reactions, known as "shake-n-bake". This is the method-of-choice for most commercial syntheses of ceramics, consisting of a high-temperature multi-step process involving[69]:

(i) *Grinding/milling*—silicate/aluminosilicate minerals (e.g., various clays, silica) and bauxite (hydrous aluminum oxide ore) are among the most abundant materials in Earth's crust, constituting the most common raw materials for ceramics processing. These minerals are pulverized into a fine powder, often alongside *calcination*—thermal treatment to remove volatile impurities (e.g., organics, waters of hydration, CO_2 from carbonates, etc.). Powders must be finely divided in order to maximize their surface area and rate of diffusion. As an alternative to "top-down" pulverization, "bottom-up" methods such as co-precipitation from solution (e.g., sol–gel), gas-phase pyrolysis, spray-drying, or freeze-drying are also commonly used to prepare ceramic precursor powders.

(ii) *Mixing and forming*—the powder is mixed with water to semi-bind the particles together, and is cast, pressed, or extruded into the desired shape. Sometimes, an organic polymer binder such as polyvinyl alcohol, $(\text{-}CH_2\text{-}CH(OH)\text{-})_n$, is added to enhance the adhesion of the granules. Organic lubricants may also be added during pressing to increase densification. This step is critical to enhance the intimate contact of reactant particulates while minimizing direct contact with the crucible, which may introduce impurities.

(iii) *Drying*—the material is heated at temperatures >200 °C to remove the water or organic binder(s) and lubricant(s) from the formed material, resulting in a *green body*. Care must be used to prevent rapid heating, to prevent cracking and other surface defects. During heating, shrinkage will occur resulting in a material that is denser (by a factor of 2–3) than the green precursor—especially for syntheses using polymeric precursors (e.g., organosilicon polymers for SiC or Si_3N_4 ceramics). Gas evolution may also occur that will introduce porosity into the ceramic.

(iv) *Firing/sintering*—quite often, precursor compounds at earlier stages of ceramic processing are at least partially amorphous. The final firing/sintering stage is used to fuse the particles together and convert the material into a (poly)crystalline product, which has the bulk form and physical properties desired for a particular application. Firing is usually performed at a temperature below the melting point of the ceramic. Most importantly, as we will see in the next chapter (powder metallurgy), the microstructure of the final product is strongly related to the morphology of the green body. Sample crucibles or sealed tubes may be used, with its selection based on the desired sintering temperature. For instance, ceramic refractories such as silica (1160 °C), alumina (2200 °C), or zirconia (2300 °C), as well as noble metals such as Pt (1770 °C), Au (1063 °C), Ir (2450 °C), Ta (3020 °C), or W (3422 ºC) are often used.

This method is slow and intrisically inefficient, since even well-mixed reactants are very inhomogeneous on the molecular scale. Nevertheless, crystals of desired compositions may be obtained in high purity and yield through nucleation (oftentimes onto the surface of a seed crystal) and growth. High temperatures and long reaction times are required to enable the diffusion of ions into preferred lattice positions and growth of large crystallites.[70] For instance, the synthesis of spinel crystals requires heating precursor powders for at least one week at temperatures on the order of 1500 °C! As you might imagine, impurities may be incorporated into the product either from the ambient atmosphere (usually performed in Ar or N_2), or from the ceramic crucibles used to hold the precursor pellets. Hence, extremely pure precursor powders, gases, and crucibles are all required for solid-state reactions, which adds to the expense of this procedure.

Most often, resistance heating within furnaces with SiC or $MoSi_2$ heating elements is used for sintering. This gives rise to temperatures up to 2000 °C. An electrical arc may be used to generate temperatures up to 3000 °C, whereas a CO_2 laser can yield temperatures up to 4000 ºC. However, it is becoming more common to utilize *microwave radiation* to sinter ceramic precursor powders.[71] Microwave processing offers many advantages relative to convential heating such as rapid heating rates (>400 ºC min^{-1}) and reduced processing times (<30–120 min. *vs.* 12–24 h). Due to rapid firing, the resultant ceramics are typically denser, with smaller grain sizes and a higher strength and hardness, than those produced using traditional solid-state routes.[72]

In contrast to conventional heating, in which heat is transferred to objects by conduction, radiation, and convection, microwave heating occurs through the absorption of electromagnetic (EM) radiation. However, not all objects will absorb microwave radiation. For instance, metallic conductors will reflect microwaves, and materials with low electrical conductivities such as glass will be transparent to microwave radiation. The absorption of microwaves occurs by two mechanisms: ionic conduction and dipole rotation. Sample ions may move under the influence of the applied electric field of magnetic EM radiation, resulting in an oscillating electric current. Resistance to this movement results in the conversion of electric energy to thermal energy (heat)—a process known as *conduction heating.* In contrast, neutral molecules may align their dipole moments, if present, with the electric field of microwave radiation. This causes molecular rotation as the dipole moments of discrete molecules attempt to realign with the applied electric field. However, the oscillating electric field changes its direction much more quickly (5 billion times per second at 2.5 GHz!) than dipole realignment can occur. This lag time is related to the absorption efficiency of a material, tan δ (2.30), which is affected by the dielectric and magnetic properties of the sample.

$$\tan \delta = \frac{\varepsilon''}{\varepsilon'} \tag{2.30}$$

where tan δ is the *dielectric loss tangent*, ε" is the *loss factor*—a measure of the ability of a material to convert EM energy to thermal energy, and ε' is the *dielectric constant*—the ability of a material to store microwave energy.

Ceramics with loss factors between $10^{-2} < \varepsilon'' < 5$ are good candidates for microwave heating. In comparison, materials with $\varepsilon'' < 10^{-2}$ would be difficult to heat, while those with $\varepsilon'' > 5$ would absorb mostly on the surface and not in the bulk.[lxxi] Whereas pure alumina and silicon nitride have small tan δ values and low microwave absorption properties, solvents such as water, ethylene glycol, or 2-propanol, and substances such as carbides, borides, oxides (e.g., CuO, ZnO, PbO_2, NiO, WO_3), graphite, and intermetallic compounds have values up to 1 and higher. However, even for weakly absorbing samples, the experimental conditions may be fine-tuned to improve their absorption properties. For instance, the loss factor of ceramic materials generally increases with temperature. Additionally, the frequency of incident radiation may be increased, or an absorbent (e.g., SiC or carbon) may be added to increase the absorption efficiency of the sample.

As another synthetic option, the use of *hydrothermal* conditions is often referred to as *chimie douce* (soft chemistry), since it requires relatively low reaction temperatures. Hydrothermal syntheses involve heating reactant precursors in water at high pressures and temperatures, though much less than that required for solid-state reactions—typically between 100–500 °C. Reactants and water are added to a sealed Teflon™-lined cylinder, referred to as a "bomb", and placed within an oven or within a microwave reactor ("microwave-hydrothermal syntheses"). The pressure inside the bomb is controlled by the extent of filling and reaction temperature.

More recently, Schaak and coworkers have pioneered the use of nanoparticulate precursors (see Chap. 6) to synthesize a variety of intermetallic alloys and ceramics in liquid phase at low temperatures, without the need for top-down grinding/pulverization and high-temperature sintering of precursor powders.[73] Since solid-state reactions are diffusion controlled, in accord with Fick's 1st Law (2.31), the use of nanosized particulates dramatically reduces the diffusion distances among precursor species, thus enabling the use of relatively mild reaction temperatures.

$$J = -D\left(\frac{dc}{dx}\right) \tag{2.31}$$

where J is the diffusion flux (moles cm^{-2} s^{-1}), D is the diffusion coefficient (cm^2 s^{-1}), x is the position (cm), and c is the concentration (cm^3).[74]

Porcelain, used for applications that range from toilets to decorative plates, is formed by firing the green ceramic comprised of the clay mineral *kaolinite* ($Al_2Si_2O_5(OH)_4$), and a variety of other crystalline and amorphous materials such as *feldspars* ($KAlSi_3O_8$/$NaAlSi_3O_8$/$CaAl_2Si_2O_8$), glass, ash, and quartz. At a temperature of *ca.* 1200–1400 °C, glass and an aluminosilicate mineral known as *mullite* (or porcelainite) are formed, resulting in the familiar high strength and translucence of porcelain.

Non-oxide ceramics are typically synthesized via high-temperature routes, which convert molecular precursors into the desired structures. For instance, SiC (carborundum) may be produced from the direct reaction of silica sand with carbon in an electric furnace (2.32). Industrially, a mixture of 50 wt% SiO_2, 40 wt% coke, 7 wt% sawdust, and 3 wt% NaCl is heated together at ca. 2700 ºC—known as the *Acheson process*. The purpose of the salt is to remove metallic impurities via formation of volatile metal chlorides (*e.g.*, $FeCl_3$, $MgCl_2$, etc.). To yield highly crystalline SiC, the *Lely process* uses the sublimation of SiC powder or lumps at 2500 ºC under argon at atmospheric pressure.[75]

$$SiO_2 + 2\,C \xrightarrow[\text{(electric fumace)}]{2000\text{–}2500\ ^\circ C} SiC + CO_2 \tag{2.32}$$

A lower-temperature route involves the reduction of dichlorodimethylsilane with Na or Na/K alloys in an organic solvent (2.33 and 2.34),[76] or the use of microwave-facilitated methods.[77]

$$x\,(Me)_2SiCl_2 + 2x\,Na \xrightarrow{350\ ^\circ C,\ \text{toluene (autoclave)}} (Me_2Si)_x + 2x\,NaCl \tag{2.33}$$

$$(Me_2Si)_x \xrightarrow{800\ ^\circ C} SiC_{amorphous} \xrightarrow{1500\ ^\circ C} b - SiC \tag{2.34}$$

Chlorinated silanes may also be used to synthesize silicon nitride (Si_3N_4) ceramics, via reaction with an amine (2.35)[78]:

$$x\ MeSiCl_3 + 3x\ NH_3 \xrightarrow{500^\circ C} (MeSiN_3)_x + 3x\ HCl \quad (2.35)$$

Of course, the "brute force" method of reacting silica with ammonia or N_2/H_2 gases at temperatures in excess of 1200 °C will also yield crystalline silicon nitride ceramics. Another route that does not involve chlorinated precursors consists of sintering a polymeric precursor such as poly[(methylvinyl)silazane], $[(CH_3SiHNH)_{0.8}(CH_3SiCH = CH_2NH)_{0.2}]_n$.[79]

2.5.1 Sol–Gel Processing

In contrast to the high-temperature routes discussed previously, the *sol–gel* (solution-gelation) process is a versatile solution-based technique for fabricating ceramic materials. In general, sol–gel involves the formation of a *sol* (colloidal suspension of *ca.* ≥ 200 nm solid particles) and subsequent crosslinking to form a viscous *gel*. Although this technique has been in practice since the late 1800s, the complex mechanisms that are involved during sol–gel processing were not understood and fully exploited until the 1950s–60 s.[80] The most common starting materials, or *precursors*, used in the preparation of the sol are water-sensitive metal alkoxide complexes ($M(OR)_x$, where R = alkyl group (e.g., CH_3, C_2H_5, CF_3, etc.)). Although original formulations used sodium silicates, the use of alkoxide precursors avoids undesirable salt byproducts that may only be removed through long, repetitive washing procedures. In addition, the nature of the metal and associated R groups may be altered to affect the rate and properties of the ultimate oxide material.

Sol–gel syntheses are typically carried out in the presence of polar solvents such as alcohol or water media, which facilitate the two primary reactions of *hydrolysis* and *condensation* (2.36 and 2.37, respectively). During sol–gel processing, the molecular weight of the oxide product continuously increases, eventually forming a highly viscous three-dimensional network.

$$M - OR + H_2O \rightarrow M - OR + ROH \quad (2.36)$$

$$n\ M - OR + n\ M - OH \rightarrow [M - O - M]_n + n\ ROH \quad (2.37)$$

The most widely used metal alkoxides are $Si(OR)_4$ compounds, such as tetramethoxysilane (TMOS) and tetraethoxysilane (TEOS). However, other alkoxides of Al, Ti, and B are also commonly used in the sol–gel process, often mixed with TEOS. For instance, aluminum silicates may be generated through hydrolysis/condensation of siloxides (2.38), which proceed through an intermediate Al–O–Al network known as *alumoxanes*. Alumoxanes are important for applications in antiperspirants, catalysts, and paint additives. Their structure is best described as Al(O)(OH) particles containing a core structure analogous to the minerals boehmite and diaspore (Fig. 2.83), with organic substituents on the surface.

$$Al(OSiR_3)_3 + H_2O \rightarrow [Al(O)(OH)_x - (OSiR_3)_{1-x}]_n(gel) \xrightarrow{\Delta} (Al_2O_3)_m(SiO_2)_n \tag{2.38}$$

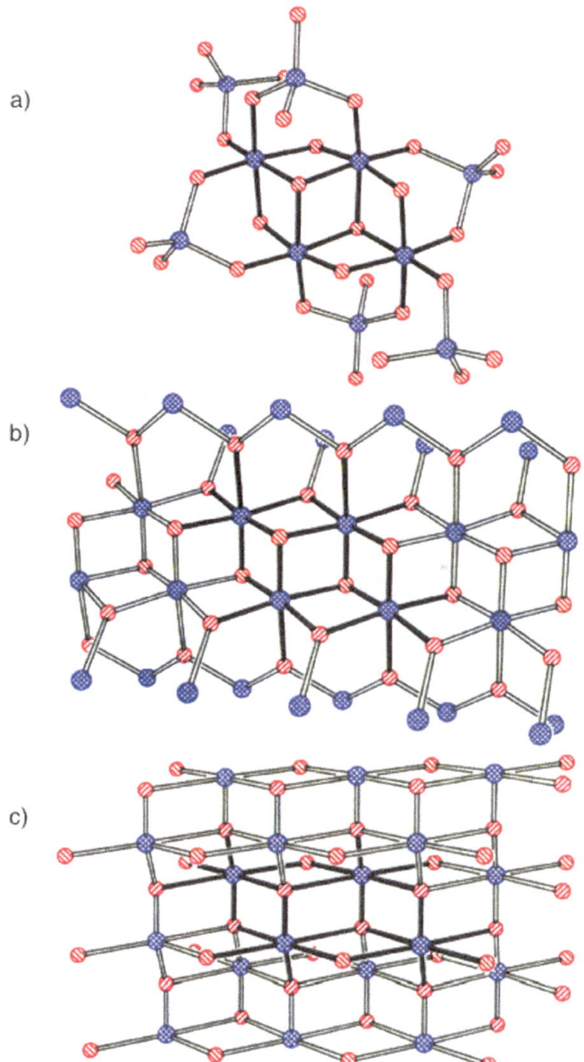

Fig. 2.83 Comparison of the core structure of **a** siloxy-substituted alumoxane gels with **b** diaspore and **c** boehmite minerals. The aluminum and oxygen atoms are shown in blue and red, respectively. Reproduced with permission from *Chem. Mater.* **1992**, *4*, 167. © 1992 American Chemical Society

As one would expect from the similar electronegativities of Si and O, the hydrolysis of silicon alkoxides are significantly slower than other metal analogues. For identical metal coordination spheres and reaction conditions, the general order or reactivity for some common alkoxides is: $Si(OR)_4 \ll Sn(OR)_4 \sim Al(OR)_3 < Zr(OR)_4 < Ti(OR)_4$. That is, larger and more electropositive metals are more susceptible to nucleophilic attack by water. As a result, the hydrolysis of most metal alkoxides is too rapid, leading to uncontrolled precipitation. Although the ratio of $H_2O/M(OR)_n$ may be tuned to control the hydrolysis rate, the nature of the metal alkoxide (e.g., altering the OR groups or metal coordination number) is the most powerful way to control the rate of hydrolysis. It is also possible to control the stepwise hydrolytic pathway, which governs the ultimate three-dimensional structure of the gel (Fig. 2.84).[81]

In stark contrast to other metal alkoxides, the kinetics for the hydrolysis of $Si(OR)_4$ compounds often require several days for completion. As a result, acid (e.g., HCl, HF) or base (e.g., KOH, amines, NH_3) catalysts are generally added to the mixture, which also greatly affects the physical properties of the final product. Under most conditions, condensation reactions begin while the hydrolytic processes are underway. However, altering the pH, $[H_2O/M(OR)_n]$ molar ratio, and catalyst may force the completion of hydrolysis prior to condensation.

A likely mechanism for an acid-catalyzed system is shown in Fig. 2.85. The protonation of the alkoxide group causes electron density to be withdrawn from Si, allowing the nucleophilic attack from water. In contrast, the base-catalyzed hydrolysis of silicon alkoxides proceeds through the attack of a nucleophilic deprotonated silanol on a neutral silicic acid (Fig. 2.86). In general, silicon oxide networks obtained via acid-catalyzed conditions consist of linear or randomly branched polymers; by contrast, base-catalyzed systems result in highly branched clusters (Fig. 2.87).

As condensation reactions proceed, the sol will set into a rigid gel. Since the reactions occur within a liquid alcoholic solvent, condensation reactions result in a three-dimensional oxide network $[M–O–M]_n$ that contains solvent molecules within its pores. The product at its gel point is often termed an *alcogel*, recognizing the trapped solvent. At this stage, the gel is typically removed from its original container, but remains saturated with the solvent to prevent damage to the gel through evaporation. It is worthwhile to note that at the gel point, the –O–Si–O– framework still contains a number of unreacted alkoxide moieties. Hence, sufficient time (typically 48 h+) must be allotted to allow for complete hydrolysis and polycondensation, so the network is suitably strengthened to prevent cracking—a process referred to as *aging*, or *syneresis*.

Depending on the post-treatment used for the sol, a wide variety of materials may be synthesized: ultra-fine powders, thin film coatings, ceramic fibers, microporous inorganic membranes, ceramics and glasses, or extremely porous materials (Fig. 2.88). Thin films are easily generated on a substrate through simple spincoating or dip-coating of the gel, followed by slow evaporation to prevent extensive cracking. Alternatively, the gel may be retained in a mold and heat-treated to convert the material into a dense ceramic or glass. If the solvent of an alcogel is removed through slow evaporation, a porous material known as a *xerogel* is formed. By contrast, if supercritical CO_2[82] is used to remove the solvent, a foam-like, porous, transparent

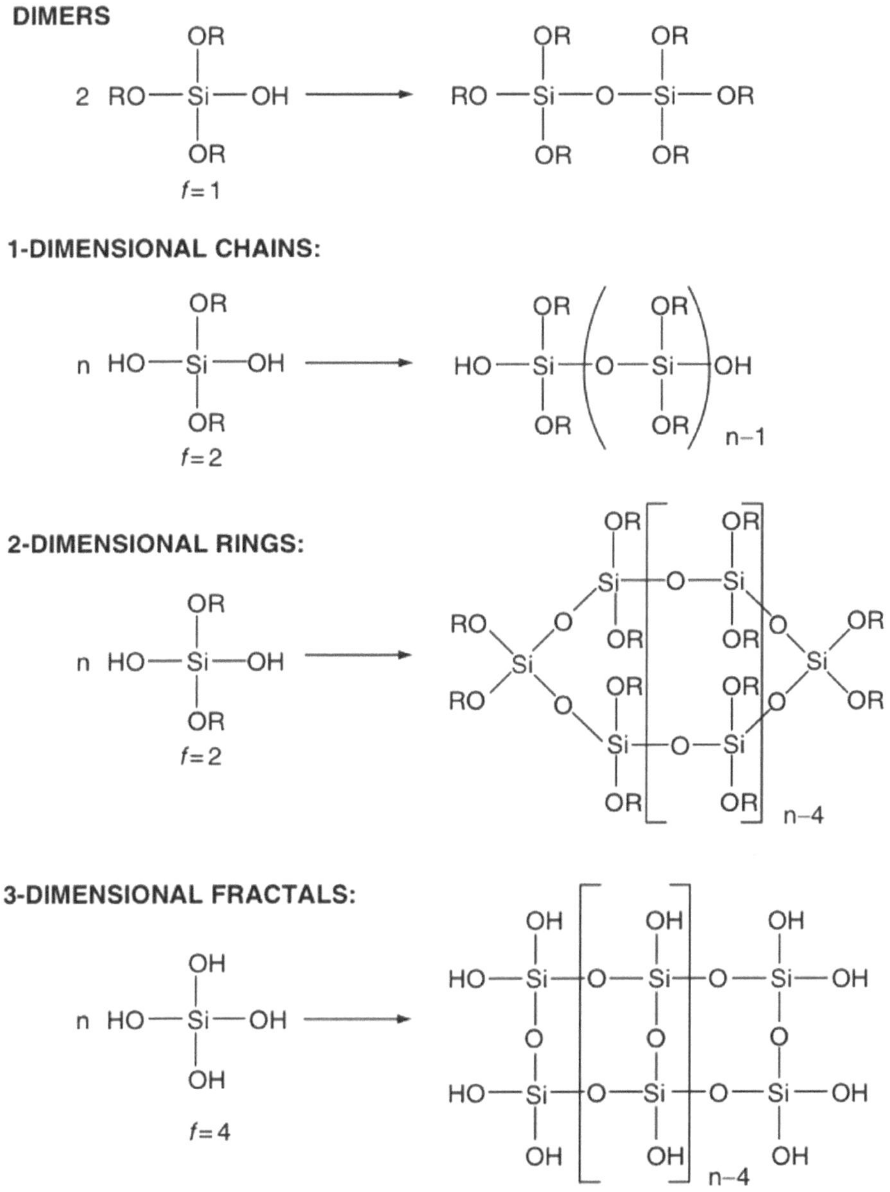

Fig. 2.84 Relationship between a siloxy precursor and its control over the three-dimensional shape of a gel. Reproduced with permission from *Chem. Rev.* **2004**, *104*, 3893. © 2004 American Chemical Society

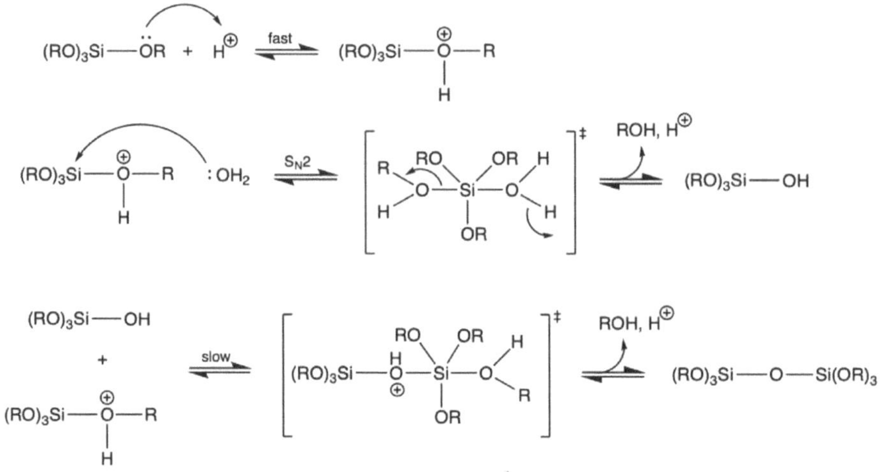

Fig. 2.85 Reaction schemes for the acid-catalyzed hydrolysis and condensation of a silicon alkoxide precursor

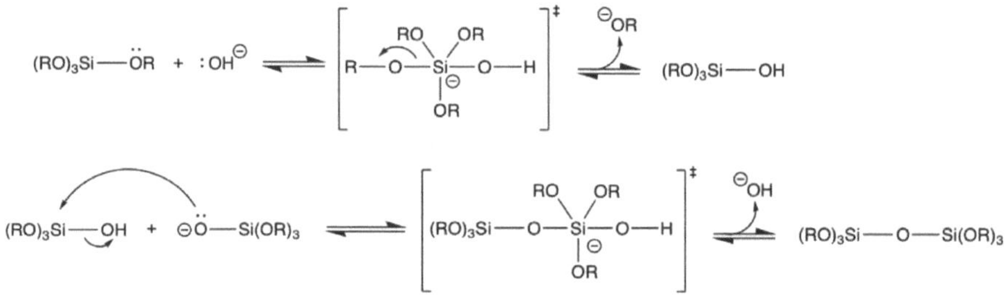

Fig. 2.86 Reaction schemes for the base-catalyzed hydrolysis and condensation of a silicon alkoxide precursor

material called an *aerogel* is formed. Silica aerogels consist of 99.8% air, and have a density of 1.9 g L^{-1} and a thermal conductivity of 1.7×10^{-2} W m^{-1} K^{-1}. The properties of aerogels afford a range of applications, among which include sound dampening, catalysis, desiccation, and thermal insulating (e.g., windows, refrigerators, walls).[83] Due to the complex, crosslinked structure of aerogels, the insulating ability is an order of magnitude greater than commonly used fiberglass. As a testament to the unique properties of silica aerogels, the Guinness Book of World Records recognizes this material as the best insulator and least-dense solid.

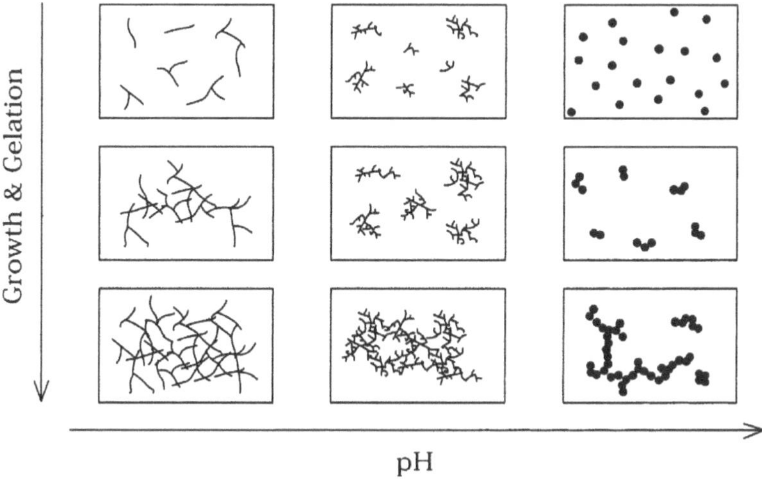

Fig. 2.87 Comparison of the morphology with the pH of the sol–gel process. Reproduced with permission from *Chem. Rev.* **2004**, *104*, 3893. © 2004 American Chemical Society

Aerogels retain the original shape and volume of the alcogel, typically >85% of the original volume. By contrast, xerogels exhibit significant shrinking and cracking during drying, even under room-temperature conditions (Fig. 2.89). It is important that the water be removed prior to the drying event. This is easily accomplished through soaking the alcogel in pure alcohol. The soaking time is dependent on the thickness of the gel. Any water left in the gel will not be removed by supercritical drying, and will lead to a dense, opaque aerogel. Similarly, water will not be removed as readily as alcohol by simple evaporation; hence, water-containing gels will result in heavily cracked and heterogeneous xerogels.

Inorganic gels are rarely used in their as-dried state. The gel is first dehydrated through thermal removal of surface –OH groups, thus preventing rehydration reactions. Most often, this step is followed by high-temperature annealing (T > 800 °C), in order to convert the amorphous material into a desired crystalline phase. As an example for alumoxanes, it has been shown that the gel is initially transformed to g-Al_2O_3, en route to its highest crystalline form, a-Al_2O_3.69 During the sintering (or firing) process, the pore size is also diminished and organic moieties are removed. Hence, this results in significantly better mechanical properties, as desired for dense ceramics and glasses.

2.5.2 Cementitious Materials

The use of cementitious materials for structural applications dates back to ancient Egypt. A type of cement was used to hold together the limestone blocks of the great pyramids that still stand today. During the time of the Roman Empire, an improvement of cement formulations was developed, which used a finely divided

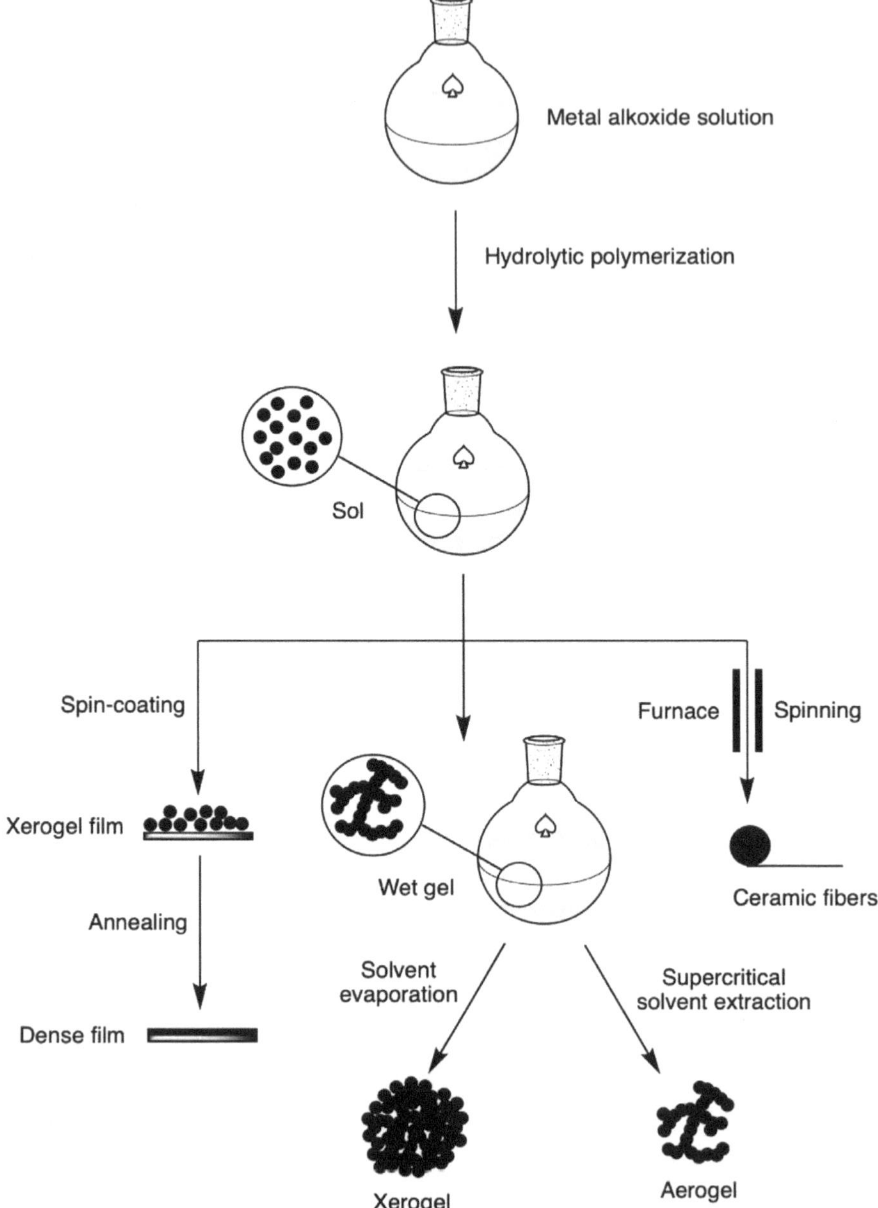

Fig. 2.88 Illustration of the products obtained through sol–gel processing

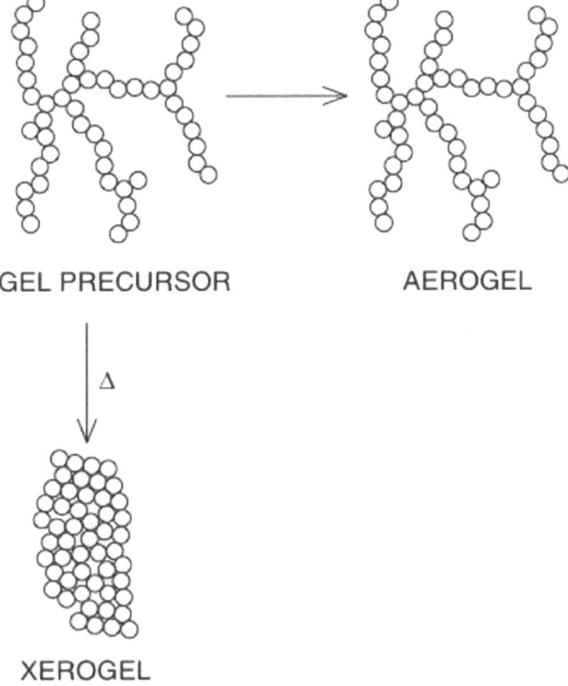

Fig. 2.89 Comparison of the three-dimensional shape of an aerogel and xerogel formed from a gel. Reproduced with permission from *Chem. Rev.* **2004**, *104*, 3893. © 2004 American Chemical Society

volcanic ash, known as Pozzolana, found in various parts of Italy. Although they did not realize it at the time, the hardening process occurred due to the reaction of the aluminosilicate-based ash with $Ca(OH)_2$ in the presence of water to yield a calcium-silicate-hydrate (CSH) rigid gel. Amazingly, thousands of years later, the CSH structure is not yet completely understood; it is likely a disordered form of the hydrated calcium silicate mineral tobermorite (Fig. 2.90).[84]

The last major development in cement technology occurred in the early nineteenth century in England. Bricklayer Joseph Aspdin first made a variety of cement known as *Portland cement*—not in a laboratory, but on his kitchen stove! His patent in 1824 changed the world forever, as this form of cement is the basic ingredient in concrete—essential for the erection of virtually all buildings and many roads throughout the world. In fact, concrete is the most heavily used man-made material. As of 2017, it is estimated that the worldwide annual production of concrete amounts to 3 tons for every man, woman, and child on earth—second only to water in terms of human consumption!

It is interesting to note the developmental timeline for cement/concrete, which has addressed many important societal needs. For example, if we consider road construction, stones were used as early as 4000 BC, and were still prevalent in early America—still evident in some historical cities such as Boston, MA. As motorcars became more abundant and faster in the early twentieth century, replacements were sought for dirt, gravel, and stone roads. Asphalt[85] and simple road tarring were the

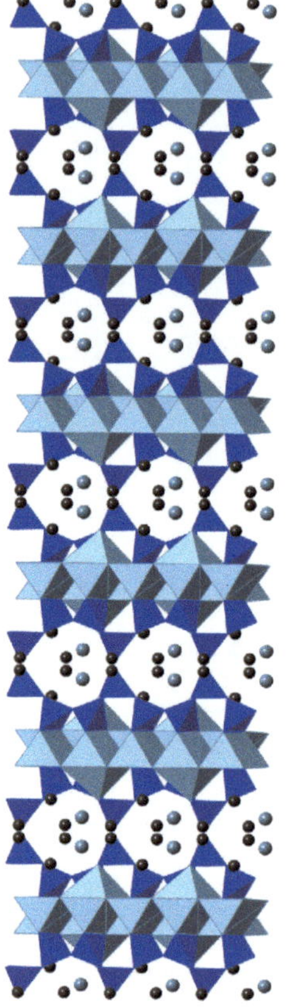

Fig. 2.90 The crystal structure of tobermorite, viewed along the **bc** plane. The amorphous gel produced during cement formation likely contains defects such as missing/disordered silicate tetrahedra and/or water sites

first alternatives that addressed the environmental problem of dust as well as road smoothness, and most highways across the world are still composed of asphalt pavement. Comparatively, a large proportion of city streets and highways, especially in the warmer locales, are made from concrete. Since concrete is more apt to crack relative to asphalt under significant temperature changes, this material is not greatly used to construct roadways in northern climates (e.g., North Dakota, Montana, Minnesota, Michigan, Canadian Provinces).

Portland cement is produced from the sintering of minerals containing $CaCO_3$, SiO_2, Al_2O_3, Fe_2O_3, and MgO in a ceramic kiln, held at a temperature of *ca.* 1500 °C. Equations 2.39–2.42 show the reactions that occur during the processing of cement. The resulting complex material is referred to as *clinker* (Fig. 2.91) and may be stored for many years under anhydrous conditions before its use in concrete. As one can see from (2.39), this process releases the greenhouse gas CO_2, causing environmental concerns over the world's increasing production of cement. It is estimated that Portland cement manufacturing accounts for over 5% of the world's total emission of CO_2. As a result, there is an increasing focus on using additives, such as fly ash (a by-product from coal-fired power plants).[86]

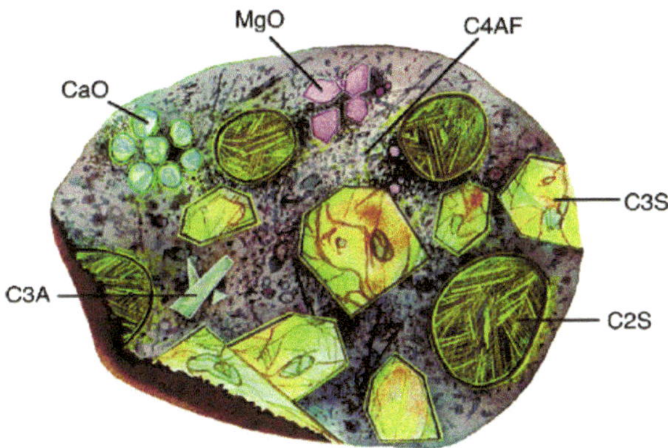

Fig. 2.91 Cross-section representation of a powdered cement particle. Dicalcium silicate (Ca_2SiO_4), tricalcium silicate (Ca_3SiO_5), tricalcium aluminate ($Ca_3Al_2O_6$), and tetracalcium aluminoferrite ($Ca_4Al_nFe_{(2-n)}O_7$) crystallites are abbreviated as C2S, C3S, C3A, and C4AF, respectively. Reproduced with permission from *Chem. Mater.* **2003**, *15*, 3074. © 2003 American Chemical Society

When water is mixed with Portland cement, the product sets in a few hours and hardens over a period of 3–4 weeks. The initial setting process is caused by the reaction between water, gypsum ($CaSO_4{\cdot}2H_2O$, added to clinker to control the hardening rate), and C3A forming calcium and aluminum hydroxides. The crystallization of these hydroxides results in an observable hardening within the first 24 h. The subsequent reactions between the formed hydroxides and C3S result in formation of a crosslinked CSH gel (vide supra); this provides further strengthening over the first week. The reaction between hydroxides and C2S proceeds the slowest, and results in hardening/strengthening of the material in latter stages of setting.[87] The hardening of concrete is a consequence of individual grains of *aggregates* being cemented together by the clinker-water byproducts. Aggregates that are present in concrete are of two or more size distributions—typically, coarse gravel/stones and fine sand. As a general rule, 1 m^3 of concrete contains over 4400 lb of gravel and sand.

$$CaCO_3 \rightarrow CaO + CO_2 \tag{2.39}$$

$$3\,CaO + SiO_2 \rightarrow Ca_3(SiO_5)\ \text{“C3S”} \tag{2.40}$$

$$2\,CaO + SiO_2 \rightarrow Ca_2(SiO_4)\,''C2S'' \tag{2.40}$$

$$3\,CaO + Al_2O_3 \rightarrow Ca_3Al_2O_6''\ C3A'' \tag{2.42}$$

$$4\,CaO + Al_2O_3 + Fe_2O_3 \rightarrow Ca_4Al_2Fe_2O_{10}\ \text{“}C4AF'' \tag{2.43}$$

2.5.3 Biomaterials Applications

The applications for ceramics span virtually every commercial sector, from porcelain dishes and sinks to uses in cutting tools, ball bearings, electronics (e.g., capacitors, insulators, integrated circuit packaging, piezoelectrics, superconductors), coatings (e.g., engine components, drill bits), filters, membranes, and catalyst support materials. It is estimated that the current annual world demand for advanced ceramics is over $50 billion. One interesting futuristic application for ceramics is a "smart ski" that uses a piezoceramic material embedded in the ski to reduce vibrations and increase stability and control at higher speeds. An emerging area for ceramic applications is the field of *biomaterials*, a $70 billion industry.

By definition, a biomaterial is a biocompatible material or device that is placed within a living creature in order to perform, augment, or replace a natural function. Throughout this textbook, we will discuss a variety of such materials, which span applications from dentistry (e.g., implants such as crowns and dentures), orthopedic (e.g., artificial limbs, joint and bone repair/replacement), optometric (contact lenses), and medicinal (e.g., soluble sutures, coronary stents, artificial organs, drug delivery agents used to deliver a chemical compound directly to the site of treatment). Such medical breakthroughs have not only extended the life expectancy of humans (currently 76 years for men and 81.2 years for women in the U.S.), but have resulted in a way of life that would have seemed impossible just a few decades ago.

Ceramics and glasses are widely used to repair or replace joints or tissue, or serve as a coating to improve the biocompatibility of metallic-based implants. Before placing a biomaterial within a body, it must be non-toxic, bioinert (i.e., able to withstand corrosion in a biological environment without causing damage to surrounding tissues), bioactive (i.e., able to undergo interfacial interactions with surrounding tissues), and biodegradable/resorbable (i.e., eventually replaced or incorporated within growing tissue). For instance, when a bioactive glass is placed in a physiological environment, there is an ion exchange among cations in the glass and hydronium ions within the surrounding solution. Hydrolysis then takes place, wherein the glass network is disrupted resulting in a change in its morphology to a porous, gel-like surface layer. Precipitation of a calcium phosphate mineral ensues, followed by further mineralization to a crystalline substance that mimics the structure of bone.

Placing a biomaterial within living tissue will always render a tissue response at the implant/tissue interface. In particular, the following four responses may result from implantation, which govern the degree of medical complications and ultimate lifetime of an implant[88]:

i. The surrounding tissue will die if the material is toxic;
ii. A fibrous tissue of varying thickness will form if the material is nontoxic and biologically inert;
iii. An interfacial bond will form if the material is nontoxic and biologically active;
iv. The surrounding tissue will replace the implant if the material is nontoxic and soluble.

The encapsulation of soft tissue (ii above), causes deleterious wear and abrasion in metallic implants. Other contributing factors in decreasing the lifetime of an implant include infection, inflammation, and lack of prolonged bonding between the implant and surrounding bone. Consequently, the current average lifetime of a titanium orthopedic implant is on the order of 10–15 years. However, there are many efforts devoted to improving the longevity of implants. One such approach features modifying the titanium surface via *anodization* (see Chap. 6) to increase the surface area and mimic the roughness of natural bone, which facilitates the adsorption of a larger number of bone-forming cells known as *osteoblasts*. To address other harmful factors, the implant is coated with anti-infection and anti-inflammation drugs (penicillin/streptomycin and dexamethasome, respectively).[89]

There are three primary methods used for bone substitution, required for applications such as spinal deformities and "non-unions"(fractures that do not heal within 9 months). *Autografting* consists of transplanting a bone from one region of the patient's body (usually the pelvic region) to the desired location. This procedure is often preferred, as it precludes immunogenicity problems; however, there may be complications and additional pain at the harvesting site, as well as the additional surgical costs for the combined harvesting/transplanting procedures. In contrast, *allografting* consists of harvesting bone from a live or deceased donor for the transplanting procedure. These implants are much less successful than autograft implants due to immuogenicity, transmitted diseases, and the absence of viable osteoblasts. Due to these limitations, it is becoming increasingly more popular to use synthetic materials as bone substitutes. Whereas refractory ceramics such as alumina (Al_2O_3) and zirconia (ZrO_2) are used in high-wear applications such as joint replacements, calcium phosphate/sulfate based ceramics are frequently used for bone regeneration applications.

The most common synthetic bone grafting ceramic is $Ca_{10}(PO_4)_6(OH)_2$, known as *hydroxyapatite* (Fig. 2.92). Unlike other calcium phosphates, hydroxyapatite is stable under physiological conditions, which features a pH range of 5.5–7.2. Hydroxyapatite serves as an osteoconductive scaffold to which proteins and cells may nucleate, and within which bone-forming cells known as *osteoblasts* are generated (Fig. 2.93). In addition to using hydroxyapatite as a filler material, it may also be used as a coating material. That is, while its mechanical properties are too brittle to withstand load-bearing applications, hydroxyapatite may be used as a coating on metallic alloys to impart a greater biocompatibility of joint replacements, while reducing the release of metal ions from the implant to the surrounding biological fluid.

Interestingly, the composition, phase, morphology, and placement of hydroxyapatite will influence the speed and extent of bone growth. Since the resorption process is surface-driven by the adsorption of osteoblasts, the ultimate solubility of a ceramic will be directly related to its surface area—i.e., crystal size and density. In addition, careful control of processing parameters is necessary to prevent thermal decomposition of hydroxyapatite into other soluble calcium phosphate phases (e.g., tricalcium phosphate, $Ca_3(PO_4)_2$, tetracalcium phosphate, $Ca_4(PO_4)_2O$, and CaO), which dissolve more rapidly in body fluid than crystalline hydroxyapatite.[90] It has

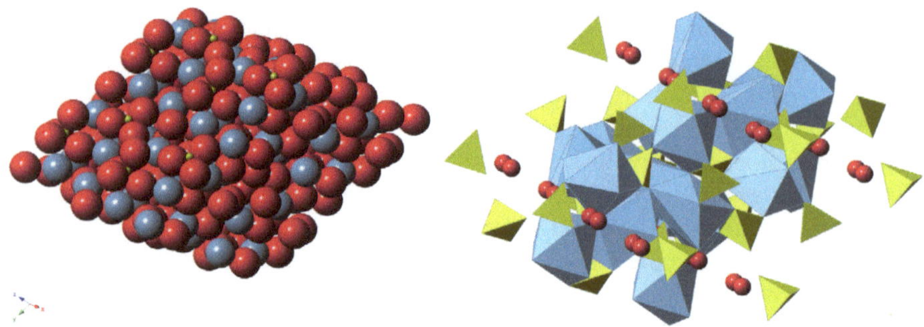

Fig. 2.92 Crystal structure of hydroxyapatite, illustrated by space-filling and polyhedral representations (yellow = P, red = O, blue = Ca). The red spheres in the polyhedral representation show the positions of hydroxyl groups that surround CaO_6 and PO_4 polyhedra

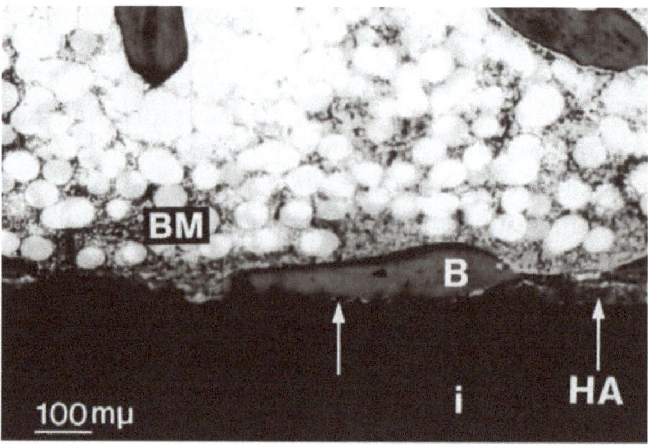

Fig. 2.93 Photomicrograph of an HA-coated implant. Areas with complete resorption of the coating (arrow) were covered by bone marrow (BM) and bone (B) ongrowth (Light Green and basic fuchsin × 75). Reproduced with permission from *J. Bone Joint Surg.* **1997**, *79-B*, 654. © 1997 British Editorial Society of Bone and Joint Surgery

been shown that the plasma spraying procedure used to deposit hydroxyapatite coatings onto orthopedic and dental implants may also result in co-deposits of amorphous calcium phosphate and other resorbable phosphates. Increased concentrations of these other phosphates are thought to result in premature resorption of the coating before the bone may attach to the implant, drastically shortening its lifetime.[91]

Due to its inherent brittleness, hydroxyapatite is often combined with calcium sulfate ($CaSO_4 \cdot 2H_2O$, denoted as *plaster of Paris*) to form a more durable ceramic material. Calcium sulfate is biocompatible and bioactive, being resorbed after 30–60 days. Calcium sulfate is mostly used in its partially hydrated form; when mixed with water, an exothermic reaction leads to recrystallization into the dihydrate final form shared by the mineral *gypsum* (2.44):

$$CaSO_4 \cdot (1/2\ H_2O) + 3/2\ H_2O \rightarrow CaSO_4 \cdot 2\ H_2O + \text{ heat} \tag{2.44}$$

Analogous to cement, plaster is used as a building material, by hydrating a finely-divided powder. However, unlike cement, plaster remains relatively soft following drying, which limits its utility for structural applications. The hardening mechanism of plaster is due to the recrystallization process.[92] That is, when the hemihydrate powder is mixed with water, the evolved heat evaporates the water, forming a network of interlocked dihydrate needle-like crystals. For identical chemical compositions, the size and morphology of the starting powder will determine the amount of water required to make a sufficiently strong ceramic material. Whereas plaster crystals are irregular in size/morphology, formed by heating gypsum in air at 115 °C, *dental stone* crystals feature a much greater uniformity—a consequence of being fabricated by heating gypsum under an applied pressure. The less uniform crystals of plaster will pack relatively inefficiently, requiring more water for mixing relative to stone. As a result, the evaporation of larger volumes of water will result in a greater porosity, causing the hardness of set plaster to be much less than dried stone.

2.6 Glasses

Thus far, we have focused on solids that have well-ordered crystalline structures. It is now time to turn our attention to those solids composed of disordered arrays of species, known as *amorphous* solids. Earlier in the chapter, we discussed the synthesis of amorphous metals—those obtained through fast nonequilibrium conditions. However, there is a more pervasive class of materials that exhibits an amorphous structure, which our society is indebted to for countless applications: silica-based glasses. Further, although most ceramic materials exhibit a crystalline structure, these materials are typically composed of polycrystals alongside embedded amorphous structures. In fact, ceramics may also have amorphous structures when synthesized at low temperatures, with the conversion to crystalline phases as their temperature is increased, a process referred to as *sintering*, *firing*, or *annealing*. This thermal processing results in the familiar properties of ceramics such as significant hardness and high melting points, which are desirable for structural applications or those occurring within extreme environments such as high temperatures and/or pressures. In this section, we will describe the structure and properties of some important classes of amorphous glasses, as well as partially-amorphous and/or polycrystalline ceramics and cementitious materials.

From drinking vessels and windows to eyeglass lenses, materials comprising glass have long played an important role in our society. In fact, it is estimated that applications for glass date back to Egypt in *ca.* 3500 BC. Although we are most familiar with transparent silica-based (SiO_2) glass (Fig. 2.94), there are many other types of glass that may be fabricated for various applications. For instance, infrared-transmitting chalcogenide glasses such as As_2E_3 (E = S, Se, Te) are suitable for specialized applications such as optical storage, sensors, and infrared lasers.[93] As discussed earlier in the chapter, even metals may be suitably synthesized to possess a bulk-disordered glassy structure. By definition, the term *glass* refers not a specific material, but a general architectural type—an amorphous solid that has cooled to rigidity without crystallizing. Glasses are most commonly made by rapidly quenching

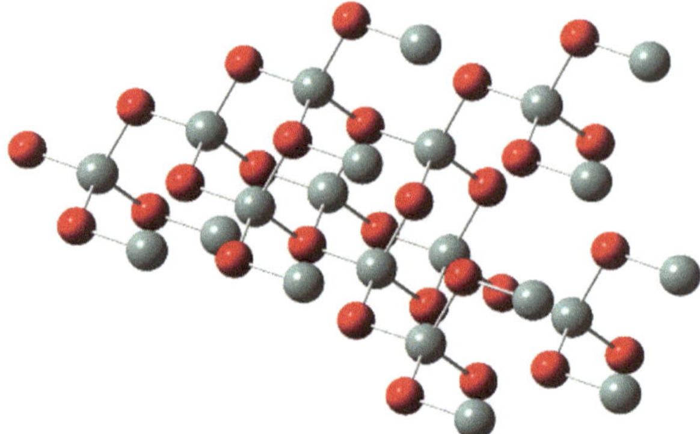

Fig. 2.94 Molecular structure of amorphous SiO_2, comprised of randomly corner-linked SiO_4 tetrahedra

a melt; accordingly, the constituent atoms are not allowed to migrate into regular crystalline lattice positions.[94]

It is noteworthy to point out why a material as disordered as glass is transparent. That is, one would think that the amorphous structure of glass should facilitate *opacity*, which is the extent to which visible radiation is blocked by the material it is passing through. There are two primary rationales for the transparency of glass: electronic and structural. First, as we will see shortly, glass may contain a variety of dopants that will afford particular colors (via electronic transitions) or physical properties (e.g., enhanced hardness, thermal/electrical conductivity, reflectivity, etc.). However, these impurities are only present in sufficient quantity to cause only partial absorption of the electromagnetic spectrum, resulting in observable transparency—although less pronounced relative to undoped glass.

Second, unlike metals, glasses are held together by covalent and/or ionic bonding and do not contain free electrons in their structure. Accordingly, the incident wavelengths are not perturbed into destructive waves and are free to transmit through the material. Additionally, the degree of disorder within glasses is the same order of magnitude as the incident radiation, allowing the light to pass through relatively unattenuated.[95] However, it should be noted that if glass contains imperfections, and/or inclusions of metals or larger particles with dimensions greater than the wavelength of indicent light, the material will become increasingly opaque due to *Rayleigh scattering* (2.45)[96]:

$$\text{scattering} \approx \frac{(\Delta\eta)\left(\mathrm{d}^3\right)}{\lambda^4} \tag{2.45}$$

where $\Delta\eta$ is the change in the refractive index and d is the spatial distance covered by the disorder.

Glasses and ceramics are largely based on a covalently bound network that is comprised of an infinite array of silicate (SiO_4^{4-}) tetrahedra.[97] As shown in Fig. 2.95, a variety of structures are possible by Si–O-Si linkages among adjacent tetrahedra. Since the silicate sub-units carry an overall −4 charge, alkali or alkaline earth metal ions are commonly present in order to afford charge neutrality, and link adjacent silicate tetrahedra via ionic bonding (Fig. 2.96). In addition to random or crystalline 3-D structures, silicates may also assemble into chain-like arrays; for instance, the large family of hydrous magnesium silicates (e.g., chrysotile ($Mg_3Si_2O_5(OH)_2$), pyroxenes ($XSiO_3$, X = Mg, Na, etc.) shown in Fig. 2.97a–b, better known as *asbestos*. Layered-sheet arrays are also well known, especially in combination with aluminum oxide such as aluminosilicate clays (e.g., kaolinite clay, Fig. 2.97c). For these latter structures, there is only weak van der Waal attraction between adjacent layers, which governs their overall physical properties. For instance, talc (layered $Mg_3Si_4O_{10}(OH)_2$) is one of the softest minerals (Mohs hardeness of 1), and may be used as a lubricant due to facile slippage of neighboring layers.

The most straightforward method to make silica (SiO_2) glass, known as *fused silica* or *quartz glass*, is through melting sand at a temperature of 1800–2000 °C followed by very slow cooling. Unlike other glasses that require a rapid quenching event, quartz will automatically form a glassy solid at all but the slowest cooling rates—a consequence of its complex crystal structure (Fig. 2.98). For example, it is estimated to have taken 100,000 years to form natural crystalline quartz! Crystalline silica exists as three varieties, with each form having slightly differing crystal structures and physical properties[98]:

$$\text{quartz} \xleftrightarrow{8700\ ^\circ\text{C}} \text{tridymite} \xleftrightarrow{1470\ ^\circ\text{C}} \text{cristobalite} \xleftrightarrow{1710\ ^\circ\text{C}} \text{liquid SiO}_2$$

Two methods commonly used to synthesize quartz are hydrothermal (autoclave at high temperature/pressure, containing water and seed crystals) and flux growth. For this latter technique, LiO, MoO, PbF_2 and silica powders are added to a crucible; the ionic compounds serve as a molten solvent to dissolve materials with a high melting point, facilitating crystallization at lower pressures/temperatures. Fused silica is thermally stable at temperatures up to *ca.* 1665 °C. Further, the coefficient of linear expansion is 5.5×10^{-7} cm cm^{-1} K^{-1}; by comparison, the softening point and coefficient of linear expansion for normal window-pane glass are *ca.* 500 °C and 9.0×10^{-6} cm cm^{-1} K^{-1}, respectively.

Although the mechanical properties of quartz are desirable for high-temperature applications, this glass is relatively difficult to mold into desired shapes through conventional glass-blowing techniques. Quartz glass is transparent toward ultraviolet radiation ($\lambda = 190 - 300$ nm), indicating that the spatial range of structural disorder is less relative to other glasses that contain additional additives (vide infra). As a result, quartz windows are used for ultraviolet lamps that are employed in a number of important applications in chemistry, biology, engineering, and materials science.

The chemistry of glass making is now a mature field, with many types available for a variety of applications. In order to decrease the prohibitively high melting point of SiO_2, *ca.* 18% of sodium carbonate ("soda," Na_2CO_3) is often added to sand, resulting in a silica framework doped with Na^+ ions.[99] The resultant glass is more easily workable than fused silica due to interruption of the silicate network. However,

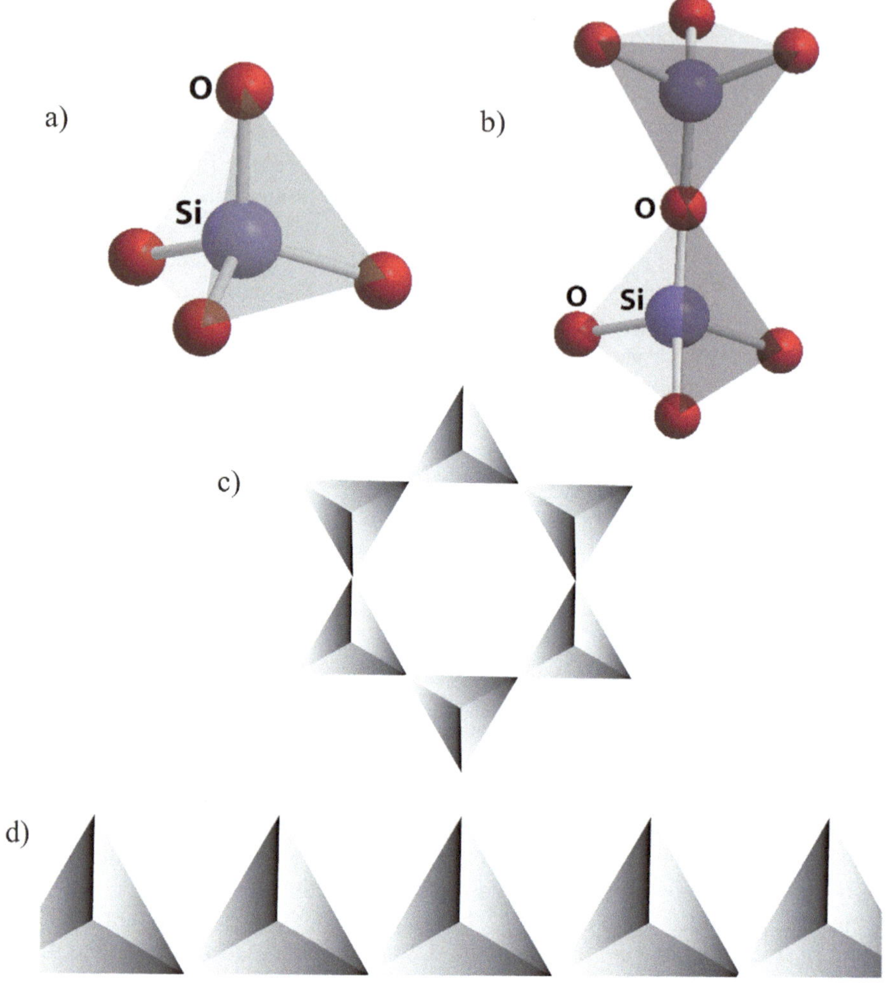

Fig. 2.95 Molecular structures of common silicate anions. Shown are: **a** $SiO_4{}^{4-}$, **b** $Si_2O_7{}^{6-}$, ◦ $Si_6O_{18}{}^{12-}$, and **d** a metasilicate polymer chain $\{SiO_3{}^{2-}\}_\infty$

the sodium ions are detrimental since they are easily solvated by water, which leads to corrosion. To prevent such weathering, *ca.* 10% of limestone ($CaCO_3$) is added to effectively replace the Na^+ ions with Ca^{2+}. When this mixture is heated to its melting point (*ca.* 1000 °C), a mixture of calcium silicate ($CaSiO_3$) and sodium silicate (Na_2SiO_3) results. Upon cooling, the most prevalent type of glass, called "crown glass" or *soda–lime* glass, is generated. This type of glass accounts for over 90% of the glass used worldwide. Interestingly, our current synthetic procedure has not deviated from the earliest glassmakers' recipes, dating back to *ca.* 1450 in Venice, which also used white stone pebbles (quartz, SiO_2) and plant ash containing sodium- and

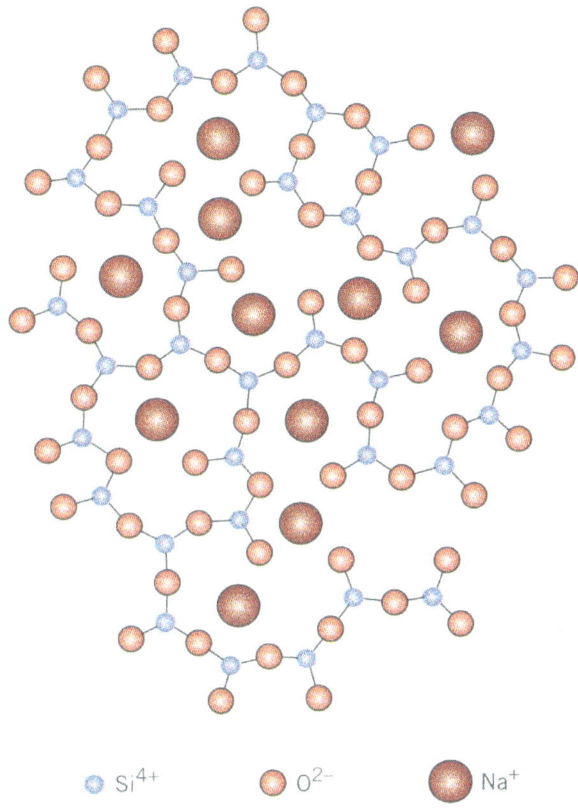

Fig. 2.96 Schematic representation of ionic positions within soda glass. Reproduced with permission from Callister, W. D. *Materials Science and Engineering: An Introduction*, 7th ed., Wiley: New York, 2007. © 2007 John Wiley & Sons, Inc

calcium-based additives (Na_2O and lime (CaO), respectively). It should be noted that glass with a molar concentration of Na_2O:CaO:SiO_2 = 16:10:74 can form crystals of *devitrite* (Na_2O•3 CaO•6 SiO_2), at a rate of $17 \mu m\, min^{-1}$ at a temperature of 995 °C—especially if the molten glass is cooled too slowly.[100] Such devitrification will alter the physical properties of glass (e.g., transparency, strength) in the area surrounding crystal growth; this occurs much less readily in ancient glasses due to their very complex compositions.

There are a number of other glass recipe variations that may be used to yield desired properties. Most likely, these formulations were discovered by accident or in a trial-and-error manner, using materials from their locale and measuring the resultant properties. For instance, the Europeans were the first to discover that K_2O, obtained locally from plant ash, could also be combined with lime and quartz to yield a potash-lime glass, later exploited for stained-glass windows. Another popular variation substitutes boric oxide (B_2O_3) for lime and soda to yield *borosilicate glass*. The physical properties of this glass resemble fused silica (e.g., coefficient of thermal expansion: $3.3 \times 10^{-7}\, cm\, cm^{-1}\, K^{-1}$), except that its softening temperature is only *ca.* 700 °C. Borosilicate glass is the variety that is sold in stores as Pyrex™ cookware

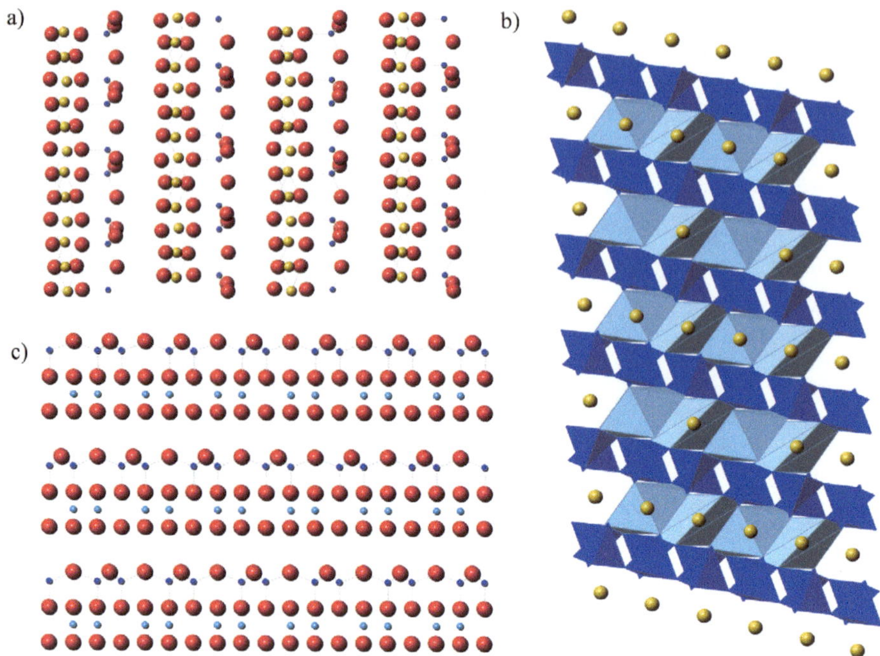

Fig. 2.97 Molecular structures of silicate-based minerals. Shown are: **a** chrysotile as viewed along [100] (blue = Si, red = O, yellow = Mg), **b** pyroxene polyhedra representation as viewed along [010] (dark blue = SiO_x polyhedra, light blue = CaO_x polyhedra, yellow = Mg^{2+}), and **c** the layered structure of kaolinite along [001], showing the layered array of SiO_4 tetrahedra (light blue = Al^{3+}; the oxygen atoms located between the SiO_4 layers are from trapped water molecules)

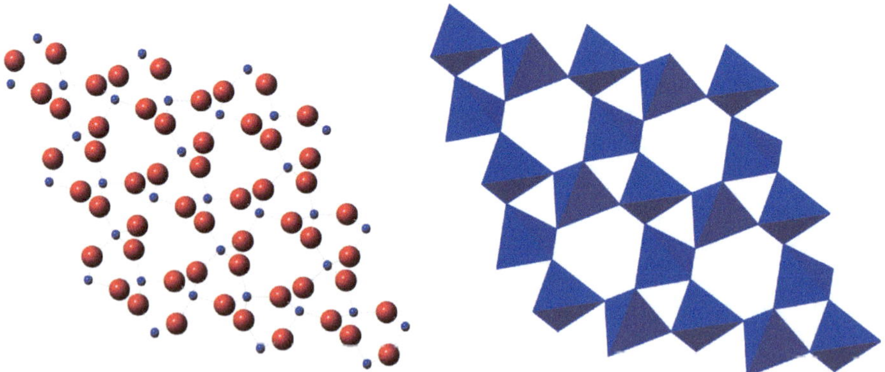

Fig. 2.98 Unit cell of the α-quartz crystal lattice

and laboratory equipment. These applications demand a glass that resists thermal expansion (i.e., cracking) as a result of significant changes in temperature.

Many electronic devices now contain a shatter-proof glass known as Gorilla glass™ (Fig. 2.99).[101] The high-strength property of this glass is due to ion-exchange of larger K^+ ions with the Na^+ ions originally present in glass. Since the larger ions take up more space in the solid, they are pressed together when the glass cools, which creates a layer of compressive stress on the surface. The degree of surface toughness is related to the diffusion distance of K^+ ions from the surface. Interestingly, the strengthening of glass by inducing compressive stresses during cooling has been known for many years, dating back to at least the mid-seventeenth century with "Prince Rupert's drops".[102]

It was not until the late seventeenth century that PbO was substituted for lime in glass formulations. This "soda-lead" glass is what we know as *crystal* (referred to as *flint glass* in pre-Civil War America) and has always been a symbol of wealth and extravagance such as expensive glassware and chandeliers. In order for crystal to be legally given the "full lead" designation, at least 24% of lead oxide must be present in its structure. The addition of the heavy element Pb adds significant weight to the glass, while increasing its refractive index. This latter property results in the familiar clear, sparkling appearance of crystal glassware. The presence of lead also makes the glass softer than regular types that must be cut with a diamond saw. Black crystal is truly one of the most fabulous materials for modern artistic design. The lack of transparency is caused by a combination of additives—typically Fe_2O_3, CuO, MnO_2, and Co_2O_3.

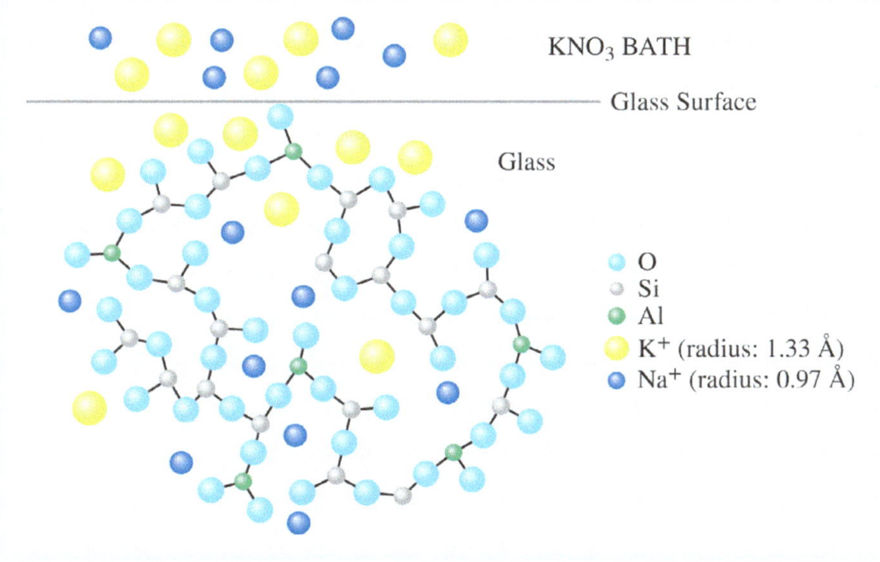

Fig. 2.99 Schematic for the formation of Gorilla Glass™. Reproduced from Corning Glass

Colored glass has been used since the construction of the first churches, prior to the tenth century.[103] Although decorative applications represent the majority of uses for colored glass, there are some other recent functional applications such as traffic light signals. The colors imparted by glass are a result of dopant species that are added during its fabrication (Table 2.12). Both transition metal ions and colloidal suspensions yield an observable color, with the hue dependent on the concentration used. Variation of the color and intensity is also extremely sensitive toward the heating regime (both temperatures and exposure times) used during the glassmaking process. In general, the observed color is the complement of the color that is absorbed by the ion. That is, the absorption of short wavelengths will result in an observable red color. Decolorizing agents may also be added; for instance, to remove a yellowish color (e.g., from the presence of Fe^{3+} impurities), a slight excess of manganese may be added that will yield the complementary color of pale purple, effectively neutralizing the glass to a colorless state.

For colloidal dopants, the particle size must be smaller than a wavelength of visible light, or an opaque glass will result. If one would prefer a cloudy glass, a number of additives (e.g., SnO_2, TiO_2, CaF_2) may be used that result in a suspension that changes the overall index of refraction. Colloidal metals yield a deep red color, with colloidal gold first used in the late seventeenth century. Alternatively, a metal salt such as $AuCl_3$ may be added to glass followed by thermal or chemical (e.g., using $NaBH_4$) reduction to metallic Au. It is important to note that a red color will only result if an agent is also added to prevent particle agglomeration. In general,

Table 2.12 Colors of glass resulting from doping

Additive	Color
Co_2O_3	Blue
Fe_2O_3	Yellow-green
FeO	Bluish-green
Colloidal Se[a]	Red
Colloidal Au[a]	Red
Colloidal Cu[a]	Red
CuO	Turquoise
NiO	Blue/violet/black
SnO_2	White
Sb_2O_3, As_2O_3	White; oxidizing agents
TiO_2	Yellow–brown
UO_2	Fluorescent yellow/green
$AgNO_3$	Orange-red
PbO/Sb_2O_3	Yellow (opaque); oxidizing agents
$K_2Cr_2O_7$	Dark green/black
Mn_2O_3	Purple

[a] With average particle diameters of *ca.* 50–100 nm

the observed color will shift toward the blue portion of the spectrum as the average particle size decreases (e.g., blue color results from diameters of <50 nm). Chapter 6 will provide more details related to the scattering properties and other applications of nanoparticles.

In comparison, transition metals are added to a molten glass matrix as soluble oxides. As you may see from Table 2.12, the observed color is a consequence of the metal ion type/concentration, as well as its oxidation state. To obtain a desired color, oxidizing agents such as $NaNO_3$, or reducing agents such as carbon powder may be added to afford the desired oxidation state. An intriguing form of glass, referred to as *vaseline glass* contains UO_2 and is slightly radioactive. Since UV radiation is sufficient to excite the weakly bound outer electrons of uranium, this additive results in a fluorescent green color. Although this is observable under normal light, it is most pronounced upon irradiation with a UV lamp. Interestingly, UO_2 was also added to ceramic glaze to yield bright orange dinner plates and tableware in the 1930s. However, it was later discovered that heat and acidic foods caused uranium to leach from the glaze, resulting in an immediate disband of this application. As one might expect, UO_2-doped materials are not currently manufactured for decorative applications, making such acquisitions a collector's item.[104]

To achieve opacity, tiny bubbles may be purposely introduced within the viscous melt—a process that dates back to ancient preparations. The resultant dispersion of light gives rise to an opalescent glass; however, it is now more prevalent to use *opalizing agents*. Earliest examples, dating back to 1400 BC, used $M_2Sb_2O_7$ (M = Pb, Ca) for opaque white glass; mixtures of Cu/Cu_2O are used to yield opaque red glass, and opaque white/blue glass often uses $CaF/CaF_3 + NaF/SnO_2$ combinations.

Thus far, we have considered the varying chemical compositions and properties of glasses. In this last section, we will examine an important architecture—glass fibers, of paramount importance in our society. The synthesis of glass fibers dates back to the early eighteenth century; applications for surgical lamps were prevalent as early as the nineteenth century. We are all familiar with the bright pink bags of fiberglass insulation that may be purchased from home improvement stores. In fact, insulation represents the leading application for fibrous glass materials.

In contrast to fiberglass, which consists of a disordered array of needle-like fibers, extremely long, one-dimensional cylindrical glass structures may be carefully fabricated to allow the transmission of light from one end to the other. Although the diameters of optical fibers are less than a human hair (i.e., 8 − 10mm), these fibers are stronger than steel and can withstand being buried underground. The fiber consists of a *core* surrounded by a *cladding* layer (Fig. 2.100). The silica core is doped with other oxides such as B_2O_3, GeO_2, and P_2O_5, resulting in a slightly higher (*ca.* 0.4%) refractive index than the cladding. The boundary between the core and cladding may either be abrupt (*step-index fiber*), or gradual (*graded-index fiber*).

There are two types of optical fibers: single-mode or multiple-mode, referring to the simultaneous transmission of single or multiple light rays, respectively. Single-mode fibers have much smaller diameters than multimode analogues (Fig. 2.100a–c), resulting in a simpler pathway for light through the fiber. Whereas multimode fibers are used to transmit information short distances (e.g., LAN applications),

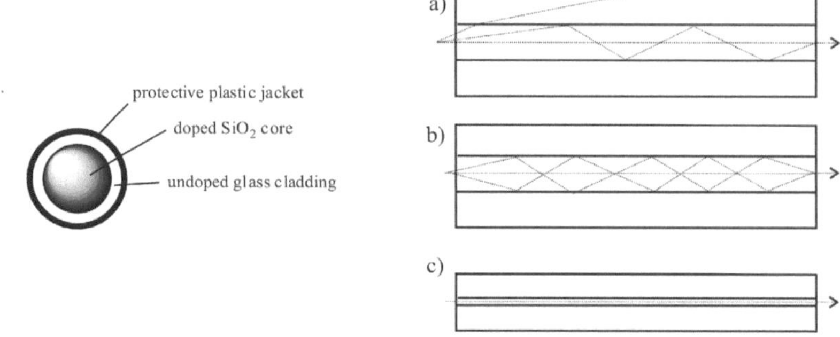

Fig. 2.100 Cross-section image of an optical fiber. Shown is the propagation of light waves through an **a** step-index multimode fiber, **b** graded-index multimode fiber, and **c** single-mode fiber

single-mode fibers are used for the long-distance transmission of cable television and telephone signals. It is not hard to see why fiber optics are desirable for transmission applications. The amount of information transmitted through 0.25 lb of optical fiber would take 33 tons of copper wire. Further, optical fibers can transmit signals at lightning speeds—unmatched by any other material. For example, three half-hour T.V. episodes may be transmitted in only 1 s!

In a step-index fiber, wavelengths of light are transmitted through the fiber core by total internal reflection (Fig. 2.90a). Depending on the angle of incidence of the light, the rays may either be transmitted through the fiber to the detector, or refracted from the core into the cladding resulting in signal loss. The desired angle (*acceptance angle*) required for total internal reflection is determined by the difference in index of refraction between the core and cladding materials. In a graded-index fiber, the refractive index in the core decreases continuously between the axis and the cladding. This causes the light rays to bend smoothly as they approach the cladding, rather than reflect abruptly from the core-cladding boundary. Light may be lost by attenuation due to absorption by impurities and scattering from microscopic density variations in the glass. To achieve sufficient transparency, the concentration of impurities such as iron and hydroxyl ions (OH^-) must be reduced to less than 1 and 10 ppb, respectively.

A fiber optic system typically consists of a transmitting device that generates the light signal, a fiber cable that transmits the light, and a receiver. The information (voice, data, or video) is encoded into electrical signals. At the light source, these electrical signals are converted into either digital or analog light signals. Once the signals are converted to light, they travel down the fiber until they reach a detector, which changes the light signals back into electrical signals. Finally, the electrical signals are decoded into the original voice, data, and/or video information.

The most common method to make optical fibers is heating a rod (preform), of the desired refractive index, to temperatures of *ca.* 2000 °C. The preform is made from the high-temperature (2000–2300 °C) reaction of $SiCl_4$ in the presence of dopant gases such as BCl_3, $GeCl_4$.[105] Once the tip of the preform is melted, it falls by gravity to form a thin strand. This wire is threaded through a coating reel, and then pulled into an optical fiber of the desired diameter. The draw towers used for this process are impressive buildings, often 8–10 stories in height. The speed of the pulling process (typically 10–20 m s^{-1}) governs the ultimate diameter of the fibers. For subsequent applications, the fiber is spooled onto shipping reels and cut to the desired length.

Another interesting application for glasses is for light control, referred to as "smart glass." We are all familiar with movie scenes where a top-secret meeting takes place, and a flip of the switch instantly darkens or clouds the windows. More routinely, it is now commonplace to have self-dimming mirrors that react to trailing vehicle headlights. Three main technologies are responsible for these intriguing materials applications: *photochromic glasses*, *electrochromic devices* (ECDs), and *suspended-particle devices* (SPDs).

Photochromic glasses exhibit a darkening effect upon exposure to particular wavelengths (usually in the UV regime) of light, and date back to the work of Corning in the 1960's. Transitions™ lenses that have appeared in television commercials use this technology, effectively protecting eyes from harmful UV irradiation. The darkening effect results from redox reactions involving microcrystalline metal halides (*e.g.*, AgCl,[106] CuCl[107]) that are present within the glass. As one would expect, the size of these dopants must be controlled to prevent reduced transmittance due to scattering before photochromic darkening may take place. However, it has been proposed that the photo-induced formation of *nanoparticles* (see Chap. 6) may also contribute to the observable darkening effect (e.g., 2.46)[108]:

$$Cu^{+} + \xrightarrow{h\nu} Cu^{2+} + Cu^{0}_{nanoparticle} \tag{2.46}$$

To synthesize photochromic glass, silica and metal halide powders are placed into a platinum crucible and heated in air to 1400 °C, followed by pouring into slabs and annealing at *ca.* 400 °C overnight. Another heat treatment at a temperature around 600–650 °C (*ca.* 1 h) is then performed to control the size of the inclusions, required for high transmittance and spectral response. We will see examples of organic molecules that also give rise to photochromism in Chap. 5 for plastic lenses, CD-R memory, and molecular switch applications.

The composition of the glass is directly related to the observed photochromic response. In general, as the silica concentration is increased, the maximum photochromic response is observed as the alkali:B_2O_3 ratio is decreased (where alkali = Na_2O:Li_2O:K_2O ratio). Likely, this delicate balance is related to governing the necessary oxidation state of the metal, and size of metal halide and/or colloidal metals precipitates formed during heat treatment. That is, metal halide solubility is related to the number of non-bridging oxygens present in the host glass, which is influenced by the concentrations of B and alkali metal ions, via formation of M^{+}–O—B bonds during heating.[cvi] Salts containing fluoride, tungstate or molybdate anions are also often added to alter the photochromic response. These additives likely serve as effective nucleation agents that facilitate precipitation of metal halide crystallites of the appropriate size during the heat treatment.

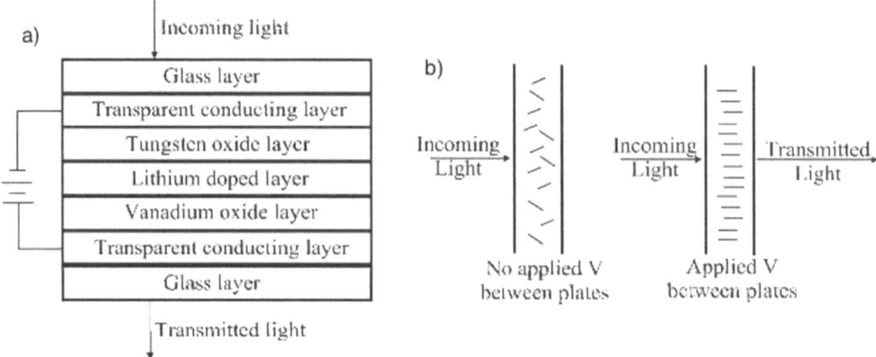

Fig. 2.101 Cross-section schematic of an **a** electrochromic device and **b** suspended-particle device

As their name implies, electrochromic materials change color due to an injection of electrons. The typical ECD has a variety of layers, sandwiched between glass (Fig. 2.101a). When no voltage is applied to the device, the incoming light will pass through undisturbed (*ca.* 70–80% transmittance). However, when a negative voltage is applied, the positive Li^+ ions are injected into the WO_3 layer of the distorted perovskite structure. A redox reaction takes place, in which some of the tungsten sites are reduced from W^{6+} to W^{5+}, and an electron is placed into the conduction band. Charge-balance is maintained through the interstitial placement of Li^+ ions in the lattice. Since the electron becomes delocalized, metallic behavior is induced in the tungsten oxide layer changing the transparent layer to a dark reflective color (*ca.* 10% transmittance of incoming light through the device). The dark color will remain even if the applied voltage is removed, since the reverse reactions are not spontaneous. However, if a reverse bias is applied to the device, lithium ions flow from the WO_3 layer reoxidizing the W^{5+} ions and restoring transparency. We will discuss more details regarding the electrical properties and band structure of semiconductive oxides such as TiO_2, SnO_2, and WO_3 in Chap. 4.

More recently, thin films of Ni/Mg hydride alloys have also been developed for light attenuation using electrochromic or gas-chromic (injection of H_2 and O_2 gases) technology.[109] Although they can technically be classified as electrochromic materials, the new reflective hydrides that are being developed behave in a noticeably different way. Instead of absorbing light, they reflect it. Thin films made of nickel-magnesium alloy can switch back and forth from a transparent to a reflective state. The switch can be powered by low voltage electricity (electrochromic technology) or by the injection of hydrogen and oxygen gases (gas-chromic technology). Furthermore, this material has the potential to be even more energy efficient than other electrochromic materials.

By comparison, SPDs operate through the behavior of rod-like particles (e.g., liquid crystals, see Appendix C.3) toward an applied voltage (Fig. 2.100b). When no voltage is applied, the particles are randomly aligned, and do not allow light to pass through the device. However, an electric charge will polarize the particles to align with the field. We will describe the molecular behavior of polarizable particles in more detail later (Chap. 4), related to dielectric materials placed in a parallel plate capacitor.

2.7 Solid-State Case Study I: Solid Electrolytes for Energy Storage Applications

With growing concerns over climate change, the switch from fossil fuels to alternative sources is growing at an unprecedented rate. In May 2017, solar panels set a one-day record for electricity generation in the U.K., demonstrated to provide almost 25% of demand—surpassing the output from nuclear power stations for the first time.[110] Wind farms now permeate the landscape in mid-Michigan and most other states—a technology once largely limited to California and Texas.

On the transportation side, bike-sharing services such as Mobike and Ofo are now commonplace and heavily used throughout China (Fig. 2.102). Electric vehicle (EV) sales jumped 37% in the U.S. and 27% in China from 2015–16; currently, China accounts for 46% of all EVs sold worldwide. After many years of development, the first commercially produced hydrogen fuel cell vehicles were released in 2015 by Hyundai and Toyota. There are currently 23 hydrogen fueling stations located in the U.S., with many more planned. As additional fueling and charging stations are added to the grid, the popularity of these modes of transportation will continue to rise. However, there are still concerns related to the safety, cycle life, and usable range of these technologies. In this final section, we will discuss the role of solid-state ceramics in fuel cells and batteries to improve their longevity and prevent safety issues.

The three primary types of energy storage devices are supercapacitors, batteries, and fuel cells. As illustrated in Fig. 2.103, supercapacitors feature high *power densities* and may quickly release their stored energy. In contrast, both fuel cells[111] and batteries feature relatively high *energy densities* that generate electricity more slowly via chemical reactions that occur at the positively charged (cathode) and negatively charged (anode) electrodes. As opposed to batteries that store a limited amount of energy, fuel cells operate with a continuous fuel flow that allows prolonged periods of electricity generation. In addition, these systems may be easily scaled-up to power large electrical grids.

An electrolyte is an essential component within energy-storage devices, used to facilitate the selective migration of ions between the electrodes. For batteries, the electrolyte is liquid-based, composed of organic solvents such as ethyl carbonate, methyl carbonate, or propylene carbonate. Although liquid electrolytes offer benefits of high conductivity and electrode wetting, they often suffer from thermal and electrochemical instabilities. In particular, news reports of battery fires, such as those recently occurring with the Samsung Galaxy Note 7,[112] have caused great concern with consumers regarding the safety of this technology. To offset this potential

Fig. 2.102 Photograph of shared bikes parked along a roadside in Tianjin, China. Photo taken by B. D. Fahlman

issue, there has been renewed interest in the development of solid-state electrolytes, which are composed of inorganic compounds, instead of flammable organic-based liquids.[113]

In general, solid electrolytes must have the following characteristics:

1. High ion conductivity
2. Low volatility and flammability
3. Stability in both oxidizing and reducing environments
4. Chemical compatibility with other cell components
5. High density to prevent mixing of fuel and oxidant gases (fuel cells)
6. Desirable thermal expansion properties, to prevent cracking of the device at high temperatures.

For proton-exchange membrane (PEM) fuel cells, the solid electrolyte is a polymer (typically Nafion, see Chap. 5) that selectively governs the flow of positively charged protons between anode and cathode compartments. For solid-oxide fuel cells (SOFCs), a material that satisfies all the above requirements is yttrium-stabilized ZrO_2. In these devices, the electrolyte must allow oxide-ion transport between the anode and cathode (2.47 and 2.48).[114] In its high-temperature cubic form, zirconia is not able to conduct oxide ions. That is, there are no suitable interstitial sites in the lattice to trap oxide ions. However, when a lower-valent metal oxide

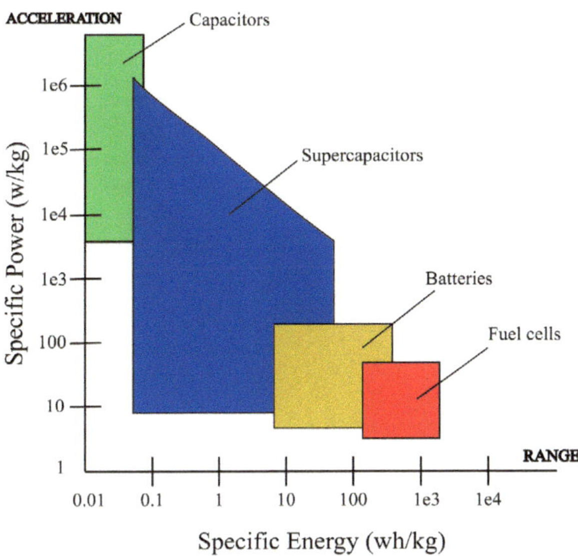

Fig. 2.103 A comparison of the relative power and energy densities for supercapacitors, batteries, and fuel cells. If we consider a plastic water bottle as analogous to an automotive energy storage device, its size would represent the energy density (vehicle range); in contrast, the size of its opening would denote its power density (vehicle acceleration)

such as Y_2O_3 (i.e., Y^{3+}) is substituted for ZrO_2 (i.e., Zr^{4+}), a vacancy is created in the unit cell of the extended lattice (Fig. 2.104). This site may accept an O^{2-} ion generated at the cathode and deliver it to the anode where it is transformed to H_2O. As illustrated in Fig. 2.98, the oxide ion is transferred among adjacent zirconia unit cells *en route* from the cathode to anode compartments of the fuel cell. By contrast, the "holes" created by Y^{3+} substitution may be envisioned to move in the opposite direction, from the anode to cathode.

$$H_2 + O^{2-} \rightarrow H_2O + 2\,e^-(\text{at anode}) \tag{2.47}$$

$$1/2\,O_2 + 2\,e^- \rightarrow O^{2-}\ (\text{at cathode}) \tag{2.48}$$

The design of solid-state electrolytes for rechargeable batteries began in the 1980s, with the development of a variety of Li-ion conductive polymers such as poly(acrylonitrile) (PAN) and poly(vinylidene fluoride) (PVDF). Inorganic-based solid electrolytes such as lithium phosphorus oxynitride (LiPON) were developed in the 1990s, which showed better ion conductivity and less issues with potential flammability. Today, most work focused on the development of inorganic-based solid electrolytes, which may be grouped into five main categories[115]:

1. Perovskite-type ($Pm\overline{3}m$ space group; e.g., $Li_{3x}La_{2/3-x}\square_{1/3-2x}TiO_3$, where □ are vacancies created by aliovalent doping of Li^+ ions into La^{3+} sites).
2. Lithium superionic conductor (LISICON)-type (Pnma space group; e.g., $Li_{3+x}(P_{1-x}Si_x)O_4$, $Li_{10}MP_2S_{12}$ (M = Si, Ge, Sn), $Li_{11}Si_2PS_{12}$).

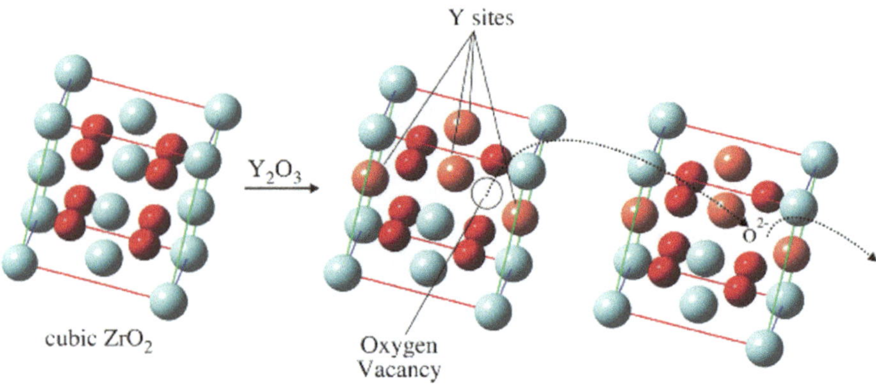

Fig. 2.104 Comparison of unit cells of cubic zirconia and yttrium stabilized zirconia (YSZ), showing transport of oxide ions through a lattice via vacant sites in neighboring YSZ unit cells

3. Sodium superionic conductor (NASICON)-type (R $\overline{3}$ c space group; e.g., $L_{1+6x}M^{4+}{}_{2-x}M'{}_{3+x}(PO_4)_3$ (L = Li, Na; M = Ti, Ge, Sn, Hf, Zr; M' = Cr, Al, Ga, Sc, Y, In, La)).
4. Argyrodite-type (F $\overline{4}$ 3 m space group; e.g., Li_6PS_5X (X = Cl, Br, I)).
5. Garnet-type (Ia $\overline{3}$ d space group; e.g., $Li_5La_3M_2O_{12}$ (M = Nb, Ta), $Li_6ALa_2M_2O_{12}$ (A = Ca, Sr, Ba; M = Nb, Ta), $Li_{5.5}La_3M_{1.75}B_{0.25}O_{12}$ (M = Nb, Ta; B = In, Zr), $Li_7La_3Zr_2O_{12}$, $Li_{7.06}M_3Y_{0.06}Zr_{1.94}O_{12}$ (M = La, Nb, Ta).

The diffusion of an ion[116] within a crystalline solid may only proceed if sufficient energy is provided to overcome the activation energy (also known as the "migration energy") that separates two crystallographic sites (revisit Fig. 2.55)—typically interstial or vacant sites in the lattice. As the valency of the migrating cation increases, the activation energy of its migration will increase due to enhanced electrostatic attractions between the surrounding lattice counterions (e.g., O^{2-}, S^{2-}, Cl^-, $PO_4{}^{3-}$, etc.). Hence, as a general rule, monovalent cations (e.g., Li^+, Na^+, K^+, etc.) will exhibit the highest diffusion coefficients and lowest migration energies than divalent (e.g., Mg^{2+}, Ca^{2+}, Zn^{2+}, etc.) or trivalent (e.g., Al^{3+}) species.

For ions of the same valency, such as Li^+ and Na^+, the smaller ion (Li^+) will generally move more efficiently through a given crystal lattice. However, the framework lattice also plays a governing role in ionic migration. If the diffusing cation is too small for a given lattice site, it will occupy a large "electrostatic well", formed from close interactions between the surrounding lattice counterions This will result in slow diffusion due to a high activation energy for its migration. On the other hand, if the ion is too large for a lattice position, the cation will experience slower diffusion as it attempts to navigate through the bottlenecks of the lattice framework.

For satisfactory ion conduction, the crystal lattice of the solid electrolyte must be designed to enhance the diffusion coefficient of the migrating cation. Not only should there be an excess of equivalent (or near-equivalent) sites relative to the number of migrating ions, but these sites must be connected to form continuous diffusion

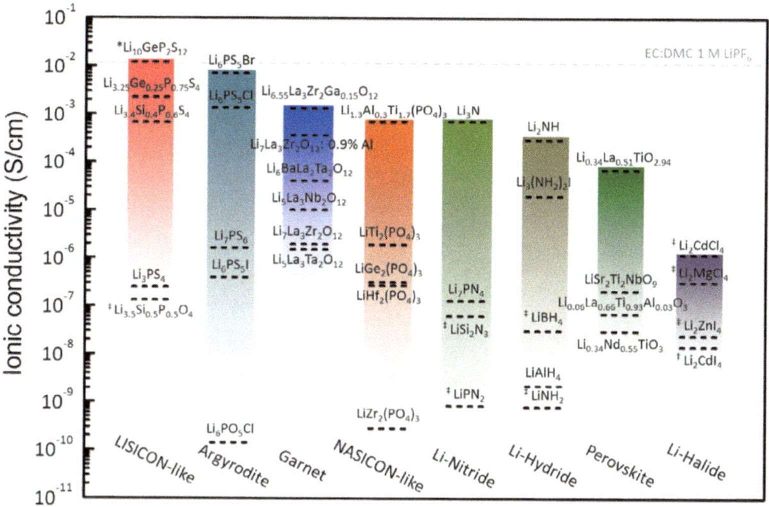

Fig. 2.105 Comparative ionic conductivities for various families of inorganic-based solid electrolytes, relative to PF_6 in 1:1 EC:DMC—the most popular liquid electrolyte for Li-ion batteries. Reproduced with permission from *Chem. Rev.* **2016**, *116*, 140. © 2016 American Chemical Society

channels. Although the diffusion of oxygen ions has been increased by inducing a mechanical strain to the crystal lattice,[117] this has not yet been shown to significantly enhance cation mobilities. Hence, increasing the lattice volume by framework-ion substitution appears to be the only viable means to dramatically improve the mobility of cations within solid electrolytes. For a given crystal structure, the concentration of migrating ion, as well as framework-substitution with ions of differing sizes and valencies will vary the ionic conductivity of the solid by as much as 5–6 orders of magnitude (Fig. 2.105). For instance, the ionic conductivity of perovskite-based solid electrolytes is increased from 10^{-7} to 10^{-3} S cm^{-1} as the rare-earth metal is changed from Sm^{3+} to the larger La^{3+}.

For many solid electrolytes such as garnet-based structures, ionic conductivity is not only dependent on the concentration of Li ions within the lattice, but which positions the ions occupy. For garnets with the general formula $A_3B_2(XO_4)_3$, A-sites have eightfold coordination (antiprismatic), B-sites are octahedral, and X-sites are tetrahedrally coordinated. In garnets such as $Li_3Nd_3Te_2O_{12}$, Li^+ ions occupy tetrahedral sites exclusively, which limits the ionic conductivity of the solid (*ca.* 10^{-6} S cm^{-1}). However, aliovalent metal doping to yield structures such as $Li_5La_3M_2O_{12}$ (M = Nb, Ta, Sb), $Li_6ALa_2M_2O_{12}$ (A = Mg, Ca, Sr, Ba; M = Nb, Ta), $Li_7La_3M_2O_{12}$ (M = Zr, Sn) feature a much higher Li content, with Li^+ ions in both distorted octahedral and tetrahedral sites. Since the Li^+ sites are very well connected within the garnet structure, an increasing Li content results in faster Li-ion mobility, up to 10^{-3} S cm^{-1} (Fig. 2.106).

Structures based on the LISICON framework has provided some of the highest Li-ion conductivities to date. In fact, the $Li_{10}MP_2S_{12}$ (M = Si, Ge, Sn) and $Li_{11}Si_2PS_{12}$

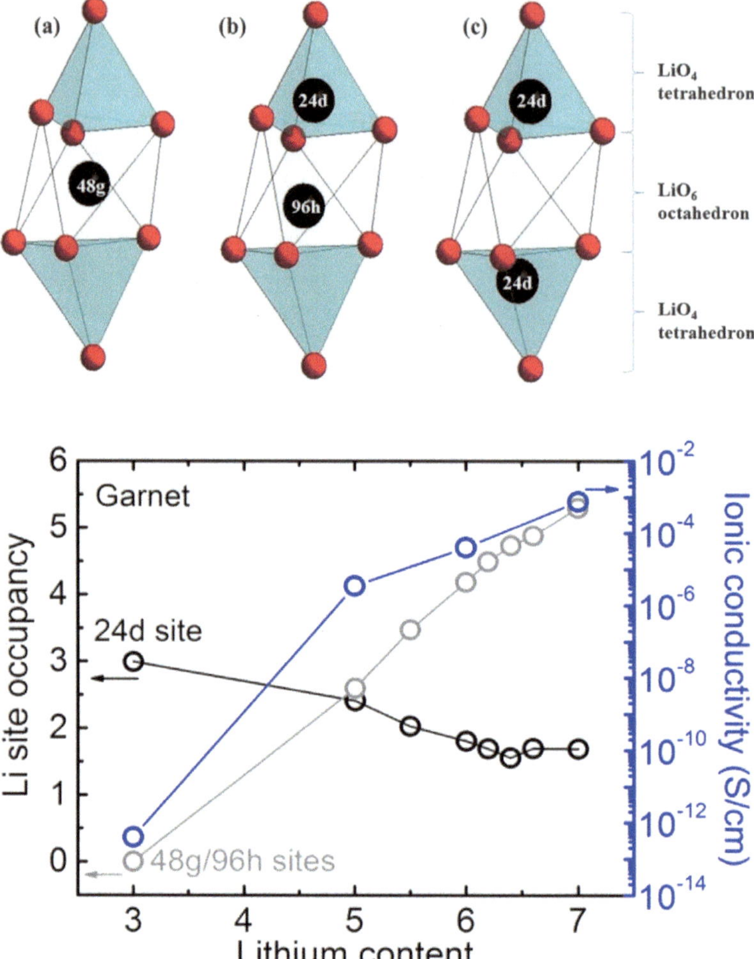

Fig. 2.106 (top) Lithium occupancy in the garnet-type $Li_{5+x}La_{3-x}A_xM_{2-y}B_yO_{12}$ where A = divalent, B = tri- or tetravalent and M = pentavalent ions, showing three possibilities for Li-ion distribution. Reproduced with permission from *Chem. Soc. Rev.* **2014**, *43*, 4714. © 2014 The Royal Society of Chemistry. **(bottom)** The importance in Li-ion placement within octahedral sites (48 g/96 h sites), relative to tetrahedral sites (24d), for enhanced ionic conductivity (bottom). Reproduced with permission from *Chem. Rev.* **2016**, *116*, 140. © 2016 American Chemical Society

structures have ionic conductivities above 10^{-2} S cm^{-1} at room temperature—of the same order of magnitude as $LiPF_6$ dissolved in a 1:1 mixture of ethylene carbonate (EC):dimethyl carbonate (DMC), the liquid electrolyte of choice for Li-ion batteries. As shown in Fig. 2.107, the structure of $Li_{10}GeP_2S_{12}$ features channels along the **c**-axis, which can accomodate significant numbers of Li^+ ions.[118] Interestingly, as Ge is replaced with Sn, the ionic conductivity decreases from 12 to 4 mS cm^{-1}—even

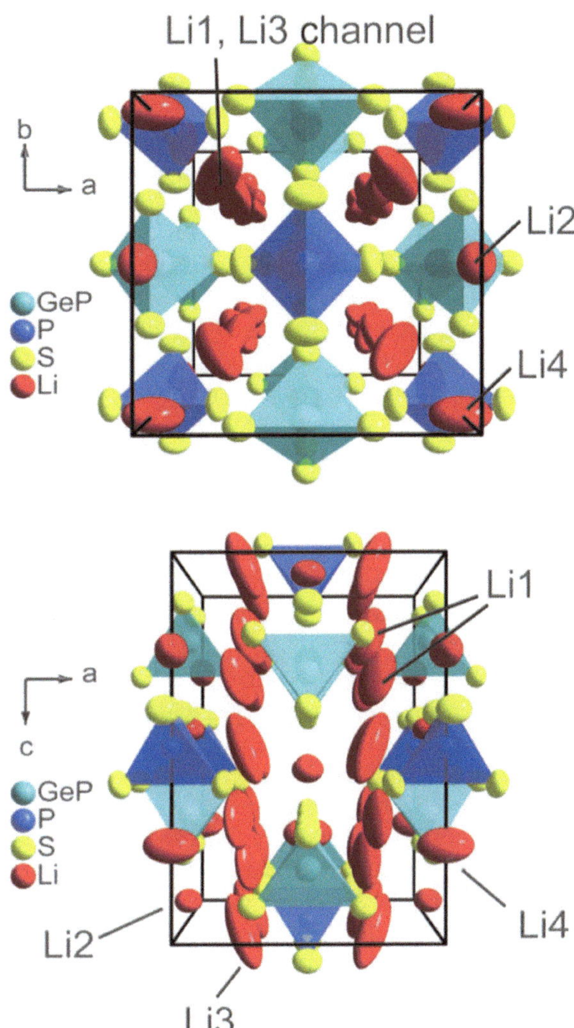

Fig. 2.107 Unit cell representation for $Li_{10}GeP_2S_{12}$ LISICON solid electrolye, as viewed along the [001] and [010] directions. Reproduced with permission from **Phys. Chem. Chem. Phys. 2013**, *15*, 11620. © the Owner Societies 2013

though this substitution results in a larger cell volume. Hence, the channel size, and not overall cell dimensions, appears to be most important for Li^+ migration in this system.

This section featured a number of structural motifs that are currently being investigated as solid electrolytes for battery and fuel cell applications. There is still much work to be done, since the best candidates are still less conductive at room temperature than their liquid counterparts. This will continue to be a fruitful area of investigation for many years; especially as the research community begins to shift their focus beyond Li-ion batteries to sodium, magnesium, and even aluminum systems.

2.8 Solid-State Case Study II: Porous Materials: Zeolites And Metal–Organic Frameworks (MOFs)

Due to the extensive number of structures and applications covered in this key chapter, we present another case study, related to porous materials typically synthesized via sol–gel routes. As illustrated in Fig. 2.108, *Zeolites* are materials best described as hydrated aluminosilicate minerals that consist of interlocking SiO_4^{4-} and AlO_4^{5-} tetrahedra, of general formula $M_{x/n}[(AlO_2)_x(SiO_2)_y] \cdot zH_2O$ (M = cation of valence n). By definition, the (Si + Al)/O ratio in zeolites must equal 0.5, resulting in a negatively charged polyhedron structure. Therefore, other cations (e.g., M = Na^+, K^+, Ca^{2+}, Mg^{2+}, etc.) must occupy the large spaces or cages of the zeolite structure in order to maintain overall charge neutrality. Though not immediately apparent from their general formulae, zeolites contain a number of reactive Br∅nsted acidic (e.g., aluminol (−AlOH), silanol (−SiOH), and bridging Si–O(H)–Al groups), Lewis acidic (e.g., framework Al^{3+} ions, charge-balancing cations), and Lewis basic (e.g., framework oxygen) sites that have important consequences in catalytic reactions. In general, the number of Br∅nsted acid sites increases linearly with the charge density of the charge-balancing cation(s), due to a greater polarization of adsorbed water. The strength of Lewis acidity is inversely proportional to the Si/Al ratio of the zeolite framework.

There are 48 naturally occurring zeolites (some even found on Mars!), and more than 150 synthetic varieties.[119] The natural varieties are mostly used in applications that are not affected by their high levels of compositional and structural heterogeneity. Such examples include concrete, soil treatment ("zeoponics"; e.g., the controlled release of fertilizer or nutrients such as K^+ or NH_4^+), and even kitty litter. However, synthetic zeolites possess a uniform and precise composition and structure, which is suited for applications such as catalysis, molecular sieves, photovoltaics, sensors, laundry detergents, and water purification. Both natural and synthetic zeolites are being explored for an intriguing adsorptive application as a blood clotting facilitator (e.g., QuikClot™ and Hemosorb™).[120] More recently, there is interest in exploiting zeolite thin films as low-dielectric constant (low-κ) materials for future computer chips and biocompatible/antifouling coatings for fuel cells and water desalination, to name only a few of the plethora of possible applications.[121]

The three-dimensional structure of zeolites is characterized by a complex system of interconnected channels, resulting in low density.[122] The structural rigidity and channel system of zeolites allow allows facile movement of ions and molecules into and out of the structure. Hence, zeolites are able to adsorb and desorb water without damage to the crystal structure, and ion-exchange takes place readily among the chelated cations. The pervasive use of zeolites in the petrochemical industry for oil refining and organic syntheses is related to their very high surface area and tunable hydrophobicity/hydrophilicity, which govern their adsorption capacity. Further, the tunable sizes and geometry of the channels and cavities provide efficient sequestering of guest molecules, facilitating shape-selectivity of catalytic reactions to increase the residence time of reactants and improve the overall yield/efficiency.

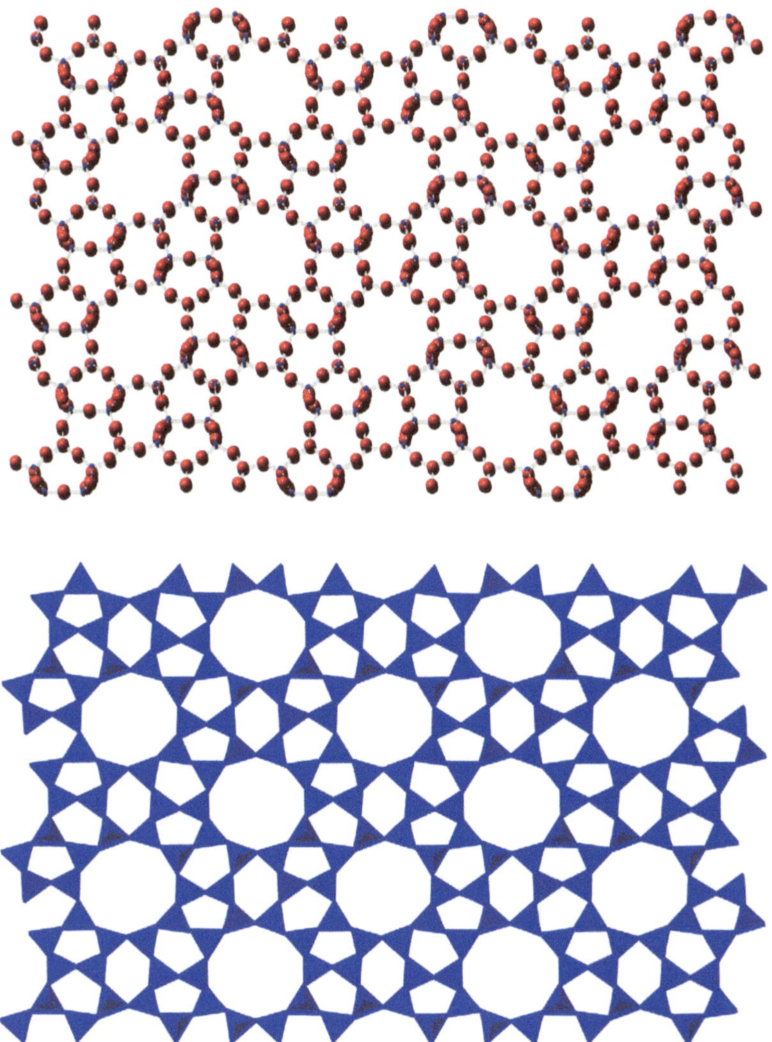

Fig. 2.108 The molecular and polyhedral representations of the ZSM-5 zeolite crystal structure formed from interlocking SiO_4 and AlO_4 polyhedra, as viewed along [010]. This zeolite belongs to the Pnma space group and has a composition of $[Na_n(H_2O)_{16}]$ $[Al_nSi_{96-n}O_{192}]$, n < 27

Depending on the type of zeolite, the interior voidspaces may also be large enough to accommodate larger molecular species such as organic molecules, water, ammonia, and anions such as CO_3^{2-} and NO_3^{-}. By definition, *microporous* materials have pore diameters of < 2 nm and *macroporous* materials have pore diameters > 50 nm. In contrast, *mesoporous* materials have intermediate pore diameters of 2–50 nm. Whereas the earliest examples of zeolites featured pore sizes in the 2–15 Å, more recent precedents have been focused on mesoporous structures with pore sizes up to 200 Å (e.g., M41S) that extend the range of applications for chemical syntheses, electronic arrays, and biomaterials.[123]

Zeolites are usually synthesized under hydrothermal conditions, from alkaline solutions of sodium aluminate and sodium silicate at temperatures between *ca.* 80–200 °C (Fig. 2.109).[124] Although zeolites exhibit a 3-D crystalline structure, they are obtained from amorphous alkaline gel precursors. The as-prepared zeolitic structure will vary depending on the reactants used and the experimental conditions (i.e., temperature, time, and pH). Particularly important is the presence of a templating ion, which directs the 3-D structure of the aluminosilicate lattice (Fig. 2.110). In alkaline media, an organic ammonium cation is used (e.g., propylammonium); neutral framework zeolites such as $AlPO_4$ require the use of amines rather than quaternary salts. Both require the removal of the organic template via calcination in order to yield open-pore zeolitic structures.

Zeolite synthesis may also be carried out at a lower pH using fluoride-containing media, wherein F^- ions are thought to act as structure directors via strong interactions with framework Si atoms. Consequently, the nucleation rate is decreased, which yields larger crystals relative to standard alkaline hydrothermal routes.[125] The fluoride route under neutral/acidic pH conditions is also extremely useful to synthesize zeolite-like materials called *zeotypes*, which contain elements other than silicon and aluminum (e.g., titanosilicates, zirconosilicates, etc.).[126]

Similar to zeolites, metal–organic frameworks (MOFs) are a class of porous crystalline[127] materials with very high porosities (up to 90% free volume), high surface areas (>6000 m^2/g), and tunable structures (Fig. 2.111).[128] These materials have risen in popularity to become one of the leading candidates for applications that span gas storage/separation, catalysis, magnetism, sensing, and drug delivery. However, whereas zeolites are purely inorganic in composition, MOFs feature two types of secondary building units (SBUs): (i) metal atoms or polyatomic metal clusters, and (ii) di- or polytopic organic ligands that serve as 3-D linkers. Due to the seemingly endless choices of ligand/metal combinations, MOFs are much more tunable than their zeolitic counterparts, which further extends the possible range of applications. The choice of metal ions and organic ligands will govern the resulting 3-D network (Fig. 2.112).[129]

Not unlike zeolites, MOFs may be synthesized by a variety of techniques to yield many possible morphologies and final products.[130] Conventional syntheses take place in a solvent at temperatures ranging from room temperature to *ca.* 250 °C. Although energy is typically introduced via electrical heating, other energy sources such as microwave-assisted, sonochemical, electrochemical, or mechanochemical (e.g., ball-milling) may be used to fabricate MOF arrays. The most important parameters for MOF syntheses are:

- Solvent—structure is governed by polarity of the solvent and solubility of the organic linker; solvent molecules may also act as ligands or structure-directing agents during the synthesis—often through use of an ionic liquid (IL), which is known as *ionothermal synthesis*.

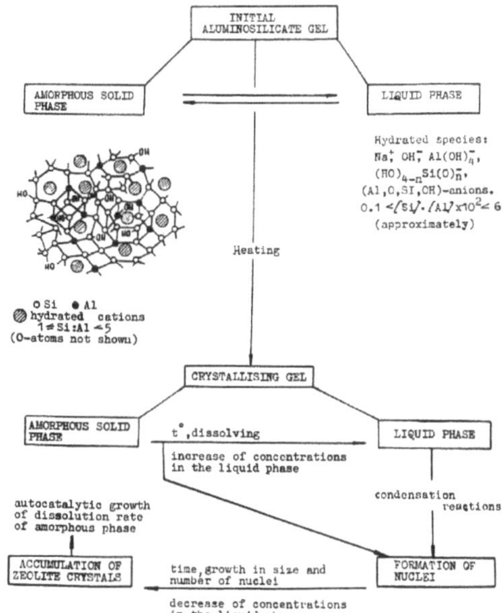

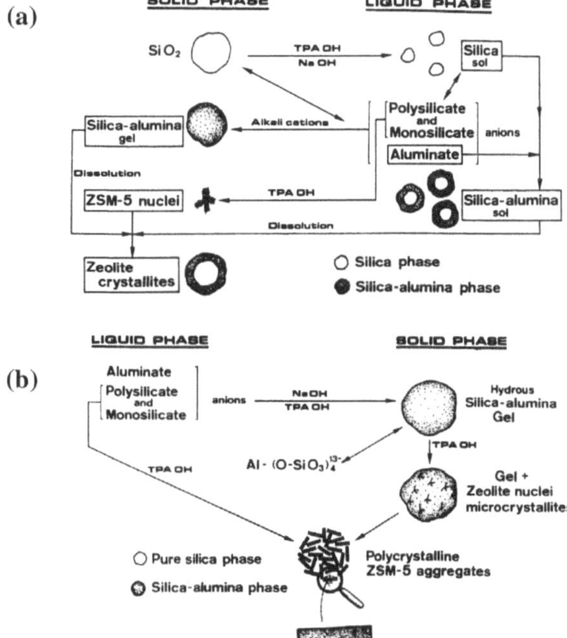

Fig. 2.109 Top: Schematic of aluminosilica gel crystallization. Reproduced with permission from *ACS Adv. Chem. Ser.* **1971**, *101*, 20. © 1971 American Chemical Society. Bottom: Two synthetic routes for ZSM-5 zeolites. Reproduced with permission from *ACS Symp. Ser.* **1984**, *248*, 219. © 1984 American Chemical Society

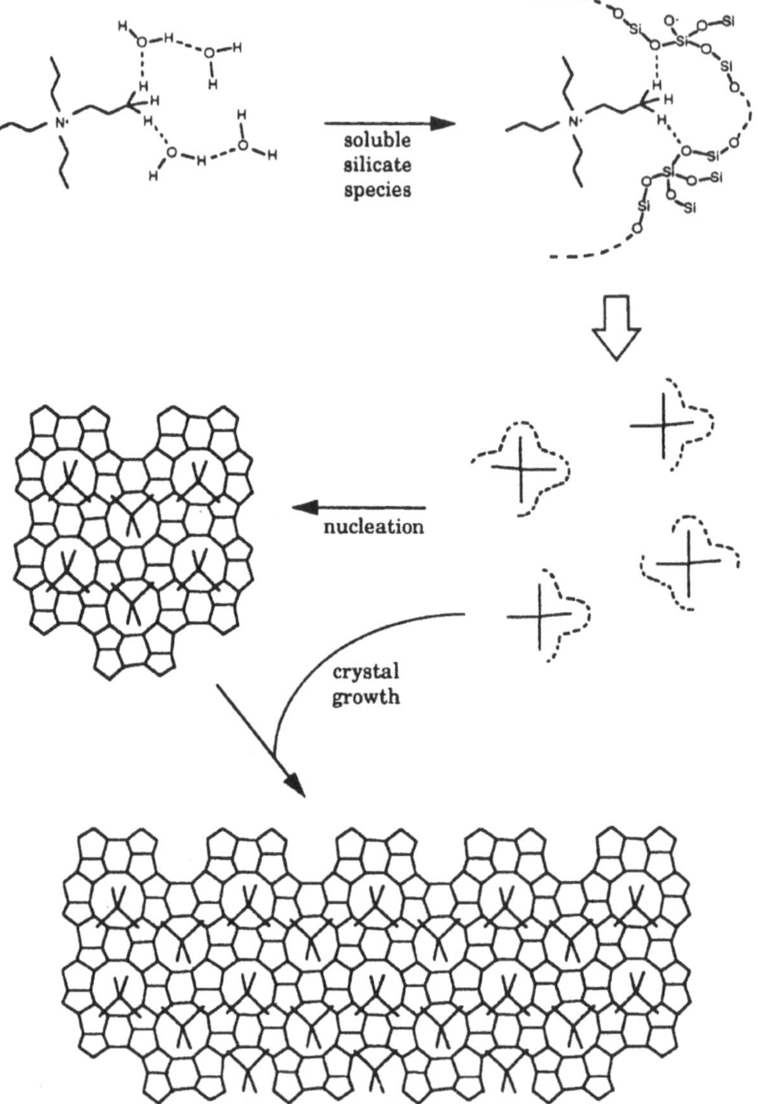

Fig. 2.110 Mechanism of template-directed crystal growth for the synthesis of a zeolite. Reproduced with permission from *J. Phys. Chem.* **1994**, *98*, 4647. © 1994 American Chemical Society

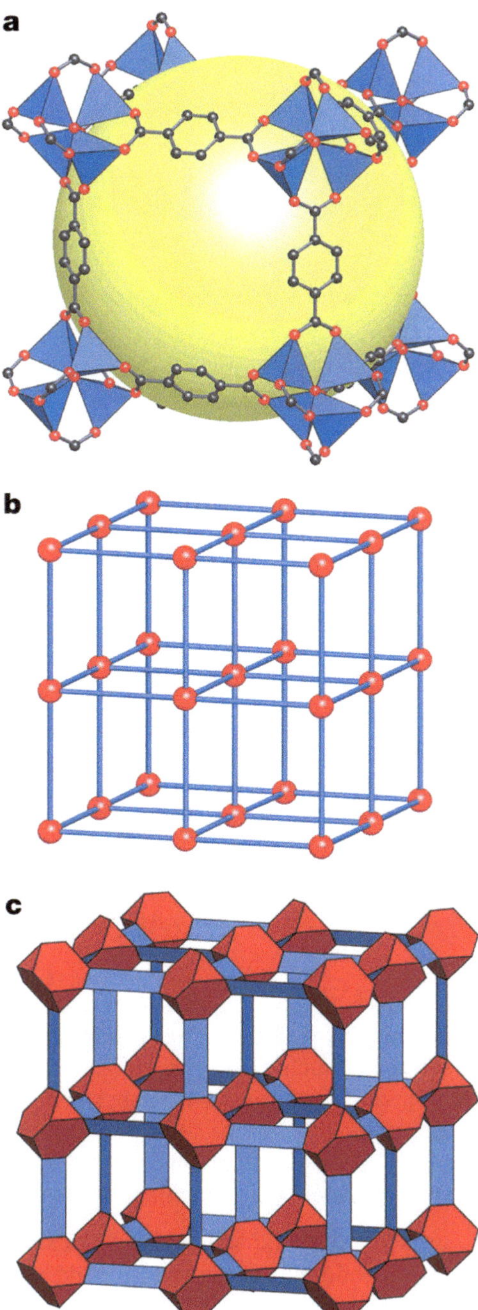

Fig. 2.111 **a** The MOF-5 structure shown as ZnO_4 tetrahedra (blue polyhedra) joined by benzene dicarboxylate linkers (O = red, C = black). The 3-D cubic framework exhibits interconnected pores of 8 Å width and 12 Å pore diameter. The yellow sphere represents the largest sphere that can occupy the pores without coming within the van der Waals size of the framework. **b** The topology of the structure (primitive cubic) shown as a ball-and-stick model. **c** The structure shown as $(OZn_4)O_{12}$ clusters (red truncated tetrahedra) and benzene dicarboxylate ions (shown as blue lines). Reproduced with permission from *Nature* **2003**, *423*, 705. © 2003 Macmillan Publishers Limited

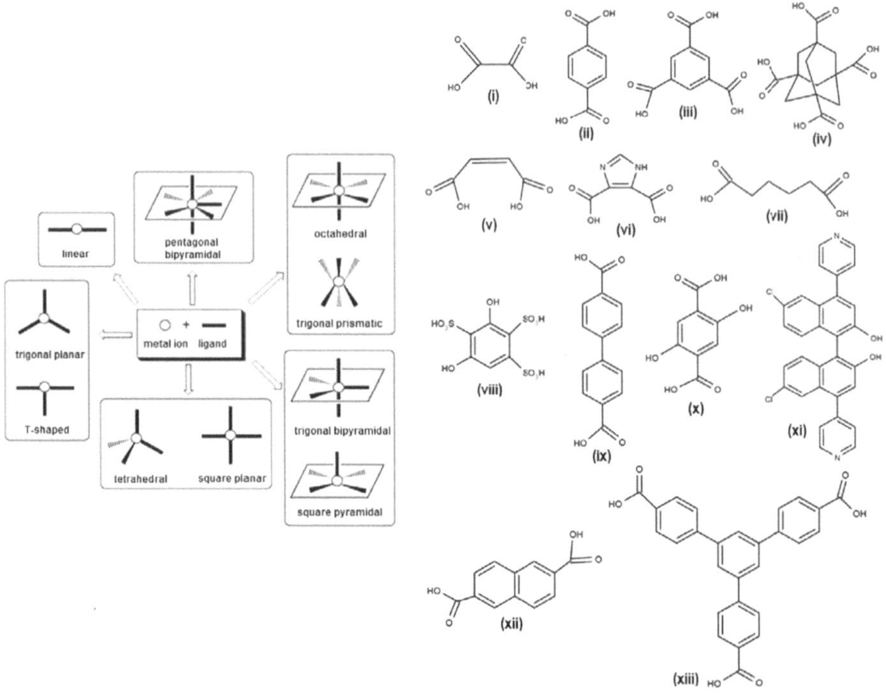

Fig. 2.112 (Left) Coordination geometries of transition metal ions used in MOFs. (Right) Some examples of polytopic organic acids used as linkers in MOFs. Shown are: (i) oxalic acid, (ii) terephthalic acid, (iii) trimesic acid, (iv) adamantane-1,3,5,7-tetracarboxylic acid, (v) fumaric acid, (vi) 4,5-imidazoledicarboxylic acid, (vii) 1,4-butanedicarboxylic acid, (viii) 4,6-dihydroxy benzene-1,2,5-trisulfonic acid, (ix) biphenyl-4,4'-dicarboxylic acid, (x) 2,5-dihydroxyterephthalic acid, (xi) 6,6'-dichloro-4,4'-di(pyridin-4-yl)-1,1'-binaphthyl-2,2'-diol, (xii) 2,6-naphthalenedicarboxylic acid, and (xiii) 1,3,5-benzenetribenzoic acid. Reproduced with permission from *Inorg. Chim. Acta* **2016**, *446*, 61. © 2016 Elsevier B. V

- Structure-directing agents—pore dimensions are often defined by unreacted organic linker molecules or solvent; the 3-D structure is altered by the presence of molecules such as benzene, amines, surfactants (e.g., cetyltrimethylammonium bromide, CTAB; 1,3,5-trimethyl benzene, TMB) or inorganic templates (e.g., Keggin polyoxometallate anion, **I**).
- Mineralizers—analogous to zeolites, ions such as F^- are used to help solubilize the starting materials and facilitate the formation of well-defined crystalline phases.

In addition to the self-assembly of metal ions and organic linkers in suitable reaction conditions, a "molecular building block" (MBB) approach may be used. This strategy employs precursors that contain the specific metal coordination and functional groups that are desired in the final MOF product. For instance, metal–organic cubes (MOCs) or supertetrahedron building units may be used to design

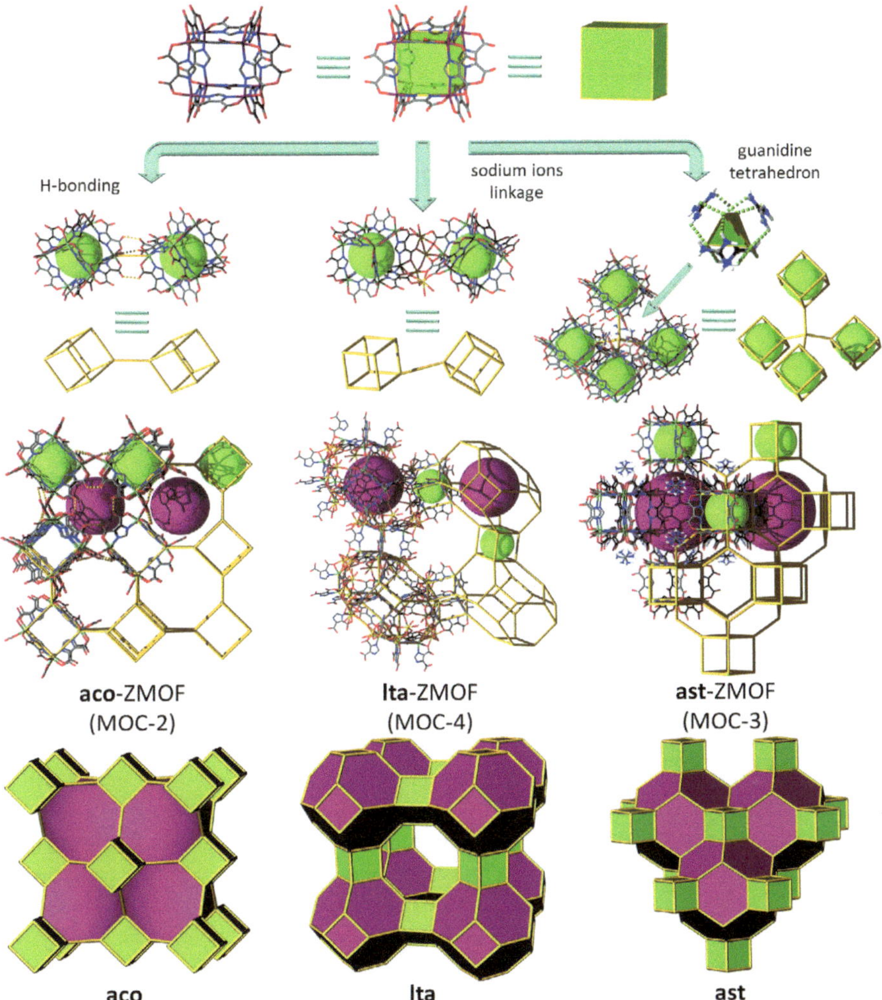

Fig. 2.113 Illustration of the design of zeolite-like MOFs from metal–organic cube precursors. Reproduced with permission from *Chem. Soc. Rev.* **2015**, *44*, 228

zeolitic-like MOFs with a variety of 3-D structures, based on the choice of linker units employed in their synthesis (Fig. 2.113).[131]

Topics for Further Discussion

1. Considering that Si has the zincblende crystal structure, draw the (111), (110), and (100) planes of Si. Place these planes in order of highest atomic density,

from least to greatest. What impact would the structures of these planes have on their relative surface reactivities?

2. Show the relationship between 4_1 and 4_3 crystallographic point group operations.
3. A metal alloy consists of 5 at% Au and 95% at% Pt; calculate the composition in terms of wt%. How many atoms of gold will be present per cubic meter of the alloy?
4. What are some differences between amorphous and crystalline materials? Cite examples of each.
5. Prove that the theoretical packing densities for BCC, FCC, and HCP are 68%, 74%, and 74%, respectively.
6. Describe some techniques used to fabricate amorphous and crystalline materials.
7. What is the difference between point groups and space groups?
8. Is it possible to have a material that is both an ionic solid and a molecular solid? Explain your reasoning and cite examples of such hybrids (if possible).
9. Consulting a periodic table and table of atomic radii, what atoms would be suitable (a) interstitial dopants, and (b) substitutional dopants within a Mn lattice? Show your calculations and rationale.
10. Using diagrams, determine whether (111) or (110) planes are more densely packed with atoms for FCC crystals.
11. Explain why the dislocation density of a single crystal is 6 times greater within 200μ m from the surface, relative to its bulk structure. Would a surface oxide layer induce more dislocations to form in a metal crystal, or insulate against dislocation formation? Explain.
12. Based on lattice parameters, explain why tetragonal, trigonal, and hexagonal crystals are often dichroic, whereas orthorhombic, monoclinic, and triclinic crystals may exhibit trichroism.
13. What are the benefits of fuel cells, as compared to batteries and other "standard" power sources such as fossil fuel power plants?
14. Why do acid-catalyzed and base-catalyzed sol–gel processes result in linear and branched metal oxides, respectively?
15. Describe the unit cell symmetry/centering and crystallographic symmetry operations (including translational operations such as glide planes and screw axes) for each of the following space groups: (a) $P6_3/m$, (b) $Fd\overline{3}$, (c) Imma, (d) P6mm, (e) Fddd, (f) $P2_1/m$.
16. For each of the crystallographic point groups represented in #15 above, determine whether the crystal is *piezoelectric* (change in voltage (ΔV) in response to a change in pressure) and/or *pyroelectric* (ΔV in response to a change in temperature). What are the governing symmetry elements that determine whether a crystal will exhibit these properties?

17. For each of the following: (i) indicate the positions of ALL symmetry elements, (ii) list the coordinates of all points **within** the unit cell, and (iii) determine the space group symbol.

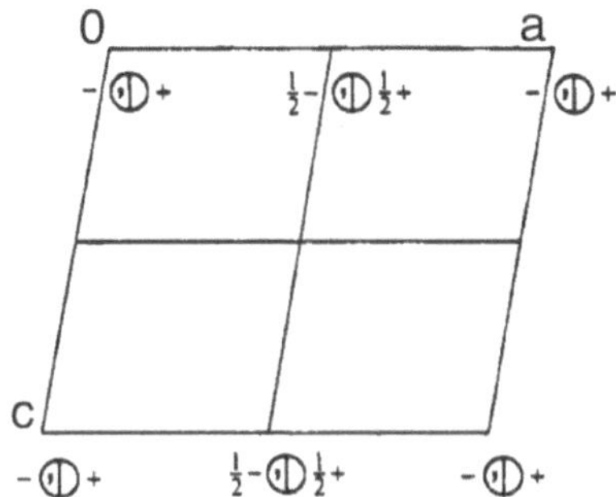

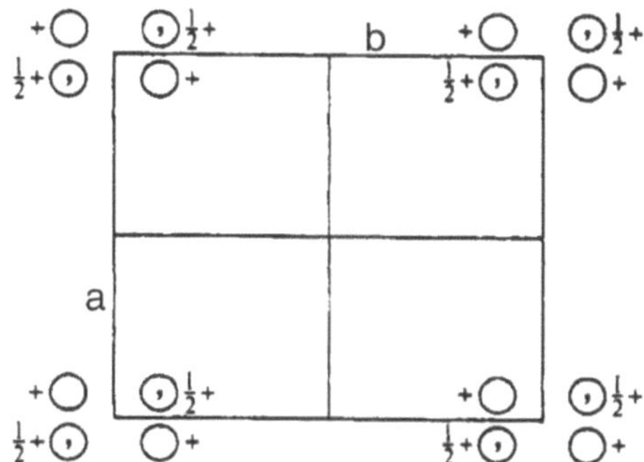

18. Provide precedents for sol–gel syntheses using tetramethoxysilane, with references. Describe the differences in experimental conditions and control over product morphology relative to using TEOS.
19. We have all seen cement trucks that have a distinctive rotating tank. Why is this constant rotation necessary, and what would happen if the tank stopped rotating?
20. Why are coarse and fine aggregates needed for the overall strength of concrete?
21. Explain why the reaction of C2S with hydroxides occurs much slower than the C3S/hydroxide reaction during cement hardening.
22. Suggest low- and high-temperature synthetic routes for AlN and BN ceramics.
23. How would you design a sensor based on smart glass, which would change from colorless to bright red upon sensing an intruder?

24. What are 'chalcogenide glasses'? Describe some recent applications for these materials.
25. How would you design a smart glass windshield, which would self-repair itself after an automobile crash?
26. Ancient colored glasses contained nanoparticles of Au and Cu. How were these synthesized (*i.e.*, what metal salts were used, along with other experimental conditions that were required to yield nanosized metal clusters)?
27. Explain why borosilicate glass is transparent to visible radiation, while quartz is transparent to both visible and UV radiation.
28. For metal oxide-based heterogeneous catalyst supports, describe the role of (a) surface hydroxyl groups (concentration and speciation), and (b) the Lewis acidity of the metal center on the chemisorption of transition metal catalysts. Cite various examples of surface-bound M^0 and M^{n+} catalytic species.
29. Both homogeneous and heterogeneous catalysis offer distinct advantages and disadvantages. Although a particular system is usually designed to be either homo- or heterogeneous in nature, there is sometimes ambiguity regarding the exact nature of the active catalytic species. For instance, there may be leaching of the transition metal species into solution, which may also involve "release and capture" mechanisms. For a model system of your choosing (where leaching is a possibility), how would you unambiguously determine whether the active catalytic species was heterogeneous or homogeneous in nature?
30. For each of the following, determine the molecular and corresponding crystallographic point groups: (a) a piece of 8.5" × 11" paper (no lines), (b) staggered ferrocene, (c) eclipsed ferrocene, (d) hydrogen peroxide, (e) cyclopropane.
31. You are given a piece of golden-colored glass and must add appropriate dopant(s) to yield a colorless material. What dopant(s) are likely present, and which would you add? Explain.
32. Explain the role of point and/or bulk defects in the origin of color for blue, pink, and black diamonds. Why are blue diamonds electrically conductive at room temperature, whereas yellow diamonds are insulating?
33. Using diagrams, explain the color origin for smokey quartz and amethyst. What factors govern the hue and intensity of color in these crystals?
34. The lattice parameter of Ag is 0.4084 nm. Calculate the planar concentration (in units atoms nm^{-2}) of atoms on the planes {100}, {110} and {111}.
35. The lattice parameter of Ge is 0.5659 nm. Calculate the number of atoms per unit length, that is the linear atomic concentration (atoms per nm of length) along <100> and <110> directions.
36. γ -Al_2O_3, commonly used as a heterogeneous catalyst support, crystallizes in the Fd-3 m space group. Describe the multiplicity and fractional coordinates of O and Al ions in the crystal lattice.
37. Solid electrolytes are being investigated for battery applications, to minimize the potential for dangerous fires. What electrolytes were featured inside the Samsung Note 7 smartphones and why did these cell phone batteries ignite? Would this problem have occurred if solid electrolytes were used? Explain your rationale.

38. What are some functional disadvantages of using a solid electrolyte in energy storage devices?
39. Have metal–organic frameworks (MOFs) been used as an active material in energy storage devices? Describe some recent precedents and some structural traits of MOFs that enable this functionality.
40. Are there any literature precedents for MOF/zeolite hybrid materials? If so, describe their fabrication and potential applications.
41. Elaborate on the 'molecular building block' (MBB) approach to MOF synthesis. What benefits would this strategy present relative to traditional synthetic routes?
42. The following data was obtained from a YBCO bulk sample. Calculate the resistance for each trial given that a constant current of 100 mA was flowing through the sample. Using this data, determine the critical temperature for YBCO. How does this compare to the literature value?

Voltage	T (K)	Voltage	T (K)
0.0010370	118.2	0.008440	93.5
0.0010270	116.1	0.007830	93.2
0.0010600	114.8	0.006390	93
0.0010490	112.9	0.0005050	92.6
0.0010350	110.9	0.0003790	92.3
0.0010220	109.1	0.0002430	92.1
0.0010090	106.9	0.0000930	91.7
0.0010010	105	0.0000100	91.4
0.0009890	103.5	0.0000030	91
0.0009750	102.2	0.0000002	90.8
0.0009670	100	−0.0000002	90.1
0.0009510	97.9	−0.0000001	89.9
0.0009440	95.8	0.0000003	89.5
0.0009180	95	−0.0000001	88.8
0.0009110	94.3	0.0000001	88.5
0.0008920	93.8		

References and Notes

1. Zallen, R. *The Physics of Amorphous Solids*, John Wiley and Sons: New York. 1983.
2. There is no empirical evidence that supports the flow of glass over time. That is, there is no analogous glass thickening observed in ancient Roman or Egyptian objects, no degradation in the optical performance of antique telescopes/microscopes, and no change in physical properties of prehistoric artifacts (e.g., dulling of obsidian swords).

3. Gases most closely approximate ideal gas behavior at high temperatures and low pressures, where the kinetic energies of gaseous molecules are sufficient to overcome all intermolecular forces. Temperature and/or pressure deviations result in real gas behavior, with intermolecular forces becoming more pronounced between gaseous molecules. For more information, see: https://courses.lumenlearning.com/boundless-chemistry/chapter/deviation-of-gas-from-ideal-behavior/.
4. For a description of the Born exponent, see: https://doi.org/10.1063/1.3684608, and references therein.
5. Electrons in an s-orbital have a finite probability of being found at the nucleus. As the principal quantum number increases, the s-orbitals become more diffuse, leading to electrons being found at distances further from the nucleus. With less attraction toward the nucleus, these electrons may orbit the nucleus at speeds approaching the speed of light. When objects move at such high speeds, an increase in relativistic mass occurs, whereby the s-electrons behave as though they were more massive than electrons moving at slower speeds. This mass increase causes the orbiting electrons to be slightly contracted toward the nucleus, decreasing their availability to participate in chemical reactions.
6. The Pauli exclusion principle states that each electron must possess a different set of four quantum numbers; that is, two electrons housed in the same orbital (identical n, l, m_l) must be of opposite spin ($m_s = \pm 1/2$).
7. The work function is the solid-state analogy of *ionization energy*, defined as removing the outermost electron from a gaseous atom. In general, the work function is *ca.* 1/2 the value of the ionization energy of its corresponding gaseous atoms.
8. (a) Reif, F. *Fundamentals of Statistical and Thermal Physics.* McGraw–Hill: New York, 1965. (b) Blakemore, J. S. *Semiconductor Statistics.* Dover: Canada, 2002.
9. For AC current, the velocity would be identical to DC, but the electrons would travel back/forth, resulting in a much smaller drift velocity.
10. It should be noted that multiple forms of an *amorphous* material are denoted as *polyamorphs*, as illustrated by the polyamorphism exhibited by silica at elevated pressures. For instance, the high-pressure polyamorphism of silica is described in: (a) https://doi.org/10.1103/PhysRevLett.80.2145. (b) https://doi.org/10.1038/nmat1458.
11. https://doi.org/10.1002/3527607889.ch2.
12. (a) https://doi.org/10.1038/s41467-018-07265-z, and references therein. (b) https://doi.org/10.1103/PhysRevB.78.161202. (c) https://doi.org/10.1103/PhysRevB.56.6662.

13. There is an ongoing debate whether the term "pseudopolymorph" should be abandoned, instead designating these compounds as "solvates". Two viewpoints may be found at: (a) https://doi.org/10.1021/cg050343e. (b) https://doi.org/10.1021/cg1013335.
14. For a nice summary and literature survey regarding the implications of polymorphism toward drug design, see: http://www.allfordrugs.com/polymorphism/.
15. For instance, see: (a) https://doi.org/10.1093/petrology/44.4.773. (b) https://doi.org/10.3390/cryst10010040.
16. The Miller indices for the (211) plane may also be visualized by extending the unit cell beyond a cell volume of 1 cubic unit. For instance, this plane would also pass through (2,0,0), (0,2,0), and (0,0,2), as well as other extended coordinates. For the (001) plane, the zeroes indicate that the plane does not intercept either the *a* or *b* axes).
17. Cullity, B. D. *Elements of X-ray Diffraction*, 2nd ed., Addison-Wesley: Reading, Massachusetts, 1978.
18. For more details about the structure and applications of SiC, see: (a) http://www.ioffe.ru/SVA/NSM/Semicond/SiC/bandstr.html#Band. (b) http://www.tf.uni-kiel.de/matwis/amat/semi_en/kap_a/backbone/ra_1_1.html.
19. There are precedents for crystals with five-fold rotation axes. These crystals are known as quasicrystals since they exhibit long-range orientational order but are not consistent with lattice translations. For example, see: https://doi.org/10.1103/PhysRevLett.53.1951. Another nice online summary of quasicrystals is: http://www.tau.ac.il/~ronlif/quasicrystals.html.
20. The *d* notation indicates a diamond glide plane, found in diamond or zinc blende extended crystal structures. Whereas glide planes are found in many inorganic-based crystals, screw axes are commonly present in protein structures.
21. $CdCl_2$ may also exist as a hcp lattice, analogous to CdI_2.
22. This is the first of five rules that govern the geometric stability of ionic packing, as proposed by Nobel Laureate Linus Pauling (*J. Am. Chem. Soc.* **1929**, *51*, 1010). For more details, see: http://positron.physik.uni-halle.de/talks/CERAMIC1.pdf.
23. (a) Honle, W. *J. Solid State Chem.* **1983**, *49*, 157. (b) Perrin, et al. *Acta Crystallogr.* **1983**, *C39*, 415.
24. For a recent discovery of a safer cathode alternative, $LiFeO_4$, see: https://doi.org/10.1038/nature07853. Information regarding a proposed intercalation mechanism for $LiFeO_4$ may be found in: http://doi.org/10.1038/nmat2230. For a method to study the thermal stability of a variety of oxide cathode materials for Li-ion battery applications, see: https://doi.org/10.1021/cm0620943.
25. Qi-Hui, W. *Chinese Phys. Lett.* **2006**, *23*, 2202.
26. For example, see: Ferracin, L. C. et al. *Solid State Ionics* **2002**, *130*, 215.

27. Willard, M. A.; Nakamura, Y.; Laughlin, D. E.; McHenry, M. E. *J. Am. Ceram. Soc.* **1999**, *82*, 3342.
28. Nakamura, Y.; Smith, P. A.; Laughlin, D. E.; De Graef, M.; McHenry, M. E. *IEEE Trans. Magn.* **1995**, *31*, 4154.
29. For background information regarding NTC thermistors, see: https://www.ametherm.com/thermistor/thermistor-applications.
30. It should be noted that one nitride also exhibits the rutile structure (ε-Ti_2N); however, this is known as *antirutile* since the position of nitride and titanium ions are reversed.
31. Another well-known polymorph of TiO_2 is *brookite*, which features an orthorhombic crystal structure. It has a larger unit cell volume than either rutile or anatase, with 8 TiO_2 units per unit cell. This polymorph is converted to rutile at temperatures greater than 750 ºC. Brookite is relatively difficult to synthesize, which has hampered its feasibility assessment for photocatalytic applications relative to rutile or anatase. However, recent studies indicate that this polymorph may exhibit a greater photocatalytic activity than anatase, under certain conditions, for the degradation of organic compounds in water: e.g., Li, Z.; Cong, S.; Xu. Y. *ACS Catal.* **2014**, *4*, 3273; Kandiel, T. A.; Robben, L.; Alkaim, A.; Bahnemann, D. *Photochem. Photobiol. Sci.* **2013**, *12*, 602.
32. For a thorough review of the anatase to rutile phase transformation, see: Hanaor, D. A. H.; Sorrell, C. C. *J. Mater. Sci.* **2011**, *46*, 855.
33. Bourikas, K.; Kordulis, C.; Lycourghiotis, A. *Chem. Rev.* **2014**, *114*, 9754.
34. (a) Yang, M.; Oro-Sole J.; Rodgers, J. A.; Belen, A.; Fuertes, J. A.; Attfield, J. P. *Nature Chem.* **2011**, *3*, 47. (b) Pan, C.; Takata, T.; Nakabayashi, M.; Matsumoto, T.; Shibata, N.; Ikuhara, Y.; Domen, K. *Angew. Chem. Int. Ed.* **2015**, *54*, 2955. (c) Jansen, M.; Letschert, H. P. *Nature* **2002**, *404*, 980. (d) Hitoki, G.; Takata, T.; Kondo, J. N.; Hara, M.; Kobayashi, H.; Domen, K. *Chem. Commun.* **2002**, 1698.
35. Pena, M. A.; Fierro, J. L. G. *Chem. Rev.* **2001**, *101*, 1981.
36. (a) Honle, W. *J. Solid State Chem.* **1983**, *49*, 157. (b) Perrin, et al. *Acta Crystallogr.* **1983**, *C39*, 415.
37. All superconducting transition temperatures are at ambient (1 atm) pressure. Although there are claims for T_c values as high as 310 K on http://superconductors.org, these findings have not yet been confirmed. To date, there are no confirmed reports of room-temperature superconductors.
38. Drozdov, A. P.; Eremets, M. I.; Troyan, I. A.; Ksenofontov, V.; Shylin, S. I. *Nature* **2015**, *525*, 73.
39. For example, see: (a) Emery, V. J.; Kivelson, S. A.; Tranquada, J. M. *Proc. Natl. Acad. Sci.* **1999**, *96*, 8814. (b) J. M. Tranquada, H. Woo, T. G. Perring, H. Goka, G. D. Gu, G. Xu, M. Fujita and K. Yamada *Nature* **2004**, *429*, 534. (c) Lee, K. H.; Hoffmann, R. *J. Phys. Chem. A* **2006**, *110*, 609.

40. For instance, see: (a) Plakida, N. M.; Oudovenko, V. S. *J. Supercond. Nov. Magn.* **2016**, *29*, 1037. (b) Lanzara, A.; Bogdanov, P. V.; Zhou, X. J.; Kellar, S. A.; Feng, D. L.; Lu, E. D.; Yoshida, T.; Eisaki, H.; Fujimori, A.; Kishio, K.; Shimoyama, J. -I.; Noda, T.; Uchida, S.; Hussain, Z.; Shen, Z. -X. *Nature* **2001**, *412*, 510. (c) Shchetkin, I. S.; Osmanov, T. S. *Powder Metall. Metal Ceram.* **1993**, *32*, 1068. (d) Abd-Shukor, R. *Solid State Commun.* **2007**, *142*, 587.
41. For instance, see: (a) Mazurenko, A.; Chiu, C. S.; Ji, G.; Parsons, M. F.; Kanasz-Nagy, M.; Schmidt, R.; Grusdt, F.; Demier, E.; Greif, D.; Greiner, M. *Nature* **2017**, *545*, 462. (b) Cheuk, L. W.; Nichols, M. A.; Lawrence, K. R.; Okan, M.; Zhang, H.; Khatami, E.; Trivedi, N.; Paiva, T.; Rigol, M.; Zwierlein, M. W. *Science* **2016**, *353*, 1260.
42. One model used to explain superconductivity in YBCO is the reduction of unstable Cu^{3+} sites by redox reactions initiated by electrons passing through the solid in adjacent planes. As an electron passes by a Cu^{3+} ion, it causes an electron to be injected from a neighboring Cu^{2+} ion resulting in a lattice distortion, & hole propagation in the opposite direction. This concept is illustrated at: http://www.chm.bris.ac.uk/webprojects2000/igrant/hightctheory.html.
43. There are two ways to account for charge neutrality in p-type superconductors. First, $La_{2-x}Sr_xCuO_4$ may be formally written as: $La_{2-x}Sr_xCu^{2+}{}_{1-x}Cu^{3+}{}_xO_4$ where one Cu^{3+} (or a Cu^{2+} with a trapped hole, $Cu^{2+}(h^+)$) forms for each Sr^{2+} added. Alternatively, the formula may be written as $La_{2-x}Sr_xCu^{2+}O_{4-(x/2)}$, where one oxygen vacancy is formed for every two Sr^{2+} ions added to the lattice.
44. For recent reviews regarding iron-based superconductors, see: (a) Si, Q.; Yu, R.; Abrahams, E. *Nature Rev. Mater.* **2016**, *1*, 16,017. (b) Hosono, H.; Kuroki, K. *Physica C* **2015**, *514*, 399.
45. An issue of the New Journal of Physics was devoted to discoveries related to iron-based HTS: http://www.iop.org/EJ/abstract/1367-2630/11/2/025003.
46. The most promising processing method for YBCO applications involves deposition onto a flexible metal tape coated with buffering metal oxides. Crystal-plane alignment can be introduced into the metal tape itself (via the RABiTS process) or a textured ceramic buffer layer can be deposited on an untextured alloy substrate (the IBAD process). Subsequent oxide layers prevent diffusion of the metal from the tape into the superconductor while transferring the template for texturing the superconducting layer.
47. https://energy.gov/sites/prod/files/oeprod/DocumentsandMedia/LIPA__5_16_08.pdf.
48. For a recent review of HTS power installations, and future considerations (including public perception/acceptance), see: Thomas, H.; Marian, A.; Chervyakov, A.; Stuckrad, S.; Salmien, D.; Rubbia, C. *Renew. Sust. Energ. Rev.* **2016**, *55*, 59.

49. For a listing of some recent HTS wire installation projects in the U.S., and DoE goals related to HTS-based electrical applications, see: https://www.energy.gov/sites/prod/files/oeprod/DocumentsandMedia/Supercon_Overview_Fact_Sheet_7_14_09.pdf.
50. For instance, see: Mukoseev, A. G.; Shabashov, V. A.; Pilugin, V. P.; Sagaradze, V. V. *Nanostruct. Mater.* **1998**, *10*, 273.
51. For a nice review of defect equilibria, see: https://www.uio.no/studier/emner/matnat/kjemi/nedlagte-emner/KJM4120/v04/undervisningsmateriale/KJM4120-Ch3-Defect-equilibria.pdf.
52. Suss, R.; van der Lingen, E.; Glaner, L.; du Toit, M. Gold Bull. 2004, 37, 196 (may be accessed online at https://link.springer.com/article/10.1007/BF03215213).
53. (a) Yasuda, K.; Ohta, M. *J. Dental Res.* **1982**, *61*, 473. (b) Uzuka, T.; Kanzawa, Y.; Yasuda, K. *J. Dent. Res.* **1981**, *60*, 883.
54. For more details related to Fick's first and second laws as they pertain to dopant diffusion, see: https://www.engineeringenotes.com/metallurgy/diffusion/diffusion-coefficient-and-laws-ficks-laws-metallurgy/41803.
55. For example, see: http://xrayweb.chem.ou.edu/notes/twin.html.
56. For example, see: Lakes, R. S. *Science* 1987, *235,* 1038.
57. (a) Carneiro, V. H.; Meireles, J.; Puga, H. *Mater. Sci.-Poland* **2013**, *31*, 561. (b) https://physicsworld.com/a/new-auxetic-material-stretches-the-limits/.
58. The yield point for metals with gradual elastic–plastic transitions is constructed by drawing a straight line parallel to the elastic portion of the stress vs. strain curve at a specific strain offset, usually 0.002. The intersection of that line and the stress vs. strain curve gives rise to the yield strength, σ_y, of the material.
59. Garlick, G. D.; Kamb, W. B. *J. Geol. Educ.* **1991**, 39, 398.
60. For a thorough treatment of crystal field theory, see Cotton, F. A.. Wilkinson, G.; Murillo, C. A.; Bochmann, M. *Advanced Inorganic Chemistry*, 6th ed., Wiley: New York, 1999.
61. The ground-state atomic term symbol for Cr^{3+} is ^4F, which splits into $^4T_2$, $^4T_1$ and $^4A_2$ for an octahedral transition metal complex. For more information regarding term symbol notation and absorption spectra for transition metal complexes, see: Shriver et al. *Inorganic Chemistry*, 4th ed., W. H. Freeman: New York, 2006.
62. Excited electrons give off their energy via infrared emission and thermal interactions with the corundum crystal lattice, referred to as electron–phonon (lattice vibrations) interactions.
63. For an explanation of the electronic transitions underlying ruby lasers, as well as tunable lasers such as alexandrite, Ti:sapphire, and Nd:YAG, see: Thyagarajan, K. *Lasers, Theory and Applications*, Plenum Press: New York, 1981.
64. Shriver et al. *Inorganic Chemistry*, 4th ed., W. H. Freeman: New York, 2006.

65. The name F-center defect is derived from the German word for color, *farbe*, since these point defects give rise to highly colored crystals.
66. For example, see: Fritsch, E.; Massi, L.; Rossman, G. R.; Hainschwang, T.; Jobic, S.; Dessapt, R. Diam. Rel. Mater. 2007, 16, 401.
67. For a nice review related to the use of antiferroelectric ceramics for high-energy density capacitor applications, see: Chauhan, A.; Patel, S.; Vaish, R.; Bowen, C. R. Materials 2015, 8, 8009.
68. For instance, see: (a) https://www.nature.com/subjects/ferroelectrics-and-multiferroics for a list of recent research articles related to multiferroic materials. (b) Scott, J. F. ISRN Mater. Sci. 2013, 2013, 187,313; available online at https://www.hindawi.com/journals/isrn/2013/187313/.
69. A nice summary of ceramic processing is found online at: https://www3.epa.gov/ttnchie1/ap42/ch11/final/c11s07.pdf.
70. *Tamman's Rule* states that a temperature of at least 2/3 the melting point of the lower-melting reactant is needed for the reaction to occur in a reasonable timeframe.
71. (a) https://www.nap.edu/read/2266/chapter/7. (b) https://www.mri.psu.edu/sites/default/files/file_attach/025.pdf. (c) https://arxiv.org/pdf/1401.1965.pdf.
72. http://cdn.intechopen.com/pdfs/29775.pdf.
73. For instance, see: (a) Sra, A. K.; Schaak, R. E. *J. Am. Chem. Soc.* **2004**, *126*, 6667. (b) Schaak, R. E.; Sra, A. K.; Leonard, B. M.; Cable, R. E.; Bauer, J. C.; Han, Y. –F.; Means, J.; Teizer, W.; Vasquez, Y.; Funck, E. S. *J. Am. Chem. Soc.* **2005**, *127*, 3506. (c) Henkes, A. E.; Bauer, J. C.; Sra, A. K.; Johnson, R. D.; Cable, R. E.; Schaak, R. E. *Chem. Mater.* **2006**, *18*, 567. (d) Wiltrout, A. M.; Read, C. G.; Spencer, E. M.; Schaak R. E. *Inorg. Chem.* **2016**, *55*, 221.
74. The negative sign in Fick's Law indicates that diffusion always takes place from regions of higher concentration to lower concentration; i.e., down the concentration gradient. The diffusion coefficient, also referred to as diffusivity, is predicted by the Arrhenius equation, which takes into account the activation energy and temperature of diffusion.
75. http://www.google.com/patents/US2854364.
76. Hu, J.; Lu, Q.; Tang, K.; Qian, Y.; Zhou, G.; Liu, X.; Wu, J. *Chem. Mater.* **1999**, *11*, 2369.
77. For instance, see: http://theses.gla.ac.uk/3032/.
78. Birot, M.; Pillot, J. P.; Dunogues, J. *Chem. Rev.* **1995**, *95*, 1443.
79. Nghiem, Q. D.; Kim, D. P.; Kim, S. O. *J. Nanosci. Nanotechnol.* **2008**, *8*, 5527.
80. For sol–gel reviews and resources, see: (a) Levy, D.; Zayat, M. *The Sol–Gel Handbook: Synthesis, Characterization and Applications*, Wiley: New York, Nov. 2015 (ISBN: 978–3-527–33,486-5). (b) Cheetham, A. K.; Mellot, C. F. *Chem. Mater.* **1997**, *9*, 2269. (c) Hench, L. L.; West, J. K. Chem. Rev. **1990**, *90*, 33.

81. It should be noted that a sol–gel process may also take place through nonhydrolytic pathways. In these systems, a metal halide reacts with an oxygen donor such as ethers, alkoxides, etc. to yield a crosslinked metal oxide product.
82. A supercritical fluid has intermediate properties of liquid and gas. Typically, the alcogel is placed in an autoclave filled with ethanol. The system is pressurized to 750–850 psi with CO_2 and cooled to 5–10 °C. Liquid CO_2 is then flushed through the vessel until all the ethanol has been removed from the vessel and from within the gels. When the gels are ethanol-free, the vessel is heated to a temperature above the critical temperature of CO_2 (31 °C). As the vessel is heated, the pressure of the system rises. The pressure of CO_2 is carefully monitored to maintain a pressure slightly above the critical pressure of CO_2 (1050 psi). The system is held at these conditions for a short time, followed by the slow, controlled release of CO_2 to ambient pressure. The length of time required for this process is dependent on the thickness of the gels; this process may last anywhere from 12 h to 6 days.

83. For a recent review on the synthesis, properties, and applications of aerogels see: Pierre, A. C.; Pajonk, G. M. "Chemistry of Aerogels and Their Applications", *Chem. Rev.* **2002**, *102*, 4243.
84. (a) Pellenq, R. J. M.; Kushima, A.; Shahsavari, R.; Van Vliet, K. J.; Buehler, M. J.; Yip, S.; Ulm, F. J. *Proc. Nat. Acad. Sci.* **2009**, *106*, 16,102. (b) Skinner, L. B.; Chae, S. R.; Benmore, C. J.; Wenk, H. R.; Monteiro, P. J. M. *Phys. Rev. Lett.* **2010**, *104*, 195,502. (c) Myers, R. J.; Hopital, E. L.; Provis, J. L.; Lothenbach, B. *Dalton Trans.* **2015**, *44*, 13,530.

85. Asphalt is a black, sticky, viscous liquid that is obtained from crude petroleum. It comprises almost entirely a form of tar called bitumen. The structure of asphalt is actually a colloidal suspension, with small particulates called asphaltenes dispersed through the petroleum matrix. More environmentally friendly aqueous-based asphalt emulsions are currently being used for road repair applications.

86. (a) https://www3.epa.gov/warm/pdfs/Fly_Ash.pdf. (b) Miller, S. A.; Horvath, A.; Monteiro, P. J. M. *Environ. Res. Lett.* **2016**, *11*, 1.
87. For more details regarding the role of C4AF in the hardening mechanisms of Portland cement, see: Meller, N.; Hall, C.; Jupe, A. C.; Colston, S. L.; Jacques, S. D. M.; Barnes, P.; Phipps, J. *J. Mater. Chem.* **2004**, 14, 428, and references therein.
88. Ratner, B. D.; Hoffman, A. S.; Schoen, F. J.; Lemons, J. E. *Biomaterials Science*, 2nd ed., Academic Press: New York, 2004.
89. Aninwene, G. E.; Yao, C.; Webster, T. J. *Int. J. of Nanomed.* **2008**, *3*, 257.
90. For instance, see: (a) Klein, C. *Biomaterials* **1990**, *11*, 509. (b) LeGeros, R. Z.; Lin, S.; Rohanizadeh, R.; Mijares, D.; LeGeros, J. P. *J. Mater. Sci.: Mater. Med.* **2003**, *14*, 201. (c) Yamada, S.; Heymann, D.; Bouler, J. M.; Daculsi, G. *Biomaterials* **1997**, *18*, 1037.

91. LeGeros, R. Z.; Ito, A.; Ishikawa, K.; Sakae, T.; LeGeros, J. P. *Fundamentals of Hydroxyapatite and Related Calcium Phosphates* in *Advanced Biomaterials: Fundamentals, Processing, and Applications*, Wiley: New York, 2010.
92. Weiser, H. B.; Moreland, F. B. J. Phys. Chem. 1932, 36, 1.
93. Adam, J. L.; Zhang, X. *Chalcogenide Glasses*, Elsevier: New York, 2013 (ISBN: 9,780,857,093,455).
94. For more information regarding the mechanism for glass formation, see: Royall, C. P.; Williams, S. R.; Ohtsuka, T.; Tanaka, H. *Nature Mater.* **2008**, *7*, 556.
95. It should be noted that glass is not 100% transparent; that is, some incident light is reflected-even in glass that is free from dopant or other inclusion impurities.
96. In addition to scattering processes, a better rationale for the transparency of glass is due to its electronic *band structure*, in which the HOMO/LUMO gap is too large to absorb visible light.
97. It should be known that other oxides are capable of glass network formation, such as B_2O_3, GeO_2, P_2O_5, As_2O_5, As_2O_3, Sb_2O_3, and to a limited degree V_2O_5, ZrO_2 and Bi_2O_3. The oxides of Te, Mo, W, Bi, Ba, Nd, Ti, Zn, Pb, Al, Th and Be are known as conditional glass formers. These may be included in varying concentrations, but will not on their own, yield a glass. These, and other oxides that will not form a glass (including Sc, La, Y, Sn, Ga, In, Mg, Li, Sr, Cd, Rb, Hg, and Cs) are used as network modifiers, to vary the melt viscosity and afford varying properties to the glass.
98. For more information regarding other crystalline forms, see: Douglas, B. E.; Ho, S. -M. *Structure and Chemistry of Crystalline Solids*, Springer: New York, 2006.
99. This is a useful exploitation of the freezing-point depression *colligative property*, as taught in introductory physical chemistry (e.g., adding salt to icy roads in the winter).
100. *The Complete Book on Glass and Ceramics Technology*, NIIR Board of Consultants and Engineers, Asia Pacific Business Press, Inc., 2005.
101. For details regarding how Gorilla glass is fabricated, see: https://www.corning.com/gorillaglass/worldwide/en/technology/how-it-s-made.html.
102. http://richannel.org/mystery-of-prince-ruperts-drop-at-130000-fps.
103. For a history of stained glass and glassmaking, see: (a) http://stainedglass.org/?page_id=169. (b) Rehren, Th.; Freestone, I. C. *J. Archaeol. Sci.* 2015, 56, 233. (c) Boyd, D. C.; Danielson, P. S.; Thompson, D. A.; Velez, M.; Reis, S. T.; Brow, R. K. *Encyclopedia of Chemical Technology*, 2004.
104. Some remaining stock of safe, weakly radioactive glass items such as ceramic plates, ore, marbles, etc. may still be acquired online from http://www.united nuclear.com/.
105. For more details on the history, properties, and fabrication of fiber optics, see: Glass, A. M.; DiGiovanni, D. J.; Strasser, T. A.; Stentz, A. J.; Slusher,

R. E.; White, A. E.; Kortan, A. R.; Eggleton, B. J. *Bell Labs Techn. J.* **2000**, Jan.–Mar., 168 s.

106. For example, see: Armistead, W. H.; Stookey, S. D. *Science* **1964**, *144*, 15.
107. For example, see: Morse, D. L. *Inorg. Chem.* **1981**, *20*, 777, and references therein.
108. For instance, see: (a) Trotter, D. M.; Schreurs, J. W. H.; Tick, P. A. *J. Appl. Phys.* **1982**, *53*, 4657. (b) Isaji, T.; Wakasugi, T.; Fukumi, K.; Kadono, K. *Chem. Phys. Lett.* **2012**, *522*, 72.
109. Richardson, T. J.; Slack, J. L.; Armitage, R. D.; Kostecki, R.; Farangis, B.; Rubin, M. D.; *Appl. Phys. Lett.* **2001,** *78*, 3047.
110. http://www.independent.co.uk/environment/solar-power-electricity-record-uk-energy-quarter-demand-supply-renewables-green-a7757836.html.
111. Fuel cells are typically classified according to the electrolytes used: alkaline fuel cell (AFC), polymer electrolyte (or proton exchange membrane) fuel cell (PEMFC), phosphoric acid fuel cell (PAFC), molten carbonate fuel cell (MCFC), and solid oxide fuel cell (SOFC). The high efficiencies and low emissions of fuel cells are extremely intriguing, but problems with emission-free production, and safe storage of hydrogen gas remain the primary stumbling blocks for widespread incorporation of this technology. Alkaline fuel cells have been used the longest, since the 1960s by NASA for space shuttles. In fact, this application illustrates the utility of fuel cells. Hydrogen and oxygen gases are used to power the fuel cell, which powered the electrical components of the space shuttle. Water, the only byproduct of the reaction, was used as onboard drinking water for the crew. Although AFCs are the most inexpensive to produce and have some of the highest efficiencies, they require high-purity oxygen to prevent catalyst poisoning by carbon dioxide. PEMFC designs also suffer from catalyst poisoning. The presence of small concentrations of CO (>1 ppm) in the reformate of fuels drastically alters the performance of the anodic catalyst (e.g., Pt, Pt/Ru).
112. (a) http://money.cnn.com/2016/09/02/technology/samsung-galaxy-note-7-recall/index.html (b) https://www.cnet.com/news/why-is-samsung-galaxy-note-7-exploding-overheating/.
113. (a) http://cen.acs.org/articles/94/i45/Making-batteries-fire-resistant-solid.html. (b) https://www.electrochem.org/dl/interface/sum/sum12/sum12_p045_049.pdf.
114. For solid-oxide fuel cells (SOFCs), the anode consists of a porous mixture of a Ni or Co catalyst on yttria-stabilized zirconia. Such a mixture of metal and ceramic is referred to as a *cermet*. The zirconia acts to inhibit grain growth of the catalyst particles of nickel or cobalt and protects against thermal expansion. The cathode is generally a Sr-doped $LaMnO_3$ perovskite. The Sr dopant provides for oxygen transfer to the cathode–electrolyte interface.

115. There are other types such as nitrides, halides, argyrodites, and hydrides. For recent reviews regarding inorganic-based solid electrolytes, see: (a) Manthiram, A.; Yu, X.; Wang, S. *Nature Rev., Mater.* **2017**, *2*, 16,103. (b) Bachman, J. C.; Muy, S.; Grimaud, A.; Chang, H. H.; Pour, N.; Lux, S. F.; Paschos, O.; Maglia, F.; Lupart, S.; Lamp, P.; Giordano, L.; Shao-Horn, Y. *Chem. Rev.* **2016**, *116*, 140.
116. Although we are focusing the discussion on the migration of cations, the same principles apply for the migration of anions such as O^{2-} through a solid electrolyte, as applicable for solid-oxide fuel cell (SOFC) applications.
117. For instance, see: (a) Schichtel, N.; Korte, C.; Hesse, D.; Zakharov, N.; Butz, B.; Gerthsen, D.; Janek, J. *Phys. Chem. Chem. Phys.* **2010**, *12*, 14,596. (b) Jiang, J.; Hu, X.; Shen, W.; Ni, C.; Hertz, J. L. *Appl. Phys. Lett.* **2013**, *102*, 143,901. (c) Aydin, H.; Korte, C.; Rohnke, M.; Janek, J. *Phys. Chem. Chem. Phys.* **2013**, *15*, 1944.
118. Kuhn, A.; Kohler, J.; Lotsch, B. V. *Phys. Chem. Chem. Phys.* **2013**, *15*, 11,620.
119. For a comprehensive database of zeolite structures refer to: http://www.iza-structure.org/databases/.
120. Li, J.; Cao, W.; Lu, X. X.; Jiang, L.; Li, Y. J.; Chen, S. Z.; Li, X. Y. Acta Pharmacol. Sin. 2013, 34, 367.
121. For a review of applications for zeolite thin films, see: Lew, C. M.; Cai, R.; Yan, Y. *Acc. Chem. Res.* **2010**, *43*, 210.
122. For detailed schemes of zeolite frameworks, see: http://www.iza-structure.org/databases/ModelBuilding/Introduction.pdf.
123. For a review of applications for mesoporous zeolites, see: (a) Garcia-Martinez, J.; Li, K.; Davis, M. E. Mesoporous Zeolites: Preparation, Characterization and Applications, Wiley: New York, 2015 (ISBN: 978–3-527–33,574-9). (b) Mesoporous nanofiber review: Anis, S. F.; Khali, A.; Singaravel, G.; Hashakeh, R. *Micropor. Mesopor. Mater.* **2016**, *236*, 176. (c) Corma, A. *Chem. Rev.* **1997**, *97*, 2373.
124. For a comprehensive review of hydrothermal methods used to synthesize zeolites, see: Cundy, C. S.; Cox, P. A. *Chem. Rev.* **2003**, *103*, 663. A laboratory protocol for the synthesis and characterization of the ZSM-5 zeolite may be found online at: http://www.rsc.org/suppdata/ra/c3/c3ra42199f/c3ra42199f.pdf.
125. For F-based zeolite syntheses, see: (a) Koller, H.; Wolker, A.; Eckert, H.; Panz, C.; Behrens, P. *Angew. Chem. Int. Ed. Engl.* **1997**, *36*, 2823. (b) Koller, H.; Wolker, A.; Villaescusa, L. A.; Dıaz-Cabanas, M. J.; Valencia, S.; Camblor, M. A. *J. Am. Chem. Soc.* **1999**, *121*, 3368.
126. (a) Comyns, A. E. *Focus on Catalysts* **2009**, *4*, 1. (b) Zeolitic-like metal–organic frameworks: Eddaoudi, M.; Sava, D. F.; Eubank, J. F.; Adil, K.; Guillerm, V. *Chem. Soc. Rev.* **2015**, *44*, 228.

127. Although most zeolites and MOFs are crystalline, it is possible to fabricate amorphous powders by heating, pressure, or ball-milling that features a disordered structure but still retains the basic building blocks and connectivity of their crystalline counterparts. For instance, see: (a) Amorphous MOFs: Bennett, T. D.; Cheetham, A. K. *Acc. Chem. Res.* **2014**, *47*, 1555. (b) Amorphous zeolites: Thomas, J. M.; Bursill, L. A. *Angew. Chem. Int. Ed.* **1980**, *19*, 745.
128. A special edition of *Chemical Reviews* was devoted to metal–organic frameworks: *Chem. Rev.* **2012**, 112(2).
129. For a recent review of MOFs, see: Gangu, K. K.; Maddila, S.; Mukkamala, S. B.; Jonnalagadda, S. B. *Inorg. Chim. Acta* **2016**, *446*, 61.
130. An interesting website that shows the 3-D structure of various MOFs, as well as other solid-state crystal structures discussed earlier in this chapter: http://chemtube3d.com/solidstate/MOF-home.html.
131. Eddaoudi, M.; Sava, D. F.; Eubank, J. F.; Adil, K.; Guillerm, V. *Chem. Soc. Rev.* **2015**, *44*, 228.

Further Reading

1. R.E. Hummel, *Electronic Properties of Materials* (Springer, New York, 2013)
2. A.R. Barron, *Chemistry of Electronic Materials* (MiDAS Green Innovations, 2021)
3. J.P. Glusker, M. Lewis, M. Rossi, *Crystal Structure Analysis for Chemists and Biologists* (VCH, New York, 1994)
4. G. Ferey, *Crystal Chemistry: From Basics to Tools for Materials Creation* (World Scientific Publishing, Hackensack, NJ, 2017)
5. W. Massa, R.O. Gould, *Crystal Structure Determination*, 2nd edn. (Springer, New York, 2011)
6. R.E. Newnham, *Properties of Materials: Anisotropy, Symmetry, Structure* (Oxford Univ, Press, 2005)
7. D.A.J. Rand, A. Dicks, *Fuel Cell Systems Explained*, 3rd edn. (Wiley, New York, 2018)
8. A.R. West, *Solid State Chemistry and Its Applications, 2nd edn.* (Wiley, New York, 2022)
9. N. Brandon, *Solid Oxide Fuel Cell Lifetime and Reliability: Critical Challenges in Fuel Cells* (Academic Press, New York, 2017)
10. A.R. West, *Basic Solid State Chemistry*, 2nd edn. (Wiley, New York, 1999)
11. E. Moore, L. Smart, *Solid State Chemistry: An Introduction*, 5th edn. (CRC Press, New York, 2020)
12. C.B. Carter, M.G. Norton, *Ceramic Materials: Science and Engineering*, 2nd edn. (Springer, New York, 2013)
13. P.M. Woodward, P. Karen, J.S.O. Evans, T. Vogt, *Solid State Materials Chemistry* (Cambridge University Press, 2021)
14. S. Elliott, *The Physics and Chemistry of Solids* (Wiley, New York, 1998)
15. L.R. MacGillivray, *Metal-Organic Frameworks: Design and Application* (Wiley, New York, 2010)
16. S. Kaskel, *The Chemistry of Metal-Organic Frameworks: Synthesis, Characterization, and Applications* (Wiley, New York, 2016)

17. B. Sels, L. Kustov, *Zeolites and Zeolite-Like Materials* (Elsevier, New York, 2016)
18. J. Garcia-Martinez, K. Li, M.E. Davis, *Mesoporous Zeolites: Preparation, Characterization and Applications* (Wiley, New York, 2015)
19. S.C. Pillai, S. Hehir, *Sol-Gel Materials for Energy, Environment and Electronic Applications (Advances in Sol-Gel Derived Materials and Technologies)* (Springer, New York, 2017)

Chapter 3
Metals

Of the 118 elements listed in the periodic table, 80% are metals. Since the discovery of copper and bronze by early civilizations, the study of metals (i.e., metallurgy) contributed to most of the early investigations related to materials science. Whereas iron-based alloys have long been exploited for a variety of applications, there is a constant search for new metallic compositions that have increasing structural durability, but also possess sufficiently less density. The recent exploitation of titanium-based alloys results from this effort, and has resulted in very useful materials that are used for applications ranging from aircraft bodies to hip replacements and golf clubs. Indeed, there are many yet undiscovered metallic compositions that will undoubtedly prove invaluable for future applications.

In Chap. 2, you learned how individual atoms pack in crystal lattices. Moreover, the nature of metallic bonding was described, which is responsible for the characteristic physical properties of these materials. This chapter will continue this discussion, focusing on the structure–property relationship of metallic structures.

3.1 Mining and Processing of Metals

Before we examine the structures and properties of various metals, we should first consider the natural sources of the metals—generally as oxide and/or silicate-based mineral formations. If the mineral deposit contains an economically recoverable amount of a metal, it is referred to as an *ore*. The waste material of the rock formation is known as *gangue*, which must be separated from the desired portion of the ore through a variety of processing steps.

There are three main types of rocks, grouped according to their form of origin. *Igneous* rocks are those formed from the solidification of molten mass following volcanic activity. Common examples include granite, feldspar, mica, and quartz; metals such as the alkali and alkaline earths, gold, platinum, and chromium are isolated from these formations. *Sedimentary* rocks are those formed through

B. D. Fahlman, *Materials Chemistry*,
https://doi.org/10.1007/978-3-031-18784-1_3

compaction of small grains deposited as sediment in a riverbed or sea. Common examples include shale, limestone, sandstone, and dolomite. Metals such as copper, iron, zinc, lead, nickel, molybdenum, and gold may all be found together within sulfur-based sedimentary deposits. *Metamorphic* rocks are those formations that have changed composition and morphology over time through the influence of temperature and pressure. Examples of these rocks include marble, slate, and gneiss, and yield similar metals as igneous formations.

Most metals are obtained through physical and chemical modification of ores that contain a sufficiently high concentration of the desired metal. Table 3.1 lists some metals, regions of their primary occurrence, and some applications for the metallic materials. Quite often, the processing steps used to isolate and purify metals from ores have a significant impact on their overall microstructure, which will affect the physical properties of the materials fabricated from them.

To obtain the ores in sufficient quantities needed to recover a desirable amount of metal, either open pit or shaft mining is used depending on how accessible the ore is located. We are all familiar with the picture of treacherous mine shafts that have even been featured in movies such as *The 33*. However, such "brute force" is sometimes not required. For instance, to recover coal that is used in power plants to generate electricity, there are often sufficient resources near the surface that may be obtained using large cranes. Such *pit- mining* techniques are commonly used throughout the world to extract ores of copper, iron, nickel, and a variety of precious metals.

Of course, once the ore is obtained from its deposit, the laborious task of extracting the desired metal has yet to be accomplished. In addition to metals, a variety of other substances comprise natural minerals. Since aluminum and silicon are the most prevalent elements in Earth's crust, most metals exist naturally as aluminates, silicates,[1] or aluminosilicates. The most common minerals are feldspars (comprising 60% of Earth's crust) and clays. These materials have been used since ancient times for the production of materials such as pottery, brick, and china. An example of a feldspar is $K_2Al_2Si_6O_{16}$, which corresponds to a mixture of potassium superoxide, alumina, and silica ($K_2O{\cdot}Al_2O_3{\cdot}6\ SiO_2$). Upon contact with water and carbon dioxide, feldspar is weathered to produce kaolinite, an aluminosilicate clay (3.1). However, in addition to these oxidized sources of metals, there are substances such as alkaline carbonates, sulfates, phosphates, as well as organic matter that need to be removed to yield the desired metal. As you would expect, the yield for this process is quite low; ores typically possess less than 1% of the desired metal—on the order of ppm-levels for noble metals such as gold!

$$\underset{\text{(feldspar)}}{K_2Al_2Si_6O_{12}} + 3\ H_2O + 2\ CO_2 \rightarrow \underset{\text{(kaolinite)}}{Al_2Si_2O_7} + 2\ H_2O + 2\ KHCO_3 + 4\ SiO_2 \tag{3.1}$$

The most common mineral classes from which metals are extracted include:

(i) *Silicates*—e.g., feldspars, quartz, olivines (Mg/Fe), pyroxenes ($XY(Si,Al)_2O_6$), garnets ($X_3Y_2(SiO_4)_3$), and micas ($X_2Y_{4\text{-}6}Z_8O_{20}(OH,F)_4$)

Table 3.1 Natural occurrence of selected metals and materials-based applications

Metal	Selected applications[a]	Primary locations[b]
Li	Batteries, glasses, aerospace components	US, Chile, China, Argentina
Na	Dyes, pigments, lamps, photoelectric cells	Chile, Peru, US
K	Nuclear reactors, respiratory equipment	Ger., Can., France, Spain
Be	X-ray tubes, ship/aircraft navig. systems	US, Africa, Brazil, Rs
Mg	Auto parts, coatings, photoengraving	US, Can., S. Amer., Asia
Ca	Lighter flints, lights, plastics stabilizer	US, Can., S. Amer., Asia
Sr	Glasses, ceramics, paints, TV tubes	Mexico
Sc	High-intensity lamps, nuclear reactors	China, US, Aus., Ind
Y	Coatings, capacitors, superconductors, lasers	China, US, Aus., Ind
Ti	Golf clubs, surgical implants, jet engines	US, Aus., S.A., Norway
Zr	Prosthetic devices, nuclear reactors	Ind., US, Aus., S.A
V	Tools, ceramics, batteries, magnets	US, Mex., Peru, Rs., S.A
Nb	Heat shields, electromagnets, spaceships	Japan, Rs., Can., Brazil
Ta	Aircraft turbines, rocket nozzles, HT vessels	Aus., Brazil, Thailand, Can
Cr	Corrosion-resistant coatings, tools	S.A., Turk., Rs., Cuba, Phil
Mo	Lubricants, lamp filaments, integ. circuits	US, Aus., Sweden
W	Filaments, coatings, missiles, tools, paints	US, Asia, S. Amer., Rs
Mn	Steels, paints, batteries	US, Can., France, Ger., Ind
Fe	Autos, tools, structural materials	US, Can., China, Brazil, Rs
Ru	Coatings, superconductors, electrodes	S.A., Rs., US
Os	Fountain pen tips, pacemakers, forensics	S.A., Rs., US
Co	Magnets, batteries, recording media	Can., N. Africa, Swe
Rh	Catalytic converters, coatings, crucibles	S.A., Rs., US
Ir	HT crucibles, spark plugs	S.A., Rs., US
Ni	Batteries, coatings, currency	Can., Rs., Aus
Pd	Jewelry, dental crowns, surgical instruments	S.A., Rs., US
Pt	Jewelry, coatings, surgical implants, lighters	S.A., Rs., US
Cu	Conductive wire, pipes, currency	Chile, US, Can
Ag	Jewelry, dental fillings, photography	Mex., Peru, US, Can
Au	Jewelry, integrated circuits, heat shields, coatings	S.A., US, Aus., Can
Zn	Coatings, pipes, highway guard rails	US, Can., Aus., China
Cd	Alloys, batteries, nuclear reactors	Canada, Aus., Mexico
Al	Auto parts, ceramics, coatings, paints	Brazil, Jamaica, Asia
Sn	Alloys, plating, flat glass	Bolivia, Brazil, Malaysia

(continued)

Table 3.1 (continued)

Metal	Selected applications[a]	Primary locations[b]
Pb	HT vessels, pipes, sound abs. materials	Aus., US, China, Peru
Ce	Lighter flints, ceramic capacitors, magnets	China, US, Aus., Ind
Eu	Television screens, nuclear reactors	China, US, Aus., Ind
Er	Lasers, catalysts, phosphors	China, US, Aus., Ind

[a]These selected applications may be for the pure metal or other compounds/alloys
[b]Can. = Canada, Rs. = Russia, Aus. = Australia, Ind. = India, S.A. = South Africa

(ii) *Carbonates*—e.g., calcite and aragonite (both $CaSO_4$), dolomite (Mg/Ca), siderite (Fe)
(iii) *Sulfates*—e.g., anhydrite ($CaSO_4$), celestine ($SrSO_4$), barite, gypsum (hydrated $CaSO_4$), chromate, molybdate, selenate, sulfite, tellurite, tungstate
(iv) *Halides*—e.g., fluorite (CaF_2), halite (NaCl), sylvite (KCl), sal ammoniac (NH_4Cl)
(v) *Oxides*—e.g., hematite (Fe_2O_3), magnetite (Fe_3O_4), chromite (Fe/Cr), spinel (Mg/Al), ilmenite (Fe/Ti), rutile (TiO_2)
(vi) *Sulfide*—e.g., pyrite (FeS), chalcopyrite (Cu/Fe), pentlandite (Ni/Fe), galena (PbS), sulfosalt (S/As), selenide/telluride/arsenide/antimonide minerals
(vii) *Phosphates*—e.g., AO_4 (A = P, As, Sb, V), apatite ($Ca_5(PO_4)_3(OH, F, Cl)$)

The variety of procedures that are used to obtain metals from their ores are known as *extractive metallurgy*. Once the ore is acquired from the deposit, a *flotation* process is first used to separate the metal from the gangue. This consists of grinding the ore into a powder and mixing it with water, oil, and surfactants (i.e., collectors) to yield hydrophobic surface charges on the particulates. The most common collector molecules are alkali salts of the *O*-alkyl dithiocarbonates (**I**)

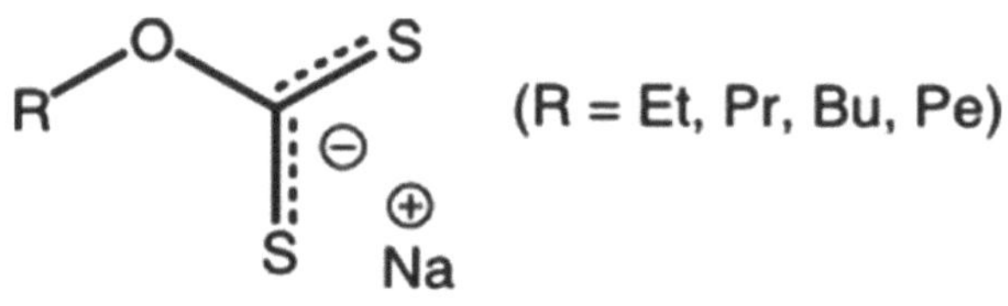

(R = Et, Pr, Bu, Pe)

I

commonly known as xanthates, which are widely used for the concentration and separation of metals from sulfide-based ores. The mixture is vigorously stirred into a frothy mixture; the coated grains attach themselves to air bubbles that float to the solution surface, where they are removed through a skimming/filtration process. This technique is very useful in the separation of lead and zinc components, often found together in natural sulfide deposits.

Once the bulk ore is concentrated from its many impurities, the metal is reduced from its natural oxides. Most often this is performed by *smelting*, where the ore is heated in the presence of a reducing agent such as carbon (e.g., coal, coke), or carbon monoxide (3.2 and 3.3, respectively). Although carbonaceous materials were used exclusively by early civilizations, modern extractive methods commonly employ other reducing agents such as Na, Mg, Ca, Al, or alloys such as Zn/Mg. However, with the advent of biomass as an alternative fuel, organic-based waste materials are now being used as effective reducing agents.[2] Often, the ore is not pure following flotation; therefore, a flux such as limestone or dolomite is added to remove the gangue as *slag*—a mixture of metal oxides (e.g., SiO_2, Al_2O_3, CaO, MgO, etc.), sulfides and impurity metals.

$$M_xO_y + y\,C \xrightarrow{1200--1500\,^\circ C} x\,M + y\,CO \tag{3.2}$$

$$M_xO_y + y\,CO \xrightarrow{1500-2000\,^\circ C} x\,M + y\,CO_2 \tag{3.3}$$

If the ore consists of metal sulfides, it must first be converted to the oxide through reaction with oxygen at high temperature (3.4):

$$2\,MS + 3\,O_2 \xrightarrow{1500-2000^\circ C} 2\,MO + 2\,SO_2 \tag{3.4}$$

The temperature required for tin and lead smelting is only on the order of 300–400 °C; hence, it is no surprise that early civilizations discovered these metals from heating local minerals over campfires, which yields a maximum temperature of *ca.* 600 °C. However, this temperature regime is not sufficient for copper smelting, which requires temperatures in excess of 1100 °C. Consequently, copper metal was most likely discovered serendipitously by using higher temperature coal-fired pottery kilns that contained the copper-rich mineral malachite.

Let's consider some tangible examples of metal extraction/refining, starting with vanadium. There are approximately 50 types of minerals that will yield a viable amount of V metal, such as vanadinite, patronite, bauxite, and carbon deposits such as crude oil, coal and butiminous matter. The oxide obtained from these deposits via flotation is first mixed with NaCl or Na_2CO_3 at 900 °C to yield sodium orthovanadate, $NaVO_3$. This salt is dissolved in water and acidified with sulfuric acid to a pH between 2–3 to yield a red solid, which is hydrated V_2O_5.[3] The vanadium(V) oxide product is then melted at 690 °C in the presence of a reducing agent (either C or Ca under an inert atmosphere) to yield pure vanadium metal.

Titanium is refined using the Kroll process, which is effective in separating Ti from Zr, which is almost always present in significant concentrations in the particular ore. A mineral such as rutile is mixed with coke and subjected to 1000 °C in a fluidized bed reactor to yield an impure form of titanium metal. This is then reacted with chlorine gas at elevated temperatures to yield $TiCl_4$, which is separated from other volatile chlorides by fractional distillation. Magnesium metal is then added to $TiCl_4$ at 800–850 °C under argon gas to yield Ti metal, as a porous metallic sponge. Aqueous HCl or liquid sodium leaching is then used to drive off the $MgCl_2$ byproduct. To refine Zr metal rather than Ti, baddeleyite ore (ZrO_2) is used, which contains significant concentrations of Fe. In this process, $FeCl_3$ must be separated from $ZrCl_4$ by fractional distillation, and 1100 °C is used instead of *ca.* 800 °C for the magnesium reduction reaction above.

The above techniques that involve high-temperature processes are known as *pyrometallurgy*. Another common technique involves the electrolytic reduction of metal compounds, often referred to as *hydrometallurgy* or *electrorefining*, depending on whether the procedure is carried out before or after the metal has already been separated from its ore, respectively. Hydrometallurgy consists of three steps:

(i) *Leaching*—a caustic or acidic solution is added to the ore to yield a concentrated solution containing the metal of interest. This may be done in situ (e.g., solution added directly to uranium deposits), or in heaps or vats of the ore. For these latter methods, the crushed ore is placed over a storage pond and irrigated with the leaching solution, known as a *lixiviant*. The solution collected following slow gravity-controlled leaching is referred to as the pregnant solution. Equations 3.5–3.9 illustrate some example lixiviants that are used to leach a variety of ores. It should be noted that refractory ores such as ZnS require high pressures to achieve effective leaching (3.7).

$$Al_2O_3 + 3\ H_2O + 2\ NaOH \rightarrow 2\ NaAl(OH)_4 \tag{3.5}$$

$$ZnO\ (\text{also}\ CuO,\ NiO) + H_2SO_4 \rightarrow ZnSO_4 + H_2O \tag{3.6}$$

$$ZnS + O_2 + 2\ H_2SO_4 \xrightarrow{\text{autoclave}\ (>0.6\ \text{MPa})} 2\ ZnSO_4 + 2\ H_2O + 2\ S \tag{3.7}$$

$$Au\ (\text{in ore}) + \frac{4}{3}\ O_3 + 4\ HCl \rightarrow AuCl_4^- + O_2 + 2\ H_2O \tag{3.8}$$

$$2\ Au(\text{in ore}) + 4\ CN^- \rightarrow 2\ Au(CN)_2^- + Zn \rightarrow Zn(CN)_4^- + 2\ Au \tag{3.9}$$

(ii) *Solution Concentration*—two methods are used to separate the metallic compound from the many impurities generated during leaching: solvent extraction and chemical precipitation (e.g., 3.10 and 3.11). Whereas the former is used to concentration metals from dilute low-value leaching solutions, the latter is used for more concentrated solutions.

$$H_2S(aq) \rightleftharpoons HS^-(aq) + H^+(aq)\ (pK_1 = 6.99) \tag{3.10}$$

$$Ni^{2+}(aq) + HS^-(aq) \rightarrow NiS(s) + H^+(aq) \tag{3.11}$$

(iii) *Metal Recovery*—the metal is recovered using electrolytic reduction of the concentrated solution, which is known as *electrowinning*. This is suitable for most metals, but is most commonly used for the alkali metals, rare earths, Pb, Cu, Zn, Ni, Co, Au, Ag, Al, Cr, and Mn.

Figure 3.1 illustrates the electrorefining of sodium and aluminum metals, after their separation from their native ores. For aluminum, the natural bauxite ore is first refined via the Bayer process, which consists of digesting bauxite with a hot solution of NaOH at 175 °C (3.12 and 3.13). The impurity oxides of Si, Fe, and Ti are not soluble and are effectively separated from the ore by filtration.

$$Al_2O_3 \cdot H_2O\ (\text{or } 2\ AlO(OH)) + 2\ OH^- \xrightarrow{\Delta} 2[AlO_2]^- + 2\ H_2O \xrightarrow{\Delta,\ 2H_2O} 2\left[Al(OH)_4\right]^- \tag{3.12}$$

$$2\left[Al(OH)_4\right]^- + CO_2 \xrightarrow{-\Delta} 2\ Al(OH)_3(s)\ (\text{or } Al_2O_3 \cdot 3\ H_2O(s)) + H_2O + CO_3^{2-} \tag{3.13}$$

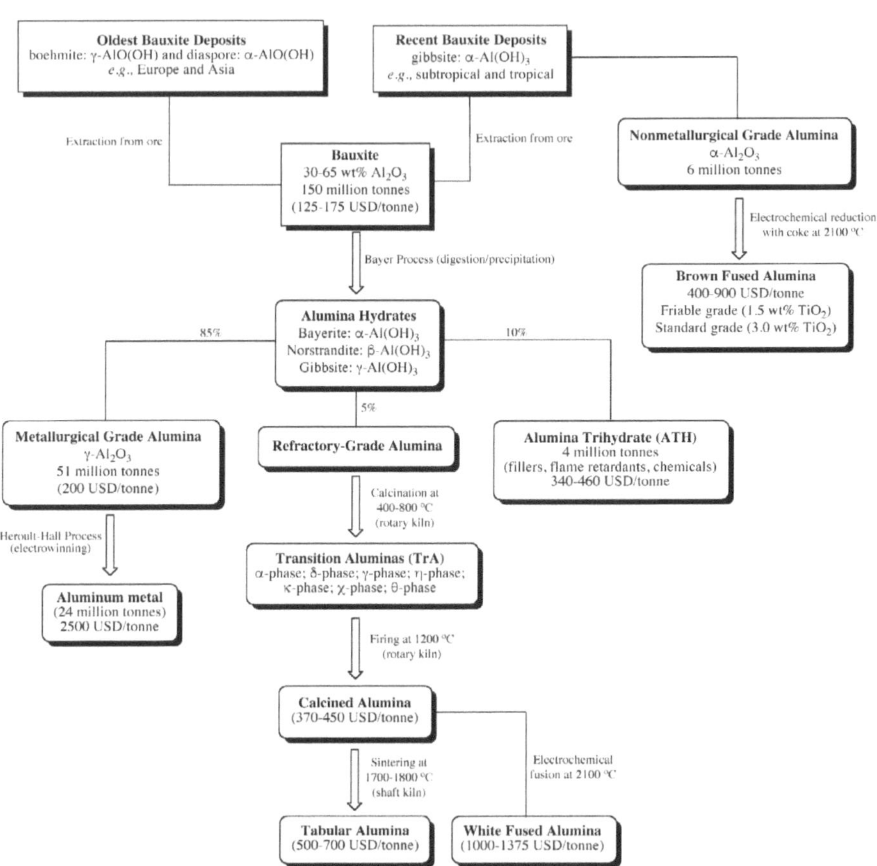

Fig. 3.1 Processing routes for the conversion of bauxite into alumina and aluminum metal Reproduced with permission from *Chemistry: The Central Science*, 8th ed., Brown, LeMay, Bursten. © 2002 Prentice-Hall

Approximately 85% of the resulting hydrated alumina is converted into aluminum metal, with the remaining converted to various forms of alumina (Fig. 3.2), for applications ranging from flame-retardants to polishing media and sapphire production.

For the purification of sodium, a fused salt is used at high temperatures. As is often necessary for ionic salts, a solid solution is necessary to reduce the melting point of the salt. The addition of calcium chloride effectively reduces the melting point of the solution from 800 °C (for NaCl alone) to 600 °C. To circumvent the high melting point of aluminum (2030 °C), anhydrous aluminum oxide (i.e., alumina, m.p. 2050 °C) is dissolved in molten cryolite (Na_3AlF_6, m.p. 1009 °C) and excess AlF_3, resulting in a more reasonable operating temperature (920–980 °C) for electrolysis, known as the Heroult-Hall process.

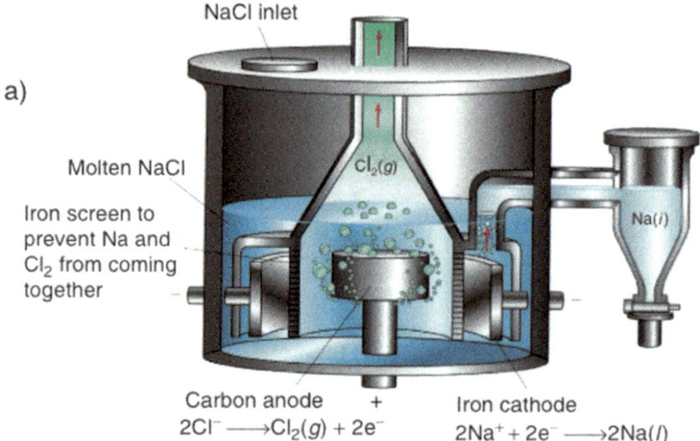

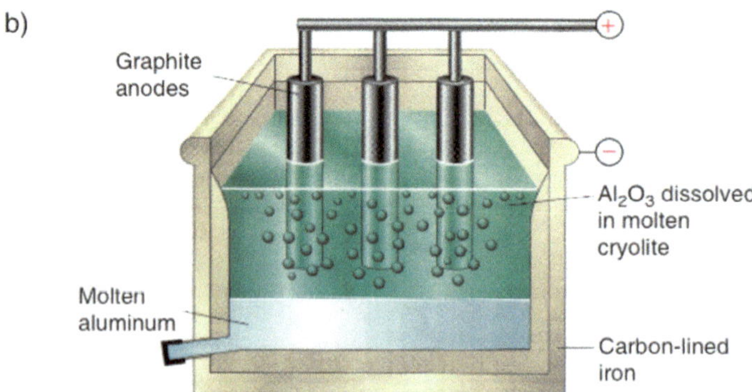

Fig. 3.2 Schematic of electrometallurgical processes used to purify metals. Shown is the **a** Downs cell used to purify sodium metal and **b** Hall cell used to purify aluminum metal. Reproduced with permission from *Chemistry: The Central Science*, 8th ed., Brown, LeMay, Bursten. © 2002 Prentice-Hall

It is estimated that iron constitutes 90% of all applications that involve metals. Hence, it is not surprising that the purification and post-processing of iron is the most widely practiced. The most primitive method used to purify iron from its ore is called *bloomery,* which dates back to the nineteenth century. This method uses pure carbon in the form of charcoal to reduce the metal (3.2). In this process, the temperature is not sufficient to completely melt the iron, so a spongy mass consisting of iron and silicates are formed. Through heat/hammering treatments, the silicates mix into the iron lattice, creating *wrought iron*. This form of iron was used exclusively by early blacksmiths, since the heating of wrought iron yields a malleable, bendable, and extremely easy compound to work with.

Most modern applications for metallic iron are steel-related, exploiting its high hardness, ductility and tensile strength. Figure 3.3 shows a flowchart for the various procedures that are used for modern steelmaking. The first step uses a *blast furnace*[4] that is composed of a massive, refractory-lined steel column wherein pelletized iron ore, charcoal, and calcium sources (from limestone and dolomite) are poured into the top, and a large jet of pre-heated (*ca.* 1050 ºC) air is blown in from the bottom. As mixing of the components occurs at various temperature regimes, the various oxides present in the ore are reduced to metallic iron. From the coolest–hottest portions of the blast oven, corresponding to the highest–lowest regions, respectively, the following oxides are reduced (Fig. 3.4):

- 500–600 °C: Hematite (Fe_2O_3)
- 600–900 °C: Magnetite (Fe_3O_4)
- 900–1100 °C: Wustite (FeO)
- >1100 °C: $FeO_{0.5}$

Since iron ore is largely composed of aluminosilicate minerals, a byproduct is also formed within the blast furnace, known as *slag* (*ca.* 30–40 wt% SiO_2, 5–10 wt% Al_2O_3, 35–45 wt% CaO, 5–15 wt% MgO, and 5–10 wt% CaS).

It should be noted that it takes 6–8 h for the native iron ore to descend toward the bottom of the blast furnace, but only *ca.* 8 s for the pre-heated air to reach the top of the furnace. Oftentimes, a fused solid known as *sinter* is also added to the blast furnace, which is composed of fine particulates of iron ore, coke, limestone and other steel plant waste materials that contain iron. The reducing agent within the blast furnace (*coke*) is composed of 90–93% carbon and is formed by heating coal to remove the volatile components such as oil and tar. The coke is ignited at the bottom of the blast furnace immediately upon contact with the air blast. Since there is excess carbon in the furnace, the combustion product is CO rather than CO_2.

The molten iron collects at the bottom of the blast furnace and is cooled into a form known as *pig iron*. This form of iron is composed of *ca.* 93.5–95% Fe, 0.3–0.9% Si, 0.025–0.05% S, 0.55–0.75% Mn, 0.03–0.09% P, 0.02–0.06% Ti, and 4.1–5% C. Due to the relatively high carbon concentration, this form of iron is too brittle and hard for any structural applications. It should be noted that re-heating pig iron with slag and hammering to remove most of the carbon will yield *wrought iron*. However, it is

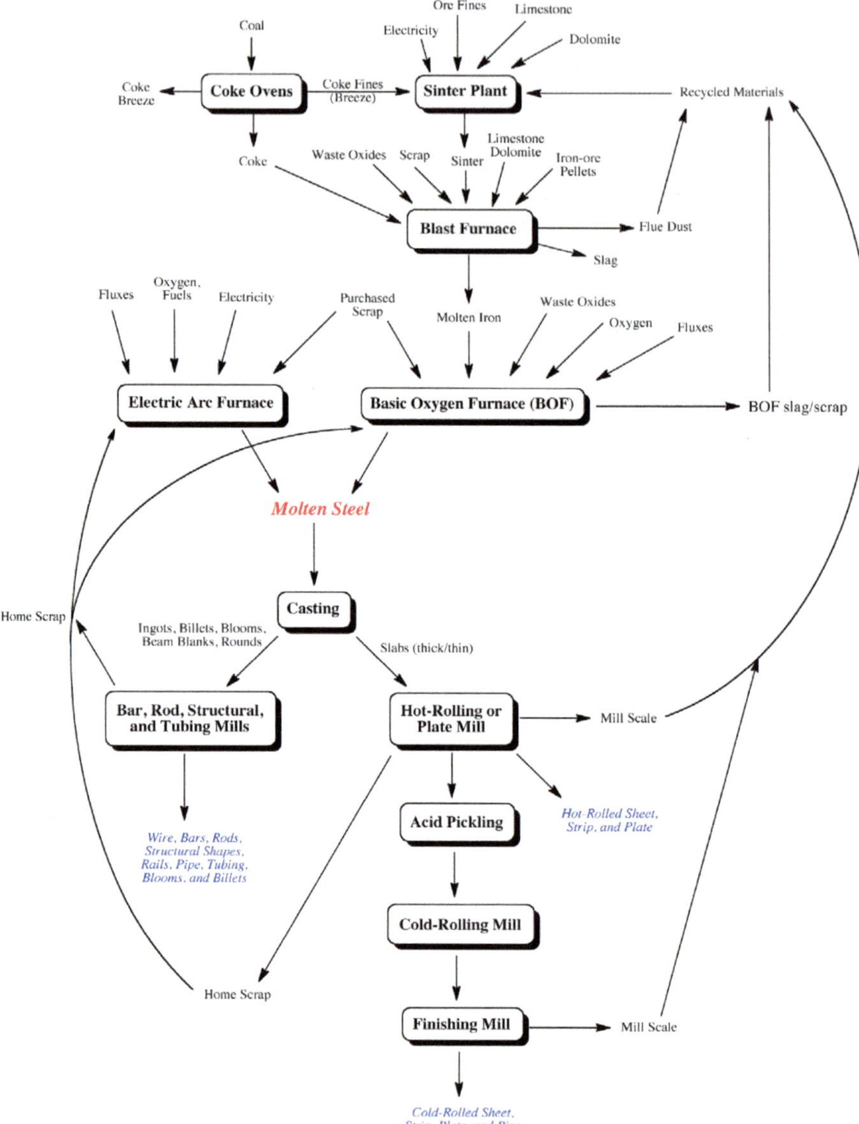

Fig. 3.3 Flowchart illustrating the various steps involved in steelmaking and processing

often most useful to convert this form into high-strength steel. While impurities such as silicon, calcium, aluminum, and magnesium were largely removed in the slag, phosphorus and sulfur impurities remain behind, and are dissolved into the molten iron.

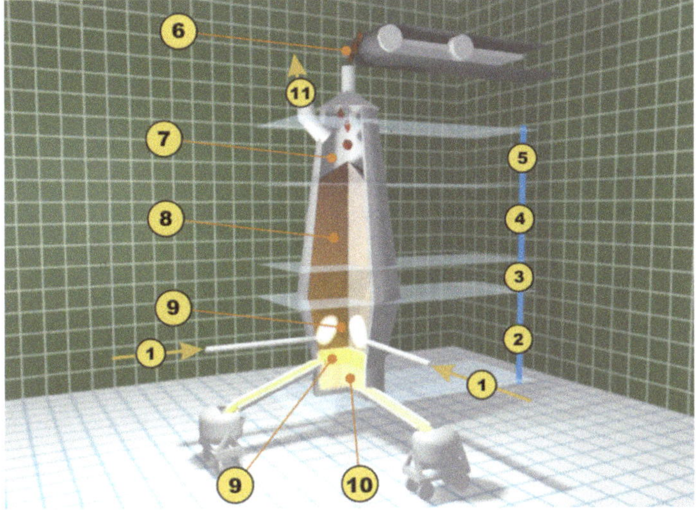

Fig. 3.4 Schematic of a blast iron. Shown are: (1) incoming pre-heated air from the Cowper stove; (2) melting zone; (3) reduction zone of FeO; (4) reduction zone of Fe_2O_3; (5) pre-heating zone; (6) feed of iron ore, limestone, and coke; (7) exhaust gases; (8) column of ore, coke, and limestone; (9) removal of slag; (10) tapping of molten pig iron; (11) collection of waste gases. Image courtesy of Robert Blazek

The next step involved in steelmaking is reducing the carbon concentration of the pig iron. The molten iron formed in the blast furnace is transported to the basic oxygen furnace (BOF) via transfer ladles (Fig. 3.5a). Lime (CaO) is added directly to the transfer ladle at 1350–1400 °C (Fig. 3.5b) in order to facilitate removal of sulfur (3.14):

$$CaO + FeS + C \rightarrow CaS + FeO + CO \tag{3.14}$$

The resultant slag is raked from the ladle and the remaining molten iron is charged into the BOF along with steel scraps (Fig. 3.4c). A supersonic jet of oxygen is used within the BOF to yield a temperature of 1620–1720 °C. Under these extreme oxidizing conditions, most of the remaining C, P, Si and Mn impurities are depleted from the hot metal, resulting in a steel with a concentration of 0.03–0.07 wt% carbon. The entire oxidation process within the BOF lasts only 20–25 min (Fig. 3.6).

The next process of steelmaking consists of fine-tuning the composition of the steel generated from the BOF. If low-carbon steel is desired, the molten steel is degassed to yield a product concentration of <0.005 wt% C (Fig. 3.7). Alternatively, desired dopants (to alter overall properties—see Sect. 3.2) may be added directly to the molten iron within the alloying ladle as a metal wire or powder. Alloying is afforded by lowering arc furnace electrodes into the ladle; argon gas is injected at the bottom of the ladle to prevent oxidation and induce stirring that is required for homogenization of the steel.

Fig. 3.5 Aa Transfer of pig iron from the blast furnace to basic oxygen furnace (BOF) in transfer ladles, **b** sulfur removal via CaO addition, and c) charging the BOF with pig iron and steel scraps. Images courtesy of Severstal Steel, Dearborn, MI

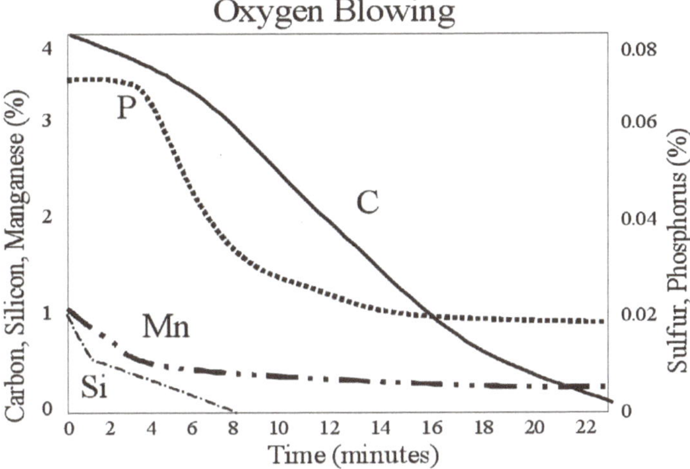

Fig. 3.6 Melt composition within the basic oxygen furnace. Data courtesy of Severstal Steel, Dearborn, MI

Fig. 3.7 Degassing unit used to remove carbon under vacuum. Image courtesy of Severstal Steel, Dearborn, MI

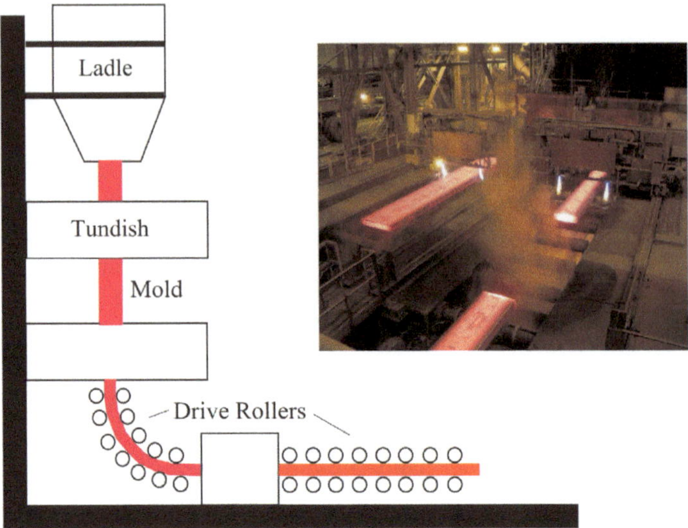

Fig. 3.8 Schematic of continuous casting, with a photograph of the resulting iron slabs. Image courtesy of Severstal Steel, Dearborn, MI

Once the desired composition of the steel has been achieved, the processes of bulk forming may begin. Long blocks of steel (*ca.* 30') are formed by *slab casting*, whereby the molten steel is poured into a water-cooled copper mold via a *tundish*. As shown in Fig. 3.8, a series of rollers slowly draw the solidified steel downward from the mold, allowing one to control the number of strains and dislocations formed during solidification. After the slabs are cut to the desired lengths, they are transferred to the *hot rolling* facility (Fig. 3.9), where the original *ca.* 8" × 30' slabs are reduced to 0.1" × 3000'. During hot rolling, refinement of the microstructure (e.g., phase transitions, precipitate formation, grain size alteration, etc.) takes place, which governs the ultimate properties of the steel such as yield/tensile strengths.

The final processing steps consist of *cold strip milling*, which consists of:

1. *Pickling*: uses hydrochloric acid to remove the oxide coating, formed during hot strip milling under ambient conditions;
2. *Cold-rolling*: reduces the gauge from 0.1" to a thickness as small as 0.017";
3. *Annealing*: relieves stresses induced during cold rolling, and develops the microstructure to improve the formability of the steel;
4. *Temper-rolling*: improves the surface finish and flatness, oil is applied to prevent rust formation.

If corrosion resistance is desired, the ribbon may be *galvanized*—i.e., coated with a protective layer of zinc. Upon exposure to the atmosphere, the Zn coating sacrificially oxidizes to form a protective layer of zinc oxide. Further interaction with moisture

Fig. 3.9 Photographs of the hot-rolling process. Images courtesy of Severstal Steel, Dearborn, MI

results in the formation of zinc hydroxide (3.15), which may also react with carbon dioxide in the atmosphere to form a thin, impermeable, and water-insoluble coating of zinc carbonate (3.16).

$$ZnO + 2\,H_2O \rightarrow Zn(OH)_2 + H_2 \tag{3.15}$$

$$Zn(OH)_2 + CO_2 \rightarrow ZnCO_3 + H_2O \tag{3.16}$$

The industry standard for galvanized coatings is a minimum thickness of 70 microns, or 505 g of Zn/m^2. A zinc coating may be applied using either *electro-galvanization* or a *hot-dip galvanization* process. Whereas the former applies a thin layer of metallic zinc, hot-dipping deposits a thicker coating that is more desirable for the undercarriage of automobiles or building nails, for instance. *Thermal-diffusion galvanizing* is a new process that applies Zn powder to the desired part within a slowly rotating sealed drum, heated to temperatures of *ca.* 600–850 °C. The Zn/Fe alloying takes place at a lower temperature relative to hot-dipping, resulting in a more uniform and wear-resistant coating. This process also eliminates the need for caustic, acidic, and flux baths required to prepare parts for hot-dipping. A coating of Zn may also be deposited by *mechanical galvanization*, in which zinc powder is pressed onto the surface of steel via the interaction of sand or glass beads within a rotating drum at elevated temperatures (*ca.* 300–350 °C). It should be noted that no galvanization process is sufficient to protect the steel in highly corrosive environments (e.g., seawater). For applications within this media, *stainless steel* is preferred wherein the chemical composition of the steel is appropriately doped with Cr to attain corrosion resistance (see Sect. 3.2).

3.1.1 Powder Metallurgy

Although the origin of fabricating metallic materials through flame sintering dates back to *ca.* 3000 BC, this method was not widely applied until the late eighteenth century. The earliest foundations of metallurgy focused on doping and strengthening bulk metallic materials; however, powders are now frequently used as precursors for metallic materials. For instance, tantalum powder is used in the fabrication of capacitors for electronics and telecommunications, including cellular phones and computer chips. Iron powder is used as a carrier for toner in electrostatic copying machines; also, over 2 million pounds of iron powder is incorporated each year in iron-enriched cereals! Copper powder is used in antifouling paints for boat hulls and in metallic pigmented inks for printing and packaging. Indeed, the list of applications for metal powders goes on and on and must constantly be updated as new applications arise.

Modern *powder metallurgy* consists of placing a metal powder(s) into a closed metal cavity, or die, compacting under high pressure (typically 200–300 + MPa), and sintering in a furnace to yield a metal with the desired porosity and hardness. The *sintering* process effectively results in the welding together of granules to form a mechanically strong material.

Metal and alloy powders may be produced through the following routes, with the last three accounting for the most common methods currently employed:

1. Grinding and pulverization of a metallic solid or oxide-based ore
2. Reductive precipitation from a salt solution
3. Thermal decomposition of a chemical compound, or precursor
4. Electrodeposition
5. Atomization of molten metal

For relatively brittle materials such as intermetallic compound and ferro-alloys, mechanical pulverization is sufficient to produce metallic powders. This process uses a ball or rod mill, a cylindrical-shaped steel container filled with ceramic balls or rods, respectively. As the grinding mill is rotated, the grinding media collides with the ore/metallic compound effectively grinding the material into a fine powder. Either alumina or zirconia represents the most common ceramic material used within grinding mills. This procedure is also commonplace for refining iron powder from the co-grinding and post-annealing of the ore with carbon (3.2). Refractory metals are normally refined through the reduction of oxides with hydrogen gas.

Chemical precipitation of metal from a solution of a soluble salt may also be used to form metallic powders. In this procedure, a reducing agent such as sodium borohydride is added to an aqueous metal salt, MX (3.17). A mixture of aqueous products will be produced in addition to the reduced metal, since sodium borohydride also reacts exothermically with water to yield *borax*, $Na_2B_2O_7$. As we will see in Chap. 6, this is the most widely used procedure for the synthesis of nanoparticulate metals, from the reduction of metal salts confined within nanosized entrainer molecules.

$$MX + 3\,NaBH_4 + 10\,H_2O \rightarrow M^0 + NaX + B(OH)_3 + Na_2B_2O_7 + \frac{19}{2}\,H_2 \quad (3.17)$$

Another useful means of producing metal powders is through thermolysis of a chemical precursor, such as metal carbonyl complexes. This process was originally developed to refine nickel from the crude product extracted from its ore. Carbon monoxide gas readily reacts with late transition metals, due to the synergistic effects of σ-electron donation from the ligand to metal, and π -back donation from the metal to the ligand (Fig. 3.10). Hence, by passing CO gas over impure nickel at 50 °C, $Ni(CO)_4$ gas is formed, leaving the impurities behind. The carbonyl decomposes upon heating at *ca.* 250 °C, forming pure nickel powder. Industrially, the *Mond process* uses the same chemistry to reduce nickel oxides extracted from an ore. Upon reaction with a mixture of H_2 and CO gases, the nickel is first reduced to form an impure product, followed by conversion to ultra-high purity Ni through the $Ni(CO)_4$ intermediate.

Electrolysis may also be used to produce metallic powders, through redox reactions at electrode surfaces. By choosing suitable reaction conditions—composition

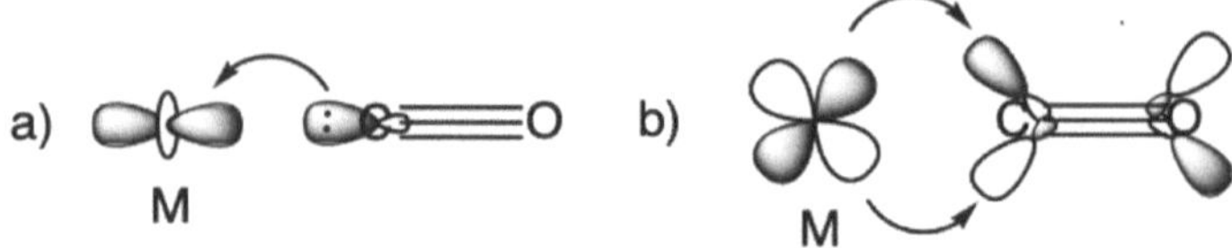

Fig. 3.10 The synergistic stabilizing effect of metal carbonyl complexes. Shown is **a** ligand-to-metal σ donation from the carbon lone pair to the metal $d_{z}{}^{2}$ orbital and **b** metal-to-ligand back-donation from the d_{x2-y2} orbital to the empty π^* orbital on CO. This weakens the C–O bond, while concomitantly strengthening the M–C interaction

and strength of the electrolyte, temperature, current density, etc., many metals can be deposited in a spongy or powdery state. However, most often a brittle deposit is formed, requiring extensive post-processing such as washing/drying, reduction, annealing, and crushing. Although this technique could be used for virtually all metals, it has been replaced with other less expensive methods such as solution reduction. Nevertheless, metallic powders of copper, chromium, and manganese are still mostly produced through electrolytic means. Interestingly, toward the ongoing search for structures at the nanoregime (Chap. 6), *electrodeposition* has recently been applied for the intriguing synthesis of metal nanoparticles and nanowires (Fig. 3.11).

The last method for generation of metallic powders that we will consider is atomization. In this high-temperature process, molten metal is broken up into small droplets and rapidly quenched to prevent wide-scale agglomeration (Fig. 3.12). The atomization process occurs through the bombardment of a stream of molten metal with a high-energy jet of gas (e.g., air, N_2, Ar) or liquid (e.g., H_2O, hydrocarbons). Argon gas is used extensively to prevent the oxidation of reactive metals and alloys such as chromium or tungsten. Atomization is very different than ionization. Whereas the former consists of gaseous ground-state and excited-state metallic atoms, the latter contains electrons and metallic ions that are much more reactive.

By varying parameters such as jet design, pressure/volume of the atomizing fluid, and density of the liquid metal stream, it is possible to control the overall particle size and shape. In principle, atomization is applicable to all metals that can be melted, and is commercially used for the production of iron, steels, alloy steels, copper, brass, bronze, and other low-melting-point metals such as aluminum, tin, lead, zinc, and cadmium.

Atomization is particularly useful for the production of homogeneous powdered alloys, since the constituent metals are intimately mixed in the molten state. Further, this process is also useful to produce powders of difficult compositions. For instance, copper-lead powders may not be formed through simple precipitation from liquid solutions. Upon solidification, the lead will preferentially precipitate, resulting in a copper-rich metallic powder. By comparison, atomization of a Cu/Pb molten solution results in a copper powder containing a very fine and uniform distribution of lead inclusions within each particulate.

For powder metallurgy, the density of the powder strongly influences the strength of the material obtained from compaction. As one would expect, the density of the

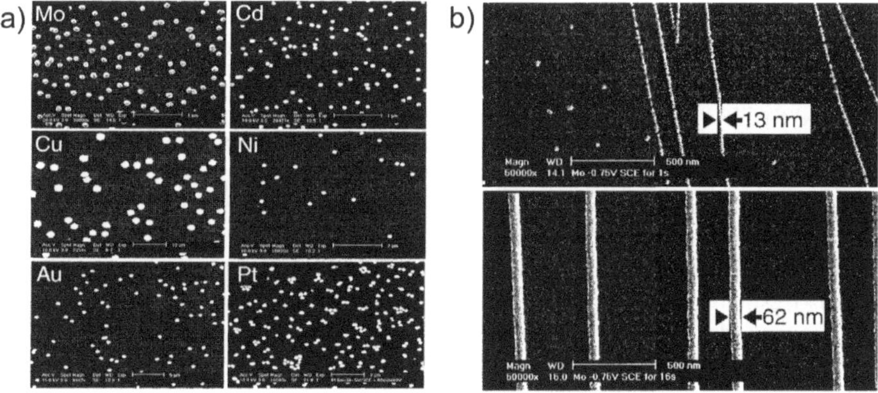

Fig. 3.11 Electrodeposition of **a** silver nanoparticles and **b** silver nanowires. The co-evolution of hydrogen gas during electrodeposition is thought to assist in monodisperse nanocluster growth by interrupting interparticle coupling via convection effects at the electrode surface and surface mobilization of growing nanoclusters. Reproduced with permission from *J. Phys. Chem. B.* **2002**, *106*, 3339. © 2002 American Chemical Society

powder depends on both the shape and porosity of individual micron-sized particulates. We saw in Chapter 2 that close-packed metals will have higher densities than simple cubic materials (Fig. 3.13). Among the close-packed metals, the theoretical percentages of total space occupied by atoms, relative to voidspace for bcc (coordination number 8), fcc (coordination number 12), and hcp (coordination number 12) unit cells are 68%, 74%, and 74%, respectively. Even if the metal particulates have the same diameter and are completely spherical, the actual packing density is typically on the order of 55–60%. This value may be improved by introducing nanosized particles that will fill the voids among the larger particles. During subsequent high-temperature sintering, the larger particles will grow at the expense of the nanoparticles, leaving behind relatively small voids that are closed during the thermal treatment.

A lubricant is also typically added during powder compaction. The most common lubricants are stearic acid (octadecanoic acid), stearin, zinc stearate, and other waxy organic compounds (e.g., palmates). The name stearate should be vaguely familiar, as the sodium salt is often employed as the active ingredient in soap. The primary use for the lubricant is to reduce friction between the powder mass and the surface of the die walls. For this purpose, it is often sufficient to apply lubrication to the walls of the die, rather than introducing the organic compound to the metallic powders. If a significant amount of organic residue is left following compaction, it will be removed upon sintering, leaving behind large voids that will greatly detract from the overall strength of the finished material.

The density of the bulk material following the pressing event is referred to as the *green density*, coined more frequently for ceramic processing. It is most desirable to have a powder with lower density, as this will undergo a greater change in volume during compaction (Fig. 3.14). The intimate pressing together, or alloying, of metals during the pressing process is known as *cold-welding*. Sometimes, the powder is too dense for efficient cold-welding; for these samples, such as heavy metal alloys, a greater pressure is required necessitating the use of larger presses and stronger dies. It should be noted that powders under pressure do not behave as liquids; the pressure is not uniformly transmitted and very little lateral flow takes place within the die.

The compacted powder will only be as pure as the initial components. The addition of small impurities will cause dramatic differences in the resultant metallic material following the pressing and sintering steps. The presence of bound *vs.* free impurities may also result in observable differences in the compaction behavior for powders. For iron powders, the presence of iron carbide (Fe_3C) will increase the hardness of the matrix, requiring higher pressures for compaction. However, free graphite particles will act as a lubricant, increasing the pressing efficiency at lower pressures.

Unless handled under an inert atmosphere, metal powder grains will be coated with a thin oxide film. Unless excessively strong SiO_2 or Al_2O_3 films are produced,

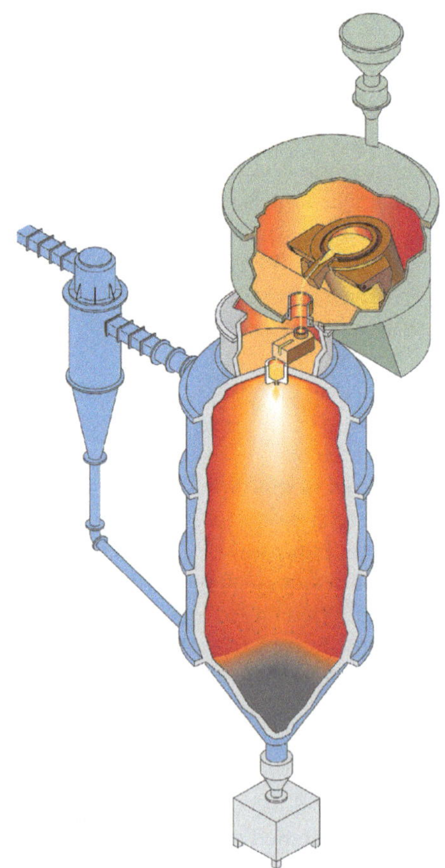

Fig. 3.12 Illustration of an atomizer for the production of metallic powders. The molten metal/alloy is sprayed into a cooling tower under the flow of an atomizing gas. The particulates are allowed to cool as they descend downward, and are collected in a hopper at the bottom of the tower. Reproduced with permission from Crucible Materials Corporation

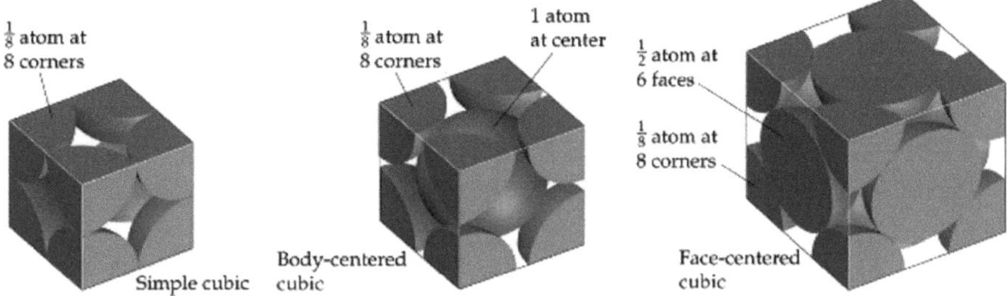

Fig. 3.13 Space-filling models showing the occupancy and available interstitial sites within cubic unit cells. Reproduced with permission from *Chemistry: The Central Science*, 8th ed., Brown, LeMay, Bursten. © 2002 Prentice-Hall

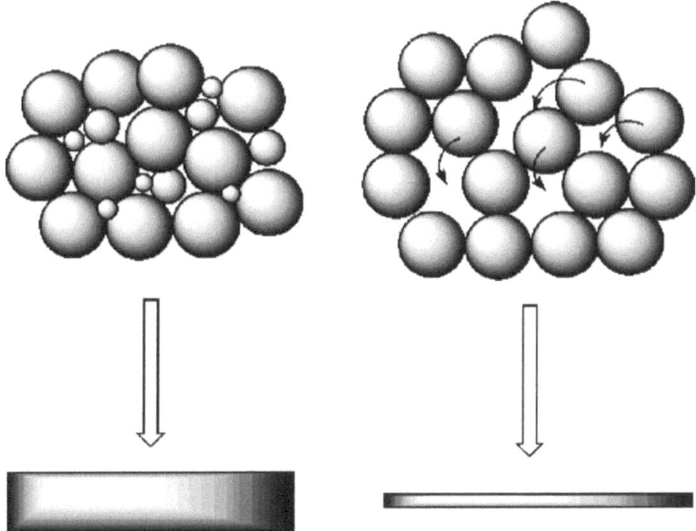

Fig. 3.14 The effect of matrix density on the compaction volume yielded from pressing

the coating will rupture during the pressing process, exposing the clean underlying metal surfaces. It should be noted that for alloying metals such as copper and zinc, pressing is not necessary. The powders are simply placed in a mold and sintered, a process aptly referred to as loose-powder sintering.

The major applications for powder metallurgical products center around the automotive industry, specifically for engines, transmissions, and brake/steering systems. Following pressing, the compacted metals may be injected into a mold, or pressed under vacuum at high temperature within a hot isostatic press (HIP). A great deal of consumer products are fabricated using these techniques; high-tech plastics and other composite parts represent a significant market share for powder metallurgical materials. This is especially the case since the soaring gas prices and stringent environmental regulations dictate the design of lighter vehicles, to improve gas consumption. Outside of automotive and aerospace[5] applications, other uses are prevalent such as parts for air-conditioner and refrigerator compressors, permanent magnets,[6] and even monetary coinage.[7]

There are numerous attractive benefits for powder metallurgy:

- Lack of machining eliminates scrap losses
- Facile alloying of metals
- In situ heat treatment is useful for increasing the wear resistance of the finished material
- Facile control over porosity and density of the green and sintered materials
- Fabrication of complex/unique shapes which would be impractical/impossible using other metalworking processes
- The rapid solidification process extends solubility limits, often resulting in novel phases.

Although we have described powder metallurgy as being an ideal process without disadvantages or hazards, it does pose some serious safety risks and limitations. Most metallic powders and other finely divided solids are *pyrophoric*, meaning that they will spontaneously ignite in air at temperatures below 55 °C. Unlike black powder, which contains both the fuel (C and S) and oxidizer (potassium nitrate), it is not immediately apparent why metallic powders would ignite, since both key components are not present within the powder matrix.

There are two primary reasons for this pronounced reactivity. The extremely large exposed surface area of powders relative to the bulk results in rapid oxidation upon exposure to air, especially for metals that form stable oxides such as aluminum, potassium, zirconium, etc. Also, there is enhanced internal friction among the individual micron- or nanosized individual particulates comprising the powder. Simply pouring the powder onto a table will yield sparks that may or may not be visible to the naked eye. Indeed, if one does not physically see the spark, he or she will soon know that there was one! As you would imagine, both pulverizing and pressing steps in powder metallurgy are especially dangerous, as the particles are forced into contact with one another and the equipment surfaces (another purpose for an added lubricant during compaction). NASA has published a technical paper that describes the production of rocket propellants[8]; this is definitely worth a read, to find out how one prepares mixtures of such reactive components.

3.1.2 Additive Manufacturing (AM)

In addition to pressing or using injection molding to fabricate a metal part from powders, a variety of other advanced fabrication techniques may be used. Additive manufacturing, also known as *3-D printing*, involves the spreading of a metal powder into layers, with high-energy sources used to melt and form the desired part.[9]

Although 3-D printing is most often associated with plastics, this fabrication technique is also widely used for ceramics and metals. The use of AM allows for the creation of parts via computer-aided design (CAD) software and hardware that forms the material layer-by-layer. This results in faster fabrication of finished parts with complex, customizable designs. By contrast, traditional powder metallurgy (PM) requires removal of material through machining, milling, carving, or shaping. An interesting case study for AM is the fabrication of rocket engine components by NASA and Aerojet Rocketdyne in 2013. Such a complex injector design would have taken more than a year to make by traditional processing, but was fabricated in less than four months via additive manufacturing with a 70% reduction in cost.[10] However, it should be noted that AM systems are not yet widespread in industry since they are typically much more expensive than PM and traditional machining processes.[11]

The primary consideration in AM techniques is the flowability of metal powders, which governs the precision and repeatability of parts fabrication. Three factors that control powder flow are the morphology, size, and size distribution of the metallic particles. As the average particle size decreases, the specific surface area increases. Hence, interparticle friction becomes significant, which lowers the overall flowability of the metallic powder. In contrast, the particle size distribution is related to the packing efficiency. However, as the particle size and distribution increase, the packing efficiency of the particles increases, which also lowers the overall flowability of the powder. Hence, the best powders for AM are composed of spherical particles with small dimensions and uniform size distribution.

Figure 3.15 provides a comparison between AM and PM for metal parts fabrication. Powder bed fusion (PBF) is the most common method for AM, which first consists of the deposition of a powder onto a starting surface known as the build platform. An inert gaseous atmosphere such as argon is used in the fabrication chamber to prevent oxidation of the starting metal powders. The deposited powder is subsequently melted or sintered using either a laser (laser sintering/melting, Fig. 3.16) or electron beam (electron beam melting. This process of powder deposition and melting/sintering continues until the desired part is formed from the bottom-up via a layer-by-layer process.[12] Complex shapes may easily be formed through this process, as demonstrated by Bondareva et al. for the fabrication of biodegradable FeSi bone-grafting scaffolds.[13] As we will discuss in Chap. 5, it is also possible to combine the metal powder with a polymer via inkjet 3-D printing,[14] photopolymerization, or extrusion process to form the desired composite part.

3.2 Metallic Structures and Properties

We are now ready to investigate a question that will be posed throughout this textbook: *What is the relationship between the microstructure of a material, and its overall properties*? If our world wishes to stay on its current path of unprecedented growth in areas of electronics, building materials, homeland security devices, and future

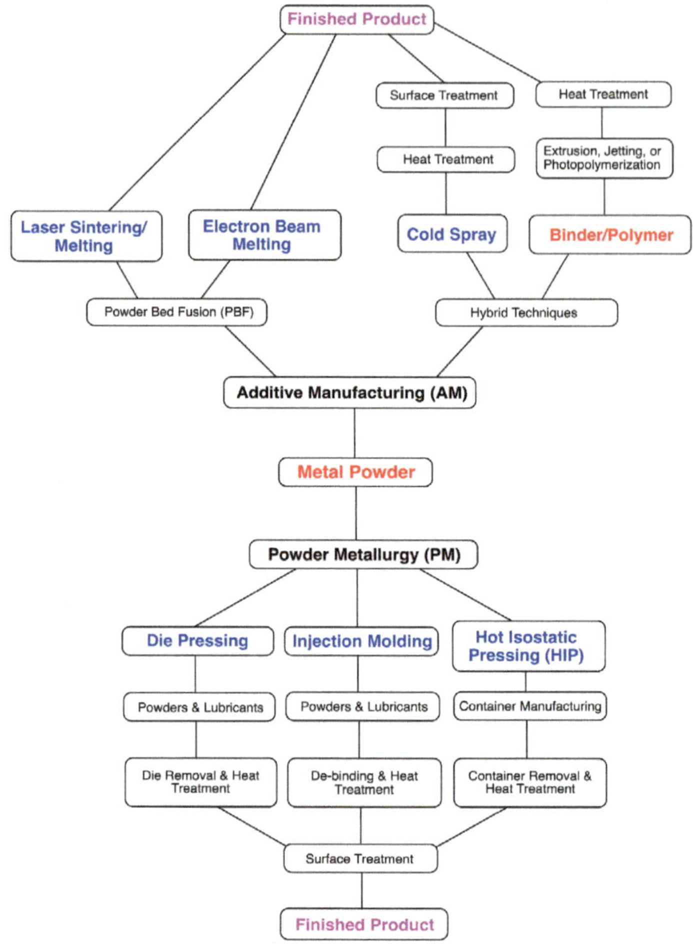

Fig. 3.15 Comparative schematic for powder metallurgy (PM) and additive manufacturing (AM)

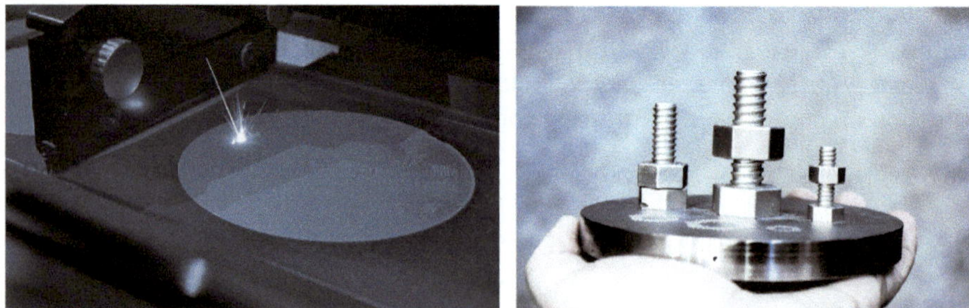

Fig. 3.16 Fabrication of metal parts via additive manufacturing. Shown is laser melting of a metallic powder bed (left) and finished parts that were fabricated via layer-by-layer laser melting (right). © MarinaGrigorivna/Shutterstock.com

"smart" materials, it is essential that we become familiar with the properties of individual classes of materials and current applications. Only then will we be able to extrapolate these properties into new and exciting applications for the future.

In Sect. 3.1, we saw that a wide variety of applications employ metallic substances (revisit Table 3.1). In this section, we will examine the various classes of metals and alloys in more detail, focusing on phase transitions, changes in the microstructure, and atomic packing of the materials. With this insight, you will be in a good position to evaluate why a particular metal is more suited than others for an existing or future application. It should be noted that certain organic polymers may also exhibit high electrical conductivities. However, this chapter will only discuss inorganic-based metallic classes; organic-based electrical conductors will be detailed in Chap. 5.

3.2.1 Phase Behavior of Iron-Carbon Alloys

In general, for a mixture of two or more pure elements, there are two types of solid-solution alloys that may be obtained. Type I alloys are completely miscible with one another in both liquid and solid states. As long as the Hume-Rothery rules are satisfied, a random or ordered substitutional alloy will be produced. We will see many examples of these alloys for a variety of metal dopants in stainless steels. By comparison, type II alloys are only miscible in the molten state, and will separate from one another upon cooling. These alloys are usually associated with compound formation from the alloying of metals or metals/nonmetals that are too dissimilar in their reactivities (e.g., Cu and Al to form $CuAl_2$ precipitates). The *eutectic* composition represents the lowest melting point of type II alloys.

Type I alloys contain two types of atoms that are arranged within a single lattice. When solidification of the solution begins, the temperature may be higher or lower than the freezing point of the pure solvent. Unlike a pure molten metal, most solid solutions will solidify over a temperature range due to differing diffusion rates of the metals *en route* toward their preferred crystal arrangement (Fig. 3.17).

Pure iron exists as a variety of allotropes depending on the external temperature or pressure. As the temperature is increased, iron undergoes allotropic transformations from α-Fe (ferrite, bcc) to γ -Fe (austenite, fcc), and finally to a narrow region of δ -Fe (bcc) before melting. As the temperature of the standard bcc crystal lattice is increased, thermally induced atomic motion increases, and it becomes more energetically favorable for atoms in the center of lattice unit cells to migrate into face-centered positions of neighboring unit cells (Fig. 3.18). However, as the magnitude of lattice vibrations continue to increase toward the melting point, the bcc structure is favored. This is due to the more open bcc structure being able to accommodate a larger range of vibrational motion than a relatively dense fcc array.

As seen earlier, the steps used to purify iron involves a carbonaceous material to remove the oxide-based impurities via exothermic formation of CO and CO_2. Hence, carbon will be pervasive in a variety of concentrations throughout all phases of iron and steels, present as an interstitial dopant within these lattices. Experimental

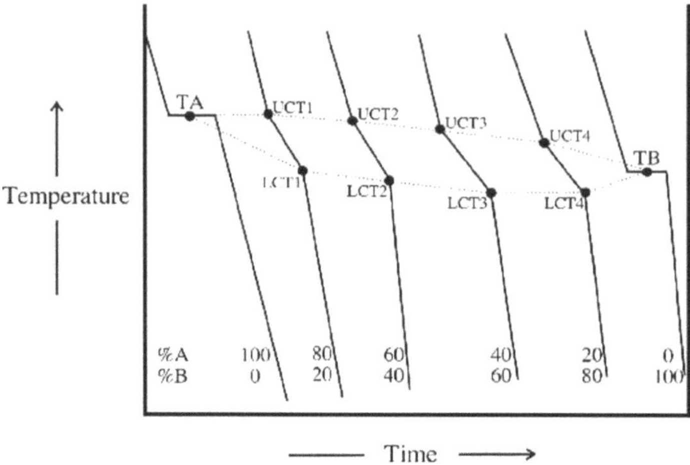

Fig. 3.17 Representative cooling curves for an alloy, A–B. The terms LCT and UCT refer to lower and upper critical temperatures, respectively. TA and TB designate the melting points of pure A and pure B, respectively

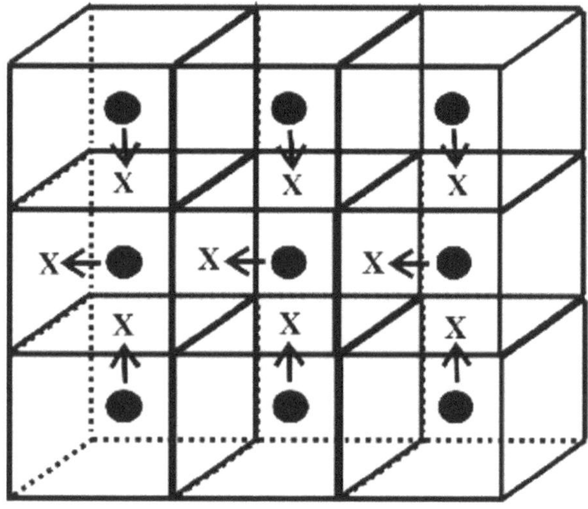

Fig. 3.18 Simplified schematic of the transformation from BCC to FCC, exhibited between the three allotropes of iron. Corner atoms have been omitted for clarity

evidence shows that carbon-doped iron polymorphs are indeed interstitial solid solutions. For instance, the carbon atoms in bcc ferrite are located only on empty face-centered positions. However, very few of these positions are occupied throughout the lattice, as the maximum solubility of carbon in α-Fe is only between 0.01 and 0.02 wt%. From a metallic-bonding standpoint, the addition of carbon in the lattice acts as an "electron sink," that may accept some of the delocalized electron density from the metallic lattice. This results in a stronger interaction among all atoms in the lattice

Table 3.2 Unit Cell Dimensions of Iron Allotropes and Fe–C Alloys[a]

Fe–C composition (crystal structure)	Unit cell parameters (Å)
α-Fe (ferrite, BCC)	$a = 2.8665$
γ-Fe (austenite, FCC)	$a = 3.555 + 0.044x$[b]
δ-Fe (BCC)	$a = 2.9323$
Martensite (tetragonal)	$a = 2.867 - 0.013x$[b]
	$c = 2.867 + 0.116x$[b]
Cementite (orthorhombic)	$a = 4.525$
	$b = 5.088$
	$c = 6.740$

[a]Values taken from Cullity, B. D. *Elements of X-Ray Diffraction*, 2nd ed., Addison Wesley: Reading, MA, 1978
[b]x = wt% C in interstitial sites of the iron lattice

that adds to physical hardness, but detracts from the overall electrical conductivity, relative to pure (undoped) iron.

Table 3.2 compares unit cell dimensions for the various allotropes and Fe–C alloys. Since the dopant species are entrained within individual unit cells, the volume of each unit cell will increase concomitantly with the concentration of carbon. Although the δ -Fe lattice is isomorphous with ferrite, the difference in volume between these allotropes corresponds to different concentrations of carbon in each solid solution. That is, δ-Fe contains an order of magnitude greater concentration of carbon than ferrite. Since a greater number of interstitial sites may be occupied by fcc unit cells relative to bcc, austenite may contain an even greater concentration of C in the lattice, up to 2.1 wt%. It must be noted that the trend of increasing volume with dopant concentration is not only exhibited by the Fe–C system, but is also followed by all other interstitial alloys that we will examine later.

In general, the density of interstitial solid solutions is given by (3.18). Since the change in volume is usually more significant than the increase in number of unit cell atoms, interstitial solids usually exhibit a decrease in density, relative to the pure allotrope. For instance, the density of pure iron (7874 kg m^{-3}) shows a significant decrease upon interstitial placement of carbon in cast irons (*ca.* 7400 kg m^{-3}).

$$\rho = \frac{1.6604 \sum(\mathrm{n}_1 \mathrm{A}_1 + \mathrm{n}_i \mathrm{A}_i)}{\mathrm{V}} \tag{3.18}$$

where n_1, n_i are the number of regular lattice and dopant atoms, respectively, per unit cell and A_1, A_i are the atomic weights of the regular lattice and dopant atoms, respectively.

The complex binary phase diagram for the Fe–C system is shown in Fig. 3.19, and illustrates a number of important transitions. In particular, as the temperature is increased from ambient to its melting point, pure iron exhibits a variety of allotropic changes. At room temperature, the ferrite form is most stable; conversion to austenite occurs at 910 °C. Austenite is much softer, being more easily formed into desired shapes, relative to ferrite. At a temperature of 1403 °C, the fcc γ -Fe allotrope converts back to the bcc form of δ-Fe before melting at 1539 °C. This behavior is quite atypical; most metals do not alter their crystal structure *en route* toward their melting points.

By definition, iron containing between 0.15 and 1.4 wt% C is typically referred to as *steel*.[15] Hence, although we typically think of steel as containing chromium and other metal dopants, some γ - and δ-phases of pure iron could also be considered forms of steel. Steels with a carbon concentration of 0.83 wt% undergo a transformation from austenite into two intimately mixed solid phases at a temperature of 723 °C. Although this phase transformation looks like a eutectic, the material above this point is a solid rather than a liquid. Hence, the transition is referred to as the *eutectoid*. As its name implies, alloys with carbon concentrations greater or less than the eutectoid are known as *hypereutectoid* or *hypoeutectoid* steels, respectively.

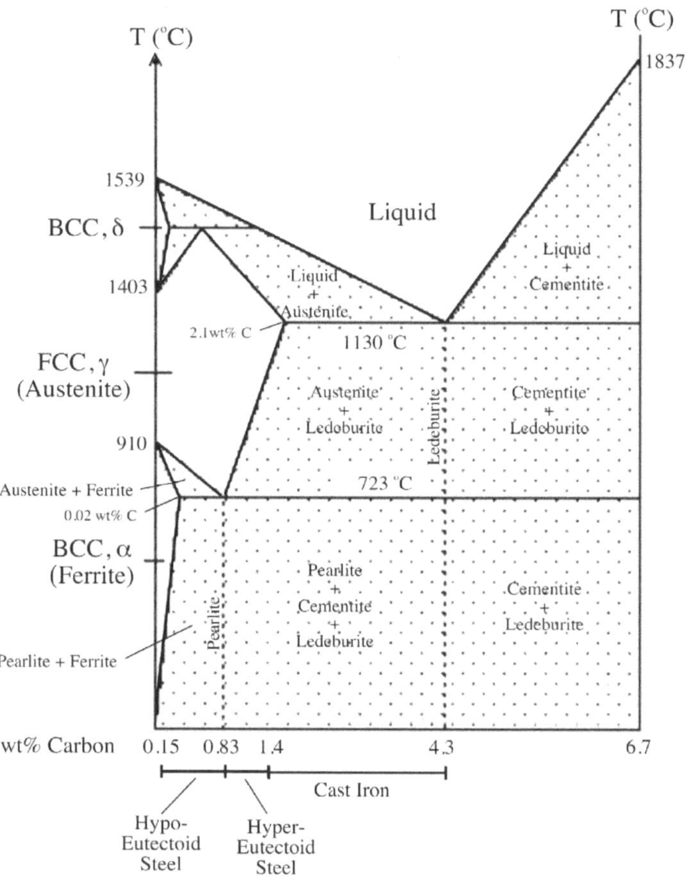

Fig. 3.19 The equilibrium phase diagram for the iron–carbon system

The eutectoid mixture of steel consists of a lamellar microstructure of soft/ductile ferrite and hard/brittle *cementite* (Fe_3C). Accordingly, this phase is known as *pearlite* since the interaction of light gives rise to a mother-of-pearl multicolored pattern when viewed through a light microscope. Cooling of austenitic steel at a higher cooling rate will yield a ferrite/cementite mixture in the form of needles or plates, known as *bainite*. Although the composition is identical to pearlite, the microstructure is markedly different (non-lamellar), which yields a stronger, more ductile alloy that is used for applications such as shovels, garden tools, etc.

Figure 3.20 illustrates the microstructural changes when austenitic steel is slowly cooled. For hypoeutectoid steel, ferrite begins to form along the austenite grain boundaries. Further cooling results in a ferrite-rich phase, with some remaining austenite crystals. At the eutectoid point of 723 °C, the residual austenite is converted to pearlite, yielding a phase that contains both ferrite and pearlite crystals upon further cooling.

By comparison, hypereutectoid steel contains significantly greater carbon concentrations; cooling results in the precipitation of the excess carbon in the form of cementite nuclei that form along austenite grain boundaries. In Chap. 2, we showed how polycrystalline aggregates, always found in pure metals and alloys, form grain boundaries due to misaligned crystallites. Since the bonding character of neighboring atoms is broken across the grain boundary, the diffusion of impurities occurs more readily in these areas, thus explaining the preferential nucleation and growth of cementite in these regions.

Considering the atomic weights of Fe and C, pure cementite corresponds to 6.7 wt% carbon. It has been determined experimentally that the strength of steel increases with carbon content up to the eutectoid composition, and then begins to drop as cementite nuclei are formed in the material. It should be noted that other Fe–C phases exist with greater carbon concentrations than cementite. However, Fig. 3.17 shows only the phases to the left of cementite that are technically useful for materials applications. Cementite is actually a metastable phase, with graphite representing the most stable form of carbon at equilibrium. However, it is difficult to obtain stable graphitic nuclei in steels due to the low concentration of carbon.

As a final note regarding the Fe–C phase diagram, the eutectic temperature corresponding to the minimum melting point of the Fe–C system is 1130 °C. As the liquid is cooled at the eutectic temperature, solidification of *ledeburite* will occur. The microstructure of ledeburite consists of tiny austenite crystals embedded in a matrix of cementite. At carbon concentrations less than the eutectic (i.e., 4.3 wt% C), ledeburite and austenite will form a solid solution. By contrast, increasing carbon concentrations will result in ledeburite/cementite solutions. At temperatures lower than the eutectoid (and carbon concentrations greater than the eutectoid), ledeburite will still be present alongside cementite or pearlite.

The incorporation of carbon into an iron lattice affects the interactions between neighboring iron atoms. As carbon is introduced at relatively low concentrations, the carbon atoms rearrange themselves within interstitial sites of the iron lattice, creating a strengthening effect. That is, metal atoms have a lesser range of movement due to the "glue" formed by interstitial carbon atoms. As a result, external forces such as temperature and pressure will not as readily cause atomic movement and surface/bulk deformation or fracturing. This is the reason why pure iron is not particularly hard or physically durable, but steels are significantly improved in these properties. Slow cooling of carbon-rich iron will yield a supersaturated solid solution. The carbon solubility in austenite decreases from about 1.7% at 1150 °C to about 0.7% at 715 °C, causing the precipitation of the excess carbon in the form of microscopic carbides or graphitic nuclei.

Supersaturated iron lattices yield a material known as *cast iron*, a ternary Fe–C–Si alloy containing much higher carbon than steel, typically around 3–5 wt% C. As the name implies, these materials are cast from their molten states into molds to yield the desired shapes. Due to the oversaturation of carbon present in these solids, cast iron is not suitable for structural applications. However, cast iron is extremely inexpensive to produce, making this material one of the most heavily used materials in industry for the manufacture of tools, valves, and automotive parts. Cast iron cookware has been employed for culinary applications since the late nineteenth century. However, with the advent of nonstick coatings such as Teflon™ in the 1940s, this application has largely been abandoned in favor of coated aluminum pans. A useful form of cast iron known as "Duriron," features a high silicon concentration (13–16 wt% Si, relative to standard cast irons with 1–3 wt% Si), which is resistant to strong acids and high temperatures.

There are a variety of cast irons, each differing with respect to the carbon impurity associated with austenite within the iron lattice. For instance, white and gray cast irons contain cementite and graphite nuclei within the microstructure, respectively. The graphitic suspensions may be present as flakes (gray cast iron), or as spheres (ductile and malleable cast iron) depending on the cooling conditions employed. For gray cast irons, the formation of iron carbide must be minimized to prevent localized hard spots that would degrade ductility and machinability.

A number of dopants may be added to facilitate the preferential formation of graphite rather than cementite. As we have discussed earlier, the excess carbon precipitated from supersaturated iron will most often yield cementite. This is especially intriguing, since the formation of graphite actually represents the lowest-energy alternative for the Fe–C system. However, as a carbon-rich pure Fe/C alloy is cooled, the localized density of carbon atoms is never enough to serve as a nucleus for graphite formation. Rather, since the carbon is distributed throughout the lattice, the intimate combination of iron and carbon atoms makes it relatively easy to form Fe_3C nuclei, relieving the supersaturation and lowering the overall energy of the system. On the other hand, if a dopant is added to serve as a nucleation site, the formation of graphite will occur due to more favorable thermodynamics.

It is proposed that the major nucleation mechanism in cast iron doping, known as *inoculation*, is the formation of sulfide species upon the addition of strong sulfide formers such as calcium, barium, cerium, or strontium. These sulfides possess lattice parameters very similar to the graphite crystal structure, serving as substrates for nucleation/epitaxial growth of graphite. Other common graphitizer dopants are Si (typically added as metal ferrosilicon compounds), Ni, and Cu; by contrast, Cr, Mo, V, and W are antigraphitizers, promoting the formation of carbides ($Cr_7C_3/Cr_{23}C_6$, Mo_2C/Fe_3Mo_3C, $VC–V_4C_3$, W_2C/Fe_3W_3C, respectively).

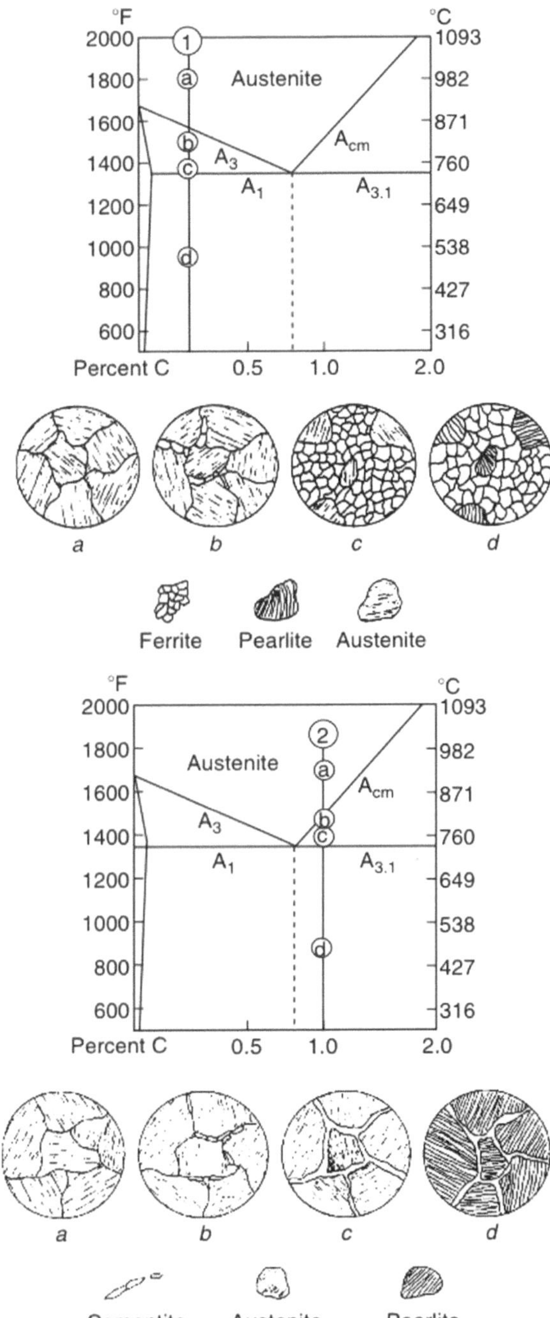

Fig. 3.20 Comparison of the microstructural changes upon very slow cooling of hypoeutectoid (upper) and hypereutectoid (lower) steel. Reproduced with permission from *Machine Tools and Machining Practices*, White, W.; Wiley: New Jersey, 1977. © John Wiley & Sons Limited

3.2.2 Hardening Mechanisms of Steels

In this section, we will describe the primary techniques that may be used to strengthen a metal. It should be noted that these methods are applicable to all metal classes, not just the iron alloys predominantly described herein.

Strain Hardening

An external pressure (stress) that is exerted on a material will cause its thickness to decrease. A *shear stress* is applied parallel to the surface of a material and may cause the sliding of atomic layers over one another. The resultant deformation in the size/shape of the material is referred to as *strain*, related to the bonding scheme of the atoms comprising the solid. For example, a rubbery material will exhibit a greater strain than a covalently bound solid such as diamond. Since steels contain similar atoms, most will behave similarly as a result of an applied stress. If a stress causes a material to bend, the resultant flex is referred to as *shear strain*. For small shear stresses, steel deforms elastically, involving no permanent displacement of atoms. The deformation vanishes when shear stress is removed. However, for a large shear stress, steel will deform plastically, involving the permanent displacement of atoms, known as *slip*.

Dislocation defects and thermal energy assist slip, allowing a sheet of atoms to slip gradually past one another. To stop slip, one can either lower the temperature, or spoil the crystal structure. A process referred to as *work hardening* (bending/hammering the cold material) is used to break up crystallites, introducing dislocations in the material. The pinning or entanglement of dislocations impedes their movement, which serves to strengthen the material. This is done at low temperature ($T \leq 0.5T_m$, where T_m = melting point), so the metal atoms cannot rearrange themselves, which would negate the effect. Consequently, low-melting metals such as Sn and Pb may not be cold-worked at room temperature. In contrast, hot-working is performed at temperatures above the recrystallization temperature of the metal. Although hot-working requires less energy and induces larger deformations than cold-working, most metals exhibit surface oxidation that may deleteriously affect its overall properties and applications.

Grain Size Hardening

As we have seen, it is not simply the carbon concentration, but rather the microstructure of Fe–C alloys that governs its physical properties. The size of the individual microcrystals (or grains) that comprise these aggregates greatly affects many properties of the bulk crystal. Both optical microscopy and X-ray diffraction are used to determine the grain sizes; most commercial metals and alloys consist of individual crystallites with diameters ranging from 10 to 100 μm, each corresponding to millions of individual metal atoms. A decrease in the size of these microscopic particles results in an increase in both strength and hardness of the bulk material. This can be understood by the tighter packing of smaller spheres relative to larger ones, effectively resisting atomic repositioning as a result of an external stress such as bending.

Heat treatment of iron alloys will affect the slip characteristics of the material through changes in crystallite sizes. As the grain diameters become larger through annealing, more grain boundaries will tend to form, resulting in a greater proclivity for slip deformations. In general, as the grain size is decreased, the strength of the alloy will increase. As an illustration, consider a bag filled with sand *vs.* a bag filled with marbles. The extremely small granules of sand will be packed more efficiently with respect to one another, providing a solid with significantly greater density and less opportunity for individual granules to slide past one another, or change their positions in response to a bending force (Fig. 3.21).

Precipitation Hardening

We are all familiar with the picture of a blacksmith withdrawing a red-hot iron from a furnace and hammering it into the desired shape. Although these early laborers were not familiar with the microcrystalline changes they were imposing, they knew through experience that heating/cooling regimes were effective means to improve many properties of steel such as hardness, toughness, ductility, machinability, and wear/stress resistance. The softening of steel at elevated temperatures is due to the formation of large iron carbide crystallites that may undergo facile slip deformations. The prolonged high temperature environment affects the microstructure through interrupting Fe–Fe and Fe–C bonds of either pearlitic or bainitic steel. This allows for

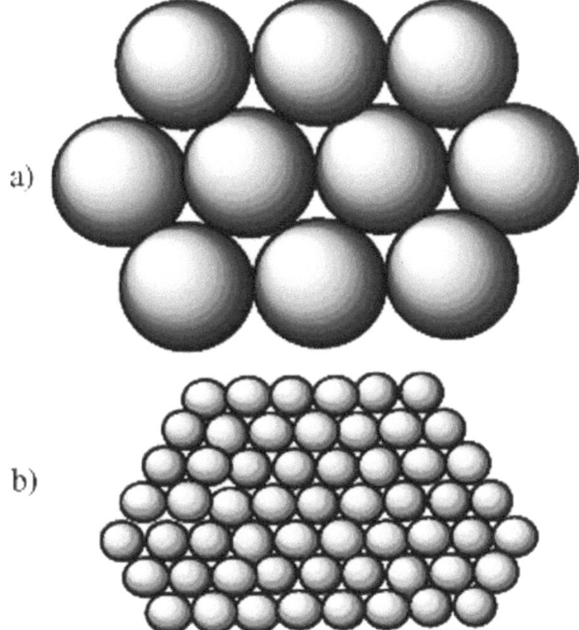

Fig. 3.21 Schematic of the effect of grain size on ease of atomic movement. If a metal consists of large grains shown in (**a**), much less external force would be necessary to cause atomic movement (*i.e.*, bending), relative to a metal comprising small grains shown in (**b**)

cementite (Fe_3C) regions to agglomerate into spheres, which are dispersed within a ferrite matrix—aptly referred to as spheroidite (Fig. 3.22). This process is an example of *precipitation hardening*, in which grain slippage is prevented through the formation of homogeneous suspensions of small, finely dispersed particles in a matrix (Fig. 3.23). Since this process occurs as the alloy ages, this is also referred to as *age hardening*. Examples of alloys that are hardened via precipitation formation include Al/Cu, Cu/Be, Cu/Sn, and Mg/Al. It should be noted that the heating of iron surfaces is now often achieved by more energetic sources such as electron beams or lasers. Such a focused thermal treatment allows for localized *surface hardening* to improve its wear resistance.

The phases of austenite, pearlite, and ferrite are relatively soft; hence, the observed high hardness of steels is obtained through processing of these materials. For instance, hypoeutectoid steel may be heated to form austenite and then slowly cooled so the cementite/ferrite phases may be worked into desired shapes. If the material is re-austenized and quickly quenched to room temperature, a very hard phase known as *martensite* is formed. Some of the remaining pearlite and ferrite phases (if present) would still remain in the matrix. Hence, quenching will result in pure martensitic steel only when the steel has been heated to temperatures sufficiently high to convert all of the ferrite into austenite. It should be noted that the martensite phase does not appear in the above Fe–C phase diagram since it is a non-equilibrium phase.

The temperature range is not the only crucial variable affecting the properties of the material, but rather the *rate* of heating and cooling. For example, when austenitic steel is very slowly cooled to room temperature, the resulting solid will be soft

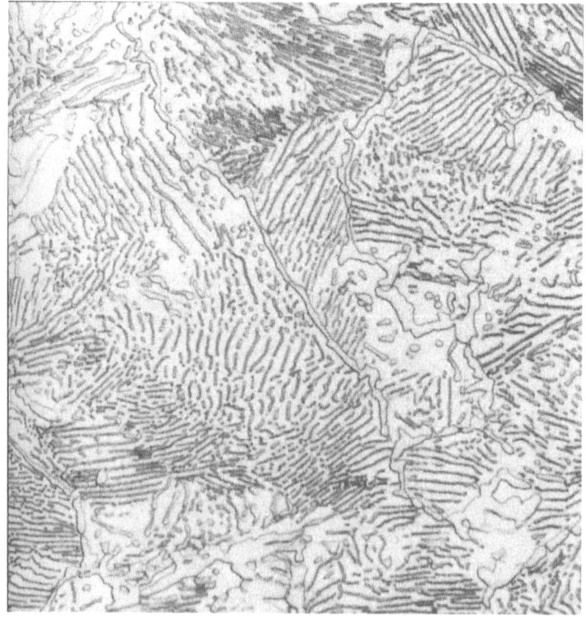

Fig. 3.22 Optical micrograph of a pearlitic steel (note the lamellar regions) that has been partially transformed to spheroidite. Image taken at 2000 × resolution. Photograph courtesy of U.S. Steel Corporation

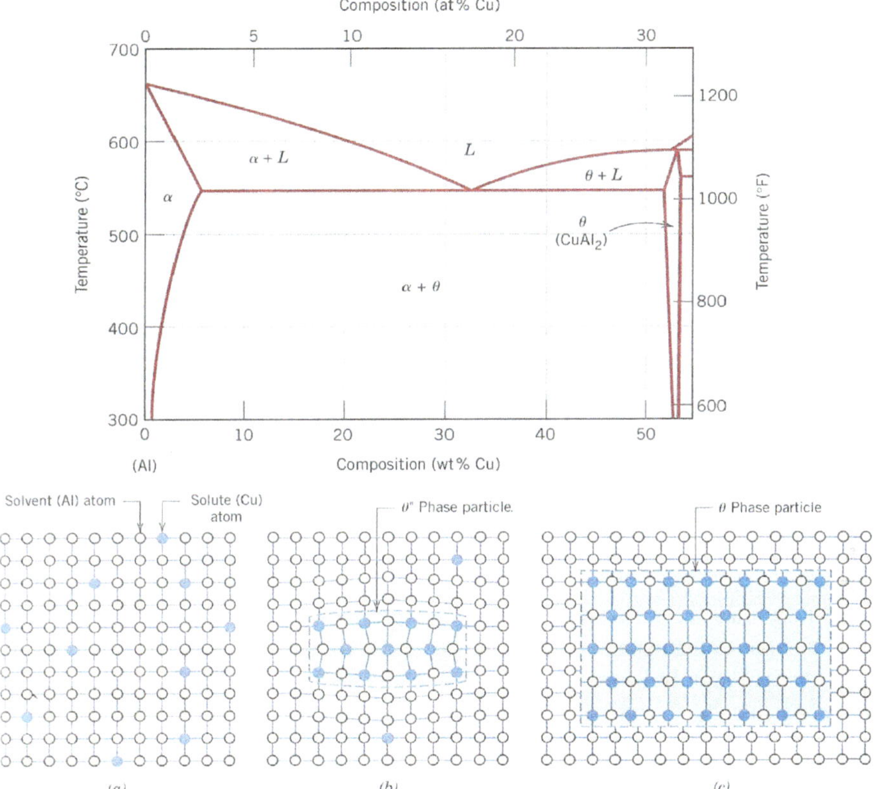

Fig. 3.23 Top: The Al-rich side of the Al-Cu phase diagram. Bottom: schematic of the formation of the equilibrium precipitate s phase: **a** a supersaturated α solid solution, **b** a transition, θ ", precipitate phase, and **c** the precipitate θ phase within the -matrix phase. Reproduced with permission from Callister, W. D. *Materials Science and Engineering: An Introduction*, 7th ed., Wiley: New York, 2007. © 2007 John Wiley & Sons, Inc

and malleable. However, when the same steel is rapidly quenched in cold water to temperatures less than 250 °C, the normal phase transformations to pearlite/ferrite or pearlite/cementite (depending on %C present) are suppressed. Rather, the γ -Fe phase is converted to *martensite*, the hardest and strongest of all possible Fe–C microstructures. Interestingly, the formation of martensite may also occur through fast-quenching of other austenitic mixtures, such as ferrite- or cementite-rich austenite (3.19–3.20):

$$\{\gamma\text{–Fe} + \alpha - \text{Fe}\} \xrightarrow{\text{rapid cooling}} \text{martensite} + \alpha\text{–Fe} \quad (3.19)$$

$$\{\gamma - \text{Fe} + \text{Fe}_3\text{C}\} \xrightarrow{\text{rapid cooling}} \text{martensite} + \text{Fe}_3\text{C} \quad (3.20)$$

To summarize, the comparative Brinell hardness values of common Fe phases are: martensite (460–710) > cementite (*ca.* 550) > tempered martensite (300–450) > bainite (*ca.* 400) > fine pearlite (100–300) > coarse pearlite (100–220) > spheroidite (90–180) > ferrite (80–90). Whereas the hardness and brittleness of cementite is much greater than ferrite, the latter has significantly greater ductility.

In order to predict the resultant phase arising from varying the cooling rate of austenitic steel, one would use a *time–temperature-transformation* (TTT) curve (e.g., Fig. 3.24). To generate a TTT diagram, thin metal specimens are annealed at a sufficient temperature to form austenitic steel. This temperature is then held for a specified duration to ensure full conversion of the microstructure to austenite. This isothermal equilibrium step allows the metal carbides to fully dissolve in austenite; incomplete conversion of carbides will result in ferrite grains that will ultimately weaken the material. The austenized specimens are then removed at specific times and quenched in cold water. Using optical and electron microscopies, the microstructure of the products is determined (*i.e.*, pure martensite, ferrite, pearlite, etc.). As one would expect, a large number of samples are needed to determine the time intervals required for the initial and full transformation of austenite to other phases, making this process an extremely labor-intensive exercise.

As shown in Fig. 3.22a, cooling austenitic steel at a rate fast enough to avoid the nose of the transformation curve (i), will result in 100% martensite. Cooling curve (ii) is tangent to the nose of the TTT plot. This will also result in 100% martensite, but with significantly lower internal stresses and distortions than (i). Such a tangent line represents the slowest cooling rate that prevents formation of the non-martensitic decomposition products of pearlite or bainite, referred to as the *critical cooling rate* (CCR). A tangent to the 50% transformation curve (cooling curve (iii)) will result in a mixture of 50% pearlite and 50% martensite. Cooling curve (iv) represents relatively slow cooling to *ca.* 600 °C, where 50% of the austenitic steel is converted to pearlite (i.e., still above the pearlite/bainite threshold temperature of 538 °C). Subsequent fast cooling to *ca.* 450 °C, and isothermal equilibration for a few hundred seconds converts the remaining austenite into bainite. Hence, the final product from (iv) will be 50% pearlite and 50% bainite. Cooling curve (v) will result in 100% bainite, since fast cooling to *ca.* 500 °C avoids the nose of the TTT diagram and is held isothermally below the pearlite/bainite boundary. Finally, cooling curve (vi) illustrates very slow cooling (*e.g.*, furnace cooling), which will result in 100% pearlite.

Solute Hardening

The introduction of dopants into a metallic lattice will induce the formation of alloy crystallites. These species will impede slip by increasing the lattice energy in the vicinity of the dopant atoms. Strengthening occurs since more work is required to propagate a dislocation through these areas. In particular, if the alloying agent is carbon, hard crystallites of iron carbide may form, thereby changing its microstructure. By comparison, austenite usually does not contain iron carbide, and is quite susceptible to slip.

From an analysis of various types of steels, only the following carbides will be present: Fe_3C, Mn_3C, $Cr_{23}C_6$, Cr_7C_3, Fe_3Mo_3C, Fe_3W_3C, Mo_2C, W_2C, WC, VC,

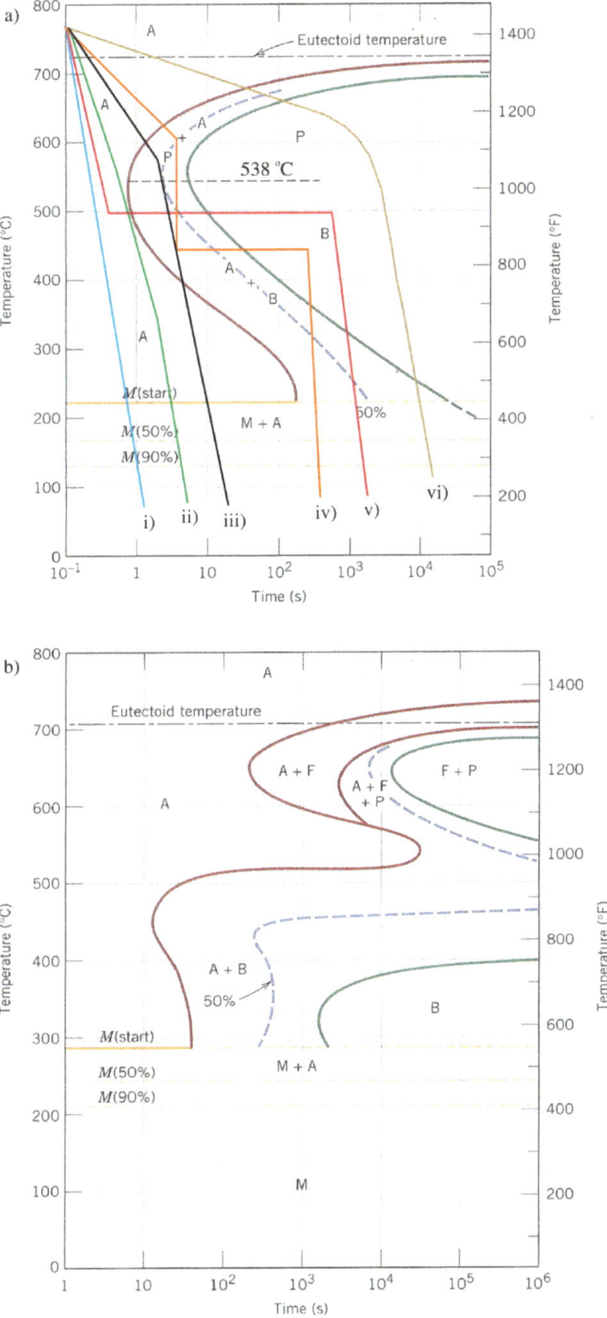

Fig. 3.24 Time–temperature-transformation (TTT) diagrams for **a** austenitic steel, and **b** an alloy steel (type 4340); A = austenite, B = bainite, P = pearlite, M = martensite, F = proeutectoid ferrite. Reproduced with permission from Callister, W. D. *Materials Science and Engineering: An Introduction*, 7th ed., Wiley: New York, 2007. © 2007 John Wiley & Sons, Inc

TiC, NbC, TaC, Ta_2C, and ZrC. The occurrence of these species will depend on the type and concentration of the transition metal dopants within the iron lattice. For interstitial carbides, the size of the metal atoms will govern the type of carbide formed. In general, the metal radius must be > 1.35 Å (*e.g.*, Ti, Zr, Hf, V, Nb, Ta, Mo, and W) to generate an interstitial vacancy large enough to accommodate C atoms. Metals with smaller radii (e.g., Cr, Mn, Fe, Co, Ni) do not form M_xC_y species, and form carbides with relatively complex crystal structures (Fig. 3.25). It should be noted that metal carbides do not generally exist as isolated pure species. That is, carbides of all alloying elements will exist as clusters that also contain iron. Further, when several carbide-forming dopants are present that share the same crystal structure, the resultant carbide will exist as a combination of those elements. As an example, steel containing Cr and Mn dopants will contain particulates of the complex carbide $(Cr, Mn, Fe)_{23}C_6$, rather than isolated $Cr_{23}C_6$ and Mn_3C species.

We have seen that only certain transition metals will form stable carbides; as a relevant digression, let us consider the chemical rationale behind such reactivity. The general trend for increasing carbide-forming ability of transition metals is:

$$\text{Fe} < \text{Mn} < \text{Cr} < \text{Mo} < \text{W} < \text{V} < \text{Nb} < \text{Ta} < \text{Ti} < \text{Zr} < \text{Hf}$$

If one follows this sequence using the Periodic Table, a preference is identified for early transition metals that are relatively electron deficient. As you may recall, the valence shell of zero-valent transition metals in a crystal lattice is $[(ns^2)((n-1)d^x))]$. In the bulk solid state, the outer *s* electrons are completely delocalized, whereas the wave functions of the *d* electrons remain localized on the respective metal atoms. When a carbon atom enters the crystal field it behaves as a ligand toward the metal, with the ligand and metal electrons electrostatically interacting causing the d orbitals to lose their original degeneracy.[16] Since this is an electrostatic effect, stronger metal–carbon bonds will result from more diffuse metal d orbitals (5d *vs.* 4d *vs.* 3d), and metals with fewer *d* electrons (i.e., both corresponding to less electron–electron repulsions between ligands and the metal).

The transference of electron density from the metal to carbon will result in the formation of a strongly polar covalent bond between carbide ions ($C_x{}^{n-}$) and transition metal ions. It should be noted that only the portions of the alloying elements and carbon that cannot be dissolved in austenite at a given temperature may be used in carbide formation. Two types of carbides are possible. If the atomic radius ratio of carbon:metal is <0.59, an interstitial phase will result; otherwise, carbides of complex compositions will form, having a different crystal lattice from the host metal. The latter carbides are characteristic of Cr, Mn, and Fe, which have five or more *d* electrons and a correspondingly weaker interaction with ligands such as carbon. Hence, such complex carbides (e.g., Fig. 3.25a–e) will have lower melting points and hardness than analogous interstitial carbides of the 5d and/or early transition metals (e.g., W_2C used as cutting blades).

It is interesting to note that alloying metals present in steel will cause stark changes in TTT diagrams (Fig. 3.24b). In particular, the nose of the austenite/pearlite transformation will be compressed, allowing for pearlite (and martensite) formation at slower

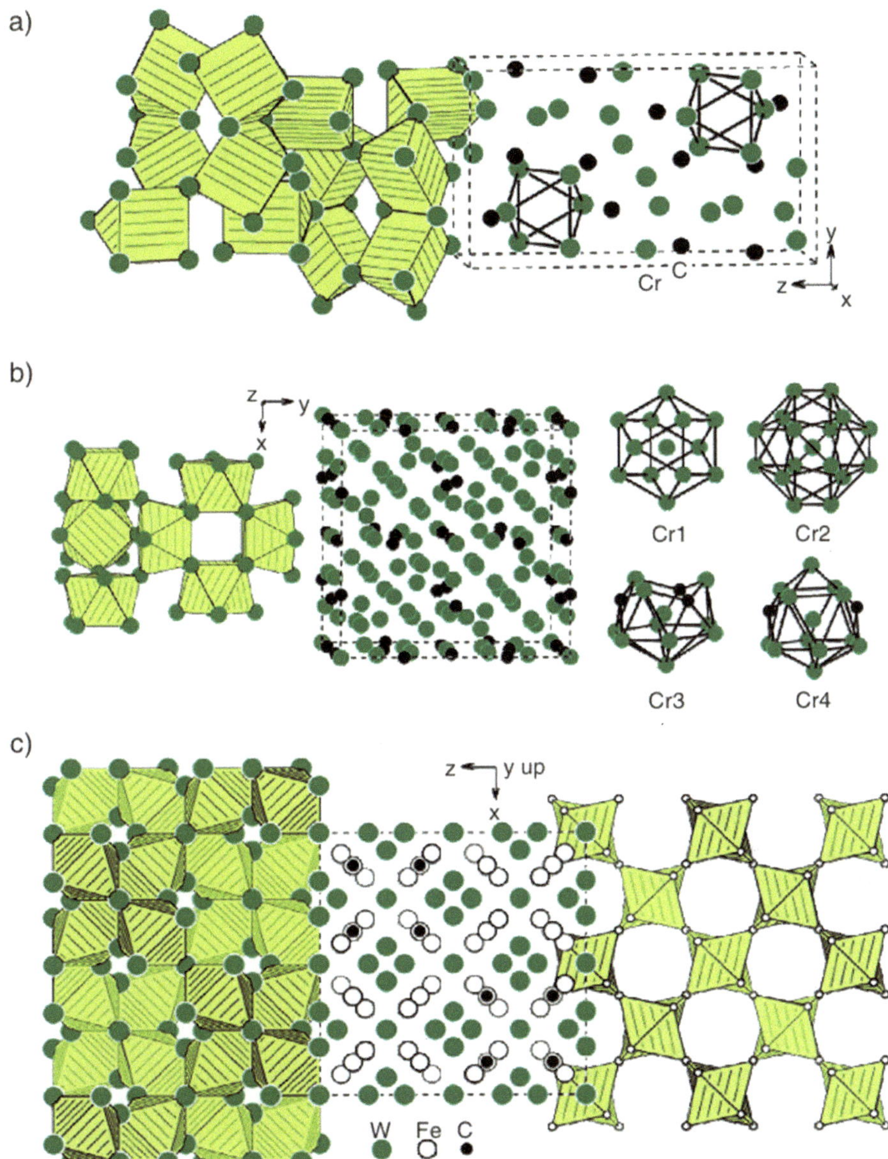

Fig. 3.25 Crystal structures of typical carbides in steels. Shown are **a** orthorhombic Cr_7C_3 (space group: *Pnma*), **b** cubic $Cr_{23}C_6$ (space group: *Fm3m*), **c** cubic Fe_3W_3C (space group: *Fd3m*), **d** orthorhombic Fe_3C (space group: *Pnma*), and **e** cubic TiC (space group: NaCl) and hexagonal W_2C (space group: CdI_2). Reproduced with permission from *Handbook of Ceramic Hard Materials*, Riedel, R. ed., Vol. 1. © 2000 Wiley–VCH (http://www.hardmaterials.de/html/_crystal_structures.html)

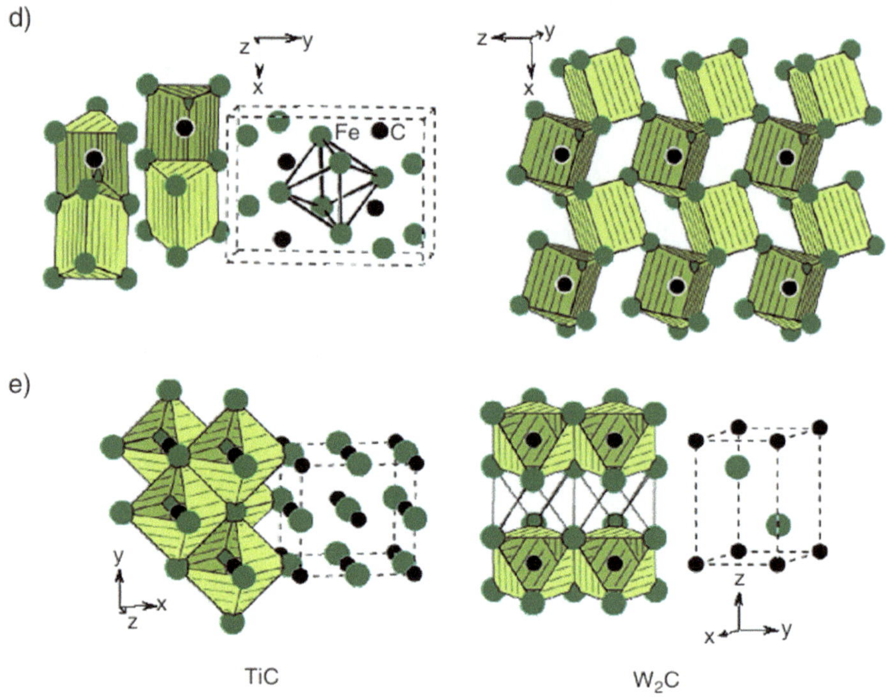

Fig. 3.25 (continued)

cooling rates. In addition, a second bainitic nose will appear in the TTT diagram. For a given steel with constant %C, as the concentration of metallic dopants increase, there will be a decrease in the temperature required for the onset of martensite formation, M_s (3.21):

$$M_s(°C) = 539 - 423(\%\,C) - 30.4(\%\,Mn) - 17.7(\%\,Ni) - 12.1(\%\,Cr) - 7.5(\%\,Mo) \tag{3.21}$$

The greatest effects are seen for the austenite-forming elements of C, Mn, and Ni, where even small concentrations result in a sharp decrease in M_s. Whereas pure γ -iron may be converted to martensite at temperatures > 500 °C, hypereutectoid steel is not transformed to martensite until a temperature of *ca.* 160 °C is reached during quenching. For steels with carbon concentrations > 0.7% and/or high dopant concentrations, martensite may be formed at temperatures well below 0 °C. Hence, high-carbon steels must be quenched in low-temperature media (e.g., dry ice/acetone, liquid nitrogen) to ensure full conversion of austenite to martensite.

The effect of alloying elements may be understood by examining the austenite/ferrite regions of the Fe–C phase diagram with respect to the alloy concentrations (Fig. 3.26). Ferrite-forming elements such as Al, Si, W, Cr, and Mo result in a contraction of the austenite region, forming a gamma loop. By contrast, austenite-stabilizing elements such as C, N, Mn, Ni, and Cu cause an expansion of the austenitic phase boundary. Austenite stabilizers inhibit the nucleation and growth of ferrite and pearlite/bainite phases, assisting in the formation of pure martensite upon quenching. In general, since bcc ferrite contains more voidspace than fcc austenite, larger interstitial dopants may be incorporated into these lattices. Hence, ferrite stabilizers tend to be larger in contrast to the smaller size of austenite stabilizers. Most importantly, in accord with the "like dissolves like" principle, bcc and fcc dopants will tend to stabilize ferrite and austenite, respectively.

A slow cooling rate will give the greatest opportunity for controlled atomic migration within the lattice, and growth of large ordered crystallites. However, in the rapid non-equilibrium conditions used to form martensite, there is no time for carbon diffusion to occur; this yields a supersaturated solution, with > 2 wt% C present as an interstitial impurity. During this phase transition, the fcc lattice of austenite is transformed to a distorted bcc lattice, commonly referred to as body-centered tetragonal bct (Fig. 3.27). The degree of distortion from a perfect bcc lattice (cubic lattice axes ratio, $c/a = 1$) is amplified with increasing carbon concentration (3.22):

$$\frac{\mathrm{c}}{\mathrm{a}} = 1 + 0.045(\mathrm{wt}\,\%\,\mathrm{C}) \tag{3.22}$$

The fcc–bct conversion, known as the *Bain transformation*, is a diffusionless process. That is, unlike the previous high-temperature conversions we discussed earlier (*e.g.*, austenite to ferrite), martensite can form at temperatures significantly below room temperature, within 1×10^{-7} s. Such a fast growth rate precludes the deciphering of the exact mechanism for the nucleation and growth of martensite. However, leading theories suggest that the growth initiates from dislocations in the solid.[17]

Tempering

Even though martensite exhibits a high hardness, the as-quenched material is much too brittle and highly stressed for structural applications. The ductility and toughness of martensite is greatly improved through post-annealing, a process known as *tempering*. This process relieves stresses in the solid through conversion of bct martensite into bcc ferrite, with precipitation of iron carbide particulates. It is important to note that the annealed structure is not simply pearlite/ferrite, but is best referred to as *tempered martensite*. During the annealing process for martensite, a number of key transformations occur:

(i) 50–250 °C: Interstitial carbon atoms in martensite begin to diffuse within the bct lattice. This results in precipitation hardening from the formation of hexagonal ε -iron carbide crystallites, Fe_xC ($2 < x < 3$). The martensite bct crystal lattice begins to lose its tetragonality.

(ii) 250–350 °C: Decomposition of retained austenite to fine aggregates of ferrite plates (acicular ferrite) and cementite crystallites. This intermediate microstructure is referred to as *bainite*, nucleating from the surface of ferrite

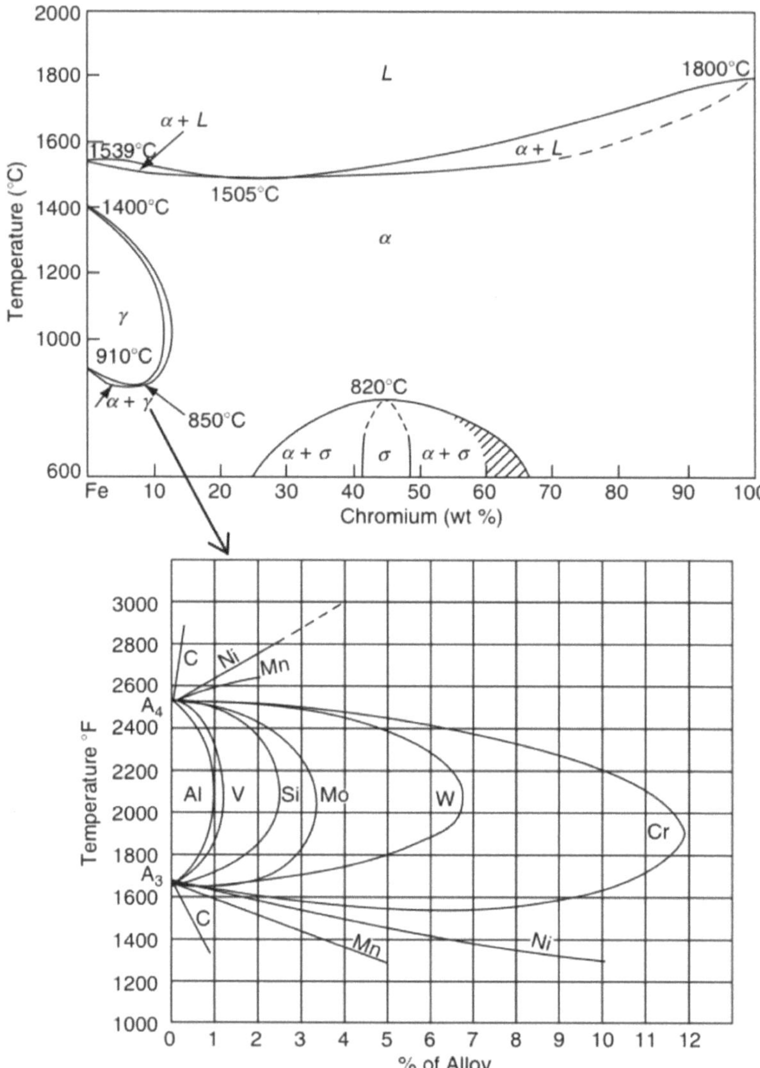

Fig. 3.26 The phase diagram for the Fe–Cr system, illustrating the expanded ferrite region, relative to the normal Fe–C diagram. Below is shown the "gamma loop" illustrating the size of the austenite region from the presence of various dopants. Reproduced with permission from (top) *Steels: Microstructure and Properties*, Honeycombe, R. W. K.; Bhadeshia, H. K. D. H.; 2nd ed.; Copyright 1995 Elsevier, and (bottom) *Basic Metallurgy: Volume I, Principles*, Grosvenor, A. W.; 3rd ed.; American Society for Metals: Cleveland, OH, 1958. (http://www.asminternational.org)

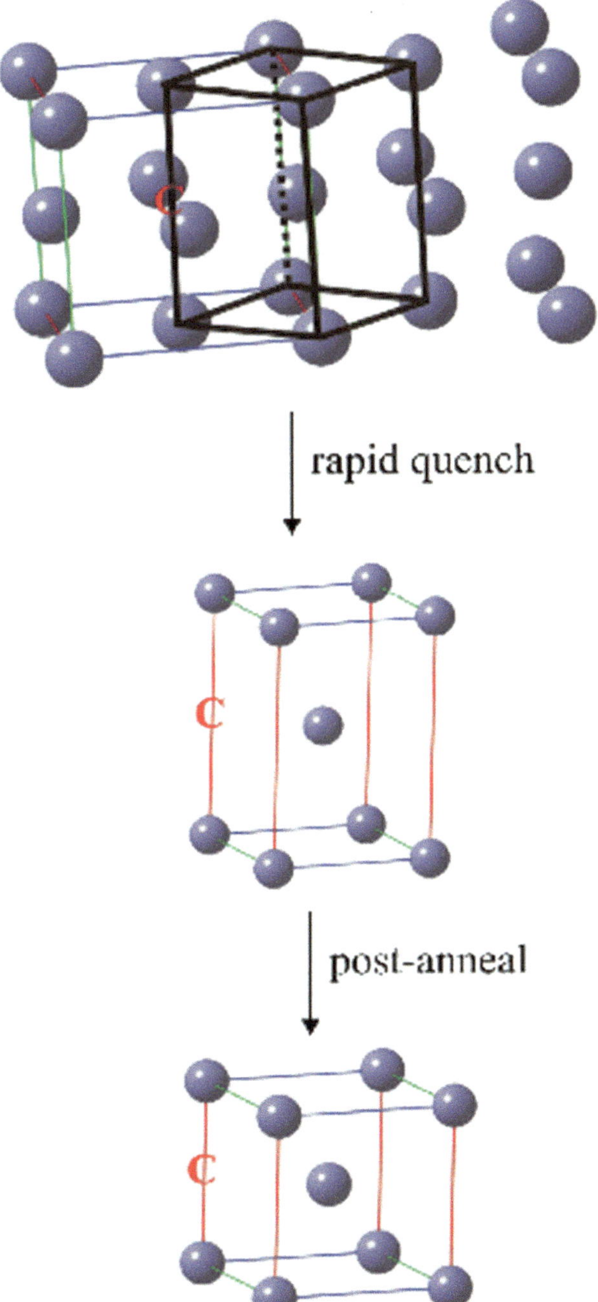

Fig. 3.27 Crystallographic representation of the phase transformation from austenite to martensite. Two neighboring fcc unit cells of austenite associate, resulting in a body-centered tetragonal (bct) unit cell. A postanneal known as tempering converts the bct structure to α -Fe. Also shown is the placement of an interstitial carbon atom, remaining in an octahedral site among lattice iron atoms

crystallites. As the temperature approaches 300 °C, cementite begins to dominate the microstructure at the expense of the ε -Fe_xC particles. At temperatures approaching 350 °C, the bct lattice is transformed to ferritic bcc.

(iii) 350–700 °C: The cementite particles undergo a coarsening process (between 300 and 400 °C), and spheroidization (near 700 °C). These processes drastically reduce the hardness of the material, but improve overall ductility and brittleness characteristics (more desirable for particular applications).

Hence, to obtain high-strength tempered steels, it is essential to anneal at low temperatures (< 350 °C). As one might expect, the presence of alloying elements will have a dramatic effect on the microstructure during annealing. For instance, if 1–2 wt% Si is present, ε -iron carbide particles are stabilized up to temperatures of 400 °C, yielding a much harder material at elevated temperatures. Further, transition metals such as Cr, Mo, V, W, and Ti will form stable carbides with higher enthalpies of formation than Fe_3C, typically at temperatures between 500 and 600 °C. A high temperature is required due to the relatively low diffusivity of the alloying elements that must substitutionally diffuse through the iron lattice. By contrast, interstitial dopants such as C, N, and B move between the iron lattice sites with a much greater diffusivity. It is important to note that the alloy carbides remain as fine suspensions even after prolonged tempering. This results in substantial strengthening referred to as *secondary hardening*.

One serious drawback of the above austenization/rapid quench method for martensite formation is the possibility of distorting and cracking the metal due to the rapid cooling event. During the quenching process, thermal stresses arise from the varying cooling rates experienced by outer and interior areas of the steel. In addition, there is a volume change when austenite is transformed to martensite. Two methods that have been used to reduce quenching stresses are *martempering* and *austempering* (Fig. 3.28). Martempering allows the transformation of austenite to martensite to take place at the same time throughout the structure of the metal part. By using interrupted quench, the cooling is stopped at a point above the martensite transformation region to allow sufficient time for the center to cool to the same temperature as the surface. Then cooling is continued through the martensite region, followed by the usual tempering process. By comparison, in austempering, the austenized steel is quenched at a rate faster than that required for pearlite formation, but above the temperature required for martensite growth. Hence, rather than transforming to martensite, the center and surface are converted to bainite—a strong material that shares the hardness of martensite with the toughness of pearlite.

Surface Hardening

The above changes in the microstructure upon annealing do not only apply to the bulk material, but also for the surface. If an iron material is placed at high temperature in the presence of carbon vapor, a procedure known as *carburization* occurs, where carbon atoms diffuse into the surface of the steel, increasing the surface hardness. There must be careful control of the annealing atmosphere; if the steel is brought

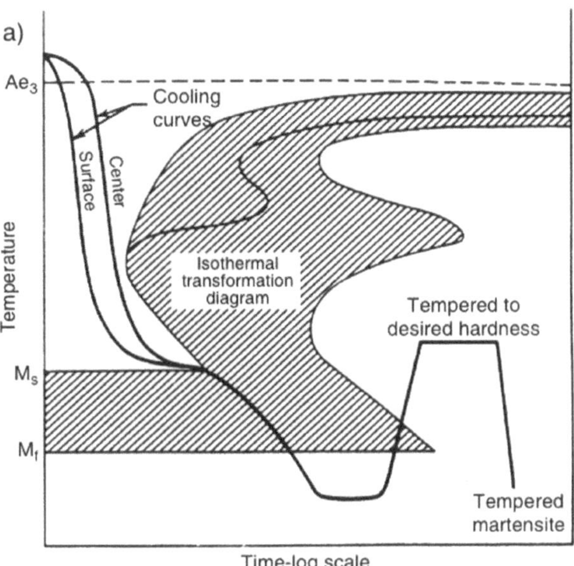

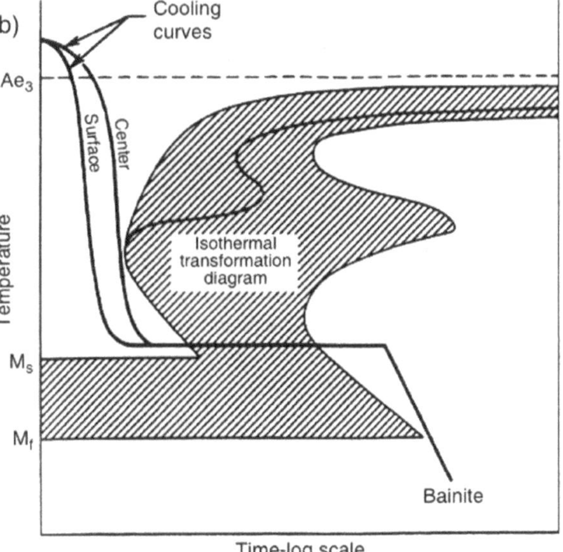

Fig. 3.28 Schematic of **a** martempering and **b** austempering processes for steel. Courtesy of the International Steel Group and Mittal Steel Company (http://www.mittalsteel.com)

into contact with an oxidizing atmosphere, decarburization of the surface will occur through preferential formation of CO_x.

Other surface hardening techniques introduce nitrogen at elevated temperature to form stable nitrides; for instance, in steels that contain metals such as Al, Cr, or V (3.23). Surface hardening techniques add a variety of attractive properties to steel components such as increasing its resistance to wear, stress, fracture, and corrosion.

$$4\,Cr + 3\,NH_3 \xrightarrow{600\,^\circ C} Cr_4N_3 + \frac{9}{2}H_2 \tag{3.23}$$

Strengthening of the exterior of a material may be achieved through either diffusional incorporation of dopants (e.g., B, C, N), or selective portions of the surface using flame, induction, laser, electron beam, or ion bombardment. Strengthening occurs through either grain size reduction or solute hardening, but occurs only on the periphery of the material due to the controlled, limited exposure.

The typical method used to assess the strength of a metal is a *tensile test*, where the metal is clamped to upper and lower jaws and pulled until it fractures. Figure 3.29 illustrates typical stress *vs.* strain curves for iron, steel, and hardened steel. The elastic limit (EL) is the greatest stress that a material may withstand and still revert back to its original shape when the stress is removed. In general, the closer a material is to its elastic limit, the longer it will take for the material to subsequently return to its original size/shape. For a metal within its EL the crystal lattice will be elastically lengthened, while becoming thinner at right angles to the applied stress. The ratio of lateral change to the change in length is referred to as *Poisson's ratio*.

When the elastic limit of a metal has been exceeded, it will undergo plastic flow beginning at the yield point. It is this property of metals that is exploited for cold- and hot- working into desired shapes. When a metal is deformed permanently from the tension force, it exhibits a property known as *ductility*. By comparison, the term *malleability* refers to the permanent deformation of a metal under a compression force (*e.g.*, hammering, cold-rolling, etc.). Although most ductile metals are also malleable, the reverse is not always true. For example, lead is extremely malleable but is not easily drawn into a wire without the use of die-extrusion techniques. It should be noted that excessive cold-working of metals may cause brittleness, where the metal will fracture rather than exhibiting plastic flow under stress. Brittleness may also be caused by a high concentration of carbon in the metal lattice (*e.g.*, cast iron), which explains the lack of structural applications for these metals.

As a metal transitions from ductility to brittleness, dire consequences may occur. From 1940–45, the U.S. started an emergency ship-building program to supply over 2,700 ships to European countries fighting Nazi Germany during WWII. However, over 1,500 of these ships sank due to sudden fractures, with 19 ships broken in half (Fig. 3.30)! A number of factors are thought to be responsible for the catastrophic failure of the so-called "Liberty Ships".[18] The low temperatures of the Atlantic Ocean likely produced the ductile-to-brittle transition (DBT, also known as "cold shortness"). As a consequence, the steel was not able to undergo dislocation gliding

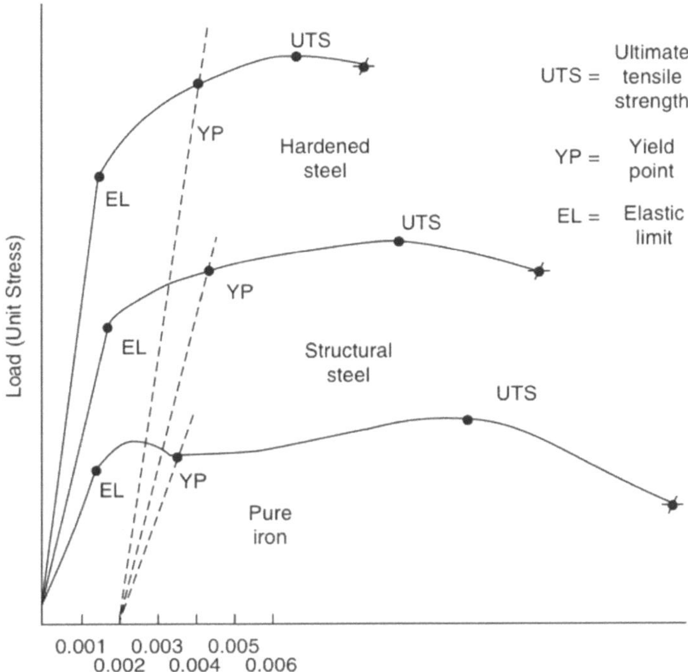

Fig. 3.29 Comparative *stress vs. strain* curves for pure iron, steel, and hardened steel. Reproduced with permission from *Practical Metallurgy and Materials of Industry*, Neely, J. E.; Bertone, T. J., 5th ed., Prentice-Hall: New Jersey, 2000

to react to stresses, which caused crack formation. The wartime steel shortage also likely played a role with low-quality steel primarily used for shipbuilding during that time. Furthermore, the weld quality was probably suspect, with large panels welded together instead of using rivets. This allowed for the propagation of cracks along welded joints; in contrast, rivets would have been more effective in stopping the spread of cracks at the joints between two plates. Although DBTs are most common for iron alloys, but can also occur in all bcc, and some hcp metals as well as non-metallic crystals and polymers. Interestingly, there is no DBT that has been observed in fcc metals, which stay ductile even at very low temperatures. The mechanism of the DBT is still largely unclear despite numerous experimental and theoretical studies.[19]

3.2.3 Stainless Steels

Technological advancements in iron ore processing and metal doping have resulted in the fabrication of many types of high-strength steels for diverse applications.

Fig. 3.30 The S.S. Schenectady on the morning of Jan. 17, 1943 after suddenly cracking while moored at the fitting dock at Swan Island. © U.S. GPO

By contrast, earlier generations worked exclusively with wrought iron, an inferior material containing >20 wt% C—formed through simple annealing of the ore with coal. We have seen that the concentration and form (e.g., Fe_3C, graphite, etc.) of dopant species will alter the physical properties of the material. Hence, the bulk and/or surface thermal and physical processing of steels is the most important consideration for ultimate material performance, as these treatments greatly affect its microstructure.

Thus far, we have focused primarily on Fe–C alloys, with carbon atoms positioned within vacant interstitial sites within the iron lattice. As you may expect, a variety of other elements may also be present in steel that will alter its overall physical properties. For example, all steels contain manganese that assists in hardening mechanisms, as well as facilitating the removal of sulfur and oxygen atoms in the matrix. This prevents FeS formation and removes bubbles in the molten state of steels, both of which would greatly contribute to brittleness of the final product.

Typically, large transition metal dopants will exist as substitutional alloys, randomly replacing iron sites throughout the lattice. Steel containing < 0.30 wt% C and chromium concentrations >10.5 wt% are referred to as *stainless steels*. The addition of Cr results in the formation of a native layer of Cr_2O_3, providing corrosion resistance. As the concentration of Cr is increased, the material is concomitantly less predisposed to rust. Such protection occurs from the comparative oxidation potentials between Cr and Fe (3.24–3.25). For redox processes, the spontaneity (Gibbs free energy) is governed by (3.26), where a negative ΔG indicates a spontaneous reaction at equilibrium. If both chromium and iron are present together, chromium atoms are considered as a *sacrificial anode*, being preferentially oxidized leaving the iron untouched. Other metals with large positive oxidation potentials such as Zn (+0.763 V), Al (+1.68 V), Ni (+0.257 V), and Ti (+2.00 V) are also useful additives that serve as corrosion barriers.

$$Fe \rightarrow Fe^{3+} + 3e^- \; E^\circ = -0.331\,V \tag{3.24}$$

$$Cr \rightarrow Cr^{3+} + 3e^- \; E^\circ = +1.32\,V \tag{3.25}$$

$$\Delta G^\circ = -nFE^\circ \tag{3.26}$$

where n is the number of electrons involved in the redox process, F is Faraday's constant (9.64853×10^4 C mol^{-1}) and E° is the reaction potential, measured at STP.

A major difference between stainless and plated steels is the former will actually self-repair itself when scratched. Since the chromium is homogeneously dispersed throughout stainless steel, a scratch will serve to expose additional Cr sites forming additional layers of the protective oxide. By contrast, the application of a protective coating over steel will only be an effective barrier as long as it remains intact. When this coating is penetrated by a scratch/crack, the bare steel is exposed to the surrounding environment allowing the possibility for corrosion. Often, aluminum and silicon are also added to steel that also form native oxides, which are effective in preventing surface corrosion of the underlying metal.

When some stainless steels are overheated (*ca.* 400–800 °C) for a prolonged period, there exists the possibility for chromium carbide formation. Most often this results from an attempt to weld steels that are not suitable for such high-temperature treatment. If such a precipitous reaction causes the bulk Cr concentration to fall below 10.5 wt%, corrosion protection is drastically reduced. To make the situation worse, the carbide usually forms at grain boundaries, leading to intergranular corrosion and stress cracking. Amazingly, this process is reversible, by reheating the steel to temperatures in excess of 1000 °C for a period long enough to redissolve the chromium carbide particles and form a homogeneous solid solution. Rapid cooling must then be introduced to suppress the reformation of carbide. Hence, if one wishes to use a stainless steel at high temperatures, either low C compositions must be used, or doping with carbide-forming metals such as V, Ti, or Ta that are more easily oxidized than Cr. As a general rule of thumb, more chromium must be added as the concentration of carbon is increased to ensure effective corrosion resistance.

There are currently over 200 commercially available types of stainless steels. Hence, there is an exact composition of stainless steel for virtually any application. As we have already seen, a tremendous number of substitutional and interstitial dopants may be alloyed with iron, resulting in significant changes in their physical

Table 3.3 General types and properties of stainless steels

Type	Concentration	Properties/applications
Martensitic	11–20 wt% Cr 0.15–0.75 wt% C	High hardness, magnetic/cutlery, blades, surgical instruments, valves, springs
Austenitic	16–26 wt% Cr 35 wt% Ni 20 wt% Mn	High and low temperature resistance, ductility, superior corrosion resistance/kitchen sinks, ovens, reaction vessels, food processors, gutters
Ferritic	10.5–30 wt% Cr <1 wt% C, N, Ni	Magnetic, inexpensive/automotive exhaust and fuel lines, cooking utensils, bank vaults, washing machines, dishwashers
Duplex (austenitic–ferritic)	18–26 wt% Cr 4–7 wt% Ni 2–3 wt% Mo	Weldable, high tensile strength, Cl^- ion resistance (acidic environments)/desalination plants, food pickling plants, petrochemical plants, pulp and paper industries

properties. In addition, varying the heat treatment of the bulk or surface of steels will change these properties even further. It is truly mind-boggling to think of all the combinations of dopant composition/postprocessing that are possible! Fortunately, all of these combinations fall under the umbrella of four general types of stainless steels, classified according to their microstructural phases/compositions (Table 3.3).

The industrial applications for austenitic stainless steels far outweigh the other types due to their facile work-hardenability and high corrosion resistance. As we have seen, the fcc austenite phase is not stable at temperatures below 723 °C; however, austenite-stabilizers such as Ni, Mn, Cu, C, or N may be added to extend the stability of this phase down to room temperature. Figure 3.31 shows the stable phases that exist at room temperature, as a function of the Cr and Ni concentrations. An easy way to think about the effect upon Ni alloying is the replacement of an increasing number of iron atoms in the lattice with Ni, which exhibits a fcc lattice at room temperature. This substitution results in the solid solution being "fooled" into crystallizing in an fcc array rather than bcc ferrite. Due to high concentrations of easily oxidizable elements such as Cr, Ni, and Mn, the corrosion resistance is the greatest for austenitic stainless steels. However, their Achilles' heel is their reaction with chloride ions. Due to the large concentration of transition metals, Cl^- will preferentially react with the metal centers, forming MCl_x rather than a protective coating of M_xO_y.

Since ferritic stainless steels contain more carbon than other classes, they are relatively harder to weld and shape than other varieties, which have historically limited their applications. However, since the 1960s, processes such as *argon oxygen decarburization* (AOD) have resulted in steels with less carbon, allowing for smaller concentrations of chromium to be used.[20] As a result, the price for ferritic stainless steel has dropped dramatically, and a number of applications now employ these materials—more than 2/3 of which include automotive exhaust systems.

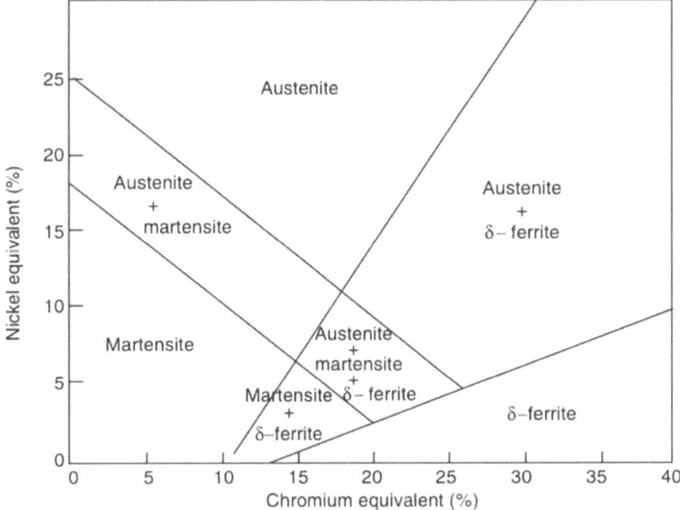

Fig. 3.31 Relative phase stabilities of Ni–Cr steels. Reproduced with permission from *Steels: Microstructure and Properties*, Honeycombe, R. W. K.; Bhadeshia, H. K. D. H.; 2nd ed., Wiley: New York, 1995

Duplex stainless steels feature the "best of both worlds" since they contain both ferritic and austenitic phases. The ferritic phase helps to circumvent the problems associated with stress-corrosion cracking in chloride environments, while the austenitic component helps to improve the generally low strength and ductility of purely ferritic steels. This biphasic steel is generated through careful heating/cooling of the Fe–Cr–Ni alloy. When these materials initially solidify, they are 100% ferritic in nature; subsequent cooling causes the precipitation of austenite within the ferrite matrix.

Strengthening of stainless steels is carried out through cold-working processes—rolling into sheets or drawing into wires/rods at temperatures around 25 °C. This generates a strong material through formation of a distorted bcc lattice that is roughly analogous to martensite. Although we indicated that austenite may be stabilized at room temperature through alloying, this is only a metastable phase. In fact, if the steel is cooled to subzero temperatures or cold-worked, the ferrite phase will be generated. Since austenite would have contained at least a small amount of carbon, the resulting ferrite phase will also contain carbonaceous suspensions—i.e., resulting in a martensite-like structure that will possess high hardness.

From a re-examination of the gamma loop in Fig. 3.24, one can see that heating stainless steel with typical Cr concentrations >11 wt% will not yield austenite upon heating, as this region is outside the loop. Hence, it is not possible to quench-harden these materials through transformation to the martensite phase. For example, austenitic, ferritic, and duplex stainless steels may not be hardened through heat treatment due to their high Cr/C ratios. Interestingly, increasing the carbon concentration

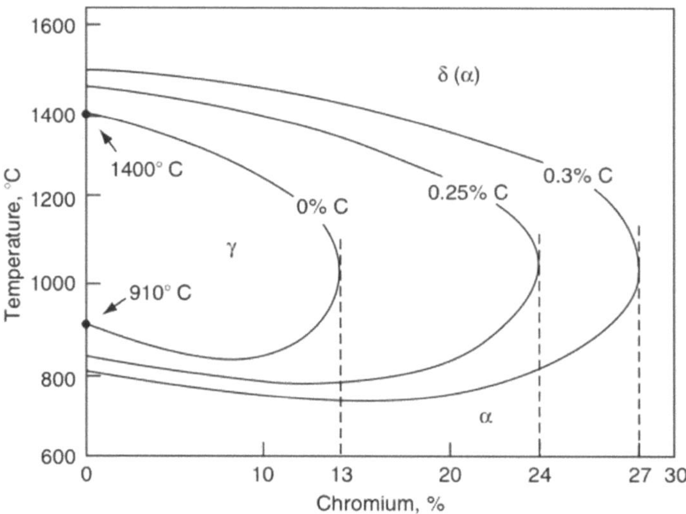

Fig. 3.32 The effect of carbon on the gamma loop. Reproduced with permission from *Stainless Steels*, Lacombe, P.; Baroux, B.; Berauger, G. eds. © 1993 EDP Sciences, Les Ulis, France

will extend the gamma loop (Fig. 3.32), allowing one to austenize the stainless steel and harden through fast-quenching to martensite.

3.2.4 Nonferrous Metals and Alloys

Although we have focused on iron for the majority of this chapter, many of the materials we use on a daily basis comprise other metals. Applications that are particularly suitable for other metals include those that require more lightweight, highly conductive, and/or corrosion-resistant materials (often at a lower cost), relative to iron-based materials. Since metals comprise only one portion of this textbook, we do not intend on discussing all of the other metal classes in as much detail as the iron system. We may now expand our discussion a bit since the general structures and mechanisms involved for alloying, surface/bulk hardening, annealing, etc. also apply for other metals. For more information about the structure and processing of other metal classes that are not discussed herein, refer to the Further Reading section at the end of this chapter.

The coinage metals

Let us begin our survey of other nonferrous metals with "*show me the money*!" The coinage metals consisting of copper, silver, and gold represent the first metals known to man. The first reports for copper purification date back to 3500 BC in the Middle East; bronze alloys were first introduced in *ca.* 3000 BC in India and

Greece. However, it is likely that earliest use for copper may have been much earlier for weaponry applications. As a testament to the durability of the coinage metals, 5000 years after an Egyptian Pharaoh had copper pipes installed in his bath, those same pipes were discovered, dug up, and were still in sufficient shape to carry water![21]

The first application for gold currency dates back to around 3400 BC in Egypt; however, gold was probably employed for decorative applications much earlier—before 9000 BC. Although gold comprises an insignificant 0.004 ppm of Earth's crust, its early widespread use was a consequence of its availability as the uncomplexed element. Hence, simple techniques such as panning along riverbeds were necessary to isolate the gold, requiring no previously developed knowledge of refining. Discovering any of the three coinage metals was as simple as noticing colors in rocks!

Needless to say, the surface deposits of coinage metals have long been harvested, and the metals must now be isolated in small quantities from sulfide-based ores. As early as 3000 BC, a process known as *cupellation* was used to isolate the precious metals from their ores—a method still in use today. In this technique, lead was added to the ore and heated to a temperature of *ca.* 800 °C in air. Any gold or silver in the sample was dissolved in the liquid metal, separating it from the undissolved matter. The insoluble material—primarily containing iron, copper, and zinc compounds—was discarded and the remaining precious metals were brought to a higher temperature by blowing the fire with bellows. This raised the temperature to the point where lead oxide formed rapidly. This was usually performed in a hearth comprising clay or crushed bones; the PbO would be absorbed into the hearth, while the precious metal deposited at the surface.

Another early process known as *amalgamation* was used by the Romans in the Middle Ages. This simple procedure consisted of combining a precious metal ore in mercury; the gold and silver content dissolved, forming liquid alloys. The metals were then obtained through simple distillation of mercury. However, the method of choice for gold production is cyanidation that usually follows a froth flotation process. The crushed ore is treated with an aqueous NaCN solution, along with enough CaO to neutralize any acid present in the rock that would generate highly toxic HCN. This results in the formation of a cyanoaurate complex (3.27); any silver that is present also forms the analogous cyanoargentate complex. Finely divided zinc metal is then added to reduce the metal ions (3.28). The addition of base regenerates cyanide through formation of zinc hydroxide, which is more stable than $Zn(CN)_4$ (3.29).

$$4\,Au + O_2 + 2\,H_2O + 8\,CN^- \rightarrow 4\,Au(CN)_2^- + 4\,OH^- \tag{3.27}$$

$$2\,Au(CN)_2^- + Zn \rightarrow Zn(CN)_4^{2-} + 2\,Au \tag{3.28}$$

$$Zn(CN)_4^{2-} + 4\,OH^- \rightarrow Zn(OH)_4^{2-} + 4\,CN^- \tag{3.29}$$

The electronic configuration of the coinage metals is $nd^{10}\,(n+1)s^1$. Hence, one may suspect that Cu, Ag, and Au would share similar properties to the isoelectronic alkali metals. However, it should be noted that a filled d-shell is far less effective at shielding an outer s electron from nuclear attraction than the less diffuse p-shell. As a result, the 1st ionization energy of the coinage metals is much higher than the alkali metals, and their bonding is significantly more covalent in nature. This explains their relatively higher melting points, hardness, density, and inertness relative to the alkali metals.

The unreactivity of the coinage metals increases dramatically from Cu to Au; whereas copper and silver readily react with sulfur and halogens, gold is completely unreactive to all reagents except very strong oxidizing acids such as *aqua regia* (3:1 $HCl:HNO_3$). Acid rain produced from gaseous sulfur compounds (SO_x) react with copper surfaces, eventually resulting in the formation of a basic copper sulfate film, as evidenced by a green–blue-patina color. A useful property of copper is its antifungal behavior; fine granules of ceramic-coated copper oxide are now placed within specialized asphalt shingles to prevent discoloration by algae. The activity can last as long as the shingles themselves, 25–30 years, until all of the Cu^{2+} ions are leached from the porous ceramic granules.

Although all metals possess metallic luster, the only metals that exhibit colors in their bulk state are copper and gold. The familiar reddish and golden colors of these elements arise from the filled d-shell near the top of the conduction band of the solid. By definition, the highest-occupied energy level at 0 K is referred to as the *Fermi level*. For copper, the gap from the top of the 3d-band to the Fermi surface is *ca.* 544 nm (Fig. 3.33). Hence, energy in the green/blue region of the spectrum may be absorbed, resulting in an observed red/orange color. For gold, the gap is *ca.* 400 nm, corresponding to absorption in the blue region of the spectrum and an observed golden color. By contrast, the analogous energy gap for silver is 311 nm, resulting in UV absorption and an observed white/silver color (i.e., equal reflection of all visible wavelengths).

In addition to high inertness, the coinage metals exhibit other desirable properties that are of tremendous use for materials applications. Silver and copper have the highest electrical and thermal conductivities of all metals in the periodic table. Gold is the most electronegative metal, behaving as a halogen through formation of Au^- ions in the presence of strongly electropositive metals such as cesium. These properties may all be rationalized by the band structures of the metals. As a consequence of the filled d-shells of the coinage metals, there is facile thermal promotion of valence electrons into the delocalized s/p conduction band. In contrast, other metals such as iron do not have a filled d-shell (Fig. 3.30), and electrons remain within the partially filled d-band. This shell has more localized character than the s/p band, which results in a lower electrical conductivity of the bulk solid relative to Cu, Ag, or Au. In general, the population of the electronic energy levels (i.e., *density of states*) immediately surrounding the Fermi level, E_f, is most important, as typical thermal/electrical energy supplied to a solid is only sufficient to interact with a small fraction (*ca.* 0.4%) of electrons. That is, the overwhelming majority of the electrons are separated from the top of the Fermi surface by much more than thermal energy.

If you have purchased jewelry, you will have heard the terms "carat," "karat," and "fineness." Whereas "carat" refers to the weight of precious stones (1 carat =

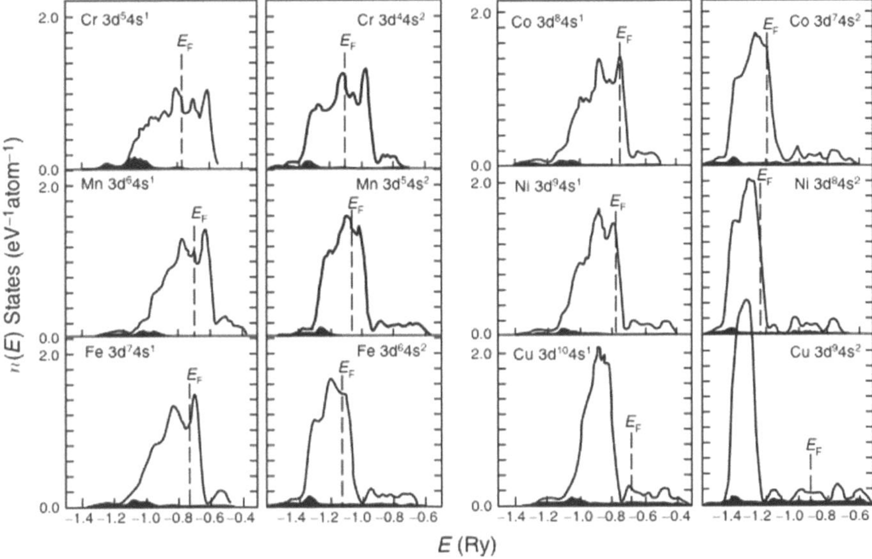

Fig. 3.33 Calculated density of states (DOS) for the fcc phases of various transition metals. Reprinted from Snow, E. C.; Waber, J. T. *Acta Metall.* **1969**, *17*, 623. © 1969, with permission of Elsevier

200 mg), the term "karat" is used to describe the purity of metals. For example, 24 karats is the pure, unalloyed metal that is almost never used for applications due to its softness. The term "fineness" refers to the weight portion of the precious metal in the alloy; 24 karats represents 100% purity, and a fineness of 1000, whereas 18 karats represents a purity of 75% and a fineness of 750. Various alloying agents may be added to gold such as Ag, Cu, Zn, Ni, Pt, and Pd. As required for solid solutions, each of these dopants is of a similar size, and has an fcc crystal lattice that matches that of gold. In addition to improving the strength, these dopants also impart colors to gold (Table 3.4).

Interestingly, "white gold" was developed in the 1920s as a substitute for platinum jewelry. Nickel and palladium are the most common dopants used for white golds, although copper may also be added to improve the strength and decrease the price. In order to retain a grayish-white color, these alloys are only available up to a purity of 21 karats. However, due to cases of skin irritation through contact with nickel, European countries have already phased out the nickel whites from jewelry—not yet a policy in the U.S. It should also be noted that the bright white color is actually an artifact of the rhodium plating that is usually applied to white golds; this film will wear off over time diminishing the color and requiring reapplication of the coating.

Intermetallics

As a more stringent application of the Hume-Rothery rules that govern the alloying of metals, if the difference in radii is less than 8%, the metals will be soluble throughout

Table 3.4 Commonly Used Alloys for Colored Golds

Composition (karat)	%Au	%Ag	%Cu	%Zn	%Ni	%Ti	%Pd	%Fe	%Si	%Co
Yellow gold										
23	99.0						0.9			
18	75.0	13.0	12.0							
18	75.0	15.0	10.0							
14	58.3	4.0	31.2	6.4	0.1			0.05	0.01	
14	58.3	24.8	26.8	0.14						
10	41.7	11.7	40.8	5.83					0.03	
10	41.7	5.5	43.8	9.0						
White gold										
18	75.0		2.23	5.47	17.8					
18	75.0	15.0					10.0			
14	58.3			28.3	4.8	8.6				
14	58.3	32.2					9.5			
10	41.7		29.2	12.1	15.1					2.0
10	41.7	47.4		0.9			10.0			
Rose gold										
18	75.0	5.0	20.0							
14	58.3	2.1	39.6							
10	41.7	2.8	55.5							

the full range of compositions. This is the case for nickel and copper, whose radii are 1.49 and 1.45 Å, respectively. Hence, there are over 20 different alloys that are used in industry based on the mutual solubility of copper and nickel in all proportions. Monel (68% Ni, 32% Cu) is used to handle corrosive materials such as F_2 or HF. U.S. currency coinage such as quarters and nickels are alloys of Ni/Cu, containing 25% Ni and 75% Cu, respectively. By comparison, Canadian quarters and nickels are predominantly steel, with only *ca.* 3.5% Cu and 2% Ni plating. Pennies are predominantly zinc (97.5%), with the remainder consisting of a thin layer of copper deposited by electroplating. High-strength Cu/Ni alloys are produced with the addition of 1.5–2.5 wt% Al, which causes precipitation-hardening through formation of Ni_3Al crystallites.

The atomic radii of Cu, Sn (1.45 Å), and Zn (1.42 Å) are nearly identical, allowing for a full gamut of Cu/Sn and Cu/Zn alloy concentrations to be produced, known as *bronze* and *brass*, respectively. Although the use of bronze dates back to at least 3000 BC, there are also early examples of brass artifacts that date back to *ca.* 2200 BC in India. Most likely, the discovery of bronze resulted from the inquisitive mixing of available metals at the time, only to discover that Au/Sn alloys possessed a greater strength than iron; steels were not developed until thousands of years later. Since zinc metal was not available until the mid-eighteenth century but tin was readily obtained, the widespread production of bronzes occurred at the expense of brasses. In the absence of pure zinc, early formulations of Cu/Zn alloys were most likely made through heating a mixture containing zinc oxide, copper metal, and a reducing agent such as charcoal in a closed crucible. A temperature in excess of 950 °C was necessary to reduce ZnO, and even a trace amount of oxygen would preclude the formation of the alloy.

The strength of a bronze increases with its tin content; however, its toughness and malleability decreases. The maximum strength of bronze occurs at *ca.* 30% Sn, but at this concentration the alloy is much too brittle for most applications due to the formation of Cu_3Sn particles. Recall that this phenomenon also occurred for the formation of Fe_3C in iron–carbon alloys, which also involved a transition metal and Group 14 dopant. If more than 15% Sn is used, the alloy is called "bell metal," due to its resonating sound when tolled.

Although the radii of Cu and Zn satisfy the Hume-Rothery constraints for solid solutions, these metals do not share the same crystal lattice. Whereas the coinage metals are fcc, zinc crystallizes in a hcp array. Hence, as we introduce more Zn into the Cu lattice, there will be a shift in the overall structure. We may think of this structural change as a consequence of the varying electron concentration of the solid. For instance, each Cu atom contributes one 4 s electron to the valence shell of the extended lattice; by contrast, each Zn atom contributes two 4 s electrons. For small concentrations of Zn, the fcc α -brass structure is formed. However, as the Zn concentration reaches 50%, the bcc β -brass (CuZn) phase predominates. The average electron concentration, n, for the β -brass structure is 1.5 (i.e., 1 e^- (for Cu) + 2 e^- (for Zn) $\div$ 2 atoms). Due to the 1:1 combination of Cu and Zn, and the overpowering 2:1 electronic effect of Zn:Cu, the bcc structure becomes more stable. Other intermetallics that exhibit the β -brass structure are AgZn, AuZn, AgCd, Cu_3Al, Cu_5Sn, CoAl, FeAl, and NiAl.

A further increase in the Zn concentration results in the complex γ -brass structure (Fig. 3.34), Cu_5Zn_8, with $n = 1.615$ (i.e., [5 $\times$ 1 e^- (for Cu) + 8 $\times$ 2 e^- (for Zn)] $\div$ 13 atoms). Other intermetallic compounds with this structure include Ag_5Zn_8, Cu_9Al_4, $Cu_{31}Sn_8$, $Na_{31}Pb_8$, Rh_5Zn_{21}, and Pt_5Zn_{21}. Additional zinc may continue to dissolve in this phase until a concentration of *ca.* 75% Zn is reached, which results in the final hcp phase referred to as ε-brass, $CuZn_3$ ($n = 1.75$). Other intermetallic compounds that share this structure include $AgZn_3$, Ag_5Al_3, Cu_3Sn, and Cu_3Si. Beyond this concentration, additional zinc results in the hcp η-brass phase which is no longer considered an alloy, but pure Zn with $n = 2$. Only the α and β-phases are useful alloys; the others are too hard and/or brittle. These various intermetallic structures are often called *electron compounds* or *Hume-Rothery phases* since they are governed by the ratio of electrons:atoms.

Laves phases represent the largest group of intermetallic compounds, whose structures are governed by a combination of atomic sizes, electronegativities, and valence electron numbers of the metals. These compounds are of form "AB_2", where the larger A atoms are positioned in the center of a 16-atom polyhedron and the smaller B atoms in centers of icosahedra. There are three primary motifs for Laves phases:

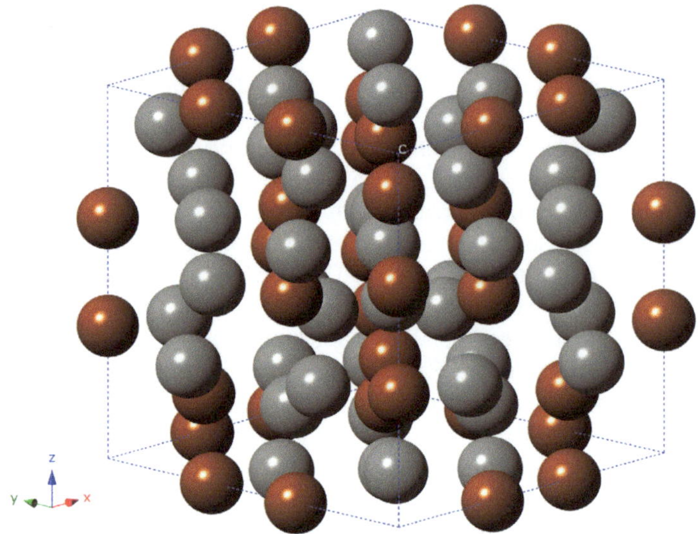

Fig. 3.34 The complex unit cell of γ -brass, Cu_5Zn_8 (space group $I\bar{4}3$ m)

- $MgCu_2$-type (Fig. 3.35): Space group = Fd $\overline{3}$ m; A:B radius ratio = 1.25; e.g., $CaPt_2$, $HfCo_2$, $CeCo_2$, $BaPt_2$, $RbBi_2$, $CsBi_2$, $PbAu_2$, $LaPt_2$, VIr_2, $ZrFe_2$
- $MgZn_2$-type (Fig. 3.36): Space group = $P6_3/mmc$; A:B radius ratio = 1.17; e.g., $CaMg_2$, $BaMg_2$, KNa_2, β -$FeBe_2$, $TaFe_2$, $MoFe_2$, WBe_2, $ZrRu_2$, $TiCr_2$, $LiOs_2$, $TiFe_2$
- $MgNi_2$-type (Fig. 3.37): Space group = $P6_3/mmc$; A:B radius ratio of 1.28; e.g., $NbZn_2$, $ScFe_2$, $ThMg_2$, $HfCr_2$, UPt_2

There also exist mixed Laves phases of type A(M'M"), where M' and M" are Cu, Ag, Zn, Al, or Si. An increasing electron concentration favors the Laves phases in the order $MgCu_2$ (1.33–1.8 e^-/atom), $MgNi_2$ (1.7–1.9 e^-/atom), and $MgZn_2$ (*ca.* > 1.9 e^-/atom). AB_5 intermetallic structures such as $AuBe_5$, $LuMn_5$, and $MgSnCu_4$ are also Laves phases related to either $MgCu_2$ or $MgZn_2$, with two arrays of Mg sites occupied equally by the two metals.

A structural motif for intermetallic compounds that contain a Group I/II and late transition metal are known as *Zintl phases*. Unlike other metallic alloys, these compounds are typically diamagnetic insulators, with a high degree of brittleness. Although late transition metals have similar electronegativities to those of the late main group elements, it has only recently been accepted that transition metal atoms present in alloys such as CsAu, $K_{34}In_{96.19}Au_{8.81}$, Yb_3Ag_2, Ca_5Au_4, and Ca_3Hg_2 exist as Zintl anions of form M^-, $[M\text{-}M]^{4-}$, etc.[22] However, the anions of late transition metal elements behave differently than their late main group counterparts (e.g., halides), exhibiting covalent bonding within alloys when their p-shells are partially filled.

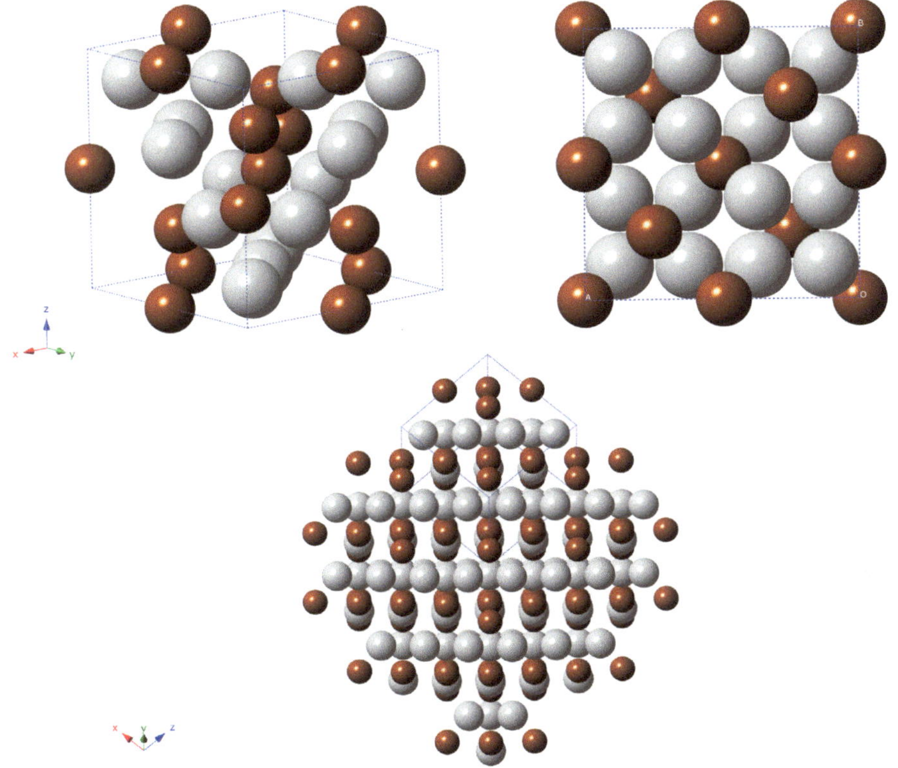

Fig. 3.35 Unit cell representations of the cubic $MgCu_2$ laves phase. Shown are (top, left–right): the unit cell as viewed along [110] and [001], respectively. The bottom image displays an expanded unit cell representation as viewed along [010], which shows the planes of Mg atoms (silver) between adjacent layers of Cu/Mg units. The original unit cell is superimposed in blue

The Zintl compound[23] NaTl features a diamond lattice of Tl^- ions with Na^+ ions in tetrahedral interstitial sites (Fig. 3.38). However, unlike most other Zintl compounds that are insulators, NaTl does not have a bandgap and exhibits metallic conductivity.[24] Insulating *half-Heusler* compounds of form AML (Fig. 3.39; where A = Grp. 2–4 such as Mg, Sc, Y, Ti; M = late transition metal such as Ni, Pd, Ag, Au; L = heavy main group atom of Grp. 14/15 such as As, Sn, Sb, Pb, Bi) are structurally related to NaTl, with M and L forming a zincblende lattice and A atoms occupying 1/2 of the 10-coordinate sites defined by a M_4 tetrahedron and L_6 octahedron.[25]

Aluminum and magnesium alloys

It is hard to imagine a world without aluminum-based materials. From the foil that we use to wrap leftovers, to the variety of beverage and deodorant cans, our world is inundated with applications for aluminum. The widespread use of this metal is a direct result of its availability—8.3 wt% in Earth's crust—making it the most naturally

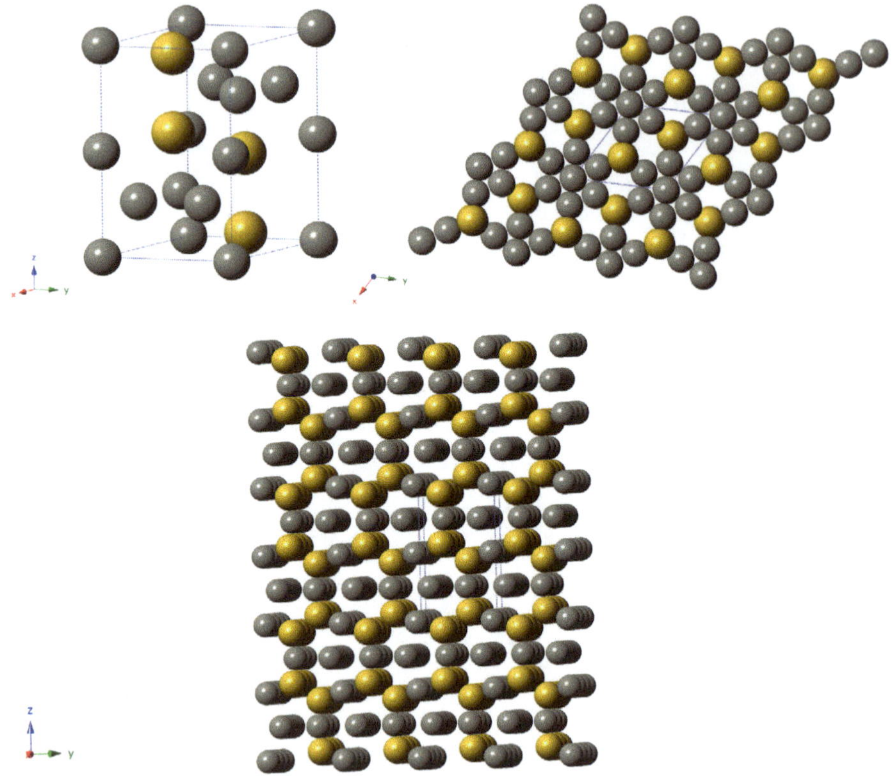

Fig. 3.36 Unit cell representations of the hexagonal $MgZn_2$ laves phase. Shown are (top, left–right): the unit cell as viewed along [110] and an expanded unit cell along [001], illustrating its hexagonal symmetry of Mg (orange) and Zn (grey) atoms. The bottom image is an expanded lattice representation as viewed along [100], which displays planes of Zn atoms between layers of Zn/Mg units. The original unit cell is superimposed in blue

abundant metal. The malleability of Al is second only to gold, and it possesses other desirable characteristics such as non-sparking, high thermal/electrical conductivity, corrosion resistance, and high ductility.

With growing demand for fuel-efficient vehicles, the automotive industry has largely shifted from steel and cast irons to more lightweight metals. As a rule of thumb, a 10% weight reduction of the vehicle results in a 5.5% improvement in fuel economy. The high stiffness-to-weight ratio, formability, corrosion resistance, and recyclability of aluminum make this metal an ideal candidate to replace heavier metals in vehicles. The first sports car featuring an aluminum body was revealed at the Berlin International Motor Show in 1899. Carl Benz began to use Al for engine components shortly thereafter, as the price for the metal dropped after World War II. Now widespread, the use of aluminum in vehicle components and frames has grown more than 80% over the past 20 years. For instance, Ford has incorporated an all-aluminum frame into their popular Ford F-150 trucks since 2015—a weight savings

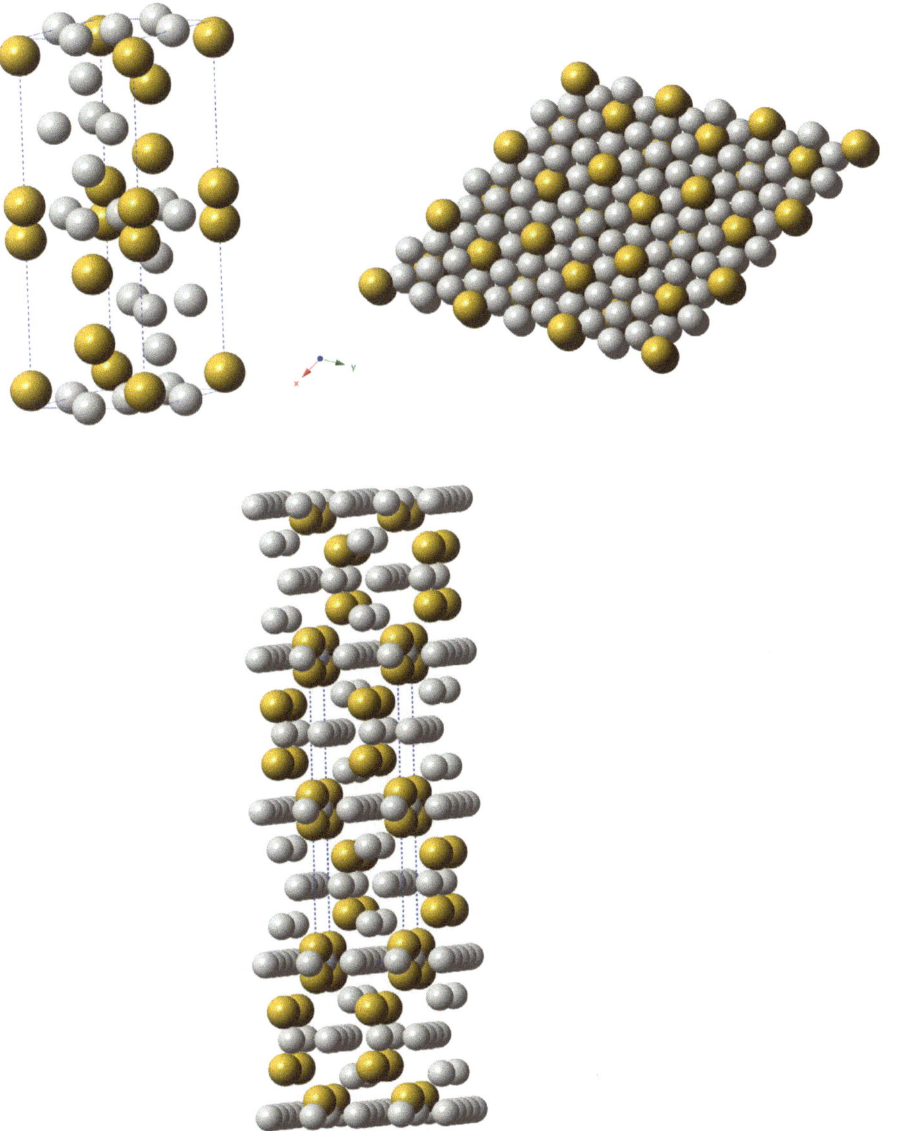

Fig. 3.37 Unit cell representations of the hexagonal $MgNi_2$ laves phase. Shown are (top, left–right): the unit cell as viewed along [110] and an expanded unit cell along [001], illustrating its high density and hexagonal symmetry of Mg (orange) and Ni (grey) atoms. The bottom expanded lattice representation is viewed along [100], which displays its layered structure composed of Ni atoms and Ni/Mg units. The original unit cell is superimposed in blue

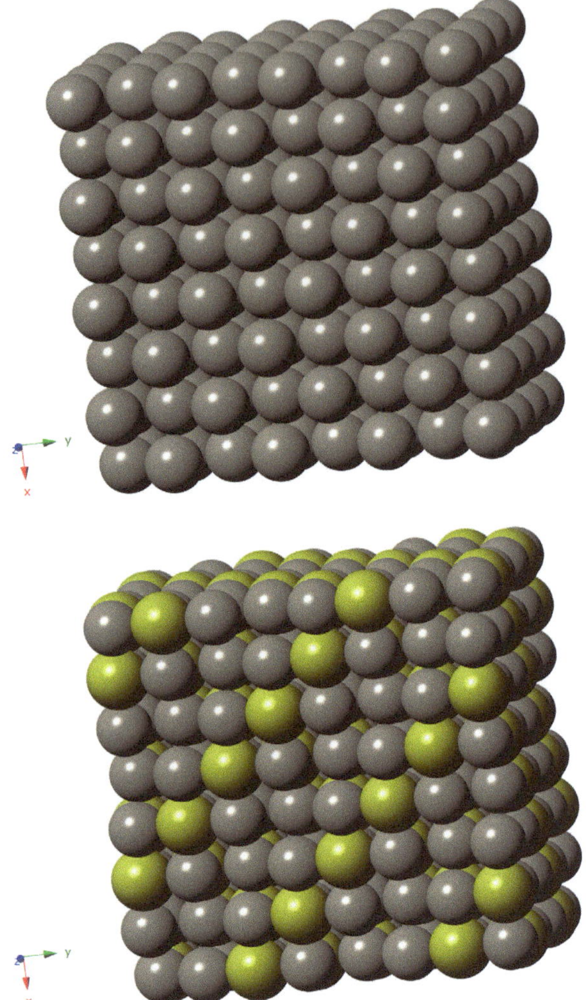

Fig. 3.38 Space-filling representations of the the NaTl Zintl phase (space group Fd $\overline{3}$ m). The Tl^- ions (grey) form a diamond lattice, with Na^+ ions (yellow) occupying interstitial holes

of 700 pounds relative to its steel version—and began selling an all-aluminum bodied Expedition SUV in 2018. It should be noted that the use of Al is not limited to frames and body components; this metal is now utilized for a variety of components such as cylinder heads, pistons, radiators, and wheel rims.

However, automotive manufacturers are constantly searching for metal alloys even lighter than aluminum. With a weight 30% less than aluminum and 60% that of steel, magnesium has been of increasing interest for transportation applications. Magnesium is the lightest of the structural metals, and exhibits a high specific strength, good castability, and superior electromagnetic shielding.[26] Table 3.5 provides a comparison of some physical properties between Mg, Al, and Fe. The specific strength and stiffness of a material is paramount for the design of weight-saving components in

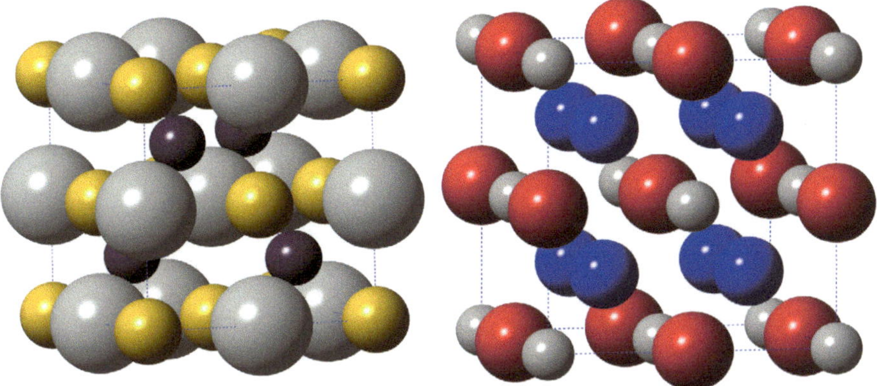

Fig. 3.39 Unit cell representations of the half-Heusler alloy AgAsMg (space group: F $\overline{4}$ 3 m; Ag = silver; Mg = orange; As = purple), as compared to the Heusler alloy $AlCu_2Mn$ (space group: Fm $\overline{3}$ m; Al = grey; Cu = dark blue; Mn = red). As shown in these images, the half-heusler structure is identical to Heusler alloys with half of the Cu atoms missing. For each structure, all atoms are located on the sites of a body-centered cubic lattice. For the AgAsMg alloy, the Mg and Ag atoms form a rocksalt structure, whereas the As and either Mg or Ag atoms form a zincblende structure

structural applications. Whereas the specific stiffness between Mg, Al, and Fe are similar, magnesium exhibits a higher specific strength than either Al or Fe.

Magnesium is easier to machine than aluminum, requiring 55% less power while generating 160% greater cutting speeds than Al. However, Mg presents a significant fire risk since the small chips and fine dust produced during cutting are highly flammable. An oil-based coolant can minimize the fire hazard, but would require extensive cleaning of the machined parts to remove the oily residue. Fortunately, machining can be minimized or eliminated on many magnesium castings because the die caster is able to hold tighter tolerances than aluminum metal.[27]

The use of Mg in vehicles dates back to race cars of the 1920s; however, this metal was not used in commercial vehicle components until the mid-1930s with Volkswagen and Porsche leading the developmental efforts. Magnesium is now used in a variety of automotive components such as crankcases and manifolds; however, its lower creep resistance relative to Al has precluded its use as the engine block and cylinder head, the two heaviest components of the engine. Porsche experimented with an all-magnesium engine block from 1968–1977, which dramatically lowered the engine weight relative to aluminum. However, the Mg cases were prone to warping and head studs pulled out after only a few years of use. The BMW N52 water-cooled engine incorporated a Mg shell with an Al block core to overcome the limitations of magnesium alloys. Produced in 2004–2015, this lightweight engine design was touted to avoid potential problems with high-temperature creep and moisture corrosion of Mg. However, there were many reports of block failures and oil leaks. As a result, BMW reverted back to a more conventional Al block from MY07 to the present.

Table 3.5 Common magnesium and aluminum Alloys

Alloy designation	Alloying components	Basic properties and applications
Magnesium		
AZ91	8–9% Al, 0.3–1% Zn, 0.13% Mn, 0.1–0.35% Cu, 0.2–0.5% Si	Good castability, good mechanical properties and corrosion resistance at T < 150 °C; used for die-casts
AM60	5.5–6.5% Al, 0.13–0.25% Mn, 0.1–0.5% Si, 0.22% Zn, 0.01–0.35% Cu	Greater toughness and ductility than AZ91, slightly lower strength; preferred for automotive structural applications
AE42	2% Al, 2% Ce, 1% La	Low castability, good creep behavior; used for die-casts
AE44	4% Al, 2% Ce, 1% La	Better creep behavior and castability than AE42; used for automotive structural parts working at elevated temperatures
AS41	4.2% Al, 1.0% Si	Better creep resistance than AZ91 at elevated temperatures, but lower strength; used for die-casting
ZE41	4.2% Zn, 0.7% Zr, 1.2% rare earth	Good creep resistance at elevated temperatures and pressures; used for aircraft components, military equipment, and video cameras
AZ31	2.5–3.5% Al, 0.7–1.3% Zn, 0.2% Mn	Good strength, ductility, and corrosion resistance; good extrusion properties; used in the aircraft industry to produce flat parts with ribs such as brackets
AMCa602	6% Al, 0.1% Zn, 0.35% Mn, 2% Ca	Non-combustible Mg alloy
WE43	4% Y, 2.25% Nd, 0.15% Zr	High tensile strength (250 MPa); used in aircraft and high-performance vehicles
AJ62	5.6–6.6% Al, 0.2% Zn, 0.08% Si, 0.26–0.5% Mn, 2.1–2.8% Sr	Good thermal and mechanical strength, superior castability, corrosion resistance, and creep behavior; good high-temperature engine Mg alloy
Aluminum[a]		
1100-X	0.12% Cu	Resistant to chemical attack and weathering; used for chemical equipment, fan blades, sheet metal

(continued)

Table 3.5 (continued)

Alloy designation	Alloying components	Basic properties and applications
2024-X	4.4% Cu, 0.6% Mn, 1.5% Mg	Heat treatable, high strength-to-weight ratio, limited corrosion resistance and weldability; used for auto wheels and suspension components, aircraft fuselage, power lines
3003-X	1.2% Mn, 0.12% Cu	Moderate strength without heat treating, high workability; used for beverage cans, cooking utensils, heat exchangers, storage tanks, furniture, highway signs, roofing, side panels
3105-X	0.55% Mn, 0.5% Mg	Good corrosion resistance and strength; used for siding, sheet metal, auto parts, building products, electronics, furniture
5005-X	0.8% Mg	Used for appliances, utensils, hardware, marine applications
5052-X	2.5% Mg, 0.25% Cr	Good weldability and strength, good corrosion resistance; used for ornamental trim, cans, household appliances, boats and ships, bridge railings, race cars
6061-X	0.6% Si, 1% Mg, 0.28% Cu	Low thermal expansion and m.p., high wear resistance; used for forged engine pistons, welding rod, brazing alloys, architectural products
7050-X	6.2% Zn, 2.3% Mg, 2.3% Cu, 0.12% Zr	High-strength alloy used for aircraft and structures, recreation equipment

[a] The "X" designation is a suffix that varies depending on how the alloy was tempered (e.g., -H = strain-hardened (cold-worked), -O = annealed, -T = heat-treated, etc.)

The strength of aluminum and magnesium metals improve upon alloying, which extends their range of applications (Table 3.6). The addition of Zn improves the strength as well as heat and creep resistance of a Mg/Al alloy. Whereas Al, Zn, and Mn are common additives for room-temperature applications of Mg alloys, Th, Ce, and Zr are commonly added for high-temperature applications. Only a maximum of 5 wt% of Mg may be dissolved in Al to provide solid-state strengthening. However, only 1.5 wt% of magnesium may be dissolved at room temperature, implying that supersaturation will often occur, with precipitation of Mg species. This is reminiscent of the Fe–C system, with precipitation hardening through dispersion of metal carbides. Frequently, iron and silicon will be present as processing impurities (or deliberately added), which will strengthen the alloy due to the formation of Mg_2Si and Fe_3Si precipitates upon cooling. In the presence of Mn, the hardening effect is even more pronounced, due to the formation of $FeMnAl_6$ crystallites.

Table 3.6 Comparison of selected physical properties for Mg, Al, and Fe

Property	Mg	Al	Fe
Crystal structure	hcp	fcc	bcc
Density at 20°C	1.7	2.7	7.9
Melting point (°C)	650	660	1540
Coeff. of linear thermal expansion (10^{-6} K^{-1})	26.0	23.1	11.8
Elastic modulus[a] (10^6 Mpa)	44.1	68.9	207
Tensile strength (Mpa)	240	320	350
Specific stiffness[b]	25.9	25.5	26.2
Specific strength[c]	141	118	44.3

[a] Also known as Young's modulus of elasticity
[b] Also known as *specific modulus.* Defined as Young's modulus/density
[c] Defined as tensile strength/density

Among the various Al-alloys at our disposal, most may not be heat-treated. In particular, alloys such as pure Al (i.e., containing trace dopants), Al–Mn, Al–Si, and Al–Mg alloys deleteriously form precipitates along grain boundaries. However, Cu–Al and Al–Zn–Mg alloys are greatly strengthened by heat treatment, through formation of $CuAl_2$ and $MgZn_2$ precipitates, respectively. Even lithium may be added as a hardening agent, via the formation of Al_3Li precipitates. As with all age-hardening techniques, the size and dispersion of the crystallites must be carefully controlled through the heating/cooling regime.

Whereas the solubility of Cu in aluminum metal is *ca.* 5 wt% at temperatures in excess of 500 °C, the solubility drops to *ca.* 0.1 wt% at room temperature. Hence, a metastable alloy is present when the high-temperature alloy is rapidly quenched. Subsequent annealing will result in further strengthening, similar to what we discussed for martensite. The strengthening effect is thought to occur due to the formation of Cu-rich discs (approximate diameter of 100 atoms, and thickness of *ca.* 4 atoms) that align themselves preferentially with selected planes of the host Al lattice, causing coherency strains within the solid-state structure.

Refractory metals

By definition, refractory metals exhibit low thermal and electrical conductivities, with equally low thermal expansion properties. As a relative benchmark to the values listed in Table 3.7, common metals such as iron and copper have coefficients of linear thermal expansion on the order of 12.1 and 17.7 $\mu m\ m^{-1}\ K^{-1}$, respectively. Also for comparative purposes, the electrical/thermal conductivities for Fe and Cu are 9.71 $\mu\Omega\ cm^{-1}$/78.2 $W\ m^{-1}\ K^{-1}$ and 1.67 $\mu\Omega\ cm^{-1}$/397 $W\ m^{-1}\ K^{-1}$, respectively.

However, it is the extremely high melting points (>1650 °C) of the refractory metals that separate this class from the others, which has enabled the widespread use of tungsten and molybdenum metals in lighting applications (Fig. 3.40). As a result, these metals may not be processed through cold- or hot-working; powder metallurgy must be used to form the metals into desired shapes. Although we also generally

Table 3.7 Properties of the Refractory Metals

Metal (Lattice)	Density (kg m^{-3})	Melting point (°C)	Resistivity ($\mu\Omega$ cm^{-1})	Thermal conduct (W m^{-1} K^{-1})	CLTE[a] (μm m^{-1} K^{-1})
Ti (HCP)	4,540	1,668	42.0	21.9	8.35
Zr (HCP)	6,506	1,852	42.1	22.6	5.78
Hf (HCP)	13,310	2,233	35.5	22.3	5.90
V (BCC)	6,110	1,915	25.0	30.7	8.40
Nb (BCC)	8,570	2,230	15.2	53.7	7.10
Ta (BCC)	16,654	2,996	13.2	57.5	6.60
Cr (BCC)	7,140	1,900	13.0	93.9	4.90
Mo (BCC)	10,220	2,610	5.70	139	5.43
W (BCC)	19,300	3,407	5.65	174	4.59
Re (HCP)	21,010	3,270	13.5	48.0	6.70
Ir (FCC)	22,650	2,410	5.30	146	6.40
Os (HCP)	22,590	3,054	8.12	87.6	4.57

[a] Coefficient of linear thermal expansion

associate the refractories with high hardness, it is worthwhile to point out that these metals are all soft and ductile in their pure states. However, the metals are rarely obtained in their pure forms, since they spontaneously react with C, B, O, N, and other nonmetals to form stable interstitial compounds. The incorporation of small main-group elements within the large lattices of electron-rich refractory metals results in solute hardening via localized covalent bonding. An example is the formation of the refractory ceramic OsB_2, which is possible since boron is significantly smaller than Os (0.87 Å *vs.* 1.85 Å, respectively).[28]

The incompressibility (bulk modulus) of a material is directly related to its valence electron density, in units of electrons $Å^{-3}$. For example, diamond, the hardest known substance, has a high valence electron density (0.705 electrons $Å^{-3}$) and an exceptionally high bulk modulus (442 GPa). By comparison, osmium has one of the highest valence electron densities for a pure metal (0.572 electrons $Å^{-3}$), resulting in an accompanying large bulk modulus (*ca.* 400 GPa). However, while the bulk moduli of diamond and osmium are equivalent, the hardness of diamond is more than an order of magnitude larger than Os—a consequence of covalent *vs.* metallic bonding. Upon doping of Os with boron, the hardness improves dramatically, while retaining a high valence electron density (0.511 electrons $Å^{-3}$). There are current studies underway that are evaluating the hardness and bulk moduli of mixed solutions such as $Os_{1-x}M_xB_2$, which are likely harder than either OsB_2 or MB_2 compounds alone.[29]

As we saw earlier, the facile reaction of Cr with oxygen is the acting principle behind the anticorrosive property of Cr-containing steels. This analogous reactivity

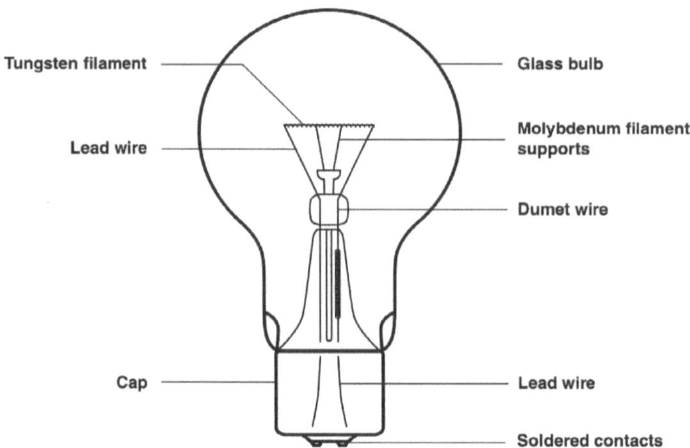

Fig. 3.40 Illustration of an incandescent lightbulb

also explains the high corrosion resistance exhibited by the refractory metals, especially those of Group 4 (Ti, Zr, Hf—often termed the "reactive metals"). Since refractories show a high reactivity toward constituent gases of the atmosphere, the metals in their finely divided forms are highly pyrophoric and must be handled within an inert-atmosphere glove box. In general, all metals that form stable oxides or nitrides (e.g., iron, zinc, nickel, etc.) are dangerous as finely divided powders. The relatively large surface area of individual crystallites leads to oxide formation (an exothermic process) that releases enough heat to spontaneously catch the material on fire.

As one moves across the groups of refractory metals, a number of trends are noteworthy and greatly affect their applications. For instance, moving left to right causes a decrease in atomic sizes due to ineffective shielding of the nuclear charge by d-electrons. As additional d-electrons are added to the valence shell, stronger metal–metal bonds are formed and the metals become increasingly dense/harder, with higher melting points (Table 3.7). Also noteworthy is the equal atomic sizes of 4d and 5d congeners due to the lanthanide contraction effect.

Among the refractory metals, perhaps none are as widely exploited for commercial applications as titanium. From golf clubs and shavers, to knee and hip replacements, titanium is now pervasive throughout our modern world. Although some of the proposed applications may be suitably classified as "hype," the broad appeal for titanium alloys is due to its favorable properties such as high strength:weight ratio and superior corrosion resistance. Titanium is also readily available from a variety of mineral sources; it is the sixth most abundant metal, behind Al, Fe, Cu, Zn, and Mg.

The density of titanium is significantly less than the other refractories (midway between aluminum and iron); even so, its yield strength ranges up to 1800 MPa. To put this in perspective, this strength is of the same magnitude as Ni-doped ultrahigh strength stainless steels—at a fraction of the weight. The low density of Ti (and other

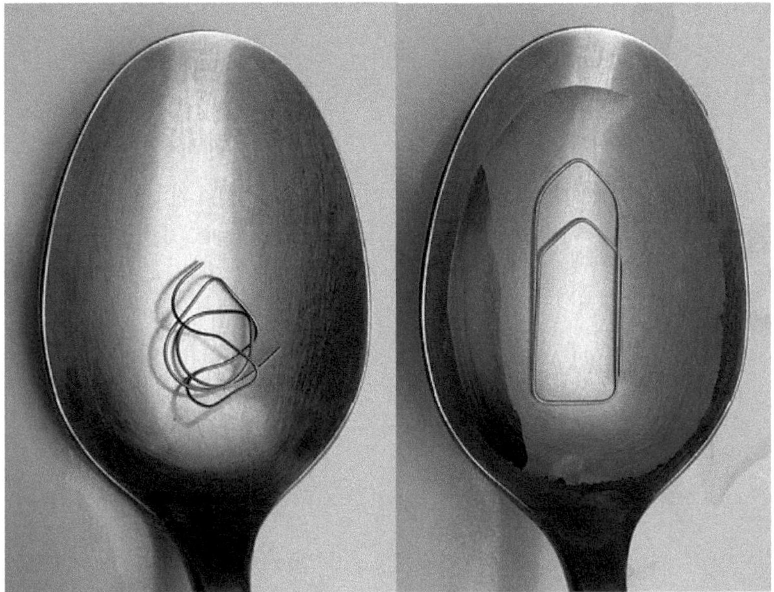

Fig. 3.41 A Ni–Ti paperclip deformed (left) and recovered (right) after being placed in hot water. Images from Peter Maerki, available via Creative Commons CC BY-SA 3.0

metals such as Be and Mg) is due to the hexagonal close-packed crystal structure, which is much less dense than bcc or fcc arrays (revisit Chap. 2).

Shape-memory alloys

As their name implies, shape-memory alloys are able to revert back to their original shape, even if significantly deformed (Fig. 3.41).[30] This effect was first discovered in 1932 for Au–Cd alloys. However, there were no applications for these materials until the discovery of Ni–Ti alloys (e.g., NiTi, known as *nitinol*) in the late 1960s. As significant research has been devoted to the study of these materials, there are now over 20 different binary, ternary, and quaternary alloys that also exhibit this property. Other than the most common Ni–Ti system, other classes include Co–Ni-Al, Ni–Mn–Ga, Au–Cu–Zn, Cu–Al–Ni, Cu–Zn–Al, and Fe–Mn–Si alloys.[31]

As illustrated in Fig. 3.42, the shape-memory effect is observed when the temperature of a piece of alloy is cooled to below that required to form the martensite phase: M_s (initial martensite formation) until M_f (martensite formation complete). Upon heating the martensitic material, reformation of austenite begins to occur at A_s until a final temperature, A_f, is reached. Since the martensite phase is a highly distorted structure relative to fcc austenite (Fig. 3.43), the alloy is highly soft and ductile and may be easily deformed while in its low-temperature phase. You may be thinking that this is the opposite of the Fe–C system that was previously discussed. That is, the martensite phase that was generated through fast-quenching austenite resulted in an extremely hard material—much stronger than the native austenite phase. However,

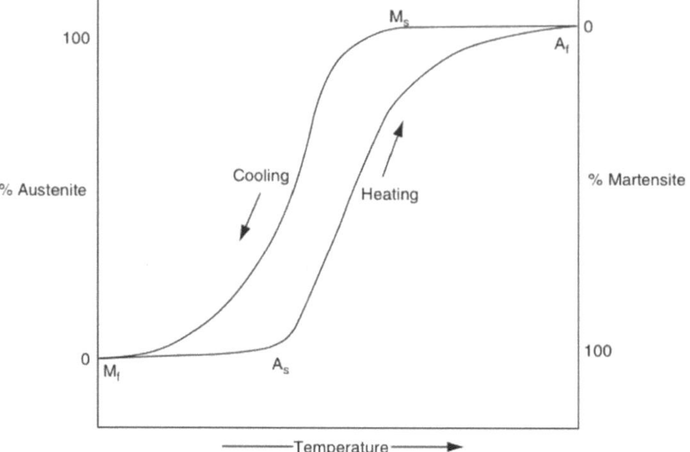

Fig. 3.42 Hysteresis loop associated with the phase transitions of shape-memory alloys

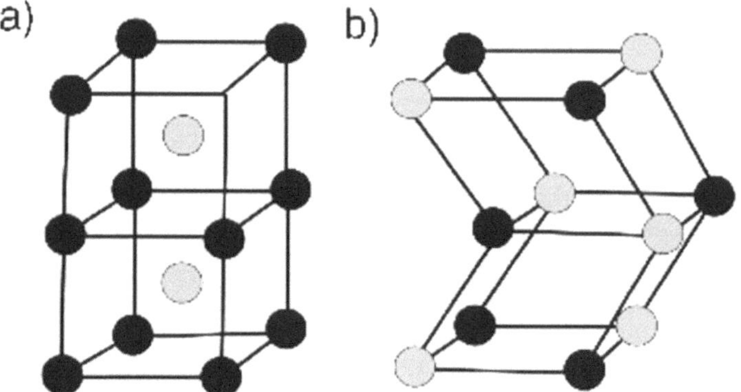

Fig. 3.43 Unit cell representations of (**a**) the CsCl structure of austenitic TiNi and (**b**) the monoclinic structure of martensitic TiNi

in that system, martensite is associated with interstitially dissolved carbon that adds strength through solute-hardening, but also brittleness due to the introduction of additional grain boundaries. It should be noted that there are 24 possible ways of accomplishing the austenite–martensite transformation; austenite has six equivalent facial planes, and each of these may shear along two perpendicular axes.

When austenite is cooled in the absence of applied stress, the material transforms into a twinned form of martensite (Fig. 3.44a). Since both austenite and twinned martensite have the same macroscopic shape/size, reheating the material will not result in any observable shape change. However, if the material is plastically deformed through bending, etc. at low temperature, it will become de-twinned and the new shape will prevail. The Ni–Ti alloys are preferred since they have a

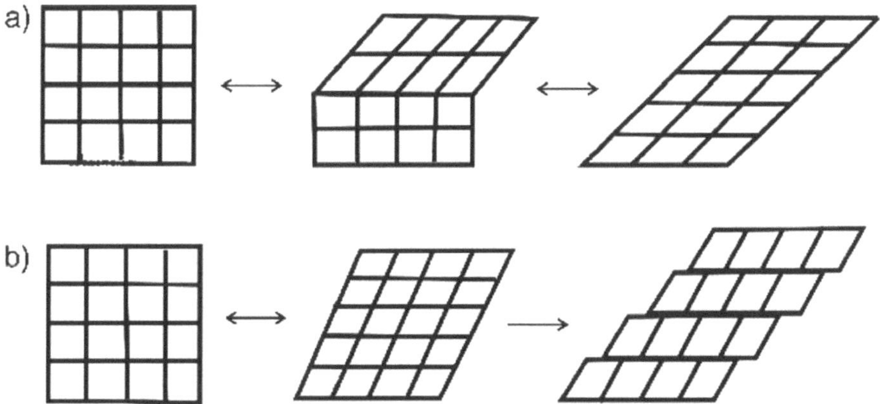

Fig. 3.44 Deformation pathways for (**a**) shape-memory alloys, showing the reversible movement of twin boundaries. Shown in (**b**) is the irreversible slip deformation of other alloys, such as carbon steels

greater range of deformation (up to 8%), relative to other Cu-based alloys (4–5%). When the material is reheated, the deformed martensite structure will be converted to the original austenite phase with a different macroscopic structure. For comparative purposes, Fig. 3.44b illustrates the irreversible slip deformation that other types of metals such as steel undergo as a result of the same stresses. Since these latter materials do not have suitable twin planes, shape-memory transitions are not possible resulting in a permanent shape alteration of the metal.

It is also possible to apply a stress to the material in its high-temperature austenitic phase. However, since the temperature is above A_f, the original shape will be reformed immediately after the load is removed. Such an immediate shape change is referred to as *pseudoelasticity* (or *superelasticity*), and is the active principle underlying cellular phone antennae that may be greatly distorted only to immediately return to their original shapes.

In addition to temperature- or stress-induced transitions, there are now many *ferromagnetic* shape-memory alloys that alter their shapes in response to a magnetic field. Examples of these systems include Fe–Pd, Fe–Pt, Co–Ni–Al, Co–Ni–Ga, and Ni–Mn–Ga. These materials are of great interest since the magnetic response time is faster and more reliable than temperature-based transitions. Whereas traditional alloys alter their structures due to the martensite–austenite transition, magnetic analogues exhibit a change in structure while remaining in the martensite phase. The change in shape is a result of the de-twinning of preferred planes based on their orientations with the applied magnetic field.

As one would expect, there are a variety of applications that currently use shape-memory alloy materials, with many more are projected for the future. The earliest application was for greenhouse window openers, with the metal serving as an actuator to provide temperature-sensitive ventilation. Some commercial faucets/showerheads are already equipped with this material that shuts off the water if a certain temperature is reached, which effectively prevents scalding. An intriguing future application will be for automobile frames; however, as we will discuss in Chap. 5, some plastics may also be designed with shape memory. Someday soon, your car may reshape itself in front of your eyes within minutes after an accident!

Since NiTi alloys have been shown to be biocompatible, perhaps the greatest use for these alloys has been for medical applications. In particular, for minimally invasive surgery where a metal wire is inserted through tiny incisions, and then reshaped into the original form (e.g., tweezers, specialized probes, etc.) while *inside* the patient due to an increase in temperature! Likewise, metal probes may be bent into specific shapes for open surgeries and then returned to their original positions afterward through the heating/sterilization process. Other widespread applications include nitinol filters that are designed to trap blood clots in arteries, as well as suture anchors that are inserted directly into bone to facilitate the attachment of soft tissue such as ligaments and tendons.[32]

Although this textbook has not yet introduced the topic of nanotechnology, it is worthwhile to point out that shape-memory behavior has also been discovered in single-crystalline Cu nanowires.[33] This is interesting since this effect does not occur in bulk copper metal. Due to the extremely high surface/volume ratio of nanowires relative to bulk structures, these materials show reversible strains of *ca.* 50%—an order of magnitude larger than bulk shape-memory alloys. Further, the martensite–austenite transformation temperature changes dramatically with minute changes in nanowire diameter. For instance, increasing the diameter from 1.76 to 3.39 nm causes a *ca.* 800 K increase in the transition temperature! This will allow for the design of nanoscale components of varying sizes that will be functional over an extremely wide temperature range—not possible with bulk alloy systems. More importantly, the response time for these materials is also orders of magnitude faster than bulk alloys due to the extremely small dimensions of the nanoscale. We will see more of the "nanoworld" in Chap. 6; this intriguing example was inserted to wet your appetite a bit …

Bulk Metallic Glasses (BMGs)

Metallic glasses combine properties of both plastics and metals, and are currently used within electric transformers, armor-piercing projectiles, and even sports equipment. The media have recently been focused on this latter application, for Liquid-Metal™ sporting goods, sensors, gears, and surgical instruments.[34] Unlike window glass, metallic glass is not brittle. Many traditional metals are relatively easy to deform, or bend permanently out of shape, because their crystal lattices are riddled with defects. A metallic glass, in contrast, will spring back to its original shape much more readily.

For most metals, the phase change from liquid to solid occurs at the freezing point (T_f). However, as shown in Fig. 2.2, the solid may be cooled below its freezing point at a rate fast enough to prevent crystallization. This supercooled solid nough to cause crystallization, the solid is supercooled. In the absence of supercooling, a liquid will tend to crystallize; however, if cooling occurs sufficiently slow to prevent crystallization, supercooling will occur which will result in an amorphous structure.

Amorphous metallic materials may be produced through a variety of procedures. Pure individual metal powders may be melted at high temperatures in an arc furnace. Depending on the composition of the melt, supercooling may be possible, resulting in a vitreous solid rather than a crystalline form. Although facile glass-forming solids such as B_2O_3 will form amorphous solids even upon relatively slow cooling (e.g., 1 K s^{-1}), metals generally require very high rates (>10^4 K s^{-1}) to prevent crystal growth. These high rates are accomplished by placing a material with high thermal conductivity (e.g., copper) in contact with a molten metal or alloy. This method is referred to as "melt spinning" or "melt extraction," and results in metallic ribbons up to 15 cm wide and 30 μm thick.

Another common procedure uses a vapor deposition technique to form amorphous metallic thin films (see Chap. 4.). Upon thermal annealing, the irregularly deposited atoms in the film have an opportunity to assemble into a crystalline array. A procedure referred to as *ball-milling* may also be used to create amorphous metal alloy powders. This method uses a mixture of crystalline powders that is vigorously agitated with ball bearings within a milling vessel. This results in regions of high pressure that cause local melting of the crystalline powders, breaking apart metallic bonds, and facilitating atomic diffusion along preferential crystallite interfaces. Interestingly, long durations of ball-milling have also been reported to induce crystallization of amorphous metal alloys.[35]

3.3 Metal Surface Treatments for Corrosion Resistance

The corrosive deterioration of metal surfaces incurs a great cost to the worldwide economy. Accordingly, there have been many research efforts devoted to understanding the surface chemistry behind these reactions. As we have already seen, this has led to the development of many useful alloys that are sufficiently resistant to corrosion via spontaneous formation of a protective oxide layer. However, for other less resistant metals such as carbon steels, a protective layer must be post-deposited onto a metal surface in an effort to prevent corrosion. In this section, we will discuss three strategies that may be used to protect a metal surface from its environment: inert-layer passivation—either native (e.g., Cr_2O_3 on Cr) or purposefully deposited, sacrificial metal coatings (e.g., galvanized steel), and organic-based coatings. It should be noted that a pretreatment process is often required for metal surfaces to allow the coatings to be strongly adsorbed to the surface. These steps effectively remove organic components such as oils, as well as inorganic species such as welding flux. The most common methods are either mechanical descaling (e.g., abrasive blast techniques) or chemical pickling (i.e., acid treatment).

The primary corrosive agents are oxidizing agents (e.g., moist air, HNO_3, H_2SO_4), or halogenated species (e.g., Cl_2, HCl, HF, CFCs). These agents degrade the metal by forming oxides, hydroxides, or halides that introduce embrittling grain boundaries on the surface. Other particularly detrimental gases for corrosion of metal surfaces are CO and H_2S. Both of these gases exhibit dissociative adsorption on metal surfaces, resulting in carbide or sulfide formation and concomitant embrittlement of the metal. In the presence of H_2S (or H_2O) at elevated temperatures, the hydrogen atoms may also interact with surface metal sites and cause surface cracking. Hydrogen-induced cracking is especially detrimental for iron surfaces.

For carbonaceous gases such as CO and CH_4 at relatively high temperatures (*ca.* > 800 °C), carburization of steel surfaces takes place in the form of brittle interstitial carbides that may cause surface cracking. Cementite may also form on the surface of steel; since its melting point is lower than the underlying metal, it may cause melting of the steel surface that is subsequently eroded by the gas stream.

One simple method used to deter the onset of corrosion is *phosphating*. This process is often used to chemically passivate a metal surface with a crystalline coating of zinc phosphate. The phosphating bath is an aqueous solution of dilute phosphoric acid, containing anionic and cationic elements that are capable of reacting with the metallic surface to yield a crystalline film on this surface. Other components of phosphating baths, known as accelerators, influence the kinetics of the reaction process and permit the control of redox reactions at the interface. The most common surface species present after phosphating are vivianite [$Fe_3(PO_4)_2{\cdot}4H_2O$], hopeite [$Zn_3(PO_4)_2{\cdot}4H_2O$], and phosphophyllite [$Zn_2Fe(PO_4)_2{\cdot}4H_2O$]. For more complex substrates such as steel coated with Zn/Fe, Zn/Ni, Zn/Al, and Zn/Cr alloys, tricationic phosphatings have been developed. For these systems, the surface is coated with crystalline pseudophosphophyllite [$(Zn, M, Ni)_3(PO_4)_2{\cdot}4H_2O$], where M = Fe, Al, Cr, etc. For aluminum-containing alloys, or hot-dipped galvanized steel (containing Al_2O_3 on the surface), it is necessary to use fluoride-based additives to cause surface crystallization.

Another useful passivation technique is *anodizing* or *anodic oxidation*. In this method, the metallic surface acts as an anode—being oxidized during an electrochemical event. The most common metals/alloys are those containing aluminum, magnesium, and zinc. However, it is also possible to anodize other metals such as copper, steel, and cadmium for protective and decorative applications. The anodizing electrolytic solution consists of strong acids, generally combinations of chromic, sulfuric, oxalic, or boric acids. The anodic layer formed in sulfuric acid baths consists of a relatively thin barrier layer, overlaid with a porous array. The density and morphology of the pores may be varied through manipulation of the electrical current or nature of the electrolytic solution. However, for protective applications, hydration sealing is often required that consists of steam treatment in the presence of chromate or Ni/Co salts. In the case of aluminum, this post-treatment results in boehmite, AlO(OH), which sufficiently seals the pores.

A variety of other nonmetallic coatings may be used to impart corrosion resistance to the underlying metal surface. Common inorganic-based coatings include vitreous enamels, ceramics, glass, cements, carbides, and nitrides. By contrast, organic-based protectants are paint coatings, plastic coatings, adhesive tapes, and sheet linings. Whereas the inorganic layers are often used to coat internal surfaces of piping and reactors, organic films are most often used for external surface protection. Refractory coatings such as carbides (e.g., TiC, B_4C, WC, and WCO), nitrides (e.g., AlN and BN), oxides (e.g., Al_2O_3, BeO, Cr_2O_3, ThO_2, and ZrO_2), silicides (e.g., $NbSi_2$, WSi_2,

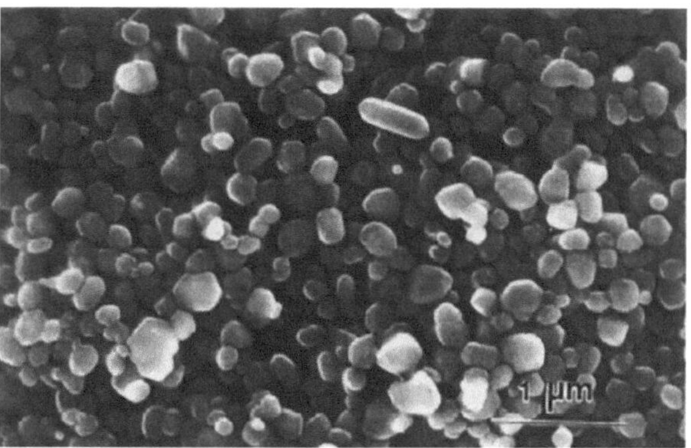

Fig. 3.45 Scanning electron micrograph of TiO_2 particles, used as white pigments in paint formulations

and $MoSi_2$), and borides (e.g., ZrB_2 and TiB_2) impart both corrosion/abrasive wear and temperature resistance to the underlying substrate.

More recently, an even greater corrosion resistance has been generated through use of composite coatings, comprising the above refractory compounds in association with a metal powder (e.g., Cr + ZrB_2, Cr/SiC + HfO_2, Al + SiC, Ti + TiB_2, and PtRh + ZrB_2).[36] For these coatings, the ceramic and metal powders are suspended in an aqueous solution with the assistance of surfactants, and sprayed onto the metal surface. This film is then allowed to dry at a temperature of *ca.* 70–90 °C, and annealed with an energetic laser source (e.g., Nd:YAG), resulting in formation of an interwoven matrix of metal and ceramic species.

Without question, the easiest and most inexpensive method to protect metal surfaces from corrosion is through simple painting. Paints are composed of finely divided solid inorganic or organic pigments (Fig. 3.45) that are suspended in a volatile solvent, alongside binder molecules. In paints and varnishes, the nature of the binder defines the type of paint system, such as oil or water based, epoxy, etc. Since varnishes do not contain light-scattering pigments their application results in a transparent coating. A volatile solvent is used to solubilize the binder and dilutants and is normally composed of organic compounds such as hydrocarbons, alcohols, ketones, ethers, and esters. Additives such as antifungal agents, driers, etc. are often used to broaden the application of the coating.

Contrary to popular belief, a paint coating will not be completely impervious to environmental agents surrounding the material. However, the integrity of the coating may be improved by using structurally complex pigments such as graphite, mica, aluminum oxide, etc. Due to the incomplete blockage of corrosive agents, sacrificially active paint pigments are often chosen based on the corrosive agents with which they will be in contact. For instance, impeding the corrosive ability of oxidizing agents can be achieved through use of Zn-rich pigments; likewise, pigments consisting of strong oxidizing agents such as chromates or manganates will offset reductive corrosion pathways.

Passivation and painting are only effective when the coating completely encapsulates the metal surface. Sacrificial metal coatings are also problematic, as the

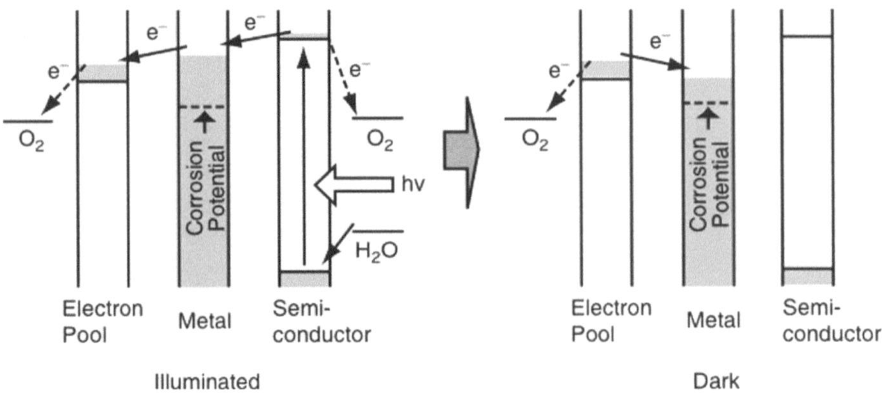

Fig. 3.46 Schematic of an anticorrosion system containing both semiconductor and "electron pool" storage components. Reproduced with permission from *Chem. Mater.* **2001**, *13*(9), 2838. © 2001 American Chemical Society

protective metal will eventually degrade exposing the underlying substrate. Most recently, an alternative strategy has been developed, referred to as *photoelectrochemical* protection by semiconductor coatings.[37] When photoactive semiconductor particles such as TiO_2 are deposited onto steel or copper substrates, the underlying metal exhibits significant corrosion resistance upon exposure to UV irradiation. Although the details of the photoactivity of semiconductor films will be discussed in Chap. 4, it is important to note here that exposure to light of the appropriate energy causes electrons in the semiconductor to be excited. These electrons are transferred to the underlying metal, which creates a potential that is more negative than its corrosion potential.

Although photoelectrochemical effects will usually only occur in the presence of UV light, there are reports of using complex multilayered or composite films to yield anticorrosion properties even in the dark.[38] Although the mechanisms are not presently well known, the general principle involves the use of films that contain an intimate connection between semiconductor and electron-storage particles. After the UV light has been turned off, the stored electrons are injected to the metal so it is still protected from corrosion (Fig. 3.46).

3.4 Magnetism

Even as young children, we are introduced to the intriguing magnetic properties of iron compounds, based on their strong attraction toward a permanent magnet. For instance, one of the most popular toys, MagnaDoodle™, operates solely through the attraction of magnetic iron oxide (magnetite, Fe_3O_4) particles toward a handheld stylus magnet. In the case of elemental iron, not all allotropes are magnetic; among the pure iron forms, only ferrite (α, bcc) is magnetic. This is intriguing, as the δ-Fe form also exhibits a body-centered cubic crystal structure. This must indicate that in addition to the simple 3-D arrangement of lattice iron atoms, their individual magnetic dipoles must also be suitably aligned in order to yield a particular magnetic behavior.

In contrast to *diamagnetism*, where all valence electrons of each atom are spin-paired, *paramagnetism* is found in solids where the constituent atoms contain an unpaired valence electron(s). In a simple paramagnetic substance, the unpaired electrons' spins are randomly oriented within the solid. Upon exposure to an external magnetic field, the spins become collectively oriented along the direction to the applied field. However, the dipoles re-randomize when the field is removed. In contrast, when a diamagnetic material is placed into an external field, the induced dipoles become aligned opposite to the field direction. This results in a very weak effect that is of little practical importance. It should be noted that both diamagnetic and paramagnetic materials are considered to be nonmagnetic since they exhibit magnetization only in the presence of an external field.

The magnetic responses of diamagnetic and simple paramagnetic substances are small enough that a special instrument called a magnetic susceptibility balance is required to measure these effects. This technique measures the amount of repulsive (for diamagnetic) or attractive (for paramagnetic) force between the sample and a permanent magnet within the instrument. For paramagnetic substances, the magnitude of the attractive response is proportional to the number of unpaired electrons present in the sample. Hence, this technique provides an efficient means to determine the ground-state electronic configuration of transition metal complexes.

As temperature increases, the magnetic susceptibility, χ, of a paramagnetic substance decreases. The increasing thermal motion of atoms comprising the solid disrupts the ordering among neighboring magnetic dipoles. Most often, the *effective magnetic moment*, μ_{eff}, is used to describe paramagnetic behavior, since this quantity is independent of both the temperature and the magnitude of the external field. Qualitatively, the macroscopic magnetic moment of a solid may be thought of as a vector summation of all the microscopic magnetic dipole moments of each atom.

In a *ferromagnetic* material, the magnetic dipoles generated from unpaired electrons tend to align in the same direction, even in the absence of an external magnetic field. This phenomenon is aptly referred to as ferromagnetic coupling. It should be noted that ferromagnetism is the direct opposite of superconductivity, where all electron spins pair to form a perfectly diamagnetic material. Examples of ferromagnetic behavior may be observed in bulk iron, cobalt, nickel, and some rare-earth elements (e.g., Gd). The regions containing parallel-aligned magnetic spins are known as ferroelectric or *Weiss domains*, with a Bloch wall providing an interface between two adjacent domains. When an external magnetic field is applied, those domains that are aligned parallel to the direction of the field are energetically favored over those magnetized in opposing directions. The favored domain walls then expand at the expense of the unfavored, resulting in a net magnetization (Fig. 3.47). Whereas the magnetic susceptibility of paramagnetic materials is on the order of 10^{-5}–10^{-2}, ferromagnetic materials exhibit values of *ca.* 10^{6}.

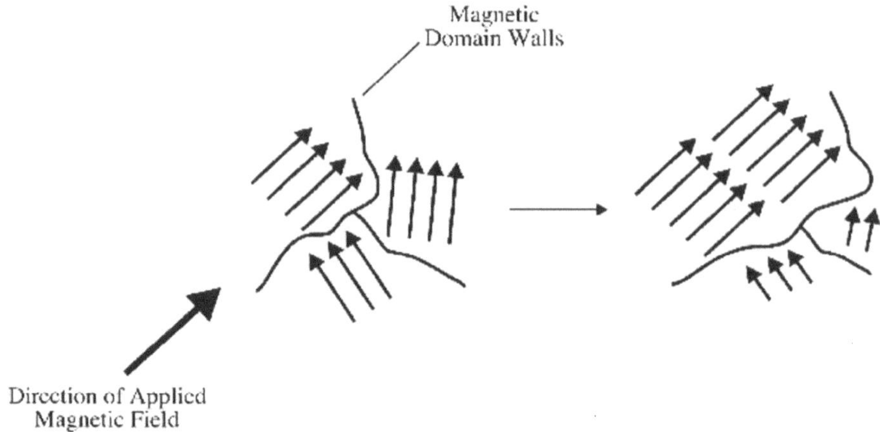

Fig. 3.47 Representation of a Bloch wall expansion resulting from an applied magnetic field impinging on a ferromagnetic material

The ground-state electronic configuration of iron in the solid-state metal lattice is [Ar] $4s^23d^6$. The magnetic moment per iron atom in a cm^3 solid is 2.2×10^{-20} emu, or approximately 2 Bohr magnetons ($\mu_B = 1 \times 10^{-20}$ emu).[39] Therefore, there are two unpaired electrons in each iron atom throughout the lattice. By comparison, the μ_B values for cobalt ([Ar] $4s^23d^7$) and nickel ([Ar] $4s^23d^8$) are 1.72 and 0.61, respectively. Since there are five d orbitals and six d-electrons for iron, two separate d orbitals must house the unpaired electrons. In the bcc array of iron, two orbitals (d_{z^2} and $d_{x^2-y^2}$, positioned *along* the Cartesian axes), are not directed toward neighboring atoms in the lattice.[40] Hence, these orbitals will have nonbonding character, and may therefore accommodate two unpaired electrons. The remaining four electrons within d_{xy}, d_{xz}, and d_{yz} orbitals (having lobes directed *between* the Cartesian axes) participate in metallic bonding between neighboring iron atoms, forming a valence band of paired electrons.

The magnetic response of a ferromagnetic material exposed to an external field is typically represented by an S-shaped field-dependent *magnetization curve* (Fig. 3.48). By definition, the applied and induced magnetic fields are given the symbols H and B_o, respectively. When the external magnetic field reaches a maximum value (H_s), the material will form a single domain with a net *saturation* magnetization (B_s) in a direction parallel to the applied field. Once the external field is removed from the material, the domain walls spring back toward their original positions, and the magnetization decreases to a value referred to as the *remanence* (B_R). This is the operating principle of magnetic storage devices such as credit card magnetic stripes and computer hard drives. To remove the induced magnetism of the material, an opposite magnetic field is applied, known as the *coercive magnetic field* (H_c). If the magnitude of the opposed field is increased to a maximum value ($-H_s$), a saturation magnetization will again be found in the material, parallel to the new direction of the applied field.

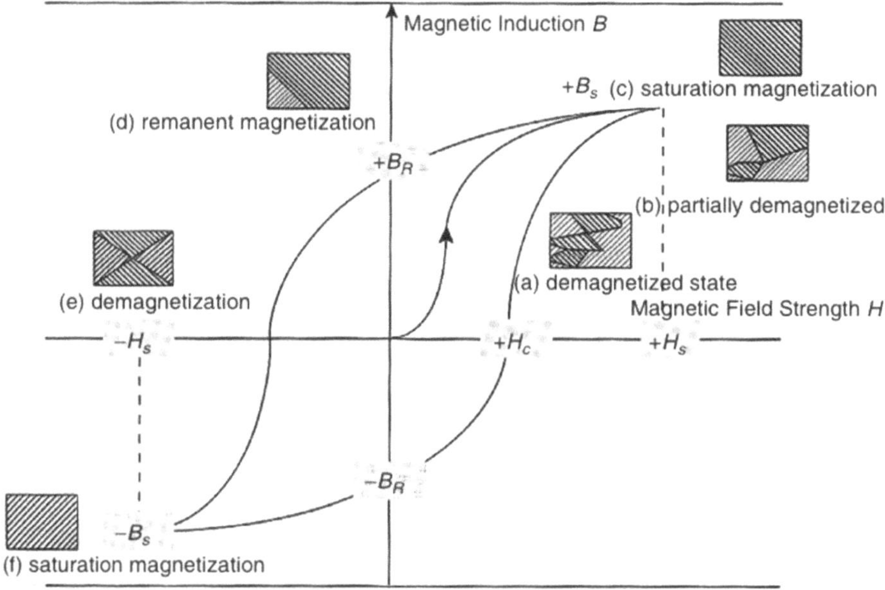

Fig. 3.48 A *B–H* magnetization hysteresis curve for a ferromagnetic material. Reproduced with permission from Cardarelli, F. *Materials Handbook*, Springer: New York. © 2000 Springer Science and Business Media

It is noteworthy that the bulk size and shape of a ferromagnetic metal may also change, due to reversible magnetization. This phenomenon, referred to as *magnetostriction*, is due to the coupling of electron spins between neighboring atoms, which affects the delocalized electrons involved in metallic bonding. This effect is responsible for the familiar hum of power line transformers and fluorescent lights, due to the vibration of the iron components within these materials.

At low temperatures, spontaneous *antiferromagnetic coupling* between neighboring atoms results in an equal number of magnetic dipoles in opposing directions. However, as the temperature is increased, the dipoles are randomized resulting in paramagnetic behavior. Primary examples of antiferromagnetic behavior include transition metal compounds such as MnO, NiO, MnS, $FeCO_3$, MnF_2, as well as certain metal clusters (e.g., Fig. 3.49). In contrast to ferromagnetic behavior, *ferrimagnetic* coupling results in magnetic spins in two opposite orientations, with more in one direction than in the other. Commonly, ferrimagnetic materials crystallize in a spinel lattice such as magnetite, Fe_3O_4. Magnetite is comprised of a Fe^{2+}:Fe^{3+} ratio of 1:2, with spin magnetic moments of 4 and 5 Bohr magnetons, respectively. As shown in Fig. 3.50, the spins of Fe^{3+} ions in octahedral and tetrahedral sites cancel each other; hence, the resultant magnetization is due solely to the Fe^{2+} ions in tetrahedral sites.

Vibrational motion of the molecules can disrupt the domain structure. Hence, the magnetic properties of ferro-, antiferro-, and ferrimagnetic materials are strongest at low temperatures. At sufficiently high temperatures, no domain structure may

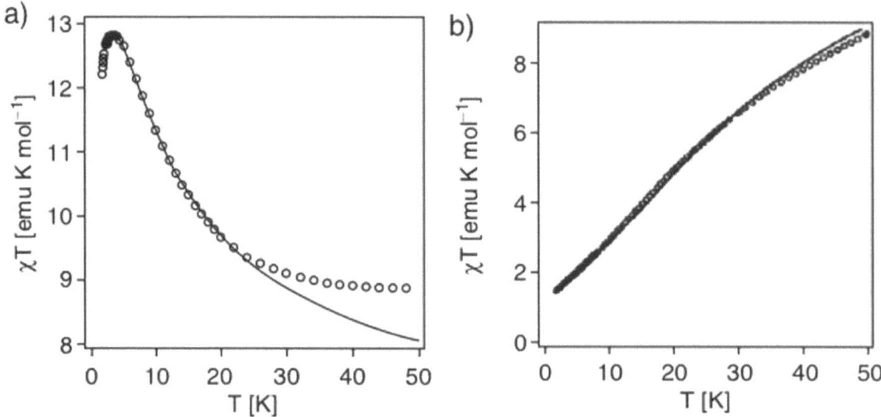

Fig. 3.49 Comparison of magnetic susceptibility profiles for **a** $Na_{12}[Co_3\ W(H_2O)_2(ZnW_9O_{34})_2] \cdot 46H_2O$ and **b** $Na_{12}[Co_3\ W(D_2O)_2(CoW_9O_{34})_2] \cdot 46D_2O$. Whereas the profile of the Co_3 cluster is indicative of ferromagnetic coupling between Co(II) ions, the trend for the Co_5 cluster is representative of antiferromagnetic coupling between Co(II) centers. Reproduced with permission from *Inorg. Chem.* **2001**, *40*, 1943. © 2001 American Chemical Society

form, resulting in paramagnetic behavior. The cutoff temperature for the onset of paramagnetic behavior is referred to as the *Curie* (T_c) or *Neel* (T_N) temperature for ferro-/ferrimagnetic or antiferromagnetic materials, respectively. Curie temperatures range from 16 °C for Gd, to 770 °C and 1120 °C for Fe and Co, respectively. By contrast, Neel temperatures range from −271 °C for $MnCl_2 \cdot 4H_2O$ to 680 °C for α-Fe_2O_3.

The larger the gap between B_o and B_R, the more effective the material will be for magnetic storage applications. Magnetically "soft" materials such as pure iron, low carbon steels, and alloys of ferromagnetic elements (e.g., Fe, Co, Ni), consist of an ordered array of ferromagnetic atoms that easily revert back to their original domain structures following the removal of the external field. In contrast, a large remanence value indicates that the domain walls are not transformed back to their original position—known as "hard" magnetic materials. As one would expect, the microstructure of the solid is paramount to the relaxation efficiency of the Bloch walls. That is, the domain walls will be less likely to relax to their original positions if the lattice contains trace amounts of interstitial dopants such as Si, C, O, or N, or dispersed particles from precipitation hardening processes (i.e., "domain-wall pinning"). Such hard-magnetic materials are known as *permanent magnets*, retaining their magnetism over prolonged periods of time after the external field is removed. In addition to high B_R, the materials also exhibit a relatively high coercive field, H_c.

Common examples of magnetically hard materials are high-carbon steels, precipitation- hardened alloys (e.g., Alnico), and sintered or bonded fine-particle alloys (e.g., ferrites, rare-earth alloys). The earliest examples of rare-earth magnets are $SmCo_5$ and Sm_2Co_{17}, with recent developments focused on the incorporation of Fe rather than other costly transition metals. Rare-earth alloys that incorporate Fe

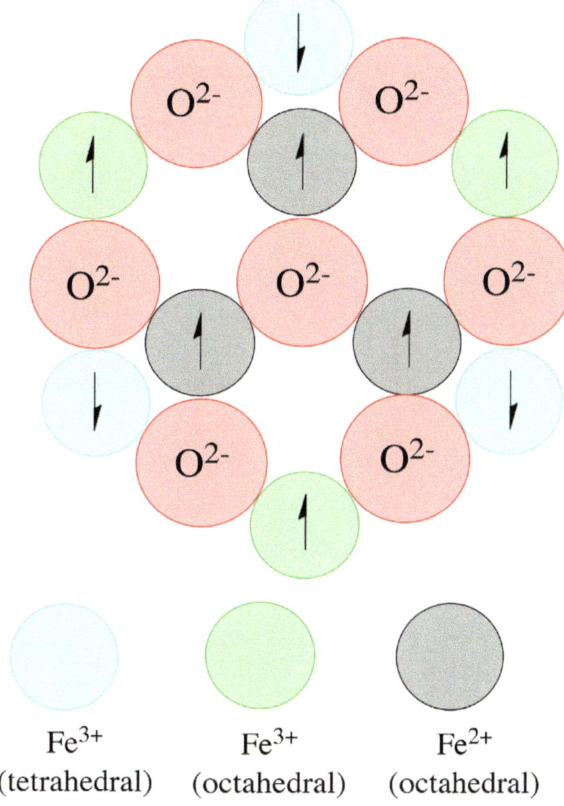

Fig. 3.50 Schematic showing the spin magnetic moment configuration for Fe^{2+} and Fe^{3+} ions in Fe_3O_4

such as R_2Fe_{17} typically have low operating temperatures; however, the addition of boron results in the ternary compound $Nd_2Fe_{14}B$ with strong uniaxial magnetocrystalline anisotropy, and a higher operating temperature. The absorption of nitrogen to yield $Sm_2Fe_{17}N_3$ causes a further improvement in the magnetic properties; however, the applications are limited by its complex synthesis. It should be noted that the partial substitution of Co for Fe reduces the surface oxidation. This is important, since the processing of rare-earth magnets involves the compaction of finely divided powders, which are more difficult to obtain through ball-milling if a hard oxide layer is present. Interestingly, since the Curie temperature of Co greater than Fe, the T_c of the ternary alloy $Nd_2(Co_xFe_{1-x})B$ increases at a rate of *ca.* 10 °C per at% Fe.

The origination of the desirable magnetic properties within the complex structures of rare-earth alloys is not completely understood. Likely, the interaction of the 4f electrons with neighboring lattice atoms results in a preferential alignment of the rare-earth magnetic moments along specific lattice directions. This allows for saturation magnetization to be achieved with only relatively small applied fields. A high intrinsic coercivity will also result, since significant energy is required to disrupt the preferential alignment, known as the *magnetocrystalline anisotropy energy*. It should be noted that such magnetic anisotropy is also found in single crystals of pure metals such as Fe, Ni, or Co (Fig. 3.51).

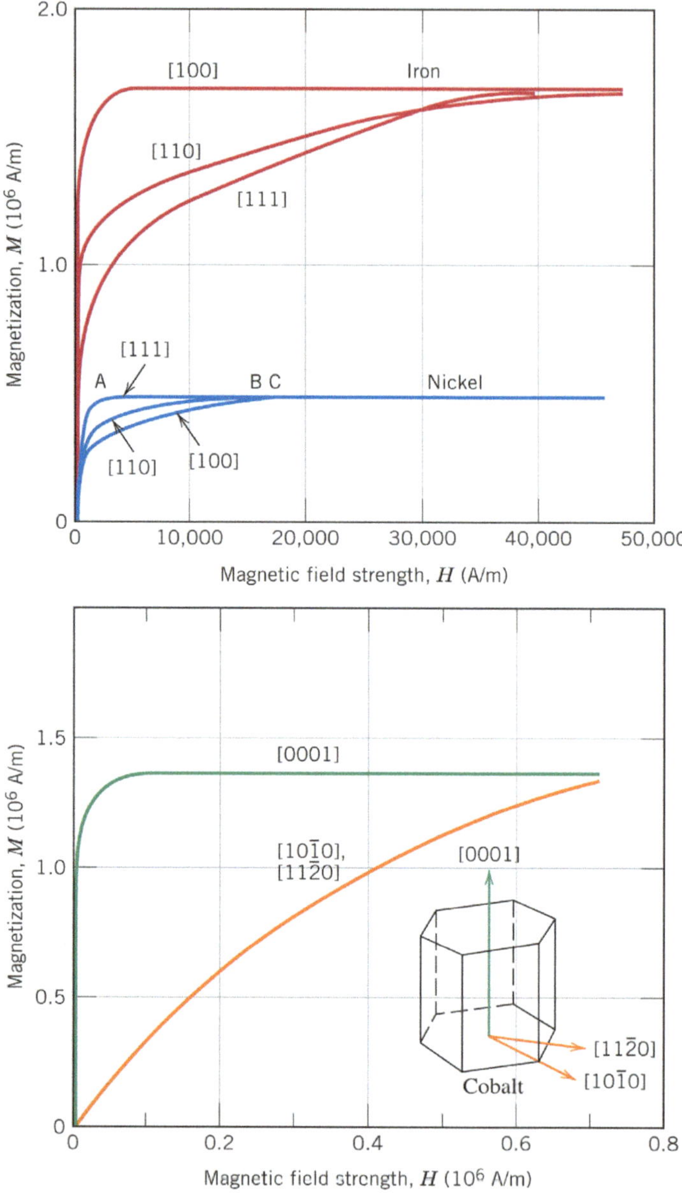

Fig. 3.51 Top: Magnetization curves for single-crystal Fe and Ni, with varying directions of the applied magnetic field. Bottom: Magnetization curves for single-crystal Co. Reproduced with permission from Callister, W. D. *Materials Science and Engineering: An Introduction*, 7th ed., Wiley: New York, 2007. © 2007 John Wiley & Sons, Inc

Table 3.8 Common applications for the rare-earth elements

Application	Rare-earth element
Magnetics (computer hard drives, disk drive motors, wind turbines, vehicle anti-lock brakes, automotive parts, frictionless bearings, magnetic refrigeration, power generation, microphones and speakers, MRI, communication systems)	Nd, Dy, Pr, Sm
Phosphors (LED and fluorescent lighting, lasers, medical imaging, light bulbs, fiber optics)	Nd, Eu, Tb, Y, Er, Gd, Ce, Pr
Metal alloys (Ni-metal hydride batteries, fuel cells, sports equipment, steel, Al/Mg alloys)	Nd, Y, La, Ce, Pr
Catalysts (petroleum refining, vehicle catalytic converters, fuel additives, chemical processing, air pollution controls)	Nd, La, Ce, Pr
Ceramics (capacitors, sensors, colorants, scintillators, refractories)	Nd, Y, Eu, Gd, Lu, Dy, La, Ce, Pr
Glass (digital camera lenses, polishing compounds, pigments and coatings, UV-resistant glass, photo-optical glass, X-ray imaging)	Nd, Gd, Er, Ho, La, Ce, Pr
Homeland security (satellite communications, guidance systems, aircraft structures, smart missiles)	Nd, Eu, Tb, Dy, Y, Lu, Sm, Pr, La

Rare-earth metals: function versus availability

The use of rare-earth (RE) metals in magnets represents only one application for these elements. Table 3.8 lists some representative applications for the collective group of rare-earth elements, composed of the lanthanides, yttrium, and scandium. The world's demand for the rare earths began to flourish in the mid-1960s, with the use of europium in color televisions. Currently, the largest use of these elements is for catalysis (62%), followed by structural alloys (13%), glass (9%), and magnets (7%).

Contrary to popular belief, the rare-earth elements are not abundantly rare; other than promethium, the rare earths have a greater natural abundance than silver, gold, or platinum. In fact, cerium is more abundant in Earth's crust than copper or lead! However, it is generally difficult to find these elements in concentrations large enough to allow an economical extraction from their ores. Typical RE deposits include carbonatites (e.g., bastnasite, a carbonate-fluorine mineral), monazite (a thorium-phosphate mineral), and clays.[41]

The hydrometallurgy processes used to mine and extract REs from their ores are environmentally hazardous and dangerous to human health. For instance, natural deposits typically contain radioactive elements such as thorium or uranium, which may be released to the environment or impart radiological effects to workers. The extraction of 1 ton of RE generates 1000 tons of wastewater and 2000 tons of toxic and radioactive mine tailings, which may pollute the surrounding soil and nearby water supplies. Other effects of RE mining include soil erosion, loss of biodiversity, flooding, pollution, and crop uptake of REs, which may lead to human health issues.[42]

As shown in Fig. 3.52, China currently produces 84% of the world's supply of the RE metals, but also accounts for 70% of the global demand. A global wakeup call occurred in 2010 when China controlled about 95% of the world's RE production and imposed significant restrictions on their exports. As a result, the prices for rare-earth oxides rose over 500% in just a few years, and nations became dependent on China for virtually all high-tech sectors, including advanced weaponry and defense systems. The U.S. Department of Energy has defined a criticality matrix (Fig. 3.53), which lists five REs (Nd, Dy, Eu, Y, and Tb) that are important for many advanced technologies, but pose a serious near- and long-term supply risk.

Countries such as Brazil, Vietnam, Russia, and India have since identified significant reserves. Hence, although China is still the world leader in RE production, they now only control about 36% of the world's total reserves. These factors have caused the price of rare-earth metals to decline in recent years, which has led some to conclude that the "rare-earth crisis" is over. However, many geopolitical factors could still lead to large variances in the price and availability of these critical metals.

Due to supply and cost instabilities, as well as unsustainable mining practices, manufacturers have begun to reduce the amount of REs in their devices. For instance, dysprosium may be replaced with cerium in Nd-Fe-B high-performance magnets[43]; Nissan has decreased the amount of Dy in the motor of the Leaf electric car by 40% through more efficient processing. The shift from fluorescent lighting to LEDs has also greatly reduced the amount of REs used in the lighting industry. Although many companies such as Apple have vowed to eliminate the use of REs altogether, it is not an easy proposition. Many of the unique magnetic, electronic, and optical properties of these elements, which made them so desirable in the first place, have not been reproduced by any natural or synthetic substitutes investigated thus far. However, some progress has been made; for instance:

- Iron nitride permanent magnets are reported to outperform neodymium analogues[44];
- Hitachi has introduced RE-free motors and pumps[45];
- Tesla has created a copper-rotor induction motor for its Model S electric vehicles, which eliminates the use of RE magnets.[46]

With only minor resources at their disposal, some nations (e.g., U.S., Canada, or E.U.) may always be dependent on foreign producers for these metals, which are essential for their technological development and military prowess. To expand their production capabilities, some mining companies are now exploring sites such as the ocean floor,[47] or even the surface of the moon or asteroids![48] However, recycling efforts may offer the best opportunity to decrease a country's reliance on foreign suppliers.[49] Currently, only about 1% of rare-earth metals are recycled. This is largely due to the very low concentrations of these metals in electronics. For instance, a smartphone only contains 50 mg of Nd and 10 mg of Pr. Furthermore, the pulverizing and shredding of electronic devices to recover these metals would result in very little recovery. Recently, Apple has introduced the Liam robot to carefully disassemble iPhones and improve the recovery of rare-earth and other valuable metals such as Au, Ag, and Li.[50]

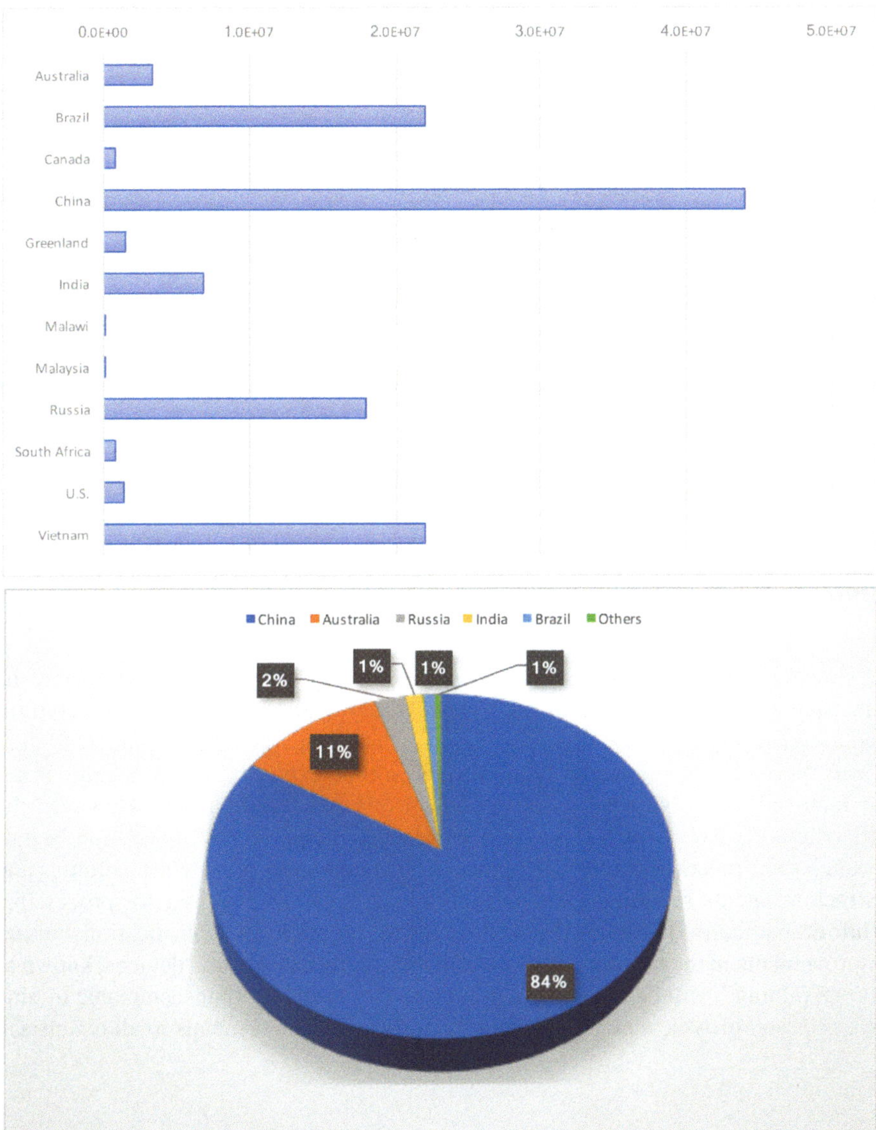

Fig. 3.52 Top: world reserves of rare-earth elements by country (2021 estimates, metric tons). Bottom: relative percentage of rare-earth element production by country. Data taken from https://geology.com/articles/rare-earth-elements/

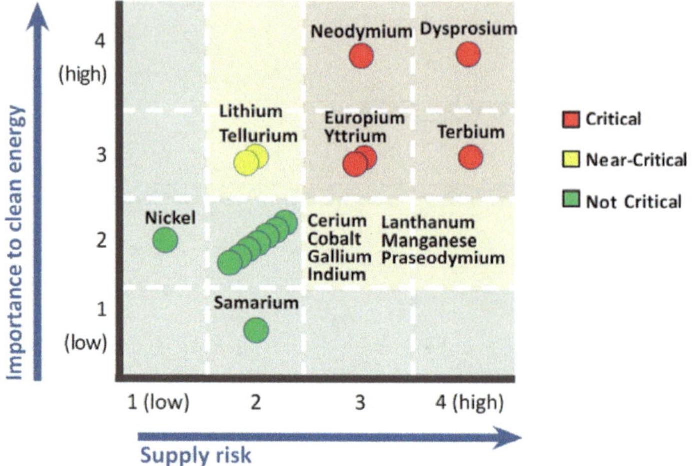

Fig. 3.53 Criticality matrix for elements of importance for clean energy (wind turbines, electric vehicles, solar cells, and fluorescent lighting. Image taken from *Critical Materials Strategy*, U.S. Department of Energy, Dec. 2011. May be accessed online at: https://www.energy.gov/sites/prod/files/DOE_CMS2011_FINAL_Full.pdf

However, once the RE-containing components are removed from the device, the subsequent extraction and purification processes typically require high temperatures and use of strong acids and organic solvents, which generate large amounts of liquid waste. Notwithstanding the problem of radioisotope exposure, this procedure is not far removed from that used to extract REs from their original ores. However, the efficiency of solvent extraction has been recently improved by Innovation Metals through their patented RapidSX™ process.[51] Their use of proprietary columns and extraction agents reduce the amount of organic solvents required for processing, while dramatically shortening the time needed for separation and purification. Further improvements in the recovery of rare-earth metals from electronic devices, known as "urban mining", will be essential to minimize the effects from unsustainable mining practices and provide an equitable supply of these essential metals to all countries.

3.5 Metals Case Study: Hydrogen Storage

With cyclic gasoline prices and increasing awareness of climate change, mankind is in constant search for renewable fuel sources. Hydrogen is the most attractive alternative due to its nonpolluting nature, only yielding water as a byproduct of its combustion. However, before widespread utilization of this medium is possible, two key issues must be solved: "green" hydrogen generation and storage. Although H_2 contains three times more energy per mass than gasoline (i.e., comparative specific capacities: 33.3 kWh kg^{-1} for H_2 vs. 12 kWh kg^{-1} for gasoline), a relatively large volume must be used due to its extremely low density (i.e., comparative volumetric capacities: 2.4 kWh L^{-1} for H_2 vs. 8.8 kWh L^{-1} for gasoline).

Although liquid hydrogen carries more energy density than gaseous H_2, there are prohibitive costs involved in H_2 liquefaction and cryogenic tank production. Furthermore, liquid hydrogen must be stored at temperatures below 20 K, which presents more safety issues than compressed hydrogen tanks in the event of an accident. For instance, rapid tank pressurization due to a temperature increase, cold embrittlement of construction materials, and line blockage due to solidification of moisture. As a result, compressed hydrogen gas (700 bars, 70 MPa, or 10,000 psi) is used in current commercial fuel cell vehicles (FCVs), such as the Hyundai Nexo and Toyota Mirai II. However, in order to increase the energy density, there continues to be much interest in the search for solid-state materials that can reversibly and safely store energetic fuels such as hydrogen.[52]

High surface-area carbonaceous materials such as activated carbons, aerogels, carbon nanotubes, and graphene have been shown to store significant concentrations of H_2, but are limited by their low storage packing densities (SPDs).[53] More recently, *metal–organic frameworks* (MOFs) have been shown to adsorb and pack more H_2 in its cavities at 77 K than any unpressurized structure to date, likely due to the presence of unsaturated metal centers.[54] The maximum adsorptive H_2 storage capacity of porous solids is directly related to the surface area accessible by the hydrogen molecules. Hydrogen adsorption follows Chahine's rule, which states that hydrogen adsorption is *ca.* 1 wt% H_2 per 500 m^2 g^{-1} of surface area at 77 K and pressures above 20 bar.[55]

Hydrogen also combines with many elements to form binary hydrides, MH_n. There are three general classes of hydrides:

1. Saline or binary (involving Group 1 and 2 metals; may be envisioned as an ionic lattice consisting of M^{n+} and H^- ions, e.g., LiH, NaH, BaH_2).
2. Covalent (molecules containing covalently bound hydrogen to nonmetals, with individual molecules held together by intermolecular forces, e.g., CH_4, SiH_4).
3. Metallic/interstitial (hydrogen molecules are contained in vacant interstitial sites of a transition-metal lattice, e.g., $PdH_{0.6}$).

For a material to be useful for hydrogen storage in energy devices, the following five targets have been established by the U.S. Department of Energy[56]:

1. The solid material must be able to adsorb/desorb at least 6.5 wt% (*gravimetric capacity*, known as "specific energy") and 50 g L^{-1} (*volumetric capacity*, known as "energy density") of hydrogen gas;
2. The storage system cost should be less than \$8 kWh^{-1} (\$266 kg^{-1} H_2);
3. The decomposition temperature necessary for generation of hydrogen from the material should be in the range of 60–90 °C;
4. The absorption/desorption of hydrogen from the material should be reversible, lasting a minimum of 1500 fueling/use cycles;
5. The material should be low cost, precluding the use of noble-metal alloys;

6. The storage solid should be nontoxic and inert under environmental conditions. That is, the solid should not react with water, oxygen, nitrogen, etc.

Table 3.9 lists some important metals and alloys that have been studied for hydrogen storage applications. To date, no material (neither metals nor nonmetals) has been discovered that satisfies all the above six constraints. Although hydrides exist for most elements of the periodic table, only the light elements (e.g., Li, Mg, Al) are able to meet criterion 1 above.

The most widespread application for hydrogen storage materials continues to be for the negative electrode (cathode) in rechargeable alkaline nickel–metal hydride (Ni–MH) batteries—used extensively in portable electronic devices and electric vehicles. The most common metal hydrides used for battery applications are intermetallic species such as AB2, AB5, AB, and A2B where Group A elements are La, Ti, Zr, Mg, etc. and B are Ni, Co, Fe, Mn, etc. (see Table 3.9).[57] The development of complex alloys was necessary to circumvent the high equilibrium pressures that early batteries exhibited at room temperature. The composition of metal hydrides may now be fine-tuned to offer low operating pressures, corrosion resistance, and reversible H_2 storage.

The decomposition temperature, T_{dec}, for binary metal hydrides (MH_x) is found to correlate strongly with the standard reduction potential, $E°$ (Fig. 3.54). In particular, the easier it is to reduce the metal (i.e., a larger reduction potential), the lower the temperature that is required to decompose the solid into a metal and hydrogen gas (3.30–3.32):

$$M^{n+} + n\,e^- \rightarrow M^0 \tag{3.30}$$

$$2n\,H^- \rightarrow 2n\,e^- + n\,H_2 \tag{3.31}$$

$$\text{Overall: } 2\,M^{n+} + 2n\,H^- \rightarrow 2\,M^0 + n\,H_2 \tag{3.32}$$

Ternary hydrides of the general formula $(MH_x)_a(EH_y)_b$, where E is either a metal or nonmetal, are also important candidates for hydrogen storage applications. For example, some of the highest wt% storage values are exhibited by reducing agents such as sodium metal or lithium borohydride—$NaBH_4$ and $LiBH_4$, respectively. Relevant for materials design, the decomposition temperature of ternary hydrides may be altered through choice of E^{y+}. For example, the T_{dec} of $LiGaH_4$ is *ca.* 50 °C higher than that of the binary GaH_3; on the other hand, the T_{dec} of BeH_2 is *ca.* 225 °C greater than that of $Be(BH_4)_2$.

The trend in T_{dec} may be rationalized by the relative difference in electronegativities of M^{x+} and E^{y+} species. For $LiGaH_4$, Ga^{3+} is a stronger Lewis base than Li^+, indicating that electron density will preferentially flow away from the gallium center, forming ionic Ga–H bonds. This causes a strengthening of the Li^+–H^- interactions through donation of H^- to Li^+, resulting in a higher overall T_{dec}. In contrast

Table 3.9 Comparison of metals and alloys for hydrogen storage

Metal/alloy ($MH_{x(ads)}$ compound)	Gravimetric capacity (wt%)	Decomposition temp. (°C)	Reversible H_2 adsorption/desorption?
Pd ($PdH_{0.6}$)	0.6	25	Yes
"A_2B_7"[a]	1.6	65	Yes
$La_{3-x}Mg_xNi_9$ (x = 1.0, 1.1)	1.6	<100	Yes
"AB_2"[b]	1.7–3.3	<100	Yes
$LaNi_5$($LaNi_5H_6$)	2.5	25	Yes
FeTi ($FeTiH_{1.7}$)	2.5	25	Yes
BaRe ($BaReH_9$)	3.5	<100	Yes
Mg_2Ni(Mg_2NiH_4)	3.6	25	Yes
Na (NaH)	4.2	425	Yes
$LaMg_{12}$(LaH_3, MgH_2)	4.6	290	Yes
Ca (CaH_2)	4.8	600	Yes
NaAl:Ti ($NaAlH_4$: TiO_2)	5.5	125	Yes
Li_2N(Li_2NH)	6.7	285	Yes
Mg (MgH_2)	7.6	330	No
LiAl ($LiAlH_4$)	8.0	180	No
Li_3Be_2($Li_3Be_2H_7$)	8.7	300	Yes
LiB:Si ($LiBH_4$: SiO_2)	9.0	200–400	No
NaB ($NaBH_4$: H_2O)	9.2	25	No
CaB_2 ($Ca(BH_4)_2$)	9.6	400	Yes
Al (AlH_3)	10.0	150	No
Al:N ((NH_3)AlH_3)	12	150	No
Li (LiH)	12.6	720	No
NaB ($NaBH_4$)	13.0	400	No
LiB:N ($LiBH_4$: NH_4F)	13.6	25	No
MgB_2 ($Mg(BH_4)_2$)	14.9	350	No
AlB_3 ($Al(BH_4)_3$)	16.8	158	No
Be (BeH_2)	18.2	250	No
LiB ($LiBH_4$)	18.4	380	No
BeB_2 ($Be(BH_4)_2$)	20.6	40	No

Note This table does not include important nonmetals such as carbon allotropes, metal organic frameworks (MOFs), or boron nitride compounds

[a] A = La, Mg; B = Ni. For instance, the (La, Mg)$_2$Ni$_7$ and (La, Mg)$_5$Ni$_{19}$ phases have been identified in $La_{0.75}Mg_{0.25}Ni_{3.05}Co_{0.2}Al_{0.05}Mo_{0.2}$ alloys[64]

[b] A = V, Ti; B = Zr, Ni. Also includes complex combinations (*e.g.*, $ZrNi_{1.2}Mn_{0.48}Cr_{0.28}V_{0.13}$)

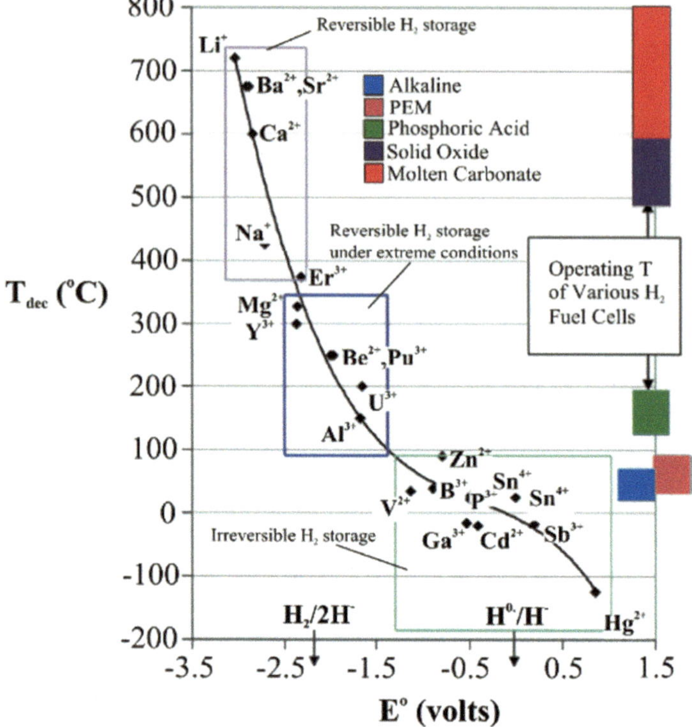

Fig. 3.54 Relationship between the reduction potential and decomposition temperature for binary hydrides. Reproduced with permission from *Chem. Rev.* **2004**, *104*, 1283. © 2004 American Chemical Society

for $Be(BH_4)_2$, the B–H bonds are covalent in nature, which causes H^- to be withdrawn from Be^{2+}. This results in a relatively low T_{dec} value that approaches room temperature.

A variety of molecular transformations take place during the formation of metal hydrides. Once hydrogen gas is adsorbed on the metal surface, the diatomic hydrogen molecule is dissociated—a process that requires a great deal of energy. At that point, individual hydrogen atoms migrate from the surface to the bulk of the material where nucleation/growth of the hydride phase begins.

Among the possible ternary hydrides, $NaAlH_4$ is most attractive for hydrogen storage applications due to its relatively low H_2 desorption temperature (80 °C *vs.* > 300 °C for magnesium compounds). The reactions involved in the thermal decomposition of complex hydrides of the general formula $MAlH_4$ (M = Li, Na) are shown by . 3.33–3.35. Whereas the first two reactions occur at temperatures around 200 °C, (3.30) only occurs at very high temperatures and is thus not considered a useful route for H_2 generation.

$$3\,MAlH_4 \rightarrow M_3AlH_6 + 2\,Al + 3\,H_2 \tag{3.33}$$

$$M_3AlH_6 \rightarrow 3\,MH + Al + \frac{3}{2}H_2 \tag{3.34}$$

$$3\,MH \rightarrow 3\,M + \frac{3}{2}H_2 \tag{3.35}$$

There continues to be significant research efforts devoted to the design of suitable catalysts that will improve the relatively slow kinetics associated with the reversible H_2 storage of $MAlH_4$ compounds. A new high-pressure polymorph of the boron analogue $LiBH_4$ has also been investigated, which appears to be more successful in low-temperature release of hydrogen.[58] Intermetallic compounds and metal aluminum hydrides doped with Ti and Zr have been successfully used to improve the rate of hydride formation/release in these compounds[59]; typically, levels between 2–4 mol% Ti is sufficient to facilitate reversibility. The catalytic mechanism may be rationalized by the donation of H_2 s-electron density to empty d orbitals on the Ti, along with synergistic donation of electrons from filled Ti d orbitals to the σ^* orbital of H_2. This two-way electron donation weakens the H–H bond, while strengthening the H–Ti interaction (Fig. 3.55).

Sometimes catalytic dopants do not alloy with the host metal; for example, the addition of Nb and V to Mg form heterogeneous mixtures rather than intermetallic compounds. A new generation of catalysts is being developed that deliver hydrogen to the metal surface as $H^\bullet$ radicals rather than H_2.[60] It may then be possible to combine a high level of hydrogen storage with low T_{dec}, both associated with desirable adsorption/desorption kinetics.

It has been discovered that decreasing the particle size of metal alloy particles through ball-milling processes will increase the adsorption kinetics by an order of magnitude.[61] This enhanced activity is due to the increased surface area of the ground particulates, and decreased surface reaction path length. For $LiAlH_4$, only prolonged milling is of sufficient energy to desorb H_2. When grinding is coupled with catalytic dopants, the H_2 storage kinetics increases even further. Upon milling bulk Mg_2NiH_4, the T_{dec} decreases by *ca.* 40 °C. Such mechanical processing has also been used to synthesize H_2-storage compounds (e.g., $La_{1.8}Ca_{0.2}Mg_{14}Ni_3$, $Li_xBe_yH_{x+2y}$, $MAlH_4$; M = Mg, Ca, Sr) that consist of a metastable amorphous or nanocrystalline structure.[62]

In order to improve the hydrogen storage capacity of solid-state materials, one must first understand the interfacial chemical reactions that occur upon hydrogen adsorption/desorption. Ray and coworkers at the Lawrence Livermore National Laboratory (LLNL) have recently investigated the hydrogenation of magnesium diboride (MgB_2).[63] Their calculations indicated that H_2 molecules likely attack highly reactive (10 $\bar{1}$ 0) edge planes. Charge polarization by Mg^{2+} ions drive the binding of hydrogen anions to form the $Mg(BH_4)_2$ phase, but this transformation slows the hydrogenation process. The boron centers within MgB_2 hexagonal sheets are more reactive toward hydrogen when the local environment is Mg-poor. However, as the $Mg(BH_4)_2$ phase is formed, the surfaces of the remaining MgB_2 planes become more Mg-rich, which slows hydrogenation in the absence of higher temperatures or hydrogen pressures.

Fig. 3.55 The influence of Ti doping on the dissociation H_2 on a (100) Al surface. For clarity, Al atoms from the layers below are shown as crosses. Reproduced with permission from *J. Phys. Chem. B* **2005**, *109*, 6952. © 2005 American Chemical Society

Topics For Further Discussion

1. For iron allotropes, why is the solubility of carbon greater in the austenite phase, relative to the ferrite phase?
2. Although iron is most stable in its bcc form, why are heavier Group 8 congeners Ru and Os most stable as hcp?
3. What is the difference between substitutional and interstitial dopants? Provide examples for each type.

4. Calculate the number of vacancies per cubic meter in gold at 900 °C. The energy for vacancy formation is 0.98 eV/atom. The density of gold at this temperature is 18.63 g/cm^3.
5. What is the composition, in weight percent, of an alloy that consists of 5 at% Cu and 95 at% Pt?
6. For each of the following metals, provide a description of the extractive metallurgy used to isolate the metal from specific ores (describe the minerals present in the ore), and post-treatment methods used to purify the metal: (a) Mo, (b) Rh, (c) Sn, (d) W.
7. For iron allotropes, provide a rationale for: (a) Why iron converts between BCC (α) & FCC (γ) lattices at 910 °C, and back to BCC (δ) at 1403 °C; (b) Why α -Fe loses its ferromagnetism at 769 °C (both magnetic (α-Fe) and nonmagnetic (β-Fe) allotropes are BCC, with identical lattice parameters and densities!).
8. Briefly compare the following metallurgical processing techniques: forging, casting, drawing, and extrusion.
9. Explain the atomic diffusion processes that occur when steel is heated and subsequently quenched by cold water. Use diagrams to illustrate your rationale.
10. Name the three types of hydrogen-storage metals/alloys and describe the placement of hydrogen within each lattice.
11. Consider the sintering process of compacted metal powders. Would the resulting sintered material be more or less desirable (from a mechanical standpoint), if an excessive amount of metal oxides were present in the pre-sintered matrix? How would you design the sintering conditions (co-reactant gases, temperature, etc.) for these matrices?
12. How does precipitation hardening work to strengthen the material?
13. For surface phosphating of aluminum-containing metals, fluoride-based additives are needed to cause crystallization. Why?
14. Using redox potentials, explain the frequent occurrence of perforations in domestic hot tap water pipes manufactured from galvanized steel.
15. Explain how shape-memory metals are able to manipulate their shapes in response to temperature fluctuations. Are there other alloy candidates for this type of behavior?
16. Classify the various phases in the Fe–C system as type I or II alloys (or both).
17. For the density of states for transition metals (e.g., Fig. 3.30), the d-band is much narrower than the overlapping s/p band. Why is this so, and what physical properties does this govern?
18. Explain why many ferrimagnetic materials crystallize in a spinel lattice.
19. Why are finely divided metals pyrophoric?
20. You have been awarded $2.3 million dollars to yield ultrahigh purity germanium from an ore that contains high concentrations of GeS_2, as well as Zn and Pb silicates. Outline a strategy that you will use to accomplish this goal.
21. Compare martempering and austempering of steel. What products do you obtain following these treatments?
22. Describe the processes involved during the extraction and purification of rare-earth metals from their ores.

23. List some environmental and health hazards associated with rare-earth metal mining. What strategies have been introduced to reduce these problems?
24. Using the TTT diagram below for an iron-carbon alloy of eutectoid composition, specify the nature of the final microstructure (in terms of microconstituents present and approximate percentages) of a small specimen that has been subjected to the following treatments. In each case, assume that the specimen begins at 760 °C and that it has been held at that temperature long enough to have achieved a complete and homogeneous austenitic structure.

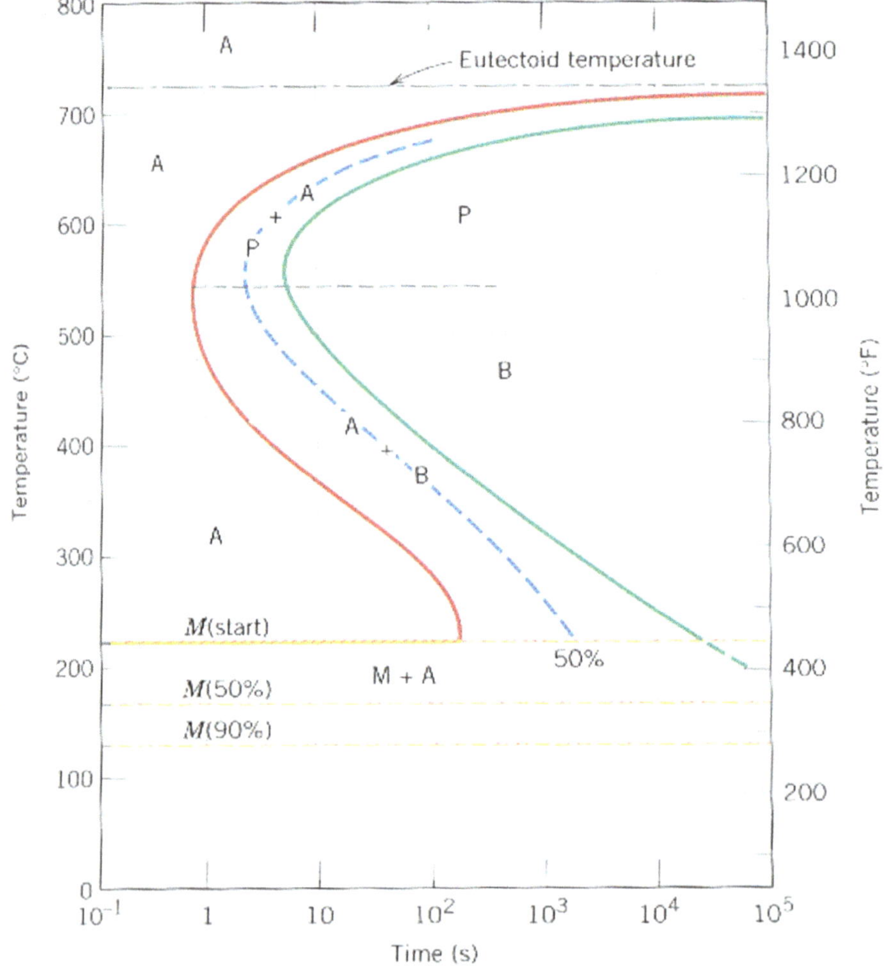

(a) Rapidly cool to 350 °C, hold for 10^4 s, and quench to room temperature.
(b) Rapidly cool to 250 °C, hold for 100 s, and quench to room temperature.

(c) Rapidly cool to 650 °C, hold for 20 s, rapidly cool to 400 °C, hold for 10^3 s, and quench to room temperature.
(d) Draw a cooling curve that corresponds to the "critical cooling rate".

References and Notes

1. Most rocks are composed of >95% silicates.
2. For instance, see: Srivastava, U.; Kawatra, S.; Eisele, T. Internat. *J. Mineral Proc.* **2013**, *119*, 51, and references therein.
3. The main use for V_2O_5 is for the catalysis of $2SO_2 + O_2 \Leftrightarrow 2SO_3$, used in sulfuric acid production, corresponding to annual production of 165 million tons, and over $8 billion.
4. Also known as a *blast oven.*
5. For instance, see: http://papers.sae.org/2006-01-2851/
6. For example, see: http://www.pm-review.com/powder-metallurgy-review-archive/powder-metallurgy-review-winter-2017-vol-6-no-4/
7. Canadian nickels were once made from strips rolled from pure nickel powder, but are now fabricated from steel (3.5% Cu) and contain only about 2% Ni, applied as an electrolytic coating.
8. Hohmann, C.; Tipton Jr., B.; Dutton, M. *Propellant for the NASA Standard Initiator* October 2000 (NASA/TP-2000–210,186).
9. For more background on additive manufacturing, see: (a) https://www.ge.com/additive/additive-manufacturing. (b) https://www.linde-engineering.com/en/images/21430_MPG_MI_Additive_Manufacturing_brochure_final_VIEW_tcm19-269682.pdf. (c) https://www.metal-am.com/introduction-to-metal-additive-manufacturing-and-3d-printing/metal-additive-manufacturing-processes/. (d) https://www.twi-global.com/technical-knowledge/faqs/what-is-additive-manufacturing
10. http://www.nasa.gov/exploration/systems/sls/3dprinting.html
11. https://www.intechopen.com/chapters/46882
12. https://www.youtube.com/watch?v=XsEPGVD1qdY&t=165s
13. Bondareva, J. V.; Dubinin, O. N.; Kuzminova, Y. O.; Shpichka, A. I.; Kosheleva, N. V.; Lychagin, A. V.; Shibalova, A. A.; Pozdnyakov, A. A.; Akhatov, S.; Timashev, P. S.; Evlashin, S. A. *Biomed. Mater.* **2022**, *17*, 1.
14. https://www.youtube.com/watch?v=UdiwBiw5dyo&t=105s
15. Note: these cutoff values for steels are arbitrary. Iron at the lower end of this range is referred to either mild steel, or low-carbon steel.
16. A good reference for "crystal field theory" is Cotton, F. A.; Wilkinson, G.; Gaus, P. L. *Basic Inorganic Chemistry*, 3rd ed., Wiley: New York, 1994.
17. (a) Olson, G. B.; Cohen, M. *Metallurg. Mater. Trans. A* **1976**, *7*, 1897. (b) http://hal.archives-ouvertes.fr/docs/00/25/56/55/PDF/ajp-jp4199707C558.pdf
18. Zhang, W. *Challenges* **2016**, *7*, 20 (DOI: 10.3390/challe7020020).
19. https://www.tf.uni-kiel.de/matwis/amat/iss/kap_9/illustr/s9_1_1.html
20. http://www.cfd.com.au/cfd_conf03/papers/063Tan.pdf

21. Unfortunately, we now know that copper may be leached from pipes and pollute the drinking water supply. The use of lead pipes dates back to the Roman Empire; however, this material was still employed for plumbing in the early twentieth century. As we are familiar from the recent reports of water pollution in Flint, Michigan and many other places across the United States, lead, copper, and iron may leach into water supplies. The U.S. established the "Lead and Copper Rule (LCR) in 1991, which monitors the levels of heavy metals in drinking water. Since the replacement of all pipes across the nation would be too cost prohibitive, there are numerous corrosion control strategies that can be used to limit or prevent the leaching of these heavy metals into water:
http://www.waterrf.org/resources/StateOfTheScienceReports/LeadCorrosion.pdf
22. Kohler, J.; Whangbo, M -H. *Chem. Mater.* **2008**, *20*, 2751, and references therein.
23. For more details about the structure and composition of Zintl phases, see: http://www3.nd.edu/~sevovlab/articles/SlaviChapter.pdf
24. Kohler, J.; Deng, S.; Lee, C.; Whangbo, M. -H. *Inorg. Chem.* **2007**, *46*, 1957, and references therein.
25. For a discussion regarding the bandgap of half-Heusler alloys, see: Kohler, J.; Deng, S. *Inorg. Chem.* **2007**, *46*, 1957, and references therein.
26. https://www.intechopen.com/chapters/46861
27. For a nice review of aluminum and magnesium metal for automotive applications, see Musfirah, A. H.; Jaharah, A. G. *J. Appl. Sci. Res.* **2012**, *8*, 4865.
28. Cumberland, R. W.; Weinberger, M. B.; Gilman, J. J.; Clark, S. M.; Tolbert, S. H.; Kaner, R. B. *J. Am. Chem. Soc.* **2005**, *127*, 7264.
29. Gu, Q.; Krauss, G.; Steurer, W. *Adv. Mater.* **2008**, *20*, 3620.
30. http://web.archive.org/web/20030605085042/http://www.sma-inc.com/SMAPaper.html
31. For instance, see: (a) Co-Ni-Al: Dilbal, S.; Sehitoglu, H.; Hamilton, R. F.; Maier, H. J.; Chumlyakov, Y. *Mater. Sci. Eng. A* **2011**, *528*, 2875. (b) b) Ni-Mn-Ga: Hamilton, R. F.; Dilbal, S.; Sehitoglu, H.; Maier, H. J. *Mater. Sci. Eng. A* **2011**, *528*, 1877. (c) Fe-Mn-Si-Cr-Ni-Sm: Shakoor, R. A.; Khalid, F. A. *Mater. Sci. Eng. A* **2009**, *499*, 411. (d) Fe–Mn-Si-Ni-Co: Wang, X. -X.; Zhang, C. -Y. *J. Mater. Sci. Lett.* **1998**, *17*, 1795. (e) Cu–Zn-Al: Lin, G. M.; Lai, J. K. L.; Chung, C. Y. *Scripta Metallurgica Mater.* **1995**, *32*, 1865. (f) Cu–Zn-Al-Mn: Gil, F. J.; Guilemany, J. M.; Sanchiz, I. *J. Mater. Sci.* **1993**, *28*, 1542.
32. For other biomedical applications for shape-memory alloys, see: (a) Lendlein, A.; Langer, R. *Science* **2002**, *296*, 1673. (b) El Feninat, F.; Laroche, G.; Fiset, M.; Mantovani, D. *Adv. Engin. Mater.* **2002**, *4*, 91. (c) http://www.scielo.br/pdf/bjmbr/v36n6/4720.pdf
33. Liang, W.; Zhou, M.; Ke, F. *Nano Lett.* **2005**, *5*, 2039.
34. http://liquidmetal.com
35. Dednarcik, J.; Kovac, J.; Kollar, P.; Roth, S.; Sovak, P.; Balcerski, J.; Polanski, K.; Svec, T. J. Non-Cryst. Solids 2004, 337, 42.

36. For example, see: (a) Xu, J.; Liu, W. *Wear* **2006**, *260*, 486. (b) Mergia, K.; Liedtke, V.; Speliotis, T.; Apostolopoulos, G.; Messoloras, S. *Adv. Mater. Res.* **2009**, *59*, 87. (c) Benea, L.; Bonora, P. L.; Borello, A.; Martelli, S. *Wear* **2001**, *249*, 995.
37. Park, H.; Kim, K. Y.; Choi, W. *J. Phys. Chem. B* **2002**, *106*, 4775.
38. For instance, see Liang, Y.; Guan, Z. C.; Wang, H. P.; Du, R. G. *Electrochem. Communic.* **2017**, *77*, 120.
39. The unit emu is short for "electromagnetic unit", which is equivalent to erg G^{-1}, where 1 erg = 1×10^{-7} J and G = Gauss (1×10^{-4} T, T).
40. Think of the atom in the middle of the bcc unit cell, at lattice position (1/2, 1/2, 1/2). Since there are no atoms on the unit cell faces in a bcc array, no atoms will be positioned directly along the *x*, *y*, and *z* axes emanating from this central atom.
41. For more information and statistics related to rare-earth metals, see: https://minerals.usgs.gov/minerals/pubs/commodity/rare_earths/
42. (a) Dutta, T.; Kim, K. H.; Uchimiya, M.; Kwon, E. E.; Jeon, B. H.; Deep, A.; Yun, S. T. Environ. Res. 2016, 150, 182, and references therein. (b) McLellan, B. C.; Corder, G. D.; Ali, S. H. Minerals 2013, 3, 304.
43. Pathak, A. K.; Khan, M.; Gschneidner, K. A.; McCallum, R. W.; Zhou, L.; Sun, K.; Dennis, K. W.; Zhou, C.; Pinkerton, F. E.; Kramer, M. J.; Pecharsky, V. K. *Adv. Mater.* **2015**, *27*, 2663 (https://doi.org/10.1002/adma.201404892).
44. http://license.umn.edu/technologies/20120016_iron-nitride-permanent-magnet-alternative-to-rare-earth-and-neodymium-magnets
45. http://www.ims.org/wp-content/uploads/2012/10/Session7a_01_Mikami_Hiroyuki_Technologies_to_replace_rare_earth_elements.pdf
46. Widmer, J. D.; Martin, R.; Kimiabeigi, M. Sustain. Mater. Technol. 2015, 3, 7, and references therein (https://doi.org/10.1016/j.susmat.2015.02.001).
47. http://www.pbs.org/wgbh/nova/next/earth/deep-sea-mining/
48. http://www.planetaryresources.com/#home-intro
49. For example, see: Sprecher, B.; Xiao, Y.; Walton, A.; Speight, J.; Harris, R.; Kleijn, R.; Visser, G.; Kramer, G. J. *Environ. Sci. Technol.* **2014**, *48*, 3951 (https://doi.org/10.1021/es404596q).
50. http://www.businessinsider.com/apple-liam-iphone-recycling-robot-photos-video-2017-4
51. http://www.innovationmetals.com/the-rapidsx-process/
52. For a comprehensive review of materials, techniques, and methods for hydrogen storage, see: (a) https://doi.org/10.1016/j.jallcom.2019.153548 (b) https://doi.org/10.3390/en14185917
53. For instance, see:
a) https://doi.org/10.1016/j.ijhydene.2019.06.093
b) Wong-Foy, A. G.; Matzger, A. J.; Yaghi, O. M. *J. Am. Chem. Soc.* **2006**, *128*, 3494.
c) Latroche, M.; Surble´, S.; Serre, C.; Mellot-Draznieks, C.; Llewellyn, P. L.; Lee, J. H.; Chang, J. S.; Jhung, S. H., Fe´rey, G. *Angew. Chem., Int. Ed.* **2006**, *45*, 8227.

d) Chahine, R.; Benard, P. In *Advances in cryogenic engineering* Kittel, P., Ed.; Plenum Press: New, York, 1998.

e) Kabbour, H.; Baumann, T. F.; Satcher, J. H., Jr., Saulnier, A.; Ahn, C. C. *Chem. Mater.* **2006**, *18*, 6085.

54. For a recent review of MOFs for hydrogen storage: https://doi.org/10.1038/s41467-019-09365-w
55. a) https://doi.org/10.1016/S0360-3199(01)00014-3
    b) https://doi.org/10.1016/j.scriptamat.2007.01.008
56. a) https://www.energy.gov/eere/fuelcells/hydrogen-storage
    b) https://www.energy.gov/eere/fuelcells/doe-technical-targets-onboard-hydrogen-storage-light-duty-vehicles
57. Schlapbach, L.; Zuttel, A. *Nature*, **2001**, *414*, 353.
58. Filinchuk, Y.; Chernyshov, D.; Nevidomskyy, A.; Dmitriev, V. *Angew. Chem. Int. Ed. Eng.* **2008**, *47*, 529.
59. For instance, see: Bogdanovi, B.; Schwickardi, M. *J. Alloys Compounds* **1997**, *253–254*, 1.
60. For example, see: Maiti, A.; Gee, R. H.; Maxwell, R.; Saab, A. P. *Chem. Phys. Lett.* **2007**, *440*, 244.
61. For instance, see: (a) Moller, K. T.; Grinderslev, J. B.; Jensen, T. R. *J. Alloys Compounds* **2017**, *720*, 497 (https://doi.org/10.1016/j.jallcom.2017.05.264). (b) Schimmel, H. G.; Huot, J.; Chapon, L. C.; Tichelaar, F. D.; Mulder, F. M. *J. Am. Chem. Soc.* **2005**, *127*, 14,348.
62. For instance, see: (a) Poletaev, A. A.; Denys, R. V.; Haehlen, J. P.; Solbrg, J. K.; Tarasov, B. P.; Yartys, V. A. *Internat. J. Hydrogen Energy* **2012**, *37*, 3548. (b) Zhang, Q. A.; Jiang, C. J.; Liu, D. D. *Internat. J. Hydrogen Energy* **2012**, *37*, 10,709. (c) Yuan, H. J.; An, Y. Xu, G. H.; Chen, C. P. *Mater. Chem. Phys.* **2004**, *83*, 340. (d) Wang, Y.; Wang, X.; Li, C. M. *Internat. J. Hydrogen Energy* **2010**, *35*, 3550. (e) Zhang, Y. H.; Yuan, Z. M.; Yang, T.; Bu, W. G.; Hou, Z. H.; Zhao, D. L. *J. Central South Univ.* **2017**, *24*, 773.
63. https://doi.org/10.1021/acsami.1c23524

Further Reading

1. W.D. Callister, *Materials Science and Engineering: An Introduction*, 10th edn. (Wiley, New York, 2018)
2. R.C. Hibbeler, *Mechanics of Materials*, 10th edn. (Pearson, New York, 2016)
3. D.A. Porter, K.E. Easterling, M.Y. Sherif, *Phase Transformations in Metals and Alloys*, 4th edn. (CRC Press, New York, 2022)
4. R.W.K. Honeycombe, H.K.D.H. Bhadeshia, *Steels: Microstructure and Properties*, 4th edn. (Wiley, New York, 2017)
5. A.W. Grosvenor, *Basic Metallurgy: Volume I, Principles* (American Society for Metals, Cleveland, OH, 1962)
6. J. Beddoes, J.G. Parr, *Introduction to Stainless Steels*, 3rd ed. (ASM International: Materials Park, OH, 1999)

7. E.T. Lacheisserie, D. Gignoux, M. Schlenker (eds.), *Magnetism: Fundamentals* (Springer, New York, 2004)
8. M. Reis, *Fundamentals of Magnetism* (Academic Press, New York, 2013)
9. D.C. Mattis, *The Theory of Magnetism Made Simple: An Introduction to Physical Concepts and to Some Useful Mathematical Methods* (World Scientific Publishing Company, New York, 2006)
10. C. Lefteri, R. Arad, *Metals: Materials for Inspirational Design* (Rotovision, London, 2004)
11. A.J. Sedriks, *Corrosion of Stainless Steel* (Wiley, New York, 1996)
12. *Damping Structural Vibrations with Shape-Memory Metals* (NASA Publication, University Press of the Pacific, 2004)
13. L. Lecce, A. Concilio, *Shape Memory Alloy Engineering: For Aerospace, Structural and Biomedical Applications, 2nd edn.* (Butterworth-Heinemann, New York, 2021)
14. D.I. Arun, P. Chakravarthy, B. Santhosh, R. Arockiakumar, *Shape Memory Materials* (CRC Press, New York, 2018)

Chapter 4
Semiconductors

Our technologically advanced way of life would not be possible without the semiconductor industry. The first semiconductor device, known as a *transistor,* was discovered at Bell Labs in the late 1940s and was widely used shortly thereafter for radio electronics. Today, transistors are still pervasive in every chip that lies at the heart of portable electronic devices, modes of transportation, and computers. In fact, modern computer chips now contain over 10 billion individual transistors—all on a surface that is smaller than a fingernail!

This chapter will investigate various types of semiconducting materials, focusing on the influence of structure and composition on their overall properties. We will also detail the many applications for semiconductors, especially within the framework of microelectronic circuitry. It should be noted that nanostructural materials represent the most recent realm of semiconducting materials, of use for next-generation solar devices, sensors, batteries, etc. However, these materials will not be considered in this chapter, but will instead be detailed in Chap. 6 that focuses solely on nanotechnology.

4.1 Properties and Types of Semiconductors

As their name implies, semiconductors possess an electrical conductivity between that of conductors such as metals, and insulators such as ceramics. In Chap. 2, we introduced the concept of band theory to describe the molecular orbitals of infinite lattices. For metals, the lack of an energy gap between the filled (valence) and empty (conduction) bands indicated that no thermal energy was required to facilitate electrical conductivity. As a consequence, metals are able to transport electrons through their lattices even at absolute zero. By contrast, semiconductors possess a *bandgap* between valence and conduction bands (Fig. 4.1). Hence, an appropriate amount of energy must be supplied to a semiconductor in order to promote an electron from the valence band to the conduction band, wherein electron transport may occur. For semiconductors, any temperature greater than 0 K is sufficient for such electron

B. D. Fahlman, *Materials Chemistry*,
https://doi.org/10.1007/978-3-031-18784-1_4

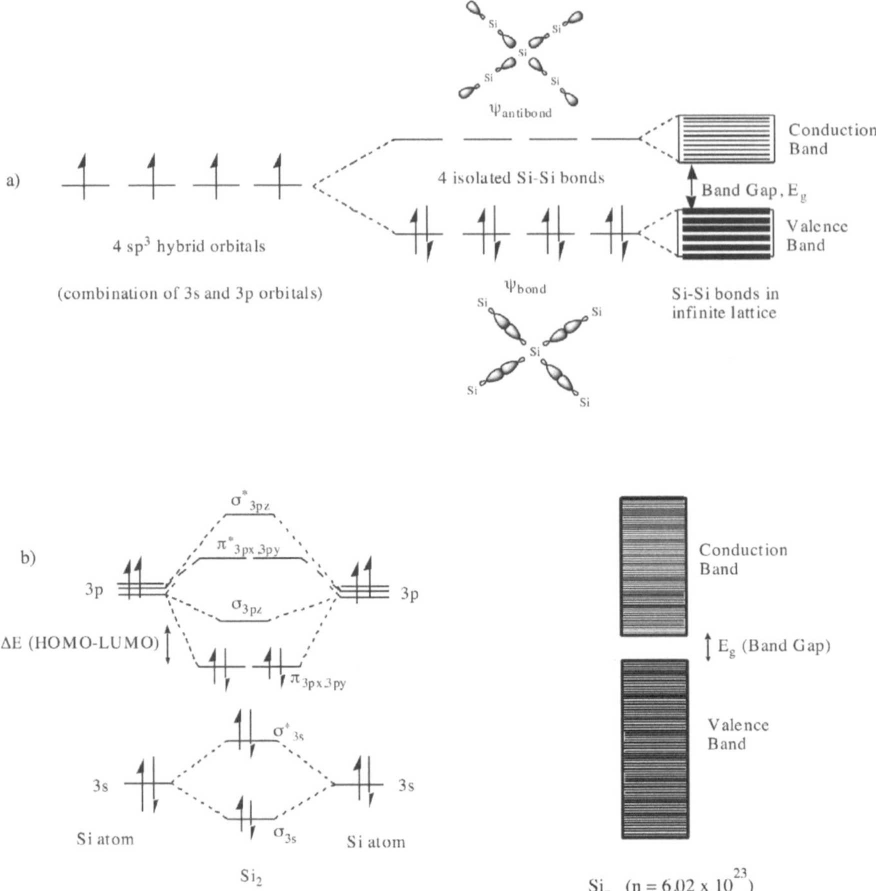

Fig. 4.1 Electronic band diagrams for silicon. Shown is (a) bands resulting from overlap of sp^3 hybrid orbitals and (b) bands resulting from overlap of molecular orbitals

promotion. If the bandgap is too large, the material is unable to conduct electricity, and is known as an *insulator*.

The bandgap of semiconductors is typically in the range 150–290 kJ mol^{-1} (*i.e.*, 1–4 eV). Since the bandgap of Si is 1.12 eV, only radiation with wavelengths ca. $\leq$ 1100 nm (near IR, visible, UV, X-ray, etc.) will be absorbed by silicon semiconductors, as the absorption of less-energetic radiation will place the electron in the bandgap, where there are no available energy states. The absorption of visible light by silicon results in its opacity, relative to transparent glasses that do not absorb radiation within this wavelength regime.

Group 14 provides an interesting case-study for the most dramatic change in electrical conductivity among its congeners. In fact, all types of conductors are present in this group, from insulating carbon (diamond) to metallic tin and lead. Since the electronic configurations of these atoms are all (ns^2 np^2), the change in electrical conductivities is related to the bond strengths among the atoms comprising the crystal lattice. Recall that the individual atoms in diamond, Si, Ge, Sn, and Pb are tetrahedrally linked through the crystal lattice due to sp^3 hybridization. As the individual s and p orbitals that comprise the hybrid orbital become more diffuse (e.g., Sn, Pb), the bonding electrons are more weakly bound to individual atoms and become more polarizable. This results in delocalized metallic bonding in Sn and Pb relative to very strong localized bonding in diamond. For intermediate Si and Ge, the bonding between individual atoms is weaker than C, allowing the possibility for thermal motion to break bonds in the solid-state lattice by promoting bonding electrons into the conduction band, thereby facilitating electrical conductivity.

There are two types of semiconductors: *intrinsic* and *extrinsic* (Fig. 4.2). Intrinsic semiconductors contain the same numbers of free bonding electrons (e^-) and holes (h^+), created from the migration of electrons from the valence to conduction bands. The temperature-dependent concentration of e^-/h^+ carriers is known as the *intrinsic concentration*, n_i (Fig. 4.3). As we discussed in Chap. 2, the number of electrons that may traverse the bandgap is governed by the *Fermi function*, which gives the probability of an electron occupying an available energy state. The *density of states* (DOS), or number of available energy levels, is also paramount for the promotion of electrons from valence-conduction bands. The DOS begin at the bottom of the valence band and continue to the *Fermi level*—the highest occupied state at absolute zero (*i.e.*, −273.15 °C). For metals that readily exhibit electrical conductivity, the Fermi level lies within the conduction band due to the lack of a bandgap. However, unlike metals, the DOS for conduction electrons in a semiconductor do not coincide with the Fermi level, but rather begin at the top of the bandgap. Hence, the placement of the *Fermi level* represents the relative ease at which an electron is promoted from the valence band to the conduction band in a bulk semiconductor. Quantitatively, the density of occupied states per unit volume is the product of the DOS and the Fermi function, $f(E)$; the density of holes is the product of the DOS and $[1 - f(E)]$.

For *extrinsic semiconductors*, the Fermi level corresponds to a level slightly above or below conduction or valence bands, depending on the type of dopant atoms introduced into the lattice—either an excess of electrons or holes (Fig. 4.2b and c). Since our frame of reference for semiconducting ability is the Group 14 element of Si, the terms *electron-deficient* and *electron-rich* dopants refer to atoms possessing <4 valence electrons (e.g., B, Al) and >4 electrons (e.g., N, P), respectively. By convention, if additional electrons are introduced into the lattice, the semiconductor is designated as *n-type*, whereas the doping of additional holes yields a *p-type* semiconductor designation. From the convenient "n" and "p" notation, one may immediately recognize whether there are excess negative or positive carriers in the lattice, respectively.

Figure 4.4 illustrates the effect of temperature on the carrier concentration of a semiconductor. At low temperatures, the e^-/h^+ concentration of extrinsic semiconductors is governed by the ionization of the dopants (e.g., B^-/h^+, Al^-/h^+ for p-type; P^+/e^-, As^+/e^- for n-type, see Fig. 4.4a). At the *saturation temperature*, T_s, all donors or acceptors have been ionized resulting in a constant carrier concentration (Fig. 4.4b). As the temperature is increased beyond T_s, the *intrinsic temperature*, T_i, is reached when the thermal promotion of electrons across the bandgap exceeds the concentration of acceptors/donors. Hence, at temperatures above T_i, an

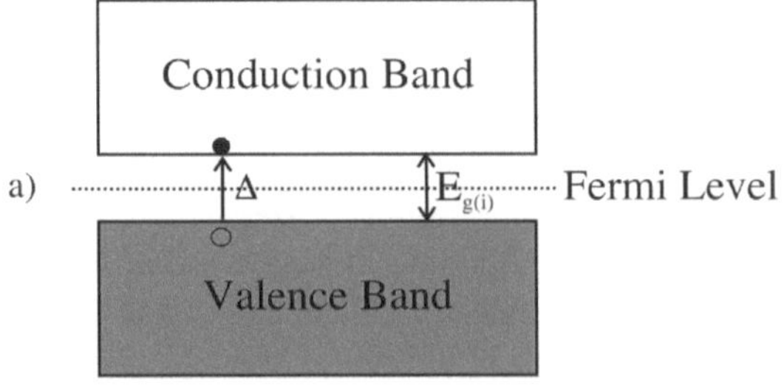

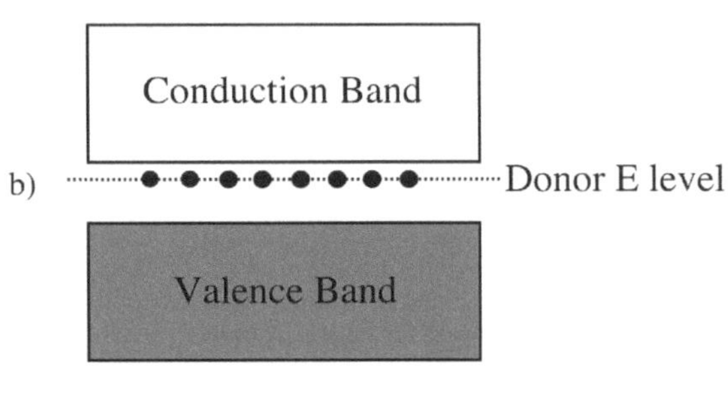

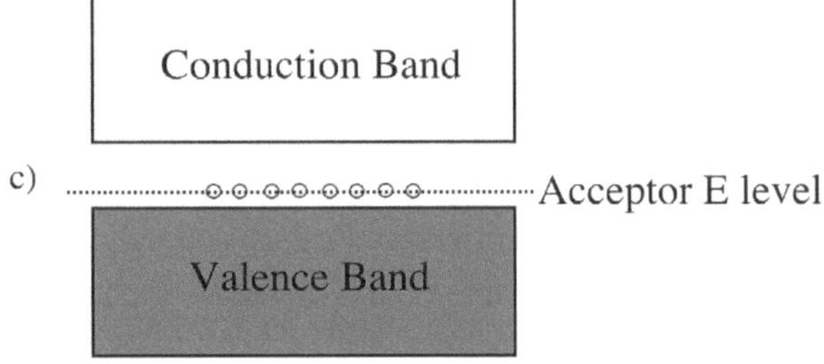

Fig. 4.2 Band diagrams for intrinsic and extrinsic semiconductors. Shown are **a** an intrinsic semiconductor, with an equivalent number of free electrons and holes, **b** an n-type extrinsic semiconductor, with a greater number of electrons, and **c** a p-type extrinsic semiconductor, with an excess of holes

extrinsic semiconductor will exhibit an intrinsic carrier concentration, with $[e^-] = [h^+]$ (Fig. 4.4c). For a given carrier concentration, the lattice constant will increase concomitantly with temperature. Accordingly, this will decrease the energy required to break bonds thereby lowering the bandgap (4.1—known as the Varshni equation, Fig. 4.5).

$$E_g = E_{g'} - \frac{AT^2}{B + T}, \tag{4.1}$$

where Eg′ is the bandgap at T = 0 K (GaAs = 1.519 eV, Si = 1.7 eV), A and B are material-specific constants (e.g., Si: $A = 4.73 \times 10^{-4}$ eV K^{-1}, B = 636 K; GaAs: $A = 5.405 \times 10^{-4}$ eV K^{-1}, B = 204 K), and T is the temperature (in K).

The temperature dependence is quite different for metals and semiconductors. At a temperature of 0 K, a semiconductor will behave as a perfect insulator. However, a metal will exhibit electrical conductivity at absolute zero due to its delocalized electron density and lattice structure described in Chap. 3. Interestingly, as the temperature is increased, the respective conductivities of these materials will be reversed, with metals showing a decrease and semiconductors an increase in conductivity (Fig. 4.6). The thermal motion of metal atoms causes less efficient electron mobility through the lattice, whereas a temperature increase causes the bandgap to narrow for semiconductors, resulting in more effective electrical conductivity. As the temperature continues to increase for semiconductors, the linear relationship does not continue to hold, and the resistance begins to increase analogous to metals. Instead, the atomic vibrations caused by the elevated temperature begin to outweigh the thermally induced decrease of the bandgap.

As Fig. 4.7 illustrates, when thermal energy promotes a bonding electron from the valence band to the conduction band, the released electrons are free to migrate throughout the lattice. However, the vacancies (i.e., holes) left behind are also free to move—in the opposite direction as electrons. One may consider these holes as positively charged species formed from loss of an electron. Thus, electrons and

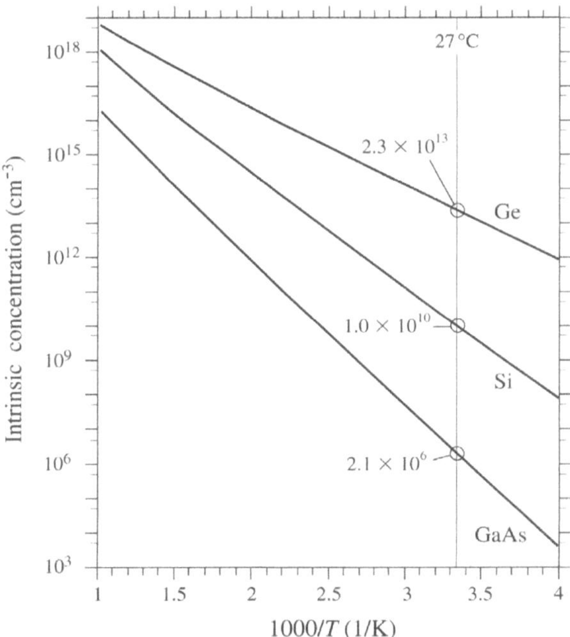

Fig. 4.3 The intrinsic carrier concentration *vs.* temperature for Ge, Si, and GaAs. Reproduced with permission from Kasap, S. O. *Principles of Electronic Materials and Devices*, 3rd ed., McGraw-Hill: New York, 2007. Copyright 2006 The McGraw-Hill Companies

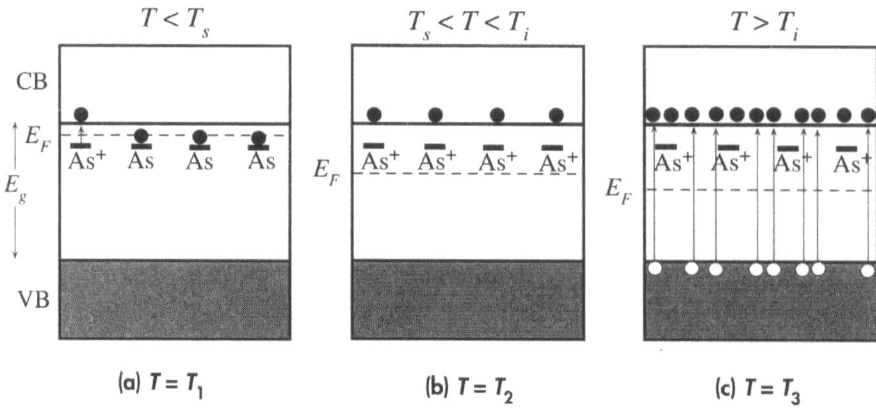

Fig. 4.4 Position of the Fermi level (E_F) and relative ionization of n-type Si with increasing temperature. Below the saturation tempature (T_s), only a few As dopants are ionized; however, at increasing temperatures, more As atoms become ionized until the intrinsic temperature is reached (T_i). At that temperature, all dopants have been ionized, and the electrical conductivity results from promotion of electrons across the bandgap. Reproduced with permission from Kasap, S. O. *Principles of Electronic Materials and Devices*, 3rd ed., McGraw-Hill: New York, 2007. Copyright 2006 The McGraw-Hill Companies

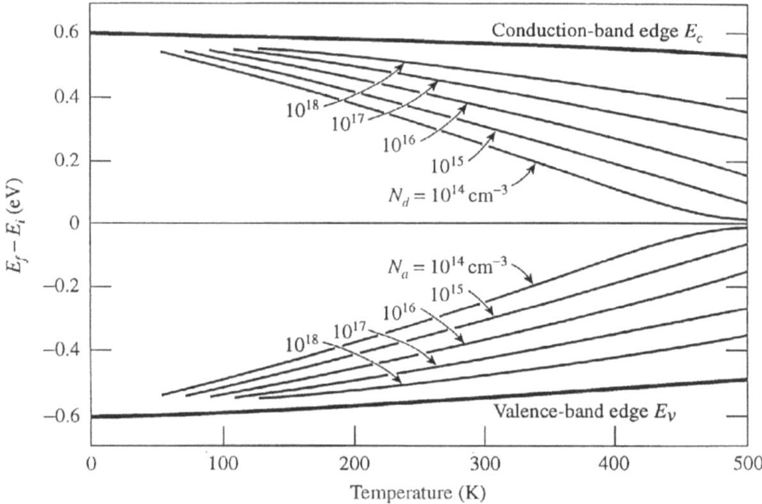

Fig. 4.5 The effect of increasing temperature on the position of the donor/acceptor energy levels within the bandgap for various doping levels, N_D or N_A. Reproduced with permission from Kasap, S. O. *Principles of Electronic Materials and Devices*, 3rd ed., McGraw-Hill: New York, 2007. Copyright 2006 The McGraw-Hill Companies

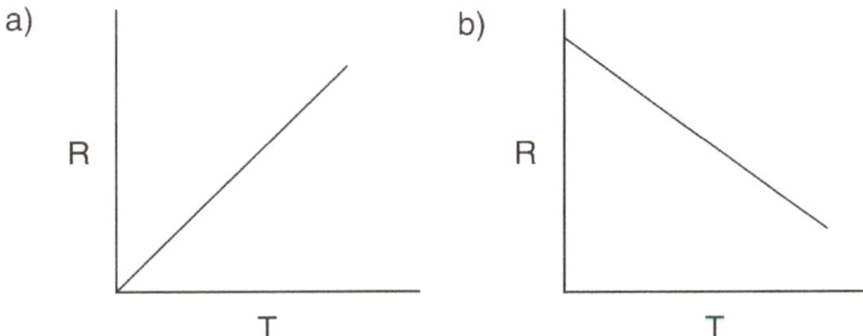

Fig. 4.6 Resistivity–temperature relationships for **a** metals and **b** semiconductors

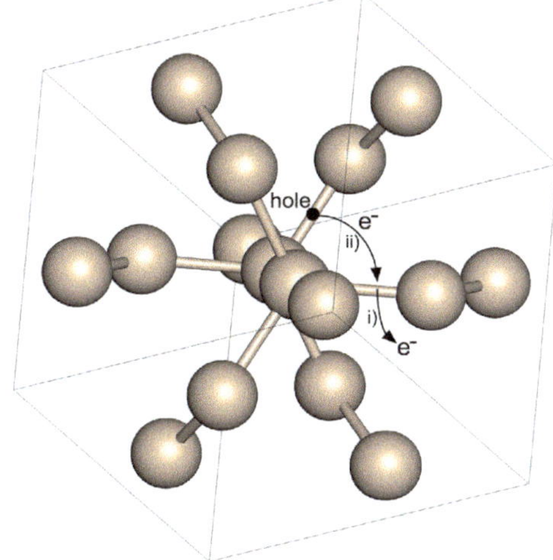

Fig. 4.7 Illustration of creation/migration of electrons and holes created through Si–Si bond thermolysis. Shown are (a) the release of an electron and concomitant formation of a hole and (b) the migration of an electron from a nearby bond to fill the vacancy

holes represent the two types of carriers that correspond to electrical conductivity in semiconductors.

Since an electron that has been promoted to the conduction band will have a greater energy than those left in the valence band, there is a possibility for the electron to lose this excess energy. The spontaneous return of conduction-band electrons to the valence band is known as *recombination* and is usually accompanied by light emission and heat (Fig. 4.8). This phenomenon happens frequently for excited-state molecules. For instance, consider what happens when one supplies sodium atoms with sufficient energy to promote an electron from the 3s energy level into an empty 3p, or 4s orbital. The electron does not remain in the higher energy level very long (typically ca. 10^{-6} s) before it spontaneously returns to its original ground state—accompanied by the release of heat and/or light. There are three types of recombination[1]:

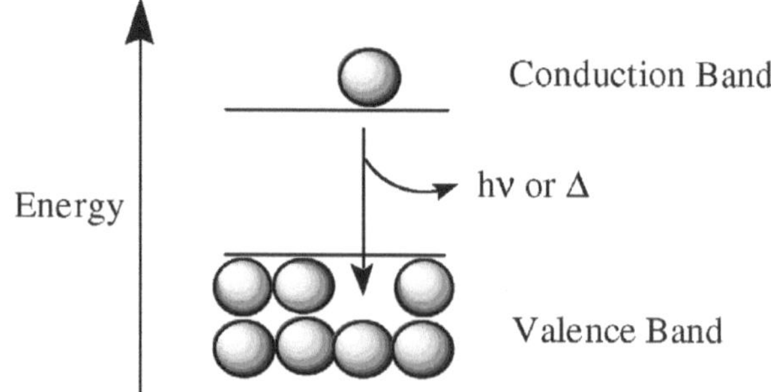

Fig. 4.8 Schematic of recombination of electron–hole pairs generating either a photon of energy or heat

(i) *Band-to-band*: an electron moves from the conduction band into an empty state in the valence band associated with the hole. This yields a radiative emission, whose wavelength is inversely proportional to the bandgap.
(ii) *Trap-assisted*: an electron falls into an energy level within the bandgap that is associated with a dopant atom or structural defect.
(iii) *Auger recombination*: a conduction-band electron and valence-band hole may recombine. Rather than emitting light, the energy may release an outer-shell electron known as an Auger electron.

Some basic concepts of solid-state physics are needed to understand the recombination process in more detail. Recall that matter and light exhibit both wave- and particle-like behavior. This duality may be expressed by the de Broglie equation, which equates the wavelength and momentum, p, of a particle (4.2). The potential energy of an electron in a crystal lattice depends on its location and will be periodic due to the regular array of lattice atoms. The periodic wavefunctions that result from solving the Schrödinger equation are referred to as *Bloch wavefunctions*. Each wavefunction represents the energy of an electron at a specific location in the lattice, referred to as k-space. Hence, an E-**k** diagram may be constructed, with the potential energy of the electron on the y-axis, and the wavevector, **k**, on the x-axis. The wavevector represents lattice directions of the semiconductor crystal; changing values of **k** represent a change in momentum of the electron. A comparison of the simplified E-**k** diagrams for Si or Ge and a compound semiconductor such as GaAs is shown in Fig. 4.9. For GaAs, the minimum of the conduction band (CB) is directly above the maximum of the valence band (VB). Accordingly, we refer to such solids as *direct bandgap* materials. In contrast, for Si and Ge, the CB minimum and VB maximum are offset, resulting in an *indirect bandgap.*

$$\lambda = \frac{\mathrm{h}}{p} \tag{4.2}$$

For Si, in order for an electron at the bottom of the CB to recombine with a hole from the top of the VB, the momentum of the electron must shift from $\mathbf{k}_{cb}$ to $\mathbf{k}_{vb}$ (Fig. 4.9b). However, this is not allowed by the Law of Conservation of Momentum. Instead, an indirect recombination mechanism must take place, wherein

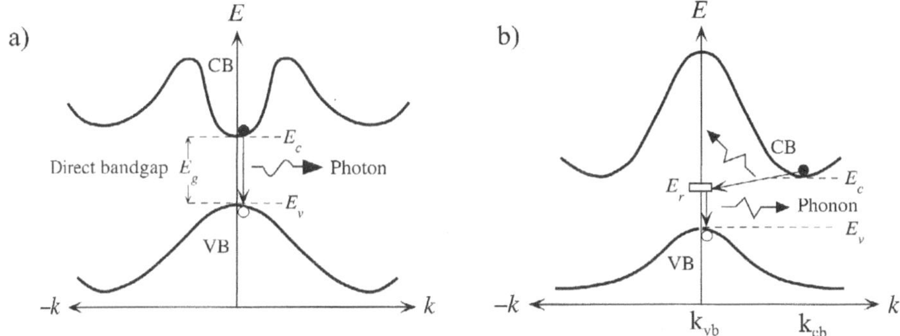

Fig. 4.9 Comparison of **a** direct bandgap (e.g., GaAs) and **b** indirect bandgap (e.g., Si, Ge) materials. Reproduced with permission from Kasap, S. O. *Principles of Electronic Materials and Devices*, 2nd ed., McGraw-Hill: New York, 2002

the electron is captured by an interstitial defect with energy E_r, which facilitates its relaxation to the top of the VB. This process is accompanied by the emission of phonons, or lattice vibrations, rather than light emission. In contrast, efficient electron–hole recombination may occur directly without any change in momentum for direct bandgap materials, resulting in the emission of photons. We will describe some important applications for direct bandgap semiconductors later in this chapter.

4.2 Silicon-Based Applications

Silicon is the second-most abundant element in Earth's crust, next to oxygen. Due to the stability of silicon oxide compounds, elemental silicon does not occur in its free state in nature, but occurs as oxide (e.g., sand, quartz, amethyst, flint, opal, etc.) and silicate (e.g., granite, asbestos, feldspar, clay, mica, etc.) minerals. The widespread availability of silicon-containing minerals is responsible for the ubiquitous use of pottery, bricks, glass, and cement since the days of the earliest civilizations.

However, most applications for silicon now involve electronic devices. A silicon-based device is found in almost every consumer product available in our world today. Even refrigerators have extensive microprocessor controls, with some even fitted with television screens! In addition to electronics, as the world looks for alternative sources of energy due to fluctuating petroleum prices and concerns over global climate change, silicon-based photovoltaic devices will represent an increasingly important application for our society.

4.2.1 Silicon Wafer Production

The silicon employed for microelectronic and photovoltaic applications must go through extensive processing to ensure that the material is of utmost purity. This section will describe these steps, with a discussion of perhaps the most intriguing conversion in the realm of materials science: the fabrication of ultrahigh purity polished silicon wafers from a naturally occurring form of silicon—sand.

Sea sand is primarily comprised of silicon dioxide (silica), which may be converted to elemental silicon (96–99% purity) through reduction by a carbon source such as charcoal or coal (4.3). A slight excess of SiO_2 is used to prevent silicon carbide (SiC) formation, which is a stable product at high reaction temperatures. Scrap iron is often present during this transformation in order to yield silicon-doped steel as a useful by-product.

The purity of silicon in this first step is only ca. 98% and is referred to as *metallurgical grade* silicon (MG-Si). In order for the silicon to be used for electronics applications, additional steps are necessary to decrease the number of impurities. Reaction of MG-Si with hydrogen chloride gas at a moderate temperature converts the silicon to trichlorosilane gas (4.4). When $SiHCl_3$ is heated to a temperature of ca. 1150 °C, it decomposes into high-purity silicon and gaseous by-products (4.5). This reaction is typically performed in a bell-shaped Siemens-type reactor, where Si is deposited onto heated U-rod silicon filaments. Excess hydrogen gas is also fed into the reactor, which prevents homogeneous nucleation of Si dust within the reactor.[2]

$$SiO_2(s) + 2C(s) \xrightarrow{2000-2500\,^\circ C} Si(s) + 2CO(g) \tag{4.3}$$

$$Si(s) + 3\,HCl(g) \xrightarrow{300\,^\circ C} SiHCl_3(g) + H_2(g) \tag{4.4}$$

$$2SiHCl_3(g) \xrightarrow{1150\,^\circ C} Si(s) + 2HCl(g) + SiCl_4(g) \tag{4.5}$$

Recently, a *fluidized-bed* approach has been developed wherein SiH_4 and H_2 gases are fed into the bottom of a vertical reactor held at a temperature >600 °C.[3] Silicon seed crystals of ca. 100 μm diameter are suspended in the chamber, and decomposition of the gaseous precursor causes the nucleation and *homoepitaxial* growth[4] of silicon on the surface of the seeds. When grown to large sizes (ca. 1 mm diameters), the particles no longer remain suspended and are collected on a filter at the bottom of the reactor. The decomposition temperature of SiH_4 is ca. 2/3 that of $SiHCl_3$, but is significantly more pyrophoric and air-sensitive. Relative to the Siemens process, the fluidized-bed approach represents a smaller system for an equivalent throughput, while using much less energy. Further, fluidized-bed systems exhibit greater heat-transfer efficiencies since the gases are heated; in contrast, Siemens reactors heat only the electrode/growing rod to prevent homogeneous nucleation of Si on the reactor walls.

These chemical processes result in *electronic-grade* polysilicon (EG-Si), with a purity of 99.9999999%; that is, only one out of every billion atoms in the solid is something other than silicon! To put this into perspective, imagine stacking yellow tennis balls from the earth's surface to the moon; replacing only one of these with a blue ball would represent the level of impurities in EG-Si. Every year, ca. 100,000 to 200,000 metric tonnes of EG-Si is manufactured throughout the world for an ever-increasing number of applications.

Although the purity of EG-Si is suitable for electronics applications, the atomic structure must first be converted from its polycrystalline structure (*polysilicon*) to a single crystal. There are two main techniques used for this conversion: Czochralski (CZ; Fig. 4.10) and float-zone (FZ; Fig. 4.11). For CZ growth, silicon and any desired dopants are molten together in a crucible at a temperature above the melting point of silicon, 1414 °C. A rod fastened with a single crystal of silicon is positioned on the surface of the melt and is pulled upward while rotating. This results in a long cylinder of Si that is referred to as an *ingot*. Due to the surface tension of the liquid, a thin film of silicon first forms on the seed crystal surface. Additional silicon atoms orient themselves to the seed crystal; hence, the final ingot has the same crystal lattice as the original seed. The diameter of the resulting ingot may be finely controlled by manipulating the temperature and rate of pulling/rotation.

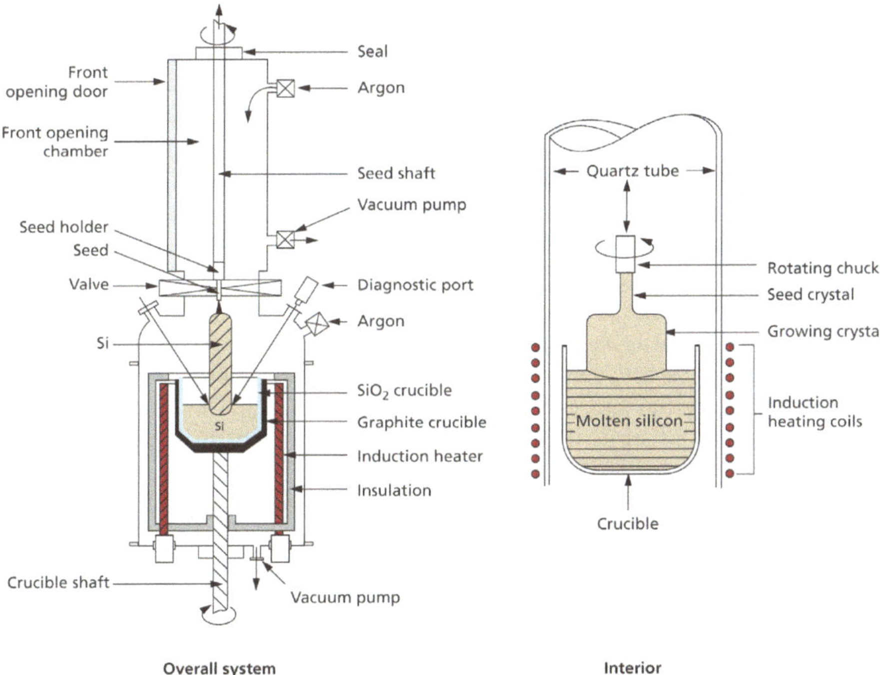

Fig. 4.10 Schematic of a Czochralski apparatus used to grow single-crystal Si ingots. Reproduced with permission from Clemens, J. T. *Bell Labs Tech. J.* **1997**, 76. Copyright 1997 Wiley–VCH

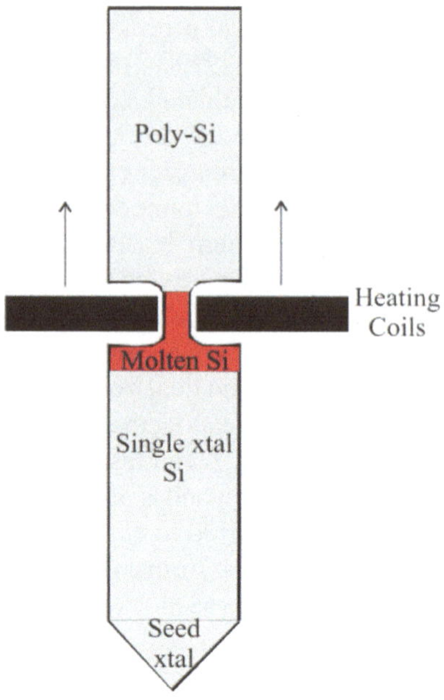

Fig. 4.11 Illustration of the float-zone (FZ) method used to convert a polysilicon ingot into single-crystal silicon

In order to ensure high purity of the ingot, this process is normally performed *in vacuo* (ca. 10^{-6} Torr), or under an inert atmosphere (e.g., 99.999% Ar) using an unreactive chamber material such as ceramic or quartz. Even under these reaction conditions, the CZ method technique suffers from O and C impurities that arise predominantly from the crucible walls. It should be noted that small quantities of O impurities are actually desirable, as they may trap unwanted transition metal impurities, a process referred to as *gettering*. The CZ technique is especially useful to yield doped semiconductors. For example, to yield p-doped Si, the desired concentration of pure Ga metal is added to molten Si within the crucible.

The float-zone technique uses inductive heating to convert poly-Si into single-crystal ingots of EG-Si. As a heating coil is slowly passed from one end to the other, the impurities become concentrated in the moving molten zone, becoming concentrated at the finished end of the cylinder (Fig. 4.11). The resultant cylinder is typically of greater purity than those using CZ growth, making this technique the most heavily used for modern semiconductor processing. The single-crystal ingot is subsequently sliced into thin wafers using a diamond saw, with typical thicknesses of ca. 0.7 mm, and diameters of 300 mm. The crystal orientation and doping (n- or p-type) of the wafer are designated by the location of primary and secondary flats (Fig. 4.12).

The wafer surface must be free of topographic defects, microcracks, scratches, and other residual surface imperfections. To ensure that the wafers remain free from dust and other airborne contaminants, the finished wafers are only handled within a *clean room*. To control the particulate contamination, clean rooms exhibit extensive use of stainless steel, perforated floors and ceiling tiles to promote air circulation,

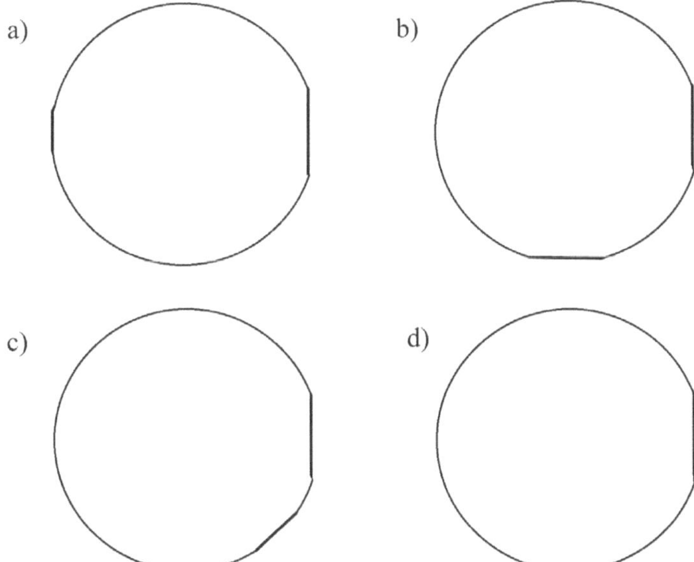

Fig. 4.12 Position of flats on Si wafers, used to identify the crystalline plane and type of dopants (n- or p-type). Shown are (a) n-type (100)—1° and 2° flats 180° apart, (b) p-type (100)—1° and 2° flats 90° apart, (c) n-type (111)—1° and 2° flats 45° apart, and (d) p-type (111)—1° flat only

and sloped surfaces to avoid dust accumulation. Yellow lighting is used to prevent light-sensitive material from being exposed to UV light, until desired as a part of chip fabrication (vide infra). Prior to entering the clean room, personnel must cover their clothing with a white "bunny suit" made of Tyvek®, a polymeric nonwoven fabric that exhibits non-lint and antistatic properties and contains few surface sites for particulate adhesion (Fig. 4.13). *En route* to the clean room, the worker must also walk over a sticky pad and pass through an air shower to remove dust particles from shoes and clothing. The ratings of clean rooms range from Class 1 to Class 10,000—an indication of the number of particles per cubic foot. As a familiar reference, in uncontrolled environments such as a typical home or office, the particle count is approximately 5 million per cubic foot!

4.2.2 *Integrated Circuits*

Integrated circuits (ICs), or chips, are the brains of all electronic devices. Each IC is a compilation of billions of complex subunits such as transistors, resistors, capacitors, and diodes that are all interconnected in a specific manner depending on the desired application. Incredibly, this minute wonder of microcircuitry is fitted onto the surface of a thin silicon substrate smaller than an individual fingernail (Fig. 4.14). Although

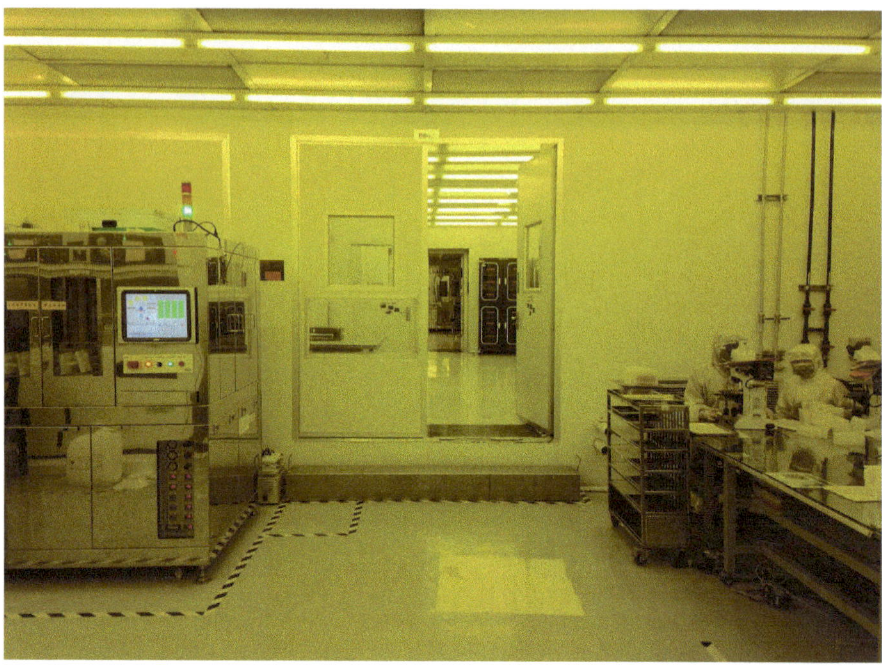

Fig. 4.13 A clean room used for chip processing. Workers must be covered in lint-free "bunny suits" to prevent contamination of microelectronic circuitry. Stainless steel surfaces and efficient laminar-flow air circulation systems are used to maintain a very low level of particulates in the air. Photoresists do not react to yellow light; hence, this lighting is used to prevent light-induced side reactions of photoresist films. Photo taken by B. D. Fahlman

ICs have only been used since the early 1970s, consider some of the ways these devices have changed our lives:

Lightweight laptop computers and tablets allow for productivity while traveling;

Vehicle emissions are continuously monitored and engine conditions adjusted for maximum gasoline efficiency;

Autonomous vehicles from Tesla and other manufacturers are in advanced stages of development;

Medical doctors are able to operate on patients from remote locations;

Answers to virtually all questions we may pose are available within seconds via the Internet;

Televisions are most commonly mounted on walls; 4 K/8 K clarities rival viewing the world through a window;

Phones may be used from virtually anywhere on Earth to keep in touch or check email;

GPS systems tell us how to best arrive at our desired destination based on current traffic conditions…

… What a world we live in—what will the next 5–10 years bring?…

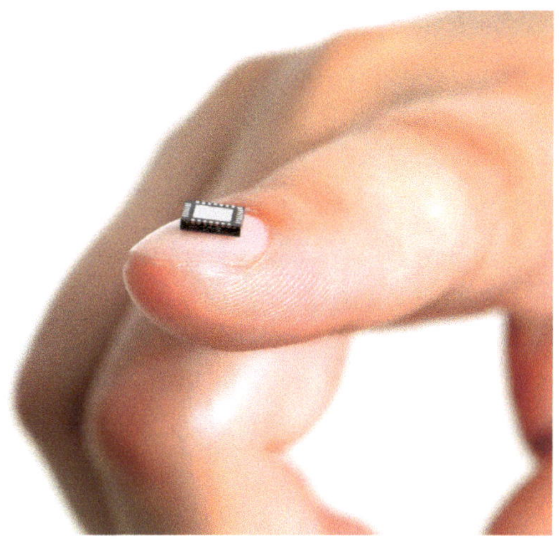

Fig. 4.14 A computer chip on a fingernail. © LeoSad/Shutterstock.com

4.2.3 Field-Effect Transistors (FETs)

The workhorses of ICs are *transistors*, which act as electronic switches in digital circuitry. Transistors were discovered by Bell Labs in the late 1940s as a replacement for vacuum tubes, which were much larger and consumed significantly more power. The earliest ICs utilized individual transistors; however, these circuits quickly became too large and complex to assemble for all but the simplest applications. In particular, computations were slow due to the long distances traversed by the electrical signals. It was clear that the only way to make ICs behave faster was by increasing the density of transistors. In 1965, Gordon Moore (a co-founder of Intel) predicted that the number of transistors on a chip would double every 2 years. *Moore's law* has been upheld since this early prediction, with an impressive rise in the number of transistors used in ICs from the first commercial release in the early 1970s, to the latest multi-core processors that run both PC and Mac systems (Fig. 4.15).

The first commercial applications for transistors were radios, which utilized a small number (i.e., <10) of transistors. From the first microprocessor chip introduced in 1974—the Texas Instruments TMS 1000 with 8,000 transistors—it is hard to believe that current microprocessor chips now contain over 100 million transistors/mm^2 (Fig. 4.16)! The M1 Ultra microprocessor from Apple currently contains the largest number of transistors—114 billion on a chip area of 840 mm^2! The latest graphics processing units (GPUs) also contain an extraordinary number of transistors; for example, AMD's Instinct MI250X chip contains 59.4 billion transistors on a chip size of 1480 mm^2. It should be noted that flash-memory chips have long featured over 1 billion transistors. Since 2005, 8 GB memory chips (146 mm^2) have featured over 4 billion transistors; Samsung's 1 TB eUFS 3-D stacked V-NAND flash memory chip contains over 2 trillion floating-gate MOSFETs! Such a technological advancement is only possible through an exponential decrease in the price of single

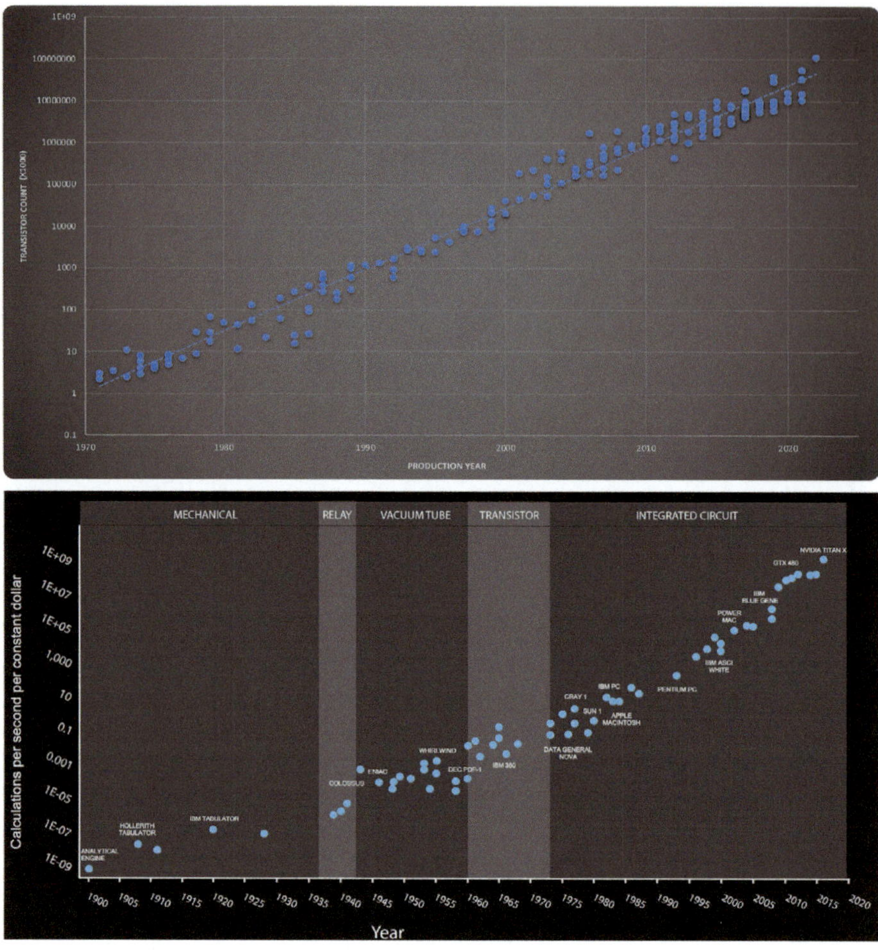

Fig. 4.15 The number of transistors employed within memory and processor chips (top). The trend of increasing computational speed over time, which has increased most rapidly with the advent of integrated circuits (bottom). © Steve Jurvetson (https://creativecommons.org/licenses/by/2.0/)

transistors. For instance, the first commercially available transistor was the Raytheon CK703, priced at \$18 in 1951. Taking inflation into account, the price of a single transistor would be ca. \$199 today, resulting in computers that would be worth over \$22 trillion! Fortunately, developments in chip fabrication processes have made it possible to reduce the effective cost of a single transistor to less than \$0.000000003 (3 nanocents!).[5]

The number of transistors within an IC is typically used to classify the complexity of the circuits. For instance, the earliest ICs of the early 1960s were considered *small-scale integration* (SSI), since they contained fewer than 100 transistors. In the late 1960s, ICs contained between 100 and 500 transistors, referred to as *medium-scale*

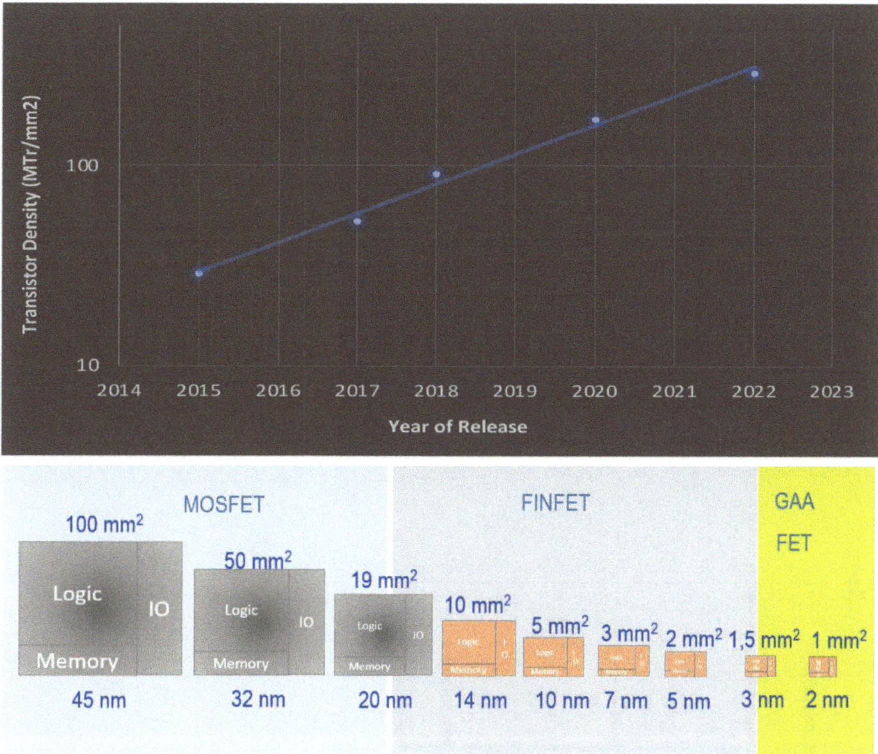

Fig. 4.16 (Top) Trend in improving logic transistor density in units of million transistors per mm^2 (MTr/mm^2) *vs.* start date of high-volume manufacturing. (Bottom) The area scaling trend of a hypothetical 100 mm^2 microprocessor from the 45-nm node down to the 2-nm node. © Intel Corporation (http://www.intel.com)

integration (MSI). Large-scale integrated circuits (LSI), containing 500–20,000 transistors, were prevalent in the early 1970s, responsible for the first handheld calculators. The 1980s brought about *very large-scale integration* (VLSI), in which ICs contained more than 20,000 transistors. Although the current architectures with greater than 1 million transistors are often referred to as *ultralarge-scale integration* (ULSI), the semiconductor industry still commonly classifies the latest ICs as being VLSI. Even though integrated circuits are our current means for calculations, this paradigm will likely not be the last. As developments in nanotechnology continue, alternative technologies will likely replace transistor-based ICs (e.g., optical computing, quantum-well transistors, neuromorphic computing, etc.[6]), allowing for computational improvements well into the future.

The most important component within all transistors is the *p–n junction*, referring to the interface between p- and n-type Si. This contact results in a concentration gradient being established, with a greater electron density in the n-doped side relative to the p-doped region (vice versa for the hole concentration). As a result, free electrons from the n-type Si readily diffuse to the p-doped Si nearest the p–n junction. A significant number of free electrons will recombine with holes in the p-Si region, leaving behind positively charged n-dopant ions (e.g., As^+, P^+). Likewise, negatively charged p-dopant ions (e.g., B^-) are also formed on the p-Si side (Fig. 4.17a). Since the region in the vicinity of the p–n junction contains ions rather than free electrons/holes, this is referred to as the *depletion zone* (Fig. 4.17b).

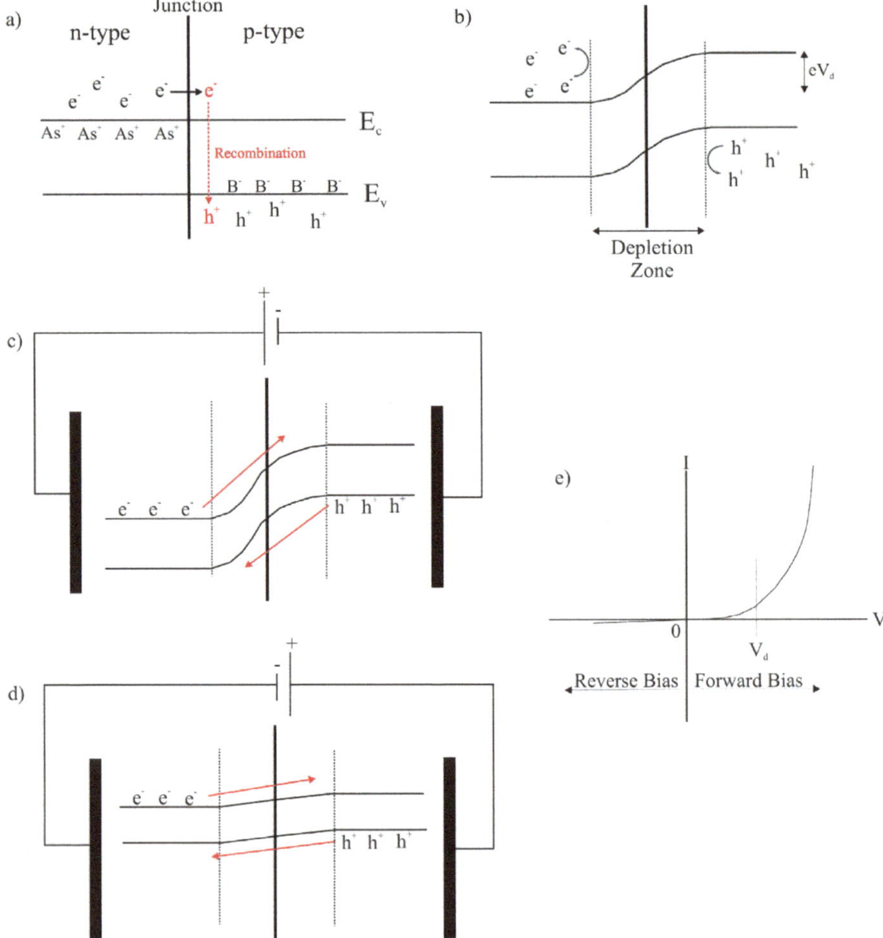

Fig. 4.17 Schematic of carrier flow within a p–n junction. Shown are: **a** the diffusion of free electrons from n- to p-type Si resulting in recombination with holes, **b** formation of a depletion zone that repels the flow of electrons and holes from n- and p-type regions, respectively, **c** application of a reverse bias, resulting in widening the junction potential, eV_O, **d** application of a forward bias, which lowers eV_O and results in an exponential voltage spike once a threshold bias of V_D is applied, **e**

The presence of oppositely charged ions within the depletion region forms an electric field (junction potential, eV_o) that opposes, and eventually ceases, the spontaneous movement of electrons/holes from their respective n- and p-type Si regions. An external voltage may be applied that causes further band bending within the depletion zone, referred to as *reverse bias* (Fig. 4.17c). This will increase eV_o, making it harder to push electrons and holes toward the junction, which results in little/no current.[7] However, if an external voltage is applied in the opposite direction (*forward bias*), electrons and holes may easily diffuse across the p–n junction (Fig. 4.17d).[8] This electrically induced flow of electrons in only one direction is referred to as a *diode* (Fig. 4.17e).

If a current is introduced to a simple configuration containing islands of n- or p-type Si within a matrix of an oppositely doped Si substrate (*i.e.*, p- or n-type Si, respectively—Fig. 4.18a), no electron flow will occur. This is because there is no conduit between the electron-rich and electron-deficient regions. Electron flow will occur if a metal is allowed to contact the doped-Si regions (Fig. 4.18b); however, there will be no control over the electrical current (analogous to the flow of water through a fully opened water faucet). Hence, the most effective arrangement that will allow for controllable electrical conductivity is shown in Fig. 4.18c. This architecture allows one to control the electron flow through a *gate* by varying the magnitude of the applied external voltage on the metal (or polysilicon—the modern composition of complementary metal–oxide–semiconductor (CMOS) gates). Such a configuration is known as a metal oxide semiconductor field-effect transistor (MOSFET), where an insulating layer known as the *gate oxide* is used to prevent uncontrollable tunneling of electrons between the gate and doped-Si substrate. The term "field effect" refers to using an electrical field to control the flow of electrons or positively charged holes.

Let us now consider how an electrical charge may be controlled using a MOSFET. To simplify the discussion, we will first consider a metal-on-semiconductor (nMOS) capacitor—an nMOSFET without source and drain contacts (Fig. 4.19a). When a negative bias is applied to the gate (V_{gate}), the positive holes are attracted from the substrate, forming an *accumulation layer* of positive charge on the SiO_2/Si interface (Fig. 4.19b). The increase in hole concentration on the Si surface causes the conduction and valence bands to bend upward, increasing the gap between the Fermi level and conduction band.

By comparison, if a positive bias is applied to the gate, the positive holes will be repelled from the SiO_2/Si interface leaving behind a *depleted region* (Fig. 4.19c). Since the number of holes and electrons is constant, the electron concentration increases near the silicon surface resulting in a downward bending of the conduction and valence bands (*i.e.*, narrowing the gap between E_F and E_c). However, for p-type silicon, the electron density is negligible for small positive biases. As a result, the gate charges are balanced not by electrons, but by negative acceptor ions (e.g., B^-) in the depletion layer. An increase in the gate bias causes the depletion layer to widen, yielding more acceptor ions in order to balance the gate charge. The degree of E_c and E_v band bending is directly related to the intensity of the positive gate bias. At large positive gate biases, an *inversion layer* will form from the conduction band lying closer to the Fermi level than E_v. That is, at large applied voltages the surface

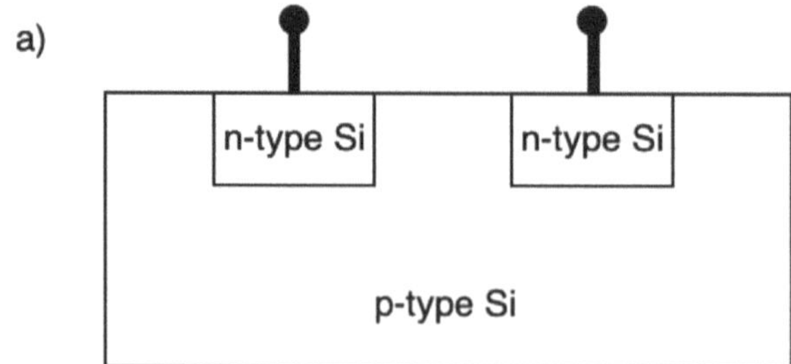

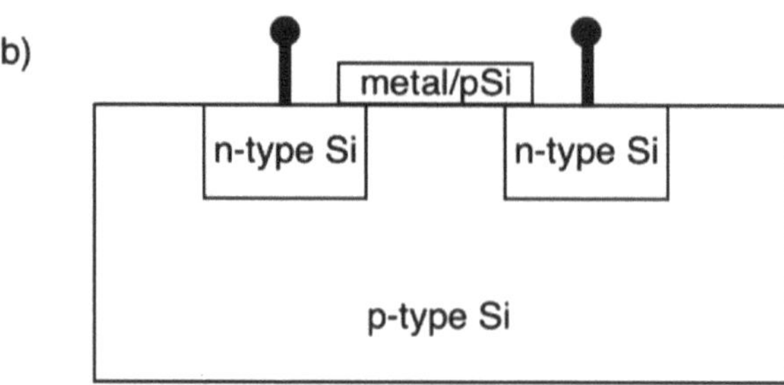

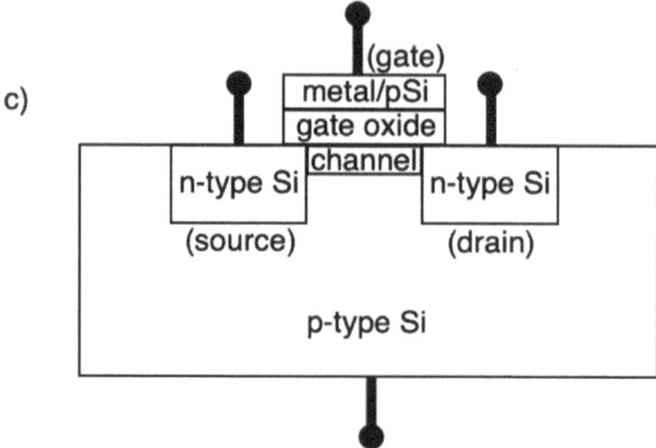

Fig. 4.18 Schematic of possible n–p transistor architectures. Shown are: **a** n–p junction not capable of transporting an electrical current, **b** metallic connectivity between electron-rich reservoirs, and **c** an n-channel MOSFET, which features controllable electrical conductivity

of the Si is inverted from p-type to n-type due to the generation of free electrons that form a channel. In a MOSFET, the inversion-induced channel is located between the *source* and *drain*. Accordingly, there are two types of MOSFETs: an n-channel (NMOS or nMOSFET) and p-channel (PMOS or pMOSFET), where the substrate comprises p- and n-type silicon, respectively.

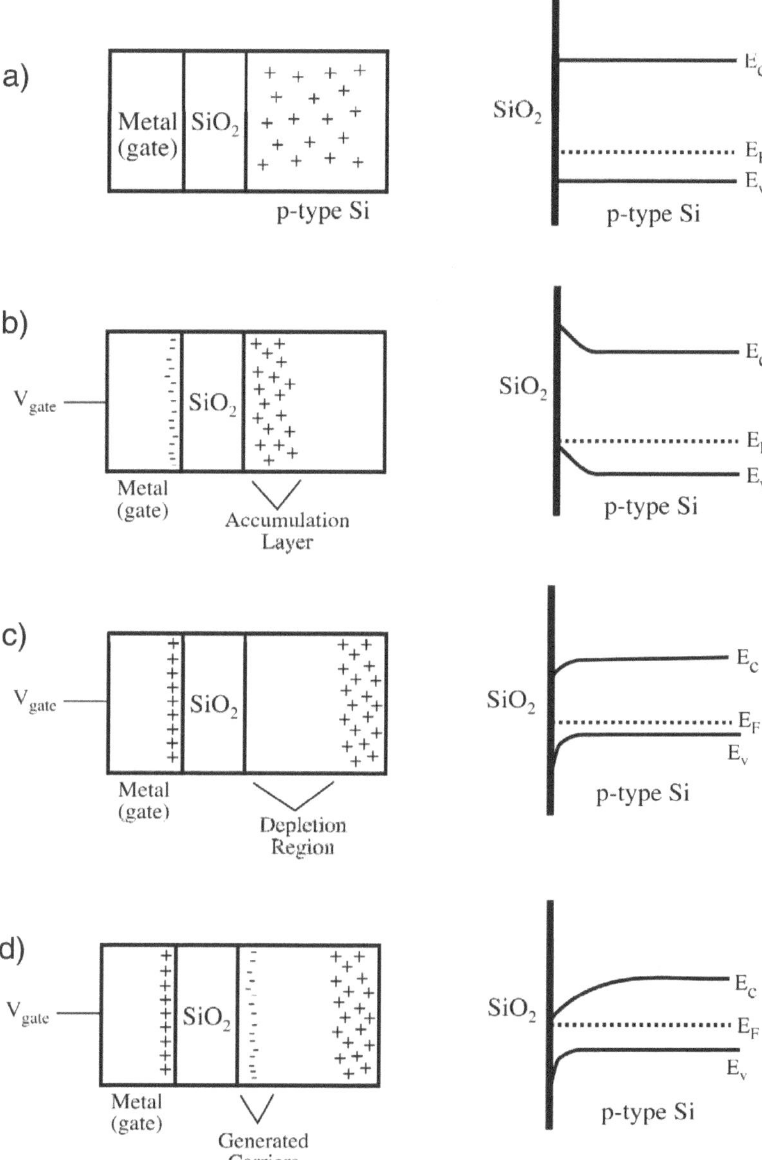

Fig. 4.19 Charge distributions (left) and band diagrams (right) for an n-type metal–oxide–semiconductor (nMOS) capacitor, with different gate bias modes. Shown are: **a** $V_{gate} = V_{FB}$, **b** accumulation ($V_{gate} < V_{FB}$), **c** depletion ($V_{FB} < V_{gate} < V_{T}$), and **d** inversion ($V_{T} < V_{gate}$). V_{T} and V_{FB} are threshold and flat-band voltages, respectively. The flat-band voltage is obtained when the applied gate voltage equals the work function difference between the gate and the semiconductor

4.2.4 Transistor Scaling: Advantages and Limitations

Over the past 60 years, the feature size of transistors has steadily decreased each year in accord with Moore's law. The continual miniaturization of MOSFETs is desirable for many reasons. Most obviously, smaller transistors may be more densely packed on ICs. The shrinking dimensions of individual transistors will shorten the distance from source–drain, resulting in higher computational speeds. The high density of transistors per chip is also essential to maintaining affordable ICs, since the cost of their fabrication is directly related to the number of chips that may be contained on each wafer. For instance, it is estimated that a 15% decrease in transistor size reduces fabrication costs by 50% per device. In addition to the density advantage, there is a significant operational benefit of smaller MOSFETs. Since transistors may be considered as resistors in the on-state, a smaller transistor will have less electrical resistance as well as lower gate capacitance, allowing more current to pass through in a shorter timeframe (i.e., higher processing speeds and less power consumption). Equation 4.6 shows the relationship between various transistor properties and its observed current.

$$I_{DS} = \frac{W}{2Lt_ox}\mu C(V_{GS} - V_T)^2, \tag{4.6}$$

where I_{DS} is the drain current, W is the channel width, L is the channel length, t_{ox} is the gate oxide thickness, μ is the channel carrier mobility, C is the capacitance density of the gate oxide, V_{GS} is the gate voltage, and V_T is the threshold voltage.[9]

The empirical scaling law of MOSFETs is provided in 4.7, which describes how the channel length, $\mathrm{L_{min}}$, is influenced by geometrical parameters ($\mathrm{r_j}$ and t_{ox}) as well as voltage-dependent parameters (w_{s} and w_{D}):

$$L_{\mathrm{min}}[\mu\mathrm{m}] = 0.41\left[r_j[\mu\mathrm{m}] \cdot t_{ox}\left[\mathrm{\mathring{A}}\right] \cdot (W_s[\mu\mathrm{m}] + W_D[\mu\mathrm{m}])^2\right]^{\frac{1}{3}}, \tag{4.7}$$

where r_{j} is the source/drain junction depths, t_{ox} is the oxide thickness, and w_{s} and w_{D} are the widths of source and drain depletion layers, respectively.

There is no electrical lower limit for L_{min} if the junction depths, oxide thicknesses, and doping are properly scaled down. However, in recent years, the channel length has been scaled more aggresively than t_{ox} or $\mathrm{r_j}$—to the point where the gate length has become comparable to the width of source and drain depletion zones (Fig. 4.20). This results in *short-channel effects* (SCEs), in which depletion is influenced by drain and source biases instead of only the gate voltage. Since less gate voltage is needed for depletion, the barrier for electron injection from source to drain (i.e., threshold voltage, V_{T}) decreases—an effect known as *drain-induced barrier lowering* (DIBL). As V_{T} is reduced, a weak inversion layer is formed even when $V_{\mathrm{gate}} = 0$, referred to as *sub-threshold leakage*. When magnified over billions of transistors, this leakage results in high power consumption for the chip—especially relevant for applications that rely on low power dissipation to prolong battery life. If the channel is shrunk to a minimum length in which the source and drain depletion regions coincide, an

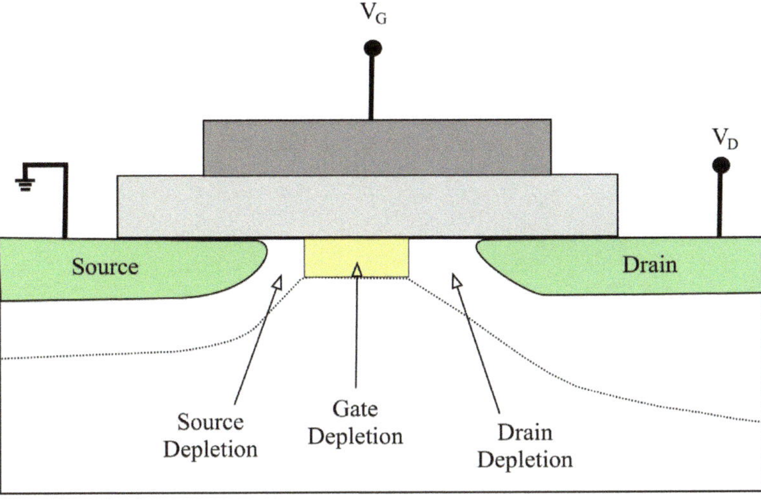

Fig. 4.20 Schematic illustrating short-channel effects (SCEs), in which the length of source/drain depletion zones are no longer small in comparison to the channel length. These depletion charges are not gate-controlled, but serve to change the gate-defined threshold voltage

effect known as *punchthrough* is observed, which deleteriously affects the current saturation of the MOSFET device.[10]

Since the gate oxide serves as the insulator between the gate and channel, its thickness directly influences channel conductivity when the transistor is on, and sub-threshold leakage when the transistor is off. As one can imagine, such a thin insulating layer is prone to dielectric breakdown—electron tunneling between the polysilicon gate and channel, which leads to increased power consumption and heat buildup. For instance, a SiO_2 thickness of 1.2 nm would result in a leakage current on the order of 10 A mm^{-2}, which would be sufficient to drain a cell phone battery in minutes! Furthermore, as the oxide thickness decreases, other issues arise such as boron diffusion from the polysilicon gate into the channel.[11] This significantly alters the threshold voltage, and ultimate drain current, of the transistor. Nitrogen has been incorporated within the oxide layer in an attempt to prevent B-diffusion; however, nitrogen also diffuses into SiO_2 and Si regions resulting in surface "roughness" and a decrease in mobility and on-current.

The 45-nm technology node[12] in 2007 first introduced the replacement of SiO_2 with a *high-k dielectric* gate oxide (Fig. 4.21; *k* refers to its *dielectric constant*).[13] These materials have a larger dielectric constant than previously used SiO_2 (k = 3.82), which allowed one to use thicker oxide layers for an identical gate capacitance (4.8):

$$C_{vol} \propto \frac{k}{t^2}, \tag{4.8}$$

where C_{vol} is the capacitance per unit volume, k is the dielectric constant (or, *relative permittivity*), and t is the thickness of the dielectric medium.

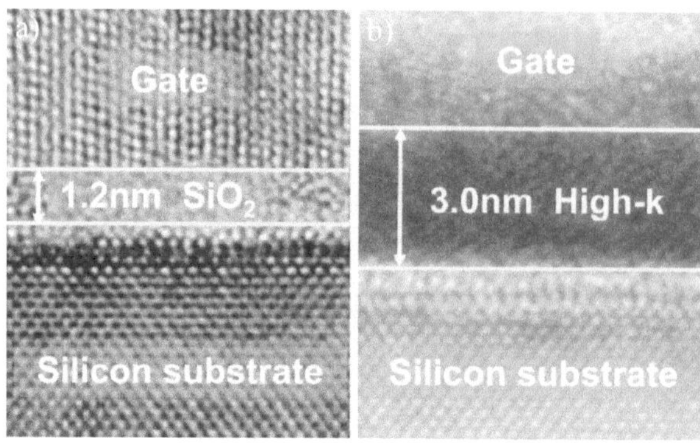

Fig. 4.21 Cross-section high-resolution transmission electron microscopy (HRTEM) images of gate oxides for MOSFETs. Provided is an illustration of film thicknesses that result in identical capacitance for SiO_2 ($k = 3.8$), relative to a high-k dielectric ($k = 23.9$) gate oxide. Hence, increased gate capacitance will result from thinner films comprising high-κ dielectric materials. © Intel Corporation (http://www.intel.com)

Materials that exhibit a high dielectric constant are strongly polarizable. That is, when exposed to an electric field, the dipole moments of the constituent molecules and atoms of the dielectric material will become aligned. This results in the buildup of electrical charges on the surface of the capacitor plates (Fig. 4.22). In general, inorganic materials with ionic or partial ionic interactions will possess a relatively large dielectric constant (high-κ), whereas organic molecules with nonpolar bonds are typically low-κ materials. Materials containing relatively large atoms will be most useful as dielectrics, since they exhibit greater polarizabilities due to valence electrons being situated farther from the nucleus.

In addition to being polarizable, it is also paramount that the dielectric medium exhibits a low electrical conductivity (i.e., a large bandgap—an offset with Si of >1 eV), to minimize carrier injection into its bands. For example, although the dielectric constants for Ta_2O_5 and TiO_2 are sufficiently large for high-κ gate oxide applications (see below), both of these materials have small bandgaps (4.5 and 3.5 eV; relative to 8.9 eV for SiO_2), which would result in high leakage currents. On the other hand, more promising candidates include Al_2O_3, HfO_2, and ZrO_2, with bandgaps of 8.7, 5.7, and 7.8 eV, respectively.

Other properties that must be inherent in the high-κ dielectric layer include thermodynamic and kinetic stabilities with the Si channel and gate.[14] Early MOSFETs used polysilicon gates since it is chemically identical to the underlying substrate. However, during the high-temperature processing conditions of IC design, interfacial silicides are formed through reactions with the high-κ dielectric layer (e.g., $HfSiO_4$ for a HfO_2 gate oxide), which leads to dielectric breakdown. A solution to this problem is the use of a metal gate stacks such as Ti/Hf, WN, or TaN instead of polysilicon. This also circumvents problems associated with boron diffusion, since metal gates do not contain boron dopant species.

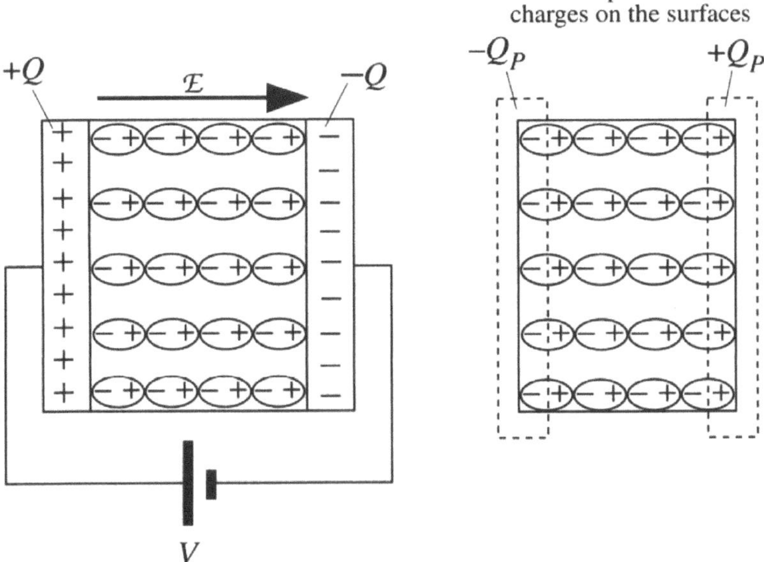

Fig. 4.22 Schematic of bound polarization occurring on capacitor plates due to polarization of the dielectric medium

The longevity of the dielectric material is dependent on the exposed voltage. When too high a voltage is introduced between the electrodes, dielectric breakdown of the solid may ensue. The solid-state structure, presence of impurities or microstructural defects (e.g., cracks/voids), and ambient conditions (e.g., temperature, humidity) are all paramount in assessing the *dielectric strength* (i.e., resistance to breakdown) of an insulating material. As you might imagine, these factors become even more relevant when considering the 0.8–1 nm SiO_2 layers found in modern MOSFETs.

Hafnium(IV) oxide is the current high-κ material of choice due to its mature deposition technologies and lack of silicide formation with suitable gate stacks. However, there are three types of high-κ dielectrics, which have been investigated as gate oxide layers (measured dielectric constants in parentheses):

i. Materials with $4 < \kappa < 10$ (*i.e.*, relatively weak polarizability) such as: Al_2O_3 (7.5–10), Si_3N_4 (7.9–8.1), sodalime glass (7.2), $Al_6Si_2O_{13}$ (mullite, 6.3), $MgAl_2O_4$ (spinel, 8.6), $MgSiO_4$ (forsterite, 6.2), MgO (9.65), ZnO (8.15), BN (4.15), BeO (6.8), CaF_2 (6.8), $BaCO_3$ (8.5), $SrCO_3$ (8.85), ($KAl[Si_3O_{10}(OH)_2]$) (mica muscovite, 6.5–8.7), and organic polymers (polyurethane: 7.1, polyvinyl fluoride: 7.4, polyvinylidene fluoride: 6.4).
ii. Materials with $10 < \kappa < 100$ such as: MnO_2 (pyrolussite, 12.8), ThO_2 (18.9), Ta_2O_5 (27.6), UO_2 (24), Nb_2O_5 (67), HfO_2 (25), ZrO_2 (12.5–24.7), $Zr_xHf_{1-x}O_2$ (25), Y_2O_3 (15), La_2O_3 (25), Gd_2O_3 (23), Pr_2O_3 (25), and $BaZrO_3$ (43).

iii. Materials with $\kappa > 100$ (*i.e.*, strongly polarizable) such as: TiO_2 (rutile, 85–170), $PbZrO_3$ (200), $Pb(Zr_{0.52}Ti_{0.48})O_3$ ("PZT," 1800), $SrTiO_3$ (2080), $Ba(Sr_{0.52}Ti_{0.48})O_3$ (3000), $Pb(Ni_{0.33}Nb_{0.67})O_3$ (5500), and $BaTiO_3$ (6000).

More recently, the use of high-κ ferroelectric materials in the gate stack has been investigated to prevent gate leakage via *negative capacitance*, denoted as negative capacitance field effect transistors (NCFETs).[15] To date, the materials evaluated for NCFETs have consisted of organic ferroelectric-poly(vinylidene difluoride-trifluoroethylene), [P(VDF-TrFE)],[16] titanate and niobate compounds such as lead zirconate titanate (PZT),[17] and $Hf_xZr_{1-x}O_2$ (Fig. 4.23)[18] (4.9) provides the definition for the subthreshold swing (SS) of an intrinsic MOSFET.

$$SS = \frac{\partial V_G}{\partial \Psi_S} = \left(\frac{\partial V_G}{\partial \Psi_S}\right)\left(\frac{\partial \Psi_S}{\partial \log_{10} I}\right) \tag{4.9}$$

where Ψ_S is the surface potential of the semiconductor channel, V_G is the gate voltage, and I is the drain current.

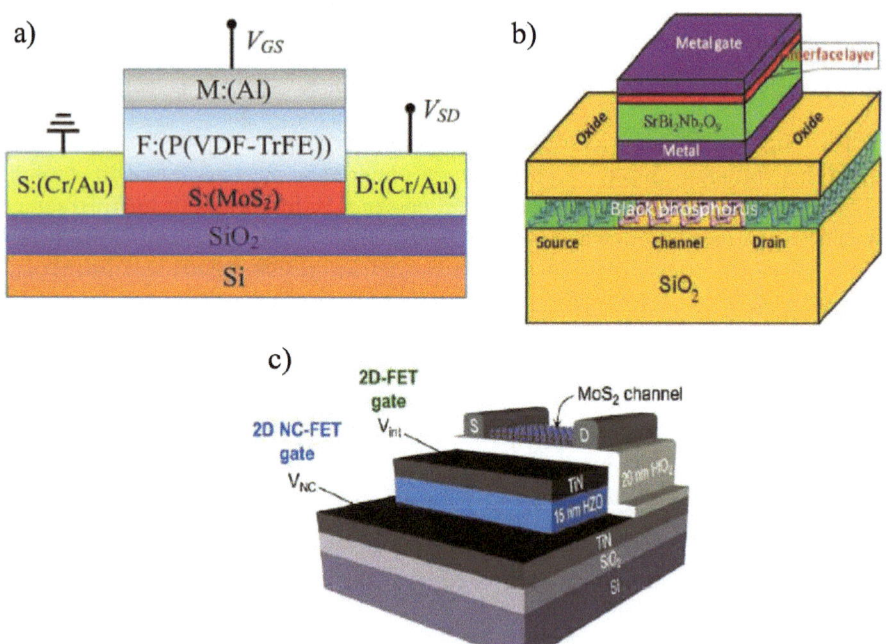

Fig. 4.23 Cross-section illustrations of next-generation NCFETs currently in development. Shown are **a** organic poly(vinylidene difluoride-trifluoroethylene), [P(VDF-TrFE)], **b** inorganic perovskite-type, and **c** Hf-doped ferroelectrics. Reproduced with permission from *Adv. Electron. Mater.* **2018**, *4*, 1,800,231. © Wiley–VCH

The first term of 4.9, $\left(\frac{\partial V_G}{\partial \Psi_S}\right)$, denotes the control of the gate voltage to surface potential, whereas the second term represents the current generated by the surface potential. Within the subthreshold region of a MOSFET, the channel current is dominated by minority diffusion; that is, at room temperature[19]:

$$\frac{\partial \Psi_S}{\partial \log_{10} I} = 60\text{mV dec}^{-1} \tag{4.10}$$

Since $\left(\frac{\partial V_G}{\partial \Psi_S}\right)$ >1 for traditional MOSFET devices, the minimum value for SS is 60 mV dec^{-1}. This implies that for a tenfold increase in drain current, I_D, the gate voltage (V_G) must be increased by at least 60 mV. In order to reduce SS below the lower limit, a variety of approaches have been investigated, including the use of suspended-gate MOSFETs.[20] The use of NCFETs allows one to decrease $\left(\frac{\partial V_G}{\partial \Psi_S}\right)$ to a value less than 1 via a voltage amplification mechanism. These devices are capable of achieving low-power dissipation and high on-state current with off-state current effectively eliminated. Furthermore, the fabrication of NCFETs is straightforward and is compatible with existing techniques used for nonvolatile memory chips.

In addition to shrinking T_{ox}, the thicknesses of source/drain regions (r_j) have also decreased in recent years. However, this scaling leads to increased resistance and slower carrier mobilities. One strategy to circumvent these issues is the use of raised source/drain regions. In these designs, the electrons have more room to maneuver thus resulting in less electron–electron repulsions and lower overall resistance. This translates to a lower operating voltage for the transistor, which results in low power consumption with less heat generation during its operation.

As another strategy to improve carrier mobilities, Intel first reported the use of "strained Si" in their 90-nm technology node (2002) in which the silicon lattice in the channel region is strained through compression or expansion. It has been shown that electrons flow through strained silicon 70% faster than in unstrained silicon. For ICs, this translates to a 17% increase in speed at the same power, or 35% reduction in power consumption at the same speed.

To understand the effect of incorporating strained-Si channels, we need to consider Si–Si bonding within the bulk crystalline solid. As we discussed earlier, electrons are promoted from valence to conduction bands due to thermal excitation. The valence band of the extended solid is formed from the overlap of sp^3 hybridized orbitals residing on each Si atom. When an electron migrates from valence to conduction bands under normal circumstances, there is no directional preference. However, when a strain is introduced along a specific direction of the lattice, the energies of the hybrid orbitals along this direction are altered.

A stretching, or tensile, strain will cause the orbitals between neighboring Si atoms to be expanded, which weakens Si–Si bonds along the strain direction. Hence, it becomes easier for an electron to be thermally promoted from the valence to conduction band, corresponding to greater electron mobility. This effect is more pronounced for n-channel MOSFETs; the strain-expanded hybrid orbitals between neighboring Si atoms may accept electron density with less electron repulsion. By contrast, compressive strain is typically used for p-channel MOSFETs, where the presence of holes may offset the electron–electron repulsion within the compressed hybrid orbitals. More simplistically, crystal strain increases the free mean path of electrons (*i.e.*, less scattering through electron–lattice interactions), which increase their velocity under a given field.

Compression strain of p-doped Si is most often accomplished by carving trenches on opposite ends, and filling these with a material with greater lattice spacing than Si. This is typically a material such as $Si_{1-x}Ge_x(x = 0.1–0.3)$ solid solutions (Fig. 4.24); since the lattice constants of Si and Ge are 5.43 and 5.66 Å, respectively, the lattice constant of the solution is [$5.43(1 - x) + 5.66 \times$ Å]. A common method used to induce tensile strain is the deposition of a dense ceramic Si_3N_4 film on top of the channel. These methods result in *uniaxial strain*, which is preferred since localized regions may be altered rather than the entire wafer. In fact, these techniques may be used to modify both nMOSFETs and pMOSFETs on a single chip. By contrast, *biaxial strain* is induced through the heteroepitaxial deposition of a Si thin film on top of a SiGe substrate (Fig. 4.25). A number of precursors such as germane (GeH_4; deposition T, T_{dep}, of 400–700 °C), germanium tetrachloride ($GeCl_4$; T_{dep} > 850 °C), isobutylgermane ($C_4H_9GeH_3$; T_{dep} 400–500 °C) and methylgermanium chloride ($(CH_3)GeCl_3$; T_{dep} 750–1000 °C) may be used to deposit the SiGe base layer using chemical vapor deposition (a technique described in more detail later in this chapter). The amount of strain and mobility enhancement depends on the germanium content of the SiGe layer.

Dramatic increases in the number of transistors and the growing market for low-power applications require transistors with increasingly lower leakage currents ($<10^{-5}$ A cm^{-2}). As gate lengths shrunk to dimensions <10 nm, traditional MOSFETs behave as resistors, in which the drain and gate compete to control the channel barrier. A higher threshold voltage is able to offset this effect; however, increasing V_T

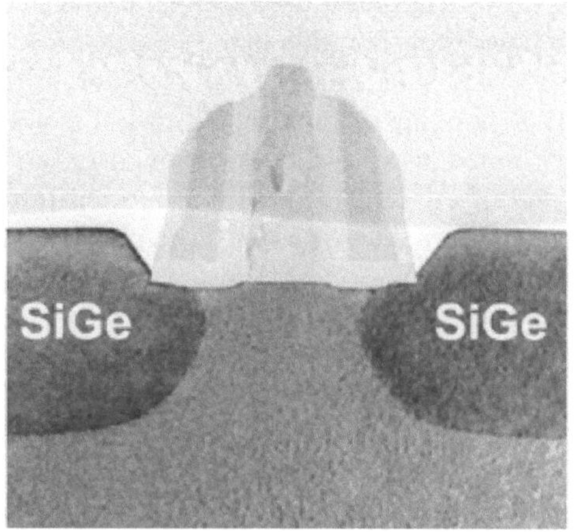

Fig. 4.24 Cross-section HRTEM image of a transistor containing a compression-strained channel. © Intel Corporation (http://www.intel.com)

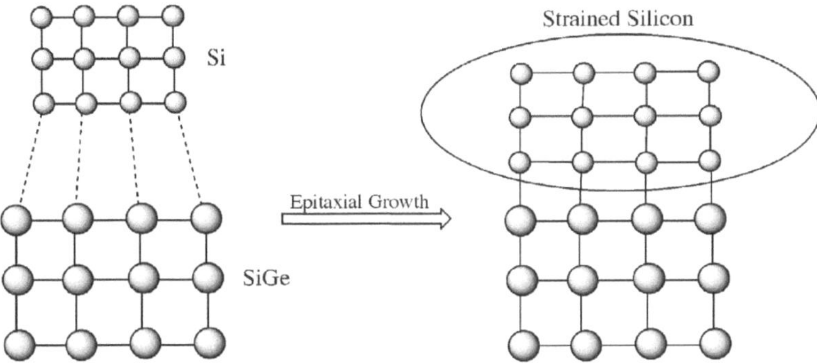

Fig. 4.25 Formation of strained silicon from the epitaxial growth of lattice mismatched Si onto a SiGe substrate

results in a higher output resistance and lowers I_{on}, which increases power consumption. Furthermore, increasing V_T requires selective dopant ion implantation into the channel region, which adds additional fabrication cost. As a consequence, the use of III-V semiconductors, Ge, SiGe, and 2-D transition metal dichalcogenides (e.g., MoS_2, WS_2, $MoTe_2$) are already in use, or have been investigated, as alternative channel materials due to their high carrier mobilities, high on/off ratios, and large intrinsic bandgaps.[21]

Although reducing T_{ox} has alleviated short channel effects, the gate is not able to control the leakage current pathways that occur far from the gate region. One way to address these issues is through the use of silicon-on-insulator (SOI) substrates (Fig. 4.26).[22] The presence of an insulating layer directly beneath a thin Si channel is reported to reduce off-state leakage from source to drain regions by more than two orders of magnitude.[23] More recently, ultra-thin-body SOIs (UTB-SOIs) with silicon-channel film thicknesses of $<L_g/3$ have been shown to effectively reduce subthreshold leakage (Fig. 4.27).[24] The insulation between source and drain cavities allows the threshold voltage (also decreasing the supply voltage) of the SOI CMOS device to be extremely low, with minimal off-state (*i.e.*, sub-threshold) leakage. When the top of the buried oxide layer (BO_x) coincides with the bottom of source/drain, the architecture is referred to as a *fully depleted SOI*; a small interfacial gap is known as partially depleted SOI, which will result in greater off-state leakage current. Since depleted substrates feature a nonconductive barrier, the transistor is able to switch on/off faster due to more effective drive current while in the on-state.

Silicon-on-insulator technology is not new; in fact, SOI ICs have been used since the 1960s for military and space applications. It is reported that this simple substrate alteration results in chips that operate more than 30% faster and use ca. 20% less power than standard CMOS analogs. The use of less power per chip is important for extending the battery life of portable electronics, as well as dealing with the heat generated by ICs. If one extrapolates the enormous heat generated from high-density bulk Si chips into the near-future, it may be necessary to use liquid coolants within the IC package—clearly not a desirable (or feasible) option.

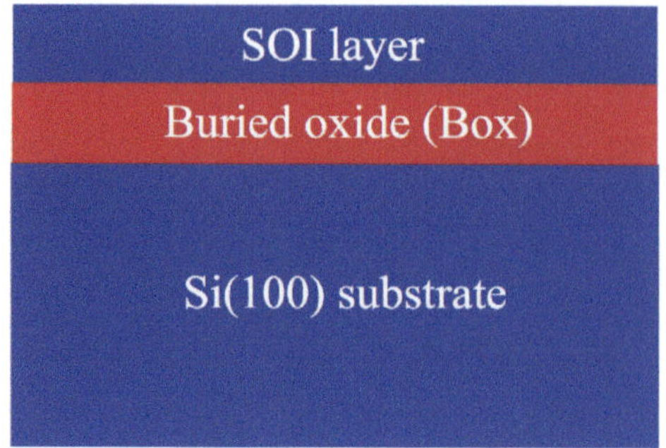

Fig. 4.26 Cross-section schematic of a SOI substrate used in modern transistor designs

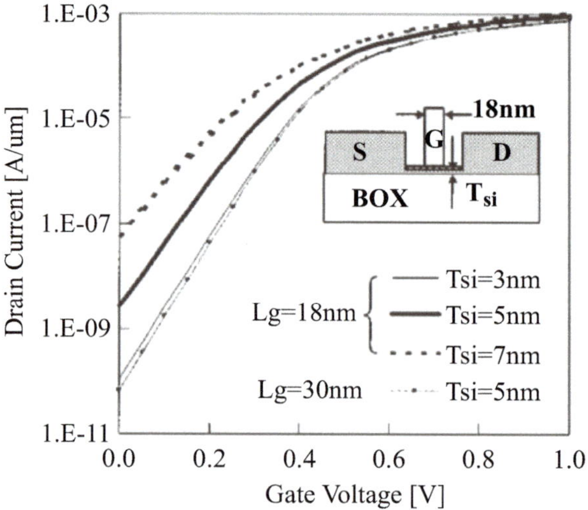

Fig. 4.27 Impact of body thickness on the I_{DS}-V_{GS} characteristics of an ultrathin body (UTB) SOI device ($T_{ox} = 1.5$ nm, $N_{sub} = 1 \times 10^{15}$ [cm^{-3}], $V_{DS} = 1$ V). Calculations show that nearly all of the leakage current at $V_g = 0$ V with a Si body thickness of 7 nm flows along the bottom 2-nm of the body, which is only weakly controlled by the gate. Elimination of this 2 nm of the body (i.e., reduction to T_{Si} from 7 to 5 nm) results in a 30-fold reduction in the leakage current. Reproduced with permission from *IEEE Electron. Device Lett.* **2000**, *21*, 254. © 2000 IEEE

Figure 4.28 summarizes the features of a MOSFET that can be optimized to improve overall chip efficiency. A further decrease in off-state leakage currents has been realized through the use of nonplanar tri-gate transistors, known as FinFETs—employed for the 14-nm (2015), 10-nm (2017), 7-nm (2019), 5 nm (2021) and 3 nm (2023) nodes (Fig. 4.29).[25] This unique design provides much better control of the channel entrained within the fin. Since electrical signals must move up and over the vertical gate, the effective pathway for electrical current transmission is tripled relative to today's planar MOSFETs. This leads to higher device drive currents than planar transistors at the same footprint dimensions. The leakage current in FinFETs is well suppressed as long as the fin thickness (currently ca. 5 nm) is less than

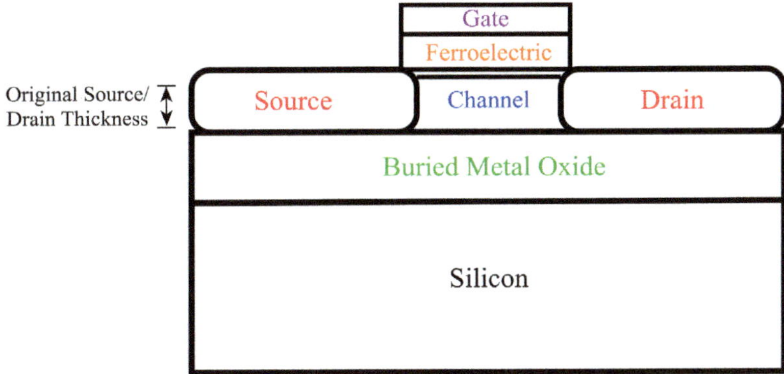

Fig. 4.28 Schematic summary of transistor design elements that have been incorporated, or are being investigated, for planar MOSFETs. Shown is the use of (i) a SOI buried oxide substrate (green), (ii) raised source/drain cavities (red), (iii) alternative channel materials (blue), (iv) a high-k dielectric gate insulating layer (yellow), (v) a ferroelectric material in the gate stack (orange), and alternative gate stack materials (purple)

the gate length (currently ca. 14 nm). Further improvements in fabrication techniques should allow for gate lengths to be scaled to <5 nm. The drive current of the FinFET devide may be optimized by increasing the width of the channel and interconnecting multiple fins. However, it should be noted that not all modern chips utilize FinFET architectures. For instance, planar CMOS chips are still employed in radio frequency (RF) wireless telecommunication technology (GPS, bluetooth, Wi-Fi, near-field communication (NFC), and mobile networks (4G, 5G)).

FinFETs have been critical in maintaining Moore's law through the 3-nm node. However, with the increasing use of machine learning and AI, transistors will need to transition to new designs within the next 4–5 years. Alternative designs that will move transistors into the Angstrom-node era will be based on next-generation 3-D FETs with further optimization of power delivery to the transistor (Fig. 4.30). However, instead of vertical fins, stacked nanosheets or nanowires/nanotubes will likely be employed, denoted as gate-all-around transistors (GAAFETs). In a FinFET device, opposite ends of an individual fin act as source and drain regions and the gate is in contact with the fin on all three sides (Fig. 4.31). In contrast, the gate in a GAAFET is in contact with nanowires or nanosheets on all four sides, providing better control over channel characteristics and producing a larger drive current at much lower supply voltage. Further down the line, forksheet FETs (Fig. 4.32) are in development, which combine both nFET and pFET in the same structure, relative to GAAFETs that use different devices for nFETs and pFETs.

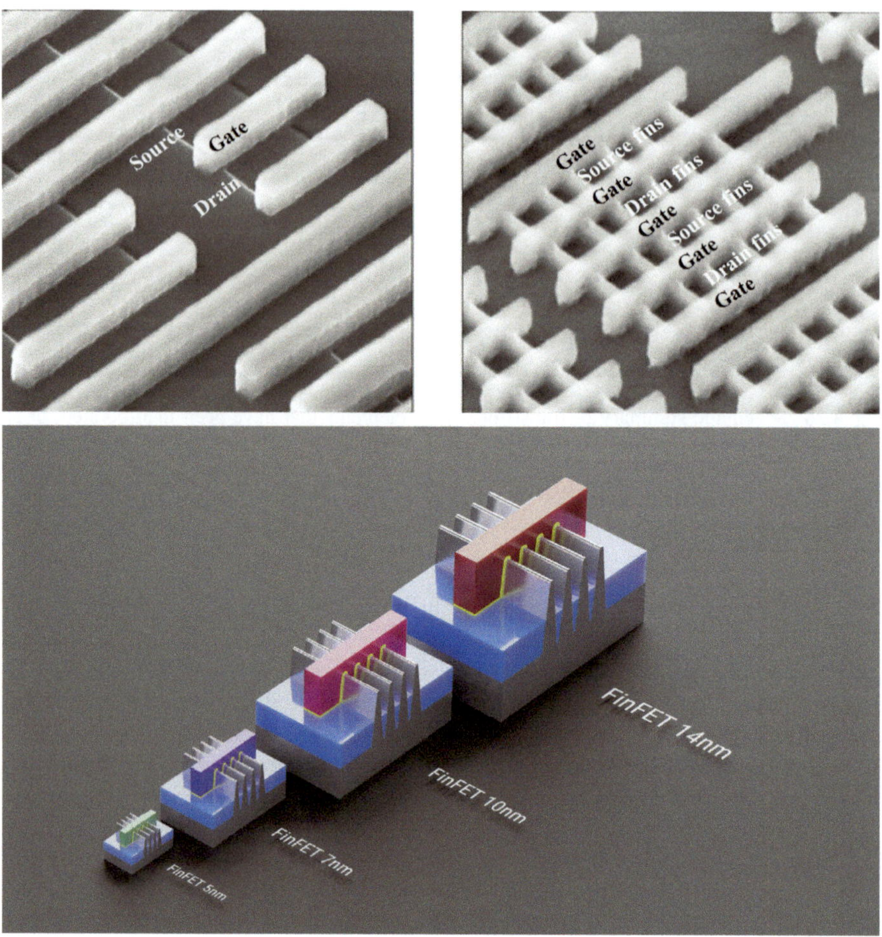

Fig. 4.29 (Top) SEM images of a planar transistor (left) *vs.* a FinFET transistor (right). © Intel Corporation (http://www.intel.com). (Bottom) comparison of FinFET transistors from the 14-nm to 5-nm technology nodes. © Mundissima/Alamy Stock Photo

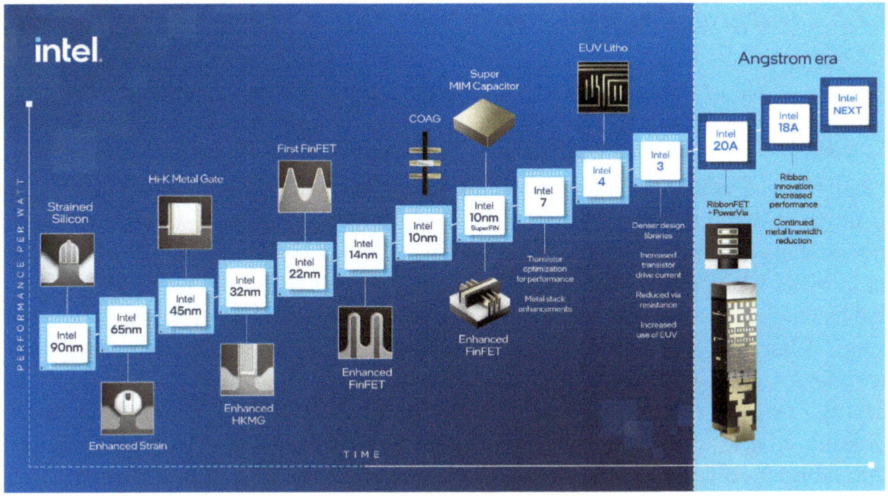

Fig. 4.30 Transistor design timeline used by Intel to continue the progression of Moore's Law. © Intel Corporation (http://www.intel.com)

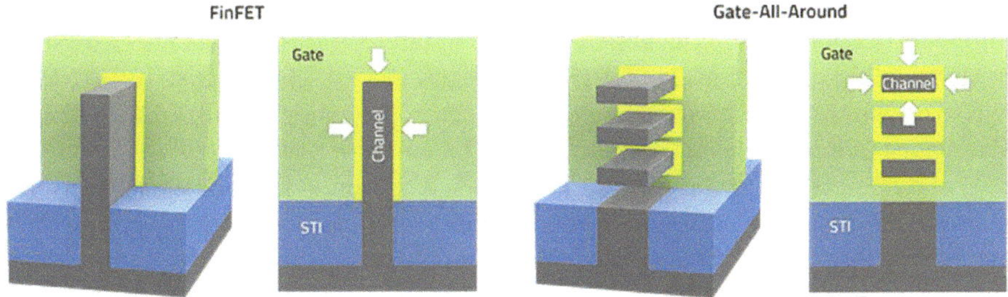

Fig. 4.31 Comparison of FinFET transistor designs (left) and gate-all-around transistors (GAAFETs), illustrating the full encapsulation of channel by the gate in GAAFETs

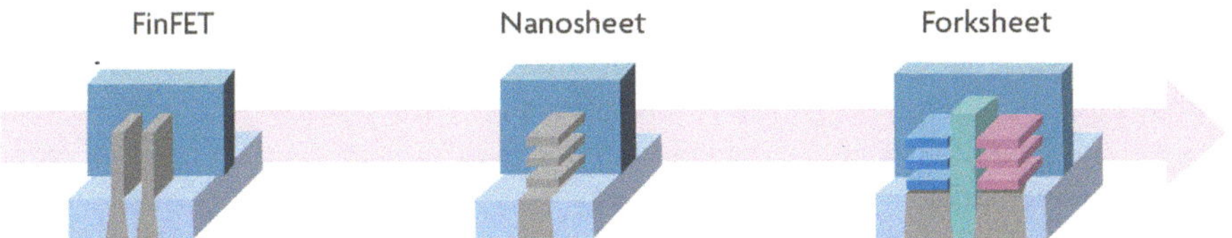

Fig. 4.32 Evolution of advanced transistors. Shown are FinFETs, gate-all-around nanosheets, and forksheet FETs

4.2.5 Integrated Circuit Fabrication

From the dawn of civilization, no other field has progressed as rapidly as microelectronics. Gordon Moore, an integrated circuit pioneer and co-founder of Intel stated (in 1998): "If the auto industry advanced as rapidly as the semiconductor industry, a Rolls Royce would get a half a million miles per gallon, and it would be cheaper to throw it away than to park it." The computers we purchase today are almost immediately out-dated, due to constant improvements in microprocessor speeds and memory/storage capabilities. A tiny integrated circuit (IC) lies beneath the protective packaging material of virtually any modern electronic device, acting as its nerve center. Like any other advanced technology, we take these feats of modern engineering for granted. Beneath the familiar black plastic casing of a computer chip are billions of complex transistor subunits, along with additional components and circuitry, to offer precise control over the flow of electrons. As you might imagine, chip fabrication is an extremely expensive endeavor. The cost of designing a 28-nm chip is ca. $40 M; this cost rises to $217 M for a 7-nm chip, and $416 M for a 5-nm chip. A 3-nm chip is estimated to cost up to $600 M!

The major class of digital ICs is the CMOS. The most familiar CMOS applications include microprocessor and RAM chips. As its name implies, CMOS technology involves the complimentary operation of interconnected n-channel MOSFET (NMOS) and p-channel MOSFET (PMOS) pairs (Fig. 4.33). Relative to NMOS or PMOS circuitry, CMOS is the design-of-choice for the majority of ICs due to distinct advantages in overall design simplicity and reduction in power.

Without question, the fabrication of modern ICs from grains of sand is one of the greatest feats ever accomplished. More than 300 carefully designed processing steps are used to form a dense packing of nanoscale objects onto a thin wafer of silicon. The silicon wafers used for IC fabrication are obtained through precise slicing of high-purity ingots using a diamond saw. The wafer edges are ground to a round finish polished until they have a defect-free mirrored finish. In order to reduce the cost/transistor, there has been a shift to larger wafers that may hold a greater number of ICs, from diameters of 200 to 300 mm (*i.e.*, 12″). Should a SOI substrate be desired, two methods are typically used for its fabrication (Fig. 4.34). A bulk Si wafer may be exposed to high-energy beam of oxygen atoms that diffuse below the surface. Also, two wafers—a bulk Si and one coated with a SiO_2 film—are bonded together using specialized techniques.

Whether the substrate is bulk Si or SOI, the most critical step in IC fabrication is the initial cleaning of the wafer. The RCA clean procedure is the industry standard, consisting of the following complex treatment to remove both organic and inorganic contaminants. It should be noted that these procedures are performed within a clean room, to avoid particulate contamination:

(i) *SC-1 clean*

- The wafer is degreased using acetone, isopropyl alcohol, and ultrapure water (UPW) rinses.
- The native SiO_2 layer is removed from the surface by exposing the wafer to a 50:1 UPW:HF solution for 30 s, followed by a prolonged UPW rinse.

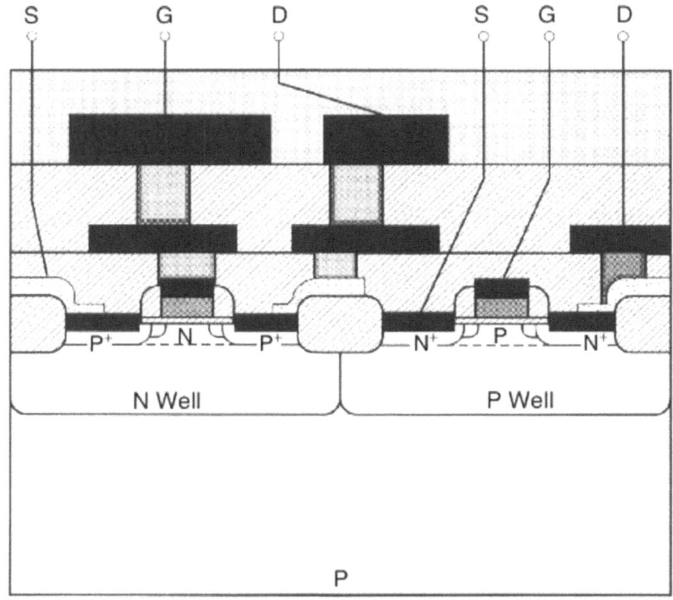

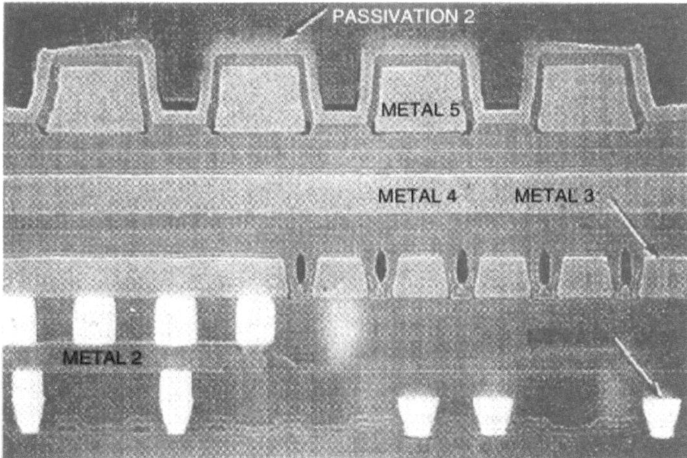

Fig. 4.33 Cross-section illustration and SEM image of an integrated circuit. Reproduced with permission from Plummer, J. D.; Deal, M. D.; Griffin, P. B. *Silicon VLSI Technology*, Prentice-Hall: New York, 2000. SEM image Courtesy of Chipworks, Inc., Ottawa, Ontario, Canada

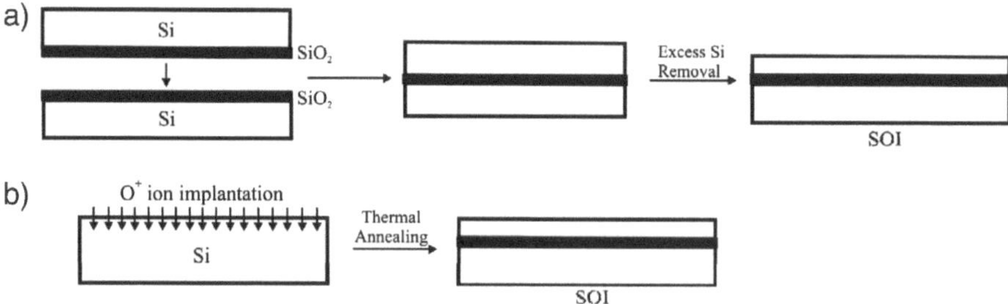

Fig. 4.34 Two methods used to fabricate silicon-on-insulator (SOI) wafers. Illustrated are the **a** Smartcut procedure where two oxidized wafers are bonded together and **b** SIMOX procedure where oxygen ions are implanted into a bulk Si wafer

- The wafer is exposed to a 10:2:1 DIW:H_2O_2: NH_4OH solution at 75 °C for 10 min, followed by a final UPW rinse. This effectively removes particulate contamination from the surface.
- Nirile gloves and Teflon tweezers are used to handle the wafer during treatments.

(ii) *SC-2 clean*

- Remaining surface oxides and hydroxides are removed by exposing the wafer to a 50:1 UPW:HF etching solution for 15 s, followed by a prolonged UPW rinse.
- The wafer is exposed to a 10:2:1 UPW:H_2O_2: H_2SO_4 solution at 75 °C, followed by a UPW rinse. This effectively removes metals/ions from the surface.
- The wafer is rinsed with DIW for ca. 20 min., followed by drying under N_2 flow.
- When finished, the polished wafer surface should be reflective with no observable residues.

Since we are considering the starting wafer, it should be noted that often the substrate consists of an epitaxially grown doped Si film on the surface of a more heavily doped bulk Si wafer. If the film is of the same composition of the underlying substrate, this process is referred to as *homoepitaxy*, or simply *epi*. By contrast, *heteroepitaxy* refers to a Si film grown onto a different substrate such as sapphire (α-alumina). Using epi wafer substrates is generally more expensive than bulk analogues, but offers the advantage of fine-tuning the conductivity of the channel region for optimal operation of the CMOS IC. Through careful film deposition methodology, it is also possible to eliminate O and C impurities resulting in further improvements in device performance.

It should be noted that there are benefits of using GaAs instead of Si for IC applications such as:

i. Faster discrete components and ICs due larger low-field electron mobility (9200 vs. 1600 $cm^2 V^{-1} s^{-1}$, respectively) and lower saturation field.
ii. The bandgap of GaAs is greater than Si (1.4 eV vs. 1.12 eV, respectively), which results in semi-insulating properties and reduced parasitic capacitance (and speed).
iii. Direct bandgap properties result in light emission for laser, LED, and microwave emitters for cellular phone applications.

However, GaAs has not been able to replace Si for microelectronic applications since Ga and As have much lower natural abundances, and As has a high toxicity. In addition, no native oxide forms on GaAs, which increases the complexity and cost of production. Lastly, GaAs has a much lower thermal conductivity relative to Si, resulting in lower packing densities at a time when Moore's Law demands significant increases. Germanium and other III-V such as InP, InAs, or InSb have also been suggested to replace Si in future technology nodes due to their higher electron mobilities (3900, 5400, 40,000, and 77,000 $cm^2V^{-1} s^{-1}$, respectively), of interest for nMOS device applications. Germanium also has a very high hole mobility (1900 $cm^2V^{-1} s^{-1}$ vs. 430 (Si), 400 (GaAs), 200 (InP), 500 (InAs), and 850 (InSb)), which could be useful for pMOS devices.[26]

A substrate candidate that is used for high-speed optical networking and inexpensive, lightweight personal communication devices is SiGe ("siggie"). Whereas Si does not operate at frequencies above a few GHz, SiGe semiconductors have speeds up to 120 GHz, which increases current speeds by up to a factor of four.[27] Although the added processing cost will not replace Si for CMOS applications, but will replace III-V semiconductors (e.g., InP, GaAs) for niche future applications such as watch-size cellular phones, collision-avoidance radar systems, wireless ICs, and low-power radio-frequency (RF) chips.

Regardless of bulk Si, epi, or SOI substrates, a key consideration that is paramount toward subsequent processing steps is the crystal orientation employed. Three important crystal planes for silicon include Si(100), Si(110), and Si(111)—Fig. 4.35. The surface atomic densities increase in the order Si(100) < Si(110) < Si(111). Empirically, this translates to available Si–Si bond densities of 6.77×10^{14}, 9.59×10^{14}, and 11.76×10^{14} cm^{-2}, respectively.[28] Hence, the rates required to remove (etch), or react with, surface atoms (e.g., thin-film deposition) should follow the reverse order as above. However, the Si(110) orientation etches fastest due to its more corrugated structure, relative to the other atomically flat surfaces (Fig. 4.36). To illustrate this effect, exposure of a wafer to a 40% alcoholic KOH solution results in Si etching at rates of 13,000, 6000, and 90 Å min^{-1}, for the Si(110), Si(100), and Si(111) planes, respectively.

Since bonds are broken during Si wafer formation from the bulk ingot, atoms at the surface will have a lower coordination number, resulting in a higher surface energy and greater reactivity (Fig. 4.37). One way to remove these surface defects is through passivation. For example, a large number of dangling bonds are tied up by H

Fig. 4.35 Unit cell representations of silicon showing the (100), (110), and (111) planes, respectively

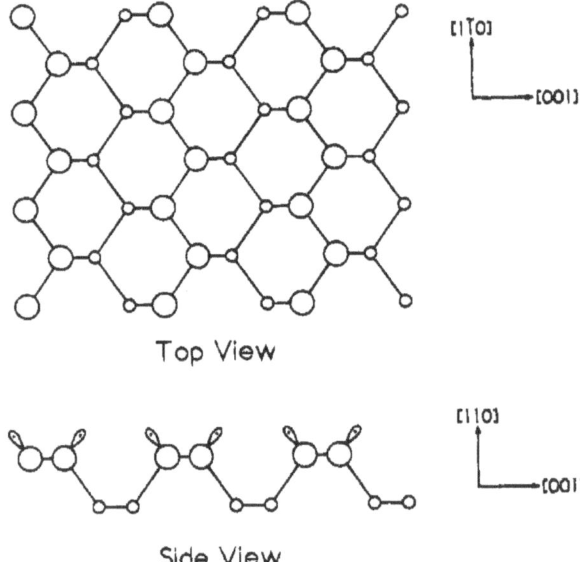

Fig. 4.36 Top and side views of the unreconstructed Si(110) surface. Each of the upper surface atoms has one dangling bond, as illustrated in the side view. Reproduced with permission from Waltenburg, H. N.; Yates, J. T. *Chem. Rev.* **1995**, *95*, 1589. © 1995 American Chemical Society

atoms during HF etching to remove native SiO_2 (the RCA clean procedure above). The unfavorable positive free energy of the surface may also be minimized by using solid surfaces with a high atomic density (*i.e.*, stability: Si(111) > Si(110) > Si(100)).

Neighboring unsaturated Si atoms may also self-assemble in a variety of ways, referred to as *surface reconstruction*, which minimizes the overall energy of the surface. For instance, the formation of a Si(100) surface leaves two dangling bonds per surface Si atom; dimerization reduces the coordinative unsaturation by 50% (Fig. 4.38a). A much more complex multi-step reconstruction for annealed Si(111) surfaces may also be observed directly by scanning tunneling microscopy (STM, Fig. 4.38b).[29] In addition to reconstructed surfaces, native surfaces that contain adsorbates are also denoted by the *Woods notation*, which provides the crystal orientation

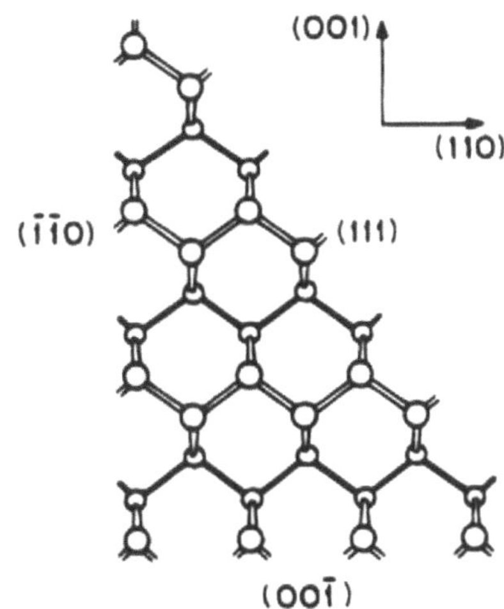

Fig. 4.37 Dangling bond formation on reconstructed surfaces of (110), (111), and (100) planes of Si. The sizes of silicon atoms are shown to decrease away from the page. Reproduced with permission from Waltenburg, H. N.; Yates, J. T. *Chem. Rev.* **1995**, *95*, 1589. © 1995 American Chemical Society

of the surface, along with the dimensions/alignment of the adsorbate units in comparison to the underlying bulk substrate (Fig. 4.38c).[30] For instance, a nitrogen atom adsorbed onto a 2 × 2 reconstructed Cu(100) surface would be designated as:

Cu	(100)	c	(2x2)	-	N
substrate	*plane*	*centering*	*mesh ratio*		*adsorbate*

The fabrication of a CMOS IC involves a repeating sequence of film deposition, patterning, and dopant implantation procedures.[31] FinFETs may be fabricated on SOI or bulk Si substrates; the first step involves e-beam lithographic patterning and etching of thin fins on the substrate using a hard mask (Fig. 4.39). In the "gate-first" process, fabrication steps after fin formation are similar to that used in conventional bulk MOSFET fabrication. In the gate-last process, the source and drain cavities are formed immediately after fin patterning. Since the introduction of FinFETs in the 22-nm node, the height and spacing (pitch) of fins have been scaled to become taller and narrower to increase transistor density and limit sub-threshold leakage (Fig. 4.40).

Silicon nitride, Si_3N_4, is typically used as a hard mask for fin fabrication and subsequent doping of source/drain cavities. The nitride film is generated through a high-temperature reaction between ammonia and silane gases (4.11). The thickness of the Si_3N_4 film is typically on the order of 75 nm. Since the density of silicon nitride is sufficiently greater than SiO_2 (3.3 g cm^{-3} *vs.* 2.6 g cm^{-3}, respectively), Si_3N_4 is used as an effective passivating layer to prevent oxidation of underlying Si regions.

$$3\ SiH_4 + 4\ NH_3 \xrightarrow{800\,^{\circ}C} Si_3N_4 + 12\ H_2 \tag{4.11}$$

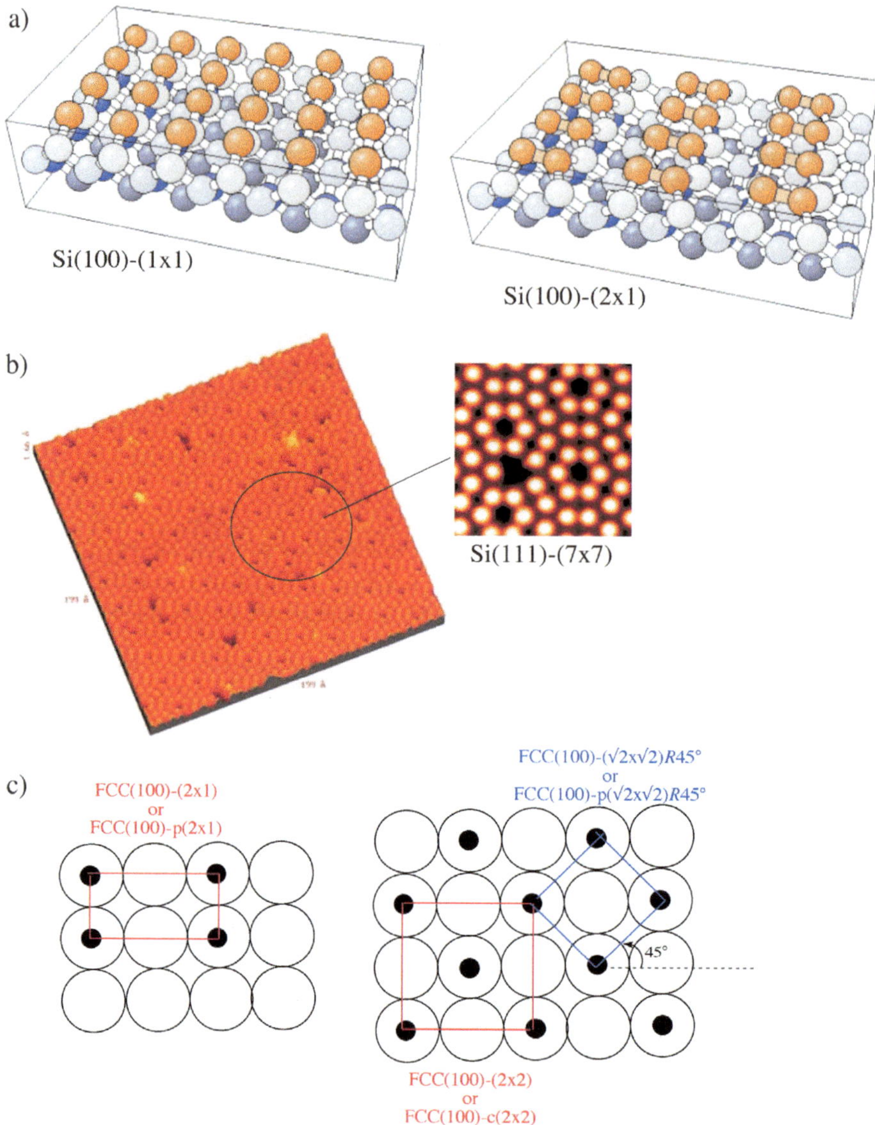

Fig. 4.38 Surface reconstruction of Si planes. Shown are **a** fcc-(100)-(2 × 2), also referred to as fcc-(100)-p(2 × 2) referring to a primitive unit, **b** an STM image of a Si(111) reconstructed surface, and **c** the Woods notation of reconstructed fcc(100) surfaces

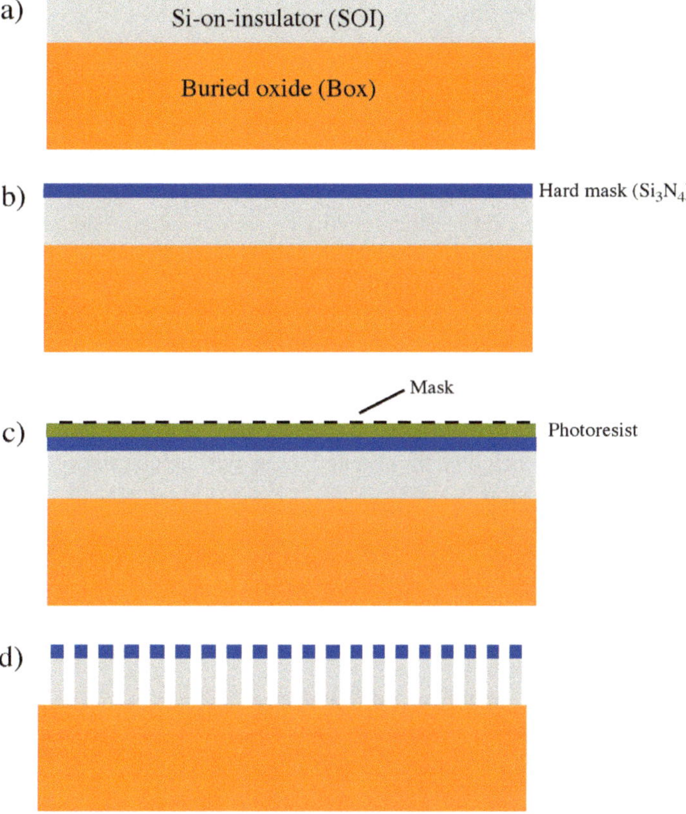

Fig. 4.39 Illustration of fin formation via e-beam lithographic patterning of a hard mask and subsequent plasma/wet etching

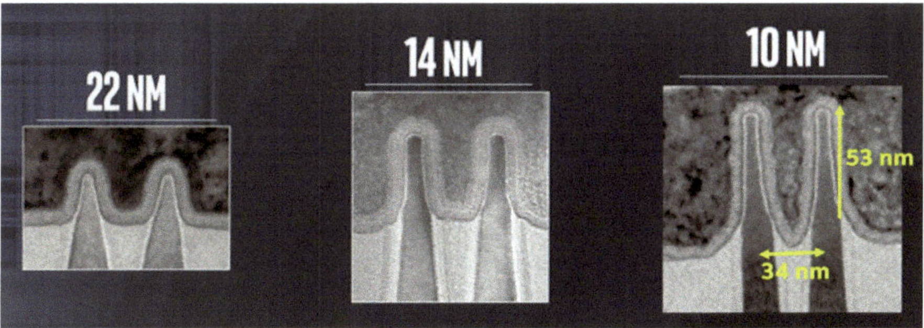

Fig. 4.40 Three generations of fin profiles on Intel technology. The fin pitch scaled from 60 to 42 nm to 34 nm. © Intel Corporation (http://www.intel.com)

(i) *IC Fabrication: Patterning*

Patterning of fins and other layers in IC fabrication are based on optical lithography, which accounts for 90% of the overall production cost. In this procedure, a photosensitive compound known as a *photoresist* is first spin-coated onto the surface of the wafer. Frequently, the wafer is pretreated by a dehydration bake (to remove adsorbed water) and application of an adhesion promotor (e.g., hexamethyl disilazane, HMDS—Fig. 4.41; others include trichlorophenylsilane ($C_6H_5SiCl_3$) and bis(trimethylsilyl)acetamide ($(CH_3)_3SiNCH_3COSi(CH_3)_3$) for Si substrates[32]). These "primers" chemically bond to the Si substrate, thereby generating a polar electrostatic surface that will more efficiently bond the photoresist.

In order to achieve the best possible line resolution, the photoresist should possess the following requirements:

i. High sensitivity (results in less exposure time and lower cost).
ii. High contrast (only brightly illuminated areas will be chemically changed).
iii. Strong surface adhesion.
iv. Resistance to etching conditions.

A circuit pattern of the microscopic IC known as a *mask*[33] is placed over the wafer, and the uncovered molecules of the photoresist that are exposed to high-energy UV

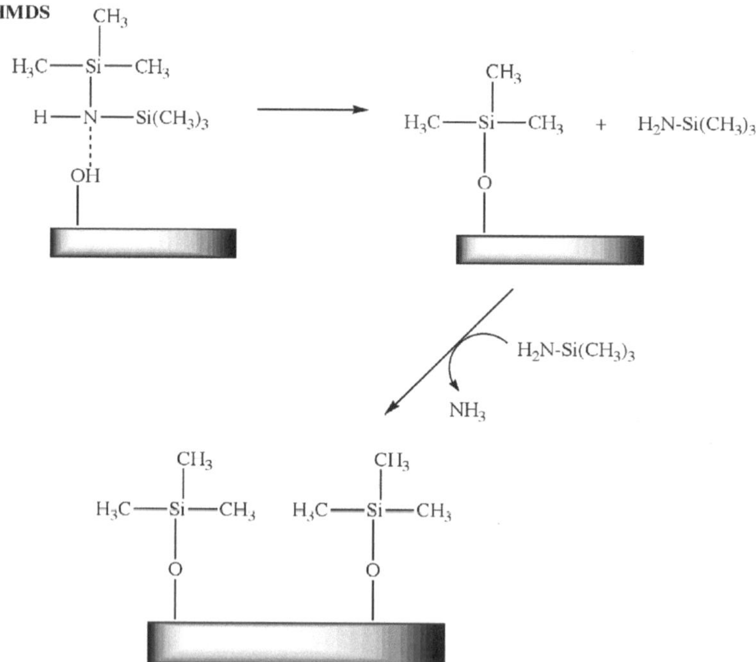

Fig. 4.41 The reactions involved in the adhesion of HMDS to a SiO_2 surface. Attachment occurs through available surface hydroxyl groups, with the release of ammonia

light will exhibit a chemical change (Fig. 4.42). *Negative tone* photoresists undergo photo-induced crosslinking mechanisms upon UV exposure. The ensuing polymerization reaction is due to the presence of photosensitive groups such as epoxy, vinyl, or aryl halides on the photoresist backbone. Accordingly, they become insoluble during subsequent contact with a basic developing solution (e.g., MOH, tetramethyl ammonium hydroxide). In contrast, use of *positive tone* photoresists results in the exposed-polymer regions becoming preferentially soluble in the developing solution. Although early photolithographic applications used negative tone photoresists exclusively, the organic matrices of these materials caused swelling, which results in pattern distortion during development. To reproduce line features below 3 μm present in today's electronic devices, aqueous base-soluble positive photoresists are most commonly employed. However, aqueous negative tone photoresists have now been developed with line resolutions below 1 μm.[34]

As one can see from (4.12), decreasing the wavelength of exposure for photolithography will directly improve the resulting line resolution. Table 4.1 lists the wavelengths and optimum resolution values for various lithographic techniques. Since 2002, IC fabrication has used 193-nm UV irradiation (ArF source) for patterning. Photolithography using 157-nm (F_2 laser source) was to be instituted few years later to further progress the rapid miniaturization of ICs; however, it was deemed too costly. That is, in addition to switching from atmospheric-pressure to high-vacuum environments, new photoresists and masks would also need to be designed.

$$R = \frac{0.61\lambda}{\mathrm{NA}}, \tag{4.12}$$

where R is the *resolution limit* (line spacing capable of being resolved; smaller R is better) and NA is the *numerical aperture* of the exposure tool (the light-gathering power of a lens, discussed in more detail in Chap. 7).

The use of extreme UV light (EUV), within the range of soft X-rays ($\lambda = 13.5$ nm; 92 eV), is now in use for IC fabrication.[35] Although EUV was once predicted to replace 193-nm photolithography in the 7-nm node, the use of phase-shift masks (Fig. 4.43) and double-/multiple-exposure techniques[36] has extended the 193-nm exposure beyond these original predictions. The EUV source is based on a plasma generated from an IR laser that impinges upon gas-phase Xe clusters expanding at supersonic speeds. Rather than conventional lenses, EUV also dictates the use of Si/Mo multilayered mirrors (Fig. 4.44) to focus the plasma radiation and reduce the size of the projected image from the mask.[37]

Although decreasing the exposure wavelength improves line resolution, issues associated with absorption has become increasingly problematic. Hence, new types of photoresists are needed that will be stable upon contact with higher energy radiation. Whereas the DNQ photoresist system is effective in the MUV range, photoresists that feature *chemical amplification* (CAM) moieties are used exclusively for DUV and shorter wavelengths.[39] The enhancement of quantum efficiency and sensitivity results from a photogenerated acid that catalyzes either crosslinking (negative photoresists),

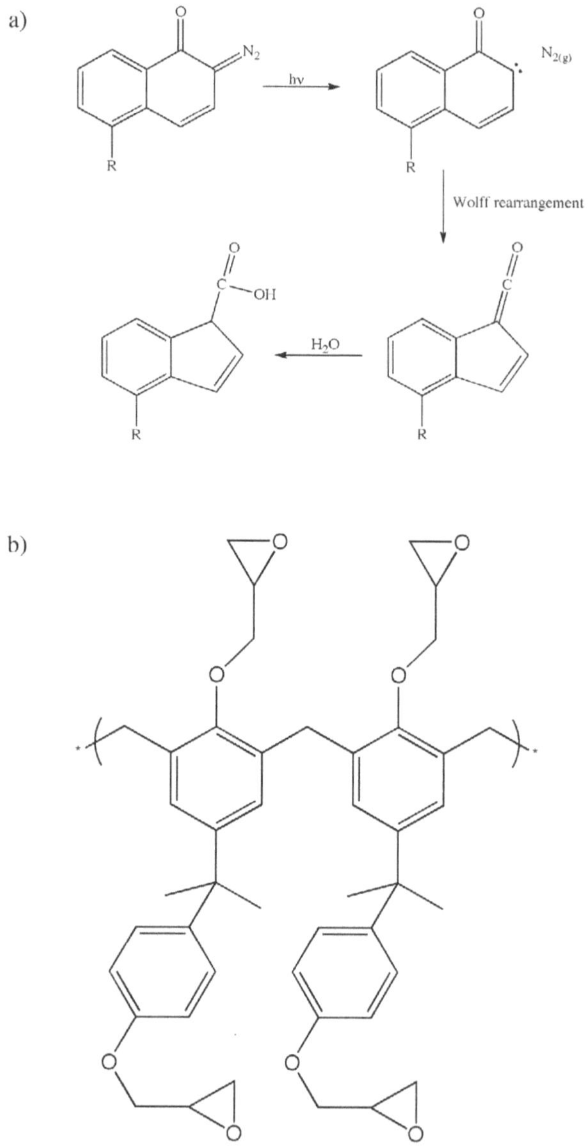

Fig. 4.42 Molecular structures and photoinduced reactions of common photoresists. Shown are **a** the diazonaphthoquinone (DNQ) positive tone photoresist, and **b** the SU-8 epoxy-based negative tone photoresist

a)

N_2 hν $N_{2(g)}$ R R Wolff rearrangement C—OH H_2O C R R

b)

or deprotection (positive photoresists) reactions. Since the diffusion of acid during the post-exposure bake cycle sparks reactions with numerous functional groups of the surrounding photoresist, fewer photons are required during UV exposure. In order to facilitate water solubility, modern photoacid generators (PAGs) are typically aryl sulfonium triflate salts (Fig. 4.45). However, compounds such as *o*-nitrobenzyl tosylate or organic/inorganic onium salts may also be used as PAGs within organic solvents.[40]

Within the 13.5-nm wavelength regime of EUV, many elements exhibit strong absorption (Fig. 4.46); this is in contrast to X-ray lithography (0.8–1.4 nm), which

Table 4.1 Resolution limits for various lithographic techniques

Lithographic technique	Exposure wavelength (nm)	Resolution
Photolithography[a] mid-UV (MUV)	350–450	0.35–3 μm
Photolithography deep-UV (DUV)	248	≥0.25 μm
Photolithography deep-UV (DUV)	193	<30 nm[b]
Photolithography extreme-UV (EUV)	13.5	<20 nm
Electron-beam lithography[c]	ca. 1	<10 nm
X-ray lithography[d]	0.4–20	<30 nm

[a] Using a standard chrome-on-glass photomask
[b] Using advanced photomask techniques such as phase-shift masks (PSMs, Fig. 4.32)
[c] Using a photomask of Si_3N_4 membrane and Cr/W patterned regions
[d] Using a photomask of $Si/Si_3N_4/SiC/BN$ membrane and Au/W patterned regions

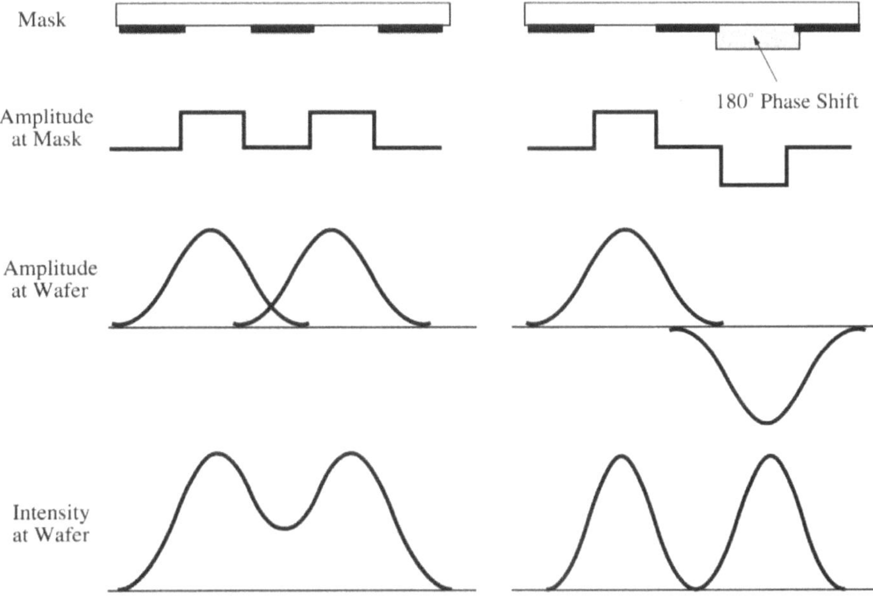

Fig. 4.43 The benefits of using phase-shift masks for photolithography. Reproduced with permission from Plummer, J. D.; Deal, M. D.; Griffin, P. B. *Silicon VLSI Technology*, Prentice-Hall: New York, 2000

is largely transparent to most elements. However, strong absorption by EUV resists is not as problematic as it is for 193 nm and 157 nm lithographies.[41] For EUV photolithography, a weak absorbance from resist materials generally results in low sensitivity that can lead to problems with line resolution. A good measure of comparing the overall performance of a photoresist is the Z-factor (4.13).[42]

$$Z\text{-factor} = (\text{Sensitivity}) \times (\text{Line-Edge Resolution})^2 \times (\text{Half-Pitch})^3 \quad (4.13)$$

Hence, the incorporation of heavier elements such as Sn or I with larger absorption cross-sections, have been shown to improve the sensitivity of EUV resists.[43] However, it is more important to consider the types of bonds present in the photoresist molecules than simply the atomic composition. For instance, the photoelectron intensity of iodine-substituted compounds have been shown to be much larger than analogues substituted with lighter elements (Fig. 4.47). The enhanced intensity of low-energy photoelectrons for compounds substituted with heavier halogens is due to core electrons residing within energy levels that preferentially interact with the ionizing radiation.[44]

In general, traditional CAM photoresists may not be used for EUV photolithography since they are comprised of C, H, and O atoms, which are largely transparent in this wavelength regime. Instead, the most widely developed photoresists with increased EUV sensitivity feature fluorinated PAGs (Fig. 4.48). Other polymer-bound PAGs such as poly(hydroxystyrene), poly(trimethylsilylstyrene-co-chloromethylstyrene) or silicon-containing polymers have also been evaluated as EUV resists.[45] However, varying combinations and degrees of polymer swelling, low sensitivities, and low etch resistances have plagued most EUV photoresists developed to date. Perhaps the most attractive candidates for EUV photoresists are inorganic molecules featuring Hf, Zr, Zn, Al, and Sn atoms due to their enhanced photoelectron production upon EUV irradiation. Oftentimes, these elements are introduced into the resist formulation as oxide nanoparticles or metal–oxygen clusters in tandem with organic ligands.[46]

In addition to decreasing the exposure wavelength, an increase in the numerical aperture of the exposure system will also yield improvements in the resultant line

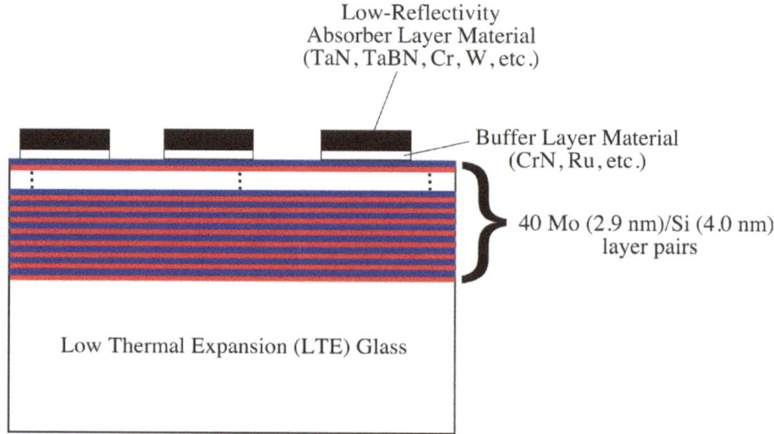

Fig. 4.44 Schematic of an extreme UV (EUV) mask[38]

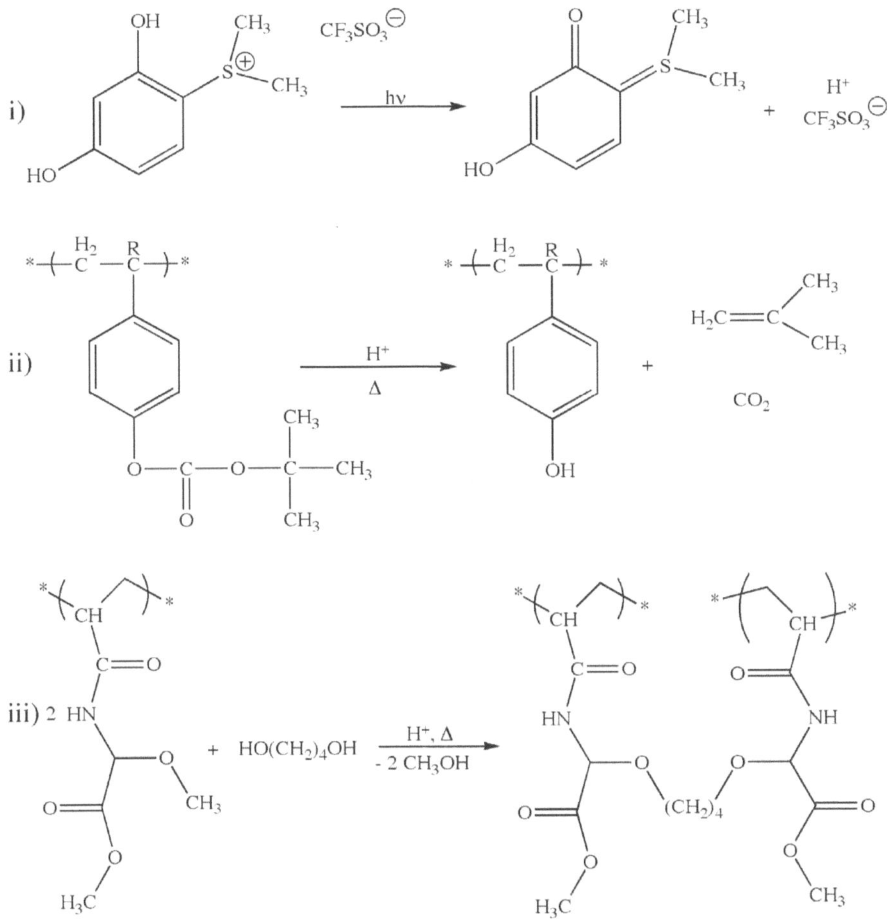

Fig. 4.45 Chemical amplification (CAM). Reaction (i) represents photoinduced acid generation (PAG); step (ii) is an acid-catalyzed deprotection mechanism (positive tone resist); and step (iii) is an acid-catalyzed crosslinking mechanism (negative tone resist)

resolution (4.13).[47] Since the refractive index of the medium is directly proportional to the numerical aperture (NA), one simply needs to surround the lithographic lens with a fluid of greater refractive index (e.g., water: $\eta = 1.47$ at 193 nm; air: $\eta = 1.0$). Although such *immersion lithography* has been suggested as a candidate for sub-50-nm line resolution, there are significant technical challenges (e.g., CaF_2-based optics, which are hygroscopic) that must first be addressed. There are current research efforts devoted to utilizing higher-index fluids; thus far, a variety of candidates have been identified that exhibit a refractive index close to 1.7 (Fig. 4.49). However, as one would expect, immersion lithography suffers from a variety of technical issues such as[33]:

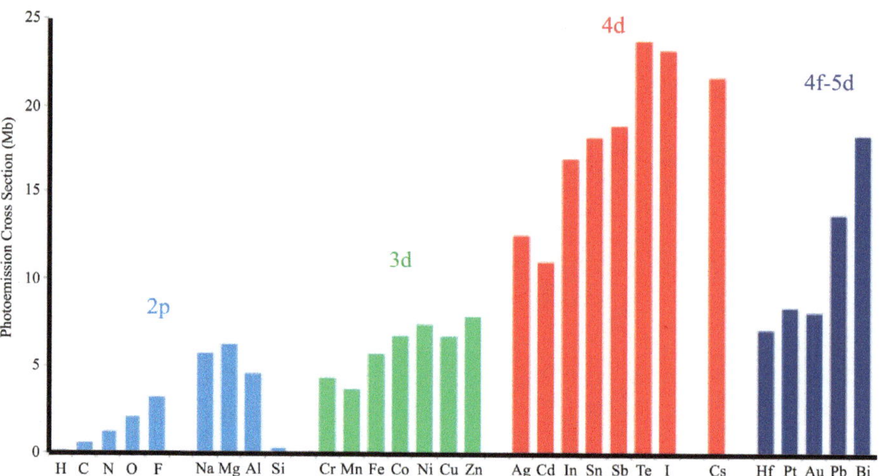

Fig. 4.46 Photoemission cross-sections at 92 eV (13.5 nm) of selected atoms. Figure adapted from *Nanomaterials* **2020**, *10*, 1593. © MDPI

i. Micro-/nano-sized bubble formation (scattering the incident radiation, causing significant losses in line resolution.
ii. Contamination of the immersion fluid by the resist (altering the refractive index, again hurting line resolution).
iii. Heating the immersion fluid during exposure (resulting in the same effect as ii)).

However, it should be noted that in addition to improving resolution via tweaking the NA, the *depth of focus* (DOF $= \lambda/(\mathrm{NA})^2$) is also paramount for the patterning of sharp features. As the depth of focus decreases with resolution, it will become increasingly more difficult to define features simultaneously at the top and bottom surfaces. Consequently, chemical mechanical polishing (CMP) is used to planarize the wafer prior to high-resolution photolithography, and as thin a layer as possible of photoresist must be applied to the wafer. As its name implies, CMP entails the use of an abrasive, corrosive slurry to physically grind flat and chemically remove the microscopic topographic features on a wafer so that subsequent processes can begin from a flat surface.

(ii) IC Fabrication: Etching

Once the pattern in the photoresist layer is formed, a variety of chemical or high-energy plasma techniques may be used to remove exposed underlying layer(s)—a process referred to as *etching*.[49] It should be noted that “wet” etching, or dipping the wafer in an acidic solution, typically results in *isotropic etching* (Fig. 4.50a), and a line resolution of >2 μm. By contrast, “dry” etching that uses a plasma[50] generally results in *anisotropic* conditions (Fig. 4.50b), and line resolutions <0.25 μm. It has been shown that the pre-application of an adhesion promoter results in greater

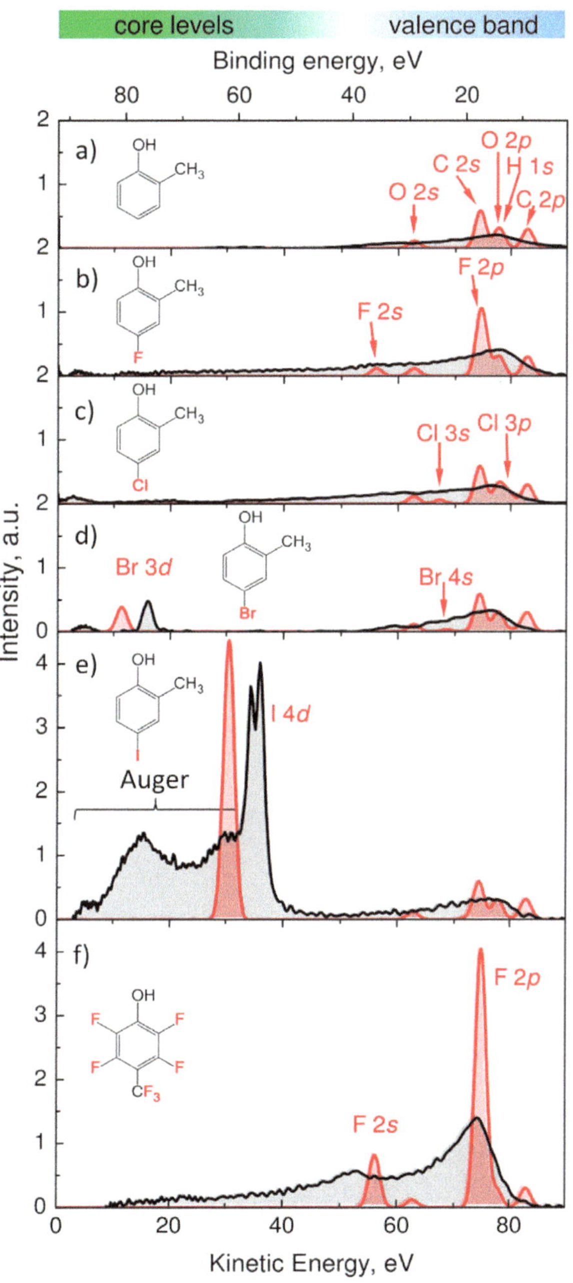

Fig. 4.47 Photoelectron spectra of gas-phase molecules irradiated by 13.5 nm EUV radiation. The black line corresponds to experimental data; the red line refers to calculated values. Shown are **a** 2-methylphenol, **b** 4-fluoro-2-methylphenol, **c**) 4-chloro-2-methylphenol, **d** 4-bromo-2-methylphenol, **e** 4-iodo-2-methylphenol, and **f** 2,3,5,6-tetrafluoro-4-(trifluoromethyl)phenol. These data indicate the utility of I in the resist composition, which greatly enhances its absorption under EUV radiation. Figure adapted from *Nanomaterials* **2020**, *10*, 1593. © MDPI

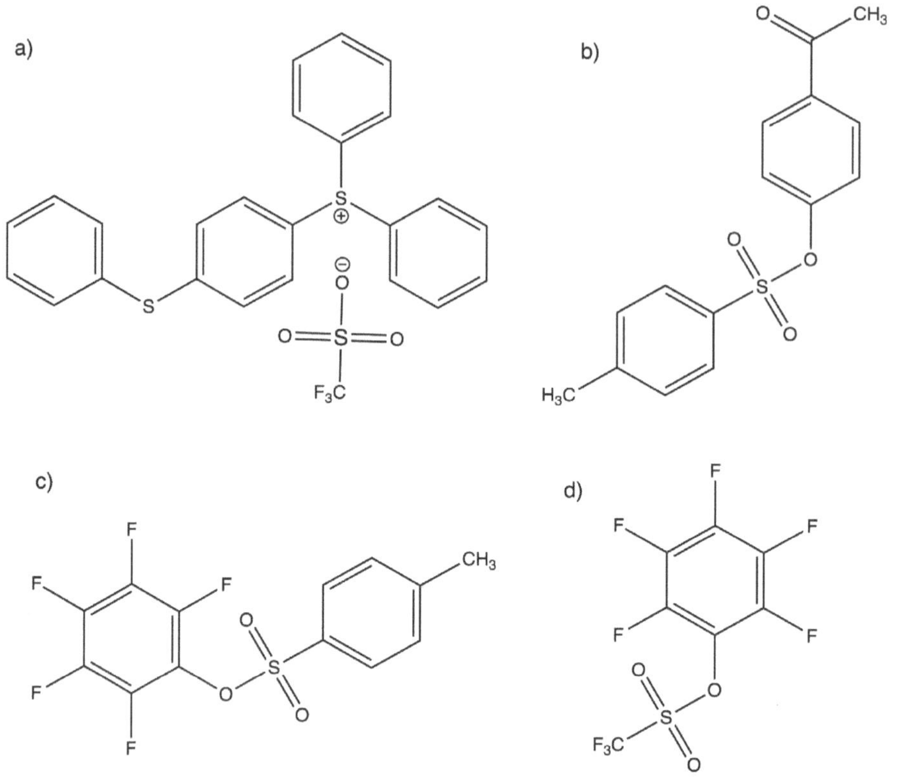

Fig. 4.48 Comparison of traditional PAGs (top) with fluorinated analogues used for EUV photoresists

anisotropic etching (less undercutting), since the photoresist and wafer surface are in closer contact with one another.

The specific etching conditions used will depend on the surface to be removed. Tables 4.2 and 4.3 provide a comprehensive list of wet and dry (plasma) etching recipes that may be used to selectively remove a variety of materials. Other plasmas used for selective SiO_2 removal include CF_4/H_2 and C_2F_6/CHF_3. Whereas a plasma of Cl_2/C_2F_6 may be used for n-type Si, Cl_2/Argon is used for undoped Si. The addition of oxygen to a fluorocarbon plasma increases the amount of free F through the formation of oxyfluorides from fluorocarbons (4.14).[51] However, if more than 10% O_2 is added to the plasma, a decrease in the Si etching rate will result. Most likely, this is due to the competition between O and F atoms for active sites on the silicon surface. Recipes used to remove Si_3N_4 layers include CF_4/H_2 or SF_6/O_2 combinations, which generate reactive fluorine species within the plasma (4.15). It should be noted that O_2/H_2O and O_2/CF_4 plasmas, or high-temperature annealing, are typically used to remove the photoresist that remains following the wet/dry etching. However, in order to remove non-postbaked photoresists, simple washing with organic solvents is often

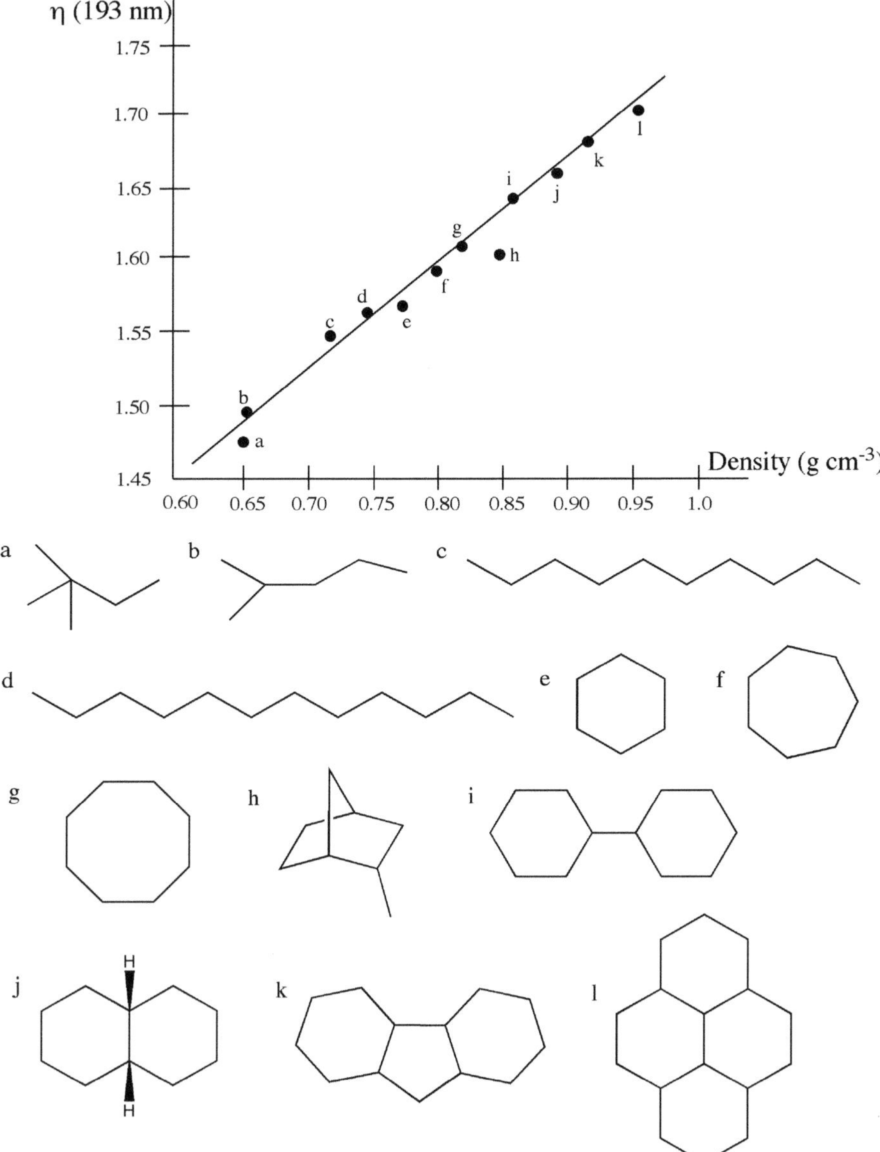

Fig. 4.49 Indices of refraction, densities, and molecular structures of high-η fluids proposed for immersion-lithography applications. Adapted with permission from *Proc. SPIE* **2008**, 6923, 69230A-1[48]

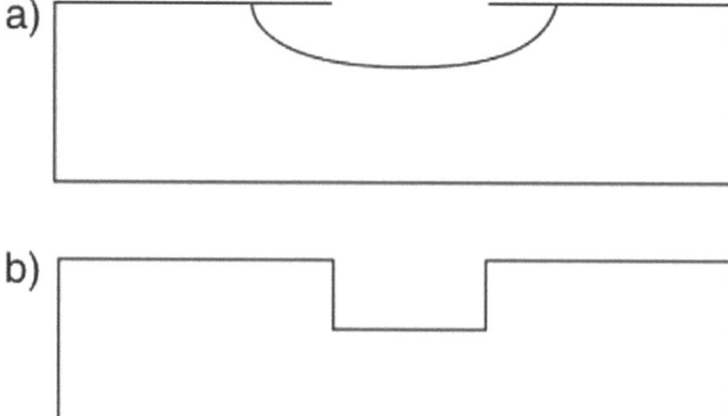

Fig. 4.50 Illustration of **a** isotropic etching evidenced by severe undercutting and **b** anisotropic etching with 90° sidewall channels (or *vias*)

sufficient (e.g., positive tone photoresists: acetone, trichloroethylene (TCE); negative tone photoresists: methyl ethyl ketone (MEK, $CH_3C(O)C_2H_5$), methyl isobutyl ketone (MIBK, $CH_3C(O)C_4H_9$).

$$CF_4 + O \xrightarrow{\text{plasma}} COF_2 + 2F \tag{4.14}$$

$$Si_3N_4 + 12\ F \xrightarrow{\text{plasma}} 3\ SiF_4 + 2\ N_2 \tag{4.15}$$

Figure 4.51 illustrates the lithographic/etching steps required to fabricate a dense array of Si fins. The narrow regions between adjacent fins are back-filled with a dielectric such as SiO_2 or a high-κ dielectric, which serves to electrically isolate individual devices on the chip from one another. This process, referred to as *shallow trench isolation* (STI), is the method-of-choice for modern CMOS fabrication since it allows the deposition of smaller insulating regions resulting in greater device density, in accord with Moore's Law. Once the thick oxide is deposited into the trench, *chemical mechanical polishing* (CMP) is used for *planarization* of the surface prior to subsequent layering steps.[52]

The STI regions are then recessed to a lower level to define individual fins. A high-κ gate dielectric thin film (ca. 10–20 nm) such as HfO_2 is then deposited onto the sidewalls and bottom of the STI fins, followed by the deposition of a gate electrode such as doped polysilicon, metals, or metal nitrides. The specific methodologies used for thin-film deposition will be described later in the chapter. The gate electrode and dielectric layers are then patterned to form gate stacks (Fig. 4.52a). Depending on the circuitry, adjacent fins may serve as NMOS and PMOS device regions. The formation of strained-Si channel regions is accomplished by removing portions of the exposed fins, followed by deposition of an appropriate stressor material (e.g., SiC for NMOS or SiGe for PMOS).

Table 4.2 Wet etching recipes

Material to be etched	Recipe	Comments
Al	16:2:1:1 H_3PO_4:water:acetic acid:HNO_3	PAN etch; 200 nm/min At 25 °C; 600 nm/min At 25 °C
Al	Conc. H_3PO_4	At 120 °C
Ag	1:1 NH_4OH:H_2O_2	
Au	3:1 HCl:HNO_3	Aqua regia
Cr	3:1 HCl:water	
Cu	5:1 HNO_3:water	
Fe	1:1 HCl:water	
Fe	1:1 HNO_3:water	
Mo	1:1 HCl:H_2O_2	
Mo	1:1:1 H_2SO_4:HNO_3:water	
Ni	5:1 HCl:HNO_3	
Ni	1:1 HF:HNO_3	
Pb	2:2:5 acetic acid:H_2O_2:water	
Pb	1:1 acetic acid:H_2O_2	
Pd	3:1 HCl:HNO_3	
Pt	3:1:4 HCl:HNO_3:water	At 95 °C
Pt	8:1 HCl:HNO_3	At 70 °C
PolySi	3:1 HNO_3:HF	4.2 μm/min at 25 °C
PolySi	50:20:1 HNO_3:water:HF	540 nm/min at 25 °C
Si	2:2:1 HF:HNO_3:water	
Si	5:3:3 HNO_3:HF:acetic acid	
SiO_2	6:1 NH_4F:HF	120 nm/min at 25 °C
SiO_2	1:10 HF:water	30 nm/min at 25 °C
SiO_2	1:100 HF:water	1.8 nm/min at 25 °C
SiO_2	Conc. HF	1.8 μm/min at 25 °C
Si_3N_4	Refluxing H_3PO_4	At 180 °C
Sn	1:1 HF:HNO_3	
Ta	2:2:5 HF:HNO_3:water	
Ti	1:30:69 HF:H_2SO_4:water	At 70 °C
Ti/W alloy	H_2O_2	5 nm/min at 25 °C
W	1:1 HF:HNO_3	

(iii) IC Fabrication: Ion Implantation

Figure 4.52b–f illustrate the steps involved for the placement of n- and p-type dopants into NMOS and PMOS fins. A hard mask such as Si_3N_4, followed by anti-reflective and photoresist coatings, are deposited onto the fins. Photolithography and etching are used to selectively expose the NMOS region, while leaving the PMOS still covered.

Table 4.3 Dry (plasma) etching recipes

Material to be etched	Recipe	Comments
Al	10 sccm Cl_2	BCl3 (30 sccm) is added to scavenge O in the native oxide layer; 180 mTorr (30 mTorr results in anisotropic etch); 200 W; 0.5 $\mu m\ min^{-1}$
PolySi	30 sccm Cl_2 (isotropic); 5:25 sccm Cl_2:HBr	180 mTorr (isotropic); 30 mTorr (anisotropic); 100 W; 0.5 $\mu m\ min^{-1}$ (isotropic); 0.3 $\mu m\ min^{-1}$ (anisotropic)
SiO_2	45:5 sccm CF_4:O_2	150 mTorr; 100 W; 0.15 $\mu m\ min^{-1}$
Si_3N_4	45:5 sccm SF_6:O_2	150 mTorr; 100 W; 0.2 $\mu m\ min^{-1}$
Polyimide	47:3 sccm O_2:CF_4	200 mTorr; 20 W; 1 $\mu m\ min^{-1}$

A p-type impurity (e.g., B) is introduced into the NMOS; subsequent removal of the photoresist and anti-reflective coatings from the PMOS allow for introduction of an n-type impurity (e.g., P, As).

The technique of impurity doping is known as *ion implantation*, which involves the collision of Si lattice atoms with p-type or n-type ions that have been accelerated to very high kinetic energies (ca. 50–200 keV). The chosen energy must be sufficient to penetrate through the capping Si_3N_4 layer to a desired fin depth, but not the exposed photoresist layer that protects the neighboring fin. For the same accelerating voltage, ions of smaller mass will penetrate to deeper/broader regions of the solid relative to heavier ions (Fig. 4.53), due to differences in their kinetic energies and concomitant collisions with Si atoms.

An energy on the order of 15 eV is enough to dislodge a Si atom from its crystalline lattice, forming a Frenkel defect. However, the energies of the impinging dopant ions are an order of magnitude larger, which will cause significant damage to the Si crystal structure. For instance, bombardment of a Si surface with a single arsenic ion with an energy of 30 keV will displace 1000 Si atoms—all on a timescale of ca. 10^{-13} s! Multiply this by thousands of bombarding ions during the implantation process, and one has a picture of a highly perturbed Si crystal. Amazingly, thermal annealing is able to cause the dislodged Si atoms to find vacant sites, removing the Frenckel defects and thus repair the damage. Prolonged high-temperature annealing at temperatures >750 °C is subsequently used to further restore the periodicity of the Si crystal lattice and diffuse the n- and p-dopant wells (*i.e.*, the substrates for PMOS and NMOS devices, respectively) to their desired Si depths. Alternatively, hot-implantation may be used, which uses low-energy bombardment (e.g., 2–5 keV) onto heated wafers. This induces self-annealing, which improves localized crystallization and lowers the rate of defect generation. It should be noted that the above ion bombardment processes are repeated using an appropriate masking photoresist to introduce a higher concentration of dopants near the surface of the n and p wells. This is done to adjust the threshold voltages (V_T) of the NMOS and PMOS devices.

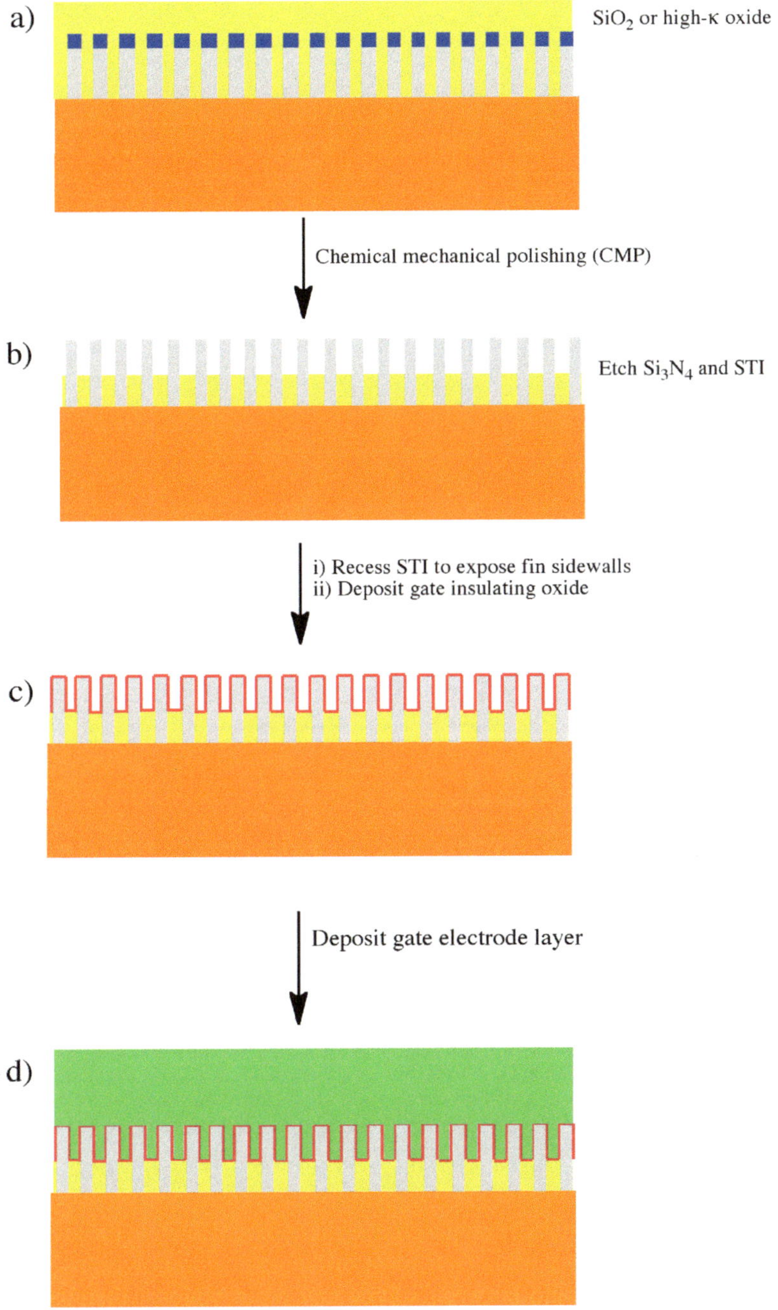

Fig. 4.51 Processing steps involved for the fabrication of Si fins and deposition of a gate electrode layer

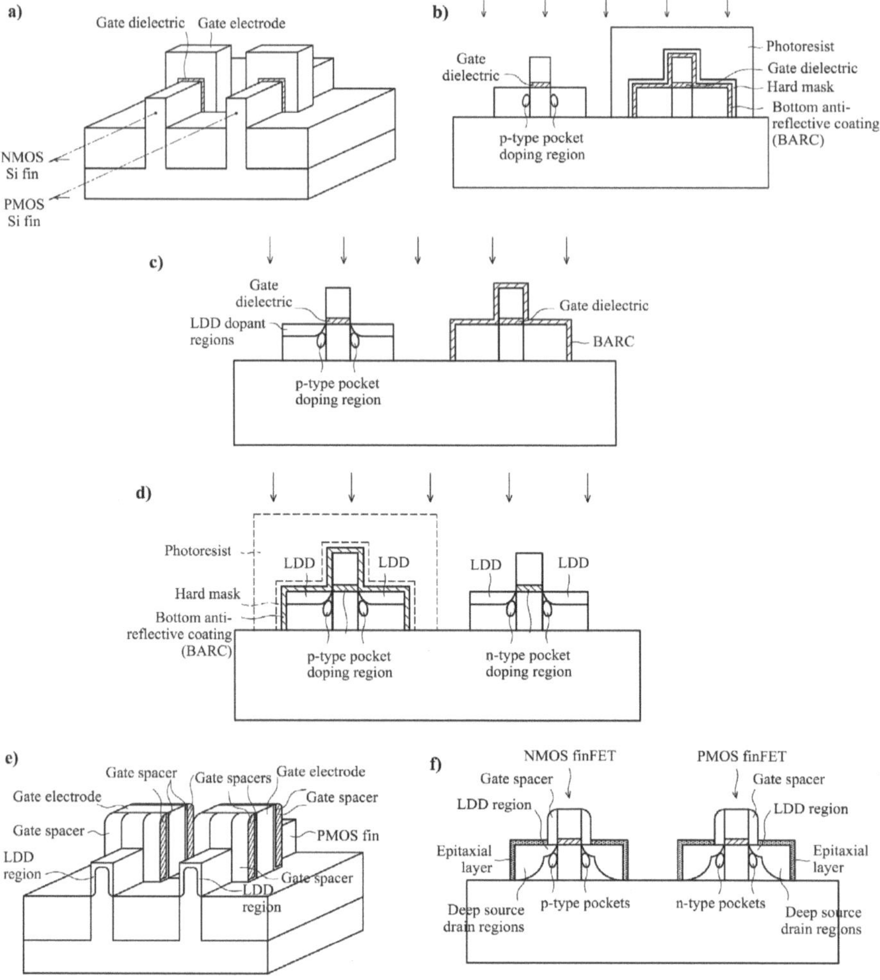

Fig. 4.52 Fabrication steps for the formation of NMOS and PMOS fins. Shown are: **a** gate electrode patterning to form gate stacks, **b** and **c** light doped drain (LDD) implantation of p-type dopants, **d** LDD implantation of n-type dopants, **e** formation of gate spacers and LDD annealing (900–1100 °C), and **f** formation of epitaxial layers on the exposed top surfaces and sidewalls of fins, followed by deep implantation of n- and p-type dopants to form deep source/drain regions. For more details, see: US Patent #20110171795A1

Two doping levels are used, composed of light doping (LDD) in the area adjacent to the channel and heavy doping over most of the source/drain regions. The LDD structure decreases the field between source and drain regions, thereby reducing injection into the gate oxide, resulting in *hot-carrier effects.* A gate spacer composed of SiO_2 or a high-κ dielectric are formed onto the fin sidewalls after LDD formation

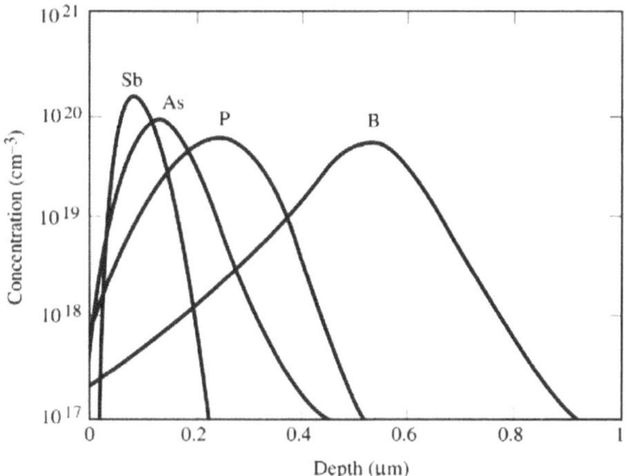

Fig. 4.53 Comparison of the Si depth penetration of various n- and p-dopant ions. Reproduced with permission from Plummer, J. D.; Deal, M. D.; Griffin, P. B. *Silicon VLSI Technology*, Prentice-Hall: New York, 2000

to prevent part of the source and drain cavities next to the channel from becoming heavily doped.

(iv) IC Fabrication: Interconnect Fabrication

The remaining steps involve *metalation*—the selective deposition of metals that form the interconnection between the active devices on the IC (Fig. 4.54). In order to reduce interconnect resistivities, metal silicides such as $TiSi_2$, WSi_2, or $TaSi_2$ are first deposited onto the surfaces of source/drain cavities and the top of the gate using photolithography/etching. A thick layer (ca. 1 μm) of insulating oxide is then deposited onto the evolving IC, which serves as the interconnect dielectric medium. Although SiO_2 was typically used for this purpose, the latest ICs utilize a low-κ dielectric such as carbon-doped SiO_2 that is reported to increase the capacitance by up to 20% relative to undoped media. This improvement translates to higher performance at lower power consumption—paramount for high-density modern ICs.

Following planarization of the oxide layer, selected regions are patterned/etched using our workhorse of photolithography. The etched regions are then back-filled with the interconnect metal; copper is now the metal-of-choice due to its more desirable electrical resistivity, relative to Al (1.7 μΩ cm *vs.* 2.7 μΩ cm, respectively) that was exclusively used in earlier ICs. Due to its low resistivity and high density, titanium nitride (TiN) is an efficient barrier level that prevents surface oxidation of Cu, as well as the interdiffusion of Cu into adjacent layers. Subsequent planarization results in a flat surface containing isolated regions of copper—a process known as the *Damascene process.*[53] Additional layers of the insulating low-κ dielectric and patterned copper interconnects continues until the desired circuitry is completed.

In ICs, insulating dielectrics separate the conductive portions (e.g., wire interconnects and individual transistors) from one another. In order to fabricate chips with higher speeds, the transistors must be placed closer and closer together, thus resulting in a thinner insulating layer. This leads to charge buildup and crosstalk, which adversely affects the maximum operating speed and performance of the chip.

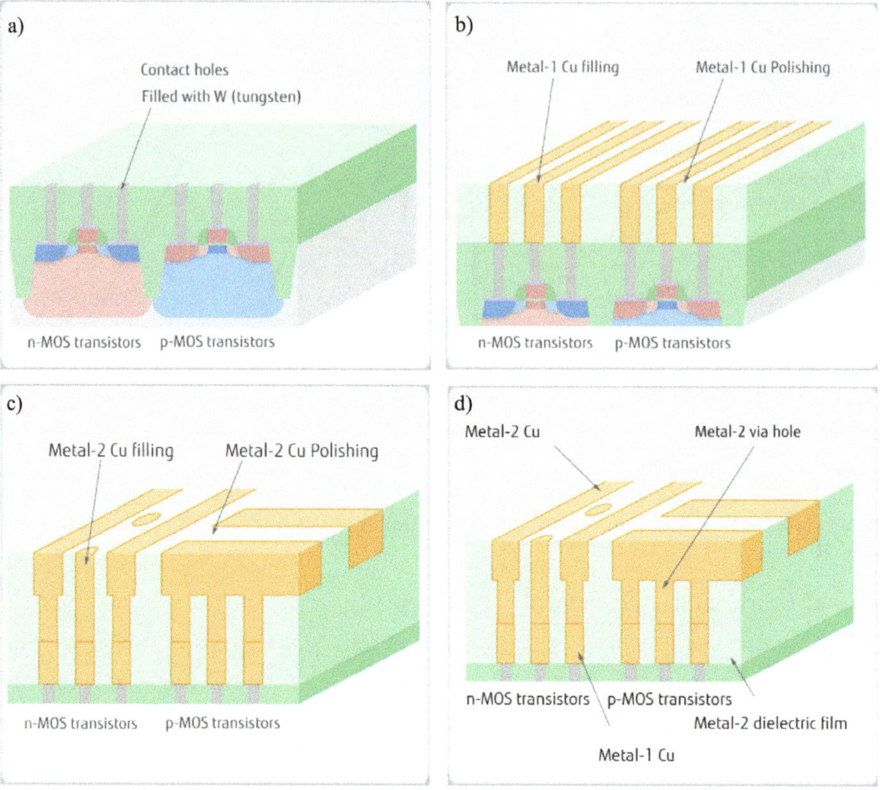

Fig. 4.54 Illustration of back end of line (BEOL) processes for interconnect fabrication. Shown are: **a** contact holes are first patterned into the dielectric film using a photoresist/etching treatment; the holes are then filled by the deposition of a tungsten film using CVD, followed by polishing the surface to remove the excess W film; **b** a thick dielectric film (e.g., SiO_2) is deposited onto the surface using CVD, followed by patterning/etching of trenches. A copper film is then deposited by electroplating, which fills the trenches, with the excess Cu removed by polishing; **c** a thick dielectric layer is deposited and patterned to form via holes and trenches, which are then filled with Cu using electroplating; **d** cross-section illustration showing the connection of two metal layers that are isolated by a dielectric layer. Integrated circuits typically contain 8–10 layers of copper interconnects, each embedded in a low-κ dielectric. © Mie Fujitsu Semiconductor Limited (http://www.fujitsu.com/jp/group/mifs/en/resources/tech-intro/process/index.html)

The use of low-κ dielectrics is important to reduce the parasitic capacitance, hence enabling faster switching speeds and lower heat dissipation. Doping SiO_2 with fluorine reduces the κ from 3.9 to 3.5; other approaches that yield values of κ <1.5 involve the use of nanoporous dielectrics based on silsesquioxane polymers (**I**) or poly(amidoamine-organosilicon) (PAMAMOS) films (see Chap. 5). The current 14-nm technology node utilizes 8–15 layers of interconnects that are embedded within a low-κ dielectric. Air gaps ($\kappa = 1$) are also purposely placed within the dielectric medium in order to reduce the dielectric constant even further (Fig. 4.55).

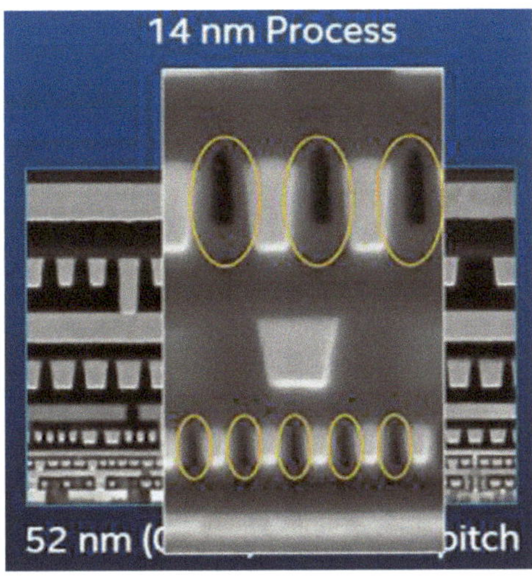

Fig. 4.55 Cross-section SEM images of the copper interconnect stack used on Intel 22-nm and 14-nm logic technologies. The 14-nm image includes a bow-up of the air-gap interlevel dielectric used on two of the interconnect layers for lower capacitance. © Intel Corporation (http://www.intel.com)

One has to remember that these multistep processes were not performed from start to finish, one IC at a time. If that were the case, each chip would cost millions of dollars! Instead, a large polished wafer that is currently 300 mm in diameter is used as the base to assemble hundreds of ICs (called *dies* when they are perforated from the large wafer). It is easy to see why the semiconductor industry has shifted to larger-diameter wafers; though the fabrication facility (known as a *fab*) costs billions of dollars to establish, the price/chip is miniscule since one is essentially able to assemble hundreds of chips at once (ca. 700 chips/wafer using 300 mm wafers (4.16—an increase of over 200% relative to 200 mm). It has been proposed that the industry will shift to 450 mm wafers in the near future; however, this will dramatically increase the costs associated with single-crystal ingot processing.

$$\text{Dies/wafer} = \frac{\pi(\text{wafer diameter} \div 2)^2}{\text{die area}} - \frac{\pi(\text{wafer diameter})}{\sqrt{2(\text{die area})}} \quad (4.16)$$

It is interesting to note that although each wafer goes through the entire CMOS fabrication pathway, all of the completed chips/dies may not be equal. That is, microprocessor chips running at 1.5 MHz and 2.3 GHz may have been produced on the same wafer! From our detailed discussion of CMOS fabrication, it is not hard to see how neighboring chips may have slight variations in layer thicknesses/composition, contaminants, etc. that will significantly alter their ultimate performance. As you might expect, it is typical for companies to sell the fastest of the fabricated chips at a premium price and decrease the price accordingly for slower ones.

4.2.6 Thin-Film Deposition Methodologies

The previous section described the deposition of a number of films onto the growing IC, from insulating HfO_2 layers to the copper interconnects. As you might imagine, a number of growth strategies are employed to yield the most desirable films for the particular application. In IC fabrication, individual layers are deposited with varying levels of thicknesses. For instance, relatively thick layers such as photoresist and interconnect dielectrics may be deposited with a higher variability in film thickness. However, for layers such as gate oxide and $TiSi_2$, techniques that are able to deposit films a monolayer at a time may be required.

In addition to the application of adhesion promotors and photoresists during IC fabrication, "bulk" deposition techniques such as dip- or spin-coating are commonly used for decorative and/or protective coating applications. Films of organic or inorganic (e.g., sol–gels) materials are also spin-coated onto a desired substrate, air-dried to remove the solvent, and postannealed (if desired) to yield the appropriate morphology/porosity of the final film.[54] It is possible to control the film thickness during spin-coating through varying the solvent, spin rate, drop height, etc. However, this technique is not suitable for the growth of thin films where control over film-thickness homogeneity, morphology, composition, conformality, and selectivity are paramount to resultant performance. As we mentioned in the Introduction, the "bottom-up" approach to materials design, or building the structure one molecule/atom at a time, provides the ultimate in control over the final properties of the material. For thin-film growth, this corresponds to *vapor deposition* techniques, rather than the "top-down" approaches of dip- and spin-coating.

Vapor deposition techniques feature the introduction of gaseous molecular/atomic subunits that self-assemble on the surface of the substrate to yield the desired film. The rate of deposition is on the order of Å min^{-1}, which allows for intimate control over the properties of the growing film. There are two types of vapor deposition methods: *physical vapor deposition* (PVD) and *chemical vapor deposition* (CVD). Both methods may be used to grow thin films of metals, alloys, oxides, nitrides, carbides, silicon, or amorphous/graphitic carbon. PVD may occur through the evaporation of atoms/molecules from a precursor solid *in vacuo* (*evaporation*, Fig. 4.56a), or through use of a high-energy Ar plasma source that causes the vaporization of atoms from a solid target (*sputtering*, Fig. 4.56b).[55] Since all atoms of the solid in both techniques will enter the gas phase, the resultant film is only as pure as the solid precursor that is used. As a result, high-purity precursors must be used for PVD; for example, a piece of gold foil of purity >99.99999% is typically used as a target for Au sputtering.

In general, coatings produced by the PVD process are hard, with a high atomic density due to slow and efficient nucleation/growth. Depending on the exposure time of the substrate with the plasma, the thickness of PVD coatings ranges from a few angstroms to $> 30\ \mu m$. Since the substrate is maintained at room temperature, there are no limitations related to the thermal stability of the substrate. This is an important

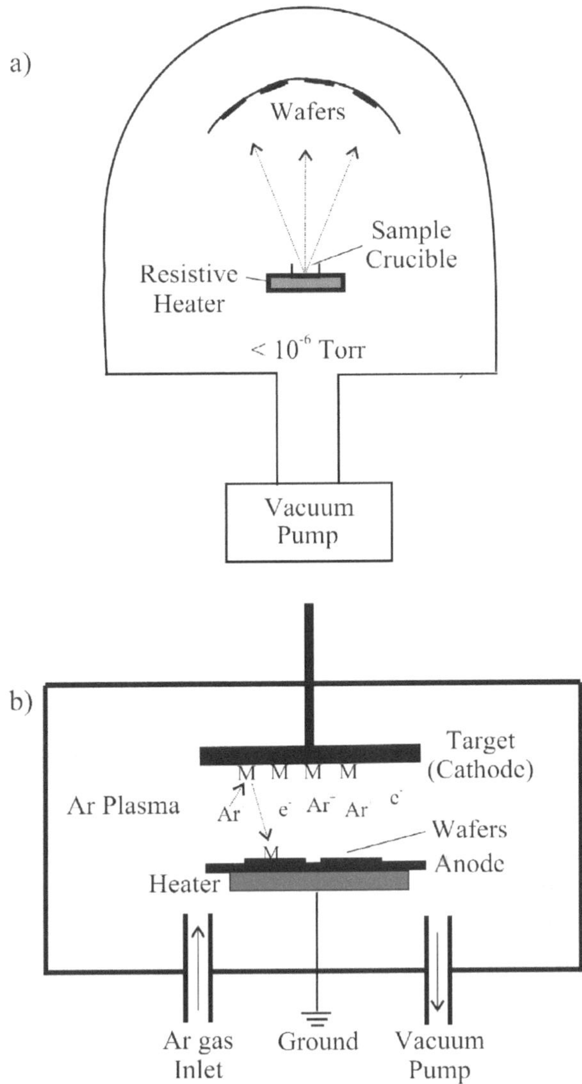

Fig. 4.56 Schematic of physical vapor deposition apparati. Shown are: **a** an evaporation system and **b** a sputtering system

consideration for coating materials that will decompose at high temperatures such as plastics, paper, etc.

Two methods for the evaporation of precursors may be employed—resistance heating and electron beam collision. The first method employs a simple alumina crucible that is heated by a W filament. Temperatures as high as 1800 °C may be reached inside the chamber, which is enough for some metals or metal salts to vaporize. Deposition rates for this method are 1–20 Å s^{-1}. The use of an electron beam to assist in the precursor evaporation results in temperatures on the order of 3000 °C, being more suited for the deposition of refractory metals/alloys and metal oxides such as alumina, titania, and zirconia. Since the temperature of the chamber interior is much higher than the walls, the gas-phase ions/atoms/molecules condense on the sidewalls as well as the substrate. This may lead to film contamination as the nonselective coating flakes off the chamber walls.

For high-purity metal or carbon films, an ultrahigh vacuum (UHV) environment (typically $<1 \times 10^{-6}$ Torr) must be used during PVD. This is necessary in order to prevent the gas-phase reaction of metal/carbon atoms with atmospheric gases (e.g., H_2O, O_2, N_2) that would preferentially form metal oxides, hydroxides, or nitrides rather than the desired film. If mixed phases such as nitrides or oxides are desired, purified nitrogen or oxygen is introduced into the chamber, respectively. For carbide films, targets of both the metal and carbon are placed together within the vacuum chamber. As one would expect, since such a high-purity metal and high vacuum chamber must be used, PVD is relatively quite expensive. Another limitation is the selectivity and conformality of the procedure. While PVD works well to deposit material on surfaces in line-of-sight of the source, nonconformal deposition of complex or rough surfaces (e.g., fibers) is a critical limitation. This issue becomes more important for integrated circuits with feature sizes less than 100 nm, often with high *aspect ratios* (height/width ratio of the surface feature, Fig. 4.57).

By contrast, CVD is a process in which gaseous precursors are *reactively* transformed into a thin film, coating or other solid-state material on the surface of a catalyst or substrate. It should be stressed that CVD is no longer limited to thin film growth; this method is now the preferred route to generating fiber-optic preforms,[56] as well as an increasingly diverse nature of nanostructural architectures, especially carbon nanotubes (CNTs) that will be detailed in Chap. 6. The CVD procedure is often denoted as *metal–organic* CVD (MOCVD), which more accurately specifies the use of an organometallic precursor, containing a central metal and ancillary ligands.

The steps involved in the growth of thin films by CVD are shown in Fig. 4.58. Once the gas-phase precursor molecules enter the deposition zone, (a), they are *physisorbed*

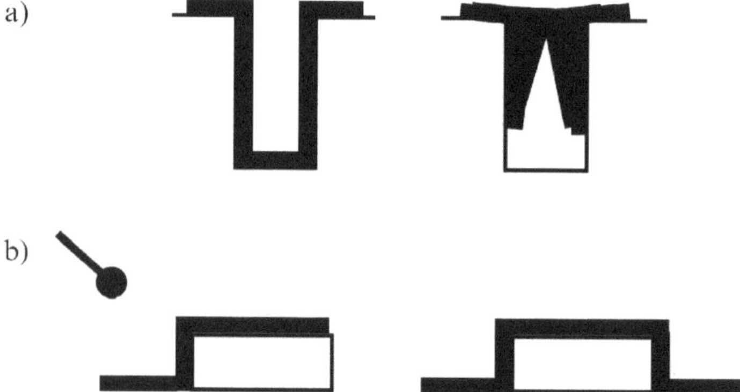

Fig. 4.57 Illustration of conformal thin-film growth for **a** trench filling and **b** step coverage. The line-of-sight limitation of PVD, relative to a conformal CVD technique, is shown in (**b**)

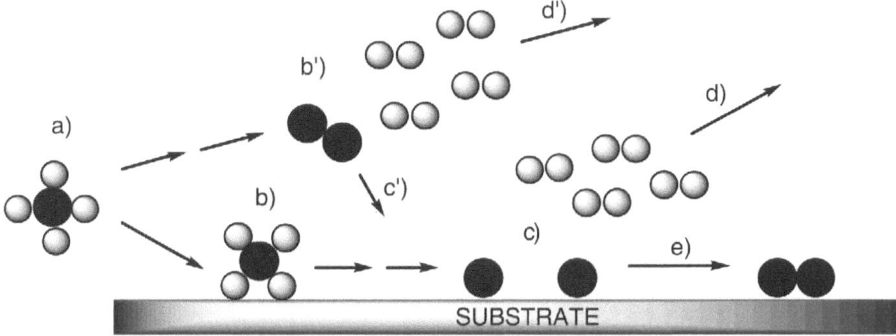

Fig. 4.58 The important steps involved in CVD using the thermolysis of precursor molecules (see text for details)

to the substrate surface through weak van der Waals interactions, (b). The ancillary ligands are removed through thermolysis, leaving the desired residual species on the surface, (c), and ligands being removed from the deposition chamber via the carrier gas flow, (d). Strong covalent interactions are then formed between the surface and adsorbed species, resulting in *chemisorption.* The surface-bound species may migrate along the surface and/or react with other surface species, eventually nucleating on thermodynamically favored positions, (e) *en route* toward thin-film growth. Since film growth occurs through surface migration of intermediate species, CVD is the method-of-choice for depositions onto irregular surfaces where conformality is not possible using PVD techniques (Fig. 4.59).

In addition to the surface governed reactions described above, there are also gas-phase reactions that may take place between precursor/intermediate molecules. Although these reactions may also be important in the growth mechanism (e.g., for plasma-enhanced CVD), significant gas-phase reactions will result in less desirable impure, granular, non-adhering, and non-conformal films. At relatively high temperatures, the gas-phase precursors may preferentially react with one another rather than adsorbing to the substrate surface (b′, d′). The gas-phase nucleation causes granules to form and fall to the substrate surface due to gravitational forces (c′). Typically, these reactions may be minimized by lowering the deposition temperature, which will facilitate the surface-bound growth route. Though gas-phase growth is generally not desired for thin-film applications, this technique has also been applied for the synthesis of nanoparticles.[57]

By definition, CVD is a non-equilibrium process. Although thermodynamics may provide useful information about the overall energetics of the growth process, kinetics must be used to provide information regarding reaction pathways or the transformation rates of the gaseous precursors. The kinetic description of CVD is divided into two parts: mass transport and the rates of the specific chemical reactions involving the precursor and intermediate species. Since the growth rate for CVD is relatively slow, the differences in forward and reverse reaction rates for interfacial events are much smaller than the absolute rates themselves.[58] Thus, CVD is frequently treated

Fig. 4.59 SEM images of an electroluminescent phosphor particle, ZnS (used in backlight displays for cell phones, watches, etc.), before (**a**) and after (**b**) the deposition of an aluminum oxide thin film. This film is a transparent coating that prevents the phosphor particle from undergoing humidity-accelerated decay. A technique known as *fluidized-bed CVD* was used, where a carrier gas both delivered the precursors to a vertically aligned CVD chamber and dispersed the powdery sample in order to expose all surface regions to the precursor vapors. Reproduced with permission from *Adv. Mater. Opt. Electron.* **2000**, *10*, 135

as a pseudo-equilibrium system, considering only the vapor and solid immediately adjacent to the interface, the boundary layer.

As illustrated in Fig. 4.60, the deposition may be controlled by surface-reaction kinetics (e.g., high gas velocity at low temperatures/pressures), or by diffusion/mass transport (e.g., low gas velocity at elevated temperatures or pressures (such as atmospheric)).[59] Whereas the deposition rate in the former system is dependent on the concentration/reactivity of the precursor gases, the deposition rate in the latter is dependent on the diffusion rates of reactants and byproducts. Although the substrate is often placed horizontally in the CVD chamber, it is more desirable to tilt the substrate in order to increase the deposition rate and film-thickness homogeneity. For horizontally-positioned substrates, the velocity of the precursor vapor does not remain constant across the substrate surface, but will decrease downstream. Accordingly, the thickness of the boundary layer will increase at downstream substrate positions, giving rise to depressed thicknesses of the deposited film along the flow direction. By tilting the substrate, the gas velocity over the wafer surface will increase, leading to a uniform boundary layer thickness. This will also enhance the residence time of precursor vapors over the substrate surface, leading to increased growth rates.

At the microscopic level, there are a few key stages of thin-film growth. From adsorbed monomers and surface migration, the first stage is formation of subcritical embryos of varying sizes. These particles will further nucleate while taking

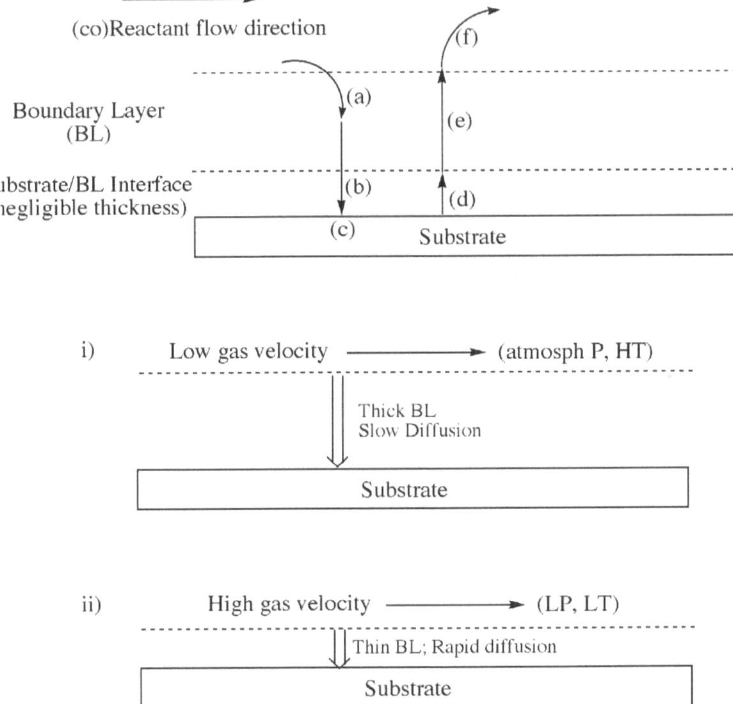

Fig. 4.60 Schematic of the kinetically controlled processes involved in MOCVD. When molecular precursors are introduced into the reaction chamber, they must first diffuse through the boundary layer, (a). The precursors or reactive intermediates/radicals are then adsorbed onto the substrate surface, (b), where surface chemical reactions take place, (c). The ancillary ligands and organic residues from the precursor are then desorbed from the substrate surface, (d), and must diffuse through the BL, (e), en route from the deposition chamber as gaseous by-products, (f). Also shown are the conditions that give rise to surface-reaction or diffusion/mass-transport controlled kinetics

on more precursor adsorption, forming a supercritical sized cluster. These clusters will coalesce into growth *islands*, which will expose new regions of the substrate that serve to adsorb additional precursor species. Isolated surface islands eventually grow together, leaving holes and channels that are filled by adsorbing precursor molecules to form a continuous thin film.

Two general types of reactors are used in CVD processes. In a *hot-walled* reactor (Fig. 4.61), a tube furnace completely surrounds the deposition chamber containing the substrate. The desired deposition temperature depends on the thermal stabilities of the substrate to be coated, and the reactor (*i.e.*, glass ($T < 600$ °C) or quartz ($T < 1100$ °C)). Hot-walled reactors are used extensively for laboratory studies, and also in industry for the CVD of semiconductors and oxides. These reactors are often preferred due to the simplicity of setup, while being able to maintain a uniform temperature over a large number of substrates. However, secondary coating of the reactor walls is unavoidable; frequent cleaning of these reactors is therefore necessary, as deposits may easily flake off these surfaces and contaminate the growing film. To prevent such problems, *cold-walled* reactors may also be used for a CVD process, wherein only the substrate is heated. This focuses the surface and gas-phase reactions to a region immediately surrounding the substrate surface. However, this reactor type generally results in narrower deposition zones and slower growth rates than hot-walled analogues.

Perhaps more than any other materials synthesis technique, semantics becomes a challenge due to the plethora of acronyms that are used to describe a specific CVD process. In particular, it is not sufficient to simply cite "CVD" alone in an article title; a more explicit acronym must be used that states the type of deposition chamber and precursor decomposition methodology employed. If no prefix is affixed to CVD, it usually denotes simple thermolysis of a precursor within a standard cold- or hot-wall reactor. However, if more energetic sources of energy are used to degrade the precursor such as laser, plasma, or microwave plasma, the acronyms laser-assisted

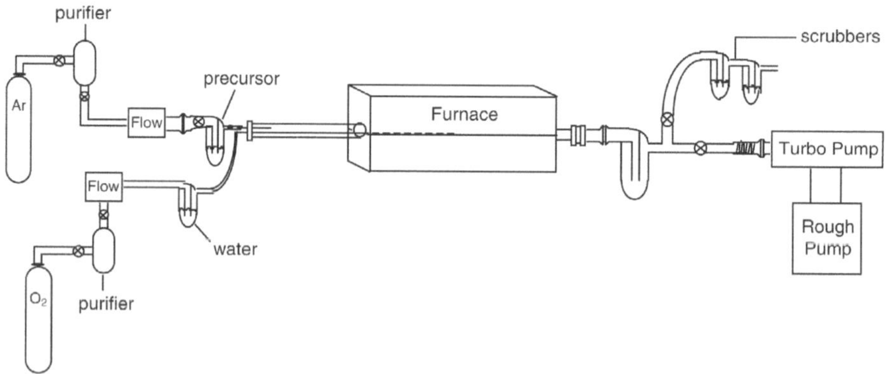

Fig. 4.61 Schematic of a horizontal hot-walled CVD reactor. Shown is a two-precursor system, where the water sensitive precursor contacts water vapor directly over the heated substrate

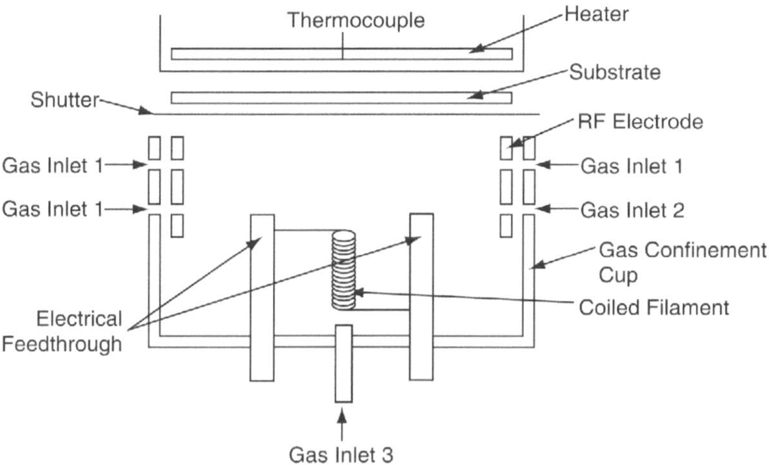

Fig. 4.62 Schematic of a hot-wire CVD system. Reprinted from Povolny, H. S.; Deng, X. *Thin Solid Films* **2003**, *430*, 125. © 2003, with permission from Elsevier

CVD (LACVD), plasma-enhanced CVD (PECVD), and microwave plasma CVD (MPCVD) are used, respectively. Rather than using high-energy plasma sources, a relatively new CVD technique uses a heated filament to degrade precursor gases and form reactive intermediates. This technique is referred to as either catalyzed CVD (Cat-CVD) or hot-wire CVD (HWCVD, Fig. 4.62).[60]

The pressure of the reactor must also be defined, and acronyms APCVD, LPCVD, and SCF-CVD are used to denote ambient, *in vacuo*, and supercritical pressure conditions within the deposition chamber, respectively. In general, the resultant film conformality of a CVD process follows the order LPCVD > APCVD, due to the relative migration rates of intermediate species along the surface of the substrate. However, it has been shown that conformal thin films of metal oxides may be deposited on nonplanar substrates such as carbon fibers under APCVD, even at low temperature.[61]

While plasma-enhanced methods are very useful to lower the substrate temperature, the as-deposited films are typically less conformal and often contain more surface impurities than competing methods. In this method, reactive radicals, ions, and atoms/molecules are formed in the gas phase that interact with the relatively low-temperature substrate to generate a film. Some of the more recent applications for plasma CVD include growth of cubic boron nitride (c-BN)[62] and diamond[63] thin films.

A complementary process known as *atomic layer deposition* (ALD) may also be utilized for more controllable thin film growth.[64] In contrast to CVD, ALD features the *sequential* exposure of a substrate to two or more precursors. The precursor vapors are pulsed into the reactor one at a time, being separated by inert-gas purges or evacuation (Fig. 4.63). This establishes user-controlled deposition cycles, each comprising surface saturation by one precursor, followed by the self-limiting surface reaction of a second precursor. Consequently, deposition is unaffected by varying

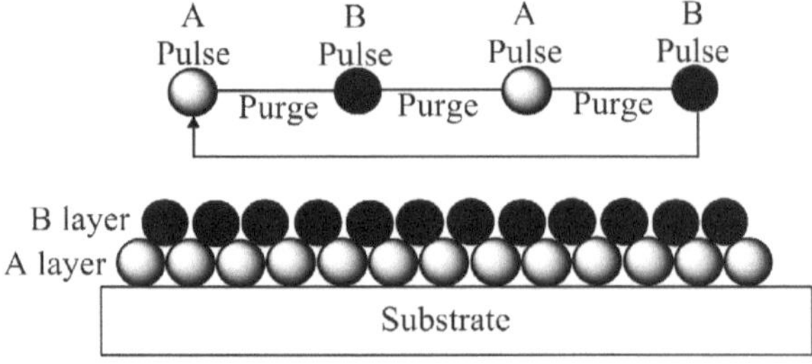

Fig. 4.63 Illustration of atomic layer deposition (ALD)

vaporization rates of solid precursors—especially problematic for CVD. Further, ALD is generally carried out at lower temperatures than CVD since co-reactants are highly reactive toward each other (e.g., trimethylaluminum and water vapor). In contrast, using the same co-reactants for CVD results in hard-to-control gas-phase reactions, even at low temperatures.[65] Since film growth occurs sub-monolayer at a time, ALD allows for an unprecedented control over the resultant film thickness, conformality, homogeneity, and stoichiometry.[66]

Ideally, the film thickness (*i.e.,* growth rate) will be uniform across the entire substrate surface. The flux of gas molecules that impinge on a substrate surface may be expressed as 4.17:

$$J = \frac{P}{\sqrt{2\pi mkT}}, \tag{4.17}$$

where J is the rate at which gas-phase molecules hit a surface (in units $m^{-2}\ s^{-1}$), m is the mass of a gas molecule, k is the Boltzmann's constant (1.38×10^{-23} J K^{-1}), and T is the temperature in K. Hence, the growth mechanism of ALD is unaffected by the presence of an inert carrier gas, as long as it exhibits a purity level of at least the parts-per-billion (ppb) level.

In theory, ALD growth proceeds by one atomic layer per cycle; however, due to steric hindrances and limited number of reactive surface sites, the growth rate per cycle is a fraction of a monolayer (ML) thickness—typically less than 0.9 ML. The *sticking coefficient*, S_c, is often used to describe the fraction of incident molecules that adsorb upon the substrate surface. In general, S_c is dependent upon the degree of coverage, temperature, and crystal structure/reactivity of the substrate surface. It should be noted that S_c is generally lower for CVD/ALD relative to PVD techniques. Consequently, the lower S_c results in being able to coat complex topographies (conformal growth), as well as selected areas of the substrate.

A number of optimization runs must be performed in order to achieve self-limited growth, known as the *ALD window*. If the thickness is greater at the inlet end, it may be due to insufficient purging between pulses or too low a deposition temperature that results in condensation of the precursor (i.e., physisorption is occurring rather than chemisorption). If the thickness increases toward the outlet end, it is usually a sign of too high a deposition temperature that is causing surface decomposition of the precursor. If the purge/evacuation cycles are too long, the precursor may be desorbed from the substrate. This would result in film thickness that decreases along the inlet–outlet direction. Finally, if the growth rate is too slow, the system is activation-energy limited; that is, the temperature must be increased in order to facilitate a suitable level of precursor reactivity.

Oftentimes, the deposition mechanisms between ALD and CVD are quite different, even for identical precursor combinations. In particular, it is well known that thin-film growth by CVD is heavily influenced by side reactions (Fig. 4.64)—not as problematic for ALD (Fig. 4.65). A CVD process often generates reaction products such as CO, RH, or HCl that may be preferentially adsorbed onto the substrate surface. This will cause film-growth termination unless the competitive adsorbents are removed through purging with inert gas or introduction of a reducing gas such as H_2.

The chemical nature of the precursor represents the most critical component of a CVD/ALD process. Generally speaking, the choice of a particular precursor is governed by the relative stabilities of the precursor and substrate, as well as the volatility, cost, and hazards of the precursor. The coordination sphere of ligands surrounding the central metal is extremely important; the organic ligands in these precursors may lead to contamination of the films if they are not completely removed through a combination of pyrolysis, reduction, or oxidation processes. Some precursors pose a high risk when being used. For example, $Ni(CO)_4$ has a very high toxicity, $Al(Me)_3$ is pyrophoric, SiH_4 and B_2H_6 are explosive, and chloride-containing species are corrosive. In general, organometallic precursors pose lower hazards than hydrides and halides, but are much more costly.

Although it was once essential that volatile precursors be used, this is no longer a synthetic limitation. Within the last decade, the gas/liquid properties of supercritical fluids (e.g., CO_2) have been used to solvate certain precursors, facilitating their use for CVD.[67] Two variations of this technique may be used; supercritical fluid transport (SFT), using the fluid as an aerosol-like delivery vehicle, or in situ thin-film growth within a high-pressure reactor, known as supercritical fluid deposition (SFD). In these methods the precursor must be soluble in CO_2, which is analogous in solvating ability to hexane with an enhanced fluorophilic character.

A CVD process may involve the use of either single or mixed precursors. The best precursor is a molecule that has sufficient volatility (or CO_2 solubility) and contains labile ligands that will leave no organic residue behind during its surface-catalyzed decomposition. In general, volatile liquids and oils are attractive CVD precursors due to the relative ease of vapor transport through simple carrier-gas bubbling, vaporization, or *direct liquid injection* (DLI) techniques (Fig. 4.66). However, low-melting powders with high volatilities and low decomposition temperatures are regarded as the best candidates for CVD, as they are often less susceptible to premature hydrolysis during transfer to the CVD system. Co-reactant gases such as NH_3, H_2, H_2S, H_2O, etc. are often added in order to assist the removal of organic moieties from the organometallic precursor molecule, preventing their incorporation in the growing film.

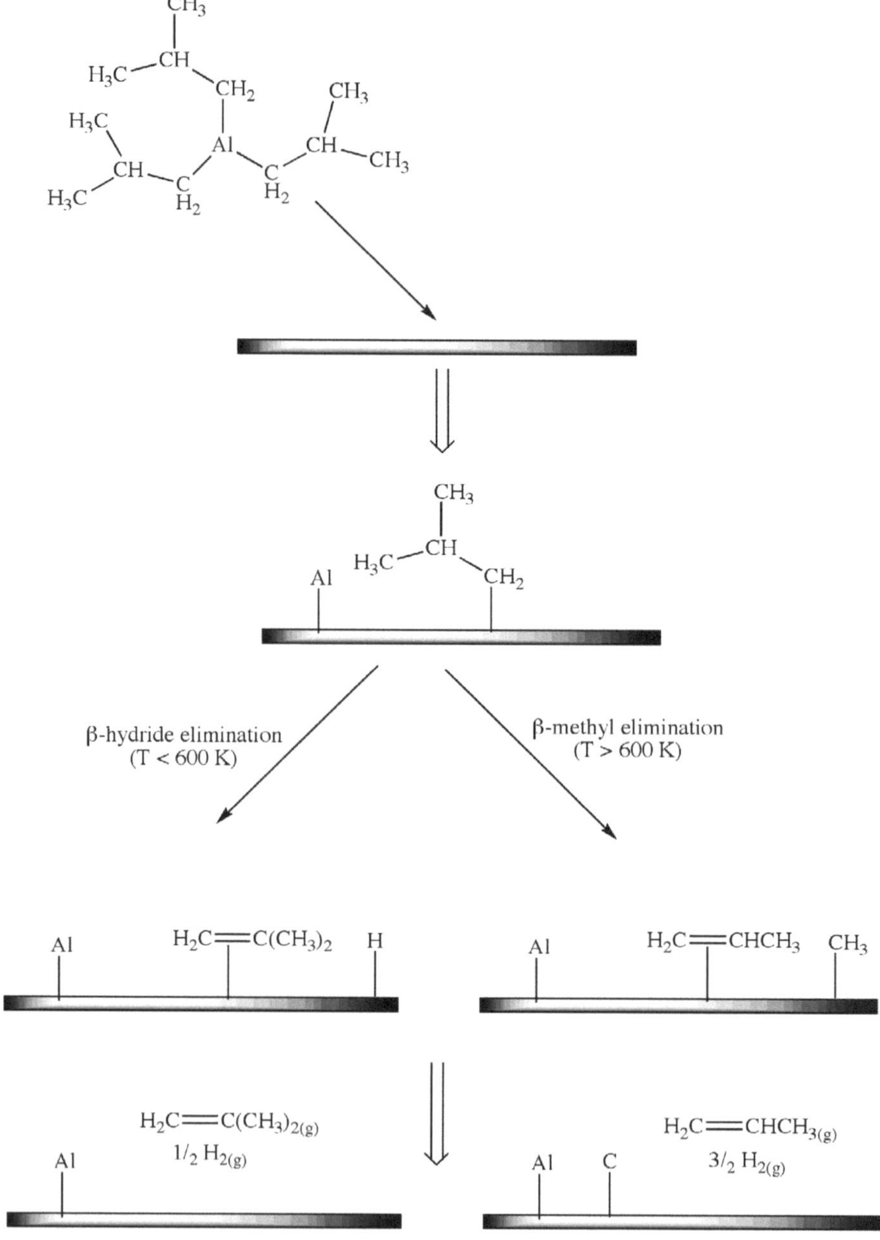

Fig. 4.64 Illustration of chemical vapor deposition (CVD) of an aluminum film from $Al(O^iPr)_3$, showing the influence of temperature (decomposition route) on the resultant film purity

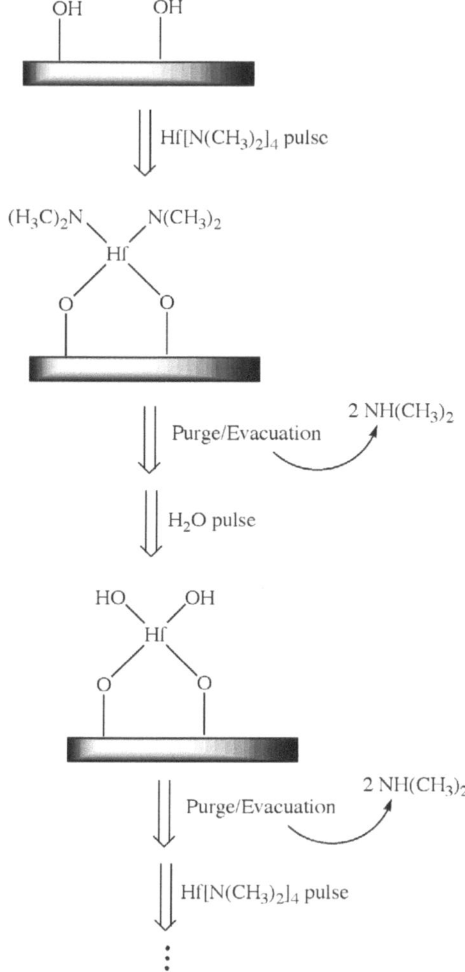

Fig. 4.65 Stepwise scheme for the atomic layer deposition (ALD) of a HfO_2 thin film from $Hf(NMe_2)_4/H_2O$ sequential pulsing

The use of mixed precursors adds complexity to the control over film stoichiometry. The different volatilities of the precursors often lead to irreproducible vapor-phase concentrations. As a result, flow rates have to be carefully controlled to give films of a desired composition. Another problem that is posed by the use of mixed precursors is the variable rate of hydrolysis/oxidation, resulting in morphologically varied films. Hence, it is most preferable to use a precursor that contains as many desired building blocks of the thin film as possible. For the films deposited during IC fabrication, common precursors include silane (SiH_4) with co-reactant gases of H_2 (for polysilicon films), NH_3 (for Si_3N_4), and O_2 (for SiO_2). The deposition of TiN films commonly employs co-reactants of $TiCl_4$ and NH_3 (4.18):

$$6\ TiCl_4 + 8\ NH_3 \xrightarrow{500\,^{\circ}C} 6\ TiN + 24\ HCl + N_2 \qquad (4.18)$$

a)

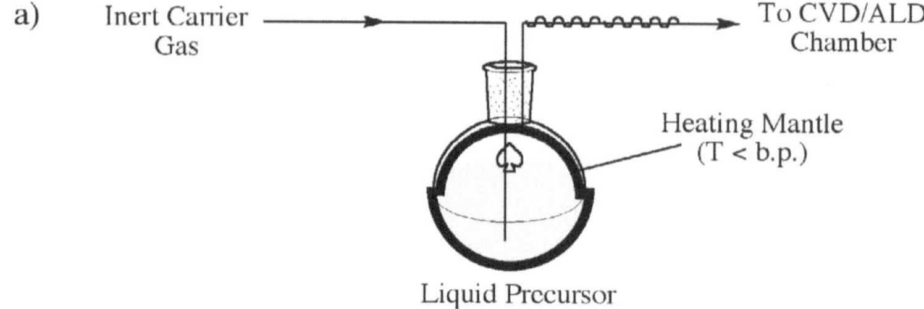

b)

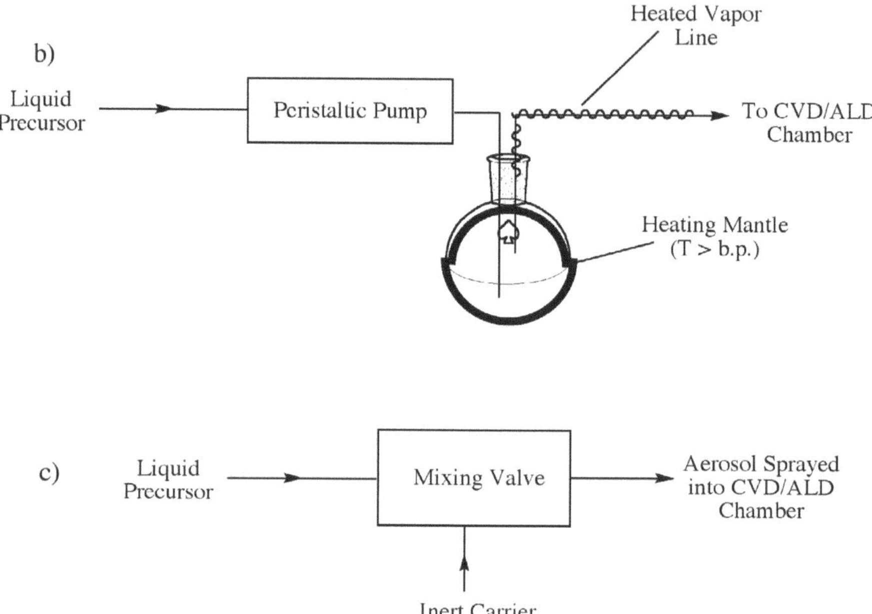

Fig. 4.66 CVD/ALD delivery methods for liquid precursors. Shown are **a** carrier gas bubbling of a volatile precursor, **b** vaporization of a volatile precursor, and **c** direct liquid injection (DLI) of a volatile/non-volatile precursor

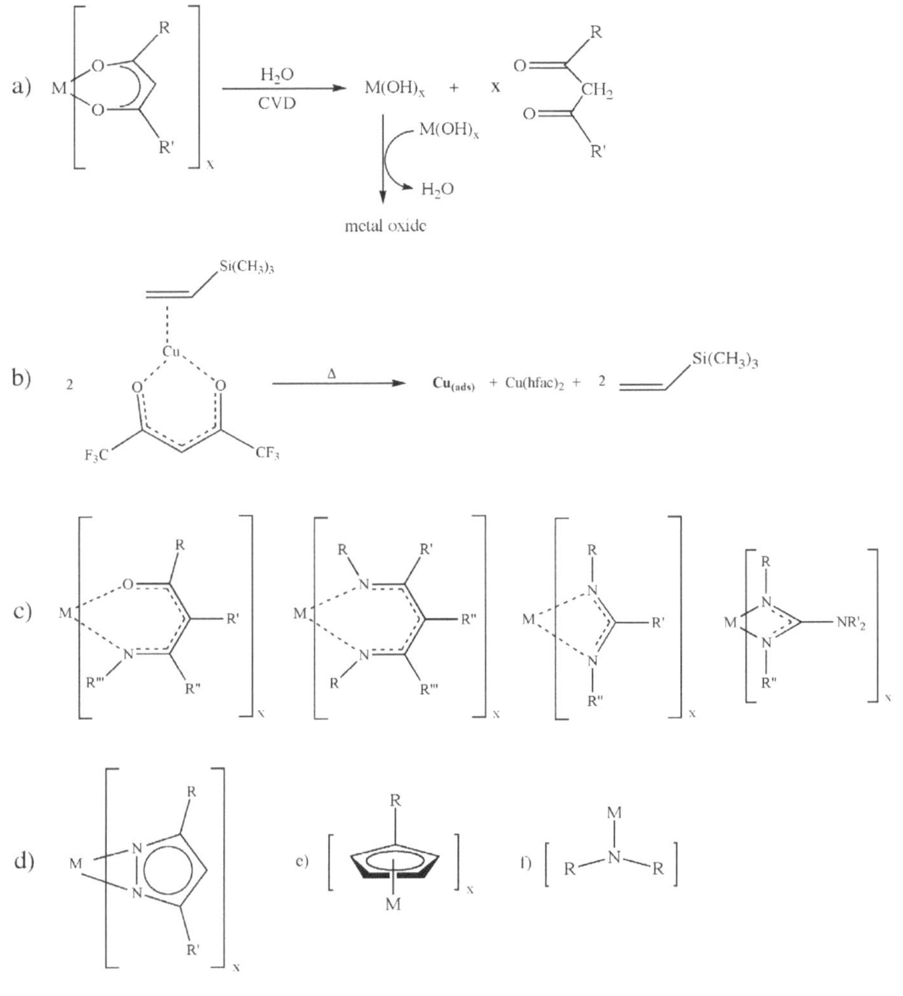

Fig. 4.67 Molecular structures of commonly used CVD precursor classes. Shown are: **a** metal β-diketonate (acetylacetonate, acac) complex to grow a metal oxide film (H_2 as the coreactant gas yields a metal film); **b** a heteroleptic (more than one type of ligand bound to the metal) β-diketonate complex to yield a Cu film; the ancillary ligand helps prevent oligomerization, enhancing volatility; **c** various types of complexes to deposit metallic, oxide, nitride, or oxynitride films (depending on coreactant gas(es) used—respective ligands are β-ketoiminato, β-diketiminato, amidinato, and guanidinato; **d** a metal azolato complex commonly used to deposit lanthanide metal thin films

Figure 4.67 illustrates the chemical structures of a variety of precursors that may be used for CVD/ALD applications.[68] Metal alkoxides ($M(OR)_x$) and halides represent the earliest known examples; mixed metal alkoxide precursors (e.g., for bimetallic thin film growth) may be synthesized by the following route (4.19–4.20; A = Group 1 metal):

$$M(OR)_x + AOR \rightarrow AM(OR)_{x+1} \tag{4.19}$$

$$y\ AM(OR)_{x+1} + M'Cl_y \rightarrow M'[M(OR)_{x+1}]_y + y\ ACl \tag{4.20}$$

Metal β-diketonate complexes have been used extensively to deposit thin films of most metals and their oxides. Their low cost and high volatility make them good candidates for precursors, but may yield significant carbon contamination. Replacing one oxygen with-NR substituents results in the β-ketoiminato ligand system. The greater basicity of the nitrogen-donor ligands relative to oxygen analogues has been reported to increase the volatility through enhanced metal–ligand bonding.[69] Replacing both oxygen groups with NR moieties results in the β-diketiminato ligand, which provides further thermal stability for the resultant metal complex while offering a tunable volatility by varying the imine substituents.

The volatility of a precursor is related to its molecular weight (*i.e.,* London dispersion intermolecular forces), as well as its degree of oligomerization. As such, it is also common to have a metal complex that features bulky ligands[70] or the use of an ancillary Lewis base (e.g., alkene, alkyne, triorganophosphine, vinylsilane, etc.) to prevent oligomerization. Such a complex with more than a single type of ligand surrounding the central metal is known as a *heteroleptic* complex; in contrast, *homoleptic* complexes have only a single ligand type, such as $Al(C_2H_5)_3$). Another effective method to enhance precursor volatility is to incorporate partially and fully fluorinated ligands. The enhancement may be rationalized either by an increased amount of intermolecular repulsion due to the additional lone pairs, or that the reduced polarizability of fluorine (relative to hydrogen) causes fluorinated ligands to have less intermolecular attractive interactions.

Prior to CVD studies, it is essential that the potential precursor undergo thermogravimetric analysis (TGA) and differential scanning calorimetry (DSC) to determine the most suitable temperature regime for thin-film growth (Fig. 4.68). A sharp TGA curve with no remaining residue indicates that the precursor vaporizes without significant ligand decomposition—most desirable for CVD applications. Further, if the mass loss via TGA occurs concomitantly with a single endotherm via DSC, then sublimation of the precursor is occurring without decomposition (Fig. 4.68, bottom)—ideal behavior for CVD applications.

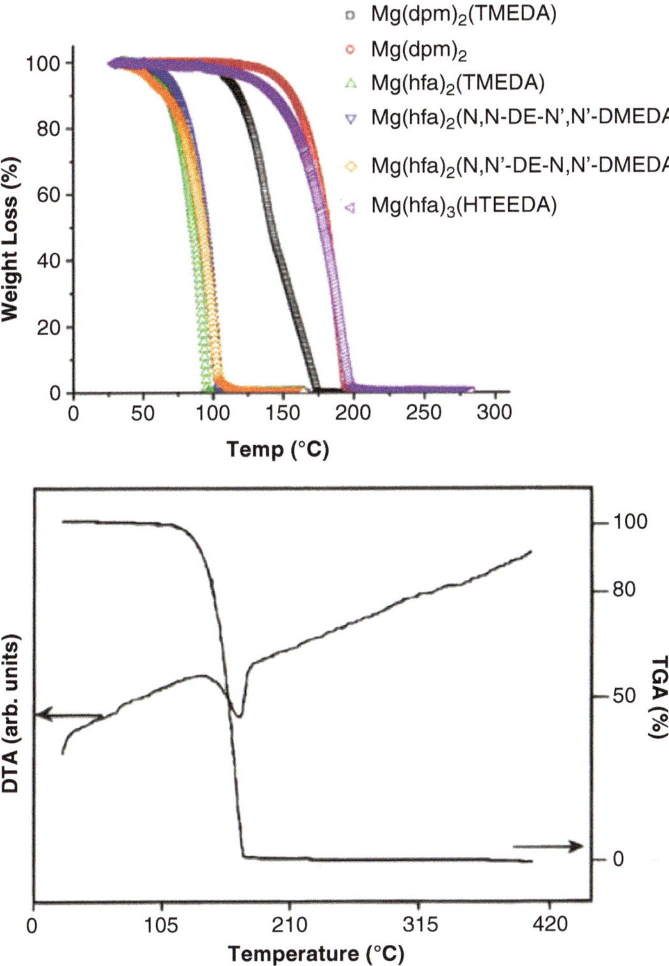

Fig. 4.68 Thermogravimetric analysis of organomagnesium CVD precursors (top), indicating the dependence of molecular structure on its decomposition temperature. Tandem TGA/DSC analyses showing a single endothermic event (sublimation) that coincides with the mass loss (bottom). The ligand abbreviations are: dpm = 2,2,6,6-tetramethyl-3,5-heptanedionate, TMEDA = *N,N,N′,N′*-tetramethylethylenediamine, hfa = 1,1,1,5,5,5-hexafluoro-2,4-pentanedione, HTEEDA = H(*N*, *N,N′,N′*-tetraethylethylene diamine). Reproduced with permission from *Chem. Mater.* **2005**, *17*, 5697. © 2005 American Chemical Society

TGA may also be applied to accurately determine the enthalpy of sublimation (ΔH_{sub}), as well as the sublimation temperature (T_{sub}) of a solid.[71] In these investigations, the mass loss through sublimation (m_{sub}) will be constant at a given temperature, as long as the phase change occurs without appreciable decomposition. Hence, by measuring the mass loss over a variety of isothermal regions, a plot of $\log(m_{sub}T^{1/2})$ *vs.* $1/T$ may be generated, which readily yields values of ΔH_{sub} and T_{sub} from the slope and *y*-intercept, respectively (4.21):

$$\log(m_{sub}\sqrt{T}) = \frac{-0.0522(\Delta H_{sub})}{T} + \left[\frac{0.0522(\Delta H_{sub})}{T_{sub}} - \frac{1}{2}\log\left(\frac{1306}{M_w}\right)\right] \quad (4.21)$$

It is often desirable to deposit films that possess more complex stoichiometries. These films may include mixed-metal, and binary/ternary metal oxides and sulfides that are of importance for emerging applications for catalysis and microelectronics. Although these materials may be generated through use of individual species, such as Y, Ba, and Cu β-diketonates for $YBa_2Cu_3O_{7-\delta}$ superconductor films,[72] it is most advantageous to use a single-source precursor for these applications. The fixed ratio of component metals in the individual precursor molecules offers a unique route toward stoichiometric control over thin-film and nanoparticulate growth. Further, lower deposition or post-annealing temperatures are often required for single-source precursors relative to their co-reactant analogs.

Perhaps the greatest "Achilles heel" of using single-source precursors is the complex and expensive synthetic procedures that are often required. Although this is usually not an issue for laboratory-scale processes, this represents a major limitation for industrial scale-up considerations. However, there are precedents for one-step reaction schemes, such as that reported for a variety of single-source precursor molecules suitable for CVD of I–III–VI semiconductors (Fig. 4.69).[73] Subsequent CVD studies using the ternary sulfide precursors were performed using an aerosol delivery methodology, with films of comparable purity to those generated by highly expensive procedures such as PVD from co-evaporation of source metals.

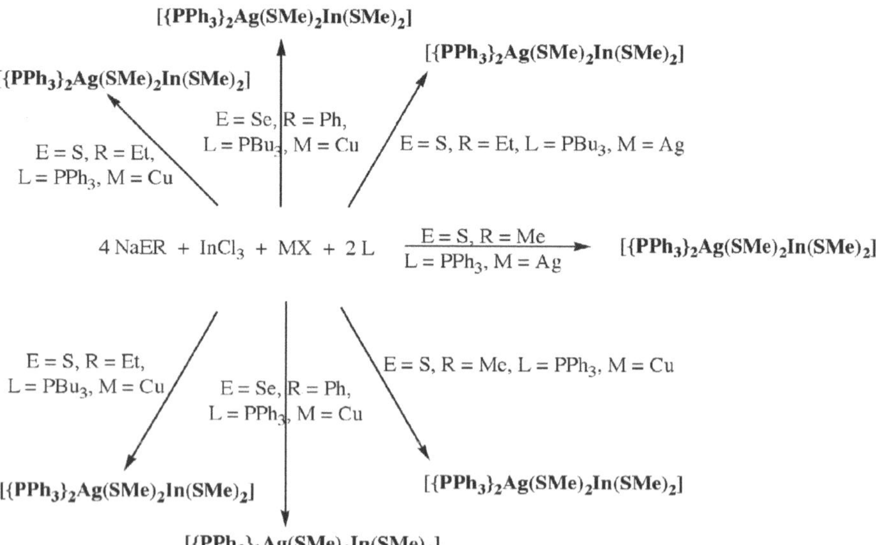

Fig. 4.69 Synthetic pathway for single-source precursor design. Reproduced with permission from *Inorg. Chem.* **2003**, *42*, 7713. © 2003 American Chemical Society

4.3 Light-Emitting Diodes: There is Life Outside of Silicon!

Thus far, we have considered the structure and applications of silicon—the most heavily employed semiconductor. It should be noted that gallium arsenide (GaAs) is also widely used for FET applications; since GaAs does not form a natural protective layer analogous to SiO_2 onto Si, CVD is used to grow films such as GaS for surface passivation. Many other direct bandgap semiconductors such as II–VI (e.g., ZnS, ZnSe) and III–V (e.g., GaN, GaP) are widely used for optoelectronic (light emission) and photonic (light detection) applications. The most important applications for non-Si semiconductors are light-emitting diodes (LEDs) and solid-state lasers. Unlike Si-based devices, the bandgap of compound semiconductors may be significantly altered by varying the stoichiometry of the composite elements.

A light-emitting diode is a p–n junction device that is comprised of a direct bandgap semiconductor(s). The recombination of an electron–hole pair (EHP) results in photon emission, with energy equivalent to the bandgap, E_g (Fig. 4.70). The LED may be structured as a simple multilayer (Fig. 4.71a), or as a heterojunction that is comprised of two different bandgap semiconductors (Fig. 4.71b). The latter is preferred for high-intensity LED applications, since the emission is confined to certain regions of the device.

Historical applications for LEDs have been centered on commercial electronic displays such as clock radios, microwave ovens, watches, etc. However, the current availability of LEDs in a spectrum of colors has opened the floodgates for new applications. The greatest breakthrough was realized in the 1990s with the discovery of wide bandgap blue LEDs, making it possible to create any color of light (Table 4.4). More than half of the traffic lights in the United States have now been replaced with

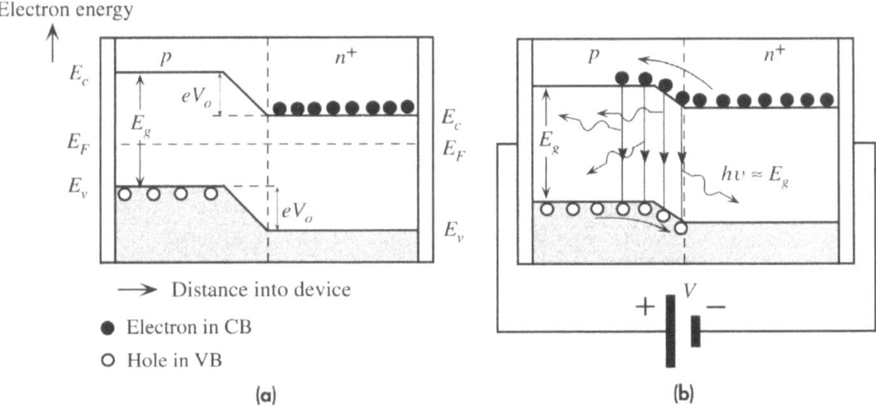

Fig. 4.70 Band diagram of a p–n junction with **a** no bias voltage, and **b** with forward bias, V, resulting in photon emission. Reproduced with permission from Kasap, S. O. *Principles of Electronic Materials and Devices*, 3rd ed., McGraw-Hill: New York, 2007. © 2007 The McGraw-Hill Companies

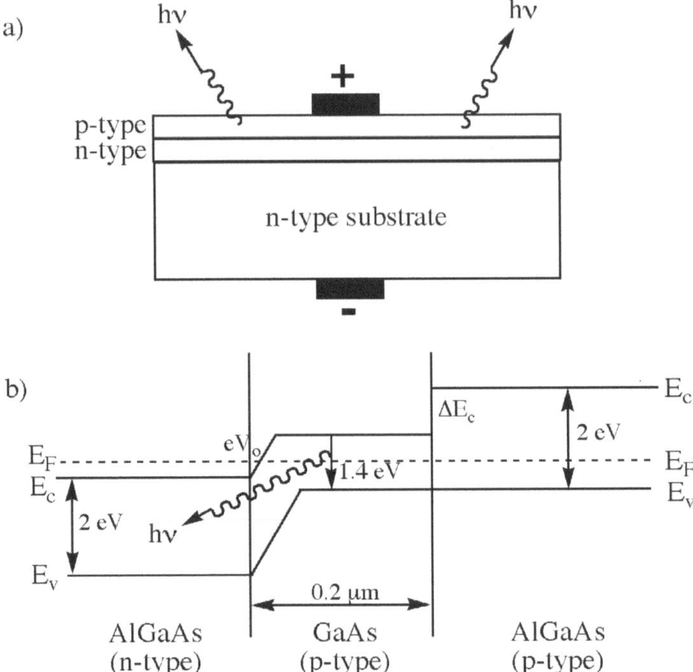

Fig. 4.71 Light-emitting diode (LED) structures. Shown are **a** a simple multilayer structure of epitaxial p- and n-type layers, and **b** a double heterostructure with accompanying band diagram, illustrating light emission from the p-type region due to confinement between wide-bandgap surrounding layers

LED-based lamps. New automobiles also utilize this lighting for ultrabright brake and turn-signal lighting. The higher initial cost of the LEDs is quickly recovered due to their greater efficiency in converting electrical current to light emission relative to incandescent lighting. That is, whereas LEDs consume 20–1000 mW, an incandescent bulb of similar brightness consumes ca. 50–150 W. Furthermore, LED bulbs can last between 50,000–100,000 h; in contrast, incandescent bulbs are typically limited to <1200 h. Although ca. 5 times more expensive, the increasing availability of soft-white LED light bulbs are enabling our society to begin the shift from long-standing reliance on inefficient and short-lived incandescent bulbs. It is estimated that if half of all lighting would shift to solid-state, it would cut worldwide power use by 120 GW. This would save over \$100 billion yr^{-1}, while reducing the carbon dioxide emitted by electrical power plants by 350 Mt yr^{-1}.

The light emission from LEDs is classified as a type of *luminescence*, which is different from *incandescence*—the generation of light from a material as a result of its high temperature. Since LEDs glow as a result of an electrical current, this emission is referred to as *electroluminescence*. Two other common types of luminescence include *chemoluminescence* (induced by a chemical reaction(s); e.g., “glow sticks”), and

Table 4.4 Comparison of the Observed Colors of LEDs

Observed color	Wavelength of emission (nm)	Semiconductor
(Infrared)	880	GaAlAs/GaAs
Red	660	GaAlAs/GaAlAs
Red	633	AlGaInP
Orange	612	AlGaInP
Orange	605	GaAsP/GaP
Yellow	585	GaAsP/GaP
Green	555	GaP
Blue	470	GaN/SiC
Ultraviolet	395	InGaN/SiC
White	–	InGaN/SiC

photoluminescence (induced through photon excitation; e.g., vaseline glass discussed in Chap. 2). If the emission is prolonged, lasting long after the stimulation source is removed, it is known as *phosphorescence*; otherwise, the short-lived process is termed *fluorescence.*

As we saw earlier in this chapter, the wavelength (and color) of light emitted by a direct bandgap material through electron–hole recombination is influenced by its bandgap. In order to change the wavelength of emitted radiation, the bandgap of the semiconducting material utilized to fabricate the LED must be changed. For instance, gallium arsenide has a bandgap of 1.35 eV (Table 4.5) and emits in the infrared (ca. 900 nm). In order to decrease the wavelength of emission into the visible red region (ca. 700 nm), the bandgap must be increased to ca. 1.9 eV. This may be achieved by mixing GaAs with a material with a larger bandgap, such as GaP ($E_g = 2.35$ eV). Hence, LEDs of the chemical composition $GaAs_xP_{1-x}$ may be used to produce bandgaps from 1.4 to 2.3 eV (and varying colors), through adjustment of the As:P ratio.

Table 4.5 Bandgaps of III–V semiconductors

Semiconductor	Bandgap (eV)
AlN	6.02
AlP	2.45
GaN	3.50
GaP	2.35
GaAs	1.35
GaSb	0.67
InN	1.95
InP	1.27
InAs	0.36

The bandgap and concomitant wavelength of light that is emitted from LEDs is related to the bond strength between atoms in the lattice. For these compounds, as the bond strength increases, there is more efficient overlap between molecular orbitals that gives rise to a larger bandgap between bonding and antibonding MOs (*i.e.*, valence and conduction bands of the infinite lattice, respectively). For a particular Group 13 metal, as one moves down the Group 15 congeners, the bonding interaction between III–V elements will become weaker through the interaction of more diffuse atomic orbitals. For instance, the bond strengths of Ga–N and Ga–As bonds are 98.8 and 50.1 kcal mol^{-1}, respectively. The larger bandgap for GaN relative to GaAs translates to a short wavelength (blue color) of emitted light that is observed.

The ideal white light source should emit a continuous broad spectrum analogous to natural sunlight. There are four primary modes of generating white light from LEDs (Fig. 4.72). Most white LEDs employ a semiconductor chip emitting at a short wavelength (blue), and a wavelength converter that absorbs light from the diode and undergoes secondary emission at a longer wavelength (Fig. 4.72a and b). Such diodes emit light of two or more wavelengths, that when combined, appear as white. The most common wavelength converter materials are termed *phosphors* (e.g., ZnS, Fig. 4.59), which exhibit luminescence when they absorb energy from another radiation source (Table 4.6).[74] Typical LED phosphors are present as a coating on the outside of the bulb and are composed of an inorganic host substance (e.g., yttrium aluminum garnet, YAG) containing an optically active dopant (e.g., Ce). Use of a single-crystal phosphor produces a yellow light, upon combination with blue light gives the appearance of white. For instance, phosphors such as $Ba_3Si_6O_{12}N_2$:Eu^{2+} (yellow) and $SrAlS_4N_7$:Eu^{2+} (orange-red) may be combined with YAG:Ce and blue LEDs to produce warm white lighting that is comparable to incandescent. Nitride-based phosphors are desired since they do not degrade under high temperature/humidity conditions—a critical limitation of sulfide-based materials.[75] A similar result has recently been produced through use of CdSe *nanoparticles*.[76] White LEDs may also be made by coating near ultraviolet (NUV) emitting LEDs with a mixture of europium-based red and blue emitting phosphors, plus green emitting copper- and aluminium-doped zinc sulfide (ZnS:Cu,Al). It is also possible for LEDs to emit white light without the use of phosphors. For instance, homoepitaxially grown ZnSe crystals simultaneously emit blue light from the film, and yellow light from the ZnSe substrate.

More recently, it is possible to generate white electroluminescence (EL) without the use of phosphors. Multiple LED chips or RGB emitters may be combined to generate white light emission (Fig. 4.72c). In contrast, a single chip may emit white EL directly (Fig. 4.72d). These two modes would exhibit a higher energy conversion efficiency relative to phosphor-based white LEDs. Futhermore, these designs would generate less heat during use. Halide perovskites are extremely promising materials for next-generation LEDs due to their tunable emission profile, high spectral purity and luminescence efficiency, and low cost.[77] Perovskites have the general formula ABX_3, where A^+ = monovalent cation (e.g., Cs^+, methylammonium, or formamidinium), B^{2+} = divalent metal cation (e.g., Pb^{2+}, Sn^{2+}, Ge^{2+}, Mn^{2+}, Ni^{2+}, Zn^{2+}, or Cu^{2+}) and X^- = single or mixed halide anion (e.g., Cl^-, Br^-, or I^-). For instance,

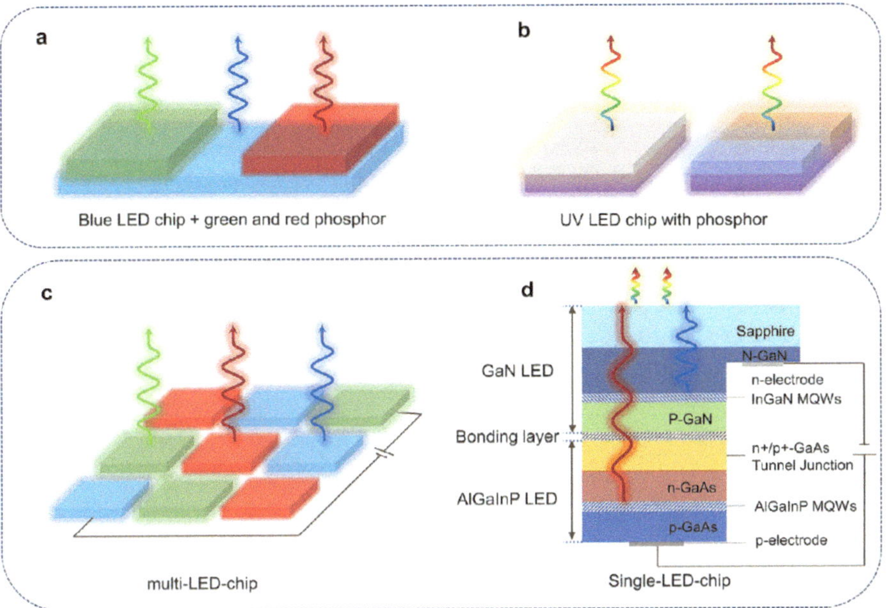

Fig. 4.72 General approaches to generate white light from LEDs. **a** Multicolor phosphors are excited by and blended with the EL of blue LED chip. **b**) UV LED light induces a direct photoluminescent emission of white light from the emissive conversion layer. **c** Multicolor LED chip emitting RGB colors are optically blended to generate white EL. **d** Single LED chip with variety of band gaps showing white EL. Reproduced with permission from *ACS Nano* **2021**, *15*, 17150. © 2021 American Chemical Society

LEDs composed of guanidinium (GA) doped formamidinium (FA) lead bromide perovskite nanocrystals exhibit an external quantum efficiency of 23.4%, the highest to date. Most desirable for future lighting applications, the optoelectronic properties of halide perovskite crystals may be fine-tuned by varying their composition or morphology (e.g., quantum dots, nanowires, nanosheets, etc.).

There has also been much interest in organic light-emitting diodes (OLEDs). Flat-panel televisions, cellular phones, and digital cameras already employ this technology; it is only a matter of time before the "holy grail" of *flexible* display screens and luminous fabrics are commercially available. Samsung has already demonstrated a Galaxy Note smartphone with a foldable screen, with plans to soon introduce this technology into their laptops. OLED displays offer many benefits relative to standard CRTs and LCDs such as enhanced brightness, lower power consumption, and wider viewing angles.

The multilayered structure and electroluminescent mechanism of OLEDs is illustrated in Fig. 4.73. As a frame of reference, the diameter of a human hair is ca. 200 times greater than the thickness of individual layers (80–200 nm) in an OLED. Depending on whether small organic molecules or long repeating-unit polymers are used (Fig. 4.74), the diodes are referred to as OLEDs or PLEDs, respectively.[78] Under

Table 4.6 Emission wavelengths of some selected inorganic phosphors

Phosphor	Emission wavelength (nm)
$Y_3Al_5O_{12}:Ce^{3+}$ (YAG:Ce)	460
$BaSi_2O_2N_2:Eu^{2+}$	498
$CaSi_2O_2N_2:Eu^{2+}$	560
$CaAlSiN_3:Eu^{2+}$	650
$Sr_5Al_{5+x}Si_{21-x}N_{35-x}O_{2+x}:Eu^{2+}$	550
$SrAlSi_4N_7:Eu^{2+}$	700
$Ba_2AlSi_5N_9:Eu^{2+}$	620
$BaSi_6O_{12}N_2:Eu^{2+}$	580
$(Sr, Ca, Ba)_3SiO_5:Eu^{2+}$	720
$Sr_2LaAlO_5:Ce^{3+}$	700
$LaAl(Si_{6-x}Al_x)N_{10-x}O_x:Ce^{3+}$	475
a-Sialon $(M_xSi_{12-m-n}Al_{m+n}O_nN_{16-n}):Ce^{3+}$	495
$La_{2.82}Ce_{0.18}Si_8O_4N_{11}$	425
β-Sialon:Eu^{2+}	538
$BaYSi_4N_7:Eu^{2+}$	510
$SrYSi_4N_7:Eu^{2+}$	550
$Ca_2Si_5N_8:Eu^{2+}$	623
$Sr_2Si_5N_8:Eu^{2+}$	640
$Ba_2Si_5N_8:Eu^{2+}$	650
β-SiAlON:Ce^{3+}	520
Ca-α-SiAlON:Eu^{2+}	570
$(Sr, Ca)_2SiO_4:Eu^{2+}$	590

positive current, electrons and holes are injected into the emissive layer from opposite directions—from the cathode and anode, respectively. The metal cathode is usually an alkaline earth or Al, which readily release a valence electron (*i.e.*, possess a low work function). The holes that migrate from the anode are blocked from further transport to the cathode by an organic layer of 2,9-dimethyl-4,7-diphenyl-1,10-phenanthroline (BCP, Fig. 4.74c), which has a relatively low-lying HOMO.

In contrast, the anode is usually tin-doped In_2O_3 (ITO, typically 1:9 Sn:In), since this material is transparent and highly conductive (ca. 1×10^4 W^{-1} cm^{-1}).[79] The substitutional replacement of In^{3+} ions with Sn^{4+} ions results in n-doping of the lattice, injecting electrons into the conduction band. The doping of Sn into the In_2O_3 lattice may also result in SnO sites; this introduces holes in the lattice that reduces its conductivity. The conductivity of ITO is due to both Sn dopants and O vacancies, as represented by the general formula $In_{2-x}Sn_xO_{3-2x}$.[80] Since the organic hole-transport polymer is in direct contact with ITO in an OLED, the surface properties of ITO is important for effective hole injection. It has been shown that surface treatments such as UV ozone cleaning or Ar/O_2 plasma treatments, result in an increase in the work

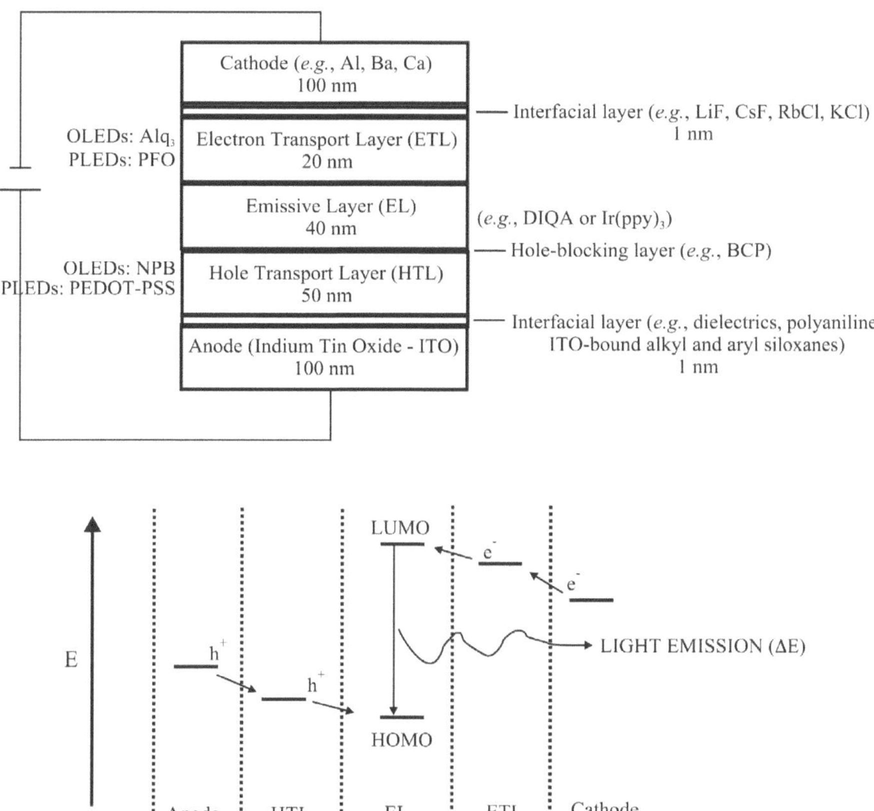

Fig. 4.73 Multilayered structure of OLEDs/PLEDs. Also shown are the relative energy levels for individual layers; light is emitted as a result of the radiative recombination of electron–hole pairs

function of the ITO surface as the Sn:In ratio is decreased (and oxygen concentration is increased).

The recombination of a hole and electron creates a quasiparticle known as an *exciton*, which releases a photon of energy. Organic molecules in the emissive layer facilitate exciton formation—likely through hole-trapping, followed by Coulombic attraction to the free electrons. A challenge in OLED design is to ensure that an equal number of holes and electrons meet in the emissive layer. This is not trivial since holes migrate much slower than electrons in conductive organic molecules. The organic molecules used in OLEDs/PLEDs are π-conjugated, meaning that p_z orbitals on adjacent $(\text{–C= C–C= })_n$ units overlap, resulting in π valence and π^* conduction bands. The observed color will depend on the HOMO–LUMO gap of the molecule, which may be fine-tuned by altering the length of the conjugation, or nature of the molecular backbone (e.g., substitution of electron withdrawing/donating groups, etc.). A wide variety of light-emitting small molecules and polymers are now available that are able to emit light across the entire visible wavelength range.[81]

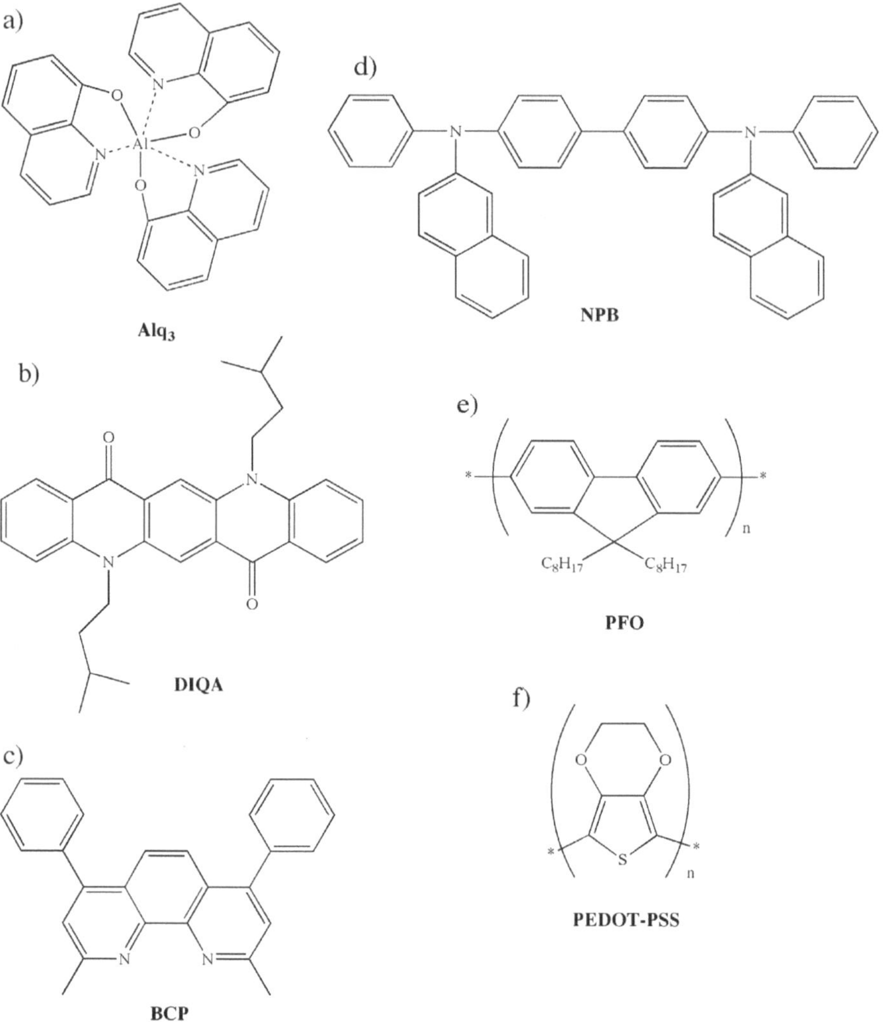

Fig. 4.74 Molecular structures of commonly used OLED/PLED materials. Shown are: **a** Alq_3 (tris(quinoxalinato)Al (III)) used as an electron-transport material; **b** DIQA (diisoamylquinacridone) used as an emissive dopant; **c** BCP (2,9-dimethyl-4,7-diphenyl-1,10-phenanthroline) used as an exciton/hole blocking agent; **d** NPB (1,4-bis(1-napthylphenyl amino)biphenyl); **e** PFO (9,9-dioctylfluorene) used as an emissive polymer in PLEDs; (f) PEDOT–PSS (poly-3,4-ethylenedioxythiophene–polystyrene sulfonate) used as a hole transport material in PLEDs

A limitation of traditional OLEDs/PLEDs is their relatively poor quantum efficiency. An exciton may exist in either a singlet (total spin, $S = 0$) or triplet state ($S = 1$), with the triplet outweighing the singlet by a 3:1 ratio. The molecules in the emissive layer are referred to as fluorophores, which yield fluorescent radiation only when a singlet exciton is formed. Hence, the theoretical maximum efficiency that is possible is only 25%. In an effort to improve the efficiency, a number of recent studies have been devoted to using organometallic compounds (transition metals with organic ligands), which allow for rapid intersystem crossing (ISC) from the excited singlet to light-emitting triplet states (Fig. 4.75). The process of collecting the excitons in the lowest excited triplet state is referred to as *triplet harvesting*, and results in a 100% theoretical quantum efficiency of the OLED.[82] This effect is most pronounced for complexes of 4d and 5d transition metals with well-shielded valence electrons, which exhibit large *spin–orbit coupling*[83] and concomitant mixing of singlet and triplet states.[84]

Organic LEDs may be passive or active, based on what governs the on/off switching of individual pixels: either external circuitry or a thin-film transistor (TFT), respectively (Fig. 4.76).[85] Analogous to MOSFETs (or, more generally, MISFETs—metal–insulator field-effect transistor) discussed earlier, thin-film transistors consist of source, drain, and gate electrodes alongside a dielectric layer and active semiconductor.[86] Whereas passive-matrix OLEDs (PMOLEDs) are easy and inexpensive to produce, active-matrix OLEDs (AMOLEDs)—most prevalent for television and cell phone displays—consume less power, have faster refresh rates, and are more amenable for larger displays with higher resolutions. The heart of TFTs has long featured amorphous Si (a-Si)[87]; however, the ultrafast frame rates (>240 Hz) that current LCD or OLED displays require necessitate the use of alternate materials with a higher electron mobility (>3 $cm^2\ V^{-1}\ s^{-1}$) such as amorphous oxide semiconductors (e.g., In_2O_3, Ga_2O_3, ZnO, SnO_2; either as discrete oxide layers or as multilayer heterojunctions and quasi-superlattices).[88] The threshold voltage that is required to turn on an a-Si TFT tends to drift over time, which worsens if deposited onto a flexible plastic substrate. In comparison, transparent amorphous oxide semiconductors exhibit much higher electron mobilities (3–50 $cm^2\ V^{-1}\ s^{-1}$ *vs.* 1 $cm^2\ V^{-1}\ s^{-1}$ for a-Si), more stable currents, and lower-temperature deposition methods than a-Si based TFTs.[89]

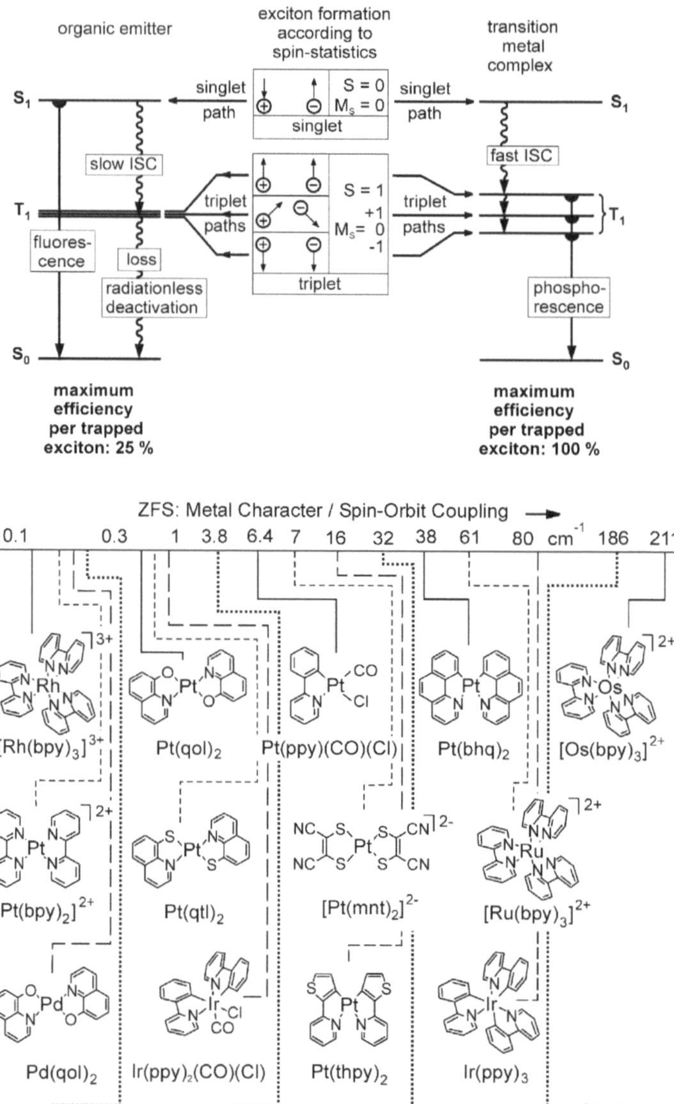

Fig. 4.75 Illustration of triplet harvesting. In the absence of a triplet emitter, the triplet excitation energy is converted to heat, losing 75% of the quantum efficiency. Also shown are a variety of organometallic complexes, and their relative spin–orbit coupling values—directly proportional to their use in phosphorescent OLEDs. Reproduced with permission from Yersin, H. *Top. Curr. Chem.* **2004**, *241*, 1. © 2004 Springer Science and Business Media

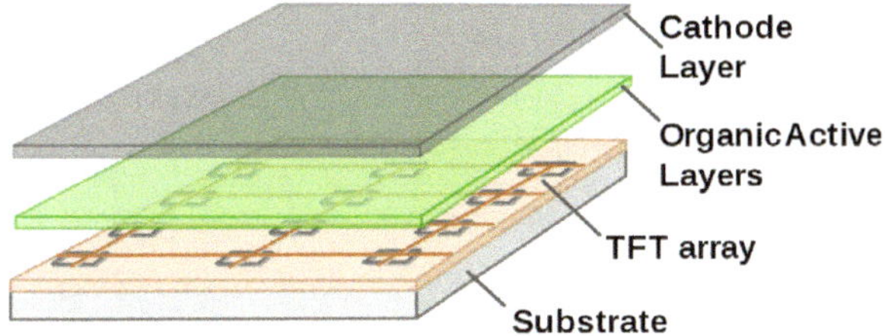

Fig. 4.76 Schematic cross-section of an organic light-emitting diode (OLED) panel, driven by an active matrix, thin film transistor (TFT) array

4.4 Thermoelectric (TE) Materials

As our society attempts to wean itself from a dependence on fossil fuels, many alternative energy sources are being investigated. One interesting potential is from the conversion of waste heat (e.g., vehicle engine heat) into electricity. This conversion is due to the *thermoelectric effect*, first discovered for junctions (or *thermocouple*) of two dissimilar metals. When metals are welded together, free electrons are able to drift across the junction in a preferential direction, based on the different atomic lattice structures of each metal. The migration of electrons effectively leaves one metal electron-deficient (positively charged), and the other metal negatively charged.

If the two metals are held at different temperatures, a voltage will result that is proportional to the ΔT. The ratio ($\Delta V/\Delta T$) is referred to as the *Seebeck coefficient*, α, related to the band structure of the materials involved. The α value for semiconductors is at least two orders of magnitude larger for semiconductors relative to metals, giving rise to much greater voltages. Contrary to the Seebeck effect, when a current is passed through a closed circuit of the two metals, heat is produced at one of the junctions and is absorbed at the other. Since the latter junction is cooled, this *Peltier effect* has been largely exploited for thermoelectric cooling applications (e.g., auto seat coolers, computer component cooling).

A TE device consists of a heat source and sink, joined together via n-type and p-type semiconductor materials (Fig. 4.77).[90] The *figure of merit*, ZT (4.22), of the device is useful to determine its suitability for power generation or refrigeration—a means to describe the transport properties of the TE material. An effective TE material should have a high Seebeck coefficient (heat conversion efficiency) and electrical conductivity, as well as low thermal conductivity to maintain thermal isolation in the device. The thermal conductivity is related to the transfer of heat through a material through either electron transport or quantized lattice vibrations (phonons). Hence, the ideal TE material has been described as a phonon-glass/electron-crystal, having the electrical properties of a crystalline lattice, and thermal properties of an amorphous/glass-like solid.

$$\mathrm{ZT} = \frac{\alpha^2\sigma}{\lambda}, \tag{4.22}$$

where α is the Seebeck coefficient (μV K^{-1}; 1–10 for metals, 150–250 for semiconductors), σ is the electrical conductivity, and λ is the thermal conductivity (electronic + lattice terms).

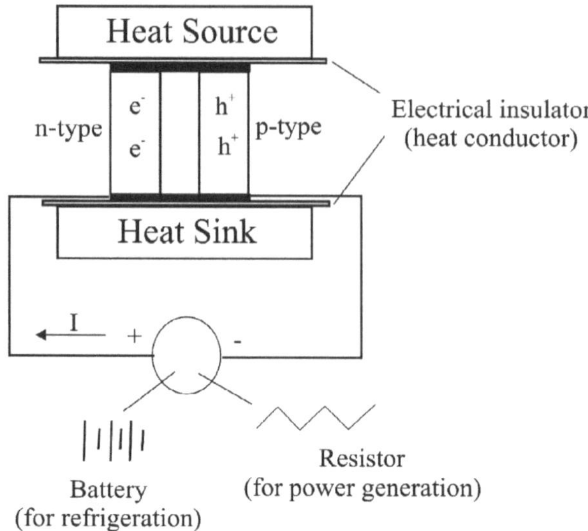

Fig. 4.77 Schematic of a thermoelectric device that may be used for power generation or refrigeration

Semiconductors are much more effective TE materials (greater ZT values) than metals due to their significantly greater Seebeck coefficients and lower thermal conductivities. A barrier toward higher ZT values for semiconductors is their relatively small electrical conductivity, especially at low temperatures. In order to overcome this limitation, the chemical composition of the semiconductor may be fine-tuned to yield a small bandgap material ($\geq 6\ k_BT$; k_B = Boltzmann cst; *i.e.*, ≥ 0.16 eV at 300 K), or one that is sufficiently doped with an intermediate concentration (ca. 10^{19}–10^{21} cm^3) of electronic/thermal carriers exhibiting high mobility through the lattice.

The most widely studied TE material to date is Bi_2Te_3, consisting of a hexagonal unit cell with repeating…[Te–Bi–Te–Bi–Te… Te–Bi–Te–Bi–Te]… units (Fig. 4.78). Whereas Te–Bi layers are bound by strong covalent interactions, the bonding between adjacent Te layers is through weak van der Waals interactions. This results in bulk anisotropic electrical and thermal conductivity, being most pronounced along planes that are perpendicular to the *c*-axis of the unit cell. To further improve ZT values, Bi_2Te_3 crystals may be doped with n- or p-type dopants, with most desired compositions of $Bi_2Te_{2.7}Se_{0.3}$ (n-type) and $Bi_{0.5}Sb_{1.5}Te_3$ (p-type), yielding ZT = 1 at room temperature. This improvement results from a decrease in thermal conductivity of the lattice, brought about by phonon scattering by the dopant atoms (*i.e.*, perturbing the symmetry of the lattice, affecting the organized lattice vibration modes).

The following list represents the primary materials classes that have been designed in an effort to optimize ZT for thermoelectric applications.[91] The general strategy is to dope the lattice with sufficient carriers (n- or p-dopants), while also interrupting the phonon transport through the solid through the introduction of large interstitial atoms.

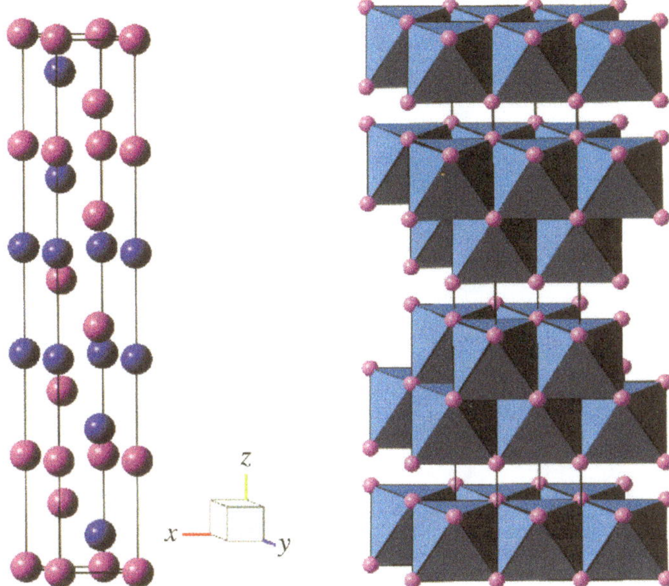

Fig. 4.78 Unit cell of Bi_2Te_3, a widely studied thermoelectric material. The blue atoms are Bi, and the pink atoms are Te. Reproduced with permission from Tritt, T. M.; Subramanian, M. A. *MRS Bull.* **2006**, *31*, 188. © 2006 Materials Research Society

i. *Complex Solid-State Inorganic Lattices* (e.g., $CsBi_4Te_6$, $Bi_{2-x}Sb_xTe_{3-y}Se_y$, $ZrTe_5$, $Ag_nPb_mM_nTe_{m+2n}$ (M = Sb, Bi), $AgPb_{10}SbTe_{12}$, "Half-Heusler alloys" MNiSn (M = Zr, Hf, Ti), and $Zr_{0.5}Hf_{0.5}Ni_{0.5}Pd_{0.5}Sn_{0.99}Sb_{0.01}$). Halide perovskites, previously discussed for LED applications, have also become recently relevant for TE applications.[92] For half-Heusler alloys, the unit cell is a combination of a NaCl lattice of two metals, with the third metal occupying tetrahedral interstitial sites. For instance, for TiNiSn, the Ti and Sn form the NaCl lattice, with Ni occupying 1/2 of the available tetrahedral interstitial sites (*i.e.*, Ti:Ni:Sn = 4:4:4 atoms per unit cell). This combination offers a great deal of control over electronic/thermal conductivity of the solid. While the introduction of Sn (p-doping) increases electrical conductivity, heavy metal atoms such as Ti and Ni cause a decrease in thermal conductivity through phonon scattering.
ii. *Crystal Structures with "Rattlers"* (e.g., rare-earth[93] (e.g., La, Ce, Nd, Sm, Eu, etc.) doped AB_3-based (A = Co, Fe, Ru, Os, Ir; B = P, As, Sb), Fig. 4.79),[94] or complex antimonide skutterudites (e.g., β-Zn_4Sb_3, $Yb_{14}MnSb_{11}$, $A_yMo_3Sb_{7-x}Te_x$),[95] metal organic frameworks, or clathrates (e.g., $Ba_8Ga_{16}E_{30}$, E = Ge, Si).[81] In these structures, dopant atoms are weakly bound to the cage, and "rattle" in response to increasing temperature. As the atom within the cages becomes smaller/heavier, the amount of structural disorder will increase causing a larger decrease in lattice thermal conductivity.

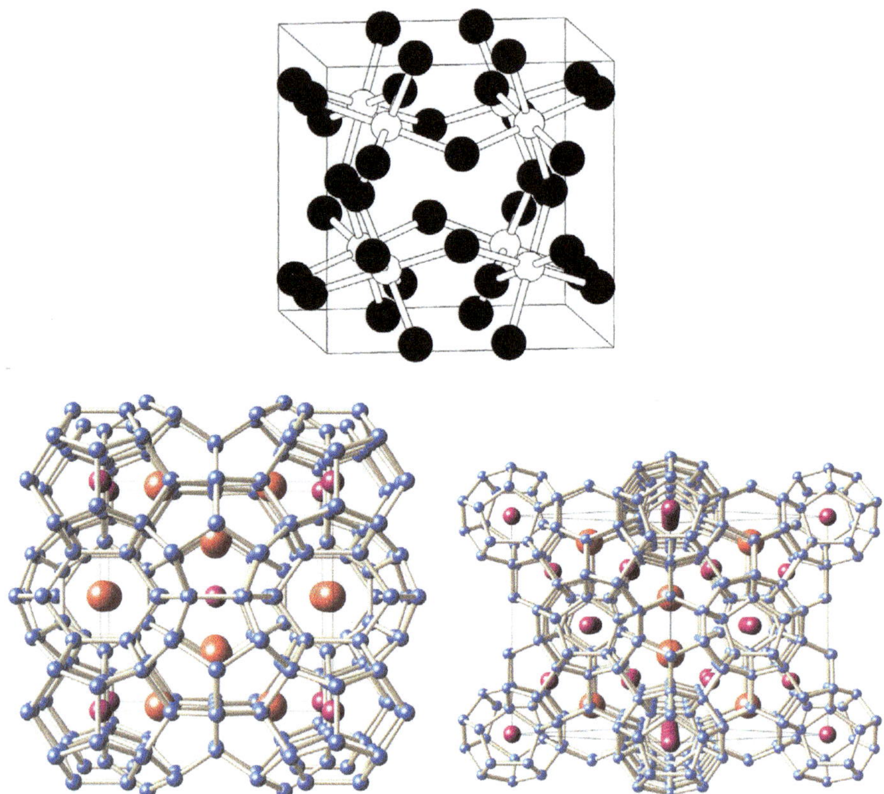

Fig. 4.79 Crystal structures of an unfilled $IrSb_3$ skutterudite (Sb atoms shown in black, top) and Type I/Type II clathrates (bottom). In the skutterudite structure, a void is present in the center of the unit cell, surrounded by 12 Sb atoms. For the clathrate structures, tetrahedrally bound framework atoms (e.g., Ge, Sn, Si) are illustrated in blue, and guest atoms within the various cages are shown in orange and pink. Reproduced with permission from (i) Nolas, G. S.; Poon, J.; Kanatzidis, M. *MRS Bull.* **2006**, *31*, 199. © 2006 Materials Research Society and (ii) *Chem. Mater.* **2000**, *12*, 697. © 2000 American Chemical Society

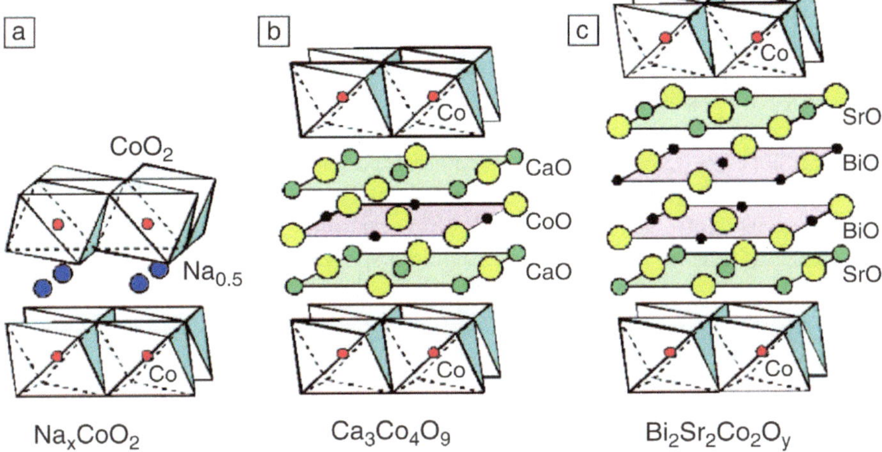

Fig. 4.80 Crystal structure schematic of CoO_2-based TE oxides. Shown are **a** Na_xCoO_2, **b** $Ca_3Co_4O_9$, and **c** $Bi_2Sr_2Co_2O_y$. Reproduced with permission from Koumoto, K.; Terasaki, I.; Funahashi, R. *MRS Bull.* **2006**, *31*, 206. © 2006 Materials Research Society

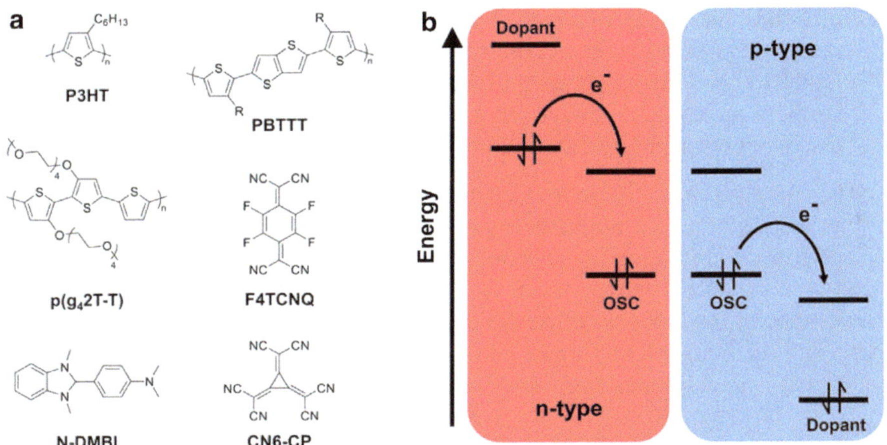

Fig. 4.81 Some examples of semiconducting polymers and dopants investigated for TE applications. Also shown are energy-level diagrams illustrating n- and p-type doping of an organic semiconductor by charge-transfer from a dopant molecule. Reproduced with permission from *Front. Electron. Mater.* **2021**, *19*, 1

iii. *Oxides* (e.g., $NaCo_2O_4$, $Ca_3Co_4O_9$,[96] $Al_{0.02}Zn_{0.98}O$, β-$SrRh_2O_4$[97]). These structures consist of CoO_x layers, which serve as effective electronic transport media within the lattice. A variety of metal/metal oxide layers (e.g., Na, Ca, Sr, Bi) are placed in between the metal oxide layers to serve as phonon scatterers (Fig. 4.80).

iv. *Tin and Lead Chalcogenides* (e.g., PbSe, PbS, PbTe, SnSe).[98] These materials exhibit high figure-of-merits for both n-type and p-type compositions. Tin(II) selenide (SnSe) is arguably the least thermally conductive crystalline material, due to its "soft", accordion-like layered structure. This results in poor phonon transport, but very effective electronic transport.[99]

v. *Nanoarchitectures* (e.g., PbTeSeTe/PbTe quantum dot superlattices, PbSe nanostructures,[100] Bi_2Te_3 nanowires,[101] nanoporous silicon,[102] Na-doped PbTe-PbS nanoparticles,[103] endotaxial nanostructural precipitates,[104] carbon nanotubes,[105] and graphene[106]). The electrical and thermal conductivities of a bulk material are strongly coupled, which makes it difficult to increasingly improve ZT values. However, in a solid with at least one dimension in the nanoregime, one may dramatically alter the carrier-transport properties by simply varying its dimensions due the confinement of the carriers inside the material.[107] We will discuss quantum confinement within nanostructures in more detail in Chap. 6.

vi. Organic semiconductors, including semiconducting polymers and small molecules (Fig. 4.81).[108] The complexity of polymer structures and phases, especially in composites, have greatly limited the development of this class of materials for TE applications. There is still a need to understand the role of the conjugated polymer backbone, side chains, and overall molecular weight on the thermoelectric properties of the polymer. Since polymer processing also strongly affects final properties, there is a need for more work in this area to discover promising candidates for future TE applications.

The comparison of ZT values for a host of the above bulk materials is displayed in Fig. 4.82, along with the variation of ZT with temperature. The search for new TE materials continues to be an extremely active area of research; recent work has resulted in suitable TE components at virtually all temperature regimes. However, the search continues for materials with ZT values in excess of 2.0 at RT. Although most bulk materials exhibit figure-of-merits ≤ 1.5, certain types of nanomaterials are the only candidates to date with ZT values >2.0.[109]

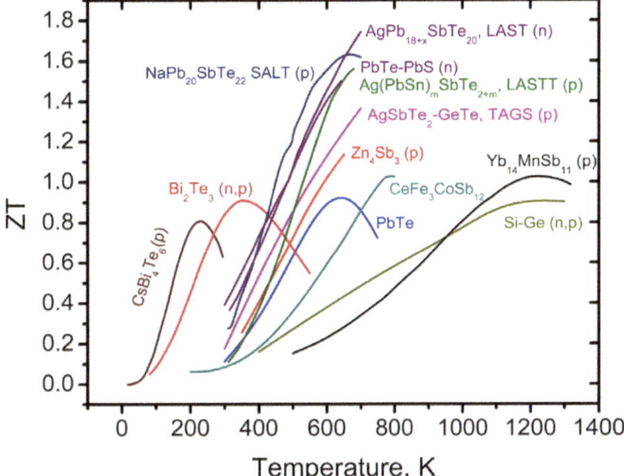

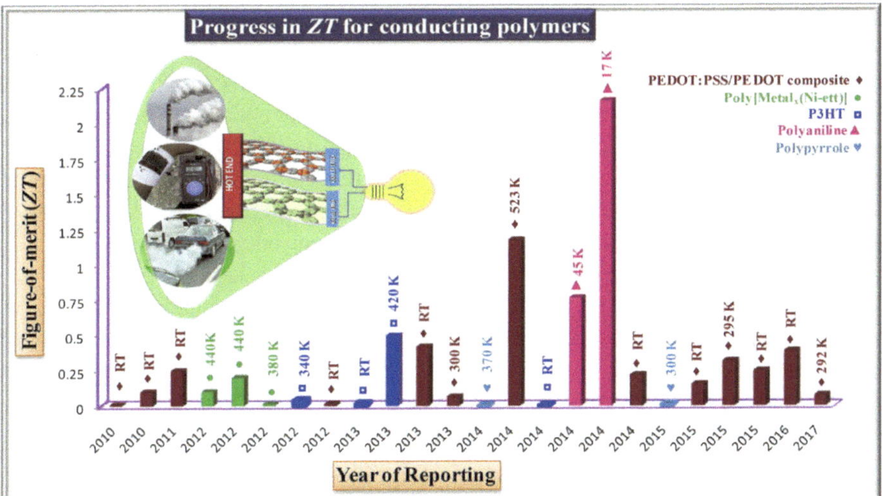

Fig. 4.82 The temperature dependence of ZT values for various TE materials. Shown are: (Top) inorganic-based crystals. Reproduced from (http://chemgroups.northwestern.edu/kanatzidis/greatthermo.html) © Kanatzidis Research Group. (Bottom) organic semiconductors. Reproduced with permission from https://doi.org/10.1016/j.pmatsci.2017.09.004 © 2018 Elsevier

4.5 Semiconductors Case Study: Photovoltaic (Solar) Cells

The increasingly volatile prices of gasoline and natural gas due to natural disasters and geopolitical factors have brought about a heightened awareness of our dependence on nonrenewable resources. Currently, ca. 80% of the total energy demand in the U.S. is powered by the combustion of fossil fuels (*i.e.*, coal, oil, and natural gas).

These fuels provide electricity and heat to homes and businesses, as well as power our vehicles and other modes of public transportation.

While the U.S. coal reserve is estimated at 252 billion short tons[110] (enough to last >250 years at current production and consumption levels), there is an increased concern about the adverse environmental effects associated with its combustion, namely SO_x, NO_x, and CO_2 emissions. With the energy needs of our world likely to double within our generation, there continues to be active development of alternative renewable energy sources. In an earlier chapter, we dealt with hydrogen-based fuel cells; we now will describe another extremely attractive option—harvesting the unlimited power from the sun.

The majority of commercial photovoltaic cells utilize silicon-based technology. When sunlight comes in contact with a p–n diode, the absorbed energy causes promotion of electrons from valence to conduction bands, generating additional electron–hole pairs (EHPs). Since the n-side is much thinner than the p-side, most of the photons are absorbed within the depletion region and p-side. The most noticeable result of this excitation is an effective increase in the number of electrons in the conduction band of p-Si (or holes in n-Si, referred to as the *minority-carrier concentration*). As a result, electrons in the p region will diffuse into the depletion region, where the junction potential (Fig. 4.83, E_o) propels them back into n-Si (vice versa for holes). It is important to note that only EHPs that are generated within the depletion region or within a certain distance from this region (Fig. 4.86; electrons within diffusion length L_e, or holes within diffusion length L_h) may contribute to photocurrent. If the minority carriers are further away from the depletion zone, recombination will occur rather than migration to their respective electrodes (*i.e.*, electrons in n-Si flowing through the external circuit to neutralize the excess holes in p-Si).

In the design of solar cells, a small bandgap semiconductor is desirable since it requires less energy to promote electrons from valence to conduction bands. Although the resultant photovoltage of the solar cell is directly proportional to the bandgap, higher energy photons would be converted to heat rather than electrical energy. Hence, the use of a higher bandgap material would result in a lower photocurrent since only high-energy photons ($E \geq E_g$) would be absorbed. The bandgaps of Si and GaAs (1.1 eV (1127 nm) and 1.4 eV (886 nm), respectively) are able to absorb a broad range of electromagnetic radiation. However, single-junction Si or GaAs solar cells, with commercial efficiencies typically in the 22–26% range, may only have a maximum power conversion (PCE) efficiency of 31%—known as the Shockley/Queisser limit.[111] Thin-film solar cells that utilize amorphous silicon (a-Si) are also currently under investigation. Since Si atoms do not form a continuous crystalline array, surface sites contain significant numbers of dangling bonds, which are easily passivated through reaction with hydrogen (yielding hydrogenated amorphous silicon, a-Si:H). The larger bandgap of a-Si (1.7 eV) is more effective at capturing broadband solar energy; however, the best efficiencies to date are less than 15% for this technology.[112]

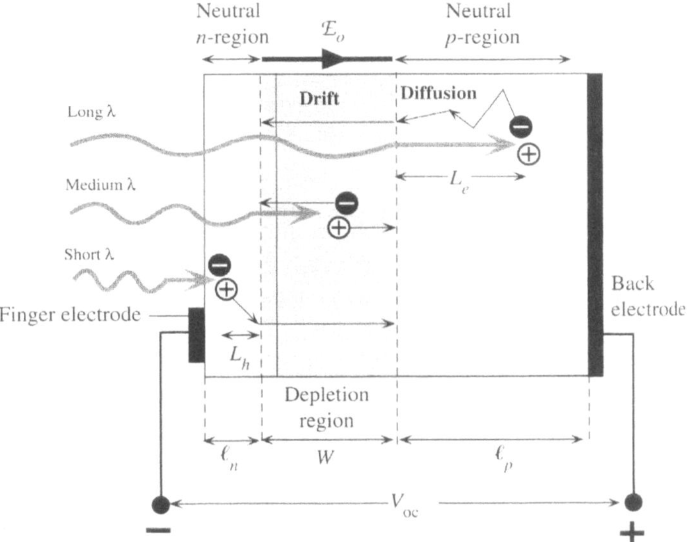

Fig. 4.83 Schematic of a single-junction photovoltaic cell. Reproduced with permission from Kasap, S. O. *Principles of Electronic Materials and Devices*, 3rd ed., McGraw-Hill: New York, 2007. © 2007 The McGraw-Hill Companies

In order to capture a larger range of photon energies, multijunction photocells have recently been introduced. This method consists of a stack of semiconductor layers with decreasing bandgaps (Fig. 4.84). Top layers have higher bandgaps and are able to absorb higher energy photons; lower layers of the cell absorb the transmitted lower energy wavelengths. A 6-junction solar cell (Fig. 4.85) developed by the National Renewable Energy Laboratory (NREL) has resulted in an efficiency of 47.1%, the highest value reported for all photovoltaic cells to date.[113] It should be noted that multijunction solar cells are limited by a number of intrinsic and extrinsic losses. Extrinsic sources include reflection, series resistance, absorption within interlayers, nonradiative recombination, and many others. Intrinsic sources include inefficient collection of solar photon energies by each layer, and radiative recombination.[114] It has been proposed that replacement of GaAs with a material with an E_g of 1.25 eV would result in a higher overall efficiency, since more photons would be collected by the second layer and fewer would be transmitted to the bottom Ge substrate. With additional tweaking such as altering the number of layers, individual layer thicknesses/stoichiometries, it is expected that multijunction photocells with efficiencies of >50% may soon be realized.

While it is important to control the stoichiometry of each layer to adjust their bandgaps, it is equally important to have as few interfacial mismatches as possible. That is, each layer must be epitaxially grown to ensure that the lattice constants are perfectly matched. It has been shown that a lattice mismatch of only ca. 0.01% is

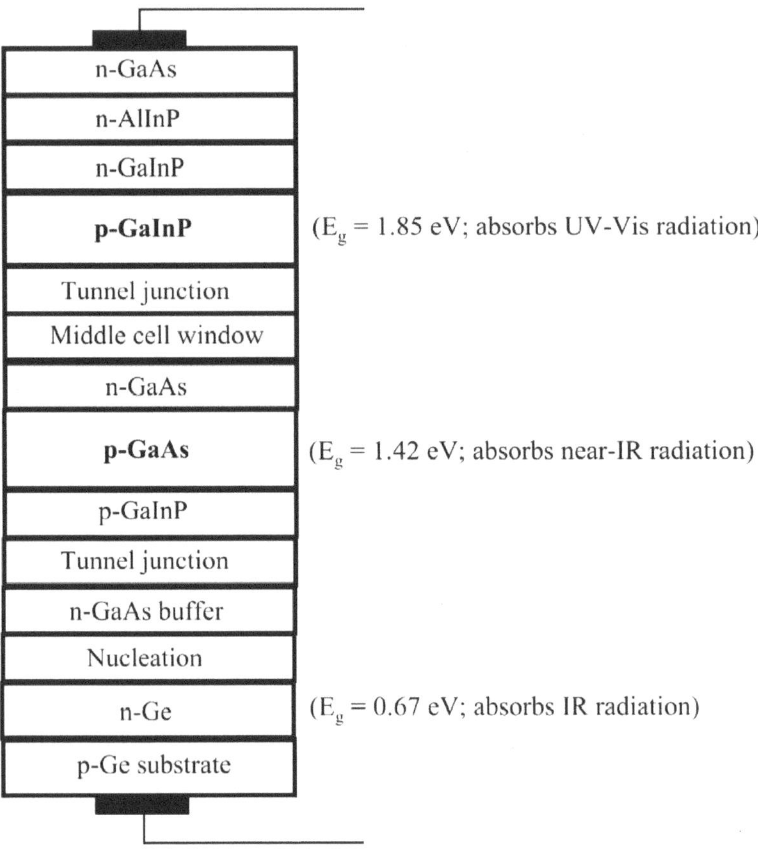

Fig. 4.84 Schematic of a multi-junction photovoltaic cell

enough to cause significant electron–hole recombinations, resulting in lower cell efficiency. CVD is the method of choice for the fabrication of these multilayer devices; as you might expect, cells of this variety are relatively quite expensive.

In an effort to decrease the cost of production for solar cells, *dye-sensitized solar cells* (DSSCs) have been widely investigated since the early 1990s.[115] Instead of light being absorbed by inorganic semiconductor layers, absorption occurs in an organic-based film that comprises light-harvesting dye molecules bound to a TiO_2 surface (Fig. 4.86). When the dye absorbs a photon, the resultant excitation injects electrons into the titanium, which transports them to the negative electrode, with the positive electrode attached to the electrolyte (e.g., aqueous iodide, aqueous polypyridyl copper complexes,[116] or sodium sulfide solutions[117]) acting as the hole-scavenging medium.[118] To date, the maximum efficiency of DSSCs is only 13.5% using a copper-based electrolyte and molecular photosensitizers (Fig. 4.87).[119]

The TiO_2 used in DSSCs is not present as a smooth film, but as a porous nanostructural sponge, with a very high surface area. This improves the number of active sites

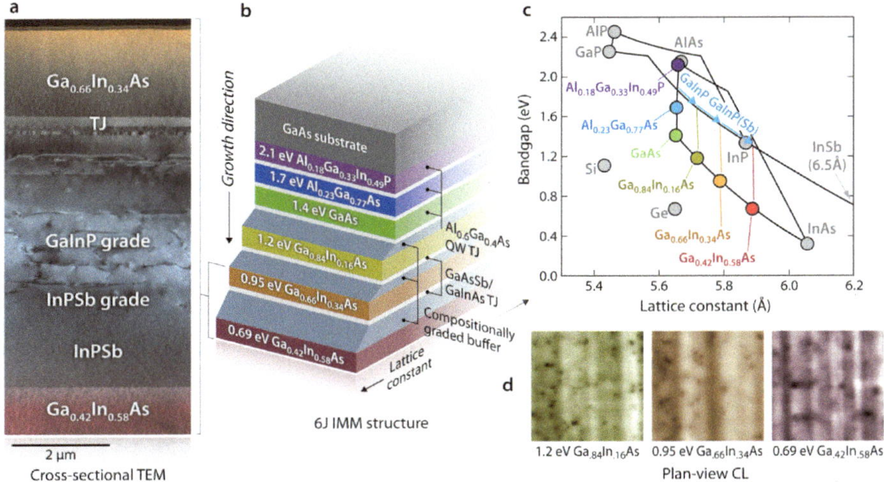

Fig. 4.85 A six-junction III-V solar cell with 47.1% conversion efficiency. Shown are: **a** Cross-section TEM image of selected layers, **b** simplified schematic of the multijunction solar cell, **c** bandgap vs. lattice constant in multilayer solar cell design, **d** cathodoluminescence images of GaInAs junctions. CL images are 65 × 65 μm. Reproduced with permission from *Nature Energy* **2020**, *5*, 326. © 2020 Springer Nature

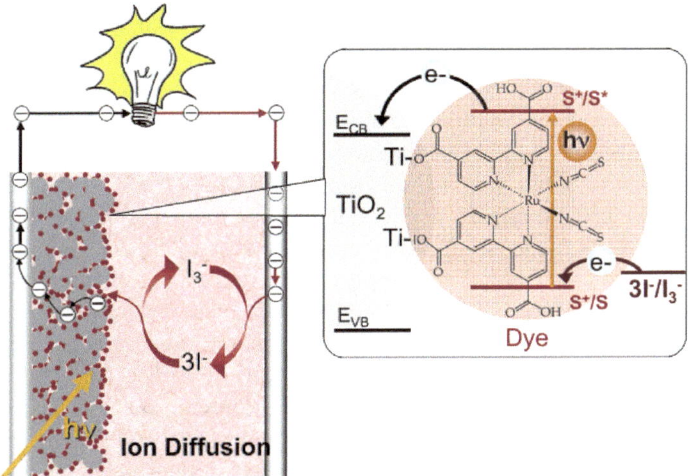

Fig. 4.86 Schematic of a dye-sensitized photovoltaic cell. The TiO_2-bound dye molecules act as the light harvester. Electrons are injected into TiO_2, flow to the collector electrode and through the circuit to the counter electrode. The dye is regenerated by electron donation from the $I_3^-/3I^-$ couple (0.536 V). Reproduced with permission from *Inorg. Chem.* **2005**, *44*, 6841. © 2005 American Chemical Society

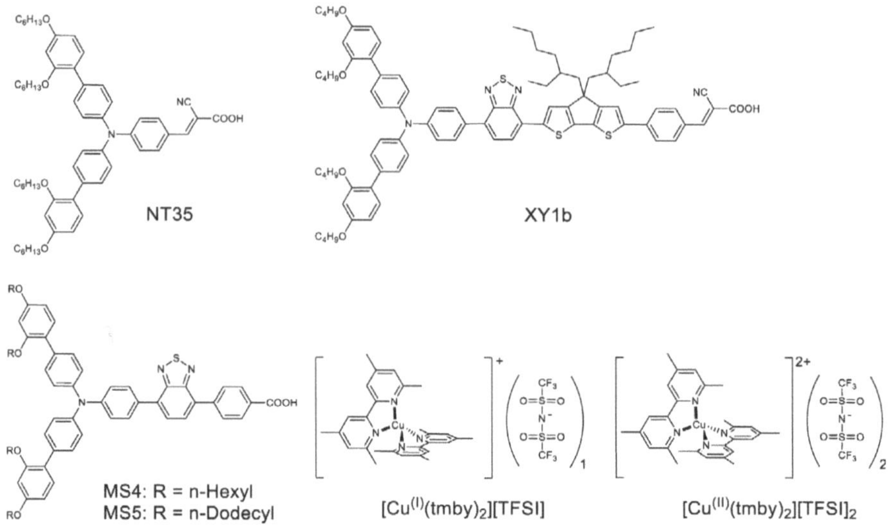

Fig. 4.87 Molecular structures of dyes (NT35, XY1b, MS4, and MS5), and $[Cu^{II}(tmby)_2][TFSI]_2$, tmby = 4,4′,6,6′-tetramethyl-2,2′-bipyridine; TFSI = bis(trifluoromethylsulfonyl)imide). Reproduced with permission from *Nature Communic.* **2021**, *12*, 1. © 2021 Springer Nature

that may absorb light, analogous to the stacked structue of leaves that convert light to energy using the light-harvesting molecule chlorophyll. The oxide nanoparticles are carefully prepared to preferentially expose the (101) face of anatase, which is the lowest energy surface—providing the lowest LUMO level to accept electron density from the sensitizer. It should be noted that other wide bandgap semiconductor oxides such as SnO_2 and ZnO have also been employed within DSSCs.

Not unlike their inorganic semiconductor counterparts, the light-harvesting molecule is perhaps the most important component of a DSSC. The dyes employed for this application are ruthenium complexes that feature highly conjugated ligands similar to those employed for OLED applications. The absorption of UV–Visible radiation causes $\pi \rightarrow \pi^*$ and metal-to-ligand charge transfer (MLCT) electronic transitions. The latter transition is quite pronounced for ruthenium, since its valence d-electrons are heavily shielded from the nuclear charge and are more easily promoted to ligand unoccupied orbitals. The donation of electron density from the bound sensitizer is due to MLCT which results in the oxidation of the Ru(II) metal center. The design of sensitizer ligands is not trivial; the π^* LUMO of the ligand must be of appropriate energy to allow electron transport to flow to the TiO_2 surface from the metal ion. However, it is also possible for an electron to be injected back to the sensitizer from the conduction band of the TiO_2 (Fig. 4.88a). By designing the sensitizer appropriately, the rate of electron flow from the dye to TiO_2 will be much faster than the reverse reaction. An intriguing way to accomplish this is through the incorporation of a hole-trapping group attached to the ligands (Fig. 4.88b). After MLCT takes place and the electron is injected into the TiO_2 layer, the positive charge is transferred to the hole-trapping moiety. This increases the distance of the hole from the surface, which slows the rate of TiO_2-sensitizer back-donation considerably.

Although there is ongoing research related to DSSCs,[120] interest in this technology is waning since conversion efficiencies have been greatly surpassed by quantum dot, organic, and perovskite solar cells, which have shown the largest efficiency improvements in recent years (Fig. 4.89). For instance, *quantum dot solar cells* were only first developed in 2010 by the National Renewable Energy Laboratory (NREL). ZnO/PbS quantum dots resulted in an efficiency of <2%; however, current examples now exhibit efficiencies of >15%.[121] The ability of quantum dots to absorb different regions of the UV-Visible range of electromagnetic radiation, based solely on the size of nanoparticle, is extremely desirable for PV efficiency. The fundamentals of quantum dot solar cells will be described in Chap. 6.

Organic solar cells rely on the use of conductive organic polymers or small molecules for light absorption and charge transport.[122] With current efficiencies of

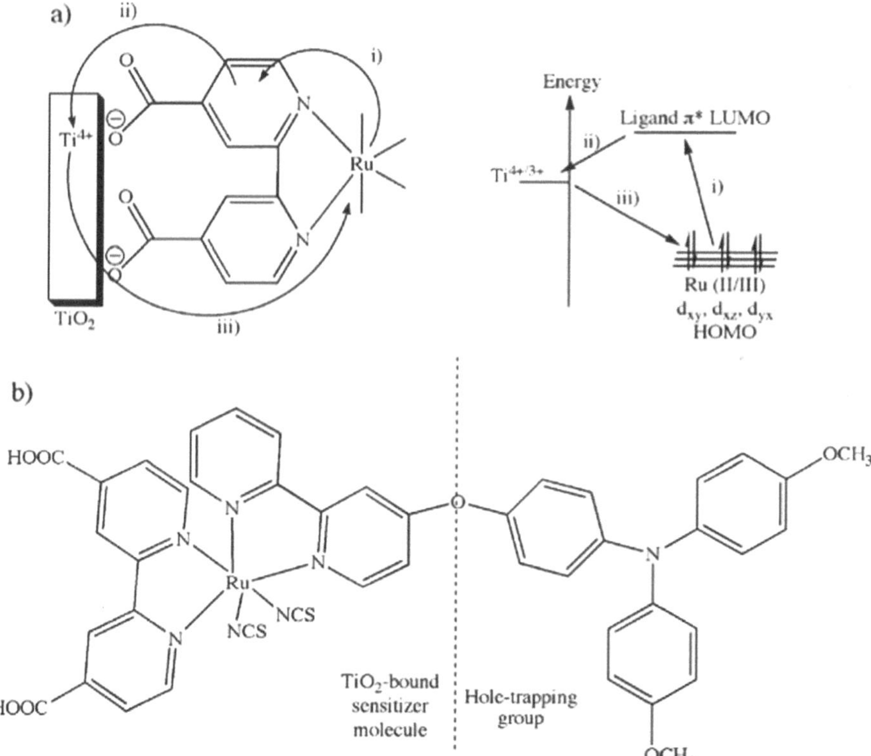

Fig. 4.88 **a** Detailed schematic of the forward/backward electron donation between the sensitizer and TiO_2 surface. A carboxylate group is shown on the sensitizer ligand—important for Lewis acid/base binding to the oxide surface. Shown in **b** is an example of a hole-trapping group that prevents back-donation from the TiO_2 surface to the sensitizer (improving the photocurrent generated by the DSC)

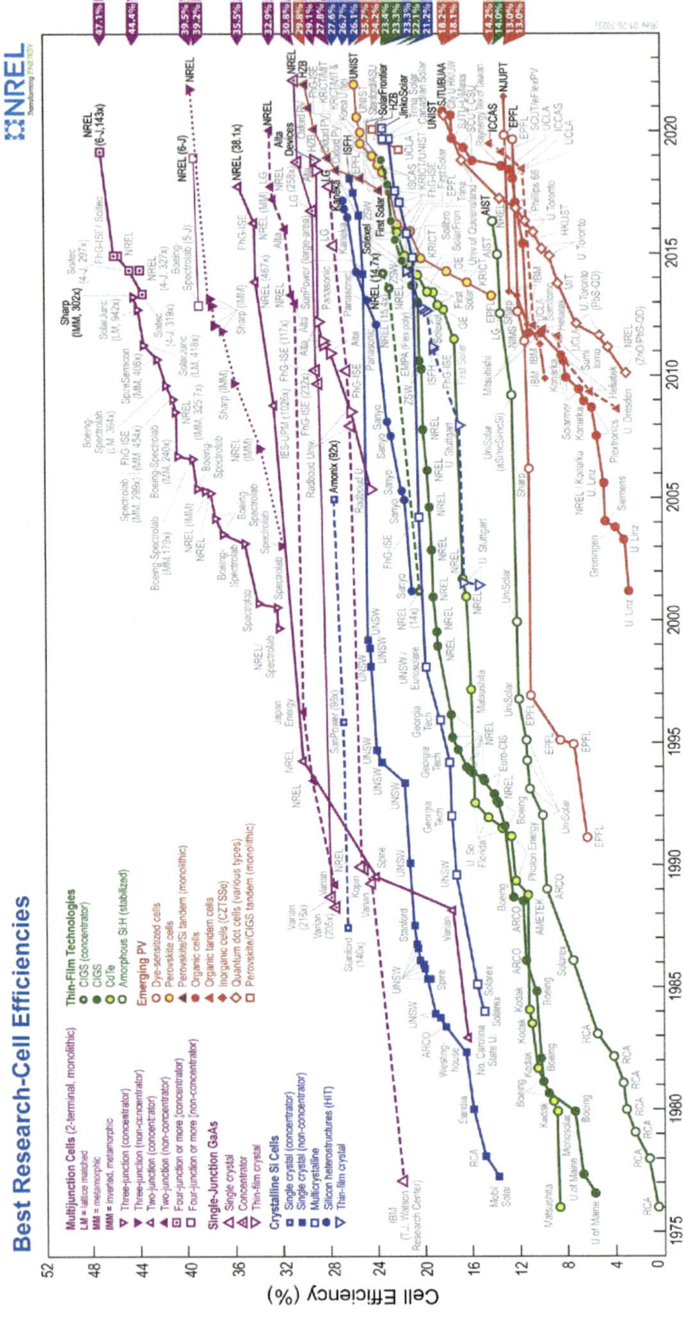

Fig. 4.89 Comparison of various cell efficiencies. © 2022 National Renewable Energy Laboratory (NREL)

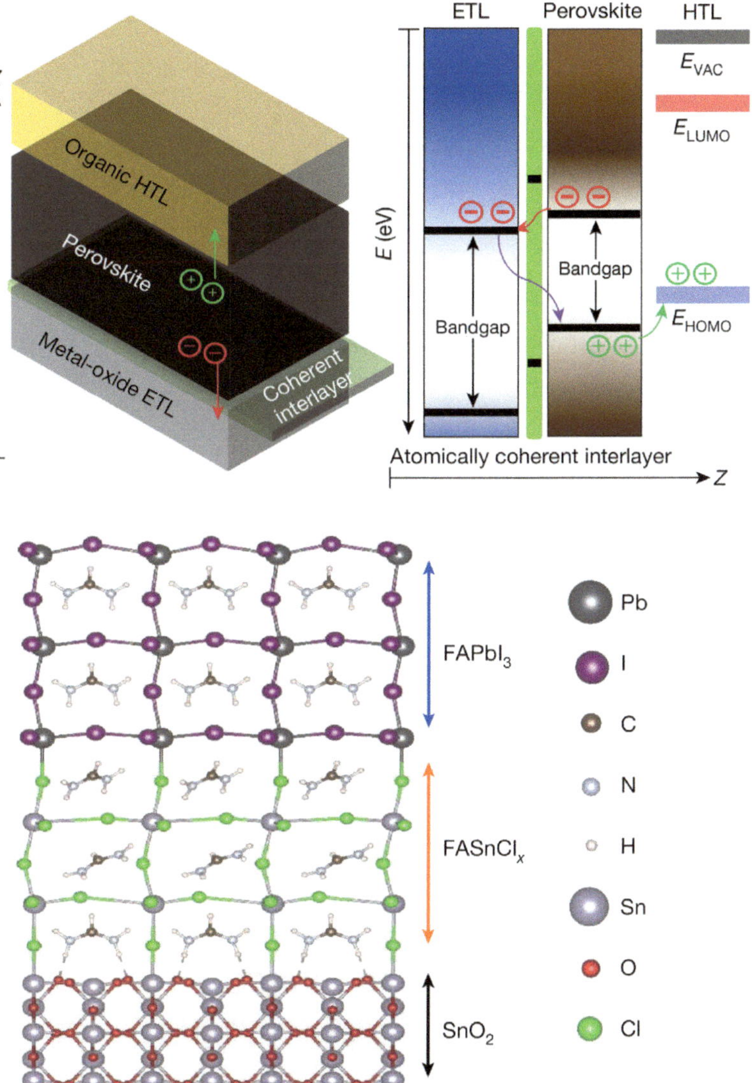

Fig. 4.90 (Top) Schematic and energy level diagram of a perovskite solar cell on a SnO_2 electrode. (Bottom) Simulation of the interlayer formed between the perovskite layer and SnO_2 substrate. Reproduced with permission from *Nature* **2021**, *598*, 444. © 2021 Springer Nature

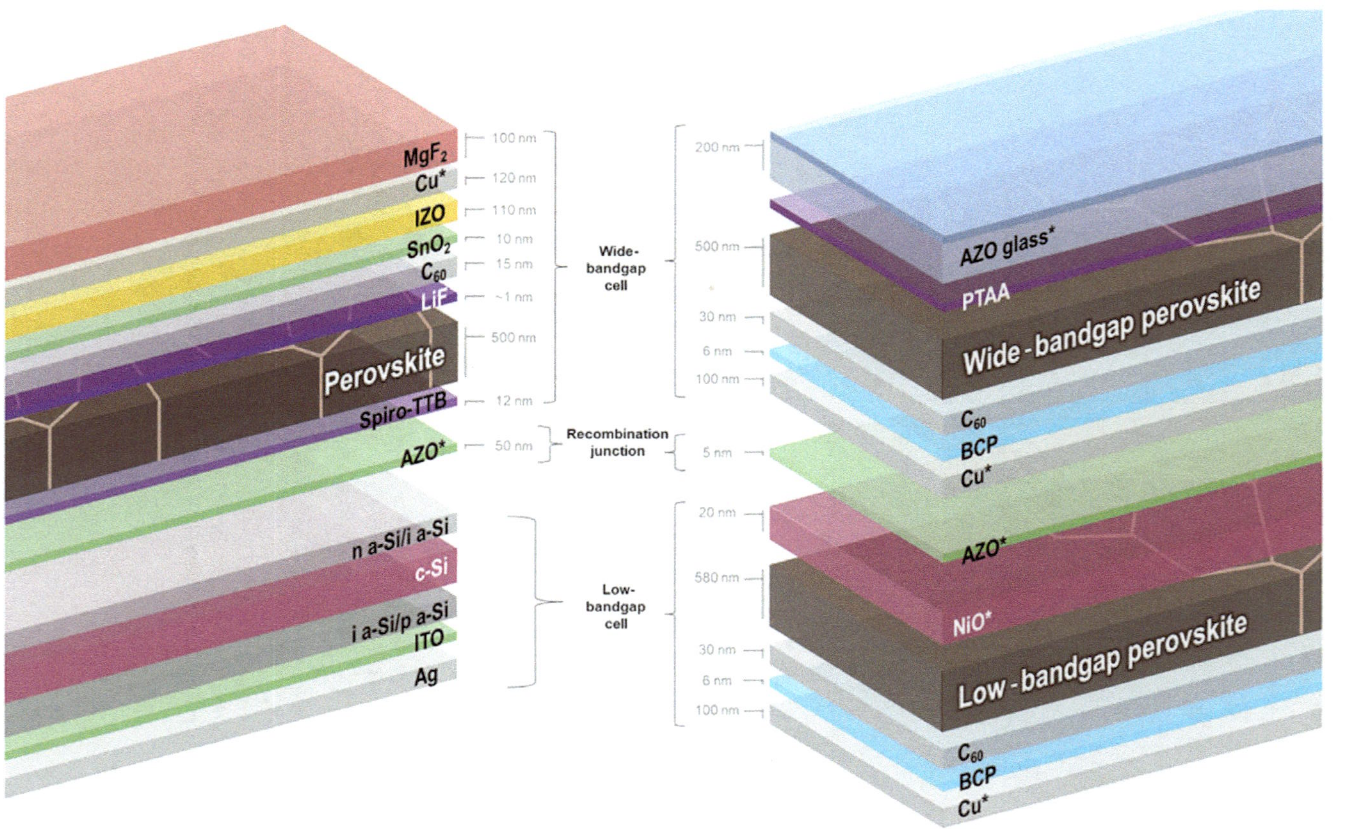

Fig. 4.91 Comparison of a perovskite-silicon tandem solar cell (left) and all-perovskite tandem solar cell (right). Reproduced with permission from *Sci. Adv.* **2020**, *6*, 1

18%,[123] the benefits of organic solar cells relative to Si-based photovoltaic devices is the low cost of fabrication, as well as being lightweight, flexible, and transparent. These properties will allow for these devices to be used in windows, walls, and portable/wearable electronics. The conjugated ring structure and delocalized π-electrons of the organic molecules allow for the formation and transportation of photogenerated charge carriers, with mobilities comparable to amorphous silicon.

With low processing costs and single-junction power conversion efficiencies of 25.5% after just a few years of investigation,[124] *perovskite solar cells* are perhaps the most favored emerging technology for photovoltaic applications (Fig. 4.90). Typical compositions feature ABX_3 perovskites where A is Cs^+, Rb^+, methylammonium ($CH_3NH_3{}^+$) or formamidinium ($HC(NH_2)_2{}^+$), B is commonly Sn^{2+}, Ge^{2+}, or Pb^{2+},[125] and X is a halide (Cl^-, Br^-, or I^-). Since the bandgap of metal halide perovskites may be tuned from 1.5–1.8 eV, one application is to design tandem junction solar cells, which place a wide bandgap material above a Si cell,[126] or use two perovskite materials with different bandgaps (Fig. 4.91).[127] This approach has been reported to increase the efficiency of the solar cell to ca. 30%.[128] Although the environmental implications of perovskite tandem solar cells have been controversial,[129] the greenhouse gas emission factor and energy payback time for all-perovskite tandem solar cells are reported to outpace those of crystalline silicon solar cells.[130]

Topics for Further Discussion

1. If a circular silicon wafer, with a diameter of 300 mm and thickness of 0.5 mm, is 99.9999999% pure, how many impurity atoms are present in the wafer?
2. Describe the steps involved in transforming sand to UHP silicon. What techniques are used to fabricate silicon wafers from the larger pieces of purified silicon sold by a company such as Hemlock Semiconductor?
3. What is meant by the current '10 nm' technology node (i.e., what distance does 10 nm refer to)? What is the next IC technology node, and when is this scheduled to be initiated?
4. What technology node is proposed as the 'end of the roadmap'? Beyond this limit, what can be done to further extend Moore's Law?
5. The cleanrooms that house photolithographic equipment must have yellow lighting. Why?
6. Can one over-expose a photoresist? If so, what will happen?
7. Provide examples (with literature references) for surface patterning techniques that do not use organic photoresists. Be sure to also include techniques other than "soft lithography".
8. Do any commercial processes currently utilize soft lithography techniques? If so, describe these applications.
9. Explain the benefits and disadvantages of using EUV as a development source for photolithography.
10. Describe the influence of the incident exposure wavelength on the resulting line resolution in photolithography. Would you expect the resolution to be better by using X-rays ($\lambda = 1.5$ Å) instead of EUV? Explain.

11. What is meant by "density of states," and how does this influence the electrical conductivity of a semiconductor?
12. Your supervisor has asked you to deposit the following thin films; what transition metal/main group complexes would you employ that would impart a high volatility of the overall complex, and low carbon incorporation in the growing film? Also, think of what coreactant gas(es) would be most appropriate for CVD. (a) ZrO_2 (b) Zr_3N_4 (c) $Zr_xHf_{1-x}O_2$ (d) PZT (e) Mg (f) Alq_3.
13. Explain how LEDs/OLEDs operate and think of some intriguing applications for these types of materials beyond smartphone and flat-panel displays.
14. Why is TiO_2 used in dye-sensitized solar cells (DSCs)? What other oxides would possibly be useful?
15. Explain how triplet harvesting improves the efficiency of OLEDs.
16. As discussed in this chapter, current CMOS ICs employ high-κ dielectric films as the gate insulator. However, a major roadblock to this replacement was the incompatibility with gate stacks—explain this problem and how it was circumvented.
17. Explain the differences between ALD, CVD, and PVD. Which one would be the method of choice to deposit a conformal silica (SiO_2) film of 1 nm (± 0.1 nm) onto a carbon fiber with a diameter of 200 microns? Rationalize your choice.
18. What are some benefits of finFET designs relative to traditional planar MOSFETs?
19. What are some differences between 'gate first' and 'gate last' process flows for finFET fabrication? Explain which route is preferred and why.
20. Describe how fin width, height, and pitch affect the performance of the finFET transistor.
21. How would you design a CVD system (reactor type/material; nature of the precursor, co-reactant, and carrier gas; precursor decomposition methodology; etc.) to deposit a Si_3N_4 film onto the surface of fumed alumina powder (particulate diameters of ca. 70 nm)? Use diagrams with full explanation of your design choices.
22. Explain the relative effect of growth rate *vs.* temperature for surface-reaction and mass-transport kinetically controlled thin film MOCVD processes. Why does the growth rate decrease for depositions taking place at very high temperatures?
23. In this chapter, there was a lot of discussion related to silicon-based devices and applications. As we attempt to continue Moore's Law into the future, how likely is it that Si will be replaced entirely? If this paradigm shift *was* proposed, what would be some negative consequences associated with such a move?
24. Why are chlorinated co-reactants used during the initial oxidation step of CMOS fabrication, and what are the mechanisms involved with its reactive participation?
25. How are the bandgap (E_g) and dielectric constant (κ) of a thin film determined experimentally?
26. Heusler alloys are intermetallics with the general formula M_2NiSn ($M = Zr$, Ti), with the unit cell being a NaCl lattice of two metals, with the third metal occupying tetrahedral interstitial sites. In contrast, "half-Heusler alloys" have

the structure MNiSn, with the third metal occupying only one-half of the interstitial sites. With this in mind, would "Heusler" or "half-Heusler" alloys be a more efficient thermoelectric material? Explain your choice.

27. Your stellar academic performance has resulted in a job interview for a very lucrative materials scientist position. As a test, they have given you the following LEDs: GaP, $Al_{0.5}Ga_{0.5}As$, GaAs, GaN, GaSb. Arrange these in order of observed color, from red - blue. Explain your rationale.
28. How are the bandgap and dielectric constant of a thin film determined experimentally?
29. Whereas the positive effects of tensile strain of a MOSFET channel is easy to visualize (increases the lattice parameter & the free mean path of electrons), the effects of channel compression strain seem counterintuitive. Explain how compression strain increases the conductivity of the channel region. Why is this type of strain not preferred for n-type Si?
30. The military has designed a futuristic soldier protective suit that would provide various functionalities such as: artificial muscles, vision enhancement, chemical/biological warfare agent protection, and temperature modulation. What strategy(ies) would you use to provide long-term electrical power to this suit, other than battery power? Note that the first missions to use this suit will be in the caves of Afghanistan.
31. What are the controversies surrounding perovskite tandem solar cells? What are some design strategies that could be used address these concerns?

References and Notes

1. http://ece-www.colorado.edu/~bart/book/book/chapter2/ch2_8.htm.
2. Hemlock Semiconductor homepage: http://www.hscpoly.com.
3. For example, see: (a) Sanjurjo, A. U.S. Patent 5006317. (b) Yoon, P.; Song, Y. U.S. Patent 4786477. (c) Boone, J. E.; Owens, D. W.; Farritor, R. E.; Blank, W. D. U.S. Patent 4806317.
4. Homoepitaxial growth refers to growing a film onto an atomically-flat substrate, wherein both film and substrate are compositionally equivalent. In contrast, heteroepitaxial growth would refer to growing a thin film of different composition onto a substrate (*e.g.*, GaAs thin film on a Si substrate). Such film growth occurs usually by vapor-phase techniques, which facilitates exact lattice matching of the crystal orientation and spacing of the growing thin film with the underlying substrate.
5. The cost of a single transistor is not possible to determine since multiple chips are produced at one type during IC manufacturing. This value is calculated based on the price of a M1 Ultra chip in 2022 ($350) divided by 114 billion (the number of transistors on the chip).
6. An article that describes some possible strategies beyond Moore's Law: http://www.economist.com/technology-quarterly/2016-03-12/after-moores-law.

7. Although the current should, in theory, be zero for a reverse-bias diode, there will still be a very small number of electrons/holes with enough energy to overcome the large junction potential, resulting in a very small current.
8. Whereas it is easy to conceptualize the uphill movement of electrons due to an applied voltage that exceeds the junction potential, it is not as straight-forward to rationalize hole migration. That is, they will move downhill from p-n regions during forward bias. A picture that may help visualize this is to think of holes as helium-filled balloons that are adhered to a ceiling. Energy would be required in order to pull them down; having a larger balloon with more helium would require even more energy (analogous to reverse-bias), whereas a small balloon would be easier to pull down (forward bias).
9. The V_T is related to the difference in the *work function* (*i.e.*, ϕ—the energy required to remove an electron from the surface of a material) between the gate and channel regions.
10. For a thorough description of the advantages and disadvantages of transistor scaling, see: (a) http://www-inst.eecs.berkeley.edu/~ee130/sp06/chp7full.pdf. (b) https://people.eecs.berkeley.edu/~tking/theses/bsriram.pdf.
11. Due to its high solubility in silicon lattices, boron was commonly used as a dopant for the p-MOSFET, being implanted as B^+ or $BF_2{}^+$ ions. However, boron has a greater solubility in SiO_2, which results in diffusion from the polysilicon gate into the SiO_2 gate insulating layer and even into the silicon channel. These positive charges within the oxide and channel regions causes an increase in scattering, thus reducing the carrier mobility and on-current.
12. It should be noted that the "technology node" no longer refers to physical dimensions of the transistor, but is rather an industry term related to a new fabrication process every 2 years, in accord with Moore's Law. By definition, the technology node refers to half the distance (half-pitch) between cells in a DRAM memory chip. It is noteworthy that the gate length, *L*, officially moved into the nanoregime in the year 2000. For more details about node nomenclature, see: (a) https://en.wikichip.org/wiki/technology_node. (b) https://www.theverge.com/2021/7/26/22594074/intel-acclerated-new-architecture-roadmap-naming-7nm-2025.
13. For instance, see: (a) Jones, A. C. et al. *J. Mater. Chem.*, **2004**, *14,* 3101. (b) Musgrave, C.; Gordon, R. G. *Future Fab International* **2005**, *18*, 126. (c) Vincenzini, P.; Marletta, G. *Adv. Sci. Technol.* **2006**, *51*, 156.
14. Locquet, J. P.; Marchiori, C.; Sousa, M.; Fompeyrine, J. *J. Appl. Phys.* **2006**, *100*, 051610.
15. (a) https://doi.org/10.1002/aelm.201800231. (b) https://doi.org/10.48550/arXiv.1809.07964. (c) https://doi.org/10.1088/1361-6528/aab9e6.
16. (a) G. A. Salvatore, D. Bouvet, A. M. Ionescu, *Tech. Dig.-Int. Electron. Devices Meet.* **2008**, 1. (b) A. Rusu, G. A. Salvatore, D. Jiménez, A. M. Ionescu, *Tech. Dig. Int. Electron. Devices Meet.* **2010**, 16.
17. S. Dasgupta, A. Rajashekhar, K. Majumdar, N. Agrawal, A. Razavieh, S. Trolier-Mckinstry, S. Datta, *IEEE J. Explor. Solid-State Comput. Devices Circuits* **2015**, *1*, 43.

18. (a) M. Lee, P. G. Chen, C. Liu, K. Chu, C. C. Cheng, M. J. Xie, S. N. Liu, J. W. Lee, S. J. Huang, M. H. Liao, *Tech. Dig. - Int. Electron. Devices. Meet.* **2015**, 22. (b) C . H. Cheng, A. Chin, *IEEE Electron Device Lett.* **2014**, *35*, 274.
19. C. Hsing, D. Kennedy, A. Sutherland, K. Van Vliet, *Phys. Status Solidi A* **1979**, *56*, 129.
20. (a) A. M. Ionescu, V. Pott, R. Fritschi, K. Banerjee, M. J. Declercq, P. Renaud, C. Hibert, P. Fluckiger, G. A. Racine, *Proc. ISQED* **2002**, 496. (b) K. Akarvardar, C. Eggimann, D. Tsamados, Y. S. Chauhan, G. C. Wan, A. M. Ionescu, R. T. Howe, H. S. P. Wong, *IEEE Trans. Electron Devices* **2008**, *55*, 48.
21. (a) https://doi.org/10.1038/s41467-020-20732-w. (b) https://doi.org/10.1063/1.5143939.
22. (a) Sadana, D. K.; Current, M. "Fabrication of Silicon-on-Insulator (SOI) Wafers Using Ion Implantation", in *Ion Implantation Science and Technology*, Ziegler, J. F. ed., Edgewater: Ion Implantation Technology Co., 2000. (b) Colinge, J. P. *Silicon-on-Insulator Technology: Materials to VLSI*, 2nd ed., Dordrecht: Kluwer Academic Publishers, 1997. (c) Marshall, A.; Natarajan, S. *SOI Design: Analog, Memory, and Digital Techniques*, Boston: Kluwer Academic Publishers, 2002.
23. (a) http://www.soiconsortium.org/pdf/Consortium_9april09_final.pdf. (b) http://www.advancedsubstratenews.com/.
24. Choi, Y. K. Et al. *IEEE Electron Device Lett.* **2000**, 254.
25. https://hal.archives-ouvertes.fr/hal-03254444/document.
26. An issue of the *MRS Bulletin* was recently devoted to scaling future CMOS logic devices with Ge and III-V materials: *MRS Bull.* **2009**, *34*, 485.
27. (a) Nihei, R.; Usami, N.; Nakajima, K. *Jpn. J. Appl. Phys.* **2009**, *48*, 115507, and references therein. (b) Jankovic, N. D.; O'Neill, A. *Semicond. Sci. Technol.* **2003**, *18*, 901.
28. Plummer, J. D.; Deal, M. D.; Griffin, P. B. *Silicon VLSI Technology: Fundamentals, Practice, and Modeling*, Prentice Hall: New Jersey, 2000.
29. A detailed video of the Si(111) 7 × 7 reconstruction may be found at: http://www.vimeo.com/1086112.
30. For more details, see: http://www.chem.qmul.ac.uk/surfaces/scc/.
31. For a video that shows FinFET processing developed for the 22-nm node, see: https://youtu.be/W3rfVpkNquA.
32. For GaAs substrates, primers such as xylene or trichlorobenzene are used, as GaAs is already a polar surface.
33. Masks are either comprised of soda-lime glass (coated with either a photographic emulsion, Fe_2O_3, or Cr films), or quartz (with a Cr film). Due to the absorption of UV light by glass, the latter is required for deep UV (DUV) photolithography. Masks may be classified as either "light-field" or "dark-field"; whereas the former is mostly clear with opaque patterns, the latter is an opaque mask, with transparent features.
34. For example: (a) Nonogaki, S. *Polymer J.* **1987**, *19*, 99. (b) Liu, H. -H.; Chen, W. -T.; Wu, F. -C. *J. Polym. Res.* **2002**, *9*, 251.

35. http://www.anandtech.com/show/10097/euv-lithography-makes-good-progress-still-not-ready-for-prime-time.
36. The use of double- and triple-patterning has extended the 193-nm photolithography timeline beyond that originally anticipated. For more details on this technology, see: (a) http://www.semi.org/cms/groups/public/documents/web_content/ctr_030805.pdf. (b) Wu, B.; Singh, A. K. *Extreme Ultraviolet Lithography*, McGraw-Hill: New York, 2009 (page 5 of the Introduction has a lithography roadmap for the extension of 193-nm photolithography, and EUV not likely being instituted until the 22 or 16 nm node (*ca.* 2015+).
37. For more details regarding EUV, see: Hutcheson, G. D. et al. *Scientific American* **2004**, *290*, 76.
38. For information regarding EUV mask design may be found online at: http://www.sematech.org/meetings/archives/litho/euvl/20030930/presentations/2C%20Shoki%20EUV%20Symp.pdf.
39. Hiroshi, I. *Adv. Polym. Sci.* **2005**, *172*, 37, and references therein.
40. (a) http://people.ccmr.cornell.edu/~cober/MiniPresentations/PAG_RBSPC.pdf. (b) Kim, K. -M.; Ayothi, R.; Ober, C. K. *Polym. Bull.* **2005**, *55*, 333.
41. *Irie, S.; Endo, M.; Sasago, M.; Kandaka, N.; Kondo, H.; Murakami, K. Study of Transmittance of Polymers and Influence of Photoacid Generator on Resist Transmittance at Extreme Ultraviolet Wavelength. Jpn. J. Appl. Phys. 2002, 41, 5864–5867.*
42. *T. Wallow, C. Higgins, R. Brainard, K. Petrillo, W. Montgomery, C.-S. Koay, G. Denbeaux, O. Wood, and Y. Wei, [Evaluation of EUV resist materials for use at the 32 nm half-pitch node] SPIE, AL (2008).*
43. For a nice review of EUV photoresists, see: http://doi.org/10.3390/nano10081593.
44. *Kostko, O.; Xu, B.; Ahmed, M.; Slaughter, D.S.; Ogletree, D.F.; Closser, K.D.; Prendergast, D.G.; Naulleau, P.; Olynick, D.L.; Ashby, P.D.;* et al. *Fundamental understanding of chemical processes in extreme ultraviolet resist materials. J. Chem. Phys. 2018, 149, 154305.*
45. For instance, see: (a) Dai, J.; Ober, C. K.; Wang, L.; Cerrina, F.; Nealey, P. F. *Proc. SPIE - Int. Soc. Opt. Eng.* **2002**, *4690*, 1193. (b) Kessel, C. R.; Boardman, L. D.; Rhyner, S. J.; Cobb, J. L.; Henderson, C. C.; Rao, V.; Okoroanyanwu, U. *Proc. SPIE - Int. Soc. Opt. Eng.* **1999**, *3678*, 214. (c) Bratton, D.; Yang, D.; Dai, J.; Ober, C. K. *Polym. Adv. Technol.* **2006**, *17*, 94, and references therein.
46. For example, see: (a) https://doi.org/10.1117/1.JMM.18.1.011007. (b) https://doi.org/10.1039/C9RA08977B.

47. The depth of focus (DOF $= \lambda/(\mathrm{NA})^2$) is also paramount toward resolution, as wafers are not atomically flat. Though it may be possible to adjust the wavelength and NA to achieve better resolution, the depth of field will decrease, making it difficult to define features simultaneously at the top and bottom surfaces. Consequently, chemical mechanical polishing (CMP) is used to planarize the wafer prior to high-resolution photolithography, and as thin a layer as possible of photoresist is applied to the wafer.

48. Article may be accessed online at: http://turroserver.chem.columbia.edu/PDF_db/publications_801_850/NJT849.pdf.
49. There are two methods used to remove the patterned material. *Etching* is where the photoresist is developed on top of the deposited layer. The underlying material is then removed by etching through openings in the mask. In contrast, *lift-off* is used when the material is deposited on top of the developed photoresist. The material is then lifted off when the resist is removed. For a nice summary of wet/dry etching, as well as etching vs. lift-off, see: http://www.mrsec.harvard.edu/education/ap298r2004/Erli%20chen%20Fabrication%20III%20-%20Etching.pdf.
50. A plasma is considered the fourth class of matter, in addition to solids, liquids, and gases. A plasma contains a mixture of ground-state and excited-state atoms, as well as ions.
51. C. J. Mogab, A. C. Adams, and D. L. Flamm, *J. Appl. Phys.* **1978**, *49*(7), 3796.
52. As its name applies, chemical mechanical polishing/planarization utilizes a hybrid of chemical and mechanical forces to yield a flat surface. It should be noted that using mechanical force alone (*e.g.,* grinding) would successfully planarize a surface; however, this would cause too much surface degradation. For more information about this process, see: http://maltiel-consulting.com/CMP-Chemical-mechanical_planarization_maltiel_semiconductor.pdf.
53. For more information regarding interconnect fabrication, see: (a) http://web.stanford.edu/class/ee311/NOTES/Interconnect_Cu.pdf. (b) https://dokumente.unibw.de/pub/bscw.cgi/d9299614/04_Short-Channel%20MOSFET.pdf.
54. This is typically performed through use of templates that are sacrificially removed following film deposition. For example, see: (a) Fuertes, M. C.; Soler-Illia, G. J. A. A. *Chem. Mater.* **2006**, *18*, 2109. (b) Xiao, L.; Zhang, H.; Scanlon, E.; Ramanathan, L. S.; Choe, E.-W.; Rogers, D.; Apple, T.; Benicewicz, B. C. *Chem. Mater.* **2005**, *17*, 5328. (c) Kanungo, M.; Deepa, P. N.; Collinson, M. M. *Chem. Mater.* **2004**, *16*, 5535. (d) Li, X. S.; Fryxell, G. E.; Birnbaum, J. C.; Wang, C. *Langmuir* **2004**, *20*, 9095.
55. For an animated website to illustrate the DC-diode and magnetron sputtering processes, see: http://www.ajaint.com/whatis.htm.
56. For example, Sigel, G. H.; Homa, D. S. U.S. Patent 7181116.
57. (a) Nasibulin, A. G.; Shurygina, L. I.; Kauppinen, E. I. *Colloid J.* **2005**, *67*, 1, and references therein. (b) Suzuki, K.; Kijima, K. *Jpn. J. Appl. Phys.* **2005**, *44*, 2081. (c) Chen, R. S.; Huang, Y. S.; Liang, Y. M.; Tsai, D. S.; Tiong, K. K. *J. Alloys Compd.* **2004**, *383*, 273. (d) Wang, Y. Q.; Chen, J. H.; Yoo, W. J.; Yeo, Y. -C. *Mat. Res. Soc. Symp. Proc.* **2005**, *830*, 269.
58. Barron, A. R. in *CVD of Nonmetals*, Rees, W. S. ed., Wiley: New York, 1996.
59. For a thorough description of kinetic and mass-transport mechanisms involved in CVD, as well as dependent variables, see: Pierson, H. O. *Handbook of Chemical Vapor Deposition*, 2nd ed., William Andrew: Norwich, NY, 1999.
60. (a) Schropp, R. E. *Mater. Res. Soc. Symp. Proc.* **2003**, *762*, 479. (b) Schropp, R. E. *Thin Solid Films* **2004**, *451*, 455. (c) Lau, K. K. S.; Murthy, S. K.; Lewis, H. G.; Pryce, C.; Jeffrey, A.; Gleason, K. K. *J. Fluorine Chem.* **2003**, *122*, 93. (d)

Mahan, A. H. *Solar Energy Mater. Solar Cells* **2003**, *78*, 299. (e) Stannowski, B.; Rath, J. K.; Schropp, R. E. *Thin Solid Films* **2003**, *430*, 220. (f) Schroeder, B. *Thin Solid Films* **2003**, *430*, 1. (g) Duan, H. L.; Zaharias, G. A.; Bent, S. E. *Curr. Opin. Solid State Mater. Sci.* **2002**, *6*, 471. (h) Mahan, A. H. *Solar Energy* **2004**, *77*, 931. (i) Matsumura, H.; Umemoto, H.; Masuda, A. *J. Non-Cryst. Solids* **2004**, *338*, 19.

61. (a) Fahlman, B. D.; Barron, A. R. *Adv. Mater. Opt. Electron.* **2000**, *10*, 135. (b) Richards, V. N.; Vohs, J. K.; Williams, G. L.; Fahlman, B. D. *J. Am. Ceram. Soc.* **2005**, *88*, 1973.
62. Zhang, W. J.; Bello, I.; Lifshitz, Y.; Chan, K. M.; Meng, X. M.; Wu, Y.; Chan, C. Y.; Lee, S. T. *Adv. Mater.* **2004**, *16*, 1405.
63. http://www.thindiamond.com/wp-content/uploads/2015/06/201406-MRS-CVD-Diamond-Research-Applications-and-Challenges.pdf.
64. For instance, see: (a) Xiong, G.; Elam, J. W.; Feng, H.; Han, C. Y.; Wang, H.-H.; Iton, L. E.; Curtiss, L. A.; Pellin, M. J.; Kung, M.; Kung, H.; Stair, P. C. *J. Phys. Chem. B.* **2005**, *109*, 14059. (b) Niinisto, J.; Rahtu, A.; Putkonen, M.; Ritala, M.; Leskela, M.; Niinisto, L. *Langmuir* **2005**, *21*, 7321. (c) Sechrist, Z. A.; Fabreguette, F. H.; Heintz, O.; Phung, T. M.; Johnson, D. C.; George, S. M. *Chem. Mater.* **2005**, *17*, 3475. (d) Reijnen, L.; Meester, B.; de Lange, F.; Schoonman, J.; Goossens, A. *Chem. Mater.* **2005**, *17*, 2724. (e) Matero, R.; Rahtu, A.; Ritala, M. *Langmuir* **2005**, *21*, 3498. (f) Min, Y.-S.; Cho, Y. J.; Hwang, C. S. *Chem. Mater.* **2005**, *17*, 626. (g) Gu, W.; Tripp, C. P. *Langmuir* **2005**, *21*, 211. (h) https://pure.tue.nl/ws/files/3625583/671193474080774.pdf. (i) http://www.cambridgenanotechald.com/pdf/Li-ion-battery-web.pdf.
65. Fahlman, B. D.; Barron, A. R. *Adv. Mater. Opt. Electron.* **2000**, 10(3–5), 135.
66. Although commercial ALD systems are priced well over $100,000 USD, effective laboratory-scale systems may be fabricated at much lower cost: Lubitz, M.; Medina, P. A.; Antic, A.; Rosin, J. T.; Fahlman, B. D. *J. Chem. Ed.* **2014**, *91*, 1022.
67. (a) Hansen, B. N.; Hybertson, B. M.; Barkley, R. M.; Sievers, R. E. *Chem. Mater.* **1992**, *4*, 749. (b) Lagalante, A. F.; Hansen, B. N.; Bruno, T. J.; Sievers, R. E. *Inorg. Chem.* **1995**, *34*, 5781. (c) Fernandes, N. E.; Fisher, S. M.; Poshusta, J. C.; Vlachos, D. G.; Tsapatsis, M.; Watkins, J. J. *Chem. Mater.* **2001**, *13*, 2023. (d) Cabanas, A.; Long, D. P.; Watkins, J. J. *Chem. Mater.* **2004**, *16*, 2028. (e) Blackburn, J. M.; Long, D. P.; Watkins, J. J. *Chem. Mater.* **2000**, *12*, 2625. (f) Blackburn, J. M.; Long, D. P.; Cabanas, A.; Watkins, J. J. *Science* **2001**, *294*, 141. (g) Cabanas, A.; Blackburn, J. M.; Watkins, J. *J. Microelectron. Eng.* **2002**, *64*, 53. (h) Ohde, H.; Kramer, S.; Moore, S.; Wai, C. M. *Chem. Mater.* **2004**, *16*, 4028.
68. For a thorough review of CVD/ALD precursors, see: Fahlman, B. D. *Curr. Org. Chem.* **2006**, *10*, 1021.
69. Gardiner, R. A.; Gordon, D. C.; Stauf, G. T.; Vaarstra, B. A.; Ostrander, R. L.; Rheingold, L. *Chem. Mater.* **1994**, *6*, 1967.
70. For an example of fluorine-free polyether ligands used to successfully prevent oligomerization of barium complexes (particularly problematic for heavy

Group II complexes due to the large ionic radius of the metal), see: Studebaker, D. B.; Neumayer, D. A.; Hinds, B. J.; Stern, C. L.; Marks, T. J. *Inorg. Chem.* **2000**, *39*, 3148.

71. For example, see: Gillan, E. G.; Bott, S. G.; Barron, A. R. *Chem. Mater*. **1997**, *9*, 796.
72. Hansen, B. N.; Brooks, M. H.; Barkley, R. M.; Sievers, R. E. *Chem. Mater.* **1992**, *4*, 749.
73. (a) Banger, K. K.; Jin, M. H.-C.; Harris, J. D.; Fanwick, P. E.; Hepp, A. F. *Inorg. Chem.* **2003**, *42*, 7713. (b) Castro, S. L.; Bailey, S. G.; Raffaelle, R. P.; Banger, K. K.; Hepp, A. F. *Chem. Mater*. **2003**, *15*, 3142.
74. For a nice summary of phosphor classes, see: (a) https://www.electrochem.org/dl/interface/wtr/wtr09/wtr09_p032-036.pdf. (b) Xie, R. J.; Hirosaki, N. *Sci. Technol. Adv. Mater.* **2007**, *8*, 588.
75. For a review of nitride- and oxynitride-based phosphors, see: Xie, R. J.; Hirosaki, N. *Sci. Technol. Adv. Mater.* **2007**, *8*, 588.
76. Bowers, M. J.; McBride, J. R.; Rosenthal, S. J. *J. Am. Chem. Soc.* **2005**, *127*, 15,378.
77. For more details regarding perovskite LEDs, see: (a) https://doi.org/10.1038/s41467-021-24616-5. (b) https://doi.org/10.1038/s41566-021-00857-0. (c) https://arxiv.org/pdf/2106.01924.pdf. (d) https://pubs.acs.org/doi/10.1021/acsnano.1c06849?ref=PDF.
78. The fabrication steps for polymer OLED displays include: (i) deposit and pattern the ITO anode, (ii) pattern the conducting and emissive polymer layers (via spin-coating, ink jet printing, screen printing, or web coating methods), (iii) pattern and vacuum deposit the cathode (Ba, Ca/Al, etc.). For small-molecule OLEDs, the multilayer structure is fabricated *in vacuo*, involving a variety of patterning steps via shadow masking, followed by thermal evaporation (PVD) of layers using appropriate source materials.
79. For a review of transparent conductive films (fabrication and applications), see: Gordon, R. G. *MRS Bull.* **2000**, *8*, 52.
80. For a review of the band structure of doped tin oxide, see: Batzill, M.; Diebold, U. *Prog. Surf. Sci.* **2005**, *79*, 47.
81. http://www.qcrsolutions.com/Site/OLED_and_PLED_QCR_Solutions_Corp.html.
82. For a review of exciton formation and OLEDs, see: Yersin, H. *Top. Curr. Chem.* **2004**, *241*, 1.
83. Note: spin–orbit coupling refers to the interaction of the spin magnetic moment of an electron with the magnetic moment arising from the orbital motion of the electron.
84. For a nice summary of triplet emitters for OLED applications, see: Yersin, H.; Finkenzeller, W. J. in *Highly Efficient OLEDs with Phosphorescent Materials*, Yersin, H. ed., Wiley–VCH: Weinheim, 2008.
85. https://electronics.howstuffworks.com/oled3.htm.
86. Vidor, F. F.; Meyers, T.; Hilleringmann, U. *Electronics* **2015**, *4*, 480.

87. For a historical review of TFT technology, see: http://www.electrochem.org/dl/interface/spr/spr13/spr13_p055_061.pdf.
88. For more details regarding amorphous oxide semiconductors and thin-film transistor designs, see: (a) Fortunato, E.; Barquinha, P.; Martins, R. *Adv. Mater.* **2012**, *24*, 2945. (b) Park, J. S.; Maeng, W. J.; Kim, H. S.; Park, J. S. *Thin Solid Films* **2012**, *520*, 1679. (c) Lin, Y. H.; Faber, H.; Labram, J. G.; Stratakis, E.; Sygellou, L.; Kymakis, E.; Hastas, N. A.; Li, R.; Zhao, K.; Armassian, A.; Treat, N. D.; McLachlan, M.; Anthopoulos, T. D. *Adv. Sci.* **2015**, *2*, 1500058. (d) Kwon, J. Y.; Lee, D. J.; Kim, K. B. *Electron. Mater. Lett.* **2011**, *7*, 1. (e) Hosono, H.; Kim, J.; Toda, Y.; Kamiya, T.; Watanabe, S. *Proc. Nat. Acad. Sci.* **2016**, *114*, 233.
89. https://spectrum.ieee.org/semiconductors/materials/thin-fast-and-flexible-semiconductors.
90. A nice brief overview of thermoelectricity may be found online: https://www.technologyreview.com/s/401415/thermoelectric-materials/.
91. For details regarding all aspects of thermoelectric materials, refer to the March 31, 2006 issue of the MRS Bulletin, which is devoted entirely to this topic. More information is also found in: Uher, C. *Materials Aspect of Thermoelectricity*, CRC Press: New York, 2016 (ISBN: 9781498754903). For a recent review of TE materials, see: https://doi.org/10.3389/femat.2021.677845.
92. (a) http://doi.org/10.1002/advs.201903389. (b) http://doi.org/10.1016/j.matchemphys.2020.123389.
93. (a) http://doi.org/10.1002/adma.201501078. (b) http://doi.org/10.1016/j.joule.2017.07.018.
94. For example, see: Mangersnes, K.; Lovvik, O. M.; Prytz, O. *New J. Phys.* **2008**, *10*, 1.
95. For example, see: Kleinke, H. *J. Appl. Phys.* **2009**, *105*, 053,703.
96. For example, see: Hebert, S.; Lambert, S.; Pelloquin, D.; Maignan, A. *Phys. Rev. B* **2001**, *64*, 172,101, and references therein.
97. For example, see: Wilson-Short, G. B.; Singh, D. J.; Fornari, M.; Suewattana, M. *Phys. Rev. B* **2007**, *75*, 035,121, and references therein.
98. For example: (a) Johnsen, S.; He, J.; Androulakis, J.; Dravid, V. P.; Todorov, L.; Chung, D. Y.; Kanatzidis, M. G. *J. Am. Chem. Soc.* **2011**, *133*, 3460. (b) Maltsev, Y. V.; Nensberg, E. D.; Petrov, A. V.; Semiletov, S. A.; Ukhanov, Y. I. *Phys. Solid State* **1967**, *8*, 1713. (c) Sootsman, J. R.; Pcionek, R. J.; Kong, H. J.; Uher, C.; Kanatzidis, M. G. *Chem. Mater.* **2006**, *18*, 4993.
99. Lo, S. H.; Zhang, Y.; Sun, H.; Tan, G.; Uher, C.; Wolverton, C.; Dravid, V. P.; Katatzidis, M. G. *Nature* **2014**, *508*, 373.
100. For example, see: Liang, W.; Hochbaum, A. I.; Fardy, M.; Rabin, O.; Zhang, M.; Yang, P. *Nano Lett.* **2009**, *9*, 1689, and references therein.
101. For instance, see: Ham, J.; Shim, W.; Kim, D. H.; Lee, S.; Roh, J.; Sohn, S. W.; Oh, K. H.; Voorhees, P. W.; Lee, W. *Nano Lett.* **2009**, *9*, 2867, and references therein.
102. Lee, J.-H.; Galli, G. A.; Grossman, J. C. *Nano Lett.* **2008**, *8*, 3750.

103. Girard, S. N.; He, J.; Zhou, X.; Shoemaker, D.; Jaworski, C. M.; Uher, C.; Dravid, V. P.; Heremans, J. P.; Kanatzidis, M. G. *J. Am. Chem. Soc.* **2011**, *133*, 16588.
104. (a) Biswas, K.; He, J.; Blum, I. D.; Wu, C. I.; Hogan, T. P.; Seidman, D. N.; Dravid, V. P.; Kanatzidis, M. G. *Nature* **2012**, *489*, 414. (b) Zhao, L. D.; Zhang, X.; Wu, H.; Tan, G.; Pei, Y.; Xiao, Y.; Chang, C.; Wu, D.; Chi, H.; Zheng, L.; Gong, S.; Uher, C.; He, J.; Kanatzidis, M. G. *J. Am. Chem. Soc.* **2016**, *138*, 2366. (c) Lee, Y.; Lo, S. H.; Chen, C.; Sun, H.; Chung, D. Y.; Chasapis, T. C.; Uher, C.; Dravid, V. P.; Kanatzidis, M. G. *Nature Commun.* **2014**, *5*, 3640. (d) Zhao, L. D.; Lo, S. H.; He, J.; Li, H.; Biswas, K.; Androulakis, J.; Wu, C. I.; Hogan, T. P.; Chung, D. Y.; Dravid, V. P.; Kanatzidis, M. G. *J. Am. Chem. Soc.* **2011**, *133*, 20476.
105. http://doi.org/10.1021/acsnano.0c06181 and references therein.
106. For example, see: http://doi.org/10.1007/s40820-020-0374-x.
107. More details regarding the benefits of nanostructures for thermoelectric applications may be found at: http://www.cs.duke.edu/~reif/NSF.NanoEnergy/Report/.
108. (a) http://doi.org/10.1021/acs.chemmater.0c00229. (b) https://doi.org/10.1016/j.progpolymsci.2022.101548. (c) https://doi.org/10.3389/femat.2021.677845.
109. For a nice summary of the development of nanomaterials with high ZT, see: Li, J. F.; Liu, W. S.; Zhao, L. D.; Zhou, M. NPG Asia Mater. 2010, 2, 152 (available online: https://www.nature.com/am/journal/v2/n4/pdf/am201012a.pdf).
110. https://www.eia.gov/energyexplained/coal/how-much-coal-is-left.php..
111. Shockley, W.; Queisser, H. J. *J. Appl. Phys.* **1961**, *32*, 510. The remaining photon energy is primarily lost by reflection or transmission through the cell, conversion to heat, or exciton recombination.
112. https://iopscience.iop.org/article/10.1088/1755-1315/726/1/012001/pdf
113. https://doi.org/10.1038/s41560-020-0598-5.
114. For details on these limitations, see: Henry, C. H. *J. Appl. Phys.* **1980**, *51*, 4494.
115. For a review of the processes involved in DSSCs, see: Peter, L. *Acc. Chem. Res.* **2009**, *42*, 1839.
116. https://doi.org/10.1021/acsaem.1c02789.
117. There have been recent efforts toward designing solid-state DSSCs, which represent a more commercially-viable device. Previous designs featured liquid electrolytes; for example: Yanagida, S.; Yu, Y.; Manseki, K. *Acc. Chem. Res.* **2009**, *42*, 1827. However, iodine/iodide-based electrolytes have certain disadvantages. Other designs have featured solid electrolytes poly(N-alkyl-4-vinylpyridine) iodide/N-methyl pyridine iodide and 2,2′,7,7′-tetrakis(N,N-di-p-methoxyphenylamine)-9,9'-spiro-bifluorene (Spiro-OMeTAD). For example, see: (a) Wu, J.; Hao, S.; Lan, Z.; Lin, J.; Huang, M.; Huang, Y.; Li, P.; Yin, S.; Sato, T. *J. Am. Chem. Soc.* **2008**, *130*, 11568. (b) Moon, S. -J.; Yum, Y -H.; Humphry-Baker, R.; Karlsson, K. M.; Hagberg, D. P.; Marinado, T.; Hagfeldt, A.; Sun, L.; Gratzel, M.; Nazeeruddin, M. K. *J. Phys. Chem. C* **2009**, *113*,

16816. (c) Cappel, U. B.; Karlsson, M. H.; Pschirer, N. G.; Eickemeyer, F.; Schoneboom, J.; Erk, P.; Boschloo, G.; Gagfeldt, A. *J. Phys. Chem. C* **2009**, *113*, 14595.
118. For a thorough review of DSSCs, see: (a) Gratzel, M. *Inorg. Chem.* **2005**, *44*, 6841. (b) Hagfeldt, A.; Boschloo, G.; Sun, L.; Kloo, L.; Pettersson, H. *Chem. Rev.* **2010**, *110*, 6595.
119. http://doi.org/10.1038/s41467-021-21945-3.
120. https://doi.org/10.3389/fchem.2019.00077.
121. (a) https://doi.org/10.1038/s41560-019-0535-7. (b) https://doi.org/10.1038/s41467-020-20749-1.
122. For a review of polymer-based organic solar cells, see: (a) Gunes, S.; Neugebauer, H.; Sariciftci, N. S. *Chem. Rev.* **2007**, *107*, 1324. (b) Kim, H.; Nam, S.; Jeong, J.; Lee, S.; Seo, J.; Han, H.; Kim, Y. *Kor. J. Chem. Eng.* **2014**, *31*, 1095.
123. https://doi.org/10.1038/s41467-020-20580-8.
124. https://doi.org/10.1038/s41586-021-03964-8.
125. Due to the toxicity of lead, there is interest in replacing Pb with Sn for perovskite solar cells. For instance, see: Hao, F.; Stoumpos, C. C.; Cao, D. H.; Chang, R. P. H.; Kanatzidis, M. G. *Nat. Photonics* **2014**, *8*, 489.
126. McMeekin, D. P.; Sadoughi, G.; Rehman, W.; Eperon, G. E.; Saliba, M.; Horantner, M. T.; Haghighirad, A.; Sakai, N.; Korte, L.; Rech, B.; Johnston, M. B.; Herz, L. M.; Snaith, H. J. *Science* **2016**, *351*, 151.
127. https://doi.org/10.1021/acs.chemrev.9b00780.
128. (a) http://doi.org/10.1126/science.abd4016. (b) https://doi.org/10.1038/s41598-021-94848-4. (c) Sivaram, V.; Stranks, S. D.; Snaith, H. J. *Sci. Am.* **2015**, *313*, 54. (d) Green, M. A.; Emery, K.; Hishikawa, Y.; Warta, W.; Dunlop, E. D. *Prog. Photovolt. Res. Appl.* **2015**, *23*, 805.
129. Serrano-Lujan, L.; Espinosa, N.; Larsen-Olsen, T. T.; Abad, J.; Urbina, A.; Krebs, F. C. *Adv. Energy Mater.* **2015**, *5*, 01119.
130. http://doi.org/10.1126/sciadv.abb0055.

Further Reading

1. S.O. Kasap, *Principles of Electronic Materials and Devices*, 4th edn. (McGraw Hill, New York, 2017)
2. J.D. Plummer, M.D. Deal, P.B. Griffin, *Silicon VLSI Technology: Fundamentals, Practice, and Modeling* (Prentice Hall, New Jersey, 2000)
3. J.M. Rabaey, A. Chandrakasan, B. Nikolic, *Digital Integrated Circuits*, 2nd edn. (Pearson, New York, 2003)
4. S.A. Campbell, *Fabrication Engineering at the Micro- and Nanoscale*, 4th edn. (Oxford Univ. Press, New York, 2012)
5. N. Weste, D. Harris, *CMOS VLSI Design: A Circuits and Systems Perspective*, 4th edn. (Pearson, New York, 2010)
6. *Physics and Chemistry of III–V Compound Semiconductor Interfaces*, ed. by C. Wilmsen (Springer, New York, 1985).

7. A. Misra, J.D. Hogan, R.A. Chorush, *Handbook of Chemicals and Gases for the Semiconductor Industry* (Wiley, New York, 2002)
8. S.J. Moss, A. Ledwith, *The Chemistry of the Semiconductor Industry* (Chapman and Hall, New York, 1987)
9. J. Turley, *The Essential Guide to Semiconductors* (Prentice Hall, New Jersey, 2003)
10. R.F. Pierret, *Semiconductor Fundamentals,* vol. I, 2nd edn. (Prentice Hall, New Jersey, 1988)
11. B. Streetman, S. Banerjee, *Solid State Electronic Devices*, 7th edn. (Prentice Hall, New Jersey, 2014)
12. L. Chen, R. Liu, X. Shi, *Thermoelectric Materials and Devices* (Elsevier: New York, 2020)
13. H.N. Waltenburg, J.T. Yates, Surface chemistry of silicon. Chem. Rev. **95**, 1589 (1995)
14. Reviews of high-κ dielectric materials: (a) J.-P. Locquet, C. Marchiori, M. Sousa, J. Fompeyrine, J.W. Seo. J. Appl. Phys., **100**, 051610; (b) Robertson. J. Rep. Prog. Phys. **2006**(69), 327 (2006)
15. http://smithsonianchips.si.edu/index2.htm
16. S.T. Barry, *Chemistry of Atomic Layer Deposition* (De Gruyter, Boston, 2022)
17. M.J. Madou, *Fundamentals of Microfabrication and Nanotechnology*, 3rd edn. (CRC Press, New York, 2011)
18. A.C. Jones, M.L. Hitchman, *Chemical Vapor Deposition : Precursors, Processes and Applications* (RSC, London, 2009)
19. D.M. Dobkin, M.K. Zuraw, *Principles of Chemical Vapor Deposition* (Kluwer, New York, 2003)

Chapter 5
Polymeric Materials

Take a minute to look at the room and furnishings around you. Virtually everything you see is at least partially composed of organic-based building blocks. From plastic packaging materials to individual carpet fibers, no other type of material is as heavily utilized in our society as polymers. Currently, over 367 million metric tons of plastics are produced each year—a total of more than 8 billion metric tons since the early 1950s! The average person in Western Europe or North America consumes 100 kg (220 lb) of plastic each year, mostly in the form of consumer product packaging. Unfortunately, recovery and recycling efforts remain insufficient; it is estimated that 60% of all plastics ever produced have accumulated in landfills and the environment (Fig. 5.1).[1]

As first defined by Staudinger in 1920, a *polymer* is any material that is composed of an extended structure of small chemical repeat units, known as *monomers*.[2] For simplicity, the monomeric unit is almost always clearly identified within the polymer name (*i.e.*, "poly(*monomeric unit*)", Fig. 5.2). A polymer is generally made up of more than 100 repeat units; structures with lower numbers of chemical repeat units are known as *oligomers*. Strictly speaking, all solid-state materials with an infinite structural array are classified as polymers—even inorganic structures such as metals, ceramics, and glasses. However, since we have described inorganic-based materials in previous chapters, we will focus our present discussion on polymeric materials that feature a carbon-containing backbone.

Organic-based materials are generally associated with "soft" characteristics—*i.e.*, low-melting and facile plastic deformation. Although this is generally the case for popular commercial polymers such as plastics and rubbers, there are numerous other polymer classes that exhibit a hardness and thermal stability that rival inorganic ceramics. Polymeric materials have long been used in packaging and other consumer product applications; however, recent developments have extended the utility range to now include microelectronics, photovoltaics, self-healing materials, and drug-delivery agents. As a testament to the importance of polymers in our world, the current volume of polymers used for commercial and industrial applications already exceeds the volume of steel and aluminum combined. In our era of soaring gas prices,

B. D. Fahlman, *Materials Chemistry*,
https://doi.org/10.1007/978-3-031-18784-1_5

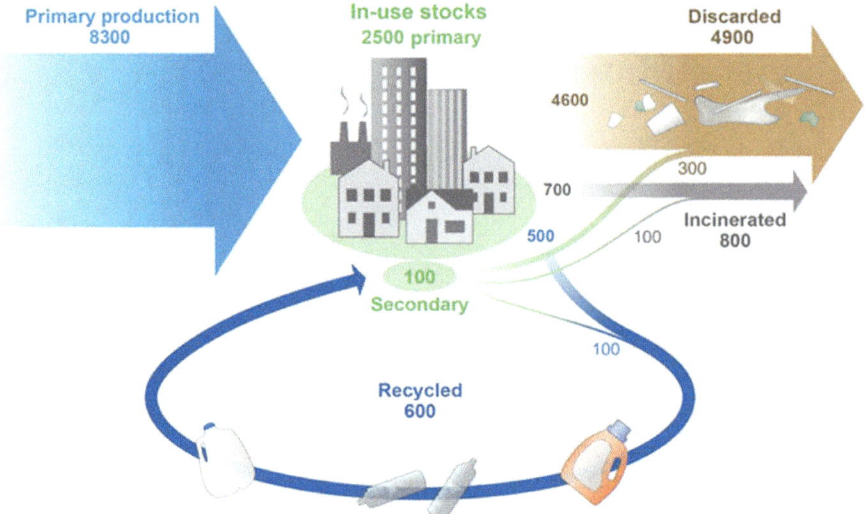

Fig. 5.1 Global production, use, and fate of polymer resins, synthetic fibers, and polymer additives from 1950 to 2015 (in million metric tons). Reproduced with permission from *Sci. Adv.* **2017**, *3*, e1700782. © 2017 American Association for the Advancement of Science

automobile manufacturers are also scrambling to add more plastics to their vehicles.[3] This chapter will delve into various polymer classes, with a detailed examination of the structural influence on resultant properties.

5.1 Polymer Classifications and Nomenclature

The most fundamental classification of polymers is whether they are synthetic or naturally occurring. Common synthetic polymers (Fig. 5.2) are pervasive in commercial applications; for example, Table 5.1 lists some synthetic polymers used for automotive applications. In contrast, natural polymers include macromolecules such as polysaccharides (*e.g.*, starches, sugars, cellulose, gums, etc.), proteins (*e.g.*, enzymes), fibers (*e.g.*, wool, silk, cotton), polyisoprenes (*e.g.*, natural rubber), and nucleic acids (*e.g.*, RNA, DNA). Accordingly, these polymer classes are often referred to as *biopolymers*, of which some recent materials applications will be discussed later in this chapter.

Synthetic polymers may be classified under two general umbrellas: *thermoplastics* and *thermosets*. As their names imply, this definition is illustrative of the properties exhibited by these materials under elevated temperatures. Thermoplastics consist of long molecules with side chains or groups that are not connected to neighboring molecules (*i.e.*, not *crosslinked*). Hence, both amorphous and crystalline thermoplastics are glasses at low temperature and transform to a rubbery *elastomer* or

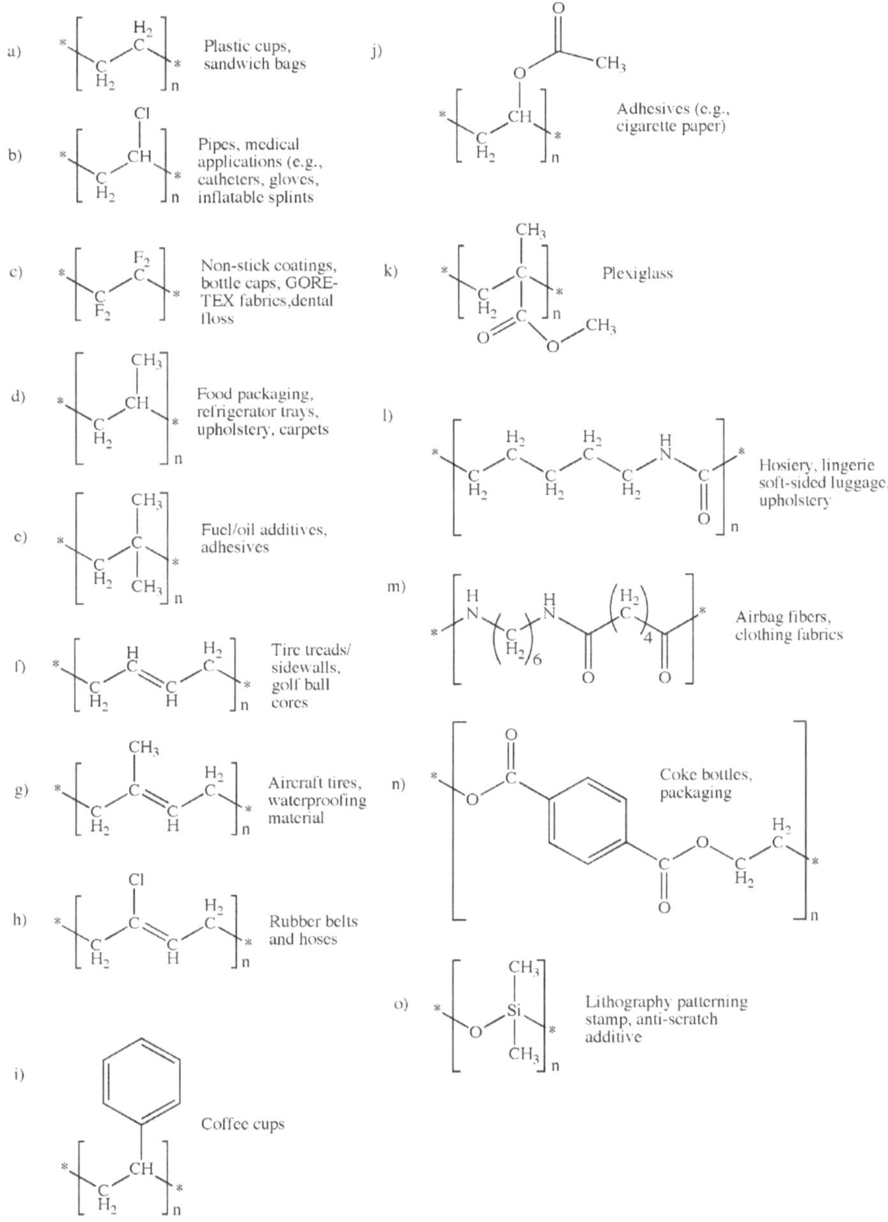

Fig. 5.2 Molecular structures of the chemical repeat units for common polymers. Shown are **a** polyethylene (PE), **b** poly(vinyl chloride) (PVC), **c** polytetrafluoroethylene (PTFE), **d** polypropylene (PP), **e** polyisobutylene (PIB), **f** polybutadiene (PBD), **g** *cis*-polyisoprene (natural rubber), **h** *trans*-polychloroprene (Neoprene® rubber), **i** polystyrene (PS), **j** poly(vinyl acetate) (PVAc), **k** poly(methyl methacrylate) (PMMA), **l** polycaprolactam (polyamide—nylon 6), **m** nylon 6,6, **n** poly(ethylene teraphthalate), **o** poly(dimethyl siloxane) (PDMS)

Table 5.1 Polymers used for automotive applications

Polymer	Application
Poly(ethylene), PE	Fuel tanks, windshield washer bottles
Poly(propylene), PP	Bumpers, external trim
Poly(vinyl chloride), PVC	Interior trim
Poly(acrylonitrile) (PAN) + poly(styrene) (PS) blend + Poly(butadiene) = ABS	Exterior and Interior trim, wheel covers
Nylon-6,6	Intake manifolds, rocker cover/air cleaner, hubcaps
Polyester	Grill opening panel, sunroof frame, passenger-side airbag doors
Poly(methylmethacrylate), PMMA	Lenses
Polycarbonate, PC	Headlamp lenses, trim
Polyurethane, PU	Foam, bumpers
Poly(butylene terephthalate), PBT	Headlamp bezel
Poly(vinyl butyral), PVB	Laminated safety glass
Poly(ethylene terephthalate), PET	Windshield wiper brackets

flexible plastic at an elevated temperature—known as the *glass-transition temperature* (T_g; Fig. 5.3). The T_g is the most important property of polymers, being analogous to the melting point of low molecular weight compounds.[4] In contrast to thermoplastics, thermosets are initially liquids and become hardened by a thermally induced crosslinking process known as *curing*. Also, unlike thermoplastics, since the crosslinking process yields a stable 3-D network, thermosets may not be re-melted/re-processed. Thermoset polymers are usually synthesized within a mold to yield a desired shape; once the polymer cures, the only way to reshape the material is through machining processes (*e.g.*, drilling, grinding).[5] The most common type of thermosetting polymer is epoxy resin, widely used for adhesives and paints/coatings, which harden through crosslinking reactions with a curing agent such as primary and secondary polyamines (*i.e.*, containing reactive $–NH_2$ groups; Fig. 5.4).

Regarding the overall structure of polymers, there are five classes of macromolecular architectures, ranging from simple linear arrays to *megamers*—complex structures built from ordered hyperbranched (dendritic) polymers (Fig. 5.5). As you might imagine, linear chains are best able to pack into a regular crystalline array; however, as the degree of chain branching increases, only amorphous phases are formed. Due to the pronounced length of polymer chains, it should be noted that regions of crystallinity generally form where only a portion of the polymer chains are regularly organized, while others remain disordered. This structural diversity directly affects physical properties such as tensile strength, flexibility, and opaqueness of the bulk polymer.

As a more general structural definition, a polymer synthesized from only one type of monomer is referred to as a *homopolymer*. In contrast, a polymer that is formed from more than one type of monomer is known as a *copolymer*. The terms

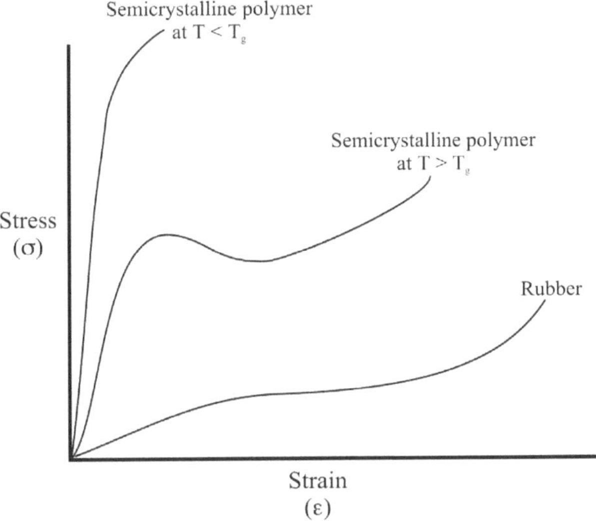

Fig. 5.3 Stress *versus* strain curves for various polymers around its glass-transition temperature. The maximum in the curve that occurs at T_g is referred to as the *yield point* (onset of plastic deformation)

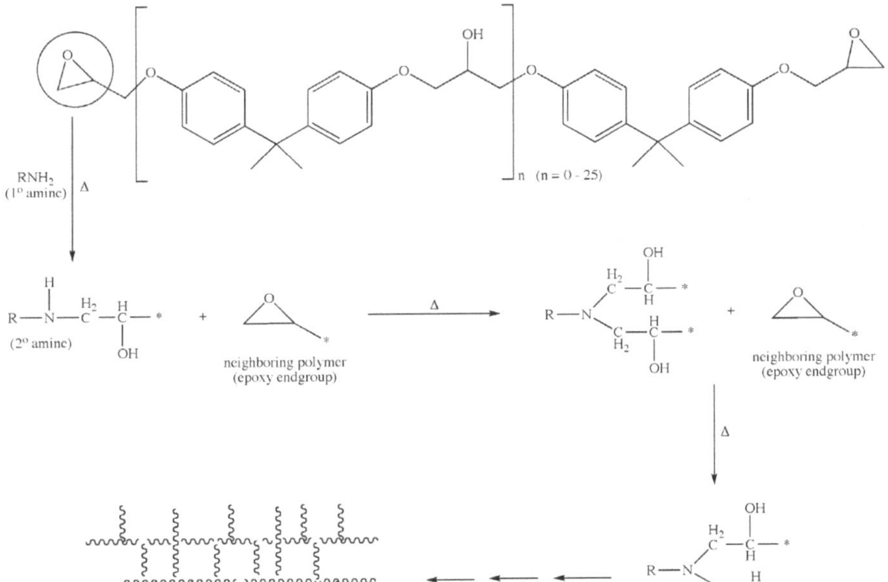

RNH_2 (1° amine) Δ

n (n = 0 - 25)

(2° amine)

neighboring polymer (epoxy endgroup)

Highly crosslinked 3D network; tough, insoluble solid

Fig. 5.4 Illustration of a hardening mechanism responsible for epoxy resin curing

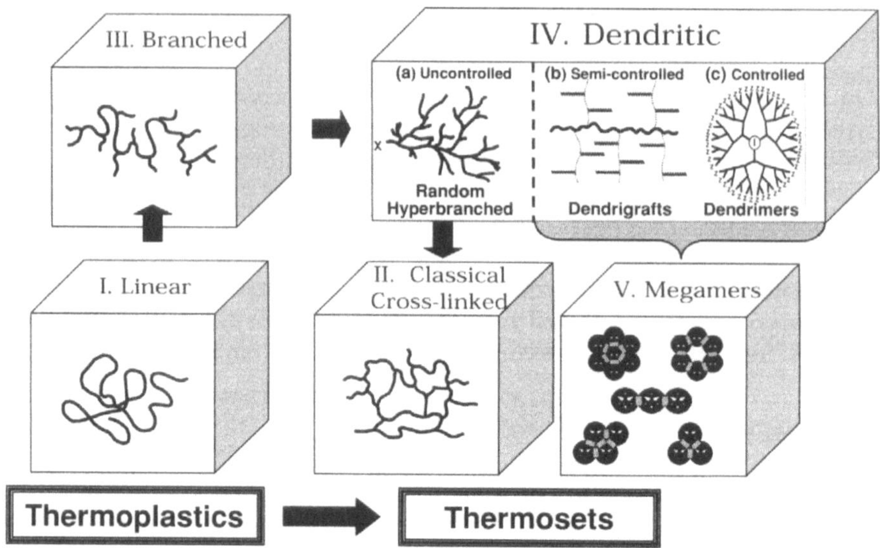

Fig. 5.5 The five major structural classes of polymers. Reproduced with permission from Frechet, J. M. J.; Tomalia, D. A. *Dendrimers and Dendritic Polymers*, Wiley: New York, 2001

terpolymer, tetrapolymer, pentapolymer, etc. are used to designate a polymer derived from three, four, five, etc. co-monomers. Although homopolymers contain only one repeating chemical unit in their structure, this distinction is not always clear. For instance, polyethylene polymers often contain short-chain branched impurities as a consequence of the polymerization process. However, since this unintentional structural deviation results from a polymerization process involving only one type of monomer, the term homopolymer is still most appropriate.

There are four types of copolymers: *random*, *alternating*, *block*, and *graft*, each exhibiting varied physical properties. In contrast to block copolymers, which contain long adjacent sequences of A and B monomers, the chain of an alternating copolymer is composed of the sequence $(–A–B–)_n$. A random copolymer is formed from the random linkages of A and B monomers along the chain (*e.g.*, –AAABAABBBAAB–). Graft copolymers may be considered as another type of block copolymer, which feature one long chain of one monomer with offshoots of a second monomer emanating along its length (Fig. 5.6). The structure, length, and placement of the copolymer units directly affect physical properties such as crystallinity, density, strength, brittleness, melting point, and electrical conductivity (for conductive polymers).

For polymers that contain different substituents, each carbon in the chain may be considered as a chiral center. As a result, a parameter known as *tacticity* may be defined that describes the stereoregularity of adjacent carbon centers (Fig. 5.7). A polymer that contains substituents on the same side of the polymer chain is referred to as *isotactic*, and often exhibits some degree of crystallinity. In contrast, if adjacent

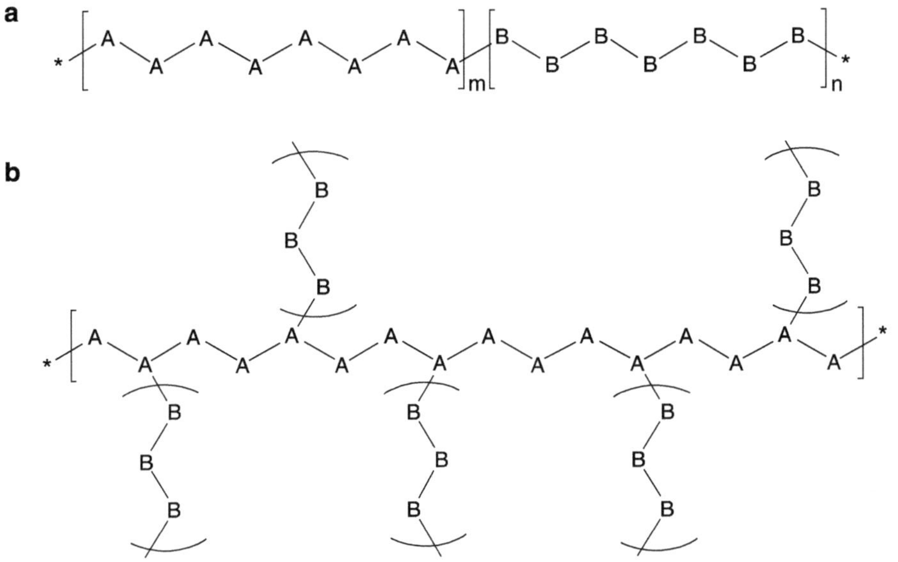

Fig. 5.6 Illustration of **a** block copolymers and **b** graft copolymers

substituents are arranged on alternating sides of the polymer chain, a *syndiotactic* polymer is formed. Unless a polymerization scheme uses a catalyst that confines the nucleation/propagation site, the substituents will be randomly organized—either disordered throughout the entire polymer chain (*atactic*) or along every other repeat unit (*hemiisotactic*). Due to their structural randomness, atactic polymers are almost always amorphous. Not surprisingly, the tacticity of a polymer significantly affects its mechanical properties. For instance, isotactic polypropylene has an elastic modulus and hardness of 1.09 GPa and 125 MPa, whereas atactic polypropylene has values of 0.15 GPa and 1.4 MPa, respectively.

Another form of isomerism that exists for polymers is the arrangement of adjacent repeat units. For an individual monomeric unit, the subsequent monomer may add in a head-to-tail or head-to-head/tail-to-tail fashion (Fig. 5.8). The sequence is often determined by the substituents; for instance, sterically bulky groups (*e.g.*, phenyl) dictate a head-to-tail array for polystyrene. On the other hand, polymers that feature halide substituents (*e.g.*, poly(vinyl fluoride)) contain significant numbers of head-to-head/tail-to-tail sequences.

Thus far, we have considered the organization and composition of the main polymer chain, held together through covalent bonding of neighboring carbons. However, the physical properties of a particular polymer are more strongly affected by *intermolecular forces* that exist between individual polymer chains. For instance, the crosslinking of adjacent polymer chains through sulfide linkages is responsible for the vulcanization of rubber (Fig. 5.9a), yielding a material with a high hardness and rigidity. In contrast, the pronounced flexibility and high tensile strength of nylon

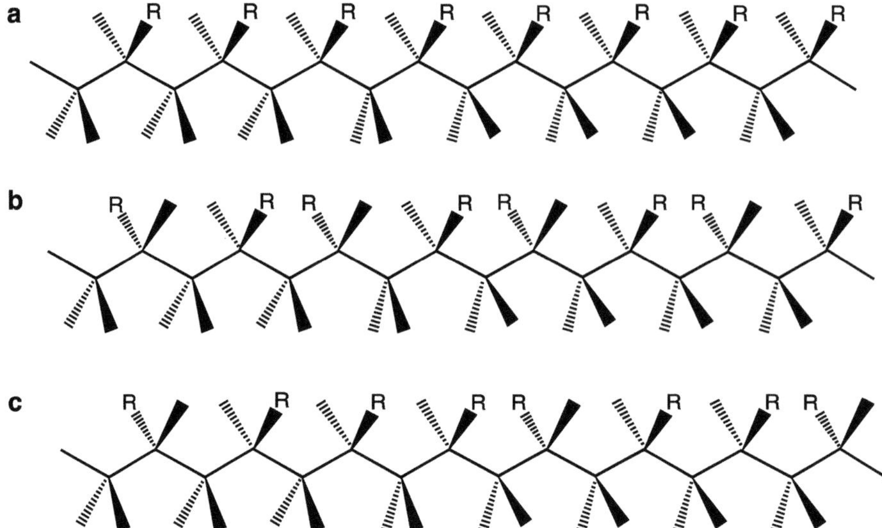

Fig. 5.7 The tacticity of polymer chains. Illustrated are **a** isotactic, **b** syndiotactic, and **c** atactic polymers

Fig. 5.8 Illustration of sequence isomerization exhibited by polymer chains. Shown are **a** head-to-tail and **b** head-to-head/tail-to-tail sequencing

are due to inter-chain hydrogen bonding interactions (Fig. 5.9b). Other types of inter- or intra-chain interactions include dipole–dipole and van der Waal (induced dipole) forces (Fig. 5.9c, d).

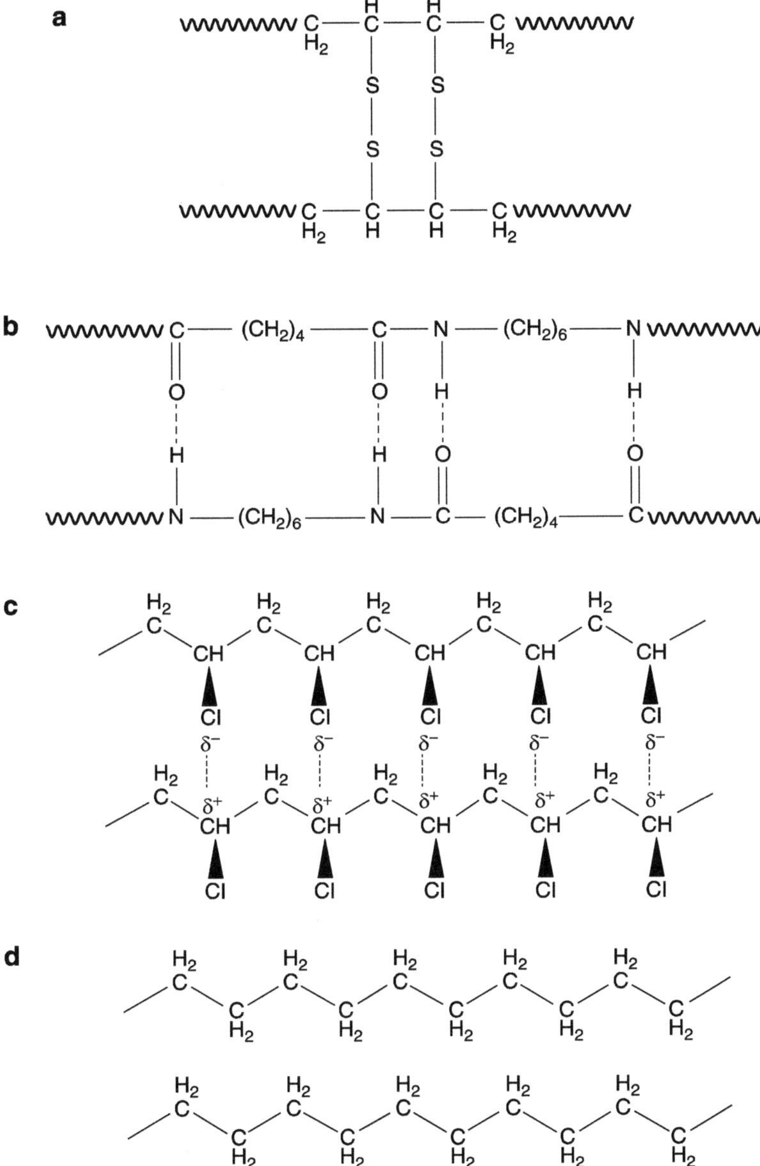

Fig. 5.9 The intermolecular forces involved in adjacent polymer chains. Shown are **a** covalent crosslinking (vulcanized rubber), **b** hydrogen bonding (nylon 6,6), **c** dipole–dipole (PVC), and **d** van der Waal interactions (polyethylene)

5.2 Polymerization Mechanisms

Polymers may be synthesized by two general mechanisms: *step-growth* (condensation) or *chain* (addition) polymerization. Step-growth is often the most preferred designation relative to condensation since it is also appropriate for polymerizations that do not eliminate water during the reaction (5.1). Table 5.2 lists the important distinctions between addition and step-growth polymerization schemes. As its name implies, step-growth involves the reactions of functionalized monomers to build up polymer chains through dimers, trimers, etc., *en route* toward oligomers, and polymers. The acid/base-catalyzed synthesis of SiO_2 networks by sol–gel (Sect. 2.5.1) is a widely used application of step-growth polymerization. In contrast, addition polymerization involves the activation/addition of unsaturated monomers, yielding a much higher MW polymer in a relatively short period of time. Since step-growth proceeds through the reaction of neighboring complementary functional groups, small molecular byproducts such as HX, H_2O, etc. are generated; addition reactions do not yield such elimination products.

$$\text{n X-R-X} + \text{n H-Y-R}'\text{-Y-H} \rightarrow \text{X} - (\text{R-Y-R}')_{\text{n}} - \text{Y-H} + \text{n HX}, \quad (5.1)$$

where X = OH, Cl, etc. and Y = NH or O

The molecular weight of a polymer may be generally described as the molecular weight of the monomer(s) multiplied by the *degree of polymerization* (DoP, 5.2)—the number of repeat units. Whereas oligomers have DoP values between 2–10 (*e.g.,* polysaccharides, polypeptides), most commercial plastics are high-mass polymers

Table 5.2 Features of addition and condensation polymerization schemes

Addition (chain growth)	Condensation (step-growth)
1. Unsaturated monomers	Monomers contain ≥2 functional groups
2. No products are eliminated	Elimination of H_2O, HCl, etc.
3. Only monomer and polymer are present during polymerization	Monomers and polymer are accompanied by dimers, trimers, and oligomeric species
4. Only monomers add to the growing polymer	All intermediate species are reactive, and contribute to the growing polymer
5. Mechanism involves reacting with double bond by active species like free radicals or ions	Involves simple elimination reaction between monomer functional groups
6. Rapidly yields a high MW polymer; crosslinking is achieved through use of monomers with two double bonds (*e.g.,* divinylbenzene)	Molecular weight is typically lower than addition polymerization. The presence of small amounts of multifunctional monomers results in extensive crosslinking (gels)
7. Examples: polyolefins, polydienes, vinyl polymers, acrylic polymers	Examples: polyesters, polyamides, polycarbonates,[a] epoxies

[a](Fun fact) a polycarbonate layer is used between glass panels to absorb the energy of a bullet blast—"bullet-proof" glass

with have DoP values >1000. If all individual polymer chains were equivalent in length and composition, this would yield an exact molecular weight value analogous to molecular or ionic solids. However, due to the variance in polymer chain lengths, the best we can attain is to assign a molecular weight *distribution*, characterized by using an empirical technique known as *gel-permeation chromatography* (GPC).[6] A broad peak indicates a greater variance in molecular weight, which is not desirable for most applications.

$$\mathrm{DoP} = \frac{\mathrm{M_r}}{\mathrm{M_R}}, \tag{5.2}$$

where $\mathrm{M_r}$ = the relative molar mass of the polymer and $\mathrm{M_R}$ = the relative molar mass of the monomer.

The range of the molecular weight distribution is referred to as the *polydispersity*, given by the polydispersity index (PDI). This value is calculated from the weight average molecular weight, $\overline{\mathrm{M}}_\mathrm{w}$, divided by the number average molecular weight, $\overline{\mathrm{M}}_\mathrm{n}$ (5.3). Whereas the number average molecular weight is experimentally determined by colligative properties (*e.g.*, elevation of boiling point, depression of freezing point, osmotic pressure), the weight average molecular weight is determined by light scattering, neutron scattering, or ultracentrifugation.

The PDI for man-made polymers will always be >1.0; however, as the polymer chains approach a uniform length, the PDI will approach unity. The type of polymerization used, as well as experimental conditions (*e.g.*, temperature and nature of the catalyst), are paramount in generating a polymer with a narrow polydispersity. For instance, typical addition polymerization, results in PDI values of *ca.* 10–20, compared to 2–3 for step-growth polymerization. However, using precise temperature control to limit termination mechanisms, PDIs of <1.5 may be generated for both techniques. Not surprisingly, nature is far ahead of human ingenuity; biopolymers have PDI values very close or equal to one, indicating that only one length of polymer is present—directly responsible for the high specificity and efficiency of complex living systems.

$$\mathrm{PDI} = \frac{\overline{\mathrm{M}}_\mathrm{w}}{\overline{\mathrm{M}}_\mathrm{n}}, \tag{5.3}$$

where

$$\overline{\mathrm{M}}_\mathrm{w} = \frac{\sum_i \mathrm{N}_i \mathrm{M}_i^2}{\sum_i \mathrm{N}_i \mathrm{M}_i} \text{ and } \overline{\mathrm{M}}_\mathrm{n} = \frac{\sum_i \mathrm{N}_i \mathrm{M}_i}{\sum_i \mathrm{N}_i}$$

(N_i is the number of molecules with molecular weight, M_i).

Addition polymerization

Addition polymerization involves three steps: initiation, propagation, and termination. During initiation, either radicals (Fig. 5.9) or ionic species are generated from the controlled decomposition of an initiator molecule. The reactive intermediates are then sequentially added to the C=C bonds of monomers to propagate the growing polymer chain. Free-radical polymerization is the most common method currently used to synthesize polymers from vinyl-based monomers.

Due to the high reactivity of the radical fragments, facile *chain-transfer* may occur (Fig. 5.10d, i, ii), whereby the radical end of the growing chain abstracts an atom from another molecule/chain. The second molecule may be monomer, solvent, initiator, or other polymers that exist in solution. As a result, the growth of the primary chain is terminated, and a new radical capable of propagation/polymerization is generated. Oftentimes, chain-transfer results in hydrogen abstraction from the second molecule, which causes branching (Fig. 5.10d, ii). The second molecule may be totally unreactive or moderately reactive with the monomer (*i.e.*, relative to normal propagating radicals). Hence, a deliberate addition of these molecules may be used to inhibit or slow the polymerization process.

Although radical polymerization is highly susceptible to chain-transfer, it is possible to suppress these side reactions, resulting in *living polymerization*.[7] By definition, a living process will result in a linear increase in the polymer molecular weight with monomer consumption. In order to gain strict control over the polymerization process, a number of approaches have been designed:

i. *Nitroxide-Mediated Polymerization* (NMP). A stable free radical (*e.g.*, **I**, 2,2,6,6,-tetramethyl-1-piperidinyloxyl (TEMPO)) is added to the solution, acting as a radical scavenger. The TEMPO radical is exceedingly stable due to the nearby methyl groups that supply electron density and help stabilize the unpaired electron that is delocalized over the N–O bond. The coupling of TEMPO with the polymeric radical is reversible, which allows one to control the molecular weight and polydispersity of the resultant polymer by varying the monomer:TEMPO ratio.

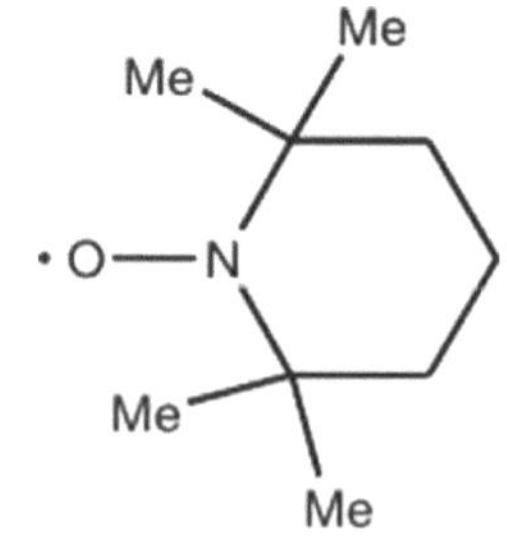

Me
Me
•O—N
Me
Me

I

ii. *Atom-Transfer Radical Polymerization* (ATRP). A halogenated organic and metal complex[8] are added to the solution, which generate a radical initiator (Fig. 5.11). As the monomer reacts with the initiator, the halide moiety preferentially terminates the chain. This process is considered redox-controlled polymer growth, since the polymer chain may grow only as additional atom-transfer reagents are added to the solution.

iii. *Reversible Addition Fragmentation Chain-Transfer Polymerization* (RAFT).[9] This process features the addition of RAFT agents, containing thiocarbonyl-thio groups. The reaction of radicals with the C=S bond forms a stabilized radical intermediate. In an ideal system, these intermediates do not undergo termination reactions, but rather reintroduce a radical capable of initiation or propagation with monomer, while they themselves reform their C=S bond (Fig. 5.12). The cycle of addition to the C=S bond, followed by fragmentation of a radical,

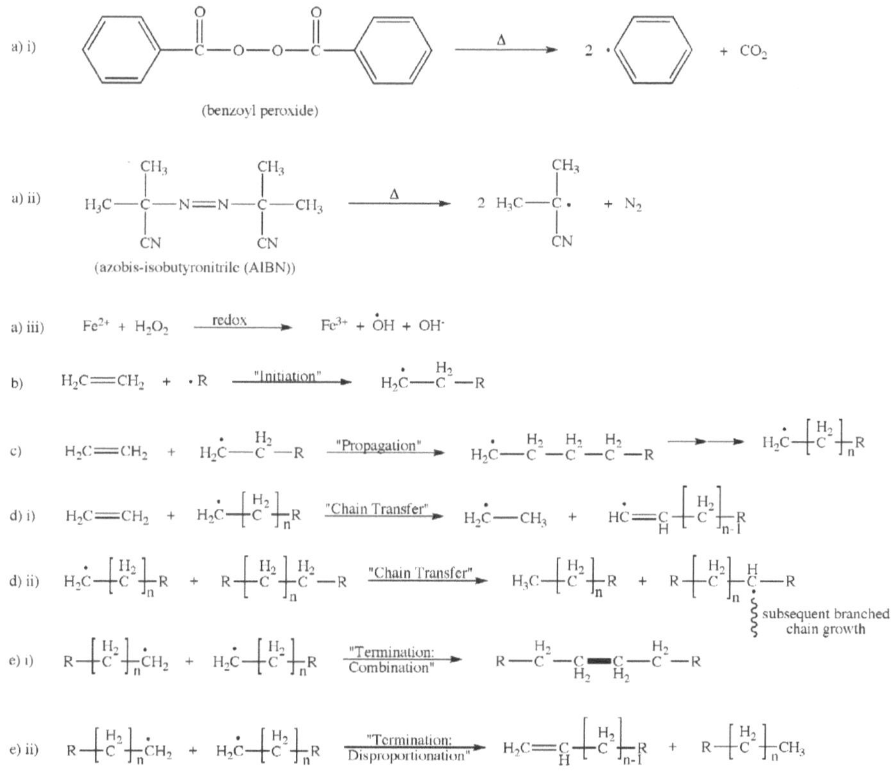

a) i) $C_6H_5-C(=O)-O-O-C(=O)-C_6H_5 \xrightarrow{\Delta} 2\ {\cdot}C_6H_5 + CO_2$

(benzoyl peroxide)

a) ii) $H_3C-C(CH_3)(CN)-N{=}N-C(CH_3)(CN)-CH_3 \xrightarrow{\Delta} 2\ H_3C-\dot{C}(CH_3)(CN) + N_2$

(azobis-isobutyronitrile (AIBN))

a) iii) $Fe^{2+} + H_2O_2 \xrightarrow{\text{redox}} Fe^{3+} + \dot{O}H + OH^-$

b) $H_2C{=}CH_2 + {\cdot}R \xrightarrow{\text{"Initiation"}} H_2\dot{C}-CH_2-R$

c) $H_2C{=}CH_2 + H_2\dot{C}-CH_2-R \xrightarrow{\text{"Propagation"}} H_2\dot{C}-CH_2-CH_2-CH_2-R \longrightarrow H_2\dot{C}-[CH_2]_n-R$

d) i) $H_2C{=}CH_2 + H_2\dot{C}-[CH_2]_n-R \xrightarrow{\text{"Chain Transfer"}} H_2\dot{C}-CH_3 + H\dot{C}{=}CH-[CH_2]_{n-1}-R$

d) ii) $H_2\dot{C}-[CH_2]_n-R + R-[CH_2]_n-CH_2-R \xrightarrow{\text{"Chain Transfer"}} H_3C-[CH_2]_n-R + R-[CH_2]_n-\dot{C}H-R$

subsequent branched chain growth

e) i) $R-[CH_2]_n-\dot{C}H_2 + H_2\dot{C}-[CH_2]_n-R \xrightarrow{\text{"Termination: Combination"}} R-CH_2-CH_2-CH_2-CH_2-R$

e) ii) $R-[CH_2]_n-\dot{C}H_2 + H_2\dot{C}-[CH_2]_n-R \xrightarrow{\text{"Termination: Disproportionation"}} H_2C{=}CH-[CH_2]_{n-1}-R + R-[CH_2]_n-CH_3$

Fig. 5.10 Reactions involved in free-radical addition polymerization. Shown are **a** (i)–(iii) generation of free radicals from a variety of initiators, **b** initiation of polymer chain growth through the combination of a free radical and unsaturated monomer, **c** propagation of the polymer chain through the combination of growing radical chains, **d** chain-transfer of free radicals between the primary and neighboring chains, and **e** termination of the polymer growth through either combination (i) or disproportionation (ii) routes

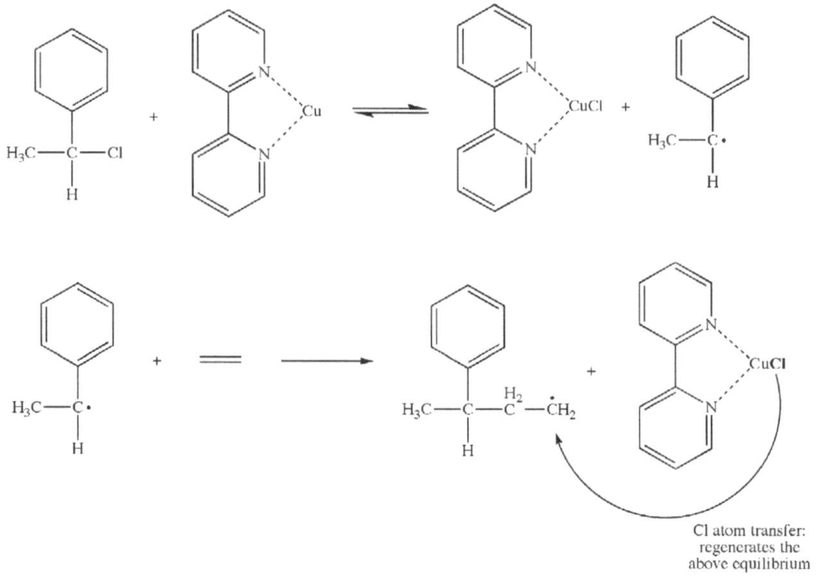

H_3C—C—Cl, H; Cu; CuCl; H_3C—C•, H; H_3C—C—C^{H_2}—$\dot{C}H_2$; Cl atom transfer: regenerates the above equilibrium

Fig. 5.11 Mechanism of atom-transfer radical living polymerization. In this process, addition of atom-transfer agents results in initiating radicals that react with monomers. Rather than terminating polymer growth, halogenated end units are formed that are capable of propagating chain growth when additional monomer is added

continues until all monomer is consumed. Termination is limited in this system by the low concentration of active radicals.

A more recent development in living free-radical polymerization is the use of *iniferters*—a single molecule that is capable of *ini*tiating, trans*fer*ring, and *ter*minating the radical polymerization process. Figure 5.13 illustrates the polymerization scheme exhibited by the most common type of iniferter. In these systems, UV absorption generates a carbon radical and sulfur-based dithiocarbamyl radical. Whereas the carbon radicals are extremely reactive toward the monomer, the dithiocarbamyl radical is not sufficiently reactive toward propagation. Termination in

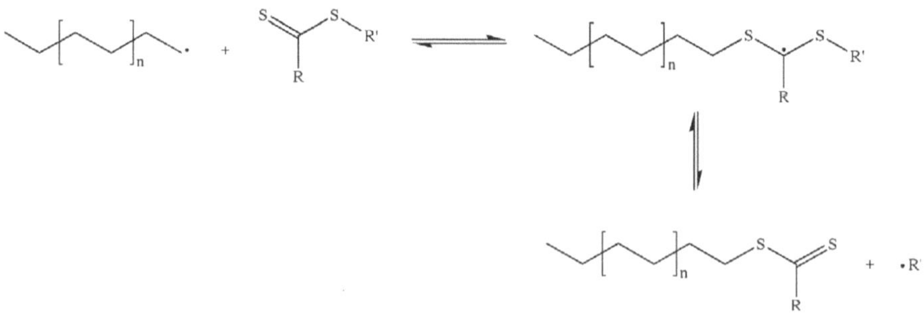

Fig. 5.12 Illustration of the general reaction scheme responsible for RAFT polymerization

iniferter systems may take place through either carbon–carbon or carbon–dithiocarbamyl bimolecular radical termination. The former route results in a dead unreactive polymer, whereas the latter route forms another iniferter species that may reinitiate upon UV light irradiation.

As previously mentioned, addition polymerization may also be initiated by cations/anions. In these ionic systems, propagation occurs through the combination of additional monomer with carbocation/carbanion intermediate species. In cationic polymerization, a Lewis acid (*e.g.*, $AlCl_3$, BF_3, etc.) may be used in isolation, or accompanied by a protic Lewis base (*e.g.*, NH_3, H_2O), which renders the proton as the actual initiator (Fig. 5.14). For cationic polymerization, termination may occur through proton, halide, or hydroxyl abstraction from the counteranion. For instance, $AlCl_3$ serves as the initiator for the cationic polymerization of isobutylene ($(CH_3)_2{=}CH_2$) to yield butyl rubber—used for the inner tube linings of automobile tires. Figure 5.15 illustrates an example of using BCl_3/2-chloroisopropyl benzene as an iniferter for the living cationic polymerization of isobutylene. There are also reports of alkylaluminum chlorides being used for living cationic polymerization,[10] as well as other base-assisted routes.[11]

Another example of cationic addition polymerization is the ring-opening polymerization of hexachlorocyclotriphosphazene to yield polydichlorophosphazene (Fig. 5.16). The polyphosphazenes represent one of the largest classes of polymers that are used for applications such as fuel cell membranes, flame-retardants, lubricants, and biomedical-related (*e.g.,* microencapsulating agents, biodegradable materials, tissue engineering scaffolds, biocompatible coatings, etc.). A living polymerization route toward polyphosphazenes is also possible using N-(trimethylsilyl)-trichlorophosphoranimine in the presence of trace amounts of PCl_5 (Fig. 5.17); the PDI and molecular weight may be controlled by varying the ratio of monomer:initiator.

In contrast to cationic routes, anionic addition polymerizations are initiated by using organolithium compounds (*e.g.*, butyllithium) or alkali metal amides (*e.g.*, $NaNH_2$, Fig. 5.18), amines, alkoxides, hydroxides, or phosphines as initiators. Many anionic polymerizations occur via a living process; that is, termination does not occur spontaneously and is controlled through the addition of a Lewis base. However, termination may also occur through unintentional quenching due to trace impurities such as oxygen, CO_2, or moisture. Polymers that are synthesized via this route include polydiene synthetic rubbers, polymethacrylates, and polystyrene.

Heterogeneous catalysis

We have considered a variety of precedents for radical/ionic addition polymerization of unsaturated monomers to yield polymeric materials. However, none of the aforementioned techniques offer stereoselective control over the growing polymer chain, resulting in purely atactic polymers. In order to introduce such control, it is necessary to spatially confine the reactive site to control the direction of incoming monomer/growing polymer.[12] The most common method used to control the tacticity of the resulting polymer is *Ziegler–Natta polymerization.*

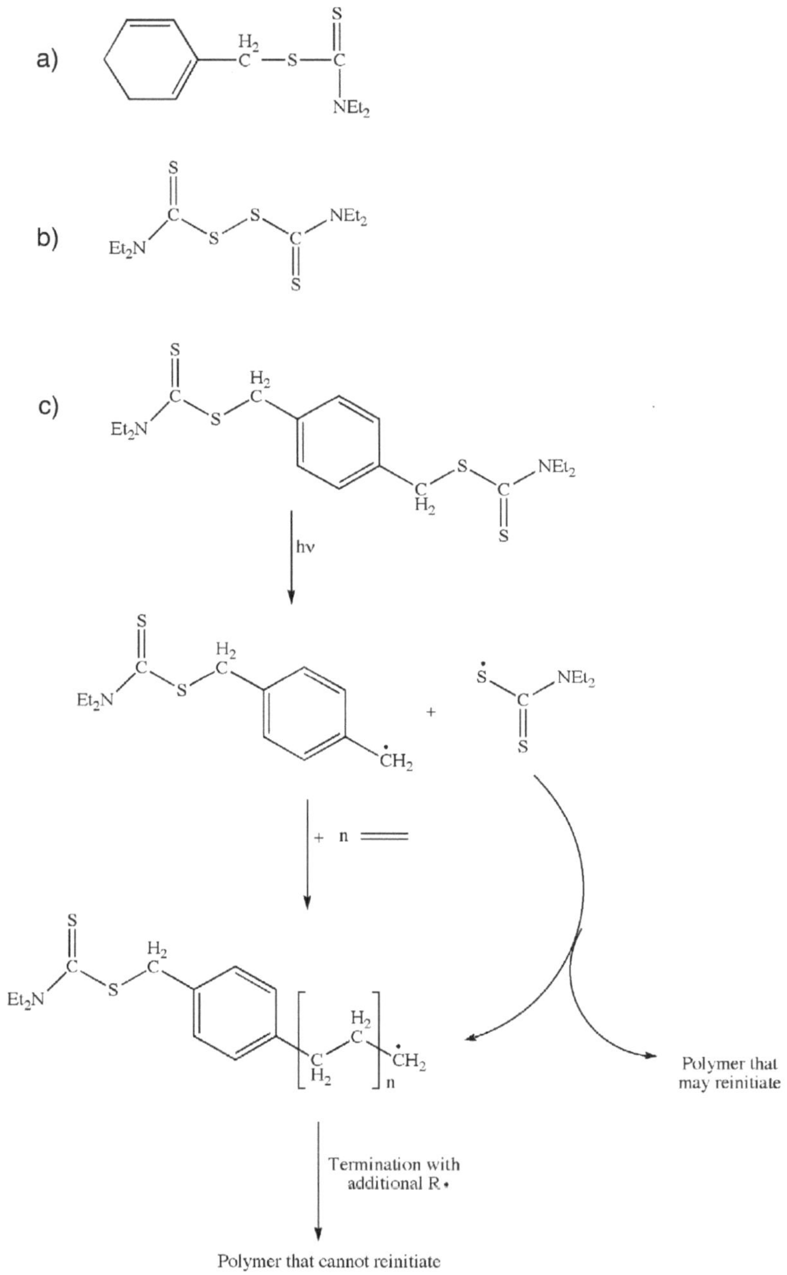

a)

b)

c)

Fig. 5.13 Molecular structures of common iniferters **a**–**c**, and illustration of polymerization using an iniferter

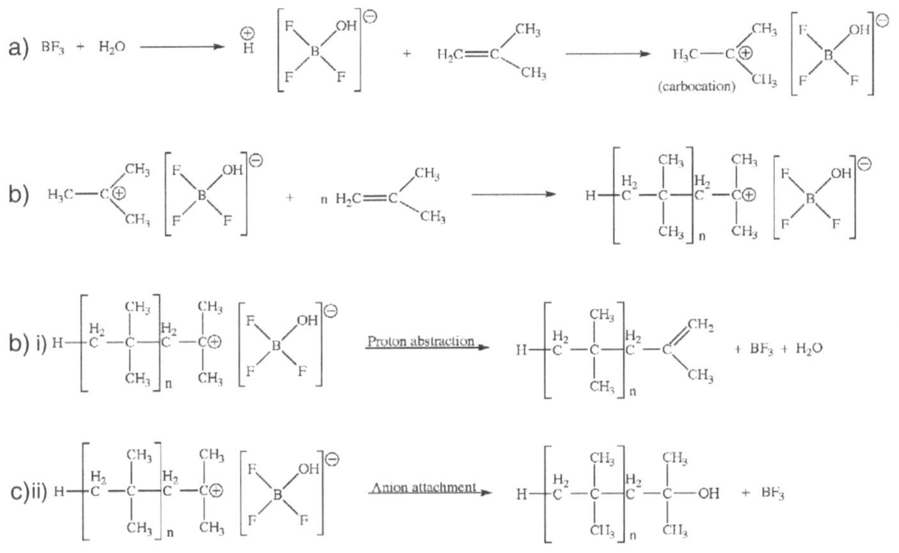

Fig. 5.14 Reactions involved in cationic addition polymerization. Shown are **a** generation of a carbocation intermediate from a Lewis acid initiator, **b** propagation of the polymer chain through the combination of the carbocationic polymer chain and additional monomers, and **c** termination of the polymer growth through either proton abstraction (i) or anionic attachment (ii) routes

Ziegler–Natta polymerization is an example of heterogeneous catalysis—a dual-phase system where polymerization occurs on the surface of the catalyst. A crystal of $TiCl_3$ or $TiCl_4$ is used in association with an aluminum alkyl co-catalyst (Fig. 5.19). Since the Ti sites on the surface are coordinatively unsaturated, monomer may attach along a controlled direction. The aluminum (Lewis acidic) complex acts as an initiator, facilitating monomer coordination through abstraction of a Cl group from the Ti coordination sphere. The Lewis acidic Al site also assists in the intramolecular rearrangements that are essential for chain propagation.

If $TiCl_3$ is used as the catalyst surface, an isotactic polymer is formed. However, changing the surface to VCl_3 yields a syndiotactic product. This difference may be explained by looking at the relative sizes of Ti^{3+} and V^{3+}. Due to an increase in the effective nuclear charge (Z_{eff}) as one moves from left to right of the Periodic Table, the V^{3+} center is smaller which creates more steric hindrance among the coordinated polymer and incoming monomer. As a result, there is less room for the ligands to undergo an equatorial–axial shift for V^{3+}, and the incoming monomer may approach in two directions. Since this spatial ligand shift readily occurs for the larger Ti^{3+}, the monomer approaches along a single direction, resulting in an isotactic polymer. The decrease in reaction rate and polymerization efficiency upon substitution of Ti with V is likely an artifact of electronic differences. Since V^{3+} has an extra d-electron relative to Ti^{3+} (d^2 *versus* d^1), there is added Coulombic repulsion with the π-electrons of the incoming olefin.

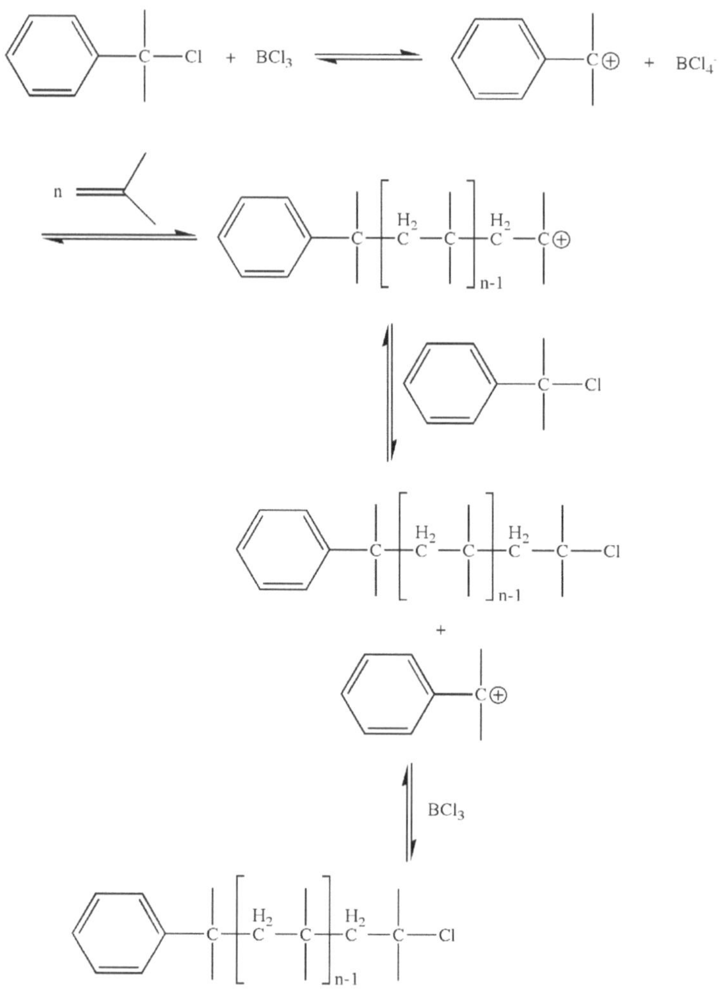

Cl
BCl3
BCl4-
n
H2
n-1
BCl3

Fig. 5.15 Iniferter-based living cationic polymerization of isobutylene

Homogeneous catalysis

Although separation of products from catalyst is easily accomplished within a heterogeneous system, the polydispersity of the product will be relatively high due to multiple reaction sites on the catalyst surface. In order to improve the overall selectivity, a variety of single-phase *homogeneous* catalytic routes have also been developed.[13] Although the polydispersity of the products are much narrower using homogeneous catalysts, the primary limitation is the difficulty in separating products from reactants/catalyst. However, supercritical fluids (*e.g.*, CO_2 at *ca.* 40 °C and 2000 psi) are now commonly used as the solvent in these systems in order to facilitate product separation.[14] Supercritical fluids have properties intermediate between gases and

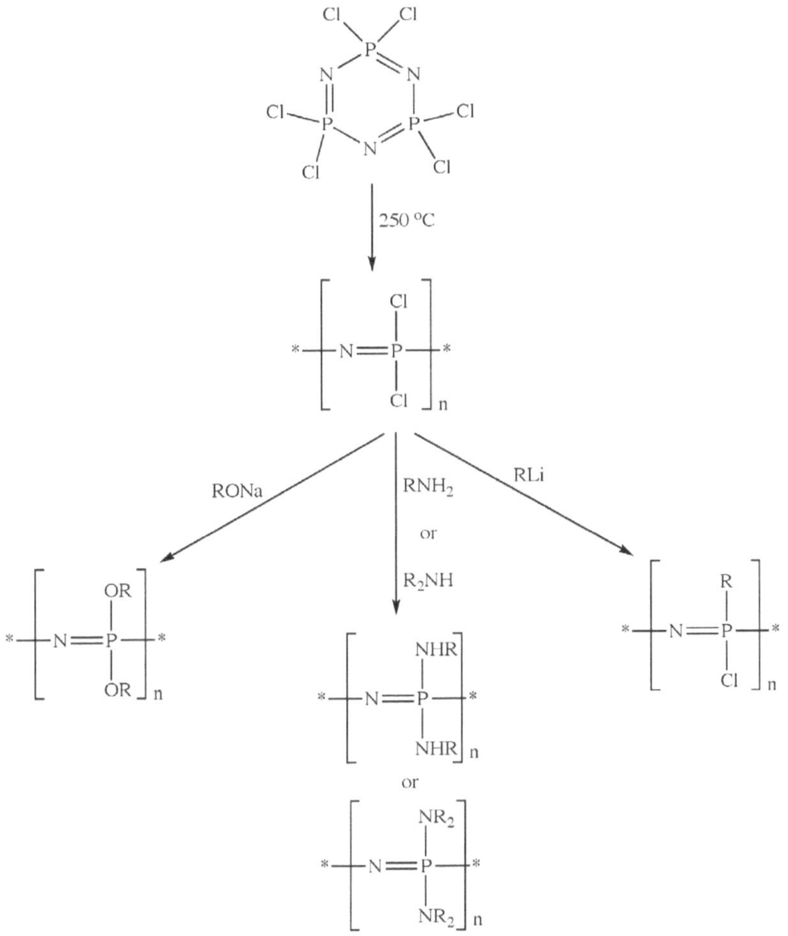

Fig. 5.16 Ring-opening polymerization of hexachlorocyclotriphosphazene, and subsequent reactions to yield functionalized polyphosphazenes

liquids; minute changes in pressure/temperature of the fluid near its critical point result in dramatic changes in its density and solubility characteristics.

Homogeneous Ziegler–Natta polymerization catalysts are typically of the *Kaminsky-type*, based on Group 4 metallocene molecules that contain bulky ligands (Fig. 5.20).[15] Based on their sterically-encumbered metal site, it is easy to see how stereocontrol over the polymer is obtained—through limiting the path of approach for the incoming olefin. A more recent technique to offer further stereoselectivity is "heterogenizing" the catalyst by anchoring the metallocene structure onto a solid support such as silica or alumina.[16]

In order to obtain catalytic activity in the system, a co-catalyst featuring an electron-deficient center (*e.g.*, B, Al) must also be present in the solution. The mechanism for activity of aluminum oxide co-catalysts, referred to as *methyl alumoxanes*

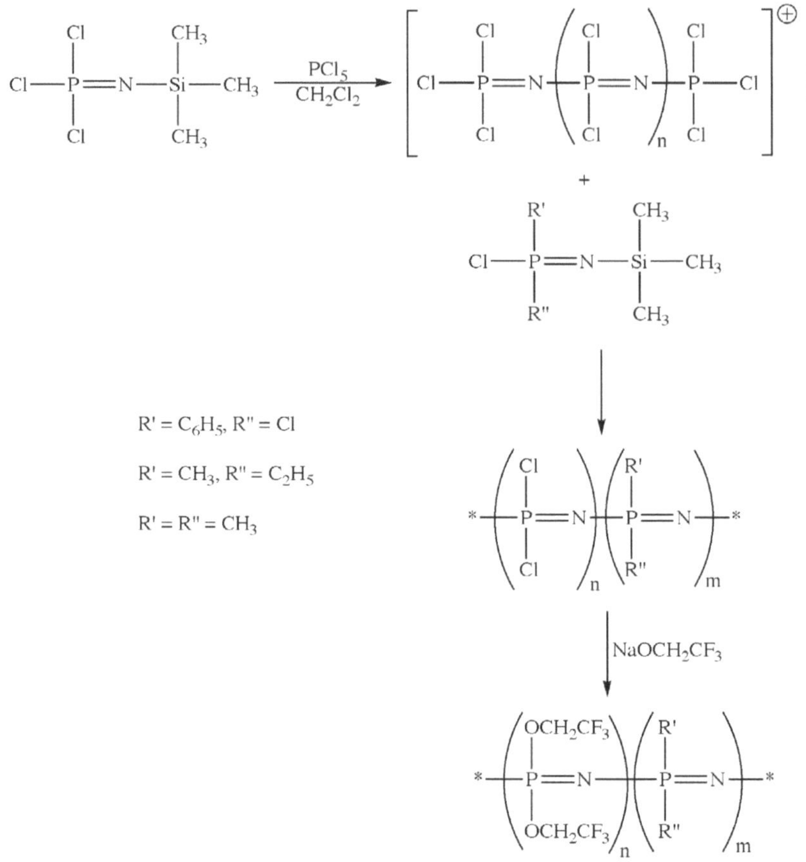
PCl5
CH2Cl2
[PCl6]
R' = C6H5, R" = Cl
R' = CH3, R" = C2H5
R' = R" = CH3
NaOCH2CF3

Fig. 5.17 An example of a living polymerization route for polyphosphazenes

(MAOs, formed from the controlled hydrolysis of $AlMe_3$[17]), has been shown to involve the abstraction of an alkyl or chloro group from the metallocene structure.[18] For aluminum co-catalysts, this yields a $[(Cp)_2MX]^+$ $[(^tBu)_2AlX]^-$ ion pair (X = R, Cl). This activates the metallocene structure toward polymerization, since the M^+ site more readily accepts electron density from the alkene monomer. In order for the ion pair to be active toward polymerization, the anion must be non-coordinating to ensure a high Lewis acidity toward the incoming olefin. Accordingly, one of the most effective co-catalysts is $B(C_6F_5)_3$, a very strong Lewis acid due to highly electronegative fluorinated substituents that pull electron density away from the central B atom. Upon abstraction of Me or Cl groups, the resulting anion $[BX(C_6F_5)_3]^-$ (**II**) is non-coordinating toward the metallocene structure, which freely allows the catalyst to accept electron density from the incoming monomer.

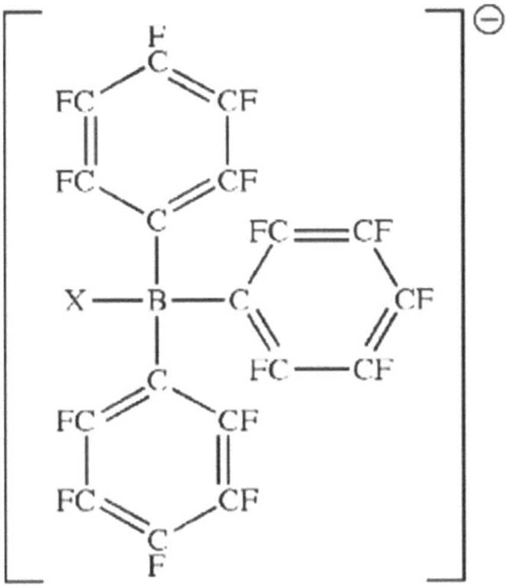

$$\left[X{-}B(C_6F_5)_3 \right]^{\ominus}$$

II

Since the Lewis acidity of the co-catalyst is most important toward its activity, it was first thought that three-coordinate Al complexes (Fig. 5.21) would be most effective for activating the metallocene structure, relative to coordinatively saturated four-coordinate Al compounds. However, it has been shown that the latter structures $\{(CH_3)Al(\mu{-}O)\}_n$, $n = 6$–$12\}$, which exist as complex caged structures, are most effective for activating the metallocene catalyst. The term "latent Lewis acids" has been coined for these structures by Barron,[19] indicating that the isolated caged structures do not themselves possess acidity, but become activated due to a cage-opening

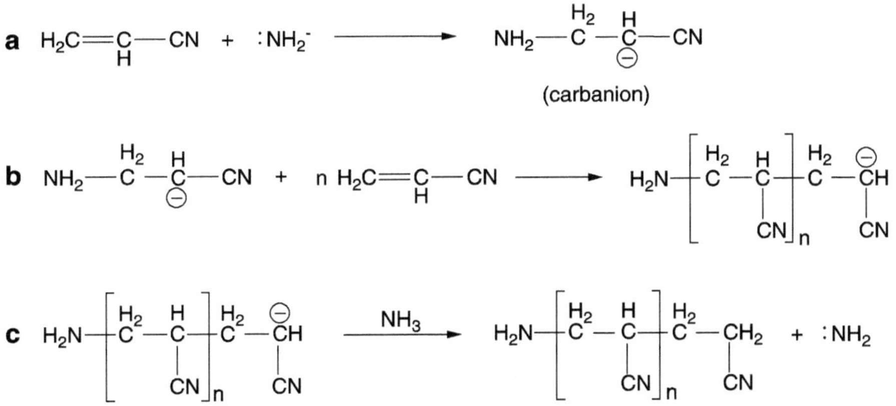

a $H_2C{=}CH{-}CN + :NH_2^- \longrightarrow NH_2{-}CH_2{-}\overset{\ominus}{C}H{-}CN$ (carbanion)

b $NH_2{-}CH_2{-}\overset{\ominus}{C}H{-}CN + n\,H_2C{=}CH{-}CN \longrightarrow H_2N{-}[CH_2{-}CH(CN)]_n{-}CH_2{-}\overset{\ominus}{C}H(CN)$

c $H_2N{-}[CH_2{-}CH(CN)]_n{-}CH_2{-}\overset{\ominus}{C}H(CN) \xrightarrow{NH_3} H_2N{-}[CH_2{-}CH(CN)]_n{-}CH_2{-}CH_2(CN) + :NH_2$

Fig. 5.18 Reactions involved in anionic addition polymerization. Shown are **a** generation of a carbanion from a Lewis basic initiator, **b** propagation of the polymer chain through the combination of the carbanionic polymer chain and additional monomers, and **c** termination of the polymer growth through the addition of a Lewis base. Unlike the other addition polymerization schemes, termination does not occur in situ, but must be initiated deliberately

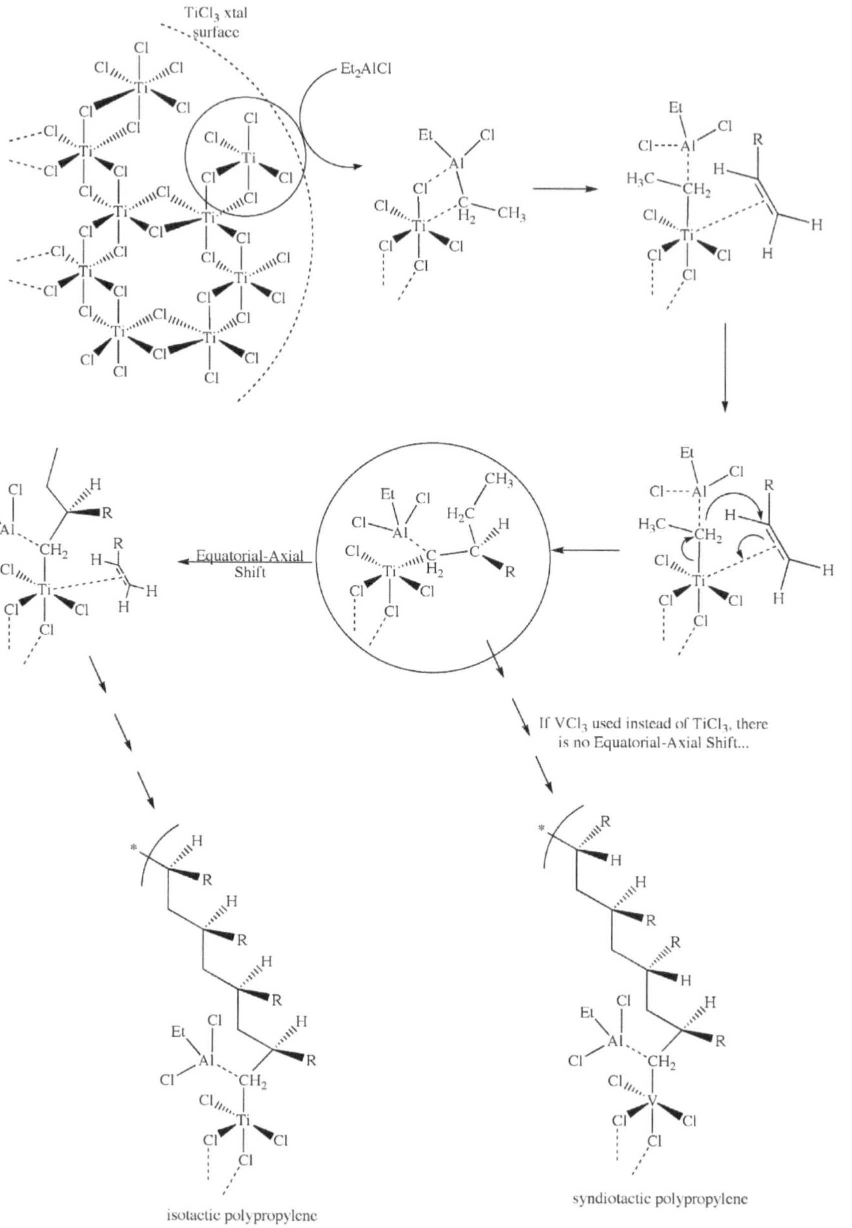

Fig. 5.19 Mechanism involved in Ziegler–Natta polymerization for tacticity control over the growing polymer chain

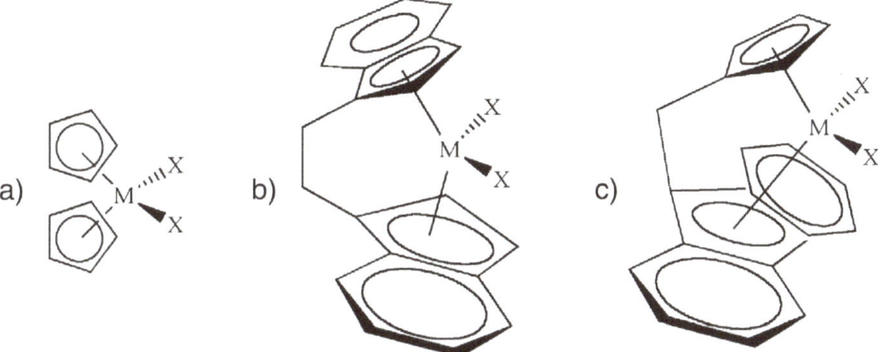

Fig. 5.20 Molecular structures of common metallocene homogeneous polymerization catalysts (M=Zr, Hf, X=Me, Cl). The cyclopentadienyl ligands may be abbreviated as "cp" in the molecular formula (*e.g.*, $(cp)_2HfMe_2$)

mechanism upon contact with the metallocene catalyst (Fig. 5.22). The magnitude of Lewis acidity is directly related to the cage strain; as one would expect, metallocene activation is more pronounced with increasingly smaller cages.

Three mechanisms have been proposed to explain metallocene-based homogeneous and Ziegler–Natta polymerization schemes. The Cossee–Arlman mechanism (Fig. 5.23a) proposes the coordination of the olefin to the vacant metal site, with

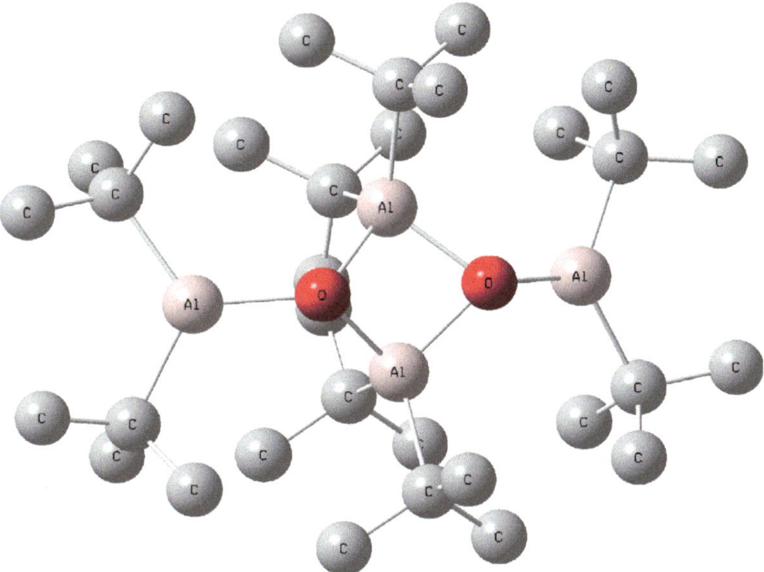

Fig. 5.21 Molecular structure of $[(^tBu)_2Al(\mu\text{-}O)Al(^tBu)_2]_2$, featuring three-coordinate, unsaturated Al centers

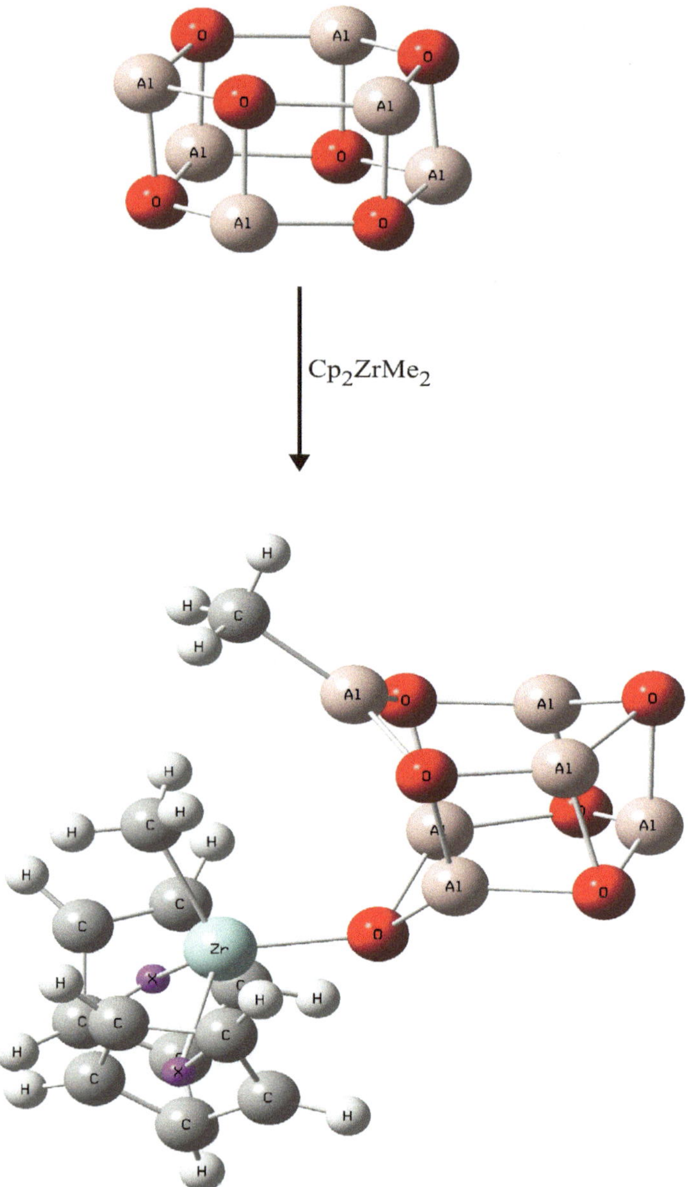

Fig. 5.22 Molecular structure of $[Me_2AlO]_6$, and schematic of the cage-opening mechanism of the alumoxane co-catalyst during metallocene-catalyzed polymerization

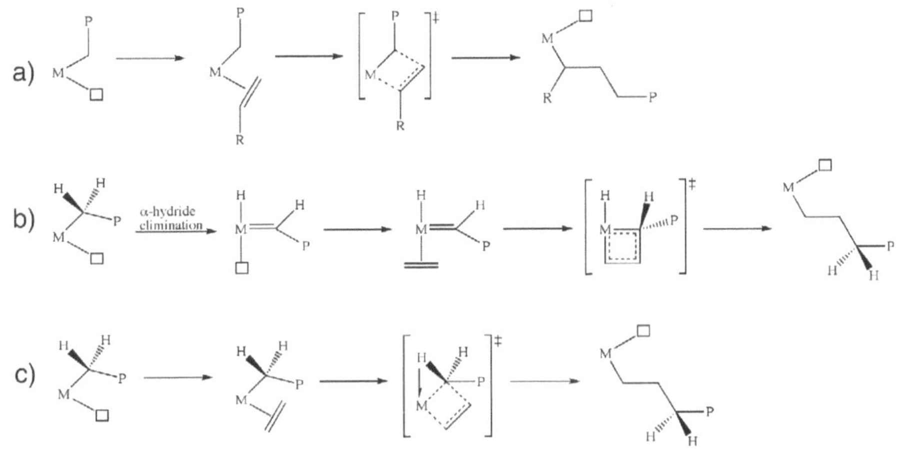

Fig. 5.23 Proposed mechanisms for transition metal-catalyzed olefin polymerization

its subsequent migratory insertion into the M–P bond (P = growing polymer). The insertion step likely proceeds through a four-membered cyclic transition state. Another mechanism proposes an initial α-hydride shift, resulting in a metal hydride species that may add an olefin and form a growing polymer chain through migratory insertion (Fig. 5.23b). The last mechanism that involves α-agostic Ti–H interactions (Fig. 5.23c) is generally accepted as being an important part of the Cossee–Arlman mechanism, serving to slow down the insertion reaction and favor the approach of incoming ligand.

It should be noted that early transition metals (Group 4 or 5) are most favorable for transition metal catalyzed living polymerization. Since these metals have few d-electrons in their valence shell, there is a lesser chance for *β*-hydride elimination (polymer termination) to occur.[20] In order to purposely terminate chain growth, either the M–C bond may be broken through reaction with hydrogen gas (Fig. 5.24a) or thermally induced *β*-hydride elimination. Whereas the first termination route results in low polydispersity, the latter method results in a terminal olefin (Fig. 5.24b) that may insert into a neighboring polymer chain yielding a branched, low-density product.

Step-growth polymerization

Thus far, we have considered addition polymerization routes—either catalyzed or uncatalyzed. Although this is sufficient to describe the synthesis of common packaging materials such as polyethylene, polypropylene, polystyrene, etc., other common classes of polymers such as nylon, PETE, and polyacrylamide are generated through step-growth mechanisms. Although the synthetic pathway for these polymers is more straightforward than addition polymerization, there are many intricate considerations that affect overall polymer properties.

The general types of step-growth polymerizations are shown in Fig. 5.25. Since the resulting polymers contain reactive functional groups on either/both ends, it is

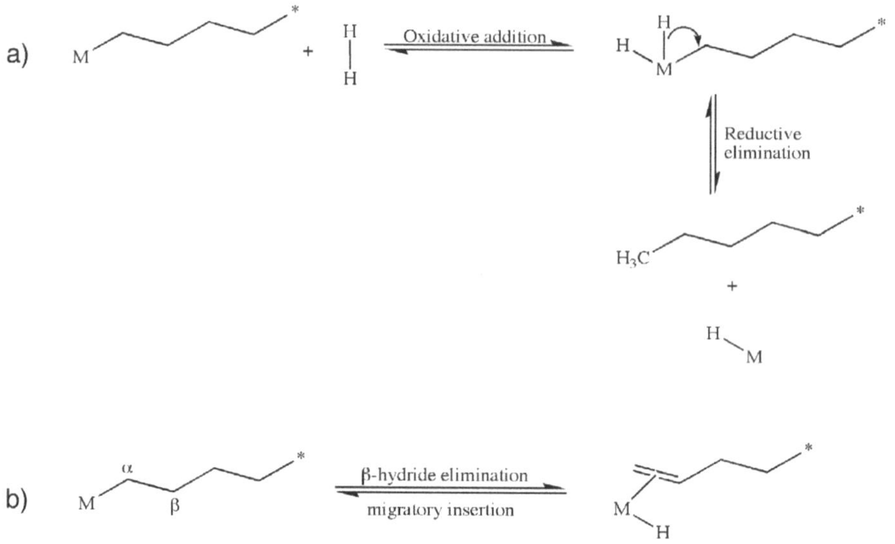

Fig. 5.24 Schematic of **a** hydrogenolysis termination of a polymer chain and **b** β-hydride elimination termination, which results in an olefin-terminated polymer chain

referred to as a *telechelic* macromolecule. Although water is typically released as a byproduct of these polymerizations, other small molecules such as alcohols and alkyl halides may also be generated based on the monomers that co-condense. In its simplest form, condensation reactions result in linear polymers from the reaction of bifunctional monomers. For example, one of the most common linear step-growth polymers is poly(ethylene terephalate) (PET; commonly denoted by the trademarks Dacron or Mylar), used for soft drink bottles due to its impermeability toward liquids and gases. This polymer is obtained through a two-step reaction between ethylene glycol (HO–C_2H_4–OH) and the dimethyl ester of terephthalic acid (CH_3O–C(O)–benz–C(O)–OCH_3). Both methanol and ethylene glycol byproducts are evolved during the condensation reaction.

In contrast, if the monomers are multifunctional rather than difunctional, a branched polymeric structure will result (Fig. 5.26). For instance, consider either the "$AB_x/AB_x + B_y$" (Fig. 5.26a) or the "$A_2 + B_y$" (Fig. 5.26b) systems, where A and B refer to the reactive functional groups of the monomers (subscripts indicate the number of reactive functional groups).[21] In describing branched step-growth polymers, it is generally assumed that:

(i) A and B endgroups react only with each other
(ii) All endgroups of a given type are equally reactive
(iii) Intramolecular cyclization reactions do not occur
(iv) Probability of A–B reactions is independent of molecular size.

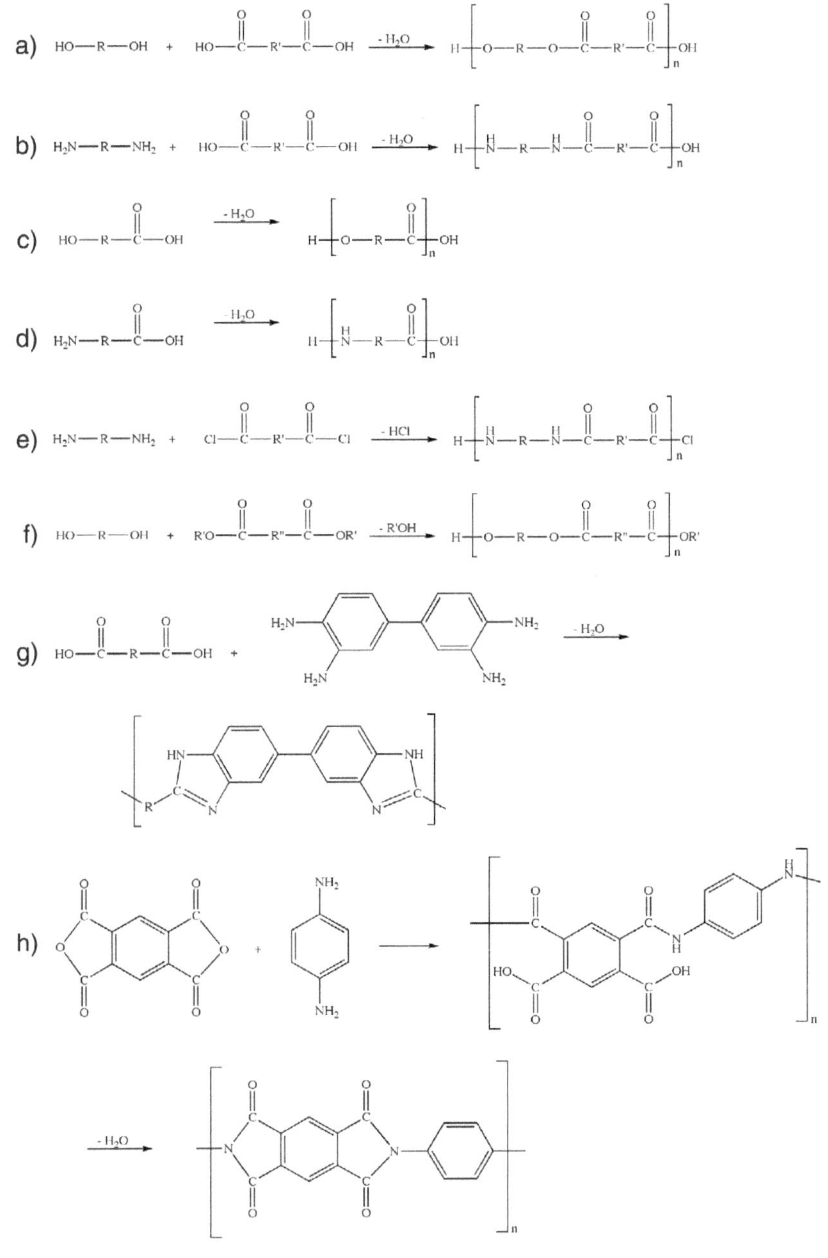

Fig. 5.25 Reaction schemes for the most common types of step-growth polymerization. Shown are **a/c** polyester formation, **b/d** polyamide formation, **e** polyamide formation through reaction of an acid chloride with a diamine, **f** transesterification involving a carboxylic acid ester and an alcohol, **g** polybenzimidazole formation through condensation of a dicarboxylic acid and aromatic tetramines, and **h** polyimide formation from the reaction of dianhydrides and diamines

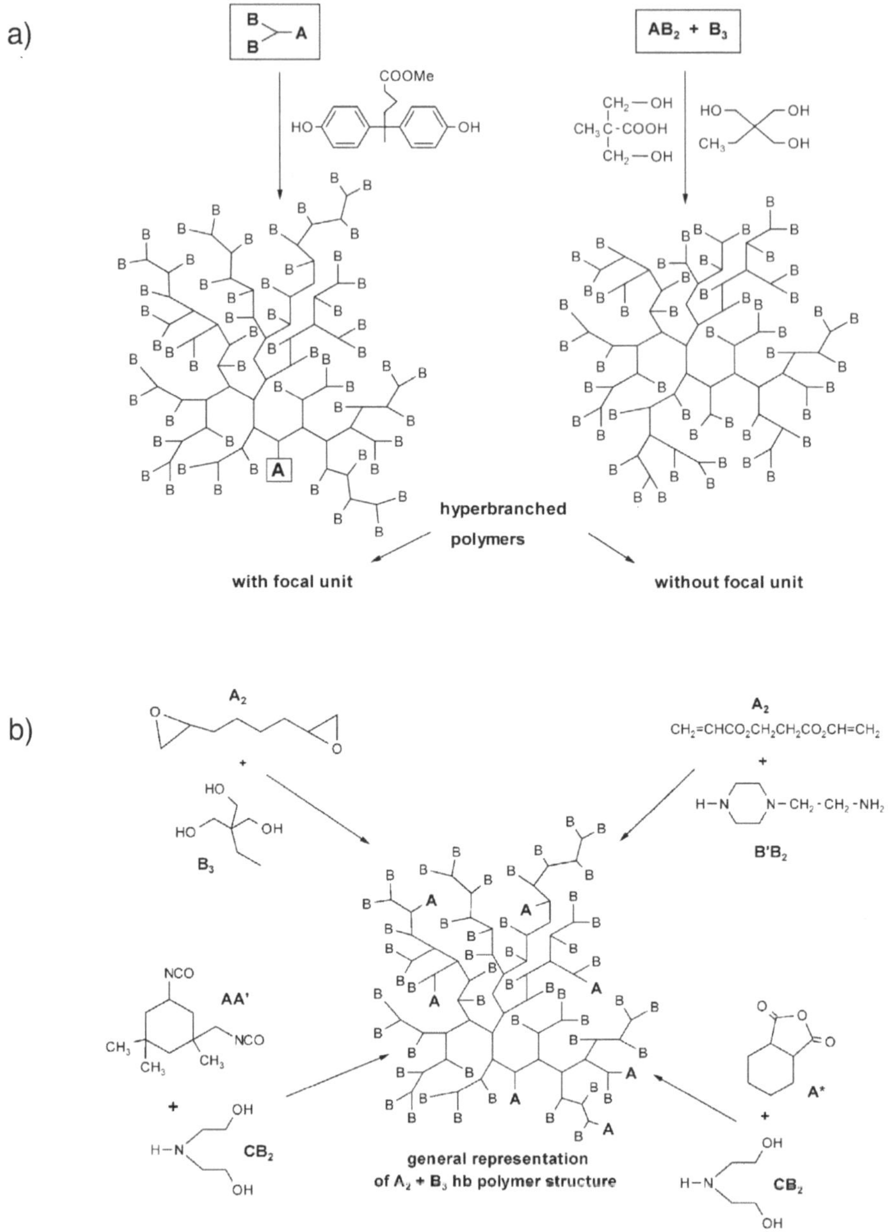

Fig. 5.26 **a** Scheme of the formation of hyperbranched polymers via the AB_x and $AB_x + B_y$ approach ($x \geq 2$; here, 2; $y \geq 3$; here, 3). **b** Scheme of the synthesis of hyperbranched polymers by various $A_2 + B_y$ ($y \geq 3$) approaches, with examples of monomer combinations ($A*$ indicates cyclic monomer; AA' indicates differences in the reactivity). Reproduced with permission from *Chem. Rev.* **2009**, *109*, 5924. © 2009 American Chemical Society

It is interesting to note that whereas $A_x + B_y$ hyperbranched systems have both A and B endgroups, AB_x-based hyperbranched polymers contain only B endgroups. This leads to the discrepancy between the two systems in forming infinite networks (gels). The $A_x + B_y$ system results in gel formation if $x \geq 2$ and $y \geq 3$, whereas it is not possible to form a gel via the AB_x system. Although the polydispersity of hyperbranched polymers is typically quite large, there is increasing interest for their use in a number of blends/coatings applications due to the high density of peripheral groups and resultant enhanced solubility/surface adhesion, relative to linear analogues. As a more recent extension, network structures that consist of interwoven hyperbranched polymers may also be synthesized (Fig. 5.27), which have been proven useful for lithographic applications.

Silicones

Although the backbone units of polymers are often assumed to be carbonaceous, there is one major class of polymers that are composed of crosslinked $[O–SiR_2–]_n$ units. These polymers are known as *silicones* or *polyorganosiloxanes*—infamously popular due to the bad press over a decade ago concerning silicone breast implants. By varying the –Si–O– chain lengths, Si-alkyl groups, and extent of crosslinking, the resultant polymer may exist as a viscous liquid (*e.g.*, vacuum pump oil), a gel (*e.g.*, silicone grease), or rubbery material (*e.g.*, used for remote control keypads). Figure 5.28 illustrates the diverse use of silicone-based devices within the body. Their extensive use for biomedical applications is due to their biocompatibility, sterilizability, surface adhesion, oxygen permeability, and resistance to attack from body defense systems. The most widely used silicones are known as *polydimethylsiloxanes* (PDMS), formed from the hydrolysis of chlorosilanes in the presence of water (5.4). In order to increase the polymer size to that required for most applications, the linear and cyclic oligomeric units are subsequently condensed and polymerized, respectively.[22]

$$\mathrm{x\ Me_2SiCl_2} \xrightarrow{\mathrm{H_2O}} \mathrm{y}\ \underset{\text{(linear)}}{\mathrm{HO(Me_2SiO)_nH}} + \mathrm{z}\ \underset{\text{(cyclic)}}{\mathrm{(Me_2SiO)_m}} \tag{5.4}$$

Dendritic Polymers

Thus far, we have only considered step-growth polymers obtained through random condensation reactions of multifunctional monomers. The pioneering work of Nobel Laureate Flory[23] in the 1940s helped the polymer community understand the kinetics of branched polymer growth, suggesting that control over sequential step-growth should be possible. However, this was not proven empirically until the work of Vögtle[24] and Tomalia[25] in the late 1970s–early 1980s. Vögtle developed a repeatable "cascade" route to produce low molecular weight amines (Fig. 5.29). However, due to cyclization side reactions, successful polymer growth via the Vögtle approach was not possible—finally being realized in 1993 for the synthesis of poly(propyleneimine) (PPI). The work of the Tomalia group at Dow Chemical was the first to yield a

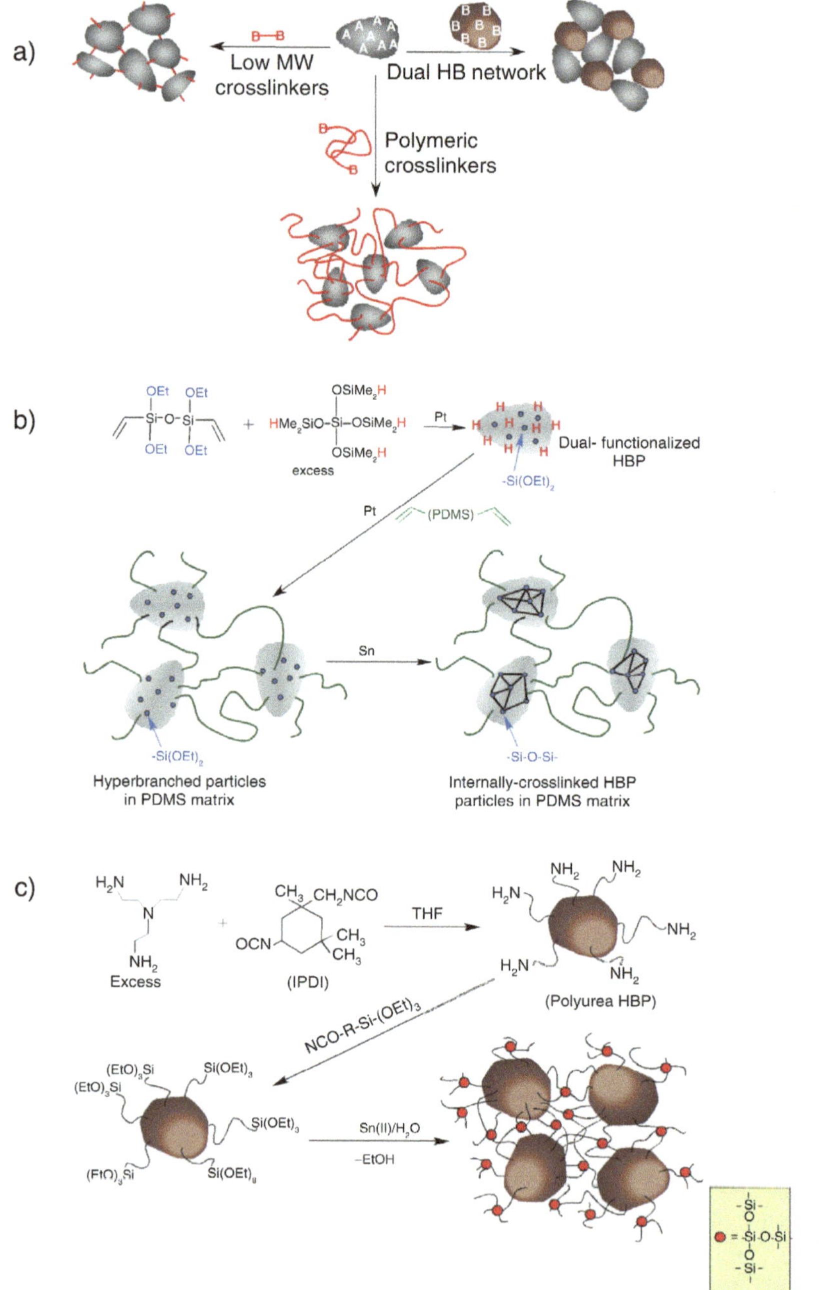

Fig. 5.27 The general strategy **a** and examples **b–c** of hyperbranched polymer network structures featuring a siloxane crosslinkage. Reproduced with permission from Meijer, D.; Dvornic, P. R. Fall 2005 ACS meeting, Midland, MI

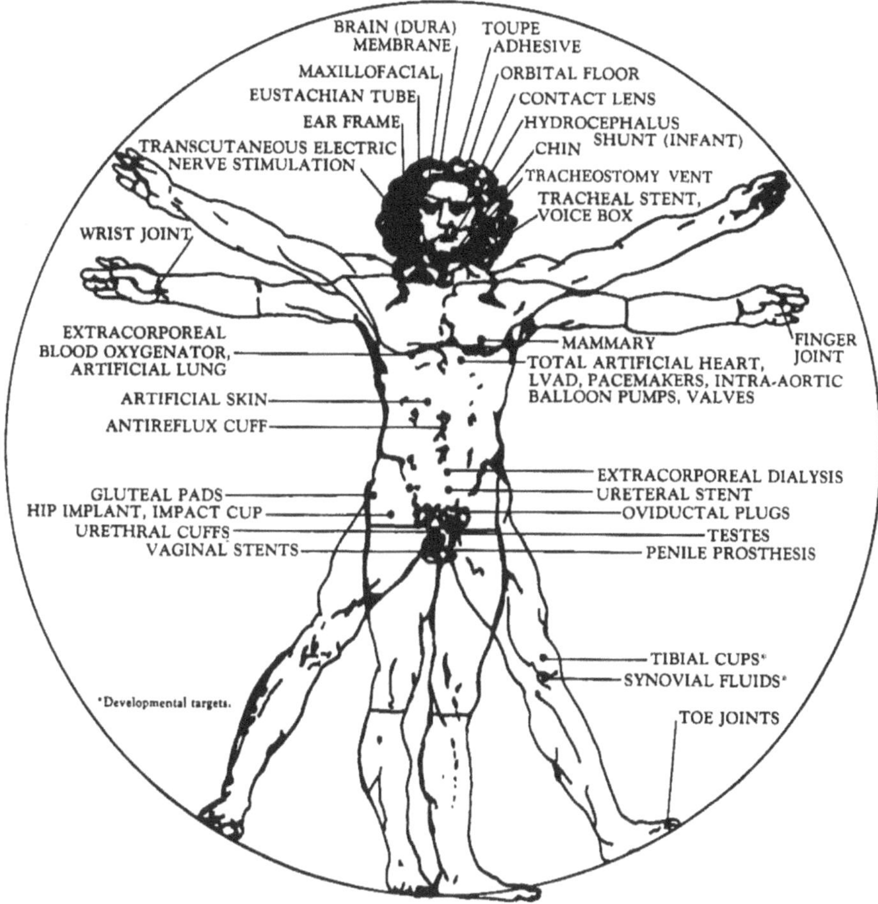

Fig. 5.28 Placement of silicone-based medical devices in the body. Reproduced with permission from *CHEMTECH* **1983,** *13*, 542

perfectly defined dendritic polymer structure, with an extremely low polydispersity (*ca.* PDI of 1.00–1.05; Fig. 5.30).[26] These polymers were coined *starburst dendrimers*, referring to the star-branched architecture and the Greek word "dendra" for tree. Not unlike other major scientific discoveries, the first report of the dendritic architecture was riddled with skepticism by the scientific community. Shortly thereafter, Newkome helped silence the critics with his publication of branched dendrimers that he called an *arborol* (Fig. 5.31).[27] To illustrate the novelty of these structures, Table 5.3 lists the comparative properties of dendrimers and linear polymers.

The earliest syntheses of dendritic polymers were divergent in nature; with growth initiating from a core and outward propagation. The terminal groups of the core react with complementary groups on the monomer, which forms a new branching point for

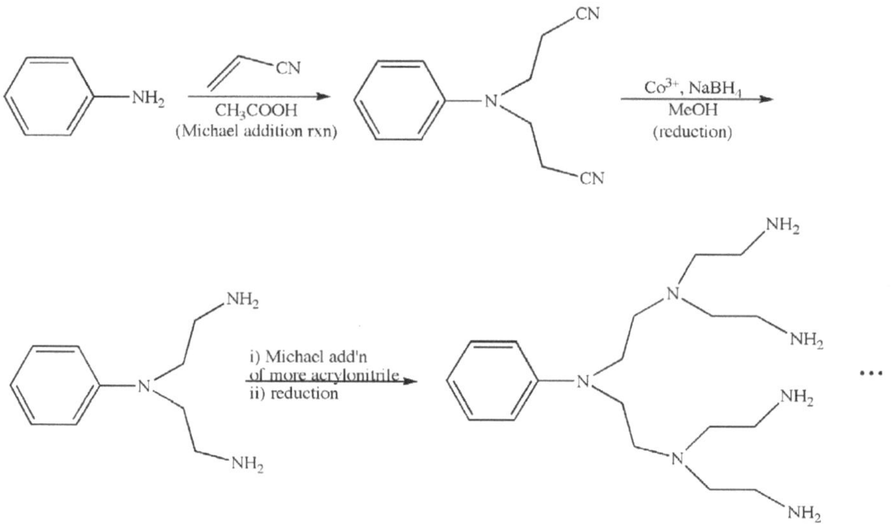

Fig. 5.29 The Vögtle approach to yield low molecular weight amines via controlled sequential synthesis

subsequent branching reactions (Fig. 5.32a). Most importantly, the terminal functional groups on the monomers are designed to be reactive with only the outwardly growing polymer, which prevents random hyperbranched growth. Such a repetitive procedure results in an exponential increase in reactions that occur on the periphery of the growing polymer, requiring a large excess of reagents. Although this technique is used for the large-scale synthesis of many dendrimers (*e.g.*, poly(amidoamine), PAMAM), a leading drawback is the relatively high number of defect structures—especially for higher *generations* (a term used to describe the sequential branches emanating from the core of the dendritic structure). With each generation, there is an increased probability for incomplete/side reactions, which are not easily removed from the solution due to their structural similarity to the final product.

In order to circumvent the purity issues associated with divergent syntheses, Frechet and coworkers designed a convergent approach in the late 1980s.[28] In contrast to the divergent approach, growth initiates from the exterior of the molecule progressing inwardly by coupling endgroups to each branch of the monomer (Fig. 5.32b). The functional group at the focal point of the wedge-shaped dendritic fragment (known as a *dendron*) may react with additional monomers to build up higher-generation dendrons. When the desired generation dendron is attained, these units are then attached to a polyfunctional core to form the final dendrimer. This route drastically increases the purity of higher-generation dendrimers relative to divergent syntheses, as there are many fewer reactions per molecule.

In addition, the reactions only require a slight excess of reagent, in contrast to the large excesses that are needed for divergent growth. However, this technique is not useful for commercial large-scale dendrimer synthesis, as the mass of the

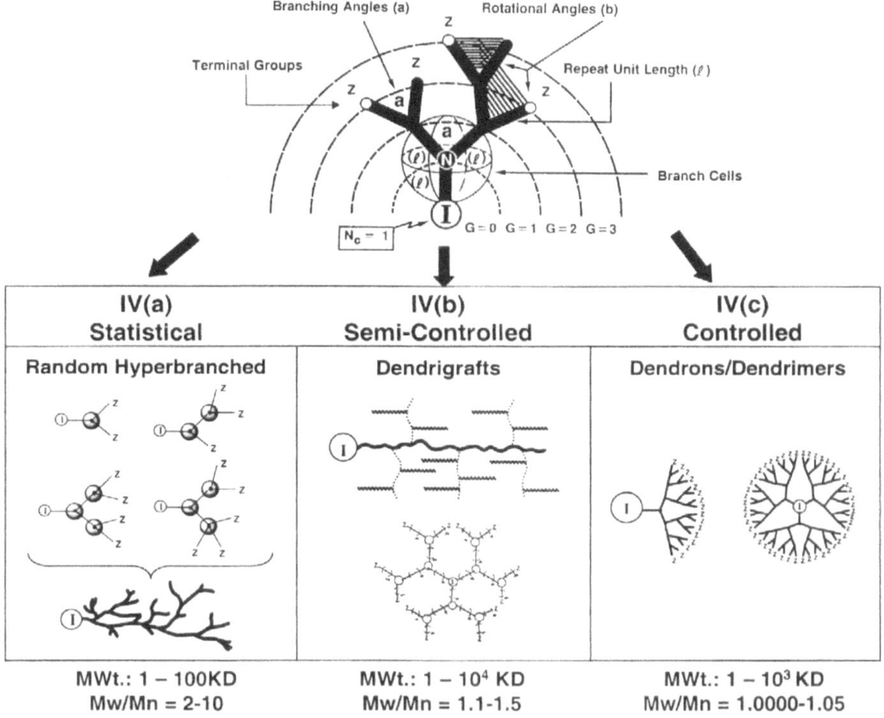

Fig. 5.30 Illustration of resultant polymers through varying the degree of control of step-growth polymerization. Each successive growth layer is referred to as a *generation* (G). Reproduced with permission from *Dendrimers and Dendritic Polymers*, Wiley: New York, 2001

sample decreases with additional generation growth, resulting in low yields of higher-generation dendrimers due to steric crowding around the focal point of the growing dendron. Nevertheless, this is the only route that offers precise structural control over the growing dendrimer, such as being able to modify the focal point/chain ends to yield well-defined unsymmetrical dendrimers. This strategy is being developed to synthesize "bowtie" dendrimers containing both target and drug-delivery agents (Fig. 5.33). As an alternative strategy for drug delivery, surface modification of PAMAM with cancer targets and anticancer drugs (Fig. 5.34) has also been proven successful in preliminary trials. We will further discuss advances in drug-delivery agents later in this chapter.

To date, the PAMAM dendrimer remains the most heavily utilized for applications, due to its facile scale-up and commercial availability. In Chap. 6, we will discuss its use as a nanoreactor/nanocapsule stabilizing agent for nanoparticle growth within both aqueous or organic solvent media.[29] Not only can one alter its solubility characteristics by changing the peripheral moieties from hydrophilic/hydrophobic character, but also its overall properties. For instance, Starpharma in association with Dendritic Nanotechnologies, Inc. have developed a HIV/AIDS drug that is based on

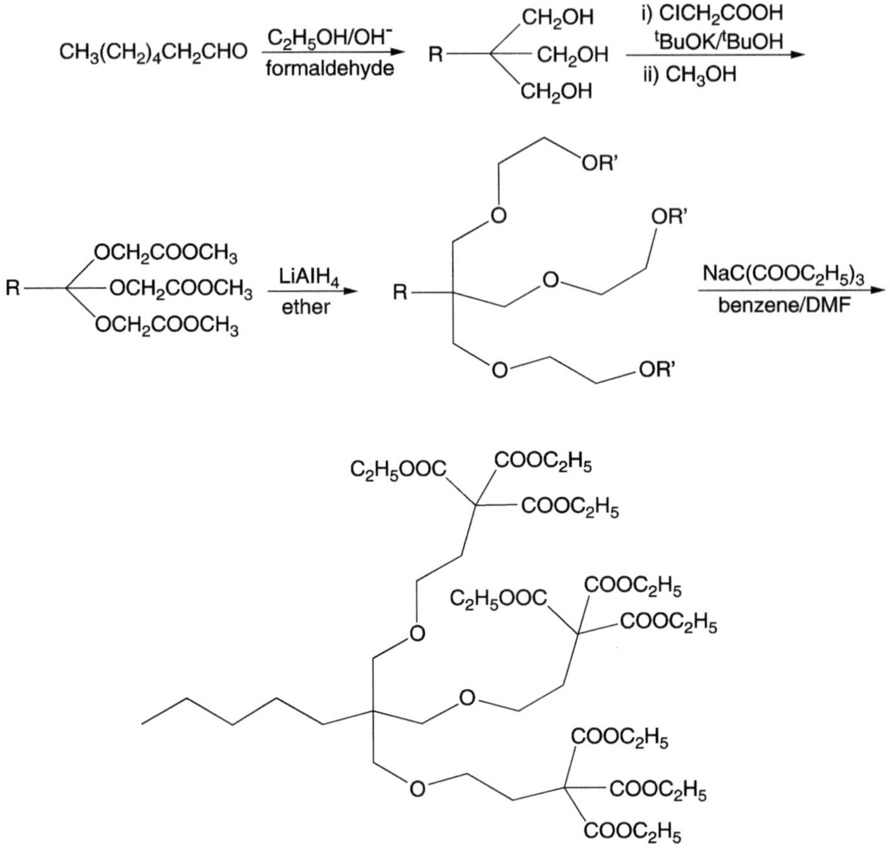

Fig. 5.31 The Newkome approach for the sequential step-growth of arborols

Table 5.3 Comparative properties of dendrimers and linear polymers

Property	Dendrimers	Linear polymers
Structure	Compact, globular	Not compact
Synthesis	Controlled, stepwise growth	Single-step polycondensation
Structural control	Very high	Low
Architecture	Regular	Irregular
Shape	Spherical	Random coil
Crystallinity	Non-crystalline, amorphous	Semi-crystalline/crystalline
T_g	Low	High
Aqueous solubility	High	Low
Nonpolar solubility	High	Low
Viscosity	Non-linear relationship w/M_w	Linear relationship w/M_w
Compressibility	Low	High
Polydispersity	Monodisperse	Polydisperse

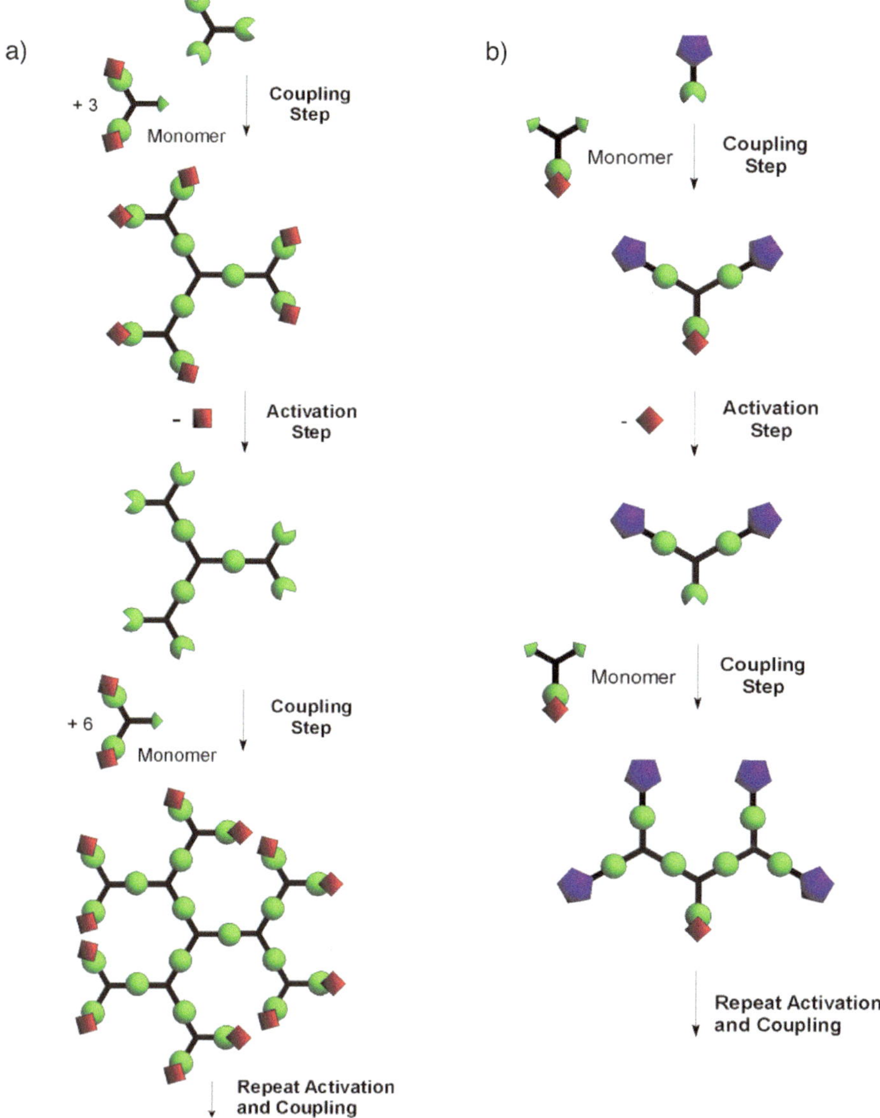

Fig. 5.32 Schematic comparison of **a** divergent and **b** convergent dendrimer synthetic routes. Reproduced with permission from *Chem. Rev.*, **2001**, *101*, 3819. © 2001 American Chemical Society

a PAMAM architecture functionalized with sulfonic acid end groups.[30] However, if the terminal groups are changed to oligo(ethylene glycol), the dendrimer may be used as a pore generating agent in the development of dielectric thin films for microelectronic devices.[31] Poly(lysine) dendrimers modified with sulfonated napthyl groups have shown activity as antiviral drugs against the herpes simplex virus.[32] Svenson and

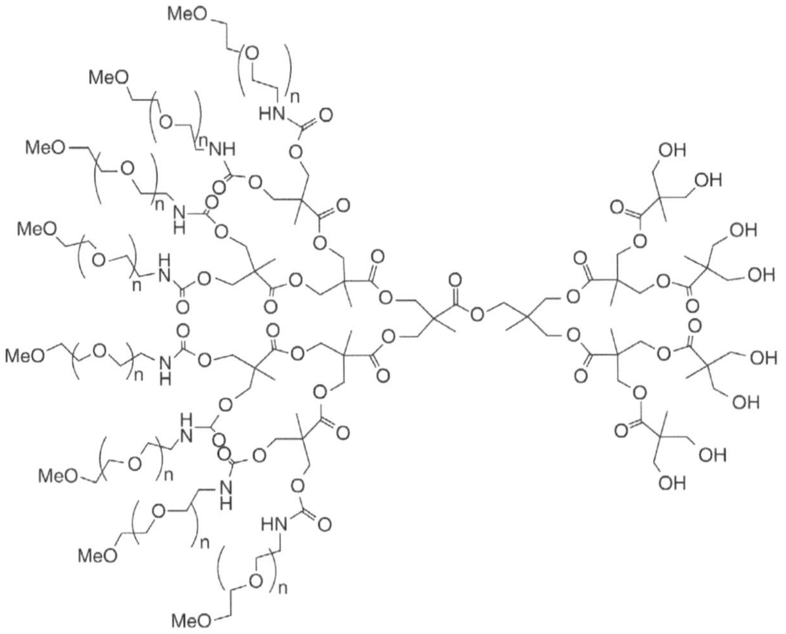

Fig. 5.33 Bowtie dendrimer synthesized via the convergent approach, for drug delivery of anti-cancer drugs to target organs. Reproduced with permission from *Mol. Pharm.*, **2005**, 2, 129. © 2005 American Chemical Society

Tomalia provide a nice review of the multifaceted use of dendrimers for biomedical applications[33]; Li and Aida provide a review of dendrimer porphyrins and phthalocyanines, which have attracted recent interest as sensitizers for photodynamic theraphy (PDT) and biosensing applications.[34]

The first "co-polymer dendrimer" was developed by Dvornic and coworkers, which feature both a hydrophilic PAMAM core and a hydrophobic organosilicon shell (Fig. 5.35).[35] This structure proves extremely useful for the encapsulation of polar species within organic solvents, for the growth of nanoparticles (Chap. 6). Due to the water-sensitive alkoxysilyl groups (*e.g.*, Si–OCH_3), facile network formation may also take place (Fig. 5.36) via the analogous hydrolysis reactions that were previously discussed for sol–gel growth of SiO_2 networks (Chap. 2). The crosslinking of dendritic units to form an extended network is sometimes referred to as a *megamer*—of increasing interest for functional coatings (*e.g.*, sensors, smart fabrics, etc.) applications. A further utility of the PAMAMOS structure is its reactivity toward a glass surface, which contains silanol (Si–OH) reactive groups (Fig. 5.37). This results in a permanent coating, with a controllable degree of surface adsorption based on the peripheral groups of the dendrimer. These properties have recently been exploited by the deposition of copper-encapsulated PAMAMOS anti-fouling coatings on ship hulls, to prevent the adhesion of zebra mussels.[36]

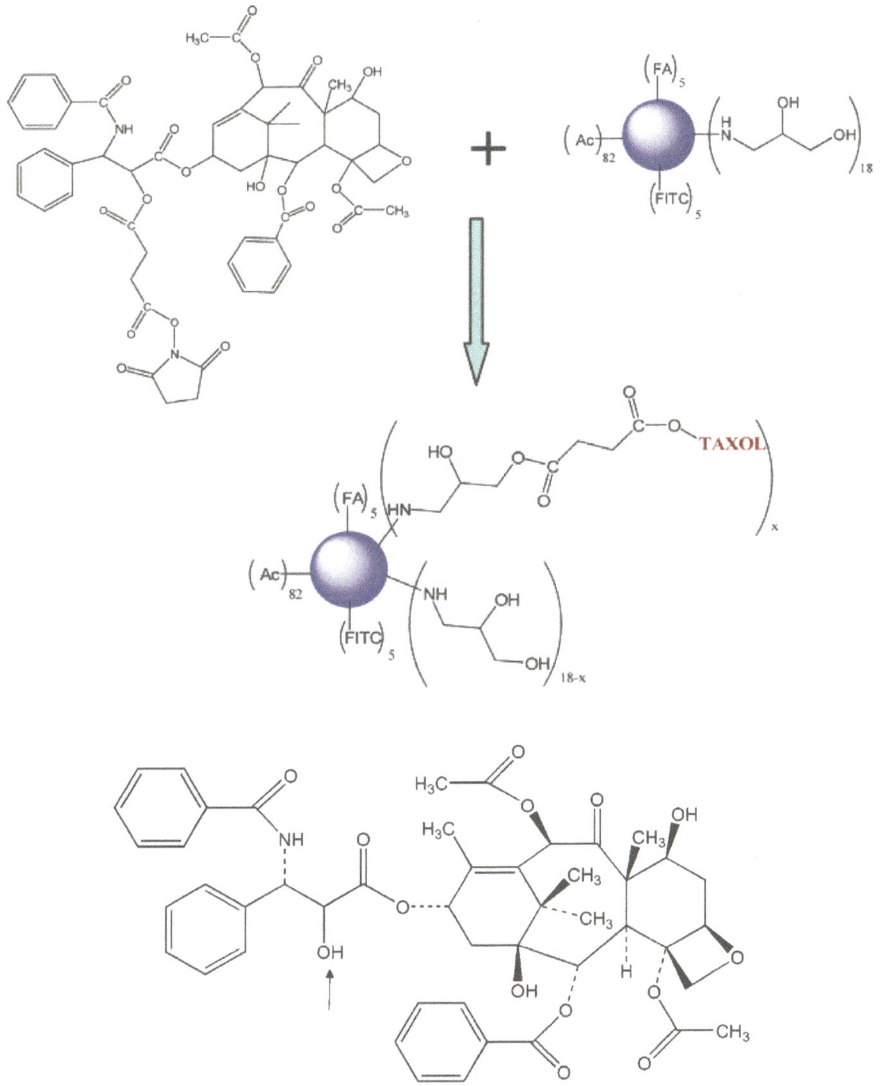

Fig. 5.34 PAMAM dendrimer multifunctional conjugates for cancer treatment. The FA group is a folic acid cancer cell target, and FITC is fluorescein isothiocyanate, used as an imaging agent. Also shown (bottom) is the molecular structure for the anticancer drug, taxol, denoting the –OH group that covalently attaches to the dendrimer. Reproduced with permission from *Biomacromolecules*, **2006**, 7, 572. © 2006 American Chemical Society

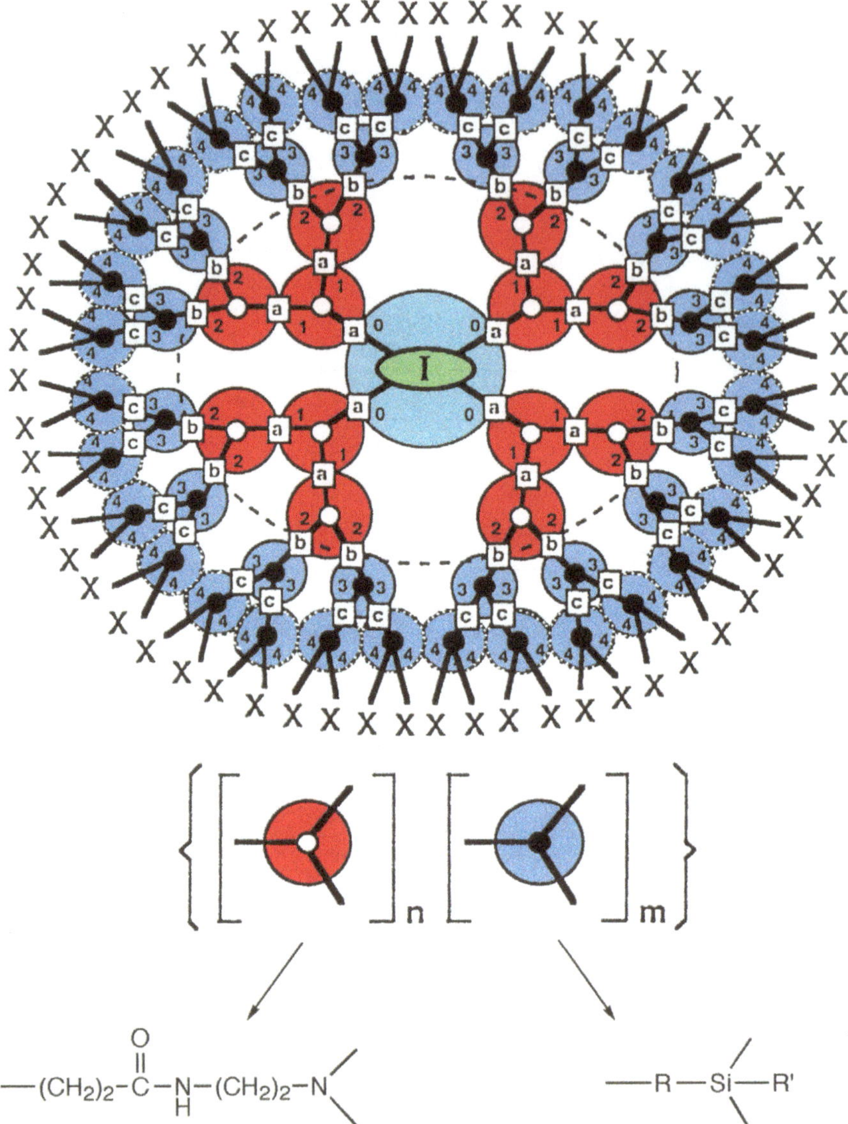

Fig. 5.35 Illustration of a poly(amidoamine-organosilicon) (PAMAMOS) dendrimer, with two generations of each PAMAM and organosilicon units. Although the PAMAMOS represents a block copolymer, an unlimited number of other variations that contain a random copolymer array, or varying dendron subunits, may also be synthesized. Reproduced with permission from *Synthesis and Properties of Silicones and Silicone-Modified Materials*, ACS Symposium Series 838, **2002**, 236. © 2002 American Chemical Society

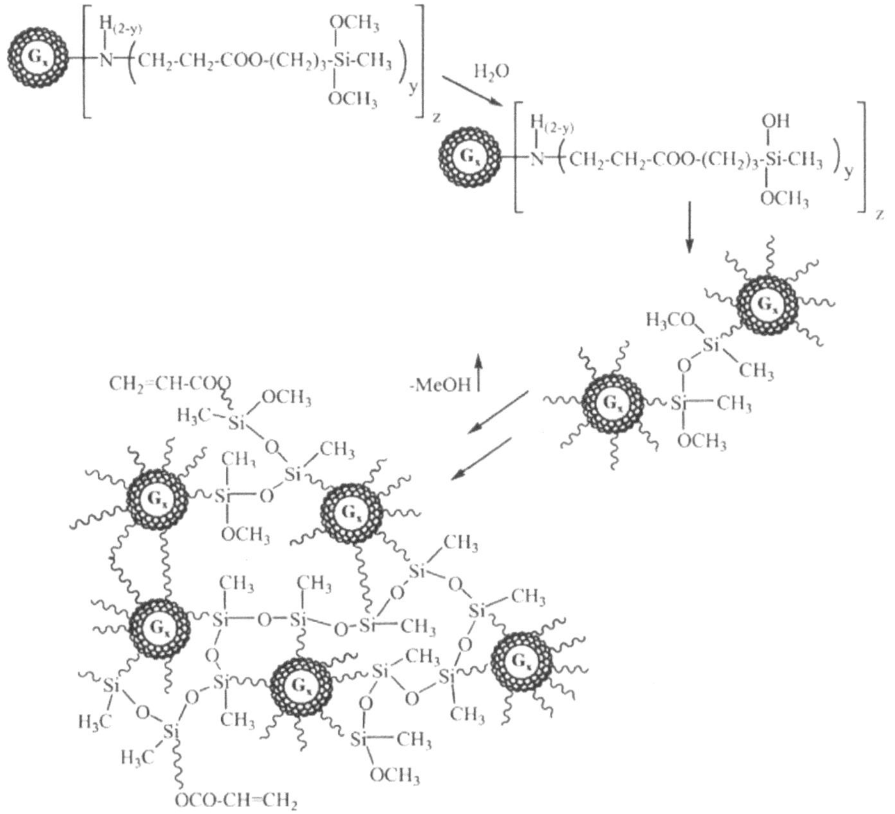

Fig. 5.36 Network (megamer) formation through the hydrolysis/crosslinking of neighboring PAMAMOS dendrimer units. Hydrolysis of the C–O–Si bond may also be exploited for the controlled-release of entrained agents (*e.g.*, cancer drugs, etc.). It should be noted that subsequent thermal annealing to remove the PAMAM cores results in a nanoporous network that has a dielectric constant (k) of *ca.* 1.5—of extreme interest for next-generation IC interconnect applications. Reproduced with permission from *Macromolecules*, **2002**, *35*, 9323. © 2002 American Chemical Society

We have described the growth of dendritic polymers as being highly controllable; however, the resultant size of the polymer is mathematically limited. This is in direct contrast to linear polymers that may increase in size to infinity (as long as they remain soluble within the solvent). As the dendritic structure grows, there become significant steric interactions among the exponentially increasing number of peripheral groups. This phenomenon is known as the *De Gennes dense packing*,[37] and results in a more structurally flawed, globular structure as the dendrimer generation increases.

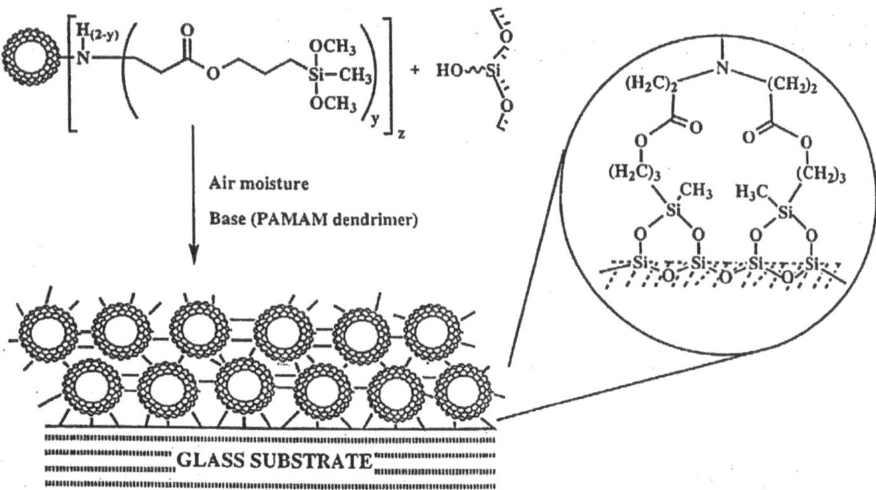

Fig. 5.37 The formation of covalently bound coatings of PAMAMOS onto a glass surface. Reproduced with permission from *Macromolecules*, **2002**, *35*, 9323. © 2002 American Chemical Society

Polymerization via *"Click" Chemistry*

As we know from experiment, any chemical reaction will result in byproducts and side-reactions that will limit the overall yield to <100%. Quantitative yields are quite rare for synthetic chemistry; that is, until the introduction of *click chemistry* by Sharpless and coworkers in 2001.[38] By definition, click chemistry involves reactions that occur by high/quantitative yield, generate few/no byproducts, and are stereospecific—setting an important precedent toward mimicking nature's synthetic efficiency. In particular, nature efficiently links small molecules together via heteroatomic C–X–C bonding to yield primary metabolites (polypeptides, polynucleotides, and polysaccharides, Fig. 5.38), which are essential for life.

Click processes occur through simple reaction conditions such as air/moisture insensitivity, solventless or aqueous media, readily available precursors and reagents, and simple product isolation—mostly precluding chromatographic separation (unlike most organic syntheses). The utilization of click chemistry in combination with combinatorial screening will speed up the discovery of new pharmaceuticals that our society will continue to rely upon.[39]

Beyond small-molecule drug discovery,[40] click chemistry may also be exploited for the synthesis of polymers and supramolecular[41] architectures.[42] Since the overall properties of the polymer are closely related to its side groups, this technique has been used to easily fine-tune polymeric structures by simple high-yield reactions of monomeric units (Fig. 5.39). One example is the coupling of two linear-chain polymers to generate a block co-polymer; of significant challenge due to the reduced reactivity of the polymeric chain ends. A strategy to accomplish this difficult chemistry is to functionalize the endgroups with alkyne or azide groups, which undergo

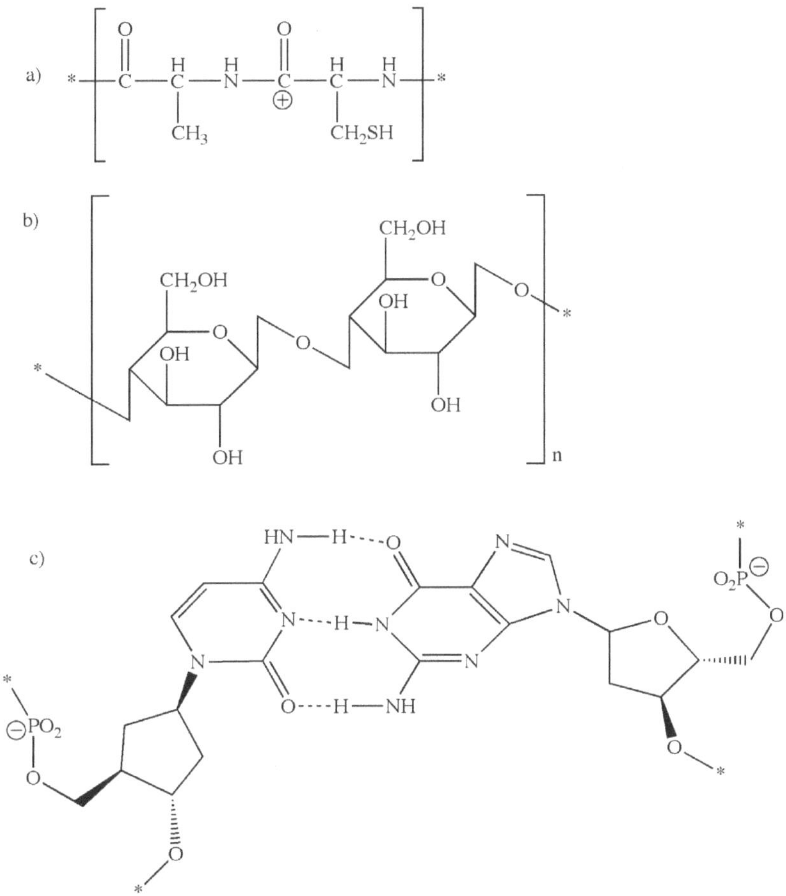

Fig. 5.38 Examples of molecular structures for **a** polypeptides, **b** polysaccharides, and **c** polynucleotides

high-yielding Cu- or Ru-catalyzed cycloaddition reactions (Fig. 5.40).[43] Metal-free click polymerization routes have also been reported, as well as new varieties such as nitrile oxide-alkyne cycloadditions and thiol-ene/-yne hydrothiolations (Fig. 5.41).[44] Even dendritic polymers may be synthesized or surface-functionalized using click chemistry (Fig. 5.42). This may be of extreme interest to the industrial and scientific community, as the current cost of PAMAM dendrimers are still rather high. The latest version of PRIOSTAR™ dendrimers developed by Tomalia and coworkers are generated via click chemistry and offer the analogous functionality and applications as PAMAM dendrimers at a fraction of the cost.[45]

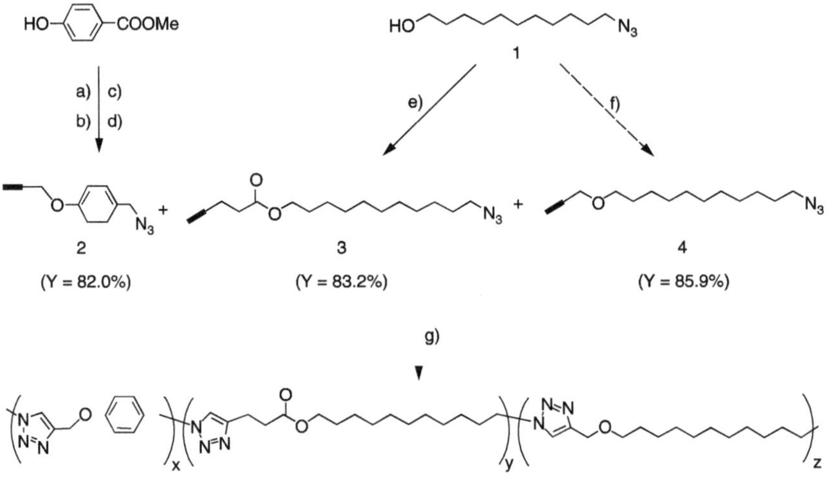

2 (Y = 82.0%) 3 (Y = 83.2%) 4 (Y = 85.9%)

5: x = 1 6: y = 1 7: z = 1

8: x = 1/2, y = 1/2 9: x = 1/2, z = 1/2 10: y = 1/2, z = 1/2 11: x = 1/3, y = 1/3, z = 1/3

a) propargyl bromide, K_2CO_3, acetone; b) $LiAlH_4$, THF; c) CBr_4, PPh_3, THF; d) NaN_3, H_2O;
e) 4-pentynoic acid, DCC, DPTS, DMAP, CH_2Cl_2; f) propargyl bromide, NaH, THF; g) $Cu(PPh_3)_3Br$, DIPEA, $CHCl_3$.

Fig. 5.39 Click chemistry step growth (co)polymerization of α-azide-ω-alkyne monomers **2–4**. Adapted with permission from *Chem. Commun.*, **2008**, 4138. © 2008 Royal Society of Chemistry

5.3 "Soft Materials" Applications: Structure *Versus* Properties

Thus far, we have considered many polymer types, which are used for a diverse range of applications. In this section, we will now consider more specific examples of applications, to show how the molecular structure of the polymer drastically affects its properties. For both synthetic (*e.g.*, nylons) and natural (*e.g.*, proteins) polymers, only slight changes in the polymer backbone, or interaction with neighboring polymer chains, will lead to very different physical properties (and resultant applications). Whereas the polymer backbone structure affects its general flexibility, the intermolecular interactions between side groups of neighboring polymer chains affect overall physical properties such as strength, solubility, crystallinity, and glass-transition temperature (T_g). The reversible self-assembly of polymeric units into macromolecular arrays is known as *supramolecular polymerization*.[46]

As you might expect, the influence of the nature and density of side groups *vs.* the backbone structure will strongly depend on the relative composition of each polymer unit, as well as how open the structure is to its environment. Regarding the relative composition of a polymer, even though a polymer may have a polar backbone structure (*e.g.*, ether and/or ester linkages), the structure may not be soluble within polar solvents. That is, if long nonpolar side chains are also present in the polymeric structure, its solubility and reactivity will be outweighed by the presence of

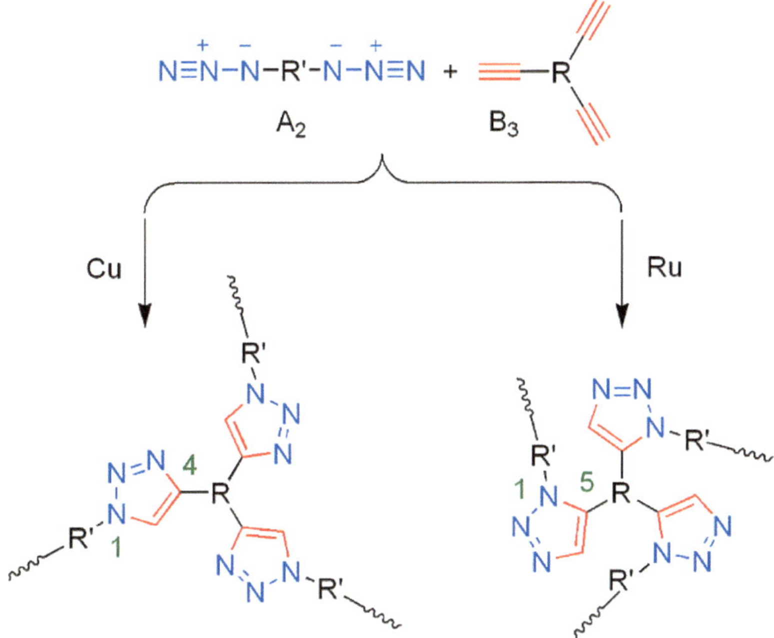

Fig. 5.40 Synthesis of Cu- and Ru-catalyzed diazide (A_2) and triyne (B_3) monomers. Reproduced with permission from *Macromolecules* **2010**, *43*, 8693. © 2010 American Chemical Society

hydrocarbon chains. A useful analogy to keep in mind is the pronounced decrease in water solubility of simple alcohols, from methanol (completely miscible) to hexanol and higher-hydrocarbon chains (complete immiscibility) as the overall ratio of non-polar:polar functionality increases.

Regarding the interaction of a polymer with its environment, consider the PAMAM dendrimers described in the previous section. For higher generations, the pronounced density of the peripheral functional groups prevents the solvent or other environmental molecules from interacting with the core groups. As a result, the solubility/reactivity of these polymers may be fine-tuned by varying the surface functional groups. As we will see in Chap. 6, the inner-core shielding exhibited by dendrimers (*e.g.*, G4–G6 for PAMAM or PPI) allows one to encapsulate a number of species within these "nanocavity reactors" for a number of useful applications for novel nanomaterials synthesis, catalysis, and drug-delivery.

Sometimes, the solubility of immiscible polymers may be improved by adding an anchoring polymer. For instance, polyethylene (PE) and isotactic polypropylene (iPP) represent the most commonly used plastics for packaging applications, but are not able to be blended due to their disparate repeat units. Melt-blending of these polymers will result in a brittle material; hence, the applications for their recycled products are limited due to the expense required to separate them from one another. However, the addition of a tetrablock copolymer, composed of PE and iPP units,

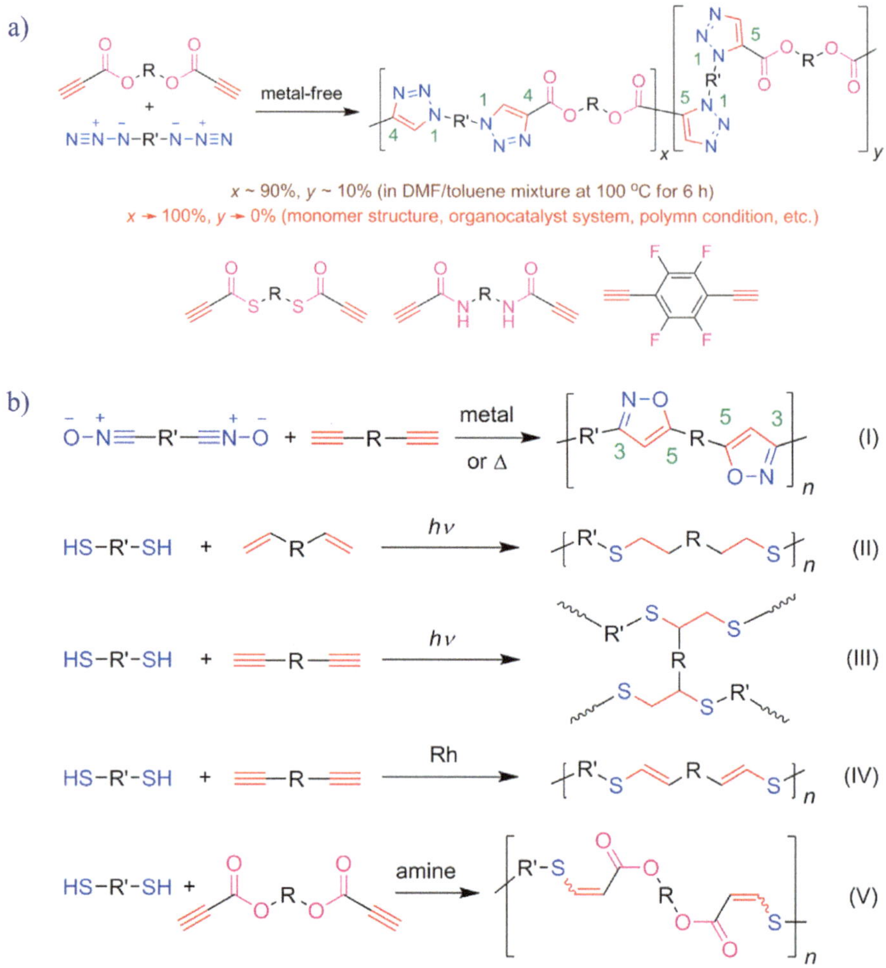

Fig. 5.41 **a** Metal-free click polymerizations of azide with propiolate and other alkynes containing electron-withdrawing groups. **b** Other examples of click polymerizations under development. Reproduced with permission from *Macromolecules* **2010**, *43*, 8693. © 2010 American Chemical Society

serves to stitch the incompatible polymers together, transforming the blend into a mechanically tough material.[47]

Table 5.4 lists the various functional groups that may be present in the polymer backbone and as side-groups, with their general effect on overall properties. It should be noted that these comparisons are over-simplications; for example, the presence of crosslinking and/or combinations of different functionalities will yield very different polymer properties. As discussed above, the general trend of chemical inertness and solubility will depend on both the openness of the structure and nature/density of the side groups. For thermal stability, conductivity, and flexibility, the polymer backbone

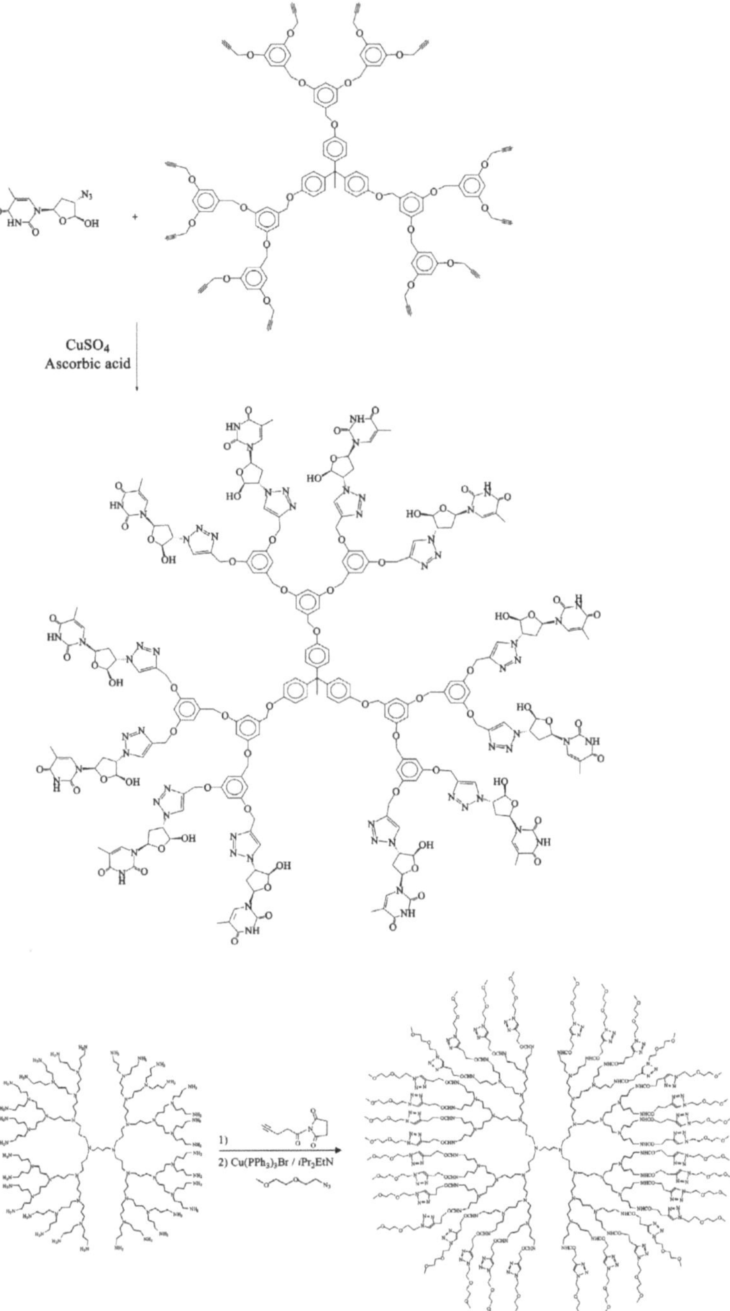

Fig. 5.42 Dendrimer synthesis using aqueous click chemistry. Reproduced with permission from *Macromolecules*, **2005**, *38*, 3663. © 2005 American Chemical Society

structure is most relevant. Changes from single to multiple bonding, and/or introduction of a delocalized resonance unit in the polymer chain will have pronounced effects on overall properties. One factor that is not explicitly mentioned in Table 5.4 is the degree of branching. However, it should be intuitive that the greater degree of interaction among adjacent polymer chains will enhance the tensile strength and density of the polymer, as well as increase the T_g. Perhaps the best example of this behavior is for highly branched *low-density polyethylene* (LDPE; used for squeeze bottles, food packaging film, plastic tubing, etc.) *versus* weakly branched *high-density polyethylene* (HDPE; used for tupperware, milk cartons, plastic bags, etc.), as illustrated in Fig. 5.43. Whereas the density and tensile strength for HDPE is 0.941–0.965 g cm^{-3} and 43 MPa, LDPE has values of 0.910–0.925 g cm^{-3} and 37 MPa, respectively.

The crystallinity of a polymer is related to the packing efficiency of individual polymer chains with respect to one another. Not unlike the formation of small molecule single crystals, processing variables such as temperature and time are paramount for influencing the degree of crystallinity. As previously mentioned, non-amorphous polymers may be considered as semicrystalline at best, with regions of crystallinity within the largely disordered polymer matrix. Typical crystalline polymers are polypropylene and polyethylene (especially isotactic and syndiotactic, Fig. 5.44), acetals, nylons, and most thermoplastic polyesters. In general, crystalline polymers have high shrinkage, low transparency, a distinct melting point, and possess good chemical and wear resistance. By contrast, polymers that contain bulkier side-groups on their chains, such as polystyrene, polycarbonate, acrylic, ABS, and polysulfone, tend to form amorphous polymers. Based on their disordered structure, amorphous polymers have low shrinkage, transparency, a broad melting point, and poor chemical and wear resistance.

5.3.1 3-D Printing

Also referred to as additive manufacturing (AM), 3-D printing can be used to replicate tools, homes, cars, and even human organs in the near future (Fig. 5.45). With applicability of materials including plastics, metals, and ceramics, perhaps no other technology has the opportunity to literally shape our future as does 3-D printing. Currently, over 100,000 Americans currently on the national organ transplant list; imagine the future possibility of printing a heart on-demand![49] The principle of 3-D printing is quite simple (Fig. 5.46); a 3-D model is first created using computer-aided design (CAD) software. The sliced model data is then transferred to a 3-D printer that deposits the material layer-by-layer through a nozzle. As shown in Fig. 5.58, objects with overhanging artifacts, such as the coffee mug handle, are designed with temporary support structures to prevent collapse during 3-D printing. A final post-processing step is often needed to remove any support structures and smooth the final surface. It should be noted that 3-D printing is not limited exclusively to plastics, but can be used for a wide variety of materials such as metals and ceramics.

Table 5.4 Influence of polymer structure on resultant properties

[b]Backbone unit/[s]substituent	Induced molecular properties
[b]Saturated carbon (–C–C–)	Chain flexibility, thermal/oxidative reactivity
[b]Unsaturated carbon (–C=C–)	Chain rigidity, high T_g, oxidative reactivity
[b]Aromatic (–C–C=C–)	Chain rigidity, colors, electrical conductivity (if aromatic rings: oxidative resistance, high strength, stacking/self-alignment (liquid crystals[48]))
[b]Ether (–C–O–C–)	Chain flexibility, oxidative/hydrolytic stability (unless in Lewis acidic media), soluble in polar solvents (for small substituents)
[b]Anhydride (–C–C(O)–O–C(O)–C–)	Chain flexibility, water sensitive (especially for short aliphatic chains)
[b]Amide (–NH–C(O)–)	Chain rigidity, crystalline, water sensitive
[b]Siloxane (–R_3Si–O–)	Chain flexibility and low T_g (especially for small substituents), stable toward oxidation, acid/base reactive
[b]Phosphazene (–R_3P=N–)	Chain flexibility, high chemical inertness
[b]Sulfur (–S–S–)	Chain flexibility, thermal/oxidative reactivity
[s]Hydrogen (–H)	Chain flexibility, low T_g; if bound to non-C atoms: H-bonding (high T_g, m.p., chemical reactivity)
[s]Alkyl (–CR_3)	Chain rigidity, hydrophobicity, chemical inertness, noncrystallinity, solubility in nonpolar solvents (properties dependent on size of R groups—most pronounced for aryl groups)
[s]Halogens (–F, –Cl)	Fluoride: extreme chemical inertness, hydrophobicity, insolubility in virtually any solvent
	Chloride: chemical inertness, chain stiffness, photolytic sensitivity, $\downarrow$solubility, fire retardancy
[s]Hydroxyl (–OH)	Hydrophilic, H-bonded rigid framework ($\uparrow T_g$)
[s]Cyano (–CN)	Hydrophilic, dipole–dipole interactions: $\uparrow T_g$ and crystalline
[s]Amide (–C(O)–NH_2)	Hydrophilic, very high T_g due to strong H-bonding
[s]Ester (–C(O)–O–)	May be hydrophilic or hydrophobic depending on alkyl substituents, T_g also varies with nature of R and tacticity of units
[s]Ether (–O–CR_3)	Hydrophilic at low temperatures, solubility and properties vary with R groups and tacticity
[s]Carboxylic acid (–C(O)–OH)	Hydrophilic and hygroscopic in solid-state

The most common techniques for polymer 3-D printing include melt extrusion, resin stereolithography, and powder (selective) laser sintering (Fig. 5.47).[50] A wide variety of polymers have been employed for 3-D printing, depending on the deposition process being used. For extrusion-based 3-D printing, a thermoplastic is heated to a semi-liquid state close to its melting point. Upon extrusion, the printed layers

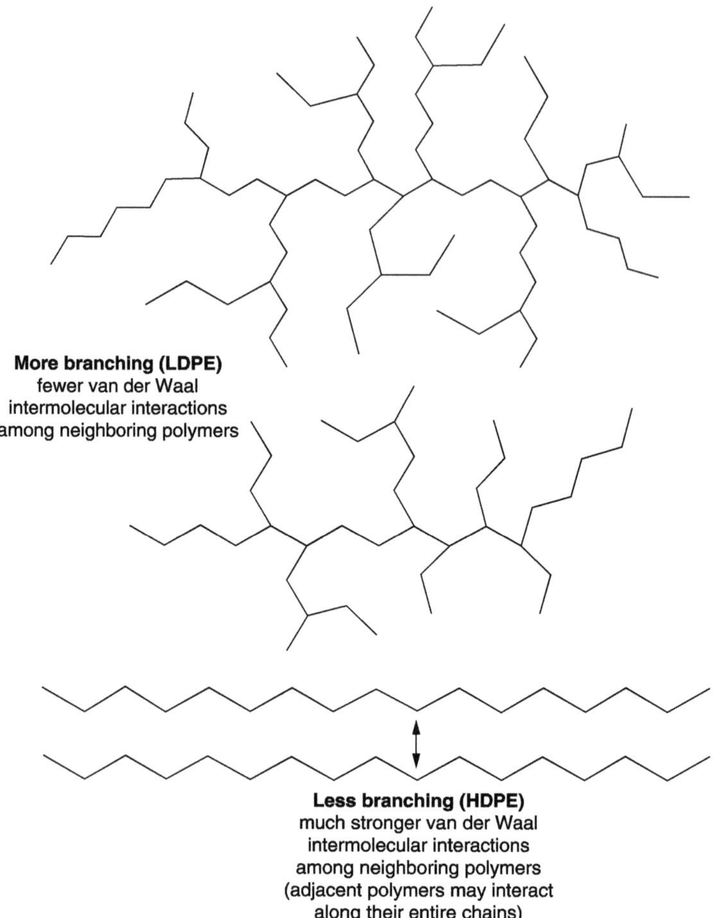

Fig. 5.43 Comparison of low-density polyethylene and high-density polyethylene backbone molecular structures. A higher degree of branching (LDPE) results in greater separation of neighboring polymer chains and fewer intermolecular interactions. By comparison, with little/no branching, adjacent chains may closely associate with one another, resulting in strong van der Waal forces along the entire length of the polymer chains

solidify into the desired shape. Some common thermoplastics used for extrusion-based 3-D printing include polycarbonate, acrylonitrile–butadiene–styrene (ABS), poly ether ester ketone (PEEK), polylactic acid (PLA), polyether imide (ULTEM), nylons, and acrylonitrile styrene acrylate (ASA). Physical properties such as density and tensile strength of the resulting ABS polymer may be tuned by varying the ratio of its monomers. While PLA is often preferred due to its biocompatibility, ASA is often chosen due to its improved heat resistance and UV stability.

The properties of elastomers such as silicones (Table 5.5) make them especially desirable for future 3-D printing applications. The ability of elastomers to restore their

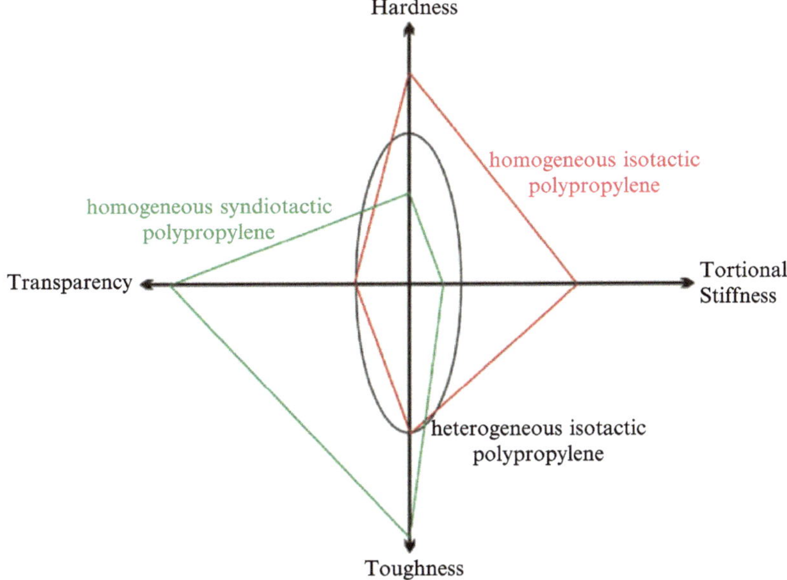

Fig. 5.44 Tacticity *versus* bulk properties for polypropylene polymers. The "homogeneous" and "heterogeneous" notations refer to whether the entire polymer chains, or only regions, are of a certain tacticity

Fig. 5.45 Examples of 3-D printed objects. Shown are (top-left to bottom-right): tools (© Materdome3d), house (© SQ4D), vehicle (© Local Motors), and jawbone and ear cartilage (© Wake Forest University)

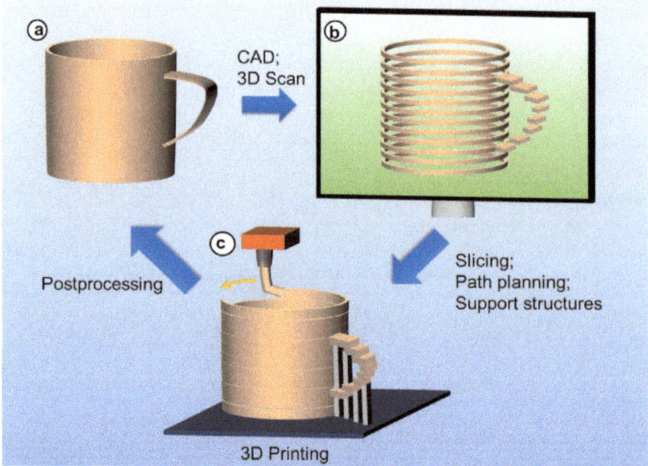

Fig. 5.46 Schematic of 3-D printing, showing the virtual slicing of the object designed by CAD and postprocessing to remove supporting structures. Reproduced with permission from *Chem Rev.*, **2017**, *117*, 10,212. © 2017 American Chemical Society

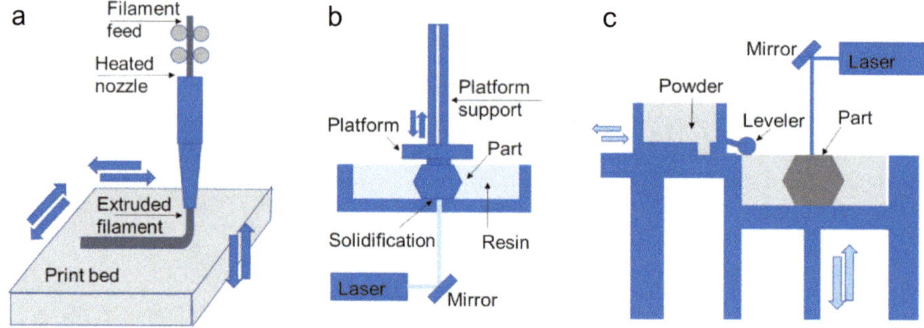

Fig. 5.47 Schematics of 3-D printing techniques. Shown are: **a** melt extrusion, **b** stereolithography, and **c** selective laser sintering. Reproduced with permission from *Polymers*, **2021**, *13*, 1499. © 2021 MDPI

shapes after significant deformations is especially relevant for actuators and flexible electronic devices. However, the low T_g and T_m of elastomers creates complexity for 3-D printing, which has necessitated the use of *thixotropic* pastes that flow like liquids above a certain yield stress but retain their shapes after the stress is released.[51] Furthermore, the pastes must be able to be quickly cross-linked (cured) in air after exiting the nozzle of an extruder. Hydrosilylation cross-linking is a desirable option since it is reproducible and may be easily controlled.[52] Dvornic and coworkers have pioneered the development of vinyl-functionalized siloxane terpolymers for "all-in-one" printing pastes that may be fine-tuned for different 3-D printing techniques (Fig. 5.48).[53] The thixotropic property of these pastes was maintained by adding trimethylsilylated (hydrophobic) fumed silica filler and repeat units specially designed to prevent crystallization, which is often problematic for silicones thus limiting their low-temperature applications. Despite exhibiting extremely fast permanent cross-linking, the pastes also have long shelf lives (>6 mo.), which is important for 3-D printing applications.[54] Interestingly, super-soft elastomers may also now be used for 3-D printing, for applications such as touch sensing, biomimetic tissue, and actuators.[55] Xie et al. have reported a strategy to incorporate bottlebrush elastomers for direct ink writing (DIW) 3-D printing, without the use of solvent or thixotropic additives.[56]

For stereolithography, thermosetting polymers are required that become solidified after curing via exposure to UV radiation. As described earlier for addition polymerization, a photoinitiator is required to generate reactive species (radical or cation) that can induce polymer growth via a radical or cationic mechanism.[57] Figure 5.49 illustrates the two types of radical photoinitiators used for photopolymerization, typically classified as Norrish Type I or Type II. The latter are two-component systems, which consist of a light-absorbing conjugated molecule and a co-initiator known as a synergist. Tertiary amines are common forms of synergists, which donate a hydrogen

Table 5.5 Physical properties of silicones of benefit for 3-D Printing

Physical Property	Details
Low-temperature flexibility	Flexible to temperatures as low as −100 °C and below, if completely amorphous
High temperature thermal and thermo-oxidative stability	To and above 400–450 °C in various atmospheres
Wide application temperature range	Some of the lowest T_g of any polymer (*e.g.*, −123 °C for PDMS; −145 °C for PDES) and very high decomposition temperatures of 425–625 °C
Unique surface properties	Some of the lowest surface energies; high hydrophobicity (contact angle of 100–109°) and pronounced non-stick properties
Excellent electrical insulation	High insulation resistance ranging from 1 TΩ to 100 TΩ, remaining stable over a wide temperature/frequency range
Biocompatibility and non-toxicity	High hydrophobicity and low surface tension results in hemocompatibility and is unaffected by host tissue attack and repeated sterilization
Low viscosity-temperature coefficients	Almost flat viscosity-temperature curves over extended intervals of operating temperatures
Pronounced resistance to atmospheric conditions	Resistance to acid rain and marine environments
Good mechanical properties	High gas permeability, optically transparent, flame resistant

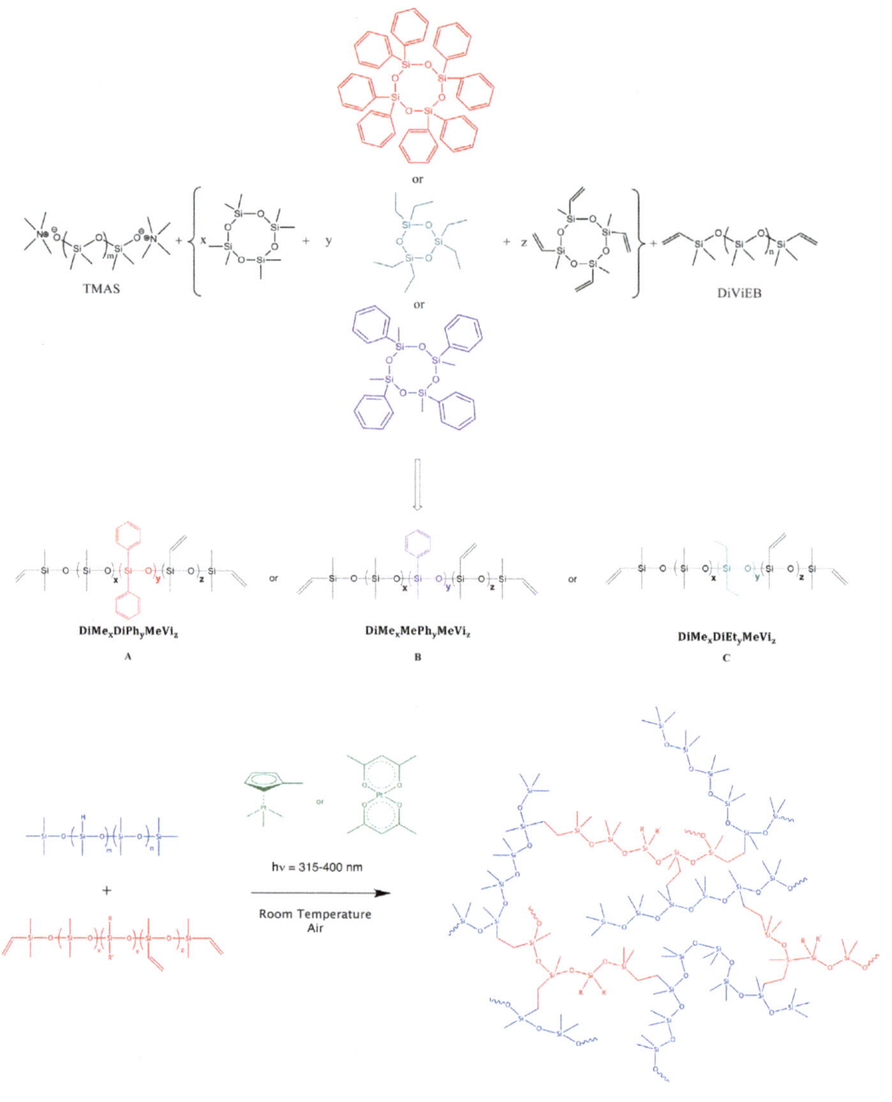

Fig. 5.48 (Top) Ring-opening polymerization of **a** diphenyl-, **b** methylphenyl, and **c** diethyl-based terpolysiloxanes. (Bottom) UV-activated cross-linking by hydrosilylation of vinyl- and hydrido-functionalized polysiloxanes. Reproduced with permission from *Macromolecules*, **2021**, *54*, 1715. © 2021 American Chemical Society

atom to the excited initiator molecule after UV irradiation (Fig. 5.49b). Some examples of Type I photoinitators used for stereolithography are shown in Fig. 5.50, which break into radical fragments when exposed to UV light.

The most common types of monomers used for radical photopolymerization in 3-D printing are (meth)acrylate-based monomers/oligomers (Fig. 5.51). These

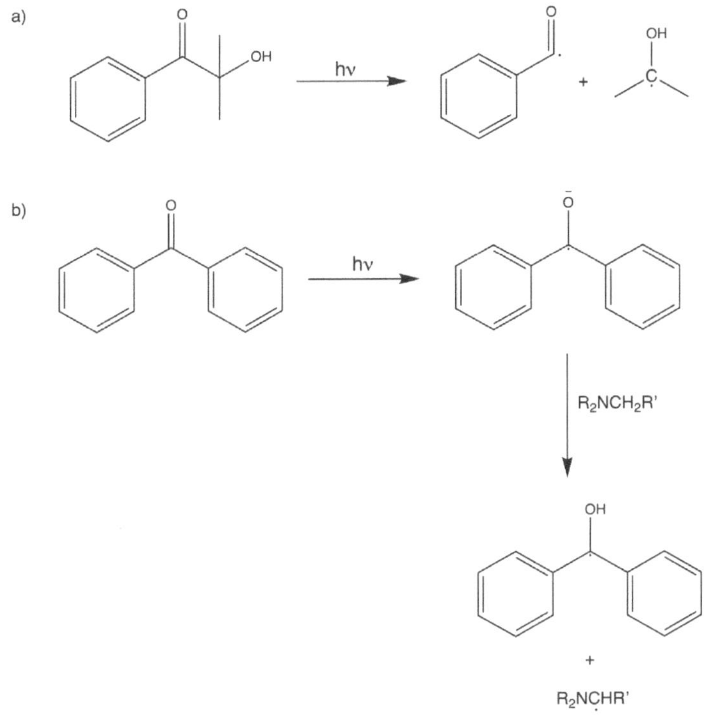

Fig. 5.49 **a** Scheme showing photocleavage of Darocur 1173 by a Norrish Type I photoinitiator. **b** Scheme of radical generation from a Norrish Type II photoinitiator

resins also have applications for shape-memory polymers, stretchable photopolymers, and a variety of bioapplications. A major limitation of acrylate resins is their propensity to undergo shrinkage during photopolymerization. In general, aromatic and cycloaliphatic acrylates show less shrinkage than aliphatic-based resins such as PEGDA or TEGDMA (Fig. 5.51a, b). Furthermore, acrylate radicals tend to be inhibited by molecular oxygen dissolved in the resin. In contrast, thiol and ene/yne monomers (Fig. 5.52) exhibit less shrinkage and can alleviate problems with oxygen inhibition by forming a peroxyl radical that generates a thiyl radical (Fig. 5.53).[58] Thiol-ene/yne resins are popular choices for 3-D printing of biodegradable hydrogels and optical waveguides. However, these resins are prone to oxidative disulfide bond formation and exhibit poor shelf life. To improve these properties, other resins that undergo RAFT polymerization have been developed, which affords the opportunity of producing 3-D materials that may be reinitiated to introduce new functionalities and properties in a living manner.[59]

Selective laser sintering (SLS) enables a broad range of materials and may employ thermoplastics (either semicrystalline or amorphous), a two-component thermoset, or an elastomeric material. The resultant properties of the 3-D printed part may be influenced by the powder particle size and size distribution, as well as processing

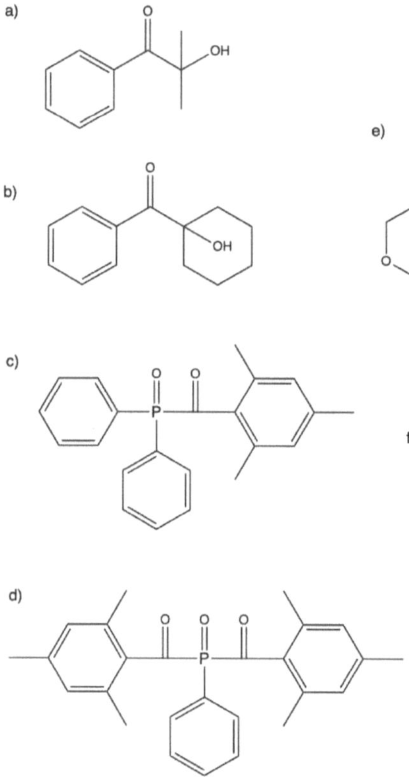

Fig. 5.50 Examples of Type I radical photoinitiators commonly used in stereolithography

parameters that may affect the microstructure of the sintered material. Figure 5.54 illustrates the mechanical properties for common polymers used for SLS, which is currently dominated by polyamides. The physical properties of the polymer are strongly affected not only by the type of polymer, but the type of filler used in the formulation. To date, the best formulations for both ductility and toughness are those that include semicrystalline polymers with fillers.[60] The effect of filler particles on resultant properties is not completely understood, since these particles affect laser absorption, thermal conductivity, melt viscosity, and surface energy of the powder. In some cases, fillers such as clay and nanofibers have been shown to improve the fire retardance properties of the final material.[61]

Beyond 3-D printing, it is possible to perform additive manufacturing with stimuli-responsive materials—known as *4-D printing*.[62] The controlled assembly of stimuli-responsive materials is of tremendous interest for applications such as sensors, robotics, flexible electronics, and biomimetics. Although all of the above techniques used for 3-D printing may be used, to date, the most common method used for 4-D printing is direct-ink writing due to its ease of use and wide applicability for material types and environmental processing conditions. The use of hydrogel

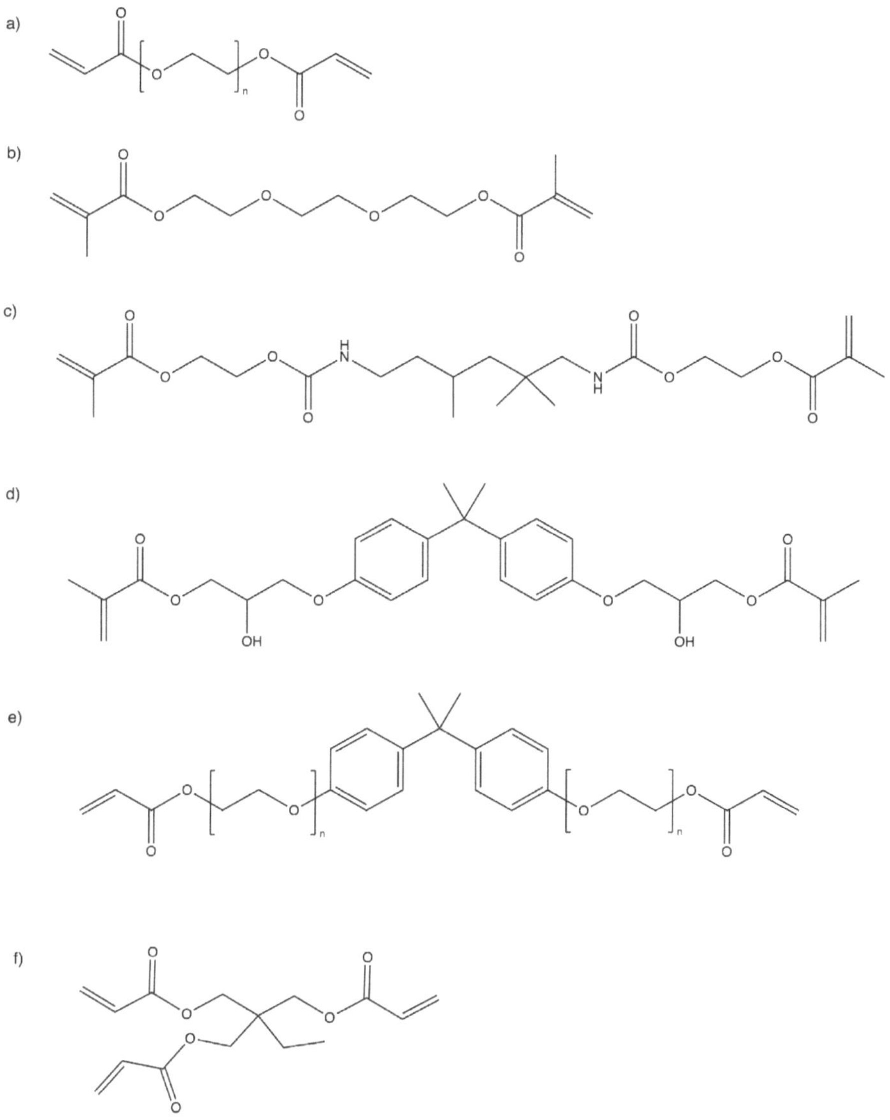

a)

b)

c)

d)

e)

f)

Fig. 5.51 Examples of (meth)acrylate monomers and oligomers commonly used in stereolithography

inks have been shown to be responsive to moisture[63]; however, there were problems with slow response speeds and instability in shape after repeated cycles. This latter limitation was addressed by incorporating a cellulose-hydrogel composite ink, with clay platelets added to enable hydrogen bonding among the constituents.[64] Another strategy to enable humidity responsive materials was reported by Debije and coworkers, who utilized a cholesteric liquid crystal oligomer ink that changed

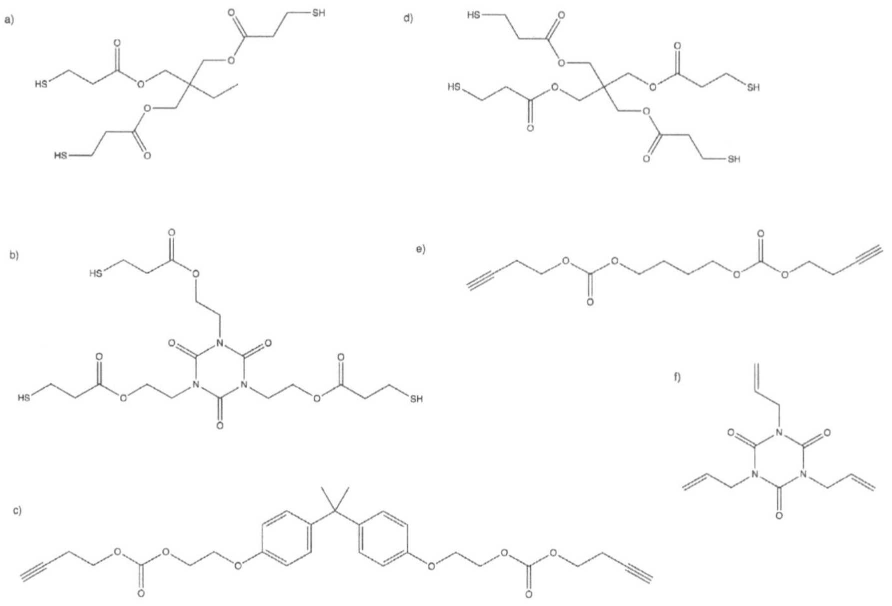

Fig. 5.52 Examples of thiols and ene/yne monomers commonly used in stereolithography

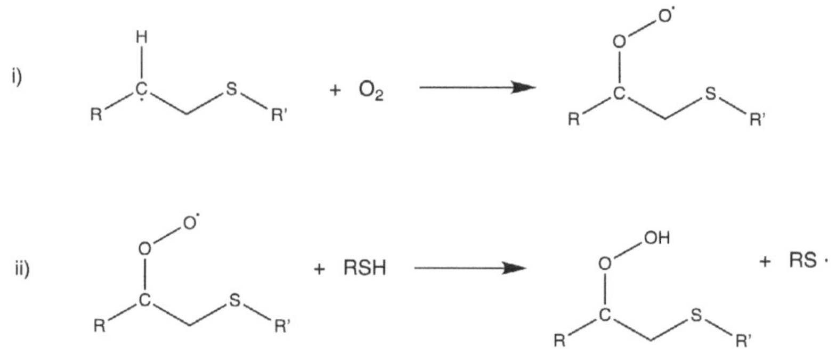

Fig. 5.53 Thiol-ene reactions with oxygen

color upon contact with moisture.[65] The use of shape-memory polymers (SMPs) and shape-memory alloys (SMAs), shape-memory ceramics (SMCs), and hybrids, have been used for the design of thermo-sensitive materials. In contrast, a chromophore must be incorporated into a polymer block to enable photo-sensitive properties. A wide variety of materials may be used to enable structural responses upon exposure to magnetic fields, pH, mechanical stress, and electrical charge (Fig. 5.55).[66]

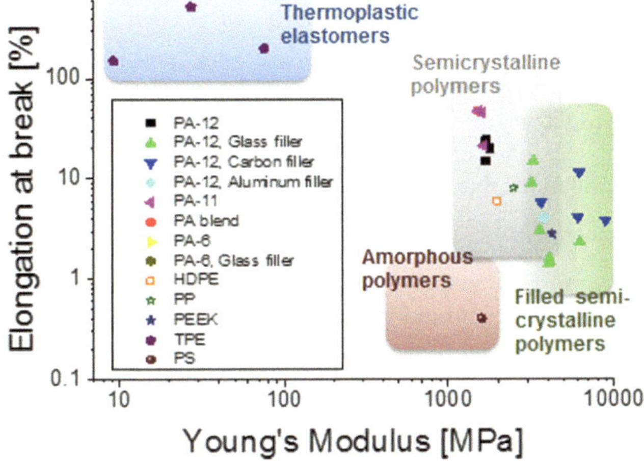

Fig. 5.54 Comparative mechanical properties of commercial SLS materials (PA-X = various polyamide polymers). Reproduced with permission from *Chem Rev.*, **2017**, *117*, 10,212. © 2017 American Chemical Society

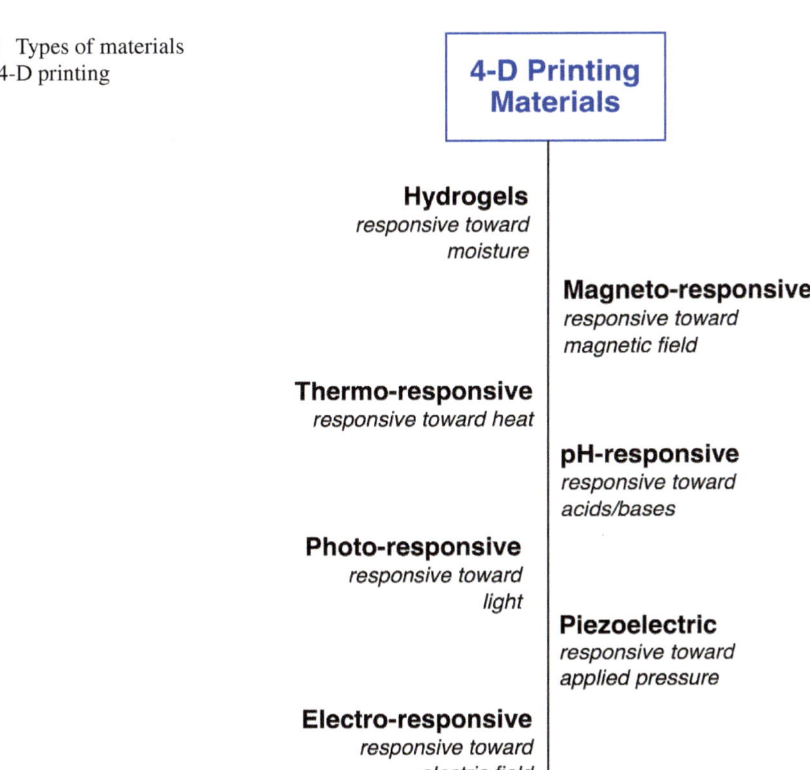

Fig. 5.55 Types of materials used for 4-D printing

5.3.2 Biomaterials Applications

The world of biomaterials is diverse, encompassing ceramics and metals used for orthopedic implants (already discussed in Chaps. 2 and 3), to artificial heart valves, blood vessel stents, and contact lenses.[67] This section will delve into those biomaterials applications that utilize "soft" polymeric-based materials.

Biodegradable polymers

In our ever-increasingly "green" society, it is becoming hard to justify the use of traditional synthetic, non-biodegradable polymers for short-lived applications such as packaging, personal hygiene, surgery, etc. In addition to the persistency of non-biodegradable polymers in the environment, groundwater/soil pollution is also possible via additive leaching. Combustion (and even recycling processes) of these polymers consumes large quantities of energy and may also yield toxic emissions such as dioxins. Anyone who has seen dramatic images of plastic bottle holders wrapped around the necks of wildlife should also be moved to evaluate viable alternatives that will break down more readily in the environment.

By definition, *biodegradable* materials exhibit chemical structures that will afford their decomposition under aerobic (*e.g.*, composting) and/or anaerobic (*e.g.*, landfill) conditions.[68] Typical degradation byproducts are CO_2, CH_4, H_2O, inorganic compounds (*e.g.*, PO_x, SiO_x), or biomass. Whereas *degradation* refers to the decomposition of a polymer by chemical means (*e.g.*, hydrolysis), *biodegradation* is carried out by the enzymatic activity of microorganisms. In this latter route, the decomposition of the plastic is afforded by the metabolism by microorganisms that generate an inert humus-like byproduct that is much less environmentally-harmful/pervasive. The terms *bioabsorption* or *bioresorption* refer to the degradation byproducts being used in cellular processes.[69]

Biodegradable polymers should exhibit the following characteristics:

i. Biocompatible (*e.g.*, elicits little/no immune response or is able to integrate with a particular tissue)
ii. Tunable and predictable degradation mechanism and rate
iii. Non-toxic degradation products and metabolites
iv. Non-toxic additives (*e.g.*, monomers, initiators, emulsifiers, etc.)
v. Easy to synthesize and process into the desired shape
vi. Long shelf-life and facile sterilization.

There are two primary classes of biodegradable polymers: naturally-occurring or generated using synthetic organic procedures (Fig. 5.56).[70] There are three origins of natural polymers: plants, animals, and microbes. Polymers such as polysaccharides may be formed from all of these sources, whereas natural polyesters (*e.g.*, poly(3-hydroxyalkanoate)s) are generated exclusively from microbial routes. It should be noted that all natural polyesters are biodegradable; however, most synthetic varieties such as those used for fabrics, bottles, tarpaulins, canoes, LCDs, filters, etc., are not.

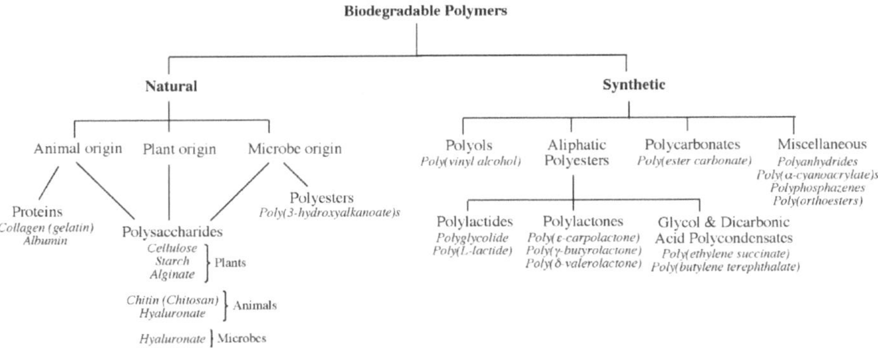

Fig. 5.56 Classification scheme for biodegradable polymers

For biomedical applications, an implanted polymer may be designed to degrade at an appropriate rate to transfer stress to surrounding tissues as they heal. This is afforded by designing the bulk geometry of the implant, the chemical stability of the polymer backbone, and the presence/concentration of additives such as plasticizers.[80] The degree of polymer crystallinity also plays a crucial role in its overall degradation kinetics. For instance, whereas amorphous poly(*dl*-lactide), is 100% degraded in 3–6 months, it may take 1–2 years for crystalline PLA to degrade under analogous conditions. Accordingly, one may design a polymer with varying degrees of crystallinity for time-delayed drug delivery applications.[81] The preferred use of PLA for implantable biomedical applications is due to its high degree of bioresorption via degradation to lactic acid, which is metabolized through the Krebs cycle. Accordingly, since its development in the 1960s, there are now more than 200 biomedical products made from PLA or PLA-based co-polymers.

To further illustrate the effect of polymer structure on its degradation—of importance for its viability for implant applications—consider silicone breast implants developed by Dow Corning in 1961 and first implanted in 1962. Due to reports of numerous health complications resulting from breast implants, countless class-action suits were brought against Dow Corning in the mid-80 s to late-90 s.[82] The ensuing multi-billion dollar settlement in 1998 nearly caused the financial ruin of Dow Corning Corporation.[83] One of the early generations of implants featured a poly(ester urethane) (**III**) coating on the silicone shell, in order to prevent *capsular contracture*—an abnormal response of the immune system to foreign materials. However, due to the large numbers of ester linkages, this polymer was shown to degrade resulting in carcinogenic aromatic amines such as 2,4-toluenediamine.[84] Limiting the number of ester groups (*e.g.*, **IV**, poly(ether urethane) or **V**, poly(ester urethane urea)) results in a significant improvement in the hydrolytic stability of the polymer. As shown by structures **III**–**V**, the polymeric coating must exhibit a tunable stiffness ("hard" and "soft" regions) that will hold the silicone gel in place while suitably interacting with the surrounding soft tissue.

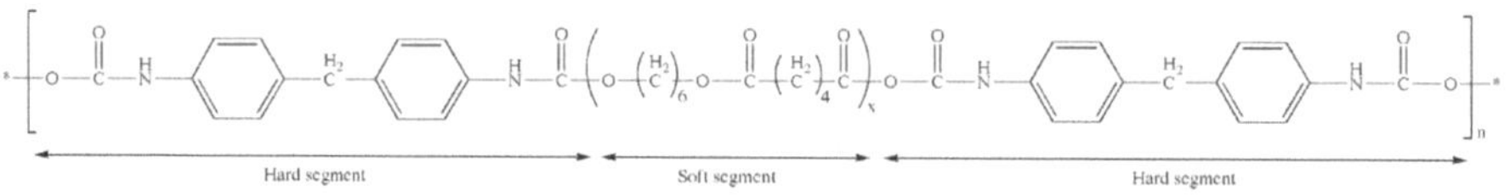

Hard segment Soft segment Hard segment

III

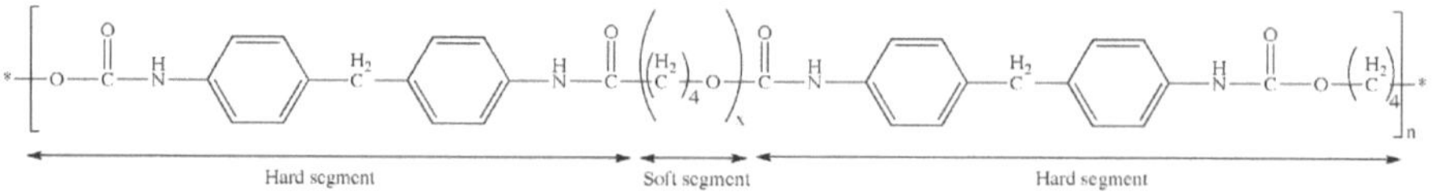

Hard segment Soft segment Hard segment

IV

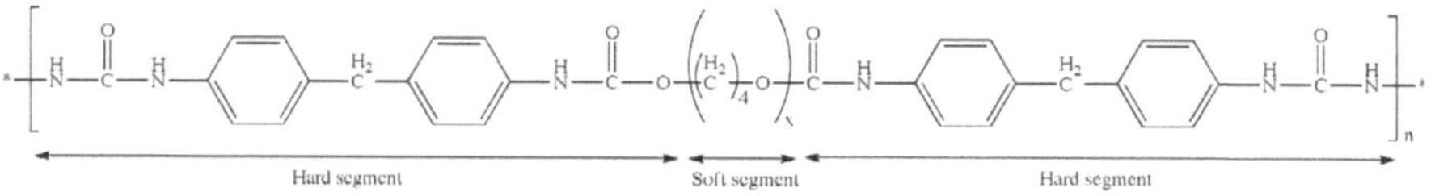

Hard segment Soft segment Hard segment

V

Figures 5.60 and 5.61 illustrate the various oxidative and hydrolytic biodegradative reactions, respectively, which are responsible for polymer degradation. As one can see, polymers with C(O)X groups such as esters, amides, urethanes, etc. are particularly hydrolytically degradable. In contrast, groups that are stable toward hydrolysis (non-biodegradable) include hydrocarbon backbones (*e.g.,* polyethylene, polypropylene), halocarbon backbones (polytetrafluoroethylene, poly(vinylidene fluoride), polychlorotrifluoroethylene, etc.), alkylsiloxanes (*e.g.,* dimethylsiloxanes, $(R_2SiO)_x$, etc.), and sulfones $((O)S(O))_x$.The majority of biodegradable polymers are synthesized by a *ring-opening polymerization* mechanism (*e.g.,* Fig. 5.57).[71] One example is the ring-opening polymerization of lactide to yield poly(lactic acid) (PLA)—one of the most widely used biodegradable polymers.[72] This route features an aluminum or tin[73] alkoxide complex that catalyzes a "coordination insertion" mechanism, which proceeds via rupture of the acyl-oxygen bond of the monomer (Fig. 5.58).[74] Monocarboxylic iron complexes have also been shown to polymerize *l*-lactide by an anionic-based insertion mechanism.[75] By definition, either cationic or anionic ring-opening polymerization may take place, wherein the reactive center of the propagating chain is a carbocation or carbanion, respectively. IBM has developed a family of organocatalysts such as N-heterocyclic carbenes, bifunctional thiourea-amines, and "superbases" (*e.g.,* guanidines, amidines) for the controlled polymerization of strained heterocyclics.[76] As Fig. 5.59 illustrates, this synthetic method is amenable for the polymerization of lactones (*e.g.,* PLA; poly(vinylalcohol)—PVA, etc.), cyclic carbonates (*e.g.,* poly(tetramethylene carbonate)—PTMC, etc.), ethers (*e.g.,* poly(ethylene oxide), PEO), and silyl ethers (*e.g.,* poly(dimethylsiloxane), PDMS, etc.). Alkoxides and thioureas have also been shown to effectively catalyze rapid and selective ring-opening polymerizations of lactones.[77] It should be noted

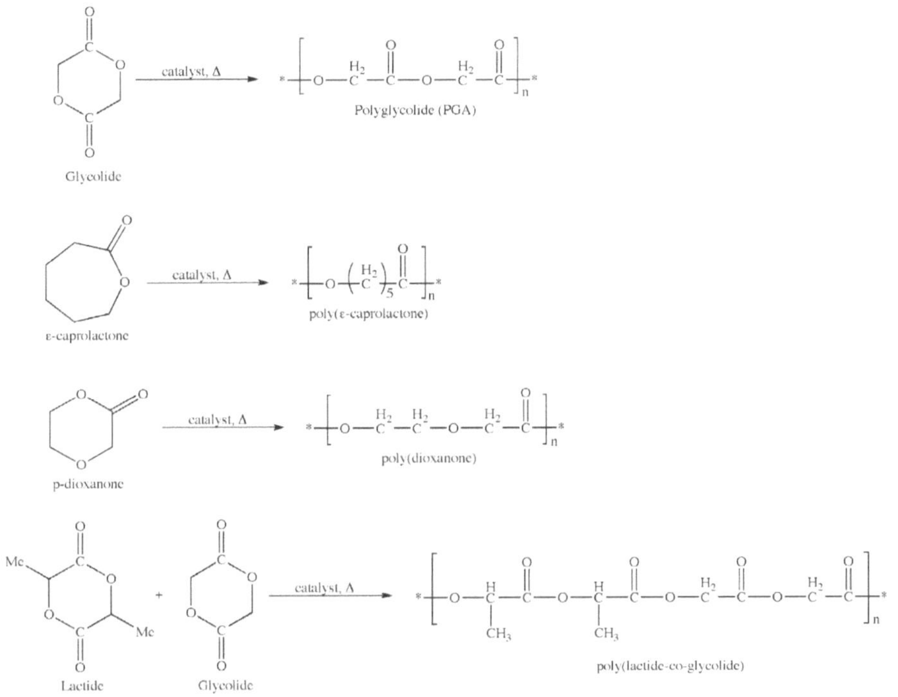

Fig. 5.57 Ring-opening polymerization of various biodegradable polymers

that enzymatic routes have also been investigated in recent years—a truly "green" alternative for polymer synthesis.[78]

The applications for biodegradable polymers span a variety of fields such as:

i. Agriculture (*e.g.,* mulch films, temporary planting pots, fertilizer/pesticide delivery)
ii. Fisheries (*e.g.,* fishing lines, nets, hooks)
iii. Sporting goods (*e.g.,* golf tees)
iv. Food packaging (*e.g.,* disposable plates, cups, bags, cutlery, bottles, retail bags, six-pack rings)
v. Hygiene (*e.g.,* feminine hygiene products, refuse bags, cups)
vi. Medical (e.g., suturing, fractured bone fixation, wound covering, skin/blood vessel/nerve reconstruction, controlled drug delivery, bioprinted artificial organs[79]).

For applications that require sealing or adhesion to tissues, liquids such as 2-cyanoacrylates are commonly used that rapidly cure to form an adhesive gel. We are all familiar with the fast-setting ethyl-2-cyanoacrylate adhesive that is sold under the trade name of Super Glue® or Krazy Glue® (ethyl-2-cyanoacrylate). Veterinary

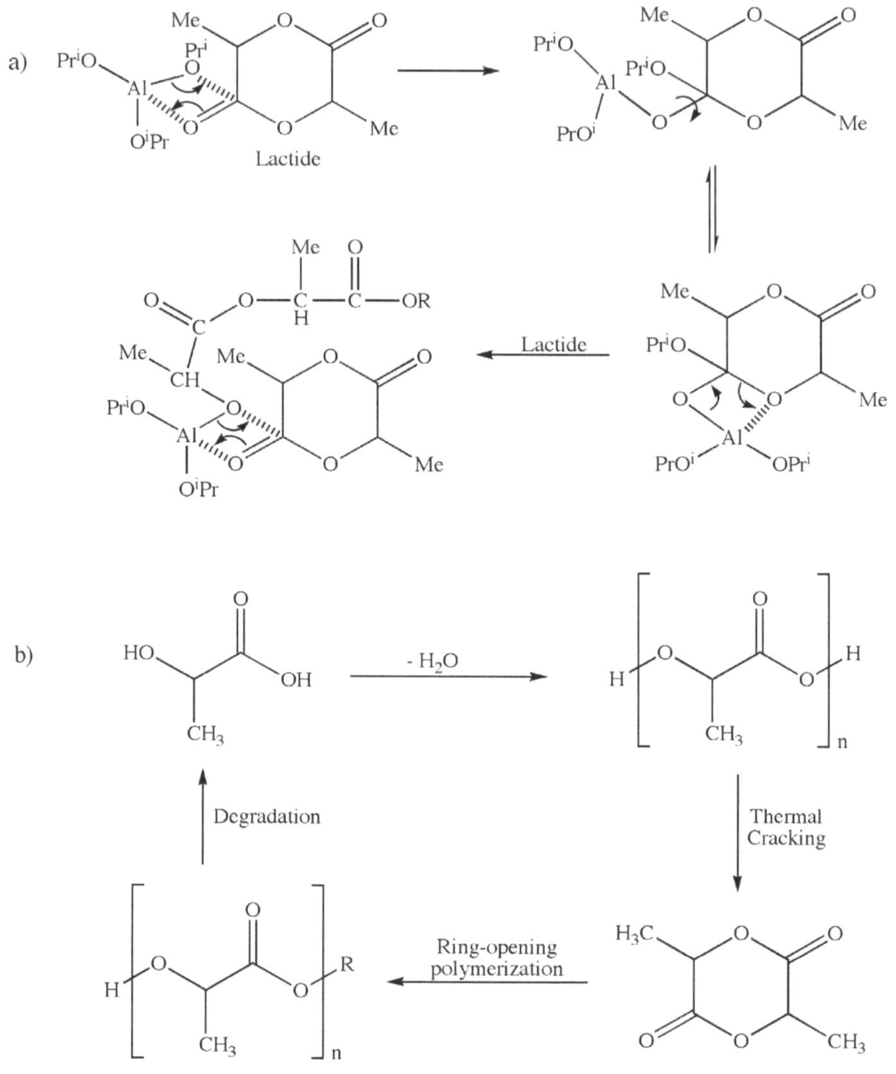

a) Lactide

b)

Fig. 5.58 Mechanism for the ring-opening polymerization of poly(lactic acid), catalyzed by an aluminum alkoxide complex

glues employ n-butyl-cyanoacrylate (Vetbond® and LiquiVet®), whereas medical-grade adhesives such as LiquiBand®, SurgiSeal®, and Dermabond® employ 2-octyl-cyanoacrylate. Polymerization is catalyzed by the presence of moisture; chain initiation occurs via the nucleophilic attack of the ethylenic carbon by hydroxide ions (Fig. 5.62). The fast-setting properties of the monomer to form a strong waterproof polymer (often in <1 min) is due to the presence of additives such as toluidine (o-methylaniline, or 2-aminotoluene; $C_6H_4(NH_2)(CH_3)$).

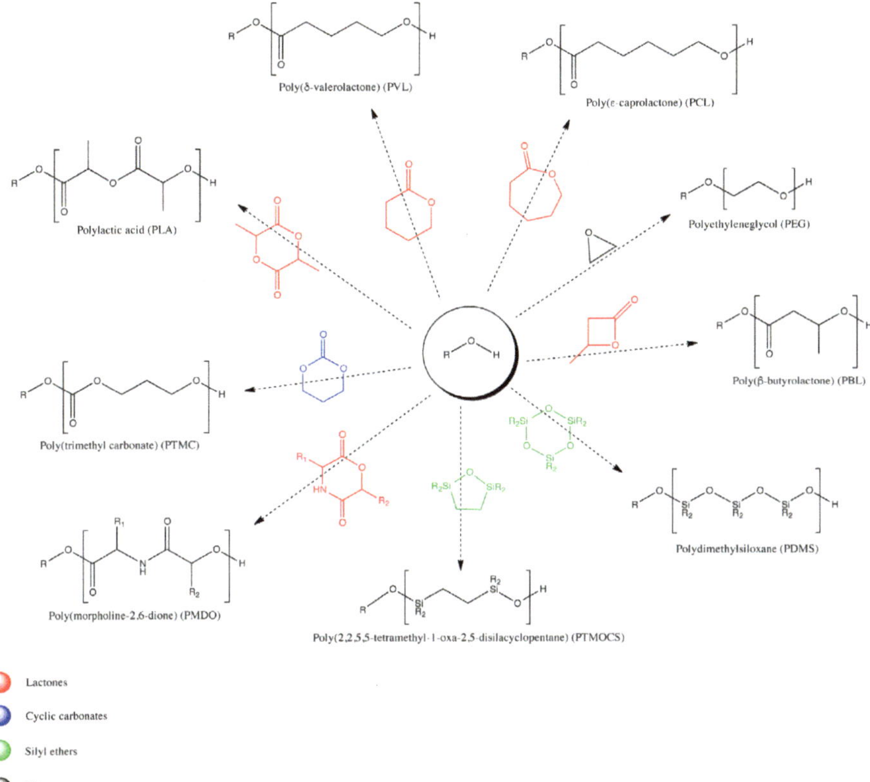

Fig. 5.59 Diverse applicability of novel N-heterocyclic carbene, bifunctional thiourea-amine, and "superbase" catalysts for the polymerization of strained heterocyclics. Reprint courtesy of International Business Machines Corporation © 2009 International Business Machines Corporation

Another important application for biodegradable polymers is the design of *arterial stents*, used to prevent or counteract restricted blood flow that may lead to heart attacks or strokes if left unaddressed. Every year, over 800,000 angioplasty procedures are performed in the U.S., consisting of expanding the lumen of the coronary artery with a balloon.[85] Although this is a more desirable option to coronary bypass surgery, *ca.* 30–50% of angioplasty patients soon develop significant narrowing of the artery through migration and nucleation of smooth muscle cells, known as restenosis.

In order to offset the high restenosis rate following angioplasty, scaffolding devices known as *endoprostheses* or *stents* have been used in recent years to facilitate fluid flow through a diseased coronary artery.[86] However, *ca.* 10–50% of patients receiving stents still develop restenosis; consequently, patients must receive close monitoring (and possible additional treatment) long after the surgery is performed.[87] Magnetic resonance imaging (MRI) is emerging as a preferred method to detect, diagnose, and monitor the formation of dangerous plaques within arteries. However, most stents are comprised of metal alloys, which produce distortion in MR images. In

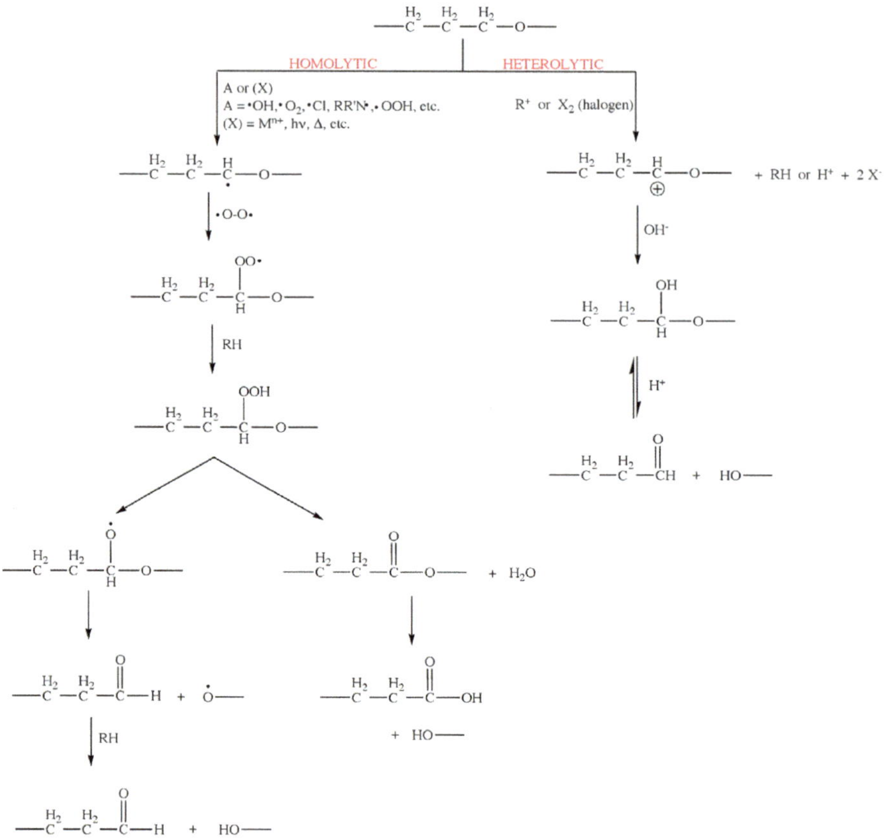

Fig. 5.60 Stepwise degradation routes for a biodegradable polymer via oxidative chemical reactions

order to circumvent these problems, DeSimone and coworkers at the University of North Carolina at Chapel Hill have developed stents that are comprised of biodegradable/erodable polymers.[88] Most importantly, these stents may contain a variety of agents (*e.g.*, sensitizers, dissolution inhibitors, photo-acid generators, thermally-, pH-, irradiation-, or light-activated catalysts, etc.) that allow one to achieve a controllable rate of degradation. In addition, the stent has also been shown to effectively elute a drug such as everolimus, an immunosuppressant used to prevent restenosis.[89] Studies showed that the stent was completely bioabsorbed in patients after two years after implantation, with full restoration of vasomotion (ability of the blood vessel to contract/expand). Due to the drug-eluting properties of the stent, patients had a 0% rate of stent thrombosis (blood clot formation). This work sets an amazing precedent for "smart" implantable devices, which perform their designed function(s) and then completely disappear without any side-effects![90]

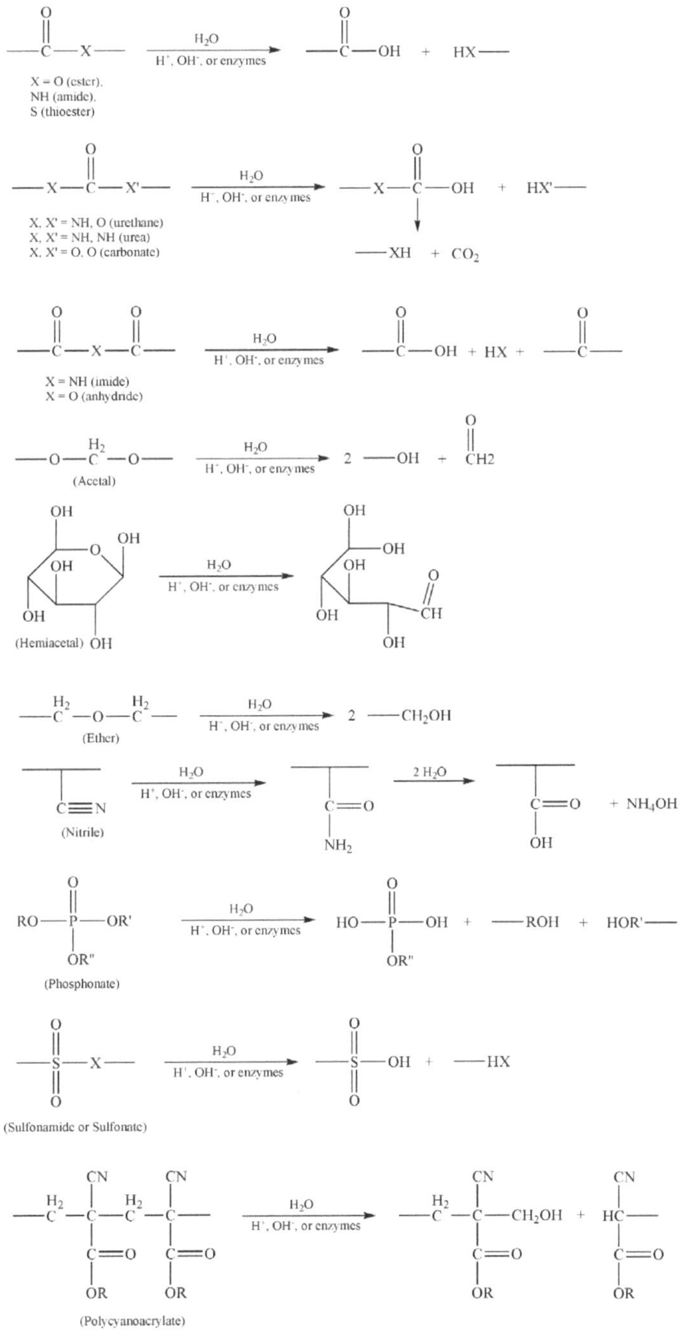

X = O (ester), NH (amide), S (thioester)

X, X' = NH, O (urethane)
X, X' = NH, NH (urea)
X, X' = O, O (carbonate)

X = NH (imide)
X = O (anhydride)

(Acetal)

(Hemiacetal)

(Ether)

(Nitrile)

(Phosphonate)

(Sulfonamide or Sulfonate)

(Polycyanoacrylate)

H_2O / H^+, OH^-, or enzymes

Fig. 5.61 Degradation of various functional groups via hydrolytic chemical reactions

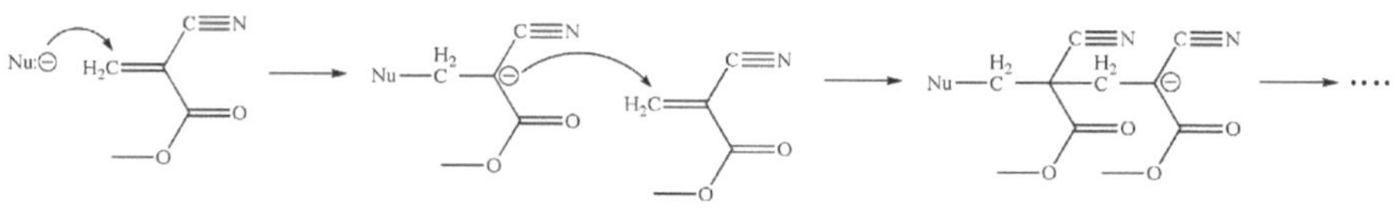

Fig. 5.62 Mechanism for the polymerization of methyl-2-cyanoacrylate, catalyzed by the nucleophilic attack of OH^-

Contact Lenses

The concept of contact lenses was first reported by Leonardo DaVinci (1508) and Rene Descartes (1632). However, it wasn't until the late 1800s that the first contact lenses were developed by Muller, Fick, and Kalt.[91] The lenses were composed of blown glass, being molded from rabbit and cadaver eyes.[92] While these early designs proved to be successful for eye protection and vision correction, you might imagine that a piece of glass covering the entire eye would be quite uncomfortable. In fact, there are reports that the pain resulting from wearing these lenses was so intense that a cocaine-based anesthetic was required![93]

It is easy to get overwhelmed when one considers the requisite properties of contact lenses. In addition to using inexpensive and reproducible fabrication methods, the lenses must be biocompatible. That is, the polymers used in modern lenses must be capable of being in direct contact with the eye for an extended duration, as well as produced in such a way that any residual monomer/additive/catalyst does not pose a health risk. Further, since the lens floats on a layer of water known as the tear film, the polymer must be hydrophilic (especially for extended-wear varieties), while resisting the deposition of other components contained in tear drops such as protein, lipids, ions (*e.g.*, Na^+, Ca^{2+}, HCO_3^-), and enzymes.[94] In addition, the lens must be lightweight with sufficient mechanical strength to avoid being torn or scratched; of course, this must be balanced with a relatively high modulus of elasticity to facilitate overall comfort when in use and ease of handling. The thickness of the lens must be optimized to afford the desired optical correction and gas permeability (sufficient O_2 supply and CO_2 removal), while not interfering with eyelid functions. Lastly, the specific weight of the lens must be of the same order of magnitude as tear drops, so it will not migrate around the corneal surface. With so many property constraints, contact lens design represents a remarkable challenge for zealous materials chemists and scientists!

The first plastic ('hard') corneal contact lens was developed in 1936, fabricated from poly(methyl methacrylate), PMMA (**VI**).[95] However, these lenses were not commercially available until the mid 1960s. In comparison, 'soft' contact lenses were first composed of poly(hydroxyethyl methacrylate), PHEMA (**VII**), and became commercially available shortly thereafter. Both types of lenses are comprised of a 3-D amorphous network of cross-linked polymer chains. However, the 'soft' *versus* 'hard' character of a lens material is directly related to the ambient operating temperature relative to its glass-transition temperature (T_g). Whereas PMMA has a T_g of 100–120 °C, the T_g of PHEMA (*ca.* 10–15 °C) lies below room temperature.[96]

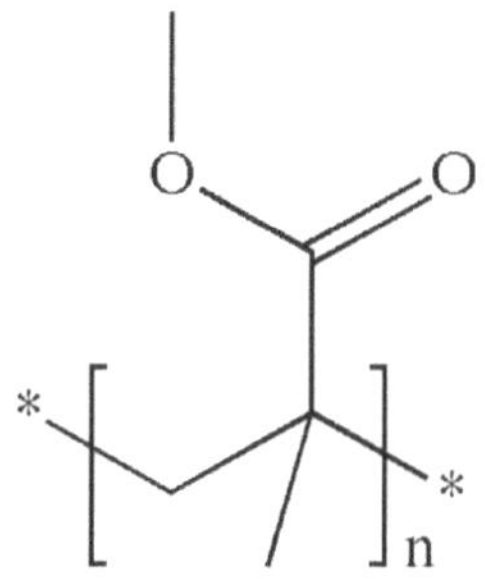

VI

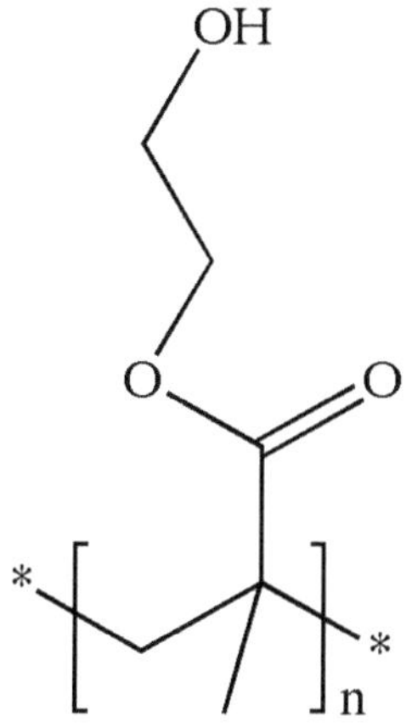

VII

The first type of hard contact lens was fabricated exclusively from PMMA. These lenses are hydrophobic, which effectively repels protein build-up; however, these materials are impermeable toward gases such as O_2 and CO_2. Due to the lack of blood vessels within the cornea, the cornea must obtain oxygen directly from the atmosphere. Hence, it is essential to afford the highest level of oxygen diffusion through the lens in order to prevent hypoxia and promote corneal health by avoiding chronic limbal inflammation and facilitating turnover of epithelial/limbal cells during wear.[97]

Although tears generated from normal blinking may facilitate sufficient oxygen transport, the impermeability of PMMA lenses may be overcome by copolymerization of MMA (to maintain its structural rigidity) with silicone monomers. Accordingly, the first rigid gas-permeable (RGP) lens was composed of methacryloxypropyl tris(trimethysiloxy) silane (TRIS, **VIII**) copolymerized with MMA. Since silicones are hydrophobic, a hydrophilic monomer such as methacrylic acid (MAA, **IX**) or N-vinylpyrrolidone (NVP, **X**) must be added to improve the wetability and overall comfort.[98] Even with these improvements, PMMA-TRIS lenses exhibit a relatively low gas permeability, which limits their use to no more than *ca.* 8 h at a time.

In order to design extended-wear RGP contacts, MMA-TRIS lenses may be doped with fluoromethacrylates, which allows greater gas permeability via increasing its free volume, while more effectively repelling proteins and lipids. However, since fluorinated chains are hydrophobic, it is also necessary to cap the chains with hydrophilic methacrylate groups.[99] With these modifications, one is able to wear such RGP lenses for up to 7 days.

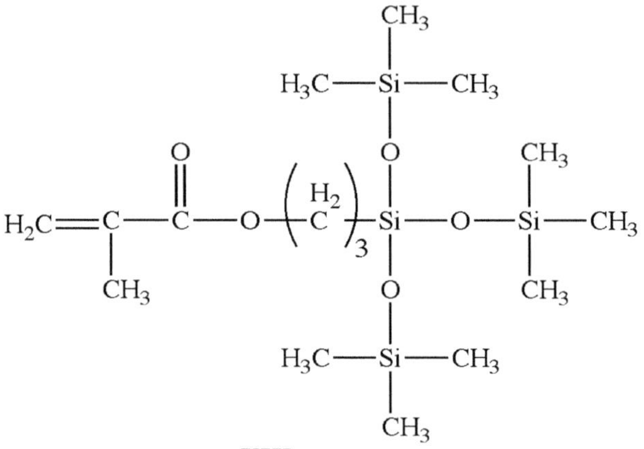

VIII

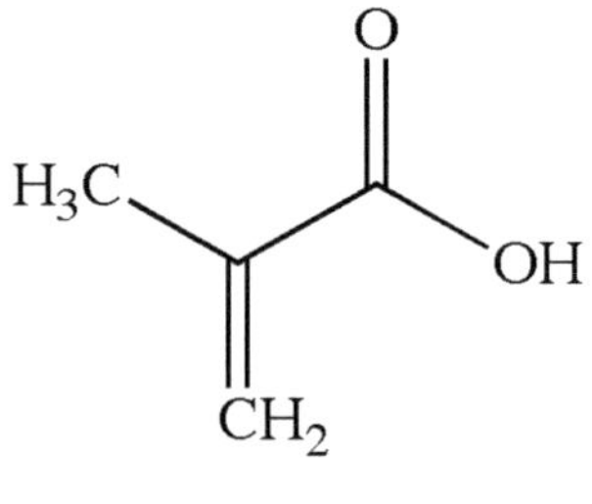

IX

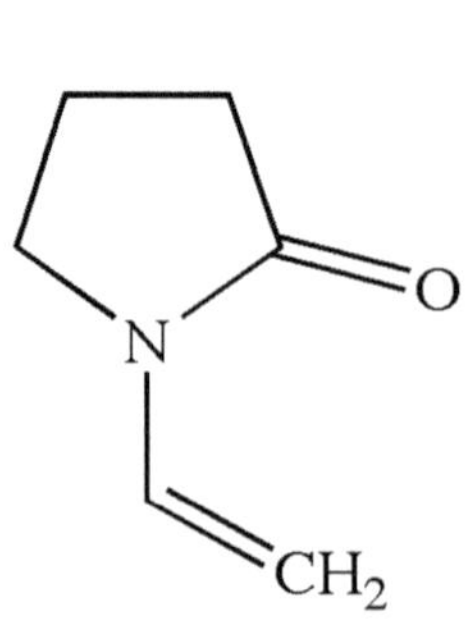

X

The earliest soft lenses were fabricated from polymers such as PHEMA, later crosslinked with monomers such as ethylene glycol dimethacrylate (EGDMA, **XI**).[100] These polymeric materials are classified as *hydrogels* since they are extremely hydrophilic, being able to absorb 10–1000 times their dry weight of water. The polymer chains of hydrogels are normally decorated with hydrophilic groups such as –OH or –COOH to facilitate H-bonding with surrounding water molecules. The copolymerization of hydrogels with MAA and NVP will yield a polymer with much greater moisture content—up to 80% for modern lens designs. Network formation may be afforded by entanglement of (co)polymer chains, covalent crosslinking, or association of polymeric units via ionic or intermolecular forces (Fig. 5.63). The water within hydrogels is either strongly bound or freely exchangeable, which is also paramount for their use as drug-delivery agents (*e.g.,* Fig. 5.64) and mineralization medium.[101]

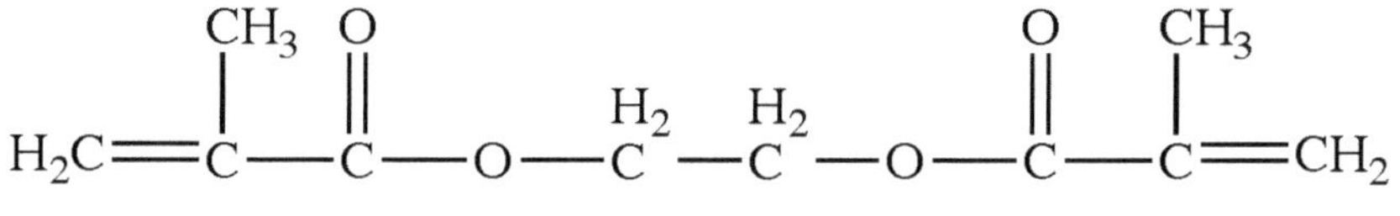

$$H_2C{=}C(CH_3){-}C({=}O){-}O{-}CH_2{-}CH_2{-}O{-}C({=}O){-}C(CH_3){=}CH_2$$

XI

The latest varieties of contact lenses are based on silicones, first brought to the commercial market by Bausch & Lomb (PureVision®)[102] and Ciba Vision (Focus Night & Day®). The primary benefit of silicone hydrogels is their significantly greater water capacity, while retaining the comfort, wetability, and biofilm resistance of non-silicon based hydrogels. However, due to their pronounced hydrophobicity, the oxygen permeability of silicone hydrogels decreases exponentially as water content increases (Fig. 5.65). Consequently, there has been much research devoted to improving their wetability via surface functionalization to replace siloxane groups. Thus far, this has been successfully accomplished by grafting polyoxyethylene to the surface, as well as using surfactants containing copolymers of lauryl-, hexyl-, and methyl-methacrylate and polyethylene glycol methacrylate.[103]

The flux of oxygen through a polymeric contact lens is governed by *Fick's Law* (5.5). In fact, Fick's law is used to predict the shelf life for food/beverages that are encased within polymeric packaging (*e.g.,* polyethylene terephthalate (PET), etc.). The units of transmissibility are 10^{-9} cm sec^{-1} mLO_2 mL^{-1} mmHg^{-1}, which is equivalent to units of Barrer cm^{-1}. The first RGP lenses exhibited Dk/t values of <12 Barrers cm^{-1}, corresponding to a lower oxygen flux than a cornea of a non-wearer (both open-eye and closed-eye).[104] Most typical modern RGP lenses have Dk/t values of 15–25 Barrers cm^{-1}, whereas soft lenses have much higher values (Table 5.6). For a lens to be approved for extended-wear applications, the Dk value must be $\geq$100 Barrers.

$$J = \left(\frac{Dk}{t}\right)\Delta P, \tag{5.5}$$

where

J = oxygen flux (units: m s^{-1}).

Dk/t = transmissibility (*i.e.,* permeability per unit thickness, t); D = diffusion coefficient, and k = solubility coefficient.

ΔP = difference in oxygen partial pressure between the front and back of the lens (P_{front}: 159 mm Hg (open eye) and 59 mm Hg (closed eye); $P_{back} \propto$ lens transmissibility).

As one would expect, there is a difficulty in determining the pressure at the posterior surface of the lens; consequently, a value known as the equivalent oxygen percentage (EOP, 5.6) may be used. The EOP provides an indirect measure of this

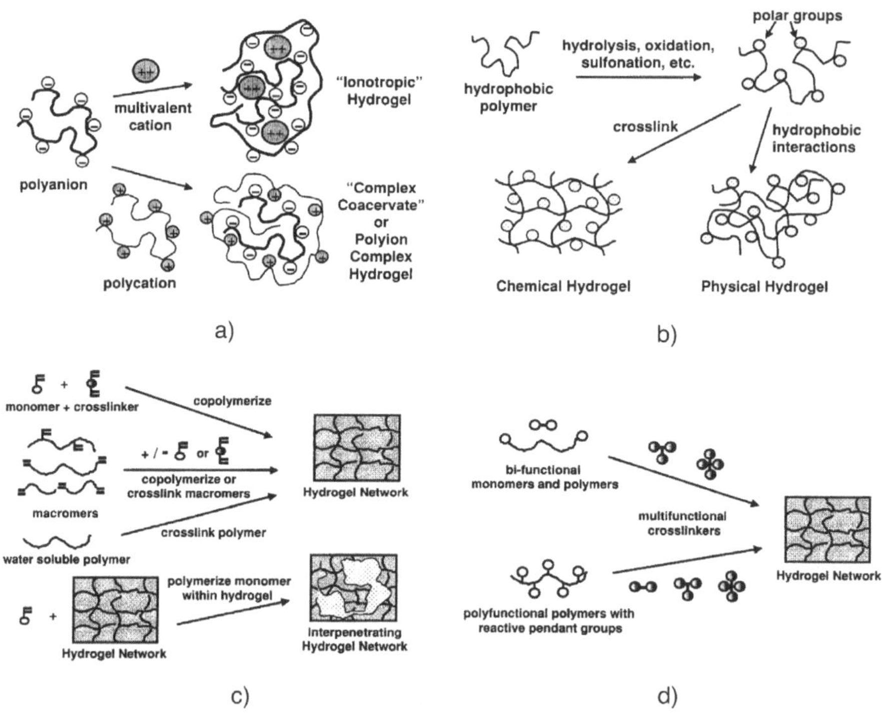

Fig. 5.63 Four techniques used to synthesize hydrogels. **a** Ionic hydrogel formation by gelling a polyelectrolyte solution with a multivalent ion of opposite charge (*e.g.,* sodium alginate + $CaCl_2$), or by mixing polyanion and polycation solutions to form a complex gel (*e.g.,* sodium alginate + polylysine); **b** chemical conversion of a hydrophobic polymer (*e.g.,* partial hydrolysis of acetate groups in PVAc to –OH groups in PVA); **c** copolymerization of a monomer and crosslinker in solution via free radical reactions (*e.g.,* crosslinked PHEMA and PEG gels); **d** crosslinking of polymers in solution (or solid state) with multi-functional reactive compounds (*e.g.,* PEG + diisocyanate to yield a PU hydrogel). Reproduced with permission from *Adv. Drug Deliv. Rev.* **2002**, *43*, 3

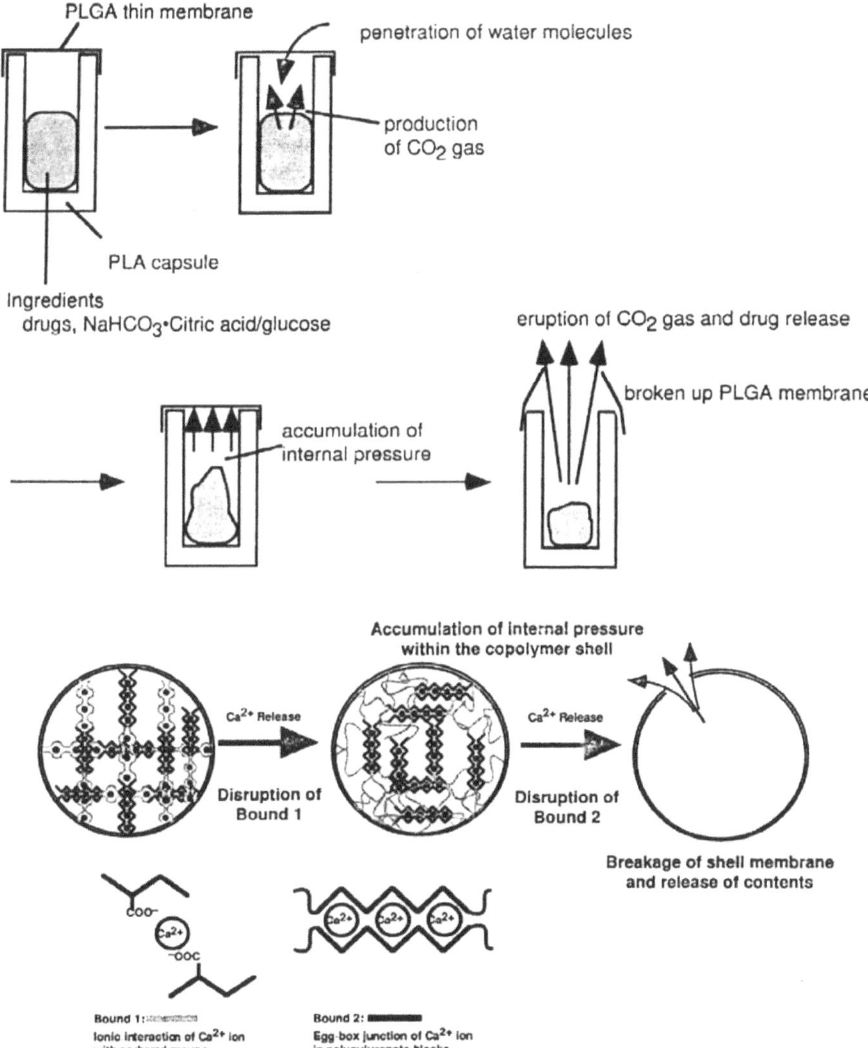

Fig. 5.64 Examples of drug delivery using hydrogel membranes. Top: carbon dioxide gas pressure causes rupture of membrane, controlled by the membrane thickness and amount of CO_2 being produced. Bottom: drug release from a polysaccharide alginate capsule, facilitated by multivalent/monovalent ion-exchange, controlled by the molecular weight of the alginate and size of the capsule. Reproduced with permission from *Adv. Drug Deliv. Rev.* **2002**, *54*, 53

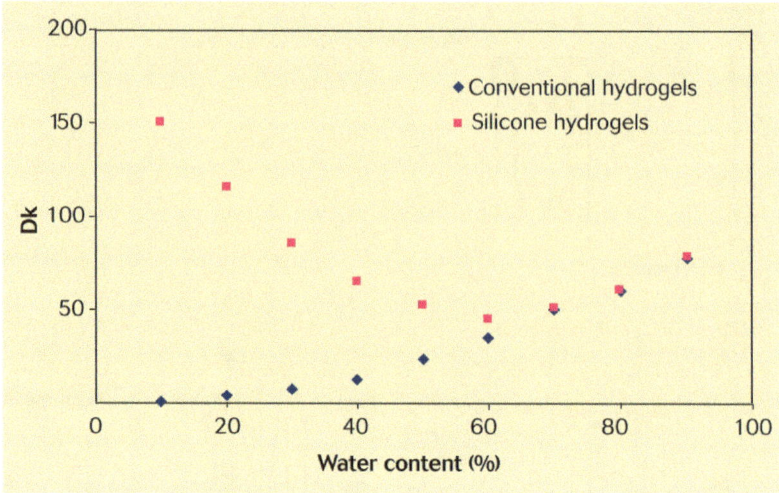

Fig. 5.65 Variation of oxygen permeability with equilibrium water content for conventional and silicone-based hydrogels. Reproduced with permission from *Optician*, **2005**, *230*, 16

Table 5.6 Oxygen transmissibility for various soft contact lenses[a]

Lens	Dk/t[b]	$J_{open\text{-}eye}$[c]	% of maximum	$J_{closed\text{-}eye}$[c]	% of maximum
100 μm HEMA	7.5	3.95	52%	1.50	25%
Acuvue 2	26	6.65	88%	4.09	68%
Acuvue advance	86	7.31	97%	5.55	92%
Purevision	110	7.37	98%	5.68	94%
O_2 optix	138	7.4	98%	5.8	96%
Acuvue oasys	147	7.4	98%	5.8	96%
Focus night and day	175	7.44	99%	5.84	97%
No lens	∞	7.54	100%	6.04	100%

[a] Data taken with permission from (a) French, K. *Optician* **2005**, *6030*, 16 and (b) Morgan, P.; Brennan, N. *Optician* **2004**, *5937*, 27
[b] Manufacturer values quoted for −3.00DS lenses
[c] Units: $\mu L\ cm^{-2}\ h^{-1}$

pressure, determined empirically by placing a membrane-covered sensor against the corneal surface and determining its oxygen consumption rate.[105] Whereas lenses with low Dk/t values such as PMMA result in very rapid O_2-uptake rates, a lens-free cornea would deplete oxygen at a much slower rate. As Fig. 5.66 illustrates, there is a large increase in the oxygen flux with increasing Dk/t until a threshold value is reached, corresponding to meeting the corneal oxygen demand.

$$EOP = 0.252(WC) - 2.912\log(L) - 20.611, \tag{5.6}$$

where

WC = water content of the lens (relevant for 'soft' hydrogel lenses).

L = thickness of the lens, in μm (typical center thicknesses: 30–50 μm for soft; *ca.* 100 μm for hard).

As one can see from 5.6, the permeability of a soft contact lens is directly proportional to its water content (Fig. 5.67). Whereas the moisture content of early PHEMA lenses was only *ca.* 35%, more recent silicone-based hydrogels contain up to 80% water. Although hydrophilic lenses are able to absorb water and permit extended wear without corneal irritation, an increase in water content also corresponds to lower mechanical strength that may result in tearing/scratching.

Drug Delivery

Anyone who has taken time-release medications (*e.g.*, nasal decongestants, allergy medications, etc.), or witnessed a loved one undergo chemotherapy has experienced a tangible application for biodegradable polymers: *drug-delivery systems* (DDS).[106] In the last few years, these materials have evolved from pills that generate systemic negative side-effects, to miniaturized devices that are able to target the region of interest and control the kinetics of drug release. Current DDS feature both spatial control (allowing for lower dosage and decreased side effects) and temporal control (to minimize fluctuations from the *therapeutic window*.[107] Typical means of control over drug release include: membrane-controlled diffusion (*e.g.*, osmosis, swelling), chemical (*e.g.*, biodegradable polymer backbones), particulate (*e.g.*, polymer-drug conjugates, liposome systems, polymeric micelle systems), and responsive (*e.g.*, temperature, solvents, pH, mechanical/magnetic/ultrasound). Future "smart" DDS

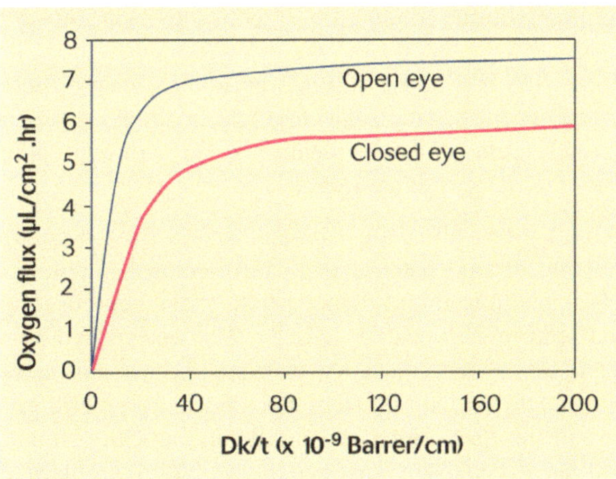

Fig. 5.66 The relationship of oxygen flux and oxygen transmissibility for open- and closed-eye wearing conditions. Reproduced with permission from *Optician*, **2005**, *230*, 16

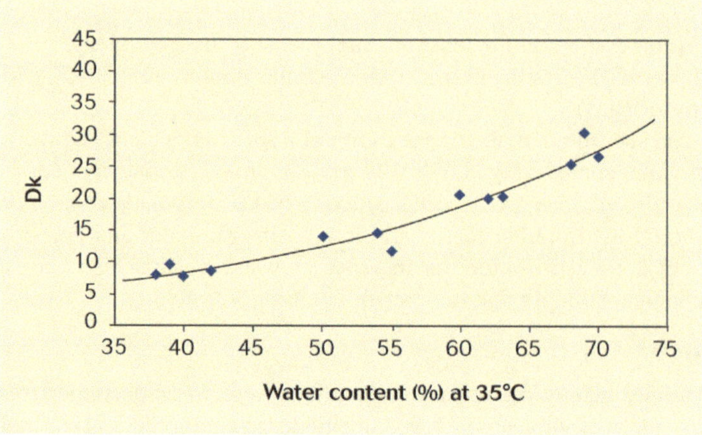

Fig. 5.67 Relationship between oxygen permeability and water content for conventional hydrogel lenses at 35 °C. Reproduced with permission from *Optician*, **2005**, *230*, 16

are currently in development, which will contain a variety of sensors for the 'on-demand' release of a drug in the proper dosage based on the patient's varying conditions. For instance, the delivery of insulin based on varying glucose levels, and growth hormones at key developmental stages to promote tissue growth.[108]

An ideal "smart" drug-delivery agent will have the following characteristics[109]:

i. Inexpensive fabrication
ii. Ease of drug loading
iii. Biocompatible—minimal trauma during both ingestion/implantation and drug delivery
iv. Facile tracking within the body, with external control via pre-/post-programming
v. Small size for ease of ingestion/injection and to allow for precise spatial positioning of drug release (<1 μm is ideal for passage through all capillaries with minimal risk of damage or embolism).[110]

No current device meets all of these parameters. Diffusion-controlled release is demonstrated by transdermal patches (*e.g.*, birth-control, nicotine, etc.), whereas particulates such as metallic micro-/nanoparticles have been used to ablate cancer cells by localized heating with near-IR light.[111] Whereas normal blood vessels have pore sizes <100 nm, the vessels adjacent to cancer cells have pore sizes between 100–200 nm. Hence, if DDS particulates (*e.g.*, liposomes, polymeric micelles, metallic nanoparticles) are of this size regime, they will automatically situate nearby cancer cells. However, it usually takes at least one day of blood circulation to accumulate sufficient amounts. In order to "hide" the particles from the immune system, the particulates must be covered with a 'stealth lipsome' such as poly(ethyleneglycol), PEG; otherwise, the particulates will be filtered out of the body via the liver as urine.

Although the above strategies are currently in use for a variety of treatment options, most efforts in the design of DDS have been focused on biodegradable reservoir systems; for example, the treatment of prostate cancer and protein hormone delivery. In these materials, degradation occurs in four steps:

i. Water flows into the device, swelling the matrix and ester bond cleavage;
ii. The interior degrades faster due to a high concentration of acid (product, catalyst);
iii. Oligomers become solubilized and diffuse out of the device;
iv. Slow erosion of the shell.

In addition to natural polymers such as arginine, chitosan, and cyclodextrin derivatives,[112] some common biodegradable polymers used for DDS include PGA, PLA and their copolymers. The rate of degradation may be controlled by its composition as well as crystallinity. However, due to the faster degradation of the interior, the localized acid concentration may become high enough to harm the therapeutic agent. A base such as $Mg(OH)_2$ may be incorporated to offset this acidity. For the treatment of malignant brain cancer, the use of polyanhydrides (*e.g.,* bis(p-carboxyphenoxy)propane—PCPP, **XII**) has proven successful; control over degradation rates may be afforded by decorating the backbone with aromatic groups, as well as copolymerization (*e.g.,* with sebacic acid—SA, **XIII**). The use of nonlinear polyanhydrides offers the benefit of enhanced mechanical properties, with the degradation rate being controlled by copolymerization (*e.g.,* an erucic acid dimer/sebacic acid copolymer, **XIV**, and a ricinoleic acid/sebacic acid copolymer, **XV**).

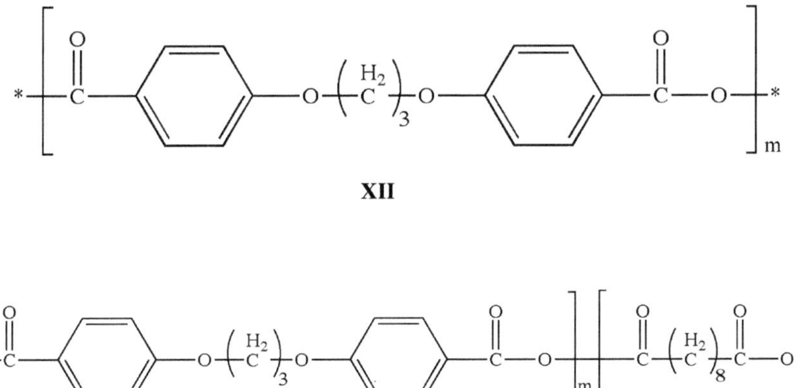

XII

XIII

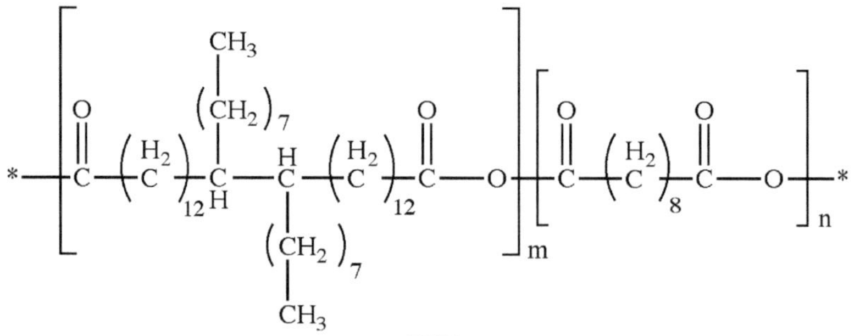

XIV

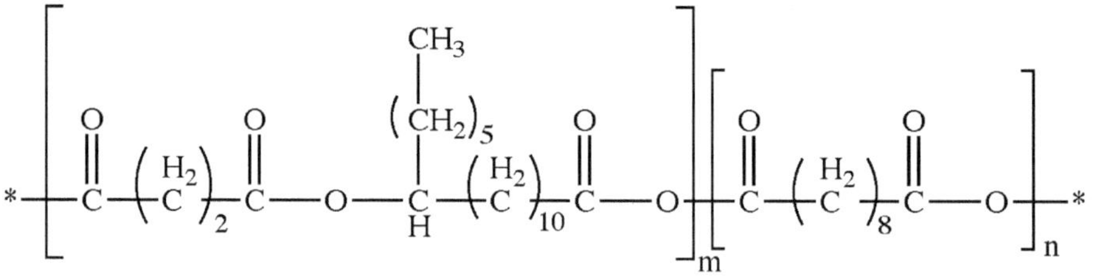

XV

Another class of biodegradable polymers used for DDS includes poly(orthoesters), POEs. The benefits of these polymers include facile synthesis and control over degradation rates and mechanical properties (*e.g.,* POE **XVI**, rigid *versus* POE **XVII**, flexible liquid/gel). Due to the highly hydrophobic nature of their backbones, POEs normally exhibit slow degradation; however, copolymerization with PLA (**XVIII**, used for periodontal disease and anesthetics) offers control over these kinetics. Though originally developed as flame retardants, poly(phosphoesters) may also be used as biodegradable DDS (*e.g.,* polylactophate (**XIX**), used for the controlled release of insulin and paclitaxel (cancer treatment)). These structures feature an extensive range of mechanical properties, while generating non-toxic degradation byproducts of phosphate ions, alcohols and diols.

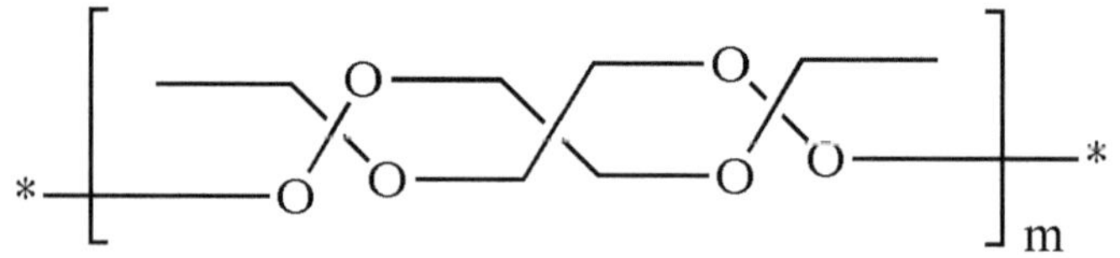

XVI

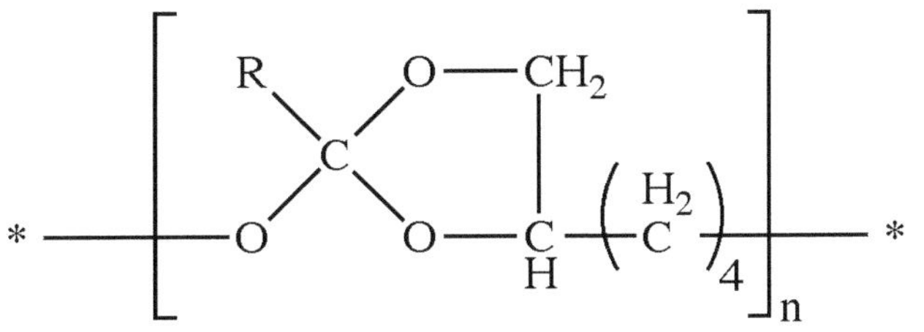

XVII

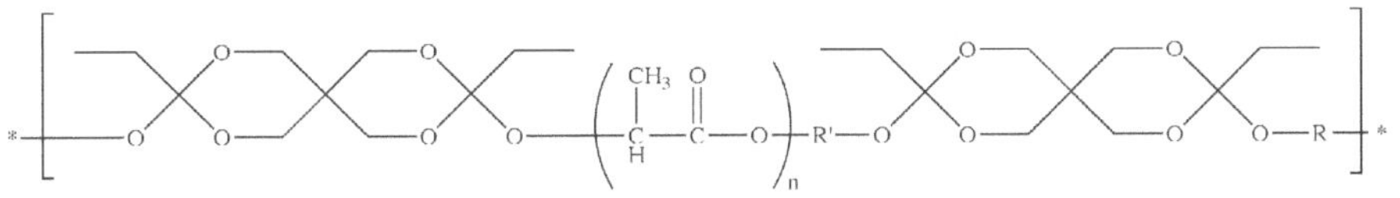

XVIII

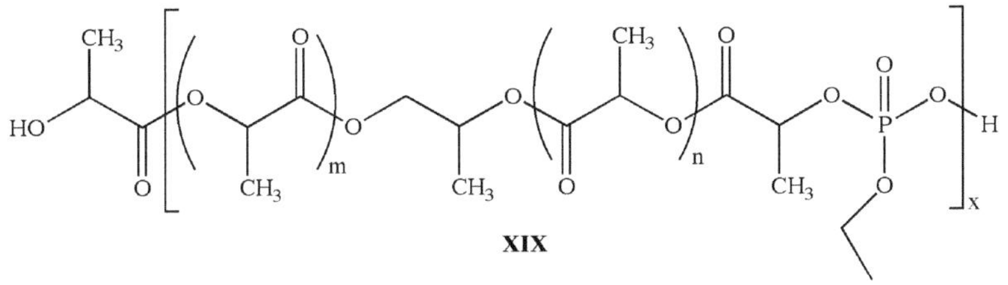

XIX

As mentioned earlier (Fig. 5.34), dendritic polymers may also be used for drug-delivery applications. Due to their unique 3-D structures and the high availability and tunability of surface functional groups, drugs may be loaded within the interior as well as attached to the periphery via electrostatic or covalent interactions. The use of PAMAM dendrimers is in development for the delivery of doxorubicin, an anti-cancer drug, as well as VivaGel®, a vaginal anti-HIV microbicide that is currently in Phase III clinical trials.[113] However, PAMAM dendrimers resist degradation within the body, which has spurred recent interest in the development of alternative dendritic polymers that are biodegradable. For example, Frechet and coworkers have developed polyester dendrimers based on the monomeric unit 2,2-bis(hydroxymethyl)propanoic acid, **XX**, for the controlled delivery of doxorubicin.[114] Biodegradable bifunctional dendrimers, **XXI**, may also be synthesized, which will offer a greater opportunity for the binding of varied functionalities (*e.g.*, drugs, targeting/imaging agents, etc.) to the periphery relative to symmetric analogues.[115]

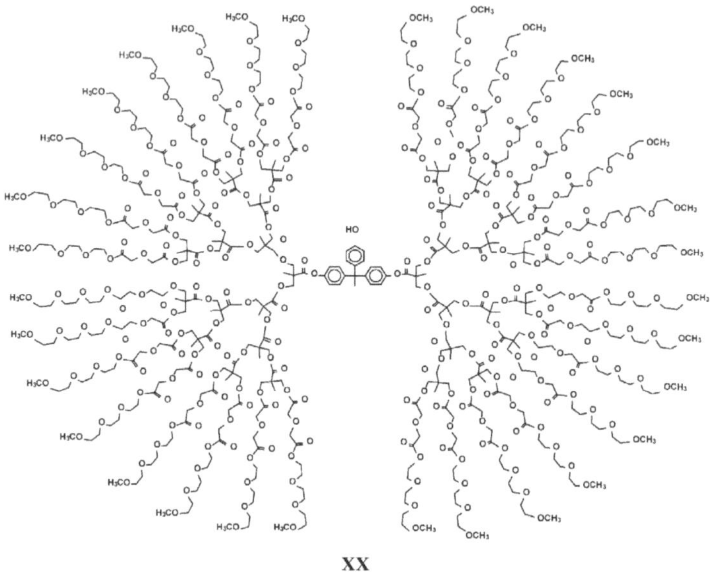

XX

XXI

Advanced DDS have also been fabricated using advanced lithographic techniques.[109] As discussed earlier, 'soft' imprint lithographic techniques such as solvent-assisted micro-molding (SAMIM), micro-molding in capillaries (MIMIC) and microcontact printing (μ-CP or MCP) most frequently employ PDMS as the material of choice for the mold. Though PDMS is used due to its UV transparency, biocompatibility, gas permeability, and low Young's modulus (allowing for conformal contact and facile release from the (nonplanar) master without cracking), its use for small-feature patterning has recently been usurped by fluorinated analogues.

5.3.3 Conductive Polymers

Although one generally considers polymeric materials as being non-conductive, there are specific structures that feature either metallic or semiconductive electrical properties. Unlike their inorganic analogues, conductive polymers generally feature high elasticity, toughness, solvent solubility, and low-temperature synthetic routes with facile structural (and property) tunability. Accordingly, these polymers are used for an increasing number of applications such as: antistatic electromagnetic shielding, conducting fibers, electrochemical batteries, electrochromic "smart" windows, light-emitting electrochemical cells, organic field-effect transistors, organic light-emitting diodes (OLEDs), photovoltaic devices,[116] nonlinear optics, and many more.[117]

Whereas traditional polymers feature a saturated sp^3-carbon framework, conductive polymers feature an unsaturated carbon backbone with 1 e^-/C atom placed in sp^2p_z (π) orbitals. As we know from resonance structures, the overlap of π-orbitals on successive C atoms leads to electron delocalization along the length of the polymeric framework. Though one may surmise that such an inter-connected 1-D array with delocalized electrons would result in metallic conductivity, this is not the case. In the 1930's, Peierls[118] asserted that a 1-D chain of equally-spaced ions is energetically unstable, and will distort via an electron–phonon coupling mechanism to lower its overall energy. One may consider the Peierls distortion for chain structures as analogous to the Jahn–Teller effect for transition metal complexes.

A prototypical example of Peierls distortion is the bond alteration exhibited by polyacetylene (PA) first synthesized by Natta and coworkers in 1958 (Fig. 5.68). Rather than considering polyacetylene as having equivalent C–C bond distances, adjacent bond lengths differ by *ca.* 6 pm (Fig. 5.69). This conformational distortion is enough to increase its bandgap, making PA a semiconductor in its pristine

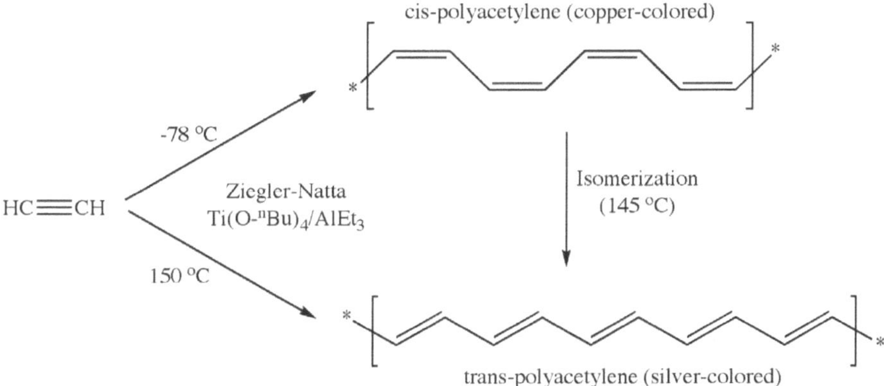

Fig. 5.68 Synthetic routes for cis- and trans-polyacetylene. It should be noted that the trans-isomer of PA is more stable than the cis-isomer since the former has two degenerate ground states (two energetically equivalent arrangements of alternating double bonds)

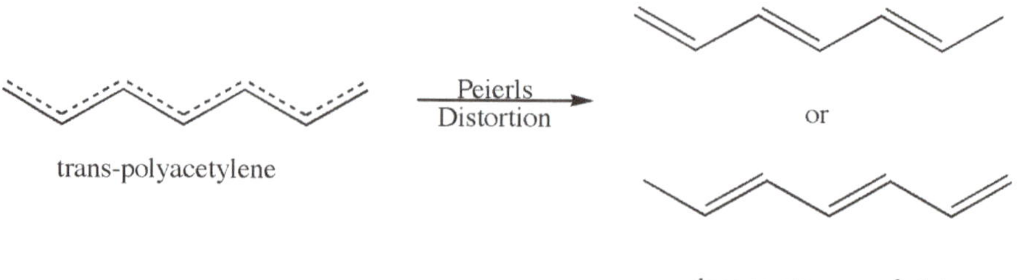

Fig. 5.69 Illustration of differing C–C bond lengths for polyacetylene

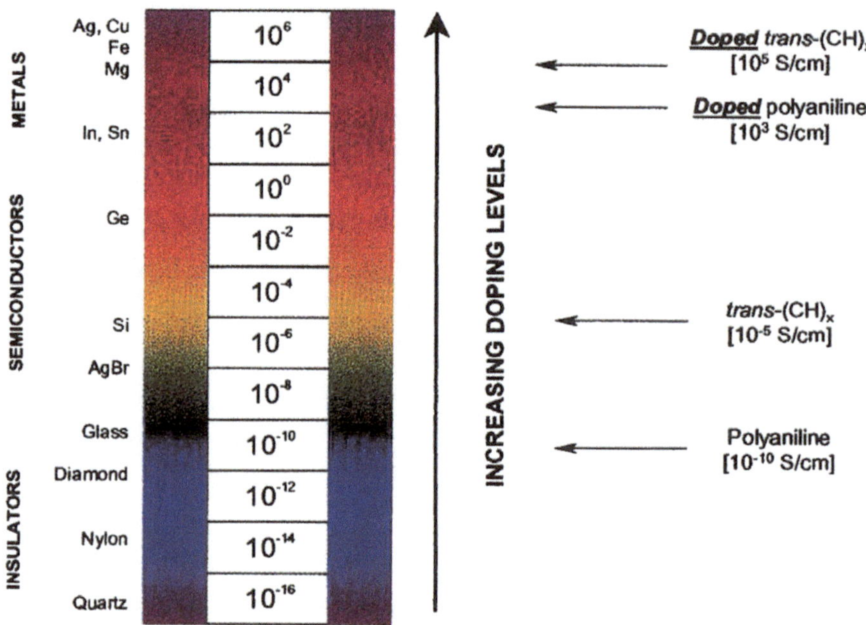

Fig. 5.70 Conductivity values for pristine (undoped) and doped conductive polymers, relative to common metals, semiconductors and insulators. Reproduced from MacDiarmid, A. G. *Synthetic Metals: A Novel Role for Organic Polymers*, Nobel Lecture, Dec. 8, 2000

(undoped) state; other pristine conductive polymers such as polyaniline are insulators (Fig. 5.70).

In order to increase the conductivity, a dopant may be added to the polymer that will either remove or add electrons (Table 5.7). The influence of doping on resultant conductivity is striking; whereas the conductivity of pristine PA is *ca.* 1×10^{-7} S cm^{-1}, doping may yield values as high as 1×10^5 S cm^{-1}—an improvement of 12 orders of magnitude! However, before discussing polymer doping in more detail, we should first point out the difference in nomenclature used by solid-state physicists and chemists. Whereas physicists refer to the generated carriers as polarons, bipolarons, and solitons, chemists are more familiar with the terms radical cations/anions, carbodications/carbodianions, and radicals/carbocations/carbanions, respectively (Fig. 5.71).

Redox doping is the first method used to enhance the conductivity of a conjugated polymer. *Oxidative doping* (p-doping) occurs when an oxidizing agent such as I_2, Br_2, ClO_4^-, AsF_5, $FeCl_3$, $NOPF_6$, etc. is added to the polymer, resulting in an abstraction of an electron and formation of a *radical cation*, also known as a *positive polaron* (5.7). In contrast, *reductive doping* (n-doping) occurs when a reducing agent such as alkali metals, LiBu, or $Li(C_{10}H_8)$ is added to yield a *radical anion*, also known as a *negative polaron* (5.8).

$$[CH]_n + \frac{3x}{2} I_2 \rightarrow [CH]_n^{x+} + x\, I_3^- \tag{5.7}$$

$$[CH]_n + x\, Na \rightarrow [CH]_n^{x-} + x\, Na^+ \tag{5.8}$$

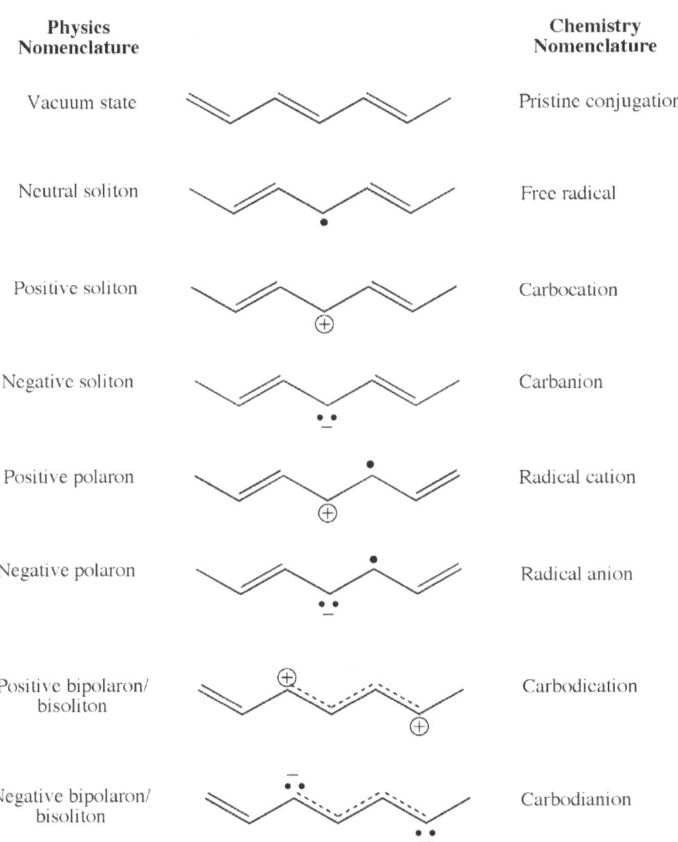

Fig. 5.71 Nomenclature of the generated carriers within conductive polymers

Table 5.7 Common conductive polymers

Polymer	Structure	Dopants	Ω^{-1} cm^{-1a}
Polyacetylene		I_2, Br_2, Li, Na, AsF_5	20,000–150,000[b]
Polyphenylene	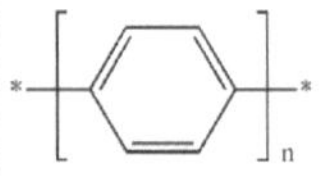	AsF_5, Li, Na	10,000
Polyparaphenylene vinylene	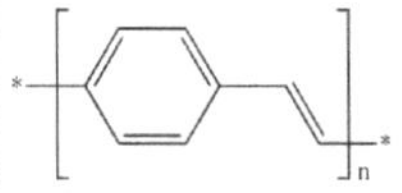	AsF_5	10,000
Polypyrrole	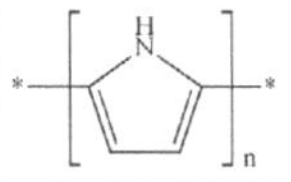	BF_4^-, ClO_4^-	7500
Polythienylvinylene	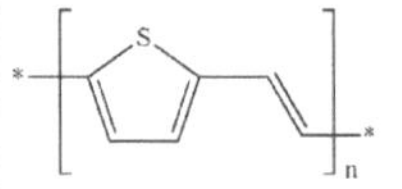	AsF_5	2700
Polyethylenedioxy thiophene (PEDOT)	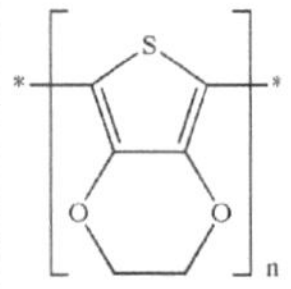	BF_4^-, ClO_4^-	550[c]
Polythiophene	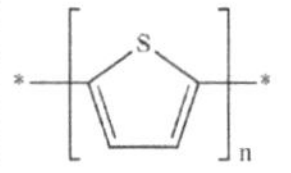	BF_4^-, ClO_4^-	500
Polyphenylene sulphide	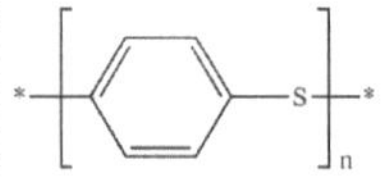	AsF_5	500
Polyaniline (PANI)	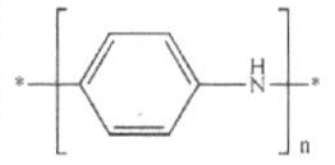	HCl	200
Polyfuran	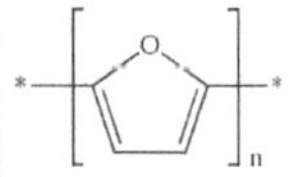	BF_4^-, ClO_4^-	100

(continued)

Table 5.7 (continued)

Polymer	Structure	Dopants	Ω^{-1} cm^{-1}[a]
Polyisothianaphthene	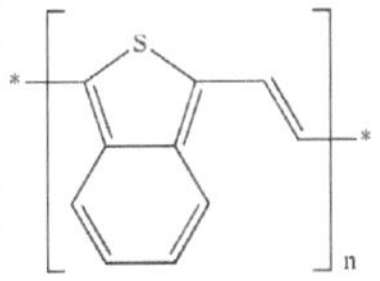	BF_4^-, ClO_4^-	100
Poly(3-alkylthiophene)	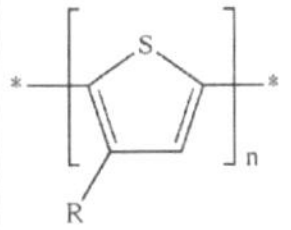	BF_4^-, ClO_4^-	100
Polyazulene	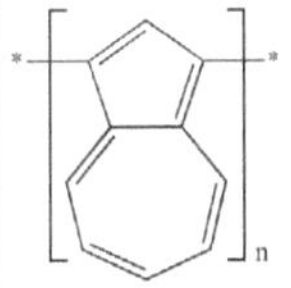	BF_4^-, ClO_4^-	2.5

[a] Maximum conductivity of doped polymers; the conductivity of pristine (undoped) polymers is *ca.* 0.001–0.1 Ω^{-1} cm^{-1}. The unit Ω^{-1} cm^{-1} is equivalent to S cm^{-1}

[b] The maximum anisotropic (stretch oriented) conductivity for polyacetylene: Naarmann, H.; Theophilou, N. *Synth. Met.* **1987**, *22*, 1

[c] Maximum conductivity of PEDOT is obtained by doping with poly(styrene sulfonate)—PSS. Accordingly, PEDOT:PSS is the industry leader in transparent conductive polymer films, with *ca.* 80% light transmission

It should be noted that in contrast to classical band theory, the doped electrons/holes generated in conductive polymers are not fully delocalized. The reason for the localization of polarons is due to Coulombic attractions to their counter ions (*e.g.*, I_3^-), which normally have very low mobilities. However, in order to achieve high electrical conductivities, the polarons must be able to migrate along the polymer chain. Figure 5.72 illustrates the migration mechanism of a polaron; for migration, a high concentration of the counter ions must be present to induce the movement of the polaron into the Coulombic field of nearby ions. Accordingly, high doping levels are required to achieve sufficient conductivity in conductive polymers. This may be compared with inorganic-based semiconductors such as Si, which require minute concentrations of dopants due to the pronounced mobility of electrons/holes through the extended solid-state lattice.

If a second electron is abstracted from a section of the polymer that is already oxidized, either a second independent polaron may be generated, or two polarons may condense to form a *bipolaron*. The two positive/negative charges of the bipolaron are not independent, but rather move as a pair similar to the Cooper pair of superconductors. To broaden the discussion beyond polyacetylene, Fig. 5.73 shows both polaron and bipolaron formation for polypyrrole.

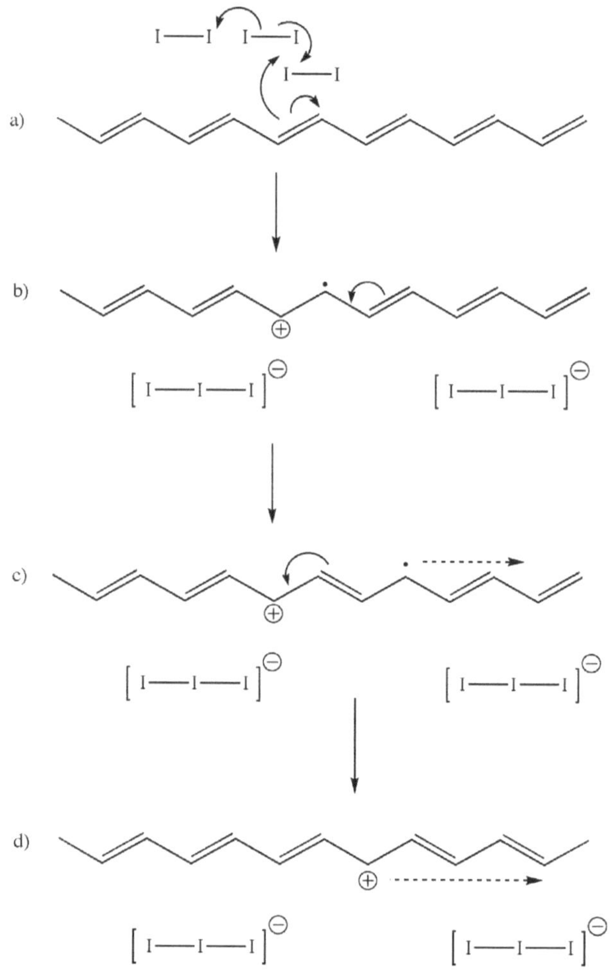

a)

b)

c)

d)

Fig. 5.72 The formation and migration mechanisms of a polaron

Another type of carrier is known as a *soliton*, formed by two primary modes: doping or bond-alternation defect formation (*e.g.*, thermal isomerization of *cis*-polyacetylene, Fig. 5.74a). Not surprisingly, solitons generated via doping are more prevalent than those formed by thermal isomerization. As illustrated in Fig. 5.74a, solitons may migrate along individual polymer chains by pairing with an adjacent electron; however, they may also contribute to charge transfer between different chains via *intersoliton hopping* (Fig. 5.74b). This latter mechanism is important for bulk conductivity, since carriers must move readily among individual chains to facilitate macroscopic electrical conductivity of the polymer. As the number of solitons increases, their spheres of influence overlap leading to metallic-like conductivity.

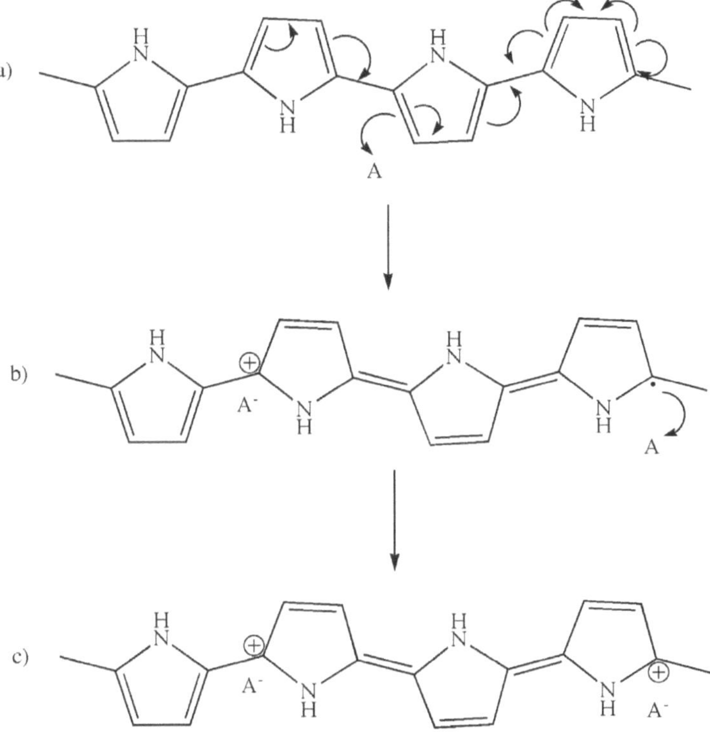

Fig. 5.73 Polaron and bipolaron formation for polypyrrole

Before closing this topic, let's consider the band structure and optical properties of doped conductive polymers, once again considering the most widely studied structure, polyacetylene. If all C–C bond distances were equivalent, one would expect a partially-filled π-band and metallic conductivity (Fig. 5.75a). However, due to Peierls distortion, the π-band splits in half, and a bandgap is now introduced at the Fermi level (Fig. 5.75b). The introduction of polarons and solitons via doping (or thermal isomerization for soliton formation) generates mid-gap states (Fig. 5.75c, d) whose optical absorption intensity grows at the expense of the E_g transition (Fig. 5.76).

The tremendous strength of conjugated polymers is the ability to fine-tune the bandgap via modification of the polymer structure. This may be achieved via sterics or electronic effects, by varying either the main chain or its pendant functional groups. Figure 5.77 lists some of the major applications for conjugated polymers, from battery to LED applications. Not unlike inorganic-based LEDs, conjugated polymers may also be designed to exhibit a full range of colors, from PPV (E_g = 2.5 eV, yellow) to MeH-PPV (E_g = 2.1 eV, red) to polyphenylene (E_g = 2.7 eV, blue).[119] A polyacetylene battery has also been reported, which employs a silver anode and I_2-doped PA as the cathode.[120] Conjugated polymers may also be exploited for the design of

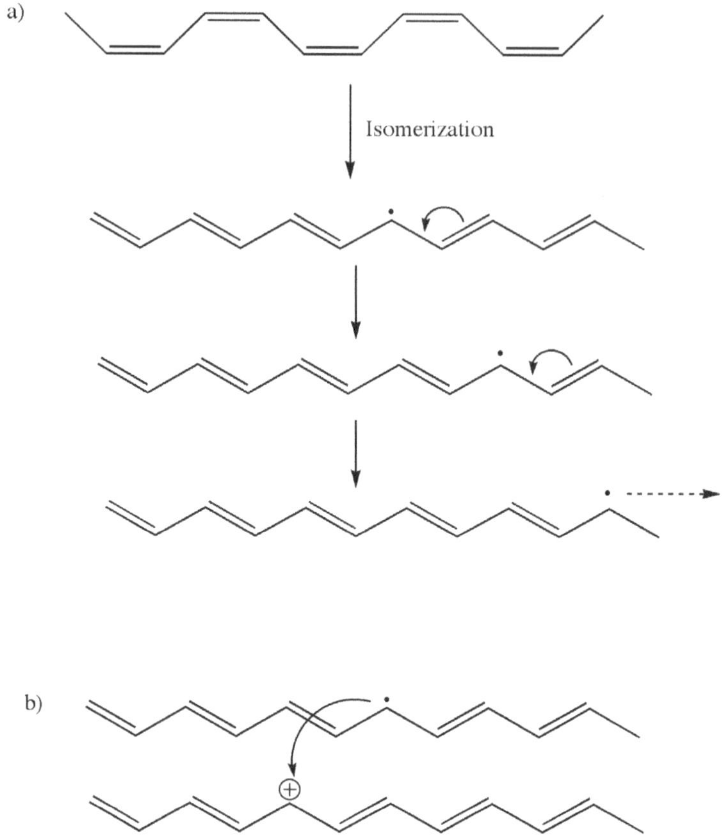

Fig. 5.74 **a** Soliton formation via isomerization of cis-polyacetylene, and migration along the polymer chain by pairing to an adjacent electron. **b** Illustration of intersoliton hopping. Charged solitons (bottom chain) are trapped by dopant counter ions, while neutral solitons (top chain) are free to move. A neutral soliton on a chain close to one with a charged soliton may interact, with the hopping of an electron from one defect site to another

wearable electronic systems that incorporate sensing, monitoring, and information processing capabilities (Fig. 5.78).[121]

Beyond polyacetylene, there are a wide variety of organic semiconducting materials that have been developed for applications in sensors, displays, and thermoelectrics (revisit Fig. 4.81).[122] The energy levels of these materials may be fine-tuned based on their backbone and substituents. However, overall transport properties are related to the orientation of the conjugated backbone, the degree of chain stacking in the π-direction, and energetic disorder due to structural defects (Fig. 5.79).[123] In general, the transport properties of solution-processed polymers are inherently anisotropic when carrier transport occurs along the chain axis and among adjacent chains across the π- or alkyl-stacking direction. The microstructure of polymer

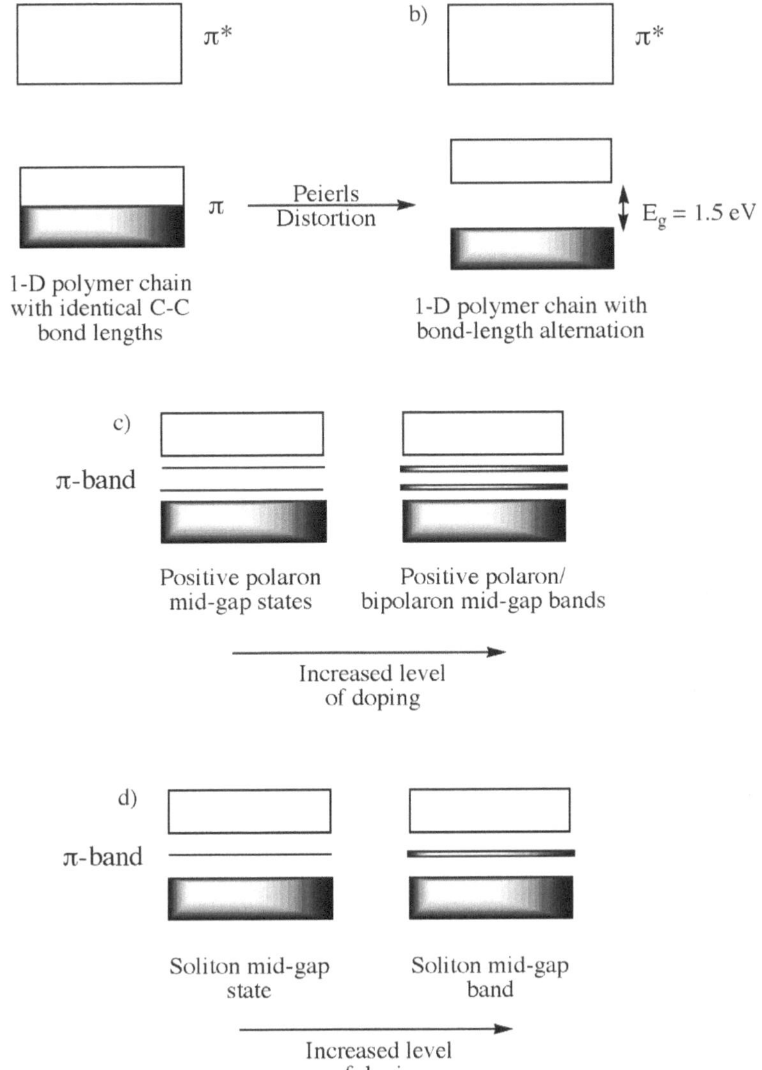

Fig. 5.75 Simplified band diagrams for a conductive polymer. Shown are: **a** polyacetylene if all C–C bond distances were equivalent (metallic), **b** actual band diagram for PA, resulting from Peierls distortion, **c** polaron mid-gap states and bands formed upon doping, and **d** soliton mid-gap state and band formed upon doping. For **c** and **d**, note that at higher levels of doping, the mid-gap bands grow at the expense of the valence and conduction bands, as midgap states are taken from band edges

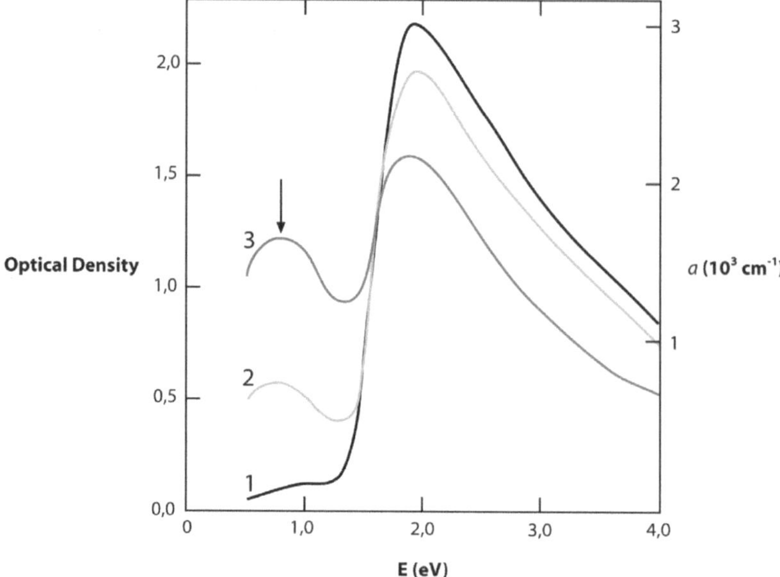

Fig. 5.76 Optical absorption of undoped polyacetylene (curve 1). Curves 2 and 3 show the absorption of polyacetylene with increasing dopant concentrations. A midgap state (arrow at 0.7 eV) emerges upon doping, becoming more intense with further doping levels, at the expense of other peaks. Results adapted from Roth, S. *One-Dimensional Metals,* Weinheim VCH, 1995

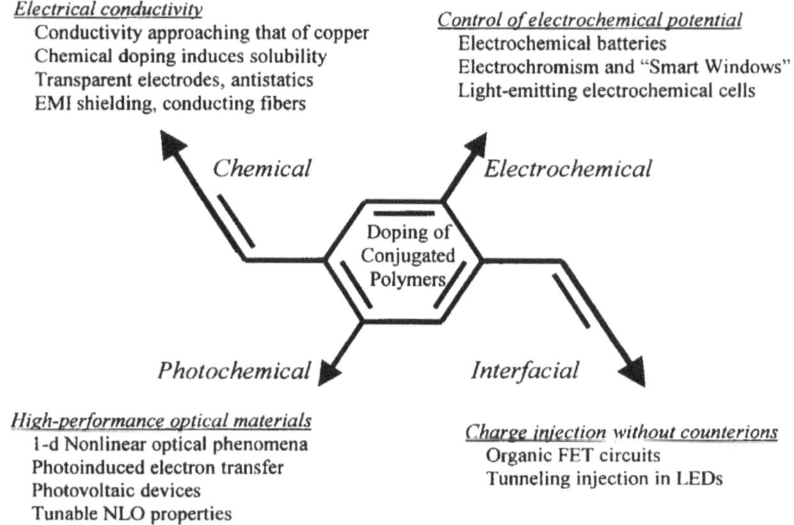

Fig. 5.77 Doping mechanisms and related applications. Reproduced from Heeger, A. J. *Semiconducting and Metallic Polymers: The Fourth Generation of Polymeric Materials*, Nobel Lecture, Dec. 8, 2000

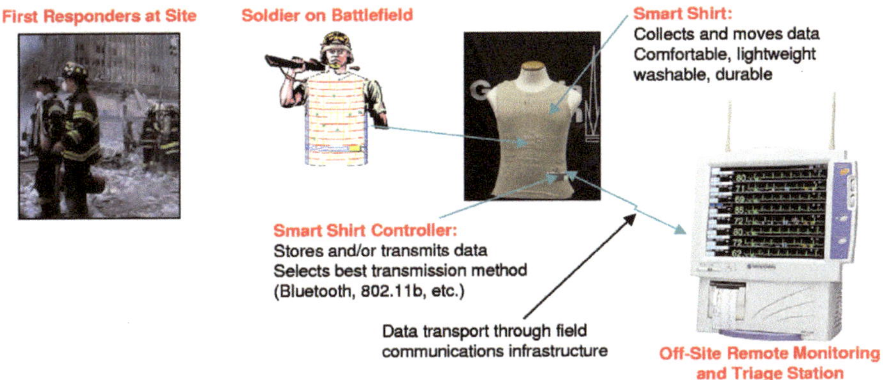

Fig. 5.78 Illustration of a 'smart shirt' for real-time monitoring of soldier vitals, so off-site medics may safely focus their recovery efforts. The use of conductive polymers would lower the weight of such wearable electronic devices

thin films affects transport between semicrystalline domains (shaded regions of Fig. 5.79c) and disordered regions. The carrier mobility of polymers may be higher for holes (p-type), electrons (n-type), or comparable for both (ambipolar).

Ambipolarity is generated from donor–acceptor co-polymers such as indacenodithiophene-co-benzothiadiazole (C16-IDTBT). The long alkyl chains in this co-polymer gives rise to a still or fused framework that withstands torsional fluctuations, thus enabling efficient transport of charge carriers along its backbone.[124] Interestingly, the electrical and thermoelectric properties of this polymer have been shown to depend on the processing conditions, which is assisted by oxygen diffusion and microstructure rearrangement upon solvent evaporation.[125]

5.3.4 "Soft" Lithography

In the previous chapter, we discussed a variety of strategies for surface patterning through the use of photolithography and variety of etching techniques. Within the last few decades, comparatively inexpensive and scaleable tehniques known as "soft lithography" have been the focus of much development. Patterning of a substrate is afforded by using a master *elastomeric* stamp that contains a nanostructured pattern, known as a *relief*, on its surface. Contrary to photolithography (Sect. 4.2.2), the resolution of the final pattern is not limited by light diffraction, but only depends on the dimensions of the relief structures—typically fabricated in the master by electron-beam lithography (Fig. 5.80). Typically, the mold (or stamp) is comprised of PDMS, which allows for intimate contact between the mold/substrate surfaces, even if nonplanar substrates are used. More recently, other polymers have been developed for this application such as polyimides, polyurethanes, and a variety of substituted siloxanes—especially fluorinated analogues—due to easy release after molding and lack of swelling by organic solvents.

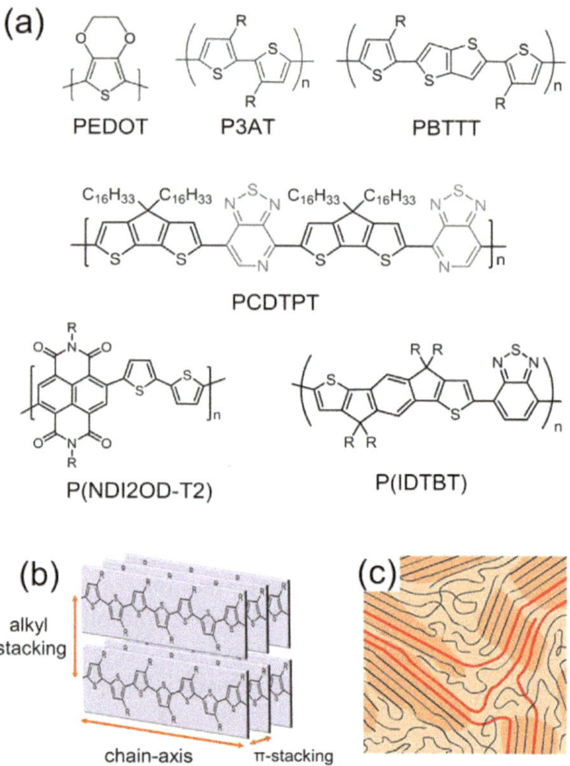

Fig. 5.79 **a** Molecular structures of common semiconducting polymers. **b** Illustration of chain stacking. **c** Illustration of thin-film microstructure of polymer films. PEDOT = poly(3,4-ethylenedioxythiophene); P3AT = poly(3-alkylthiophene); PBTTT = poly[2,5-bis(3-alkyldecylthiophen-2-yl)thieno[3,2-b]thiophene]; PCDTPT = poly[4-(4,4-dihexadecyl-4H-cyclopenta[1,2-b:5,4-b']dithiophen-2yl)-*alt*-[1,2,5]thiadiazolo[3,4-c]pyridine]; P(NDI2OD-T2) = poly{[N,N'-bis(2-octyldodecyl)-naphthalene-1,4,5,8-bis(dicarboximide)-2,6-diyl]-*alt*-5,5'-(2,2'-bithiophene)}; P(IDTBT) = indacenodithiophene-benzothiadiazole Reproduced with permission from *J. Appl. Polym.* **2017**, *134*, 44403. © 2017 Wiley

The technique of replicating a master pattern is aptly termed *replica molding*. In theory, the resolution of the replica will be identical to the master. However, due to the "soft" nature of the mold, the nanoscale features of the relief may become distorted due to polymer shrinkage (*e.g.*, solvent evaporation, in situ cross-linking, mechanical deformation), or interfacial phenomena between the mold and master surfaces (*e.g.*, differing thermal expansions, adhesive forces[126]). In contrast, a hard mold of Si or quartz exhibits significantly less distortion due to their solvent/chemical resistance, and thermal stabilities at temperatures sufficient to cause polymer cross-linking. Hard molds, used for *step-and-flash imprint lithography* (SFIL, Fig. 5.81) and *nanoimprint lithography* (NIL, Fig. 5.82),[127] are commonly used to pattern materials such as CDs, DVDs, and holographic images on the front of most credit cards.[128]

A common application for elastomeric molds is for micro- (μ-CP) or nanocontact printing, where a *self-assembled monolayer* (SAM) is placed on both planar[129] and curved[130] surfaces via contact with the reliefs on the mold (Fig. 5.83). The archetypical example of a SAM is the chemisorption of alkylthiols on a gold or silver surface, which results in self-assembly/alignment into a 3-D forest array (Fig. 5.84). Applications for SAMs span a number of fields from sensors to high-density storage; some interesting precedents include the selective adsorption and spontaneous alignment of carbon nanotubes (CNTs)[131] and directed growth of nanowires[132] from SAMs, as well as the production of graphene from SAM precursors.[133]

In order to improve the stamping resolution of the elastomeric stamp, there have been improvements in both the stamp and "molecular inks" (*e.g.*, alkylthiols, silanes). In particular, traditional PDMS exhibits a relatively high elasticity that limits possible

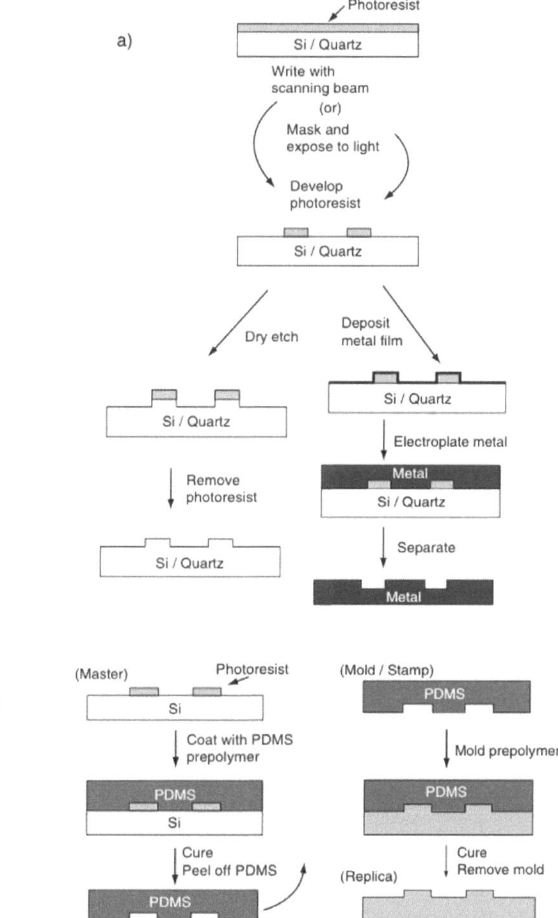

Fig. 5.80 Comparison of **a** conventional photolithography/electroplating with **b** soft lithography. Shown in **b** is replica molding, which consists of the formation of a PDMS stamp and subsequent replication of a master in a photo- or thermally curable prepolymer. Reproduced with permission from *Chem. Rev.* **2005**, *105*, 1171. © American Chemical Society

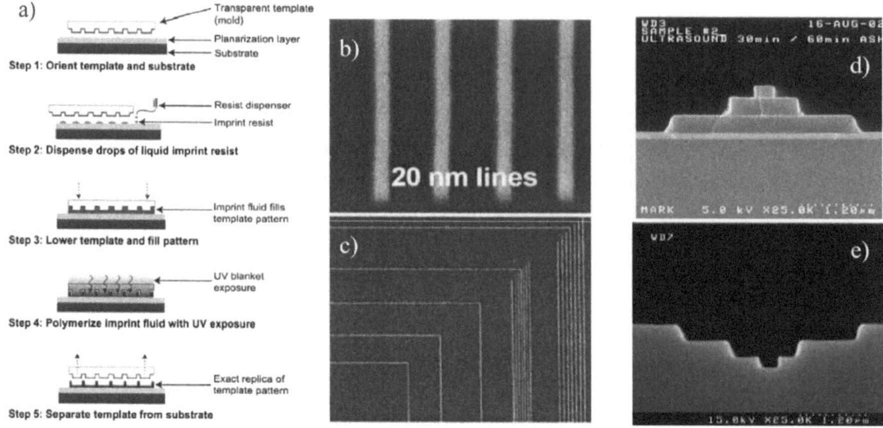

Fig. 5.81 Schematic of step-and-flash imprint lithography, SFIL. The SEM images shown in **b** and **c** represent 20- and 40-nm patterned line arrays, whereas **d** and **e** show a multi-tiered template and resultant imprint in a resist material. Reproduced with permission from *J. Microlithogr., Microfabr., Microsyst.* **2005**, *4*, 011002. © 2005 International Society of Optical Engineering

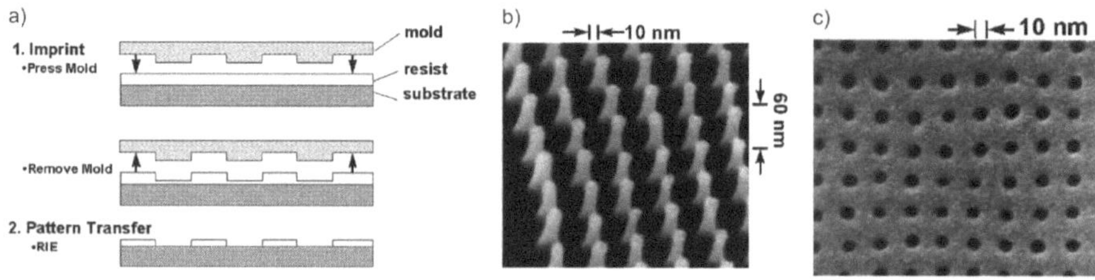

Fig. 5.82 Schematic of nanoimprint lithography (NIL) and SEM images of a nanopatterned mold and arrays imprinted in poly(methyl methacrylate). Reproduced with permission from *J. Vac. Sci. Technol. B* **1997**, *15*, 2897. © 1997 American Institute of Physics

relief linewidths; on the other hand, small molecular weight inks exhibit diffusion during patterning. Hence, the following complementary strategies have been employed:

(i) Using a composite two-layer stamp comprised of a 30 μm hardened PDMS coating on a 2–3 mm thick PDMS support[134]; especially in tandem with sharp, V shaped grooves
(ii) Using high molecular weight inks such as dendrimers[135] and biological molecules (*e.g.*, proteins[126])

Using a combination of the above modifications has now extended nanocontact printing to well below the 30 nm regime—even as low as 2 nm![136]

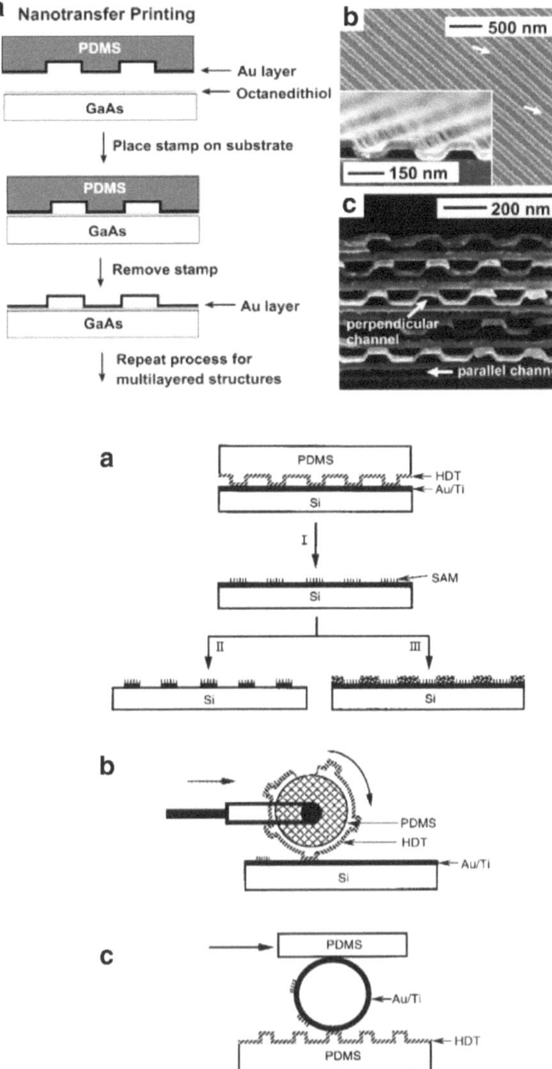

Fig. 5.83 (Top) Illustration of the general process of nanotransfer printing. SEM image B illustrates a 20-nm grooved gold layer transferred onto a GaAs surface. Image C shows a multilayered stack of 20-nm thick layers of parallel grooves; the channels in adjacent layers are aligned perpendicular to one another. Reproduced with permission from *Chem. Rev.* **2005**, *105*, 1171. © American Chemical Society. (bottom) Illustration of the diversity of microcontact printing. Shown is the printing of hexadecanethiol (HDT) via: **a** a planar surface with a planar stamp (I: SAM printing, II: etching, III: deposition), **b** large-area printing on a planar surface with a rolling stamp, and **c** printing on a nonplanar surface with a planar stamp. Reproduced with permission from *Angew. Chem. Int. Ed.* **1996**, *37*, 550. © 1996 Wiley–VCH

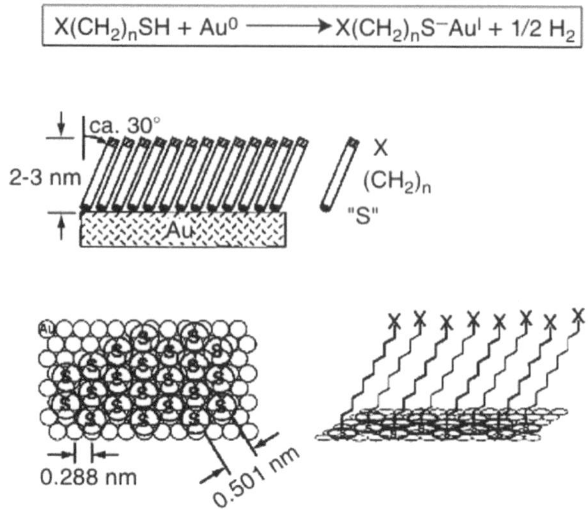

Fig. 5.84 Schematic of a self-assembled monolayer (SAM), illustrating the organization of the alkyl chains via van der Waal interactions and the close-packed array of sulfur atoms on the gold surface. Reproduced with permission from *Angew. Chem. Int. Ed.* **1996**, *37*, 550. © 1996 Wiley–VCH

Although PDMS has dominated the field of soft lithography, DeSimone and coworkers have introduced fluorinated perfluoropolyether (PFPE, **XXII**) elastomers, which offer better solvent resistance and cleaner mold-release characteristics.[137] This version of soft lithography is coined PRINT: Particle (or Pattern) Replication in Nonwetting Templates.[138] Figure 5.85 illustrates the processes involved in the PRINT technique, which allows one to fabricate surface-bound or free polymeric structures of virtually any size and shape, with feature sizes reaching well below 100 nm. A master template with the desired features is first prepared via e-beam lithography or photolithography/etching of a Si substrate. A homogeneous coating of a photocurable liquid PFPE is then applied and photochemically cross-linked to yield a mold that may be peeled away from the master. To generate a 2D array of particles or free particulates, the mold is filled with an appropriate liquid via capillary filling without wetting the land area around the cavities. The liquid in the mold is then converted to a solid using evaporation, curing, lyophilization or other mode of phase transition, and the array may then be transferred to another surface. It should be noted that this process is also amenable for the replication of naturally occurring objects such as virus particles,[139] block copolymer micelles, and protein particles.[140]

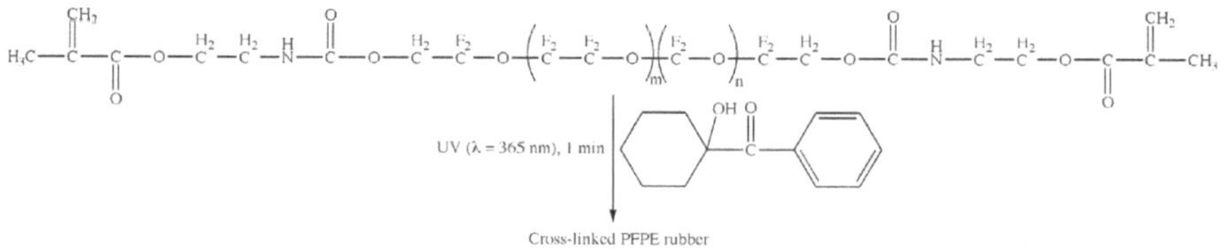

XXII

The PRINT technique is analogous to micromolding in capillaries (MMIC) or solvent-assisted micromolding (SAMIM), with the use of perfluorinated polymers rather than silicones (Fig. 5.86). The use of PFPE instead of PDMS molds offers the following benefits (Fig. 5.87)[141]:

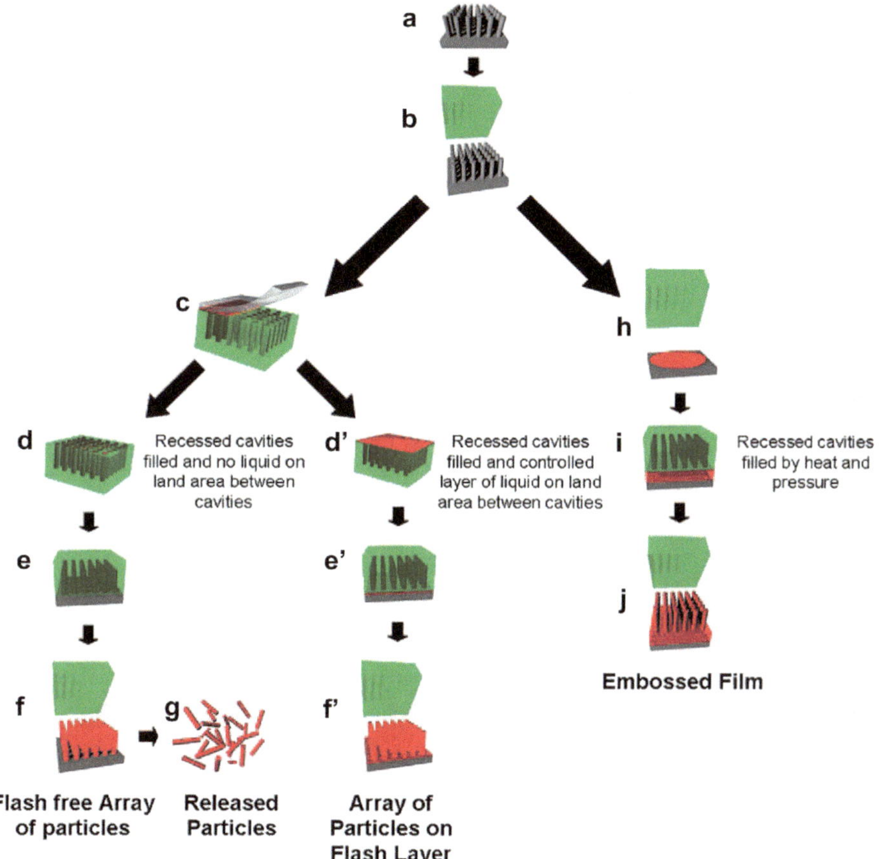

Fig. 5.85 Comparative illustrations of the particle/pattern replication in nonwetting templates (PRINT) process versus traditional embossing processes. Shown are: **a** silicon master template, **b** mold release from master template, **c** mold filling via capillary fill with counter-sheet having a higher surface energy than the PTFE mold. Depending on the exact nature of the liquid to be molded and the details of the process, one can fill the cavities only and not wet the land area around the cavities, (**d**). Alternatively, one can fill the cavities and have a thin layer of liquid on the land area around the cavities, (**d'**). The thickness of the layer of connecting flash layer liquid is determined from the principles associated with free meniscus coating processes with the resulting (**e**, **e'**) pattern transfer to substrate, (**f**, **f'**) mold release from the array of isolated features, and (**g**) dissolution of the harvesting film to yield free particles. As an alternative to PRINT, one can use PFPEs using traditional embossing processes where pressure and heat are applied (**h**, **i**) to form an embossed film (**j**) after the mold is released. Reproduced with permission from *Acc. Chem. Res.* **2008**, *41*, 1685. © 2008 American Chemical Society

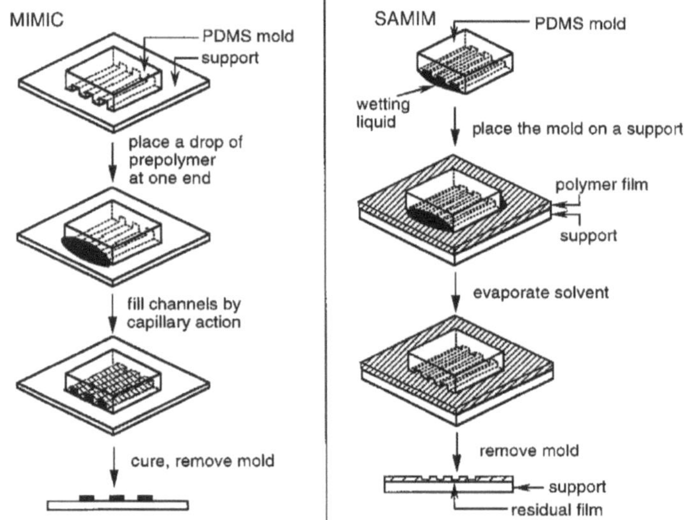

Fig. 5.86 Schematic of soft lithographic procedures for **a** replica molding (REM), **b** microtransfer molding (μTM), **c** micromolding in capillaries (MIMIC), and **d** solvent-assisted micromolding (SAMIM). Reproduced with permission from *Angew. Chem. Int. Ed.* **1998**, *37*, 550. © 1998 Wiley–VCH

(i) Lower surface energy and solvent resistance, which allows one to selectively fill the nanosized cavities of the mold without wetting the surrounding area. This results in a diverse range of particulate morphologies and surface functionalization (Fig. 5.88), of extreme importance for nanomedicine applications[142];
(ii) Significantly greater swelling resistance in organic solvents;
(iii) Greater chemical inertness, which allows the array of surface features to be easily separated from the mold to yield free-standing micro- or nanostructures;
(iv) Greater tunability of the modulus by varying the precursor molecular weight, which allows one to pattern a variety of sol–gels into <100 nm features;
(v) Non-stick properties allow the organic particles to be easily removed from the mold.

In addition to using PRINT to fabricate 3-D polymeric containers (Fig. 5.89h, i), a clever approach employs photolithography and self-assembly to generate cubic and pyramidal structures via "micro-/nano-origami (Figs. 5.87a–g and 5.90) that can be used for diverse applications such as drug delivery. In this approach, a 2-D structure is formed with temperature-sensitive hinges that fold into the desired 3-D shape upon heating.[143] As shown in Fig. 5.88 (bottom), the drug-release characteristics may be

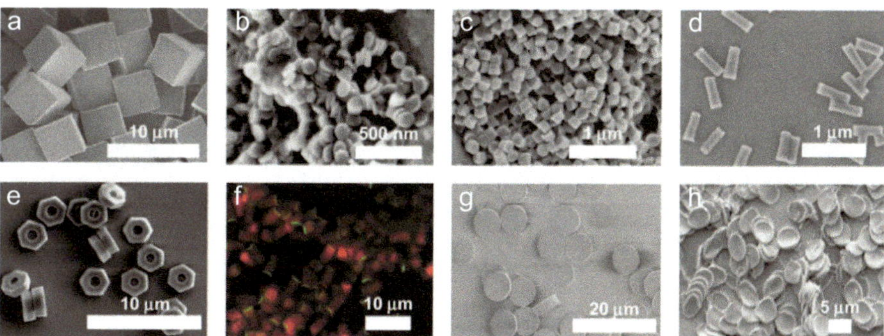

Fig. 5.87 PRINT particles varying in size, shape, surface chemistry, and deformability. The particle composition for all of these particles was approximately the same and included PEG (bulk of the matrix), a cross-linker, and a linker group for conjugation of stabilizing groups (such as PEG) or targeting ligands (such as peptides, antibodies, etc.): **a** scanning electron micrograph (SEM) of cube-shaped particles with a cube side length of 5 μm; **b** SEM of cylindrical nanoparticles having diameter) 110 nm and height) 35 nm; **c** SEM of cylindrical nanoparticles having diameter) 200 nm and height) 200 nm; **d** SEM of rod-like PRINT particles having diameter) 100 nm and height) 300 nm; **e** SEM of 3 μm "hex nut" particles; **f** cylindrical PRINT particles containing a covalently attached red fluorophore that have been functionalized on one face with a generic linker group (green fluorophore) that will allow the conjugation of targeting peptides, antibodies, and aptamers region-specifically onto the particle probes; **g**, **h** particles for mechanobiology studies having approximately the same dimensions as red blood cells (cylinders with a diameter) 7 μm and a height of 1.7 μm made from **g** a nondeformable, highly cross-linked hydrogel and **h** lightly cross-linked, deformable hydrogel. Reproduced with permission from *Acc. Chem. Res.* **2008**, *41*, 1685. © 2008 American Chemical Society

fine-tuned in these structures, from isotropic to anisotropic, based on the relative porosity/permeability of its sidewalls.

5.4 Polymer Additives

Although the properties of polymers may be fine-tuned based on the functional groups that are present in their repeat units, all commodity polymers also contain a number of components in order to impart desired properties. Some common additives include:

i. *Stabilizers* (antioxidants[144]: *e.g.,* 1,3,5-trimethyl-2,4,6-tris(3,5-di-tert-butyl-4-hydroxybenzyl) benzene, **XXIII**; UV-stabilizers: *e.g.,* TiO_2; heat-stabilizers[145]: *e.g.,* tetrabutyltin, tetraoctyltin). Used to protect the polymer from oxidation, UV light, and heat during/after processing.
ii. *Nucleating/clarifying agents* (*e.g.,* nucleation: sodium benzoate-based, nucleation/clarifying: sorbitol-based). Used to increase the crystallization rate of

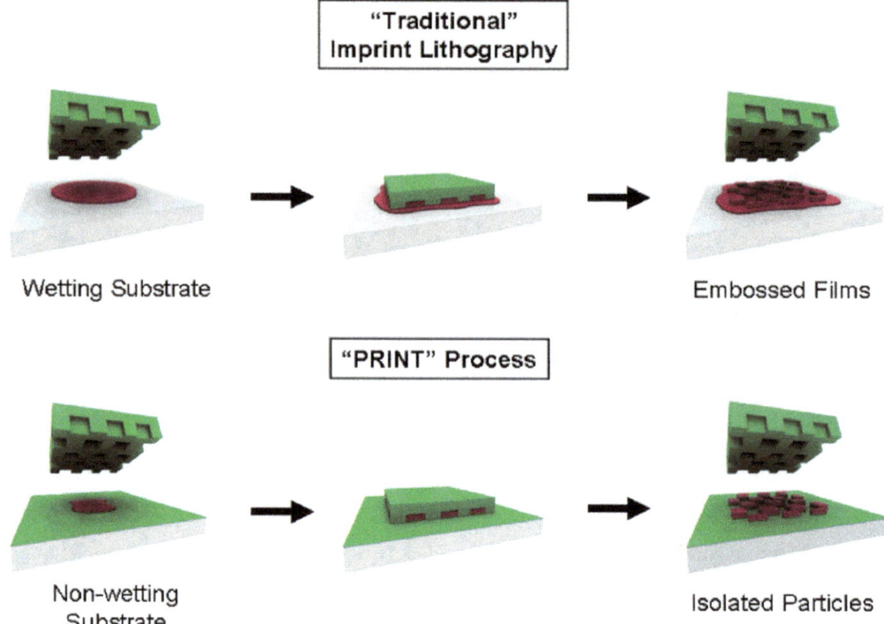

Fig. 5.88 Illustration of the PRINT process compared to traditional imprint lithography in which the affinity of the liquid precursor for the surface results in a scum layer. In PRINT, the nonwetting nature of fluorinated materials and surfaces (shown in green) confines the liquid precursor inside the features of the mold, allowing for the generation of isolated particles. Reproduced with permission from *J. Am. Chem. Soc.* **2005**, *127*, 10096. © 2005 American Chemical Society

semi-crystalline polymers such as polypropylene, polyamide, and polyester. Clarifying agents reduce haze and significantly increase polymer transparency.

iii. *Curatives* (chain extenders and crosslinkers: *e.g.,* 3,5-diethyltoluene-2,4-diamine, **XXIV**[146]; cure promoters: *e.g.,* N-(2-hydroxyethyl)-N-methyl-para-toluidine, **XXV**[147]; polymerization inhibitors: *e.g.,* tris(N-nitroso-N-phenylhydroxylamine) aluminum salt, **XXVI**). Used to promote polymer curing at lower temperatures for urethanes, epoxies, polyesters, vinyl esters, acrylates, and ureas. Also used for free-radical scavenging and metal chelation to prevent unwanted polymerization during the manufacture and storage of olefinic-type resins.

iv. *Plasticizers*[148] *or anti-plasticizers*[149] (*e.g.,* phthalates—**XXVII**[150]; trimellitates—**XXVIII**[151]; adipates—**XXIX**[152]). Used to control the flexibility/rigidity of the polymer.

v. *Coloring agents* (*e.g.,* inorganic pigments, organic dyes). Used to impart the desired color(s) to the polymer.

vi. *Flame retardants* (described later).

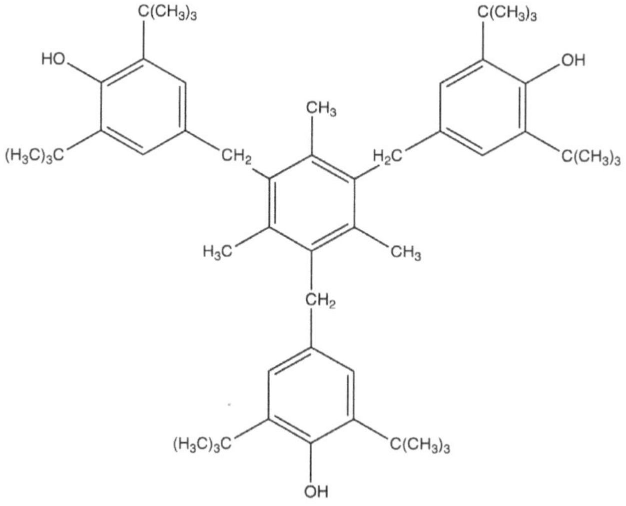

XXIII

XXIV

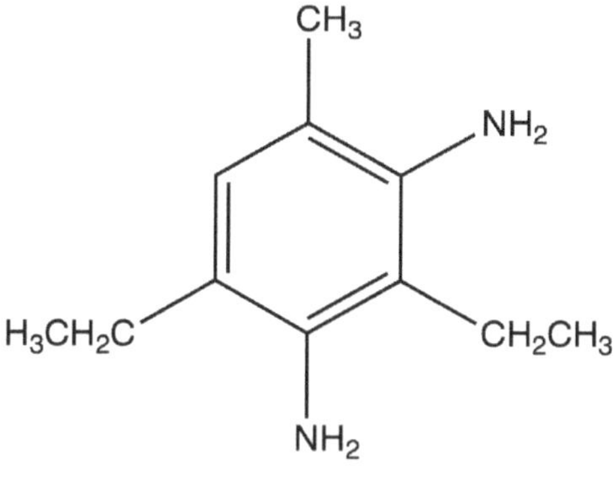

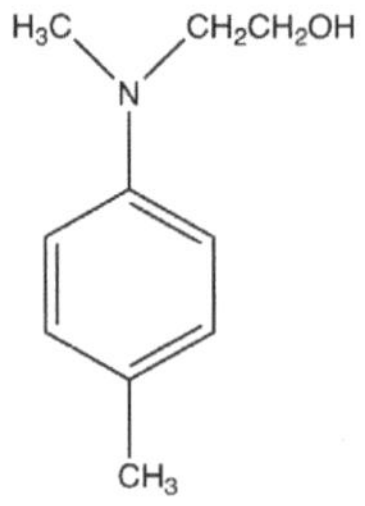

H_3C CH_2CH_2OH N CH_3

XXV

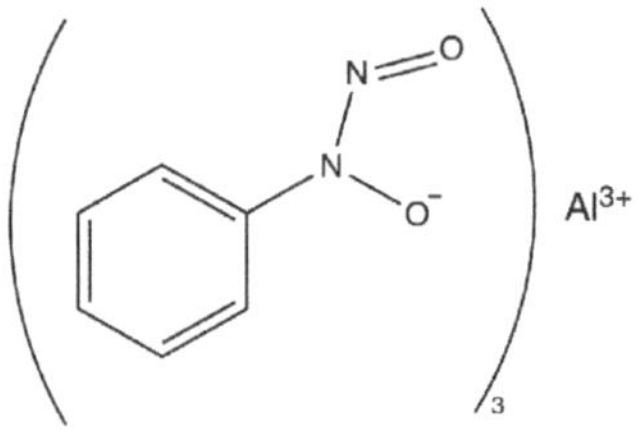

N=O N O^- Al^{3+} 3

XXVI

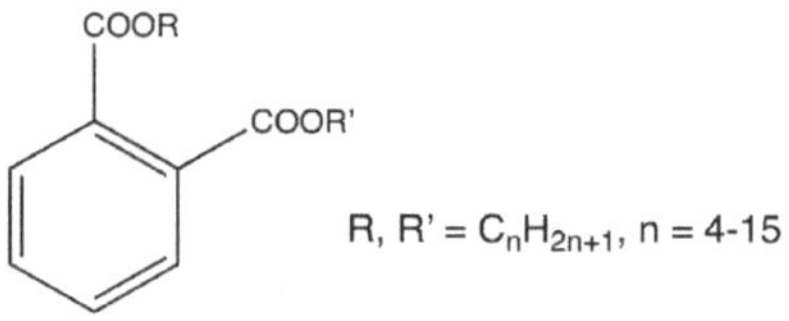

COOR COOR'

R, R' = C_nH_{2n+1}, n = 4-15

XXVII

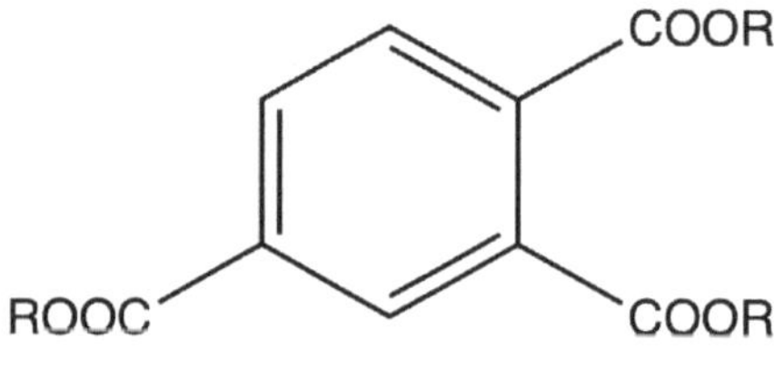

COOR ROOC COOR

XXVIII

$$R-O-\overset{O}{\overset{\|}{C}}-\left(\overset{H_2}{C}\right)_4-\overset{O}{\overset{\|}{C}}-O-R$$

XXIX

Fig. 5.89 **a** An optical image showing free-standing, self-assembled fabricated cubic containers over 2 orders of magnitude in size from 2 mm to 50 μm and **b** 15 μm along with **c–e** different shapes. **f** Scanning electron micrograph of polystyrene particles, "UFOs", created by film stretching techniques. Reproduced with permission from *Proc. Nat. Acad. Sci.* **2006**, *103*, 4930. **g** Optical micrographs of microcapsules in water, reproduced with permission from *Small* **2007**, *3*, 412. **h** SEM image of 300 nm conical-shaped poly(lactic acid) particles generated by using the Particle Replication In Nonwetting Templates (PRINT) method. **i** Manipulation of shape using PRINT: 3 μm arrow poly(ethylene glycol) particles. Images **a–c** reproduced with permission from *Langmuir* **2007**, *23*, 8747, **d–e** reproduced with permission from *J. Am. Chem. Soc.* **2006**, *128*, 11336, and images **h–i** reproduced with permission from *J. Am. Chem. Soc.* **2005**, *127*, 10096. © American Chemical Society

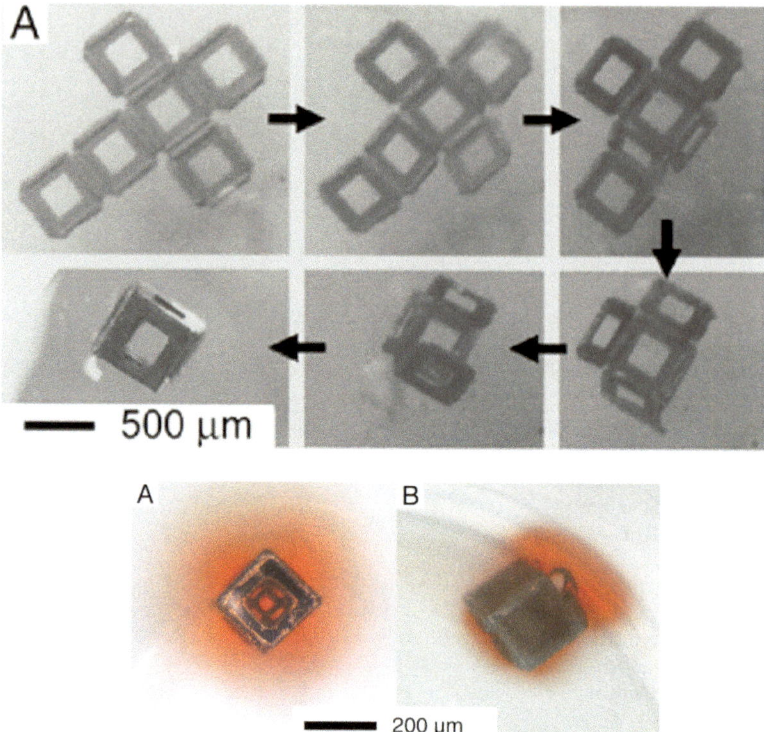

Fig. 5.90 Top: video snapshots showing the self-assembly of a lithographically fabricated template into a 3-D hollow container. Reproduced with permission from *Adv. Drug Deliv. Rev.* **2007**, *59*, 1547. Copyright 2007 Elsevier Science. Bottom: optical images of chemical release from containers: **a** isotropic release of a dye from a container with identical porosity on all faces; **b** anisotropic release of a dye from a container with varied porosity (five faces with an array of 5 μm pores; the sixth face has a 160 μm window). Reproduced with permission from *J. Am. Chem. Soc.* **2006**, *128*, 11336. © 2006 American Chemical Society

As their name implies, *plasticizers* are additives that soften a material, enhancing its flexibility. The worldwide market for plasticizers is currently over 5 million metric tons, with over 90% used to soften PVC. The most common plasticizers are phthalates; however, due to the relatively high vapor pressure of these compounds, plasticizers will evaporate from the polymer structure as evidenced by the "new car smell" of new cars, as well as the organic film that becomes deposited on the interior windshield surface. For these applications, it is best to use a plasticizer with a lower volatility such as trimellitates.[153]

There has been increased awareness regarding the health impacts of phthalate-based plasticizers, which have been identified as reproductive and developmental toxins. Since phthalates are not chemically bound to their PVC polymer host, these additives will leach out of products over time and diffuse into the environment. In 2009, the U.S. Consumer Product Safety Improvement Act restricted the most toxic di-(2-ethylhexyl) phthalate (DEHP), dibutyl phthalate (DBP), and benzyl butyl phthalate (BBP) additives in children's toys and childcare articles in concentrations that exceed 0.1%. However, though many phthalates are restricted or prohibited for used in some products in the U.S. and the E.U., these additives remain unregulated in many parts of the world such as China and India. In order to circumvent the harmful effects of phthalates, a number of alternatives have been identified such as citrates, sebacates, adipates, phosphates, and a variety of bio-based plasticizers.[154] However, these alternatives will also leach from plastic products and may present adverse health effects from excessive exposure. Hence, it is more desirable to choose another plastic that does not require a plasticizer additive. Some alternatives that have been used as substitutes for PVC include: polyurethane (PU), ethylene vinyl acetate (EVA), PET, HDPE, PP, PS, ABS, styrene butadiene styrene (SBS), ABS/PU hybrids, and bio-based plastics (*e.g.*, PLA, starch-derived plastics, polyhydroxyalkanoate (PHA), cellulose acetate).[155]

There are two leading theories concerning the mechanism of activity for plasticizer molecules. The 'lubricating theory' suggests that as the polymer is heated, the plasticizer diffuses into the polymer and disrupts the van der Waal interactions among polymer chains. Since network formation is reduced, the T_g is lowered resulting in more flexibility/softness of the bulk polymer. By contrast, the 'free volume theory' suggests that the lowering of T_g is due to the polymer chains being pushed further apart by the interdiffusion of the plasticizer molecules. Since the free volume of the polymer has been increased, the chains are free to move past one another more easily resulting in greater flexibility.

The above mechanistic explanations assume that the plasticizer molecules are not permanently bound to the polymer chains. Since these interactions are relatively weak, there is likely a dynamic adsorption/desorption at various locations among neighboring polymer chains. Accordingly, the plasticizer structure may be fine-tuned to affect its solubility/miscibility with the polymer, as well as its interactions with polymer chains and with other plasticizer molecules. As the polymer–plasticizer interactions are strengthened, the T_g will increase; at low concentrations, the rigidity of the polymer is increased due to effective rigid-network formation between the plasticizer and polymer. However, as the plasticizer concentration is increased, the additive molecules themselves interact yielding the desired softening characteristics. As illustrated by **XXVII** above, the molecular structure of a plasticizer contains both polar and nonpolar (hydrocarbon chain) units. It is usually the polar endgroups that bind reversibly with the polymer chains; the length-tunable nonpolar component affords controlled separation of neighboring polymer chains.

Flame Retardants

Since polymers exhibit a hydrocarbon-based structure, these materials pose a significant flammability threat. However, if one examines a room following a fire, it is obvious that some polymers withstand ignition much greater than others (Fig. 5.91). In fact, these polymers are not naturally fire resistant, but rather contain additives that afford this desirable property. The largest class of flame inhibiting additives is *brominated flame-retardants* (BFRs). It is estimated that bromine-containing molecules are

Fig. 5.91 Photograph of damage from a candle-ignited fire that destroyed a campus fraternity house in Amherst, MA. Flame retardants likely prevented the television and surface-treated wood cabinet from igniting. Reproduced from http://www.buildinggreen.com

added to over 2.5 million tons of polymers each year, with the electronics industry accounting for the greatest consumer market. BFRs are also used in a number of other products such as electronic equipment housing, carpets, paints/stains, fabrics, and kitchen countertops/appliances. Even the "silly string" that our children plays with contains a brominated flame retardant (hexabromobenzene) to prevent the dried-up string of poly(isobutylmethacrylate) from catching fire.

Due to the bioaccumulation of BFRs in the environment[156] and increasingly stringent environmental regulations, the use of BFRs is being dramatically reduced—especially in Europe. The most widely used alternative is organophosphorus-based (OP) flame-retardants (*e.g.*, **XXX**), which are much more expensive than organohalogen additives. However, these molecules also contribute to health and environmental hazards, being found in air samples as far away as Antarctica and in rainwater collected across European countries. Part of the problem stems from the fact that OP flame retardants are not as active as BFRs and must be present in much higher concentrations to be effective—often at the expense of altering the physical properties of the polymer.

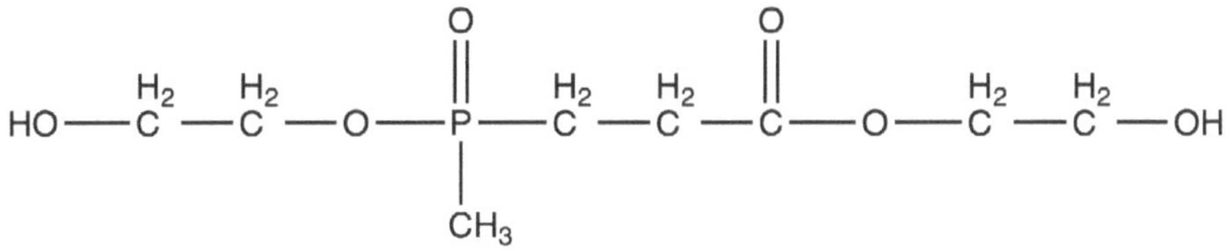

$$HO-CH_2-CH_2-O-P(=O)(CH_3)-CH_2-CH_2-C(=O)-O-CH_2-CH_2-OH$$

XXX

Not unlike plasticizers, the most common method used to impart flame retardancy is by simple mixing of the additives with the polymer during final processing. However, a growing area of development is the design of functionalized monomers that contain flame retardant groups (*e.g.*, halogens, organophosphorus, Fig. 5.92). These monomers are known as *reactive flame retardants*, which yield a polymer that has an inherent flame retardant characteristic—much more controllable than additives that are placed randomly within the polymer. However, this approach is not widely used due to its relatively high production cost. Further, the inclusion of functionalized monomeric units within the polymer chains may alter the physical properties of the bulk polymer.

There are two primary reactive modes for flame-retardants: either gas-phase or solid-state. Since a flame consists of a variety of gas-phase radicals and atoms (5.11 and 5.12), a gas-phase flame retardant is one that will scavenge the flame-propagating radicals such as $^{\bullet}OH$ and $O^{\bullet}$. In particular, if the hydroxyl radical is suppressed, the exothermic formation of CO_2 (5.13) will be prevented thereby reducing the flame temperature.

$$H\cdot + O_2 \rightleftarrows \cdot OH + O\cdot \tag{5.11}$$

$$O\cdot + H_2 \rightleftarrows \cdot OH + H\cdot \tag{5.12}$$

$$\cdot OH + CO \rightleftarrows CO_2 + H \tag{5.13}$$

Halogenated flame-retardants are active in the gas-phase by combining with radicals. The organohalogen compound (MX) first thermally decomposes to form halogen radicals that combine with hydrogen radicals or fuel to form hydrogen halide (HX) (5.14 and 5.15). It should be noted that BFRs all possess aromatic structures since the flame temperature is sufficient to break aromatic C–Br bonds (C–Cl require higher temperatures). By comparison, aliphatic C–Br bonds are broken at temperatures lower than the flame, resulting in pre-decomposition of the organobromine compound and negligent flame retardancy. Often, antimony oxide (Sb_2O_3) is used in combination with BFRs since the side-product of $SbBr_3$ is sufficiently volatile, and carries a high concentration of bromine into the gas-phase.

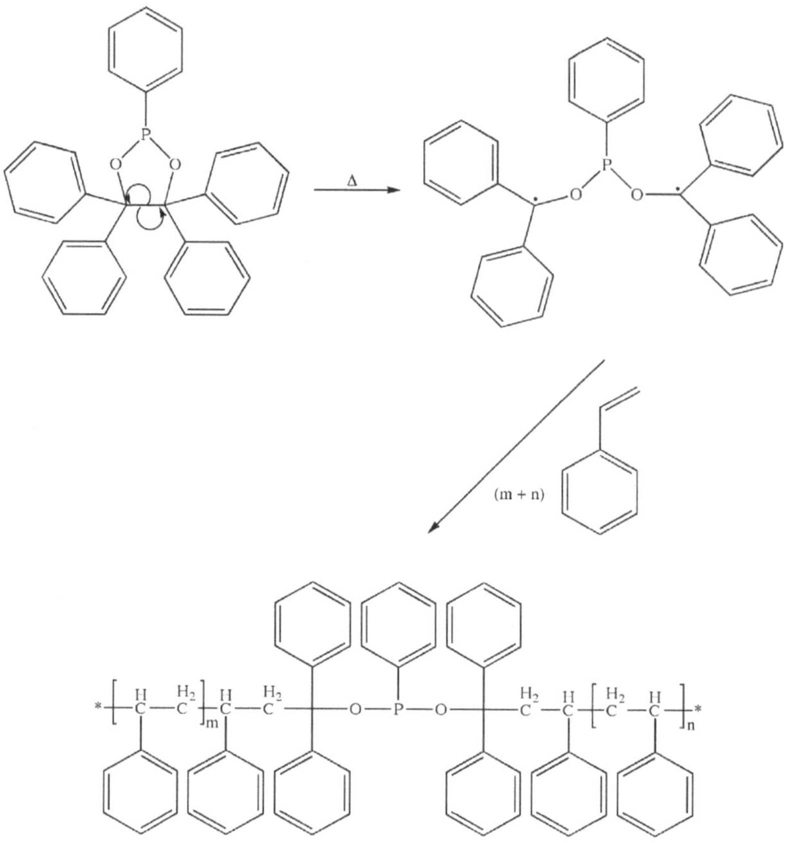

Fig. 5.92 Example of the reactive flame retardant approach through incorporation of an organophosphorus unit in the polymer main chain[157]

$$MX \rightleftarrows M\cdot + X\cdot \tag{5.14}$$

$$X\cdot + H\cdot \rightleftarrows HX \tag{5.15}$$

The thermally-generated HX species may also behave as flame inhibitors by scavenging hydrogen, hydroxyl and oxygen radicals (5.16 and 5.17):

$$H\cdot + HX \rightleftarrows H_2 + X\cdot \tag{5.16}$$

$$\cdot OH + HX \rightleftarrows H_2O + X\cdot \tag{5.17}$$

By contrast, solid-state flame-retardants such as organophosphorus agents act by passivating the surface of the polymer toward heat, flame, and oxygen through formation of a carbonaceous coating known as a *char*. When thermally decomposed, some phosphorus compounds generate acids that promote surface crosslinking reactions and resultant char formation. Quite often, phosphorus flame-retardants also have gas-phase reactivity, whereby $PO^{\bullet}$ radicals help to scavenge flame-propagating species in the flame. Though the mechanism is not currently known, the use of nitrogen compounds (or N-containing polymers such as poly(amide)s) enhances char formation. This is most likely due to a lower number of reactive C/H/O-containing units that volatilize to fuel the flame.

Another interesting area of recent development is the synthesis of dual-action flame-retardants that contain both organophosphorus and organobromine units (Fig. 5.93).[158] Since these agents are active in both gas- and condensed phases, the activity is proposed to be much higher than current additives at much lower concentrations—in accord with environmental regulations. In addition, the synthesis of these agents is relatively inexpensive, which is also a stringent guideline in the development of alternative additives for such a large market sector.

Important Polymeric Materials Applications: Self-healing Polymers

Imagine getting into an accident only to have your vehicle revert back to its original shape before your eyes! Though this certainly sounds like something out of a science fiction novel, these materials are fast becoming a reality. The general strategy for this activity is to have the healing agents as a part of the polymer, which become activated upon crack formation.[159] This was first demonstrated with microcapsules of a urea–formaldehyde shell that contained a dicyclopentadiene monomer, suspended within an epoxy polymer matrix. Of course, without a polymerization catalyst, nothing would happen; hence, crystals of Grubbs' catalyst were also dispersed in the polymer (Fig. 5.94). The encapsulation of the Grubbs' catalyst within a wax microsphere has been reported to improve the dispersity in the epoxy matrix and prevent side reactions between the catalyst and the amine-based epoxy curing agents.[160]

A technique referred to as *ring-opening metathesis polymerization* (ROMP)[161] using Grubbs' catalyst (Fig. 5.95) was chosen due to rapid polymerization under ambient conditions, in the presence of oxygen and water. Further, these conditions were also suitable for low shrinkage upon polymerization—extremely important for self-healing materials. When a crack ruptured a microcapsule, the monomer became exposed to the catalyst and a strong, highly crosslinked system was formed that bonded together adjacent crack faces within minutes. This same technology is being developed for other materials such as ceramics[162] and composites[163] for an unlimited number of applications, even including fiberglass wind turbine blades![164]

An alternative technique that does not involve microcapsules or transition metal catalyst has recently been developed. This method features a reversible polymerization using Diels–Alder reactions of furan and imide-based monomers (Fig. 5.96). When the monomers are combined, a highly crosslinked polymer is formed with properties similar to epoxy or polyester[165] composites. However, when a crack propagates through the material, the reactions are reversed generating the monomers

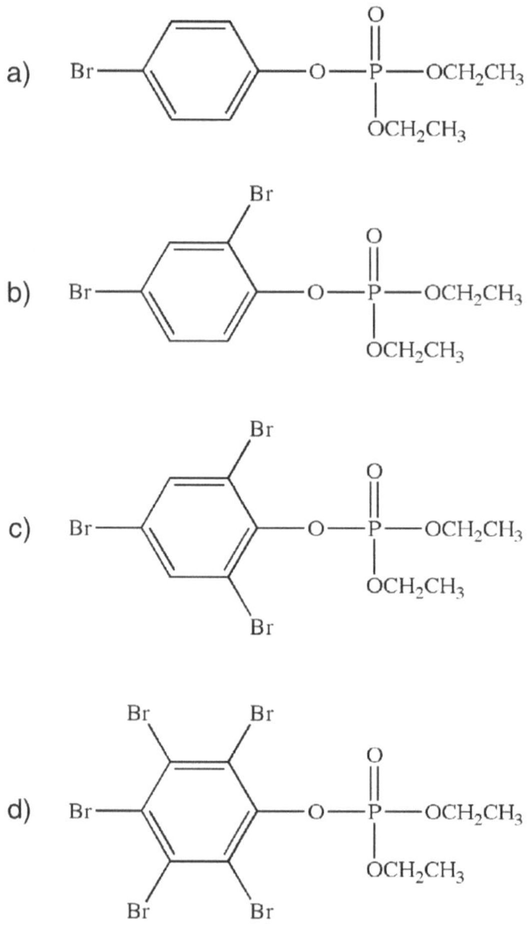
a)
Br
O
P
O
OCH2CH3
OCH2CH3
b)
Br
Br
O
P
O
OCH2CH3
OCH2CH3
c)
Br
Br
Br
O
P
O
OCH2CH3
OCH2CH3
d)
Br
Br
Br
Br
Br
O
P
O
OCH2CH3
OCH2CH3

Fig. 5.93 Molecular structures of dual-action brominated organophosphorus flame retardants. Shown are **a** (4-bromophenyl)diethylphosphate, **b** (2,4-dibromophenyl)diethylphosphate, **c** (2,4,6-tribromo phenyl)diethylphosphate, and **d** (2,3,4,5,6-pentabromophenyl)diethylphosphate

in the vicinity of the fracture. If this region is heated and clamped together, the monomers will polymerize to generate a self-healed material. The benefit of this approach is the ability of this material to undergo multiple fracture-healing cycles without a loss of activity. However, the area must be heated in order to under regeneration. Further developments in this exciting area are surfacing all the time; for instance: self-healing knee implants using encapsulated poly(methylmethacrylate) and peroxide initiators,[166] self-healing polyurethanes upon exposure to UV light (Fig. 5.97),[167] and self-healing epoxy thermosets that have repaired holes larger than 3 cm in diameter.[168]

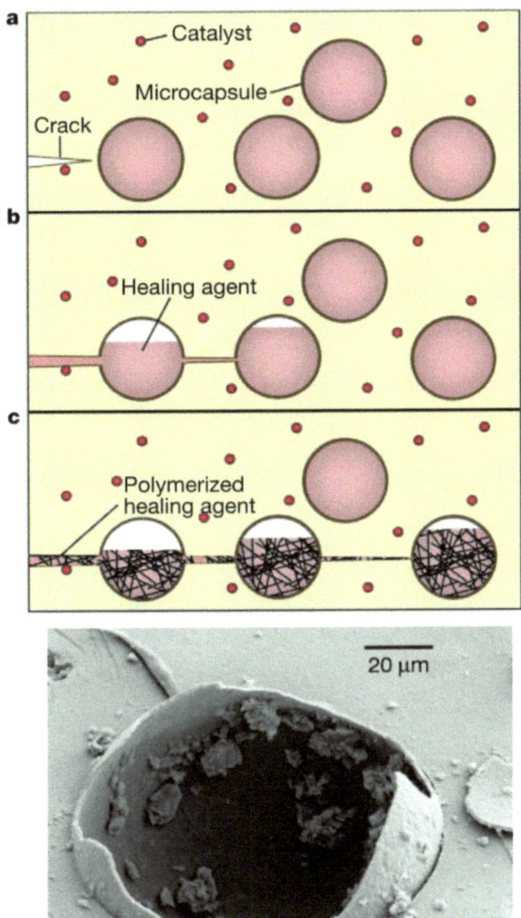

Fig. 5.94 Schematic illustrating the mode of action of self-healing polymers through embedded healing-agent microcapsules that are activated by a propagating crack. Also shown is a polymeric microcapsule following rupture. Reproduced with permission from White, S. R.; Sottos, N. R.; Geubelle, P. H.; Moore, J. S.; Kessler, M. R.; Sriram, S. R.; Brown, E. N.; Viswanathan, S. *Nature*, **2001**, *409*, 794. © 2001 Macmillan Magazines

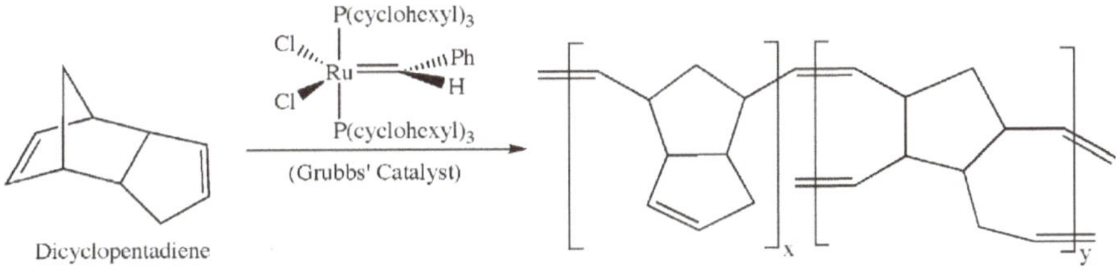

Dicyclopentadiene

$P(cyclohexyl)_3$, Cl, Cl, Ru, Ph, H, $P(cyclohexyl)_3$

(Grubbs' Catalyst)

x y

Fig. 5.95 Ring-opening metathesis polymerization (ROMP) of dicyclopentadiene catalyzed by Grubbs' catalyst

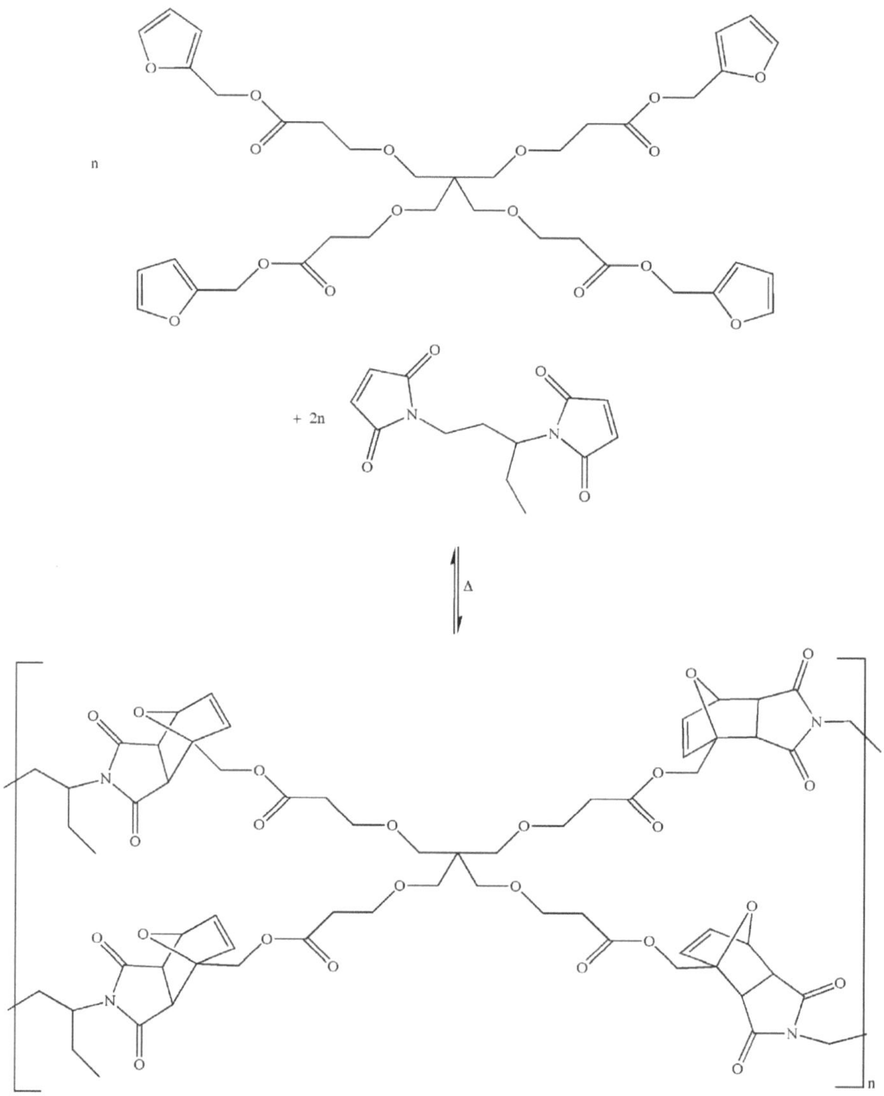

Fig. 5.96 Thermally induced self-healing through a Diels–Alder reaction[169]

Hou and coworkers have reported a terpolymerization route incorporating ethylene and two methoxyaryl-substituted propylenes, catalyzed by a half-sandwich scandium complex.[170] The glass transition temperature and mechanical properties of the polymers was altered based on the ratio of monomers, which bodes well for a wide range of possible applications. Within 1 h of rupture, polymer films exhibited 90% recovery of the original tensile strength, with >1000% elongation at the breakage point. The tensile strength of self-repaired polymers were shown to be comparable or

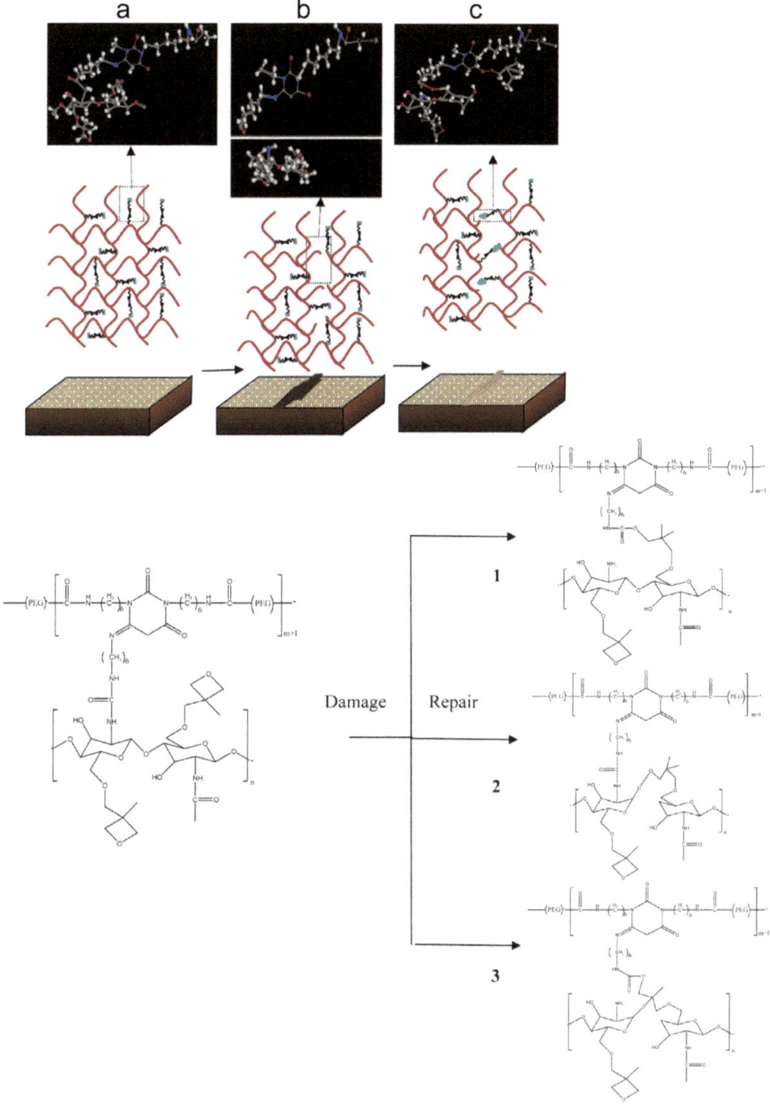

Fig. 5.97 Top: proposed mechanism for the repair of oxetane-substitute chitosan polyurethane (OXE-CHI-PUR) networks. The crosslinked network is represented by red thick lines, whereas dangling OXE entities are black thinner lines. As mechanical damage is created **(b)**, OXE rings open up. Upon exposure to UV light, crosslinking reactions of the OXE-CHI entities result in self-healing of the damaged area, **c** Bottom: illustration of chemical reactions leading to UV-induced polymer repair. Shown are (1) breaking of urea linkages as well as the ring opening of OXE and formation of urethane linkages, (2) OXE ring opening and scission of the CHI linkages and the formation of alkyl peroxide linkages, and (3) urea breakage, CHI and OXE ring opening and formation of urethane and linear –C–O–C– crosslinks. Reproduced with permission from *Science* **2009**, *323*, 1458. © 2009 American Association for the Advancement of Science

higher than traditional commercial elastomers! Such an extreme level of self-healing is thought to arise from microphase separation among the nanoscale domains of crystalline/amorphous segments of the polymer.

In a related topic, shape-morphing and shape-memory polymers are also of interest for self-repairing applications, as well as soft robotics and various types actuators for energy or sensor applications.[171] For instance, a liquid–crystal elastomer film featuring light-sensitive azobenzene functional groups (**XXXI**) exhibits photoisomerization and reversible anisotropic shape transitions upon illumination with UV light (365 nm).[172] A plethora of other polymers have been shown to exhibit shape changes in response to exposure to electric[173] or magnetic fields,[174] temperature,[175] or solvent.[176]

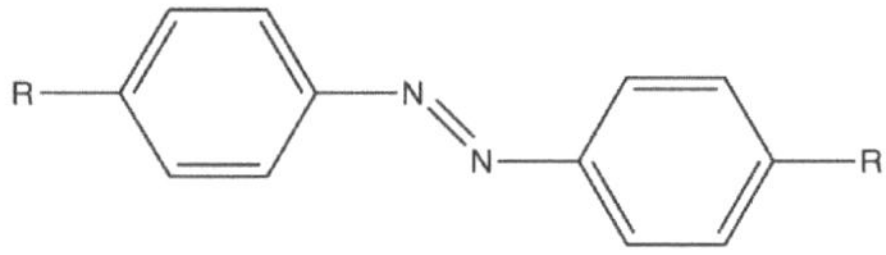

R—C6H4—N=N—C6H4—R

XXXI

Topics for Further Discussion

1. Explain the differences between a thermoplastic and thermoset.
2. Distinguish "plastics" from "polymers".
3. What is meant by "tacticity," and how does this influence overall polymer properties?
4. Draw the isotactic, syndiotactic, and atactic structures of poly(vinyl chloride).
5. Summarize the causes and ramifications of the 'gel effect' in free radical polymerization.
6. Why must water and carbon dioxide be kept out of the reaction mixture in anionic polymerization?
7. Explain the effects of the initiator concentration on the degree of polymerization (free-radical route) of styrene.
8. Why was the reported discovery of dendrimers fought with so much controversy (*i.e.*, what about the structure/route did the leading scientists not believe)?
9. What is the chemical origin of the organic film that forms on the windshield of a new car? What is responsible for the "new car smell"?
10. What is meant by the terms 'inifer' and 'iniferter' and what are some recent candidates that have been synthesized in the laboratory?
11. What is the purpose of MAO in olefin polymerizations?
12. Explain why polymers crystallize in a chain array rather than the thermodynamically favored extended chain form.
13. Earlier in this chapter, we discussed the influence of polymer structure on the efficacy for use as contact lens applications, at the molecular level. At a more bulk level, describe the industrial techniques used to inexpensively and repeatably fabricate 'hard' and 'soft' contact lenses.

14. Examine the polymeric materials in your immediate vicinity and determine whether they were formed via condensation or addition polymerization routes.
15. What are the structures of conductive polymers, and what is the mechanism for electronic transport in the solid-state?
16. Illustrate the mechanisms for polaron, bipolaron, and soliton formation for PANI, showing the mechanism for intersoliton hopping. Why is the conductivity lower for doped PANI relative to PA?
17. Which polymers may be recycled, and which ones are not? What factors govern the relative "recyclability" of a polymer?
18. What is the effect of having non-stoichiometric amounts of the two monomers in step-growth polymerization?
19. Perform a life-cycle assessment for polyethylene and PET packaging materials.
20. What are the initial components you would use to make a cross-linked polyester?
21. Give one possible monomer and initiator for free radical, anionic, and cationic polymerizations.
22. What is the Mark-Houwink equation and how is it used to determine the molecular weight of a polymer?
23. Draw a schematic that shows how a ring-opening metathesis polymerization route forms a linear polymer with a carbon–carbon double bond in the backbone.
24. What type of monomer can be polymerized using 'group transfer polymerization'?
25. What polymerization method would you use to most easily form polymers from the following monomers?

    (a) $CH_2C(CN)_2$
    (b) $CH_2CH(CH_3)$
    (c) $CH_2CHOCOCH_3$ [MMA].

26. Your supervisor has asked you to make a polymeric material with the properties listed below. What are some possible backbone and side-group structures that you would try to incorporate in the polymer? Explain your rationale.

    (a) High-melting point, flexible
    (b) Low-melting point, conductive
    (c) Able to withstand strong acids or bases
    (d) Flexible at low temperatures and becoming more rigid at elevated temperatures
    (e) Properties of (b) and biodegradable under ambient conditions.

References and Notes

1. http://advances.sciencemag.org/content/3/7/e1700782/tab-pdf.
2. Mulhaupt, R. *Angew. Chem. Int. Ed.* **2004**, *43*, 1054.
3. In addition to cost savings during automotive fabrication and enhanced design flexibility relative to steel, it is estimated that every 10% of weight reduction will yield a 5% increase in fuel economy.

4. Melting points refer to the temperature required to separate molecules from one another. By contrast, the glass-transition temperature refers to the temperature required to perturb the bonds of the polymer backbone.
5. For a very comprehensive treatise regarding epoxy-based adhesives (structure vs. properties, applications, etc.), see: Petrie, E. M. *Epoxy Adhesive Formulations*, McGraw-Hill: New York, 2006.
6. Gel-permeation chromatography (GPC) is a subdivision of size-exclusion chromatography (SEC), in which macromolecular species are separated from one another based on their size. As its name implies, GPC employs a gel (usually cross-linked polystyrene) as the stationary phase, with detection through either light-scattering or refractive index.
7. For a review of living radical polymerization (LRP) involving organotellurium, organostibine, and organobismuthine mediated routes (TERP, SBRP, and BIRP, respectively), see: Yamago, S. *Chem. Rev.* **2009**, *109*, 5051.
8. For a comprehensive review of transition-metal catalyzed living radical polymerization, see: Ouchi, M.; Terashima, T.; Sawamoto, M. *Chem. Rev.* **2009**, *109*, 4963.
9. (a) https://doi.org/10.1515/cti-2020-0026. (b) For a review of bioapplications for RAFT polymerization, see: Boyer, C.; Bulmus, V.; Davis, T. P.; Ladmiral, V.; Liu, J.; Perrier, S. *Chem. Rev.* **2009**, *109*, 5402.
10. (a) Hadjikyriacou, S.; Acar, M.; Faust, R. *Macromolecules* **2004**, *37*, 7543. (b) Aoshima, S.; Segawa, Y.; Okada, Y. *J. Polym. Sci., Part A: Polym. Chem.* **2001**, *39*, 751.
11. For a recent review of cationic living polymerization, see: Aoshima, S.; Kanaoka, S. *Chem. Rev.* **2009**, *109*, 5245.
12. For a review of stereospecific living radical polymerization (LRP), see: Satoh, K.; Kamigaito, M. *Chem. Rev.* **2009**, *109*, 5120.
13. For a thorough classic review of homogeneously-catalyzed polymerization, see: Bikales, N. M. *Adv. in Chemistry* **1968**, *70*, 233.
14. *Catalyst Separation, Recovery and Recycling: Chemistry and Process Design*, Cole-Hamilton, D. J.; Tooze, R. P., eds. Springer: New York, 2006.
15. For a review of homogeneous living Ziegler–Natta polymerization, see: http://www.scs.illinois.edu/denmark/wp-content/uploads/gp/2009/gm-2009-1-27.pdf.
16. For example, see: (a) Welborn, H. C. U.S. Patents 5,183,867, 4,808,561. (b) Bergemann, C.; Luft, G. *Chem. Eng. Technol.* **1999**, *21*, 33. (c) Broyer, J. -P.; Malinge, J.; Saudemont, T.; Spitz, R.; Verdel, N. U.S. Patent 6,239,059.
17. (a) Mason, M. R.; Smith, J. M.; Bott, S. G.; Barron, A. R. *J. Am. Chem. Soc.*, 1993, **115**, 4971. (b) Harlan, C. J.; Mason, M. R.; Barron, A. R. *Organometallics*, **1994**, *13*, 2957. (c) Landry, C. C.; Pappè, N.; Mason, M. R.; Apblett, A. W.; Barron, A. R.; *Inorganic and Organometallic Polymers*, ACS Symposium Series, Volume II, **1998**, *572*, 149. (d) Harlan, C. J.; Bott, S. G.; Barron, A. R. *J. Am. Chem. Soc.*, **1995**, *117*, 6465.
18. (a) Watanabe, M.; McMahon, C. N.; Harlan, C. J.; Barron, A. R. *Organometallics*, **2001**, *20*, 460. (b) Koide, Y.; Bott, S. G.; Barron, A. R.

Organometallics, **1996**, *15*, 2213. (c) Barron, A. R. *Macromol. Symp.*, **1995**, *97*, 15.
19. Harlan, C. J.; Bott, S. G.; Barron, A. R. *J. Am. Chem. Soc.*, **1995**, *117*, 6465.
20. Note: in general, high molecular weight polymers are typically formed from catalysts derived from early transition metals (Groups 4–6). For late transition metals, the β-hydride elimination mechanism is more preferred, leading to greater numbers of oligomers and dimers.
21. For a recent review of hyperbranched polymer architectures, see: Voit, B. I.; Lederer, A. *Chem. Rev.* **2009**, *109*, 5924.
22. A comprehensive discussion regarding siloxane polymerization routes are described within the Ph.D. dissertation (J. Daum, Univ. of Akron, Dec. 2005), found online at: http://etd.ohiolink.edu/send-pdf.cgi/Daum%20Jeremy%20L.pdf?akron1134097748.
23. Flory, P. J. *J. Am. Chem. Soc.*, **1952**, *74*, 2718.
24. (a) Buhleier, E.; Wehner, W.; Vögtle, F. *Synthesis*, **1978**, 155. (b) Moors, R.; Vögtle, F. *Chem. Ber.*, **1993**, *126*, 2133. (c) Vögtle, F.; Weber, E. *Angew. Chem. Int. Ed. Engl.*, **1974**, *13*, 814.
25. (a) Tomalia, D. A.; Baker, H.; Dewald, J.; Hall, M.; Kallos, G.; Martin, S.; Roeck, J.; Ryder, J.; Smith, P. *Polym. J.*, **1985**, *17*, 117. (b) Tomalia, D. A.; Baker, H.; Dewald, J.; Hall, M.; Kallos, G.; Martin, S.; Roeck, J.; Ryder, J.; Smith, P. *Macromolecules*, **1986**, *19*, 2466. (c) Tomalia, D. A.; Hall, M.; Hedstrand, D. *J. Am. Chem. Soc.*, **1987**, *109*, 1601. (d) Tomalia, D. A.; Berry, V.; Hall, M.; Hedstrand, D. M. *Macromolecules*, **1987**, *20*, 1164. (e) Tomalia, D. A. et al. US Patents 4,587,392 (1986); 4,558.120 (1985); 4,568,737 (1986); 4,599,400 (1986); 4,631,337 (1986); 4,507,466 (1985). (f) For an extensive review, see: Tomalia, D. A.; Naylor, A. M.; Goddard III, W. A. *Angew. Chem. Int. Ed. Engl.*, **1990**, *29*, 138.
26. Note: Robert Denkwalter and coworkers from Allied Corporation were granted the first patent for dendrimers (US Patent 4,410,688, which represents the first dendrimer-related publication. However, the term "dendrimer" can be traced back to A. J. Vogel. The generally accepted definition of dendrimers is highly branched and monodisperse polymers (particularly for convergent growth), with a degree of branching of 1.0.
27. (a) Newkome, G. R.; Yao, Z.; Baker, Z. R.; Gupta, V. K.; *Org. Chem.*, **1985**, *50*, 2004. (b) Newkome, G. R.; Yao, Z.; Baker, G. R.; Gupta, V. K.; Russo, P. S.; Saunders, M. J. *J. Am. Chem. Soc.*, **1986**, *108*, 849. (c) Newkome, G. R.; Baker, G. R.; Saunders, M. J.; Russo, P. S.; Gupta, V. K.; Yao, Z.; Miller, J. E.; Bouillion, K. *J. Chem. Soc., Chem. Commun.*, **1986**, 752.
28. (a) Hawker, C. J.; Frechet, J. M. J. *J. Am. Chem. Soc.*, **1990**, *112*, 7638. (b) Hawker, C. J.; Frechet, J. M. J. *Macromolecules*, **1990**, *23*, 4726. (c) Wooley, K. L.; Hawker, C. J.; Frechet, J. M. J. *J. Am. Chem. Soc.*, **1991**, *113*, 4252.
29. For a review of dendrimers and other (biohybrid) polymers that exhibit capsule properties, see: van Dongen, S. F. M.; de Hoog, H. -P. M.; Peters, R. J. R. W.; Nallani, M.; Nolte, R. J. M.; van Hest, J. C. M. *Chem. Rev.* **2009**, *109*, 6212.
30. Matthews, B. R.; Holan, G. U.S. Patent 6,190,650.

31. Hawker, C. J.; Hedrick, J. L.; Miller, R. D.; Volksen, W. *MRS Bull.* **2000**, *25*, 54.
32. Bourne, N.; Stanberry, L. R.; Kern, E. R.; Holan, G.; Matthews, B.; Bernstein, D. I. *Antimicrob. Agents Chemother.* **2000**, *44*, 2471.
33. Svenson, S.; Tomalia, D. A. *Adv. Drug Deliv. Reviews* **2005**, *57*, 2106.
34. Li, W. -S.; Aida, T. *Chem. Rev.* **2009**, *109*, 6047.
35. (a) Dvornic, P. R.; de Leuze-Jallouli, A. M.; Owen, M. J.; Perz, S. V. *Macromolecules*, **2000**, *33*, 5366. (b) Dvornic, P. R.; Li, J.; de Leuze-Jallouli, A. M.; Reeves, S. D.; Owen, M. J. *Macromolecules*, **2002**, *35*, 9323.
36. Sarkar, A.; Rousseau, J.; Hartmann-Thompson, C.; Maples, C.; Parker, J.; Joyce, P.; Scheide, J. I.; Dvornic, P. R. "Dendritic Polymer Networks: A New Class of Nano-Domained Environmentally Benign Antifouling Coatings", Chapter X in *"New Membranes and Advanced Materials for Wastewater Treatment"*, Mueller, A. and Sarkar, A., Eds., ACS Symposium Series 000, American Chemical Society, Washington DC, 2009, pp. 000–000.
37. De Gennes, P. G.; Hervet, H. *J. Phys. Lett.*, **1983**, *44*, 351–360.
38. Kolb, H. C.; Finn, M. B.; Sharpless, K. B. *Angew. Chem. Int. Ed.* **2001**, *40*, 2004.
39. Kolb, H. C.; Sharpless, K. B. *Drug Discov. Today* **2003**, *8*, 1128.
40. For a review of click chemistry for drug design and biological applications, see: Thirumurugan, P.; Matosiuk, D.; Jozwiak, K. *Chem. Rev.* **2013**, *113*, 4905.
41. For a review of supramolecular polymerization (non-click routes), see: De Greef, T. F. A.; Smulders, M. M. J.; Wolffs, M.; Schenning, A. P. H. J.; Sijbesma, R. P.; Meijer, E. W. *Chem. Rev.* **2009**, *109*, 5687.
42. (a) Iha, R. K.; Wooley, K. L.; Nystrom, A. M.; Burke, D. J.; Kade, M. J.; Hawker, C. J. *Chem. Rev.* **2009**, *109*, 5620. (b) Hawker, C. J.; Wooley, K. L. *Science* **2005**, *309*, 1200.
43. J. A. Opsteen, J. A.; van Hest, J. C. M. *Chem. Commun.* **2005**, 57.
44. Qin, A.; Lam, J. W. Y.; Tang, B. Z. *Macromolecules* **2010**, *43*, 8693.
45. (a) Swanson, D. R.; Huang, B.; Abdelhady, H. G.; Tomalia, D. A. *New J. Chem.* **2007**, *31*, 1368. (b) Tomalia, D. A. *Dendritic Polymers with Enhanced Amplification and Interior Functionality*, U.S. Patent 2007/0298006 A1, published Dec. 27, 2007. (c) https://news.agropages.com/News/NewsDetail---19235.htm.
46. (a) De Greef, G. F. A.; Smulders, M. M. J.; Wolffs, M.; Schenning, A. P. H. J.; Sijbesma, R. P.; Meijer, E. W. *Chem. Rev.* **2009**, *109*, 5687. (b) Smulders, M. M. J.; Nieuwenhuizen, M. M. L.; De Greef, T. F. A.; Der Schoot, P.; Schenning, A. P. H. J.; Meijer, E. W. *Chem. Eur. J.* **2010**, *16*, 362. (c) Gershberg, J.; Fennel, F.; Rehn, T. H.; Lochbrunner, S.; Wurthner, F. *Chem. Sci.* **2016**, *7*, 1729. (d) Ogi, S.; Sugiyasu, K.; Manna, S.; Samitsu, S.; Takeuchi, M. *Nature Chem.* **2014**, *6*, 188.
47. Eagan, J. M.; Xu, J.; Di Girolamo, R.; Thurber, C. M.; Macosko, C. W.; LaPointe, A. M.; Bates, F. S.; Coates, G. W. *Science* **2017**, *355*, 814.
48. For a history of liquid crystal displays (LCDs), see: Kawamoto, H. *Proc. IEEE* **2002**, *90*, 460.

49. Fully functional organs are likely decades away. For a review of 3-D printed organ and tissue development, see: https://www.frontiersin.org/articles/10.3389/fmech.2020.589171/full.
50. Arefin, A. M. E.; Khatri, N. R.; Kulkarni, N.; Egan, P. F. *Polymers* **2021**, *13*, 1499.
51. (a) Bonn, D.; Denn, M. M.; Berthier, L.; Divoux, T.; Manneville, S. *Rev. Mod. Phys. **2017**, 89*, 035005. (b) Nelson, A. Z.; Ewoldt, R. H. *Soft Matter* **2017**, *13*, 7578. (c) Lee, S.; Koo, Kang, J. S.-K.; Park, G.; Lee, Y. J.; Chen, Y.-Y.; Lim, S. A.; Lee, K.-M.; Rogers, J. A. *Mater. Today* **2018**, *21*, 207. (d) Jin, Y.; Liu, C.; Chai, W.; Compaan, A.; Huang, Y. *ACS Appl. Mater. Interfaces **2017**, 9*, 17456.
52. Dvornic, P.R. in *Silicon-Containing Polymers*, Jones, R.G., Ando, W., Chojnowski, J., (Eds.) Kluwer Academic Publishers: Dordrecht, 2000, 185 – 212.
53. (a) Zlatanic, A.; Radojcic, D.; Wan, X.; Messman, J. M.; Dvornic, P. R. *Macromolecules* **2017**, *50*, 2532. (b) Zlatanic, A.; Radojcic, D.; Wan, X.; Messman, J. M.; Bowen, D. E.; Dvornic, P. R. *J. Polym. Sci. A: Polym. Chem. Ed.* **2019**, *57*, 1122. (c) Alt, C. A.; Dvornic, P. R. *Macromol. Rapid Commun.* **2021** http://doi.org/10.1002/marc.202000692. (d) Beach, J.; Patterson, S.; Messman, J. M.; Dvornic, P. R. *MRS Commun.* **2021**, *11*, 88.
54. Beach, J.; Mann, S.; Ault, C.; Radojcic, D.; Wan, X.; Zlatanic, A.; Patterson, S.; Messman, J. M.; Dvornic, P. R. *Macromolecules* **2021**, *54*, 1715.
55. (a) Reynolds, V. G.; Mukherjee, S.; Xie, R.; Levi, A. E.; Atassi, A.; Uchiyama, T.; Wang, H.; Chabinyc, M. L.; Bates, C. M. Mater. Horizons 2020, 7, 181. (b) Varnoosfaderani, M. V.; Daniel, W. F. M.; Zhushma, A. P.; Li, Q.; Morgan, B. J.; Matyjaszewski, K.; Armstrong, D. P.; Spontak, R. J.; Dobrynin, A. V.; Sheiko, S. S. Adv. Mater. 2017, 29, 1,604,209.
56. Xie; R.; Mukherjee, S.; Levi, A. E.; Reynolds, V. G.; Wang, H.; Chabinyc, M. L.; Bates, C. M. Sci. Adv. 2020, 6, 1. http://doi.org/10.1126/sciadv.abc6900.
57. Bagheri, A.; Jin, J. *ACS Appl. Polym. Mater.* **2019**, *1*, 593.
58. Yagci, Y.; Jockusch, S.; Turro, N. J. *Macromolecules* **2010**, *43*, 6245.
59. Chen, M.; Gu, Y.; Singh, A.; Zhong, M.; Jordan, A. M.; Biswas, S.; Korley, L. T. J.; Balazs, A. C.; Johnson, J. A. *ACS Cent. Sci.* **2017**, *3*, 124.
60. Ligon, S. C.; Liska, R.; Stampfl, J.; Gurr, M.; Mulhaupt, R. *Chem. Rev.* **2017**, *117*, 10,212.
61. (a) Cheng J.; Lao S.; Nguyen K.; Ho W.; Cummings A.; Koo J. *SFF Symp. Proc.* **2005**, 141. (b) Koo J. H.; Lao S.; Ho W.; Nguyen K.; Cheng J.; Pilato L.; Wissler G.; Ervin M. *SFF Symp. Proc.* **2006**, 392.
62. https://doi.org/10.1002/adfm.202201766.
63. Naficy, S.; Gately, R.; Gorkin, R.; Xin, H.; Spinks, G. M. *Macromol. Mater. Eng.*, **2017**, *302,* 1600212.
64. Mulakkal, M. C.; Trask, R. S.; Ting, V. P.; Seddon, A. M. *Mater. Des.* **2018**, *160,* 108.
65. http://doi.org/10.1002/adfm.202201766.

66. For a review of materials used for 4-D printing, see: Ahmed, A.; Arya, S.; Gupta, V.; Furukawa, H.; Khosla, A. *Polymer* **2021**, *228*, 123,926, and references therein.
67. For a recent review of synthetic polymers used in clinical medicine, see: Maitz, M. F. *Biosurf. Biotribol.* **2015**, *1*, 161 (Open access: https://doi.org/10.1016/j.bsbt.2015.08.002).
68. Some useful references related to biodegradable polymers include: (a) Steinbuchel, A. *Biopolymers: General Aspects and Special Applications.* Wiley–VCH: Weinheim (Germany), 2003. (b) Fritz, J.; Link, U.; Braun, R. *Starch* **2001**, *53*, 105. (c) Karlsson, R. R.; Albertsson, A. -C. *Polymer Eng & Sci.* **1998**, *38*, 1251. (d) Kaplan, D. J.; Mayer, J. M..; Ball, D.; McMassie, J.; Allen, A. L.; Stenhouse, P. "Fundamentals of biodegradable polymers" in *Biodegradable polymers and packaging* Ching, C., Kaplan, D. L., Thomas, E. L. eds.; Technomic publication: Basel, 1993. (e) Van de Velde, K.; Kiekens, P. *Polym. Test.* **2002**, *21*, 433. (f) Rouilly, A.; Rigal, L. J. *Macomol. Sci.-Part C. Polymer Reviews* **2002**, *C42*, 441. (g) Chandra, R.; Rustgi, R. *Prog Polym Sci* **1998**, *23*, 1273. (h) Kaplan, D.L. *Biopolymers from renewable resources,* Springer Verlag: Berlin, 1998.
69. For more information regarding bioabsorption/bioresorption, see: Ratner, B. D.; Hoffman, A. S.; Schoen, F. J.; Lemons, J. E. *Biomaterials Science: An Introduction to Materials in Medicine*, 2nd ed., Academic Press: New York, 2004.
70. Another common classification rationale for these polymers is by application—for either biomedical or ecological use. For instance, see: Ikada, Y.; Tsuji, H. *Macromol. Rapid. Commun.* **2000**, *21*, 117.
71. For a review of biodegradable polymer syntheses using a bismuth catalyst, see: Kricheldorf, H. R. *Chem. Rev.* **2009**, *109*, 5579.
72. The most common biodegradable polymers used for medical applications (sutures, screws, pins/rods, tacks, plates, mesh, guided tissue, etc.) are poly(*d,l*-lactide) and poly(*l*-lactide), co-polymerized with polyglycolide or poly(*l*-/*d,l*-lactide). Other important varieties are poly(dioxanone)-based.
73. It should be noted that tin halides may be used to catalyze the ring-opening polymerization of lactide; however, the halide is converted to an alkoxide, which is the active catalytic species. For instance, see: Kricheldorf, H. R.; Sumbel, M. *Eur. Polym. J.* **1989**, *25*, 585.
74. Mehta, R.; Kumar, V.; Bhunia, H.; Upadhyay, S. N. *Polym. Rev.* **2005**, *45*, 325, and references therein.
75. Stolt, M.; Sodergard, A. *Macromolecules* **1999**, *32*, 6412.
76. Kamber, N. E.; Jeong, W.; Hedrick, J. L. *Chem. Rev.* **2007**, *107*, 5813.
77. Zhang, X.; Jones, G. O.; Hedrick, J. L.; Waymouth, R. M. *Nature Chem.* **2016**, *8*, 1047.
78. Kobayashi, S.; Makino, A. *Chem. Rev.* **2009**, *109*, 5288.

79. For instance, see: (a) https://3dprint.com/133438/3d-bioprinting-where-are-we/. (b) http://www.nature.com/news/the-printed-organs-coming-to-a-body-near-you-1.17320. (c) Melchels, F.; Malda, J.; Fedorovich, N. *Comprehen. Biomater.* **2011**, *5*, 587.
80. Note: the mechanical properties, crystallinity, molecular weight, and T_g/m.p. of the biodegradable polymer depends on various factors such as the monomer/initiator structure, synthetic and post-processing conditions, and the presence of additives. In particular, excessively high processing temperatures will shift the equilibrium toward monomer formation, which will affect the mechanical and degradation properties of the polymer.
81. (a) Frelberg, S.; Zhu, X. X. *Int. J. Pharmaceutics* **2004**, *282*, 1. (b) Miyajima, M.; Koshika, A.; Okada, J.; Ikeda, M.; Nishimura, K. *J. Controll. Rel.* **1997**, *49*, 207. (c) http://chemistry.creighton.edu/Opportunities/Baumann/MuszynskiBeth.pdf.
82. A nice chronology related to breast implants may be found online at: http://www.pbs.org/wgbh/pages/frontline/implants/cron.html.
83. Fortunately, Dow Corning was able to successfully emerge from the breast-implant controversy and now remains the global leader in silicon/silicone-based commercial products, as well as one of the world's leaders in the production of ultra-high purity Si, fabricated by Hemlock Semiconductor—a subsidiary of Dow Corning (http://www.hscpoly.com).
84. For a review of implant materials and their carcinogenicity, see: http://monographs.iarc.fr/ENG/Monographs/vol74/mono74-10.pdf.
85. Arteriosclerosis, Thrombosis, and Vascular Biology, Vol. 22, No. 6, 2002, p. 884.
86. Arterial stents may be self-expanding (spring-type), balloon-type, or thermal-expanding. Thermal-expanding varieties feature shape-memory alloys that expand in response to the application of heat.
87. Lowe, H. C.; Oesterle, S. N.; Khachigian, L. M. *J. Am. Coll. Cardiol.* **2002** *39*, 183.
88. (a) Williams, M. S.; DeSimone, J. M. U.S. Patent application 2006/0121087 A1. (b) Williams, M. S.; Glenn, R. A.; Smith, J. A.; Holbrook, K. D.; DeSimone, J. M. U.S. Patent 6,887,266. (c) DeSimone, J. M.; Williams, M. S. U.S. Patent 6,932,930.
89. (a) Serruys, P. W. et al. *Lancet*, **2009**, *373*, 897. (b) Ormiston, J. A. et al. *Lancet*, **2008**, *371*, 873.
90. It is preferable to use a biodegradable material as a scaffold for implanted cells during tissue formation, so the implanted material will disappear leaving behind only the generated tissue. For instance, see: Asti, A.; Gioglio, L. *Int. J. Artif. Organs* **2014**, *37*, 187.
91. In 1827, astronomer Sir John Herschel first reported the concept of making a mold of the wearer's eyes so lenses could be fabricated to perfectly conform to the eyes' surfaces. This idea was realized in 1887 by German glassblower F. A. Muller, who fabricated the first set of glass contact lenses, which were fit to adjust for nearsightedness/farsightedness by Fick and Kalt shortly

thereafter. A nice summary of the history of contact lenses may be found online at: (a) http://edwardhandmedicalheritage.org/history_of_contact_lenses.html. (b) https://www.1800contacts.com/connect/articles/da-vinci-disposable-history-contact-lenses.

92. Salvatori, P. L. *The story of contact lenses*, Obrig Laboratories: New York, 1960.
93. *Clinical Anesthesia.* Barash, P. G.; Cullen, B. F.; Stoelting, R. K.; Cahalan, M.; Stock, M. C. eds, 6th ed., Lippincott Williams and Wilkins: Philadelphia, PA, 2009.
94. Note: if the polymer contains charged monomeric units, as is common for hydrogel-based contact lenses, proteins and other charged biomolecules will be attracted resulting in biofilm formation. This will result in loss of occular properties, as well as influence the diffusivity of oxygen through the lens, requiring immediate replacement to prevent severe irritation and infection.
95. Note: PMMA is sold under a variety of trade names such as Plexiglass, Lucite, Polycast, Oroglass, Acrylite, R-Cast, Vitroflex, and many others. PMMA may be used as an alternative to glass and polycarbonate; however, it is quite brittle and has a melting point of *ca.* 130–140 °C.
96. It should be noted that the T_g may be varied by co-polymerization with other monomers or through simple intermixing of polymers; for example, the T_g of semi-interpenetrating networks of HEMA and polyurethane may be varied from −140 to 180 °C.
97. A nice article that describes the importance of oxygen diffusion for silicone-based soft contact lenses may be found online: http://www.clspectrum.com/article.aspx?article=12953.
98. Note: if the lens contains a hydrophobic surface, it will disrupt the tear flow that results in the deposition of an albumin film on the lens. Not only will this reduce the effectiveness of the lens to correct optical aberrations, but will also cause infection/irritation. For more details regarding the lens surface and eye complications, see: Rao, J. B., Saini, J. S. "Complications of Contant Lenses" in *Contact Lenses.* Aquavella, J. V., Rao, G. N., eds. Lippincott Williams and Wilkins: Philadelphia, PA, 1987.
99. Kunzler, J.; Ozark, R. *J. Appl. Polym. Sci.* **1997**, *65,* 1081.
100. For a recent summary of contact lens innovations, see: Bomgardner, M. M. *Chem. Eng. News* **2017**, *3*, 31.
101. There is an interesting recent report that delineats the 3-D structure and interactions of calcite crystallized within an agarose hydrogel, using high-resolution electron microscopy/tomography: Li, H.; Xin, H. L.; Muller, D. A.; Estroff, L. A. *Science* **2009**, *326*, 1244.
102. PureVision™ is a copolymer of tris-(trimethylsiloxy)-silyl-propylvinyl carbamate (TRIS-VC), N-vinylpyrrolidone, a vinyl carbonate functional polydimethylsiloxane (PDMS) macromer, and a vinyl carbamate derivative of alanine. For more information, see: Nicolson, P. C.; Vogt, J. *Biomaterials* **2001**, *22*, 3273.

103. (a) Toit, R.; Stern, B.; Sweeney, D. *Int. Cont. Lens Clinic* **2000**, *27*, 191. (b) http://www.siliconehydrogels.org/editorials/08_may.asp. (c) Chen, C.; Ye, H.; Manesis, N. U.S. patent 7,572,841.
104. Benjamin, W. J.; Karkkainen, T. R. "Hydrogel Hypoxia: Where We've Been, Where We're Going" *Contact Lens Spectrum*, **1996**, September issue.
105. http://www.clspectrum.com/article.aspx?article=13020.
106. For a summary of recent strategies for the design and development of polymeric materials for drug- and gene-delivery applications, see: *Adv. Drug Deliv. Rev.* **2008**, *60(9)*, 955–1094—thematic issue devoted to this topic. The thematic issue *Adv. Drug Deliv. Rev.* **2005**, *57(15)*, 2101–2286 is devoted to dendrimers as drug-delivery agents. Lastly, the thematic issue *Adv. Drug Deliv. Rev.* **2001**, *53(1)*, 1–131 also deals with polymeric materials being used for drug-delivery applications.
107. The therapeutic window is defined as the drug concentration lying between minimum-effective and toxic levels, and is different for each person based on their metabolic and circulatory systems.
108. Saltzman, W. M.; Olbricht, W. L. *Nat. Rev. Drug Discov.* **2002**, *1*, 177.
109. Randall, C. L.; Leong, T. G.; Bassik, N.; Gracias, D. H. *Adv. Drug Delivery Rev.* **2007**, *59*, 1547.
110. (a) Lehr, H. A.; Brunner, J.; Rangoonwala, R.; Kirkpatrick, C. J. *Am. J. Respir. Crit. Care Med.* **2002**, *165*, 514. (b) Martin, F. J.; Melnik, K.; West, T.; Shapiro, J.; Cohen, M.; Boiarski, A. A.; Ferrari, M. *Drugs R&D* **2005**, *6*, 71.
111. Gobin, A. M.; Lee, M. H.; Halas, N. J.; James, W. D.; Drezek, R. A.; West, J. L. *Nano Lett*. 2007, *7*, 1929.
112. Sung, Y. K.; Kim, S. W. *Biomater. Res.* **2020**, *24*, 12.
113. http://www.starpharma.com/vivagel/vivagel_clinical_trials.
114. Padilla De Jesus, O. L.; Ihre, H. R.; Gagne, L.; Frechet, J. M. J.; Szoka, F. C. *Bioconj. Chem.* **2002**, *13*, 453.
115. Goodwin, A. P.; Lam, S. S.; Frechet, J. M. J. *J. Am. Chem. Soc.* **2007**, *129*, 6994.
116. For a recent review of conjugated polymers for organic solar cell applications, see: Cheng, Y. -J.; Yang, S. -H.; Hsu, C. -S. *Chem. Rev.* **2009**, *109*, 5868.
117. For reviews of synthetic routes for conjugated polymers, see: (a) Yokozawa, T.; Yokoyama, A. *Chem. Rev.* **2009**, *109*, 5595 (chain-growth condensation polymerization).
    (b) Liu, J.; Lam, J. W. Y.; Tang, B. Z. *Chem. Rev.* **2009**, *109*, 5799 (synthesis, structure, and applications of acetylenic polymers).
118. Interesting trivia: Rudolph Peierls was one of Heisenberg's doctoral students while he was at the Universitat Leipzig; other notable students mentored by Heisenberg included Bloch, Mulliken, Slater, Teller, Wentzel and Zener!
119. http://www.qcrsolutions.com/Site/OLED_and_PLED___QCR_Solutions_Corp.html.
120. Chen, Y. -C. *Polym. Bull.* **1990**, *23*, 411.
121. Park, S.; Jayaraman, S. *MRS Bull.* **2003**, *28*, 585.

122. Patel, S. N.; Chabinyc, M. L. *J. Appl. Polym. Sci.* **2017**, *134*, 44,403. http://doi.org/10.1002/app.44403.
123. Noriega, R.; Rivnay, J.; Vandewal, K.; Koch, F. P. V.; Stingelin, N.; Smith, P.; Toney, M. F.; Salleo, A. *Nat. Mater.* **2013**, *12*, 1038.
124. (a) Lemaur, V. et al. *Chem. Mater.* **2019**, *31*, 6889. (b) Zhang, X. et al. *Nat. Commun.* **2013**, *4*, 2238. (c) Zhang, W. et al. *J. Am. Chem. Soc.* **2010**, *132*, 11437.
125. Dobryden, I.; Korolkov, V. V.; Lemaur, V.; Waldrip, M.; Un, H. L.; Simatos, D.; Spalek, L. J.; Jurchescu, O. D.; Olivier, Y.; Claesson, P. M.; Venkateshvaran, D. *Nature Communic.* **2022**, *13*, 3076.
126. In order to reduce the adhesion between a polymeric mold and a silicon/quartz master, the master surface is typically modified with a fluorosilane (*e.g.*, $CF_3(CF_2)_6(CH_2)_2SiCl_{3(g)}$). In addition, the final removal of the mold may also be carried out in the presence of a liquid with a low viscosity such as methanol (solvent-assisted micromolding (SAMIM)).
127. For a nice survey of the benefits for (nano)imprint lithography relative to photolithography, see: https://www.azonano.com/article.aspx?ArticleID=4323.
128. A recent thorough review of nanofabrication using both hard and soft molds, as well as other forms of soft lithography, see: Gates, B. D.; Xu, Q.; Stewart, M.; Ryan, D.; Willson, C. G.; Whitesides, G. M. *Chem. Rev.* **2005**, *105*, 1171; for instance, CDs are made by imprinting patterns from Ni masters in polycarbonate (a) J. S. Winslow, IEEE Trans. Consumer Electron. 1976 (Nov.), 318); holograms are made by imprinting patterns from a fused quartz master in SURPHEX photopolymer (F. P. Shvartsman in Diffractive and Miniaturized Optics (Ed.: S.-H. Lee), SPIE Optical Engineering Press, Bellingham, WA, 1993, 165).
129. For example, see: Chou, S. Y.; Krauss, P. R.; Renstrom, P. J. *Science* **1996**, *272*, 85.
130. For example, see: Jackman, R. J.; Wilbur, J. L.; Whitesides, G. M. *Science* **1995**, *269*, 664.
131. (a) Im, J.; Kang, J.; Lee, M.; Kim, B.; Hong, S. *J. Phys. Chem. B* **2006**, *110*, 12,839. (b) Cabrera, E. J.; Amade, R.; Jaller, L.; Pascual, E.; Bertran, E. *J. Nanopart. Res.* **2014**, *16*, 2172.
132. Myung, S.; Lee, M.; Kim, G. T.; Ha, J. S.; Hong, S. *Adv. Mater.* **2005**, *17*, 2361.
133. http://onlinelibrary.wiley.com/doi/10.1002/andp.201700168/full.
134. (a) Odom, T. W.; Thalladi, V. R.; Love, J. C.; Whitesides, G. M. *J. Am. Chem. Soc.* **2002**, *124*, 12,112. (b) Odom, T. W.; Love, J. C.; Wolfe, D. B.; Paul, K. E.; Whitesides, G. M. *Langmuir* **2002**, *18*, 5314.
135. Li, H.-W.; Muir, B. V. O.; Fichet, G.; Huck, W. T. S. *Langmuir* **2003**, *19*, 1963.
136. Steward, A.; Toca-Herrera, J. L.; Clarke, J. *Protein Sci.* **2002**, *11*, 2179.
137. Note: PDMS is known to swell in organic solvents and leaves a silicone residue behind during its release from the substrate; these limitations are overcome for PFPE molds; for example, see: (a) Lee, J. N.; Park, C.; Whitesides, G. M.

Anal. Chem. **2003**, *75*, 6544. (b) Rolland, J. P.; Hagberg, E. C.; Denison, G. M.; Carter, K. R.; DeSimone, J. M. *Angew. Chem., Int. Ed.* **2004**, *43*, 5796. (c) Rolland, J. P.; Van Dam, R. M.; Schorzman, D. A.; Quake, S. R.; DeSimone, J. M. *J. Am. Chem. Soc.* **2004**, *126*, 2322.

138. (a) Williams, S. S.; Hampton, M. J.; Gowrishankar, V.; Ding, I. K.; Templeton, J. L.; Samulski, E. T.; DeSimone, J. M.; McGehee, M. D. *Chem. Mater.* **2008**, *20*, 5229. (b) Rolland, J. P.; Van Dam, R. M.; Schorzman, D. A.; Quake, S. R.; DeSimone, J. M. *J. Am. Chem. Soc.* **2004**, *126*, 2322. (c) Rolland, J. P.; Hagberg, E. C.; Denison, G. M.; Carter, K. R.; De Simone, J. M. *Angew. Chem. Int. Ed. Eng.* **2004**, *43*, 5796.
139. Maynor, B. W.; Larue, I.; Hu, Z.; Rolland, J. P.; Pandya, A.; Fu, Q.; Liu, J.; Spontak, R. J.; Sheiko, S. S.; Samulski, R. J.; Samulski, E. T.; DeSimone, J. M. *Small* **2007**, *3*, 845.
140. Kelly, J. Y.; DeSimone, J. M. *J. Am. Chem. Soc.* **2008**, *130*, 5438.
141. Gratton, S. E. A.; Williams, S. S.; Napier, M. E.; Pohlhaus, P. D.; Zhou, Z.; Wiles, K. B.; Maynor, B. W.; Shen, C.; Olafsen, T.; Samulski, E. T.; Desimone, J. M. *Acc. Chem. Res.* **2008**, *41*, 1685.
142. Gratton, S. E. A.; Pohlhaus, P. D.; Lee, J.; Guo, J.; Cho, M. J.; DeSimone, J. M. *J. Controlled Release* **2007**, *121*, 10.
143. (a) In, H. J.; Arora, W.; Buchner, T.; Jurga, S. M.; Smith, H. I.; G. Barbastathis, G. *4th IEEE Conf. on Nanotech.* **2004**, 358. (b) Kubota, K.; Fleischmann, T.; Saravanan, S.; Vaccaro, P. O.; Aida, T. *Jpn. J. Appl. Phys.* **2003**, *42*, 4079. (c) Guan, J.; He, H.; Hansford, D. J.; Lee, L. J. *J. Phys. Chem., B* **2005**, *109*, 23134.
144. Examples of antioxidant additives used in polymer formulations may be found online: http://www.adeka-pa.eu/additive-formulations/antioxidant-blends.
145. Heat stabilizers are used in the processing of rigid (pipe, window profiles, siding, fencing) and some flexible (packaging) PVC applications, preventing the thermal degradation of PVC resins during elevated temperature exposure. In addition, heat stabilizers extend the lifetime of finished products.
146. A curing agent for polyurethanes and epoxies, as well as a chain extender for polyurethane and polyurea: http://www.albemarle.com/_filelib/FileCabinet/Products/Curatives/2017_New/CUR-0001_100_TDS_102417.pdf.
147. An amine cure accelerator that is used to promote free-radical formation in addition polymerizations such as unsaturated polyester resin, viny ester, and acrylate systems: http://www.albemarle.com/Products---Markets/Lithium-and-Advanced-Materials/Synthesis-and-Polymer-Solutions/Curatives/FirstCure-MHPT-curative-156C2164.html.
148. A Plasticizer Information Center with information about various types of plasticizers may be found online at: http://www.plasticisers.org.
149. Anti-plasticizers are added to rigidify polymers, and may be bulky molecules that disperse in a polymer, or "sticky" molecules that induce attractive interactions between the additive and polymer. For instance, see: Dasmahapatra, A. K.; Nanavati, H.; Kumaraswamy, G. *J. Chem. Phys.* **2009**, *131*, 074,905, and references therein.

150. Phthalates are used where good resistance to water and oils is required. For example, applications such as garden hoses, shoes, toys, food wraps, flooring materials, notebook covers, vinyl tiles, traffic cones, plastic foams, tool handles, building materials, etc.
151. Trimellitates have a low volatility, and are used in automobile interiors where resistance to high temperature is required.
152. Used for low-temperature applications or those requiring resistance to UV light.
153. For a review of plasticizers used for various applications, see: https://www.cpsc.gov/PageFiles/126379/godwin.pdf.
154. (a) Bocque, M.; Voirin, C.; Lapinte, V.; Caillol, S.; Robin, J. J. *J. Polym. Sci., Part A: Polym. Chem.* **2016**, *54*, 11. (b) http://www.phthalate-free-plasticizers.com/home.html.
155. http://sustainableproduction.org/downloads/PhthalateAlternatives-January2011.pdf.
156. https://noharm.org/sites/default/files/lib/downloads/bfrs/BFRs_Rising_Concern.pdf.
157. Howell, B. A. Presented in part at the 15th International Conference on Advances in Additives and Modifiers for Polymers and Blends, Las Vegas, NV, February, 2006.
158. For instance, see: (a) Howell, B. A.; Cho, Y. -J. "Brominated Aryl Phospholane Flame Retardants" in *ACS Symp. Series V—Fire and Polymers*, **2009**, 249. (b) Howell, B. A.; Wu, H. *J. Therm. Anal. Calorim.* **2006**, *83*, 79.
159. For a review of self-healing polymers and composites, see: (a) Blaiszik, B. J.; Kramer, S. L. B.; Olugebefola, S. C.; Moore, J. S.; Sottos, N. R.; White, S. R. *Annu. Rev. Mater. Res.* **2010**, *40*, 179 (http://sottosgroup.beckman.illinois.edu/nrs098.pdf). (b) Mauldin, T. C.; Kessler, M. R. *Int. Mater. Rev.* **2010**, *55*, 317 (https://research.libraries.wsu.edu/xmlui/bitstream/handle/2376/5545/52-self%20healing.pdf?sequence=1&isAllowed=y).
160. Rule, J. D.; Brown, E. N.; Sottos, N. R.; White, S. R.; Moore, J. S. *Adv. Mater.*, **2005**, *17*, 205.
161. *Controlled and Living Polymerizations: From Mechanisms to Applications.* Matyjaszewski, K.; Muller, A. H. E., eds., Wiley: New York, 2009.
162. (a) Feiteira, J.; Gruyaert, E.; De Belie, N. *Construct. Build. Mater.* **2016**, *102*, 671. (b) Greil, P. *J. Adv. Ceram.* **2012**, *1*, 249. (c) Escobar, M. M.; Vago, S.; Vazquez, A. *Compos. B: Eng.* **2013**, *55*, 203.
163. (a) Zhong, N.; Post, W. *Compos. A: Appl. Sci. Manuf.* **2015**, *69*, 226. (b) http://naturalprocessdesign.com/tech_polymer.htm.
164. https://www.asme.org/engineering-topics/articles/energy/selfhealing-wind-turbine-blades.
165. http://cdn.intechopen.com/pdfs/39416/InTech-Unsaturated_polyester_resin_for_specialty_applications.pdf.
166. Hasenwinkel, J. M.; Lautenschlager, E. P.; Wixson, R. L.; Gilbert, J. L. *J. Biomed. Mater. Res.*, **2002**, *59*, 411.
167. Ghosh, B.; Urban, M. W. *Science* **2009**, 323, 1458.

168. White, S. R.; Moore, J. S.; Sottos, N. R.; Krull, B. P.; Santa Cruz, W. A.; Gergely, R. C. R. *Science* **2014**, *344*, 620.
169. Chen, X.; Dam, M. A.; Ono, K.; Mal, A.; Shen, H.; Nutt, S. R.; Sheran, K.; Wudl, F. *Science*, **2002**, *295*, 1698.
170. Yang, Y.; Wang, H.; Huang, L.; Nishiura, M.; Higaki, Y.; Hou, Z. *Angew. Chem. Int. Ed.* **2021**, *60*, 26,192.
171. (a) Shim, T. S.; Kim, J. M. *Kor. J. Chem. Eng.* **2017**, *34*, 2355. (b) Manti, M.; Cacucciolo, V.; Cianchetti, M. *IEEE Robotics & Automation Mag.* **2016**, *23*, 93. (c) Meng, H.; Li, G. *Polymer* **2013**, *54*, 2199.
172. Abhoff, S. J.; Lancia, F.; Iamsaard, S.; Matt, B.; Kudernac, T.; Fletcher, S. P.; Katsonis, N. *Angew. Chem. Int. Ed.* **2017**, *56*, 3261.
173. For instance, see: https://www.techbriefs.com/component/content/article/tb/techbriefs/materials/26634.
174. For example, see: (a) Conti, S.; Lenz, M.; Rumpf, M. *J. Mechan. Phys. Solids* **2007**, *55*, 1462. (b) Zainal, M. A.; Ahmad, A.; Mohamed Ali, M. S. *Biomed. Microdev.* **2017**, *19*, 8.
175. For instance, see: Meng, Y.; Jiang, J.; Anthamatten, M. *J. Polym. Sci. B: Polym. Phys.* **2016**, *54*, 1397.
176. Leng, J.; Lv, H.; Liu, Y.; Du, S. *Appl. Phys. Lett.* **2005**, *86*, 114,105.

Further Readings

1. G.G. Wallace, G.M. Spinks, L.A.P. Kane-Maguire, P.R. Teasdale, *Conductive Electroactive Polymers: Intelligent Polymer Systems*, 3rd edn. (CRC Press, New York, 2008)
2. P.R. Dvornic, M.J. Owen (eds.), *Silicon-Containing Dendritic Polymers* (Springer, New York, 2009)
3. B.D. Ratner, A.S. Hoffman, F.J. Schoen, J.E. Lemons, *Biomaterials Science: An Introduction to Materials in Medicine*, 3rd edn. (Academic Press, New York, 2012)
4. M. Chanda, *Introduction to Polymer Science and Chemistry: A Problem-Solving Approach*, 2nd edn. (CRC Press, New York, 2013)
5. S. Chandrasekaran, *Click Reactions in Organic Synthesis* (Wiley, New York, 2016)
6. K. Matyjaszewski, A.H.E. Muller (eds.), *Controlled and Living Polymerizations: From Mechanisms to Applications* (New York, Wiley, 2009)
7. T.H. Courtney, *Mechanical Behavior of Materials*, 2nd edn. (Waveland Press, New York, 2013)
8. M. Bolgar, J. Hubball, *Handbook for the Chemical Analysis of Plastic and Polymer Additives*, 2nd edn. (CRC Press, New York, 2015)
9. G. Pritchard, *Plastics Additives: An A-Z Reference*, 1998th edn. (Springer, New York, 2012)
10. D.W. Van Krevelen, K. Nijenhuis, *Properties of Polymers: Their Correlation with Chemical Structure*, 4th edn. (Elsevier, New York, 2009)
11. H.R. Allcock, F.W. Lampe, J.E. Mark, *Contemporary Polymer Chemistry*, 3rd edn. (Prentice-Hall, New Jersey, 2003)
12. P.C. Painter, M.M. Coleman, *Fundamentals of Polymer Science*, 2nd edn. (CRC, New York, 1997)
13. P.J. Flory, *Principles of Polymer Chemistry* (Cornell University Press, Ithaca, NY, 1953)
14. G. Odian, *Principles of Polymerization*, 3rd edn. (Wiley, New York, 1991)
15. R.J. Young, P.A. Lovell, *Introduction to Polymers*, 3rd edn. (CRC, New York, 2011)
16. S. Fakirov, *Fundamentals of Polymer Science for Engineers* (Wiley, New York, 2017)

17. M.P. Stevens, *Polymer Chemistry: An Introduction*, 3rd edn. (Oxford University Press, Oxford, 1998)
18. J.B. Fortin, T.-M. Lu, *Chemical Vapor Deposition Polymerization: The Growth and Properties of Parylene Thin Films* (Springer, New York, 2003)
19. H. Hsieh, R.P. Quirk, *Anionic Polymerization* (CRC, New York, 1996)

Chapter 6
Nanomaterials

Imagine how much control over resultant properties you would have if you were able to deposit and maneuver individual atoms into predefined arrangements, *en route* toward a new material. This is fast becoming a reality and is the realization of the ultimate in "bottom-up" materials design. Thus far, one is able to easily fabricate materials comprised of a small number of atoms, with features on the nanometer scale (10^{-9} m)—one-billionth of a meter. To put this into perspective, think of a material with dimensions approximately 1000 times smaller than the diameter of a human hair follicle! As we will see, it is now even possible to push individual atoms around a surface using specialized techniques.

We are at the crossroads of unprecedented applications that will only be possible using nanoscale building blocks. More effective devices will be constructed to remove pollutants from the environment and detect/deactivate chemical and biological warfare agents. Nanomedicines composed of nanosized robots will target and kill cancer cells or human disease without any side-effects (Fig. 6.1). Integrated circuitry with the capabilities of current workstations will be the size of a grain of sand and will be able to operate for decades with the equivalent of a single wristwatch battery. Robotic spacecrafts that weigh only a few pounds will be sent out to explore the solar system, and widespread space travel will be possible for the masses. Oh, yes—one that is near to us all—inexpensive alternative energy sources will power our vehicles, rather than depending on the daily fluctuations of soaring gas prices![1]

In order to gain rapid progress toward these intriguing goals, a number of universities across the globe now offer degree programs in nanotechnology.[2] The first national network focused on the design/fabrication/testing of nanomaterials was instituted in 2004 through funding from the National Science Foundation. The *National Nanotechnology Coordinated Infrastructure* consists of a conglomerate of 16 sites across the country that are focused on all aspects of nanomaterials.[3] Since "nanotechnology" is such an interdisciplinary field, many more nano-related education programs and research centers will likely be instituted in the coming years. However, as with all scientific disciplines, a major roadblock toward research progress in the U.S. is domestic student recruitment. There are a declining number of degrees

B. D. Fahlman, *Materials Chemistry*,
https://doi.org/10.1007/978-3-031-18784-1_6

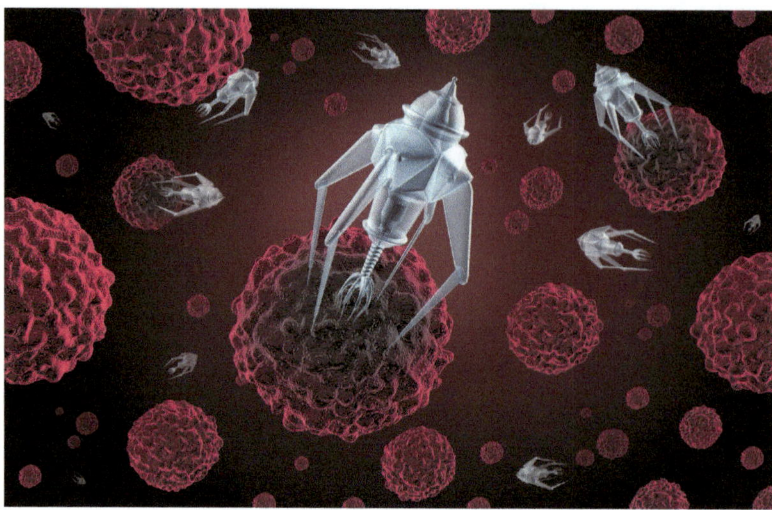

Fig. 6.1 Nanomedicine concept as a group of nanobots programed to target and kill cancer cells or human disease. © Lightspring/Shutterstock.com

awarded in the sciences within recent years (from B.S. to Ph.D. levels) in the U.S., relative to other foreign countries (e.g., China, India). This represents an ominous forecast that our advanced technology and warfare capabilities will greatly lag behind other countries, threatening our everyday way of life and the "superpower" status long enjoyed by the United States.

6.1 The Toxicity of Nanomaterials

As exciting as the futuristic applications of nanotechnology may sound, is it possible that such technological growth may be associated with dire societal consequences? In Eric Drexler's book *Engines of Creation*, it was forecasted that self-replicating nanomachines would take over all life on Earth![4] Although that notion is far from reality, there may be more serious issues that arise through introduction of nanomaterials into the biosphere. Since we haven't yet determined the extent of long-term effects that may arise from the widescale introduction of nanoscale materials, are we on the verge of upsetting the natural balance in ways that cannot be overturned?

History has revealed that chemistry is a two-edged sword, with benefits that greatly improve our lives but also negative consequences for human health and our environment. For example, think of the chlorofluorocarbons (CFCs) that were once used for refrigerants. Their discovery was heralded as one of the greatest triumphs of modern science. Alas, many years after their worldwide adoption, it was realized that they contributed to the ozone hole and likely increase in the incidences of skin cancer. Likewise, CO_2 emissions from factories and automobiles were not considered as problematic many years ago, but more active hurricane seasons, wildfires, and atypical weather patterns due to climate change are making us more aware of the destructive consequences of fossil fuel combustion.

The diversity of engineered nanostructures has expanded at an astounding pace. Nanomaterials are used in a vast range of products such as cosmetics,[5] batteries, electronics, sporting equipment, fuel catalysts, food additives, paints, drug-delivery/biomedicine,[6] ultrafiltration, stain-/wrinkle-free textiles, scratch-resistant coatings, and many more.[7] To date, more than 20 nanomaterial-based drug delivery systems have been approved by the U.S. Food and Drug Administration (FDA) for clinical use.[8] With so many stakeholders in the nanotechnology arena who stand to make a lot of money from commercial products,[9] it is no wonder why the technological innovation has far surpassed nanotoxicity studies. Most of the early work regarding nanotoxicity was reported by the European Union, with the assessment of toxicological/environmental consequences of nanoscale materials with comparisons to known contaminants such as colloids, aerosols/smoke/soot particulates, and asbestos.[10]

The life cycle and toxicity hazards of a nanostructure are governed by its physical properties such as size distribution, morphology, degree of agglomeration, specific surface area, crystallinity, and aspect ratio. The size reduction of a bulk material to the nanoregime results in an exponential increase in its surface:volume ratio. Hence, nanoparticles are generally more toxic than larger particulates of the same insoluble material (on a mass dose basis), since relatively more molecules of the material are available on its surface thus enhancing its intrinsic toxicity.[11] In addition, chemical properties that govern overall reactivity and solubility also contribute to nanotoxicity such as purity, atomic composition/structure, and surface chemistry (e.g., charge, reactive/defect sites, functionalization, protective coating).[12]

With such profound nanostructural complexity and diversity, it is not surprising that standardized *toxicokinetics* studies (i.e., absorption, distribution, metabolism, and excretion characteristics) are still greatly lacking. Furthermore, it is currently difficult to interpret and compare scientific studies—even those performed on the same type of nanostructure. Hence, there is a need for an interdisciplinary collaboration to develop databases that provide open access to toxicity data.[13] There are modeling efforts underway to translate the simplistic and vast in vitro (tissue/cell) studies into in vivo (living organism) effects, which are needed to define exposure risks for the various classes of nanomaterials.[14] More recently, Fortino et al. have utilized a multilayered omics-driven approach to identify biomarkers from both in vitro and in vivo assays to predict the hazard potential of engineered nanomaterials (Fig. 6.2).[15]

When encountering a harmful substance such as asbestos, carcinogenic solvents, or smoke particulates, we follow safety guidelines that serve to protect us from harmful side effects. In order to be useful, the suggestions provided by a *risk assessment* should consider both the exposure and toxicity of a substance.[16] Whereas toxicity is not easy to address, *exposure* deals with the concentration of the substance, the length of time of exposure, and the rate of its uptake. As shown in Fig. 6.3, human

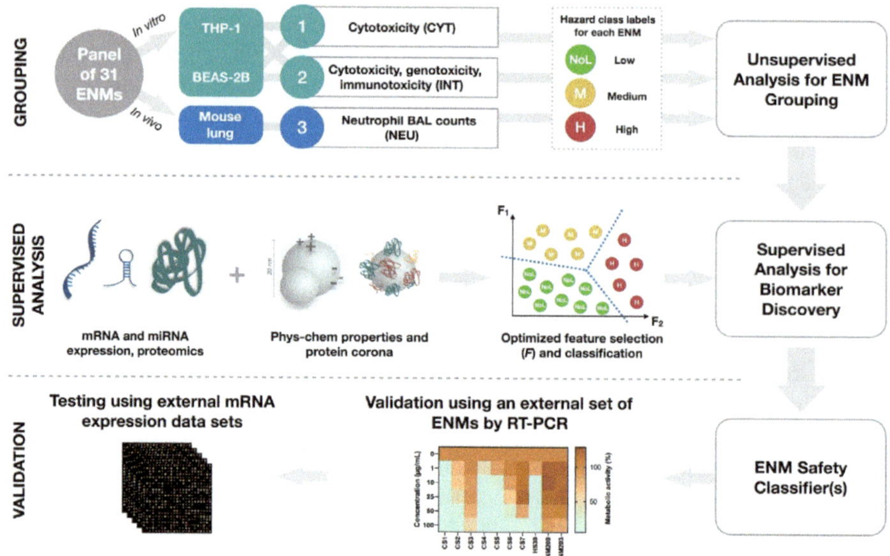

Fig. 6.2 Schematic illustrating the strategy used to evaluate the hazard potential of 31 different engineered nanomaterials (ENMs) based on their cytotoxicity (CYT), and combination of cytotoxicity, genotoxicity, and immunotoxicity (INT) in vitro, as well as neutrophil (NEU) in vivo. Reproduced with permission from *Nat. Commun.* **2022**, *13*, 3798. © 2022 Springer Nature

exposure to nanomaterials may occur at any stage of a product's life cycle: manufacturing, transportation, use, end-of-life treatment, and final disposal. The most common exposure routes for nanostructures are inhalation, ingestion, and uptake through the skin; however, intravenous injection may also be used for medical purposes (e.g., drug delivery).

Based on the current literature, we will now attempt to address some important questions related to nanotoxicity[17].

(i) *What is the danger of exposure to nanomaterials* via *skin absorption, ingestion, or inhalation routes?*

The first news report on the potential damaging effects of nanoscale materials surfaced about 15 years ago, when TiO_2/ZnO nanoparticles from sunscreen were found to cause free radicals in skin cells, damaging DNA.[18] Since then, there have been an increasing number of such reports suggesting that nanostructures are able to traverse across membranes in the body, with an increasing toxicity found with decreasing nanoparticulate dimensions. Perhaps the most widely reported study surfaced in mid-2004, where it was shown that fullerenes, a nanoscale allotrope of carbon, caused brain damage in aquatic species.[19] However, subsequent studies have shown that aqueous C_{60} solutions exhibit low toxicity in fish[20]; the harmful effects were more likely caused by the tetrahydrofuran (THF) solvent used in the synthesis than the nanoparticles themselves.[21] However, even if these nanoparticles

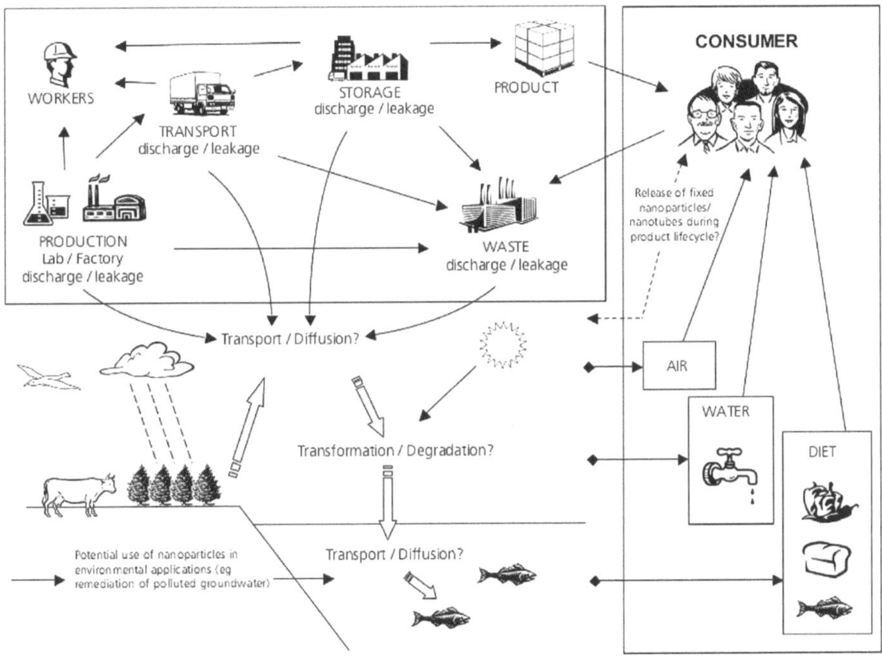

Fig. 6.3 Some possible exposure routes for nanostructures based on current/potential applications. Reproduced with permission from "Nanoscience and Nanotechnologies: Opportunities and Uncertainties", *Nanoscience and Nanotechnologies*, July 2004. © 2004 The Royal Society

are not as lethal as once thought, studies have shown that C_{60} solutions produce cytogenotoxic and reproductive effects in fish species.[22]

Skin Absorption

Skin represents the largest organ in our bodies, and serves as the first stage of protection against the uptake of airborne and/or solubilized contaminants from our surroundings. There are three layers that comprise skin: the epidermis, dermis, and hypodermis; the outermost layer of the epidermis is referred to as the *stratum corneum* (Fig. 6.4), and consists of dead cells that are strongly keratinized. However, once particulates are able to diffuse across this barrier (known as percutaneous penetration), absorption through underlying skin layers may take place. If resorption takes place, the compound is taken into the vascular system (lymph and/or blood vessels) and is delivered to organs system-wide. It has been shown that nanomaterials have a greater risk of being absorbed through the skin than larger particulates.[23] Although the therapeutic absorption of nanoparticles through skin is desired for drug-delivery applications, we need to identify the health risks posed by *unintentional* absorption of nanostructures through the skin; for instance, involving personnel active in nanomaterials synthesis/characterization.

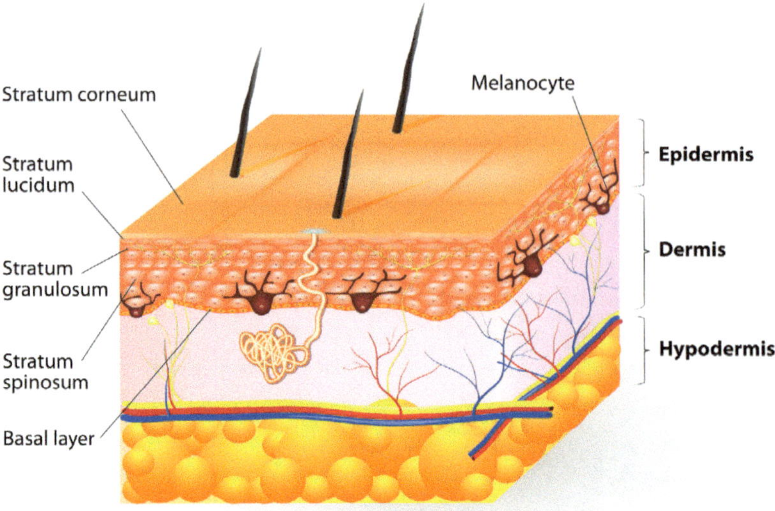

Fig. 6.4 Cross-section illustration of skin. © Designua/Shutterstock.com

The structural diversity of nanostructures equates to amazingly complex uptake/absorption profiles through skin—much more so than bulk chemical agents (e.g., organic/inorganic liquids, solutions, gases, etc.), of which have predictable diffusion/absorption profiles. Although diffusion theory suggests that only nanoparticles with dimensions below 1 nm are able to permeate intact skin,[24] other experimental studies have shown that larger sized nanoparticles may easily cross the skin barrier.[25] Not surprisingly, diffusion through skin barriers is related to the polarity of the nanoparticle. Whereas diffusion into underlying living tissue is favored for polar molecules, diffusion through the outermost stratum corneum is preferred for more lipophilic species. Hence, only substances with partition coeffients (logP) between 1 and 3 are most suited for skin absorption.[26]

As shown in Fig. 6.5, nanostructures may move between individual cells of the stratum corneum (intercellular), through cells (intracellular), or through the hair follicle or sweat ducts (transappendageal).[27] Nanoparticles may also release significant concentrations of ions relative to bulk materials due to their enhanced surface/mass ratio.[28] Prolonged release of ions from hair follicles poses an increased risk of dermatitis allergies, especially for nanoparticles containing sensitizing elements such as Ni, Pd, Co, Cd, S, Se, or As.

Thus far, in vivo diffusion-based studies have been mostly performed on pig skin, since it represents the closest match to humans. However, there are reports that have shown nanoparticle penetration through healthy human skin as a function of size,[29] as deep as the reticular dermis.[30] In vitro testing is commonly used to determine the starting dose for in vivo toxicity assays. However, what makes toxicity studies so difficult to standardize is the change in nanoparticle size, shape, surface charge, etc.

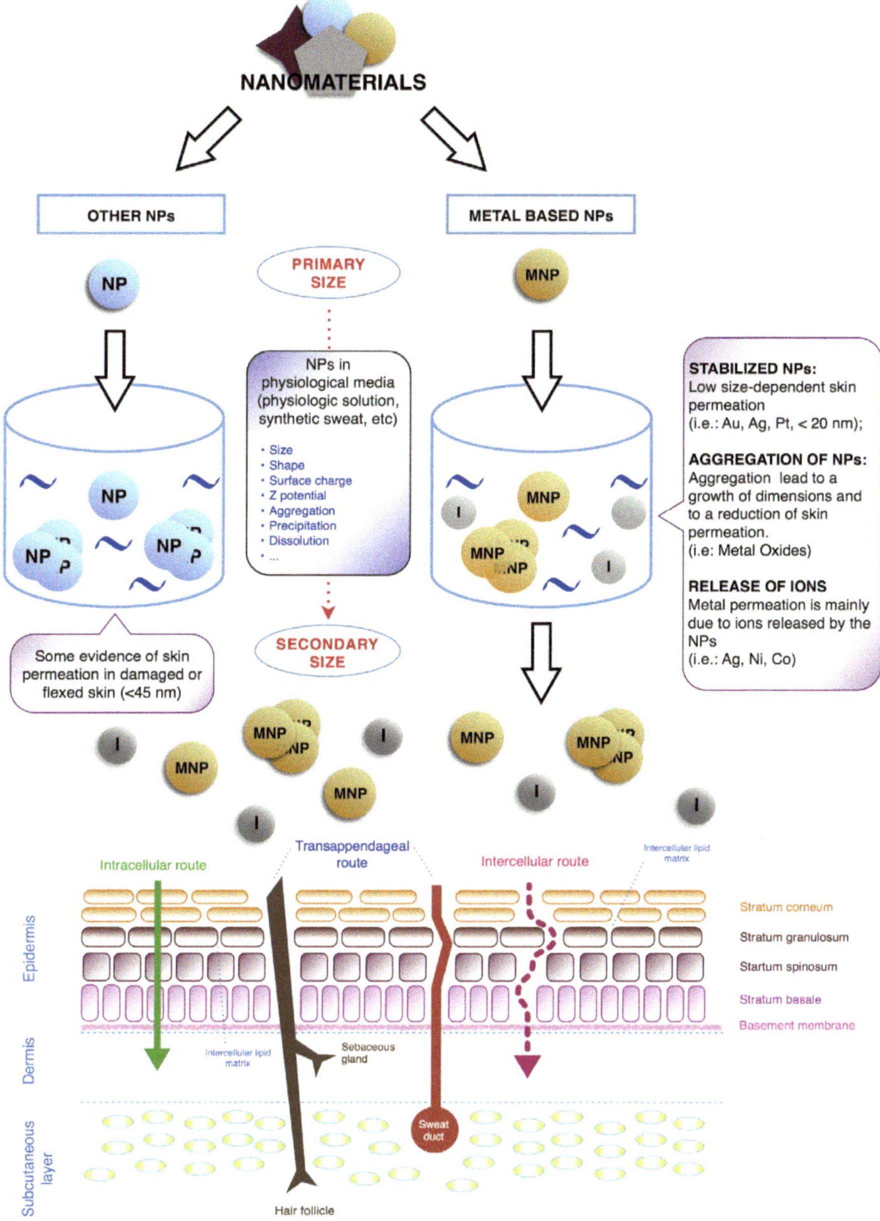

Fig. 6.5 Skin absorption of nanoparticles. NP = nanoparticle (non-metallic); MNP = metallic nanoparticle; I = ions released. Reproduced with permission from *Regulat. Toxicol. Pharmacol.* **2015**, *72*, 310. © 2015 Elsevier

that occurs in solution. Hence, physiological media or synthetic sweat are often used in absorption studies in an attempt to reproduce in vivo conditions.

To date, the greatest risk of skin exposure to nanostructures is from prolonged contact with over-the-counter pharmaceutical formulations (Table 6.1) or sunscreens. Fullerenes (carbon clusters such as C_{60}, C_{70}, etc.) are present in a number of these products, as well as other therapeutic agents that exploit their inherent phototoxic properties.[31] Water-soluble aggregates of C_{60} may exhibit significant cytotoxicity even at concentrations as low as 20 ppb; however, fullerenes with different surface functionalization (e.g., $Na^{+}{}_{2-3}[C_{60}O_{7\text{-}9}(OH)_{12-15}]^{(2-3)-}$ and $C_{60}(OH)_{24}$) are significantly less cytotoxic.[32] The toxicity of C_{60}-based nanostructures is thought to arise from the degree of oxygen radicals generated, which is less pronounced for more heavily derivatized (and water-soluble) fullerenes. Amino acid functionalized fullerenes have also been shown to penetrate the outer layers of flexed or imperfect skin (not intact skin).[33] In fact, the relationship between nanoparticulate absorption and mechanical flexing is likely responsible for the high onset of podoconiosis in unprotected feet that come into contact with certain soils that have high concentrations of Be and Zr.[34] It should be noted that high aspect ratio nanostructures such as carbon nanotubes have also been shown to be absorbed within human epidermal keratinocytes, giving rise to an irritation response.[35] However, there are currently no reports of pronounced dermal migration of these nanostructures.

Inorganic-based nanocrystals such as quantum dots (QDs)[37] and TiO_2[38] have been shown to penetrate the human stratum corneum and hair follicles, possibly allowing for accumulation in various organs[39] as well as their integration into the immune system.[40] Most importantly, it has been demonstrated that QDs are able to penetrate *intact* skin, without the assistance of any mechanical stressors.[41] The lipid layers within the cells of the stratum corneum are thought to form a pathway for the migration of nanoparticles into the skin, which may then be sequestered (phagocytized) by the Langerhans cells.[42] However, surface-functionalized QDs have also been noted to exhibit significant penetration to the highly vascularized dermis, offering a route for systemic toxicity.[43] In fact, it is believed that the dermal translocation of nanoparticulates may follow the same pathway(s) as viruses such as Herpes, which closely interact with the peripheral and central nervous systems. Based on the current literature, it can be summarized that QDs can permeate skin, mostly when damaged or pretreated with UVB light, and Cd can diffuse into internal organs. However, size, pH, temperature, and surface functionalization all influence the absorption profile and resultant toxicity of QDs.

There are many recent reports highlighting the toxicity (Fig. 6.6)[44] and pervasiveness[45] of microplastics in the environment, farm animals,[46] and even human blood.[47] However, there are far fewer studies related to the accumulation and toxicity of nanoplastics, which may be derived from microplastics (Fig. 6.7). Polystyrene nanoparticles, commonly found in paints and varnishes, have been shown to accumulate in follicular openings of skin—especially for smaller sized particulates. Lee and coworkers reported that polystyrene particles with diameters of 50 nm easily penetrated zebrafish embryos thus accumulating in lipid-rich areas of the organism.

Table 6.1 List of nanostructures commonly present in cosmetics formulations[36]

Nanostructure	Cosmetic product
Fullerenes	Anti-aging cream, around-eye cream
Vitamin nanocapsules or nanosized vitamins	After-sun products, bronzer/highlighter, sunscreen/tanning oil
Nanosized ZnO and/or TiO_2 (coated or uncoated with dimethicone)	Anti-aging cream, blush, bronzer/highlighter, concealer (also may contain Al nanoparticles), anti-itch cream, diaper cream, eye shadow, foundation, lip balm, suncreen/tanning oil
Nanosomes (e.g., sodium lactate, calendula, witch hazel, ginseng, urea, vitamins (A, E, B5), α-bisabolol, germal (II))	Anti-aging cream, body-firming lotion, concealer, body wash/cleanser
Nanosized lipsomes	Anti-aging cream, around-eye cream, facial moisturizer
Nano-delivery system	Facial moisturizer, body-firming lotion
Lyphazome nanospheres	Around-eye cream, moisturizer, nail treatment
Nanosized minerals (e.g., rose quartz, talc, topaz)	Concealer, exfoliant/scrub, facial cleanser, facial moisturizer, foundation, glitter, lip gloss, lipstick, mask, sunscreen/tanning oil

However, the presence of these nanoparticles was not shown to affect cell death or reproduction.[48]

Oxide nanoparticles such as SiO_2, which are found as an additive for foods, cosmetics, and paints, have been shown to pass through a partially damaged stratum corneum, and penetrate the skin layer. For instance, amorphous silica nanoparticles with an average diameter of 70 nm were shown to penetrate through mice skin and were transported to the lymph nodes, which could potentially be dispersed throughout the entire body via the lymphatic transport system.[49] However, other contradictory studies indicate that functionalized silica nanoparticles are not able to penetrate through either intact or damaged mice skin.[50] TiO_2 and ZnO nanoparticles, commonly found in sunscreens and cosmetics, were also found to be localized only in the outer layer of the stratum corneum—even when damaged or UV-irradiated skin were used for testing. However, there was one study that found traces of Zn in blood and urine of volunteers after 5 days of sunscreen applications.[51] Clearly, the nature and degree of surface functionalization, as well as combination with other consumer products,[52] appear to be important in the skin uptake of these nanostructures.

To summarize,[53] the data to date show that nanoparticles with dimensions $\leq$4 nm may penetrate and permeate intact skin. Nanoparticle dimensions of 4–20 nm may potentially permeate intact or damaged skin. Nanoparticles with dimensions between

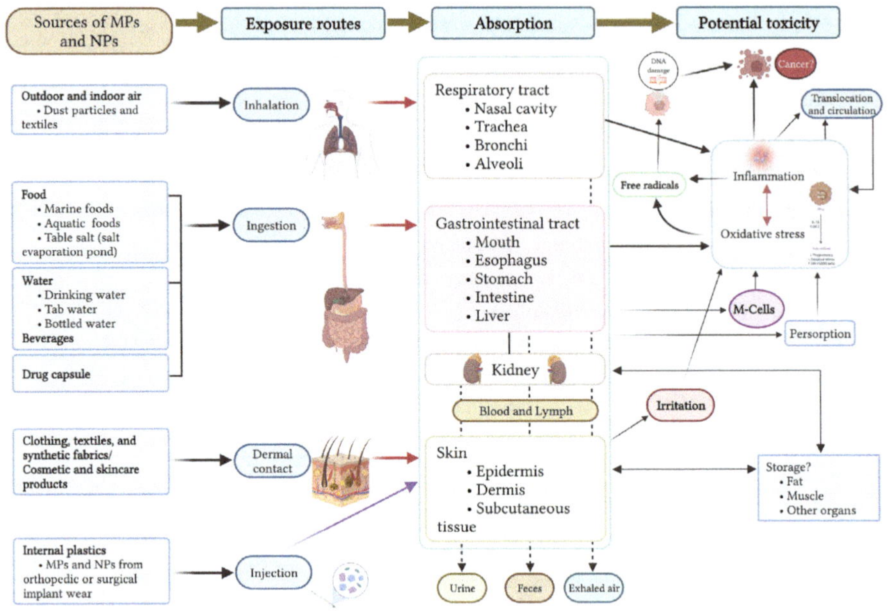

Fig. 6.6 Potential routes of exposure to micro- and nanoplastics and toxic effects on humans. Reproduced with permission from *Marine Pollut. Bull.* **2022**, *181*, 113,832. © 2022 Elsevier

21 and 45 nm may penetrate through only damaged skin, whereas nanoparticles beyond 45 nm cannot penetrate through skin. However, there are plenty of exceptions to these size-related guidelines, based on the nature and degree of surface groups used in their syntheses. Furthermore, toxic metals may be released in physiological conditions, which could lead to local or systemic effects that are not directly related to nanoparticulate dimensions. Perhaps most concerning is the use of silver nanoparticles for wound dressing[54] or engineered textiles—both of which involve contact with a large skin surface area. The data suggest that these nanoparticles are able to penetrate and permeate the skin, which could cause effects on internal organs.[55]

Inhalation or Ingestion

It has long been known that there is a significant link between exposure to inhaled particulate matter (PM) and increased incidences of cardiovascular and pulmonary diseases.[56] However, the dangers of unintentional inhalation of nanostructures is not limited to workers and researchers specifically involved in nanotechnology-related fields. In fact, there are a surprisingly large number of nanoparticles in our ambient atmosphere, reaching as high as 10^6–10^8 nanoparticles/L of air in urban (e.g., automobile exhaust, building-material particles) and rural (e.g., dust, diesel exhaust, combustion of wood, waste, etc.) areas, including indoors.[57] Diesel exhaust emissions are comprised of a wide distribution of particle sizes, with the bulk of the particulates with dimensions <50 nm—well within the nano-range. However, even

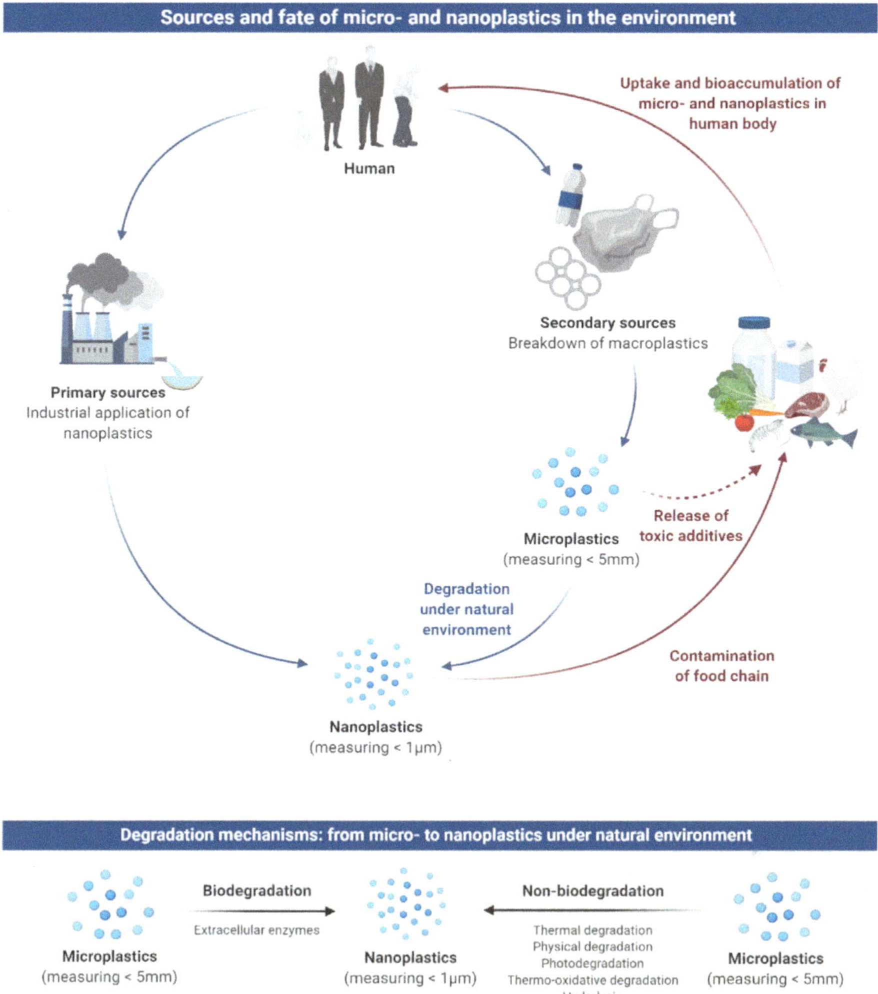

Fig. 6.7 Sources and fate of micro- and nanoplastics in the environment. Reproduced with permission from *Nanomaterials* **2021**, *11*, 496. © 2021 MDPI

gasoline exhaust gases contain nanoparticulates, ranging from 1.9×10^{14}–9.9×10^{14} particles km^{-1} and 2.2×10^{15}–1.1×10^{16} particles kg^{-1} fuel, with the greatest concentrations and smallest sizes associated with high-speed road traffic.[58] Table 6.2 compares the number and size of nanoparticles in a typical ambient environment of a variety of workplaces.

Larger particles (5–30 μm) are typically deposited in the nasopharyngeal region of the lungs, and smaller particles (1–5 μm) are deposited in the tracheobronchial region.

Table 6.2 Comparison of environmental nanoparticles in a variety of workplaces

Workplace	Nanoparticle Concentration[a] (particles.cm^{-3})	Average size (nm)
Silicon processing	Up to 100,000	280–520
Metal grinding	Up to 130,000	17–170
Soldering	Up to 400,000	36–64
Plasma cutting	Up to 500,000	120–180
Bakery	Up to 640,000	32–109
Airport field	Up to 700,000	<45
Welding	100,000–40,000,000	40–600

[a] Particles within the 14–673 nm size range only

Whereas these particles may be removed by mucociliary clearance, nanosized particles (<100 nm) may penetrate deeply into the alveolar region where removal mechanisms are insufficient (Fig. 6.8).[59] This results in a higher probability of adverse health effects due to particle-tissue and particle-cell interactions. In general, particle toxicity is governed by a number of factors such as composition, concentration, size distribution, water solubility, reactivity, frequency and duration of exposure, interaction with other airborne agents, pulmonary ventilation, and pre-existing immunological conditions.[60]

Although there are reports that challenge the notion of 'nano-specific toxicity',[61] the small size distribution of nanoparticles results in their greater propensity to cross tissue and cell membranes. Once inside the cell, either accumulation of nanoparticles or cellular injury may occur due to particle-cell interactions. Perhaps most importantly, inhaled/ingested nanoparticles have been shown to enter the blood or central nervous system, which could cause direct effects on cardiac and cerebral functions.

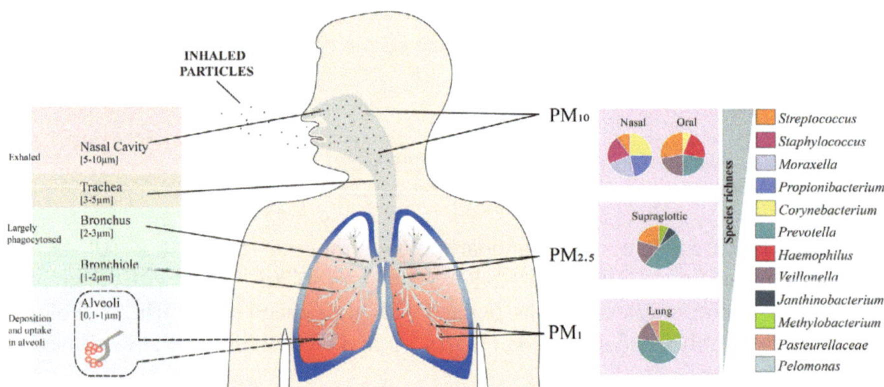

Fig. 6.8 Size-dependent deposition of inhaled nanoparticles within lung tissue and the regional variation of the top five abundant bacterial taxa. Reproduced with permission from *Partic. Fibre Toxicol.* **2018**, *15*, 46. © 2018 Biomed Central

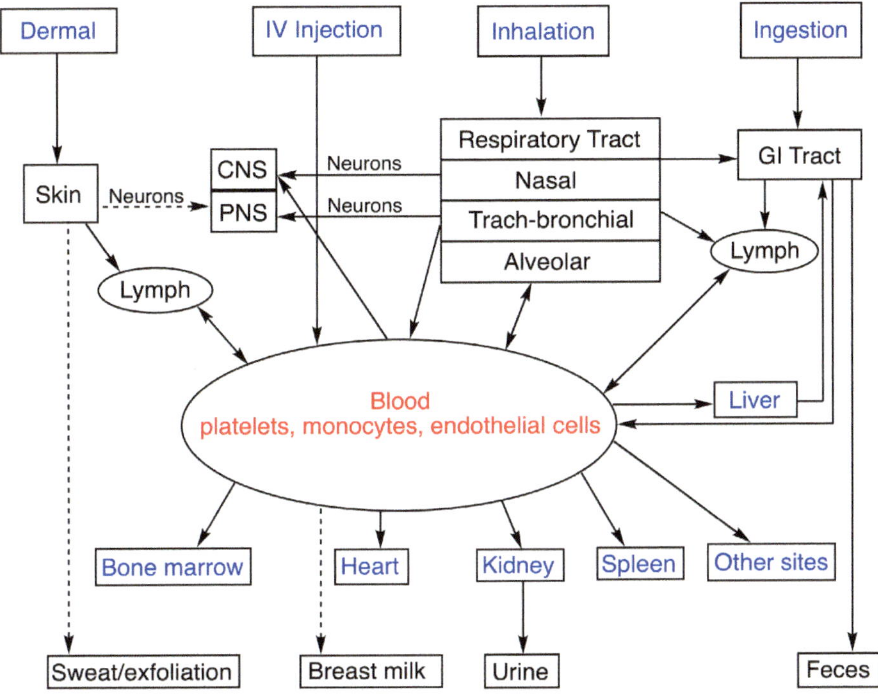

Fig. 6.9 Scheme illustrating the absorption, translocation, and excretory pathways of nanostructures within the body. The solid arrows indicate confirmed routes, whereas dashed arrows indicate potential routes. CNS = central nervous system; PNS = peripheral nervous system; GI = gastrointestinal; IV = intravenous. Adapted from *Int. J. Mol. Sci.* **2016**, *17*, 929. © 2016 MDPI

For instance, it has been reported that nanoparticles may enter the circulation system of rodents via the respiratory (inhalation) or gastro-intestinal (ingestion) tracts, which may then reach other target sites such as the liver, spleen, heart or brain (Fig. 6.9).[62]

Regarding ingestion, smaller sizes of nanoparticulates resulted in higher organ uptake, whereas larger agglomerates were found almost exclusively within the GI tract.[63] Based on the limited data available thus far, it is likely that size is the predominant factor that governs overall translocation of ingested nanostructures, which must cross the GI tract barrier in order to be distributed into the circulation system. However, dosage is also important; oral ingestion of a single dose of 500 mg kg^{-1} of TiO_2, ZnO, or Al_2O_3 nanoparticles resulted in translocation to the central nervous system of rats. Additionally, the nanoparticles were found to accumulate in the brain, ultimately leading to brain damage.[64] Food-grade nanomaterials are found in a variety of products including water bottles, taco seasoning, vitamins, meats, toothpaste, and powdered sugar (Fig. 6.10). Since these nanoparticles are ingested alongside food items, there is also a need to address the impact of nanoparticle-food interactions on nutrient uptake of the GI tract.[65]

As one would expect, the extent of circulation through the body increases when the nanostructures are functionalized with a hydrophilic and/or positively charged

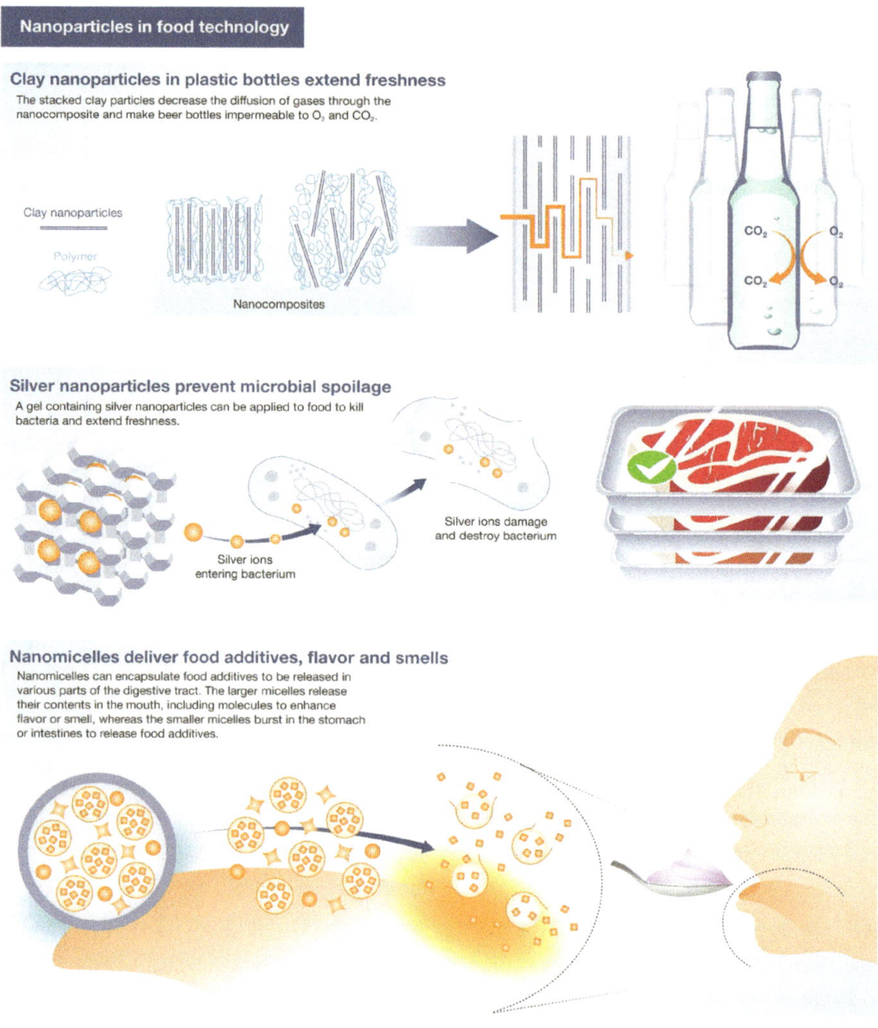

Fig. 6.10 Some uses of nanomaterials in the food industry. Reproduced with permission from *EMBO Rep.* **2014**, *15*, 638. © 2014 EMBO Press

surface. In contrast, inhaled insoluble nanoparticulates mostly remain in the lungs, which are then transported via mucociliary action to the GI tract where they leave the body as fecal excretions. Thus far, reports indicate that only about 1% of inhaled nanoparticles are translocated from the lungs across the air-blood barrier, being accumulated in secondary target organs (liver, spleen, kidneys, heart, and brain) as well as in soft tissue and bone. However, there are precedents of more extensive organ uptake (as much as 10% of the administered dose) for small-diameter nanostructures, with dimensions <10 nm.[66] In vivo studies on inhaled gold nanoparticles indicated that translocation across the air-blood barrier is most influenced by size, whereas their accumulation in secondary organs depends most strongly on specific surface area and charge.[67] However, it has been shown that rat-inhaled magnetic nanoparticles with relatively large diameters of 50 nm are still able to penetrate the blood-testes and blood–brain barriers.[68] Other direct evidence for translocation of metallic nanoparticles into the olfactory bulb from nasal exposure in rats has also been demonstrated[69]; in one case, observed in the brain up to 6 months after the inhalation exposure was administered.[70] Perhaps most interestingly, the human inhalation of diesel exhaust is reported to give rise to changes in brain activity, as demonstrated by changes in EEG signals.[71] More studies are needed to determine whether nanoparticle exposure is a contributing factor in the relatively high onset of brain diseases occurring throughout the world.

Everyone is familiar with the toxicological effects generated from the inhalation of long fibrous materials such as asbestos. The form of cancer caused by asbestos exposure is known as *mesothelioma*, responsible for *ca.* 3000 deaths per year in the U.S. and significantly more in other countries such as Britain, Australia and Belgium. Two recent reports have investigated the toxicity of analogous high aspect ratio (i.e., length:diameter ratio) nanostructures known as carbon nanotubes (CNTs). Their studies showed that nanotubes induced mesothelioma in mice, with the greatest toxicity occurring as the aspect ratio increased.[72] Another recent study by Pacurari et al. used cultured mesothelial cells to show that CNTs induced the activation of molecular-signaling pathways associated with oxidative stress, similarly to asbestos.[73] In addition, surface functionalization of carbon nanotubes with hydroxyl ($-OH$), carbonyl ($C = O$), and carboxyl (COOH) groups, as formed during acid purification steps, has been shown to generate a greater cytotoxicity than pristine analogues.[74]

In addition to size and surface-functionalization effects, there is also a compositional effect of nanoparticulates on their resultant toxicity. For instance, the cytotoxicity of carbon black has been shown to be more severe than that of titanium dioxide for identical particle sizes.[75] In addition, DuPont researchers have recently found that TiO_2 nanoparticles were not increasingly more toxic in rats than larger (micron-sized) particulates. Whereas both sizes of TiO_2 gave rise to an inflammatory response that lasted only a week, quartz nanoparticles were found to trigger a longer inflammatory response lasting for months—again, not unlike micron-sized quartz particles.[76]

What generates even more complexity is that nanoparticles may react with one another in the atmosphere, generating complex aggregates that may exhibit very different toxicity profiles than their original native compositions.[77] No wonder it is difficult to identify a standard toxicity profile for particular nanoparticles—we must consider their exact compositions, aggregate size, surface structure/charge at the time of ingestion or and/or inhalation.

(ii) *Do nanomaterials bioaccumulate?*

The pathway of nanoparticle uptake into the cells of living organisms, known as *endocytosis*, begins with the interaction of nanoparticles with the outside of the plasma membrane. This results in the formation of membrane-bound vesicles that are delivered to various intracellular compartments. Depending on the nature of the cell-nanoparticle interactions, a variety of structural changes are induced at the cell surface that facilitates nanoparticle uptake. Endocytosis is categorized into two main types: phagocytosis and pinocytosis (Fig. 6.11). Whereas larger particles (>500 nm) primarily enter via phagocytosis, pinocytosis is mainly responsible for the uptake of small nanoparticles.[78]

The use of nanomaterials for drug-delivery and cancer-targeting applications exploits their ability to localize within certain cells within the body. For instance, coated magnetite nanoparticles were found within human mammary carcinoma cells,[79] and we have already seen that functionalized carbonaceous nanostructures (fullerenes, carbon nanotubes) may become localized within human epidermal keratinocytes (HEKs).[80] Nanocrystals known as quantum dots (QDs) feature a tunable coating to improve their aqueous solubility and facilitate the attachment of antibodies or other therapeutic/diagnostic moieties for cell imaging, immunohistochemistry, and cancer-targeting applications.[81] Depending on the type of coating present, QDs have been shown to accumulate within a variety of human cells.[82] A

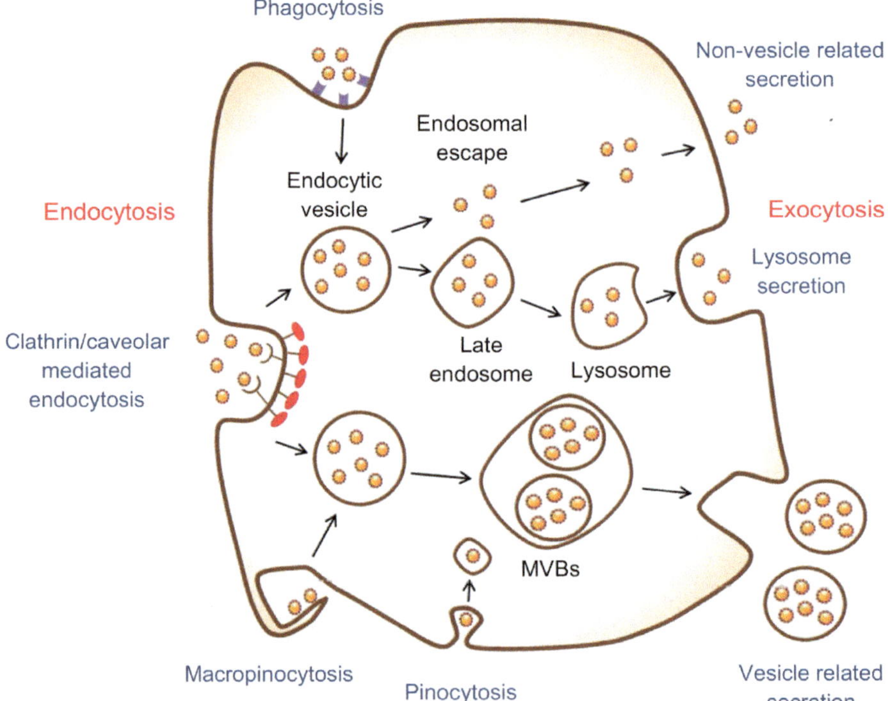

Fig. 6.11 Schematic of endocytosis and exocytosis pathways for nanoparticles in cells. MVBs = multivesicular bodies. Reproduced with permission from *Int. J. Med.* **2014**, *9*, 51. © 2014 Dove Press

recent report by Zhang and Monteiro-Riviere provides the first mechanistic insights for the uptake of quantum dots by cells, which is governed by their size, surface coating and charge.[83]

Nanoparticles administered via injection are reported to enter the systemic circulation and central nervous systems. For instance, engineered carbon nanoparticles and nanotubes were found to enhance vascular thrombosis in rat carotid arteries.[84] Rats that were injected with 20 mg kg^{-1} TiO_2 nanoparticles every two days for 20 days exhibited an accumulation of nanoparticles in the liver, lung, and brain, suggesting the ability of nanomaterials to translocate and biodistribute to various organs.[85]

It is now recognized by the breadth of research studies that materials with low solubility or degradation profiles may accumulate within biological systems and persist there for long durations. Studies have shown that bacteria and living cells are capable of taking up nanosized particulates,[86] which may lead to potential bioaccumulation in the food chain.[87] Further, the presence of nanoparticles has been shown to enhance the bioaccumulation of other toxic materials such as Cd in aquatic species.[88] Studies have also demonstrated that nanostructures may also be transferred to offspring, which may even affect future generations of species (Fig. 6.12).[89]

(iii) *What are the molecular mechanisms responsible for nanotoxicity?*

Exposure to various nanostructures may result in either direct (oxidative cellular stress) or indirect (inflammation-mediated) cytotoxicity.[90] As expected, these effects are more pronounced for susceptible populations, such as the elderly, diseased, or pregnant females.[91] The small size of nanoparticulates allow for their penetration into regions such as the mitochondria and nucleus of the cell, the latter of which would yield direct interaction with DNA. Although production of reactive oxygen species (ROS) is a normal cellular process that is used in cellular signaling and the defense

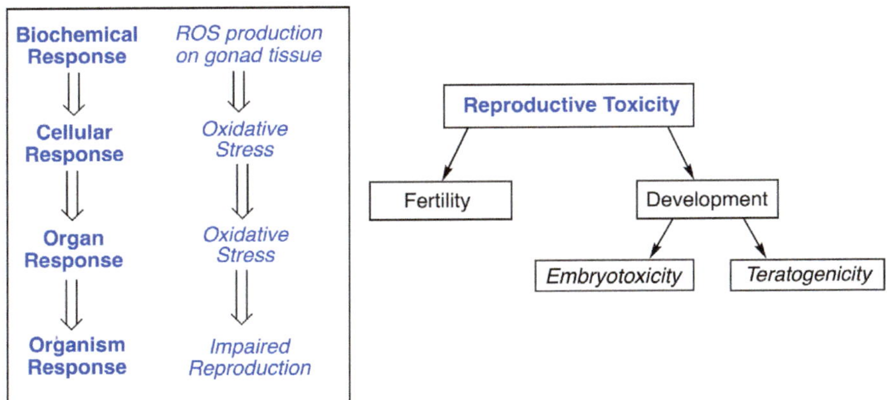

Fig. 6.12 Illustration of reproductive toxicity of nanoparticles. Adapted from *Int. J. Med.* **2021**, *148*, 103,384. © 2021 Elsevier

mechanism of the immune system, in excess it has been reported to cause severe damage to cellular macromolecules such as proteins, lipids, and DNA (Fig. 6.13).[89,92] Studies of Au nanoparticles with dimensions between 20 and 250 nm have shown that large surface area particles caused increased concentrations of ROS.[93] Both in vitro and in vivo studies with silica nanoparticles have also confirmed ROS production, which has induced pro-inflammatory responses and genotoxicity.[94] However, the degree of toxicity is affected by the type of surface functionalization.[95] Beyond silica, ROS generation has now been reported for a wide variety of nanoparticles, resulting in varying degrees of biotoxicity.[96] However, it is noteworthy that not all nanoparticles induce oxidative stress. Intracellular ROS generation depends on the physical and chemical properties of the nanoparticle, such as the degree of crystallinity and surface functionalization, which influences its adsorption profile and solubility.[97] Hence, particle size alone does not provide a direct prediction for the degree of toxicity of nanoparticles due to nanoparticle-induced oxidative stress.

Thus far, four different modes of ROS generation by nanomaterials have been identified:

- ROS generated directly from the surface of nanoparticles
- Transition metal ions catalyze oxygen-based metabolic products into more reactive hydroxyl radicals such as ·OH
- Nanoparticles trigger mitochondria dysfunction, leading to an imbalance of the respiratory chain and disturbed ROS signaling
- Macrophages and neutrophils produce ROS when activated by nanoparticles.

To avoid apoptosis (programmed cell death), cells initiate DNA repair mechanisms involving the use of enzymatic antioxidants (e.g., gluthathione reductase, catalase) and non-enzymatic antioxidants (e.g., glutathione, vitamin C, vitamin D) to scavenge ROS. However, if left unbalanced, the oxidative stress caused by nanoparticles can lead to DNA damage such as chromosomal fragmentation, strand breakage, and gene mutations.[98] For instance, it has been shown that 30–45 nm metal oxide nanoparticles such as TiO_2, ZnO, Fe_3O_4, Al_2O_3, and CrO_3 are able to induce apoptosis in mammalian cells, in which the extent of DNA damage was too severe to be overcome by inherent repair mechanisms.[99]

Since nanomaterials may be recognized by receptors as pathogens in the body, a variety of immune responses may also be triggered, leading to release of inflammatory protein mediators such as interleukins and chemokines.[100] Exposure to lipid-based nanoparticles has also been shown to lead to hypersensitivity reactions, and even anaphylaxis.[101] However, some studies have shown that even *non-metabolized* nanomaterials may result in granulomas,[102] lesions (areas of damaged cells or tissue),[103] and cancer.[104] There is also evidence from animal studies that some nanoparticles can cross the placenta, posing particularly significant risks to developing embryos.[105] However, even if the nanoparticles are not able to cross blood–brain or placental barriers, there still may be cellular damage. A mechanism elucidated by Bhabra et al. has shown that Co-Cr nanoparticles of *ca.* 30 nm diameter are able to cause damage to human fibroblast cells, even across an intact cellular barrier.[106]

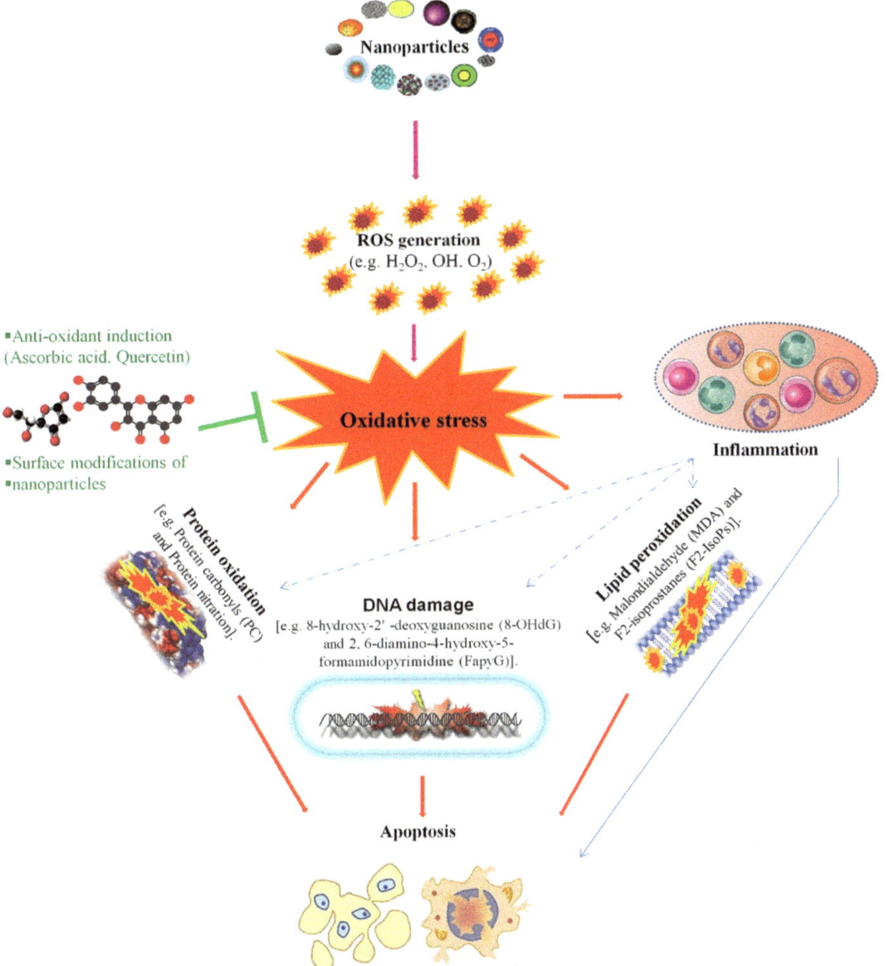

Fig. 6.13 Schematic of nanotoxicity that is caused from the generation of free radicals that induce oxidative stress. Reproduced with permission from https://doi.org/10.3389/fphar.2017.00606

(iv) *What is the fate, transport, and transformation of nanosized materials after they enter the environment?*

Unfortunately, the majority of nanostructures currently used for applications consist of non-biodegradable subunits. Further, nanostructures often exhibit a low solubility, may be transported long distances by wind, water, and wildlife, and have a strong propensity to form aggregates. All of these factors suggest that nanomaterials will have a lasting presence within the environment. In general, the fate of a nanomaterial within an environmental medium such as soil and water will be governed by the inherent properties of the nanostructure discussed above, as well as its reactivity with

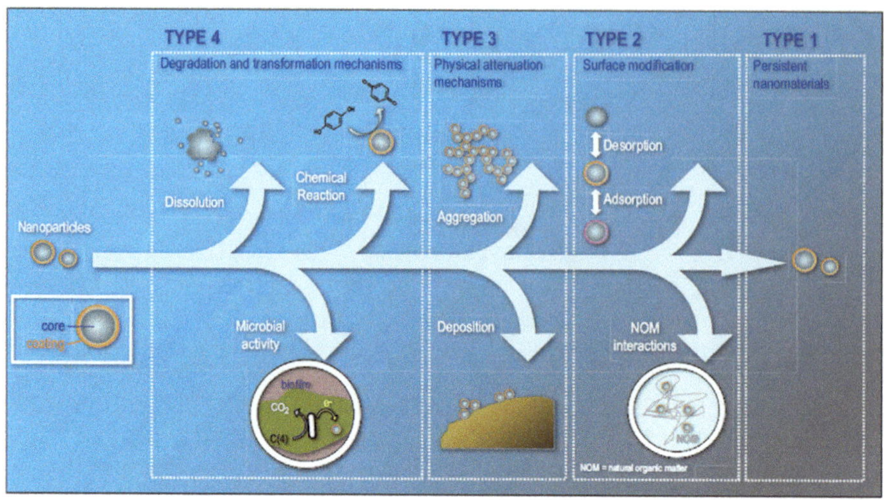

Fig. 6.14 Possible nanoparticle modifications within the environment. Reproduced with permission from *ACS Nano* **2009**, *3*, 1616. © 2009 American Chemical Society

the particular medium that might affect its degree of agglomeration transformation into a more or less stable material (Fig. 6.14).

Based on the current literature, it is clear that metal-based nanoparticles can cause toxicity to biota in the environment.[107] As shown in Fig. 6.15, exposure routes for plants include nanoparticle-embedded soils or growth media, from foliar sprays, or through the epidermis of roots (via symplastic or apoplastic regions). Not unlike human exposure, the size, degree of aggregation, and surface functionalization of nanostructures govern their uptake by plants.[108] Due to differences in plant species and growth conditions, the bioaccumulation factor (BAF; i.e., concentration of metal in plant shoot or root ÷ exposure concentration in growth medium) is often used to compare metallic nanoparticle uptake data for higher plants.

Studies have shown that metal-based nanoparticles can release ions into the substrate (soil or hydroponic growth solution), which may then undergo biotransformation to other species. For instance, Zn phosphates were formed from the exposure of wheat shoots to ZnO nanoparticles[109]; ZnS was found in wheat shoots exposed to ZnO nanoparticles.[110] However, it is not clear if nanotoxicity is due to exposure to the nanoparticles themselves, or from the release of ions and subsequent biotransformation. More studies are needed to assess whether Cu nanoparticles or ions are initially taken up by the plant species.

Morphological alterations of root systems have also been reported from exposure to metallic nanoparticles[111]—possibly more serious in plant species than ROS-related effects, since this alters nutrient transport and assimilation. Although it has been reported that nanoparticles can induce DNA damage in plants,[112] its mechanism, severity, and reversibility are still unknown.

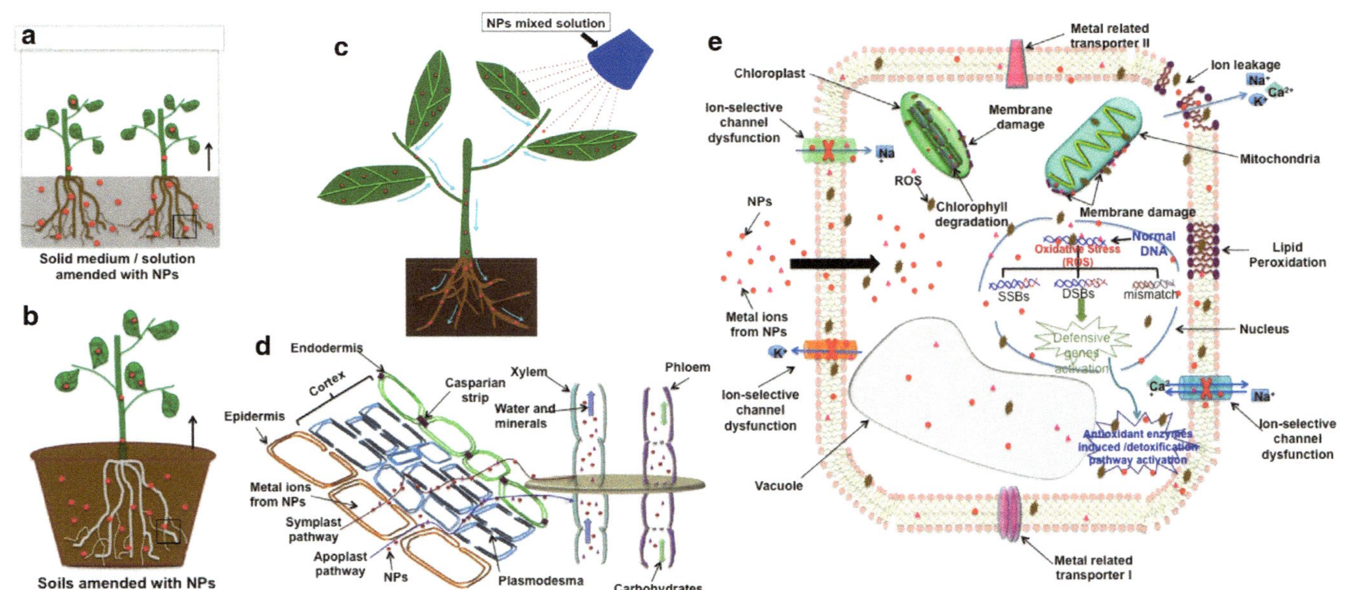

Fig. 6.15 Schematic of possible routes of nanoparticle uptake by roots and cell damage caused by nanoparticle exposure. Shown are: **a** plants grown in medium amended with nanoparticles; **b** plants grown in soils amended with nanoparticles; **c** nanoparticles introduced to plants via foliar spray; **d** nanoparticle introduction via symplast and apoplast; **e** potential damage could be caused by nanoparticles at the cellular level, but other damage at molecular/biochemical levels (e.g., DNA damage and transcription levels of genes) involved in nanoparticle detoxification pathways are also presented. Reproduced with permission from *Environ. Sci. Technol.* **2015**, *49*, 7109. © 2015 American Chemical Society

Perhaps the most compelling question is whether nanomaterials can accumulate at each level in the food chain. It is well-known that heavy metals such as Hg, Pb, or Cd may bioaccumulate; however, there are only limited studies that have investigated the effects of nanoparticles.[113] Figure 6.16 illustrates the possible trophic transfer of metal-based nanoparticles in the food chain; the majority of studies to date have focused on the aquatic ecosystem. Although evidence confirms the accumulation of nanoparticles at each trophic level, there are conflicting reports regarding whether biomagnification may occur along the food chain in either aquatic or terrestrial species.[114] Thus, more studies are needed to assess the fate and behavior of nanomaterials in the food chain.

To summarize this section, although the number of studies devoted to assessing nanotoxicity has increased by about 200% since 2000, there is still much to be learned about the possible dangers of nanostructures for human health and the environment. The data thus far are contradictory in many respects, which may depend on the processing parameters of the nanostructures used within particular in vitro/in vivo tests. In an attempt to define the risks in a more tangible and coordinated way, there has been a recent attempt at developing a risk-based classification system for nanomaterials.[115] However, research in this area must continue to expand in order for us to truly understand the risks that may accompany this exciting "nanofrontier".[116]

To complicate matters even further, in addition to the above toxicological-based questions, a number of ethical considerations must also be addressed[117]:

- Equity issues: will nanotechnology be utilized to solve third-world problems, or will it primarily be used to increase the prowess of industrially advanced countries?
- Privacy issues: imagine a world where you have invisible sensors/microphones—need we say more…
- Security: how will our country and others defend itself against invisible nanoweaponry?
- Human–machine interactions: there are many religious and philosophical issues associated with embedded nanodevices within the human body.

The introduction of a new architecture such as nanomaterials necessitates the need for new terminology and methods of classification and characterization. We must also understand the mechanisms by which individual nanostructures may assemble into larger materials, as this will greatly affect the properties of the bulk device for a particular application. This chapter will focus on all of these important issues, with an introduction to the various types of nanomaterials, laboratory techniques used for their synthesis, and (perhaps most importantly) their role in current/future applications.

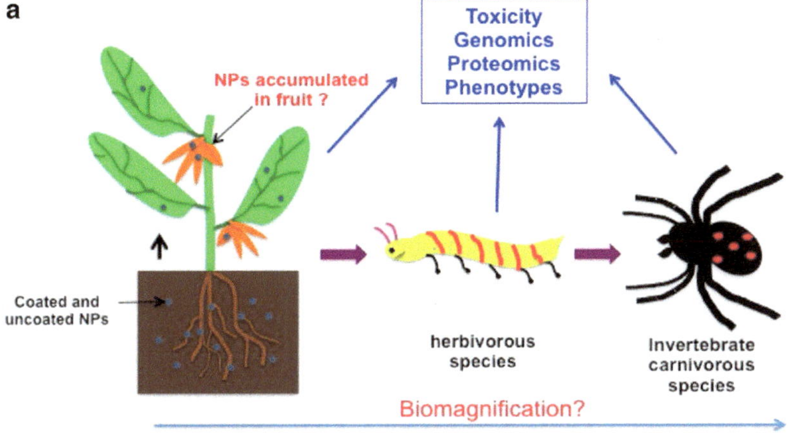

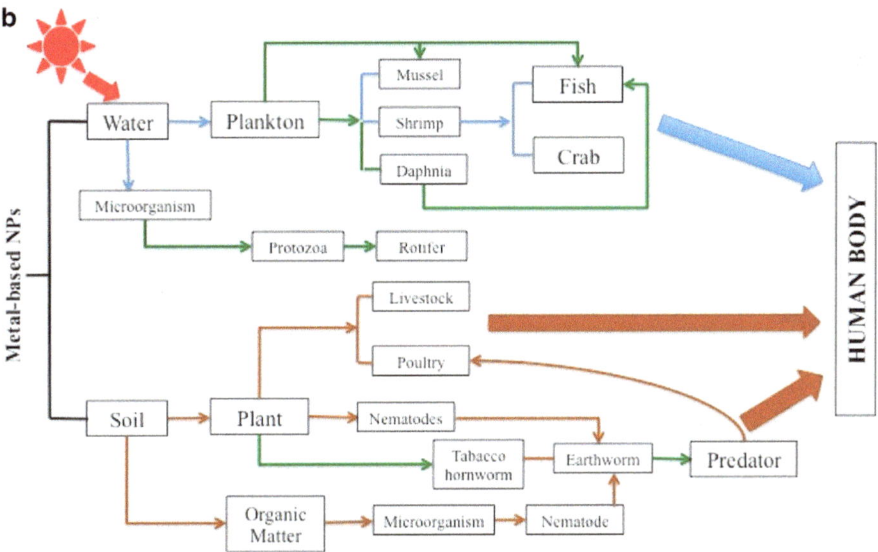

Fig. 6.16 Flowchart summarizing the possible trophic transport routes of metal-based nanoparticles in food chain ecosystems. Blue arrows show the possible food chains in the aquatic environment; brown arrows are for the possible food chains in the soil system; green arrows represent trophic transfers of nanoparticles in food chains that have already been established. Reproduced with permission from *Environ. Sci. Technol.* **2015**, *49*, 7109. © 2015 American Chemical Society

6.2 What is "Nanotechnology"?

Although there is much excitement and hype[118] regarding nanomaterials, there is really nothing new about nanoscience. In fact, the earliest civilizations used nanoscale materials for a variety of applications. For example, the Mayans used a magnesium aluminum silicate clay called palygorskite, which contained nanosized channels that were filled with water. The Mesopotamian civilizations used colored glass for decorative applications that contained embedded metallic nanoparticles.

Physics Nobel Laureate Richard Feynman gave the first lecture regarding the applications for nanoscale materials. His talk, entitled "There's Plenty of Room at the Bottom," was delivered on December 29, 1959 at the annual American Physical Society meeting on the campus of Caltech. Appendix 2 contains a transcript of his entire talk, which contains references to a future world that was never before imagined. Feynman pointed out that designing materials atom-by-atom is a real possibility, as it would not violate any physical laws. He also predicted such sci-fi accomplishments as writing 24 volumes of the Encyclopedia Brittanica on the head of a pin, and even more amazingly, the complete reproduction of *every book ever produced* in a small handheld pamphlet of less than 40 pages! To put these prophetic statements into context, at the time he delivered this speech, computers filled an entire room (Fig. 6.17) and carried a price tag of over $1 million.

The first use of the term "nanotechnology" was by Norio Taniguchi in 1974 at the International Conference on Precision Engineering (ICPE). His definition referred to "production technology to get extra high accuracy and ultra fine dimensions, i.e., the preciseness and fineness on the order of 1 nm (nanometer), 10^{-9} m in length."[119] Although many definitions for nanotechnology have been suggested, NASA has provided the most thorough description:

Fig. 6.17 Photo of the room-sized URAL-2 computer system in Tashkent, Uzbekistan, 1965. © emkaplin/Shutterstock.com

> The creation of functional materials, devices and systems through control of matter on the nanometer length scale (1–100 nm), and exploitation of novel phenomena and properties (physical, chemical, biological) at that length scale.[120]

Although Feynman and Drexler certainly popularized nanotechnology, their influence did not directly lead to the design of nanoscale materials. Rapid progress in nanotechnology could only take place after the arrival of sophisticated instrumentation, capable of viewing and manipulating materials on the nanoscale. In the 1980s, scanning probe microscopy (SPM) was developed which allowed scientists to fulfill Feynman's vision of pushing individual atoms around a surface (Fig. 6.18).

SPM was co-invented by Calvin Quate and Hemantha Kumar Wickramasinghe. Interestingly, when Quate and Binnig first submitted their work to the peer-reviewed journal *Physical Review Letters*, it was rejected due to such far-fetched claims as being able to measure forces on individual atoms. However, these results were eventually published, directly influencing the future of molecular nanotechnology. The 1986 Nobel Prize in Physics was awarded to Gerd Binnig and Heinrich Rohrer to honor their design of the scanning tunneling microscope (STM). They shared the Prize with Ernst Ruska, the inventor of the first electron microscope, another essential tool for the modern nanomaterials scientist. In fact, the resolution of modern electron microscopes are now high enough to provide images of individual atoms, and are often fitted with detectors that are capable of determining the chemical composition and/or oxidation state of the surface atoms. Chapter 7 will describe these and other instruments that are commonly used for materials-related research and development.

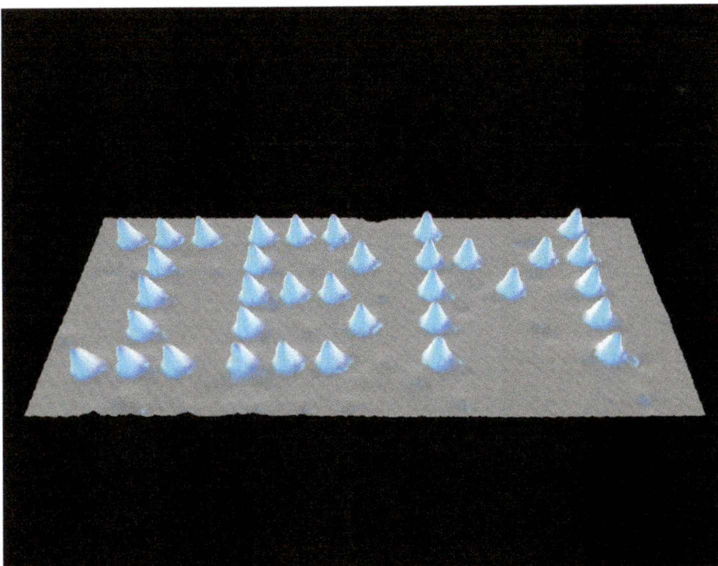

Fig. 6.18 Scanning tunneling microscope image of the placement of individual Xe atoms on a Ni(110) surface by researchers at IBM. Reproduced with permission from *Nature* **1990**, *344*, 524. © 1990 Macmillan Publishers Ltd.

6.3 Nanoscale Building Blocks and Applications

The first question almost everyone new to the nanoregime asks is "why are nanomaterials so special?" The leading advantage of this size regime is the relatively large surface area/volume ratio exhibited by nanomaterials; that is, with decreasing size, an increasingly larger percentage of (coordinatively unsaturated) atoms reside at its surface (Fig. 6.19 and Table 6.3). Accordingly, this translates to a very high surface reactivity with the surrounding surface, ideal for catalysis or sensor applications. Further, since biological systems feature the systematic organization of nanoscale materials (e.g., proteins are 1–20 nm in size, the diameter of DNA is *ca.* 2.5 nm), being able to fabricate materials in this size regime holds promise for artificial components within cells (that have *ca.* 10,000–20,000 nm diameters) to diagnose/fight diseases, illnesses, and viruses.

Another key benefit for nanomaterials is their ability to vary fundamental properties (e.g., magnetization,[121] optical properties (color),[122] melting point (Fig. 6.20),[123] hardness,[124] etc.), without a change in chemical composition. Properties such as decreased melting point, enhanced solubility, and high hardness are related to an increase of surface energy generated by the high density of surface atoms.[125] In contrast, the size-tunable electronic properties are due to quantum-confinement effects, as discussed in later in this chapter.

The melting point of nanoparticles is directly proportional to their cohesive energy, which is the sum of the bond energy of *n* atoms comprising the nanoparticle. As such, models may be constructed to predict the melting point of nanoparticles based on

Fig. 6.19 Comparison of the surface area/volume ratio of macroscopic particles (marbles) and nanoscopic aluminum oxide particles. Since nanoparticules contain a proportionately large number of surface atoms, there are a significantly greater number of adsorption/reaction sites that are available to interact with the surrounding environment. Further, whereas bending of a bulk metal occurs via movement of grains in the >100 nm size regime, metallic nanostructures will have extreme hardness, with significantly different malleability/ductility relative to the bulk material. © U.S. FDA

Table 6.3 Comparison of the % surface atoms for gold nanoparticles[a]

Particulate radius (nm)	Total # of atoms	Surface atoms (%)
0.65	79	76
0.71	116	67
0.81	140	69
0.87	201	64
0.98	225	62
1.1	309	52
1.2	459	51
1.4	807	43
1.5	976	40
1.6	1289	37
2.0	2406	31
2.2	2951	30
2.4	4033	27
2.6	4794	26
2.8	6266	23

[a] Data taken from Hostetler, M. J.; Wingate, J. E.; Zhong, C. -J.; Harris, J. E.; Vachet, R. W.; Clark, M. R.; Londono, J. D.; Green, S. J.; Stokes, J. J.; Wignall, G. D.; Glish, G. L.; Porter, M. D.; Evans, N. D.; Murray, R. W. *Langmuir* **1998**, *14*, 17. Au nanoparticles were stabilized by dodecanethiol, and sizes were controlled by varying the sodium borohydride reduction temperature and rate of addition, and the molar ratio of dodecanethiol: $HAuCl_4 \cdot xH_2O$

their size and shape by exploiting potential functions such as Lennard–Jones, Mie, Morse, and others.[126] There are three hypotheses that describe the depressed melting point of nanostructures relative to their bulk counterparts (Fig. 6.21).[127] The first, known as *homogeneous melting* (HM) assumes that no surface melting occurs. In contrast, *liquid skin melting* (LSM) assumes that a liquid layer is formed around a solid core at a relatively low temperature. The nanoparticle is thought to remain in that state until it is completely transformed into a liquid at its melting point. The third process is referred to as *liquid nucleation and growth* (LNG), which is a surface-melting phenomenon wherein a liquid layer is formed at lower temperatures and grows with increasing temperatures. Considering these three hypotheses, the general expression for the nanostructural melting temperature, $T_{m,n}$, may be expressed as:

$$\frac{T_{c,n}}{T_{c,b}} = 1 - \frac{2V(\gamma_{sv} - \gamma_{lv})\alpha}{\Delta H_f(D - 2\delta)} \tag{6.1}$$

where $T_{c,n}$ and $T_{c,b}$ = melting points of nanostructures and bulk, respectively;

γ_{sv} and γ_{lv} = surface energies of the solid–vapor and liquid–vapor interfaces of the material, respectively;

ΔH_f = bulk latent heat of fusion
D = average diameter of the nanostructures
ρ_s and ρ_l = densities of the solid and liquid, respectively
$\alpha = 2$ for LNG; $\alpha = 3$ for HM and LSM
$d > 0$ for LSM and <0 for either HM or LNG

Since we live in a macroscopic world, the next generation of materials will be of similar physical dimensions as today's consumer products. That is, we have shrunk down the size of cell phones and computers to almost their useful limits—any further, and one would need to use a sharp stylus to dial a phone number! However, as articulated in Chap. 4, although the size of electronic devices will remain somewhat constant, the speed and computational ability of these devices must continue to increase. This translates to materials that are built from the ground up, one nanoscale building block at a time. However, it is synthetically too expensive (and not industrially scaleable) to arrange such small units into their desired positions by hand. Consequently, materials chemists are largely focused on "bottom-up" techniques that afford the *self-assembly* of nanoscale species. As we will see later in this chapter, parallel efforts in "top-down" processing are being developed by materials engineers to yield nanoscale building blocks and devices through advanced milling, lithographic, ablation, or etching techniques (Fig. 6.22). In this respect, one can consider a nanoscale object as being "mesomolecular" or "mesoatomic"—an aggregate of smaller molecular/atomic subunits.

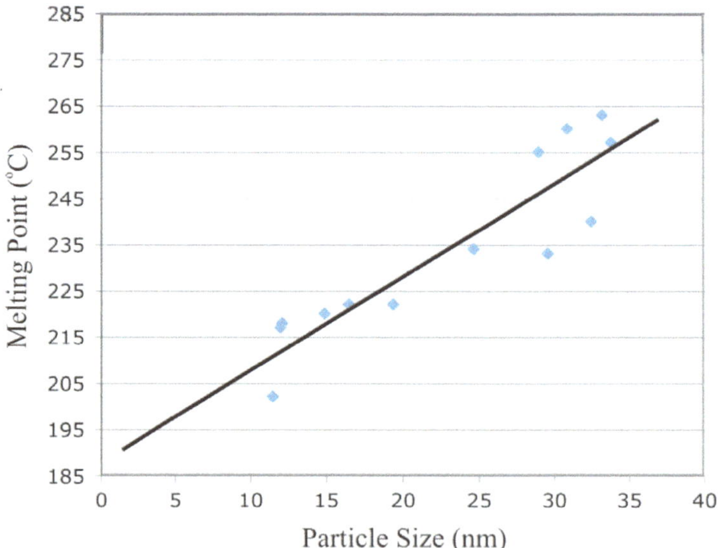

Fig. 6.20 Decrease in the melting point of gold nanoparticles with decreasing diameter. It should be noted that the melting point of bulk gold is 1064 °C! Adapted with permission from Unruh, K. M. et al. "Melting Behavior in Granular Metal Thin Films," *Materials Research Society Symposium Proceedings*, vol. 195. Materials Research Society, Apr 16–20, 1990. © 1990 Materials Research Society

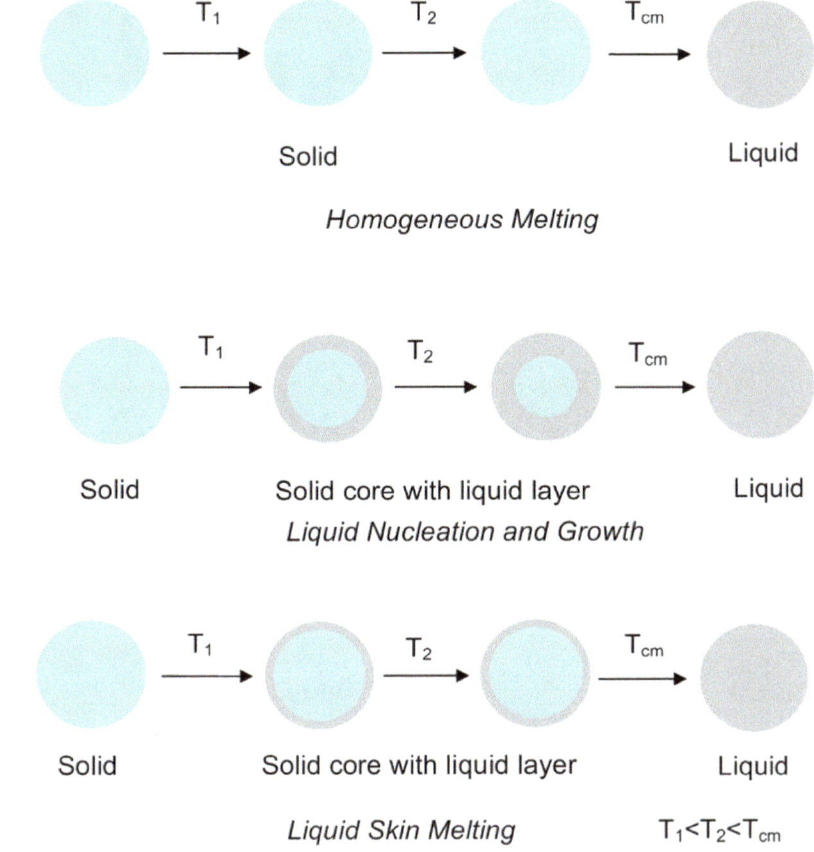

Fig. 6.21 Illustration of nanoparticle melting hypotheses. Reproduced with permission from *Pramana—J. Physics*, **2009**, *72*, 617. © 2009 Indian Academy of Sciences

There are three primary types of nanoscale building blocks that may be used for device fabrication and applications:

(i) 0-D (e.g., nanoparticles, nanoclusters, nanocrystals, quantum dots)
(ii) 1-D (e.g., nanotubes, nanofibers, nanowires, nanoribbons)
(iii) 2-D (e.g., thin films (see Chap. 4), graphene sheets).

The direct incorporation of these nanoarchitectures into existing materials to improve their properties is often referred to as *incremental nanotechnology*. However, as we will see later in this chapter, the self-assembly of nanosized building blocks into 3-D architectures may yield entirely new devices and functionalities—referred to as *evolutionary nanotechnology*.

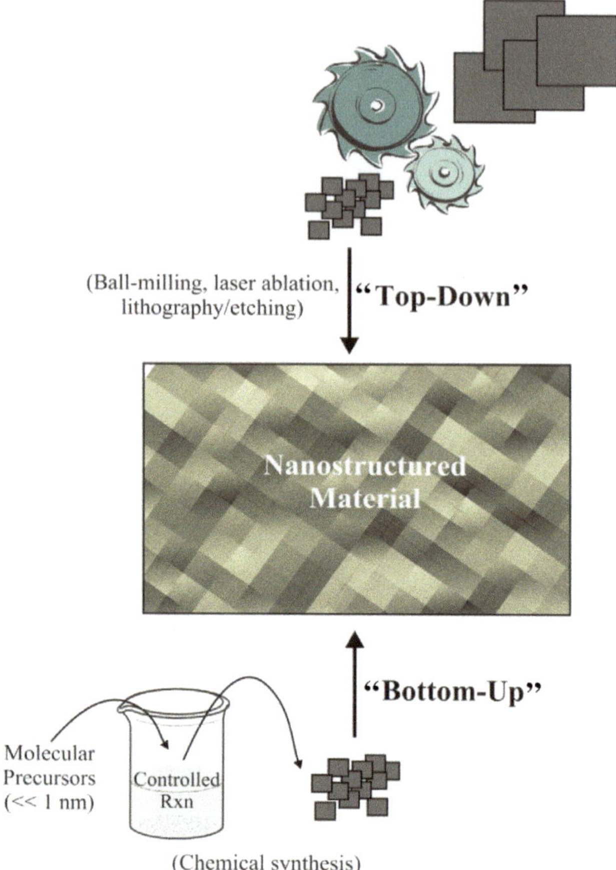

Fig. 6.22 Schematic comparison of the "top-down" and "bottom-up" approach to nanomaterials synthesis

6.3.1 Zero-Dimensional Nanomaterials

Analogous to the period in this sentence, a "zero-dimensional" structure is the simplest building block that may be used for nanomaterials design. These materials have diameters <100 nm and are denoted by nanoparticles, nanoclusters, or nanocrystals, which are used synonymously in the literature. In order to avoid nomenclature ambiguities, a primary goal of this chapter is to provide explicit definitions and examples of nanostructures (Fig. 6.23).

The term *nanoparticle* is often used to encompass all 0-D nanosized building blocks (regardless of size and morphology), or those that are amorphous and possess a relatively irregular shape. Herein, we will define nanoparticles as amorphous or

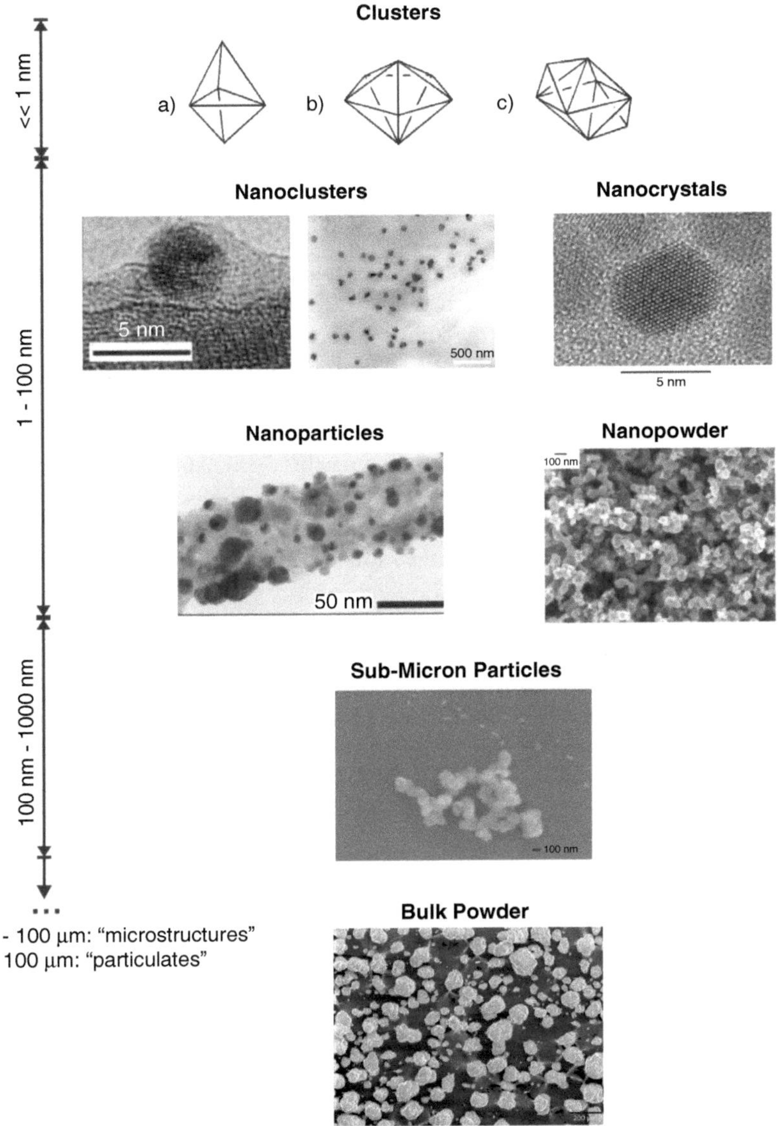

Fig. 6.23 0-D nanostructure nomenclature. Shown are the well-defined cage sizes of molecular *clusters* $(Os_5(CO)_{16})$, $(Os_6(CO)_{18})$, $[Os_8(CO)_{22}]^{2-}$.[128] Comparatively, *nanoclusters* should be used to describe 0-D nanostructures of a homogeneous size distribution.[129] By contrast, *nanoparticles* exhibit a greater range of sizes/shapes.[130] *Nanocrystals* are characterized by the presence of an ordered lattice array of the constituent subunits, as illustrated by a single nanocrystal of CdSe.[131] In stark contrast to a nanocrystal, an example of a *nanopowder* is shown that consists of microscopic grains, each comprised of nanoscale amorphous units.[132] The size regime that is intermediate between the nano- and microregimes is best referred to as *submicron*.[133] The bulk powder scale bar is 200 μm

semicrystalline 0-D nanostructures with dimensions larger than 10 nm, and a relatively large (≥15%) size dispersion. For amorphous/semicrystalline nanostructures smaller in size (i.e., 1–10 nm), with a narrow size distribution, the term *nanocluster* is more appropriate. This distinction is a simple extension of the term "cluster", which is typically used in inorganic/organometallic chemistry to indicate small molecular cages of fixed sizes.

Analogous to bulk materials, the agglomeration of noncrystalline nanostructural subunits should best be termed a *nanopowder*. Although the gross dimensions of these particulates are typically much greater than 100 nm, they will exhibit a large specific surface area (SSA), which may be easily determined by a particle-size analyzer (e.g., BET) that quantifies the surface adsorption of an inert gas such as nitrogen. Using this parameter, one can then properly define a nanopowder as having a specific surface area >60 m^2/g, which corresponds to the SSA of solid spheres of unit density with a diameter of exactly 100 nm.

It is also important here to note the difference between nanoparticles/nanoclusters and traditional *colloids*, which date back to the early 1860s (Table 6.4). We are all familiar with the term colloid, which is used to describe solid/liquid and solid/gas suspensions such as milk, paints, butter, smoke, and smog. Although colloids may feature particle dimensions within the nanoregime, particle sizes typically extend into the micron size range. Furthermore, one generally has less control over the composition and morphology of colloids relative to engineered nanostructures. As we will see shortly, in order to stabilize metal nanostructures, a stabilizing agent must be used to prevent agglomeration into a larger powder. This is also the case for colloids, which generally employ polydispersed organic polymers and other ionic species that may adsorb to the entrained colloid surface. Such a variation in the nature of the encapsulating environment leads to a large dispersity in overall morphology and properties of colloids. By contrast, in order for nanomaterials to be used for "bottom-up" design, their synthesis and resultant properties must be more reproducible. This is easily accomplished through the use of stabilizing agents with well-defined structures (e.g., dendrimers, polyoxoanions, etc.) that do not react with/surface deactivate the entrained nanostructures.

Quantum Dots: Properties and Applications

Thus far, we have defined nomenclature for amorphous 0-D nanostructures. Analogous to bulk materials, any nanomaterial that is crystalline should be referred to as a *nanocrystal*. However, this term should be reserved for those materials that are single-crystalline; if a particle exhibits only regions of crystallinity, it is better termed a nanoparticle or nanocluster depending on its dimensions. Transmission electron microscopy (TEM), especially in tandem with electron diffraction, is most useful in determining the crystallinity of any nanostructure (Fig. 6.24).

There is a special class of nanocrystals known as *quantum dots* (QDs).[136] Typically, the dimensions of these semiconductor nanostructures lie in the range 1–30 nm, based on their composition (see below). Quantum dots currently find applications as sensors, lasers, LEDs (e.g., QLED TVs, LED lighting), nanophotonic devices,[137] and fluorescent biomarkers (Fig. 6.25). In fact, high-density disks (e.g.,

Table 6.4 Comparison of 0-D nanoarchitectures with traditional colloids[134]

Nanoparticles/nanoclusters	Colloids
Size: 1–100 nm (nanoclusters: 1–10 nm)	Typically >10 nm
Homogeneous molecular composition ≤15% Size dispersion (less polydispersity for nanoclusters relative to nanoparticles)	Poorly defined compositions >15% Size dispersion
Reproducible synthesis (control over size, shape, and composition)	Nonreproducible, uncontrollable morphology/composition
Reducible physical properties and catalytic activity	Nonreproducible properties (*esp.* irreproducible catalytic activities[135])
Soluble in polar/nonpolar organic solvents (depending on stabilizing agent)	Typically only soluble in polar solvents
Contain clean surfaces	Contain surface-adsorbed species such as –OH, –X, $-OH_2$, etc.

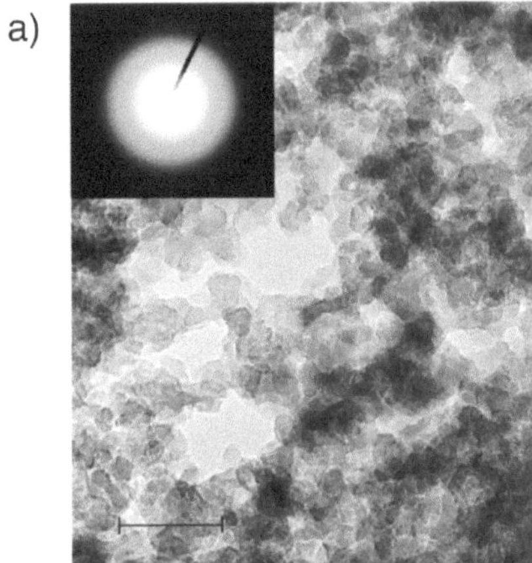

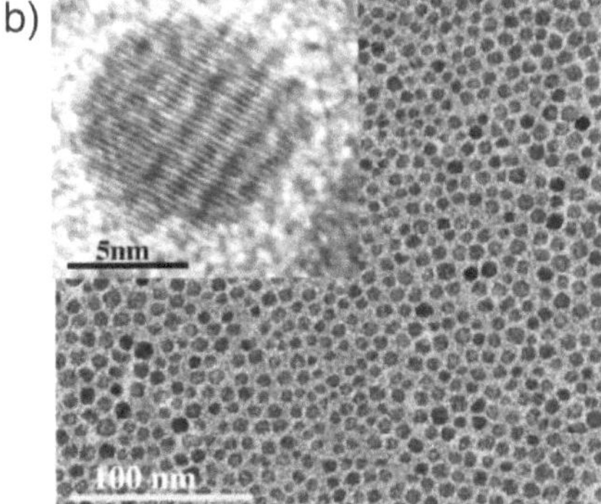

Fig. 6.24 TEM images of amorphous nanoclusters (**a**), and nanocrystals, (**b**). The inset of (**a**) shows selected area electron diffraction (SAED); the absence of a pattern indicates an amorphous structure. The scale bar is 20 nm. The high-resolution TEM image inset in (**b**) shows lattice spacings of an individual nanocrystal. Reproduced with permission from **a** *Chem. Mater.* **2003**, *15*, 2803. **b** *Chem. Mater.* **2006**, *18*, 465. © 2003 & 2005 American Chemical Society

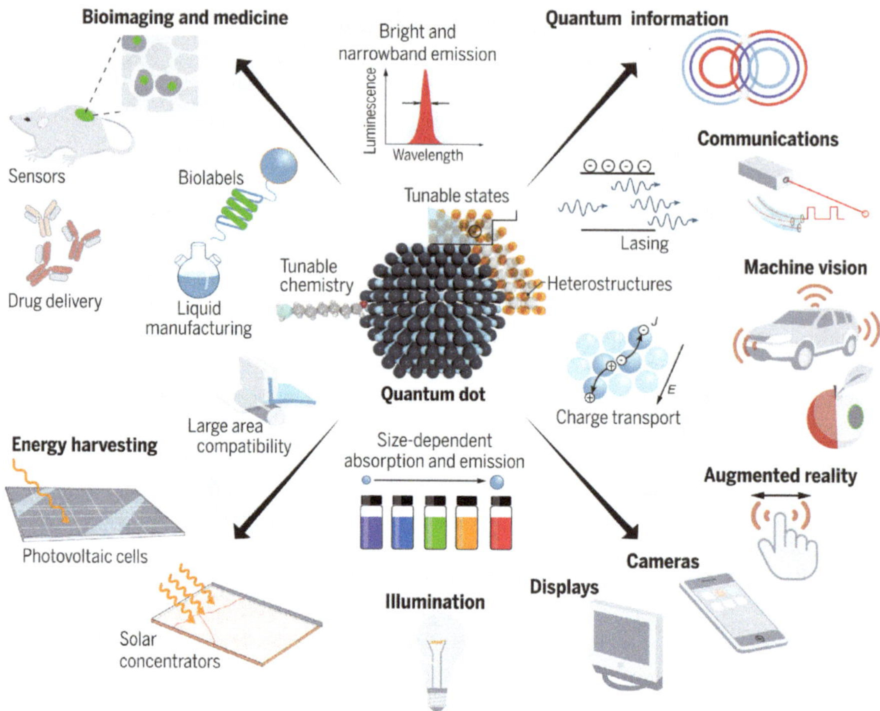

Fig. 6.25 Applications for quantum dots (QDs). Reproduced with permission from *Science* **2021**, *373*, 1. © 2021 AAAS

Blu-ray high-definition DVDs) may only be read via blue lasers, which are fabricated from quantum dots. Long-term applications for these structures will likely include quantum computing[138] and high-resolution camera sensors.[139]

As you may recall from Chap. 4, when an electron is promoted from the valence to conduction bands, an electron–hole pair known as an *exciton* is created in the bulk lattice. The physical separation between the electron and hole is referred to as the *exciton Bohr radius* (r_B) that varies depending on the semiconductor composition. In a bulk semiconductor crystal, r_B is significantly smaller than the overall size of the crystal; hence, the exciton is free to migrate throughout the lattice. However, in a quantum dot, r_B is of the same order of magnitude as the diameter (D) of the nanocrystal, giving rise to *quantum confinement* of the exciton. Empirically, this translates to the strongest exciton confinement when $D \leq 2r_B$.[140]

Analogous to the "particle-in-a-box" model from introductory physical chemistry, exciton quantum confinement results in discrete energy levels rather than the continuous bands of a bulk semiconductor crystal (Fig. 6.26). Since the gap between adjacent energy levels is infinitesimally small for a bulk semiconductor, the bandgap is considered a fixed value. In contrast, since the dimensions of a quantum dot are extremely small, the addition/subtraction of a single atom will significantly change

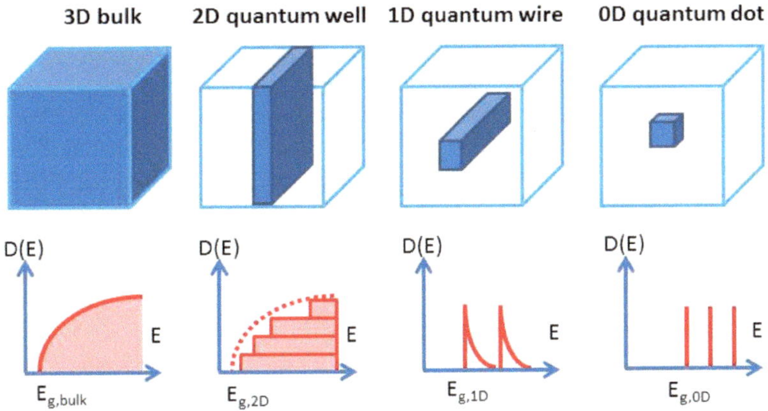

Fig. 6.26 Density of states (DOS) for electrons in 3-D bulk semiconductors, 2-D quantum wells, 1-D quantum wires, and 0-D quantum dots). Reproduced with permission from *Chem. Rev.* **2019**, *119*, 3140. © 2019 American Chemical Society

the nanocrystal dimensions and bandgap. Table 6.5 lists the r_B values for common semiconductor crystals whose bandgap may be easily fine-tuned by simply changing the diameter of the quantum dot, as long as the dimensions are smaller than r_B.

In the early 1980s, Efros described the size-dependent electronic properties of quantum dots, first delineating that the bandgap, E_n, will increase from the bulk value based on a $1/R^2$ confinement energy term (6.2). According to quantum confinement theory, electrons in the conduction band and holes in the valence band are spatially confined by the potential barrier of the surface. Due to confinement of both electrons and holes, the lowest energy optical transition from the valence to conduction band will increase in energy, effectively increasing the bandgap.

Table 6.5 Calculated exciton Bohr radii for common semiconductors

Material	r_B (Å)
Si	55
CdS	28
CdSe	61
CdTe	100
ZnO	18
ZnS	50
PbS	204
PbSe	460
InAs	340
InSb	540

$$E_n = E_g + \frac{h^2\pi^2}{2\mu R^2} \tag{6.2}$$

where E_g is the bandgap of the bulk semiconductor, h is Planck's constant, R is the radius of the quantum dot, and μ is the reduced mass of the exciton given by $m_e m_h/(m_e + m_h)$. Here m_e and m_h are masses of the electron and hole, respectively.

This model was expanded by Brus and coworkers to include Coulombic interaction of excitons and the correlation energy (6.3):

$$E_n = E_g + \frac{h^2\pi^2}{2\mu R^2} - \frac{1.786e^2}{\varepsilon R} + 0.284E_R \tag{6.3}$$

where E_R is the Rydberg (spatial correlation) energy of bulk semiconductor:

$$E_R = \frac{\mu e^4}{2\varepsilon^2 {\varepsilon_o}^2 h^2} = 13.56\left(\frac{\mu}{\varepsilon^2 m_e}\right)$$

where ε_0 is the permittivity of free space, ε is the dielectric constant of the bulk semiconductor, and m_e is the mass of the electron.

Hence, the absorption energy of quantum dots will shift to a higher frequency with decreasing radius of the dots, with a dependence of $1/R^2$. For example, CdSe QDs with a diameter of 1.2–1.5 nm have a bandgap of 3.0 eV; in contrast, 3.0–3.5 nm CdSe QDs have a bandgap of 2.3 eV, which is closer to its bulk value (1.8 eV). This is readily determined from the observed colors of quantum dots with varying radii, shifting from blue to red with increasing size (Fig. 6.27).[141]

However, due to their high surface-to-volume ratio, the surrounding environment (e.g., solvent, ligands, etc.) has a dramatic effect on the optoelectronic properties of QDs. These effects generally result from more dangling bonds at the surface of the nanocrystal, which lead to the formation of surface traps known as *mid-gap states*, which provide pathways for nonradiative exciton recombination and resultant loss of photoemission efficiency (Fig. 6.28).[142] A common method used to eliminate surface traps involves passivation of the surface by organic ligands. However, these ligands may be weakly bound, which results in a dynamic equilibrium of adsorbing/desorbing ligands. Hence, a more stable method to passivate the surface of a QD is the growth of a shell with different bandgap and/or bandgap offset relative to the core (Fig. 6.29).[143]

The choice of shell material greatly affects the optical properties of the core/shell QD (Table 6.6), which serves to move the surface defects to the outer vicinity of the inorganic shell, thereby impeding electron–hole recombination and improving the photoluminescence quantum yield (PLQY) of the quantum dot.[144] However, lattice mismatch between the core and shell materials may induce strain, which may also degrade overall optoelectronic properties.[145] For Type I core/shell nanocrystals are formed by coating the core QD with a larger bandgap material, allowing for charge confinement within the core. For Type II core/shell nanocrystals, the band alignment is designed so that one charge carrier is localized in the core and the other carrier in the shell.[146] Due to spatial separation of the charge carriers, there is reduced electron–hole overlap and longer exciton radiative lifetimes.[147] Another more recent variety

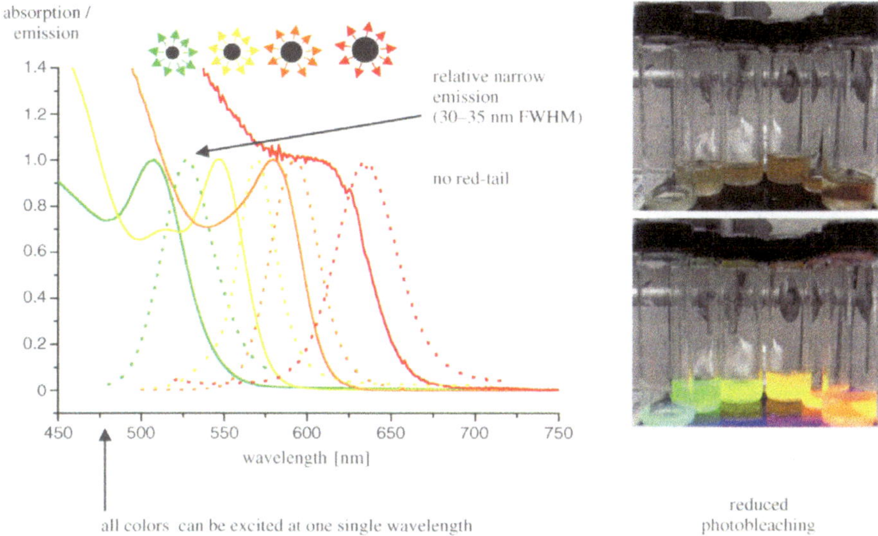

Fig. 6.27 The size-dependency on the optical properties of CdSe nanocrystals. With decreasing size, the fluorescence peak is shifted to shorter wavelengths. Also shown is the reduced photobleaching exhibited by CdSe nanocrystals—(top–bottom photos): suspension of nanocrystals in solution, before and after irradiation with UV light, respectively. Reproduced with permission from *Small*, **2005**, *1*, 48. © 2005 Wiley–VCH

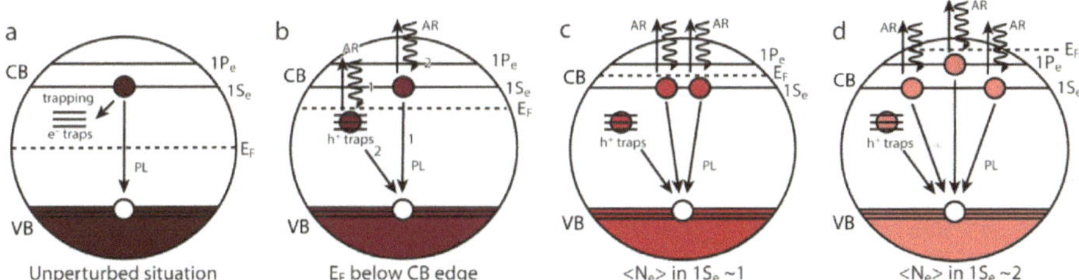

Fig. 6.28 Mechanisms for radiative and nonradiative recombination pathways in CdSe/CdS quantum dots. The illustrations of **a–d** show the different processes that occur after photoexcitation occurs when the Fermi level (E_F) is: **a** in the middle of the bandgap, **b** just above the mid-gap states within the bandgap, **c** above the $1S_e$ conduction band edge, and d) above the $1P_e$ conduction band. **a** The photogenerated electron can get trapped in states within the bandgap, which leads to nonradiative recombination or radiative recombination with the hole in the valence band. **b** An electron within a midgap state can get excited to higher energy states within the conduction band when the electron and hole recombine, without emitting a photon, through trap-assisted Auger recombination (AR)—pathway 1, or the photogenerated hole can get trapped within the mid-gap state, which promotes the photogenerated electron higher up in the conduction band through AR—pathway 2. **c** A negative *trion* (three charged particles of electrons and holes) and **d** negative *tetron* (four charged particles of electrons and holes) result in additional radiative and nonradiative recombination pathways, which compete against each other depending on the number of electrons per nanocrystal. Reproduced with permission from *Chem. Mater*, **2019**, *31*, 8484. © 2019 American Chemical Society

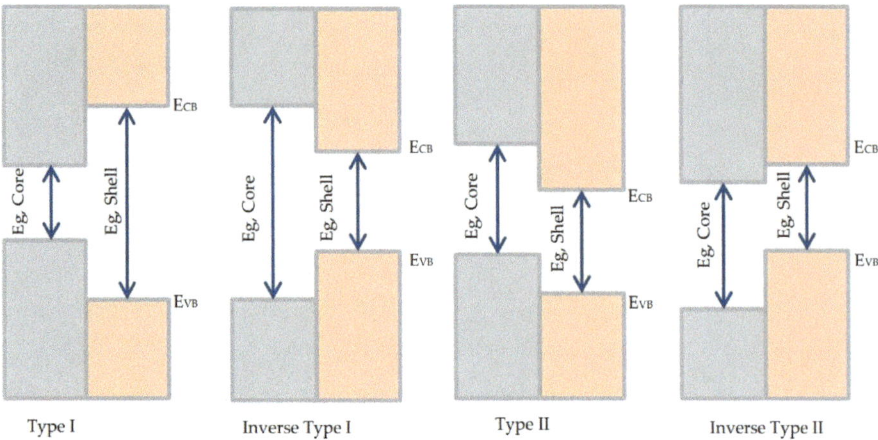

Fig. 6.29 Comparison of various core/shell quantum dots based on their relative band alignment. Reproduced with permission from *Energies* **2019**, *12*, 1. © 2019 MDPI

of core/shell nanocrystal is known as Type I–III–VI2 chalcogenides.[148] These QDs exhibit NIR emission profiles, with narrow bandgap (*ca.* 1 eV), high PLQY of 65–75%, and large Stokes shift between absorption/emission bands. A large Stokes shift is desirable in applications such as solar concentrators since this lowers the possibility of photon reabsorption.[149] However, in photovoltaics, the Stokes shift leads to a loss in the open-circuit voltage, thereby decreasing the power conversion efficiency.[150] Some examples include $CuInS_2$/MS, $CuInSe_xS_{2-x}$/MS, and $CuInS_2Se_{1-x}$/MS, and $AgInS_2$/MS, where M = Zn or Mn.

The tunability of the optical and electrical properties of QDs, as well as their solution processability and facile scale-up make them amenable for photovoltaic applications[151]; for instance, for building integrated photovoltaic (BIPV) systems.[152] The most mature solar cells composed of crystalline Si, CdTe or CIGS thin-film cells require complicated, expensive, and energy-intensive manufacturing processes. However, there is much interest in low-cost alternatives that feature inexpensive, nontoxic, abundant, and environmentally friendly materials. In Chap. 4 we discussed one variety—dye-sensitized solar cells (DSSCs), which have a current efficiency of over 13%. However, both DSSCs and other organic-based cells suffer from possible electrolyte leakage, long-term stability, and difficulties with large-scale fabrication. In an attempt to expand the limited absorption profile of organic dyes, *quantum dot sensitized solar cells* (QDSSCs) were developed that exploited the broadband absorption of QDs—even extending into the near-infrared region.[153] The decoration of QDs on the surface of a wide bandgap semiconductor (e.g., TiO_2 or ZnO) enables charge separation; with further optimization, the current maximum efficiency of QDSSCs has risen to over 14% in recent years.[154] However, this architecture has a theoretical power conversion efficiency (PCE) of up to 44%.[155] Especially problematic is the use of a liquid electrolyte that has limited commercialization success due to issues with QD corrosion and electrolyte leakage. However, solid-state varieties are currently in development to alleviate some of these issues.[156] Other limitations in efficiency are

Table 6.6 Comparison of various types of core/shell quantum dots[149]

Parameter	Type I	Inverse Type I	Type II	Inverse Type II
Bandgap of core versus shell	Bandgap of core is smaller than shell; the bandgap of core falls within bandgap of shell	Bandgap of core is greater than shell; the bandgap of shell falls within bandgap of core	Bandgaps of core and shell are identical; the valence band edge of core is within the bandgap of the shell OR conduction band edge of shell is within the bandgap of core	Bandgaps of core and shell are identical; the valence band edge of shell is within the bandgap of the core OR conduction band edge of core is within the bandgap of shell
Position of excited electrons/holes	Electrons and holes are confined in the core region	Electrons and holes are completely or partially confined in the shell, depending on shell thickness	One charge carrier is confined to the core, while the other is mostly confined to the shell	One charge carrier is delocalized in the core/shell structure; the other one is confined within the core
Quantum yield (QY)	Higher QY and long-term stability	Lower QY and poor stability	Lower QY and poor stability	Relatively higher QY and fair stability
Stokes shift	Small	Significantly large	Large	Large and tunable via controlling the size of core and thickness of shell
Avg. absorption range (nm)	400–500	400–500	600–800	300–1600
Avg. emission range (nm)	430–600	400–700	700–1000	700–1000
Limitations	The shell can trap charge carriers which reduces the fluorescence QY	Both charge carriers may leak to the surface reducing QY	One of the charge carriers may leak to the surface	The charge carriers may be absorbed resulting in reduced excited decay time
Examples	CdSe/ZnS, CdSe/CdS, CdS/ZnS	CdS/HgS, CdS/CdSe, ZnSe/CdSe	CdTe/CdSe, CdSe/ZnTe, CdSe/ZnSe	InP/CdS, PbS/CdS

related to the low QD coverage on the TiO_2 photoanode surface, which leads to low light-harvesting capacity and charge recombination in the device. A secondary deposition method was introduced by Song and coworkers to increase the QD loading, which resulted in a record PCE of 15.3% for Zn-Cu-In-S-Se (ZCISSe) QDSSCs.[157]

To circumvent problems with a liquid electrolyte, and complexities/inefficiencies associated with hybrid QD-polymer solar cells, Schottky QD solar cells were first demonstrated by Nozik et al. in 2008.[158] These designs feature thin QD layers sandwiched between an Ohmic-contact transparent electrode (ITO) and low-work-function metal electrode (e.g., Mg, Ca, Ag, or Al). Since the Shockley limit of 31% may be extended to >45% for multijunction solar cells, many groups have stacked multiple layers of QDs of varying dimensions to harvest photons at different wavelengths to improve its overall efficiency.[159] However, Fermi-level pinning at the metal-QD layer interface severely limits the open-circuit voltage that is possible from Schottky architectures. Furthermore, the thickness of the metallic layer has been reported to negatively impact the light absorption of the semiconductor.[160] This limitation has been addressed through the use of plasmonic nano-antennas such as metal nanostructures, or QD luminescent layers, which have led to enhancement of solar cells, including within the UV region.[161]

Although attempts have been made to reduce the density of electronic trap states in Schottky QD solar cells, bulk heterojunction QD architectures have been of more interest in recent years. This variety of solar cell features interpenetrating layers of QDs and a n-type wide-bandgap semiconductor, thus ensuring that photogenerated excitons are positioned close to charge-separating interfaces.[162] However, this may also lead to an increased likelihood for recombination events, which necessitates the careful engineering of the QD-metal oxide band offsets. The limitation on the built-in voltage in heterojunction QD devices has led to other more advanced architectures such as homojunction-like solar cells, in which both p- and n-doped materials are composed of QDs. However, n-type QD films is challenging since oxygen acts as a p-type dopant. A variety of synthetic strategies have been devoted to more effective passivation of the QD.[163] Methylammonium iodide (MAI) was used to effectively passivate the surface of PbS QDs, which resulted in a certified power conversion efficiency of 10.6%.[164]

Perhaps the most intriguing potential for using QDs in solar devices is the possibility for *multiple-exciton generation* (MEG).[165] When excitons have an energy greater than twice the bandgap of the semiconductor, they may relax to the band edge by exciting an additional electron–hole pair through impact ionization.[166] This would give rise to an external quantum efficiency (EQE) > 100%. Although early studies reported up to 300% MEG quantum yields for charge carrier generation in PbE (E = S, Se, Te) QDs,[167] this phenomenon has become a topic of controversy[168] since more recent studies have resulted in values significantly lower from those of earlier publications.[169] With the implementation of new designs (Fig. 6.30), it may be possible to exploit MEG to exceed the Shockley-Queisser limit for future QD-based solar cells.[170]

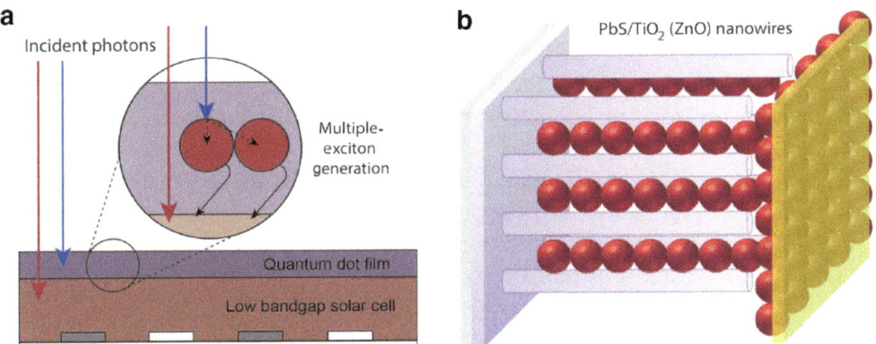

Fig. 6.30 Proposed next-generation solar cells that exploit multiple exciton generation (MEG) of quantum dots (QDs). Schematic **a** shows a bottom-contact solar cell. High-energy photons (blue arrows) are absorbed in the thin QD layer where they undergo MEG. The carriers are then transferred to the low bandgap semiconductor, thus contributing to the photocurrent. In addition, photons of lower energy are more likely to pass through the QD film due to the lower absorption cross-section and can be absorbed and converted into charge carriers within the bulk semiconductor. Schematic **b** shows a device composed of a nanostructured hetero-interface between a QD absorber and metal oxide acceptor layer. Reproduced with permission from *Nanophotonics* **2017**, *7*, 1. © 2017 De Gruyter

Optical and Magnetic Properties of Metallic Nanostructures

For anyone who has admired the bright red colors of stained-glass windows, it may be surprising to note that this too is a nanoparticle-related phenomenon. In fact, the reds and yellows observed in antiquated stained glass as far back as the early seventeenth century are caused by the presence of gold and silver nanoparticles, respectively. However, the use of gold nanoparticles may date back even further, as reports of "soluble" gold extractions date back to the fifth century BC in Egypt and China.[171] However, for metallic nanoparticles with diameters *ca.* > 2 nm, the operating principle is different from semiconductor quantum dots, since there is no bandgap between valence and conduction bands, and the energy states form a continuum analogous to bulk metal.

For these metallic nanostructures, another phenomenon known as *surface plasmon resonance* (often denoted as localized surface plasmon resonance, LSPR) is active for these structures, involving specific scattering interactions between the impinging light and the nanostructures. In particular, the oscillating electric field of the incoming light causes the coherent oscillation of the conduction electrons, resulting in a concomitant oscillation of the electron cloud surrounding the metal nuclei (Fig. 6.31).[172]

The leading theory that describes the scattering behavior of small spherical particles with light dates back to the work of Mie in the early 1900s (6.4).[173]

$$\mathrm{E}(\lambda) = \frac{24\pi \mathrm{N_A r^3} \varepsilon_\mathrm{m}{}^{3/2}}{\lambda \ln(10)} \left[\frac{\varepsilon_i(\lambda)}{(\varepsilon_r(\lambda) + 2\varepsilon_\mathrm{m})^2 + \varepsilon_\mathrm{i}{}^2(\lambda)} \right] \tag{6.4}$$

where E(λ) is the extinction (sum of absorption and scattering), N_A is the density of the nanostructures, r is the radius of the nanostructure, ε_m is the dielectric constant of the metallic nanostructure, λ is the wavelength of absorbing radiation, and ε_i, ε_r are the imaginary and real (respectively) portions of the λ-dependent dielectric function of the nanostructure.

This theory is still relevant today, alongside Raleigh scattering theory, to describe the colors of our sky, as well as the appearance of suspensions such as milk and latex paints. It should be noted that for non-spherical nanostructures, the denominator of the bracketed term above is replaced with:

$$(\varepsilon_r(\lambda) + \chi\varepsilon_m)^2$$

where χ may range from a value of 2 (perfect sphere) to 17 (e.g., for a 5:1 aspect ratio nanostructure) and beyond.[174]

Based on (6.4), the factors that govern the oscillation frequency (and the observed color) of metallic nanoparticles include: electron density (size/shape, Figs. 6.32 and 6.33), the effective nuclear charge of the nuclei, and the size/shape of the charge distribution (polarization effects, strongly affected by the dielectric constant of the metal). As you might expect, additional effects related to the resonance frequency/intensity are observed from the nanostructural composition and its surrounding solvent.[175] As such, noble metal nanoparticles have been used as tunable platforms for biological sensing (Fig. 6.34). It should be noted that LSPR contributes most strongly to the observed colors of the coinage metals (Cu, Au, Ag). Many other metals such as Pb, In, Hg, Sn, etc. also exhibit this phenomenon, but exhibit colorless solutions since either the plasmon resonance frequency lies in the UV region or the nanoparticles are readily oxidized under ambient conditions.

For the smallest of metallic nanoclusters with dimensions *ca.* <2 nm, the surface plasmon absorption disappears. Since so few atoms comprise discrete nanoclusters of this size, the spacings between adjacent energy levels (referred to as the *Kubo gap*, δ, 6.5) become comparable to the thermal energy, kT[176]—especially at lower temperatures and smaller nanocluster dimensions. This results in a shift in conductive properties of the nanocluster, from metallic to insulating, with decreasing size (Fig. 6.35).[177] For instance, photoelectron measurements of Hg_n clusters have shown

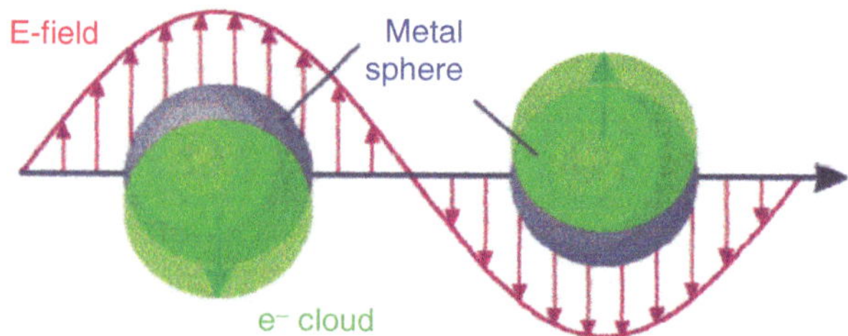

Fig. 6.31 Schematic of LSPR for a nanosphere, showing the induced oscillation of the electron cloud relative to the nuclei. Reproduced with permission from *J. Phys. Chem. B* **2003**, *107*, 668. © 2003 American Chemical Society

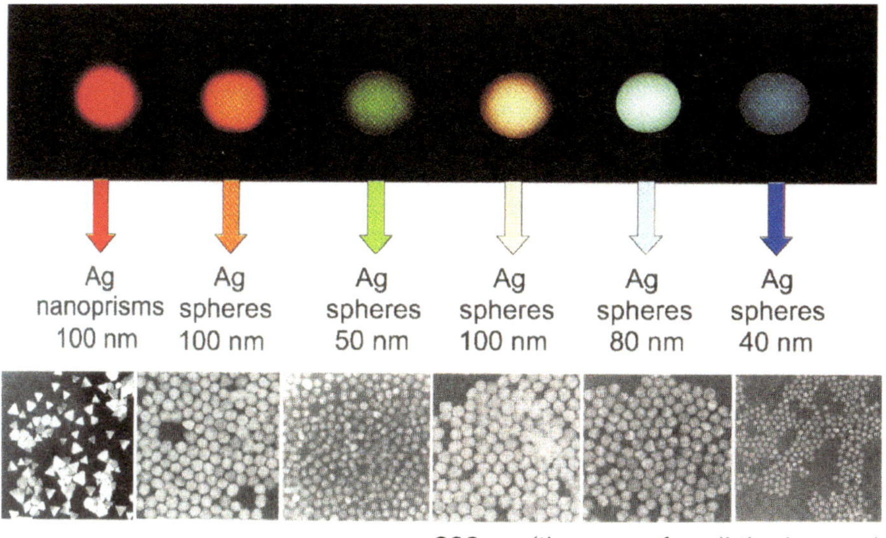

Fig. 6.32 The influence of size and shape on the light-scattering, and resultant colors, of silver nanoparticles. Reproduced with permission from *Small* **2005**, 1, 14. © 2005 Wiley–VCH

that a significant HOMO–LUMO gap of *ca.* 3.5 eV is present for Hg_3 clusters, relative to only *ca.* 0.2 eV for Hg_{250}; it is estimated that gap closure occurs for Hg clusters with $n \geq 400$ atoms.[178]

In band theory, the breadth of a band is directly related to the strength of interactions among nearest neighbors. For atoms/molecules (Fig. 6.35, far right) this interaction is weak, which corresponds to a narrow band. However, as additional atoms are added to the solid, stronger interactions will ensue, resulting in a greater density of states near the Fermi level—additional states near the band edges develop last. Hence, the insulating properties of very small nanoclusters are due to the small number of neighboring atoms, which are held together primarily by nonmetallic interactions (e.g., van der Waal forces).[179] As nanocluster size increases, the atomic s/p and d orbitals from a larger number of constituent atoms will broaden into bands, forming an energy continuum.

Quantum confinement effects also cause a change in the optical properties of metallic nanoclusters. For instance, since the spacing between interband energy levels increases with decreasing nanocluster size, the $5d^{10} \rightarrow 6(sp)^1$ band absorption for gold (Fig. 6.36) will become increasingly blue-shifted with decreasing dimensions (i.e., the observed color will be red-shifted).[180]

$$\delta(\varepsilon_F) \propto \frac{3}{2}\frac{\varepsilon_F}{N_A z} \tag{6.5}$$

where δ is the Kubo gap, energy spacing between adjacent E levels, N_A is the number of atoms in the nanocluster, z is their valence, and ε_F is the energy of the Fermi level (e.g., 8 eV for Au).

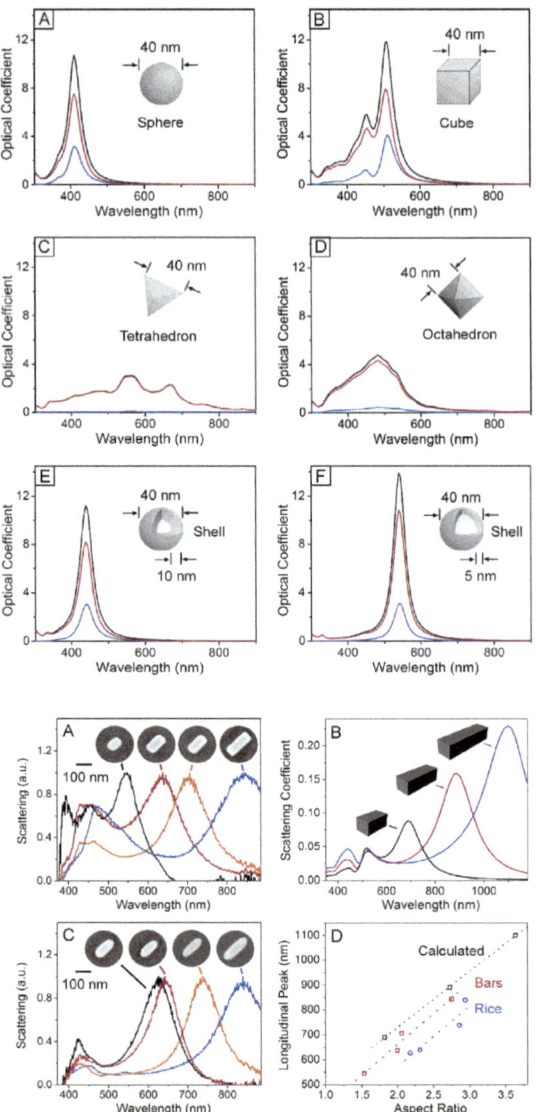

Fig. 6.33 Top: calculated UV–visible extinction (black), absorption (red), and scattering spectra (blue) of Ag nanocrystals, illustrating the effect of shape on spectral characteristics: **a** sphere, **b** cube, **c** tetrahedron, **d** octahedron, e) triangular plate, and f) circular plate. Reproduced with permission from *J. Phys. Chem. B* **2006**, *110*, 15666. Copyright 2006 American Chemical Society. Bottom: **a** SEM images of individual Ag nanobars and the corresponding normalized LSPR spectra. The longitudinal plasmon peak red-shifts with increasing aspect ratio for the nanobars; **b** calculated LSPR (scattering) spectra for Ag nanobars 100, 150, and 200 nm in length, keeping width = 55 nm and height = 50 nm; **c** SEM images of individual nanorice with the corresponding normalized LSPR spectra; **d** Plot of longitudinal plasmon peak location versus aspect ratio. The peaks of both nanobars and nanorice red-shift with increasing length, but on average the peaks of nanobars are 80 nm red-shifted from nanorice. Reproduced with permission from *Nano Lett.* **2007**, *7*, 1032. © 2007 American Chemical Society

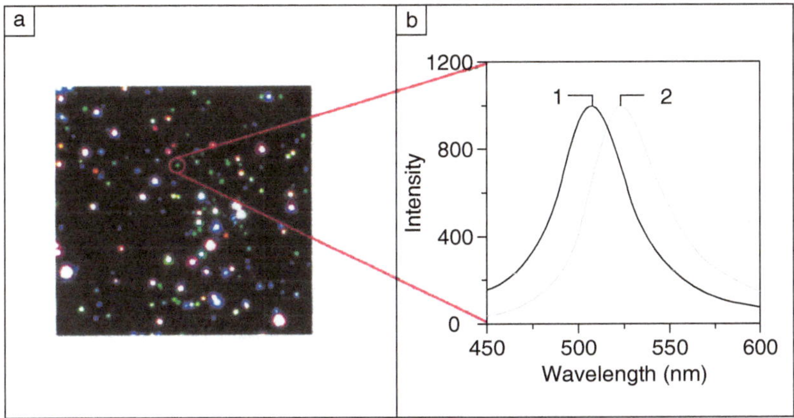

Fig. 6.34 Biological sensing using Ag nanoparticles. Shown is **a** a dark-field optical image of surface-functionalized (biotinylated) Ag nanoparticles, and **b** the shift in wavelength before (1) and after (2) exposure to 10 nM streptavidin. Reproduced with permission from *J. Fluoresc.* **2004**, *14*, 355. © 2004 Springer Science and Business Media

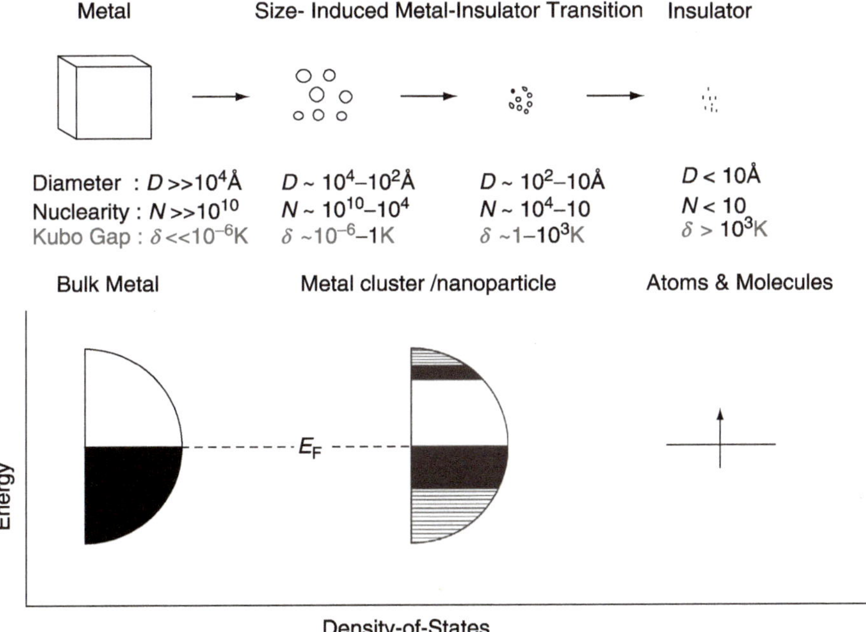

Fig. 6.35 Schematic of the density of states exhibited by bulk metal relative to increasingly smaller nanoclusters. The approximate diameter, nuclearity, and Kubo gap for each size regime are indicated. As the nanocluster size decreases, the energy continuum of the bulk metal is transformed into discrete energy levels, especially at band edges. Reproduced with permission from *Chem. Soc. Rev.* **2000**, *29*, 27. © 2000 Royal Society of Chemistry

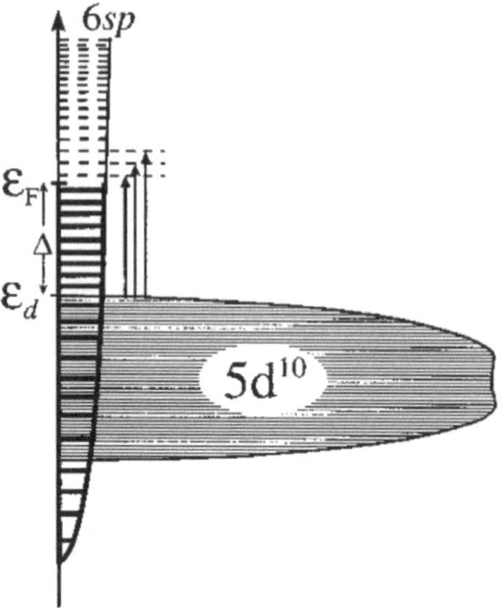

Fig. 6.36 Schematic illustration of the band structure (orbital energy vs density of states) of fcc-Au, indicating the approximately correct bulk bandwidths and displacements, and density of states (solid curve). The discrete levels are indicated by solid (broken) lines below (above) the Fermi level εF, with a (mean) level spacing appropriate to a clusters of 50 Au atoms. The vertical arrows designate optical transitions in the 1.7–2.5 eV range, originating at the top of the 5d band (εd) and terminating at each of the first few unoccupied levels. Reproduced with permission from *J. Phys. Chem. B* **1997**, *101*, 7885. © 1997 American Chemical Society

As you might expect, magnetism is also affected by decreasing particle size. Some applications that exploit the tunable magnetism of nanostructures include tunable viscous liquids (ferrofluids[181]) and biomedicine (e.g., drug-delivery, diagnostics, magnetofection, etc.). Relevant for magnetic storage applications,[182] bulk ferromagnetic materials such as α-Fe, Fe_3O_4, α-Fe_2O_3, and Ni will become paramagnetic and exhibit zero magnetization once their diameters reach critical sizes of 5 nm, 16 nm, 20 nm, and 15 nm, respectively.[183] Whereas bulk ferri- or ferromagnetic materials undergo a transition to the paramagnetic state above their Curie temperature, such *superparamagnetic* nanomaterials undergo this transition at a temperature below its Curie temperature. This occurs when the thermal Boltzmann energy at room temperature (k_BT = *ca.* 25 meV) is comparable to the magnetocrystalline energy (K_aV; K_a is the magnetocrystalline anisotropy constant and V is the magnetic nano-size volume—revisit Sect. 3.4).

As the dimensions of magnetic nanoparticles decrease, one must use a material with a large magnetic anisotropy[184] to prevent the random flipping of the magnetization direction under the influence of temperature. However, the very high coercivity of small grains with very large magnetocrystalline anisotropy make them hard to magnetize using standard writing technology.[185] Perpendicular recording devices using heat-assisted magnetic recording (HAMR) heads employ a laser to

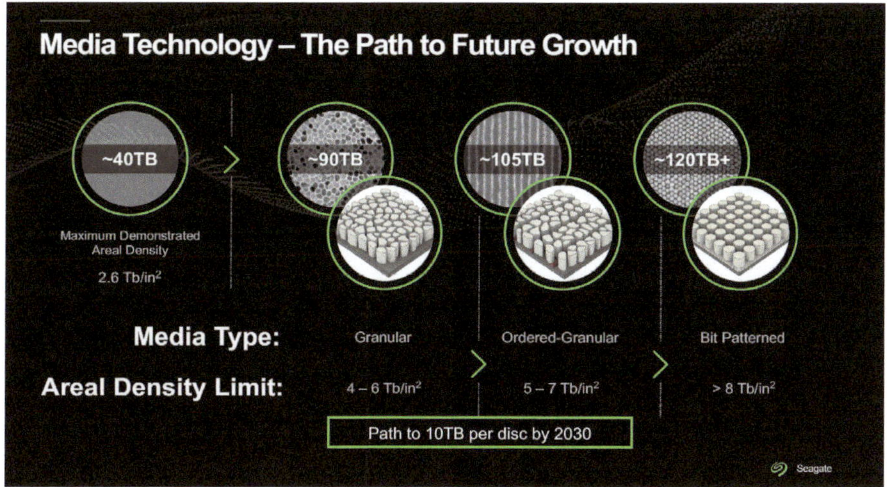

Fig. 6.37 Roadmap of magnetic recording media. © Seagate Corp

heat the part of the disk being written to, which helps to reduce the coercivity of the material.[186] However, beyond HAMR, the superparamagnetic limit may also be overcome by exploiting exchange bias effects between ferromagnetic and antiferromagnetic nanoparticles[187] or the magneto-electric effect in multiferroics.[188] The use of heated-dot magnetic recording (HDMR) or HAMR with bit-patterned media (Fig. 6.37) are being developed to extend areal densities beyond 10 Tb/in^2, which will require advanced patterning techniques to fabricate <10-nm rectangular dot media.[189]

The First 0-D Nanoarchitecture: The Fullerenes

Without question, our modern "nanotechnology revolution" was catalyzed by the mid-1980s discovery of carbon nanoclusters known as *fullerenes* (e.g., C_{60}, C_{70}, C_{84}, etc.).[190] The 1996 Nobel Prize in Chemistry was awarded to Richard Smalley, Robert Curl, and Sir Harold Kroto for this discovery, which focused the worldwide spotlight on unique nanoscale materials and their possible applications. It should be noted that the term fullerene does not simply mean the C_{60} carbon allotrope termed *Buckminsterfullerene* (or "Buckyballs"),[191] but rather the entire class of closed-cage carbon clusters that are comprised of exactly 12 pentagons, and a varying number of hexagons (e.g., 20 hexagons for C_{60}).

To date, the most common applications for fullerenes include MRI contrast agents (exploiting its container properties), drug-delivery agents (through surface functionalization), fulleride-based superconductors, and light-activated antimicrobial agents[192] (Fig. 6.38). However, Barron and coworkers at Rice University have recently shown that pyrolyzed fullerene-polyethyleneimine (PEI) composites are able to sequester carbon dioxide—of extreme importance for climate change. Their

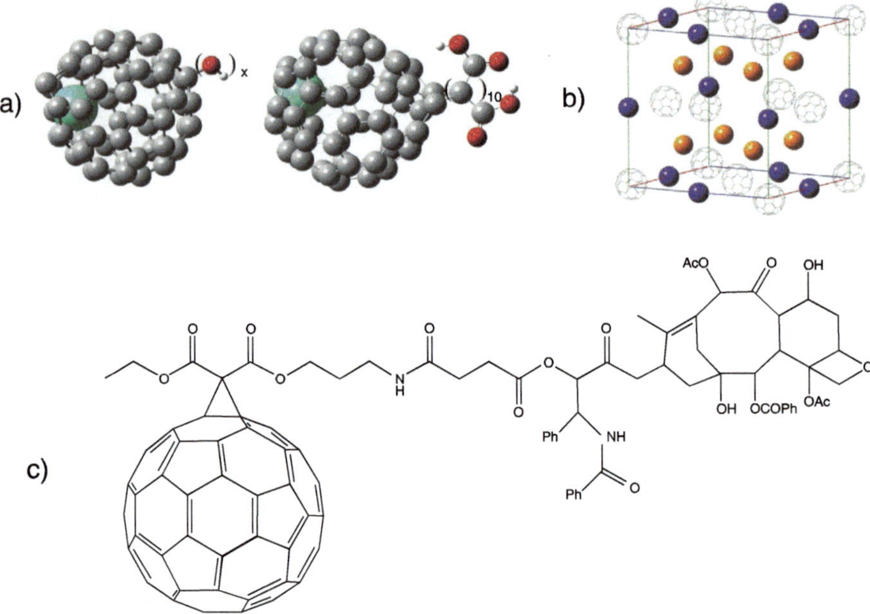

Fig. 6.38 Examples of fullerene applications. Shown are **a** Gd^{3+} @C_{60} − $[OH]_x (x \approx 27)$ and Gd^{3+} @C_{60} − $[CCOOH]_{10}$ used as MRI contrast agents,[195] **b** the ionic unit cell for the superconductive alkali metal fulleride $Cs_xRb_yC_{60}$,[196] and **c** the conjugate structure of C_{60} covalently bound to the lung cancer drug Paclitaxel[197]

work showed that the heat-treated C_{60} composite was able to absorb more than 10-times its weight in CO_2.[193] In another application related to environmental remediation, it has also been shown that hydroxyfullerenes $[C_{60}O_x(OH)_y)]^{n-}$ act as chelating agents for the preferential removal of divalent and trivalent ions in solution.[194] Some possible applications include the removal of toxic metals from acid mining waste or hydraulic fracturing fluids used in oil and natural gas production.

Another application for fullerenes that would rank high in the "cool" category is the development of surface-rolling molecules that are aptly termed *nanocars*.[198] Rather than stabilized nanoparticles, these nanovehicles are organic molecules that contain fullerenes as wheels (Fig. 6.39). The placement of the nanocars onto a gold surface is as easy as spin-casting from a toluene suspension. Due to strong adhesion forces between fullerenes and metal surfaces, the nanocars spontaneously deposit with all four "nanowheels" on the substrate. The axles consist of triple-bonded alkyne groups that allow rotation of the fullerene wheels, transporting the nanocar across a gold surface in response to changes in temperature.[199] Although only pivoting and translational motion are currently possible, this work sets an important step toward the realization of nanomachines for an endless number of possible applications.

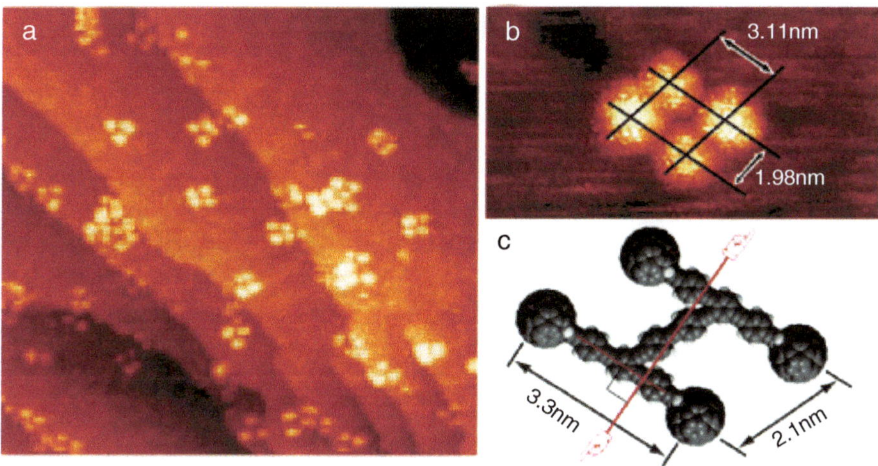

Fig. 6.39 Images and dimensions of a nanocar on a Au(111) surface. The bright features are the fullerene wheels. Reproduced with permission from *Nano Lett.* **2005**, 5, 2330. © 2005 American Chemical Society

As with other major discoveries such as Teflon and nylon, the synthesis of C_{60} was serendipitous (Fig. 6.40).[200] In fact, the experiments that led to the discovery of fullerenes were aimed at simulating the environment of a carbon-rich red giant star. As such, the discovery of C_{60} has been dubbed "the celestial sphere that fell to earth."[201] Their experimental system featured laser vaporization of a graphite target into a helium carrier gas wherein the atoms nucleated into clusters. The gas was then cooled using supersonic expansion and injected into a time-of-flight mass spectrometer for analysis (Fig. 6.41a). Reactive gases such as hydrogen or nitrogen could also be added to the carrier gas for the synthesis of other reaction products. It should be noted that this technique is now used to generate nanoclusters of a wide variety of metals, semiconductors (e.g., Si), and insulators (e.g., Al_2O_3) depending on the nature of the target and co-reactant gases employed.[202]

The first large-scale synthesis of fullerenes was discovered in 1989 by Huffman and Kratschmer[204]; this provided a macroscopic quantity of C_{60} in order to confirm the proposed icosohedral structure. Their technique consisted of the arc-evaporation of graphite electrodes via resistive heating within an atmosphere of *ca.* 100 atm helium (Fig. 6.41b).[205] Other more recent techniques such as high-temperature combustion of benzene and a benchtop graphite arc process (Fig. 6.42) were developed in order to reduce the cost and complexity associated with fullerene synthesis. Not only will such improvements allow for more widespread fabrication of fullerenes for research/applications (i.e., syntheses no longer limited to groups with supersonic laser and arc plasma systems), but may also result in lowering the cost due to a more straightforward industrial scale-up.

Buckyballs represent the smallest fullerene that obeys the *isolated pentagon rule* (IPR)—i.e., an energetic requirement that pentagons be surrounded by hexagons, so

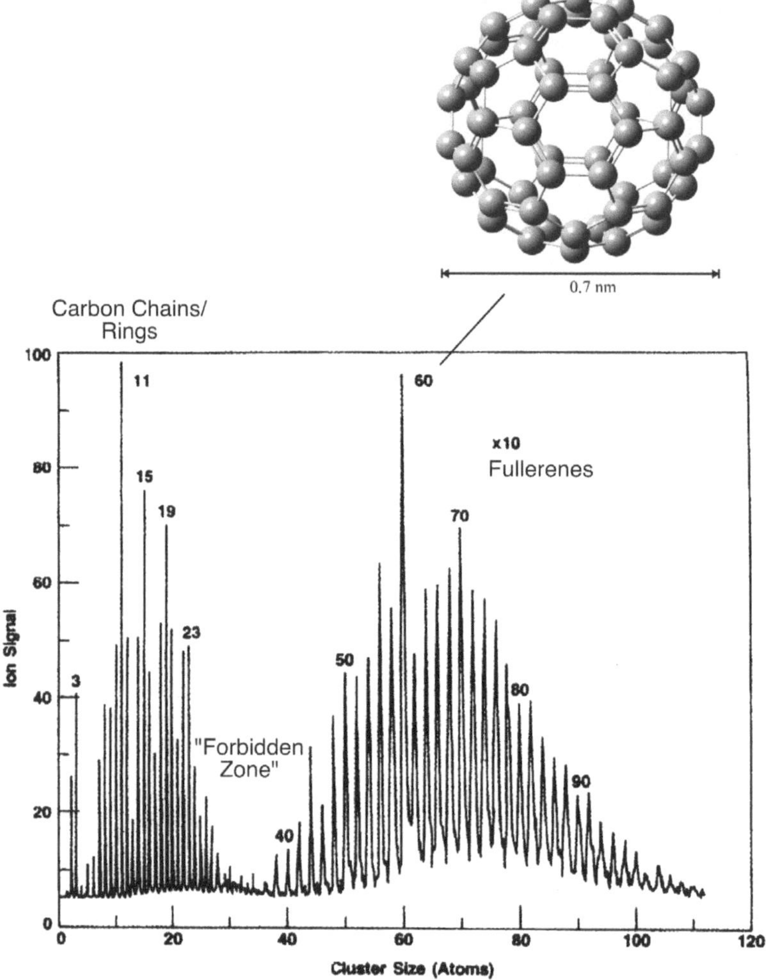

Fig. 6.40 Mass spectrum of carbon clusters in a supersonic beam, originally observed by Nobel Laureates Smalley and Curl. Reproduced with permission from *J. Chem. Phys.* **1984**, *81*, 3322. Also shown is the molecular structure of Buckminsterfullerene, C_{60}, containing alternating six- and five-membered rings of sp^2 hybridized carbon atoms. This is only one isomer for C_{60}, out of a staggering total of 1,812 possible structures[203]

that adjacent pentagons do not share an edge. Calculations show that π-bonds shared between six-membered rings have large positive bond resonance energies (BREs) and bond orders, indicating a high degree of aromaticity and stability/unreactivity. However, π-bonds between adjacent five-membered rings have large negative BREs with very small bond orders, indicating a much lower thermodynamic stability. Most likely, this difference is due to the increased ring strain that would be imposed by the fullerene structure as a result of two smaller rings directly adjacent to one

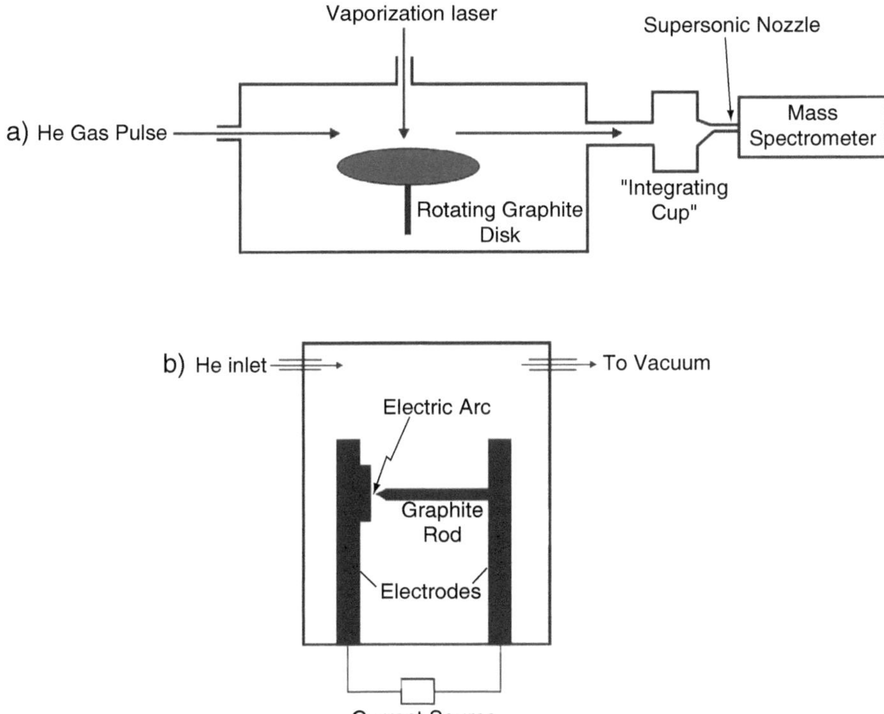

Fig. 6.41 Schematic of apparati first used to synthesize fullerenes. Illustrated are **a** the Smalley/Curl supersonic laser evaporation system, and **b** the Huffman/Kratschmer electric arc apparatus

another. Theoretical calculations indicate that the strain energy of the icosohedral Buckminsterfullerene structure is at least 2 eV lower than any other non-IPR isomer, of which there are over 1,800 possibilities.

Interestingly, it has been reported that adjacent pentagons containing at least one N atom instead of C (e.g., $C_{58}N_2$ rather than C_{60}), may actually be more stable than C_{60} (Fig. 6.43).[206] This apparent anomaly is a direct contradiction of the IPR. The most plausible explanation is the reduction of ring strain due to sp^3 hybridization of the N atom, as well as the addition of π-electron density (from the N lone-pair electrons) to the pentagons, resulting in an enhanced aromaticity/stability. To date, only short-lived azafullerenes $C_{59}N$ and $C_{58}N_2^-$ have been identified experimentally; the search continues for stable structures, since these will likely result in dramatically different properties and associated applications relative to their C-only analogues.

Although fullerenes have been actively investigated for more than three decades, there is an ongoing debate regarding the growth mechanism of these nanoclusters. Since the formation of fullerenes via laser/arc/combustion techniques occurs too rapidly to isolate intermediate species, most of the mechanistic proposals are based on theoretical techniques (quantum mechanical and molecular dynamics calculations).

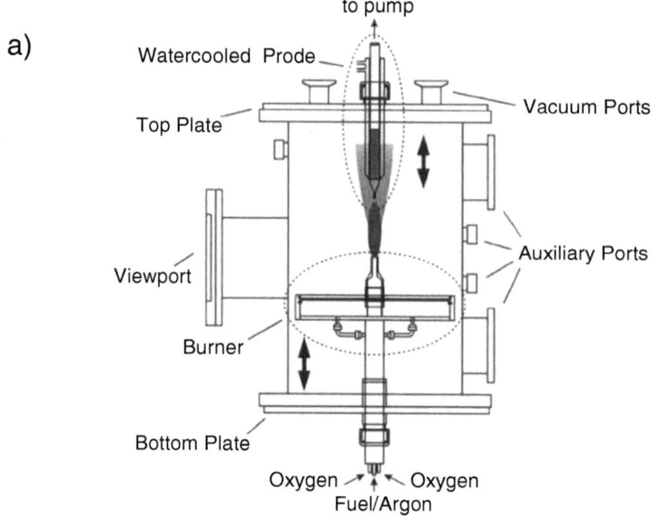

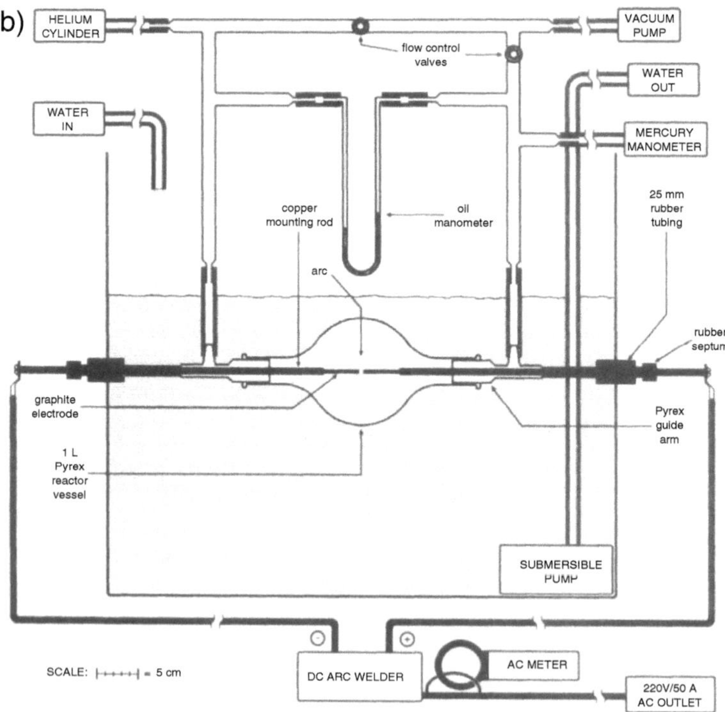

Fig. 6.42 Cross-section schematics of reactors used for fullerene synthesis. Shown are **a** a reduced-pressure fuel-rich pyrolytic chamber, and **b** a benchtop modified arc evaporation system. Reproduced with permission from **a** *Proc. Combust. Inst.* **2000**, *28*, 1397, Copyright 2000 Elsevier Ltd., and **b** *J. Org. Chem.* **1992**, *57*, 6932. © 1992 American Chemical Society

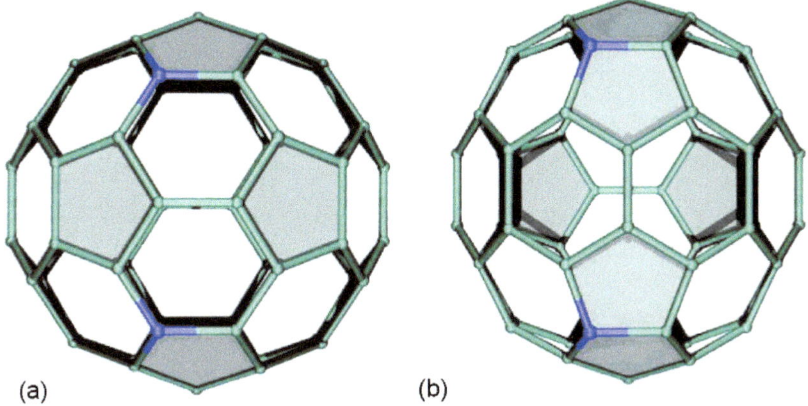

Fig. 6.43 Illustration of $C_{58}N_2$ that (a) satisfies the IPR and (b) violates the IPR with adjacent pentagons. The structure with adjacent pentagons, (**b**), is more stable than (**a**) by 12.5 kcal mol^{-1}. Reproduced with permission from *Nano Lett.* **2006**, 6, 890. © 2006 American Chemical Society

It was once thought that fullerenes were formed from the folding of preformed graphitic sheets that emanated from the target following laser ablation. However, a variety of experiments have shown that the growth process initiates from small linear chains of carbon atoms. As the number of carbon atoms increases, the chains preferentially connect into ring structures due to their greater stabilities. In particular, for C_n where $n < 10$ (with the exception of C_6 as discussed below), linear species are the preferred morphology rather than rings (Fig. 6.44). The preference for ring formation for $n = 6$, and $n \geq 10$ (especially for C_{10}, C_{14}, C_{18}, etc.) is due to the enhanced aromaticity/stability of planar rings when there are $4n + 2\pi$ electrons (where $n = 1, 2, 3, \ldots$ known as the *Hückel rule*).

Figure 6.45 illustrates a proposed mechanism for the subsequent steps of fullerene growth, involving the self-assembly of carbon rings. When $n \geq 30$ or so, the monocyclic rings are proposed to form graphitic sheets. The "pentagon road" mechanism proposed by Smalley[207] assumes that the graphitic sheets contain both hexagon and pentagon units. Closure of the sheets to form Buckyballs effectively results in growth termination. In contrast, the "fullerene road" model assumes the initial formation of smaller non-IPR fullerenes, which undergo thermal rearrangement to yield C_{60} and higher fullerenes.[208]

As discussed earlier, pentagon units are essential to the fullerene structure, since they allow the planar graphitic sheet to curl. The driving force for this rearrangement is likely the C–C bonding of edge carbons that satisfies their unfilled valences. As the prevailing mechanism points out, adequate annealing is required in order to incorporate a sufficient number of pentagons. For instance, if the cooling rate is too high, amorphous soot particles will be preferentially formed rather than fullerenes. In addition, an overall low growth temperature will not be sufficient to cause cage formation, yielding planar graphitic fragments instead of fullerenes.

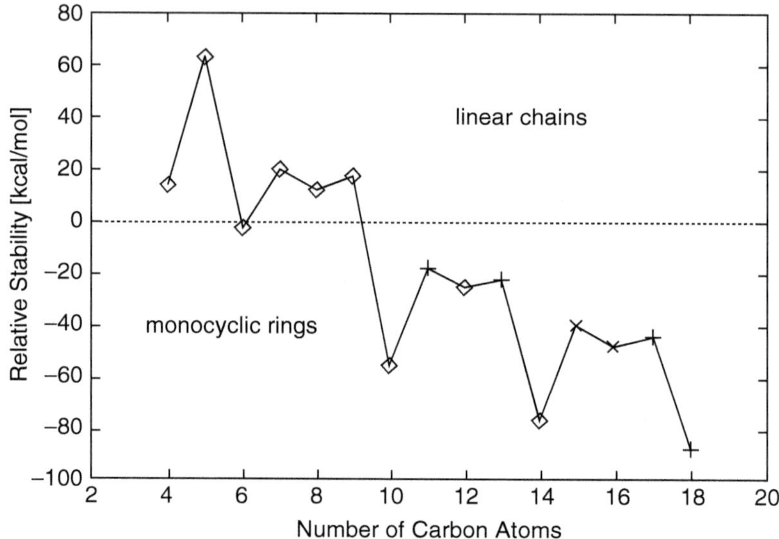

Fig. 6.44 The relative stabilities of linear-chain carbon clusters vs. monocyclic rings with changing cluster size. Reproduced with permission from *J. Am. Chem. Soc.* **1994**, *116*, 750. © 1994 American Chemical Society

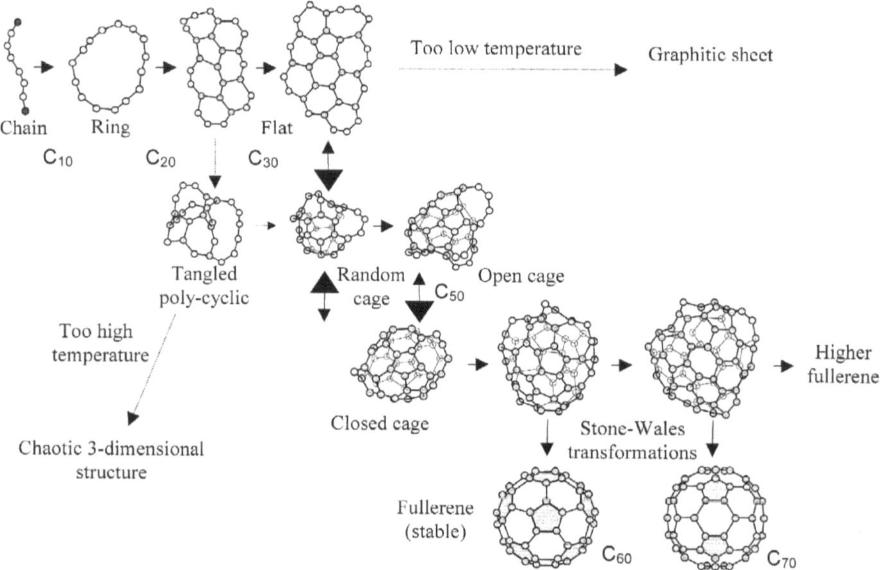

Fig. 6.45 Proposed mechanism for fullerene growth. Reprinted from *Chem. Phys. Lett.* **1998**, *286*, 343. © 1998, with permission from Elsevier Ltd.

Regardless of the proposed mechanism, a final thermal annealing step is likely required to organize the hexagon and pentagon subunits into the lowest-energy IPR arrangement. This rearrangement step is known as the *Stone–Wales (SW) transformation*, and involves a concerted reorganization of the hexagon/pentagon units. We already provided an example of a rare SW transformation where the non-IPR N-containing species was actually lowest in energy (Fig. 6.29). However, most often this rearrangement occurs in the opposite direction—transforming adjacent pentagons into a hexagon-isolated structure. It should be noted that the Stone–Wales transformation is actually thermally forbidden via the Woodward–Hoffman rules; calculations show an energy barrier of at least 5 eV for this pathway. However, it has been shown that this rearrangement may likely be catalyzed by additional carbon and/or hydrogen atoms that are present during laser/arc or thermal combustion syntheses (Fig. 6.46).

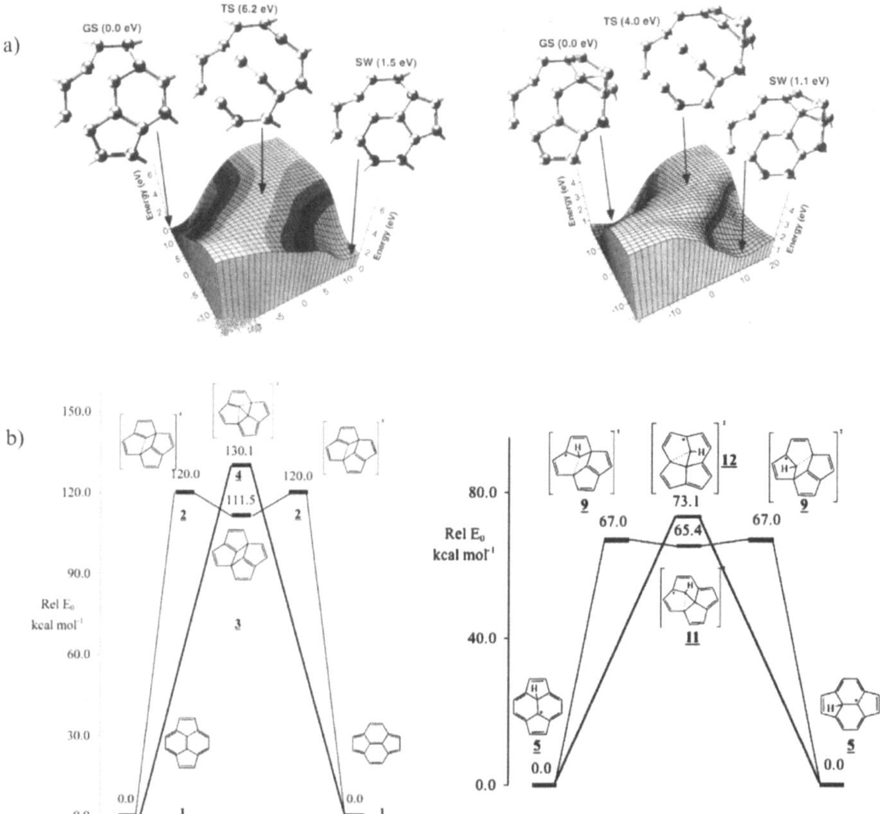

Fig. 6.46 The potential energy surface of the Stone–Wales transformation before/after the addition of catalyzing moieties such as **a** carbon, and **b** hydrogen atoms. Reproduced with permission from *Science* **1996**, *272*, 87. © 1996 AAAS; and **b** *J. Phys. Chem. A* **2005**, *109*, 9896. © 2005 American Chemical Society

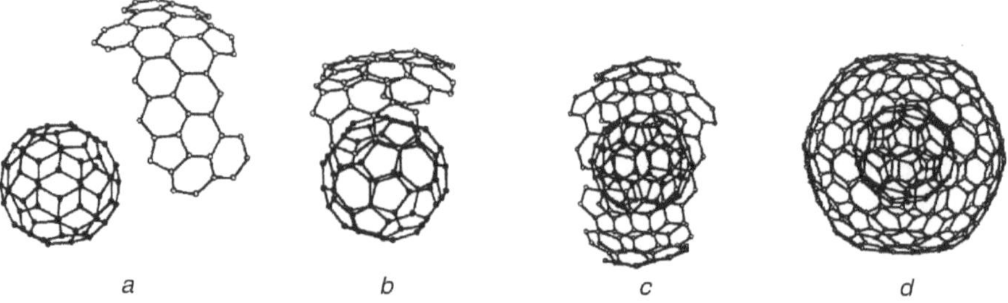

Fig. 6.47 Proposed scheme for the formation of the multishell fullerene $C_{60}@C_{240}$.[210] © Mordkovich, V. Z.; Shiratori, Y.; Hiraoka, H.; Takeuchi, Y

Interestingly, a fullerene structure may serve as a nucleation site for additional layers of graphite en route toward multishell fullerenes (Fig. 6.47). These are denoted as "$C_{60}@C_{240}$" where the @ symbol represents the encapsulated species. There are even triple-layered structures such as "$C_{60}@C_{240}@C_{560}$."[209] Although very small quantities (<0.01%) of multilayered fullerenes are found in the soot resulting from laser vaporization, the yield may be improved by *in vacuo* sublimation of the vapor phase at a high temperature (*ca.* 1200 °C).

Although the "brute force" methods of laser/arc and high-temperature pyrolysis represent the most common techniques for generating fullerenes, there is interest in developing a bottom-up synthetic approach that could produce impurity-free isomerically pure fullerenes. In 1999, a promising step in that direction was accomplished with the first nonpyrolytic synthesis of "buckybowls."[211] These structures had a bowl-shaped structure, and consisted of the hexagon-isolated pentagon backbone exhibited by fullerenes (Fig. 6.48). In early 2002, a chlorinated C_{60} precursor was reported using a traditional 12-step organic synthesis. This compound was subsequently converted to Buckyballs using high-temperature vacuum pyrolysis (Fig. 6.49).[212] Alternatively, a fluorinated precursor was also converted to C_{60} cages via laser ionization.[213] The yield of C_{60} was <1%—certainly not useful for commercial production of Buckyballs; however, the novelty of these approaches was that pyrolysis or laser ionization did not decompose the precursor into smaller units, but rather served to stitch together adjacent arms of the molecular precursor. More recently, C_{60} and the triazafullerene $C_{57}N_3$ were synthesized from aromatic precursors via surface-catalyzed cyclodehydrogenation. This route requires the deposition of a polyaromatic precursor such as $C_{60}H_{30}$ onto a Pt(111) surface, followed by annealing at 750 K.[214] Although the conversion efficiency of the precursor to fullerene product is 100%, this route is not amenable to yield bulk quantities of fullerenes.[215]

In addition to pristine fullerene structures, it has been discovered that various metal ions may be encapsulated inside the caged structure to yield *endohedral fullerenes*. Thus far, a variety of alkali and lanthanide metals, Group 5 atoms, noble gases, and neutral molecules such as H_2,[216] CO, and H_2O have been sequestered inside the C_{60} structure. Calculations have shown that the encapsulation of noble gas atoms and small ions (e.g., Li^+, F^-, Cl^-) actually stabilize the fullerene cage, whereas larger

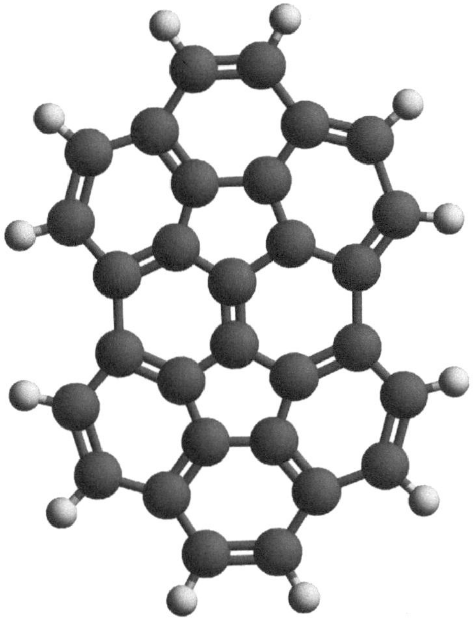

Fig. 6.48 Backbone structure of semibuckminsterfullerene, $C_{30}H_{12}$

species (e.g., Rb^+, Br^-, I^-) destabilize the cage.[217] *Metallofullerenes* ($M@C_x$) are typically grown by either laser ablation of metal-doped graphite disks at high temperature (*ca.* 1200 °C), or carbon arc techniques with metal-doped graphite rods. An example of a metallofullerene ($Gd^{3+}@C_{60}$) was shown in Fig. 6.24a; these structures are commonly employed as MRI contrast agents.

You may be wondering "*how does the ion get inside the cage?*" Does this occur during the growth of the fullerene structure itself, or does the metal ion insert after the cage is already formed? It has been shown that the latter occurs, with the exact entrance pathway dependent on the size of the dopant species. Small dopants such as He or H^+ may directly pass through either hexagon or pentagon units of the cage toward the vacant core. However, for larger ions/atoms, some framework C–C bonds must be reversibly broken in order to accommodate the incoming species—aptly referred to as a *window mechanism* (Fig. 6.50). Since non-IPR fullerenes have relatively large strain energies due to fused pentagons, this process should occur readily for these structures. Indeed, there has been much recent interest in synthesizing "unconventional" metallofullerenes such as $Sc_2@C_{66}$.[218] Unlike the empty C_{66} counterpart, these structures are stable since the incoming metal atom donates electron density to the C–C bond between adjacent pentagons, causing a decrease in the local bond strain.

As you might imagine, relatively large ions such as Cs^+, Y^{3+}, or Sc^{3+} are likely not encapsulated through a simple reversible windowing mechanism. In order for this to occur, more than one C–C bond would need to be broken (Fig. 6.36b), which increases the energetic barrier for this to occur. More recently, a "hole-repairing mechanism"

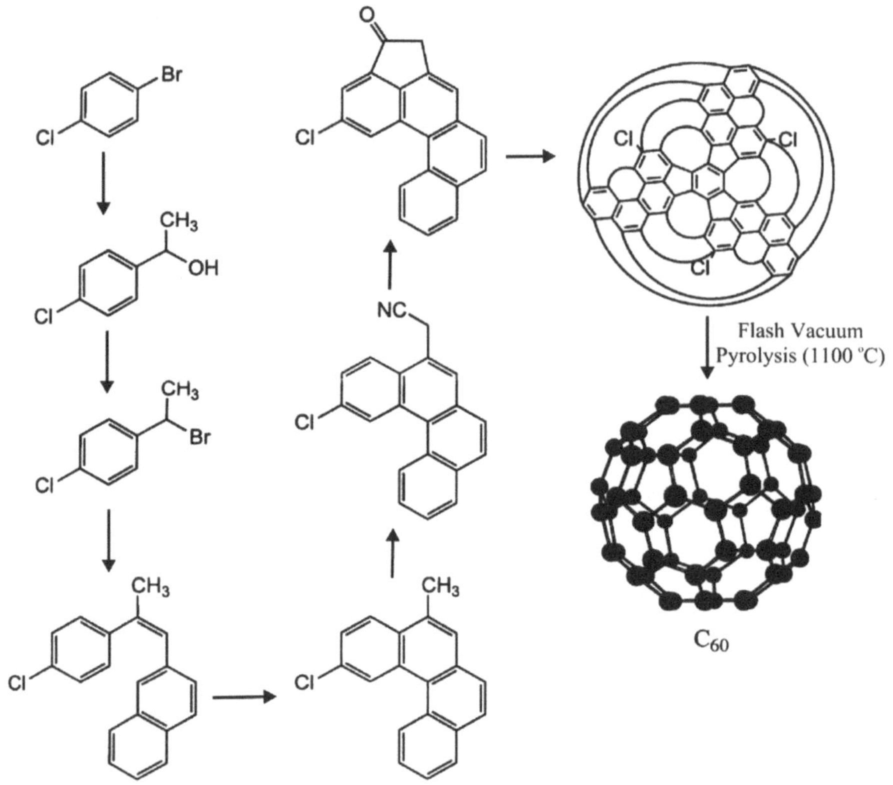

Fig. 6.49 Synthesis of a C_{60} precursor. Reproduced with permission from *Science* **2002**, *295*, 1500. © 2002 AAAS

was proposed for $Y@C_{82}$ metallofullerenes, in which calculations predict the combination of a large C_{76} open-cage fullerene and a smaller C_6Y fragment that effectively repairs the framework hole (Fig. 6.51).[219]

Endohedral guests have also been shown to stabilize Si_{20} dodecahedra—silicon analogues of fullerenes.[220] Although all-silicon fullerenes are too unstable due to the preference for Si sp^3 hybridization, $[Si_{32}Cl_{45}]^-$ clusters have been synthesized by chloride-induced disproportionation reaction of hexachlorodisilane (Si_2Cl_6). Whereas all carbon atoms in C_{60} have coordination numbers of three and form double bonds, all silicon atoms in Si_{20} have a coordination number of four and are connected via single bonds. Accordingly, the "silafullerane" structure features a dodecahedral core stabilized by an endohedral Cl^- ion; the presence of eight chloro and 12 trichlorosilyl surface groups result in a structure related to the dodecahedrane ($Si_{20}H_{20}$) molecule. Interestingly, theoretical calculations have shown that P-doped $Si_{20}H_{20}$ species should be possible and would result in tunable optical and chemical properties.[221]

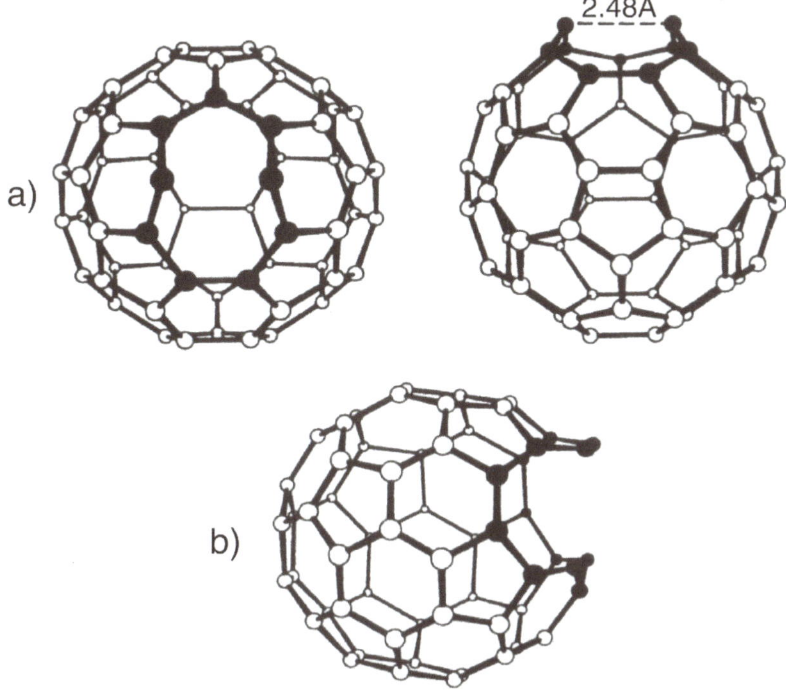

Fig. 6.50 Theoretical intermediates during endohedral fullerene formation. Image **a** shows the nine-membered ring formed by a single pentagon–hexagon bond. By comparison, image **b** illustrates the formation of a larger 13-membered ring by breaking two framework bonds. Reproduced with permission from *Science* **1994**, *263*, 791. © 1994 AAAS

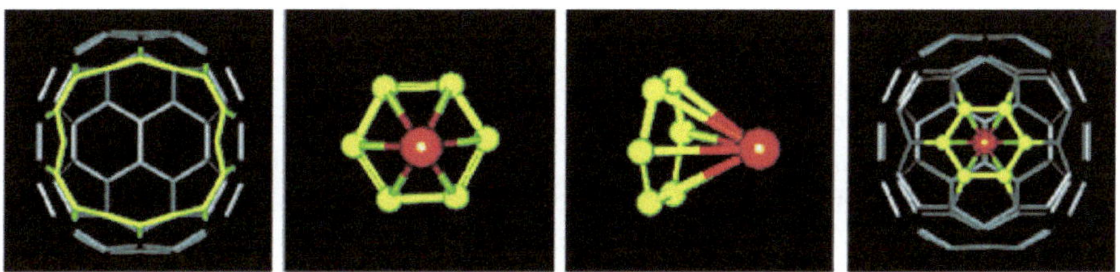

Fig. 6.51 Images of the calculated structures involved in the "hole-repairing" mechanism for endohedral fullerene growth. Shown from left to right are: the C_{76} open cage, top and side views of the C_6Y fragment, and the final $Y@C_{82}$ metallofullerene. Reproduced with permission from *J. Phys. Chem. A* **2005**, *109*, 3980. © 2005 American Chemical Society

Vapor-Phase Synthesis of 0-D Nanostructures

In addition to fullerenes, a variety of metallic and compound (oxides, nitrides, chalcogenides) nanoclusters/nanoparticles may be synthesized using vapor techniques. The primary benefits of this approach relative to solution-phase methods discussed in the next section are much higher rates of production and greater purity/compositional control. That is, gas-phase nanoparticles will be free of impurities acquired from solvents and added reagents such as surfactants, reducing agents, etc. Further, complex compositions (e.g., high-temperature superconductors, intermetallics, hydroxyapatite,[222] etc.) and metastable phases are often more easily synthesized using gas-phase decomposition techniques rather than low-temperature solvent processing.[223]

When molecular precursors are decomposed in the gas phase, they first exhibit homogenous nucleation which leads to supersaturation. Condensation of sub-nanosized particulates will ensue, followed by growth via *Ostwald ripening*—the growth of large particles at the expense of smaller ones, due to the relatively higher surface free energy (less stability) of small nanoclusters. Particle growth is quenched by shutting off the flow of precursor, or cooling the system by either cool-gas dilution or free-jet expansion through a narrow nozzle. The overall size and morphology of the nanoparticles may be controlled by favoring either nucleation or growth via sintering. That is, if sintering/annealing is faster than collisions, large spheres will be formed; in contrast, aggregates of smaller nanoparticles will be produced if nucleation processes are faster than sintering.

Figure 6.52 shows a variety of gas-phase techniques that have been used to synthesize 0-D nanoparticles. Radio frequency plasma sources have long been used for quantitative analysis by atomizing component species in liquid or solid samples—a technique referred to as inductively-coupled plasma atomic emission spectroscopy (ICP-AES). The extreme energy of an ICP may also be exploited to vaporize precursor sources to afford the growth of nanoparticles (Fig. 6.52a).[224] In this system, the nanoparticle size/morphology would be mostly controlled by the concentration of precursor in the plasma and the rate of cooling—a function of its distance from the plasma source.

Although gas-phase techniques are not generally useful for shape-control of nanoparticles, plasma techniques have been used to generate cubic nanocrystals.[232] Using a precursor gas mixture of SiH_4/H_2, cubic nanocrystals are thought to be formed due to preferential etching of spherical Si nanoparticles by H atoms formed in the plasma, which would more readily attack (100) facets than (110) or (111)—Fig. 6.53. Shape control of Fe@FeO nanoparticles is also possible using sputtering gas aggregation with varying degrees of gas-phase annealing.[233]

As one would expect, simple flame pyrolysis has been used to generate 0-D nanoparticles. In fact, millions of metric tons of carbon black and metal oxide nanoparticles are synthesized by this method each year. The complex flame chemistry is difficult to control, which often generates a broad distribution of nanoparticle sizes, compositions, and morphologies. A degree of growth control may be afforded by placing electrodes at the exit side of the flame reactor. By varying the applied field strength, one may control the sizes and degree of agglomeration of the formed nanoparticles.[234] A spark formed between two charged electrodes may also be used to form nanoparticles (Fig. 6.52b). Such spark-facilitated methods are used for solid sources with a high melting point such as Si or C, which are not easily evaporated in a furnace.

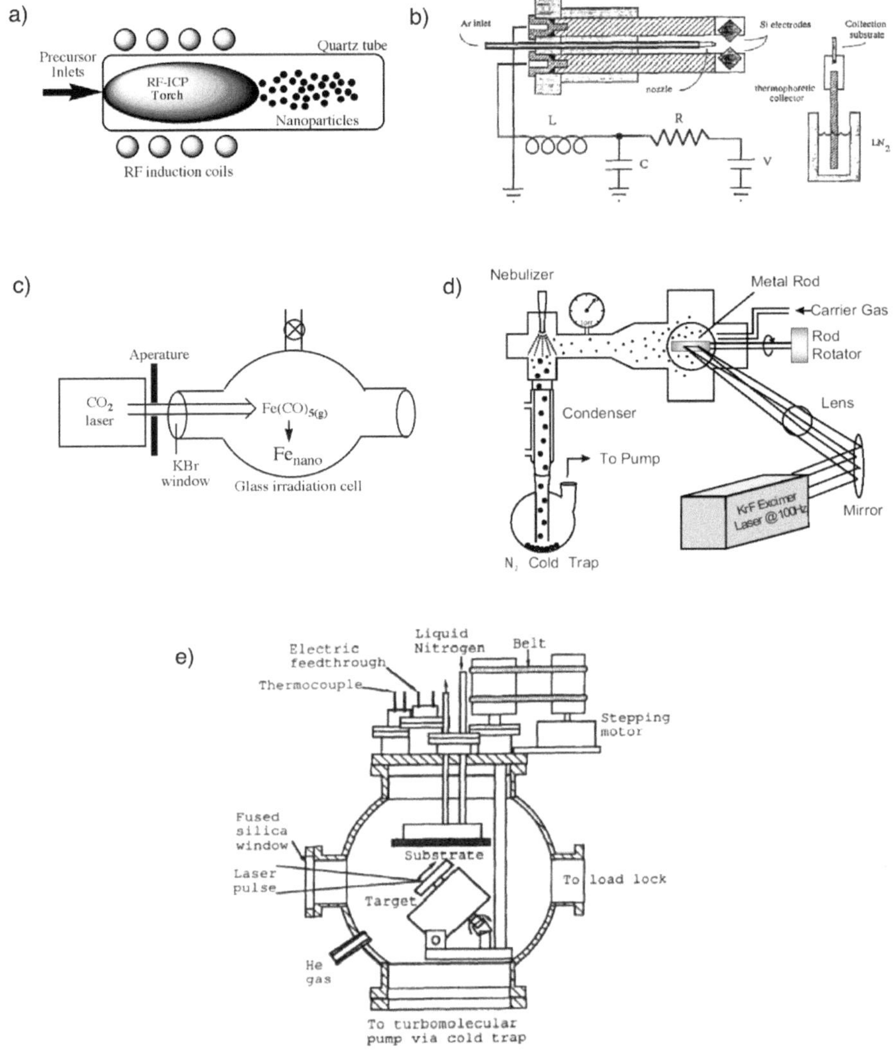

Fig. 6.52 Illustrations of apparati used in the gas-phase synthesis of 0-D nanoparticles. Shown are: **a** plasma system; **b** spark-facilitated growth[225]; **c** laser vaporization/pyrolysis[226]; **d** laser evaporation[227]; **e** laser ablation[228]; **f** inert-gas evaporation[229]; **g** electrospray system[230]; **h** spray pyrolysis[231]

Laser sources are also useful for either pyrolysis of precursor vapors (Fig. 6.52c) or vaporization of solid precursor targets (Fig. 6.52d). Using a CO_2 laser (100 W), one is able to grow nanoparticles from precursor powders, crystals, and sintered blocks of Fe_3O_4, $CaTiO_3$, Mg_2SiO_4, and metal carbides.[235] The utility of a laser source is the introduction of localized heating, which facilitates nanoparticle growth in controlled regions such as reactor flask walls or within a quenching cold trap.

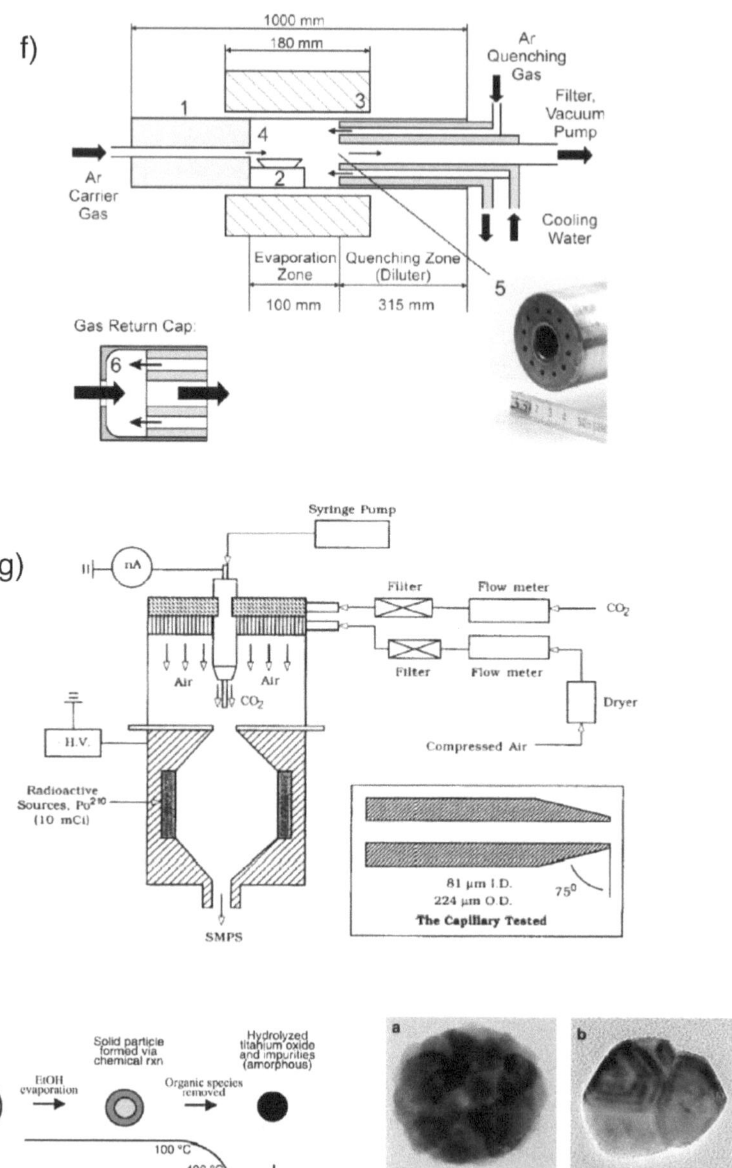

Fig. 6.52 (continued)

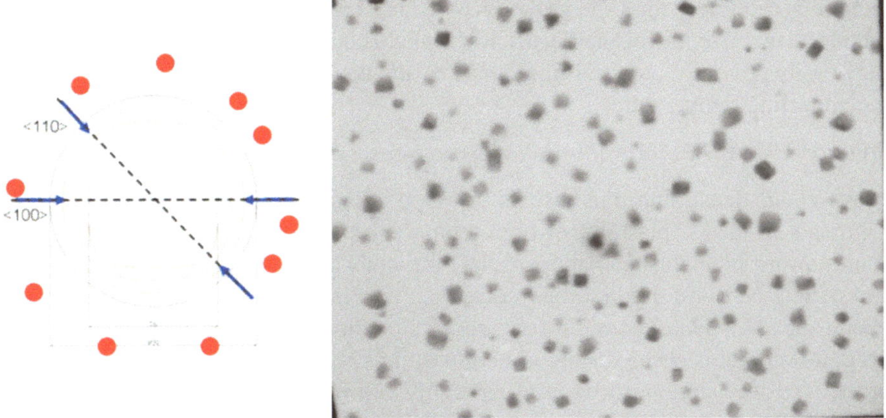

Fig. 6.53 Schematic (left) illustrating the selective etching of facets within a Si nanoparticle by hydrogen atoms present in a plasma. The TEM image (right) illustrates the resultant cubic morphology of Si nanoparticles, realized by faster etching of Si(100) facets relative to (110) planes. Reproduced with permission from *J. Phys. D* **2007**, *40*, 2247. © 2007 The Institute of Physics

Laser ablation (Fig. 6.52e) may be distinguished from laser vaporization by the type of fragments that are produced. Whereas ablation will result in both sub-nano and micron-sized fragments in the gas phase, vaporization results in purely gas-phase molecular precursors. Hence, laser ablation may be considered a type of 'top-down' approach, whereas laser vaporization would be a 'bottom-up' technique, which proceeds via homogeneous nucleation/growth.

The particle size distribution may be controlled in both laser ablation/vaporization techniques by manipulating the pulse energy/duration (usually in the timeframe of 10–50 ns/pulse). It should be noted that when laser ablation is used for thin-film deposition, the technique is referred to as pulsed-laser deposition (PLD).[236] At lower operating pressures (*ca.* < 0.1 Torr), thin films will be favored; however, at higher pressures (*ca.* > 1 Torr), nanoparticles will be formed due to a greater opportunity for gas-phase nucleation to occur.

Physical vapor deposition (PVD) techniques such as evaporation or sputtering may also be used to synthesize 0-D nanoparticles. One technique, known as inert gas evaporation (Fig. 6.52f), consists of evaporation of a precursor material within a cooling inert gas at low pressures (*ca.* 100 Pa). Vaporization may be accomplished via resistive heating, ion bombardment (sputtering), or laser irradiation. As one might expect, reactive gases such as O_2/O_3 or NH_3 may be plumbed into the system to allow for the production of oxide or nitride nanoparticles, respectively.

In addition to evaporating solid sources, one may also utilize dilute liquid suspensions of nanoparticles. However, simple evaporation of these solutions in an oven will serve to concentrate any impurities that were present in the original solvent. In order to circumvent this limitation, the nanoparticle suspensions must be sprayed into a heat source as small droplets whose uniformity may be controlled using a nebulizer or electrospray techniques (Fig. 6.52g). The growth of nanoparticles via this synthetic technique is referred to as *spray pyrolysis* and has been used to synthesize a variety of metal and metal oxide 0-D nanoparticles/nanoclusters.[237] As shown in Fig. 6.52h, one may easily control the crystallinity/phase of the developing nanoparticle by varying the annealing temperature.

Solution-Phase Synthesis of 0-D Nanostructures

Most of our discussion thus far has involved some rather extreme synthetic environments of lasers, plasmas, or high-temperature pyrolysis. However, a preferred route toward nanoclusters/nanoparticles of metals and their compounds is through use of relatively mild conditions—often taking place at room temperature on the bench-top. This is not possible for carbon nanoallotropes since the precursor (e.g., graphite) contains strong covalent bonding, thus requiring a significant amount of energy for dissociation prior to self-assembly. However, for metallic and compound nanostructural growth, the simple reduction of metal salts (usually via $NaBH_4$, H_2, or hydrazine as reducing agents), or combination of reactive molecular precursors, are amenable for mild, solution-phase growth. For instance, (6.6) and (6.7) represent the reduction of copper ions by sodium borohydride (and subsequent hydrolysis of borane); 6.8 shows the side-reaction of $NaBH_4$ with the water, indicating its relative instability in aqueous solutions. In theory, any metal with a larger standard reduction potential (E°) than the reducing agent (e.g., -1.33 V for sodium borohydride; -0.23 and -1.15 V for hydrazine in acidic and basic media, respectively) is a candidate for nanoparticle growth via reduction. This includes most of the first-row transition metal ions and many others from the main group/transition metal series. However, it should be noted that solution pH and side-reactions (e.g., metal ions being converted to borides by BH_4^- rather than reduction) often provide a barrier toward successful metal-ion reduction.

$$CuCl_2 + 2\,NaBH_4 \rightarrow Cu^0 + 2\,NaCl + B_2H_6 + H_2 \tag{6.6}$$

$$B_2H_6 + 6\,H_2O \rightarrow 2\,B(OH)_3 + 6\,H_2 \tag{6.7}$$

$$NaBH_4 + 2\,H_2O \rightarrow NaBO_2 + 4\,H_2 \tag{6.8}$$

If the above reduction were to be carried out as-is, a metallic film or bulk powder would be formed rather than nanostructures. That is, as the metal ions are reduced, they would instantly agglomerate with one another to form larger particulates. Hence, the most crucial component of nanostructure synthesis is the stabilizing agent that isolates the growing nuclei from one another. In addition, shape control of nanostructures may be afforded in solution by varying the stabilizing agent or other experimental conditions.[238] Some desirable traits of a stabilizing (or entraining) agent are:

i. Chemically unreactive toward the growing nanocluster, rendering an unpassivated nanocluster surface;

ii. Structurally well-defined (size/shape), which allows for the controlled growth of the encapsulated nanocluster;
iii. Composed of light elements (organic-based), so its structure does not interfere with the characterization of the entrained nanocluster. This will also facilitate its sacrificial removal from the nanocluster by pyrolysis at relatively low temperatures, if desired;
iv. Surface-modifiable, to allow for tunable solubility and selective interactions with external stimuli. In addition, this would afford the controllable self-assembly of entrained nanoclusters on a variety of surfaces through chemisorption, if desired.

Although some references cite the processing conditions ("continuous" gas-phase vs. "batch" solution-phase) as being superior for gas-phase techniques, the resultant batch-to-batch properties (e.g., monodispersity, morphology, purity) of solution-grown nanoparticles has greatly improved in recent years. With optimized nanoparticle synthetic recipes pervasive in the literature, it is now safe to assume that the nanoparticles generated in one batch will be virtually identical to those in subsequent batches (although one must always confirm this via appropriate characterization techniques). Furthermore, whereas the nanoparticles generated from individual gas-phase runs may exhibit very similar diameters/morphologies, their overall monodispersity will be relatively poor due to the lack of an entraining agent (Fig. 6.54). Accordingly, differential mobility analyzers (DMAs, Fig. 6.55) are often used to improve the size/morphological uniformity of gas-phase synthesized nanoparticles.[239]

In aqueous solutions, the most common method used to stabilize nanostructures is the use of organic "capping" ligands. For instance, the *Turkevich process,* which dates back to early colloidal growth of the 1950s (and refined by Frens in the 1970s), uses sodium citrate (**I**) to entrain the reduced gold nuclei obtained from aqueous tetrachloroaurate anions. Particle diameters on the order of 10–20 nm may be easily synthesized using this method. In this case, since gold is easily reduced ($E^\circ = 1.00$ V for $AuCl_4^-$), the citrate reagent acts as both the reducing and stabilizing agent.

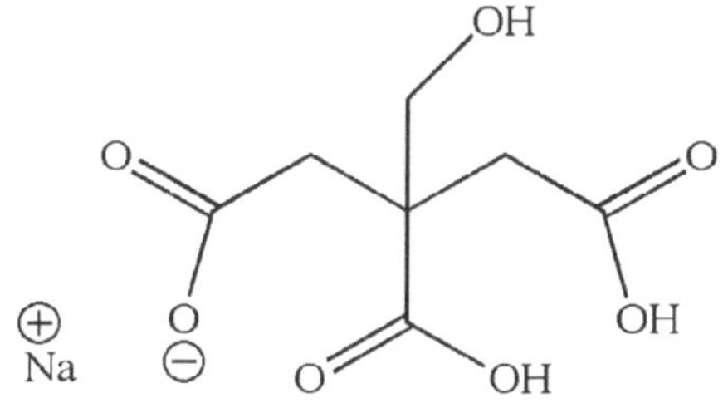

OH O O ⊕ Na O ⊖ OH O OH

I

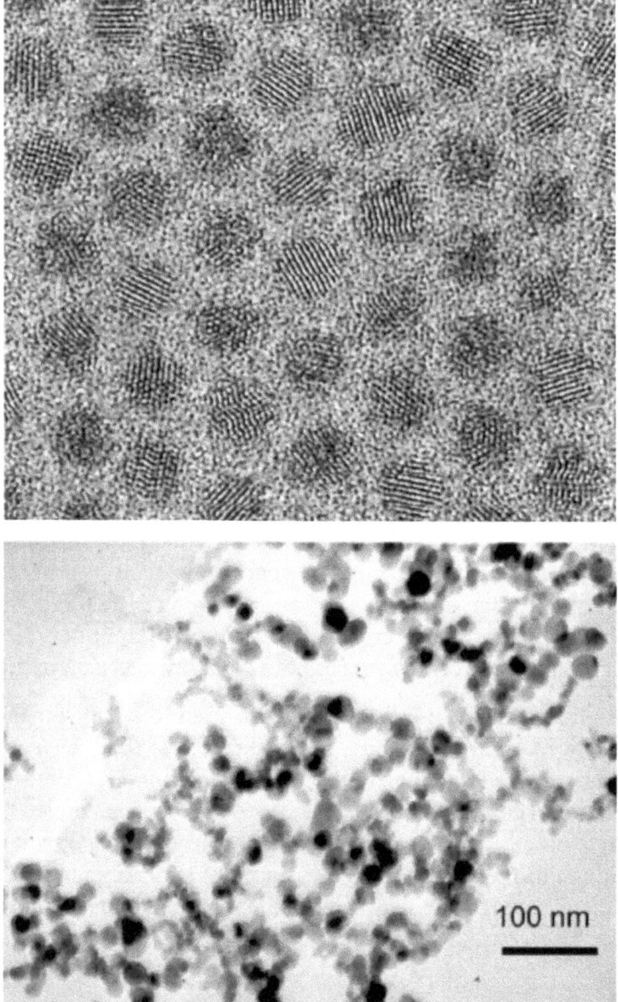

Fig. 6.54 TEM images illustrating the common relative monodispersities of solvent-based (top) and gas-phase (bottom) syntheses. The top image is of CdSe quantum dots synthesized using a solution-phase mixture of TOPO/TOP/HDA at 300 °C. Reproduced with permission from Talapin, D. V.; Rogach, A. L.; Kornowski, A.; Haase, M.; Weller, H. *Nano Lett.* **2001**, *1*, 207. Copyright 2001 American Chemical Society. The bottom TEM image shows silicon nanoparticles generated from the laser pyrolysis of silane gas. Reproduced with permission from *Curr. Opin. Coll. Interf. Sci.* **2003**, *8*, 127. © 2003 Elsevier Ltd.

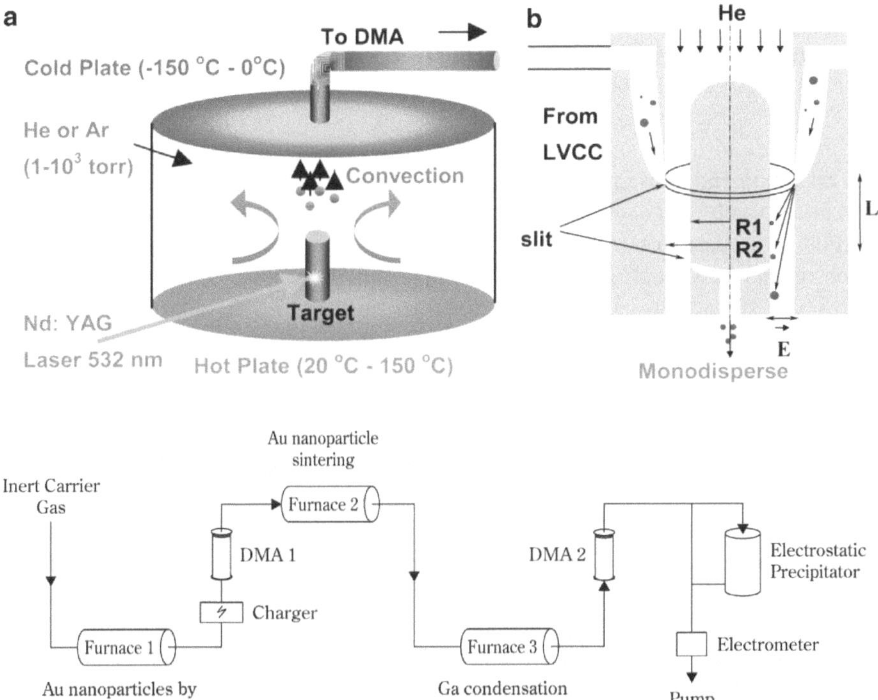

Fig. 6.55 Schematic of nanoparticle growth via pulsed-laser vaporization with controlled condensation (LVCC), coupled to a differential mobility analyzer (DMA). A DMA is used to control the size of gas-phase synthesized nanoparticles by exploiting differences in the electrical mobility of nanoparticles under a flow of an inert gas. Reproduced with permission from *Pure Appl. Chem.* **2006**, *78*, 1667. © 2006 IUPAC. The bottom image shows how gas-phase techniques may be used to synthesize Au/Ga core/shell nanoparticles with the assistance of multiple DMAs. Reproduced with permission from *Aerosol Sci. Technol.* **2004**, *38*, 948. © American Association for Aerosol Research

The cationic stabilizing agent bis(11-trimethylammoniumdecanoylaminoethyl) disulfide dibromide (TADDD, **II**) has also been utilized to grow metallic nanoclusters with diameters <10 nm. If $NaBH_4$ is used as the reducing agent, the sulfide bond is cleaved resulting in a –SH capping group. The thiol is chemisorbed to the surface of the growing nanostructure surface to prevent agglomeration (especially effective for "thiophilic" noble metals such as Pt, Ag, and Au).[240]

II

There has also been much interest in the use of structurally perfect polymeric *dendrimers* (see Chap. 5) as stabilizing templates for nanocluster growth. By varying the peripheral groups and number of repeat branching units (known as "generations"), one is able to fine-tune the size of the entrained nanocluster. Although amine-terminated dendrimers and hyperbranched polymers may be used as a template for M^{n+} chelation and subsequent chemical reduction, the size of the resultant nanoparticle is relatively large, with a greater degree of agglomeration possible. This is especially the case for hyperbranched polymers that exhibit a random structure, which results in a much greater nanoparticle polydispersity. On the other hand, if the primary surface amines ($-NH_2$) are either protonated (NH_3^+), or replaced with hydroxyl groups ($-OH$), the pre-reduced metal ions are forced further into the interior of the dendritic structure (Fig. 6.56). This results in much smaller diameters and narrow polydispersities for the reduced metal nanoclusters.[241]

In addition to controlling the surface moieties and solution pH, the generation size of the dendrimer is also paramount for successful nanocluster growth. As the degree of branching increases, so does the surface density, which prevents the incoming metal ion (or H^+ during protonation) from entering the interior of the dendritic architecture. Alternatively, for smaller generations, the entrained species becomes easily dislodged from the interior due to its open structure. Hence, the most effective size range for nanocluster growth using polyamidoamine (PAMAM) is between the fourth and sixth generations (G4–G6), which exhibit strong container properties (Fig. 6.57). As an illustration of the extreme flexibility of the dendritic architecture, the core may also be altered to change its solubility characteristics, or allow the penetration of species through the periphery at high generations (Fig. 6.58). Since dendrimers containing an almost unlimited range of cores and peripheral groups have been synthesized, it is now possible to easily control nanocluster properties such as composition, size, morphology, solubility, and encapsulation (e.g., control the release rate of entrained medicinal agents/sensors based on structural or environmental changes, for targeted drug delivery or in situ monitoring).

It should be noted that metal nanocluster growth using dendritic templates is strongly governed by the degree of complexation of the precursor metal ions. For instance, silver nanoclusters are not possible using hydroxyl-terminated PAMAM dendrimers since Ag^+ is not strongly chelated to tertiary amine groups. However, if Cu^0 nanoclusters are first generated within the structure, followed by Ag^+, a redox reaction will facilitate the displacement of Cu^0 with Ag^0 within the dendrimer interior (6.9):

$$Cu + 2\,Ag^+ \rightarrow Cu^{2+} + 2\,Ag \tag{6.9}$$

In addition to simple metallic nanostructures, more complex intermetallic species have also been synthesized through the introduction of more than one metal. For instance, bimetallic nanoclusters may be generated via three routes within a dendritic host (Fig. 6.59). In addition to already being proven for core–shell nanoclusters, this route should also be amenable for the growth of trimetallic nanostructures for interesting catalytic applications.[242]

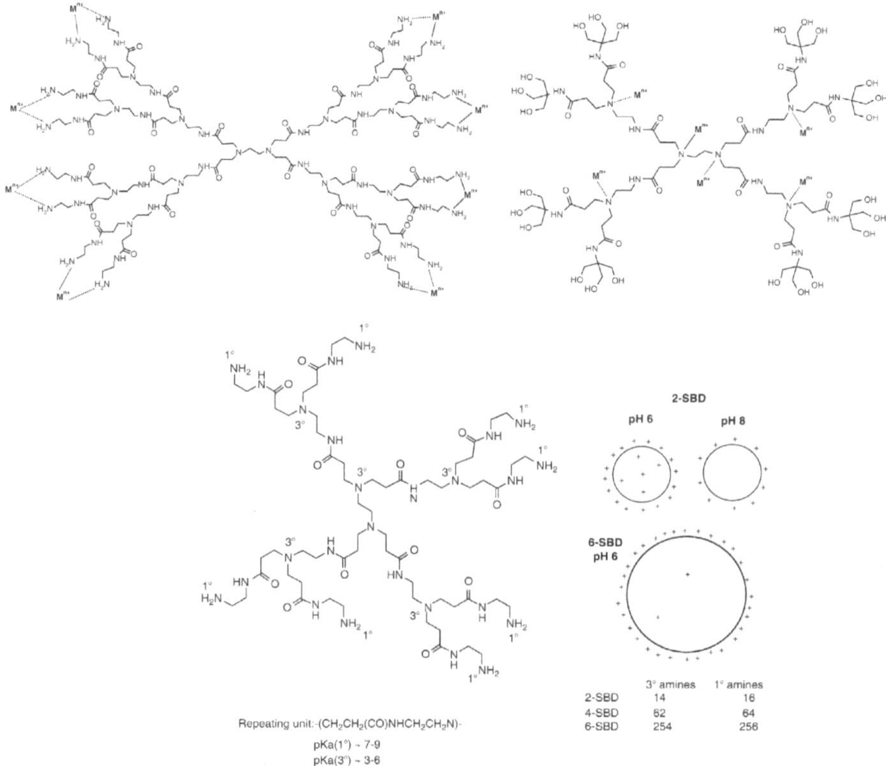

Fig. 6.56 Molecular structure of a second Generation (G2) amine-terminated PAMAM dendrimer, illustrating the positions of the metal ions chelated to the primary amine groups (prereduction). In comparison, a G2 hydroxyl-terminated PAMAM dendrimer is shown, with the metal ions now preferring to chelate to the interior tertiary amine groups. Shown on the bottom is the effect of protonation on G2/G6 amine-terminated PAMAM dendrimers. A schematic on the lower right illustrates the positions of the protonated amines at varying pH values. As the generation size increases, the surface density also increases which limits the access of protons (or chelating metal ions) to interact with the interior tertiary amine groups. Reproduced with permission from *J. Phys. Chem. B* **2000**, *104*, 11472. © 2000 American Chemical Society

Other polymers may also serve as effective stabilizing agents for nanostructural growth. For instance, poly(vinylpyrrolidone) (PVP, **III**) , poly(styrenesulfonic acid) sodium salt (PSS, **IV**) , and poly(2-ethyl-2-oxazoline) (PEO, **V**) were used to generate a number of intermetallic nanoalloys through a mild "metallurgy in a beaker" approach developed by Schaak and coworkers.[243] Since individual metal nanoparticles with high surface reactivity and low melting points are in intimate contact, the use of high-temperature annealing is not required, unlike bulk-scale alloy synthesis. The PVP architecture has also been shown to facilitate the growth of Au@Ag core–shell nanostructures, as well as Ag nanowires and nanocubes.[244] Interestingly, one is able to fine-tune the LSPR frequency of core–shell nanoclusters by varying the shell thickness or composition/size of the core.[245]

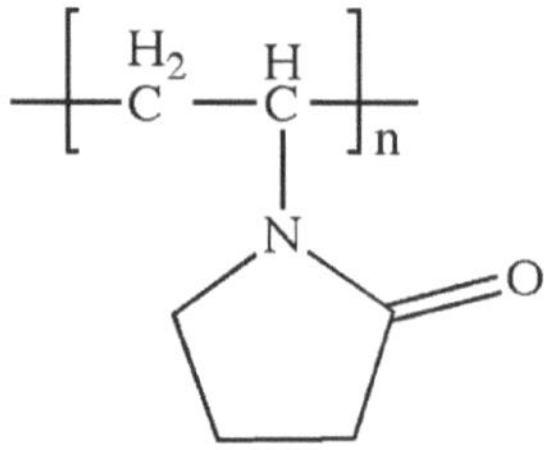

III

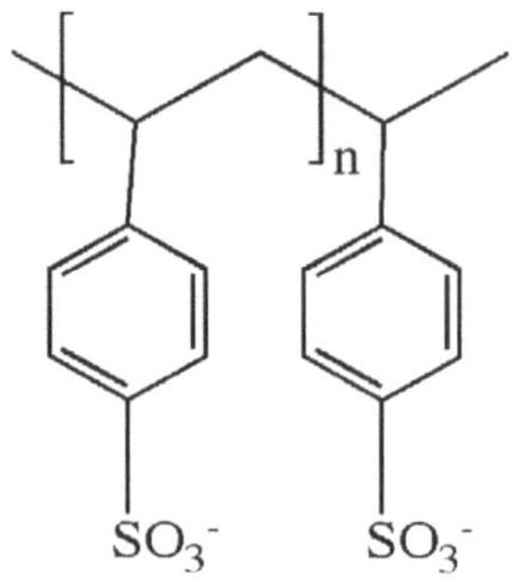

IV

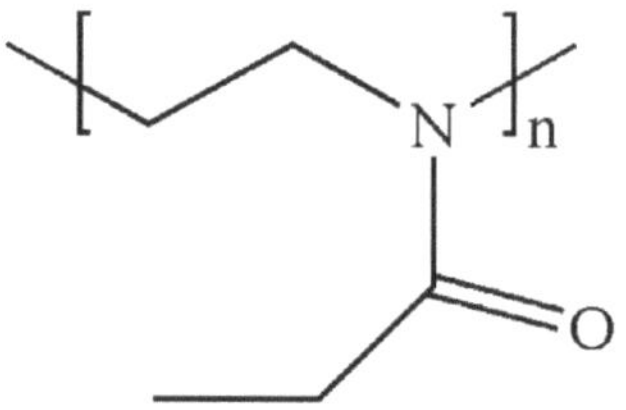

V

For the synthesis of nanostructures within nonpolar solvents, one may use a stabilizing agent that contains alkyl chains rather than –OH endgroups. One of the first capping agents to be used for noble metal colloidal growth was alkylthiols (CH_3–$[CH_2]_X$–SH).[246] In this system, the –SH end is bound to the surface of the nanostructure and the long organic tail is responsible for dispersion within the organic solvent. Although these capping stabilizers worked well for colloidal growth to prevent agglomeration—even allowing solvent removal/redispersion into organic solvents—it was relatively difficult to control the monodispersity of the nanoparticles.

Accordingly, systems that contain a nanoreactor template have been used more recently for controlled nanostructural growth. In fact, everyone who has washed

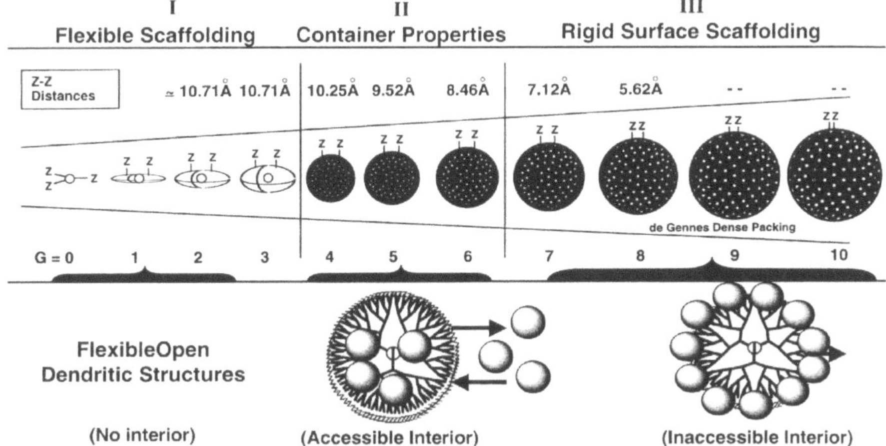

Fig. 6.57 Relative sizes and surface densities of PAMAM dendrimers, showing the most suitable range for nanocluster growth as Generation 4 to Generation 6. Reproduced with permission from *Dendrimers and other Dendritic Polymers*, Frechet, J. M. J.; Tomalia, D. A. eds. © 2001 Wiley–VCH

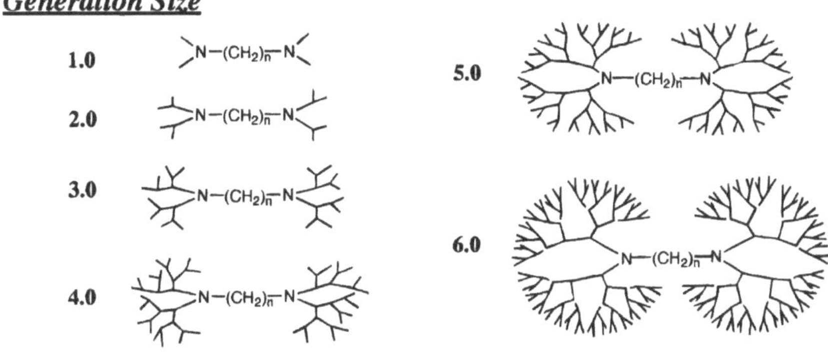

Fig. 6.58 Molecular structures of dendrimers modified with long-chain aliphatic cores. Unlike traditional dendritic structures with smaller cores, as the generation size increases, there is an available channel for external species to enter the dendrimer interior. Reproduced with permission from *Langmuir* **1997**, *13*, 3136. ©1997 American Chemical Society

dishes or laundry already has some experience with these types of stabilizing agents, known as *micelles*. These compounds contain both polar (–OH, cationic/anionic) and nonpolar (aliphatic) functional groups. Soaps and surfactants work by surrounding the dirt particle with the nonpolar ends, leaving the hydrophilic polar groups exposed to the surrounding water molecules. This serves to pull the dirt particle from the surface of the clothing fiber, forming an aqueous suspension. In a similar fashion, an oil-in-water microemulsion may be set up using common surfactants such as sodium

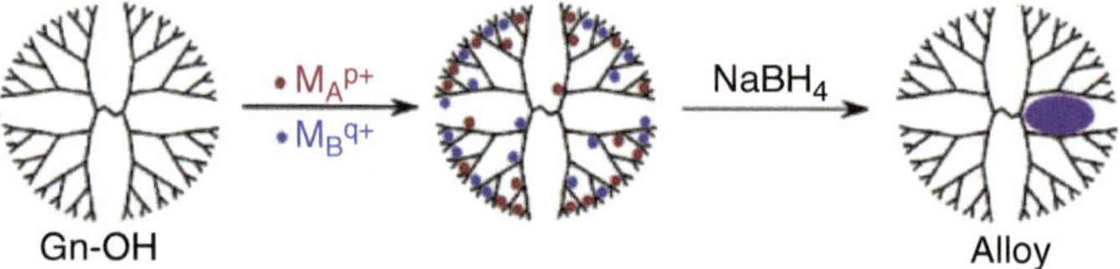

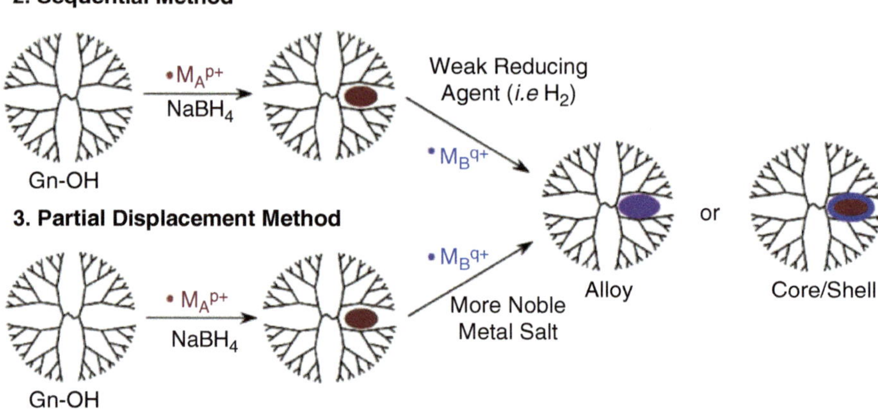

Fig. 6.59 Schematic of the three methods used to generate bimetallic nanoclusters within a dendritic host. Reproduced with permission from *J. Phys. Chem. B* **2004**, *109*, 692. © 2004 American Chemical Society

bis(2-ethylhexyl) sulfosuccinate (also referred to as Aerosol OT or AOT, **VI**), or the nonionic surfactant Triton-X (**VII**) in the presence of the biphasic oil/water mixture.[247]

VI

$$H_3C-\overset{H_2}{C}-C(CH_3)_2-C_6H_4-\left[O-\overset{H_2}{C}-\overset{H_2}{C}\right]_n-OH$$

VII

Since most precursors for solution-phase nanostructural growth are ionic metal salts, a typical micelle would not be effective for since the precursor would not be confined to the interior of the microemulsion. Hence, *reverse micelles* (or inverse micelles, Fig. 6.60) are used to confine the precursor ions to the aqueous interior, which effectively serves as a nanoreactor for subsequent reduction, oxidation, etc. en route to the final nanostructure. Not surprisingly, either PAMAMOS dendrimers (see Chap. 5) or dodecyl-terminated (hydrophobic) PAMAM dendrimers (Fig. 6.61) have been employed for this application.[248]

Increasingly complex stabilizing agents such as *cross-linked micelles* (Fig. 6.62)[249] have also been developed in recent years, which have found applications in biomedicine and environmental remediation. For drug-delivery, traditional polymeric micelles are often destabilized in the blood due to their high dilution factor.[250] However, by applying crosslinking strategies to the core, shell, or core–shell interface of the micelle, the structural integrity of the micelle may be maintained without premature drug release during systemic circulation (Fig. 6.63).

It should be noted that dendrimer-entrained nanoclusters synthesized within aqueous solutions may also be phase-transferred into an organic solvent by mixing with alkylthiols dissolved in a nonpolar solvent.[251] This also results in monodisperse nanoclusters, with much less polydispersity than early colloidal syntheses that employed thiol-based entraining agents. That is, the nanocluster size has already been controlled via intradendrimer stabilization. In contrast, the use of alkylthiols from

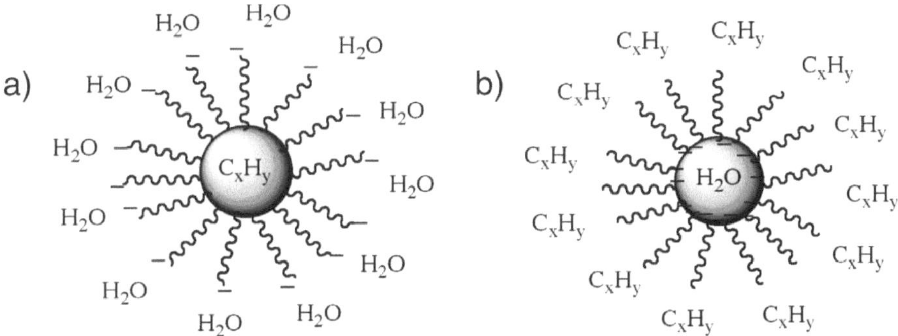

Fig. 6.60 Comparison of a traditional micelle used to entrain organic oils/dirt using an anionic surfactant, (**a**), and a reverse micelle used to stabilize aqueous nanoreactors within a nonpolar solvent, (**b**)

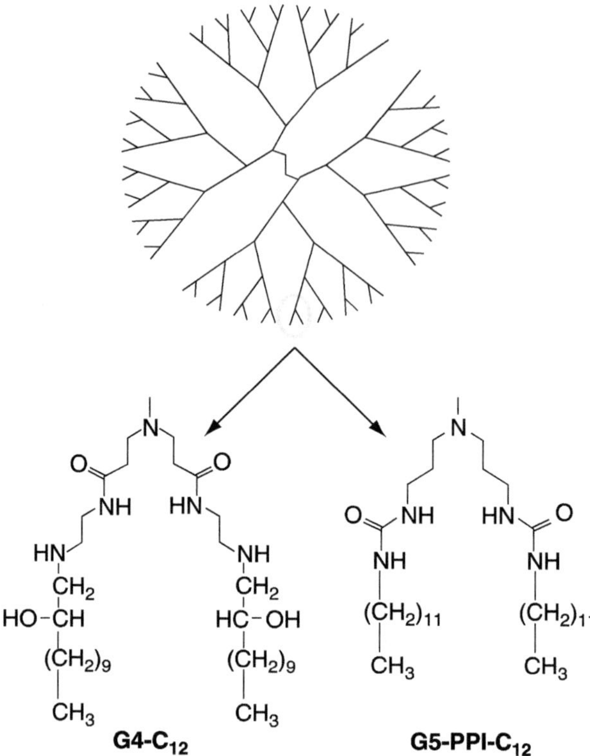

Fig. 6.61 Hydrophobic-functionalized PAMAM (G4-C_{12}) and poly(propyleneimine) (PPI) dendrimers, which serve as templates for Au nanocluster growth. Reproduced with permission from *Langmuir* **2005**, *21*, 11,981. © 2005 American Chemical Society

the initial stages of growth is not as effective in preventing agglomeration during the nucleation step.

Inorganic-based entraining agents have also been used to stabilize nanoclusters. Perhaps the earliest example is the use of polyoxometalates of general composition $M_xO_y^{n-}$ or $X_xM_yO_z^{m-}$ (where: X = P or Si; M = W, Mo, Nb or V) for the synthesis of nearly-monodisperse Ir nanoclusters (e.g., $P_2W_{15}Nb_3O_{62}^{9-}$, Fig. 6.64).[252] In addition, one may also synthesize nanoparticulates by entrapping the precursor within voids of a highly porous substrate. For instance, zeolites, layered solids, molecular sieves, gels and glasses have been used to sequester reactive precursor species wherein hydrolysis, high temperature, or other chemical/redox reactions are able to transform them into surface-immobilized nanostructures.[253] Of course, the encasing medium may then be removed by acid/base dissolution, plasma/heat treatment, etc. if discrete nanostructures are desired.

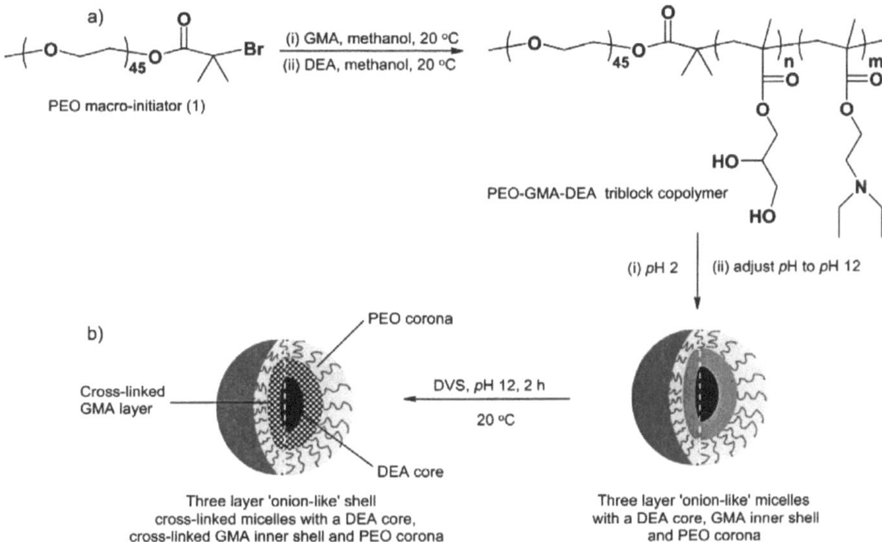

Fig. 6.62 Synthetic scheme for three-layer cross-linked micelles. Shown is the micellation of poly[(ethylene oxide)-block-glycerol monomethacrylate-block-2-(diethylamino)ethyl methacrylate] (PEO-GMA-DEA) triblock copolymers to the final "onion-like" layered nanostructure. Reproduced with permission from *Langmuir* **2002**, *18*, 8350. © 2002 American Chemical Society

It is not always necessary for the precursor metal ions to be encapsulated within a stabilizing polymer during chemical reduction. For instance, the reduced metal may be entrained by polymerization precursors, such as post-reduction living radical polymerization that takes place on the surface of gold nanoparticles (Fig. 6.65). This results in a dense "polymer brush"[254] that encapsulates the metallic nanoparticle, effectively stabilizing the structure against agglomeration. Subsequent alignment and surface reactivity of the resultant nanostructures may be fine-tuned by varying the nature of the polymer coating.

Not surprisingly, 0-D nanostructural growth need not be limited to metallic structures, but may also include other compounds such as metal oxides, sulfides, etc. There are a copious number of applications for nanostructural oxides such as high-density magnetic storage, heterogeneous catalysis, gas sensors, electrolytes for lithium batteries, and fuel cells. Sun and coworkers described a *polyol*[255] route for the synthesis of hexane-suspended Fe_3O_4 (magnetite) nanoclusters through the reaction between $Fe(acac)_3$, oleylamine, oleic acid, and 1,2-hexadecanediol at *ca.* 200 °C within phenyl ether.[256] This one-pot synthesis results in the thermal replacement of the acetylacetonate ligands (recall Chap. 4—CVD precursors) from the iron center, followed by oxide formation from reaction with the alcohol. The oleylamine and oleic acid condense in situ to form a long-chain inverse micelle that is capable of suspending the nanoclusters in nonpolar media. For purification, one simply adds an excess of polar solvent (methanol or ethanol) to precipitate the nanostructures. Following centrifugation, the nanoclusters are re-suspended in hexane, with this

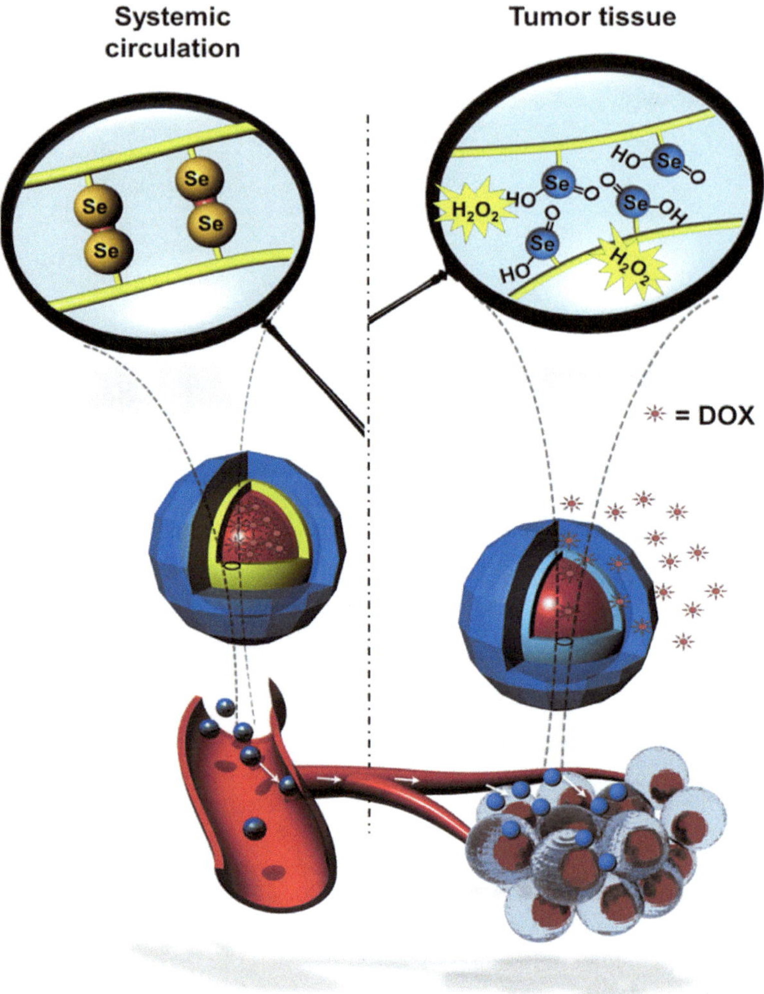

Fig. 6.63 An illustration depicting that diselenide-crosslinked micelles maintain their structural integrity during the circulation due to crosslinking. At the target site, the high level of H_2O_2 cleaves the crosslinking, which triggers drug release. Reproduced with permission from *Biomaterials* **2016**, *103*, 56. © 2016 Elsevier Ltd.

precipitation/resuspension procedure repeated 3–4 times to afford monodisperse and chemically pure Fe_3O_4 nanoclusters. Another analogous procedure uses oleic acid/trimethylamine N-oxide in trioctylamine (b.p. 365 °C) to decompose $Fe(CO)_5$ to yield monodispersed γ-Fe_2O_3 nanoclusters.[257] Interestingly, shape control of nanoparticles may be afforded by altering the nature of the solvent or injection timing of surfactants.[258]

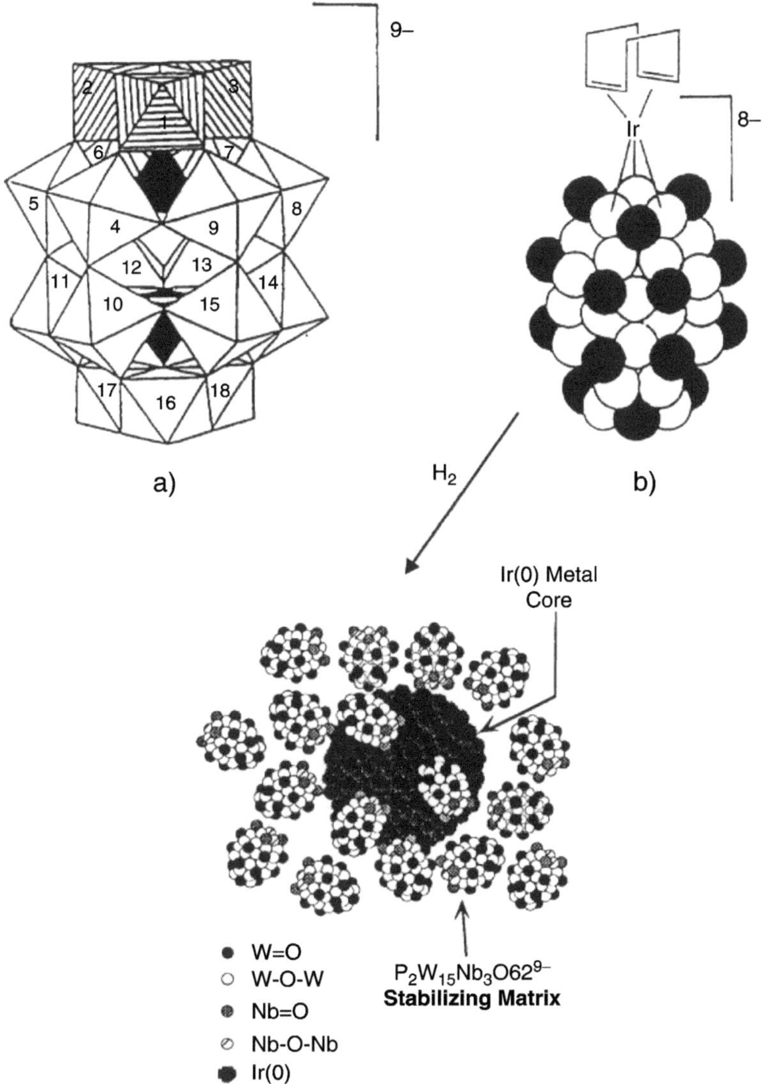

Fig. 6.64 Polyoxoanion-stabilized Ir^0 nanocluster formation. Shown is **a** Polyhedral representation of the α-1,2,3-$P_2W_{15}Nb_3O_{62}{}^{9-}$ stabilizing anion, and **b** a space-filling representation of the [(1, 5—COD)Ir($P_2W_{15}Nb_3O_{62}$)]$^{8-}$ complex. In (**a**), the three Nb atoms are indicated by the striped octahedra in positions 1–3. The WO_6 polyhedra occupy the 4–18 positions, and the internal PO_4 tetrahedral units are illustrated in black. In (**b**), the black spheres represent M–O terminal groups, and the white spheres represent M–O–M bridging groups. The Bu_4N^+ and Na^+ counterions are omitted for clarity. Images (**a**) and (**b**) reproduced with permission from *Inorg. Chem.* **1990**, *29*, 1784. © 1990 American Chemical Society. The bottom image of the stabilized nanocluster is reproduced from *Chem. Mater.* **1997**, 9, 3083. © 1997 American Chemical Society

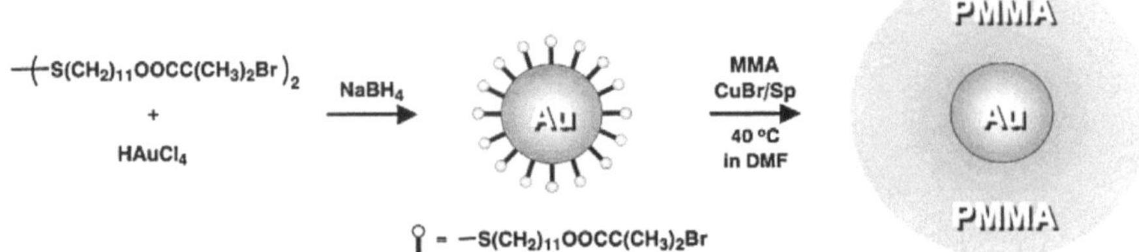

Fig. 6.65 Schematic of the formation of gold nanoparticles coated with free-radical polymerization initiators that subsequently yield Au@polymer nanostructures through a surface-controlled living polymerization process. Reproduced with permission from *Macromolecules* **2002**, *35*, 8989. © 2002 American Chemical Society

Tri-*n*-octylphosphine oxide (TOPO, **VIII**) is another common inverse-micelle stabilizing agent used to control the sizes of semiconductor nanocrystals such as PbE or CdE (E = S, Se, Te) quantum dots. The synthesis is performed at elevated temperatures since TOPO is a low-melting (53 °C), moisture-sensitive solid. A typical procedure consists of heating/degassing TOPO at *ca.* 360 °C under an inert atmosphere, quickly injecting a tri-*n*-octylphosphine (TOP) suspension of Me_2Cd and chalcogen powder at a reduced temperature (*ca.* 180 °C), and re-heating to 250–260 °C.[259] The quick injection of precursors into the hot solvent generates homogeneous nucleation of CdE nanocrystals. Subsequent re-heating affords slow growth of nanocrystals via Ostwald ripening[260] This is important in order to achieve a very high degree of monodispersity. Alternatively, PbS quantum dots have been synthesized by mixing PbO and oleic acid in a high b.p. solvent such as octadecene, followed by the injection of a solution of bis(trimethylsilyl) sulfide (TMS) in octadecene. Size control was demonstrated in this system by varying the PbO:oleic acid ratio as well as the injection and growth temperatures.[261]

O
P

VIII

Another strategy is the synthesis of metal oxide nanoparticles from the simple reaction of surfactant-stabilized basic metalate salts with CO_2 (Fig. 6.66).[262] This route should work for any metalate salt that is typically formed from the reaction of the metal hydroxide with a strong base (6.10, for sodium hafnate). Metal/nonmetal oxides may also be synthesized using a sol–gel technique, analogous to that described in Chap. 4 for thin film growth. For instance, SiO_2 nanoparticles of *ca.* 50–100 nm may be obtained through the controlled hydrolysis of a silicon alkoxide (e.g., tetraethylorthosilicate, TEOS) in a mixture of ethanol, aqueous ammonia (21%) and water. Particle sizes were controlled by the addition of an entraining silane such as 3-aminopropyltriethoxysilane that was added to TEOS prior to hydrolysis.[263] However, it should be noted that an entraining agent is not always required; for instance, iron oxide nanocrystals may be synthesized via olation/oxolation processes of Fe^{2+} ions in an acidic aqueous solution (6.11–6.13).[264]

$$2\,\mathrm{NaOH} + \mathrm{Hf(OH)_4} \rightarrow \mathrm{Na_2HfO_3} + 3\,\mathrm{H_2O} \tag{6.10}$$

$$\text{Step 1: } \left[\mathrm{Fe(H_2O)_6}\right]^{3+} \rightarrow \left[\mathrm{Fe(OH)(H_2O)_5}\right]^{2+} + \mathrm{H^+}(\mathrm{pH} = 0 - 2) \tag{6.11}$$

$$\begin{aligned}\text{Step 2: } 2\left[\mathrm{Fe(OH)(H_2O)_5}\right]^{2+} &\rightarrow \left[\mathrm{(H_2O)_4Fe(OH)_2Fe(H_2O)_4}\right]^{4+} \\ &+ 2\,\mathrm{H_2O}\,(\mathrm{pH} = 2 - 3)\end{aligned} \tag{6.12}$$

$$\text{Step 3: } \left[\mathrm{(H_2O)_4Fe(OH)_2Fe(H_2O)_4}\right]^{4+} \rightarrow \mathrm{Fe_2O_3 \cdot 9H_2O}\,(\mathrm{pH} = 3 - 5) \tag{6.13}$$

In addition to oxides, a variety of other compositions may be synthesized. In general, any stoichiometry is "fair game" through the same reactions that one does in bulk scale. For example, sulfides may be formed through reaction of precursors with H_2S,[265] nitrides through NH_3 exposure, etc.[266] Reactions with sodium chalcogenides represent a safer means to introduce chalcogens to the structure; for instance, the formation of L-cysteine stabilized Mn^{2+}-doped ZnS photoluminescent phosphor nanoparticles (6.14). As long as a suitable entraining agent is used, one is able to control the resultant size/morphology of the nanostructures. With current advances in compound and metallic nanoclusters, with the properties of each being fine-tuned via quantum confinement effects, one can begin to imagine intriguing designs for future microelectronic devices.

$$\mathrm{ZnSO_4} + \mathrm{Na_2S} + \mathrm{MnCl_2} \xrightarrow{\text{L-cysteine}} \mathrm{ZnS} : \mathrm{Mn^{2+}} + \mathrm{Na_2SO_4} + \mathrm{Cl^-} \tag{6.14}$$

Core–Shell Nanoparticles

As their name implies, core–shell nanoparticles consist of a core composition that is coated with a layer of another material.[267] However, in addition to solid bilayer nanoparticles, a hollow core–shell nanoparticle may be synthesized by sacrificially removing the core material by chemical or thermal decomposition. Such intriguing

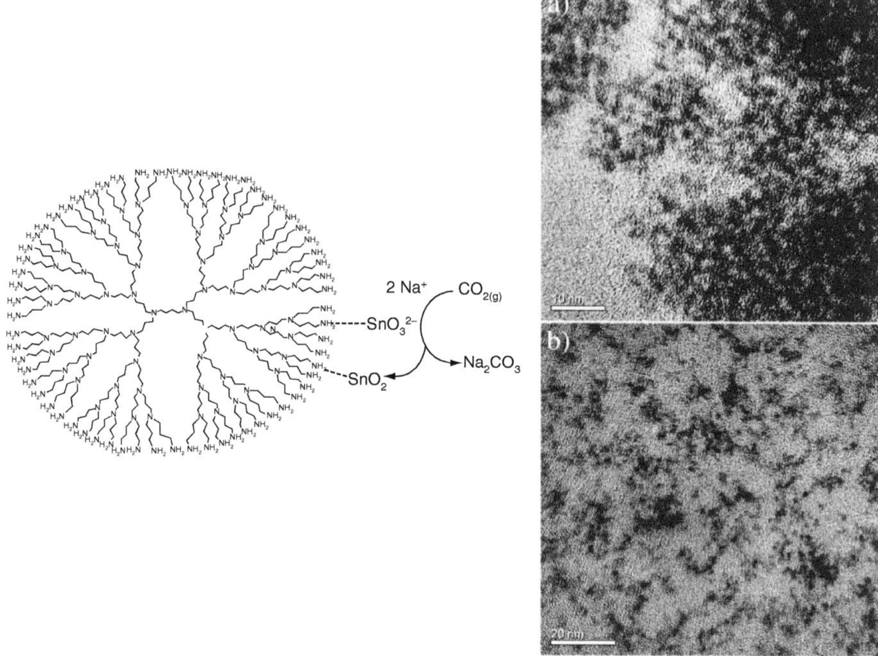

Fig. 6.66 Schematic of the growth of tin oxide nanoclusters at room temperature. The TEM images on the right illustrate interdendritic stabilized nanoclusters using **a** PAMAM and **b** amine-terminated poly(ethyleneimine) hyperbranched polymer hosts. Reproduced with permission from *J. Am. Chem. Soc.* **2006**, *128*, 420. © 2006 American Chemical Society

nanobuilding blocks will likely be of extreme importance for next-generation medical treatment/sensing, hydrogen storage, ion-exchange, energy storage, and microelectronics applications. Two types of core–shell syntheses may be employed: (i) the core nanoparticle is synthesized/purified followed by surface functionalization to afford subsequent coating of the shell material; (ii) the core nanoparticle is synthesized, followed by in situ coating of the shell material. Although the in situ route is often preferred due to its facile one-step procedure, it is possible for some impurities to be trapped between the core and shell layers.

The term *nanoshell* is often used to denote the deposition of a thin coating onto a core nanoparticle.[268] A common strategy for nanoshell growth consists of applying a thin metallic coating onto surface-functionalized silica or polystyrene templating spheres. Such gold nanoshells were pioneered by Naomi Halas at Rice University, and have far-reaching applications in photonics and biomedicine such as cancer diagnosis/treatment.[269] If desired, one may also subsequently remove the template by treating with hydrofluoric acid (HF) or toluene for silica or polystyrene cores, respectively. The reverse case of a polymer/ceramic coating onto removable metallic nanoparticles has also been applied to yield nonmetallic nanoshells.[270] However, the

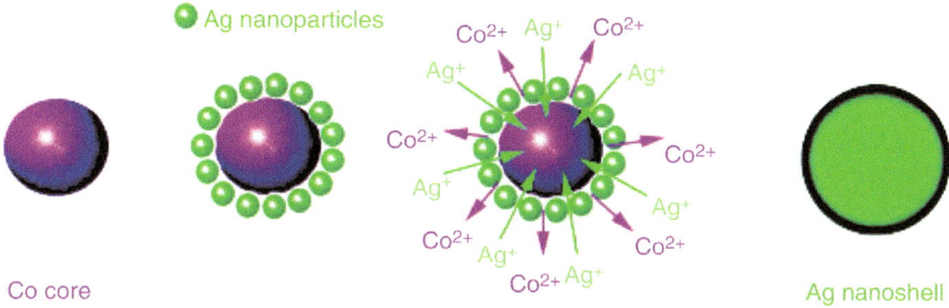

Fig. 6.67 Schematic of Ag nanoshell formation from a nanostructural Co core. Due to favorable redox couples between Ag and Co, a nanoshell of metallic silver forms at the expense of the inner Co core. Reproduced with permission from *Inorg. Chem.* **2006**, *45*, 5145. © 2006 American Chemical Society

removal of a relatively large core (i.e., typically >200 nm) from a nanoscale coating generally results in significant deformation of the resultant nanoshell.

The structural robustness of hollow core–shell nanoparticles has been recently improved through the use of sacrificial cores of smaller diameters. An example of this strategy is the growth of silver nanoshells from nanostructural Co templates (Fig. 6.67). It should be noted that $NaBH_4$ reduction of ligand-capped Co^{2+} ions preferentially yields Co_2B rather than metallic cobalt. However, a post-exposure of oxygen converts the boride into metallic cobalt nanostructures (6.15).[271] Once the Co core is formed, silver ions are introduced into the system through addition of $AgNO_3$. Since Ag^+ ($E_{red}° = 0.799$ V) is preferentially reduced relative to Co^{2+} ($E_{red}° = -0.377$ V), the exchange of Ag for Co occurs spontaneously as Ag^+/Co^0 boundaries are formed. This process continues as Ag^+ ions diffuse through the growing Ag^0 layer, until all of the metallic Co is consumed from the core.

$$4\ Co_2B + 3\ O_2 \rightarrow 8\ Co + 2\ B_2O_3 \tag{6.15}$$

Rather than redox-governed core–shell nanoparticle growth, diffusion-governed routes are also possible. For example, the exposure of cobalt nanoparticles to oxygen, sulfur, or selenium does not simply form a Co-chalcogenide coating, but rather hollow nanoparticles of the cobalt chalcogenide (Fig. 6.68).[272] This interesting result is a modern extension of an effect that has been studied since the late 1940s: the *Kirkendall effect*. This phenomenon describes the differential diffusion rates of two species in direct contact with one another at elevated temperatures (e.g., Cu and Zn in brass). During the growth of oxide or sulfide films, pores are formed in the solid due to a vacancy-exchange mechanism. That is, the outward movement of metal ions through the oxide layer is balanced with an inward migration of lattice vacancies that settle near the metal/oxide boundary. Due to the large number of defects and volume of bulk solids, the pores at the metal–oxide interface do not coalesce into an ordered array. However, for a nanostructural system that is relatively free of defects

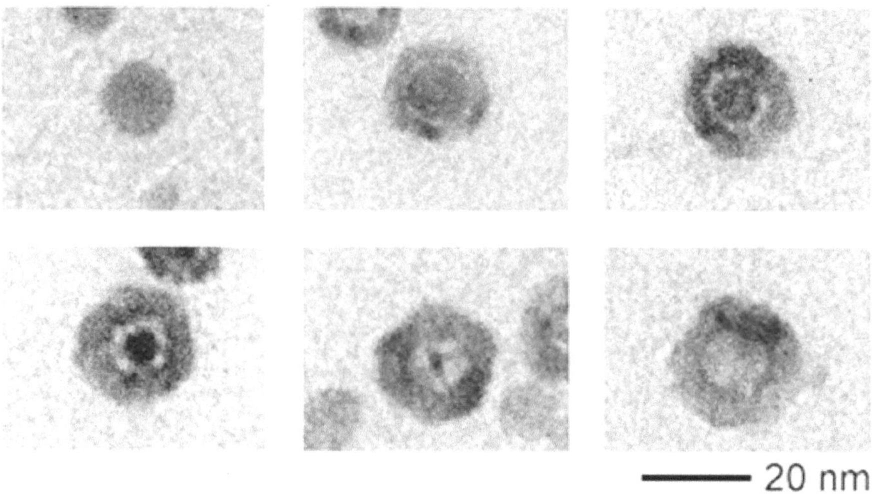

Fig. 6.68 Hollow nanoshell formation via the Kirkendall Effect. Shown is the evolution of the hollow nanoshell after reaction times of (left–right): 0, 10, 20 s, 1, 2, 30 min. An interesting feature is the formation of "bridges" that connect the core to the sulfide shell, facilitating the fast outward migration of Co. Reproduced with permission from *Science* **2004**, *304*, 711. © 2004 AAAS

and exhibits a large surface/volume ratio, the pores readily aggregate into a single hollow core.

It should be noted that organic cores or shells may also be synthesized. Organic-based polymer cores are synthesized by emulsion polymerization, followed by modification by a polyelectrolyte prior to subsequent shell polymerization.[273] For emulsion polymerization, the monomer is dispersed in water containing a surfactant. If inorganic cores are present, hydrophilic polymer coatings are often used to increase the stability and biocompatibility of the nanoparticles for medical applications. Oftentimes, the hydrophilic initiator used for polymerization also acts as the entraining agent.[274]

Growth Mechanisms of Metal Nanoclusters

In order to maintain control over the composition and morphology of a 0-D nanostructure, it is essential that we understand the growth mechanism for these structures. Only within the last decade have we figured out ways to repeatably control the morphology and composition of nanoparticles/nanoclusters. This explains why the mechanistic details of nanocluster growth have not surfaced until recently. One of the first systems to be investigated was metallic iridium nanoclusters, formed through the hydrogenation of a polyoxoanion-supported Ir complex.[275] Since Ir does not have an observable surface plasmon resonance profile, the rate of nanocluster growth was determined by following the hydrogenation of alkenes over time. This is possible since the initial complex is not active toward alkene hydrogenation; catalytic activity arises only from the reduced metal (Ir^0).

The general four-step mechanism for nanocluster growth is shown in Fig. 6.69.[276] Although this pathway is based on metallic nanocluster studies, other types of

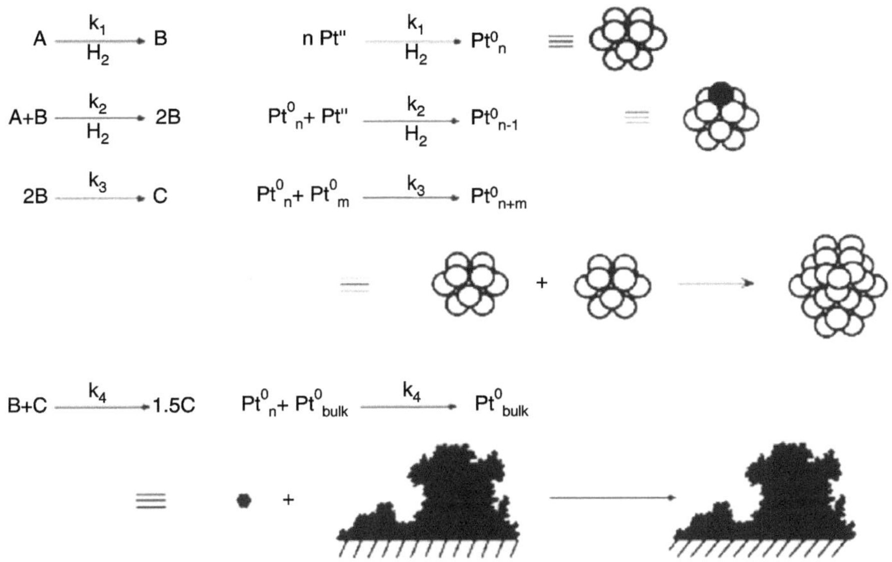

$$A \xrightarrow[H_2]{k_1} B \qquad n\,Pt'' \xrightarrow[H_2]{k_1} Pt^0_n$$

$$A+B \xrightarrow[H_2]{k_2} 2B \qquad Pt^0_n + Pt'' \xrightarrow[H_2]{k_2} Pt^0_{n-1}$$

$$2B \xrightarrow{k_3} C \qquad Pt^0_n + Pt^0_m \xrightarrow{k_3} Pt^0_{n+m}$$

$$B+C \xrightarrow{k_4} 1.5C \qquad Pt^0_n + Pt^0_{bulk} \xrightarrow{k_4} Pt^0_{bulk}$$

Fig. 6.69 Schematic of the four-step mechanism for transition metal (e.g., Pt) nanocluster formation. Shown are (i) nucleation to a desired cluster size; (ii) autocatalytic growth onto the cluster surface; (iii) diffusive agglomerative growth of two nanoclusters; and (iv) autocatalytic agglomeration into bulk metal particulates. Reproduced with permission from *J. Am. Chem. Soc.* **2005**, *127*, 8179. © 2005 American Chemical Society

nanoparticles would likely follow a similar route.[277] The first step involves the slow, continuous nucleation of clusters that are much smaller than 1 nm. When an energetically favored critical nucleus size is reached (*ca.* 15 atoms for Ir^0; varies depending on the metal), additional atoms rapidly aggregate to the surface—a process referred to as *autocatalytic surface growth.* The term "autocatalytic" is used since the nanoclusters formed from the nucleation step are also reactants for subsequent surface growth. Accordingly, the autocatalytic step will proceed faster as the reaction progresses, effectively shutting down the nucleation step—essential to achieve monodisperse nanocluster growth. That is, the size of the growing nanocluster may be controlled by varying the relative values of k_1 and k_2, as well as the availability of additional precursor molecules, both of which separate the nucleation and growth events in time. In this fashion, metal nanoclusters have been designated as "living metal polymers," analogous to organic living polymers discussed in Chap. 5.

It is widely believed that Au nanoparticles synthesized by the Turkevich method are formed via the *LaMer model*,[278] in which the concentration of metal atoms in solution constantly increases (by decomposition of a precursor compound) en route toward supersaturation. Atomic nucleation then occurs, followed by growth via the adsorption of additional gold atoms. This may be observed experimentally by measuring a sharp drop in the concentration of metal atoms in solution. Below a certain level of supersaturation, no further nucleation events will take place, which is required to effectively control the monodispersity of the resultant nanoparticles. Further growth will be accompanied by coarsening or Ostwald ripening, whereby

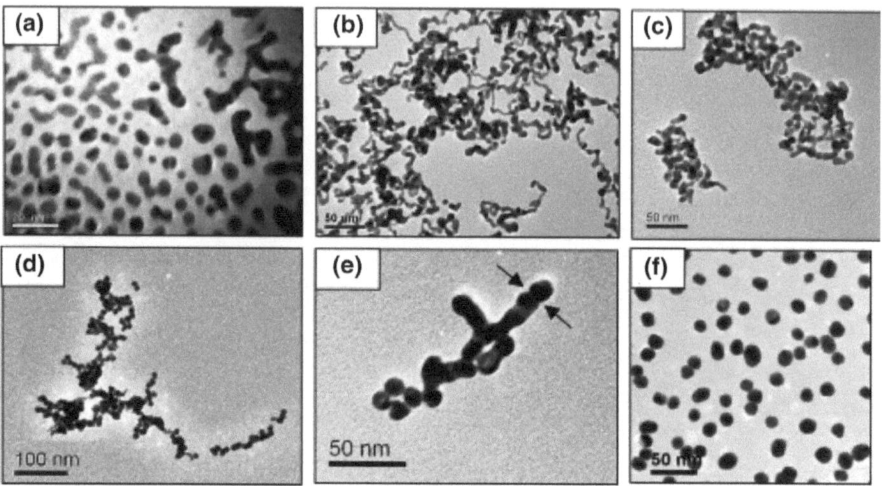

Fig. 6.70 TEM micrographs of various stages of Au nanoparticle synthesis via the citrate reduction of $AuCl_4^-$, showing nanowire intermediate formation. The colors of the solutions are: **a** colorless, **b** dark blue, **c** dark purple, **d** purple, and **e-f** ruby-red. Reproduced with permission from *J. Phys. Chem. C* **2007**, *111*, 6281. © 2007 American Chemical Society

larger particulates grow at the expense of smaller ones. Although this may yield a high degree of monodispersity, it may also yield bulk metal particles via a second autocatalytic surface-growth process (step 4, Fig. 6.78). A general prediction of the overall mechanism is that lower concentrations and higher temperatures are most conducive to yield nanoclusters rather than bulk metal (i.e., $k_1 >> k_3, k_4$). Indeed, this prediction is backed up by experimental data.

The sizes of resultant Au nanoparticles may be fine-tuned by varying the experimental conditions[279]; in fact, a recent study has even elucidated a non-LaMer growth mode via nanowire intermediates (Fig. 6.70).[280] It should be noted that in the Turkevich process, the citrate ion serves as a reducing agent by undergoing oxidative decarboxylation to yield an acetone dicarboxylic acid intermediate. This species is then proposed to chelate gold $[AuCl_x]^{n-}$ ions as they undergo stepwise reduction from $Au^{3+} \rightarrow Au^{+} \rightarrow Au^{0}$.[281] Hence, it appears as though the reduction of metal ions in solution occurs via the aggregation/nucleation of unreduced ions, rather than the nucleation of zero-valent atoms that then coalesce into larger metallic nanostructures.[282]

Interestingly, it has been shown that the growth of nanoclusters proceeds through the formation of "magic number" (or closed shell) clusters that exhibit unusual electronic stability. It should be noted that once the diameter of the cluster is <*ca.* 10 nm, a significant portion of the total atoms lie at its surface, which will result in different physical/optical properties than exhibited by the bulk metal.[283] A fcc or hcp transition metal will form stable clusters with icosohedral or truncated octahedral (TO; analogous to C_{60}) symmetry (Fig. 6.71). The number of metal atoms within an icosohedral cluster is given by the Mackay sequence of (6.16); i.e., 13, 55, 147, 309, 561, 923, 1415, 2057, 2869, etc.:

$$N_T = \frac{10}{3}n^3 + 5n^2 + \frac{11}{3}n + 1 \tag{6.16}$$

where n = packing shell (n = 1, 2, 3, 4,...). The number of surface atoms in the nth shell is given by $10n^2 + 2$.[284] For instance, the first stable shell consists of a central atom surrounded by 12 surface atoms, corresponding to 92% of the total atoms housed on the surface.

In contrast, the number of total and surface atoms within a TO array are given by (6.17) and (6.18), respectively. For instance, the first stable cluster of 38 atoms consists of 32 surface atoms surrounding a core of 6 atoms (84% of the total atoms

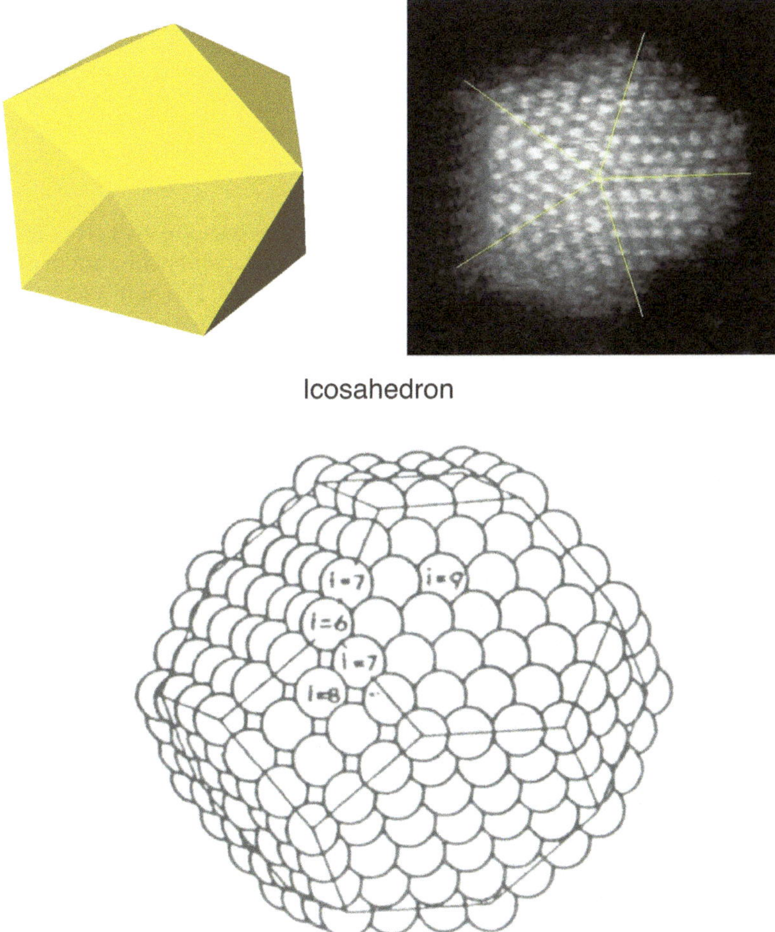

Fig. 6.71 Illustrations of icosahedral and truncated octahedral (TO) clusters. The numbers listed on the TO faces indicate atomic coordination

located on the surface). Subsequent shells yield stable clusters with 201, 586, 1289, 2406, 4033, 6266, etc. atoms. Further, the number of TO edge atoms is represented by (6.18), corresponding to the most reactive atoms that exhibit the lowest number of nearest neighbors. Hence, the first shell of a TO structure would contain 24 edge atoms—63% of the total number of atoms in the cluster. A regular TO cluster may be designated by its (n, m) indices with $n - m = 0$ $(n, m > 1)$, where n is the number of atoms on an edge adjoining (111) facets and m are atoms between adjoining (111) and (100) faces.[285] However, gold nanoclusters with $N_T > 20$ have been shown to exhibit the irregular TO^+ structure, with $0 < n - m \leq 4$, $m > 1$. Stable clusters within the TO^+ set contain 79, 140, 225, 314, 459,... metal atoms.

$$N_T = 16(n+1)^3 - 33(n+1)^2 + 24(n+1) - 6 \tag{6.17}$$

$$N_S = 30(n+1)^2 - 60(n+1) + 32 \tag{6.18}$$

$$N_E = 36(n+1) - 48 \tag{6.19}$$

It should be noted that intermediate magic-number nanoclusters represent only local minima in the potential energy surface, relative to the global minimum of a bulk metal with the lowest possible surface area (Fig. 6.72).[286] The high yield of magic number nanoclusters is a consequence of kinetically controlled surface growth. That is, once these favored intermediate structures are formed, they are less reactive toward autocatalytic surface growth relative to non-magic number clusters.

Thus far, we have considered mechanisms to explain the relative sizes of 0-D nanostructures, without consideration of their shapes. The observed shapes of nanocrystalline facets may be controlled by either thermodynamics or kinetics. For those reactions under thermodynamic control, growth will occur in such a way to minimize the total *interfacial free energy*, γ—i.e., the energy associated with the nucleation/growth of a surface of a unit area. Hence, particle growth is dependent on the surface energy of each crystallographic face. For a fcc structure, $\gamma\ \{110\} > \gamma\ \{100\} > \gamma\ \{111\}$; hence, a single crystal should prefer an octahedral structure to maximize the number of (111) facets. However, as we have seen, a truncated octahedron (TO) structure is often observed since a cube will have a lower γ than an octahedron of the same volume. It has also been suggested that one shape may evolve into another during growth, since a nanostructure will aggregate to eliminate high-energy surface facets and minimize its overall surface energy.[287]

In contrast, for kinetic-controlled growth, the concentration, reactivity, and chemical potential of the precursor monomer are governing variables to realize a particular shape. Interestingly, various shapes may be synthesized from a single set of reaction parameters by simply varying the monomer concentration. In order to achieve kinetic control over the resultant nanostructure, one must generally slow down the precursor decomposition/reduction while taking advantage of Ostwald ripening. Once the atomic nucleation and growth are slowed, the nanostructures will exhibit random hexagonal close-packing in addition to a variety of stacking faults, which will lead

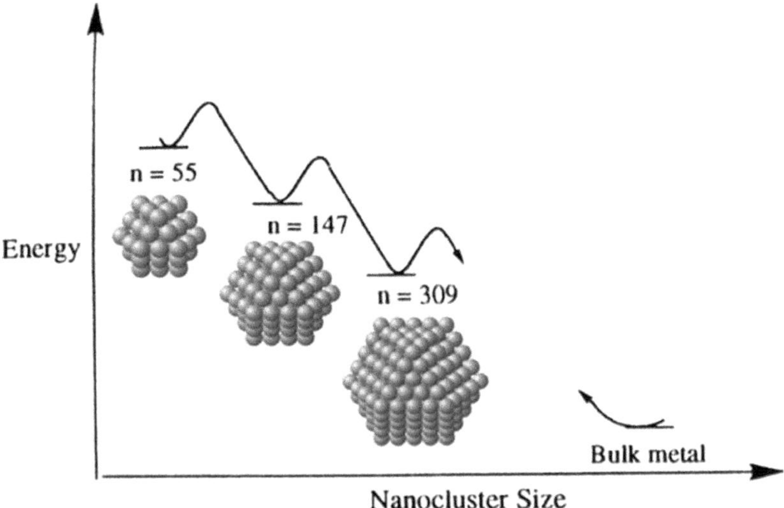

Fig. 6.72 Comparative energy levels for "magic number" noble metal nanoclusters relative to bulk metal. Also shown are the perfect fcc arrays of the nanoclusters, in which more than 75% of the atoms are located on the surface. Adapted with permission from Finke, R. G. in *Metal Nanoparticles: Synthesis, Characterization, and Applications* Feldheim, D. L.; Foss, C. A. eds., Dekker: New York, 2002. © 2002 Taylor & Francis

to thermodynamically disfavored shapes (Fig. 6.73). Further, if twinned regions are present rather than single-crystal nuclei, one is able to guide the morphology into 1-D or 2-D arrays via selective oxidation of specific crystal facets, or by the preferential adsorption of entraining agents to specific regions of the growing nanocrystal.[288]

Self-assembly of 0-D Nanostructures

Now that you understand how 0-D nanostructures are synthesized and stabilized, it is worthwhile noting how these structures are aligned into more complex arrays. As you recall, the "bottom-up" approach to materials design features the purposeful placement of individual nanoarchitectures, in order to build specific functional devices one unit at a time. Although one would have the greatest control over the device properties through the growth/arrangement of individual atoms, current synthetic methodology most easily yields nanoclusters/nanoparticles that consist of small groups of atoms (N_{atoms} ~ 50+). Nanoparticles can arrange themselves in 2-D/3-D based on interactions occurring at the liquid–liquid interface.[289] However, advanced lithographic techniques such as dip-pen nanolithography (DPN), micro/nano-contact printing, focused-ion beam (FIB), and deposition of nanoparticles within lithographically generated surface features have all been used to pattern 0-D nanostructures into regular 2-D and 3-D arrays.[290]

Rather than having to use a mold for nanocontact printing, one may also directly write features onto an appropriate substrate using a molecular ink, via the ultrafine tip of an atomic force microscope[291] (Fig. 6.74). This technique is known as

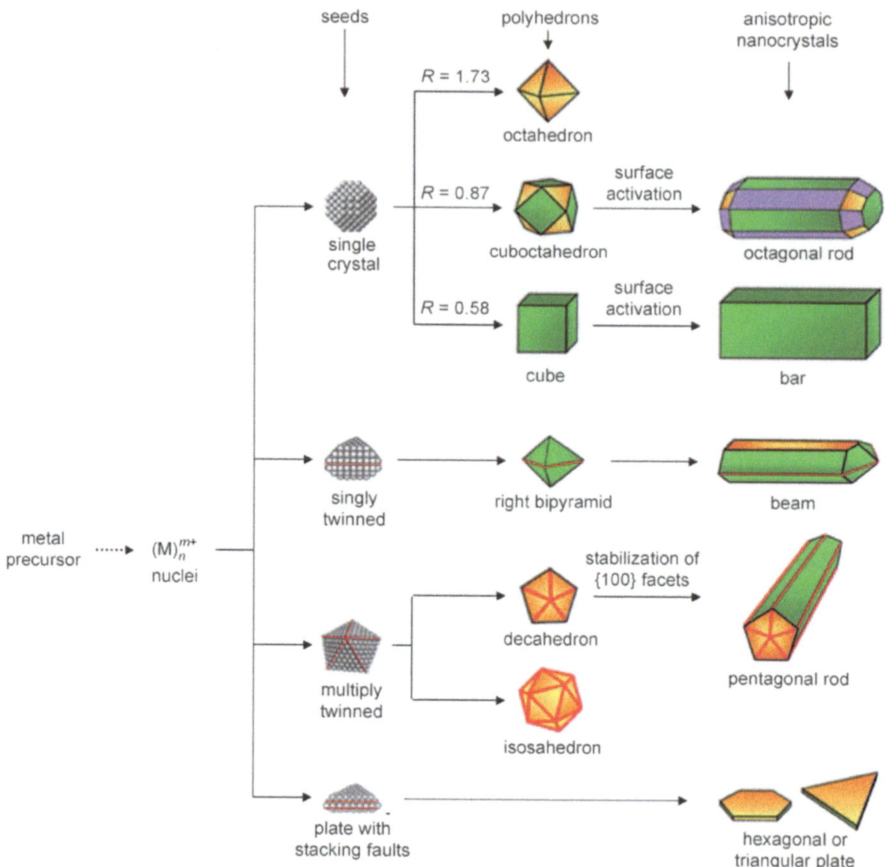

Fig. 6.73 Reaction pathways that give rise to varying shapes of fcc metal nanocrystals. First, a precursor is reduced or decomposed to form the nuclei (small clusters). Once the nuclei have grown past a certain size, they become seeds with a single-crystal, singly twinned, or multiply twinned structure. If stacking faults are introduced, then plate-like seeds will be formed. The green, orange, and purple colors represent the {100}, {111}, and {110} facets, respectively. Twin planes are delineated in the drawing with red lines. The parameter R is defined as the ratio between the growth rates along the <100> and <111> directions. Reproduced with permission from *Adv. Mater.* **2007**, *19*, 3385. © 2007 Wiley–VCH

dip-pen nanolithography (DPN), first demonstrated by Mirkin and coworkers in the late 1990s.[292] Although the earliest examples of DPN featured alkylthiols as the ink onto Au surfaces, there are now an increasingly large number of other ink/substrate combinations that have been reported for DPN (Table 6.7),[293] which have most recently been exploited for nanomaterials discovery (Fig. 6.75). A general benefit of DPN over other soft lithographic techniques is the ability to pattern nanostructures (including biological materials) by a single step without cross-contamination, since the desired chemistry occurs only in a specifically defined location of the substrate.

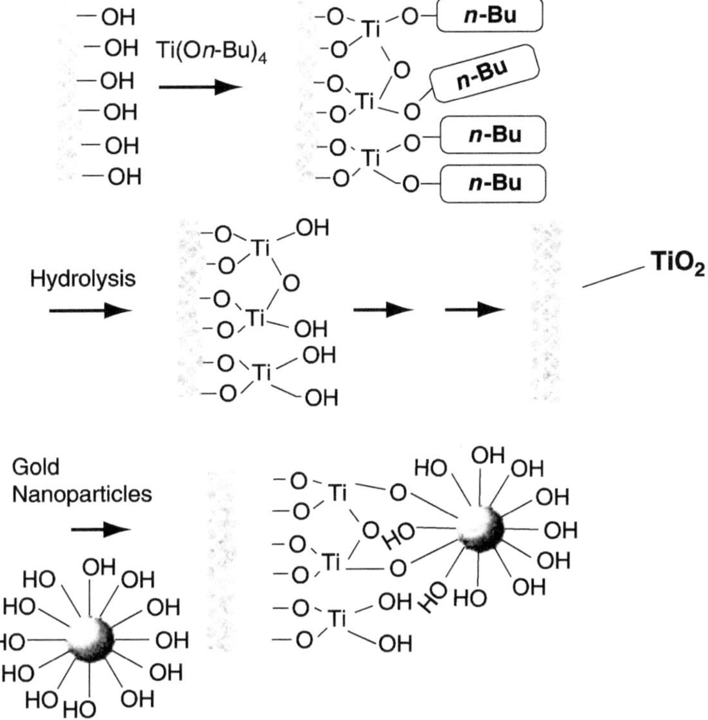

Fig. 6.74 Illustration of the chemisorption of a nanostructure onto a hydroxylated TiO_2 surface. The hydroxyl groups on the periphery of the nanostructure stabilizing agent provide the handle for surface adsorption. Reproduced with permission from *Phys. Chem. Chem. Phys.* **2002**, *4*, 6059. © 2002 PCCP Owners Societies

The mechanism of ink transport from the AFM tip to substrate is appreciably complex, being influenced by the chemical/physical nature of the ink, surface tension, interfacial wettability, tip structure, ink coverage on the tip, Laplace pressure at the meniscus, ambient temperature and humidity, and operating parameters (e.g., dwell time, contact force, lifting speed).[294] Models suggest that a meniscus forms between the tip and substrate, which aids in ink transport. As a result, the transport rate is found to increase concomitantly with the ambient humidity, but only for inks that are soluble in water. In general, the rate of ink transport is found to decrease significantly with increasing contact time, due to the changing surface energy of the substrate.[295] The resolution of DPN is primarily affected by surface spreading of the depositing ink due to interfacial wettability. To overcome this limitation and allow line resolutions of <10 nm, mechanical DPN was developed in which ink was transferred onto a diffusion-limited area that was formed using an AFM tip.[296]

Table 6.7 Summary of the ink-substrate combinations used to date for DPN

Molecular ink	Substrate
Alkylthiols (e.g., ODT[a] and MHA[b])	Au
Ferrocenylthiols	Au
Silazanes	SiO_x, GaAs
Proteins	Au, SiO_x
Conjugated polymers	SiO_x
DNA	Au, SiO_x
Fluorescent dyes	SiO_x
Sols	SiO_x
Metal salts	Si, Ge
Colloidal particles	SiO_x
Alkynes	Si
Alkoxysilanes	SiO_x
ROMP materials	SiO_x
Thioacetamide/cadmium acetate[c]	SiO_x, mica

[a] 1-Octadecanethiol
[b] 16-Mercaptohexadecanoic acid, or thiohexadecanoic acid
[c] Ding, L.; Li, Y.; Chu, H.; Li, X.; Liu, J. *J. Phys. Chem. B* **2005**, *109*, 22337

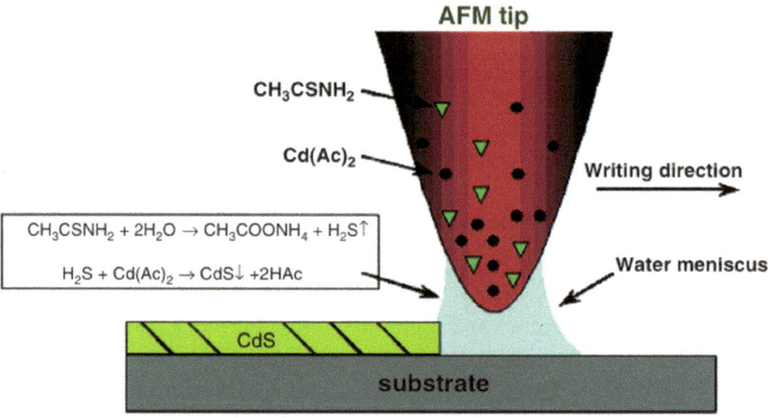

Fig. 6.75 Illustration of dip-pen nanolithography (DPN). Reproduced with permission from *Chem. Rev.* **2020**, *120*, 6009. © 2020 American Chemical Society

As we have seen, a stabilizing agent is used to prevent agglomeration of growing 0-D nanostructures. In fact, this component also provides an effective "handle" to bind the nanostructure to a particular surface. Once appropriate reactive groups (e.g., –OH, NH_2) are placed on a surface through monolayer formation, the stabilizing group surrounding the nanostructure spontaneously becomes chemisorbed (Fig. 6.76). If the nanostructures are encapsulated with stabilizing agents of well-defined sizes, then the spacing between adjacent nanostructures will also be highly ordered and predictable (Fig. 6.77). Although this most often results in a 2-D matrix of nanostructures, it is also possible to create alignment into 1-D chains. For instance, Au nanostructures that are stabilized by a long-chain thiolated poly(ethylene oxide) polymer form a linear array via interactions with sulfate/acid groups of a polysaccharide compound (chondroitin sulfate c, **IX**) (Fig. 6.78). In addition to chemisorptive methods, electrostatic forces generated from biased p–n junction-patterned substrates has also been an effective means to align metallic nanoparticles.[297]

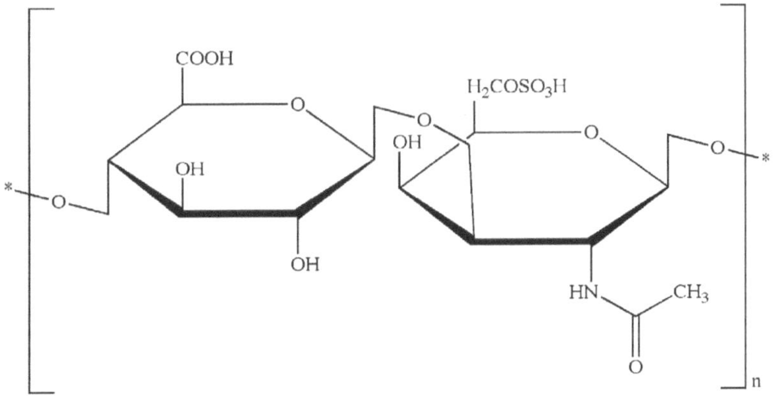

IX

An important precedent for the self-assembly of 0-D nanostructures into a 3-D superlattice array was reported by O'Brien and coworkers.[298] In this work, oleic acid stabilized γ-Fe_2O_3 nanoclusters were synthesized by the decomposition of $Fe(CO)_5$ and subsequent oxidation using trimethylamine N-oxide within a high boiling solvent (trioctylamine, b.p. 365 °C). Depending on the relative size ratios and concentrations of Fe_2O_3:PbSe QDs, AB_{13} or AB_2 superlattice arrays were generated (Fig. 6.79)—isostructural to intermetallic compounds such as $NaZn_{13}$ and AlB_2, respectively. The superlattices were made by simply placing a substrate in a suspension containing the two types of nanoparticles. Upon evaporation of the solvent in a low-pressure chamber, the nanoparticles self-assembled into the ordered arrays. Binary superlattices of this type that feature both magnetic and semiconducting QDs may have applications for data storage applications, whereas arrays of 2+ semiconductor QDs may be exploited for next-generation displays, as well as solar or thermoelectric devices.

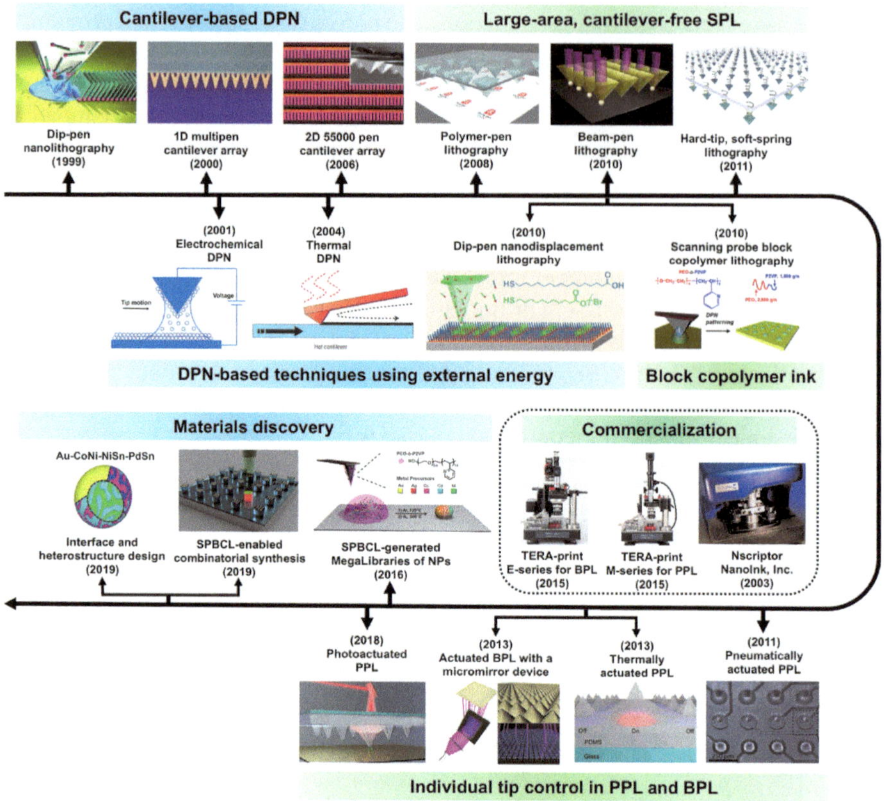

Fig. 6.76 Timeline of dip-pen nanolithography (DPN), from a cantilever-based technique to advanced lithography for heterostructure design. Reproduced with permission from *Chem. Rev.* **2020**, *120*, 6009. © 2020 American Chemical Society

The most popular method used to achieve 3-D arrays such as layered nanostructural thin films, is the *layer-by-layer* (LbL) self-assembly pioneered by Decher in the early 1990s.[299] This technique is based on the sequential adsorption of species with complimentary functional groups (e.g., ionic charges) on a variety of substrates. There are many device applications for LbL thin films, such as self-cleaning surfaces, surface deactivation of warfare agents, solar energy, drug delivery, and optoelectronics. An interesting precedent for LbL nanostructural film growth uses sequential layering of cyclodextrin-stabilized Au nanoparticles and adamantyl-terminated dendrimers (Fig. 6.80).[300] The film thickness was reported as 2 nm per bilayer, allowing for strict control over the resultant film thickness.

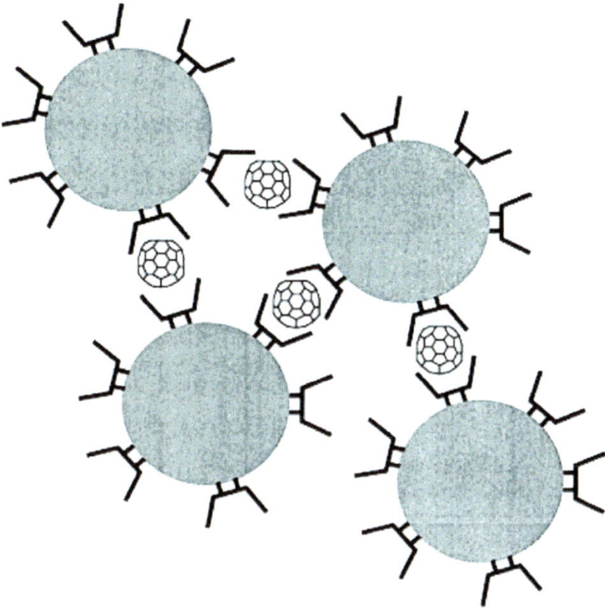

Fig. 6.77 Schematic of the controlled spacing between individually stabilized gold nanoparticles by the formation of fullerene inclusion complexes. Reproduced with permission from *Chem. Mater.* **1999**, *11*, 33. © 1999 American Chemical Society

6.3.2 One-Dimensional Nanostructures

The second class of nanoscale building blocks, referred to as 1-D nanostructures, is reserved for those materials that have nanoscale dimensions that are equivalent in all but one direction. Whereas a 0-D nanostructure is analogous to the period following this sentence (length = width); a 1-D nanostructure is analogous to the number "1" (length > width).

Since we began the discussion of 0-D nanostructures with nomenclature, we will follow suit in this section. Once again, it is easy to be confused by the common synonymous use of the terms nanotube, nanofiber, nanowire, and nanorod. However, if you think of the analogous bulk materials without the prefix "nano," there should be no ambiguity regarding the proper use of these descriptors (Fig. 6.81). The commonality among all of these structures is that their diameters must be within the 1–100 nm range; typically, their lengths are within the micron (or larger) regime. A *nanotube* is a 1-D structure that contains a hollow core, whereas the other three nanoarchitectures are solid throughout. The term *nanofiber* should be reserved for 1-D nanostructures that are amorphous (and usually nonconductive) such as polymers and other non-graphitized carbonaceous structures. By contrast, a *nanowire* designates a structure that is crystalline, with either metallic or semiconductive electrical properties.

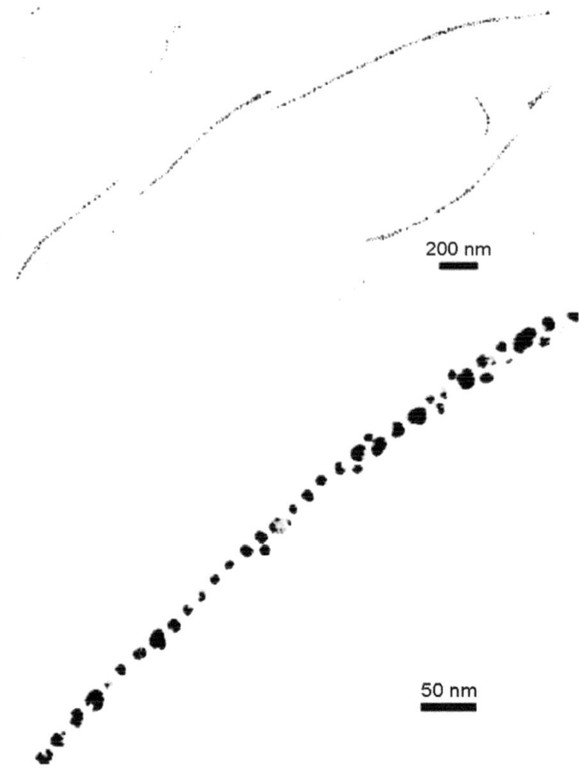

Fig. 6.78 TEM images of a 1-D array of Au nanoparticles formed through the interaction of PEO and polysaccharide chains. Reproduced with permission from *Chem. Lett.* **2002**, 990. © 2002 Chemical Society of Japan

A *nanorod* is typically a crystalline 1-D nanostructure with an overall length comparable to its width (i.e., both dimensions are <100 nm). As its name implies, another feature of nanorods is their rigid sidewall structures. However, since crystalline nanorods exhibit the same overall shape as needle-like bulk crystals, the term "nanocrystal" is probably more appropriate for these structures (or, more explicitly, "rod-like nanocrystals"). Whereas nanowires, nanofibers, and nanotubes exhibit an interwoven array, nanorods are completely linear in morphology. As such, nanorods are capable of stacking onto each other to yield interesting 2-D and 3-D arrays—not usually as easy to perform with the "spaghetti-like" morphology of the other 1-D nanostructures.

Carbon Nanotubes (CNTs)

Without question, the most widely studied 1-D nanomaterial is the *carbon nanotube* (CNT). These structures were first discovered by Iijima in 1991,[303] which may be visualized as a graphitic sheet(s) of sp^2 hybridized carbon atoms (i.e., graphene[304]) rolled into a tubular array. Based on the layers of graphene sheets that comprise the CNT, the structures are designated as single-walled, double-walled, or multiwalled

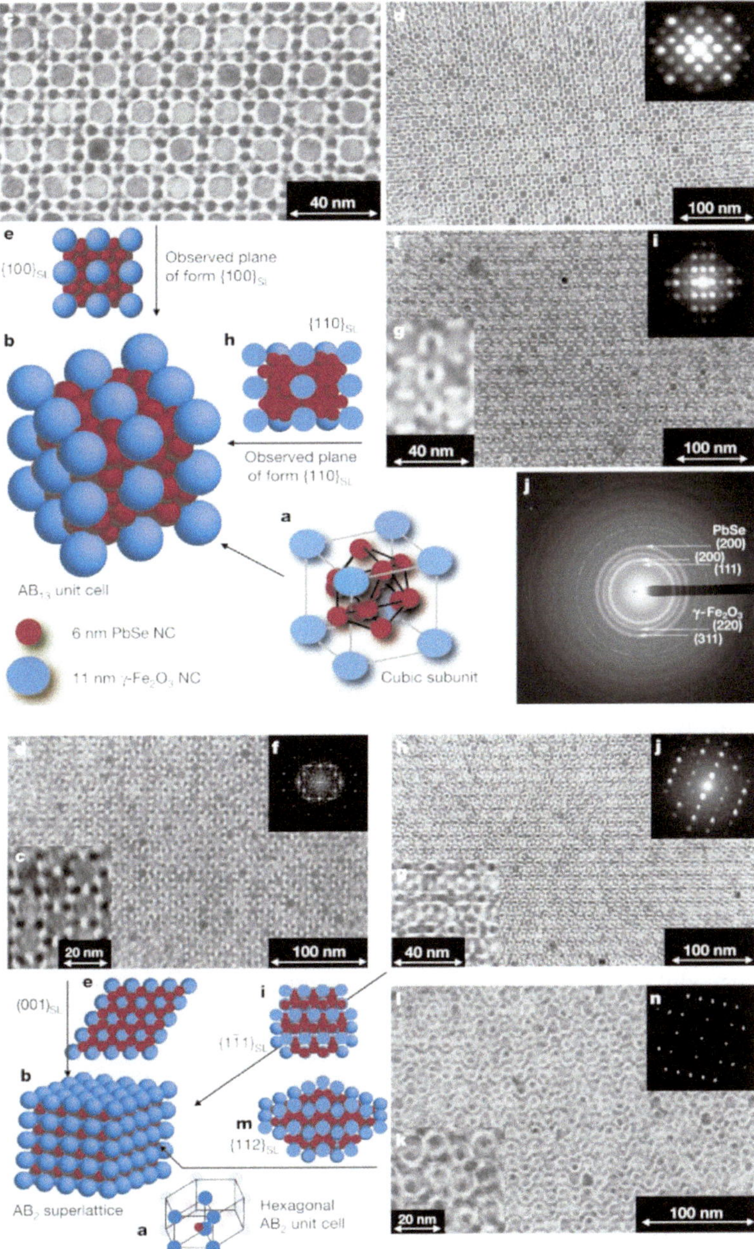

Fig. 6.79 Top: TEM micrographs and sketches of AB_{13} superlattices (isostructural with intermetallic phase $NaZn_{13}$) of 11-nm γ-Fe_2O_3 and 6-nm PbSe NCs. Bottom: TEM micrographs and sketches of AB_2 superlattices (isostructural with intermetallic phase AlB_2) of 11-nm γ-Fe_2O_3 and 6-nm PbSe NCs. Reproduced with permission from *Nature* **2003**, *423*, 968. © 2003 Nature Publishing Group

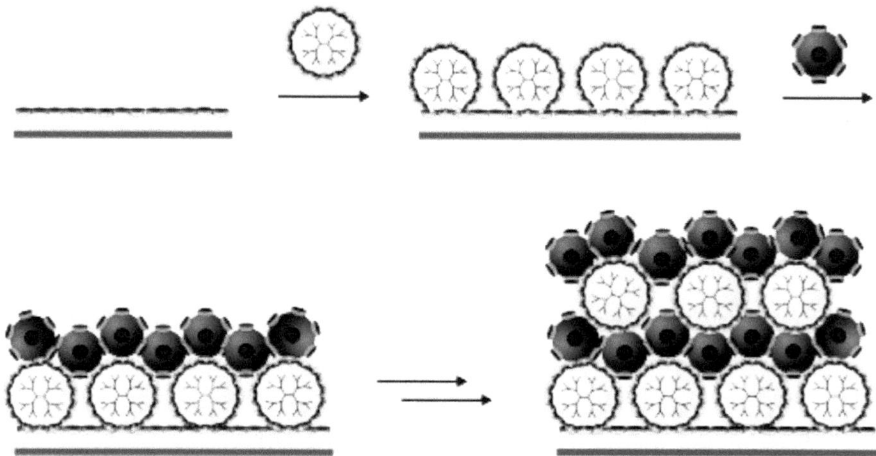

Fig. 6.80 Illustration of LbL growth of cyclodextrin-stabilized Au nanoparticles (dark spheres) and adamantyl-terminated PPI dendrimers. Reproduced with permission from *J. Am. Chem. Soc.* **2005**, *127*, 7594. © 2005 American Chemical Society

nanotubes (SWNTs, DWNTs, or MWNTs, respectively—Fig. 6.82). The diameters of CNTs range from 1 nm (SWNTs) to >30 nm (MWNTs), with aspect ratios (length:width) ranging from 100 to greater than 1×10^6.

Even though the diameters of CNTs are orders of magnitude smaller than a human hair, their theoretical tensile strength is predicted to reach *ca.* 10 times greater than steel—a property attributed to extremely strong sp^2 bonding between neighboring hexagonal units.[305] However, it should be noted that the physical or electronic properties for CNTs vary depending on their morphologies, routes of synthesis, or choice of theoretical calculations or characterization techniques employed. For instance, whereas the Young's modulus is calculated as 2.8–3.6 TPa for SWNTs, the values for MWNTs are in the range 1.7–2.4 TPa (for comparison, stainless steel has a Young's modulus of 0.19–0.21 TPa). In addition, although the tensile strength of SWNTs is predicted to be higher than 100 GPa, the measured mechanical strength is only in the range 13–52 GPa due to the formation of structural defects during their synthesis. Structural imperfections also cause thermal and electrical conductivities of CNTs to be much lower than their theoretical values.[306]

The electrical conductivity of SWNTs will vary from metallic to semiconducting, depending on the way a graphene sheet is folded (Fig. 6.83). In particular, the diameter and helicity of a SWNT are uniquely characterized by the *chirality vector* (or *Hamada vector*), **C**, which connects crystallographically equivalent graphene lattice sites. Vector indices designated by (n, m) are used to indicate the direction and length of the chirality vector (6.20). Equation 6.21 provides a formula to calculate the diameter, D, of a (n, m) nanotube, and 6.22 is the relation to determine the chirality angle, ϕ. When $n = 0$, the SWNT is denoted as the *zigzag* conformation; when $n = m$, the SWNT is in its *armchair* form. For all values in between these extremes, the nanotubes are designated simply as *chiral*.

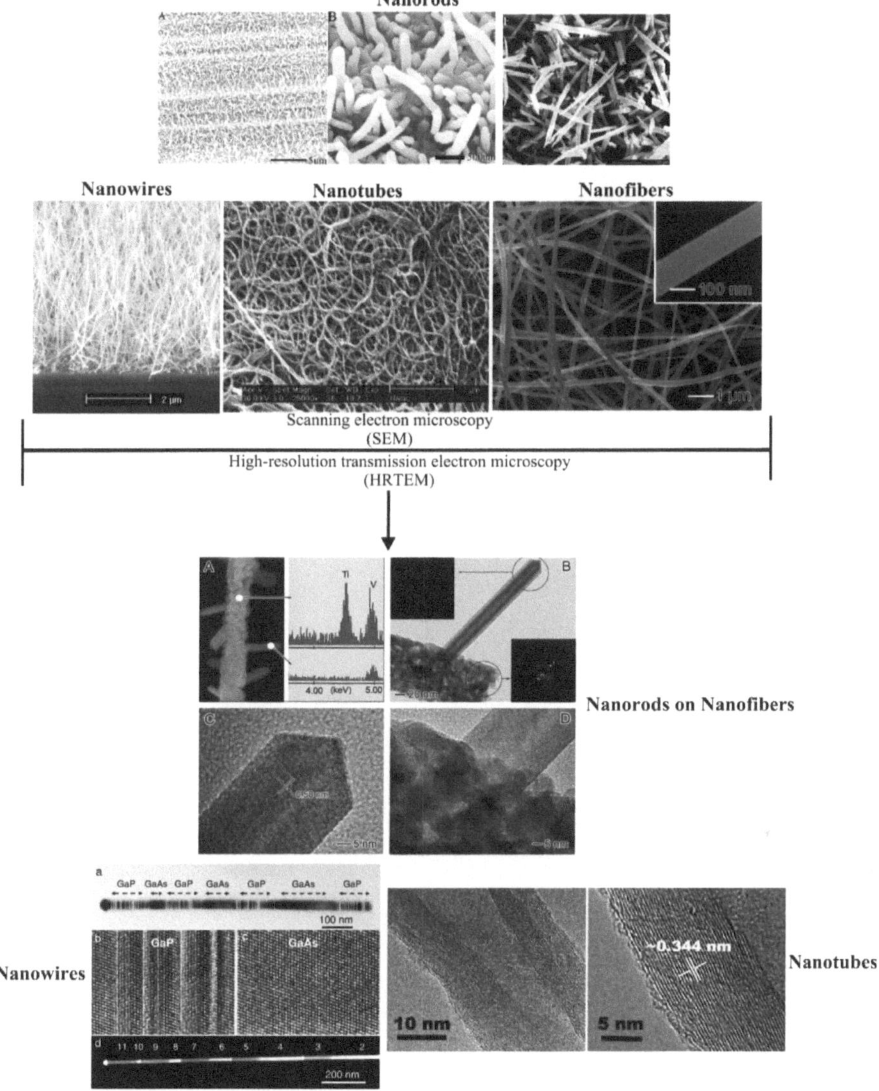

Fig. 6.81 Classifications of 1-D nanostructures. The top portion shows SEM images of various nanostructures.[301] Whereas the morphology of nanowires, nanotubes, and nanofibers look identical by SEM, nanorods are notably different, with much shorter lengths and straight sidewalls. The bottom portion shows high-resolution TEM images, which provide morphological details of the nanostructures.[302] The top TEM image shows crystalline nanorods grown on the surface of amorphous nanofibers. The bottom two TEM images illustrate the difference between nanowires/nanotubes—the latter contains a hollow core. It should be noted that crystalline nanorods may also be termed "nanocrystals," as their morphology resembles that of needle-like bulk crystallites

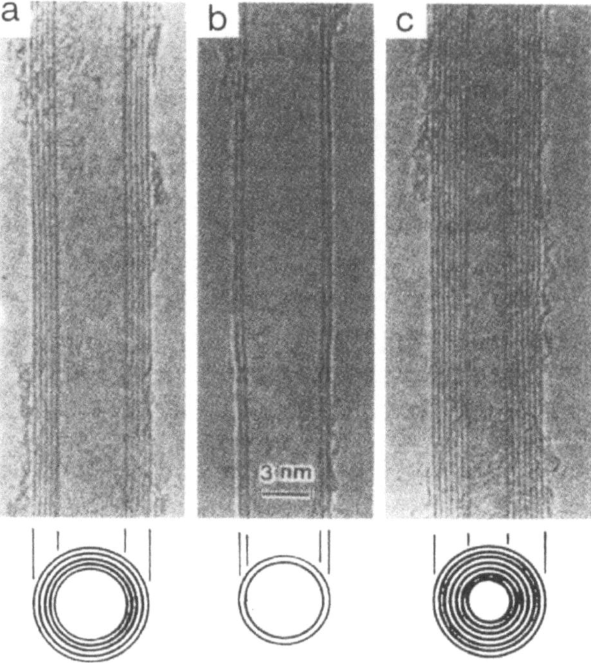

Fig. 6.82 TEM images of MWNTs, formed from the folding **a** 5-, **b** 2- (i.e., DWNT), and **c** 7-stacked graphene sheets. The spacing between individual concentric cylinders is 0.34 nm—the distance between adjacent planes in graphite. Reprinted from *Science of Fullerenes and Carbon Nanotubes*. © 1996 Elsever B.V

$$\mathbf{C} = n\boldsymbol{a}_1 + m\boldsymbol{a}_2 \tag{6.20}$$

$$\mathrm{D} = \frac{\sqrt{3}\mathrm{d}_{C-C}}{\pi}\sqrt{n^2 + nm + m^2} \tag{6.21}$$

$$\cos\phi = \frac{2n + m}{2\sqrt{n^2 + nm + m^2}} \tag{6.22}$$

where *n*, *m* are the integers denoting the number of unit cell vectors along two directions in the crystal structure of graphene, $\mathbf{a_1}$ and $\mathbf{a_2}$ are the graphene unit cell vectors, and $d_{C\text{-}C}$ is the C–C bond length (1.41–1.44 Å).

For metallic SWNTs, the electrical conductance may exceed silver or copper by three orders of magnitude. Calculations have shown that (*n*, 0) or zigzag SWNTs exhibit metallic conductivity when *n*/3 is an integer, and semiconducting properties for all other values of *n*. Similarly for chiral SWNTs, when $(2n + m)/3$ is an integer, the tubes are metallic (otherwise are semiconducting). Finally, armchair SWNTs, with $n = m$, exhibit metallic conductivity (Fig. 6.84). Most importantly, electronic band structure calculations show that metallic and semiconducting CNTs are dependent

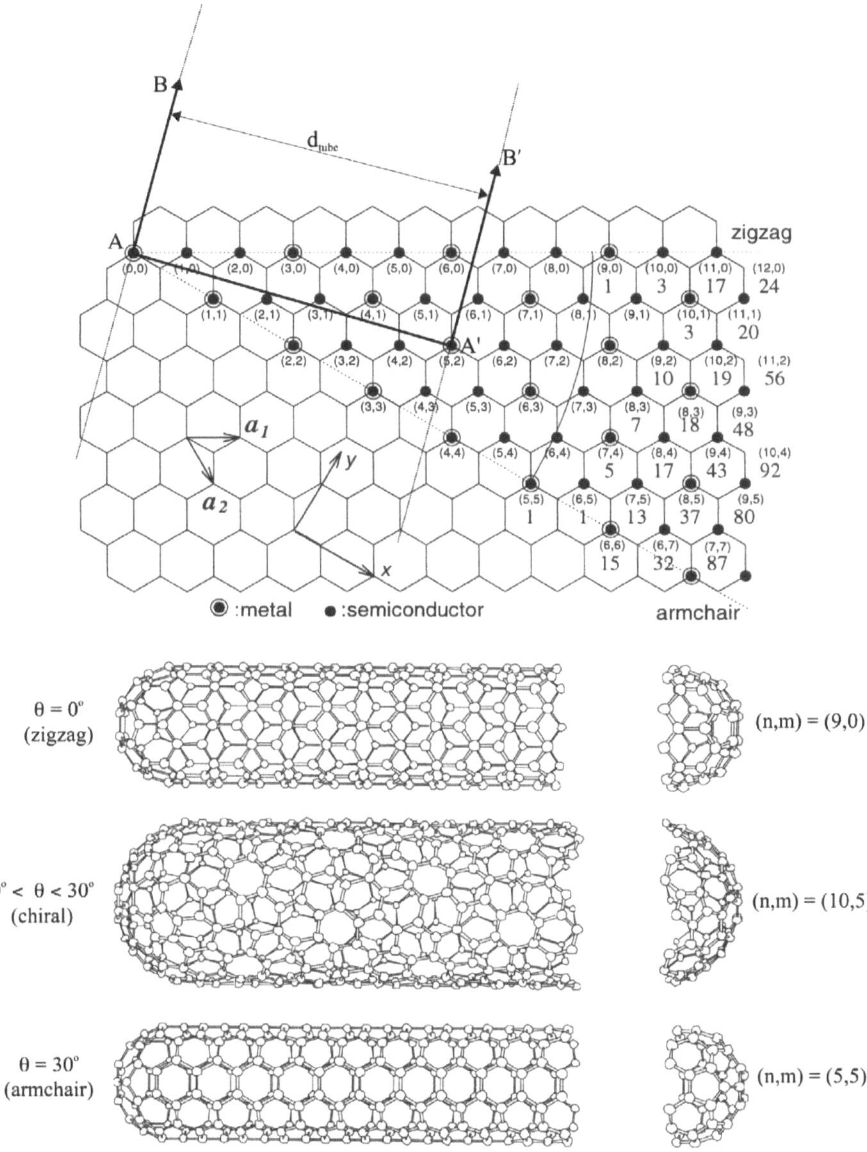

Fig. 6.83 Illustration of the honeycomb 2-D graphene network, with possible unit cell vector indices (*n*, *m*). The dotted lines indicate the chirality range of tubules, from $\theta = 0°$ (zigzag) to $\theta = 30°$ (armchair). For θ values between 0 and 30°, the formed tubules are designated as chiral SWNTs. The electrical conductivities (metallic or semiconducting) are also indicated for each chiral vector. The number appearing below some of the vector indices are the number of distinct caps that may be joined to the (*n*, *m*) SWNT. Also shown is an example of how a (5,2) SWNT is formed. The vectors AB and A′B′, which are perpendicular to the chiral vector (AA′) are superimposed by folding the graphene sheet. Hence, the diameter of the SWNT becomes the distance between AB and A′B′ axes. Reprinted from *Science of Fullerenes and Carbon Nanotubes*. © 1996 Elsevier Ltd.

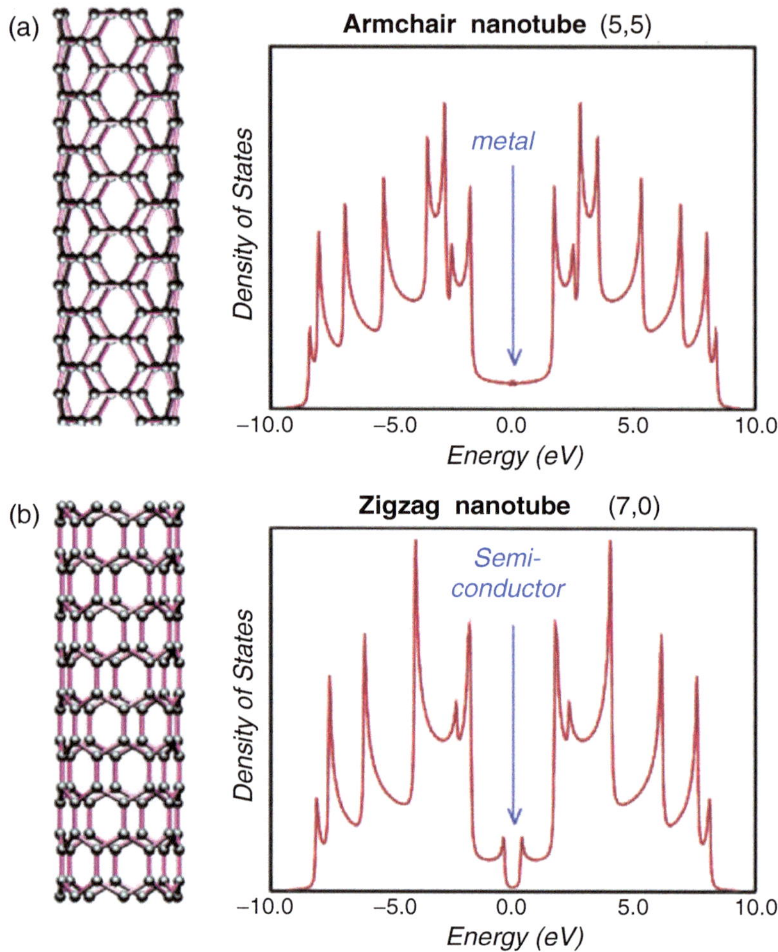

Fig. 6.84 Comparison of the density of states, and resultant electronic properties of armchair and zigzag SWNTs. Reproduced with permission from *Acc. Chem. Res.* **2002**, *35*, 1063. © 2002 American Chemical Society

only on (*n*, *m*)—that is, slight structural variations result in dramatic changes in their electronic properties. For example, the bandgap of semiconducting SWNTs may be fine-tuned from *ca.* 10 meV to 1 eV—with no required addition of dopants, unlike bulk Si counterparts (6.24). If semiconducting nanotubes exhibit the same chirality, the bandgap is inversely proportional to the diameter (e.g., E_g (7, 0) > E_g (10,0)). The density of states (DOS) for SWNTs exhibit sharp maxima, which are known as *van Hove singularities* resulting from quantum confinement in directions perpendicular to the tube axis. Optical transitions are possible between these maxima, which may be observed by Raman spectroscopy (Fig. 6.85).[307]

$$E_g = \frac{3.13\text{eV}(\text{d}_{C-C})}{\text{D}} \tag{6.24}$$

The tunable electronic properties of CNTs are being explored for next-generation IC architectures. As you may recall from Chap. 4, traditional Si-based microelectronic devices will likely reach a fundamental limit within the next decade or so, necessitating the active search for replacement materials. Accordingly, an area of intense investigation is *molecular electronics*—in which the electronic device is built from the placement of individual molecules.[308] Not surprisingly, the interconnects of these devices will likely be comprised of CNTs and other (semi)conductive 1-D nanostructures such as nanowires.

Since CNTs have a high electrical conductivity and contain sharp tips, these nanomaterials are the best-known field-emitter[309] of any material to date. In general, the smaller the radius of curvature of the tip, the more concentrated the electric field

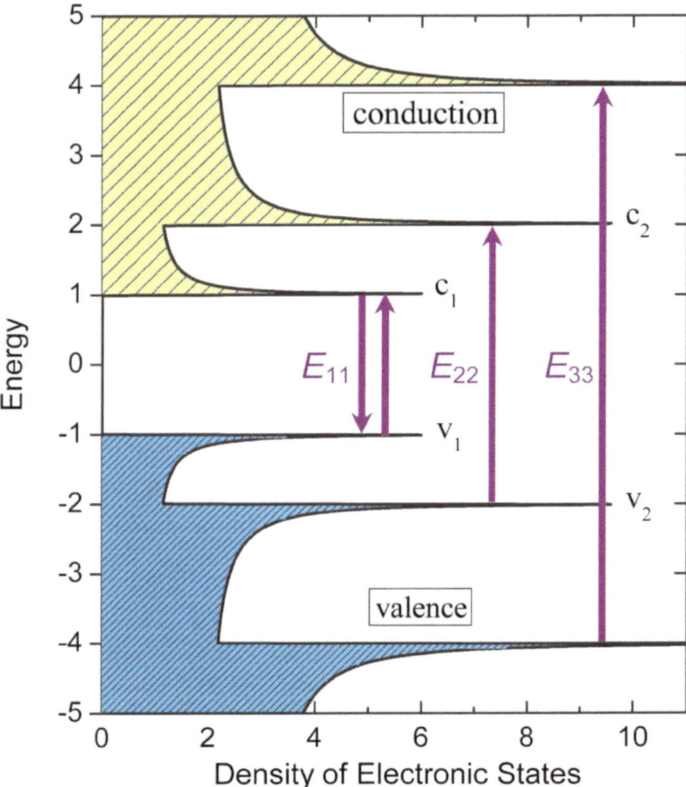

Fig. 6.85 Density of states in a semiconducting single-walled carbon nanotube. The sharp maxima are van Hove singularities. The dominant optical absorption/emission transitions are indicated by arrows. Reproduced with permission from *Electrochem. Soc. Interface*. © 2006 The Electrochemical Society

will be, which corresponds to increased field emission at low required voltages. In addition, a variety of new phosphor materials are currently being developed for the design of next-generation field emission displays.[310] Although (O)LED displays are still hot ticket items at electronics stores, they possess inherent disadvantages. LED screens are still rather expensive to produce and often lack the response time required to view fast-paced sporting events/movies without blur. In contrast to popular belief, the new 'LED TVs' available at the high end of the HDTV market are not truly LEDs. These displays are in fact LCDs that simply use LEDs—most recently mini-LEDs—instead of cold cathode fluorescent lamps as the backlight[311]; in contrast, true LED TVs would not require a backlight, since each LED pixel would be self-illuminating. Current "QLED" televisions feature photo-emissive particles, which use a layer of quantum dots to convert the backlight to emit pure basic colors, thereby improving the brightness and color range. In contrast, next-generation electro-emissive QD displays are similar to active-matrix OLEDs, in which light is produced directly by each pixel by applying electrical current to QDs. As seen earlier in this chapter, the color emitted by a QD is related to its size and composition.

As far as picture quality is concerned, traditional CRT displays are still among the finest quality. The replacement of this technology with CNTs is a logical step in the evolution of display panels. Rather than a single electron gun, CNT-based screens will contain a separate nanotube electron gun for each individual pixel in the display—dramatically enhancing the resolution and clarity of the picture. Further, in contrast to current large flat-panel televisions, the overall weight of CNT-based analogues will be significantly lower, and they will consume far less power. This concept has already been proven in prototypes, including their use in flexible displays,[312] but has largely been abandoned in recent years due to the popularity of OLEDs. The delay of large-scale field effect displays in reaching the commercial market is likely due to the complexity and cost associated with their production, which would require a high vacuum environment to operate. Hence, there is more interest in using CNTs as thin film transistor (TFT) backplanes for OLED displays (revisit Sect. 4.3 for details regarding TFTs and OLEDs).[313] Particularly attractive properties are higher mobilities relative to amorphous silicon (*ca.* 35 vs. 1 $cm^2\ V^{-1}\ s^{-1}$) and their facile deposition under low-temperature and non-vacuum conditions.[314] In addition, the flexible properties of CNTs make them ideal for elastic/wearable electronics.[315]

Another electronic application for CNTs is for next-generation field-effect transistor (FET) design.[316] The "proof-of-concept" for CNTFETs was demonstrated in the late 1990s, with a simple bridging of two noble metal electrodes with a SWNT (Fig. 6.86, top). However, the electrical characteristics of this new FET design were less than desirable, with high contact resistance (>1 MΩ) and low drive currents. Researchers at IBM have since modified the original design wherein the semiconducting SWNTs are placed directly onto an oxidized Si wafer, followed by the deposition of the source and drain (Co or Ti) electrodes (Fig. 6.86, bottom). Through subsequent annealing, a stronger interaction is afforded between the electrodes and CNT channel, which reduces the contact resistance.

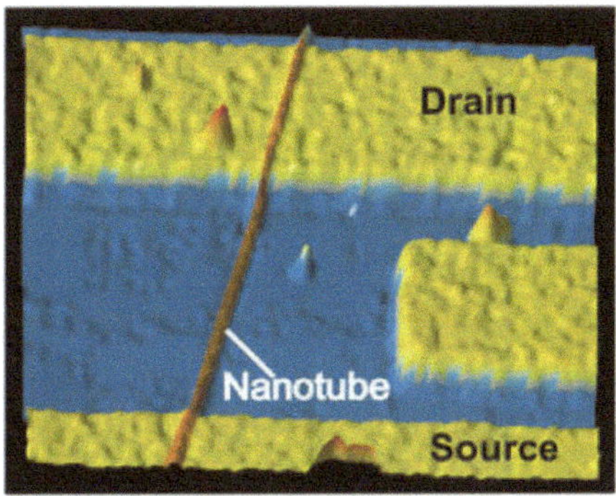

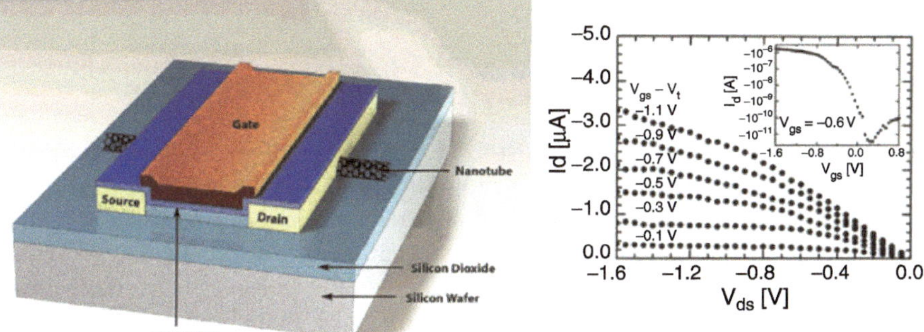

Fig. 6.86 Atomic force microscope (AFM) topographical image of an original CNTFET (top). The source/drain electrodes were fabricated on a SiO_2 thin film grown on a Si wafer. The wafer itself served as the gate electrode. Also shown (bottom) is an illustration of a top-gated CNTFET design, with an oxide thickness of 15 nm and CNT length of 300 nm. The device electrical characteristics at room temperature is also provided. The inset shows the transfer characteristic of the FET (output voltage of the device as a function of the input voltage). Reproduced with permission from *Acc. Chem. Res.* **2002**, *35*, 1026. © 2002 American Chemical Society

Avouris at IBM provided a comparison of the output from a top-gate CNTFET (Fig. 6.86, bottom) and existing Si-based FETs.[317] For CNTs with an average diameter of 1.4 nm, the on-current for a top-gated CNTFET was on the order of 2100 μA μm^{-1} at V_{DS} (drain voltage) = V_{GS} (gate voltage)–V_T (threshold voltage) = 1.3 V. In comparison, the highest drive current in a p-CMOS under the same conditions was 650 μA μm^{-1} for a gate length of 50 nm. The *transconductance* (ratio of the output current variation to the input voltage variation) of the CNTFET is 2300 μS μm^{-1}; the value for an analogous Si p-CMOS is 650 μS μm^{-1}. The values obtained thus far show that CNTFETs outperform Si-based FETs; a computer composed entirely of CNTFETs was recently demonstrated,[318] which is an exciting glimpse of what is surely on the horizon. Theoretically, carbon nanotubes could also be employed in gate-all-around FETs (GAAFETs) that are planned for the near future (Fig. 6.87). However, the reproducible production of nanotubes on a scale

Fig. 6.87 Illustration of the deformation modes of SWNTs, resulting in a high elasticity. This behavior is likely an artifact of in-plane flexibility of a graphene sheet, and facile rehybridization of carbon atoms from sp^2 to sp^3 geometries. Reproduced with permission from *Chem. Rev.* **1999**, 99, 1787. © 1999 American Chemical Society

that is sufficient to be used for transistor fabrication is still greatly lacking. Separation of semiconducting and metallic nanotubes is still difficult and time consuming. Furthermore, single-walled carbon nanotubes degrade in the atmosphere,[319] which will cause problems for chip reliability using this technology. Beyond individual transistors, it has also been suggested that CNTs could replace Cu in microelectronic interconnects, due to their high current-carrying capacity and resistance to electromigration. Lastly, the extremely efficient thermal conductivity of CNTs could also be exploited for cooling applications for future computers—of increasing concern as the chip density continues to soar.

Beyond their tunable conductive properties, CNTs are among the strongest and stiffest materials known to date (Table 6.8). The hollow, closed morphology of nanotubes results in rather intriguing deformation modes in response to a mechanical stress (Fig. 6.88). In particular, it has been suggested that Stone–Wales defects may become mobile under stress, which results in a change in tube diameter/chirality.[320] This also causes a change in the electronic properties of the CNT, opening up possibilities for sensor applications. However, for MWNTs, studies have shown that only the outer graphitic shell is able to support stress,[321] and SWNT bundles/ropes (Fig. 6.89) exhibit smaller Young's moduli, relative to isolated SWNTs, due to weak intertube cohesion.[322] Hence, although an individual SWNT has an elastic modulus of *ca.* 1.2 TPa, the value for bundles is *ca.* 100 GPa for diameters in the range of 15–20 nm. To improve the load distribution throughout the CNT and improve its mechanical strength, there are efforts to create crosslinks between individual shells of MWNTs, and among SWNTs within ropes.[323]

The desirable mechanical properties of CNTs have been exploited in recent years for the structural reinforcement of polymers.[324] This is a natural extension of traditional composite materials used for applications that require both high-strength and lightweight materials. There are currently plenty of examples of carbon fiber

Table 6.8 Specific tensile strengths of various materials[271]

Material	Tensile strength (MPa)	Density (kg dm^{-3})	Specific strength[a] (kN m kg^{-1})	Breaking length[b] (km)
Concrete	10	2.30	4.34	0.44
Rubber	15	0.92	16.3	1.66
Brass	580	8.55	67.8	6.91
Nylon	75	1.15	97.3	9.92
Polypropylene	80	0.90	88.9	9.06
Aluminum	600	2.70	222	22.7
Steel	2,000	7.86	254	25.9
Titanium	1,300	4.51	288	29.4
Silicon carbide	3,440	3.16	1,088	110
Glass fiber	3,400	2.60	1,307	133
Graphite	4,300	1.75	2,457	250
Kevlar[c]	3,620	1.44	5,246	534
CNTs	62,000	1.34	46,268	4,716

[a] The strength of a material divided by its density
[b] The length beyond which a strip of the material, of uniform width, would break under its own weight if suspended from both ends
[c] A synthetic fiber used in bullet-proof vests, comprised of poly(paraphenylene terephthalamide)

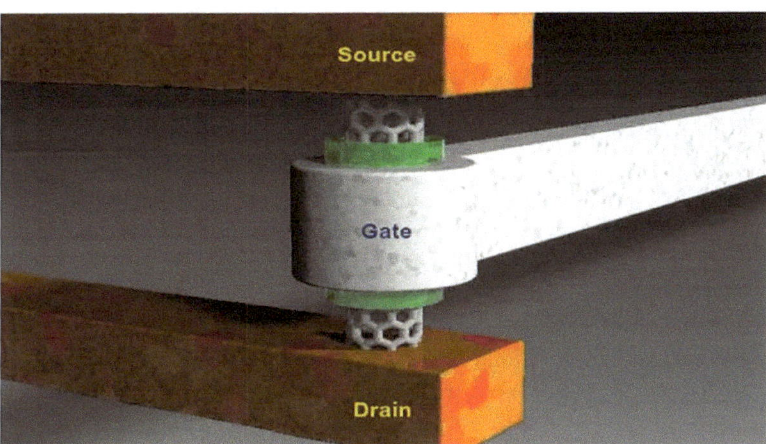

Fig. 6.88 An illustration of a vertical gate-all-around (GAA) carbon nanotube field-effect transistor. Reproduced with permission from *Mater. Today* **2004**, *7*, 46. © 2004 Elsevier

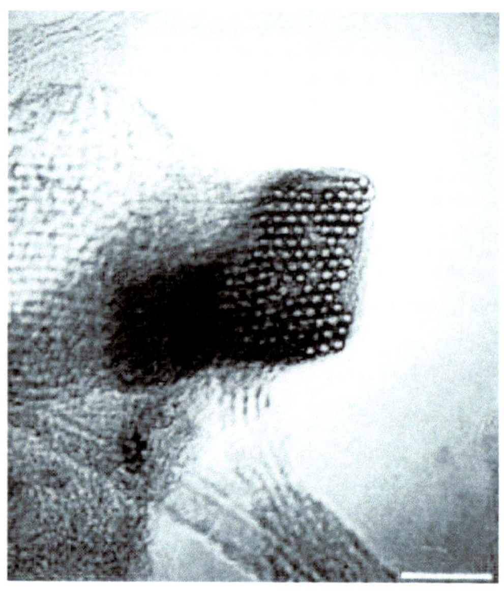

Fig. 6.89 High-resolution TEM image of a bundle/rope of single-wall carbon nanotubes, formed spontaneously through the self-assembly of individual SWNTs (scale bar is 10 nm). Reproduced with permission from *Science* **1996**, *273*, 483. © 1996 AAAS

composites—aircraft and spacecraft parts, racing car bodies, golf club shafts, bicycle frames,[325] fishing rods, automobile springs, sailboat masts, and many others—all incorporating bulk carbon fibers, *ca.* 5–10 μm in diameter. Although this technology has been around since the 1950s, the much larger aspect ratios of CNTs translates to an even greater impact on the future of materials reinforcement.

In order to broaden the scope of CNTs for textile applications, it is most desirable to convert the as-formed powders into useful fibers and yarns. An intriguing precedent consists of shrinking the yarn-spinning process, used by the earliest civilizations, to the nanoregime. In this process, MWNTs from a vertically aligned bamboo-like forest array were drawn into fibers, and weaved into yarns, that were both strong and highly flexible (Fig. 6.90). In order for spinning to be successful, the nanotube arrays must be well aligned, which is only possible through the formation of uniform and densely packed catalyst particles that are not coarsened during pre-treatment steps.[326] One possible application for these advanced nanotextiles is the design of a "supersuit" for the next generation of soldier (Fig. 6.90, bottom). In addition to exploiting the toughness and tensile strength of carbon nanotubes, a variety of other functionalities could be realized through the placement of appropriate surface coatings (e.g., adaptive camouflage, self-healing/repairing, stealth coatings, deactivation of gaseous warfare agents, exoskeletal human muscle mimicry, temperature/humidity control, etc.).[327]

In addition to increasing strength and stiffness, the incorporation of CNTs also imparts electrical/thermal conductivity to the polymer matrix. It has been shown that the addition of CNTs also enhances the solvent resistance and glass transition temperature of the native polymer. Further, due to the high aspect ratio of CNTs, much lower dopant levels are required to yield the desired properties, relative to standard additives such as carbon black and larger graphitic fibers. There are an endless number

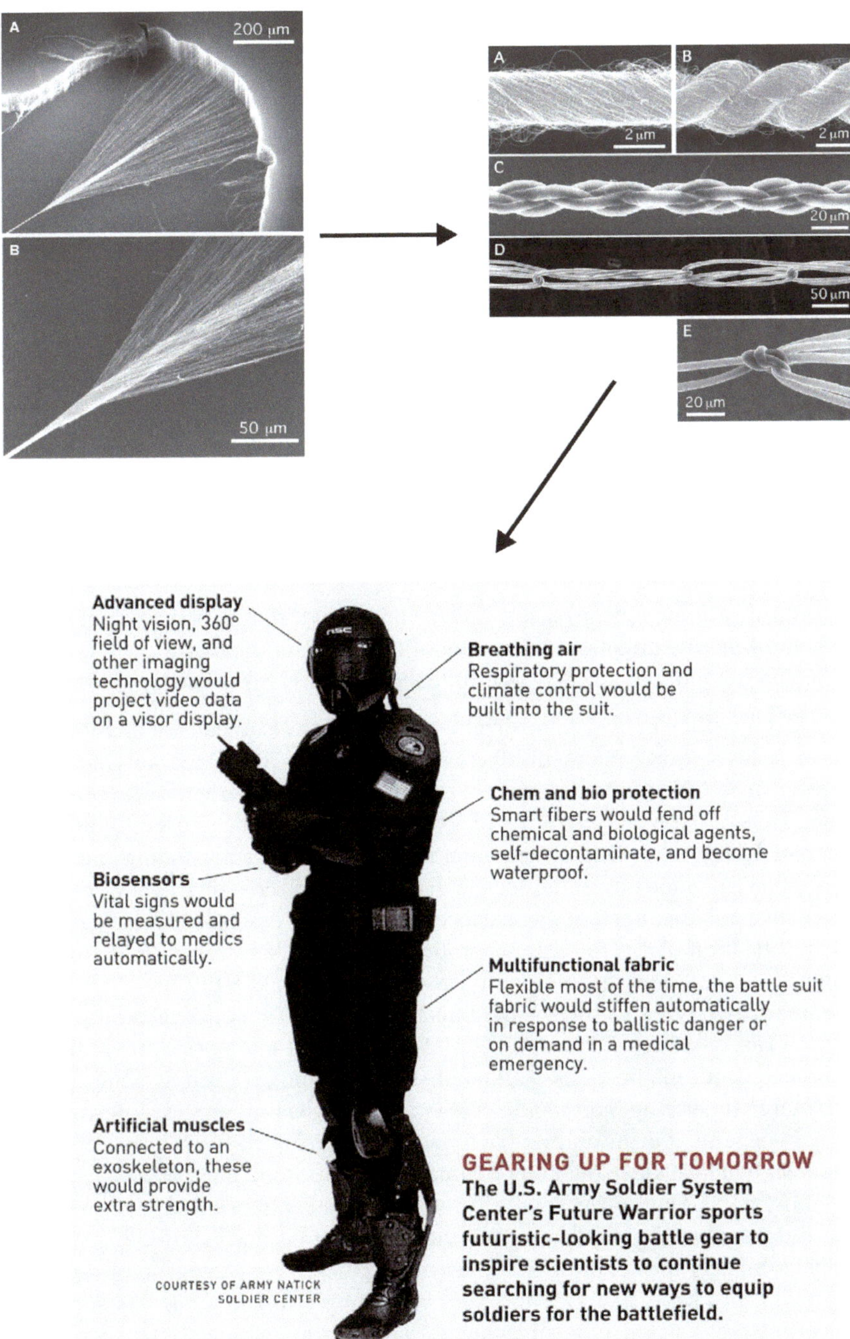

Fig. 6.90 SEM images of spinning MWNT arrays into fibers, and subsequent images of MWNT yarns. Shown below is an image of a futuristic uniform to equip the next generation of warfighter. The SEM images were reproduced with permission from *Science* **2004**, *306*, 1358, © 2004 AAAS. The image of the futuristic soldier was reproduced with permission from the Aug. 11, 2003 issue of *Chemical and Engineering News*. © American Chemical Society

of potential applications for CNT composites—imagine future flat-panel displays that are flexible, exhibiting the consistency of common fabrics, or shirts that are capable of monitoring the external temperature and automatically heating/cooling. The automotive industry currently uses CNT composites for a number of applications,[328] such as mirror housings, fenders, and door handles. Nanotubes are also frequently added to impart conductivity to plastic exterior parts that are subjected to electrostatic painting processes. In addition, CNT-doped nylon is being used for fuel lines to reduce the buildup of static electrical charges, and prevent fuel line rupturing during an accident.

It should be noted that reinforcement applications are not limited to organic polymers. Though ceramics are already hard and chemically/thermally resistant, these materials may also be doped with CNTs to improve their inherent brittleness. This has already been proven for the incorporation of 5–10% of CNTs within an alumina matrix. The resultant material exhibits five times the fracture toughness and seven times greater electrical conductivity relative to undoped Al_2O_3, with an added property of unidirectional heat conductivity.[329]

In order to continue the development of intriguing CNT-based composites, three main challenges are being addressed:

i. Ensuring a homogeneous dispersion of nanotubes throughout the polymer matrix
ii. Controlling the direction/orientation of the CNTs in the matrix
iii. Separating individual SWNTs from bundles/ropes (exfoliation).

In order for efficient load delocalization and strength enhancement of a composite, there must be a strong interaction between the dopant fibers and polymer matrix. However, since CNTs are allotropes of carbon, they are inherently insoluble in any solvent. Hence, there are a number of strategies that have been used to modify the surface of CNTs to facilitate their interactions with the surrounding matrix. Not only are these methods essential for the dispersion and self-alignment of nanotubes throughout a polymer host (or on a surface for molecular devices), but may also assist in the purification of CNTs (e.g., separation of metallic and semiconducting CNTs through selective functionalization, followed by traditional HPLC).

There are two types of surface modifications for CNTs: noncovalent interactions (Fig. 6.91) and covalent sidewall (Fig. 6.92)[330] or defect-site[331] (Fig. 6.93) functionalization.[332] Both methods, as well as physical techniques such as sonication, are successful in separating individual SWNTs from bundles—a process known as *exfoliation*. In general, it is most desirable to incorporate isolated SWNTs in a composite rather than bundles, since the latter features poor inter-tube interactions resulting in a lower overall strength—especially at low CNT concentrations.

As illustrated in Fig. 6.91, noncovalent modifications consist of either wrapping the CNT with polymers or biological macromolecules,[333] or placement of conjugated macromolecules on the surface through π-stacking π interactions. As such, these types of interactions have the advantage of not altering the electronic properties of the CNT (unlike covalent modifications). However, since covalent functionalization places specific chemical groups onto the CNT surface, this route offers a greater

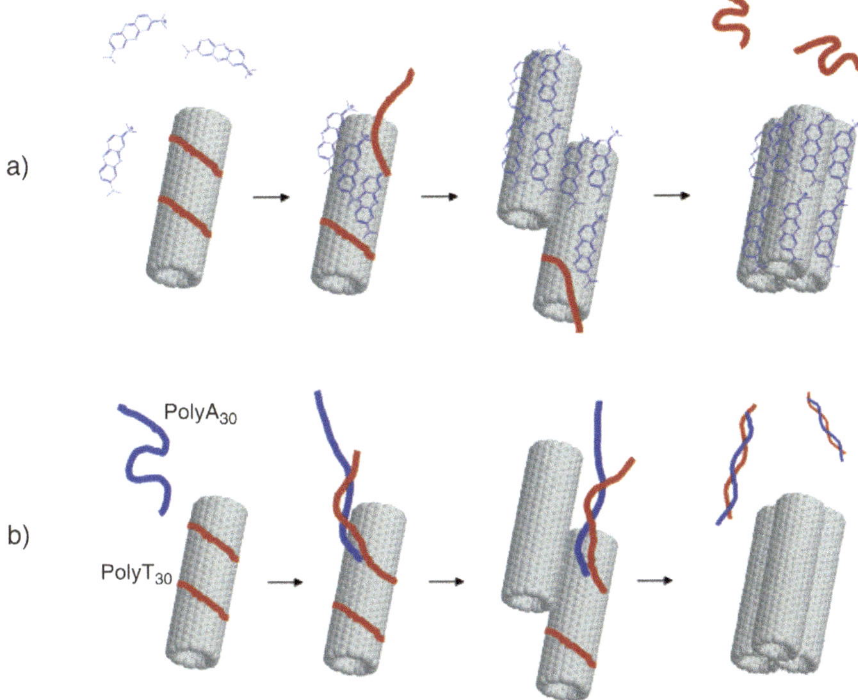

Fig. 6.91 Examples of noncovalent interactions with a macromolecule solubilizing agent and CNTs. Shown is **a** the solubilization of SWNTs with PolyT30 single-strand DNA, followed by the displacement of the biomolecule through preferred π–π stacking between the conjugated dye methylene blue and the SWNT surface—causing the controlled precipitation of SWNTs from solution. Also shown is another method for solubilization/precipitation of SWNTs through use of complimentary single-strand DNA. Reproduced with permission from *J. Phys. Chem. B* **2006**, *110*, 54. © 2006 American Chemical Society

potential for the selective tunability of CNT properties as well as molecular control over the organization of CNTs for device fabrication.

The driving force behind the endcap and sidewall functionalization of SWNTs is the strain brought about from the high degree of surface curvature. The deviation of the carbon geometry from planarity is referred to as the *pyramidalization angle*, θ_p, which is defined as 0° for the sp^2 carbon atoms in C_2H_4, and 19.5° (i.e., 109.5–90°) for the sp^3 carbon in CH_4 (Fig. 6.94). Since the θ_p of the carbon atoms in C_{60} is 11.6°, the desired geometry is closer to tetrahedral rather than trigonal planar. As a result, fullerenes readily undergo addition reactions to relieve this high degree of surface strain. Since the endcap of a SWNT may be considered as a half-fullerene structure, with a $\theta_p \geq 9.7°$ regardless of the nanotube diameter,[334] these points are highly reactive toward electrophilic addition.

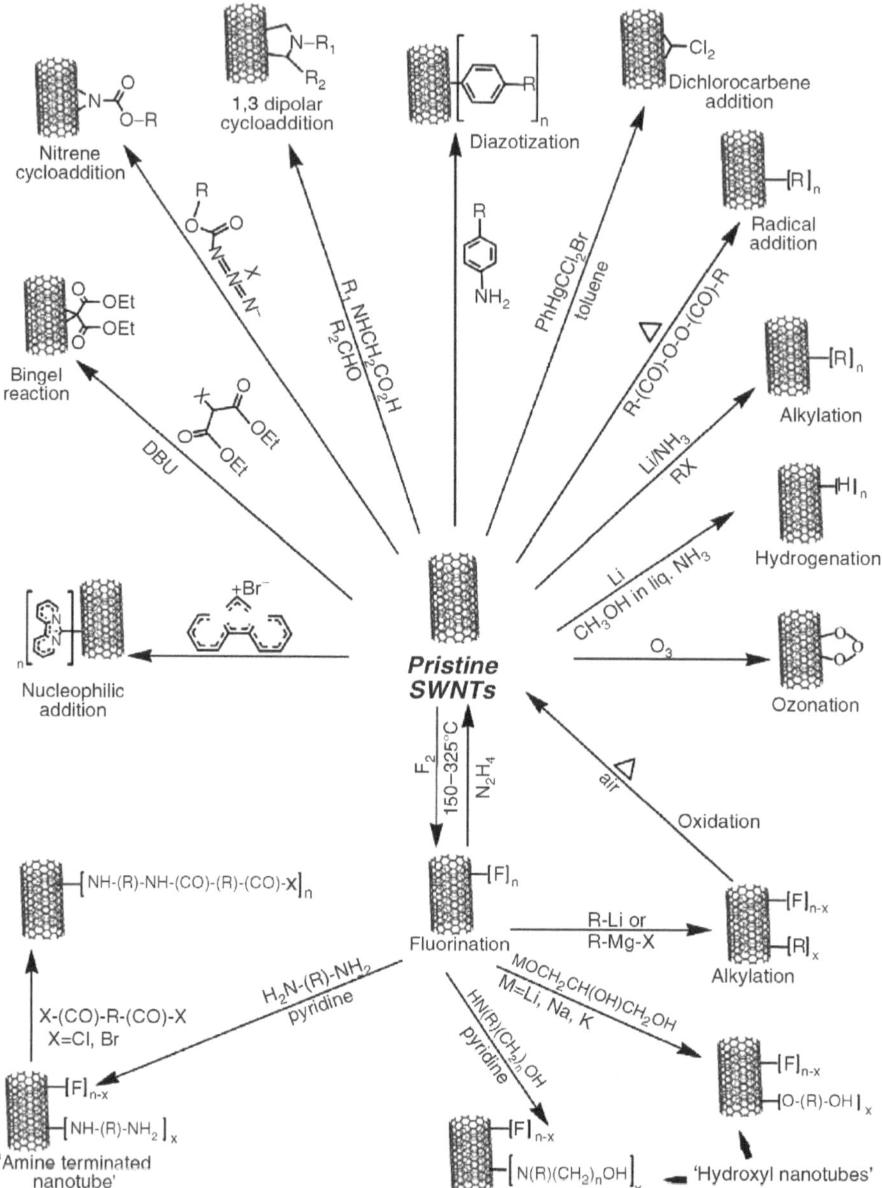

Fig. 6.92 Schematic of the various methods to perform sidewall functionalization of SWNTs. Reproduced with permission from *Adv. Mater.* **2005**, 17, 17. © 2005 Wiley–VCH

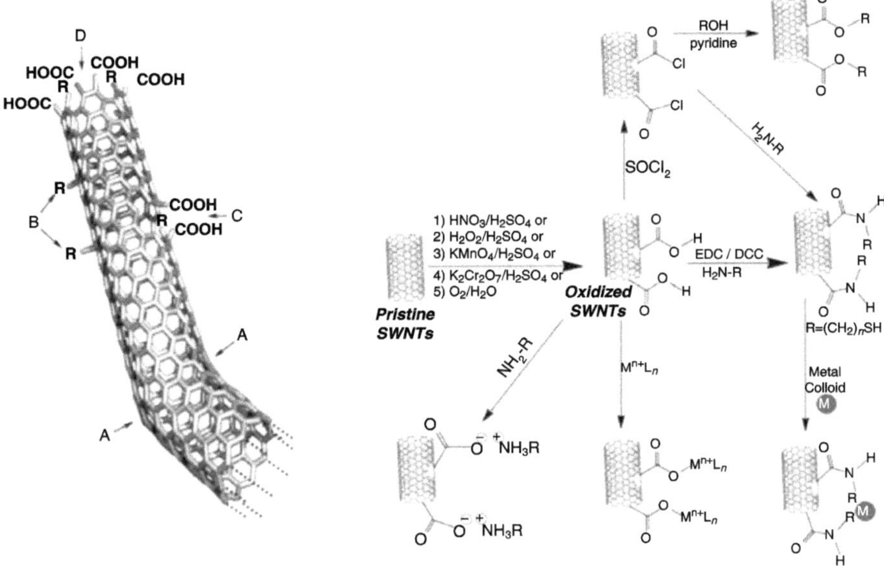

Fig. 6.93 Illustration of possible defect sites on the surface of a SWNT, along with methods used for covalent functionalization of SWNTs at defect sites. The types of defects shown are: **A** 5- or 7-membered rings in the framework, which results in bending, **B** sp^3-hybridized defects (R = –OH, –H), **C** oxidative surface degradation, with holes capped by –COOH groups, and **D** open end of the SWNT shown with—COOH capping groups (may also be terminated with –OH, = O, $-NO_2$, and –H groups). Reproduced with permission from: (**left**) *Angew. Chem. Int. Ed.* **2002**, *41*, 1853 and (**right**) *Adv. Mater.* **2005**, *17*, 17. © 2002 & 2005 Wiley–VCH

In contrast, the sidewall pyramidalization angle is much less than the endcap, on the order of 3–6°, depending on the diameter/chirality of the SWNT. For an armchair (5,5) SWNT, two types of C–C bonds are present along the sidewalls—either parallel or perpendicular to the nanotube axis. Each of these bonds exhibits a different degree of π orbital overlap with adjacent carbon atoms, as quantified by the π *orbital misalignment angle*, ϕ (Fig. 6.103). Whereas $\phi = 0°$ for fullerenes, the misalignment of neighboring π orbitals in SWNTs may be significant—as large as 21.4° for (5,5) species. In general, a SWNT of smaller diameter has a greater sidewall reactivity than larger analogues due to relatively large π orbital misalignment angles (much more influential than θ_p). It should also be noted that the overall curvature of CNTs is less than fullerenes, which make SWNTs significantly less reactive than fullerenes.

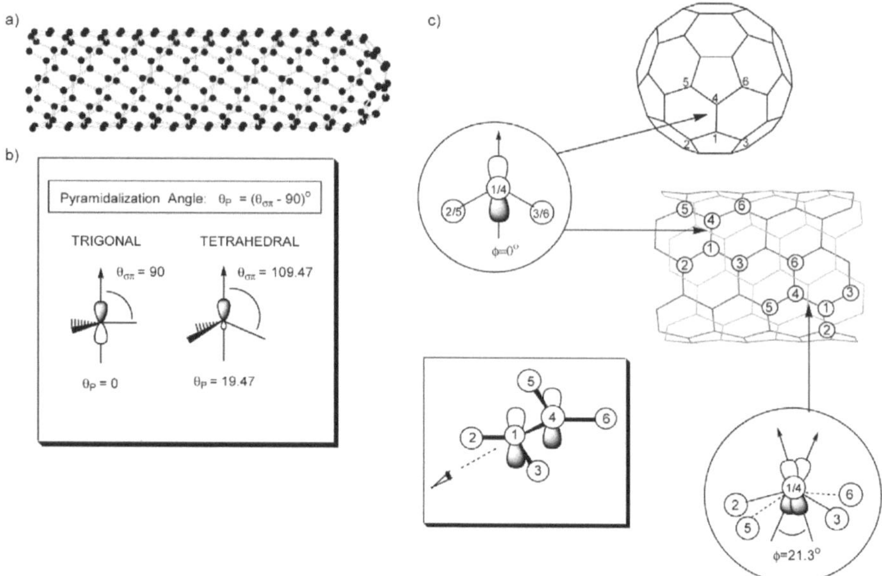

Fig. 6.94 Schematic of **a** a metallic (5,5) SWNT, **b** the pyramidalization angles of typical sp^2 and sp^3 carbon atoms, **c** the π orbital misalignment angles along C1–C4 in the (5,5) SWNT and its capping fullerene, C_{60}. Reproduced with permission from *Acc. Chem. Res.* **2002**, *35*, 1105. © 2002 American Chemical Society

Synthesis of Carbon Nanotubes

Now that you are familiar with the properties and applications of CNTs, we must now consider the techniques used for their synthesis. Although there have been reports of "bottom-up" strategies for 1-D and 2-D carbonaceous nanostructures,[335] the only routes that currently provide sufficient yield are top-down routes: laser evaporation of graphitic targets, arc-discharge methods, and chemical vapor deposition (CVD). You might recognize the first two methods as also being used for to generate fullerenes. This is not surprising, since CNTs are essentially 1-D extensions of fullerene clusters. Since both of these methods involve the sublimation of graphite followed by condensation of carbon atoms, growth occurs at an extremely high temperature—close to the m.p. of graphite (3000–4000 °C). To synthesize SWNTs, a catalytic amount of a nanosized species (e.g., metals: Mg, Ti, V, Cr, Mn, Fe, Co, Ni, Cu, Zn, Al, Ga, Zr, Mo, Ru, Rh, Pd, Ag, Sn, Pt, Au, Er[336]; semiconductors: Si, Ge; carbides: SiC, Fe_3C; oxides: SiO_2, Al_2O_3, TiO_2, lanthanide oxides)[337] must be present to prevent the leading edge from closing, which would generate spheroidal fullerenes. For MWNT growth, a catalyst is not always required[338]; additionally, the products usually contain a larger number of other carbonaceous products such as fullerenes and amorphous carbon.[339]

Whereas arc-based methods generally contain a mixture of products (SWNTs/MWNTs of varying diameters/morphologies, amorphous soot, etc.), a dual-pulsed laser approach, first used in the mid-1990s, results in SWNT ropes with >70% purity.[340] Although both of these methods may be used to generate small quantities of CNTs, they are relatively expensive and not easily scalable to generate industrial quantities. Further, the CNTs arising from vaporization methods are typically in a tangled array, with other forms of carbon intermixed with the remaining catalytic metal (especially affecting subsequent electronic applications).

Consequently, chemical vapor deposition (CVD) is now the method-of-choice for the synthesis of CNTs, which are produced in quantities of thousands of tons annually. As discussed in Chap. 4, CVD entails the volatilization and decomposition (typically thermal) of a hydrocarbon precursor on the surface of a catalytic surface. Methane, acetylene (C_2H_2), and ethylene (C_2H_4) have been used most extensively as CNT precursors; other alternatives now include CO and methanol/ethanol. As with any CVD approach, this method is easily scalable, and is used to generate kilogram quantities of CNTs for an ever-increasing laundry list of applications.

Most often, carbon nanotubes are grown as random/intertwined structures, analogous to a bowl of spaghetti. However, by varying the growth conditions, it is also possible to fabricate vertically or horizontally aligned arrays of CNTs. Vertically-aligned CNTs (VACNTs), known as "forests", are synthesized from surface-immobilized catalyst nanoclusters, which was pioneered by Dai and coworkers at Stanford.[341] Although the structure/property relationship of VACNTs is still poorly understood,[342] vertical alignment of the CNTs likely takes place through strong intratube/intertube van der Waal interactions (e.g., MWNT arrays, Fig. 6.95a) or tube-patterned substrate interactions (e.g., SWNT arrays, Fig. 6.95b). Vertically-aligned or suitably-patterned 1-D nanostructures (not limited to CNTs only) are also of use for the design of superhydrophobic coatings, which are able to levitate water droplets above a surface to preclude wetting (Fig. 6.96).[343] Beyond useful societal applications, nanotube forests may also be grown in the shape of popular media icons or even Presidential 'NanObamas' (Fig. 6.97)—a very cool artistic application for nanomaterials, indeed!

Whereas randomly and vertically aligned CNTs contain many structural defects and exhibit lengths less than millimeters, horizontally aligned (HACNTs) varieties have lengths that span into the decimeter range,[344] with aspect ratios of 10^6–10^8. Although some HACNTs are multi-walled structures with 1–5 shells, the majority are SWNTs, which are relatively free from structural defects. Growth of HACNTs is proposed to occur via gas-flow directed growth, which incorporates fast-heating of the substrate to lift the CNT tips off the surface, denoted as a "kite mechanism".[345] Morphological control of the CNTs is typically afforded by either surface-directed or electrical/magnetic field-directed alignment (Fig. 6.98). As further improvements in chirality control and length are realized, HACNTs will find many uses for a diverse range of applications such as transparent displays, micro/nanoelectronics, micro/nanoelectromechanical systems (MEMS/NEMS), energy storage, photodetectors, and functional materials.[346]

Carbon nanotube growth emanates from catalyst nanoparticles that have been deposited onto a substrate surface (Fig. 6.99), with resultant tube diameters related to the size of the seed catalyst. The morphology of discrete nanotubes is thought to originate from the incorporation of non-hexagonal defects during growth.[347] Interestingly, whereas a number of carbon sources such as methane and various alcohols may be used for SWNT growth from catalyst powders, ethylene (C_2H_4) is a key precursor for vertically-aligned SWNT growth.[348] Often, a metallic underlayer (e.g., Al/Al_2O_3, Cr, Ir, W, Ta, Ti) is used as a diffusion barrier in order to facilitate and control the growth of SWNTs on substrates such as Si, SiO_2, or metals (e.g., Au, Ag, W, steel alloys).[349] The morphology of this underlayer plays a crucial role in the overall structure (i.e., random vs. aligned arrays) of the resultant SWNTs.[350]

There are also reports of the one-step synthesis of aligned MWNT bundles and Y-junction CNTs using a controlled mixture of metallocene-based precursors ($M(C_5H_5)_2$, M = Fe, Co, Ni)—precluding the use of surface-immobilized catalyst nanoclusters.[351] Since the precursor acts as both the carbon source and catalyst, carbon growth initiates at approximately same time from multiple metal sites. Subsequent growth is self-directed into a parallel array by interactions amongst neighboring CNTs. Phosphorus- and nitrogen-doped VACNTs have also been synthesized using this route with toluene as the carbon source and ferrocene as the catalyst, alongside triphenylphosphine or ammonia as the dopant precursors.[352]

For some applications, a main drawback of CVD is that MWNTs are often generated alongside SWNTs. Since the size of the catalyst governs the diameter of resultant tubes, surface-immobilized nanoclusters of iron oxide@PAMAM dendrimer, with

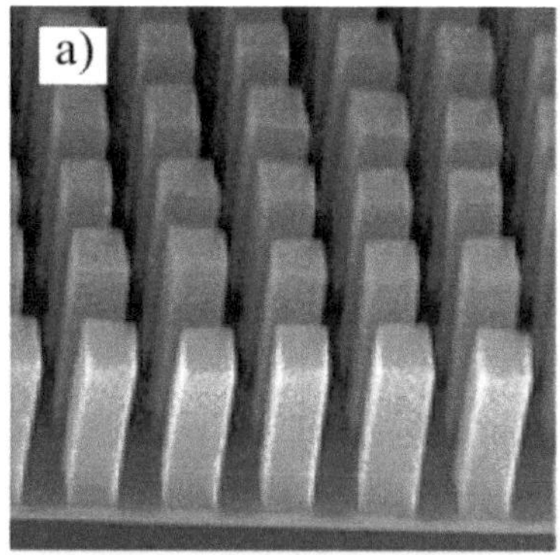

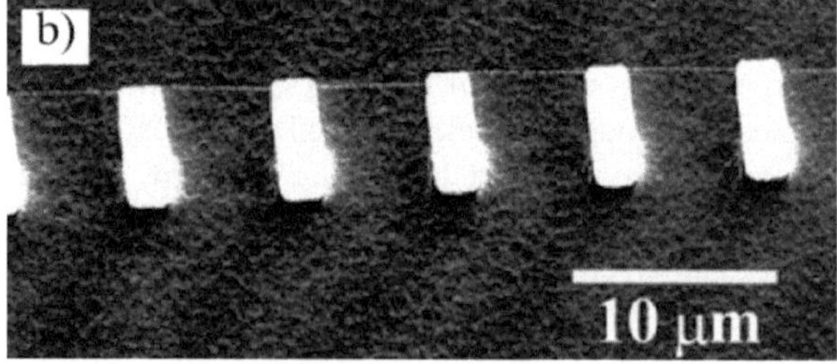

Fig. 6.95 Examples of the ordered growth of carbon nanotubes. Shown are **a** MWNT arrays grown from squared regions of iron nanoclusters, and **b** Side-view SEM image of a "SWNT power line on Si posts". Reproduced with permission from *Acc. Chem. Res.* **2002**, *35*, 1035. © 2002 American Chemical Society

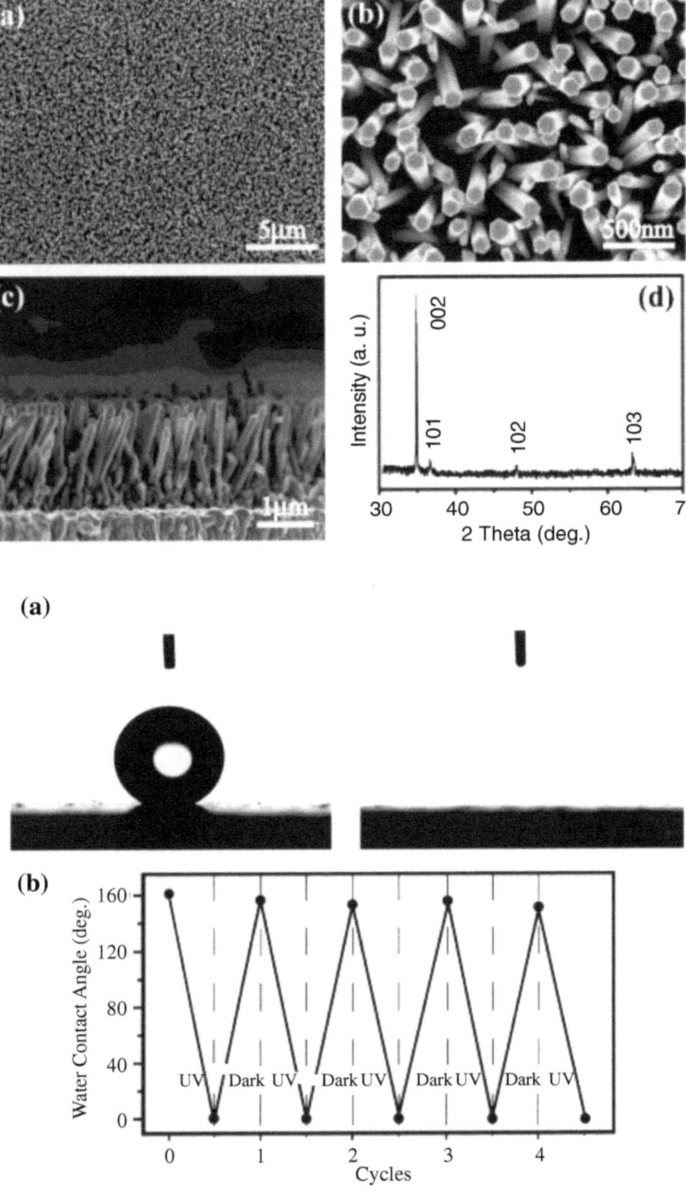

Fig. 6.96 SEM images and XRD pattern of ZnO nanorod films, and photographs of water droplet shapes on the aligned ZnO films before (left) and after (right) UV illumination. Bottom: the reversible superhydrophobicity/superhydrophilicity character of the films is demonstrated by cycling with UV irradiation/dark storage. Reproduced with permission from *J. Am. Chem. Soc.* **2004**, *126*, 62. © 2004 American Chemical Society

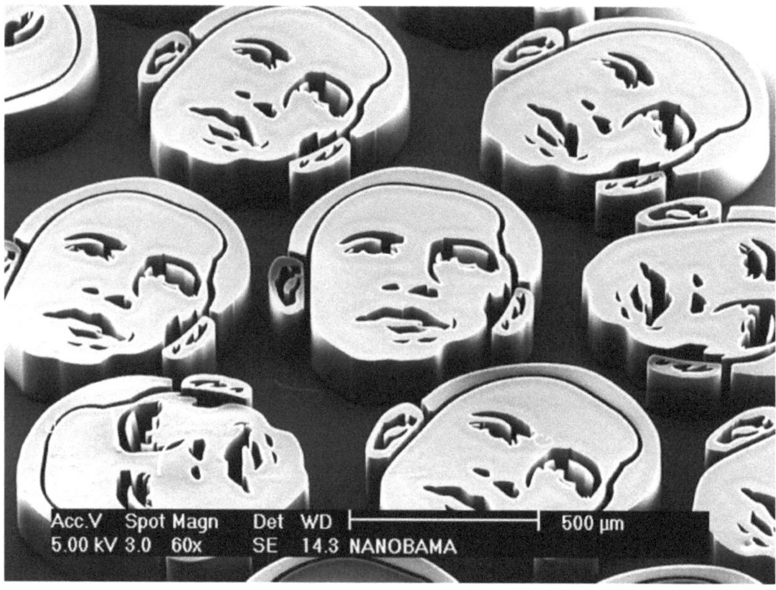

Fig. 6.97 SEM image of 'NanObamas' composed of vertically-aligned carbon nanotubes: (http://www.flickr.com/photos/ajohnhart/sets/72157608561494488/). © John Hart, Michael De Volder, and Sameh Tawfick (University of Michigan)

Fig. 6.98 SEM image of aligned SWNT growth through electric-field induction. Reproduced with permission from *Acc. Chem. Res.* **2002**, *35*, 1035. © 2002 American Chemical Society

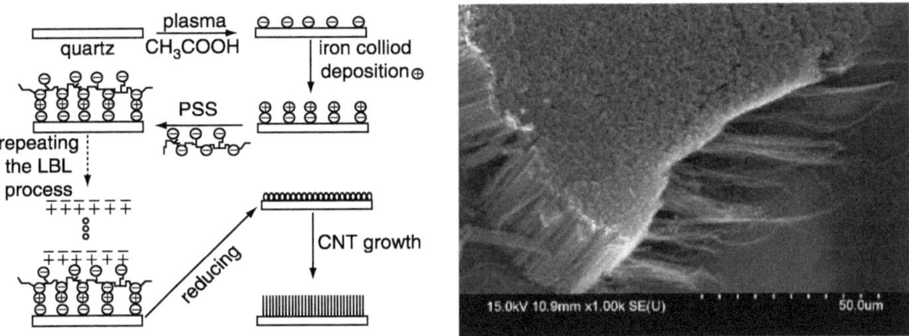

Fig. 6.99 Schematic and cross-section SEM image of a layer-by-layer (LbL) approach to deposit iron nanoclusters onto a surface to yield vertically aligned SWNTs. Reproduced with permission from *Chem. Mater.* **2005**, *17*, 6599. © 2005 American Chemical Society

well-defined sizes, has been shown to yield only SWNTs with an extremely narrow diameter distribution.[353] Two advanced CVD processes, CoMoCAT® (fluidized-bed CVD) and HiPCO® (high-pressure CO CVD), have recently been developed for the commercial production of SWNTs (Fig. 6.100). Although the experimental setup of these methods are significantly more complex than standard hot-walled CVD, these techniques are still considered an extension of CVD, as the precursor is decomposed on the surface of the catalyst. Most noteworthy about these and other recent CVD methods is that they offer the ability to fine-tune resultant lengths and diameters by varying the gas pressure and growth temperature, which affects catalyst nanoparticle agglomeration. Even chirality control is now possible; for instance, the CoMoCAT process selectively produces (6,5) and (7,5) SWCNTs.[354] It is believed that changes in the crystal structure of the catalyst, due to compositional tuning, may result in preferred chiralities due to an epitaxial growth process.[355] However, others have reported selective chiral control by simply using different carbon precursors.[356] Think of the possibilities for applications of selective CNT growth—it may not be long until you can place an order such as: "10 g of pure (5,5) SWNTs with an average diameter of 7 nm and length of 1 m." However, this degree of control will only be possible once the exact growth mechanism is known, rather than current efforts that invoke "blind" modifications of experimental variables.[357]

Growth Mechanisms of Carbon Nanotubes

Although CNT growth dates back to the early 1990s, details of the growth mechanism are still under investigation.[360] The most commonly accepted description is based on the *vapor–liquid-solid* (VLS) model, which was first proposed for the growth of semiconductor whiskers (Fig. 6.101).[361] This model assumes that the catalyst liquefies, which then acts as a preferential adsorption site for gaseous precursors. Subsequent growth of the 1-D nanostructure occurs by supersaturation of the catalyst droplet and precipitation at the liquid–solid interface.[362] However, the amount of carbon in a supersaturated catalyst nanocluster will never be sufficient to account for

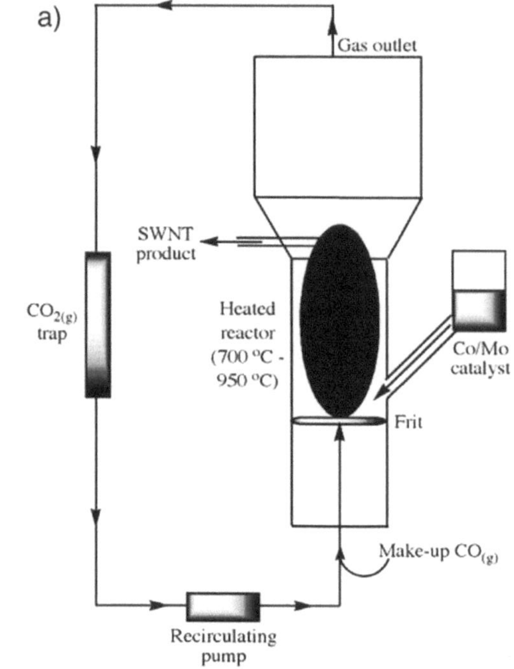

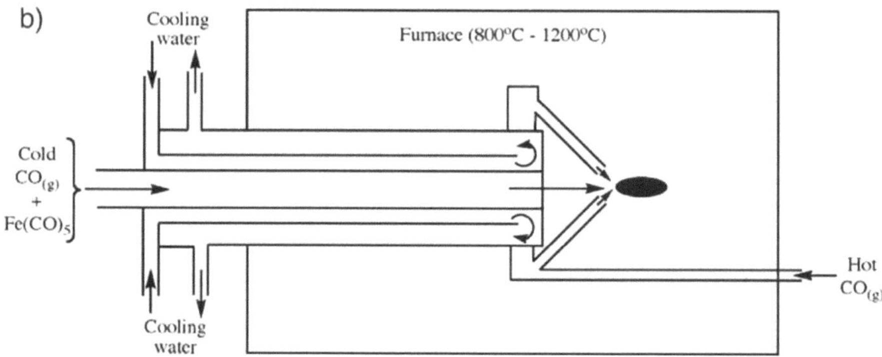

Fig. 6.100 Illustration of two methods used for the commercial production of SWNTs. Shown are **a** the CoMoCat fluidized bed method using CO as the precursor and a Co/Mo bimetallic catalyst,[358] and **b** the HiPco "floating catalyst" process using the thermal decomposition of iron pentacarbonyl at pressures of 1–10 atm[359]

the length or number of nanostructures observed in practice. As a result, additional gaseous precursor species must be supplied continuously to the catalyst nanocluster in order for sustained growth to occur.

Although VLS growth is a non-equilibrium process, the binary (i.e., precursor/catalyst) phase diagram provides important details regarding the optimal growth temperature—any value above the eutectic temperature where the catalyst

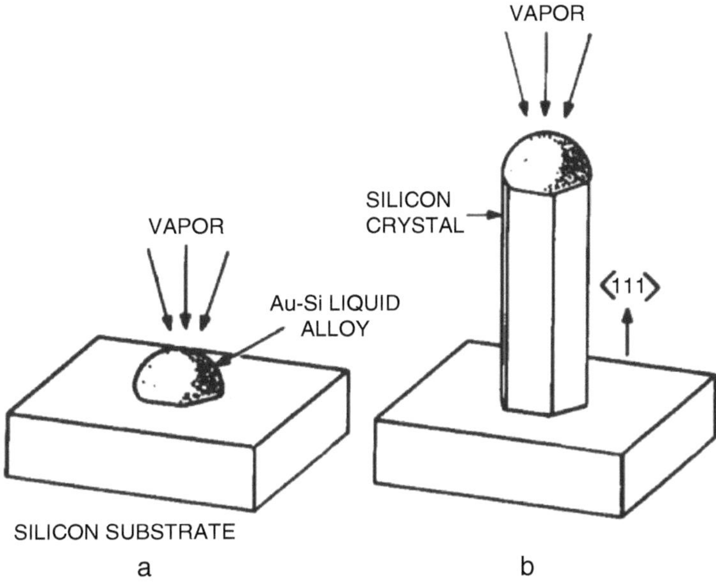

Fig. 6.101 The original schematic used to describe vapor–liquid–solid (VLS) growth of semiconductor nanowires. Reproduced with permission from *Appl. Phys. Lett.* **1964**, 4, 89. © 1964 American Institute of Physics

remains a liquid (Fig. 6.102). As we saw earlier for gold nanoparticles, as the diameter decreases, the melting point is significantly decreased relative to the bulk solids.[363] As a result, "size-corrected eutectics" must be determined for the specific nanoparticle diameter in order to determine experimental conditions for nanowire growth.[364]

The VLS mechanism is generally[365] sufficient to model the growth of nanowires[366]; however, nanowire growth may also occur in solution via *solution-liquid–solid* (SLS)[367] or via *solid–liquid-solid*[368]—the latter through simple annealing of substrate-immobilized nanoclusters (Fig. 6.103). More recently, a "nano Czochralski model" has been asserted to explain the role of the catalyst on chirality of the resultant SWNT.[369] This mechanism is proposed to be analogous to CZ growth of Si (revisit Sect. 4.2.1), which occurs by rotation of a seed catalyst from a melt (Fig. 6.104). In this model, the molten catalyst serves as a nano-crucible, in which gaseous carbon atoms continuously dissolve and supersaturate to grow a CNT cap structure. The rotation of the growing CNT relative to the catalyst particle has indeed been observed using in situ electron microscopy,[370] which provides empirical credibility to this VLS model.

Although the VLS route has been largely accepted for SWNT growth from iron-group catalytic nanoparticles,[371] it may not always be the governing mechanism.[372] For instance, there are numerous reports of CNTs being grown at temperatures below the size-corrected eutectic,[373] or from non-melting nanoclusters such as SiO_2, Al_2O_3, etc.[374] Although small nanoparticles may exhibit "fluid-like" behavior by altering

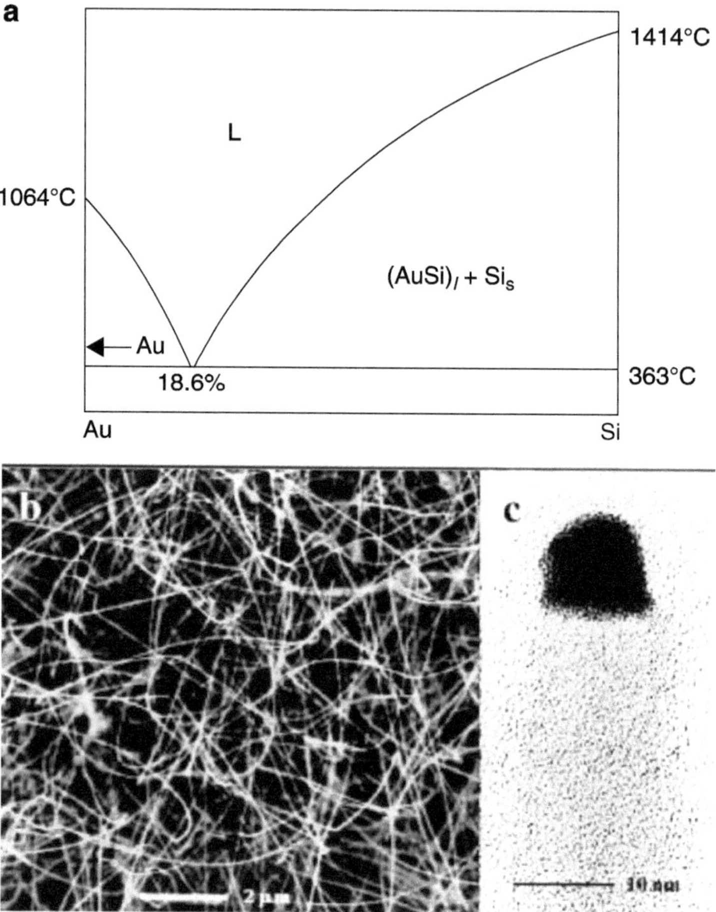

Fig. 6.102 Silicon nanowire growth from a gold nanocluster catalyst. Shown is **a** the phase diagram for the Au/Si system, showing the eutectic temperature/composition; **b** SEM image; and **c** high-resolution TEM image of the nanowires grown at a temperature of 450 °C. The dark tip of the nanowire is from the gold nanocluster. Reproduced with permission from *Acc. Chem. Res.* **1999**, *32*, 435. © 1999 American Chemical Society

their surface geometry, empirical data has shown that the catalyst particle may remain crystalline, rather than an amorphous liquid phase, during growth. This suggests that nucleation may be dominated by the catalyst *surface*—not unlike most heterogeneous catalysis processes (e.g., Ziegler–Natta polymerization).[375] This growth mechanism is referred to as *vapor–solid-solid* (VSS),[376] which proposes the formation of isolated sp^2-carbon islands on the surface of a solid catalyst particle, followed by the growth of extended polyyne chains. As additional gaseous carbon atoms are deposited, sp^2-carbon networks are formed via ring condensation, ultimately leading to a SWNT cap

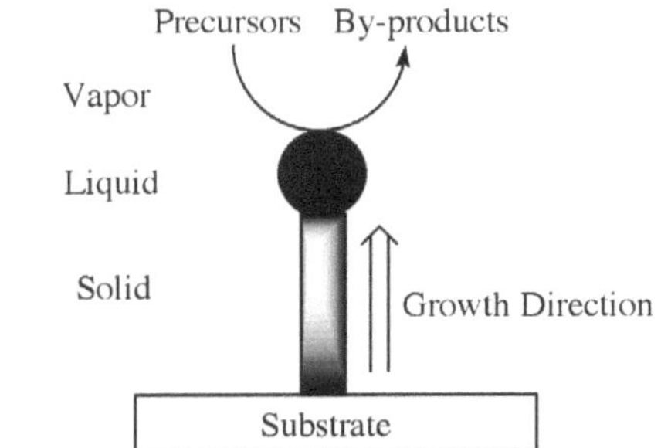

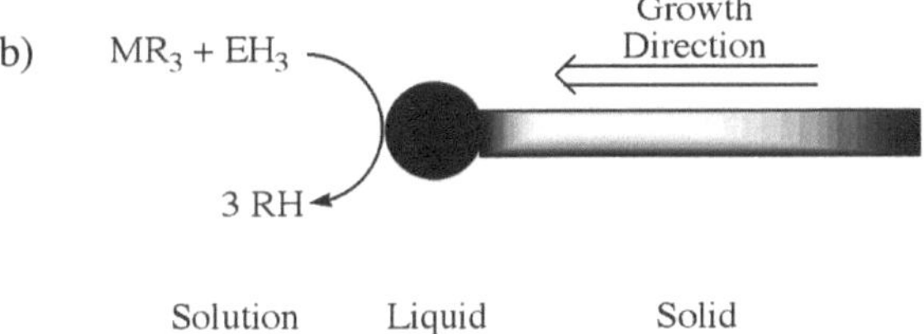

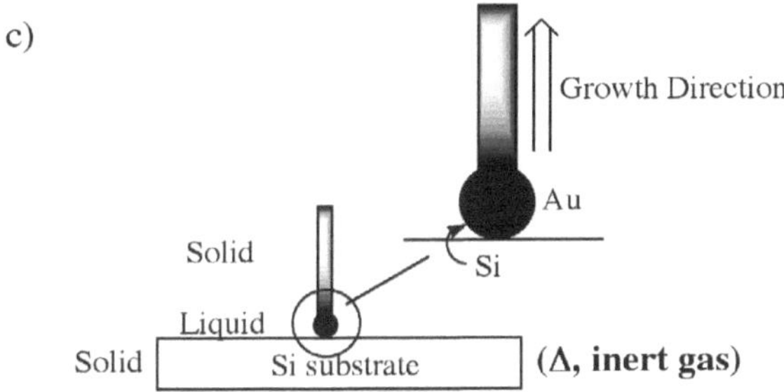

Fig. 6.103 Schematic illustrations of **a** VLS, **b** SLS: solution-liquid–solid, and **c** SLS: solid–liquid-solid 1-D nanostructural growth

structure. The VSS mechanism has also been proposed to explain nanowire growth via solid-phase diffusion, instead of the commonly accepted VLS route.[377]

As with any complicated system, it is much easier to study growth mechanisms using computational techniques. For CNT growth, it has been shown that small graphitic islands form on the supersaturated catalyst surface. When the island covers half of the catalyst particle, it lifts off and forms the SWNT—with the same diameter as the catalyst. Based on empirical data and theoretical predictions,[378] the catalytic ability of the metal/alloys for SWNT growth follows the order: Ni/Mo > Ni/Cr > Ni/Co > Ni/Pt > Ni/Rh > Ni/Fe > Ni > Fe/Mo > Fe/Cr > Fe/Co > Fe/Pt > Fe/Rh > Fe > Ni/Mo > Fe/Mo > Co/Mo > Co > Pt > Cu. Depending on whether the mechanism proceeds via VLS or VSS, the above order may be related to the ease of carbide formation and carbon diffusion through the nanocluster interior, or the morphology/composition of the catalyst surface and rate of carbon diffusion on the

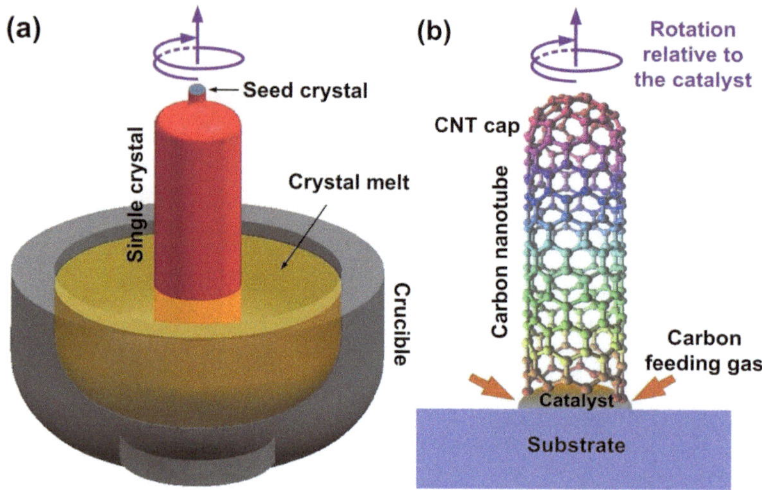

Fig. 6.104 Illustration of the conventional Czochralski (CZ) growth process and the 'nano-CZ' model for carbon nanotube growth. Reproduced with permission from Nanoscale Res. Lett. 2012, 7:356. © 2012 Lu, J. and Miao, J.; licensee Springer

catalyst surface, respectively.[379] Figure 6.105 shows some mechanistic proposals, based on experimental and theoretical data. The VLS-based proposals (Fig. 6.105, top and middle) illustrate the following basic steps:

(i) At initial growth stages, carbon dissolves in the (semi)-molten "liquid-like" catalytic nanocluster.[381] Calculations indicate that there is a dynamic process of carbon precipitation onto the catalyst surface and re-dissolution, until a highly supersaturated catalyst is obtained.
(ii) Carbon precipitates on the surface of the highly supersaturated catalyst nanoclusters, forming carbon strings/polygons. This causes a decrease in the dissolved carbon concentration.
(iii) The carbon nuclei form graphitic islands on the surface of the catalyst, which aggregate into larger graphitic clusters.
(iv) At low temperatures, the graphitic islands are not able to lift off the catalyst surface, resulting in graphite-encapsulated metal nanoclusters.[382]
(v) At relatively high temperatures (*ca.* 500 °C–1200 °C), when the diameter of the island becomes *ca.* 1/2 that of the catalyst, the graphitic nucleus lifts off the catalyst surface to form the SWNT endcap. Subsequent graphitization and growth propagation of the SWNT may occur through two routes (Fig. 6.105, middle):

   a. "Root (base) Growth" (c–d): carbon atoms migrate through the catalyst nanocluster and precipitate from the liquid-like catalyst to the open end of the growing CNT. This route has recently been confirmed for MWNT growth via in situ TEM measurements, from Fe_3C catalyst nanoclusters

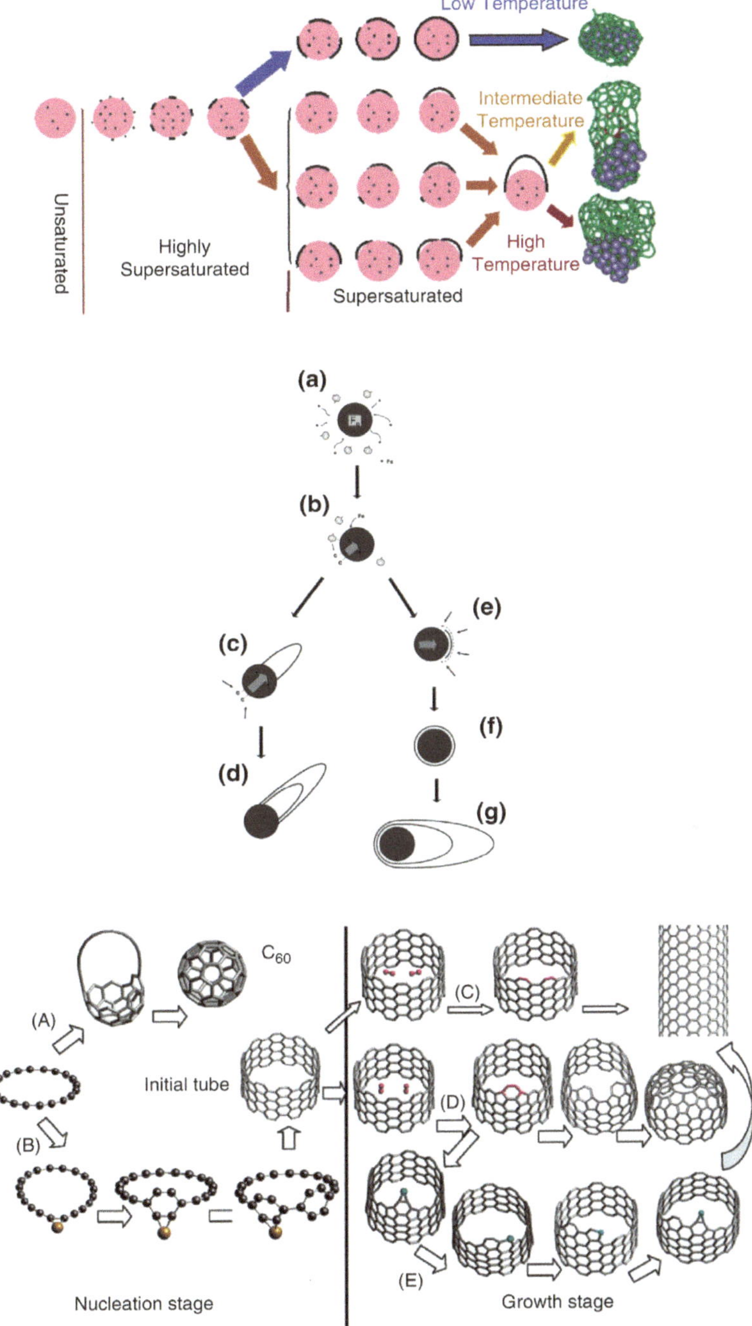

Fig. 6.105 Comparison of a VLS-based mechanism (top and middle) and a surface-mediated route[380] (see text for details)

(Fig. 6.106, top).[383] Recent theoretical[384] and empirical[385] studies have addressed the question of how growth occurs at the atomic level, described by a 'screw-dislocation-like' (SDL) mechanism (Fig. 6.107).[386]

b. "Tip (folded) Growth" (e–g): carbon atoms precipitate directly onto the catalyst surface, and are added to the graphitic endcap. This route has not been demonstrated for CNTs, but has been shown to occur for carbon nanofiber (CNF) growth using in situ TEM measurements (Fig. 6.106, bottom).[387]

In contrast to VLS, a surface-governed route is characterized by negligible carbon dissolution within the bulk of the nanocatalyst. Since the growth is not proposed to occur through supersaturation/precipitation, any species that are chemisorbed onto the catalyst surface will have a dramatic influence on SWNT growth. Figure 6.105 (bottom) shows a proposed route for a surface-based SWNT growth mechanism, which progresses through the continual addition of C_2 units onto the leading edge of the growing nanotube. As indicated in this scheme, the absence of an active catalyst will result in the formation of fullerenes rather than SWNTs. In this route, nanotube growth is proposed to initiate from surface metal atoms adding to the edge of a polyyne ring, which is subsequently able to add additional carbon atoms.

Whereas path C of Fig. 6.105 (bottom) shows the successful addition of C_2 units at the growth edge, path D shows the formation of a defect, which leads to tube closure. It should be noted that this route is 1.3 eV more favorable than route C when there is no metallic catalyst present, thus explaining the preference for closed-cage fullerenes. Most likely, this is a consequence of pentagonal defects that are formed when carbon atoms add to the open end of the growing SWNT. Hence, calculations have shown that catalyst-assisted defect repair is a crucial step in the overall mechanism (Fig. 6.105—bottom, route E). This is proposed to occur through a "scooter mechanism,"[388] whereby the catalyst atoms diffuse along the open-end surface of the SWNT, facilitating the formation of hexagons that prevent premature nanotube closure.

It should be noted that an additional variable(s) may be required to describe the full mechanism for CNT growth.[389] Studies have suggested there is likely an important role for metal oxide catalysts, or oxygenated co-reactants, in the overall growth mechanism of SWNTs. For example, the use of water vapor as a co-reactant for CVD results in the growth of a dense forest of vertically aligned SWNTs, with lengths up to 2.5 mm after only 10 min of reaction time (Fig. 6.108a).[390] The catalytic enhancement of water vapor is evidenced by the SWNT/catalyst weight ratio of 50,000%—more than 100 times greater than the HiPCO process. The effect of an oxidizing species such as water vapor seems logical from the standpoint of reducing the amount of amorphous carbon deposits, thereby controlling the degree of carbon precipitation (thus preventing MWNT growth). However, its role may be even more significant—perhaps even providing a new mode of nucleation. In particular, it is proposed that SWNT growth may be initiated from oxygen-etched graphitic shells that encapsulate the catalyst—a route recently termed "nucleation via etched carbon shells" (NECS),

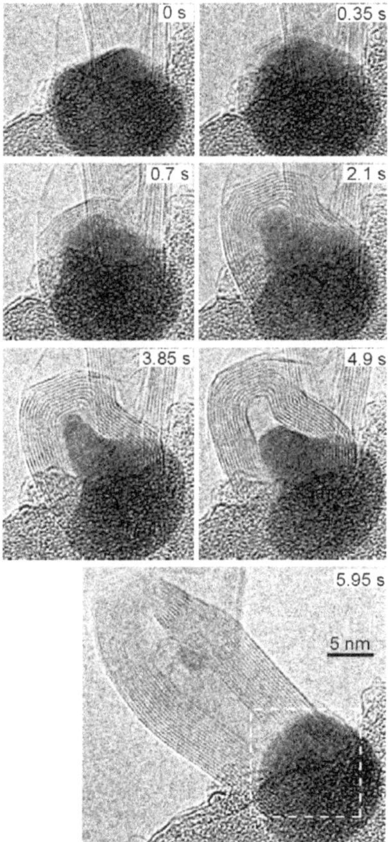

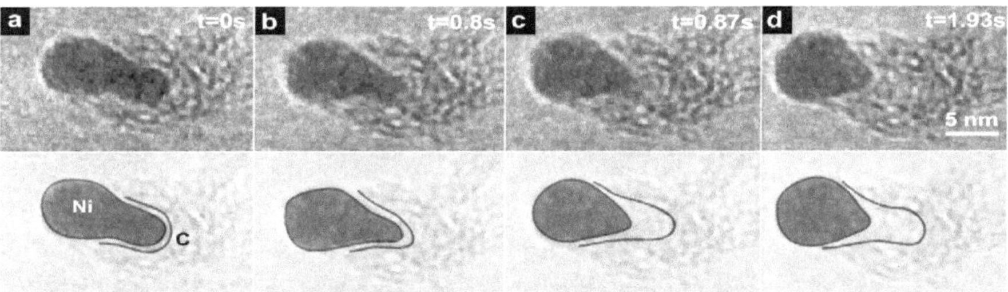

Fig. 6.106 Nucleation and growth of a MWNT from a NPC on a substrate. **a** Graphene layers are formed on a NPC and then a MWNT is suddenly expelled from the deformed NPC. The recording time is shown in images. Reproduced with permission from *Nano Lett.* **2008**, *8*, 2082 © 2008 American Chemical Society. **a**–**d** ETEM image sequence showing a growing CNF in 3:1 NH_3:C_2H_2 at 1.3 mbar and 480 °C. The video was recorded at 30 frames/s, and the time of the respective stills is indicated. Drawings (lower row) indicate schematically the Ni catalyst deformation and C-Ni interface. Reproduced with permission from *Nano Lett.* **2007**, *7*, 602. © 2007 American Chemical Society

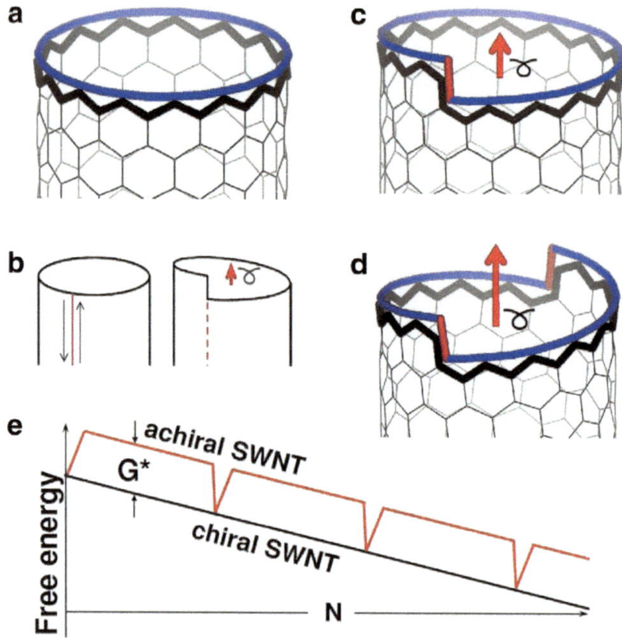

Fig. 6.107 An axial screw dislocation in the CNT. An achiral zigzag (n, 0) tube (**a**) can be viewed as a perfect crystal, and transformed into a chiral one by cutting, shifting by a Burgers vector b (red arrows in **a–d**), and resealing a tube-cylinder (**b**). The chiral (n, 1) in (**c**) and (n, 2) in (**d**) tubes contain the axial screw dislocations with a single and double value of b, accordingly; the corresponding kinks at the open tube-end are marked in red. **e** Free energy profile during the growth of a chiral or achiral nanotube. Reproduced with permission from *PNAS*, **2009**, *106*, 2506. © 2009 National Academy of Sciences

Fig. 6.108b.[391] However, other experimental optimizations have also resulted in ultra-long nanotube forest growth. Using a catalyst system of Fe/Gd/Al_xO_y in the presence of Fe and Al vapors, Sugime and coworkers were able to report the growth of VACNT forests with lengths of 14 cm and average growth rate of 1.5 μm s^{-1}.[392] The role of oxygen is also thought to be important for SWNT growth, through formation of strong M–O bonds with the catalyst nanoparticle.[393] This bonding could influence the rate of carbon adsorption, as well as Ostwald ripening of the catalyst during growth, which would affect the size/shape of the catalyst and resultant nanotube growth mechanism.

It should be noted that the formation of nanotubes is not limited to carbonaceous materials. In fact, a number of inorganic-based nanotubes (e.g., MS_2 (M = W, Mo, Nb, Ta, Zr, Hf, Au/Mo, Pb/Nb), VO_x, ZnS, TiO_2, Al_2O_3, GaN, Bi, InGaAs, GaAs, etc.) have been synthesized in recent years.[394] There are many applications for these materials, such as FETs, LEDs, photodetectors, photovoltaic devices, catalysts, advanced ceramics, and chemical sensors.[395] The most common method to synthesize these structures is by coating an anodized alumina or CNT substrate with a thin

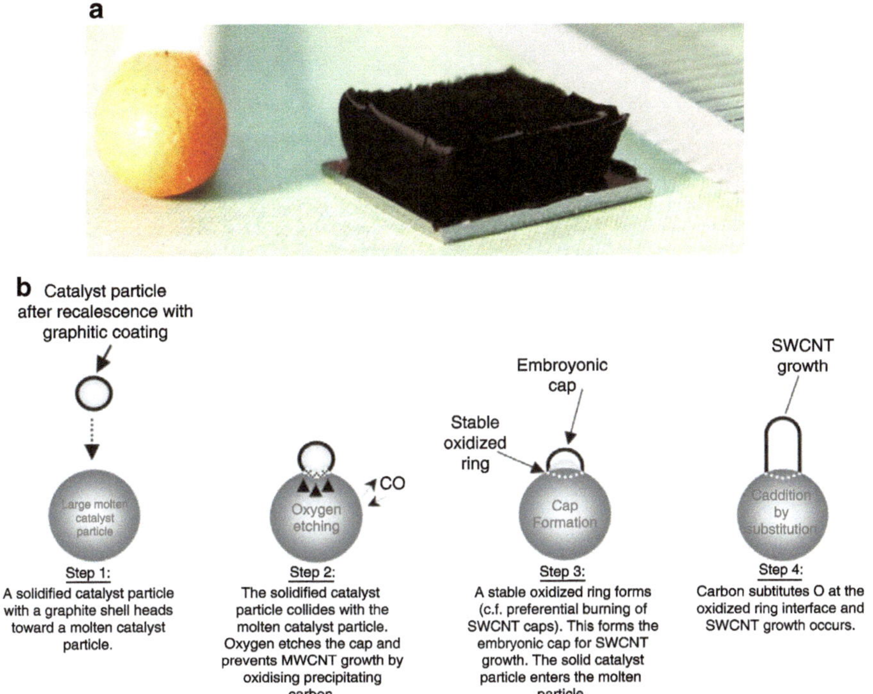

Fig. 6.108 a Photo of a cm-scale SWNT forest, shown next to a matchstick for comparative purposes. Reproduced with permission from *Science* **2004**, *306*, 1362. © 2004 AAAS **b** A proposed scheme for the laser ablative growth of SWNTs via metal oxide nanocluster catalysts. Reproduced with permission from *Nano Lett.* **2005**, 5, 1209. © 2005 American Chemical Society

film of the desired material, followed by the sacrificial removal of the template.[396] However, anodization of a substrate may also directly yield nanotubes in high yield.[397] Hydrothermal techniques with/without sacrificial organic templates have also been widely reported.[398] As we saw earlier for nanocluster growth, templating routes offer a convenient mode of nanostructural synthesis, as well as surface alignment of the nanoarchitectures on a substrate surface. In addition, self-assembly of molecular precursors in solution, analogous to sol–gel growth of thin films, may also be used to synthesize single-walled oxide nanotubes such as aluminosilicate or aluminogermate varieties.[399] Alternatively, aluminosilicate-based halloysite clay nanotubes are found naturally in Earth's crust, formed millions of years ago from the rolling of planar kaolinite at elevated temperatures and pressures (Fig. 6.109). With a cost and potential environmental impacts much less than carbonaceous nanostructures, halloysite nanotubes are are among the most popular 1-D nanostructure, with a wide array of possible applications such as drug-delivery, biomimetic reaction vessels, biocide or protective coatings, polymer additives, and cosmetic/personal care products.[400]

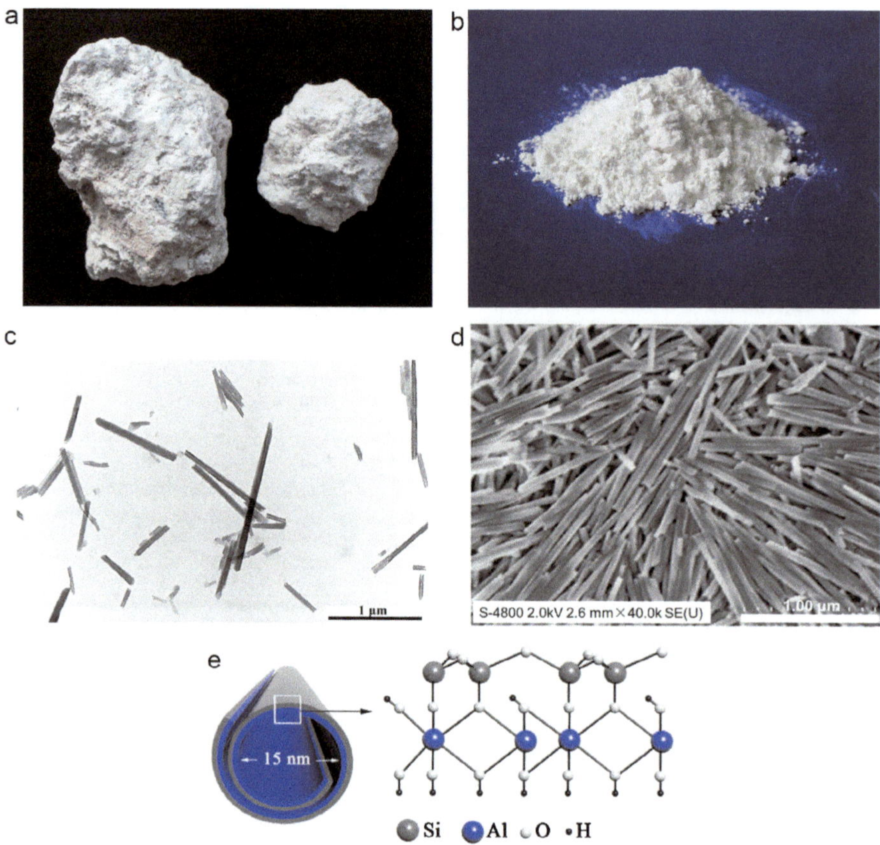

Fig. 6.109 Representations of halloysite. Shown are: **a** raw halloysite, **b** powdered halloysite obtained through grinding, **c** TEM image and **d** SEM image of halloysite nanotubes, **e** the crystal structure of halloysite. Reproduced with permission from *Prog. Polym. Sci.* **2014**, *39*, 1498. © 2014 Elsevier Ltd.

6.3.3 Two-Dimensional Nanostructures

In Chap. 4, we discussed an important 2-D nanoarchitecture—thin films. As long as the thickness is <100 nm, these are properly classified within the nanomaterials umbrella. Although spin-coating may be able to afford thin films of this overall thickness, these types of coatings are generally deposited using vapor deposition techniques (PVD, CVD, and ALD). Let's begin our discussion of these interesting nanostructures with the most well-studied 2-D nanomaterial—graphene, another nanostructural allotrope of carbon.

The term *graphene* has been used since the late 1980's to describe nanotube and graphite intercalation compounds. However, 2004 marked the earliest reported fabrication of graphene through simple exfoliation of graphite with Scotch tape(!)[401]—a technique that was recognized by the 2010 Nobel Prize in Physics.[402] Other synthetic methods have since been realized such as scratching highly-oriented pyrolytic graphite (HOPG) with a solid surface,[403] or heteroepitaxial layering of graphene onto single-crystalline SiC.[404] The exfoliation of graphite via strong mechanical shearing (stirring/sonication) at elevated temperatures will produce *expanded graphite*, which is primarily comprised of few-layer graphene sheets.[405] Graphite oxide sheets[406] that contain hydroxyl/epoxide groups on basal planes and carbonyl/carboxyl groups on sheet edges,[407] may also be reduced to pristine graphene using hydrazine solutions.[408]

More recently, "bottom-up" routes such as chemical vapor deposition has been used to grow high-yield, large area graphene films that may be post-transferred to a variety of substrates. In a typical thermal CVD process, a catalytic metal substrate (e.g., Cu) is heated to temperatures in excess of 1000 ºC in the presence of a carbon source such as methane.[409] Roll-to-roll CVD systems are now able to produce transparent graphene conductive films onto polymer substrates with lengths up to 100 m (Fig. 6.110)![410] Single and multilayer graphene films have now been reported from a variety of solid, liquid, and gas carbonaceous sources,[411] including biomass.[412]

Not unlike carbon nanotubes, solubilization and/or surface functionalization of graphene is essential to control its reactivity/self-assembly for future applications.[413] Whereas graphene oxide (GO) is soluble in polar solvents, reduced GO is insoluble. Thus far, strategies have included both non-covalent (π-stacking π) interactions with aromatic molecules,[414] and covalent interactions with amphiphilic polymers[415] as well as surface fluorination.[416]

Electronic Properties of Graphene

There has been a flurry of activity throughout the world to theoretically and empirically study the properties of graphene for applications in biomedicine, electronics, flexible transparent displays (Fig. 6.111), catalysis, spintronics, photodetectors, energy storage, sensors, environmental remediation, and a plethora of consumer products.[417] In particular, graphene exhibits the quantum Hall effect[418] at room temperature, extremely high electron/hole mobilities (>200,000 cm^2/Vs at room temperature[419]), and high thermal conductivity. It has also been reported that stacked graphene layers with small-angle twists exhibit superconductivity.[420] Graphene exhibits a similar mechanical strength as carbon nanotubes, with a specific surface area much greater than finely divided activated carbon currently used for water purification applications.[421]

The band diagram for graphene sheets features π/π^*-bands formed from the overlap of p_z atomic orbitals on neighboring C atoms, with an electronic energy dispersion described by (6.24). However, the graphene band structure is distinct from the conventional parabolic bands exhibited by metals or n-/p-type semiconductors. In particular, the E-**k** relationship of graphene is linear around the Dirac point, which directly results in relativistic behavior of the electrons (zero effective mass, referred to as *Dirac fermions*) at low energies (Fig. 6.112). Further, the Fermi surface of neutral graphene reduces to a single point (Dirac point), which defines the material as a *zero-bandgap semiconductor*.[422] Due to their structural relationships, the unique

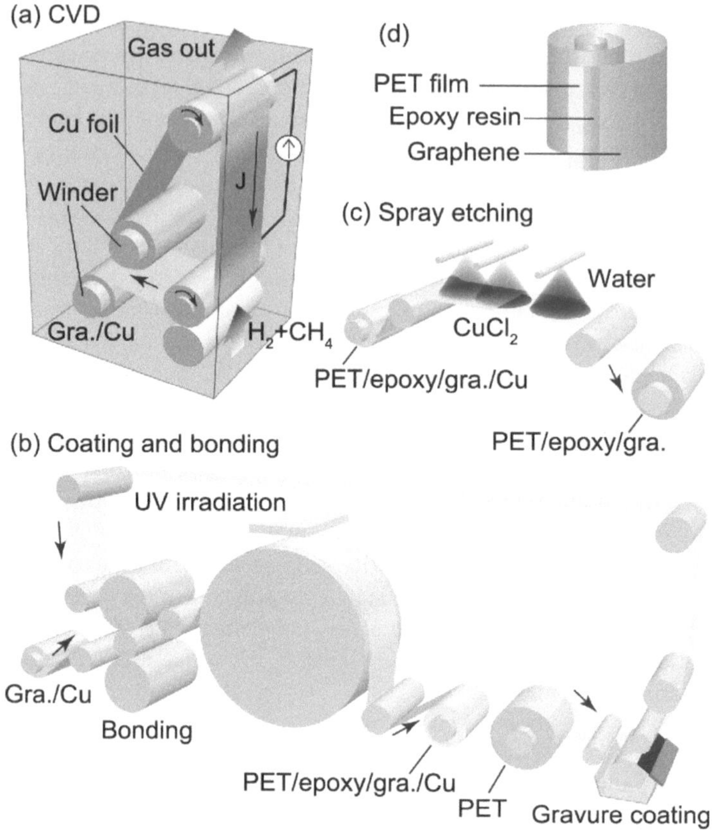

Fig. 6.110 **a** Schematic of a continuous roll-to-roll CVD system used to grow 100-m long graphene sheets. Selective Joule heating is used to heat a copper foil substrate suspended between two current-feeding electrode rollers. **b** Illustration of the reverse gravure coating of a photocurable epoxy resin onto a PET film and bonding to the graphene/copper foil, followed by curing of the epoxy resin. **c** Spray etching of the copper foil with a $CuCl_2$ solution, and **d** The structure of the fabricated graphene/epoxy/PET film. Reproduced with permission from *Appl. Phys. Lett.* **2013**, *102*, 023,112. © 2013 AIP Publishing LLC

band structure of graphene directly influences the distinct electronic properties of CNTs.

$$E(\mathrm{p}) = \pm \nu_o \left| \hbar \sqrt{k_x^2 + k_y^2} \right| \tag{6.24}$$

where ν_o is the Fermi velocity (*ca.* 1×10^6 m s^{-1}) and p is the carrier momentum.

An important consequence of the graphene band structure is its *ambipolar field effect*, in which both electron/hole carriers may exist by simply varying the applied

Fig. 6.111 Photo of a flexible, transparent sheet of graphene on a polymer substrate. © Bonnin-studio/Shutterstock.com

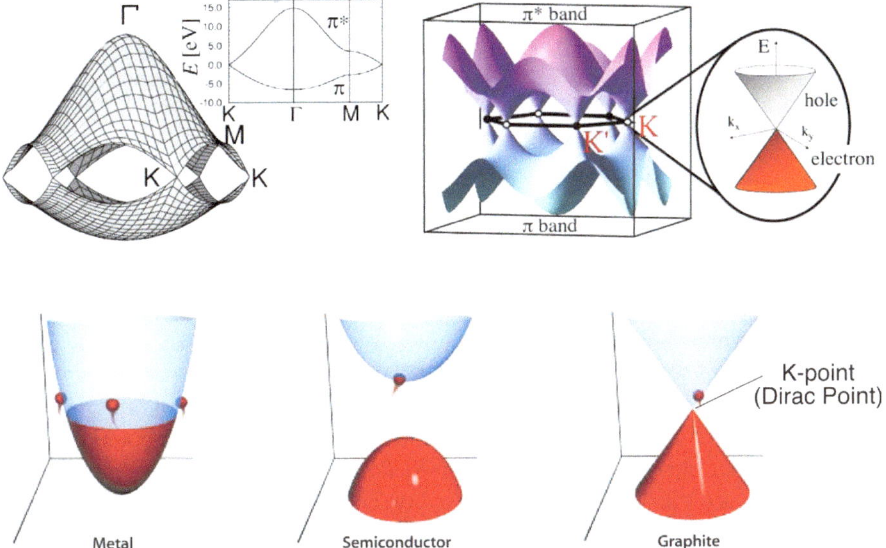

Fig. 6.112 Top: the energy dispersion relations for the π and π* bands in 2-D graphite are shown throughout the whole region of the Brillouin zone. The inset shows the energy dispersion along the high symmetry directions of the 2-D Brillouin zone. Reproduced with permission from *Phys. Rev. B* **2000**, *61*, 2981 © 2000 American Physical Society Bottom: comparison of band structures for a semiconductor, metal, and graphene sheets. Reproduced with permission from *Sci. Amer.* **2000**, *283*, 62. © 2000 Scientific American, Inc.

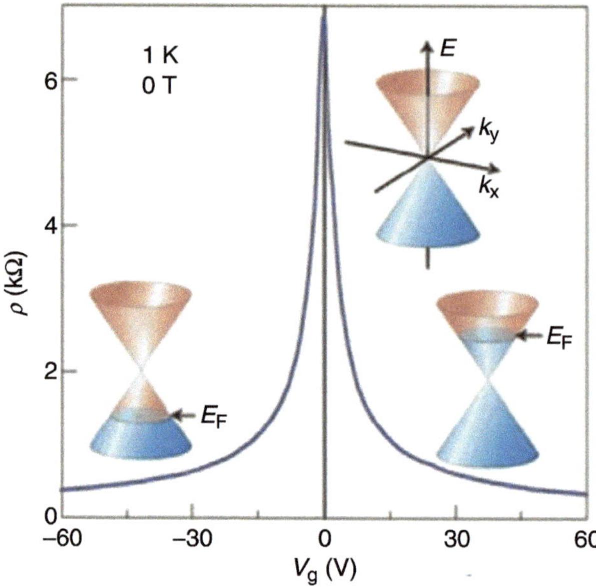

Fig. 6.113 Band structure and the ambipolar field effect in graphene. Conduction and valence bands meet at the Dirac point without an external field; in the presence of a gate bias, the Fermi level moves above/below the Dirac point to generate free carriers. Reproduced with permission from *Nature Mater.* **2007**, *6*, 183. © 2007 Nature Publishing Group

voltage (Fig. 6.113).[423] Though a single graphene sheet is a zero-bandgap semiconductor and cannot be used directly in applications such as field-effect transistors (FETs), adding another graphene layer results in a widely-tunable bandgap.[424] Accordingly, there is much current interest in *graphene bilayers* being used for FET applications, which have the potential to outperform both Si- and GaAs-based devices based on their greater carrier mobilities of electrons and holes, respectively.[425]

Another method used to overcome the zero-bandgap limitation of bulk graphene sheets for electronic applications is to cleave graphene into nanoribbons in order to induce further quantum confinement. Depending on how the *graphene nanoribbons* (GNRs) are fabricated, armchair and/or zigzag edges will be present along its sidewalls (Fig. 6.114). Analogous to carbon nanotubes, the bandgap and resultant electronic transport properties of GNRs are expected to strongly depend on the width and crystallographic orientation; however, the presence of armchair/zigzag edge configurations is also a major governing variable. Both armchair and zigzag GNRs are semiconductors with direct bandgaps that decrease with increasing ribbon widths.[426] However, zigzag GNRs are dominated by edge states, which have been demonstrated to be spin-polarized under a large electric field. This has important implications for energy storage applications as well as future *spintronic* devices, which promise to be smaller, faster, and more versatile than current technologies.[427] There are now precedents for the growth of GNRs with a variety of edge structures

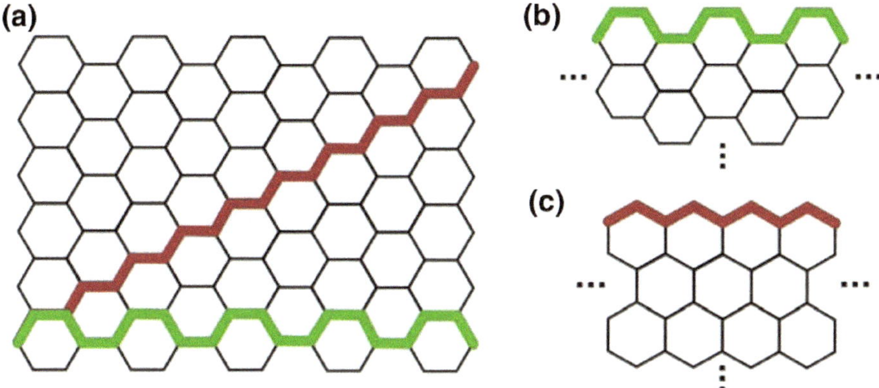

Fig. 6.114 Graphene nanoribbons with zig-zag (red) or armchair (green) edges, relative to a bulk 2-D graphene sheet

beyond zigzag and armchair, to include chevron, fjord, gulf, cove, and junction edge states (Fig. 6.115).[428]

The first reports of "top-down" syntheses of graphene nanoribbons (GNRs) via 'unzipping' carbon nanotubes were reported in mid-2009. Tour et al. report an acid-catalyzed permanganate oxidation route (Fig. 6.116, top),[429] whereas Dai et al. report a plasma-based unzipping of polymer-immobilized nanotubes (Fig. 6.116, bottom).[430] Alternatively, lithography[431] and chemical vapor deposition[432] methods have been used to afford small amounts of GNRs. The first "bottom-up" synthetic approach was reported by Mullen and coworkers using the controlled polymerization of aromatic monomers (Fig. 6.117).[433] Alternatively, surface-mediated routes have been shown to yield GNRs with high purity.[277d,e] However, the current yield of bottom-up synthesized GNRs is currently far too low for use in applications.

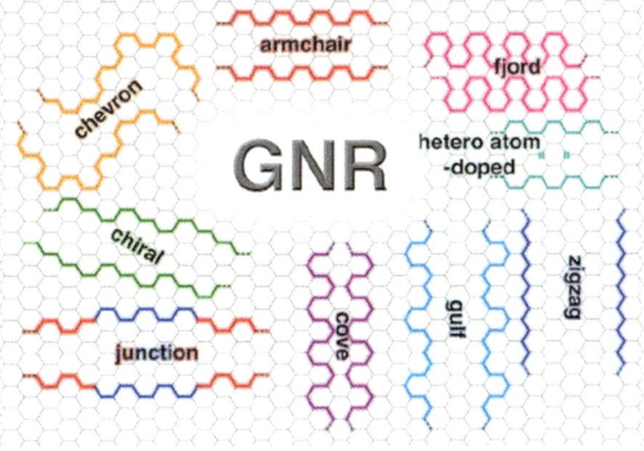

Fig. 6.115 A comparison of edge structures for graphene nanoribbons. Reproduced with permission from *J. Org. Chem.* **2020**, *85*, 4 © 2020 American Chemical Society

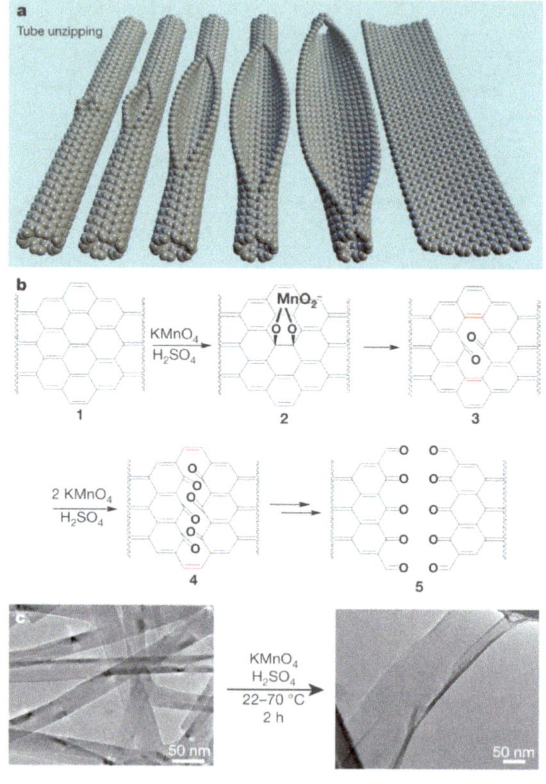

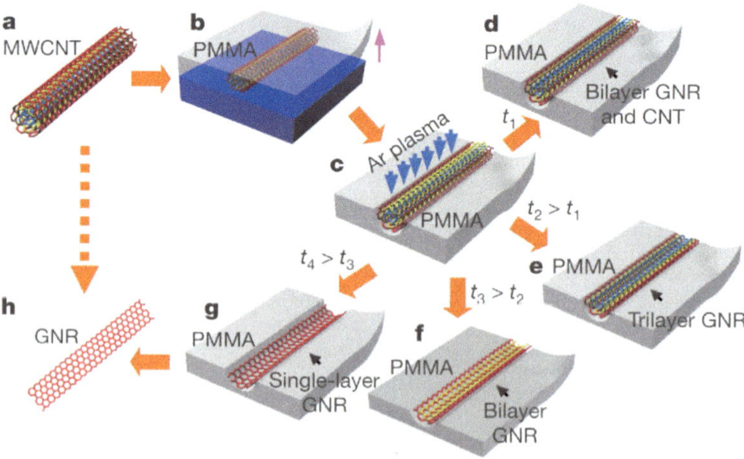

Fig. 6.116 Illustration of top-down graphene nanoribbon formation from unzipping multi-walled carbon nanotubes via a solution-based permanganate oxidation (top), and Ar plasma (bottom). Reproduced with permission from *Nature* **2009**, *458*, 872, and *Nature* **2009**, *458*, 877, respectively.

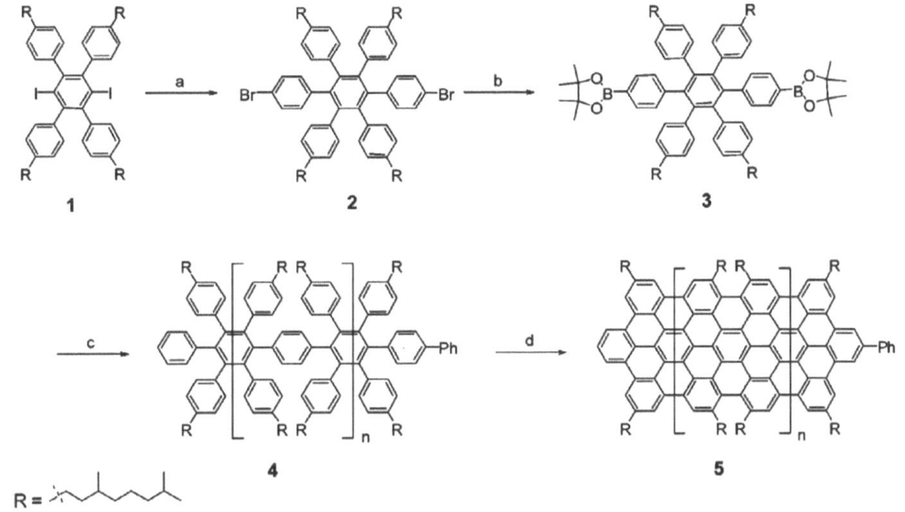

Fig. 6.117 Bottom-up synthesis of graphene nanoribbons. Reagents and conditions: (a) 4-bromophenylboronic acid, $Pd(PPh_3)_4$, K_2CO_3, toluene, 80 °C, 24 h, 93%. (b) (i) *n*-BuLi, THF, −78 °C, 1 h; (ii) 2-isopropoxy-4,4,5,5-tetramethyl[1,3,2]dioxaborolane, rt, 2 h, 82%. (c) compound **1**, $Pd(PPh_3)_4$, K_2CO_3, toluene/H_2O, reflux, 72 h, 75%. (d) $FeCl_3$, CH_2Cl_2/CH_3NO_2, 25 °C, 48 h, 65%. Reproduced with permission from *J. Am. Chem. Soc.* **2008**, *130*, 4216. © 2008 American Chemical Society

Beyond Graphene: Other 2-D Nanostructures

The excitement surrounding graphene has led to significant efforts to expand the range of 2-D materials.[434] In addition to offering a high specific surface area, 2-D nanomaterials exhibit superior mechanical, electronic, catalytic, and optical properties relative to their bulk counterparts due to their low dimensionality. For instance, MoS_2 shifts from an indirect to direct bandgap in its monolayer form, which results in a 10^4-fold enhancement in its luminescence quantum yield relative to the bulk material.[435] Furthermore, of relevance for next-generation integrated circuits, atomically thin 2-D nanomaterials are less susceptible to *short channel effects*,[436] which have threatened the survival of Moore's Law for Si-based MOSFET devices (revisit Sect. 4.2.2). To date, a variety of monolayer 2-D nanomaterials have been fabricated, such as metal oxides, carbides, nitrides, hydroxides, and dichalcogenides.[437]

Recently, *silicene*—the analogue of graphene—was fabricated,[438] which further extends the range of applications for 2-D nanomaterials.[439] Since pristine graphene does not have a bandgap, its use in transistors has been limited since this would require special bandgap engineering approaches, which would reduce carrier mobilities or would require high operating voltages.[440]

Although silicene was predicted more than two decades ago, its fabrication requires special conditions to prevent its oxidation in air. Hence, the first evidence for silicene was inside a low-temperature scanning tunneling microscope (STM) under UHV conditions (5×10^{-11} Torr), in which Si was evaporated from a heated

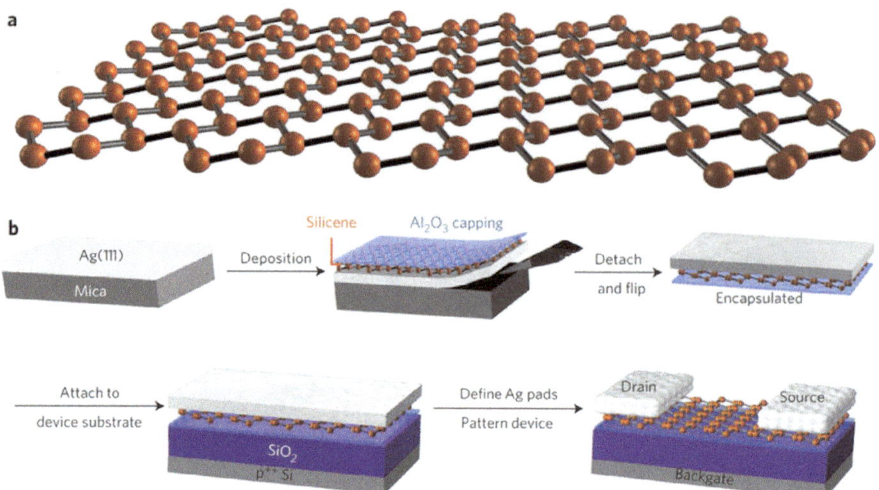

Fig. 6.118 **a** Illustration of the buckled honeycomb lattice structure of silicene. **b** Silicene encapsulated delamination with native electrode (SEDNE) process, which includes the following steps: epitaxial growth of silicene onto Ag(111); in situ alumina capping; encapsulated delamination transfer of silicene; native contact electrode formation to enable back-gated silicene transistors. Reproduced with permission from *Nat. Nanotechnol.* **2015**, *10*, 227. © 2015 Nature Publishing Group

wafer (*ca.* 1200 K) onto a heated Ag(111) surface. Likewise, single-layer Ge ("germanene") has also been fabricated using molecular beam epitaxy (MBE) onto a Au(111) surface. Interestingly, in contrast to the planar sheets of graphene, silicene layers have a buckled honeycomb morphology, which gives rise to an interesting set of tunable electronic and optoelectronic properties. Through a careful encapsulated delamination procedure, it is now possible to transfer silicene films to another surface for device fabrication, such as field-effect transistors (Fig. 6.118).[441]

In addition to the above "bottom-up" routes, 2-D nanomaterials may be fabricated using more simplistic "top-down" approaches. For instance, single-layer sheets of BN, MoS_2, $NbSe_2$, $Bi2Sr_2CaCu_2O_x$, and graphene were fabricated by the simple rubbing of a layered crystal against another surface.[442] Unexpectedly, among the thicker flakes deposited onto the surface were many single-layer films. Phosphorene films may be fabricated via mechanical exfoliation of bulk black phosphorus, which has a direct tunable bandgap depending on the number of layers produced.[443] However, in contrast to graphene or other 2-D nanomaterials such as MoS_2, single-layer or bilayer phosphorene films are extremely sensitive to oxidation and difficult to utilize for device applications. Beyond phosphorene and silicene, there are now reports of 2-D siloxene, antimonene, germanene, stanene, arsene, and bismuthene for a variety of applications such as supercapacitors.[444]

The "selective extraction" method pioneered by Gogotsi and coworkers has been used to yield 2-D nanomaterials of a variety of transition metal oxides, carbides, and carbonitrides.[445] This method employs chemical reactions involving a bulk layered structure, which leads to the formation of unique nanostructures by the selective extraction of certain elements. For instance, films of transition metal carbides and carbonitrides may be produced by the selective removal of layers of atoms within its crystal structure. Since the A layer of transition metal $M_{n+1}AX_n$ phases (known as "MAX" phases; M = early transition metal such as Ti, Nb, Mo, etc.; A = Al, Si, Sn, etc.; X = C and/or N) are most reactive, these layers may be selectively removed by using aqueous HF at room temperature (Fig. 6.119).[446] More recently, the composition range of MXenes has been expanded to include Mo, V, and Nb.[447] As a result of etching the A-layers, the metal atoms on the outer surface will become terminated with functional groups (−OH, −O, or −F), which serve to hold together adjacent layers via weak intermolecular forces. By sonicating the resultant functionalized layers, discrete films of MX compositions may be fabricated, which are denoted as "MXenes"—derived from MAX phases and analogous to graphene. In addition to HF, ammonium bifluoride (NH_4HF_2) has also been successfully utilized to etch Al from Ti_3AlC_2. Although potential applications for MXenes have yet to be widely exploited,[448] these 2-D nanomaterials show promise for energy storage applications[449] and toxic pollutant removal.[450]

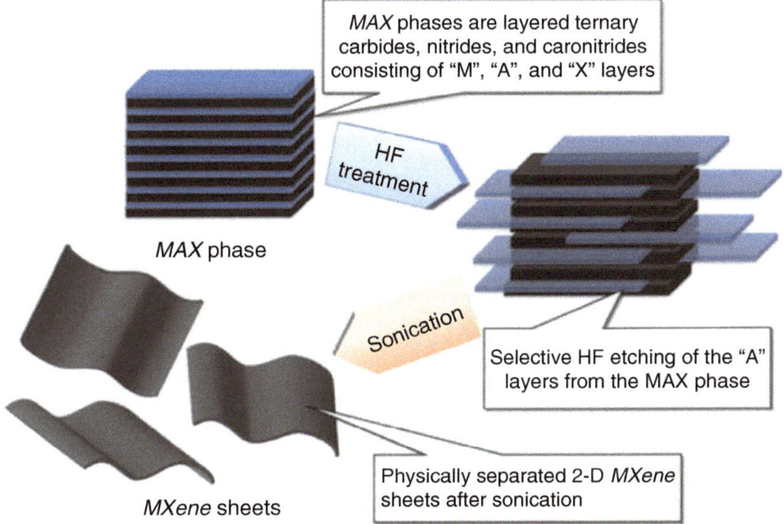

Fig. 6.119 Schematic illustrating the process of MXene formation from the selective etching of MAX phases. Reproduced with permission from *ACS Nano* **2012**, *6*, 1322. © 2012 American Chemical Society

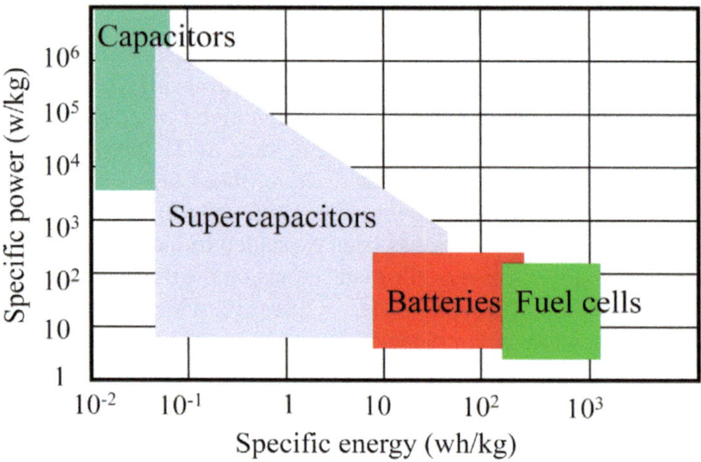

Fig. 6.120 Comparison of various energy storage technologies

6.4 Important Nanomaterials Applications I: Lithium-Ion Batteries

With cell phones and portable electronics now pervasive in our society, it is essential that our busy days of web browsing, phone conversations, texting, email contact, and—perhaps most importantly—social media posts, are not interrupted by a dead battery! In the automotive world, electric vehicles (EVs) are becoming more prevalent due to concern over global climate change. Accordingly, there has been a flurry of research effort to increase the range of vehicles through use of various energy storage technologies (Fig. 6.120). With a high specific power ("acceleration"), virtually unlimited cycle life, and ultrafast charging, *super/ultracapacitors* are good choices for high-power applications such as power tools. However, supercapacitors contain only about 5% of the energy stored by Li-ion batteries, which would limit automobiles to a few miles per charge. In addition, supercapacitors exhibit linear discharge voltages[451] and have issues with high self-discharge. In comparison, the specific energy ("range") of batteries and fuel cells are much greater than supercapacitors, but exhibit much slower charging/discharging. The combination of batteries and supercapacitors represents the best of both worlds; a capacitor could be used for fast acceleration and a battery for extended driving range. This technology is currently in use for buses in some metropolitan areas such as Shanghai, China (Fig. 6.121),[452] with expansions to many other locations across the globe in the coming years.

Table 6.9 illustrates a comparison between common battery chemistries. Despite a low specific energy (6.26),[453] large mass, and environmental hazards, lead-acid batteries continue to be the workhorse for automotive starting, lighting, and ignition (SLI) applications. The large niche market share of Pb-acid batteries is due to their low cost of production and large power-to-weight ratio. Nickel–cadmium batteries feature better energy densities, but have been largely phased out of consumer products due to the environmental impacts of Cd. Some of the few remaining applications for Ni–Cd

Fig. 6.121 Photo of a hybrid supercapacitor/battery bus in Shanghai, China. © Bradley D. Fahlman

include alarm systems, emergency lighting, and portable power tools. The energy densities of nickel-metal hydride (Ni-MH) batteries can rival that of lithium-ion, thus replacing Ni–Cd varieties for portable consumer use. Being much less expensive than Li-ion batteries, some plug-ion electric (e.g., Toyota RAV4 EV, Ford Ranger EV) and hybrid EVs (e.g., Toyota Prius, Ford Escape Hybrid, Honda Insight) have recently employed Ni-MH batteries. However, lithium-ion batteries (LiBs) currently represent the most popular choice for portable electronics and electric/hybrid vehicles.[454] This chemistry is ideal for applications that require high energy density and portability, since Li is the lightest metal in the periodic table (density $= 0.534$ g cm^{-3}) and also one of the easiest metals to oxidize (i.e., largest negative standard reduction potential, -3.04 V).

$$\text{Energy Density} = \frac{\text{Voltage} \times \#\,\text{of movable ions/species in electrodes}}{\text{Total battery weight (specific energy) or volume (energy density)}} \quad (6.25)$$

As shown in Fig. 6.122, most commercial EVs have an EPA-rated range of <300 miles per charge. Notable exceptions are the 2022 Lucid Air and selected Tesla models with ranges above 400 miles. However, the actual range of EVs is often much less based on the ambient temperature,[455] driving speeds, and use of power-consuming features (e.g., heater, A/C, stereo). Although the range of EVs is now approaching distances of relevance to commuting consumers, the battery packs in these vehicles are enormous—comprising the entire bottom of the vehicle frame![456] For instance, the 100 kWh battery pack in the Tesla P100D weighs *ca.* 1300 lb and

Table 6.9 Comparison of different battery chemistries

Properties	Pb-Acid	Ni–Cd	Ni-MH	Li-ion
Specific energy (Wh kg^{-1})	30–55	45–80	60–120	90–190
Cycle life (80% init. capacity)	200–300	1000	300–500	500–2000
Cell voltage (V)	2.0	1.2	1.2	3.3–3.8
Self-discharge/month at 25 °C	5–15%	20%	30%	<5%
Relative safety	Dangerous[a]	Safe	Safe	Dangerous[b]
Year of commercialization	1881	1950	1990	1991
Relative toxicity	High	High	Low	Low

[a] Contains *ca.* 35% sulfuric acid; risk of fire or explosion

[b] Protection circuitry is mandatory to prevent overcharging and associated risk of fire/explosion

is composed of 8256 Li-ion cells (i.e., 16 compartments, each with 516 individual "18,650" Li-ion cells[457]). In contrast, the Lucid Air Dream Edition is equipped with an impressive 118 kWh battery pack containing 22 modules with a total of 6600 "2170" Li-ion cells.[458] Hence, even though Li-ion batteries are the most lightweight option, a large number of cells are needed for usable driving ranges. As a result, the search continues for alternative electrode materials that may be used to improve the overall energy density of LiBs to further enhance the range and overall weight of the battery pack.

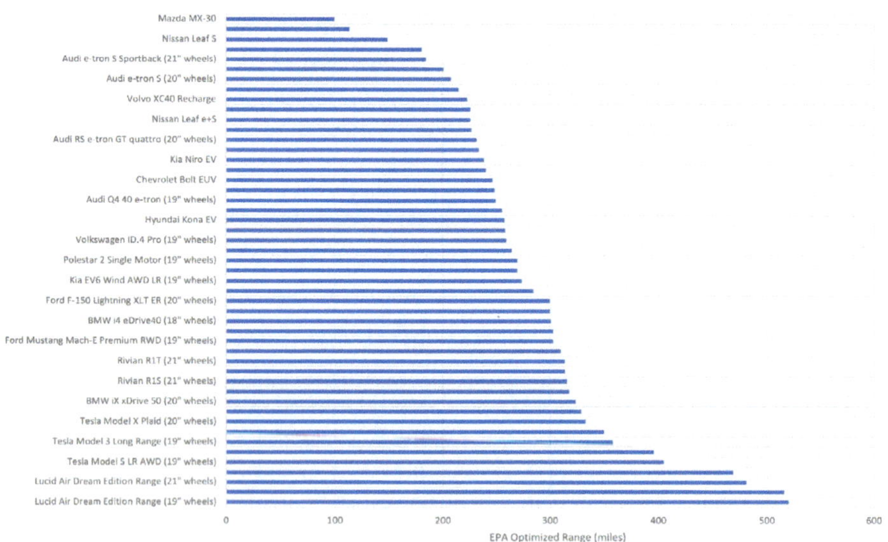

Fig. 6.122 EPA rated ranges of electric vehicles for 2022 models. Data taken from http://fuelec onomy.gov

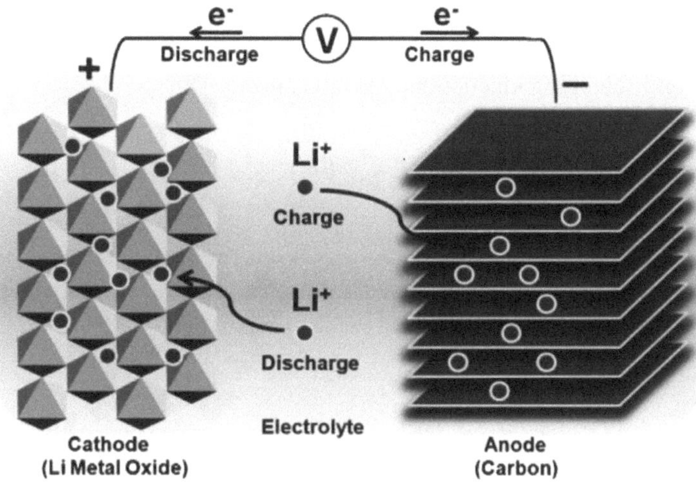

Fig. 6.123 Schematic of the charging and discharging processes occurring in a Li-ion battery

Figure 6.123 illustrates the charging/discharging mechanisms of a LiB, which proceed through simple intercalation of Li ions between layered cathode/anode electrode materials. The nominal voltage of LiBs is 3.7 V, which is much higher than other types of rechargeable batteries such as Ni–Cd or Ni-MH (1.5 V). Hence, only 1/3 the number of LiB cells are needed to generate an equivalent voltage. During charging, lithium ions move from the cathode toward the graphite anode, with one Li^+ ion becoming situated within each C_6 unit of graphene sheets. During the spontaneous discharging process, Li-ion migration is reversed, moving from the layered graphitic planes to the layered metal oxide cathode.

The specific energy density of Li-ion cells is typically 200–385 Wh kg^{-1} (720–1386 kJ kg^{-1}); in comparison, volumetric energy densities are 250–620 Wh L^{-1} (900–1900 J cm^{-3}) and specific power is in the range 250–340 W/kg. Not unlike other varieties of battery chemistries, the energy density of LiBs is governed by the electrode materials. Common cathodes are composed of $LiCoO_2$ (195 Wh kg^{-1}; 3.7 V, 150 mAh g^{-1}), $LiNi_xMn_yCo_zO_2$ (NMC; 150–300 Wh kg^{-1}; 3.7–3.8 V, 160–200 mAh g^{-1}), $LiNiCoAlO_2$ (250–300 Wh kg^{-1}; 4.0 V; 140 mAh g^{-1}), or $LiFePO_4$ (90–120 Wh kg^{-1}; 3.3 V; 170 mAh g^{-1}). Typical anode materials are composed of graphite, with a theoretical specific capacity of 372 mAh g^{-1}; however, commercial batteries with binder, conductive carbon black additive, etc. only realize a reversible capacity of 200–220 mAh g^{-1}.[459] In order to increase the capacity of LiBs—and the driving range of the EV—higher-capacity cells are needed. Hence, the capacity of both cathode and anode materials represents an intense area of research investigation.

Although much work has been performed to increase the capacity of cathodes, there has been little improvement beyond *ca.* 200 mAh g^{-1} for intercalation-type materials (Fig. 6.124).[460] In comparison, the capacity of anode materials has dramatically increased through the use of alloying and nanostructured materials

(Fig. 6.125).[461] Nanostructural electrodes feature a large contact area between the active material/electrolyte and short Li^+ diffusion pathways, which help to improve the charging/discharging kinetics and overall capacity.[462]

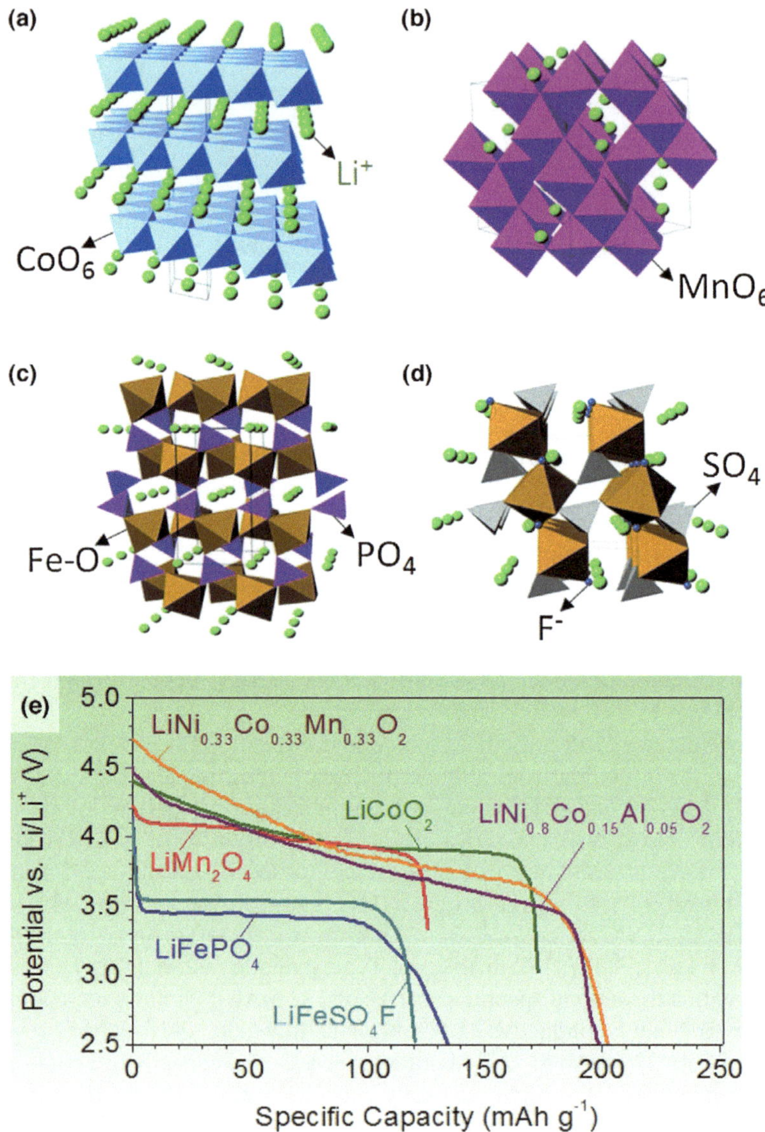

Fig. 6.124 Crystal structures of representative intercalation cathodes. Shown are **a** layered $LiCoO_2$, **b** spinel $LiMn_2O_4$, **c** olivine $LiFePO_4$, and **d** tavorite $LiFeSO_4F$. **e** Typical discharge profiles of intercalaction cathodes. Reproduced with permission from *Mater. Today* **2015**, *18*, 252. © 2015 Elevier Ltd.

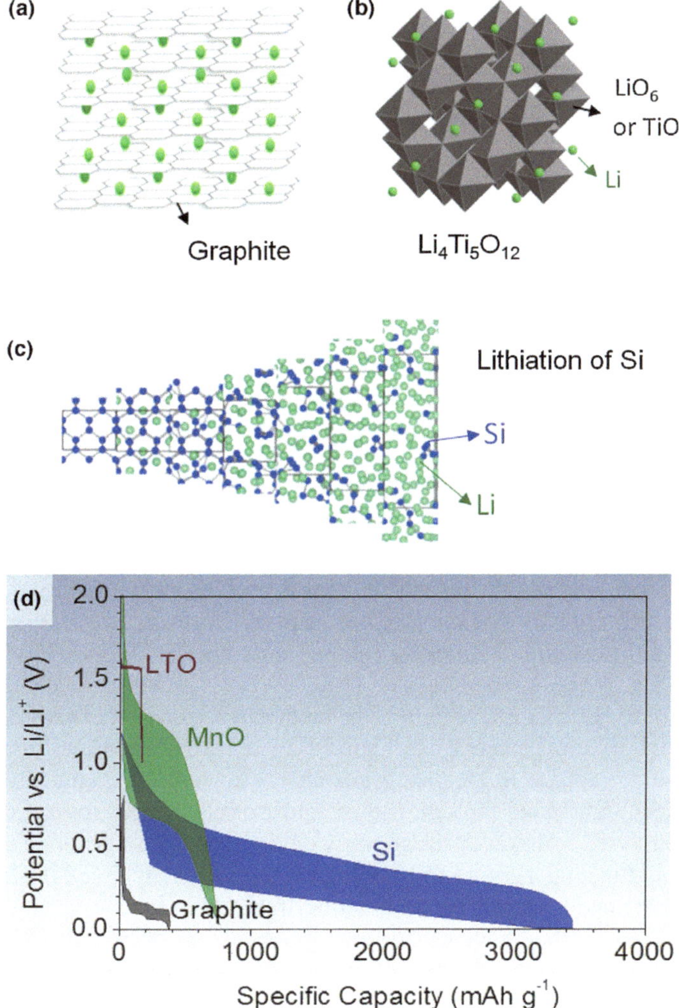

Fig. 6.125 Crystal structures of **a** lithiated graphite, **b** lithium titanate (LTO), and **c** lithiated silicon. The charge–discharge profiles at low current densities are shown in (**d**). Reproduced with permission from *Mater. Today* **2015**, *18*, 252. © 2015 Elevier Ltd.

The first nanomaterial to be investigated for LiB anode applications was CNTs, with capacities approaching 1000 mAh g^{-1} reported in the literature for ball-milled varieties.[463] However, after extended cycling (>100 cycles), the capacity was comparable to pristine graphite anodes. During the first few charging/discharging cycles, a *solid-electrolyte interface* (SEI) of complex composition[464] is formed due to the irreversible electrochemical decomposition of the electrolyte, which is not stable at the operating potential of the anode during charging.[465] The formation of this solid layer effectively removes Li^+ from the system, which causes the capacity to drop as the SEI is being formed. As such, the capacity obtained from the first few cycles is known as the “irreversible capacity”. Since only Li^+ ions are able to permeate the SEI, this layer serves to stabilize the electrode from further degradation, thus enabling extended cycling with minimal capacity fading (i.e., “reversible capacity” of the electrode material).

In order to increase the surface area of the nanostructured anode, graphene was the focus of many investigations. It has been suggested that Li ions may be adsorbed onto both sides of single-layer sheets, as well as onto edge sites of planes.[466] Although irreversible capacities >500 mAh g^{-1} have been reported, reversible capacities are typically in the range of pristine graphite after extended cycling. Interestingly, the Li-capacity of graphene sheets was reported to dramatically increase if CNTs or fullerenes were incorporated—likely due to an increase in d-spacing, which allows for more efficient adsorption of Li^+ ions.[467] More recently, Fe_3O_4 nanoparticles dispersed onto the surface of graphene nanosheets resulted in a very high reversible capacity of >900 mAh g^{-1} after 100 cycles (at 200 mA g^{-1}).[468] Even after 500 cycles at 1 A g^{-1}, the capacity of Fe_3O_4@graphene nanocomposites was still in excess of 750 mA g^{-1}!

The best capacity to date for pristine graphene was reported by Hassoun et al.[469] who showed an outstanding reversible capacity of *ca.* 650 mAh g^{-1} after 140 cycles, with a *Coulombic efficiency*[470] near 100%. However, full-battery testing of graphene nanoflakes with a $LiFePO_4$ cathode only resulted in an overall capacity of 165 mAh g^{-1} and energy density (190 Wh kg^{-1}), which is comparable to current graphite-based Li-ion batteries. The method used to synthesize the nanoflakes entailed the simple mechanical exfoliation (via ultrasonication and ultracentrifugation) of bulk graphite. In contrast, most studies have used chemically modified graphene sheets fabricated through use of acids, often referred to as the modified *Hummer's method*.[471] However, the strongly oxidizing reagents generate a variety of surface-bound functional groups (e.g., epoxides, carbonyls, hydroxyls) that decrease the conductivity and Li-ion diffusion kinetics of the cell.[472]

As described in the previous section, graphene nanoribbons may be fabricated using the Hummer's method on expanded graphite. Interestingly, theoretical studies have shown that Li atoms bind much more strongly to zigzag GNRs than armchair varieties, which exhibit similar Li-adsorption properties as bulk graphene.[473] Although the irreversible capacity of oxidized GNRs is quite high (>800 mAh g-1), the capacity fades dramatically with increasing cycling.[474] Our group evaluated the effect of surface functionalization on the Li capacity of GNRs after post-reducing oxidized GNRs (oGNRs) under a variety of conditions.[475] As shown in Fig. 6.126, the Coulombic efficiencies of both oGNRs and post-reduced GNRs were extremely low for the first cycle, but improved to >90% after 15 cycles. The best reversible capacity was obtained after mild reduction at 250 °C, which was sufficient to remove surface hydroxyl groups. Through detailed surface characterization and modeling, it was suggested that enhanced Li^+ binding occurred onto edge-bound ketone groups, which could be formed by the migration of surface-bound epoxide groups at elevated temperatures. Although this leads to very high reversible capacities (>700 mAh g^{-1} at a current density of 100 mA g^{-1}; >500 mAh g^{-1}

at 500 mA g^{-1}), the presence of these oxygen-containing surface groups is likely responsible for pronounced SEI formation and poor Coulombic efficiencies in early cycles. Hence, surface speciation plays a critical role in the Li capacity and cyclability of surface-functionalized graphenes.[476]

Although carbonaceous nanomaterials have been the focus of most research investigations for anode materials, more recent studies have been devoted to materials with Li-alloying properties such as Si, Sn, Ge, Al, Sb, Zn, Bi, or Cd—with much greater theoretical capacities (e.g., Si = 4200 mAh/g, Sn = 993 mAh/g) than insertion materials (e.g., graphite = 372 mAh/g, TiO_2 = 335 mAh/g, LTO = 175 mAh/g). Another promising candidate is tin(IV) oxide, SnO_2, which exhibits a high theoretical capacity (782 mAh/g). Its large uptake of Li^+ is most likely due to reduction of Sn^{4+} to metallic Sn (6.27), which subsequently forms stable Li_xSn ($0 \leq x \leq 4.4$) alloys.[477]

$$SnO_2 + 4\,Li^+ + 4\,e^- \rightarrow Sn + 2\,Li_2O \tag{6.26}$$

However, the higher capacities of ion-alloying materials are associated with large variations in volume upon alloying/de-alloying (e.g., 310% for $Li_{22}Si_5$, 259% for

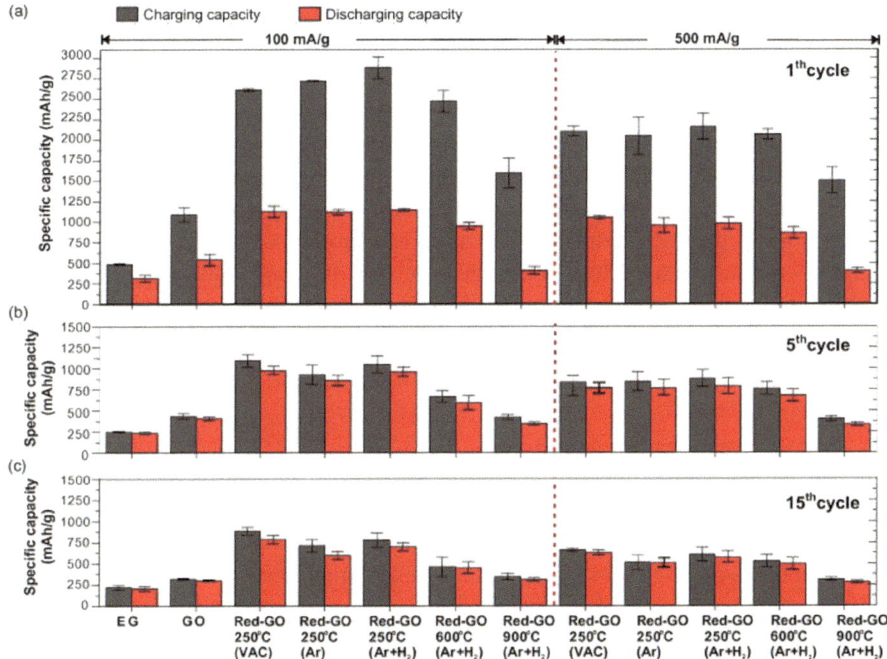

Fig. 6.126 **a** First-cycle irreversible specific capacity of expanded graphite (EG), graphene oxide (GO), and post-reduced GO. Images **b** and **c** show the effect of extended cycling with different conditions used for post-reduction of GO. Two sets of experiments at two different current densities (100 and 500 mA g^{-1}) are shown in each panel. Reproduced with permission from *Carbon* **2013**, *61*, 558. © 2013 Elsevier Ltd.

$Li_{4.4}Sn$), which limits their cyclability.[478] For silicon, there are additional limitations associated with the instability of the electrode/electrolyte interface and low inherent electronic conductivity.

A number of strategies have recently been investigated in an attempt to overcome the practical limitations with Si-based anodes.[479] In particular, the use of Si nanotubes,[480] hollow nanospheres,[481] and porous silicon[482] are able to buffer the large volume expansion that occurs upon lithiation, which has led to better cycling performance. To enhance the electrical conductivity and also restrict volume changes during cycling, carbon coatings have been used.[483] However, other additives such as carbon nanofibers,[484] carbon nanotubes,[485] graphene nanosheets,[486] metallic nanoparticles,[487] or even conductive polymers,[488] have also been shown to increase the electrochemical performance of Si anodes. With Tesla currently using Si-based nanocomposites in their batteries, this represents the most likely alternative for high-capacity EV anodes. However, the search continues to equally expand the capacity bottleneck of the cathode …

6.5 Important Nanomaterials Applications II: Nanoelectromechanical Systems (NEMS)

The past decade or so has brought about tremendous growth in the field of microelectromechanical systems (MEMS)—devices such as sensors, computers, electronics, and machines at the microscale.[489] The design of these devices utilizes a combination of traditional semiconductor processing and mechanical engineering. Typically, the output of an electromechanical device is the movement of the mechanical component; a transducer is used to convert the mechanical energy into electrical/optical signals, or vice versa. Some current MEMS applications include:

(i) Automotive (e.g., air flow and tire pressure sensors, "smart" suspension, headlight leveling, navigation, vehicle security, automatic seatbelt restraint, etc.)
(ii) Micro-nozzles for inkjet printers
(iii) Micro-tweezers
(iv) Aerospace navigational gyroscopes
(v) Disposable blood pressure transducers
(vi) Portable skin analysis sensors for cosmetics applications.

The extension of MEMS to the nanoregime is referred to as NEMS—representing the ultimate in future devices—with benefits such as lower power dissipation and ultra-sensitive and localized responses. Further, due to the size of NEMS, it will be possible to directly incorporate a number of auxiliary functionalities alongside transistors within a single chip. Indeed, the applications for NEMS will span the fields of sensors, electronics, biotechnology, affecting virtually every aspect of our lives. Though the technology is in place to fabricate NEMS, there are three primary challenges that must be overcome prior to the realization of widespread commercial applications[490]:

(i) How to communicate signals from the nanoscale to the macroscopic world
(ii) Understanding and controlling mechanical responses at the nanoscale
(iii) Developing methods for low-cost, high-volume, and *reproducible* nanofabrication.

One of the earliest applications of NEMS is for sensitive mass sensors for single-molecule mass spectrometry and single-cell characterization. To date, the most sensitive NEMS mass sensor has a resolution of 1.7 yoctograms (1 yg $= 10^{-24}$ g)—the mass of one proton![491] Incredibly, this device was able to detect the adsorption of naphthalene molecules onto the surface of a carbon nanotube, as well as measure the binding energy of a single Xe atom onto the CNT. It has been suggested that this technology could be combined with nanofluidics for the genetic analysis of the DNA present within a single cell. In comparison, current methods utilize PCR amplification, whereby small samples of DNA are repeatedly replicated in order to facilitate detection. Such analyses will be of use for the detection of genetic markers for cancer or other diseases such as HIV. The ability to identify proteins will be essential for future diagnostic and forensics applications. The use of high-throughput nanofluidic real-time PCR has already been shown to be useful to detect various serotypes of streptococcus pneumoniae, a leading cause of global mortality.[492]

Another area of NEMS that is receiving tremendous attention is the mimicry of biological systems, aptly referred to as *biomimetics*.[493] For instance, in the development of *linear molecular muscles* that undergo contraction and extension movements. Initial work in this field utilized transition metal complexes containing rotaxanes and catenanes, due to the nondestructive redox processes occurring on the metal centers.[494] Although these complexes were actuated by a chemical reaction, the movement was in a noncoherent manner. In order to better mimic skeletal muscle movement, one has to look at the mode of motion within the most efficient molecular machines—in our human bodies.

The cellular unit that is active toward the contraction of skeletal muscles, known as the sarcomere, is comprised of alternatively stacked filaments of the proteins actin and myosin. During muscle contraction, the protein filaments slide past each other as a result of a rowing action of the surface myosin heads (Fig. 6.127a).[495] Hence, an effective biomimetic approach would entail the design of a linear architecture that features sliding components that will respond to a chemical stimulus. This approach has recently been demonstrated with the design of a rotaxane molecule that exhibits redox-controlled contraction and extension of the molecular architecture, in response to a chemical or electrochemical stimulus (Fig. 6.127b).[496]

The controlled movement in this system is controlled by the interaction among redox-active units at the redox-active tetrathiafulvalene (TTF) units. In the neutral, unperturbed state, the tetracationic cyclophane units (cyclobis(paraquat-*para*-phenylene), $CBPQT^{4+}$) are most stable coordinated to the TTF moieties, due to electron donation and π-stacking π interactions. However, upon oxidation of the TTF units, the $CBPQT^{4+}$ rings become electrostatically repelled, migrating to the naphalene component (Fig. 6.127b). Hence, the "rowing" action exhibited by skeletal

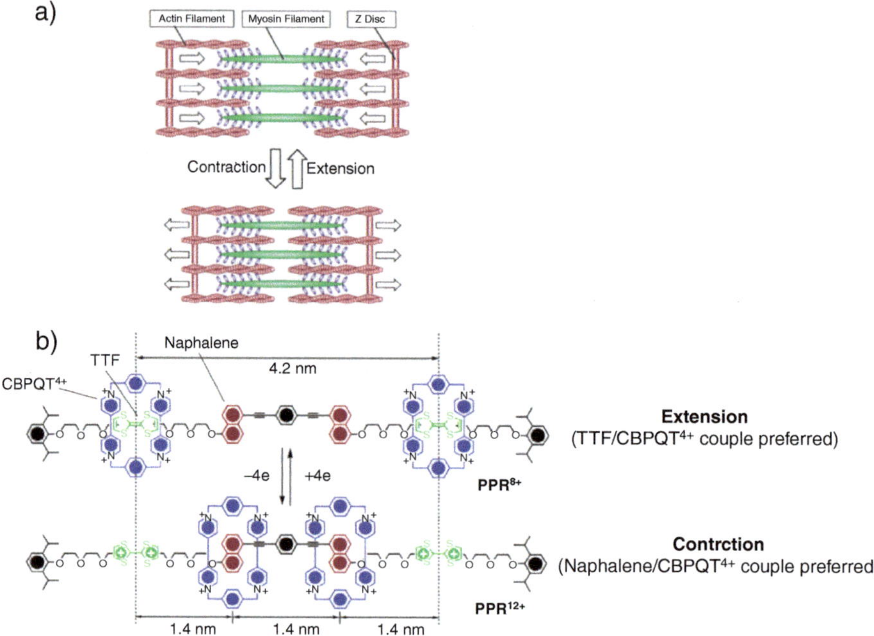

Fig. 6.127 A biomimetic approach toward skeletal muscle movement. Shown is **a** the stacked protein filaments of the sarcomere and **b** a redox-controlled molecular analogue. Adapted with permission from *J. Am. Chem. Soc.* **2005**, *127*, 9745. © 2005 American Chemical Society

muscles is emulated by the Coulombic repulsion and π-donation of the naphalene rings.[497] By attaching disulfide tethers to the $CBPQT^{4+}$ rings, the attachment of the linear muscle may be fashioned to a gold surface *en route* toward a biomimetic NEMS device.[498]

Lab-on-a-chip (LOC) technology represents another area of interest that employs MEMS- and NEMS-based devices.[499] Sensors of this variety are designed to detect specific chemical targets in bodily fluids to provide information about current health conditions. Sikdar et al. reported the use of sophisticated LOC devices composed of dielectric nanocubes as unidirectional nanoantennas and nanocantilevers for cancer detectors (Fig. 6.128).[500] Lab-on-a-chip technology is extremely powerful and virtually unlimited, since it has the potential to prevent disease before it progresses.

In the next few years, there will be many additional developments related to NEMS devices. In particular, for nanomedicine devices or replacing the current chemically driven molecular machines with those stimulated by optical or electrical pulses—dramatically extending the range of applications. With so many possible nanobuilding blocks at our disposal, the scope of devices and resultant applications is limited only by our imaginations—an exciting area of discovery awaits!

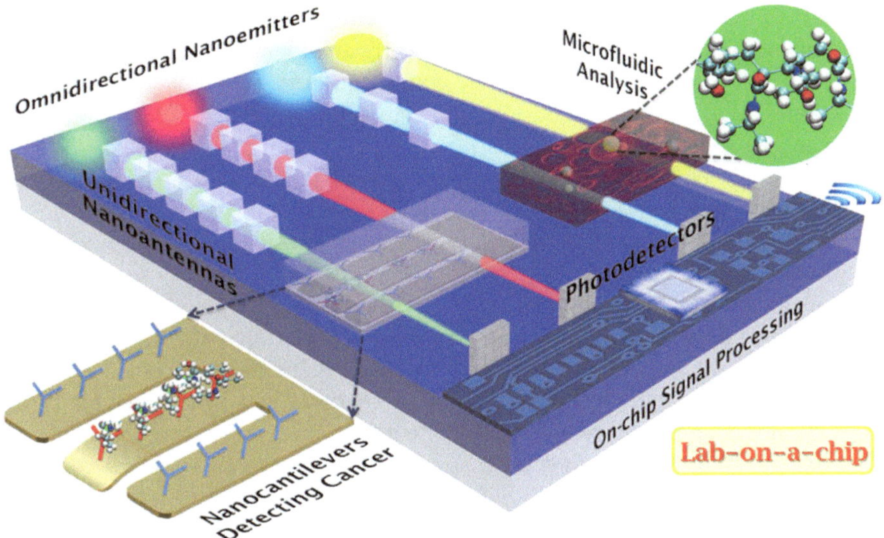

Fig. 6.128 Schematic of an advanced lab-on-a-chip (LOC) design for cancer targeting and treatment. Reproduced with permission from *J. Appl. Phys.* **2015**, *117*, 083101. © 2015 AIP Publishing

Topics for Further Discussion

1. From your knowledge of semiconductor and metallic 0-D nanostructures, think about how you would design a coating that would sense the surrounding color of a wall and adapt its color to match (e.g., color-adapting decor!).
2. Mercedes-Benz and other manufacturers feature scratch-resistant clear coats as standard on new vehicles. What are these coatings composed of, and how does this prevent surface scratching?
3. Carbon nanotubes have been touted as being useful to store large amounts of hydrogen gas for fuel cell applications. From the literature, how is this contained—within the interior or adsorbed along the sidewall?
4. What are some examples of "self-cleaning" coatings? How do these films work?
5. What factors govern the tilt angle (between the substrate and alkyl chain) of a SAM on a gold or silver surface?
6. The growth mechanism of carbonaceous nanostructural materials is generally thought to be via VLS. Cite some recent examples of carbon nanotube/nanofiber growth at temperatures far below the melting point of the nanoparticulate catalyst species.
7. Interestingly, 1-D nanostructures such as nanocrystals or nanowires may also be formed on the reactor sidewalls by laser ablation. Explain the growth mechanism of these structures, as opposed to more commonly formed 0-D nanostructures within these systems.

8. In the Finke four-step mechanism for nanocluster growth, explain why higher temperatures are most conducive for the growth of monodisperse nanoclusters.
9. Think of a new device that you could fabricate by DPN that would be composed of both nanoclusters and nanotubes/nanowires. What are some potential applications for this device?
10. How might one utilize inert gas evaporation to pattern micron-sized arrays of 0-D nanoparticles?
11. How could you synthesize free-standing nanorings using both the top-down and bottom-up approaches?
12. Calculate the total number of atoms and number of surface atoms contained within a Fe nanoparticle with a diameter of 10 nm. Calculate the number of C atoms in a (10,0) SWNT with a length of 25 μm and diameter of 1.5 nm.
13. How could you synthesize metal oxide nanotubes, using a sacrificial template? Cite any related precedents from the literature.
14. Describe how the band diagram of silicon nanowires (SiNWs) differs from bulk Si. Does this imply the use of SiNWs for applications that differ from bulk Si? Elaborate.
15. What are some examples of NEMS devices currently under development?
16. What are the major developmental efforts underway to power nanodevices?
17. How could you deposit a square grid of TiO_2 nanowires with a controllable spacing between adjacent nanowires?
18. Describe how LbL may be used to deposit coatings onto complex, nonplanar substrates. Cite examples for this strategy from the literature.
19. Cite examples of materials designs where *both* "top-down" and "bottom-up" approaches were used.
20. Explain the phenomena of "optical tweezers" and "nano pens". Describe how these techniques have been used to assemble nanostructures on various surfaces.
21. There are reports that N-doped CNTs are less toxic than pristine SWNTs or MWNTs (e.g., *Nano Lett.* **2006**, *6*, 1609). Provide some likely rationales for the varying toxicological effects for these doped nanostructures.
22. Describe a technique (with illustrations of your experimental setup, if necessary) to synthesize the following heterostructures. For each, describe how you control the diameter of the core and thickness of the shell.

    (a) 0-D quantum dots composed of a CdSe core and GaP shell.
    (b) 1-D nanowires composed of a CdSe core and GaP shell.

23. Explain how light has been used to assemble 0-D nanostructures into 2-D arrays onto planar substrates.
24. Describe a "top-down" and "bottom-up" approach to synthesizing 0-D nanostructures.
25. Describe the Vapor–Liquid-Solid, Solution-Liquid–Solid, and Solid–Liquid-Solid synthetic routes for 1-D nanostructural growth. Make sure you discuss the experimental setup and required precursor(s) for each technique, as well as the morphological control (i.e., control over thickness, length, chirality, etc.) one would have for each technique.

26. Provide some examples of nanostructures that may be used for energy storage applications, citing literature precedents. What are some benefits and challenges of using nanomaterials for this application? Do any commercial batteries currently utilize nanomaterials as electrodes?
27. Based on the structures and properties of various nanostructures detailed in this chapter, propose a lab-on-a-chip design that could detect cancer cells and monitor drug delivery treatment.

References and Notes

1. A sampling of some intriguing applications that are already possible using nanomaterials include: self-cleaning fabrics (via TiO_2 nanoparticles), automobile clearcoats that prevent scratches (PPG nanoparticle-based coatings), car wash solutions that prevent dirt from adhering to a painted surface, bandages that kill bacteria, drug-release agents and time-release biocidal coatings, and tennis balls that bounce twice as long as conventional balls.
2. https://www.nano.gov/education-training/university-college.
3. http://nnci.net.
4. Now available online at: https://www.nanowerk.com/nanotechnology/reports/reportpdf/report47.pdf.
5. Nafisi, S.; Maibach, H. I. "Nanotechnology in Cosmetics" in *Cosmetic Science and Technology: Theoretical Principles and Applications*, Elsevier: New York, pp. 337–169, 2017.
6. A review of nanoparticle-based drugs that may be inhaled for lung cancer treatment: Lee, W. H.; Loo, C. Y.; Traini, D.; Young, P. M. *Asian J. Pharm. Sci.* **2015**, *10*, 481.
7. http://www.beilstein-journals.org/bjnano/articles/6/181.
8. These include liposomes (e.g., Doxil, a PEGylated liposome), polymers (e.g., BIND-014, PEGylated polymer), metal (e.g., CYT-6091, PEGylated nano Au) and metal oxides (e.g., Ferumoxtran-10, iron oxide nanoparticles), and composite materials (e.g., Abraxane, albumin particles). For more information, see: https://www.ncbi.nlm.nih.gov/pmc/articles/PMC6764803/. (b) https://doi.org/10.1038/s41573-020-0090-8.
9. The global market for products that contain nanomaterials was valued at $8B USD in 2022 and is expected to reach $12.1B USD by 2026—an annual growth rate of 9.7%. For instance, see: https://www.prnewswire.com/news-releases/global-nanomaterials-market-to-reach-12-1-billion-by-2026--301506511.html.
10. "Discussion draft: Addressing nanomaterials as an issue of global concern", May 2009, Center for International Environmental Law (CIEL). Found online at: http://www.ciel.org/Publications/CIEL_NanoStudy_May09.pdf.
11. (a) De Jong, W. H.; Borm, P. J. A. *Int. J. Nanomed.* **2008**, *3*, 133. (b) Donaldson K, Stone V, Tran CL, Kreyling W and Borm PJA. Nanotoxicology. *Occup Environ Med* 2004, 61, 727–728. (c) http://www.ciel.org/wp-content/uploads/2015/07/Nano_ToxicRisks_Nov2014.pdf.

12. For details on the biological effects of CNTs, see: Liu, Z.; Cai, W.; He, L.; Nakayama, N.; Chen, K.; Sun, X.; Chen, X.; Dai, H. *Nat. Nanotechnol.* **2007**, 2, 47, and references therein. The biological effects of dendritic polymers is described in Boas, U.; Heegaard, P. M. H. *Chem. Soc. Rev.* **2004**, *33*, 43, and references therein. Some websites that have information regarding the toxicological effects of nanostructures include: (a) https://orau.org/ihos/tech-topics/nanotechnology-repository.html (b) http://www.cdc.gov/niosh/topics/nanotech.
13. An example of such a database: Vriens, H.; Mertens, D.; Regret, R.; Lin, P.; Locquet, J. P.; Hoet, P. in "Advances in Experimental Medicine and Biology", Springer: New York, vol. 947, 2017.
14. Shelley, M. L.; Wagner, A. J.; Hussain, S. M.; Bleckmann, C. *Int. J. Toxicol.* **2008**, *27*, 359.
15. Fortino, V. et al. *Nature Communications* **2022**, *13*, 3798.
16. For more information regarding the risk assessment of nanomaterials in a regulatory context, see: http://www.ciel.org/wp-content/uploads/2016/12/NanoRiskAssessmentFactsheetFinal_6dec2016.pdf.
17. One of the most comprehensive sources of nanotoxicity data may be found online at: https://search.data.enanomapper.net.
18. Cross, S. E.; Innes, B.; Roberts, M. S.; Tsuzuki, T.; Robertson, T. A.; McCormich, P. *Skin Pharmacol. Physiol.* **2007**, *20*, 148.
19. Zhu, S.; Oberdorster, E.; Haasch, M. L. *Marine Environ. Res.* **2006**, *62*, S5, and references therein.
20. Henry, T. B.; Petersen, E. J.; Compton, R. N. *Curr. Opin. Biotechnol.* **2011**, *22*, 533.
21. (a) Henry, T. B.; Menn, F. M.; Fleming, J. T.; Wilgus, J.; Compton, R. N.; Sayler, G. S. *Environ. Health Perspect.* **2007**, *115*, 1059. (b) Kovochich, M.; Epinasse, B.; Auffan, M.; Hotze, E. M.; Wessel, L.; Xia, T.; Nel, A. E.; Weisner, M. R. *Environ. Sci. Technol.* **2009**, *43*, 6378. (c) Spohn, P.; Hirsch, C.; Hasler, F.; Bruinink, A.; Krug, H. F.; Wick. P. *Environ. Pollut.* **2009**, *157*, 1134. (d) Zhang, B.; Cho, M.; Fortner, J. D.; Lee, J.; Huang, C. H.; Hughes, J. B.; Kim, J. H. *Environ. Sci. Technol.* **2009**, *43*, 108.
22. For example, see: (a) Sumi, N.; Chitra, K. C. *Mutation Res./Genetic Toxic. Environ. Mutagenesis* **2019**, *847*, 503104. (b) Sumi, N.; Chitra, K. C. *Environ. Sci. Pollut. Res. Int.* **2020**, *27*, 19603. (c) Holmannova, D.; Borsky, P.; Svadlakova, T.; Borska, L.; Fiala, Z. *Nanomaterials* **2022**, *12*, 1716 (a review that includes other carbon-based nanoparticles).
23. Tinkle, S. S, Antonini, J. M., Rich, B. A., Roberts, J. R., Salmen, R., DePree, K., Adkijns, E. J. *Environ. Health Perspect.* **2003**, *111*, 1202.
24. (a) Bos, J. D.; Meinardi, M. M. *Exp. Dermatol.* **2000**, *9*, 165 (the "500 Dalton Rule"). (b) Watkinson, A.C.; Bunge, A. L.; Hadgraft, J.; Lane, M. E. *Pharm. Res.* **2013**, *30*, 1943.
25. For instance, see: (a) Monteiro-Riviere, N. A.; Larese Filon, F. in "Adverse Effects of Engineered Nanoparticles", Chapter 11, Elevier: New York, 2012.

(b) Labouta, H. I.; Schneider, M. *Nanomedicine* **2013**, *9*, 39. (c) Schneider, M.; Stracke, F.; Hansen, S.; Schaefer, U. F. *Dermatoendocrinol.* **2009**, *1*, 197.

26. Potts, R. O.; Guy, R. H. *Pharm. Res.* **1992**, *9*, 663.
27. Holsapple, M. P.; Farland, W. H.; Landry, T. D.; Monteiro-Riviere, N. A.; Carter, J. M.; Walker, N. J.; Thomas, K. V. *Toxicol. Sci.* **2005**, *88*, 12.
28. (a) Sukhanova, A.; Bozrova, S.; Sokolov, P.; Berestovoy, M.; Karaulov, A.; Nabiev, I. *Nanoscale Res. Lett.* **2018**, *13*, 44. (b) Lademann, J.; Patzelt, A.; Richter, H.; Antoniou, C.; Sterry, W.; Knorr, F. *J. Biomed. Opt.* **2009**, *14*, 021,014.
29. Jung, E.; Hui, X.; Zhu, H.; Zhang, A.; Wang, W.; Buchholz, B.; Maibach, H. *Toxicol. Res. Applic.* **2019**, *3*, 1.
30. George, R.; Merten, S.; Wang, T. T.; Kennedy, P.; Maitz, P. *Australas J. Dermatol.* **2014**, *55*, 185.
31. For instance, see: Yang, X.; Fan, C.; HS, Z. *Toxicol.* In Vitro **2002**, *16*, 41, and references therein.
32. Sayes, C. M.; Fortner, J. D.; Guo, W.; Lyon, D.; Boyd, A. M.; Ausman, K. D.; Tao, Y. J.; Sitharaman, B.; Wilson, L. J.; Hughes, J. B.; West, J. L.; Colvin, V. L. *Nano Lett.* **2004**, *4*, 1881, and references therein.
33. (a) Rouse, J. G.; Yang, J.; Ryman-Rasmussen, J. P.; Barron, A. R.; Monteiro-Riviere, N. A. *Nano Lett.* **2007**, *7*, 155. (b) Rouse, J. G.; Yang, J.; Barron, A. R.; Monteiro-Riviere, N. A. *Toxicol.* In Vitro **2006**, *20*, 1313. (c) Yang, J.; Barron, A. R. *Chem. Commun.* **2004**, 2884 (Article describing the synthesis of amino-acid functionalized fullerenes).
34. (a) Blundell, G.; Henderson, W. J.; Price, E. W. *Ann. Trop. Med. Parasitol.* **1989**, *83*, 381-385. (b) Corachan, M.; Tura, J. M.; Campo, E.; Soley, M.; Traveria, A. *Trop. Geogr. Med.* **1988**, *40*, 359–364. (c) Tinkle, S. S.; Antonini, J. M.; Rich, B. A.; Roberts, J. R.; Salmen, R.; DePree, K.; Adkins, E. J. *EnViron. Health Perspect.* **2003**, *111*, 1202–1208.
35. (a) Kim, S. H. et al. *Toxics* **2020**, *8*, 122. (b) Monteiro-Riviere, N. A.; Nemanich, R. A.; Inman, A. O.; Wang, Y. Y.; Riviere, J. E. *Toxicol Lett* **2005**, *155*, 377. (c) Degim, I. T.; Burgess, D. J.; Papadimitrakopoulos, F. *J. Microencapsul.* **2010**, *27*, 669.
36. It should be noted that nanosized particulates are also present as micron-sized aggregates. As reported in the European Union's Scientific Committee on Consumer Products (SCCP), Preliminary Opinion on Safety of Nanomaterials in Cosmetic Products, June 19, 2007. May be accessed online at: http://ec.europa.eu/health/ph_risk/committees/04_sccp/docs/sccp_o_099.pdf.
37. Ryman-Rasmussen, J. P., Riviere, J. E., Monteiro-Riviere, N. A. *Toxicol. Sci.* **2006**, *91*, 159.
38. Lademann, J.; Weigmann, H.; Rickmeyer, C.; Barthelmes, H.; Schaefer, H.; Mueller, G.; Sterry, W. *Skin Pharmacol Appl Skin Physiol **1999**, 12, 247..*
39. Liang, Y.; Zhang, T.; Tang, M. *J. Appl. Toxicol.* **2021**, *42*, 1.
40. Kreilgaard, M. *Adv Drug Deliv Rev **2002**, 54, S77..*
41. Ryman-Rasmussen, J. P.; Riviere, J. E.; Monteiro-Riviere, N. A. *Toxicol. Sci.* **2006**, *91*, 159.

42. Menon, G. K.; Elias, P. M. *Skin Pharmacol* ***1997****, 10, 235.*.
43. Tang, L.; Zhang, C.; Song, G.; Jin, X.; Xu, Z. *Sci. Chin. Life Sci.* **2013**, *56*, 181.
44. (a) Gruber, E. S.; Stadlbauer, V.; Kenner, L. *Exposure and Health* **2022**, 1. (b) Hwang, J. et al. *Sci. Rep.* **2020**, *10*, 7391. (c) Sangkham, S. et al. *Marine Pollut. Bull.* **2022**, *181*, 113832.
45. (a) https://doi.org/10.1177%2F0734242X18785730 (b) https://www.nationalgeographic.com/science/article/microplastics-in-virtually-every-crevice-on-earth (c) Cozar, A.; Echevarria, F.; Gonzalez-Gordillo, J. I.; Duarte, C. M. *PNAS* **2014**, *111*, 10,239.
46. https://www.plasticsoupfoundation.org/wp-content/uploads/2022/07/Final-Report-pilot-study-plastic-particles-in-livestock-feed-milk-meat-and-blood-SIGNED.pdf.
47. Leslie, H. A. et al. *Environ. Internat.* **2022**, *163*, 107199.
48. Lee, S. W.; Kim, B.; Huh, H. Y.; Lee, J. S. *Nanoscale* **2019**, *11*, 3173.
49. Hirai, T.; Yoshikawa, T.; Nabeshi, H.; Yoshida, T.; Akase, T.; Yoshioka, Y.; Itoh, N.; Tsutsumi, Y. *Pharmazie* **2012**, *67*, 742.
50. Ostrowski, A.; Nordmeyer, D.; Boreham, A.; Brodwolf, R.; Mundhenk, L.; Fluhr, J. W.; Lademann, J.; Graf, C.; Ruhl, E.; Alexiev, U.; Gruber, A. D. *Nanomedicine* **2014**, *10*, 1571.
51. It is not known whether Zn was absorbed as ZnO or as soluble Zn. Also, hand-to-mouth contamination could also play a role in these studies. Gulson, B.; McCall, M.; Korsch, M.; Gomez, L.; Casey, P.; Oytam, Y.; Taylor, A.; McCulloch, M.; Trotter, J.; Kinsley, L.; Greenoak, G. *Toxicol. Sci.* **2010**, *118*, 140.
52. Geppert, M. et al. *Chem. Res. Toxicol.* **2020**, *33*, 1215.
53. For thorough reviews related to the skin absorption and toxicity of nanoparticles, see: (a) Filon, F. L.; Mauro, M.; Adami, G.; Bovenzi, M.; Crosera, M. *Regul. Toxic. Pharmacol.* **2015**, *72*, 310. (b) Tiwari, N. et al. *Angew. Chem. Int. Ed. Eng.* **2022**, *61*, 1.
54. Trop, M.; Novak, M.; Rodl, S.; Hellbom, B.; Kroell, W.; Goessler, W. *J. Trauma* **2006**, *60*, 648.
55. Nesporova, K. et al. *Sci. Rep.* **2020**, *10*, 15216.
56. For instance, see: (a) Frampton, M. W. *Environ. Health Perspect.* **2001**, *109*, 529. (b) Ruckerl, R.; Schneider, A.; Breitner, S.; Cyrys, J.; Peters, A. *Inhal. Toxicol.* **2011**, *23*, 555. (c) Nawrot, T. S.; Alfaro-Moreno, E.; Nemery, B. *Am. J. Respir. Crit. Care Med.* **2008**, *177*, 696. (d) Kan, H.; London, S. J.; Chen, G.; Zhang, Y.; Song, G.; Zhao, N.; Jiang, L.; Chen, B. *Environ. Health Perspect.* **2008**, *116*, 1183.
57. (a) Ronkko, T.; Timonen, H. *J. Alzheimers Dis.* **2019**, *72*, 15. b) Orru, H.; Hagenbjork, A.; Olstrup, H. *Environments* **2021**, *8*, 75. (b) Scientific Committee on Emerging and Newly Identified Health Risks (SCENIHR), March 10, 2006. May be accessed online at: http://ec.europa.eu/health/ph_risk/committees/04_scenihr/docs/scenihr_o_003b.pdf.

58. (a) Kittelson, D. B.; Watts, W. F.; Johnson, J. P. *On-Road Nanoparticle Measurements*, 8th International Conference on Environmental Science and Technology, Lemnos, Greece, September, 2003. (b) Kittelson, D. B.; Watts, W. F.; Johnson, J. P.; Zarling, D.; Schauer, J.; Kasper, A.; Baltensperger, U.; Burtscher, H. *Gasoline vehicle exhaust particle sampling study*, Proc. U. S. Department of Energy 9th Diesel Engine Emissions Reduction Conference, Newport, RI, 2003.
59. Schwotzer, D.; Ernst, H.; Schaudien, D.; Kock, H.; Pohlmann, G.; Dasenbrock, C.; Creutzenberg, O. *Particle Fibre Toxicol.* **2017**, *14*, 23.
60. Bakand, S.; Hayes, A.; Dechsakulthorn, F. *Inhal. Toxicol.* **2012**, *24*, 125.
61. Donaldson, K.; Poland, C. A. *Curr. Opin. Biotechnol.* **2013**, *24*, 724.
62. (a) Oberdörster G.; Sharp, Z.; Atudorei, V.; Elder, A. C. P.; Gelein, R.; Lunts, A.; Kreyling, W.; Cox, C. *J. Toxicol. Environ. Health* **2002**, *65A*, 1531. (b) Kreyling, W. G.; Semmler, M.; Erbe F.; Mayer, P.; Takenaka, S.; Schulz, H.; Oberdörster, G.; Ziesenis, A. *J. Toxicol. Environ. Health* **2002**, *65A*, 1513. (c) MacNee, W.; Li, X. Y.; Gilmour, P.; Donaldson K. *Inhal. Toxicol.* **2000**, *12*, 233. (d) Bakand, S.; Hayes, A. *Int. J. Mol. Sci.* **2016**, *17*, 929.
63. For example, see: (a) *Hillyer, J. F.; Albrecht, R. M. J Pharm Sci **2001**, 90, 1927. (b) Kim, Y. S.; Kim, J. S.; Cho, H. S.; Rha, D. S.; Kim, J. M.; Park, J. D. Inhal Toxicol **2008**, 20, 575.*
64. (a) Shrivastava, R.; Raza, S.; Yadav, A.; Kushwaha, P.; Flora, S. J. *Drug Chem. Toxicol.* **2014**, *37*, 336. (b) Liu, Y.; Xu, Z.; Li, X. *Brain Inj.* **2013**, *27*, 934. c) Li, T.; Shi, T.; Li, X.; Zeng, S.; Yin, L.; Pu, Y. *Int. J. Environ. Res. Public Health* **2014**, *11*, 7918.
65. Sohal, I. S.; O'Fallon, K. S.; Gaines, P.; Demokritou, P.; Bello, D. *Particle Fibre Toxicol.* **2018**, *15*, 29.
66. For instance, see: (a) Semmler-Behnke, M.; Kreyling, W. G.; Lipka, J.; Fertsch, S.; Wenk, A.; Takenaka, S. *Small **2008**, 71*, 616. (b) Takenaka, S.; Karg, E.; Kreyling, W. G.; Lentner, B.; Möller, W.; Behnke-Semmler, M. *Inhal Toxicol **2006**, 18, 733.*
67. De Barardis, B.; Marchetti, M.; Risuglia, A.; Letto, F.; Fanizza, C.; Superti, F. *J. Nanopart. Res.* **2020**, *22*, 235.
68. Kwon, J. T.; Hwang, S. K.; Jin, H.; Kim, D. S.; Minai-Tehrani, A.; Yoon, H. J. *J Occup Health **2008**, 50, 1..*
69. For example, see: Yu, L. E.; Yung, L. Y.; Ong, C. N.; Tan, Y. L.; Balasubramaniam, S.; Hartono, D. *Nanotoxicology **2007**, 1, 235..*
70. Semmler, M.; Seitz, J.; Erbe, F.; Mayer, P.; Heyder, J.; Oberörster, G. *Inhal Toxicol **2004**, 16, 453..*
71. Crüts, B.; Van Etten, L.; Törnqvist, H.; Blomberg, A.; Sandström, T.; Mills, N. L. *Part Fibre Toxicol **2008**, 5, 4..*
72. (a) Takagi, A.; Hirose, A.; Nishimura, T.; Fukumori, N.; Ogata, A.; Ohashi, N.; Kitajima, S.; Kanno, J. *J. Toxicol. Sci.* **2008**, *33*, 105. (b) Poland, C. A.; Duffin, R.; Kinloch, I.; Maynard, A.; Wallace, W. A. H.; Seaton, A.; Stone, V.; Brown, S.; MacNee, W.; Donaldson, K. *Nature Nanotechnol.* **2008**, *3*, 423.

73. *Pacurari, M.; Yin, X. J.; Zhao, J.; Ding, M.; Leonard, S. S.; Schwegler-Berry, D. Environ Health Perspect 2008, 116, 1211.*
74. Magrez, A.; Kasas, S.; Salicio, V.; Pasquier, N.; Seo, J. W.; Celio, M.; Catsicas, S.; Schwaller, B.; Forro, L. *Nano Lett.* **2006**, *6*, 1121.
75. Renwick, L. C.; Brown, D.; Clouter, A.; Donaldson, K. *Occup Environ Med* **2004**, *61*, 442.
76. Arnaud, C. H. *Chem. Eng. News* **2010**, *88*, 32.
77. Wilson, M. R.; Lightbody J. H.; Donaldson, K.; Sales, J.; Stone V. *Toxicol Appl Pharmacol* **2002**, *184*, 172.
78. (a) Oh, N.; Park, J. H. *Int. J. Nanomedicine* **2014**, *9*, 51. (b) de Almeida, M. S. et al. *Chem. Soc. Rev.* **2021**, *50*, 5397.
79. (a) Jordan, A.; Scholz, R.; Wust, P.; Schirra, H.; Thomas, S.; Schmidt, H.; Felix, R. *J. Magn. Magn. Mater.* **1999**, *194*, 185. (b) Zhang, Y., Kohler, N., and Zhang, M. *Biomaterials* **2002**, *23*, 1553.
80. For instance, see: (a) Rouse, J. G.; Yang, J.; Barron, A. R.; and Monteiro-Riviere, N. A. *Toxicol.* In Vitro. **2006**, *8*, 1313. (b) Zhang, L. W.; Zeng, L.; Barron, A. R.; Monteiro-Riviere, N. A. *Int. J. Toxicol.* **2007**, *26*, 103.
81. For example: (a) Michalet, X.; Pinaud, F. F.; Bentolila; L. A.; Tsay, J. M.; Doose, S.; Li, J. J.; Sundaresan, G.; Wu, A. M.; Gambhir, S. S.; Weiss, S. *Science* **2005**, *307*, 538. (b) Derfus, A. M.; Chan, W. C. W.; Bhatia, S. *Nano Lett.* **2004**, *4*, 11. (c) Gao, X.; Cui, Y.; Levenson, R. M.; Chung, L. W.; Nie, S. *Nat. Biotechnol.* **2004**, *22*, 969. (d) Xing, Y.; Chaudry, Q.; Shen, C.; Kong, K. Y.; Zhau, H. E.; Chung, L. W.; Petros, J. A.; O'Regan, R. M.; Yezhelyev, M. V.; Simons, J. W. *Nat. Protoc.* **2007**, *2*, 1152.
82. For instance, see: (a) A review that summarizes in vitro and in vivo testing of QDs: Yong, K. T.; Law, W. C.; Hu, R.; Ye, L.; Liu, L.; Swihart, M. T.; Prasad, P. N. *Chem. Soc. Rev.* **2013**, *42*, 1236. (b) Duan, H.; Nie, S. *J. Am. Chem. Soc.* **2007**, *129*, 3333. (c) Jaiswal, J. K.; Mattoussi, H.; Mauro, J. M.; Simon, S. M. *Nat. Biotechnol.* **2003**, *21*, 47. (d) Zhang, L. W.; Yu, W. W.; Colvin, V. L.; Monteiro-Riviere, N. A. *Toxicol. Appl. Pharmacol.* **2008**, *228*, 200. (e) Ryman-Rasmussen, J. P.; Riviere, J. E.; Monteiro-Riviere, N. A. *J. Invest. Dermatol.* **2007**, *127*, 143.
83. Zhang, L. W.; Monteiro-Riviere, N. A. *Toxicol. Sci.* **2009**, *110*, 138.
84. Radomski, A.; Jurasz, P.; Alonso-Escolano, D.; Drews, M.; Morandi, M.; Malinski, T.; Radomski, M. W. *Br. J. Pharm.* **2005**, *146*, 882.
85. (a) Younes, N. R.; Amara, S.; Mrad, I.; Ben-Slama, I.; Jeljeli, M.; Omri, K.; El Ghoul, J.; El Mir, L.; Rhouma, K. B.; Abdelmelek, H. *Environ. Sci. Pollut. Res. Int.* **2015**, *22*, 8728. (b) Almeida, J. P. M.; Chen, A. L.; Foster, A.; Drezek, R. *Nanomedicine* **2011**, *6*, 815.
86. Biswas P.; Wu, C. -Y. *J. Air & Waste Manage. Assoc.* **2005**, *55,* 708.
87. For example, see: (a) Kuehr, S.; Kosfeld, V.; Schlechtriem, C. *Environ. Sci. Eur.* **2021**, *33*, 9 (bioaccumulation in invertebrate species). (b) Handy, R. D. et al. *Environ. Sci.: Nano* **2022**, *9*, 684. (c) Murali, M. et al. *Sci. of Total Environ.* **2022**, *11*, 3, 152249 (bioaccumulation in plants). (d) Dinda, S. C.

"Bioaccumulation and Toxic Profiling of Nanostructured Particles and Materials" in *Unraveling the Safety Profile of Nanoscale Particles and Materials—From Biomedical to Environmental Applications.* Intech, 2018. Open-access: https://www.intechopen.com/chapters/59716 (e) USEPA Nanotechnology White Paper, Science Policy Council, Feb. 2007. May be accessed online at: https://www.epa.gov/sites/production/files/2015-01/documents/nanotechnology_whitepaper.pdf.

88. Zhang et al. *Chemosphere*, **2007**, 67, 160.
89. (a) Brohi, R. D. et al. *Frontiers Pharmacol.* **2017**, *8*, 606. (b) Takeda et al. *J. Health Sci.*, **2009**, 55, 1.
90. For a thorough review of the genotoxicity of nanoparticles, see: Gonzalez, L.; Lison, D.; Kirsch-Volders, M. *Nanotoxicology* **2008**, *2*, 252.
91. Li, Y.; Zhang, Y.; Yan, B. *Int. J. Mol. Sci.* **2014**, *15*, 3671.
92. Khanna, P.; Ong, C.; Bay, B. H.; Baeg, G. H. *Nanomaterials* **2015**, *5*, 1163.
93. Misawa, M.; Takahashi, J. *Nanomedicine* **2011**, *7*, 604.
94. Park, E. -J.; Park, K. *Toxicol. Lett.* **2009**, *184*, 18.
95. Inoue, M. et al. *Particle Fibre Toxicol.* **2021**, *18*, 1.
96. (a) Yu, Z.; Li, Q.; Wang, J.; Yu, Y.; Wang, Y.; Zhou, Q.; Li, P. *Nanoscale Res. Lett.* **2020**, *15*, 1. (b).
97. Horie, M.; Tabei, Y. *Free Radic. Res.* **2021**, *55*, 331.
98. (a) Singh, N.; Manshian, B.; Jenkins, G. J. S.; Griffiths, S. M.; Williams, P. M.; Maffeis, T. G. G.; Wright, C. J.; Doak, S. H. *Biomaterials* **2009**, *30*, 3891. (b) Alarifi, S.; Ali, D.; Alkahtani, S. *Int. J. Nanomed.* **2015**, *10*, 3751. (c) Al Gurabi, M. A.; Ali, D.; Alkahtani, S.; Alarifi, S. *Onco Targets Ther.* **2015**, *8*, 295. (d) Sliwinska, A.; Kwiatkowski, D.; Czarny, P.; Milczarek, J.; Toma, M.; Korycinska, A.; Szemraj, J.; Sliwinski, T. *Toxicol. Mech. Methods* **2015**, *25*, 176. (e) Lin, W.; Xu, Y.; Huang, C. C.; Ma, Y.; Shannon, K. B.; Chen, D. R.; Huang, Y. W. *J. Nanopart. Res.* **2009**, *11*, 25. (f) Karlsson, H. L.; Cronholm, P.; Gustafsson, J.; Moller, *J. Chem. Res. Toxicol.* **2008**, *21*, 1726.
99. Jeng, H. A.; Swanson, J. *J. Environ. Sci. Health A* **2006**, *41*, 2699.
100. Turabekova, M.; Rasulev, B.; Theodore, M.; Jackman, J.; Leszcynska, D.; Leszcynski, J. *Nanoscale* **2014**, *6*, 3488.
101. (a) Dobrovolskaia, M. A.; Aggarwal, P.; Hall, J. B.; McNeil, S. E. *Mol. Pharm.* **2008**, *5*, 487. (b) Chanan-Khan, A.; Szebeni, J.; Savay, S.; Liebes, L.; Rafique, N. M.; Alving, C. R.; Muggia, F. M. *Ann. Ancol.* **2003**, *14*, 1430.
102. For instance, see: Song, Y.; Li, X.; Du, X. *Eur. Respir. J.* **2009**, *34*, 559.
103. Lam, C. W.; James, J. T.; McCluskey, R.; Hunter, R. L. *Toxicol. Sci.* **2004**, *77*, 126.
104. For instance, see: Trouiller, B.; Reliene, R.; Westbrook, A.; Solaimani, P.; Schiestl, R. H. *Cancer Res.* **2009**, *69*, 8784.
105. (a) Blum, J. L.; Xiong, J. Q.; Hoffman, C.; Zelikoff, J. T. Toxicol. Sci. 2012, 126, 478. (b) Austin, C. A.; Umbreit, T. H.; Brown, K. M.; Barber, D. S.; Dair, B. J.; Francke-Carroll, S.; Feswick, A.; Saint-Louis, M. A.; Hikawa, H.; Seibein, K. N. Nanotoxicology 2012, 6, 912. (c) Pietroiusti, A.; Massimiani, M.; Fenoglio, I.; Colonna, M.; Valentini, F.; Palleschi, G.; Camaioni, A.;

Magrini, A.; Siracusa, G.; Bergamaschi, A. ACS Nano 2011, 5, 4624. (d) Zalgeviciene, V.; Kulvietis, V.; Bulotiene, D.; Didziapetriene, J.; Rotmskis, R. Medicina (Kaunas) 2012, 48, 256. (e) Noori, A.; Parivar, K.; Modaresi, M.; Messripour, M.; Yousefi, M. H.; Amiri, G. R. Afr. J. Biotechnol. 2011, 10, 1221. (f) Umezawa, M.; Kudo, S.; Yanagita, S.; Shinkai, Y.; Niki, R.; Oyabu, T.; Takeda, K.; Ihara, T.; Sugamata, M. J. Toxicol. Sci. 2011, 36, 461. (g) Jackson, P.; Hougaard, K. S.; Vogel, U.; Wu, D.; Casavant, L.; Williams, A.; Wade, M.; Yauk, C. L.; Wallin, H.; Halappanavar, S. Mutat. Res. 2012, 745, 73. (h) Barton, H. A.; Cogliano, V. J.; Flowers, L.; Valcovic, L.; Setzer, R. W.; Woodruff, T. J. Environ. Health Perspect. 2005, 113, 1125.

106. Bhabra, G.; Sood, A.; Fisher, B.; Cartwright, L.; Saunders, M.; Evans, W. H.; Surprenant, A.; Lopez-Castejon, G.; Mann, S.; Davis, S. A.; Hails, L. A.; Ingham, E.; Verkade, P.; Lane, J.; Heesom, K.; Newson, R.; Case, C. P. *Nature Nanotechnol.* **2009**, *4*, 876.
107. (a) Rajput, V. D. et al. *Toxicity of Nanoparticles in Plants*, Vol. 5, Elsevier: New York, 2022. (b) Bundschuh, M. et al. *Environ. Sci. Eur.* **2018**, *30*, 1.
108. For a thorough review of the effects of metal nanoparticles on plants and their possible effects in the food chain, see: Ma, C.; White, J. C.; Dhankher, O. M.; Xing, B. *Environ. Sci. Technol.* **2015**, *49*, 7109.
109. Dimkpa, C.; McLean, J.; Latta, D.; Manangon, E.; Britt, D.; Johnson, W.; Boyanov, M.; Anderson, A. *J. Nanopart. Res.* **2012**, *14*, 1.
110. Dimkpa, C. O.; Latta, D. E.; McLean, J. E.; Britt, D. W.; Boyanov, M. I.; Anderson, A. J. *Environ. Sci. Technol.* **2013**, *47*, 4734.
111. For example, see: (a) Ranjan, A. et al. *Environ. Nanotechnol. Monitor. Manag.* **2021**, *15*, 100457. (b) Yin, L.; Cheng, Y.; Espinasse, B.; Colman, B. P.; Auffan, M.; Wiesner, M.; Rose, J.; Liu, J.; Bernhardt, E. S. *Environ. Sci. Technol.* **2011**, *45*, 2360. (c) Priester, J. H.; Ge, Y.; Mielke, R. E.; Horst, A. M.; Moritz, S. C.; Espinosa, K.; Gelb, J.; Walker, S. L.; Nisbet, R. M.; An, Y. J.; Schimel, J. P.; Palmer, R. G.; Hernandez-Viezcas, J. A.; Zhao, L.; Gardea-Torresdey, J. L.; Holden, P. A. *Proc. Natl. Acad. Sci. U.S.A.* **2012**, *109*, E2451.
112. For instance, see: (a) Marmiroli, M.; Marmiroli, N.; Pagano, L. *Nanomaterials* **2022**, *12*, 1658. (b) Panda, K. K.; Achary, V. M. M.; Krishnaveni, R.; Padhi, B. K.; Sarangi, S. N.; Sahu, S. N.; Panda, B. B. *Toxicol.* In Vitro **2011**, *25*, 1097. (c) Faisal, M.; Saquib, Q.; Alatar, A. A.; Al-Khedhairy, A. A.; Hegazy, A. K.; Musarrat, J. *J. Hazard Mater.* **2013**, *250*, 318. (d) Ghosh, M.; Bandyopadhyay, M.; Mukherjee, A. *Chemosphere* **2010**, *81*, 1253.
113. Dang, F.; Huang, Y.; Wang, Y.; Zhou, D.; Xing, B. *Environ. Sci. Nano* **2021**, *8*, 1519.
114. Some reports for no biomagnification: (a) Mielke, R. E.; Priester, J. H.; Werlin, R. A.; Gelb, J.; Horst, A. M.; Orias, E.; Holden, P. A. *Appl. Environ. Microbiol.* **2013**, *79*, 5616. (b) Zhu, X.; Wang, J.; Zhang, X.; Chang, Y.; Chen, Y. *Chemosphere* **2010**, *79*, 928. (c) Lewinski, N. A.; Zhu, H.; Ouyang, C. R.; Conner, G. P.; Wagner, D. S.; Colvin, V. L.; Drezek, R. A. *Nanoscale* **2011**, *3*, 3080. (d) Holbrook, R. D.; Murphy, K. E.; Morrow, J. B.; Cole, K. D. *Nat. Nanotechnol.* **2008**, *3*, 352.

Some examples that show biomagnification: (a) Werlin, R.; Priester, J. H.; Mielke, R. E.; Kramer, S.; Jackson, S.; Stoimenov, P. K.; Stucky, G. D.; Cherr, G. N.; Orias, E.; Holden, P. A. *Nat. Nanotechnol.* **2011**, *6*, 65. (b) Judy, J. D.; Unrine, J. M.; Bertsch, P. M. *Environ. Sci. Technol.* **2010**, *45*, 776. (c) Unrine, J. M.; Shoults-Wilson, W. A.; Zhurbich, O.; Bertsch. P. M.; Tsyusko, O. V. *Environ. Sci. Technol.* **2012**, *46*, 9753. (d) Hawthorne, J.; De la Torre Roche, R.; Xing, B.; Newman, L. A.; Ma, X.; Majumdar, S.; Gardea-Torresdey, J.; White, J. C. *Environ. Sci. Technol.* **2014**, *48*, 13,102.

115. Tervonen, T.; Linkov, I.; Figueira, J. R.; Steevens, J.; Chappell, M.; Merad, M. *J. Nanopart. Res.* **2009**, *11*, 757.
116. (a) For a comprehensive roadmap for research strategies needed to evaluate the safety of nanomaterials, see: Holsapple, M. P.; Farland, W. H.; Landry, T. D.; Monteiro-Riviere, N. A.; Carter, J. M.; Walker, N. J.; Thomas, K. V. *Toxicol. Sci.* **2005**, *88*, 12. (b) For a review of test guidelines that should be used to assess the properties and toxicity of engineered nanostructures, see: http://www.olis.oecd.org/olis/2009doc.nsf/43bb6130e5e86e5fc12569fa005d004c/fc3f70281e7640a5c12575ef002eb46b/$FILE/JT03267900.PDF (Organisation for Economic Co-Operation and Development, Paris, 2009).
117. http://www.ethicsweb.ca/nanotechnology.
118. For example, commercial products such as Autel Robotics EVO Nano + drone, Reebok Nano X1 cross-trainer shoes, Mountain Buggy Nano V3 stroller, Flashpoint Nano light stand, and Hollywood movies (e.g., "nanomites" in G.I. Joe, "nanobots" in iRobot, etc.).
119. Taniguchi, N. On the Basic Concept of NanoTechnology. *Proc. ICPE* **1974**.
120. https://www.nasa.gov/centers/ames/news/releases/2002/02images/nanogear/nanogears.html.
121. For example, see (a) Theivasanthi, T.; Alagar, M. *Phys. Techn. Sci.* **2013**, *1*, 39. (b) Pishko, V. V.; Gnatchenko, S. L.; Tsapenko, V. V.; Kodama, R. H.; Makhlouf, S. A. *J. Appl. Phys.* **2003**, *93*, 7382.
122. http://home.iitk.ac.in/~anandh/MSE694/NPTEL_Optical%20properties%20of%20Nanomaterials.pdf.
123. Goldstein, A. N.; Echer, C. M.; Alivisatos, A. P. *Science* **1992**, *256*, 1425.
124. Wu, Q. et al. *Nanotechnol. Reviews* **2020**, *9*, 1.
125. Krahne, R. et al. *Phys. Rep.* **2011**, *501*, 75.
126. Rsheed, A. A.; Aldawood, S.; Aldossary, O. M. *Nanomaterials* **2021**, *11*, 2916, and references therein.
127. For a review of nanoparticle melting models, see: Nanda, K. K. *PRAMANA J. Phys.* **2009**, *72*, 617.
128. (a) Lewis, J. *Chem. Br.* **1988**, *24*, 795. (b) Deeming, A. J. *Adv. Organomet. Chem.* **1986**, *26*, 1.
129. Shown from left to right are (a) Pd nanoclusters supported on hydroxyapatite: Mori, K.; Hara, T.; Mizugaki, T.; Ebitani, K.; Kaneda, K. *J. Am. Chem. Soc.* **2004**, *126*, 10657. (b) Copper nanoclusters: Williams, G. L.; Vohs, J. K.; Brege, J. J.; Fahlman, B. D. *J. Chem. Ed.* **2005**, *82*, 771.
130. Huang, J.; Kunitake, T.; Onoue, S.-Y. *Chem. Commun.* **2004**, 1008.

131. Scher, E. C.; Manna, L.; Alivisatos, A. P. *Philos. Trans. R. Soc. Lond. A.* **2003**, *361*, 241.
132. Shown is a Ti/O/C nanopowder with individual nanosized grains: Leconte, Y.; Maskrot, H.; Herlin-Boime, N.; Porterat, D.; Reynaud, C.; Gierlotka, S.; Swiderska-Sroda, A.; Vicens, J. *J. Phys. Chem. B* **2006**, *110*, 158.
133. Shown are submicron particulates (with some nanoparticles also present) of aluminum oxide: Williams, G. L.; Vohs, J. K.; Brege, J. J.; Fahlman, B. D. *J. Chem. Ed.* **2005**, *82*, 771.
134. For an excellent review of transition metal nanocluster formation and nomenclature, as well as the difference between colloids and nanoclusters, see Finke, R. G. *Transition Metal Nanoclusters* in *Metal Nanoparticles: Synthesis, Characterization, and Applications*, Dekker: New York, 2002.
135. For example, there is a 500% rate difference for the photoreduction of CO_2 using 10 different samples of Pd_n colloids: Wilner, I.; Mendler, D. *J. Am. Chem. Soc.* **1989**, *111*, 1330. Also, see Kohler, J. U.; Bradley, J. S. *Catal. Lett.* **1997**, *45*, 203, wherein they describe a 670% variation in the rate of hydrogenation with PVP-protected Pt_n colloids (due to a widely dispersed composition, with varying numbers of surface Cl^- groups).
136. Garcia de Arquer, F. P.; Talapin, D. V.; Klimov, V.; Arakawa, Y.; Bayer, M.; Sargent, E. H. *Science* **2021**, *373*, 1.
137. For example, see: Dinu, M.; Rapaport, R.; Chen, G.; Stuart, H. R.; Giles, R. *Bell Labs Techn. J.* **2005**, *10*, 215.
138. Nakagawa, T. et al. *J. Appl. Phys.* **2022**, *131*, 134305.
139. Kim, J. et al. *Adv. Mater.* **2021**, *34*, 1.
140. Although quantum dots are 0-D nanostructures, quantum confinement effects are also exhibited in 1-D nanowires and nanorods. Buhro and coworkers have studied the effect on both size and shape on quantum confinement (Yu, H.; Li, J.; Loomis, R. A.; Wang, L.-W.; Buhro, W. E. *Nature Mater.* **2003**, 2, 517). Their work provides empirical data to back up the theoretical order of increasing quantum confinement effects: dots (3-D confinement) > rods > wires (2-D confinement) > wells (1-D confinement). For an example of an interesting nanostructure comprised of both a nanorod and nanodot, see: Mokari, T.; Sztrum, C. G.; Salant, A.; Rabani, E.; Banin, U. *Nature Mater.* **2005**, 4, 855.
141. It should be noted that while nanoparticulate films are generally transparent to light due to minimal scattering, films that contain nanoporous structures may not be optically transparent due to significant light scattering from interconnected/agglomerated pores.
142. For instance, see: (a) Giansante, C.; Infante, I. *J. Phys. Chem. Lett.* **2017**, *8*, 5209. (b) Mahboub, M. et al. *ACS Energy Lett.* **2018**, *3*, 767. (c) Cassidy, J.; Zamkov, M. *J. Chem. Phys.* **2020**, *152*, 110902–1.
143. For instance, see: (a) Hines, M. A.; Guyot-Sionnest, P. *J. Phys. Chem.* **1996**, *100*, 468. (b) Dabbousi, B. O. et al. *J. Phys. Chem. B* **1997**, *101*, 9463. (c) Peng, X. G.; Schlamp, M. C.; Kadavanich, A. V.; Alivisatos, A. P. *J. Am. Chem. Soc.* **1997**, *119*, 7019. (d) van der Stam, W. et al. *Chem. Mater.* **2019**, *31*, 8484. (e)

Ji, B.; Koley, S.; Slobodkin, I.; Remennik, S.; Banin, U. *Nano Lett.* **2020**, *20*, 2387.

144. (a) Reiss, P.; Protiere, M.; Li, L. *Small* **2009**, *5*, 154. (b) Toufanian, R. et al. *Front. Chem.* **2018**. https://doi.org/10.3389/fchem.2018.00567.
145. Rafipoor, M.; Dupont, D.; Tornatzky, H.; Tessier, M. D.; Maultzsch, J.; Hens, Z.; Lange, H. *Chem. Mater.* **2018**, *30*, 4393.
146. (a) Ivanov, S. A.; Piryatinski, A.; Nanda, J.; Tretiak, S.; Zavadil, K. R.; Wallace, W. O.; Werder, D.; Klimov, V. I. *J. Am. Chem. Soc.* **2007**, *129*, 11708. (b) Ning, Z.; Tian, H.; Yuan, C.; Fu, Y.; Qin, H.; Sun, L.; Agren, H. *Chem. Commun.* **2011**, *47*, 1536.
147. Carey, G. H.; Abdelhady, A. L.; Ning, Z.; Thon, S. M.; Bakr, O. M.; Sargent, E. H. *Chem. Rev.* **2015**, *115*, 12732.
148. Renuga, V.; Mohan, N. "Design, Synthesis, and Properties of I-III-VI2 Chalcogenide-Based Core-Multishell Nanocrystals" in *Core/Shell Quantum Dots*, Springer: New York, 2020.
149. Voznyy, O. et al. *Nano Lett.* **2017**, *17*, 7191.
150. Liu, Y.; Kim, D.; Morris, O. P.; Zhitomirsky, D.; Grossman, J. C. *ACS Nano* **2018**, *12*, 2838.
151. Selopal, G. S.; Zhao, H.; Wang, Z. M.; Rosei, F. *Adv. Funct. Mater.* **2020**, *30*, 1908762.
152. AbouElhamd, A. R.; Al-Sallal, K. A.; Hassan, A. *Energies* **2019**, *12*, 1058.
153. Ruhle, S.; Shalom, M.; Zaban, A. *ChemPhysChem* **2010**, *11*, 2290.
154. Pan, X.; Rao, H.; Mora-Sero, I.; Bisquert, J.; Zhong, X. *Chem. Soc. Rev.* **2018**, *47*, 7659.
155. Mora-Sero, I. *Adv. Energy Mater.* **2020**, *10*, 2001774.
156. For example, see: (a) Mao, X. et al. *New J. Chem.* **2020**, *44*, 505. (b) Duan, J.; Tang, Q.; He, B.; Chen, H. *RSC Adv.* **2015**, *5*, 33463. (c) Lee, H.; Leventis, H. C.; Moon, S. J.; Chen, P.; Ito, S.; Haque, S. A.; Torres, T.; Nuesch, F.; Geiger, T.; Zakeeruddin, S. M.; Gratzel, M.; Nazeeruddin, M. K. *Adv. Funct. Mater.* **2009**, *19*, 2735. (d) Feng, W.; Zhao, L.; Du, J.; Li, Y.; Zhong, X. *J. Mater. Chem. A* **2016**, *4*, 14849. (e) Jumabekov, A. N.; Siegler, T. D.; Cordes, N.; Medina, D. D.; Bohm, D.; Garbus, P.; Meroni, S.; Peter, L. M.; Bein, T. *J. Phys. Chem. C* **2014**, *118*, 25853.
157. Song, H. et al. *J. Am. Chem. Soc.* **2021**, *143*, 4790.
158. Luther, J. M.; Law, M.; Beard, M. C.; Song, Q.; Reese, M. O.; Ellingson, R. J.; Nozik, A. J. *Nano Lett.* **2008**, *8*, 3488.
159. Kim, M. R.; Ma, D. *J. Phys. Chem. Lett.* **2015**, *6*, 85, and references therein.
160. Dasgupta, K. et al. *Trans. Indian Natl. Acad. Eng.* **2021**, 1.
161. Nguyen, H. T. et al. *Appl. Sci.* **2022**, *12*, 83, and references therein.
162. (a) Shaheen, S. E.; Brabec, C. J.; Sariciftci, N. S.; Padinger, F.; Fromherz, T.; Hummelen, J. C. *Appl. Phys. Lett.* **2001**, *78*, 841. (b) Barkhouse, D. A.; Debnath, R.; Kramer, I. J.; Zhitomirsky, D.; Pattantyus-Abraham, A. G.; Levina, L.; Etgar, L.; Gratzel, M.; Sargent, E. H. *Adv. Mater.* **2011**, *23*, 3134.
163. Carey, G. H.; Abdelhady, A. L.; Ning, Z.; Thon, S. M.; Bakr, O. M.; Sargent, E. H. *Chem. Rev.* **2015**, *115*, 12732.

164. Lan, X.; Voznyy, O.; Garcia de Arquer, F. P.; Liu, M.; Xu, J.; Proppe, A. H.; Walters, G.; Fan, F.; Tan, H.; Liu, M.; Yang, Z.; Hoogland, S.; Sargent, E. H. *Nano Lett.* **2016**, *16*, 4630.
165. Beard, M. C. *J. Phys. Chem. Lett.* **2011**, *2*, 1282.
166. A related phenomenon to MEG, denoted as electron–hole pair multiplication (EHPM), may also occur in bulk semiconductors. However, this effect is too inefficient in bulk semiconductors since both energy and momentum must be conserved; in contrast, the conservation of momentum is relaxed in QDs. For more information, see: (a) Kershaw, S. V.; Rogach, A. L. *Materials* **2017**, *10*, 1095. (b) Beard, M. C.; Luther, J. M.; Semonin, O. E.; Nozik, A. J. *Acc. Chem. Res.* **2012**, *46*, 1252.
167. (a) The original report by NREL: https://www.nrel.gov/docs/fy06osti/38992.pdf (b) Ellingson, R. J. et al. *Nano Lett.* **2005**, *5*, 865.
168. Beard, M. C. *Phys. Chem. Lett.* **2011**, *2*, 1282.
169. For instance, see: (a) McGuire, J. A.; Sykora, M.; Joo, J.; Pietryga, J. M.; Klimov, V. I. *Nano Lett.* **2010**, *10*, 2049. (b) Nair, B.; Bawendi, M. G. *Phys. Rev. B* **2007**, 76081304R. (c) Pijpers, J. J. et al. *J. Phys. Chem. C* **2008**, *112*, 4783. (d) McGuire, J. A.; Joo, J.; Pietryga, J. M.; Schaller, R. D.; Klimov, V. I. *Acc. Chem. Res.* **2008**, *41*, 1810.
170. (a) Goodwin, H.; Jellicoe, T. C.; Davis, N.; Bohm, M. L. *Nanophotonics*, **2017**, *7*, 1. (b) Schaller, R. D.; Klimov, V. I. *Phys. Rev. Lett.* **2004**, *92*, 186,601.
171. Compton, D.; Comish, L.; van der Lingen, E. *Gold Bull.* **2003**, *36*, 10.
172. For a thorough review of surface plasmon resonance, see Kelly, K. L.; Coronado, E.; Zhao, L. L.; Schatz, G. C. *J. Phys. Chem. B* **2003**, *107*, 668.
173. Faraday was the first to assert that the red color of gold nanoparticle solutions was due to the interaction of light with particulates of varying morphologies. Mie explained the red color in 1908 by solving Maxwell's equations based on spheres with sizes comparable to the wavelength of light; however, Raleigh solved the problem for spheres smaller than wavelengths of light. For more details on the interaction of light with metallic nanoparticles and the influence of inter-particle agglomeration, see: Ghosh, S. K.; Pal, T. *Chem. Rev.* **2007**, *107*, 4797.
174. Haes, A. J.; Stuart, D. A.; Nie, S.; Duyne, R. P. V. *J. Fluoresc.* **2004**, *14*, 355.
175. For instance, one is able to vary the plasmonic properties of Au nanorods by incorporating Cu: Henkel, A.; Jakab, A.; Brunklaus, G.; Sonnichsen, C. *J. Phys. Chem. C* **2009**, *113*, 2200.
176. For a primer on "thermal energy", see: http://hyperphysics.phy-astr.gsu.edu/hbase/Kinetic/eqpar.html.
177. For a nice review of metal nanoparticles and their electronic properties, see: (a) Rao, C. N. R.; Kulkarni, G. U.; Thomas, P. J. *Chem. Soc. Rev.* **2000**, *29*, 27. (b) Eustis, S.; El-Sayed, A.; *Chem. Soc. Rev.* **2006**, *35*, 209.
178. Busani, R.; Folker, M.; Cheshnovsky, O. *Phys. Rev. Lett.* **1998**, *81,* 3836.
179. Rao, C. N. R.; Kulkarni, G. U.; Thomas, P. J.; Edwards, P. P. *Chem. Soc. Rev.* **2000**, *29*, 27.

180. For more information/precedents on quantum confinement effects for metallic nanoclusters, see: (a) Rao, C. N. R.; Kulkarni, G. U.; Thomas, P. J.; Edwards, P. P. *Chem. Soc. Rev.* **2000**, *29*, 27. (b) Mohamed, M. B.; Volkov, V.; Link, S.; El-Sayed, M. A. *Chem. Phys. Lett.* **2000**, *317*, 517. (c) Huang, T.; Murray, R. W. *J. Phys. Chem. B* **2001**, *105*, 12,498. (d) Link, S.; Beeby, A.; FitzGerald, S.; El-Sayed, M. A.; Schaaff, T. G.; Whetten, R. L. *J. Phys. Chem. B* **2002**, *106*, 3410. (e) Empedocles, S.; Bawendi, M. *Acc. Chem. Res.* **1999**, *32*, 389. (f) El-Sayed, M. A. *Acc. Chem. Res.* **2001**, *34*, 257. (g) Zheng, J.; Petty, J. T.; Dickson, R. M. *J. Am. Chem. Soc.* **2003**, *125*, 7780. (h) Schaaff, T. G.; Shafigullin, M. N.; Khoury, J. T.; Vezmar, I.; Whetten, R. L.; Cullen, W. G.; First, P. N.; Gutierrez-Wing, C.; Ascensio, J.; Jose-Yacaman, M. J. *J. Phys. Chem. B* **1997**, *101*, 7885.

181. Ferrofluids are a class of important ternary ferrite spinels are of the formula MFe_2O_4 (M = Co, Mn, Ni), synthesized as nano-sized particulates dispersed in a solvent. These solutions have applications that include electronic devices (e.g., forming seals around spinning drive shafts in computer hard drives; heat conduction in speaker tweeters), automotive (magnetorheological damping for tunable shock absorption), to medicine (hyperthermia-based cancer treatment; i.e., directing nanoparticles to a cancerous area and raising their temperature via an external magnetic field). For a review of ferrofluids, see: Odenbach, S. *J. Phys.: Condens. Matter* **2004**, *16*, R1135.

182. With areal densities in excess of 1.7 Tb in^{-2}, NAND flash memory now surpasses that of hard disk drive (HDD) technology, which are typically in the range 850 Gb in^{-2} to 1.3 Tb in^{-2} However, with developments in it is likely that areal densities greater than 3 Tb in^{-2} are possible within the next few years. A nice website that provides details about trends in the areal densities of magnetic storage devices: http://digitalpreservation.gov/meetings/documents/storage13/GaryDecad_Technology.pdf.

183. (a) Ge, J.; Hu, Y.; Diasini, M.; Beyermann, W. P.; Yin, Y. *Angew. Chem. Int. Ed.* **2007**, *46*, 4342. (b) Esteban-Cubillo, A.; Pina-Zapardiel, R.; Moya, J. S.; Pecharroman, C. *J. Nanop. Res.* **2009**, ASAP. (c) Yang, C.; Xing, J.; Guan, Y.; Liu, J.; Liu, H. *J. Alloys Compounds* **2004**, 385, 283. For more details regarding superparamagnetism, see: http://lmis1.epfl.ch/webdav/site/lmis1/shared/Files/Lectures/Nanotechnology%20for%20engineers/Archives/2004_05/Superparamagnetism.pdf.

184. For examples, see: (a) Monson, T. C.; Venturini, E. L.; Petkov, V.; Ren, Y.; Lavin, J. M.; Huber, D. L. *J. Magn. Magn. Mater.* **2013**, *331*, 156. (b) Duarte, E. L.; Itri, R.; Lima, E.; Baptista, M. S.; Berguo, T. S.; Goya, G. F. *Nanotechnology* **2006**, *17*, 1. (c) Oyarzun, S.; Tamion, A.; Tournus, F.; Dupuis, V.; Hillenkamp, M. *Sci. Reports* **2015**, *5*, 14749. (d) Gambardella, P.; Rusponi, S.; Veronese, M.; Dhesi, S. S.; Grazioli, C.; Dallmeyer, A.; Cabria, I.; Zeller, R.; Dederichs, P. H.; Kern, K.; Carbone, C.; Brune, H. *Science* **2003**, *300*, 1130.

185. For more details on next-generation materials for magnetic storage devices, see: (a) Bandic, Z. Z.; Litvinov, D.; Rooks, M. *MRS Bull.* **2008**, *33*, 831. (b)

Liao, J. W.; Zhang, H. W.; Lai, C. H. "Magnetic Nanomaterials for Data Storage" in *Magnetic Nanomaterials: Fundamentals, Synthesis, and Applications.* Wiley: New York, 2017.

186. For more information about HAMR, see: Chen, Y. J.; Yang, H. Z.; Leong, S. H.; Wu, B. L.; Asbahi, M.; YuY u Ko, H.; Yang, J. K. W.; Ng, V. *Appl Phys. Lett.* **2014**, *105*, 162402, and references therein.
187. (a) Skumryev, V.; Stoyanov, S.; Zhang, Y.; Hadjipanayis, G.; Givord, D.; Nogues, J. *Nature*, **2003**, *423*, 850. (b) Eisenmenger, J.; Schuller, I. K. *Nat. Mater.* **2003**, *2*, 437. (c) Fullerton, E. E.; Margulies, D. T.; Schabes, M. E.; Carey, M.; Gurney, B.; Moser, A.; Best, M.; Zeltzer, G.; Rubin, K.; Rosen, H.; Doerner, M. *Appl. Phys. Lett.* **2000**, *77*, 3806.
188. (a) Kief, M. T.; Victora, R. H. *MRS Bull.* **2018**, *43*, 87. (b) Vopson, M. M.; Zemaityte, E.; Spreitzer, M.; Namvar, E. *J. Appl. Phys.* **2014**, *116*, 113910. (c) Kim, H. K. D.; Schelhas, L. T.; Keller, S.; Hockel, J. L.; Tolbert, S. H.; Carman, G. P. *Nano Lett.* **2013**, *13*, 884.
189. Hono, K. et al. *MRS Bull.* **2018**, *43*, 93.
190. Kroto, H. W.; Heath, J. R.; O'Brien, S. C.; Curl, R. F.; Smalley, R. E. *Nature* **1985**, *318*, 162.
191. Smalley and Curl named this structure after Buckminster Fuller, for his discovery of geodesic domes.
192. The irradiation of C_{60} with light in the presence of O_2 causes the formation of reactive singlet oxygen ($^1O_2$); for example, see Jensen, A. W.; Daniels, C. *J. Org. Chem.* **2003**, *68*, 207.
193. Andreoli, E.; Barron, A. R. *Energy & Fuels* **2015**, *29*, 4479.
194. Heimann, J.; Morrow, L.; Anderson, R. E.; Barron, A. R. *Dalton Trans.* **2015**, *44*, 4380.
195. Sitharaman, B.; Bolskar, R. D.; Rusakova, I.; Wilson, L. J. *Nano Lett.* **2004**, 4, 2373.
196. Tanigaki, K.; Ebbesen, T. W.; Saito, S.; Mizuki, J.; Tsai, J. S.; Kubo, Y.; Kuroshima, S. *Nature* **1991**, *352*, 222.
197. Zakharian, T. Y.; Seryshev, A.; Sitharaman, B.; Gilbert, B. E.; Knight, V.; Wilson, L. J. *J. Am. Chem. Soc.* **2005**, *127*, 12508.
198. Shirai, Y.; Osgood, A. J.; Zhao, Y.; Kelly, K. F.; Tour, J. M. *Nano Lett.* **2005**, *11*, 2330. The difference between nanocars and nanotrucks has been described as the former is only able to transport itself, whereas a nanotruck is able to accommodate a load. More recently, "nanodragsters" have been synthesized that have a shorter axle with smaller wheels in the front and a larger axle with larger wheels in the back: Vives, G.; Kang, J.; Kelly, K. F.; Tour, J. M. *Org. Lett.* **2009**, *11*, 5602.
199. (a) Shirai, Y.; Osgood, A. J.; Zhao, Y.; Kelly, K. F.; Tour, J. M. *Nano Lett.* **2005**, *5*, 2330. (b) Shirai, Y.; Osgood, A. J.; Zhao, Y.; Yao, Y.; Saudan, L.; Yang, H.; Chiu, Y.-H.; Alemany, L. B.; Sasaki, T.; Morin, J.-F.; Guerrero, J. M.; Kelly, K. F.; Tour, J. M. *J. Am. Chem. Soc.* **2006**, *126*, 4854.
200. For an interesting book on the history of other serendipitous discoveries in science, see Roberts, R. M. *Serendipity: Accidental Discoveries in Science,*

Wiley: New York, 1989. Note: Exxon first obtained mass spectra of fullerenes during their analysis of coke build-up on a reforming catalyst. However, subsequent work by Smalley, Curl and Kroto yielded larger concentrations of gaseous carbon clusters, of which mass spectra more clearly showed the existence of stable, even-numbered clusters with the intensity of the C_{60} species being 20% greater than neighbors (Fig. 6.27). For more information, see: Smalley, R. E. *Discovering the Fullerenes*, Nobel Lecture, Dec. 1996, may be accessed online at: http://nobelprize.org/nobel_prizes/chemistry/laureates/1996/smalley-lecture.pdf.

201. Kroto, H. *Nanotechnology* **1992**, *3*, 111. A lecture given in the same title is also available as an audio file from: http://www.learnoutloud.com/Catalog/Science/Scientists/C60-The-Celestial-Sphere-That-Fell-to-Earth/15201.
202. For instance, see: (a) Morales, A. M.; Lieber, C. M. *Science* **1998**, *279*, 208. (b) Itina, T. E.; Sentis, M.; Marine, V. *Appl. Surf. Sci.* **2006**, *252*, 4433. (c) Saitow, K. *J. Phys. Chem. B* **2005**, *109*, 3731. (d) Marine, W.; Patrone, L.; Lukyanchuk, B.; Sentis, M. *Appl. Surf. Sci.* **2000**, *154–155*, 345. (e) Ozerov, I.; Bulgakov, A. V.; Nelson, D. K.; CasFor example, see:tell, R.; Marine, W. *Appl. Surf. Sci.* **2005**, *247*, 677.
203. Manolopoulos, D. E. *Chem. Phys. Lett.* **1992**, *192*, 330.
204. For a review of fullerene production methods, see: Lamb, L. D.; Huffman, D. R. *J. Phys. Chem. Solids* **1993**, *54*, 1635.
205. (a) Kriitschmer. W.; Lamb. L. D.; Fostiropoulos, K.; Huffman. D. R. *Nature* **1990**, *347*, 354. (b) Kratschmer, W.; Fostiropoulos, K.; Huffman, D. R. *Chem. Phys. Lett.* **1990**, *170*, 167.
206. Ewels, C. P. *Nano Lett.* **2006**, *6*, 890.
207. Smalley, R. E. *Acc. Chem. Res.* **1992**, *25*, 98.
208. Heath, J. R. *ACS Symp. Ser.* **1992**, *481*, 1.
209. (a) V. Z. Mordkovich, V. Z.; Umnov, A. G.; Inoshita, T.; Endo, M. *Carbon* **1999**, *37*, 1855. (b) Mordkovich, V. Z. *Chem. Mater.* **2000**, *12*, 2813.
210. https://www.researchgate.net/publication/226419542_Synthesis_of_multishell_fullerenes_by_laser_vaporization_of_composite_carbon_targets.
211. Sygula, A.; Rabideau, P. W. *J. Am. Chem. Soc.* **1999**, *121*, 7800.
212. (a) Scott, L. T.; Boorum, M. M.; McMahon, B. J.; Hagen, S.; Mack, J.; Blank, J.; Wegner, H.; de Meijere, A. *Science* **2002**, *295*, 1500. (b) Scott, L. T. *Angew. Chem. Int. Ed. Eng.* **2004**, *43*, 4994.
213. Kabdulov, M.; Jansen, M.; Amsharov, K. *Chem. Eur. J.* **2013**, *19*, 17,262.
214. Otero, G.; Biddau, G.; Sanchez, C.; Caillard, R.; Lopez, M. F.; Rogero, C.; Palomares, F. J.; Cabello, N.; Basanta, M. A.; Ortega, J.; Mendez, J.; Echavarren, A. M.; Perez, R.; Gomez, B.; Martin-Gago, J. A. *Nature* **2008**, *454*, 865.
215. For a review of surface-assisted syntheses of fullerenes, Buckybowls, and single-walled carbon nanotubes, see: Amsharov, K. "From Polyphenylenes to Nanographenes and Graphene Nanoribbons" in Advances in Polymer Science, vol. 278, 2017 (https://doi.org/10.1007/12_2017_7).

216. (a) Komatsu, K.; Murata, M.; Murata, Y. *Science* **2005**, *307*, 238. (b) Lopez-Gejo, J.; Marti, A. A.; Ruzzi, M.; Jockusch, S.; Komatsu, K.; Tanabe, F.; Murata, Y.; Turro, N. J. *J. Am. Chem. Soc.* **2007**, *129*, 14554.
217. Gluch, K.; Feil, S.; Matt-Laubner, S. M.; Echt, O.; Scheier, P.; Mark, T. D. *J. Phys. Chem. A* **2004**, *108*, 6990.
218. Wang, C.-R.; Shi, Z.-Q.; Wan, L.-J.; Lu, X.; Dunsch, L.; Shu, C.-Y.; Tang, Y.-L.; Shinohara, H. *J. Am. Chem. Soc.*, **2006**, *128*, 6605.
219. Gan, L.-H.; Wang, C.-R. *J. Phys. Chem. A* **2005**, *109*, 3980.
220. Tillmann, J.; Wender, J. H.; Bahr, U.; Bolte, M.; Lerner, H. W.; Holthausen, M. C.; Wagner, M. *Angew. Chem. Int. Ed. Eng.* **2015**, *54*, 5429.
221. Pichierri, F.; Kumar, V. *J. Molec. Struct.: THEOCHEM* **2009**, *900*, 71.
222. Bapat, P. V.; Kraft, R.; Camata, R. P. *J. Nanopart. Res.* **2012**, *14*, 1163.
223. Glaspell, G.; Abdelsayed, V.; Saoud, K. M.; El-Shall, M. S. *Pure Appl. Chem.* **2006**, *78*, 1667.
224. For example, see: Shimada, M.; Azuma, Y.; Okuyama, K.; Hayashi, Y.; Tanabe, E. *Jpn. J. Appl. Phys.* **2006**, *45*, 328.
225. Reproduced with permission from J. *Appl. Phys. Lett.* **1993**, *63*, 1549. © 1993 American Institute of Physics.
226. Reproduced with permission from *Jpn. J. Appl. Phys.* **1994**, *33*, 4759. © 1994 The Japan Society of Applied Physics.
227. Reproduced with permission from Ayers, T. M.; Fye, J. L.; Li, Q.; Duncan, M. A. *J. Cluster Sci.* **2003**, *14*, 97. © 2003 Springer Science and Business Media.
228. Reproduced with permission from Yamamoto, T.; Mazumder, J. *Nanostruct. Mater.* **1996**, *7*, 305. © 1996 Elsevier Ltd.
229. Cross-section illustration of an inert-gas evaporation system. Shown are an inconel pipe (1), a crucible containing the precursor (2), tube furnace (3), evaporation zone (4), and cap-free (5) or gas-return cap (6) inert-gas diluter configurations to introduce cooling carrier gas to control the size of the nanoparticles. Reproduced with permission from *ChemEng Sci.* **2002**, *57*, 1753. © 2002 Elsevier Ltd.
230. Schematic of an electrospray system. In order to prevent droplet explosion during evaporation of aerosol droplets, the highly charged aerosol is passed through a radioactive neutralizer before evaporation occurs. Reproduced with permission from *Nanostruct. Mater.* **1995**, *6*, 309. © 1995 Elsevier Ltd.
231. Schematic of the development of nanoparticulate size and microstructure. TEM images illustrate the TiO_2 nanoparticle at reactor temperatures of (a) 800 °C, (b) 1100 °C, and (c) 1300 °C. Reproduced with permission from *Aerosol Sci.* **2001**, *32*, 615. © 2001 Elsevier Ltd.
232. Bapat, A.; Gatti, M.; Ding, Y. -P.; Campbell, S. A.; Kortshagen, U. *J. Phys. D.: Appl. Phys.* **2007**, *40*, 2247.
233. Aktas, S.; Thornton, S. C.; Binns, C.; Lari, L.; Pratt, A.; Kroger, R.; Horsfield, M. A. *Mater. Res. Express* **2015**, *2*, 1.
234. Vemury, S.; Pratsinis, S. E.; Kibbey, L. *J. Mater. Res.* **1997**, *12*, 1031.
235. For example, see: (a) Shoutian, L.; Igor, N.; Germanenko, N.; El-Shall, M. S. *J. Cluster Sci.* **1999**, *10*, 533. (b) Ayers, T. M.; Fye, J. L.; Li, Q.; Duncan,

M. A. *J. Cluster Sci.* **2003**, *14*, 97. (c) El-Shall, M. S.; Si, S. T.; Graiver, D.; Pernisz, U. *Nanotech ACS Symp. Series* **1996**, *622*, 79.

236. (a) Belouet, C. *Appl. Surf. Sci.* **1996**, *96–98*, 630.
237. Ahonen, P. P.; Joutsensaari, J.; Richard O. *J. Aerosol Sci.* **2001**, *32*, 615.
238. For instance, see: (a) Lee, H.; Habas, S. E.; Kweskin, S.; Butcher, D.; Somorjai, G. A.; Yang, P. *Angew. Chem. Int. Ed.* **2006**, *45*, 7824, and references therein. (b) Ahmadi, T. S.; Wang, Z. L.; Green, T. C.; Henglein, A.; El-Sayed, M. A. *Science* **1996**, *272*, 1924.
239. Glaspell, G.; Abdelsayed, V.; Saoud, K. M.; El-Shall, M. S. *Pure Appl. Chem.* **2006**, *78*, 1667.
240. (a) Yonezawa, T.; Onoue, S.-Y.; Kimizuka, N. *Langmuir* **2000**, *16*, 5218. (b) Yonezawa, T.; Onoue, S.-Y.; Kimizuka, N. *Chem. Lett.* **2002**, 528.
241. Crooks, R. M.; Zhao, M.; Sun, L.; Chechik, V.; Yeung, L. K. *Acc. Chem. Res.* **2001**, *34*, 181, and references therein. The first precedent for the use of poly(propylene imine) (PPI) dendrimers is: Floriano, P. N.; Noble, C. O.; Schoonmaker, J. M.; Poliakoff, E. D.; McCarley, R. L. *J. Am. Chem. Soc.* **2001**, *123*, 10,545. This also contains many useful references for early precedents for metal@PAMAM nanocomposites.
242. For an example of trimetallic nanoparticle synthesis (using a nondendritic host), see: Henglein, A. *J. Phys. Chem. B* **2000**, *104*, 6683.
243. Schaak, R. E.; Sra, A. K.; Leonard, B. M.; Cable, R. E.; Bauer, J. C.; Han, Y.-F.; Means, J.; Teizer, W.; Vasquez, Y.; Funck, E. S. *J. Am. Chem. Soc.* **2005**, *127*, 3506.
244. (a) Kim, C.; Lee, H. *Catalysis Commun.* **2009**, *10*, 1305. (b) Siekkinen, A. R.; McLellan, J. M.; Chen, J.; Xia, Y. *Chem. Phys. Lett.* **2006**, *432*, 491. (c) Patel, K.; Kapoor, S.; Dave, D. P.; Mukherjee, T. *Res. Chem. Intermed.* **2006**, *32*, 103. (d) Mallik, K.; Mandal, M.; Pradhan, N.; Pal, T. *Nano Lett.* **2001**, *1*, 319. (e) Tsuji, M.; Matsuo, R.; Jiang, P.; Miyamae, N.; Hikino, S.; Kumagae, H.; Sozana, K.; Kamarudin, N.; Tang, X. -L. *Cryst. Growth Des.* **2008**, *8*, 2528.
245. Kim, J. -H.; Bryan, W. W.; Lee, T. R. *Langmuir* **2008**, *24*, 11147, and references therein.
246. Chen, L.; Zhang, D.; Chen, J.; Zhou, H.; Wan, H. *Mater. Sci. Engin. A* **2006**, *415*, 156, and references therein.
247. For example, see: (a) Saleh, N.; Sarbu, T.; Sirk, K.; Lowry, G. V.; Matyjaszewski, K.; Tilton, R. D. *Langmuir* **2005**, *21*, 9873. (b) Herrera, A. P.; Resto, O.; Briano, J. G.; Rinaldi, C. *Nanotechnol.* **2005**, *16*, S618.
248. Knecht, M. R.; Garcia-Martinez, J. C.; Crooks, R. M. *Langmuir* **2005**, *21*, 11981.
249. For instance, see: (a) Thurmond, K. B.; Kowalewski, T.; Wooley, K. L. *J. Am. Chem. Soc.* **1996**, *118*, 7239. (b) Butun, V.; Billingham, N. C.; Armes, S. P. *J. Am. Chem. Soc.* **1998**, *120*, 12135. (c) Zhou, Y.; Jiang, K.; Chen, Y.; Liu, S. *J. Polym. Sci. A* **2008**, *46*, 6518. (d) O'Reilly, R. K.; Joralemon, M. J.; Wooley, K. L.; Hawker, C. J. *Chem. Mater.* **2005**, *17*, 5976.
250. Deepagan, V. G.; Kwon, S.; You, D. G.; Nguyen, V. Q.; Um, W.; Ko, H.; Lee, H.; Jo, D. G.; Kang, Y. M.; Park, J. H. *Biomaterials* **2016**, *103*, 56.

251. (a) Garcia-Martinez, J. C.; Crooks, R. M. *J. Am. Chem. Soc.* **2004**, *126*, 16170–16178. (b) Garcia-Martinez, J. C.; Scott, R. W. J.; Crooks, R. M. *J. Am. Chem. Soc.* **2003**, *125*, 11190–11191. (c) Kim, Y.-G.; Garcia-Martinez, J. C.; Crooks, R. M. *Langmuir* **2005**, *21*, 5485–5491.
252. Finke, R. G. in *Metal Nanoparticles: Synthesis, Characterization, and Applications*, Feldheim, D. L.; Foss, C. A. ed., Marcel-Dekker: New York, 2002.
253. For instance, see: (a) Nayak, R.; Galsworthy, J.; Dobson, P.; Hutchison, J. *J. Mater. Res.* **1998**, *3*, 905ff. (b) Trindade, T. et al. *Chem. Mater.* **2001**, *13*, 3843.
254. For a review of polymer brush syntheses via surface-initiated controlled radical polymerization, see: Barbey, R.; Lavanant, L.; Paripovic, D.; Schuwer, N.; Sugnaux, C.; Tugulu, S.; Klok, H. -A. *Chem. Rev.* **2009**, *109*, 5437.
255. For examples of nanoparticles grown via the polyol route, see: (a) Elkins, K. E.; Chaubey, G. S.; Nandwana, V.; Liu, J. P. *J. Nano Res.* **2008**, *1*, 23. (b) Wiley, B.; Herricks, T.; Sun, Y.; Xia, Y. *Nano Lett.* **2004**, *4*, 2057, and references therein.
256. Sun, S.; Zeng, H. *J. Am. Chem. Soc.* **2002**, *124*, 8204. Other examples of solution-phase growth of oxide, and other compound 0D nanostructures (including quantum dots) are: (a) Strable, E.; Bulte, J. W. M.; Moskowitz, B.; Vivekanandan, K.; Allen, M.; Douglas, T.*Chem. Mater.* **2001**, *13*, 2201. (b) Frankamp, B. L.; Boal, A. K.; Tuominen, M. T.; Rotello, V. M. *J. Am. Chem. Soc.* **2005**, *127*, 9731. (c) Lemon, B. I.; Crooks, R. M. *J. Am. Chem. Soc.* **2000**, *122*, 12886. (d) Hanus, L. H.; Sooklal, K.; Murphy, C. J.; Ploehn, H. J. *Langmuir* **2000**, *16*, 2621.
257. Redi, F. X.; Cho, K. -S.; Murray, C. B.; O'Brien, S. *Nature* **2003**, *423*, 968.
258. Nandwana, V.; Elkins, K. E.; Poudyal, N.; Chaubey, G. S.; Yano, K.; Liu, J. P. *J. Phys. Chem. C* **2007**, *111*, 4185.
259. Murray, C. B.; Norris, D. J.; Bawendi, M. G. *J. Am. Chem. Soc.* **1993**, *115*, 8706.
260. Also known as 'coarsening'. For more information, see: Ratke, L.; Voorhees, P. W. *Growth and Coarsening: Ostwald Ripening in Material Processing (Engineering Materials)*, Springer: New York, 2002.
261. (a) Hines, M. A.; Scholes, G. D. *Adv. Mater.* **2003**, *15*, 1844. (b) Choi, H.; Ko, J. H.; Kim, Y. H.; Jeong, S. *J. Am. Chem. Soc.* **2013**, *135*, 5278. (c) Cademartiri, L.; Bertolotti, J.; Sapienza, R.; Wiersma, D. S.; von Freymann, G.; Ozin, G. A. *J. Phys. Chem. B* **2005**, *110*, 671.
262. Juttukonda, V.; Paddock, R. L.; Raymond, J. E.; Denomme, D.; Richardson, A. E.; Slusher, L. E.; Fahlman, B. D. *J. Am. Chem. Soc.* **2006**, *128*, 420.
263. Vacassy, R.; Flatt, R. J.; Hofmann, H.; Choi, K. S.; Singh, R. K. *J. Coll. Interf. Sci.* **2000**, *227*, 302.
264. Jolivet, J. -P.; Chaneac, C.; Trone, E. *Chem. Commun.* **2004**, 481.
265. A clever way to generate H_2S gas in situ from thiacetamide rather than having to handle a cylinder of the dangerous nerve gas: $H_3CC(S)NH_2 + H_2O \rightarrow H_2S + CH_3COO^- + NH_4^+$.

266. It should be noted that in addition to solution-phase methods, quantum dots are frequently synthesized using molecular-beam epitaxy or other vapor-phase technique. For example, see: Wang, X. Y.; Ma, W. Q.; Zhang, J. Y.; Salamo, G. J.; Xiao, M.; Shih, C. K. *Nano Lett.* **2005**, *5*, 1873, and references therein.
267. A review of core–shell nanoparticles: Chaudhuri, R. G.; Paria, S. *Chem. Rev.* **2012**, *112*, 2373.
268. For a review of nanoshell synthesis, properties, and applications, see: Kalele, S.; Gosavi, S. W.; Urban, J.; Kulkarni, S. K. *Curr. Sci.* **2006**, *91*, 1038.
269. For a review of gold nanoshells and applications, see: (a) Bardhan, R.; Chen, W.; Perez-Torres, C.; Bartels, M.; Huschka, R. M.; Zhao, L.; Morosan, E.; Pautler, R.; Joshi, A.; Halas, N. J. *Adv. Func. Mater.* **2009**, *19*, 3901. (b) Barhoumi, A.; Huschka, R. M.; Bardhan, R.; Knight, M. W.; Halas, N. J. *Chem. Phys. Lett.* **2009**, *482*, 171. (c) Lal, S.; Clare, S. E.; Halas, N. J. *Acc. Chem. Res.* **2008**, *41*, 1842. (d) Westcott, S.; Oldenburg, S.; Lee, T. R.; Halas, N. J. *Langmuir*, **1998**, *14*, 5396 (& 7378).
270. (a) Kamata, K.; Lu, Y.; Xia, Y. *J. Am. Chem. Soc.* **2003**, *125*, 2384. (b) Marinakos, S. M.; Shultz, D. A.; Feldheim, D. L. *Adv. Mater.* **1999**, *11*, 34. (c) Chah, S.; Fendler, J. H.; Yi, J. *J. Colloid Interface Sci.* **2002**, *250*, 142. (d) Marinakos, S. M.; Novak, J. P.; Brousseau, L. C., III; House, A. B.; Edeki, E. M.; Feldhaus, J. C.; Feldheim, D. L. *J. Am. Chem. Soc.* **1999**, *121*, 8518.
271. Chen, M.; Gao, L. *Inorg. Chem.* **2006**, *45*, 5145.
272. Yin, Y.; Rioux, R. M.; Erdonmez, C. K.; Hughes, S.; Somorjai, G. A.; Alivisatos, A. P. *Science* **2004**, *304*, 711.
273. Some review articles related to organic-based core–shell nanoparticles: (a) Kirsch, S.; Doerk, A.; Bartsch, E.; Sillescu, H.; Landfester, K.; Spiess, H. W.; Maechtle, W. *Macromolecules* **1999**, *32*, 4508. (b) Chan, J. M.; Zhang, L.; Yuet, K. P.; Liao, G.; Rhee, J. W.; Langer, R.; Farokhzad, O. C. *Biomaterials* **2009**, *30*, 1627. (c) Nah, J. W.; Jeong, Y. I.; Cho, C. S. *J. Polym. Sci., Polym. Phys.* **1998**, *36*, 415.
274. For example: Dresco, P. A.; Zaitsev, V. S.; Gambino, R. J.; Chu, B. *Langmuir* **1999**, *15*, 1945.
275. (a) Synthesis of the Iridium complex is reported in: Finke, R. G.; Lyon, D. K.; Nomiya, K.; Sur, S.; Mizuno, N. *Inorg. Chem.* **1990**, *29*, 1784. (b) Aiken, J. D.; Lin, Y.; Finke, R. G. *J. Mol. Catal. A* **1996**, *114*, 29.
276. For a detailed discussion of the mechanistic steps, see: Besson, C.; Finney, E. E.; Finke, R. G. *J. Am. Chem. Soc.* **2005**, *127*, 8179, and references therein.
277. For a thorough recent review on nanostructural growth via coprecipitation of multiple species (and ways to synthesize/stabilize 0D nanostructures), consult: Cushing, B. L.; Kolesnichenko, V. L.; O'Connor, C. J. *Chem. Rev.* **2004**, *104*, 3893.
278. (a) Murray, C. B.; Kagan, C. R.; Bawendi, M. G. *Annu. Rev. Mater. Sci.* **2000**, *30*, 545. (b) LaMer, V. K. *J. Am. Chem. Soc.* **1950**, *72*, 4847. (c) Saraiva, S. M.; de Oliveira, J. F. *J. Dispersion Sci. Technol.* **2002**, *23*, 837.
279. For example, see: (a) Kimling, J.; Maier, M.; Okenve, B.; Kotaidis, V.; Ballot, H.; Plech, A. *J. Phys. Chem. B* **2006**, *110*, 15700.

(b) Ji, X.; Song, X.; Li, J.; Bai, Y.; Yang, W.; Yeng, X. *J. Am. Chem. Soc.* **2007**, *129*, 13939.

280. Pong, B. -K.; Elim, H. I.; Chong, J. -X.; Ji, W.; Trout, B. L.; Lee, J. -Y. *J. Phys. Chem. C* **2007**, *111*, 6281.
281. For more details regarding proposed mechanisms for Au/Ag nanoparticle growth via citrate reduction/stabilization, see: (a) Rodriguez-Gonzalez, B.; Mulvaney, P.; Liz-Marzan, L. M. *Z. Phys. Chem.* **2007**, *221*, 415. (b) Pillai, Z. S.; Kamat, P. V. *J. Phys. Chem. B* **2004**, *108*, 945.
282. L. Colombi Ciacchi, W. Pompe, A. De Vita, J. Phys. Chem. B 2003, 107, 1755.
283. Emerson, S. C. *Synthesis of Nanometer-size Inorganic Materials for the Examination of Particle Size Effects on Heterogeneous Catalysis*, Ph.D. thesis, Worcester Polytechnic Institute, 2000. May be accessed online at: http://www.wpi.edu/Pubs/ETD/Available/etd-0503100-105634/unrestricted/emerson.pdf
284. J. M. Montejano-Carrizales and J. L. Morán-López. *Nanostructured Materials*, 1:397–409, 1992.
285. Whetten, R. L.; Khoury, J. T.; Alvarez, M. M.; Murthy, S.; Vezmar, I.; Wang, Z. L.; Stephens, P. W.; Cleveland, C. L.; Luedtke, W. D.; Landman, U. *Adv. Mater.* **1996**, *8*, 428.
286. Finke, R. G. in *Metal Nanoparticles: Synthesis, Characterization, and Applications*, Feldheim, D. L.; Foss, C. A. eds., Dekker: New York, 2002. Crooks and coworkers determined that a closed-shell metallic nanocluster of Au_{55} has a diameter of 1.2 nm: Kim, Y.-G.; Oh, S.-K.; Crooks, R. M. *Chem. Mater.* **2004**, *16*, 167.
287. (a) Yu, J. H.; Joo, J.; Park, H. M.; Baik, S. L.; Kim, Y. W.; Kim, S. C.; Hyeon, T. *J. Am. Chem. Soc.* **2005**, *127*, 5662. (b) Cho, K. S.; Talapin, D. V.; Gaschler, W.; Murray, C. B.; *J. Am. Chem. Soc.* **2005**, *127*, 7140.
288. For a thorough review of shape control for nanocrystals, see: Xie, Y.; Xiong, Y.; Lim, B.; Skrabalak, S. E. *Adv. Mater.* **2009**, *48*, 60.
289. For a review of nanoparticle self-assembly mechanisms that occur at liquid–liquid interfaces, see: Ghosh, S. K.; Boker, A. *Macromol. Chem. Phys.* **2019**, *220*, 1900196.
290. For instance, see: (a) Using focused-ion beam (FIB): McMahon, M. D.; Hmelo, A. B.; Lopez, R.; Ryle, W. T.; Newton, A. T.; Haglund, R. F.; Feldman, L. C.; Weller, R. A.; Magruder, R. H. *Mat. Res. Soc. Symp. Proc.* **2003**, *739*, H2.7.1. (b) Liu, K.; Ho, C. -L.; Aouba, S.; Zhao, Y. -Q.; Lu, Z. -H.; Petrov, S.; Coombs, N.; Dube, P.; Ruda, H. E.; Wong, W. -Y. *Angew. Chem.* **2008**, *120*, 1275. (c) Yonczawa, T.; Itoh, T.; Shirahata, N.; Masuda, Y.; Koumoto, K. *Appl. Surf. Sci.* **2007**, *254*, 621. (d) Resch, R.; Baur, C.; Bugacov, A.; Koel, B. E.; Madhukar, A.; Requicha, A. A. G.; Will, P. *Langmuir*, **1998**, *14*, 6613.
291. Note: We will discuss the operating principle of atomic force microscopy (AFM) and other scanning force microscopies in more detail in Chap. 7. At this point, simply think of this technique as analogous to an antiquated record player, in which the needle gently touches the surface of the record to produce music. Similarly, the AFM tip either gently taps, or hovers immediately above, the surface of a planar substrate.

292. For a recent review of DPN, see: Liu, G.; Petrosko, S. H.; Zheng, Z.; Mirkin, C. A. *Chem. Rev.* **2020**, *120*, 6009.
293. (a) Piner, R. D.; Zhu, J.; Xu, F.; Hong, S.; Mirkin, C. A. *Science* **1999**, *283*, 661. (b) Hong, S.; Zhu, J.; Mirkin, C. A. *Science* **1999**, *286*, 523. (c) Hong, S.; Mirkin, C. A. *Science* **2000**, *288*, 1808.
294. Urtizberea, A.; Hirtz, M.; Fuchs, H. *Nanofabrication* **2016**, *2*, 43.
295. (a) Rozhok, S.; Piner, R.; Mirkin, C. A *J. Phys. Chem. B* **2002**, *107*, 751. (b) Hampton, J. R.; Dameron, A. A.; Weiss, P. S. *J. Am. Chem. Soc.* **2006**, *128*, 1648. (c) Liu, G.; Zhou, Y.; Banga, R. S.; Boya, R.; Brown, K. A.; Chipre, A. J.; Mirkin, C. A. *Chem. Sci.* **2013**, *4*, 2093. (d) Urtizberea, A.; Hirtz, M.; Fuchs, H. *Nanofab.* **2016**, *2*, 43.
296. Amro, N. A.; Song, X.; Liu, G. Y. *Langmuir* **2000**, *16*, 3006.
297. Tsai, D. -H.; Kim, S. H.; Corrigan, T. D.; Phaneuf, R. J.; Zachariah, M. R. *Nanotechnology* **2005**, *16*, 1856.
298. Redi, F. X.; Cho, K. -S.; Murray, C. B.; O'Brien, S. *Nature* **2003**, *423*, 968.
299. (a) Decher, G.; Hong, J. D. *Makromol. Chem. Macromol. Symp.* **1991**, *46*, 321. (b) Decher, G.; Hong, J. D.; Schmitt, J. *Thin Solid Films* **1992**, *210*, 831. For a recent review of electrostatic LbL growth, see: Hammond, P. T. *Adv. Mater.* **2004**, *16*, 1271.
300. Adamantyl groups were used on the periphery of the dendrimers since they strongly interact with cyclodextrins. For example, see: Rekharsky, M. V.; Inoue, Y. *Chem. Rev.* **1998**, *98*, 1880–1901.
301. (a) The SEM image (low-resolution and high-resolution) of 9, 10-antraquinone nanorods is reproduced with permission from *J. Phys. Chem. B* **2004**, *108*, 7744. © 2004 American Chemical Society. (b) The SEM image of GaP–GaAs nanowires is reproduced with permission from *J. Am. Chem. Soc.* **2006**, *128*, 1353. © 2006 American Chemical Society. (c) The SEM image of carbon nanotubes is reproduced with permission from *J. Phys. Chem. B* **2001**, *105*, 1157. © 2001 American Chemical Society. (d) The SEM image of TiO_2 nanofibers is reproduced with permission from *Nano Lett.* **2006**, 6, 1297. © 2006 American Chemical Society.
302. (a) The HRTEM image of V_2O_5 nanorods on TiO_2 nanofibers is reproduced with permission from *J. Am. Chem. Soc.* **2005**, *127*, 9731. © 2005 American Chemical Society. (b) The HRTEM image of GaP–GaAs nanowires is reproduced with permission from reference 420b. (c) The HRTEM image of multiwall carbon nanotubes is reproduced with permission from *J. Am. Chem. Soc.* **2004**, *126*,4951. © 2004 American Chemical Society.
303. (a) Iijima, S. *Nature* **1991**, *354*, 56 (first report of MWNTs). (b) Iijima, S. *Nature* **1993**, *363*, 603 (SWNT co-precedent). (c) Bethune, D. S.; Kiang, C. H.; Devries, M. S.; Gorman, G.; Savoy, R.; Vazquez, J.; Beyers, R. *Nature* **1993**, *363*, 605 (SWNT co-precedent).
304. The term *graphene* designates a single layer of carbon atoms packed into hexagonal units. Though this structure is used to describe properties of many carbonaceous materials (e.g., CNTs, graphite, fullerenes, etc.), this planar structure is thermodynamically unstable relative to curved structures such as

fullerenes, nanotubes, and other structures found in carbon soot. As such, the isolation of single graphene sheets has only recently been reported through exfoliation from a high purity graphite crystal: Novoselov, K. S.; Geim, A. K.; Morozov, S. V.; Jiang, D.; Zhang, Y.; Dubonos, S. V.; Grigorieva, I. V.; Firsov, A. A. *Science* **2004**, *306*, 666.

305. For more information, see: (a) Ruoff, R. S.; Qian, D.; Liu, W. K. *C. R. Physique* **2003**, *4*, 993. (b) Collins, P. G.; Avouris, P. *Scientific American* **2000**, *283*, 62. (c) Ibrahim, K. S. *Carbon Lett.* **2013**, *14*, 131. (d) Dresselhaus, M. S.; Dresselhaus, G.; Avouris, P. *Carbon Nanotubes: Synthesis, Structure, Properties, and Applications*; Springer: New York, 2001. (e) De Volder, M. F. L.; Tawfick, S. H.; Baughman, R. H.; Hart, A. J. *Science* **2013**, *339*, 535.

306. (a) Yu, M. F.; Files, B. S.; Arepalli, S.; Ruoff, R. S. *Phys. Rev. Lett.* **2000**, *84*, 5552. (b) Kim, P.; Shi, L.; Majumdar, A.; McEuen, P. *Phys. Rev. Lett.* **2001**, *87*, 215502.

307. (a) Dresselhaus, M. S.; Dresselhaus, G.; Jorio, A.; Souza Filho, A. G.; Pimenta, M. A.; Saito, R. *Acc. Chem. Res.* **2002**, *35*, 1070. (b) Rao, A. M.; Richter, E.; Bandow, S.; Chase, B.; Eklund, P. C.; Williams, K. A.; Fang, S.; Subbaswamy, K. R.; Menon, M.; Thess, A.; Smalley, R. E.; Dresselhaus, G.; Dresselhaus, M. S. *Science* **1997**, *275*, 187. (c) Weisman, R. B.; Subramoney, S. *Electrochem. Soc. Interf.* **2006** (summer), 42.

308. For extensive reviews of molecular electronics see: (a) Xiang, D. et al. Chem. Rev. 2016, 116, 4318. (b) Tour, J. M. *Molecular Electronics: Commercial Insights, Chemistry, Devices, Architecture and Programming*; World Scientific: River Edge, NJ, 2003. (c) Tour, J. M.; James, D. K. in *Handbook of Nanoscience, Engineering and Technology*; Goddard, W. A., III; Brenner, D. W.; Lyshevski, S. E.; Iafrate, G. J. eds.,;RC: New York, 2003; pp. 4.1–4.28. (d) Tour, J. M. *Acc. Chem. Res.* **2000**, *33*, 791.

309. Field emission results from the tunneling of electrons from a metal tip into a vacuum, under an applied strong electric field (Chap. 7 will have more details on this phenomenon, and how it is exploited for high-resolution electron microscopy).

310. (a) Li, J.; Zhou, X.; Ding, J.; Zhou, X.; Wang, Y. *J. Mater. Chem. C* **2019**, *7*, 2257. (b) Zhou, Y.; Chen, H.; Wang, Y. *Inorg. Chem.* **2021**, *60*, 8870.

311. (a) https://www.makeuseof.com/what-is-neo-qled-better-than-qled-and-oled/. (b) https://www.cnet.com/tech/home-entertainment/qled-vs-oled-samsung-and-lg-tv-technologies-explained/.

312. (a) Wang, Q. H.; Setlur, A. A.; Lauerhaas, J. M.; Dai, J. Y.; Seelig, E. W.; Chang, R. P. H. *Appl. Phys. Lett.* **1998**, *72*, 2912. (b) http://imechanica.org/files/FinalProject_12_9_Group2.pdf. (c) Choi, Y. S.; Kang, J. H.; Park, Y. J.; Choi, W. B.; Lee, C. J.; Jo, S. H.; Lee, C. G.; You, J. H.; Jung, J. E.; Lee, N. S.; Kim, J. M. *Diam. Rel. Mater.* **2001**, *10*, 1705.

313. (a) Zhao, T. Y. et al. *ACS Appl. Mater. Interfaces* **2019**, *11*, 11,699. (b) Liang, X.; Xia, J.; Dong, G.; Tian, B.; Peng, L. *Top. Curr. Chem.* **2016**, *374*, 80. (b) https://www.chemistryworld.com/news/carbon-nanotubes-in-large-panel-displays/3000744.article.

314. (a) Sun, D. M.; Timmermans, M. Y.; Tian, Y.; Nasibulin, A. G.; Kauppinen, E. I.; Kishimoto, S.; Mizutani, T.; Ohno, Y. *Nat. Nanotechnol.* **2011**, *6*, 156. (b) McCarthy, M. A. Liu, B.; Donoghue, E. P.; Kravchenko, I.; Kim, D. Y.; So, F.; Finzler, A. G. *Science* **2011**, *332*, 570. (c) Wang, C.; Zhang, J.; Ryu, K.; Badmaev, A.; Gomez, L.; Zhou, C. *Nano Lett.* **2009**, *12*, 4285. (d) Chen, P.; Fu, Y.; Aminirad, R.; Wang, C.; Zhang, J.; Wang, K.; Galatsis, K.; Zhou, C. *Nano Lett.* **2011**, *11*, 5301.
315. https://cen.acs.org/articles/92/web/2014/01/Carbon-Nanotube-Transistors-Help-Displays.html.
316. http://www.guo.ece.ufl.edu/CNTFET_IJHES_published.pdf.
317. (a) Avouris, P. *Acc. Chem. Res.* **2002**, *35*, 1026. (b) Wind, S. J.; Appenzeller, J.; Martel, R.; Derycke, V.; Avouris, P. *Appl Phys. Lett.* **2002**, *80*, 3817. A recent strategy for the bottom-up design of CNT interconnects: Li, J.; Ye, Q.; Cassel, A.; Ng, H. T.; Stevens, R.; Han, J.; Meyyappan, M. *Appl. Phys. Lett.* **2003**, *82*, 2491.
318. Shulaker, M. M.; Hills, G.; Patil, N.; Wei, H.; Chen, H. Y.; Wong, H. S. P.; Mitra, S. *Nature* **2013**, *501*, 526.
319. Inaba, T.; Homma, Y. *Appl. Phys. Lett.* **2015**, *107*, 071907.
320. Yakabson, B. I. *Appl. Phys. Lett.* **1998**, *72*, 918.
321. Micro-Raman spectroscopy has shown that during tension, only the outer layers of MWNTs are loaded, whereas during compression, the load is transferred to all layers.
322. Salvetat, J.-P.; Briggs, G. A. D.; Bonard, J.-M.; Basca, R. R.; Kulik, A. J.; Stöckli, T.; Burnham, N. A.; Forró, L. *Phys. Rev. Lett.* **1999**, *82*, 944.
323. Filleter, T.; Espinosa, H. D. *Carbon* **2013**, *56*, 1.
324. For a review of nanotube-reinforced polymers, see: Schulte, K.; Gojny, F. H.; Fiedler, B.; Sandler, J. K. W.; Bauhofer, W. in *Polymer Composites*, Springer: New York, 2005.
325. In 2005, the winner of the Tour de France used a bicycle composed of a CNT composite: https://www.cnet.com/news/carbon-nanotubes-enter-tour-de-france/.
326. Zhang, Y.; Zou, G.; Doorn, S. K.; Htoon, H.; Stan, L.; Hawley, M. E.; Sheehan, C. J.; Zhu, Y.; Jia, Q. *ACS Nano* **2009**, *3*, 2157.
327. This work is being developed by the U.S. Army Research Laboratory, through collaboration with the MIT Institute of Soldier Nanotechnology: (a) http://isnweb.mit.edu. (b) https://www.arl.army.mil/www/pages/510/TAB_ISN-2_2012_for_print.pdf.
328. http://www.temp.speautomotive.com/SPEA_CD/SPEA2013/pdf/NN/NN4.pdf.
329. For a very nice summary of specific stiffness/specific strength regions for various materials classes see: http://www-materials.eng.cam.ac.uk/mpsite/interactive_charts/spec-spec/basic.html.
330. Fluorination has recently been used to attach amine-terminated polymers to the sidewalls of carbon nanotubes; for instance, see: (a) Dillon, E. P.; Crouse, C. A.; Barron, A. R. *ACS Nano* **2008**, *2*, 156. (b) Direct polymerization

of poly(amidoamine), PAMAM, dendrimers directly from chlorocarbonyl-functionalized (-COCl) carbon nanotubes: Pan, B.; Cui, D.; Gao, F.; He, R. *Nanotechnol.* **2006**, *17*, 2483.

331. For a nice review regarding defect sites in CNTs, see Charlier, J.-C. *Acc. Chem. Res.* **2002**, *35*, 1063.
332. For a thorough recent review of the surface chemistry (noncovalent and covalent) of CNTs, see Tasis, D.; Tagmatarchis, N.; Bianco, A.; Prato, M. *Chem. Rev.* **2006**, *106*, 1105.
333. A interesting recent precedent related to the *reversibly* tunable exfoliation of SWNTs using poly(acrylic acid) at varying pH levels is reported by Grunlan, J. C.; Liu, L.; Kim, Y. S. *Nano Lett.* **2006**, 6, 911.
334. Knupfer, M.; Reibold, M.; Bauer, H.-D.; Dunsch, L.; Golden, M. S.; Haddon, R. C.; Scuseria, G. E.; Smalley, R. E. *Chem. Phys. Lett.* **1997**, *272*, 38.
335. (a) Itami, K. *Pure Appl. Chem.* **2012**, *84*, 907, and references therein. (b) Nevius, M. S.; Wang, F.; Mathieu, C.; Barrett, N.; Sala, A.; Mentes, T. O.; Locatelli, A.; Conrad, E. H. *Nano Lett.* **2014**, *14*, 6080. (c) Han, P.; Akagi, K.; Canova, F. F.; Mutoh, H.; Shiraki, S.; Iwaya, K.; Weiss, P. S.; Asao, N.; Hitosugi, T. *ACS Nano* **2014**, *8*, 9181. (d) Cai, J.; Ruffieux, P.; Jaafar, R.; Bieri, M.; Braun, T.; Blankenburg, S.; Muoth, M.; Seitsonen, A. P.; Saleh, M.; Feng, X.; Mullen, K.; Fasel, R. *Nature* **2010**, *466*, 470. (e) Narita, A.; Verzhbitskly, I. A.; Frederickx, W.; Mall, K. S.; Jensen, S. A.; Hansen, M. R.; Bonn, M.; De Feyter, S.; Casiraghi, C.; Feng, X.; Mullen, K. *ACS Nano* **2014**, *8*, 11,622. (f) Wang, W.; Wang, Y. X.; Yang, H. B. *Org. Chem. Front.* **2014**, *1*, 1005.
336. Although these metals have been confirmed to be suitable for CNT growth, it is now accepted that most any metal should catalyze the growth of carbon nanotubes. Hong, G.; Chen, Y.; Li, P.; Zhang, J. *Carbon* **2012**, *50*, 2067.
337. For a description of methods used to form surface-bound nanosized catalysts, see: (a) Kind, H.; Bonard, J. –M.; Emmeneggar, C.; Nilsson, L. –O.; Hernadi, K.; Maillard-Schaller, E.; Schlapbach, L.; Forro, L.; Kern, K. *Adv. Mat.* **1999**, *11*, 1285. (b) Huang, Z. P.; Xu, J. W.; Ren, Z. F.; Wang, J. H.; Siegal, M. P.; Provencio, P. *Appl. Phys. Lett.* **1998**, *73*, 3845. (c) Bower, C.; Zhou, O.; Zhu, W.; Werder, D. J.; Jin, S. *Appl. Phys. Lett.* **2000**, *77*, 2767. (d) Geng, J. et al. *J. Phys. Chem. B* **2004**, *108*, 18447.
338. Rummeli, M. H.; Bachmatiuk, A.; Bornert, F.; Schaffel, F.; Ibrahim, I.; Cendrowski, K.; Simha-Martynkova, G.; Placha, D.; Borowiak-Palen, E.; Cuniberti, G.; Buchner, B. *Nanosc. Res. Lett.* **2011**, *6:303*, 1 (open access: https://nanoscalereslett.springeropen.com/articles/10.1186/1556-276X-6-303.
339. Smalley, R. E. *Discovering the Fullerenes*, Nobel Lecture, 1996. May be found online at: http://nobelprize.org/chemistry/laureates/1996/smalley-lecture.pdf (along with the Nobel lectures from Curl and Kroto).
340. The National Institute of Standards and Technology (NIST) has been recently focused on the development of standard synthesis, purification, and characterization techniques for CNTs (and other nanomaterials). To date, there are a number of competing methods for SWNTs/MWNTs, all citing percent purity

values that appear rather arbitrary. Indeed, purchasing a "90% pure SWNT" sample from multiple vendors will result in very different products! In order to continue the rapid progress in CNT synthesis/applications, it is essential that we set up a "gold standard" for CNTs that will immediately tell us what a certain purity level means. That is, if a "60% purity" value is cited, clarifying what the remaining 40% consists of (amorphous carbon, remaining catalytic metal, other nanotube diameters/morphologies, etc.).

341. (a) Dai, H. *Acc. Chem. Res.* **2002**, *35*, 1035. (b) Bourzac, K. *MIT Technol. Rev.*, Feb. 2009 issue.
342. Hajilounezhad, T. et al. *npj Comput. Mater.* **2021**, *7*, 134.
343. For a nice review of strategies used for superhydrophobic coatings, see: Ma, M.; Hill, R. M. *Curr. Opin. Coll. Interf. Sci.* **2006**, *11*, 193.
344. Zhang, R.; Zhang, Y.; Zhang, Q.; Xie, H.; Qian, W.; Wei, F. *ACS Nano* **2013**, *7*, 6156.
345. (a) Huang, S. M.; Woodson, M.; Smalley, R.; Liu, J. *Nano Lett.* **2004**, *4*, 1025. (b) Hofmann, M.; Nezich, D.; Reina, A.; Kong, J. *Nano Lett.* **2008**, *8*, 4122. (c) Huang, S.; Maynor, B.; Cai, X.; Liu, J. *Adv. Mater.* **2003**, *15*, 1651.
346. For a recent review of horizontally aligned carbon nanotube arrays, see: Zhang, R.; Zhang, Y.; Wei, F. *Chem. Soc. Rev.* **2017**, *46*, 3661.
347. Zhang, M.; Li, J. *Mater. Today* **2009**, *12*, 12.
348. (a) Zhong, G.; Hofmann, S.; Fan, F.; Telg, H.; Warner, J. H.; Eder, D.; Thomsen, C.; Milne, W. I.; Robertson, J. *J. Phys. Chem. C*, **2009**, *113*, 17321. (b) Meshot, E. R.; Plata, D. L.; Tawfick, S.; Zhang, Y.; Verploegen, E. A.; Hart, A. J. *ACS Nano*, **2009**, *3*, 2477.
349. Burt, D. P.; Whyte, W. M.; Weaver, J. M. R.; Glidle, A.; Edgeworth, J. P.; Macpherson, J. V.; Dobson, P. S. *J. Phys. Chem. C* **2009**, *113*, 15,133, and references therein.
350. Nessim, G. D.; Acquaviva, D.; Seita, M.; O'Brien, K. P.; Thompson, C. V. *Adv. Funct. Mater.* **2010**, *20*, 1306.
351. Rao, C. N. R.; Govindaraj, A. *Acc. Chem. Res.* **2002**, *35*, 998, and references therein. For recent information regarding the role of alumina on the yield/morphology of supported CNT catalysts, see: Jodin, L.; Dupuis, A.-C.; Rouviere, E.; Reiss, P. *J. Phys. Chem. B* **2006**, *110*, 7328. It should be noted that the supported nanoclusters may reside within nanochannels to facilitate 1D growth, examples of these methods, which include both template and "closed space sublimation" (CSS) are (and references therein): (a) Li, J.; Papadopoulos, C.; Xu, J. M.; Moskovits, M. *Appl. Phys. Lett.* **1999**, *75*, 367. (b) Kyotani, T.; Tsai, L. F.; Tomita, A. *Chem. Mater.* **1996**, 8, 2109. (c) Hu, Z. D.; Hu, Y. F.; Chen, Q.; Duan, X. F.; Peng, L.-M. *J. Phys. Chem. B* **2006**, *110*, 8263.
352. Yu, D.; Xue, Y.; Dai, L. *J. Phys. Chem. Lett.* **2012**, *3*, 2863.
353. Choi, H. C.; Kim, W.; Wang, D.; Dai, H. *J. Phys. Chem. B* **2002**, *106*, 12361. The first precedent for SWNT growth from gold nanoclusters has been recently reported: Bhaviripudi, S.; Mile, E.; Steiner, S. A.; Zare, A. T.; Dresselhaus, M. S.; Belcher, A. M.; Kong, J. *J. Am. Chem. Soc.* **2007**, *129*, 1516.

354. (a) Lolli, G.; Zhang, L.; Balzano, L.; Sakulchaicharoen, N.; Tan, Y.; Resasco, D. E. *J. Phys. Chem. B* **2006**, *110*, 2108. (b) Bachilo, S. M.; Balzano, L.; Herrera, J. E.; Pompeo, F.; Resasco, D. E.; Weisman, R. B. *J. Am. Chem. Soc.* **2003**, *125*, 11,186.
355. Chiang, W. H.; Sankaran, R. M. *Nat. Mater.* **2009**, *8*, 882.
356. Wang, B.; Poa, C. H. P.; Wei, L.; Li, L. J.; Yang, Y. H.; Chen, Y. *J. Am. Chem. Soc.* **2007**, *129*, 9014.
357. In 2009, researchers in China reported the longest SWNT, over 18.5 cm in length, grown using CVD from a CNT catalytic film: Wang, X.; Li, Q.; Xie, J.; Jin, Z.; Wang, J.; Li, Y.; Jiang, K.; Fan, S. *Nano Lett.* **2009**, *9*, 3137.
358. Resasco, D. E.; Alvarez, W. E.; Pompeo, F.; Balzano, L.; Herrera, J. E.; Kitiyanan, B.; Borgna, A. *J. Nanopart. Res.* **2002**, 4, 131.
359. Nikolaev, P.; Bronikowski, M. J.; Bradley, R. K.; Rohmund, F.; Colbert, D. T.; Smith, K. A.; Smalley, R. E. *Chem. Phys. Lett.* **1999**, *313*, 91. It should be noted that $Fe(CO)_5$ is not the only system in which the precursor acts as the metal catalyst and carbon source. A number of metallocenes (e.g., ferrocene, cobaltocene, and nickelocene) have also been used; however, they typically result in MWNT growth rather than SWNTs. This is most likely due to the larger number of carbon atoms from cyclopentadienyl groups that must self-assemble, relative to smaller carbon precursors (e.g., CH_4, C_2H_2, etc.) used for SWNT growth.
360. (a) Ding, L. P. et al. *J. Am. Chem. Soc.* **2022**, *144*, 5606. (b) Chufa, B. M. et al. *Green Chem. Lett. Rev.* **2021**, *14*, 647.
361. Wagner, R. S.; Ellis, W. C. *Appl. Phys. Lett.* **1964**, 4, 89. For a recent review of the solid–liquid–solid (SLS) and supercritical fluid–liquid–solid (SFLS) mechanisms for semiconductor nanowire growth, see: Wang, F.; Dong, A.; Sun, J.; Tang, R.; Yu, H.; Buhro, W. E. *Inorg. Chem.* **2006**, *45*, 7511. A recent precedent for the epitaxial growth of ZnO nanowires at the junction of nanowalls: Ng, H. T.; Li, J.; Smith, M. K.; Nguyen, P.; Cassell, A.; Han, J.; Meyyappan, M. *Science* **2003**, *300*, 1249.
362. A recent in situ TEM study of Si nanowire growth has been reported by Hofmann, S.; Sharma, R.; Wirth, C. T.; Cervantes-Sodi, F.; Ducati, C.; Kasama, T.; Dunin-Borkowski, R. E.; Drucker, J.; Bennett, P.; Robertson, J. *Nature Mater.* **2008**, *7*, 372.
363. Kharlamova, M. V. *Beilstein J. Nanotechnol.* **2017**, *8*, 826.
364. An example for Au–Ge nanoalloys: Lu, H.; Meng, X. *Sci. Rep.* **2015**, *5:11263*, 1.
365. The word "generally" is used, since there are reports of nanowire growth at temperatures below the eutectic. For example, see: Adhikari, H.; Marshall, A. F.; Chidsey, E. D.; McIntyre, P. C. *Nano Lett.* **2006**, 6, 318.
366. A 25th anniversary review article on semiconductor nanowires: Dasgupta, N. P.; Sun, J.; Liu, C.; Brittman, S.; Andrews, S. C.; Lim, J.; Gao, H.; Yan, R.; Yang, P. *Adv. Mater.* **2014**, *26*, 2137.

367. The first precedent of solution-liquid–solid (SLS) growth of nanowires: Trentler, T. J.; Hickman, K. M.; Goel, S. C.; Viano, A. M.; Gibbons, P. C.; Buhro, W. E. *Science,* **1995**, *270*, 1791.
368. For instance, see: Paulose, M.; Varghese, O. K.; Grimes, C. A. *J. Nanosci. Technol.* **2003**, *3*, 341.
369. Lu, J.; Miao, J. *Nanosc. Res. Lett.* **2012**, *7:356*, 1.
370. Marchand, M. L.; Journet, C.; Guillot, D.; Benoit, J. M.; Yakobson, B. L.; Purcell, S. T. *Nano Lett.* **2009**, *9*, 2961.
371. (a) Harris, P. J. F. *Carbon* **2007**, *45*, 229. (b) Gavillet, J.; Thibault, J.; Stephan, O.; Amara, H.; Loiseau, A.; Gaspard, J. P.; Ducastelle, F. *J. Nanosci. Nanotechnol.* **2004**, *4*, 346. (c) Irle, S.; Ohta, Y.; Okamoto, Y.; Page, A. J.; Wang, Y.; Morokuma, K. *Nano Res.* **2009**, *2*, 755.
372. A review of growth mechanisms for CNTs: https://cdn.intechopen.com/pdfs-wm/16802.pdf.
373. Hofmann, S.; Csanyi, G.; Ferrari, A. C.; Payne, M. C.; Robertson, J. *Phys. Rev. Lett.* **2005**, *95*, 036,101.
374. (a) Huang, S.; Cai, Q.; Chen, J.; Qian, Y.; Zhang, L. *J. Am. Chem. Soc.* **2009**, *131*, 2094, and references therein. (b) Liu, B.; Ren, W.; Gao, L.; Li, S.; Pei, S.; Liu, C.; Jiang, C.; Cheng, H. -M. *J. Am. Chem. Soc.* **2009**, *131*, 2082.
375. Cantoro, M.; Hofmann, S.; Pisana, S.; Scardaci, V.; Parvez, A.; Ducati, C.; Ferrari, A. C.; Blackburn, A. M.; Wang, K.-Y.; Robertson, J. *Nano Lett.* **2006**, 6, 1107.
376. Page, A. J.; Chandrakumar, K. R. S.; Irle, S.; Morokuma, K. *J. Am. Chem. Soc.* **2011**, *133*, 621.
377. Persson, A. I.; Larsson, M. W.; Stenstrom, S.; Ohlsson, B. J.; Samuelson, L.; Wallenberg, L. R. *Nat. Mater.* **2004**, *3*, 677.
378. Deng, W.-Q.; Xu, X.; Goddard, W. A. *Nano Lett.* **2004**, 4, 2331.
379. It has been shown that reducing the catalyst size causes an increase in the growth rate, whereas varying the catalyst composition affects the growth rate, activation energy, and onset temperature for CNT growth: Chiang, W. -H.; Sankaran, R. M. *Diam. Rel. Mater.* **2009**, *18*, 946.
380. The top VLS mechanism was predicted using molecular dynamics calculations. The image was reproduced with permission from *J. Phys. Chem. B* **2004**, *108*, 17,369. © 2004 American Chemical Society The middle VLS mechanism shows both "root growth" (c–d) and "folded growth" (e–g). The image was reproduced with permission from *J. Am. Chem. Soc.* **2004**, *126*, 4951. © 2004 American Chemical Society The bottom mechanism, predicted by quantum mechanics/molecular mechanics, is one of the rare examples of an atomic-level picture of CNT growth. The image was reproduced with permission from *Nano Lett.* **2004**, 4, 2331. © 2004 American Chemical Society.
381. Considering a VLS growth mechanism, the catalyst nanocluster must be molten during nucleation. However, the growth temperature of carbonaceous nanostructures is much lower than the melting point of binary C/M systems (M = catalytic metals such as Fe, Ni, Co, etc.), which lends credence to the existence of "liquid-like" deformable nanoclusters during growth.

382. Graphite-encapsulated metal nanostructures are of increasing importance for magnetic applications such as high-density magnetic recording media; for example, see: Flahaut, E.; Agnoli, F.; Sloan, J.; O'Connor, C.; Green, M. L. H. *Chem. Mater.* **2002**, *14*, 2553, and references therein. Encapsulation dominates over CNT growth at low temperatures since the kinetic energy is not sufficient for graphitic islands to lift off the catalyst surface. Hence, encapsulation may easily be limited, which enhances CNT growth, by maintaining elevated temperatures. Experimental results also show that small catalyst nanoclusters (diameters <2 nm) are free of graphite encapsulation since they do not contain a sufficient number of dissolved C atoms. However, for metal nanostructures >3 nm in diameter, calculations suggest that graphite encapsulation is thermodynamically preferred over SWNT growth. This is confirmed by the empirical observation that SWNTs form only on catalyst particles with diameters <2 nm.
383. For example, see: (a) Lin, M.; Tan, J. P. Y.; Boothroyd, C.; Loh, K. P.; Tok, E. S.; Foo, Y. -L. *Nano Lett.* **2006**, *6*, 449. (b) Yoshida, H.; Takeda, S.; Uchiyama, T.; Kohno, H.; Homma, Y. *Nano Lett.* **2008**, *8*, 2082.
384. Ding, F.; Harutyunyan, A. R.; Yakobson, B. I. *Proc. Nat'l Acad. Sci.* **2009**, *106*, 2506.
385. Using field-emission microscopy (FEM) to observe axial rotation and preferential adsorption of dimeric C_2 units during growth: Marchand, M.; Journet, C.; Guillot, D.; Benoit, J. -M.; Yakobson, B. I.; Purcell, S. T. *Nano Lett.* **2009**, *9*, 2961.
386. Jin, S.; Bierman, M. J.; Morin, S. A. *J. Phys. Chem. Lett.* **2010**, *1*, 1472, and references therein.
387. Hofmann, S.; Sharma, R.; Ducati, C.; Du, G.; Mattevi, C.; Cepek, C.; Cantoro, M.; Pisana, S.; Parvez, A.; Cervantes-Sodi, F.; Ferrari, A. C.; Dunin-Borkowski, R.; Lizzit, S.; Petaccia, L.; Goldoni, A.; Robertson, J. *Nano Lett.* **2007**, 602.
388. Lee, Y. H.; Kim, S. G.; Jund, P.; Tomanek, D. *Phys. Rev. Lett.* **1997**, *78*, 2393.
389. A recent paper by Hata et al. discusses the variables that govern highly-efficient carbon nanotube growth - an oxygen-containing "growth enhancer" (e.g., water, alcohols), and a carbon source not containing oxygen: Futaba, D. N.; Goto, J.; Yasuda, S.; Yamada, T.; Yumura, M.; Hata, K. *Adv. Mater.* **2009**, *21*, 4811.
390. Hata, K.; Futaba, D. N.; Mizuno, K.; Namai, T.; Yumura, M.; Iijima, S. *Science* **2004**, *306*, 1362.

Some additional recent precedents for millimeter-tall SWNT carpets: (a) Almkhelfe, H.; Li, X.; Rao, R.; Amama, P. B. *Carbon*, **2017**, *116*, 181. (b) Chen, G.; Sakurai, S.; Yumura, M.; Hata, K.; Futaba, D. N. *Carbon* **2016**, *107*, 433.
391. Rummeli, M. H.; Borowiak-Palen, E.; Gemming, T.; Pichler, T.; Knupfer, M.; Kalbac, M.; Dunsch, L.; Jost, O.; Silva, S. R. P.; Pompe, W.; Buchner, B. *Nano Lett.* **2005**, 5, 1209.
392. Sugime, H. et al. *Carbon* **2021**, *172*, 772.

393. Wang, Y.; Song, W.; Jiao, M.; Wu, Z.; Irle, S. *Carbon* **2017**, *121*, 292.
394. (a) Visic, B.; Panchakarla, L. S.; Tenne, R. *J. Am. Chem. Soc.* **2017**, *139*, 12,865. (b) Goldberger, J.; Fan, R.; Yang, P. *Acc. Chem. Res.* **2006**, *39*, 239, and references therein. (c) https://link.springer.com/content/pdf/10.1007/s11467-013-0326-8.pdf. (d) Remskar, M. *Adv. Mater.* **2004**, *16*, 1497.
395. Lu, J. G.; Chang, P. C.; Fan, Z. Y. *Mater. Sci. Eng. Rep.* **2006**, *52*, 49.
396. (a) Hulteen, J. C.; Martin, C. R. *J. Mater. Chem.* **1997**, *7*, 1075. (b) Sander, M. S.; Cote, M. J.; Gu, W.; Kile, B. M.; Tripp, C. P. *Adv. Mater.* **2004**, *16*, 2052.
397. (a) Kmentova, H.; Kment, S.; Wang, L.; Pausova, S.; Vaclavu, T.; Kuzel, R.; Han, H.; Hubicka, Z.; Zlamal, M.; Olejnicek, J.; Cada, M.; Krysa, J.; Zboril, R. *Catal. Today* **2017**, *287*, 130. (b) Roy, P.; Berger, S.; Schmuki, P. *Angew. Chem. Int. Ed.* **2011**, *50*, 2904.
398. For example, see: (a) Wang, J. M.; Gao, L. J. Mater. Chem. **2003**, *13*, 2551. (b) Liu, B.; Zeng, H. C. *J. Am. Chem. Soc.* **2003**, *125*, 4430. (c) Jia, C. J.; Sun, L. D.; Yan, Z. G.; You, L. P.; Luo, F.; Han, X. D.; Pang, Y. C.; Zhang, Z.; Yan, C. H. *Angew. Chem.* **2005**, *117*, 4402. (d) Sander, M. S.; Cote, M. J.; Gu, W.; Kile, B. M.; Tripp, C. P. *Adv. Mater.* **2004**, *16*, 2052.
(e) Meng, F.; Gong, J.; Fan, Z.; Li, H.; Yuan, J. *Ceram. Int.* **2016**, *42*, 4700.
(f) Niu, X.; Li, M.; Wu, B.; Li, H. *J. Mater. Sci.: Mater. Electron.* **2016**, *27*, 10198.
399. For instance, see: Levard, C.; Rose, J.; Masion, A.; Doelsch, E.; Borschneck, D.; Olivi, L.; Dominici, C.; Grauby, O.; Woicik, J. C.; Bottero, J. Y. *J. Am. Chem. Soc.* **2008**, *130*, 5862.
400. For reviews of halloysite nanotubes, see: (a) Kamble, R.; Ghag, M.; Gaikawad, S.; Panda, B. K. *J. Adv. Scient. Res.* **2012**, *3*, 25. (b) Du, M.; Guo, B.; Jia, D. *Polym. Int.* **2010**, *59*, 574. (c) http://www.tandfonline.com/doi/full/10.1080/03602559.2017.1329436. (d) Yuan, P.; Tan, D.; Annabi-Bergaya, F. *Appl. Clay Sci.* **2015**, *112*, 75.
401. (a) Novoselov, K. S.; Geim, A. K.; Morozov, S. V.; Jiang, D.; Zhang, Y.; Dubonos, S. V.; Grigorieva, I. V.; Firsov, A. A. *Science* **2004**, *306*, 666. (b) Novoselov, K. S.; Jiang, D.; Schedin, F.; Booth, T. J.; Khotkevich, V. V.; Morozov, S. V.; Geim, A. K. *Proc. Natl. Acad. Sci. U.S.A.* **2005**, *102*, 10,451.
402. http://nobelprize.org/nobel_prizes/physics/laureates/2010/
403. Novoselov, K. S.; Geim, A. K.; Morozov, S. V.; Jiang, D.; Katsnelson, M. I.; Grigorieva, I. V.; Dubonos, S. V.; Firsov, A. A *Nature* **2005**, *438*, 197.
404. Berger, C.; Song, Z.; Li, X.; Wu, X.; Brown, N.; Naud, C.; Mayou, D.; Li, T.; Hass, J.; Marchenkov, A. N.; Conrad, E. H.; First, P. N.; de Heer, W. A. *Science* **2006**, *312,* 1191.
405. *Graphite intercalation compounds and applications,* Endo, M., Ed.; Oxford University Press: Oxford, U.K., 2003.
406. Bulk graphite oxide is best prepared from purified natural graphite using the Hummers method: Hummers, W. S. & Offeman, R. E. *J. Am. Chem. Soc.* **1958**, *80*, 1339.

For the mechanical exfoliation of bulk graphene oxide into sheets within an aqueous medium, see: Stankovich, S.; Dikin, D. A.; Dommett, G. H. B.; Kohlhaas, K. M.; Zimney, E. J.; Stach, E. A.; Piner, R. D.; Nguyen, S. T.; Ruoff, R. S. *Nature* **2006**, *442*, 282.

407. (a) He, H., Klinowski, J., Forster, M. & Lerf, A. *Chem. Phys. Lett.* **1998**, *287*, 53. (b) Lerf, A.; He, H.; Forster, M.; Klinowski, J. *J. Phys. Chem B* **1998**, *102*, 4477.
408. For example, see: (a) Tung, V. C.; Allen, M. J.; Yang, Y.; Kaner, R. B. *Nat. Nanotechnol.* **2009**, *4*, 25. (b) Allen, M. J.; Fowler, J. D.; Tung, V. C.; Yang, Y.; Weiller, B. H.; Kaner, R. B. *Appl. Phys. Lett.* **2008**, *93*, 193119. (c) Fowler, J. D.; Allen, M. J.; Tung, V. C.; Yang, Y.; Kaner, R. B.; Weiller, B. H. *ACS Nano* **2009**, *3*, 301.
409. (a) Reina, A.; Jia, X.; Ho, J.; Nezich, D.; Son, H.; Bulovic, V.; Dresselhaus, M. S.; Kong, J. *Nano Lett.* **2009**, *9*, 30. (b) Kim, K. S.; Zhao, Y.; Jang, H.; Lee, S. Y.; Kim, J. M.; Kim, K. S.; Ahn, J. -H.; Kim, P.; Choi, J. -Y.; Hong, B. H. *Nature* **2009**, *457*, 706.
410. (a) Kobayashi, T.; Bando, M.; Kimura, N.; Shimizu, K.; Kadono, K.; Umezu, N.; Miyahara, K.; Hayazaki, S.; Nagai, S.; Mizuguchi, Y.; Murakami, Y.; Hobara, D. *Appl. Phys. Lett.* **2013**, *102*, 023112. (b) Bae, S.; Kim, H.; Lee, Y.; Xu, X.; Park, J. S.; Zheng, Y.; Balakrishnan, J.; Lei, T.; Kim, H. R.; Song, Y.; Kim, Y. J.; Kim, K. S.; Ozyilmaz, B.; Ahn, J. H.; Hong, B. H.; Iijima, S. *Nat. Nanotechnol.* **2010**, *5*, 574.
411. (a) Lee, H. C. et al. *RSC Adv.* **2017**, *7*, 15644. (b) Santhiran, A. et al. *J. Carbon Res.* **2021**, *7*, 76. (c) Wu, Y.; Wang, S.; Komvopoulos, K. *J. Mater. Res.* **2020**, 76.
412. Saha, J. K.; Dutta, A. *Waste Biomass & Valorization* **2022**, *13*, 1385.
413. (a) Guo, Z. et al. *Adv. Biol.* **2021**, *5*, 2100637. (b) Georgakilas, V. et al. *Chem. Rev.* **2012**, *112*, 6156. (c) Maio, A. et al. *Nanomaterials* **2021**, *11*, 1717.
414. Xu, Y.; Bai, H.; Lu, G.; Li, C.; Shi, G. *J. Am. Chem. Soc.* **2008**, *130*, 5856.
415. Stankovich, S.; Piner, R. D.; Chen, X. Q.; Wu, N. Q.; Nguyen, S. T.; Ruoff, R. S. *J. Mater. Chem.* **2006**, *16*, 155.
416. Bon, S. B.; Valentini, L.; Verdejo, R.; Garcia Fierro, J. L.; Peponi, L.; Lopez-Manchado, M. A.; Kenny, J. M. *Chem. Mater.* **2009**, *21*, 3433.
417. The October 2014 issue of *Nature Nanotechnology* was focused on the applications for graphene: http://www.nature.com/nnano/focus/graphene-applications/index.html. Graphene is even being placed in sports footwear: https://www.inov-8.com/us/?___store=us
418. The quantum Hall effect is the integer quantization of the Hall conductance for a 2-D system subjected to low temperatures and strong magnetic fields. However, for graphene, an anomalous *fractional quantization* is observed in which electrons exhibit much smaller effective masses. This corresponds to extremely facile transport through graphene, which is analogous to the mobility of relativistic particles such as photons.

For example, see: (a) http://icmt.illinois.edu/ICMT-talks/20080114-Abanin-Talk.pdf (b) Toke, C.; Lammert, P. E.; Crespi, V. H.; Jain, J. K. *Phys. Rev. B* **2006**, *74*, 235417. (c) Novoselov, K. S.; Jiang, Z.; Zhang, Y.; Morozov, S. V.; Stormer, H. L.; Zeitler, U.; Maan, J. C.; Boebinger, G. S.; Kim, P.; Geim, A. K. *Science* **2007**, *315*, 1379. (d) Ozyilmaz, B.; Jarillo-Herrero, P.; Efetov, D.; Abanin, D. A.; Levitov, L. S.; Kim, P. *Phys. Re*V. *Lett.* **2007**, *99*, 186804.
419. For instance, see: Bolotin, K. I.; Sikes, K. J.; Jiang, Z.; Klima, M.; Fudenberg, G.; Hone, J.; Kim, P.; Stormer, H. L. *Solid State Commun.* **2008**, *146*, 351.
420. (a) Park, J. M.; Cao, Y.; Xia, L. Q.; Sun, S.; Watanabe, K.; Taniguchi, T.; Jarillo-Herrero, P. *Nature Mater.* **2022**, *21*, 877. (b) Liu, X.; Zhang, N. J.; Watanabe, K.; Taniguchi, T.; Li, J. A. *Nature Phys.* **2022**, *18*, 522.
421. Li, D.; Kaner, R. B. *Science* **2008**, *320*, 1170.
422. The term 'zero-bandgap semiconductor' should be distinguished from a metal, even though both of these materials exhibit zero bandgaps. Within a graphene sheet, the Dirac point represents the sole electronic state responsible for its electrical conductivity; in contrast, electrons in many other states within the vicinity of the Fermi level may participate in electrical conductivity.
423. http://www-mtl.mit.edu/researchgroups/palacios/graphene/palacios.html.

Ambipolarity is also shared by organic semiconductors: (a) Chua, L. L.; Zaumseil, J.; Chang, J. F.; Ou, E. C. W.; Ho, P. K. H.; Sirringhaus, H.; Friend, R. H. *Nature* **2005**, *434*, 194.

(b) Meijer, E. J.; De Leeuw, D. M.; Setayesh, S.; van Veenendaal, E.; Huisman, B. H.; Blom, P. W. M.; Hummelen, J. C.; Scherf, U.; Klapwijk, T. M. *Nat. Mater.* **2003**, *2*, 678.
424. Zhang, Y.; Tang, T. -T.; Girit, C.; Hao, Z.; Martin, M. C.; Zettl, A.; Crommie, M. F.; Shen, Y. R.; Wang, F. *Nature* **2009**, *459*, 820.
425. Zhang, Y.; Tan, Y. W.; Stormer, H. L.; Kim, P. *Nature* **2005**, *438*, 201.
426. (a) Son, Y. -W.; Cohen, M. L.; Louie, S. G. *Phys. Rev. Lett.* **2006**, *97*, 216,803. (b) Han, M. Y.; Ozyilmaz, B.; Zhang, Y.; Kim, P. *Phys. Rev. Lett.* **2007**, *98*, 206,805.
427. http://www.lbl.gov/enews/back-issues/2007/Feb/SABLSelectZigzag.pdf
428. Yano, Y.; Mitoma, N.; Ito, H.; Itami, K. *J. Org. Chem.* **2020**, *85*, 4.
429. Kosynkin, D. V.; Higginbotham, A. L.; Sinitskii, A.; Lomeda, J. R.; Dimiev, A.; Price, B. K.; Tour, J. M. *Nature* **2009**, *458*, 872.
430. Jiao, L.; Zhang, L.; Wang, X.; Diankov, G.; Dai, H. *Nature* **2009**, *458*, 877.
431. (a) Han, M. Y.; Oezyilmaz, B.; Zhang, Y.; Kim, P. *Phys. Rev. Lett.* **2007**, *98*, 206805. (b) Chen, Z.; Lin, Y. -M.; Rooks, M. J.; Avouris, P. *Physica E* **2007**, *40*, 228.
432. Campos-Delgado, J. et al. *Nano Lett.* **2008**, *8*, 2773.
433. Yang, X.; Dou, X.; Rouhanipour, A.; Zhi, L.; Joachim Rader, H.; Mullen, K. *J. Am. Chem. Soc.* **2008**, *130*, 4216.
434. A special issue of Acc. Chem. Res. was devoted to 2-D nanomaterials, beyond graphene: *Acc. Chem. Res.* **2015**, *48*, 1.
435. Mak, K. F.; Lee, C.; Hone, J.; Heinz, T. F. *Phys. Rev. Lett.* **2010**, *105*, 136,805.

436. Liu, H.; Neal, A. T.; Ye, P. D. *ACS Nano* **2012**, *6*, 8563.
437. Mas-Balleste, R.; Gomez-Navarro, C.; Gomez-Herrero, J.; Zamora, F. *Nanoscale* **2011**, *3*, 20.
438. Feng, B.; Ding, Z.; Meng, S.; Yao, Y.; He, X.; Cheng, P.; Chen, L.; Wu, K. *Nano Lett.* **2012**, *12*, 3507.
439. Kharadi, M. A. et al. *ECS J. Solid State Sci. Technol.* **2020**, *9*, 115031.
440. Radisavljevic, B.; Radenovic, A.; Brivio, J.; Biacometti, V.; Kis, A. *Nature Nanotechnol.* **2011**, *6*, 147.
441. Tao, L.; Cinquanta, E.; Chiappe, D.; Grazianetti, C.; Fanciulli, M.; Dubey, M.; Molle, A.; Akinwande, D. *Nat. Nanotechnol.* **2015**, *10*, 227.
442. Novoselov, K. S.; Jiang, D.; Schedin, F.; Booth, T. J.; Khotkevich, V. V.; Morozov, S. V.; Geim, A. K. *Proc. Nat. Acad. Sci.* **2005**, *102*, 10451.
443. Liu, H.; Neal, A. T.; Zhu, Z.; Luo, Z.; Xu, X.; Tomanek, D.; Ye, P. D. *ACS Nano* **2014**, *8*, 4033.
444. Venkateshalu, S. et al. *J. Energy Storage* **2022**, *48*, 104027.
445. Naguib, M.; Gogotsi, Y. *Acc. Chem. Res.* **2015**, *48*, 128.
446. Naguib, M.; Mochalin, V. N.; Barsoum, M. W.; Gogotsi, Y. *Adv. Mater.* **2014**, *26*, 992.
447. Zhao, L.; Li, B. *Tungsten* **2020**, *2*, 176.
448. For a review of MXene applications, see: (a) Papadopoulou, K. A.; Chroneos, A.; Parfitt, D.; Christopoulos, S. R. G. *J. Appl. Phys.* **2020**, *128*, 170902. (b) Zamhuri, A.; Lim, G. P.; Ma, N. L.; Tee, K. S.; Soon, C. F. *Biomed. Eng. Online* **2021**, *20*, 33.
449. Wang, Z. *IOP Conf. Series: Earth Environ. Sci.* **2021**, *714*, 042030.
450. Kumar, J. A. et al. *Chemosphere* **2022**, *286*, 131607.
451. For example, a battery rated at 3 V, when at 50% charge would still output a voltage close to 3 V; in contrast, a supercapacitor rated at 3 V would output exactly half of its maximum charge voltage, 1.5 V at 50% charge.
452. https://sustdev.unescap.org/Files/China%20Public_Transportation_Sector_ultra-capacitor_buses.pdf.
453. *Energy density* is defined as the energy stored in a battery per unit volume. Although this term may also be used for an energy storage device per unit mass, the term *specific energy* is preferred.
454. Although LiBs are currently popular for automotive and portable electronic applications, the research community is now shifting its focus to more Earth-abundant and sustainable materials such as Na-ion (NaB), Mg-ion, Al-ion, and their metal-air varieties, to address concerns with the worldwide availability of lithium reserves. For instance, see: (a) Masse, R. C.; Uchaker, E.; Cao, G. *Sci. China Mater.* **2015**, *58*, 715. (b) Larcher, D.; Tarascon, J. M. *Nature Chem.* **2015**, *7*, 19. (c) Kim, S. W.; Seo, D. H.; Ma, X.; Ceder, G.; Kang, K. *Adv. Energy Mater.* **2012**, *2*, 710. (d) Choi, J. W.; Aurbach, D. *Nature Rev. Mater.* **2016**, *1*, 16013.
455. https://www.axios.com/2022/03/04/the-cold-hard-truth-about-electric-vehicles-in-winter

456. For detailed photos of the interior of the Tesla 100 kWh battery pack, see: https://teslamotorsclub.com/tmc/threads/pics-info-inside-the-battery-pack.34934/
457. The "18,650" (i.e., 18 × 65 mm) Li-ion cell is larger than a typical AA alkaline cell. For more details, see: https://commonsensehome.com/18650-battery/.
458. At the time of writing, the Lucid Air, Tesla Model 3/Model Y, and Rivian R1T/R1S models are employing 2170-type Li-ion cells. For a comparison of common cell types, see: https://www.torquenews.com/14093/tesla-18650-2170-and-4680-battery-cell-comparison-basics
459. It should be noted that the *current density* (mA g^{-1}) is important to compare reported capacities of electrode materials. The lower the current density, the more time the Li^+ ions have to intercalate/insert into the electrode material, resulting in a higher overall capacity. Whereas most research studies utilize current densities in the 50–500 mA g^{-1} range, commercial rechargeable batteries must typically cycle in the A g^{-1} range. The battery industry also most often refers to current densities as "C"; for instance, a 1C rate refers to the discharge current necessary to fully discharge the battery in 1 h, and so on.
460. The capacity of some alloying materials such as MX_y (M = Fe, Co, Cu, Ni, Bi; X = F, Cl; y = 2, 3), AgCl, LiCl, S, Se, Te, or I is expected to be much higher than intercalation compounds, with theoretical capacities >1000 mAh g^{-1}. For some recent reviews of materials being investigated for LiB and Na-ion battery cathodes, see: (a) Nitta, N.; Wu, F.; Lee, J. T.; Yushin, G. *Mater. Today* **2015**, *18*, 252. (b) Ellis, B. L.; Lee, K. T.; Nazar, L. F. *Chem. Mater.* **2010**, *22*, 691. (c) Demirocak, D. E.; Srinivasan, S. S.; Stafanakos, E. K. *Appl. Sci.* **2017**, *7*, 731.
461. For a review of nanostructured materials for LiB anode applications, see: Goriparti, S.; Miele, E.; De Angelis, F.; Di Fabrizio, E.; Zaccaria, R. P.; Capiglia, C. *J. Power Sources* **2014**, *257*, 421.
462. Chen, K. S.; Xu, R.; Luu, N. S.; Secor, E. B.; Hamamoto, K.; Li, Q.; Kim, S.; Sangwan, V. K.; Balla, I.; Guiney, L. M.; Seo, J. W. T.; Yu, X.; Liu, W.; Wu, J.; Wolverton, C.; Dravid, V. P.; Barnett, S. A.; Lu, J.; Amine, K.; Hersam, M. C. *Nano Lett.* **2017**, *17*, 2539.
463. Xiong, Z.; Yun, Y. S.; Jin, H. J. *Materials* **2013**, *6*, 1138.
464. For instance, see: (a) Verma, P.; Maire, P.; Novak, P. *Electrochim. Acta* **2010**, *55*, 6332. (b) Tasaki, K.; Goldberg, A.; Lian, J. J.; Walker, M.; Timmons, A.; Harris, S. J. *J. Electrochem. Soc.* **2009**, *156*, A1019.
465. (a) An, S. J.; Li, J.; Daniel, C.; Mohanty, D.; Nagpure, S.; Wood, D. L. Carbon 2016, 105, 52. (b) https://arxiv.org/pdf/1210.3672.pdf.
466. (a) Xue, J. S.; Dahn, J. R. *J. Electrochem. Soc.* **1995**, *142*, 3668. (b) Yazami, R.; Deschamps, M. *J. Power Sources* **1995**, *54*, 411.
467. Yoo, E.; Kim, J.; Hosono, E.; Zhou, H. S.; Kudo, T.; Honma, I. *Nano Lett.* **2008**, *8*, 2277.
468. Mo, Y. F.; Zhang, H. T.; Guo, Y. N. *Mater. Lett.* **2017**, *205*, 118.

469. Hassoun, J.; Bonaccorso, F.; Agostini, M.; Angelucci, M.; Betti, M. G.; Cingolani, R.; Gemmi, M.; Mariani, C.; Panero, S.; Pellegrini, V.; Scrosati, B. *Nano Lett.* **2014**, *14*, 4901.
470. The *Coulombic efficiency* is the ratio of charging and discharging capacities for a particular cycle, reflecting the amount of Li that has been lost due to SEI formation. For instance, if the cell has a charging capacity of 850 mAh g^{-1} and a discharging capacity of 620 mAh g^{-1}, the Coulombic efficiency would be 73% for that cycle.
471. Chen, J.; Yao, B.; Li, C.; Shi, G. *Carbon* **2013**, *64*, 225.
472. Mattevi, C.; Eda, G.; Agnoli, S.; Miller, S.; Mkhoyan, K. A.; Celik, O.; Mastrogiovanni, D.; Granozzi, G.; Garfunkel, E.; Chhowalla, M. *Adv. Funct. Mater.* **2009**, *19*, 2577.
473. (a) Uthaisar, C.; Barone, V.; Peralta, J. E. *J. Appl. Phys.* **2009**, *106*, 113715. (b) Uthaisar, C.; Barone, V. *Nano Lett.* **2010**, *10*, 2838.
474. Bhardwaj, T.; Antic, A.; Pavan, B.; Barone, V.; Fahlman, B. D. *J. Am. Chem. Soc.* **2010**, *132*, 12556.
475. Uthaisar, C.; Barone, V.; Fahlman, B. D. *Carbon* **2013**, *61*, 558.
476. Another recent report that shows the effect of oxygen-based functional groups on the interlayer spacing of graphene sheets, and enhanced Li capacity: Cheng, Q.; Okamoto, Y.; Tamura, N.; Tsuji, M.; Maruyama, S.; Matsuo, Y. *Sci. Rep.* **2017**, *7:14782*, 1.
477. Chen, J. S.; Archer, L.; Lou, X. W. "SnO_2 Hollow Structures and TiO_2 Nanosheets for lithium-ion batteries." *J. Mater. Chem.* **2011**, *21*, 9912.
478. McDowell, M. T.; Lee, S. W.; Nix, W. D.; Cui, Y. *Adv. Mater.* **2013**, *25*, 4966.
479. Some recent reviews: (a) Du, F. H.; Wang, K. X.; Chen, J. S.; J. *Mater. Chem. A* **2016**, *4*, 32. (b) Mazouzi, D.; Karker, Z.; Reale, C.; Hernandez, C. R.; Manero, P. J.; Guyomard, D.; Roue, L. *J. Power Sources* **2015**, *280*, 533.
480. For instance: (a) Park, M. H.; Kim, M. G.; Joo, J.; Kim, K.; Kim, J.; Ahn, S.; Cui, Y.; Cho, J. *Nano Lett.* **2009**, *9*, 3844. (b) Yoo, J. K.; Kim, J.; Jung, Y. S.; Kang, K. *Adv. Mater.* **2012**, *24*, 5452.
481. For example, see: (a) Yao, Y.; McDowell, M. T.; Ryu, I.; Wu, H.; Liu, N.; Hu, L.; Nix, W. D.; Cui, Y. *Nano Lett.* **2011**, *11*, 2949. (b) Chen, D.; Mei, X.; Ji, G.; Lu, M.; Xie, J.; Lu, J.; Lee, J. Y. *Angew. Chem. Int. Ed. Eng.* **2012**, *51*, 2409.
482. For example, see: (a) Ge, M.; Rong, J.; Fang, X.; Zhang, A.; Lu, Y.; Zhou, C. *Nano Res.* **2013**, *6*, 174. (b) Bang, B. M.; Lee, J. I.; Kim, H.; Cho, J.; Park, S. Adv. *Energy Mater.* **2012**, *2*, 878.
483. For instance, see: (a) Liu, N.; Lu, Z.; Zhao, J.; McDowell, M. T.; Lee, H. W.; Zhao, W.; Cui, Y. *Nat. Nanotechnol.* **2014**, *9*, 187. (b) Su, L.; Xie, J.; Xu, Y.; Wang, L.; Wang, Y.; Ren, M. *Phys. Chem. Chem. Phys.* **2015**, *17*, 17562. (c) Lee, W. H.; Kang, D. Y.; Kim, J. S.; Lee, J. K.; Moon, J. H. *RSC Adv.* **2015**, *5*, 17424. (d) Pan, L.; Wang, H.; Gao, D.; Chen, S.; Tan, L.; Li, L. *Chem. Commun.* **2014**, *50*, 5878. (e) Park, Y.; Choi, N. S.; Park, S.; Woo, S. H.; Sim, S.; Jang, B. Y.; Oh, S. M.; Park, S.; Cho, J.; Lee, K. T. *Adv. Energy Mater.* **2013**, *3*, 206.

484. For example: (a) Hwang, T. H.; Lee, Y. M.; Kong, B. S.; Seo, J. S.; Choi, J. W. *Nano Lett.* **2012**, *12*, 802. (b) Li, Y.; Guo, B.; Ji, L.; Lin, Z.; Xu, G.; Liang, Y.; Zhang, S.; Toprakci, O.; Hu, Y.; Alcoutlabi, M.; Zhang, X. *Carbon* **2013**, *51*, 185. (c) Lee, B. S.; Son, S. B.; Park, K. M.; Seo, J. H.; Lee, S. H.; Choi, I. S.; Oh, K. H.; Yu, W. R. *J. Power Sources* **2012**, *206*, 267.
485. For instance, see: (a) Wang, W.; Kumta, P. N. *ACS Nano* **2010**, *4*, 2233. (b) Martin, C.; Crosnier, O.; Retoux, R.; Belanger, D.; Schleich, D. M.; Brousse, T. *Adv. Funct. Mater.* **2011**, *21*, 3524. (c) Fan, Y.; Zhang, Q.; Xiao, Q.; Wang, X.; Huang, K. *Carbon*, **2013**, *59*, 264.
486. For example, see: (a) Tang, H.; Zhang, J.; Zhang, Y. J.; Xiong, Q. Q.; Tong, Y. Y.; Li, Y.; Wang, X. L.; Gu, C. D.; Tu, J. P. *J. Power Sources* **2015**, *286*, 431. (b) Xu, B.; Wu, H.; Lin, C. X.; Wang, B.; Zhang, Z.; Zhao, X. S. *RSC Adv.* **2015**, *5*, 30624. (c) Zhao, X.; Hayner, C. M.; Kung, M. C.; Kung, H. H. *Adv. Energy Mater.* **2011**, *1*, 1079. (d) Xiang, H.; Zhang, K.; Ji, G.; Lee, J. Y.; Zou, C.; Chen, X.; Wu, J. *Carbon* **2011**, *49*, 1787.
487. For instance, see: (a) Murugesan, S.; Harris, J. T.; Korgel, B. A.; Stevenson, K. J. *Chem. Mater.* **2012**, *24*, 1306. (b) Yoo, S.; Lee, J. I.; Ko, S.; Park, S. *Nano Energy* **2013**, *2*, 1271.
488. For example, see: (a) Guo, Z. P.; Wang, J. Z.; Liu, H. K.; Dou, S. X. *J. Power Sources* **2005**, *146*, 448. (b) Wu, H.; Yu, G.; Pan, L.; Liu, N.; McDowell, M. T.; Bao, Z.; Cui, Y. *Nat. Cummun.* **2013**, *4*, 1943. (c) Gu, M.; Xiao, X. C.; Liu, G.; Thevuthasan, S.; Baer, D. R.; Zhang, J. G.; Liu, J.; Browning, N. D.; Wang, C. M. *Sci. Rep.* **2014**, *4*, 3684.
489. https://doi.org/10.3389/fmech.2020.00037.
490. Bhushan, B. in *Nanotribology and Nanomechanics*: Springer: New York, 2017.
491. Chaste, J.; Eichler, A.; Moser, J.; Ceballos, G.; Rurali, R.; Bachtold, A. *Nature Nanotechnol.* **2012**, *7*, 301.
492. Downs, S. L.; Madhi, S. A.; Van deer Merwe, L.; Nunes, M. C.; Olwagen, C. P. *Sci. Reports* **2021**, *11*, 23728.
493. Simovic-Pavlovic, M.; Bokic, B.; Vasiljevic, D.; Kolaric, B. *Appl. Sci.* **2022**, *12*, 905.
494. Ginger, D. S.; Zhang, H.; Mirkin, C. A. *Angew. Chem. Int. Ed.* **2004**, *43*, 30. This review contains all of the references for the various experimental conditions.
495. Yang, Y. T.; Callegari, C.; Feng, X. L.; Ekinci, K. L.; Roukes, M. L. *Nano Lett.* **2006**, 6, 583.
496. Collin, J.-P.; Dietrich-Buchecker, C.; Gavina, P.; Jimenez-Molero, M.; Sauvage, J.-P. *Acc. Chem. Res.* **2001**, *34*, 477.
497. Geeves, M. A. *Nature* **2002**, *415*, 129.
498. Liu, Y.; Flood, A. H.; Bonvallet, P. A.; Vignon, S. A.; Northrop, B. H.; Tseng, H.-R.; Jeppesen, J. O.; Huang, T. J.; Brough, B.; Baller, M.; Magonov, S.; Solares, S. D.; Goddard, W. A.; Ho, C.-M.; Stoddart, J. F. *J. Am. Chem. Soc.* **2005**, *127*, 9745.

499. (a) Shuler, M. L. *Lab Chip* **2019**, *19*, 9. (b) Sharma, B.; Sharma, A. *Adv. Eng. Mater.* **2022**, *24*, 2100738.
500. Sikdar, D.; Cheng, W.; Premaratne, M. *J. Appl. Phys.* **2015**, *117*, 083101.

Further Reading

1. http://www.nanohub.org
2. http://www.nanowerk.com (updated news stories related to nanotechnology)
3. B. Bhushan (ed.), *Springer Handbook of Nanotechnology*, 2nd edn. (Springer, New York, 2007)
4. V.B. Sutariya, Y. Pathak, *Biointeractions of Nanomaterials* (CRC Press, New York, 2014)
5. Q. Saquib, M. Faisal, A.A. Al-Khedhairy, A.A. Alatar, *Cellular and Molecular Toxicology of Nanoparticles* (Springer, New York, 2018)
6. M.F. Bertino, *Introduction to Nanotechnology* (WSPC, U.S., 2022)
7. G. Cao, Y. Wang, *Nanostructures and Nanomaterials: Synthesis, Properties and Applications*, 2nd edn. (World Scientific Publishing, New York, 2011)
8. G.A. Ozin, A.C. Arsenault, L. Cademartiri, *Nanochemistry: A Chemical Approach to Nanomaterials*, 2nd edn. (Royal Society of Chemistry Publishing, Cambridge, U.K., 2009)
9. M.I. Mendelson, *Learning Bio-Micro-Nanotechnology* (CRC Press, New York, 2013)
10. B. Rogers, J. Adams, S. Pennathur, *Nanotechnology: Understanding Small Systems*, 3rd edn. (CRC Press, New York, 2014)
11. D. Natelson, *Nanostructures and Nanotechnology* (Cambridge University Press, UK, 2015)
12. P.N. Prasad, *Introduction to Nanomedicine and Nanobioengineering* (Wiley, New York, 2012)
13. J.F. Leary, *Fundamentals of Nanomedicine (Cambridge University Press, 2022)*
14. M.S. Dresselhaus, G. Dresselhaus, P.C. Eklund, *Science of Fullerenes and Carbon Nanotubes* (Academic, New York, 1996)
15. J.M.J. Frechet, D.A. Tomalia (eds.), *Dendrimers and Dendritic Polymers.* (Wiley, New York, 2001)
16. M. Wiesner, J.Y. Bottero, Environmental Nanotechnology: Applications and Impacts of Nanomaterials, 2nd edn. (McGraw Hill, N.Y., 2016)
17. MRS Bulletin **31**, 5 (2006). Special issue on materials for magnetic data storage
18. MRS Bulletin **30**, 12 (2005). Special issue on fabrication of sub-45-nm structures for the next generation of devices
19. M.J. Madou, *Fundamentals of Microfabrication and Nanotechnology*, 3rd edn. (CRC Press, New York, 2011)
20. C.M. Sotomayor-Torres (ed.), *Alternative Lithography: Unleashing the Potentials of Nanotechnology.* (Kluwer/Plenum, New York, 2003)
21. M. Wilson, K. Kannangara, G. Smith, M. Simmons, *Nanotechnology: Basic Science and Emerging Technologies* (CRC, Boca Raton, 2002)
22. M.J. O'Connell, *Carbon Nanotubes* (CRC, Boca Raton, 2006)
23. R.E. Smalley, M.S. Dresselhaus, G. Dresselhaus, P. Avouris, *Carbon Nanotubes: Synthesis, Structure, Properties and Applications* (Springer, New York, 2001)
24. A. Jorio, G. Dresselhaus, M.S. Dresselhaus, *Carbon Nanotubes: Advanced Topics in the Synthesis, Structure, Properties and Applications* (Springer, New York, 2008)
25. H.S. Wong, D. Akinwande, *Carbon Nanotube and Graphene Device Physics* (Cambridge University Press, UK, 2011)
26. P.J.F. Harris, *Carbon Nanotube Science: Synthesis, Properties and Applications*, 2nd edn. (Cambridge University Press, UK, 2011)
27. D. Tomanek, http://www.pa.msu.edu/cmp/csc/nanotube.html (The Nanotube Site)
28. S.E. Lyshevski, S.E. Lyshevski, *MEMS and NEMS: Systems, Devices, and Structures* (CRC, Boca Raton, 2002)

29. G. Schmid, *Nanoparticles: From Theory to Application* (Wiley, New York, 2004)
30. D. Vollath, *Nanoparticles, Nanocomposites, Nanomaterials: An Introduction for Beginners* (Wiley, New York, 2013)
31. A.S. Edelstein, R.C. Cammaratra, *Nanomaterials: Synthesis, Properties and Applications*, 2nd edn. (CRC Press, New York, 1998)
32. V. Rotello, *Nanoparticles: Building Blocks for Nanotechnology (Nanostructure Science and Technology)* (Springer, New York, 2004)
33. E.L. Wolf, *Nanophysics and Nanotechnology*, 3rd edn. (Wiley, New York, 2015)

Chapter 7
Materials Characterization

Thus far, we have focused on the relationship between the structure of a material and its properties/applications. However, we have not yet focused on how one is able to determine the structure and composition of materials. That is, when a material is fabricated in the lab, how are we able to assess whether our method was successful? Depending on the nature of the material being investigated, a suite of techniques may be utilized to assess its structure and properties. Whereas some techniques are qualitative, such as providing an image of a surface, others yield quantitative information such as the relative concentrations of atoms that comprise the material. Recent technological advances have allowed materials scientists to accomplish something that was once thought to be impossible: to obtain actual 2-D/3-D images of atomic positions in a solid, in real time. It should be noted that the sensitivity of quantitative techniques also continues to be improved, with some techniques able to measure elemental concentrations down to the parts per billion (ppb) or trillion (ppt) range.

This chapter will focus on the most widely used techniques available to characterize solid-state compounds. The primary objective of this chapter is to provide a practical description of the methods used to characterize a broad range of materials. Rather than focusing on the theoretical aspects of each technique, which may be found in many other textbooks, our treatment will focus on the suitability of various techniques, sample preparation, and anticipated results. In this manner, you will be well informed regarding the best method to use for a particular material. Since techniques such as solution-phase nuclear magnetic resonance (NMR) and infrared spectroscopy (IR) are used throughout undergraduate courses, the background of these methods will not be provided in this textbook. Likewise, it is beyond the scope of this chapter to provide detailed background in optics, electronics, and physical chemistry concepts that form the foundation of most techniques described herein. For this information, the reader is referred to the "Further Reading" section at the end of this chapter.

B. D. Fahlman, *Materials Chemistry*,
https://doi.org/10.1007/978-3-031-18784-1_7

7.1 Light Scattering

Dynamic light scattering (DLS), also known as photon correlation spectroscopy (PCS) or quasi-elastic light scattering (QELS), represents a particle size analysis technique that has many advantages relative to other characterization methods. Unlike electron microscopy, DLS does not require ultrahigh vacuum conditions and expensive optics, making this technique relatively inexpensive and facile. Furthermore, characterization is non-invasive and nondestructive, which can be performed quickly without the need of complex and time-consuming sample preparation techniques.

When electromagnetic radiation interacts with small molecules or particles, the oscillating electric field of the incoming light generates a dipole in the electric field of the illuminated particles. This induced dipole oscillates with the same frequency as that of the incident radiation, which results in the release of energy via scattered light. For metallic nanoparticles, this process is referred to as *surface plasmon resonance*, which is used to explain the observed colors of dispersed metallic nanoparticles such as those found in stained glass windows. Mie theory is used to describe scattering when the wavelength of light is comparable to the dimensions of the scattering particles. In contrast, when the wavelength of light is large relative to the particle size, scattering is approximated by Rayleigh theory—for instance, to explain why the sky is blue.

The principle of DLS is based on light scattering from suspended particulates in solution that are randomly and dynamically distributed via Brownian motion. A laser is employed as a monochromic light source, which is directed through a sample chamber, with scattered light collected by a photodetector. Due to collisions among the suspended particulates, the fluctuation of the scattered light intensity will vary over time, based on the size of particles in solution (Fig. 7.1).

The intensity versus time signal appears to be completely random over long time intervals, but can be used over extremely short timescales (typically 100–1000 ms) to determine particle sizes. This information is extracted by constructing a *correlation function,* $C(\tau)$, which measures the exponential decay in scattering intensity as a function of delay time, τ (7.1). Whereas it is not possible to determine the precise movement of individual particles in solution, it is possible to *correlate* the motion of suspended particulates with respect to each another.

$$C(\tau) = e^{-2\Gamma \tau} \tag{7.1}$$

The correlation function depends on the decay constant, Γ, which is influenced by a variety of properties of the suspended particles and solvent (7.2):

$$\Gamma = Dq^2 \tag{7.2}$$

where:

Γ is the decay constant

D is the diffusion coefficient/diffusivity of the dispersed particles.
q is the scattering vector, defined as:

$$q = \frac{4\pi n}{\lambda}\sin\left(\frac{\theta}{2}\right)$$

where:

n is the refractive index of the dispersed particles
λ is the wavelength of the incident laser.
θ is the scattering angle between sample & detector.

As shown in Fig. 7.2, decay of the scattering intensity occurs more rapidly for smaller particles due to differences in the respective diffusivities of dispersed particulates. The decay constant, Γ, may be determined from the experimentally derived correlation function (7.1). Using the decay constant shown above, the diffusivity, D,

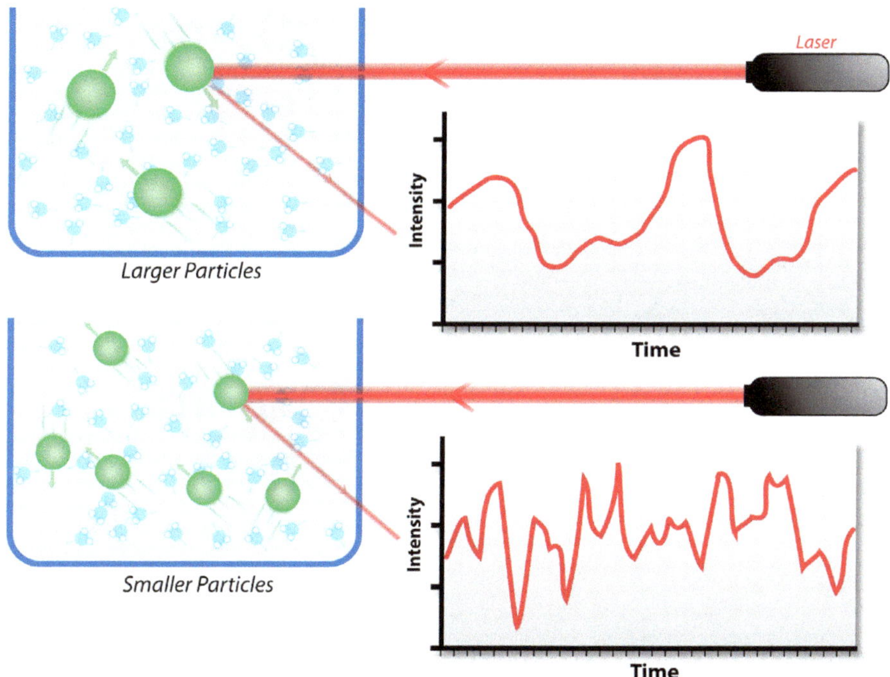

Fig. 7.1 Illustration of the fluctuation of scattering intensity of light from suspended particulates of different sizes over time. Image reproduced under the Creative Commons Attribution-Share Alike 3.0 Unported license

of the suspended nanoparticles may be determined (7.2). Since the random Brownian motion of dispersed particulates is related to size, solvent viscosity and temperature, the hydrodynamic radius, R_H, of the particles may be calculated using the Stokes–Einstein equation (7.3).

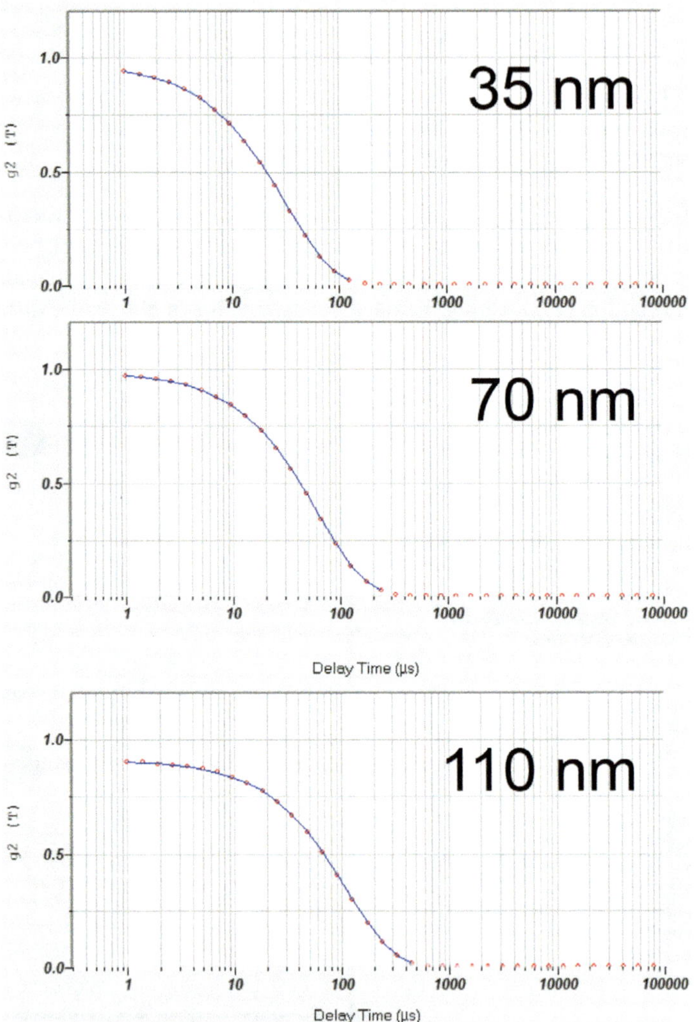

Fig. 7.2 Correlation functions for nanoparticles of varying dimensions. Reproduced with permission from Bodycomb, J. *Interpreting and Understanding Dynamic Light Scattering Data*, Horiba, 2012

$$R_H = \frac{k_B T}{6\pi \eta D} \tag{7.3}$$

where:

k_B is the Boltzmann constant (1.38×10^{-23} m^2 kg s^{-2} K^{-1}).
T is the measurement temperature (in Kelvin, K).
η is the solvent viscosity (kg m^{-1} s^{-1}).
D is the diffusivity (calculated using (7.2)) of the dispersed particles (m^2 s^{-1}).

These formulae clearly indicate that the hydrodynamic dimensions obtained from DLS are intimately related to the temperature and viscosity of the solvent. Hence, it follows that the DLS-measured sizes of suspended particulates may be altered by choosing a different solvent, and/or performing the measurements at a different temperature. This is particularly relevant if the suspended particles have varying degrees of agglomeration in different solvents.

The above formulae refer to the most simplistic example of a monodisperse nanoparticle solution. The presence of nanoparticles of differing dimensions will complicate the correlation function, which will exhibit multiple decay constants. However, the general methodology in determining the hydrodynamic radius of suspended particulates from intensity versus time data is identical. The particle dimensions obtained from DLS are commonly reported based on intensity data. However, it may be preferred to report the data in terms of the mass/volume or number distributions of the dispersed particulates. The intensity of scattered light is proportional to d^6, where d is the nanoparticle diameter. Hence, large particulates will scatter light much more strongly than smaller ones. Accordingly, the intensity-based size distribution will be biased by the presence of large-diameter particulates in solution. As illustrated in Fig. 7.3, the presence of large particulates in the 200–300 nm range alongside smaller nanoparticles will bias the intensity size distribution toward the larger particulates. However, the volume-based size distribution yields the largest peak in the 30–40 nm range, with a smaller peak >200 nm. This suggests that most of the solution was composed of small nanoparticles, with few aggregates present. This is explicitly demonstrated by examining the number-based distribution, in which the only peak present is between 30 and 40 nm. This indicates that there are essentially no larger-sized particles present in solution when normalized to the hundreds/thousands of nanosized particles.

Since the presence of large particles will strongly affect DLS results, rigorous sample preparation must be used prior to analysis. In accord with standard operating protocols, as outlined by the National Institute of Standards and Technology (NIST),[1] the solvent to be used as the dispersing medium should be filtered through a 100-nm or smaller pore size membrane prior to testing. Also, to prevent scattering from large particulates such as dust or aggregates, sample solutions should be filtered with a 200-nm or smaller pore size membrane, preferably as the solution is added to the sample cuvette. To prevent removal of sample particles, the pore size of the filter should be around 3–5 times larger than the smallest dimensions of the dispersed particles. Alternatively, centrifugation may be used instead, or in addition to filtration,

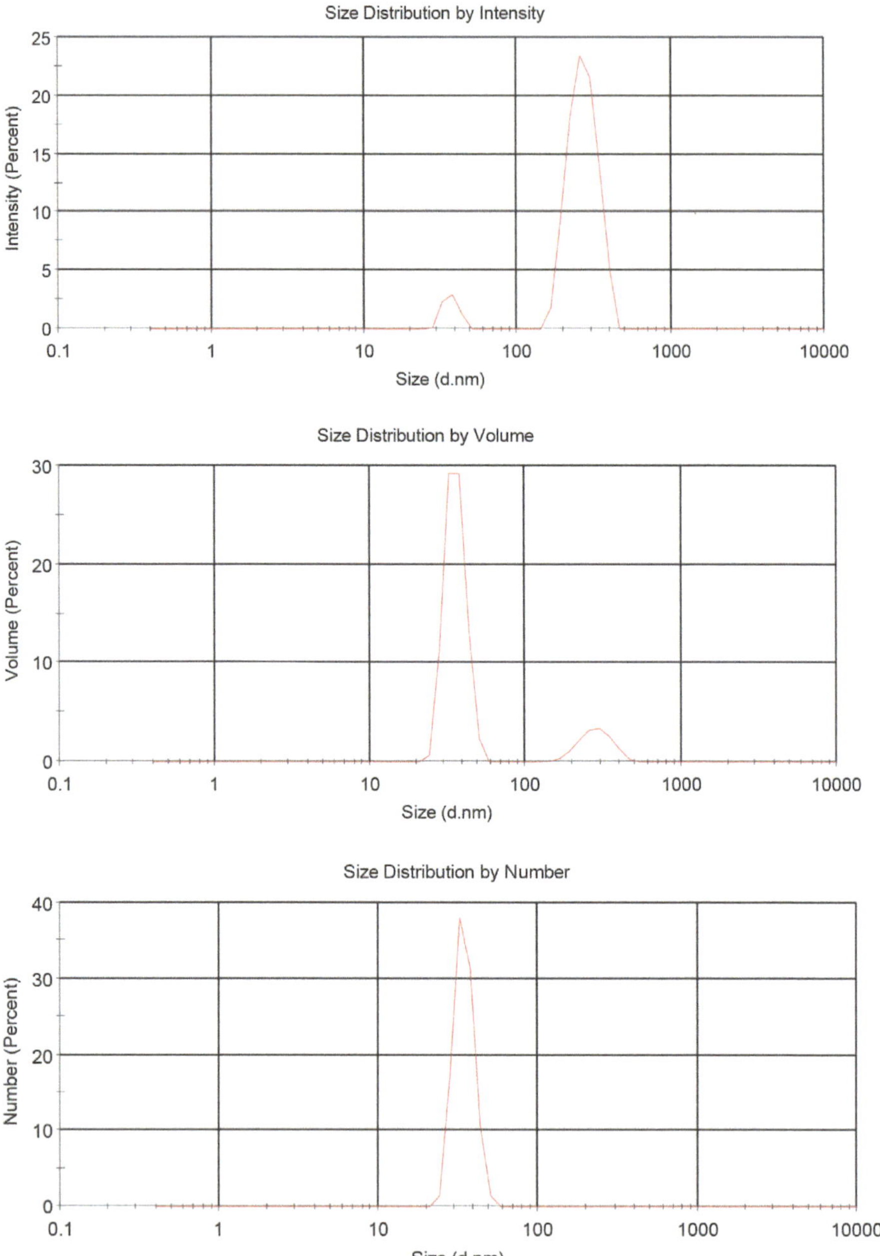

Fig. 7.3 Hydrodynamic diameter distribution of a PbS nanoparticle solution measured by DLS, reported by intensity (top), volume/mass (middle), and number (bottom)

to remove large size contaminants. Centrifugation is preferred for "soft" organic samples that may be absorbed, aggregated, or deformed by the filter media.

It should be noted that DLS is a *complementary* characterization technique and should never be used in isolation to determine nanoparticle dimensions or overall morphology. The information determined from DLS represents the *hydrodynamic* dimensions; that is, the radius/diameter of suspended particles in solution. Hence, if nanoparticles are agglomerated, or are entrained within a surfactant micelle, the measured dimensions will be of the entire aggregate, not individual nanoparticles. In contrast, transmission electron microscopy (TEM, Sect. 7.4) provides the direct imaging of nanoparticles that have been deposited onto the surface of a carbon-coated grid. Accordingly, TEM is often better to measure the dimensions of discrete nanoparticles. Even if agglomeration occurs before or after nanoparticle deposition, the resolution of TEM instruments is sufficient to observe the discrete nanoparticles that comprise the larger aggregates. Furthermore, TEM interrogates the electron-rich regions of a sample, so the observed dimensions of nanoparticles will not be influenced by the presence of capping ligands from organic-based surfactants and entraining agents.

Another common technique used to characterize semiconductor nanoparticles is UV–Visible spectroscopy. This technique illuminates a QD solution with electromagnetic radiation in the UV/Visible region, with the unattenuated light intensity measured by a photodetector. In this respect, this technique shares similar benefits as DLS, being nondestructive, non-invasive, and inexpensive. Due to quantum confinement of electron/hole carriers, as the size of semiconductor nanocrystals decreases, the optical properties will differ significantly from the bulk solid. For very small nanocrystals known as quantum dots (QDs), there will be a shift in the absorption peak toward the UV with decreasing size (Fig. 7.4). This allows for the facile determination of nanoparticle size based simply on the position of its UV–vis absorption peak.

7.2 X-ray Scattering

The wavelength range of X-rays is comparable to the size of atoms (Fig. 7.5); hence, this energetic form of radiation is ideal to probe the structural arrangement of atoms and molecules in a wide variety of solids. The generation of X-rays takes place within a vacuum tube in which electrons emitted from a heated cathode filament are accelerated toward a metal anode target by an accelerating voltage of *ca.* 50 kV. The high-energy electrons interact with the atoms in the target, and may lose energy via two processes. First, an electron may come very close to a nucleus, which causes its trajectory to be altered and release of *Bremsstrahlung* (braking) radiation (Fig. 7.6). Bremsstrahlung radiation exhibits a broad emission profile since the energy of the emitted photon can take any value up to the energy of the incident electron. Second, the high-energy incident electron may cause inner-shell electrons to be released from their orbitals. These electron vacancies are filled by electrons from orbitals farther

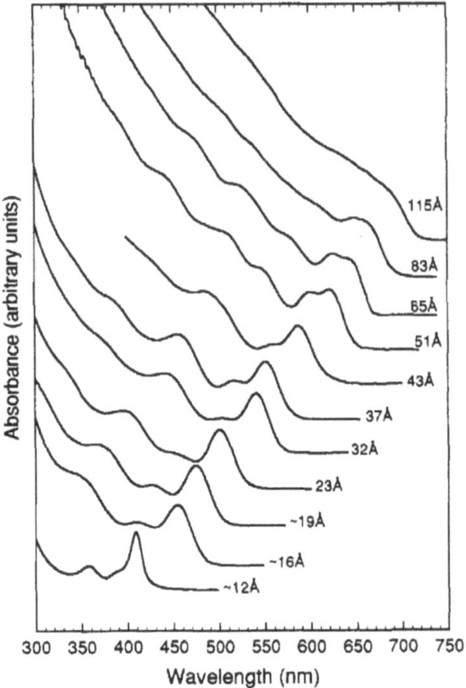

Fig. 7.4 UV–Vis absorption spectrum of CdS quantum dots of varying sizes. Reproduced with permission from *J. Am. Chem. Soc.* **1993**, *115*, 8706. © 1993 American Chemical Society

from the nucleus (Fig. 7.7, top), which result in the emission of photons of varying characteristic energies. For instance, K_α emission results from a K-shell (1s orbital) vacancy being filled by an L-shell (2s or 2p orbitals) electron; K_β radiation results from a K-shell vacancy being filled by an M-shell (3s, 3p, or 3d orbitals) electron, and so on (Fig. 7.7, bottom). In contrast to Bremsstrahlung, characteristic X-ray lines are observed as sharp peaks in the X-ray emission spectrum (Fig. 7.8).

Synchrotron facilities represent another source of X-rays, which are thousands to million times more intense than radiation emitted by X-ray tubes. The unique benefits of synchrotron radiation include its continuous spectrum, high flux and brightness,[3] and high coherence. Figure 7.9 shows an illustration of the advanced photon source (APS) at Argonne National Laboratory near Chicago, IL. Electrons are emitted from a cathode that is heated to *ca.* 1000 °C and accelerated to energies of 450 meV. At this energy, the electrons are relativistic—traveling at >99.999% of the speed of light (299,792,458 m s^{-1}). From the linear accelerator, the electrons are injected into the booster synchrotron—a circular ring of electromagnets that accelerates the electrons from 450 meV to 7 GeV in one-half second. In order to maintain the orbital path of the electrons, bending and focusing magnets increase the electron field strength in synchronization with the applied RF field. The electrons, now traveling at speeds >99.999999% of the speed of light, are injected into a larger-circumference storage ring, which consists of >1000 electromagnets. The radiation emitted from the electrons traveling in a curved path is often referred to as *magnetobremsstrahlung*,

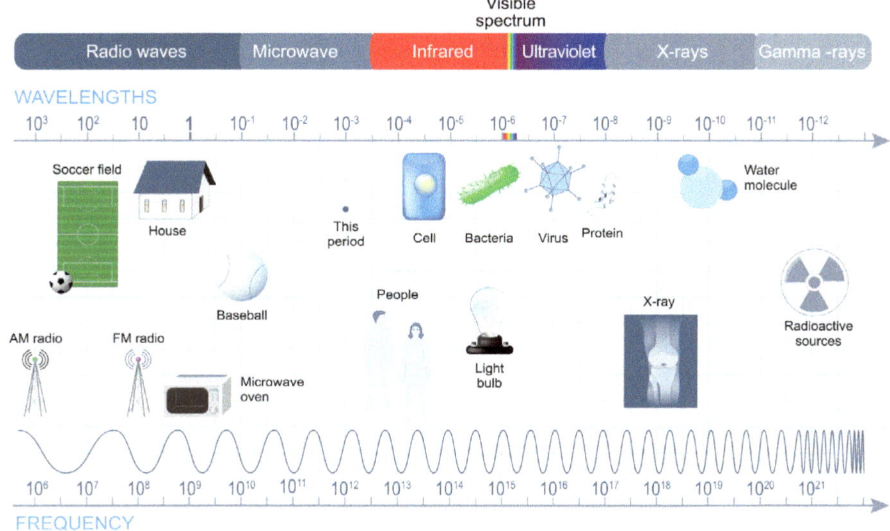

Fig. 7.5 Regions of the electromagnetic spectrum. © Designua/Shutterstock.com

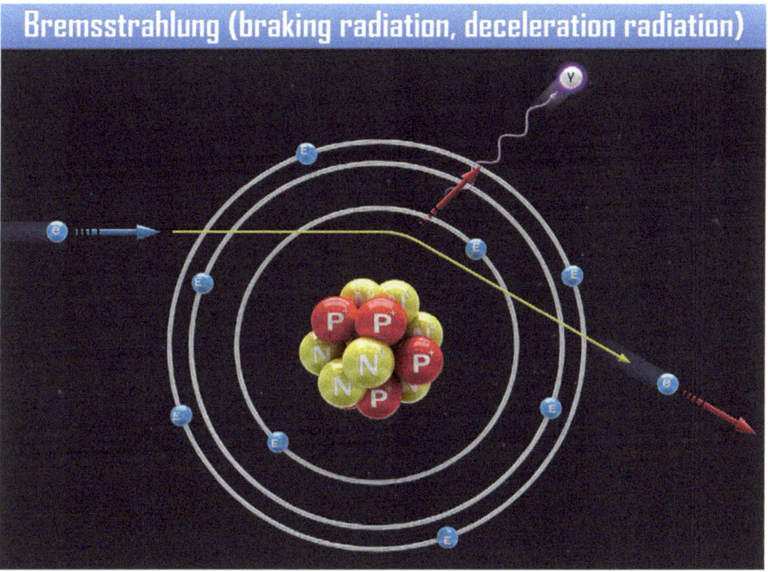

Fig. 7.6 Illustration of Bremsstrahlung radiation from the interaction of an electron with an atom. © General-fmv/Shutterstock.com

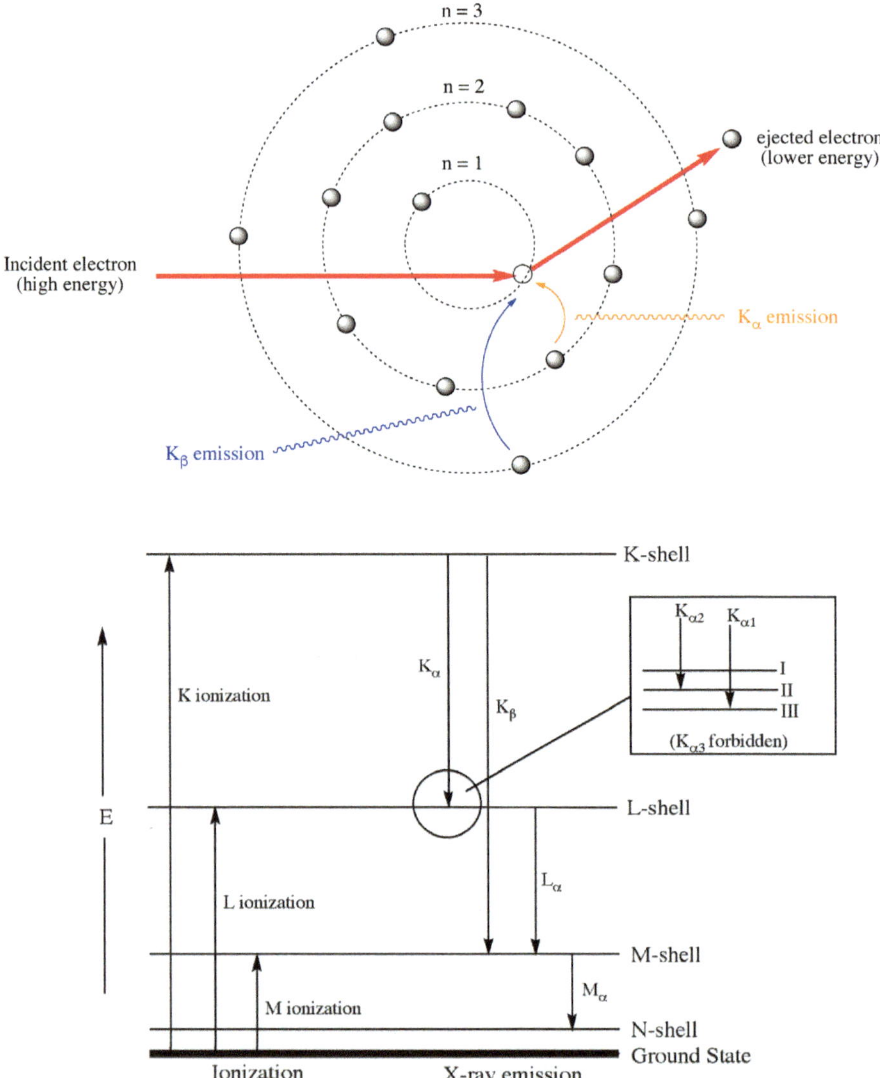

Fig. 7.7 Simplified illustrations of atomic energy levels, showing the characteristic X-rays emitted as the atomic energy returns to the ground state following ionization. The α and β designations indicate the filling of K, L, or M electron vacancies from adjacent ($n + 1 \rightarrow n$) and ($n + 2 \rightarrow n$) energy levels, respectively.[2] The inset shows transitions between the K shell and L subshells

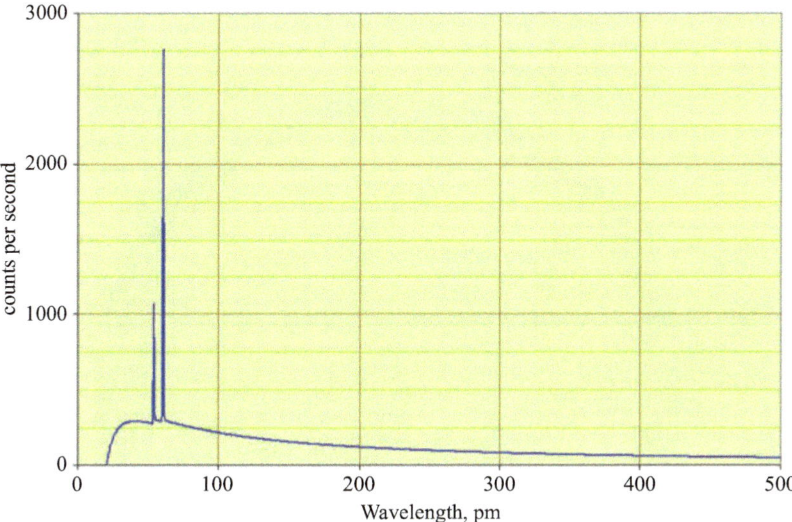

Fig. 7.8 X-ray spectrum of a Rh target operated at 60 kV. The continuous curve is from bremsstrahlung and the spikes are characteristic K-lines for rhodium. © Linguistic Demographer at English Wikipedia

which contains photon energies ranging from microwave to X-ray spectral regions. As illustrated in Fig. 7.9, a variety of straight sectors are located around the storage ring, which gives rise to varying energies of X-ray beamlines that are suitable for a myriad of experimental techniques (Fig. 7.10).[4]

7.2.1 X-ray Diffraction (XRD)

When X-rays ineteract with a crystalline solid, the incoming beam will be diffracted by the 3-D periodic array of atoms/ions to yield a characteristic diffraction pattern on a photographic film or area detector/CCD camera (Fig. 7.11). As first presented in the early 1900s, the most intense diffraction peaks correspond to scattered waves that are in-phase with one another (i.e., constructive interference), which will occur only when the conditions of the Bragg equation are satisfied (7.4 and Fig. 7.12):

$$\mathrm{n}\lambda = 2d\sin\theta, \tag{7.4}$$

where n is an integer, d is the lattice spacing of the crystal planes (e.g., d_{111} would indicate the spacing between neighboring 111 planes), λ is the wavelength of the incoming beam, and θ is the angle of the incident and diffracted beams (i.e., $\theta_{in} = \theta_{out}$).

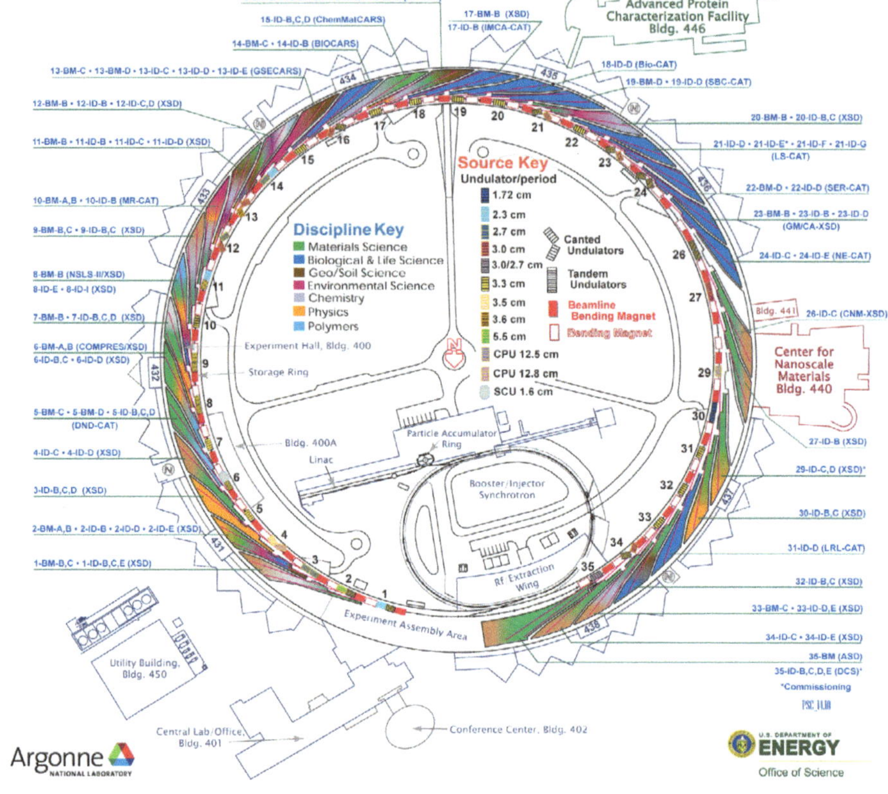

Fig. 7.9 The storage ring and beamlines available at the Advanced Photon Source, Argonne National Laboratory. © U.S. Department of Energy

Illustrated another way, if two waves differ by one whole wavelength, they will differ in phase by 360° or 2π radians. For instance, the phase difference, ϕ, of the (*hkl*) reflection resulting from a wave scattered by an arbitrary lattice atom at position (x, y, z), and another scattered from an atom at position (0, 0, 0) is:

$$\phi = 2\pi(h\mathbf{u} + k\mathbf{v} + l\mathbf{w}), \tag{7.5}$$

where the vector (**u**, **v**, **w**) corresponds to fractional coordinates of (x/**a**, y/**b**, z/**c**).

The two waves may differ in amplitude as well as phase if the two atoms are different. In particular, for scattering in the $2\theta = 0°$ (forward) direction of a wave by an atom comprised of Z electrons, the waves scattered from all of the electrons in the atom will be in phase.[5] Accordingly, the amplitude of the scattered wave is simply Z times the amplitude of the wave scattered by a single electron. The atomic scattering factor, f, is used to describe the scattering by an atom in a given direction (7.6). As just described, $f = Z$ (atomic number) for forward scattering; however, as

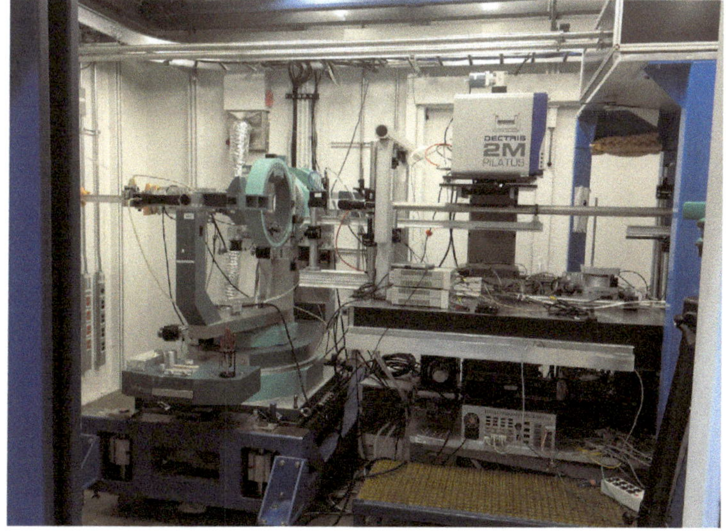

Fig. 7.10 Photos of inside the experiment hall at the Advanced Photon Source, Argonne National Laboratory. The green instrument shown in the bottom image is the sample stage and goniometer of a single-crystal X-ray diffractometer. © Bradley D. Fahlman

$(\sin\theta)/\lambda$ increases, f will decrease due to more destructive interference among the scattered waves.

$$f = \frac{\text{amplitude of wave scattered by one atom}}{\text{amplitude of wave scattered by one electron}} \tag{7.6}$$

Fig. 7.11 Example of diffraction spots from a charge-couple device (CCD) detector, from the single crystal X-ray diffraction analysis of corundum

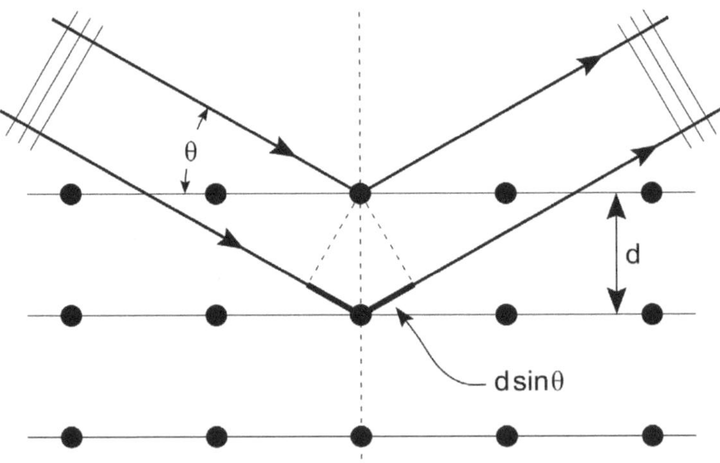

Fig. 7.12 Schematic of Bragg's Law, which governs the conditions required for the constructive interference of waves. Image from Hydrargyrum courtesy of Wikimedia Commons

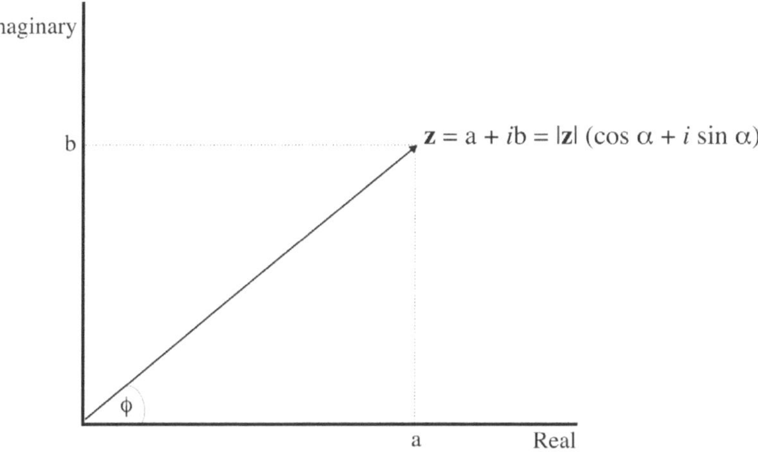

Fig. 7.13 The expression of a vector, **z**, in terms of a complex number

A description of the scattered wavefront resulting from diffraction by a unit cell of the crystal lattice is significantly more complex. That is, one would need to include the contribution of waves scattered by all atoms of the unit cell, each with differing phases and amplitudes in various directions. In order to simplify the trigonometry associated with adding two waves of varying phases/amplitudes, it is best to represent individual waves as vectors.[6] Instead of using **x** and **y** components for the vectors in 2-D real space, one may represent the vectors in *complex space*, with real and imaginary components. This greatly simplifies the system; that is, the addition of scattered waves is simply the addition of complex numbers, which completely removes trigonometry from the determination.

Complex numbers are often expressed as the sum of a real and an imaginary number of the form a + *i*b (Fig. 7.13). It should be noted that the vector length represents the wave amplitude, A; the angle the vector makes with the horizontal (real) axis represents its phase, ϕ (7.7, Euler's equation). The intensity of a wave is proportional to the square of its amplitude, which may be represented by (7.8), obtained by multiplying the complex exponential function by its complex conjugate (replacing *i* with -*i*).

$$\mathrm{A}e^{i\phi} = A\cos\phi + Ai\sin\phi \tag{7.7}$$

$$\left|\mathrm{A}e^{i\phi}\right|^2 = Ae^{i\phi}Ae^{-i\phi} \tag{7.8}$$

If we now include the phase and scattering factor expressions we used in (7.5) and (7.6), we will obtain an equation that fully describes the scattering of the wave from a lattice atom, in complex exponential form (7.9):

$$\mathrm{A}e^{i\phi} = fe^{2\pi i(h\mathbf{u}+k\mathbf{v}+l\mathbf{w})} \tag{7.9}$$

If we want to describe the resultant wave scattered by all atoms of the unit cell, we will need to replace the atomic scattering factor with the *structure factor*, *F*. This factor describes how the atomic arrangement affects the scattered beam and is obtained by adding together all waves scattered by the discrete atoms. Hence, if a unit cell contains N atoms with fractional coordinates ($\mathbf{u}_n$, $\mathbf{v}_n$, $\mathbf{w}_n$), and atomic scattering factors f_n, the structure factor for the *hkl* reflection would be given by:

$$F_{hkl} = \sum_{n=1}^{N} f_n e^{2\pi i (h\mathbf{u}_n + k\mathbf{v}_n + l\mathbf{w}_n)} \tag{7.10}$$

The intensity of the diffracted beam by all atoms of the unit cell in which Bragg's Law is upheld is proportional to $|F|^2$, which is obtained by multiplying the expression given by (7.10) by its complex conjugate. Hence, this equation is invaluable for diffraction studies, as it allows one to directly calculate the intensity of any *hkl* reflection if one knows the atomic positions within the unit cell. For instance, let's consider the structure factor for a bcc unit cell, with atoms at (0, 0, 0) and (1/2, 1/2, 1/2). The expression for the bcc structure factor is given by (7.11):

$$F_{hkl} = f e^{2\pi i (0)} + f e^{2\pi i (\frac{h}{2} + \frac{k}{2} + \frac{l}{2})} = f\left[1 + e^{\pi i (h+k+l)}\right] \tag{7.11}$$

Since $e^{n\pi i} = -1$ or $+1$, where n = odd integer or even integer, respectively, $|F|^2 = (2f)^2$ when $(h + k + l)$ is even, and $|F|^2 = 0$ when $(h + k + l)$ is odd. That is, for a bcc unit cell, all reflections with $(h + k + l) = 2n + 1$, where n = 0, 1, 2, ... such as (100), (111), (120), etc. will be absent from the diffraction pattern, whereas reflections with $(h + k + l) = 2n$ such as (110), (121), etc. may be present if the diffraction satisfies Bragg's Law. Such missing reflections are referred to as *systematic absences* and are extremely diagnostic regarding the centering present in the unit cell, as well as the presence of translational symmetry elements such as screw axes or glide planes (Table 7.1).

Another way to represent the structure factor is shown in (7.12), where $\rho(\boldsymbol{r})$ is the electron density of the atoms in the unit cell ($\boldsymbol{r}$ = the coordinates of each point in vector notation). As you may recall, this is in the form of a Fourier transform; that is, the structure factor and electron density are related to each other by Fourier/inverse Fourier transforms (7.13). Accordingly, this relation is paramount for the determination of crystal structures using X-ray diffraction analysis. This equation enables one to prepare a 3-D electron density map for the entire unit cell, in which maxima represent the positions of individual atoms.[8]

$$F_{hkl} = \int_V \rho(\boldsymbol{r}) e^{2\pi i (h\mathbf{u} + k\mathbf{v} + l\mathbf{w})} \mathrm{d}V \tag{7.12}$$

$$\rho(r) = \int_V F_{hkl} e^{-2\pi i (h\mathbf{u} + k\mathbf{v} + l\mathbf{w})} \mathrm{d}V = \frac{1}{\nabla} \sum_h \sum_k \sum_l F_{hkl} e^{-2\pi i (h\mathbf{u} + k\mathbf{v} + l\mathbf{w})} \tag{7.13}$$

Table 7.1 Systematic absences for X-ray and electron diffraction

Cause of absence	Symbol	Absences[a]
Body-centering	I	$h + k + l = 2n + 1$ (odd)
A centering	A	$l + k = 2n + 1$
B centering	B	$h + l = 2n + 1$
C centering	C	$h + k = 2n + 1$
Face centering	F	*hkl* mixed (not all even or all odd)
Glide plane ⊥ (100) (0*kl*)	b c n d	$k = 2n + 1$ $l = 2n + 1$ $k + l = 2n + 1$ $k + l = 4n + 1$
Glide plane ⊥ (010) (*h*0*l*)	a c n d	$h = 2n + 1$ $l = 2n + 1$ $h + l = 2n + 1$ $h + l = 4n + 1$
Glide plane ⊥ (001) (*hk*0)	a b n d	$h = 2n + 1$ $k = 2n + 1$ $h + k = 2n + 1$ $h + k = 4n + 1$
Glide plane ⊥ (110) (*hhl*)	b n d	$h = 2n + 1$ $h + l = 2n + 1$ $h + k + l = 4n + 1$
Screw axis // **a** (*h*00)	2_1 or 4_2 4_1 or 4_3	$h = 2n + 1$ $h = 4n + 1$
Screw axis // **b** (0*k*0)	2_1 or 4_2 4_1 or 4_3	$k = 2n + 1$ $k = 4n + 1$
Screw axis // **c** (00 *l*)	2_1 or 4_2 or 6_3 3_1, 3_2, 6_2 or 6_4 4_1 or 4_3 6_1 or 6_5	$l = 2n + 1$ $l = 3n + 1$ $l = 4n + 1$ $l = 6n + 1$
Screw axis // (110)	2_1	$h = 2n + 1$

Refers to the Miller indices ((*hkl*) values) that are absent from the diffraction pattern. For instance, a body-centered cubic lattice with no other screw axes and glide planes will have a zero intensity for all reflections where the sum of $(h + k + l)$ yields an odd #, such as (100), (111), etc.; other reflections from planes in which the sum of their Miller indices are even, such as (110), (200), (211), etc. will be present in the diffraction pattern. As these values indicate, there are three types of systematic absences: 3-D absences (true for all *hkl*) resulting from pure translations (cell centering), 2-D absences from glide planes, and 1-D absences from screw axes[7]

It is noteworthy that the units of functions related by a Fourier transform are reciprocals of one another. For instance, consider the reciprocal relationship between the period and frequency of a wave. Whereas the former is the time required for a complete wave to pass a fixed point (units of sec), the latter is the number of waves passing the point per unit time (units of s^{-1}). Similarly, the unit of the structure factor (Å) is the inverse of the electron density (Å^{-1}). Since the experimental diffraction pattern yields intensity data (equal to $|F|^2$), the spatial dimensions represented by the diffraction pattern will be inversely related to the original crystal lattice.

Accordingly, the observed diffraction pattern represents a mapping of a secondary lattice known as a *reciprocal lattice*; conversely, the "real" crystal lattice may only be directly mapped if one could obtain high enough resolution for an imaging device such as an electron microscope. The reciprocal lattice is related to the real crystalline lattice by the following (as illustrated in Fig. 7.14):

(i) For a 3-D lattice defined by vectors **a**, **b**, and **c**, the reciprocal lattice is defined by vectors **a***, **b***, and **c*** such that **a*** ⊥ **b** and **c**, **b*** ⊥ **a** and **c**, and **c*** ⊥ **a** and **b**:

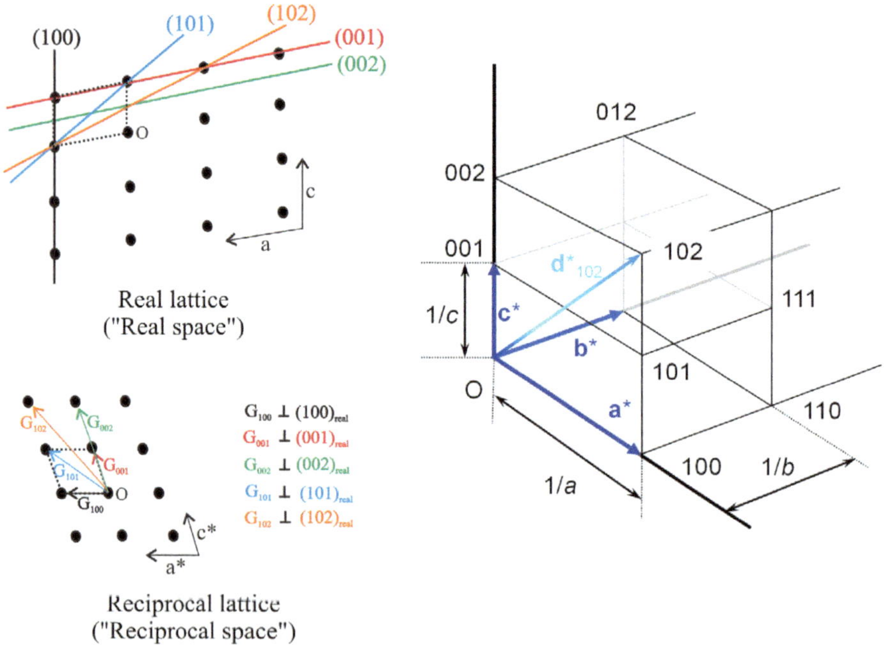

Fig. 7.14 Comparison between a real monoclinic crystal lattice (a = b ≠ c) and the corresponding reciprocal lattice. Dashed lines indicate the unit cell of each lattice. The magnitudes of the reciprocal lattice vectors are not in scale; for example, $|a^*| = 1/d_{100}$, $|c^*| = 1/d_{001}$, $|G_{101}| = d_{101}$, etc. Note that for orthogonal unit cells (cubic, tetragonal, orthorhombic), the reciprocal lattice vectors will be aligned parallel to the real lattice vectors. Reproduced with permission from *Biomolecular Crystallography: Principles, Practice, and Application to Structural Biology* by Bernard Rupp. © 2009 Taylor & Francis Group LLC

$$\mathbf{a}^* = \frac{\boldsymbol{b} \times \boldsymbol{c}}{[\boldsymbol{a} \cdot (\boldsymbol{b} \times \boldsymbol{c})]} \quad \mathbf{b}^* = \frac{\boldsymbol{a} \times \boldsymbol{c}}{[\boldsymbol{a} \cdot (\boldsymbol{b} \times \boldsymbol{c})]} \quad \mathbf{c}^* = \frac{\boldsymbol{b} \times \boldsymbol{c}}{[\boldsymbol{a} \cdot (\boldsymbol{b} \times \boldsymbol{c})]} \tag{7.14}$$

Recall that the cross-product of two vectors separated by θ results in a third vector that is aligned in a perpendicular direction as the original vectors. For instance, for $\mathbf{b} \times \mathbf{c}$, the magnitude of the resultant vector would be $|\mathbf{b}||\mathbf{c}| \sin \theta$, and would be aligned along the **a**-axis. This is in contrast to the dot-product of two vectors, which results in the scalar projection of one vector onto the other, of magnitude $|\mathbf{b}||\mathbf{c}| \cos \theta$ for $\mathbf{b} \cdot \mathbf{c}$. For instance, for a cubic unit cell, the denominator terms in (7.14), $[\mathbf{a} \cdot (\mathbf{b} \times \mathbf{c})]$, would simplify to:

$$\boldsymbol{b} \times \boldsymbol{c} = |\boldsymbol{b}||\boldsymbol{c}| \sin 90^\circ = |\boldsymbol{b}||\boldsymbol{c}|, \text{ aligned along the a axis;}$$
$$\therefore [\boldsymbol{a} \cdot \boldsymbol{b} \times \boldsymbol{c}] = |\boldsymbol{a}||\boldsymbol{b}||\boldsymbol{c}| \cos 0^\circ = |\boldsymbol{a}||\boldsymbol{b}||\boldsymbol{c}|, \text{ or the } volume\ of\ the\ unit\ cell$$

As you might have noticed, the magnitude of a reciprocal lattice vector is in units of [1/distance], relative to the basis vectors in real space that have units of distance.

(ii) A vector joining two points of the reciprocal lattice, **G**, is perpendicular to the corresponding plane of the real lattice (Fig. 7.14). For instance, a vector joining points (1, 1, 1) and (0, 0, 0) in reciprocal space will be perpendicular to the {111} planes in real space, with a magnitude of $1/d_{111}$. As you might expect, one may easily determine the interplanar spacing for sets of real lattice planes by taking the inverse of that represented in (7.14)—e.g., the distance between adjacent planes of {100} in the real lattice is:

$$\frac{[\boldsymbol{a} \cdot (\boldsymbol{b} \times \boldsymbol{c})]}{|\boldsymbol{b} \times \boldsymbol{c}|} \tag{7.15}$$

In order to more completely understand reciprocal space relative to "real" space defined by Cartesian coordinates, we need to first recall the deBroglie relationship that describes the wave-particle duality of matter:

$$\lambda = \frac{h}{\mathbf{p}} \text{ or } \mathbf{p} = \frac{h}{\lambda} \tag{7.16}$$

where h is Planck's constant (6.626×10^{-34} J s) and **p** is the momentum.

Accordingly, if we represent a wave as a vector as described previously, the *wavevector*, **k**, will have a magnitude of:

$$|\mathbf{k}| = \frac{|\mathrm{p}|}{\hbar} = \frac{2\pi}{\lambda} \tag{7.17}$$

where $\hbar$ is the reduced Planck's constant ($h/2\pi$).

As one can see from (7.17), the magnitude of the wavevector, **k**, will have units proportional to [1/distance], analogous to reciprocal lattice vectors. This is no accident; in fact, solid-state physicists refer to reciprocal space as *momentum space* (or *k-space*), where long wavevectors correspond to large momenta and energy, but small wavelengths. This definition is paramount to understanding the propagation of electrons through solids.

In order to determine which lattice planes give rise to Bragg diffraction, a geometrical construct known as an *Ewald sphere* is used. This is simply an application of the law of conservation of momentum, in which an incident wave, **k**, impinges on the crystal. The Ewald sphere (or circle in two-dimensional) shows which reciprocal lattice points, (each denoting a set of planes) which satisfy Bragg's Law for diffraction of the incident beam. A specific diffraction pattern is recorded for any **k**-vector and lattice orientation—usually projected onto a two-dimensional film or CCD camera. One may construct an Ewald sphere as follows (Fig. 7.15):

(i) Draw the reciprocal lattice from the real lattice points/spacings.
(ii) Draw a vector, **k**, which represents the incident beam. The end of this vector should touch one reciprocal lattice point, which is labeled as the origin (0, 0, 0).
(iii) Draw a sphere (circle in 2-D) of radius $|\mathbf{k}| = 2\pi/\lambda$, centered at the start of the incident beam vector.
(iv) Draw a scattered reciprocal lattice vector, **k'**, from the center of the Ewald sphere to any point where the sphere and reciprocal lattice points intersect.

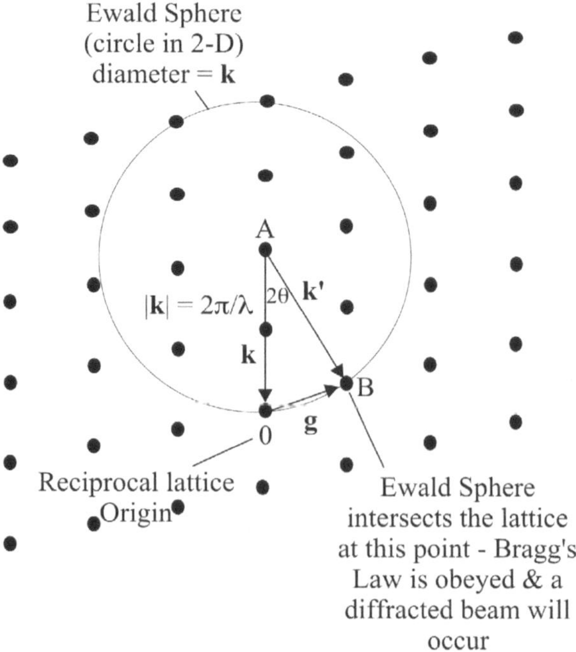

Fig. 7.15 Illustration of Ewald sphere construction, and diffraction from reciprocal lattice points. This holds for both electron and X-ray diffraction methods. The vectors AO, AB, and OB are designated as an incident beam, a diffracted beam, and a diffraction vector, respectively

(v) Diffraction will result for any reciprocal lattice point that crosses the Ewald sphere such that **k'** = **k** + **g**, where **g** is the scattering vector. Stated more succinctly, Bragg's Law will be satisfied when **g** is equivalent to a reciprocal lattice vector. Since the energy is conserved for elastic scattering of a photon, the magnitudes of **k'** and **k** will be equivalent. Hence, using the Pythagorean theorem, we can re-write the diffraction condition as:

$$(\mathbf{k}+\mathbf{g})^2 = |\mathbf{k}|^2, \text{ or } 2\mathbf{k}\cdot\mathbf{g}+\mathbf{g}^2 = 0 \tag{7.18}$$

Since $\mathbf{g} = h\mathbf{a}^* + k\mathbf{b}^* + l\mathbf{c}^*$, with a magnitude of $2\pi/d_{hkl}$ for the (*hkl*) reflection, (7.18) can be re-written in terms that will yield Bragg's Law in more familiar terms (*c.f.* (7.4)):

$$2\left(\frac{2\pi}{\lambda}\right)\left(\frac{2\pi}{d_{hkl}}\right)\sin\theta = \frac{4\pi^2}{d_{hkl}^2}, \text{ or } \frac{2\sin\theta}{\lambda} = \frac{1}{d_{hkl}} \tag{7.19}$$

In general, very few reciprocal lattice points will be intersected by the Ewald sphere,[9] which results in few sets of planes that give rise to diffracted beams. As a result, a single crystal will usually yield only a few diffraction spots. As previously seen in Fig. 7.6, a single-crystalline specimen will yield sharp diffraction spots. In contrast, a polycrystalline sample will yield many closely spaced diffraction spots, whereas an amorphous sample will give rise to diffuse rings (Fig. 7.16). Since the wavelength of an electron is much smaller than an X-ray beam, the Ewald sphere (radius: $2\pi/\lambda$) is significantly larger for electron diffraction relative to X-ray diffraction studies. As a result, electron diffraction yields much more detailed structural information of the crystal lattice.

Phase identification using X-ray diffraction primarily relies on peak positions, which may be compared to the d-spacings of pure compounds (Fig. 7.17). However, the shapes of peaks are also important to measure the degree of thermal oscillations of atoms, as well as the presence of vacancies or impurities in the sample.[10] Crystallite size also effects the broadness of peaks. The *Scherrer equation* (7.20) describes peak broadening as a function of incident beam divergence, which makes it possible to satisfy the Bragg equation for non-adjacent diffraction planes.

$$B(2\theta) = \frac{K\lambda}{L\cos\theta}, \tag{7.20}$$

where B is the peak width (corrected for instrumental broadening; units of 2θ), K is the Scherrer constant,[11] λ is the wavelength, L is crystallite size, and θ is the Bragg angle. As shown in Figs. 7.17 and 7.18, significant peak broadening occurs for crystallites with dimensions <100 nm.

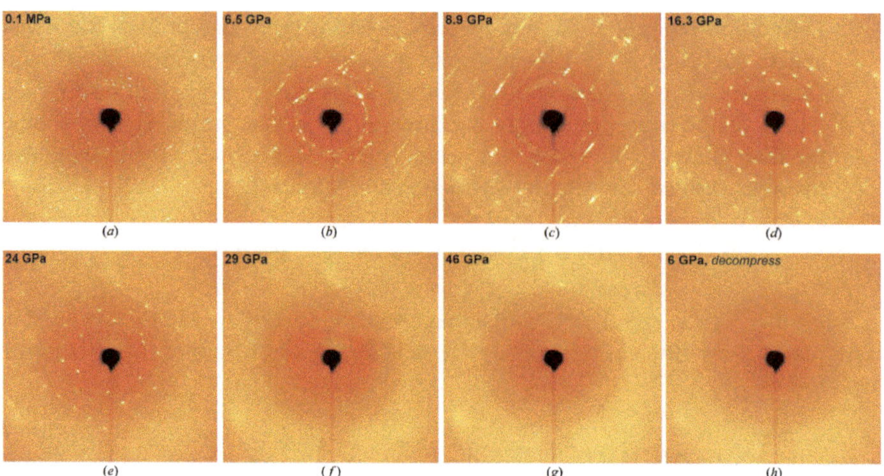

Fig. 7.16 Diffraction patterns collected from the tridymite phase of SiO_2 compressed to different pressures at room temperature. The polycrystalline sample at ambient pressure (**a**–**c**) gradually transforms to a single crystal at a pressure of 16 GPa (**d**). Further compression results in disordering of the crystal lattice (**e**–**f**), eventually resulting in diffuse scattering rings, characteristic of an amorphous morphology at pressures of 46 GPa and higher (**g**–**h**). Reproduced with permission from *J. Synchrotron Rad.* **2005**, *12*, 560. © 2005 International Union of Crystallography

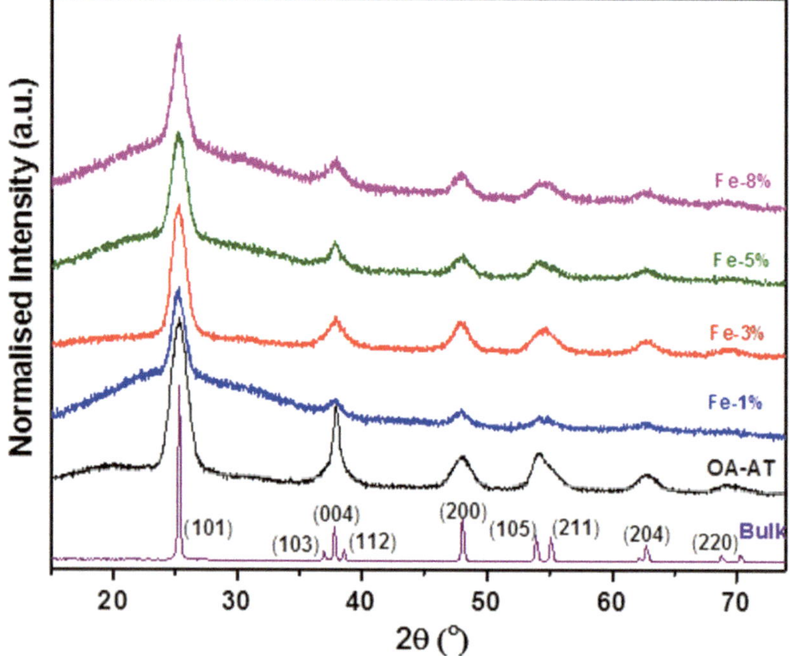

Fig. 7.17 Powder X-ray diffractograms for Fe-doped TiO_2 nanoparticles, with peaks indexed to various planes of anatase-TiO_2. © Bradley D. Fahlman/Swati Naik

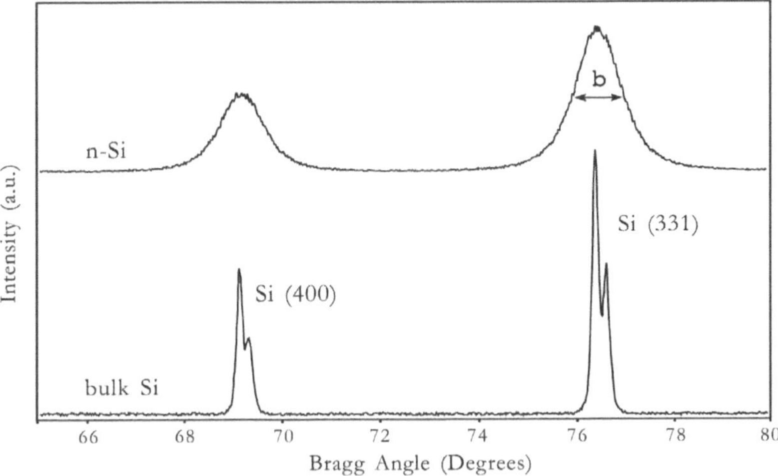

Fig. 7.18 Comparison of peak profiles for 90-nm Si nanocrystals and bulk Si, showing peak broadening for nanosized crystallites. © 2002 H & M Analytical Services, Inc.

7.2.2 Small-Angle X-ray Scattering (SAXS)

Structural information such as conformational changes of noncrystalline macromolecules may be determined by *small-angle scattering* (SAS) techniques.[12] This method examines the patterns arising from the elastic scattering of X-rays (SAXS) or neutrons (SANS) from sample components. Whereas X-ray diffraction features diffraction angles of 4–170°, small-angle scattering measurements feature scattering angles of <4°, which corresponds to structures with large d-spacings (10–100 nm). Due to the small angular separation between the incident and scattered beams, large sample-to-detector distances (0.5–10 m) and high-quality collimating optics are required to achieve good signal-to-noise.

Whereas SAXS provides information regarding the electron density distribution of the sample, SANS is sensitive toward the sample nuclei. For instance, ^{1}H and ^{2}H (deuterium) exhibit different scattering lengths; hence, many studies perform isotopic labeling to gain additional structural information about the sample.[13] The larger the scattering angle, the smaller the length scale that may be probed; hence, typical X-ray diffraction, also known as *wide-angle X-ray scattering* (WAXS) is used to determine structural information on the atomic length scale, and SAXS/SANS are used in the size regime of *ca.* 1–300 nm (Fig. 7.19). A wide variety of samples may be analyzed by SAS, most often consisting of powders or solvent suspensions of macromolecules or nanoparticles/colloids.[14]

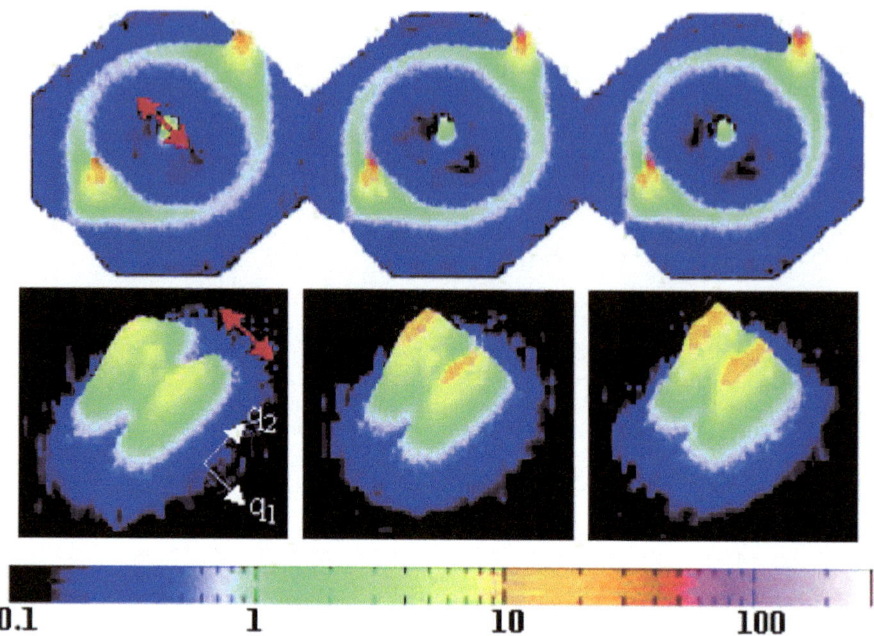

Fig. 7.19 Wide-angle (top) and small-angle (bottom) X-ray scattering patterns of drawn polyethylene fibers, annealed for 264 h at 23, 80, and 100 °C (from left to right). The red arrows indicate the drawing direction. The peaks in the WAXS pattern remain unchanged during annealing, whereas the pattern changes significantly for SAXS. This is proposed to indicate the preferential orientation of the polymeric chains along the drawing direction, whereas the lamella in the samples exhibit slight shear during annealing. Reproduced with permission from *J. Phys. Chem. B* **2005**, *109*, 16,650. © 2005 American Chemical Society

7.3 Optical Microscopy

Of the many techniques available for the analysis of solid materials, perhaps the simplest is optical microscopy. Two modes of optical microscopy are typically employed based on the measurement of transmitted (transparent sample) or reflected light (opaque sample) (Fig. 7.20). Often, a microscope is fitted with both modes, allowing one to analyze both types of samples. Most of the solid-state materials discussed in earlier chapters are nontransparent in their as-grown/as-deposited states. Further, it is usually difficult to prepare thin cross sections for transmission microscopy. Hence, materials scientists typically employ the reflection mode, also known as *episcopic light differential interference contrast (DIC) microscopy*.[15] This technique is useful for imaging of a variety of reflective samples including minerals, metals, semiconductors, glasses, polymers, and composites. The semiconductor industry relies heavily on reflective DIC imaging for quality assessment of computer chip components.

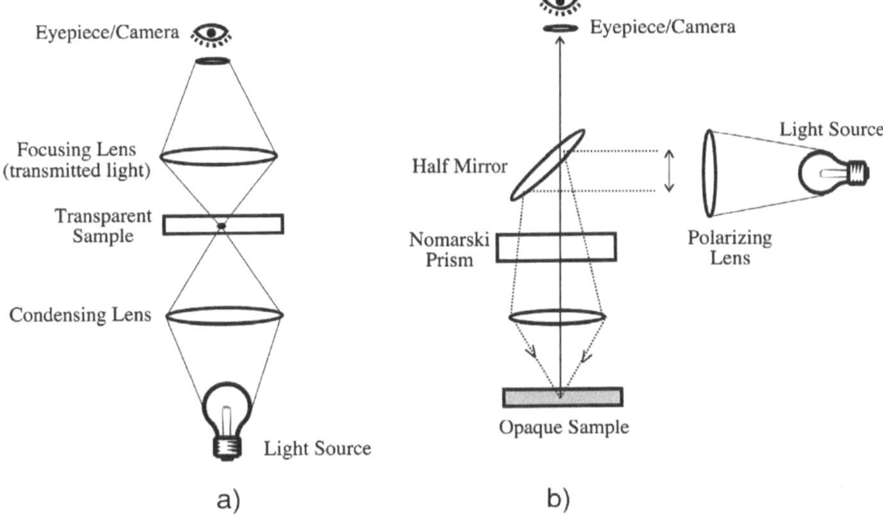

Fig. 7.20 Schematic of the components of **a** transmitted light optical microscope and **b** reflected light optical microscope

Surface artifacts such as depressions and particulates create optical path differences in the reflected beam. Formation of the final image is the result of interference between two distinct wave fronts that reach the image plane slightly out of phase with each other. Unlike the situation with transmitted light and semitransparent phase specimens, the image created in reflected light DIC can often be interpreted as a true three-dimensional representation of the surface geometry, provided a clear distinction can be realized between raised and lowered regions in the specimen. Oftentimes, the reflected wave fronts emanating from the sample are separated by only fractions of a micrometer, which is much less than the resolution of the objective.

The ability to discern fine details within a magnified image is referred to as the *resolution* of a microscope. Since light is used as the illumination source in optical microscopy, the resolution is expressed in the same units as the wavelength of light (nm). The theoretical resolution, *R*, of any optical system may be calculated using Abbe's equation (7.21):

$$R = \frac{0.61\lambda}{\eta(\sin\theta)} \tag{7.21}$$

where:

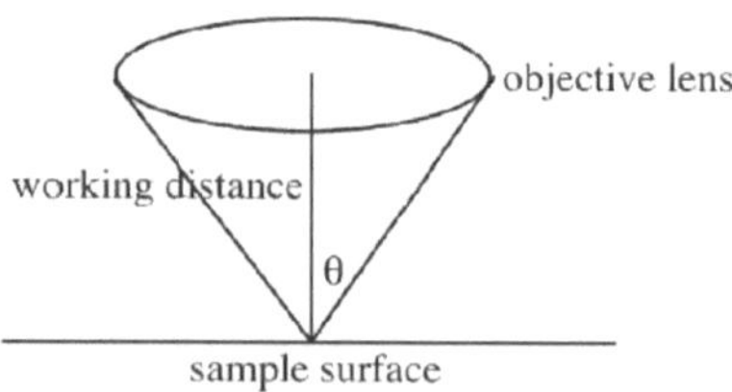

The denominator of (7.21) represents the numerical aperature (NA) of the objective lens, related to its light-gathering ability. Other primary factors that influence the resolution of a lens is the wavelength of light used, the index of refraction (η) of the environment surrounding the lens (e.g., 1.00 for air), and the angle of illumination (θ).

Examining the above equation, one can see that a resolution limit will be reached using visible light (350–700 nm) as the illumination source. That is, using high numerical apertures (e.g., 1.3–1.4 for oil-immersion lenses where $\eta = 1.5$), the theoretical resolution using polychromatic visible light (*ca.* 500 nm) is on the order of 220 nm. Hence, any sample features that are less than 220 nm apart from one another will appear blurry. It should be noted that the calculated resolution represents the best possible cutoff for clear discernment of small features. In practice, the observed resolution is often worse than the theoretical value, depending on the degree of optical aberrations that are inherent in the lens.

It may be seen by (7.21) that higher spatial resolutions (i.e., smaller R values) are possible through use of shorter wavelengths. To illustrate this concept, subsequent sections of this chapter will examine the high resolutions inherent in microscopes that use an electron beam rather than visible light. However, we first must ask ourselves whether it is possible to improve the resolution limits of optical microscopy. If this is possible, the cost of such a modification would be far less than the price of electron microscopes (currently \$300 K–\$1.5 M).

In the late 1920s, Edward H. Synge published a series of articles that conceptualized the idea of an "ultrahigh resolution" optical microscope.[16] His original idea proposed using a screen with an aperture of dimensions much smaller than wavelength of the illuminating source. Upon irradiating the screen with a high-intensity light source, the light is confined to the dimensions of the hole. If this hole is placed in close proximity (nanometer regime) to the sample surface, the light emerging from the aperture could be used to image a specimen before it had time to spread out. By comparison, the distance between the light source and sample in conventional optical microscopy is on the order of millimeters. Not unlike other visionary ideas ahead of their time, the technical difficulties associated with Synge's idea were only recently overcome at IBM in the mid-1980s, which marked the rediscovery of the technique now referred to as *near-field scanning optical microscopy* (NSOM, Fig. 7.21).[17]

A variety of apertures have been used to deliver nanometer-sized spots of light. While early NSOM tips were fabricated out of etched quartz crystals and micropipettes, tapered optical fibers with tip diameters of *ca.* 100 nm are now typically used. A metallic thin film such as aluminum is usually applied around the sides of the tapered region of the NSOM tip to focus the light toward the sample. For optical

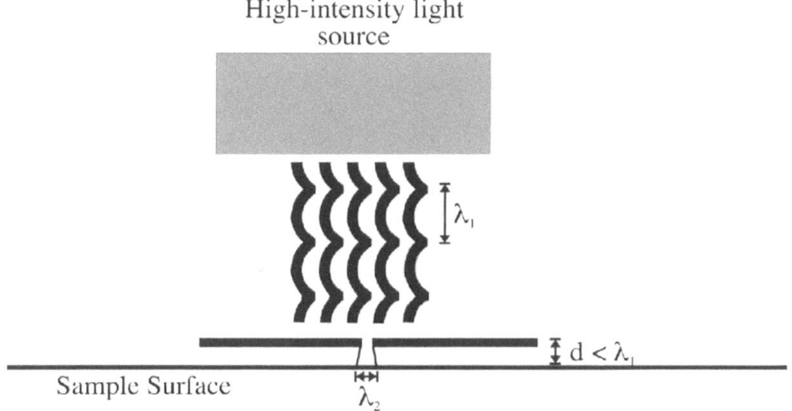

Fig. 7.21 Schematic of operating principle of near-field scanning optical microscopy (NSOM). The resulting wavelength of light impinging on the sample (λ_2) has a wavelength much smaller than the illuminating source, resulting in much higher resolution

fibers, the numerical aperture is related to the difference in the indices of refraction of the cladding and core (7.22):

$$\text{NA} \propto \sqrt{(\eta_{\text{core}})^2 - (\eta_{\text{cladding}})^2} \tag{7.22}$$

In order to obtain high-resolution optical images with NSOM, the tip must be stabilized within a few nanometers of the sample surface during scanning. Techniques to precisely position the tip above the sample were developed by researchers in 1992, resulting in more applications for both imaging and topographic profiling of surfaces. This latter capability will be discussed in more detail later in this chapter when discussing scanning probe microscopy (SPM).

7.4 Electron Microscopy

Although optical microscopy may be extended into the nanoregime, other techniques must be used to clearly discern components below 100 nm. Indeed, the current "nanotechnology revolution" that we are experiencing would not have been possible if there were not suitable techniques in order to characterize nanomaterials. As we saw in the previous section, in order to improve resolution, we must use source radiation with as small a wavelength as possible.

Louis DeBroglie was the first to assert that matter, like light, could be described as having both particle and wave characteristics (7.23). However, the application of this equation for a macroscopic item such as a golf ball in flight[18] predicts a wavelength that is too small to be meaningful (7.24). By comparison, subatomic particles have wavelengths comparable to the size of an atom. In particular, electrons that are accelerated in a potential, V, have quantifiable wavelengths[19]; for example, electrons accelerated at 100 kV have a wavelength of 0.037 Å (7.25). This is significantly shorter than the wavelengths of high-energy radiation sources such as X-rays (e.g., 1.54 Å for Cu K_α X-rays), or neutrons (*ca.* 1 nm). In addition to using electrons, it is also possible to use neutrons[20] or X-rays[21] for imaging. Although the resolution of X-ray microscopy is typically on the order of 30 nm, a hard X-ray microscope developed by Brookhaven National Laboratory and Columbia University was used to perform nondestructive 3-D imaging of nanoparticles at 7-nm resolution.[22]

$$\lambda = \frac{h}{mv} \tag{7.23}$$

$$\frac{6.626 \times 10^{-34}\ \text{J s}}{(0.045\ \text{kg})(41\ \text{m s}^{-1})} = 3.59 \times 10^{-34}\ \text{m} \tag{7.24}$$

$$\lambda = \frac{12.3\text{Å}}{\sqrt{V + \frac{V^2}{1 \times 10^6}}} \tag{7.25}$$

There are two types of scattering interactions that are possible when an electron beam impinges on atoms of the sample (Fig. 7.22). Interaction between the electron source and atomic nuclei gives rise to *elastic scattering*, which results in a large-angle deflection of the electron beam with little or no energy loss. By comparison, electron–electron interactions between the source and electron clouds of individual atoms cause small-angle deflections with a significant loss of energy, known as *inelastic scattering*. It should be noted that there is a finite probability that the electrons may continue their original trajectory through the atomic structure without scattering. In particular, this is commonly observed with molecules containing light atoms such as H, Li, B, etc. As you would expect, a solid consisting of a high density of large atoms such as metallic solids would result in shorter elastic mean free paths, and stronger elastic scattering interactions with incident electrons. By comparison, organic solids such as polyethylene would primarily exhibit inelastic scattering with an electron source.

As its name implies, *transmission electron microscopy* (TEM) measures the intensity from an electron source after it has passed through a transparent sample (Fig. 7.23a). In contrast, *scanning electron microscopy* (SEM) features the scanning of an electron beam over selected regions of an opaque sample (Fig. 7.23b). The ability to focus the extremely small incident wavelengths of energetic electrons results in an unprecedented spatial resolution compared to optical based microscopy techniques (Fig. 7.24). Since the resolution of our eyes is on the order of 0.1 mm, in order for us to observe these features, an appropriate level of magnification must be used. That is, for the increasingly greater resolution capabilities of modern SEMs and TEMs, the sample must be magnified by *ca.* 100–300 K and 1.5–2 M times, respectively (Fig. 7.25). However, it should be noted that although the wavelength of

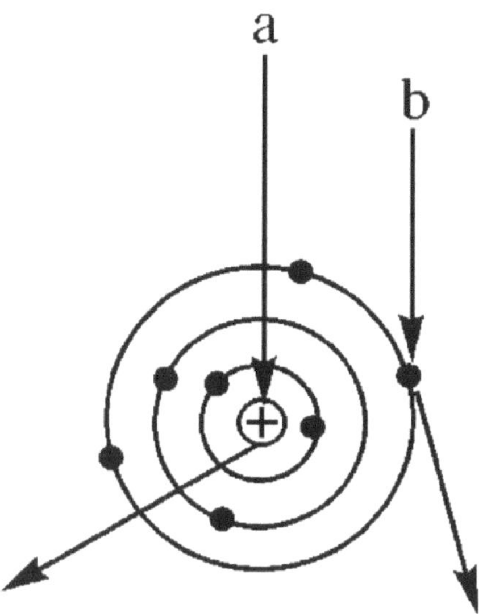

Fig. 7.22 Schematic of specific interactions between the electron beam and sample atom. Shown are **a** an elastically scattered electron and **b** an inelastically scattered electron

electrons are *ca.* 100,000 times smaller than visible light, the resolving power of electron microscopes are only on the order of 1,000 times greater than light microscopes, due to lens aberrations (especially spherical aberration).[23]

Images obtained from electron microscopy are due to the nature/degree of electron scattering from the constituent atoms of the sample. Table 7.2 provides a comparison between electron, X-ray, and neutron sources, pertaining to their utility for scattering applications. Electrons are much less penetrating than X-rays, and are readily absorbed by air, dictating the use of ultrahigh vacuum (UHV) environments. Since the mass of an electron is 1,000 times smaller than the lightest nucleus, electrons are scattered much more intensely than X-rays or highly penetrating neutrons. This results in high sensitivity from the interaction of electrons with extremely small samples.

The basic principles that govern electron microscopy are analogous to optical microscopy. Whereas optical microscopes use light and optical lenses to illuminate and magnify the sample, electron microscopes utilize high-energy electrons and electromagnetic lenses. There are two types of lenses in electron microscopes (see below); TEM has an additional lens, the projector lens, which is used to project an amplified copy of the image onto a screen:

(i) *Condenser*—used to control the illumination of the sample through concentration of the electron beam generated from the source.
(ii) *Objective*—used to magnify the sample (in TEM, the sample is inserted into the objective lens).
(iii) *Projector*—used to project an amplified copy of the image onto a screen (photographic plate, computer screen, fluorescent screen, CCD camera/monitor).

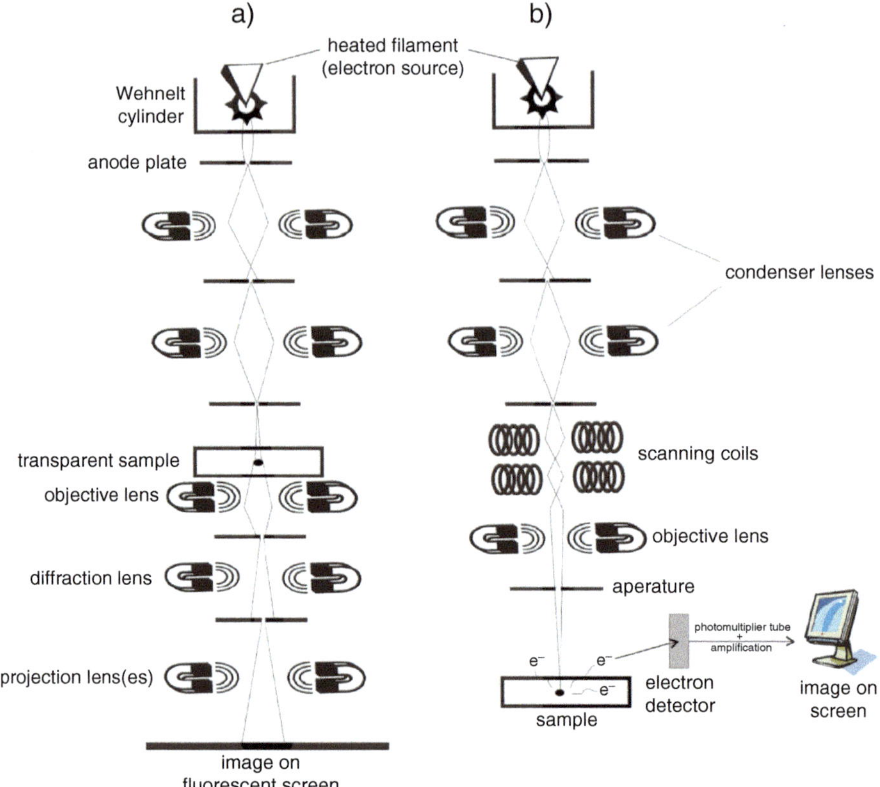

Fig. 7.23 Schematic of the major components of **a** transmission electron microscope (TEM) and **b** scanning electron microscope (SEM). *Note* stigmator coils are also associated with the lens(es) that reduce beam astigmatism, thereby improving image distortion and resolution. For TEM, the image is shown to be projected onto a viewing screen. This screen is coated with phosphorescent particles such as ZnS and CdS that emit visible light when struck by electrons (hard copies of the image may also be transferred to photographic plates via silver halide emulsions). However, modern TEM instruments do not require photographic development, as digital images are observed directly on computer monitors, in real time

Electron sources

The heart of any electron microscope is the electron source, or gun. In fact, the electron gun within microscopes is identical to that operating within modern televisions. Behind the TV monitor lays a cathode ray tube (CRT), containing a set of three tungsten filaments. The electrons generated from these filaments are accelerated toward an anode that is coated with phosphor particles, which emit either red, green, or blue light when struck with electrons. This light is focused onto the individual pixels of the monitor; if you look closely at a television screen, you may notice the millions of tiny dots that comprise individual images. The picture we see on our monitors is a result of the electron guns being scanned sequentially from left to right in a fraction

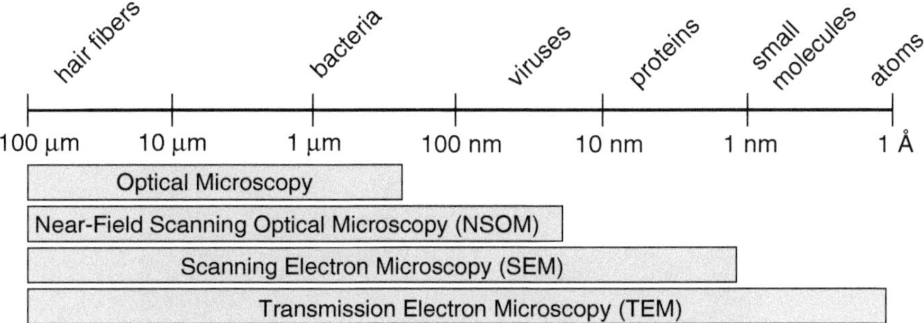

Fig. 7.24 Comparison of the characterization size regimes for optical and electron microscopy

of a second, a process known as *rastering*. This is the exact mode that is operable in SEMs and scanning transmission electron microscopes (STEMs), where the focused electron beam is allowed to scan across selected regions of a sample.

Electrons are generated by passing high currents through tungsten or crystalline LaB_6 (or CeB_6) filaments, resulting in temperatures on the order of 2700–2900 K or 1500 K, respectively. The thermal release of electrons at the Fermi level of the material is known as *thermionic emission* and may only occur once the energy exceeds the work function of the material (4.7 eV (W) or 2.7 eV (LaB_6)). Saturation of the filament occurs when further heating does not increase the number electrons being emitted. Electrons ejected from the filament cathode are focused by a negatively charged Wehnelt cylinder and drawn toward a positively charged anode plate containing a small aperture (revisit Fig. 7.23). The electrons that escape through the pinhole are directed down the column toward the sample. The voltage difference between the filament and the anode plate is referred to as the *accelerating voltage*. This value is directly proportional to the energy of the electron beam that is directed down the column. Whereas the voltage range for SEM is *ca.* 200 V–40 kV, TEMs feature much higher accelerating voltages, ranging from 60 to 400 kV. All electron microscopes must be operated under extremely high vacuum conditions (*ca.* 10^{-6}–10^{-9} Torr), which provides an insulating medium between the cathode and anode, prolongs the lifetime and efficiency of the gun, and allows for sharper beam focusing. In general, the characteristics of an electron gun may be optimized by varying four components: the distance from the filament to the opening of the Wehnelt cylinder, the temperature of the filament, the negative bias of the Wehnelt cylinder with respect to the cathode, and the brightness of the beam.

Lanthanum hexaboride crystal filaments yield a brightness *ca.* 10 times greater than tungsten, with a smaller spot size. This allows for imaging at very low accelerating voltages without a loss of resolution—of importance for beam-sensitive samples. Due to the small spot size, the beam from a LaB_6 source is highly monochromatic, which increases the overall resolution (i.e., less chromatic aberrations). Interestingly, LaB_6 filaments employ the (100) face due to its relatively low work function.[24] The preferential release of electrons from this crystal face is likely due to the

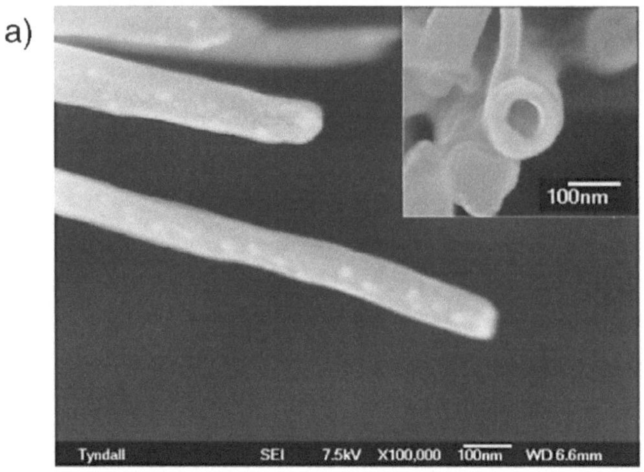

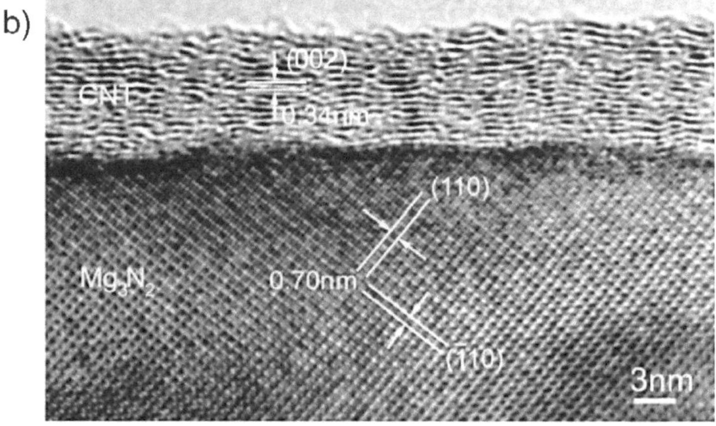

Fig. 7.25 High-resolution (HR) electron micrographs. Shown is **a** an SEM image of vanadium oxide nanotubes at a magnification of 100,000× and **b** a TEM image of the interface between a Mg_3N_2 nanowire and a graphitic carbon coating at a magnification of *ca.* 2,000,000×. Images reproduced with permission from *Chem. Mater.* **2006**, *18*, 3016 and *Nano Lett.* **2006**, *6*, 1136. © 2006 American Chemical Society

dipole existing between the positively charged metal atoms at the crystal surface, and the underlying layer of boride octahedra (Fig. 7.26). A high vacuum must be used to prevent oxidation of the crystal surface; adsorption of oxygen raises the work function of LaB_6 to 4.0 eV.[25]

It should be noted that the work function of structural alloys is always less than the work function of the isolated component metals. For instance, the work functions of B and La are 4.5 and 3.3 eV, respectively—much higher than LaB_6. The mechanisms responsible for the reduction of the work function are not completely understood, although it must depend on the three-dimensional arrangement and orbital overlap

Table 7.2 Comparative characteristics of energetic radiation sources

Source	Brightness[a] (particles cm^{-2} eV^{-1} sr^{-1})	Elastic mean free path[b] (Å)	Absorption length[c] (Å)	Minimum probe size[d] (Å)
Neutrons	1×10^{24}	1×10^{8}	1×10^{9}	1×10^{7}
X-rays	1×10^{26}	1×10^{4}	1×10^{6}	1×10^{3}
Electrons	1×10^{29}	1×10^{2}	1×10^{3}	1

[a] Related to the number of particles and the range of incident angles that irradiate a sample
[b]Related to the distance a particle travels in a solid before elastically colliding with constituent atoms
[c]Related to the distance a particle travels in a solid before its energy is absorbed through interactions with constituent atoms and electrons
[d]Related to the minimum diameter of the source that is used to generate the radiation

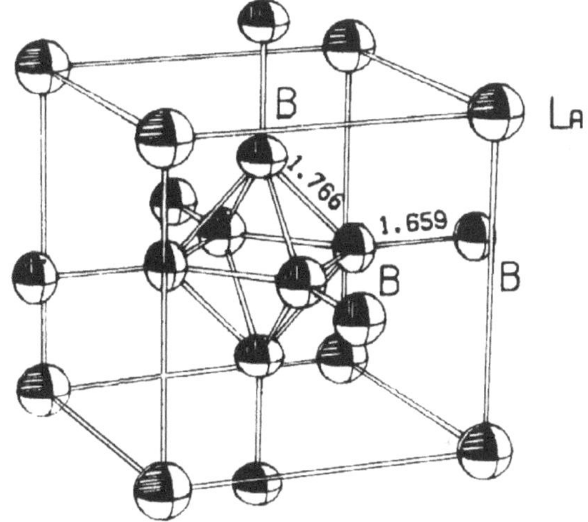

Fig. 7.26 Unit cell of LaB_6. Thermal vibration ellipsoids are also shown, illustrating a significant aniostropy in the thermal vibrations of the B atoms. Reproduced with permission from *Pure Appl. Chem.* **1985**, *57*, 1383. © 1985 IUPAC

of electron rich/deficient atoms or ions near the surface of the solid. Once this mechanism is deciphered, many more materials will be isolated as effective candidates for electron emission sources.

In order to increase the intensity and focus of electrons, a *field emission* source may be used. This consists of a single crystal tungsten or LaB_6(100) wire that is sharpened to a tip diameter of *ca.* 100 nm–1 μm. For crystalline tungsten, the axis is suitably aligned with respect to the optical axis of the microscope. For example, a beam with a diameter <5 nm is possible from alignment of the filament planes perpendicular to ⟨310⟩ and ⟨111⟩. In addition to W and LaB_6, a number of other materials are proposed for field emission applications, such as silicon, single-walled nanotubes,[26] and ultra-nanocrystalline diamond (UNCD) or Cu/Li alloy films deposited onto sharpened tips.[27]

Two classifications of field emitters are used in electron microscopes, cold and Schottky sources (Table 7.3). Whereas the former operates through Fowler–Nordheim electron tunneling[28] from a cathode wire held at room temperature, the latter features thermionic emission from a ZrO_2-coated sharpened W filament at 1800 K. In both cases, an electrical field draws electrons from the narrow filament tip into an ultrahigh vacuum (10^{-9} Pa) chamber. When the electrons are accelerated through the magnetic field and focused onto the sample, a dramatically greater resolution is obtained—*ca.* 15 times that of traditional filaments (Fig. 7.27). The magnification, M and resolution R, are given by (7.26) and (7.27)[29]:

Table 7.3 Comparison of various electron sources

	Thermionic W	Thermionic LaB_6	Cold FE W(310)	Schottky FE ZrO/W(100)
Operating temperature (K)	2800	1900	300	1800
Operating vacuum (kPa)	$\leq 1 \times 10^{-5}$	$\leq 1 \times 10^{-6}$	$\leq 1 \times 10^{-10}$	$\leq 1 \times 10^{-8}$
Cathode radius (nm)	60,000	10,000	≤100	≤1000
Effective source radius (nm)	15,000	5,000	2.5	15
Emission current density(A cm^{-2})	3	30	17,000	5,300
Total emission current (μA)	200	80	5	200
Normalized brightness (A cm^{-2} sr^{-1} kV^{-1})	1×10^4	1×10^5	2×10^7	1×10^7
Maximum probe current (nA)	1000	1000	0.2	20
Energy spread at the cathode (eV)	0.59	0.40	0.26	0.31
Energy spread at the gun exit (eV)	1.5–2.5	1.3–2.5	0.3–0.7	0.35–0.71
Beam noise (%)	1	1	5–10	1
Emission current drift (% h^{-1})	0.1	0.2	5	<0.5
Cathode lifetime (h)	200	>500	>2000	>2000
Cathode regeneration	Not required	Not required	Every 6–8 h	Not required
Sensitivity to external influences	Minimal	Minimal	High	Low

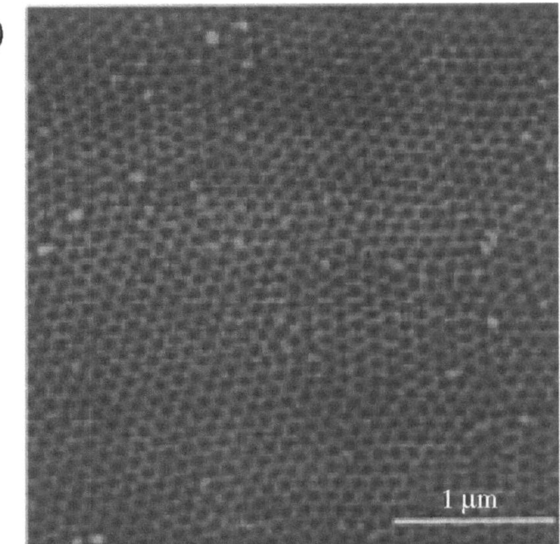

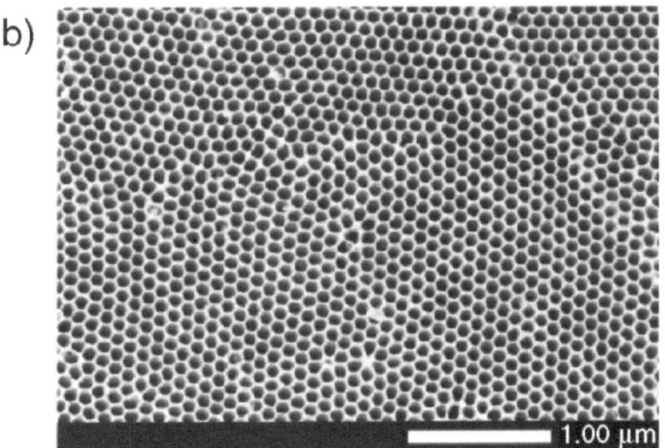

Fig. 7.27 Comparison of the resolution for **a** low-resolution SEM, with **b** high-resolution FESEM. Both images are of a nanoporous anodized aluminum oxide template at identical magnification. The low-resolution SEM image was reproduced with permission from *Chem. Mater.* **1998**, 10, 2470 © 1998 American Chemical Society. The high-resolution SEM image was reproduced with permission from Moon, J. -M.; Wei, A. *J. Phys. Chem. B* **2005**, *109*, 23,336. © 2005 American Chemical Society

$$M = \frac{L}{r}, \tag{7.26}$$

where L is the distance from the tip to the screen/detector and r is the radius of the tip apex.

$$R = \frac{M}{\tau} \tag{7.27}$$

where M is the magnification and τ is the time-of-flight of an electron from tip to the screen/detector.

Although cold FE sources feature a higher brightness and smaller spot size than a Schottky FE source, they are not as preferred due to their relatively noisy emission profile. This beam instability arises from the Brownian motion of adsorbed residual gases on the surface of the narrow tip. Even though electron microscopes are operated under UHV, these gases are always present in the column—likely originating from the sample itself (residual solvent on grid, e-beam decomposition of the sample, etc.). In addition, as electrons strike residual gas molecules in the column, positive ions are generated that may be accelerated back to the electron source. Such ion bombardment will cause a significant deformation of the filament surface, which will lead to beam instabilities. Since Schottky emitters are operated at 1800 K, these deformations are readily repaired via annealing. However, this is not possible using cold FE sources, which must be periodically "flashed" with a high-temperature pulse in order to repair surface deformations. Although this effectively improves the surface structure of the emitter, the lifetime of the emitter is shorted due to an increase in the tip radius—a consequence of thermal flashing in the absence of an electric field.

7.4.1 Transmission Electron Microscopy (TEM)

A transmission electron microscope is analogous to a slide projector, with illumination from an electron beam rather than light. When an electron beam is impinged upon a sample, a black and white TEM image is formed from the passage of some electrons through the sample untouched, alongside the combination of interactions between other electrons and sample atoms (e.g., inelastic/elastic scattering, diffraction). If the undiffracted beam is selected to form the image, it is referred to as *bright-field imaging*; in contrast, selection of strongly diffracting regions of the sample, which would appear brighter than the transmitted beam, is known as *dark-field imaging*.[30] It should be noted that electrons may also be absorbed by molecules containing large atoms, or by surface contamination (e.g., dust, grease). The absorption of a high density of electrons in a specific region will cause a buildup of heat, leading to sample destruction and poor image quality.

Analogous to throwing a baseball of varying speeds through a wall, the relative degree of penetration through a particular sample is governed by the energy of the electron source. That is, higher-energy electrons (e.g., 200 keV vs. 100 keV) will be more penetrating, allowing for the characterization of thicker and/or less transparent samples. In general, increasing the thickness of a sample, or decreasing the energy (i.e., accelerating voltage) of the electron beam, will induce more scattering events through more effective interactions between the electron beam and atoms of the sample. This effect will enhance image contrast, since there is a larger deviation between the path lengths of transmitted and scattered electrons that reach the viewing screen. However, this improvement of image quality is offset by plentiful inelastic collisions that yield a broadened wavelength distribution of the electron beam. Since individual electrons will have differing energies, they will be brought into focus at different points resulting in a blurry image (i.e., decreased resolution).

Sample Preparation Techniques

Not unlike other materials (or molecular) characterization techniques, the most important and time-consuming aspect of TEM analysis is sample preparation. Specimens for TEM analysis are placed on special micromesh grids of a conductive metal such as Cu, Au, or Ni. The typical dimensions of TEM grids are *ca.* 3 mm in diameter and 10–25 μm thick. The mesh number of a grid indicates the number of grid openings per linear inch. The smaller the grid number, the larger the hole size and the greater the ratio of open area to covered area (Fig. 7.28a). For example, a 200-mesh grid has 20 holes along its diameter; a 400-mesh grid has 40 holes. This translates to hole sizes of *ca.* 200 μm for 100-mesh, *ca.* 97 μm for 200 mesh, *ca.* 63 μm for 300 mesh, and *ca.* 42 μm for 400-mesh grids.

Most often, a grid is coated with a support film that holds the sample in place. The film must be as transparent as possible, while providing support for the sample. To prevent interference with electron-sample interactions, films containing light elements (e.g., C, Be) are used. Support films are typically deposited onto the surface of grids through a "floating" technique (Fig. 7.28b). Amorphous carbon or polymeric (e.g., Formvar,[31] Collodion (nitrocellulose),[32] Butvar B98[33]) films are first deposited onto a glass microscope slide or mica surface and floated off onto the surface of filtered water. For polymer films, the pre-cleaned glass/mica surface is simply immersed in a 0.3% ethylene dichloride solution of the polymer. By contrast, carbon films are deposited onto the glass/mica surface through sputter-coating. The

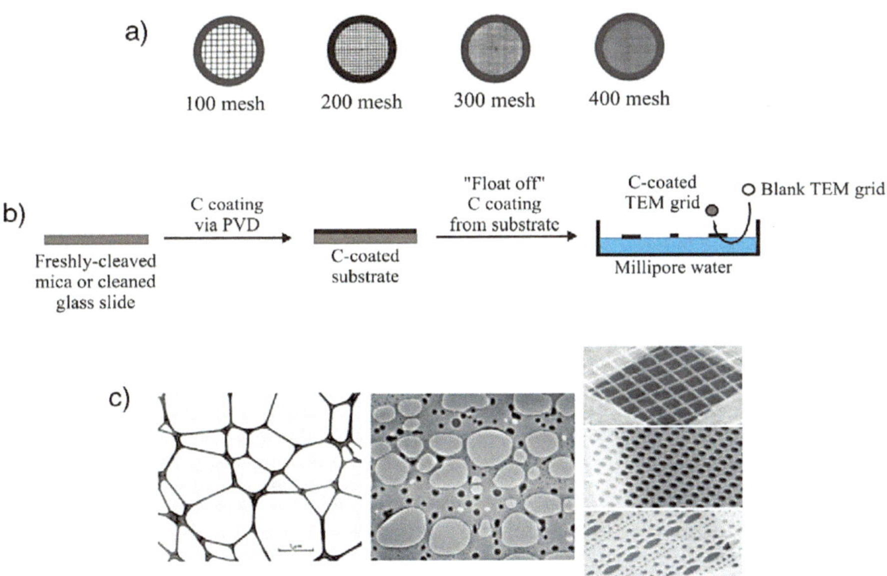

Fig. 7.28 **a** A comparison of TEM grid mesh sizes; **b** Illustration of the "float off" method to prepare carbon-coated TEM grids; **c** Comparison of a lacey carbon coating with holey carbon films deposited onto TEM grids

coated glass/mica substrate is allowed to dry completely, scored with a razor blade, and slowly immersed into a container of distilled water. When water reaches the score marks, surface tension effects cause the film to be pulled from the glass/mica, resulting in a floating film on the surface of the water. Grids are then dunked into the water and brought up underneath the floating film, lifted out of the water, and allowed to air-dry completely. The coated grids are stored in a dry, dust-free environment.

The typical thickness of amorphous carbon films is 2–5 nm, whereas polymer films have thicknesses on the order of 30 nm. As you might expect, all plastic films are subject to decomposition by the electron beam. Sometimes, this exposure also causes further crosslinking that will cause the film to shrink and become more brittle. The surface adsorption characteristics of support films vary quite significantly between polymer and carbon varieties. In general, polymer films possess hydrophilic characteristics whereas carbonaceous support films are hydrophobic.

For the imaging of nanomaterials, it is most desirable to have a support film that resembles a piece of lace, featuring many holes on the surface. In this manner, nanostructures such as nanotubes or nanowires are held on the surface of the support film, with unobstructed imaging made possible through holes in the film surface (Fig. 7.28c). Such "holey" or "lacey" grids are made from plastic film suspensions to which water or glycerin has been added. Due to the microemulsions created in the solution, when the film is dried, thousands of tiny holes are left in the support film. In this way, the coating itself acts like a mesh grid with the thousands of very tiny open viewing areas.

Due to the high vacuum environment and bombardment of the sample with high-energy electrons, not all samples are appropriate for TEM characterization. In particular, any sample comprising a volatile organic compound, or those containing residual solvents, must be dried using freeze–pump–thaw procedures, or supercritical fluid media. Failure to do so will result in collisions between the electron beam and stray gaseous molecules in the column, resulting in defocusing of the beam and a reduction in image quality. Further, the decomposition of organic matter can severely contaminate the microscope column—especially in finely machined regions such as apertures and pole pieces. Some samples may not be directly analyzed even following drying procedures, due to their instability toward the focused electron beam. For these samples (e.g., biological), the specimen is mounted within a plastic support resin, and very thin pieces of the composite are shaved off using an ultramicrotome fitted with a diamond, sapphire, or glass knife (Fig. 7.29).

TEM grids of solid samples are typically prepared by pipetting a few drops of a solution containing the suspended solid onto a grid and allowing it to air-dry. Alternatively, the suspension may be blown onto a grid using an aerosol delivery device (Fig. 7.30). If the solid is too thick, or comprises relatively large granules, the material may be embedded within a polymeric matrix and cut into thin cross sections. The best embedding medium permits thin sectioning with the least damage during preparation and gives the least interference during microscopy. While early embedding media included gelatin and paraffin, the <100 nm sections that are required for electron microscopy were not possible to cut using any of these media. Nowadays, polymers such as methacrylates, epoxy, and polyester resins are typically employed

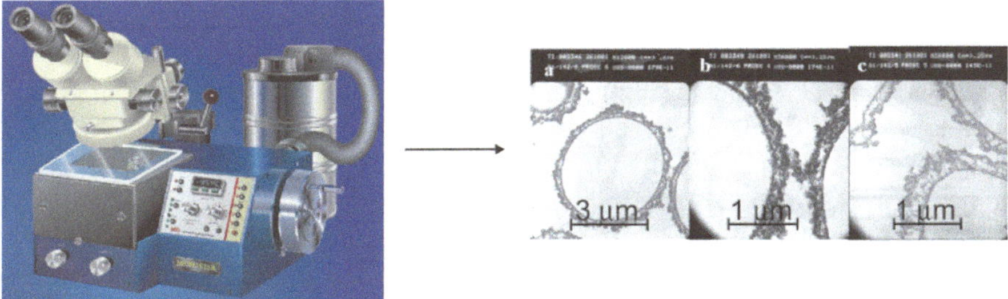

Fig. 7.29 Sectioning of a TEM sample using an ultramicrotome instrument. Shown is an ultramicrotome instrument with a built-in nitrogen cryogenic system, used to section samples with a thickness of 25 nm–5 mm.[34] Also shown is a TEM image of polymer capsules that have been sectioned using an ultramicrotome instrument. Image reproduced with permission from *Langmuir* **2002**, *18*, 9533. © 2002 American Chemical Society

for sectioning applications since they are nontoxic, inexpensive, and strong enough to withstand very thin mechanical sectioning. For methacrylates, the degree of final block hardness can be predictably controlled by mixing methyl (hard) and *n*-butyl (soft) functionalities. However, with respect to electron beam stability, epoxy resins are by far superior to both methacrylates and polyester resins.

The "best" section thickness not only depends on the electron opacity of the sample, but also on the operating conditions of the TEM. In general, the higher the accelerating voltage, the thicker the sample cross section may be. For instance, epoxy-based sections of up to 250 nm thick may be used for standard TEM analyses operating at 100 kV. However, section thicknesses in the range of 50–60 nm are typically used for TEM analysis, in order to improve the resolution through reduced electron scattering.

A recent alternative to (ultra)microtoming is the use of a *focused-ion beam* (FIB) to etch/mill away undesired portions of the sample (Fig. 7.31). The most common

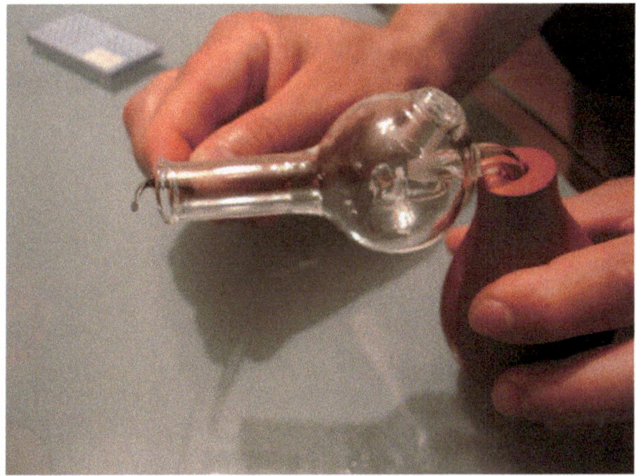

Fig. 7.30 Photo of a glass aerosol delivery system used to spray solvent-suspended particulates onto a TEM grid

beam consists of gallium ions from a liquid gallium source, with an energy of *ca.* 30 keV. Extreme control over the resultant cross section is afforded by varying the beam current/energy and scanning speed of the ion beam. A FIB/lift-out method is especially useful for the study of polymer films that have been deposited onto hard inorganic substrates.[35] Typically, a dual-beam FIB/SEM is used for sample preparation, which affords real-time monitoring of FIB sectioning (Fig. 7.31b). A more recent use of a FIB instrument is nanofabrication. In particular, in situ CVD of a thin film may be deposited through the introduction of a co-reactant precursor gas during ion bombardment (Fig. 7.31c). It should be noted that in contrast to ultramicrotomy, FIB preparation does not induce mechanical stress on the sample surface, which often leads to deformation and interfacial debonding. However, the high-energy radiation may result in other surface modifications such as amorphization, shrinkage, and bond cleavage—especially for "soft," polymeric-based samples.[36]

Oftentimes, a lack of image contrast in TEM results from the analysis of very thin sections (e.g., <30 nm), or samples comprising light elements (e.g., hydrocarbons, B, Li, etc.). In order to increase the contrast, elements of high atomic weight are introduced to selective regions of the sample—a process referred to as *staining* (Fig. 7.32). Due to their high electron density, staining agents are better able to stop or deflect the beam of electrons, whereas elements of low weight allow the beam to pass through relatively unimpeded. Electron staining falls into one of two categories:

(i) *Positive staining*—contrast is imparted to the sample.
(ii) *Negative staining*—contrast is imparted to the area surrounding the sample.

The most common staining agents are aqueous solutions (*ca.* 2 wt% concentration) of OsO_4 (osmium tetroxide), RuO_4 (ruthenium tetroxide), $UO_2(CH_3COO)_2{\cdot}2H_2O$ (uranyl acetate), $Pb(C_6H_2O_7)_2{\cdot}3H_2O$ (lead citrate), or $Na_3[PW_{12}O_{40}]{\cdot}30H_2O$ (sodium phosphotungstate). Due to the presence of heavy metals, these agents are considerably toxic and must be handled/disposed with great care. Whereas most of these agents interact most strongly with amino and phosphate groups, lead ions preferentially interact with hydroxyl groups. For biological staining, it has been widely shown that OsO_4 interacts more readily with lipids than proteins, resulting in a structure-specific staining agent. The use of lead citrate requires special considerations due to its high sensitivity to CO_2; hence, NaOH is often used as a CO_2-scavenger for the aqueous staining solution.

Non-Imaging Applications for TEM

Due to its high spatial resolution and predictive scattering modes, TEM is often employed to determine the three-dimensional crystal structure of solid-state materials. Thus, TEM may be considered as complementary to conventional crystallographic methods such as X-ray diffraction. Since the wavelength of an accelerated electron in a TEM is much smaller than an X-ray beam, the Ewald sphere (Fig. 7.15; radius $= 1/\lambda$) is significantly larger for electron diffraction relative to X-ray diffraction studies. As a result, electron diffraction yields much more detailed structural information of the crystal lattice. Information such as lattice parameters and atomic

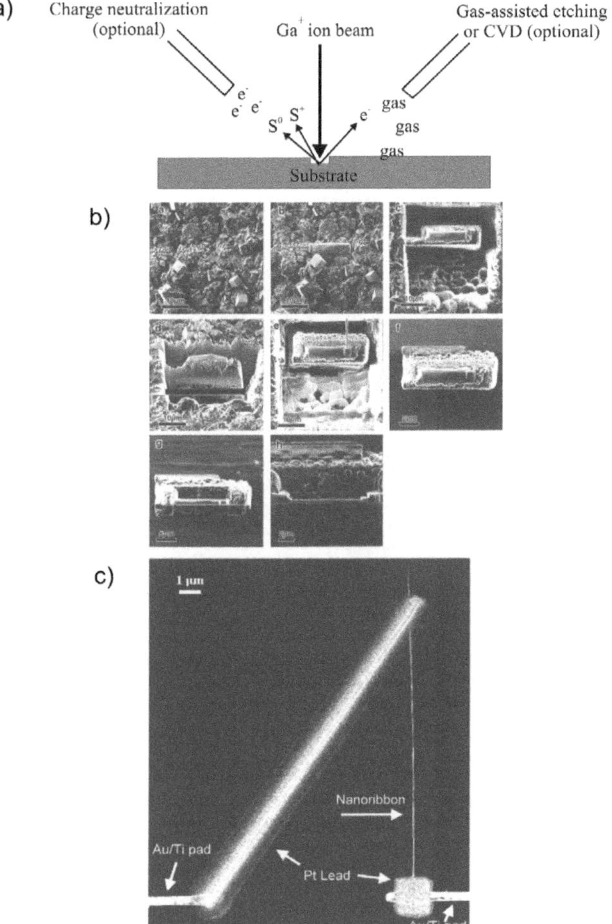

Fig. 7.31 Applications for focused-ion beam (FIB). Shown is **a** general illustration of the technique, with a gallium ion source and emission of secondary electrons, as well as charged ions and atoms from the substrate surface (S^+ and S^0, respectively). The optional "charge neutralization" electron beam is used for in situ imaging of a surface during FIB (dual-beam FIB/SEM). The images shown in (**b**) represent the sectioning of a zeolitic membrane (images c and d illustrate FIB milling to a depth of 20 μm, and cutting off the bottom surface; latter images show the manipulation of the milled target and further sectioning by FIB) Reproduced with permission from *Chem. Mater.* **2003**, *15*, 4287 © 2003 American Chemical Society. The SEM image in c illustrates the use of FIB for nanofabrication, the selective deposition of Pt contact leads onto the ends of a Ga_2O_3 nanowire. Reproduced with permission from *Chem. Mater.* **2006**, *18*, 922. © 2006 American Chemical Society

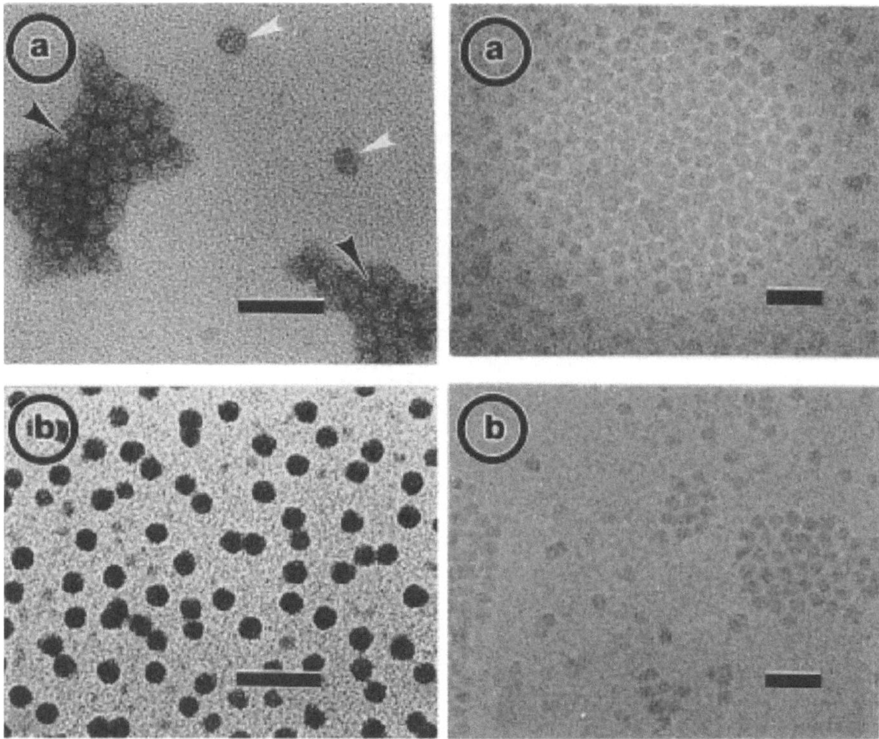

Fig. 7.32 (Left, **a**) TEM image of G10 PAMAM dendrimers stained with 2% methylamine tungstate, showing both positively (white arrows) and negatively (black arrows) stained regions. (Left, **b**) TEM image of G9 PAMAM dendrimers positively stained with sodium phosphotungstate, showing little agglomeration. (Right, **a, b**): TEM images showing the lack of contrast from unstained G10 PAMAM dendrimers, imaged in vitrified water using cryo-TEM. The scale bar for all images is 50 nm. Reproduced with permission from *Macromolecules* **1998**, *31*, 6259. © 1998 American Chemical Society

positions in a crystal may be obtained through analyzing the in situ electron diffraction pattern from a specimen size of *ca.* >400 nm (>100 nm for a field-emission source). This technique is denoted as *selected area electron diffraction* (SAED, Fig. 7.33).[37] In addition, the lenses within a TEM allow one to alter the orientation of the incoming electron beam, from a parallel beam to a cone-shaped beam. The latter orientation is known as *convergent beam electron diffraction* (CBED), and allows one to conduct a diffraction experiment over many incident angles simultaneously to reveal the full three-dimensional reciprocal lattice of the crystal.[38]

In the same manner as X-ray diffraction, by examining the systematic absences of the electron diffraction pattern, one may easily determine the appropriate Bravais lattice as well as any screw axes and glide planes that are present in the crystal lattice. Since TEM is more user-intensive, involves significant sample preparation,

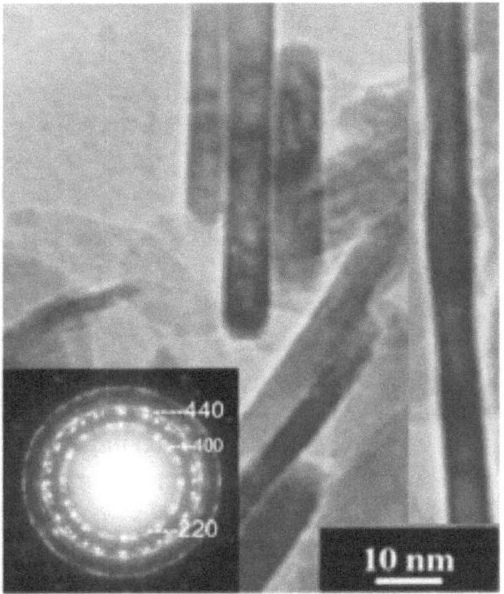

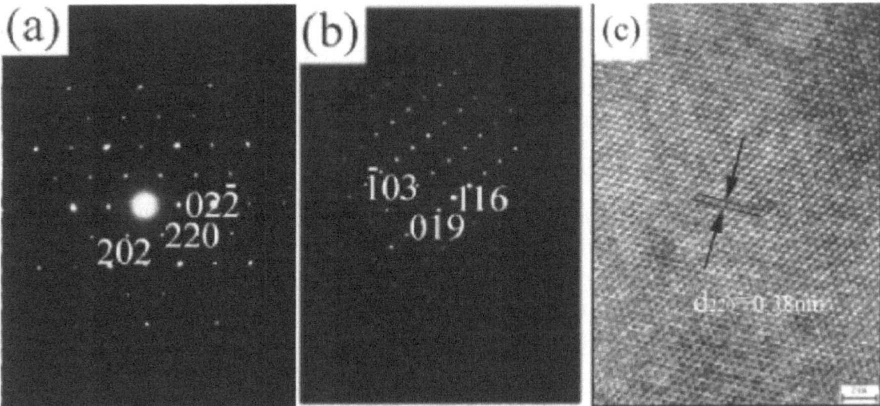

Fig. 7.33 (Top) SAED pattern from a polycrystalline $MgAl_2O_4$ nanorods. Reproduced with permission from *J. Phys. Chem. B* **2006**, *110*, 13,375. © 2006 American Chemical Society. (bottom) SAED patterns (**a, b**) and high-resolution TEM image (**c**) of a single-crystalline In_2S_3 nanoparticle. Reproduced with permission from *Cryst. Growth Design* **2006**, 6, 1304. © 2006 American Chemical Society.Images (**a**) and (**b**) illustrate the diffraction patterns for cubic and tetragonal unit cells, respectively. The TEM image shown in (**c**) illustrates the lattice fringes and interplanar spacing of 3.81 Å, corresponding to the (220) interplanar spacing of cubic In_2S_3. In high-resolution images, lattice fringes result from the interference of the transmitted and diffracted beams; these are only evident in the TEM image when the lattice spacings are larger than the resolution of the electron microscope

and often results in sample damage from the high-energy electron beam, highly automated X-ray and neutron (useful for light elements such as H) diffraction methods are the preferred methods to yield structural information from a bulk crystalline solid. However, if structural information is desired from an individual nanocrystal, TEM/SAED is the best alternative.[39]

Before leaving the subtopic of electron diffraction, it should be noted that two other techniques known as *low-energy electron diffraction* (LEED)[40] and *reflection high-energy electron diffraction* (RHEED)[41] may also be used to glean structural information from a crystalline sample. Both techniques are performed independent of a TEM instrument, and utilize an electron gun and fluorescent screen to show the structure and morphology of a crystal surface. The RHEED technique uses a high accelerating voltage and low impact angle in order to focus the electrons to the first few atomic layers of the crystal surface. This technique may be used to monitor crystal growth in real time, or to probe the surface adsorption properties for sensor development. For beam-sensitive crystalline samples (e.g., organic thin films), either LEED or microchannel plate (MCP)–RHEED may be used; the latter features the amplification of the incident electron beam via a MCP—resulting in much lower intensities of the incident electron beam.

In addition to structural information from crystalline samples, quantitative analysis may also be carried out in tandem with TEM (and SEM) analyses. The interaction among high-energy electrons and sample atoms results in a variety of emissions (Fig. 7.34) that yield important information regarding the surface morphology and elemental composition of the sample. As previously stated, if the electron beam interacts with the nucleus of surface atoms, the electrons are elastically scattered. In this case, the trajectory of the electron changes with no change in the kinetic energy or velocity, a process known as *backscattering*.

The number of *backscattered electrons* (BSE) that are produced from a given atom is proportional to the atomic number. That is, materials composed of heavy atoms will backscatter more electrons, resulting in brighter gray tones in the image relative to less dense materials. Hence, BSE produce an image that is related to material composition, providing both spatial and chemical information. Accordingly, it is often easier to accurately determine the sizes/spatial positions of high-Z nanoparticles via so-called *dark-field imaging* relative to bright-field imaging, which is more easily shaded by the varying contrast of the support material (Fig. 7.35).

The primary electron beam may also be inelastically scattered through interaction with electrons from surface atoms. In this case, the collision displaces core electrons from filled shells (e.g., ns^2 (K) or np^6 (L)); the resulting atom is left as an energetic excited state, with a missing inner shell electron(s). Since the energies of these *secondary electrons* are sufficiently low, they must be released from atoms near the surface in order to be detected. Electrons ejected from deeper regions of the sample are likely to be reabsorbed by the material before they reach the surface. As we will see in the next section (regarding SEM), as the intensity of the electron beam increases, or the density of the sample decreases, information from underlying portions of the sample may be obtained.

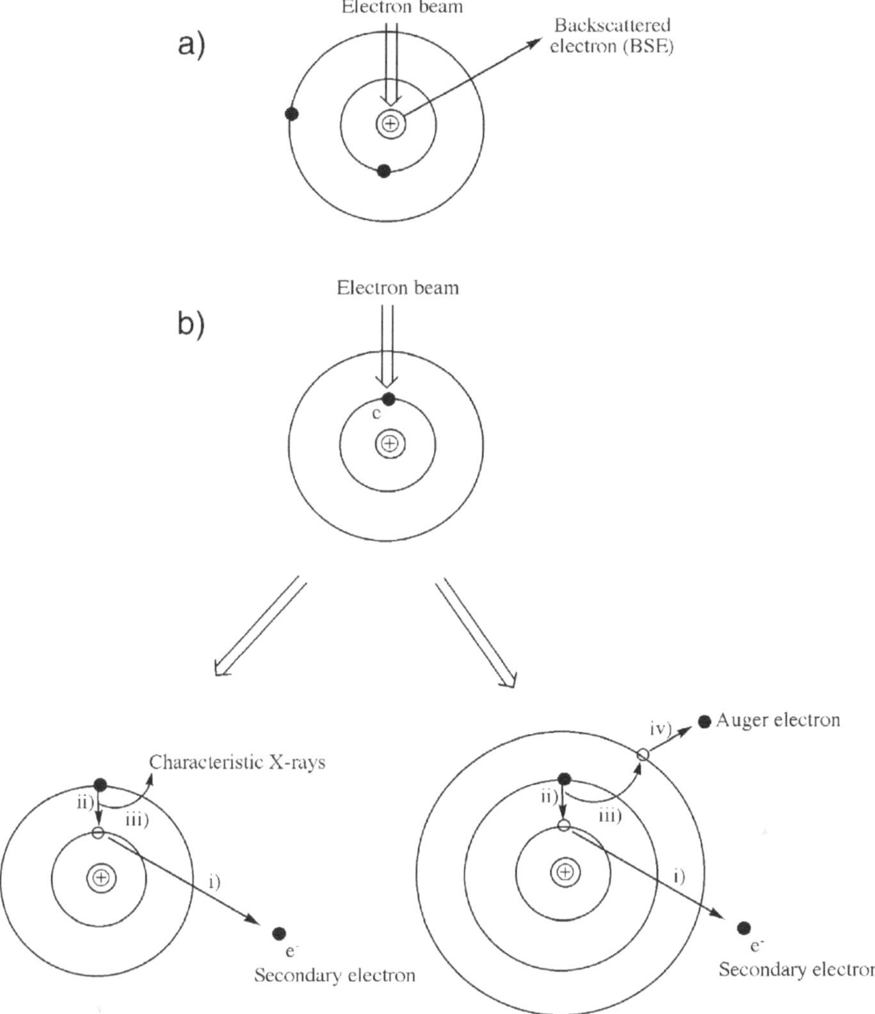

Fig. 7.34 Types of possible interactions between the primary electron beam and (**a**) the nucleus of surface atoms and (**b**) electrons of surface atoms comprising the sample. For simplicity, only K, L, and M shells (with no L and M subshells) are shown

As we saw earlier for X-ray diffraction, in order for an excited-state atom to return to its ground state (within a picosecond or so following secondary electron generation), the vacancy in an inner shell is filled with an electron from shells farther from the nucleus (e.g., L, M shells, revisit Fig. 7.7). When such post-ionization atomic relaxation occurs, the excess energy may be released as either characteristic X-rays, or through nonradiative emission of an *Auger electron*. Especially for elements with $Z > 11$ (i.e., >Na), the electronic shell structure becomes exceedingly

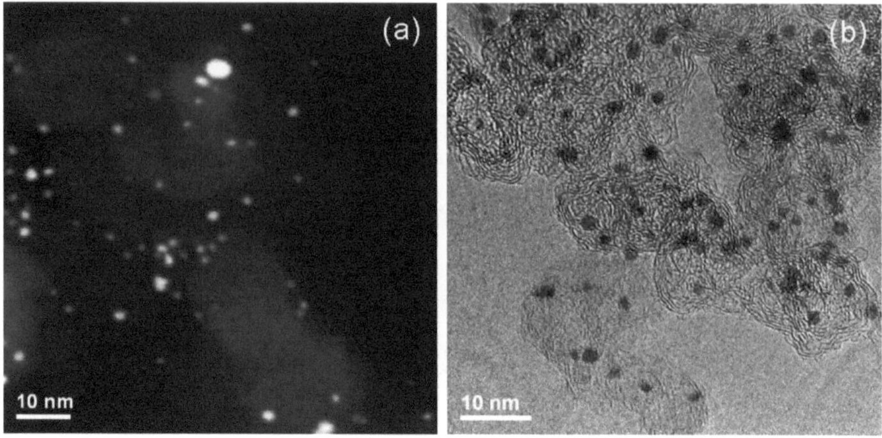

Fig. 7.35 HAADF image, **a** and bright-field image, **b** of Pt nanoparticles supported on carbon (Vulcan XC-72). Reproduced with permission from *Langmuir* **2008**, *24*, 11,350. © 2008 American Chemical Society

complex, which results in a plethora of electronic transitions that are possible from a number of outer shells.

Typically, a variety of X-rays are produced during atomic relaxation due to a cascading effect. For instance, a vacancy in the K shell may be filled with an electron in an L shell (K_α emission); the resulting vacancy in the L shell may then be filled with an electron from the M shell ($L_{\alpha 1}$ emission), and so forth. By counting the number and energies of X-rays produced from electrons interacting the sample, it is possible to both qualitatively and quantitatively (using suitable standards) determine the chemical composition of the surface being analyzed.

There are two methods used to identify and quantify the X-ray emission: *energy-dispersive X-ray spectroscopy* (EDS), and *wavelength-dispersive X-ray spectroscopy* (WDS).[42] In EDS, all the characteristic X-ray energies reaching the detector are measured simultaneously. Hence, data acquisition is very rapid across the entire spectrum. By contrast, WDS measures a single wavelength at a time through use of a detecting crystal. As the characteristic X-rays are emitted from the sample, they are diffracted in a regular manner as discussed previously. Not only does this improve the resolution of WDS to an order of magnitude greater than EDS (Fig. 7.36),[43] but also improves the count rate and deconvolution of overlapping spectral peaks. Nevertheless, due to its simplicity and speed of analysis, EDS is the standard method for chemical analysis within TEM (and SEM) instruments. Typically, if WDS is desired, an instrument known as *electron probe microanalyzer* (EPMA) is utilized, often in tandem with SEM imaging.

The spatial resolution of X-ray microanalysis may be described by (7.28). In general, prerequisites for the best spatial resolution include a high-energy electron beam and extremely thin samples. Although the resolution limits of EDS/WDS will never match the imaging resolution, it is now possible to easily determine the

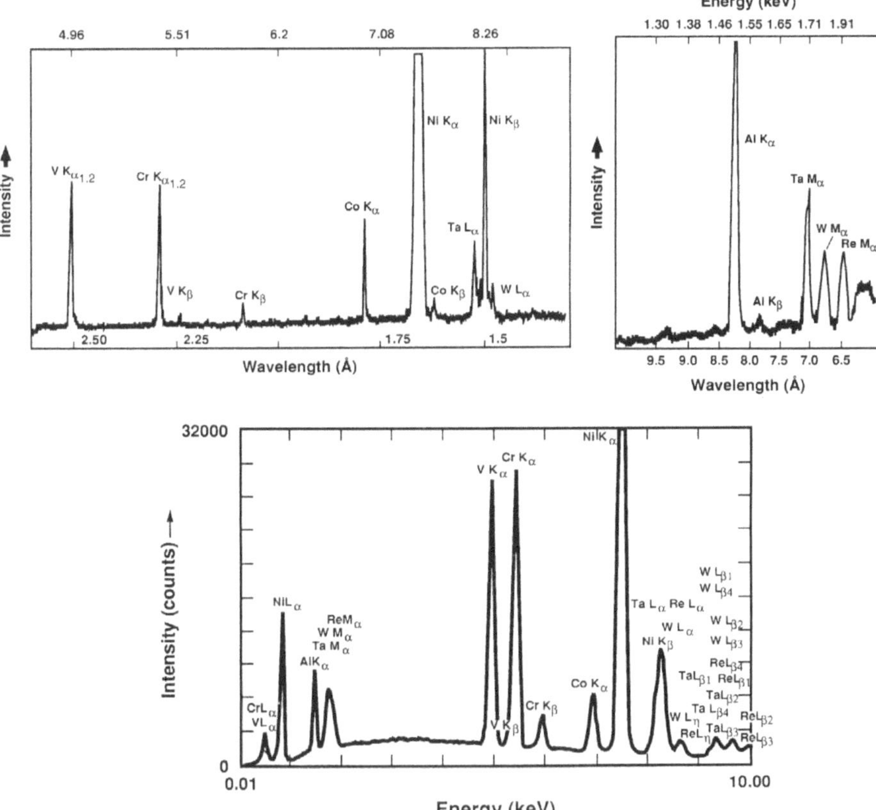

Fig. 7.36 Comparison of WDS (top, using LiF and thallium acid phthalate (TAP (1010), respectively) and EDS for the analysis of a superalloy. A number of X-ray spectral lines such as Ta L_α, Ni K_β, and W L_α are hardly discernible using EDS, but readily visible using WDS analysis. Reproduced with permission from *Scanning Electron Microscopy and X-Ray Microanalysis*, 3rd ed., Kluwer: New York. © 2003 Springer Science and Business Media

elemental composition of individual nanoclusters in the 1–5 nm size regime—as long as they are suitably dispersed with respect to one another (Fig. 7.37):

$$R = \frac{d + \sqrt{\left(7.21 \times 10^5 \frac{Z}{E_o} \left(\frac{\rho}{\mathrm{A}}\right)^{\frac{1}{2}} t^{\frac{3}{2}}\right)^2 + \mathrm{d}^2}}{2}, \tag{7.28}$$

where R is the X-ray spatial resolution, d is the beam diameter, E_o is the beam energy (eV), ρ is the specimen density, and t is the specimen thickness (cm).

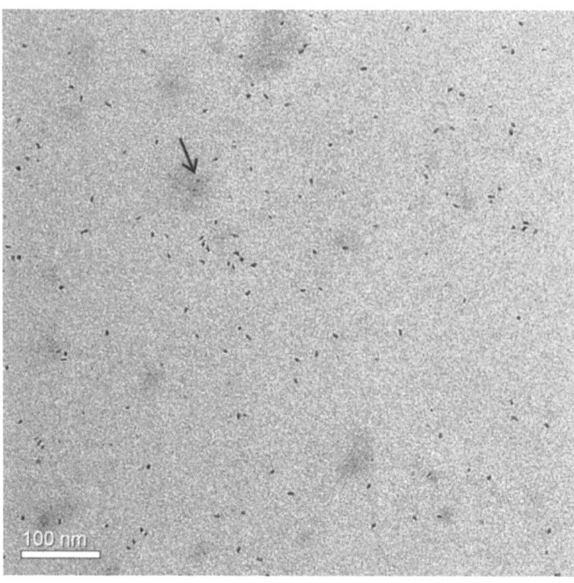

Fig. 7.37 Ambiguity in assigning a chemical composition (iron oxide) to individual nanostructures based on TEM/EDS. The small and large nanoparticles are in the same vicinity on the grid; Depending on the area selected for analysis, it may be difficult to definitely state if one (or both) comprise iron oxide (e.g., arrow region). However, by also looking at the relative contrast of the nanoparticles, it is likely that the smaller structures contain iron (higher density) and the larger nanoparticles comprise lighter elements (e.g., a carbonaceous, organic-based nanostructure). © Bradley D. Fahlman

Scanning Transmission Electron Microscopy

An imaging mode that merges both SEM and TEM is also possible on most modern TEM instruments. This method, referred to as *scanning transmission electron microscopy* (STEM), uses a LaB_6 source that produces a focused electron beam with a high current density and extremely small diameter. Instead of monitoring the transmitted electrons from a static beam as performed in standard TEM imaging, the beam within a STEM is scanned across the sample—analogous to SEM, as discussed in the next section. Due to a higher beam intensity, thicker samples may be analyzed in a STEM; furthermore, staining is generally not necessary for low-Z elements due to a higher sensitivity to sample density/composition. In particular, it is possible to overlay the image with the EDS data—a technique known as *elemental dot-mapping*, widely used for SEM/EDS analysis (Fig. 7.38).

The majority of STEM instruments are simply conventional TEMs with the addition of scanning coils. As a result, these "non-dedicated STEMs" are capable of TEM/STEM, as well as SEM imaging for thicker samples. The development of HRTEMs and 'dedicated' STEMs with lens aberration correction[44] have now pushed

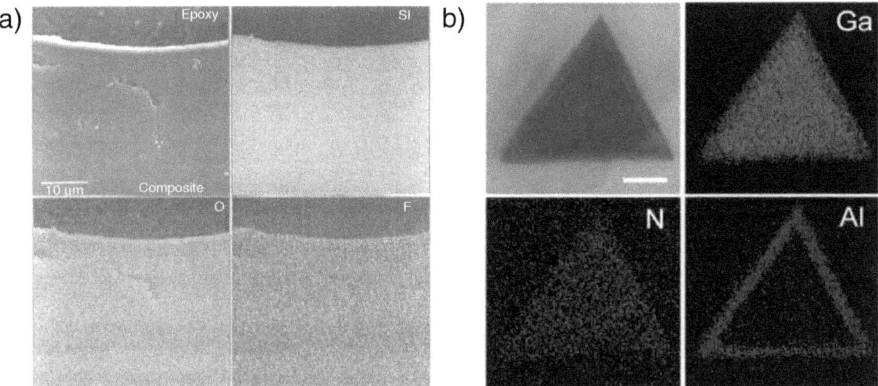

Fig. 7.38 Elemental dot-maps. Shown is **a** elemental concentrations of Si, O, and F (as white pixels) overlaid onto the SEM image of a Nafion resin/silica composite. Reproduced with permission from *J. Am. Chem. Soc.* **1996**, *118*, 7708. © 1996 American Chemical Society. **b** Bright-field STEM image of a GaN/AlN/AlGaN nanowire cross section, with elemental mapping of Ga, N, and Al (scale bar is 50 nm). Reproduced with permission from *Nano Lett.* **2006**, 6, 1468. © 2006 American Chemical Society

the resolution limits to as low as sub-50 pm, i.e., ≤0.5 Å—suitable for facile atomic-resolution imaging.[45] However, it has been shown that imaging of 2D layered materials was affected by electron beam shape, which can lead to contrast artifacts.[46] Coincidentally, the introduction of this powerful TEM in 2009 comes exactly 50 years after Feynman's famous speech (Appendix B), in which he challenged the world to increase the resolution limits of electron microscopes!

In a dedicated STEM, high-angle (elastically) scattered electrons are separated from inelastic/unscattered electrons through use of a high-angle annular dark-field (HAADF) detector (Fig. 7.39). Since there are no post-sample refocusing lenses, it is not possible to generate a bright-field image as is standard using a conventional (HR)TEM. Instead, the image is based on specific interactions between the electron beam and sample atoms (i.e., elastic/inelastic scattering, unscattered electrons). Since the incoherently scattered electrons are related to elastic scattering, they are directly related to the structure and chemical composition of the feature being imaged (Fig. 7.40).[47] In particular, the incoherent scattering of an electron is described by the Rutherford equation (7.29). This equation predicts that the intensity of the scattered electrons, and resultant image contrast, is strongly dependent on atomic number of the sample atoms. Hence, HAADF–STEM is also referred to as *Z-contrast imaging*. It should be noted that HAADF–STEM is strongly dependent on variations in sample thickness. That is, thicker regions of a sample will result in higher image intensities, which may be falsely interpreted as the presence of species with relatively high atomic numbers.

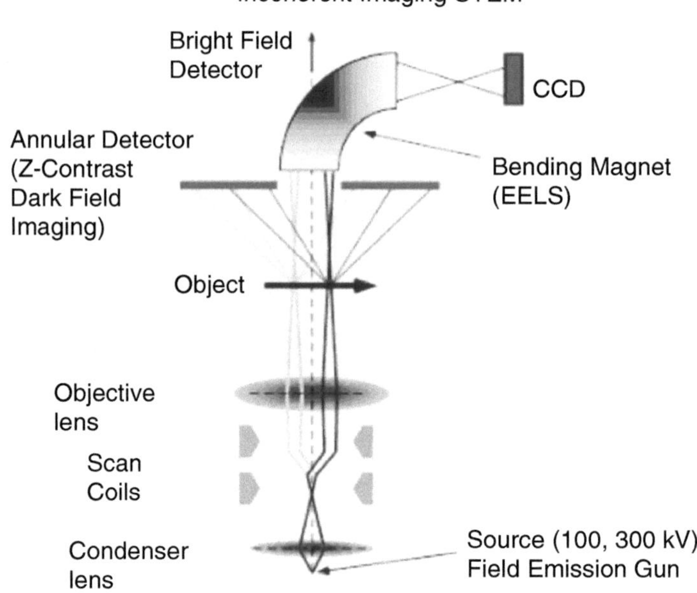

Fig. 7.39 Schematic of a dedicated HAADF–STEM. Reproduced with permission from *Nano Lett.* **2004**, 4, 1279. © 2004 American Chemical Society

$$\frac{\mathrm{d}\sigma(\theta)}{d\Omega} = \frac{e^4 Z^2}{16(E_o)^2 \sin^4(\frac{\theta}{2})}, \tag{7.29}$$

where $\mathrm{d}\sigma(\theta)/\mathrm{d}\Omega$ is the different scattering cross sections as a function of the scattering angle (θ), E_o is the incident beam energy, e is the electron charge (1.602×10^{-19} C), and Z is the atomic number of the scattering nucleus.

A primary limitation of EDS and WDS is the inability to detect light (i.e., low-Z) elements. Since atomic energy levels are closely spaced for low-Z elements, the energies of the emitted X-rays will be relatively low.[48] As a result, they are masked by the broad, continuous background spectrum (known as Bremsstrahlung[49]) that is most intense at energies below 1 keV. Furthermore, the characteristic X-ray lines are less intense for low-Z elements since they exhibit a low *X-ray fluorescence yield*—favoring non-emissive Auger[50] electron processes rather than X-ray generation (Fig. 7.41). Consequently, it becomes increasingly more difficult to observe X-ray lines from elements with $Z < 11$ (Na). Recently, there have been improvements in the design of EDS windows, which separate the detection system from the electron microscope.[51] Since low-energy X-rays are readily absorbed, new detector designs feature ultrathin windows and lightweight compositions that facilitate the detection of elements down as far as beryllium ($Z = 4$). However, the analysis of these elements by EDS/WDS is semi-quantitative, at best.

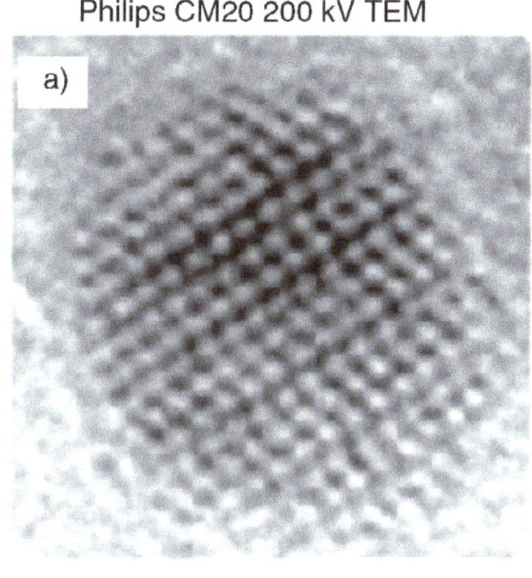

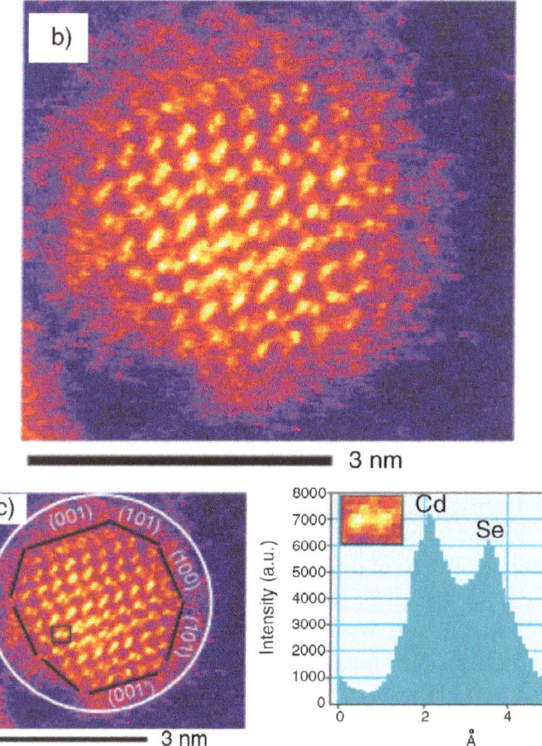

Fig. 7.40 Comparison of conventional HRTEM (**a**), with HAADF–STEM (**b**). Also shown (**c**) is the chemical analysis of an individual CdSe "dumbbell." The white circle shows the amorphous oxide region, and the surface of the nanocrystal is outlined in black. Unlike conventional HRTEM, it is also possible to label the individual nanocrystal facets, such as Cd-rich (001) and Se-rich (001′). Reproduced with permission from *Nano Lett.* **2004**, *4*, 1279. © 2004 American Chemical Society

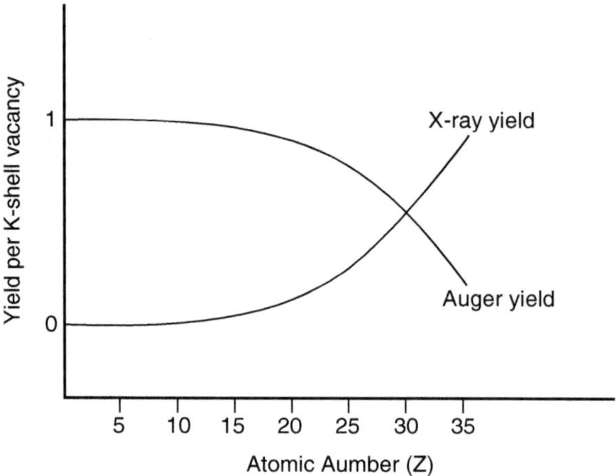

Fig. 7.41 Relative probabilities for X-ray and Auger electron emission during the decay of K-electron vacancies

Electron Energy-Loss Spectroscopy

In order to increase the sensitivity toward the detection of light elements, a technique known as *electron energy-loss spectroscopy* (EELS) may be utilized.[52] This method may be carried out within a (S)TEM and consists of monitoring the loss in energy (due to inelastic scattering) of the beam electrons as they pass through the sample. Since it is more difficult to focus X-rays relative to electrons with appropriate lenses, the collection efficiency for EELS is *ca.* 80–90%, relative to 5–6% for EDS. This leads to a greater sensitivity and spatial resolution for EELS, with elemental mapping of *ca.* ≥2 Å. The intensity is greater than EDS for light elements since the signal generated by EELS represents the total sum of the number of X-ray photons and Auger electrons emitted from the sample. This technique is useful for elements with $Z > 1$, and like EDS, is amenable for elemental mapping of a sample surface.

As previously illustrated in Fig. 7.35, a dedicated STEM is usually fitted with an EELS detector, which collects the low-angle scattered electrons that pass by the HAADF. Although the beam electrons have energies of several hundred keVs, the electrons being transmitted through the sample only have energies on the order of a few eVs. In order to perform EELS, it is therefore necessary to detect very small differences in the kinetic energies of the electrons. This is accomplished using a magnetic prism that exerts a centripetal force on each electron, causing a circular motion. While in the magnetic field, electrons move along the arc of a circle, whose radius is based on the speed and kinetic energy of the electron. In reality, there is nothing new with this concept; a magnetic prism is actually analogous to the dispersion of white light into a colored spectrum using a glass prism. However, unlike a glass prism, the magnetic field focuses the electrons as they exit the field, generating a spectrum from the grouping of electrons that exhibited identical energy losses.

In addition to detecting/quantifying particular elements in a sample, EELS also provides detailed elemental information such as the electronic structure, bonding, and nearest neighbor distribution of the atoms in the sample.[53] A representative EELS spectrum for a NiO surface is shown in Fig. 7.42. The most intense features are peaks corresponding to zero-loss—those electrons that were either unscattered, or elastically scattered, while traversing through the sample. At relatively small energy losses (*ca.* 5–25 eV), a plasmon peak is observed which corresponds to the collective oscillation of many outer-shell (valence or conduction) electrons. The most useful application for this peak is the accurate determination of the sample thickness, of up to several thousand nanometers with a precision of a few percent (7.30).[54] More recently, the plasmon region of the spectrum has been used to delineate variations in the size and geometry of metal nanoparticles.[55] As we saw in Chap. 6, the plasmon resonance frequency is directly related to the effective nuclear charge and size/shape of the charge distribution. Hence, the low-loss region of an EELS spectrum also provides information about bonding interactions and the dielectric function of the sample:

$$T \propto \log\left(\frac{\mathrm{I}_p}{\mathrm{I}_z}\right), \tag{7.30}$$

where T is the sample thickness, I_p is the intensity of the plasmon peak, and I_z is the intensity of the zero-loss peak.

At higher energy losses, an EELS spectrum will exhibit a variety of sharp features known as "edges," which are diagnostic for the presence of specific elements. The positions of the edges correspond to the binding energies of the core electrons in the sample. As shown in Fig. 7.42 (top), the K-edges for O and Ni are 530 and 860 eV, respectively. Once the background is subtracted, the area under each edge peak(s) is integrated in order to determine elemental concentrations. The shape of the peak immediately surrounding the edge is aptly referred to as the electron-loss near-edge structure (ELNES). As you might expect, these features are directly dependent on the exact band diagram and density of states (DOS) of the solid being analyzed. As such, this profile may be considered as the electron-scattering counterpart of X-ray absorption near-edge structure (XANES).[56] This region of the spectrum relates to the electronic structure, oxidation state, and bonding hybridization/symmetry of the desired element (Fig. 7.42 (bottom)).[57]

Whereas the ELNES region typically extends to *ca.* 20 eV beyond the edge, the extended energy-loss fine structure (EXELFS) provides chemical information from the scattering of electrons by neighboring electrons. Accordingly, this region of the EELS spectrum is the electron-scattering counterpart of extended X-ray absorption fine structure (XAFS, detailed later)[58]; both being used to determine nearest-neighbor distances, oxidation states, and coordination numbers of the element being probed (see Fig. 7.42 (bottom)).

The spectral fine details from EELS result from dipole-scattering of the incoming electrons. Much like IR spectroscopy, the incoming charged electron is influenced by a vibrating dipole at the sample surface. Hence, the energy loss of the electron is based on the amount of energy that was deposited into the vibrational mode.[59] The

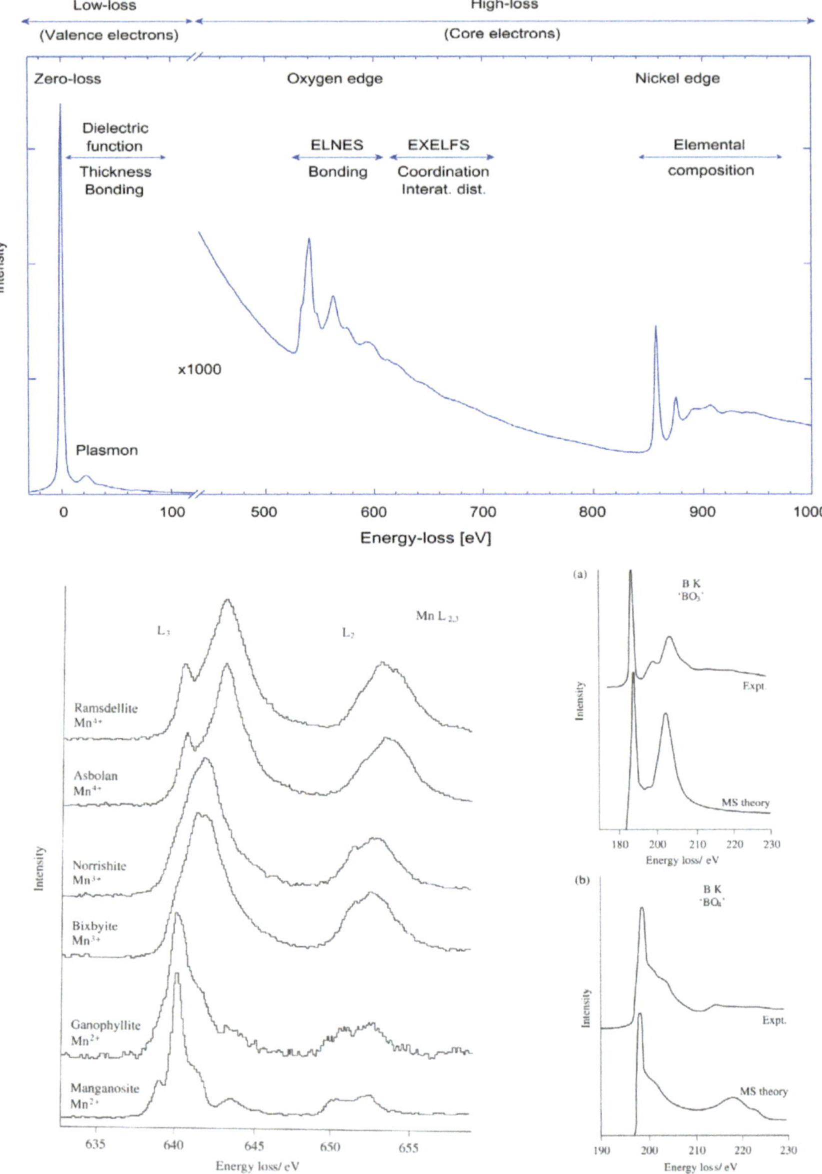

Fig. 7.42 Electron energy-loss spectroscopy (EELS) spectra. Shown (top) is a representative EELS spectrum of a nickel oxide sample. A typical EELS spectrum shows a zero-loss peak that represents the unscattered or elastically scattered electrons, the near-edge fine structure (ELNES), and extended energy-loss fine structure (EXELFS). Also shown (bottom) are the "fingerprint" regions of an EELS spectrum, just beyond the core-electron edges, which provide information regarding the detailed bonding and chemical environment of the desired element. Reproduced with permission from Brydson, R. *Electron Energy Loss Spectroscopy*, BIOS Scientific Publishers: Oxford, UK. © 2001 Taylor & Francis Group

compilation of the various regions in an EELS spectrum provides a fingerprint of the surface species being analyzed; hence, this technique is commonly used to probe the localized vibrational modes of chemisorbed molecules on a surface—relevant for the study of any heterogeneous process, including the development of novel catalysts for chemical syntheses, and gas sensors for fuel cell and homeland security applications.[60]

It should be noted that energy-filtered TEM (EF-TEM) images may be formed with electrons that have lost a specific energy, with respect to a predetermined cut-off energy of the atomic inner shell. Chemical mapping of the surface is possible in this mode, which allows one to determine the exact location of elements in the sample—most useful for surfaces that contain low-Z elements (Fig. 7.43). In addition, EF-TEM may be used to illustrate the valence-state distribution of a particular element across a surface, based on slight differences observed in the ELNES region of the EELS spectrum.[61]

To summarize the many signals one is able to detect using TEM, Fig. 7.44 illustrates all of the products arising from interactions between the incident electron beam and the sample. Not only is one able to image the surface via transmitted electrons or secondary electrons (*c.f.* SEM), but also interpret the observed image contrast in terms of the relative atomic masses of the constituent species in the sample (i.e., bright- and dark-field images, energy-filtered image). Further, quantitative elemental analysis is possible via EDS/WDS, AES, and EELS.

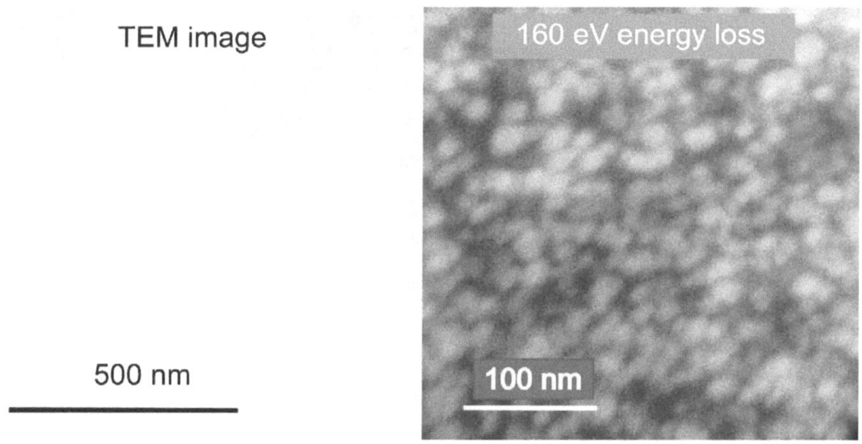

Fig. 7.43 Energy-filtered TEM showing silica nanoparticles embedded within an organic coating. The conventional bright-field TEM image (left) shows little/no contrast relative to the EF-TEM image (right)[62]

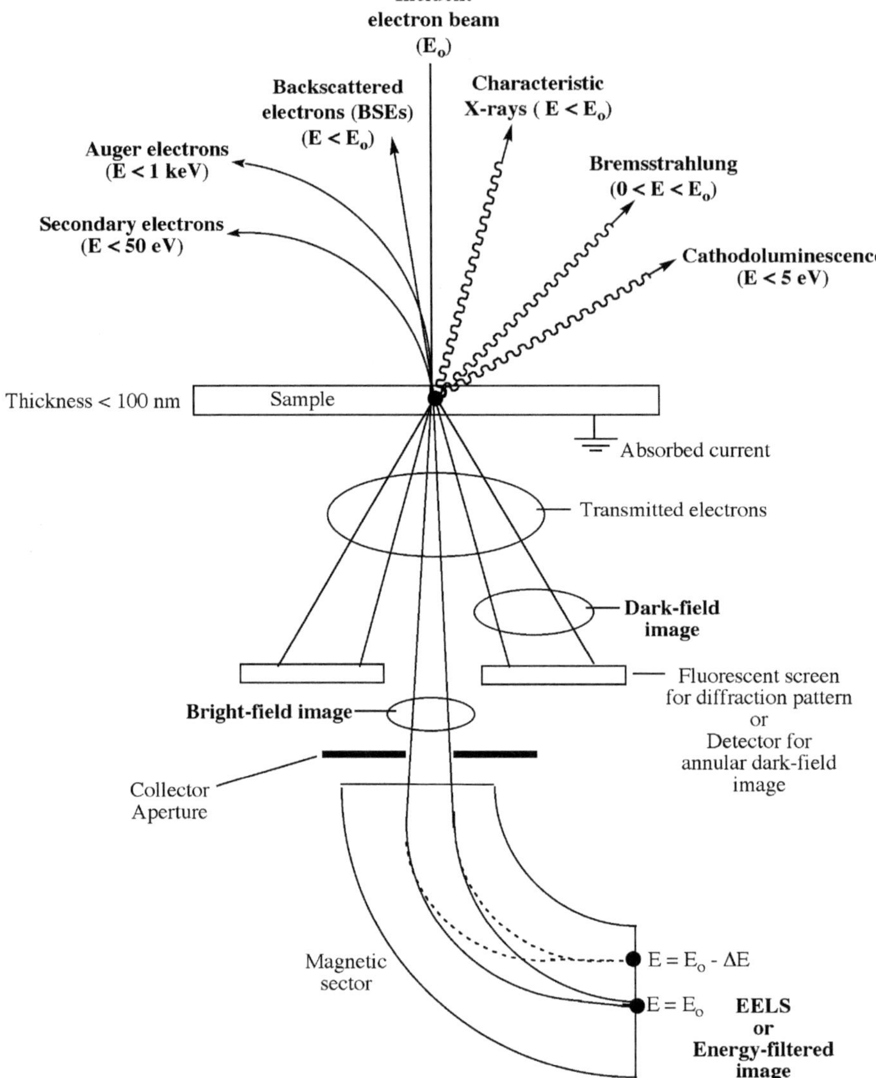

Fig. 7.44 Illustration of beam-specimen interactions within a transmission electron microscope

7.4.2 Scanning Electron Microscopy (SEM)

In contrast to TEM, with typical sample thicknesses <200 nm, the interaction volume for SEM extends into the micron range. As such, this technique is used to provide a topographic image of the sample surface as well as delineate information regarding the species present at varying depths of the sample (Fig. 7.45):

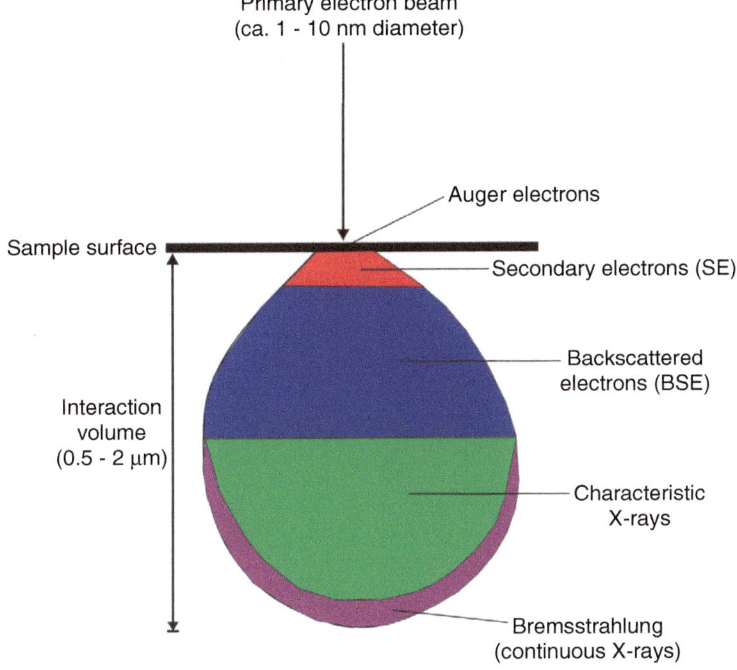

Fig. 7.45 Illustration of the sample interaction volume, and the corresponding particles emitted from various sample depths. The exact volume will depend on the accelerating voltage of the electron beam and sample composition

(i) Elastic scattering of electrons by atomic nuclei of the sample results in BSE—useful for generating images based on *Z*-contrast (Fig. 7.46).
(ii) Inelastic scattering of electrons by sample atoms results in low-energy secondary electrons—useful for providing topographic information regarding the sample surface.
(iii) Inelastic scattering of electrons by sample atoms results in X-ray generation (characteristic and Bremsstrahlung background X-rays) from lower sample depths—useful for chemical analysis of the bulk sample.
(iv) Inelastic scattering results in Auger electrons emitted from sample atoms near the sample surface—useful for surface chemical analysis.

As you will recall from Fig. 7.23b, for imaging and chemical analysis using SEM, the particles generated from the sample must reach the detector situated above the sample surface. Since the relative energies of the generated particles are in the order:

$$\text{Auger electrons} < \text{SE} < \text{BSE} < \text{X-rays}$$

The least energetic emissions will not reach the surface from lower depths of the sample. For instance, Auger electrons that are emitted from deeper regions of the

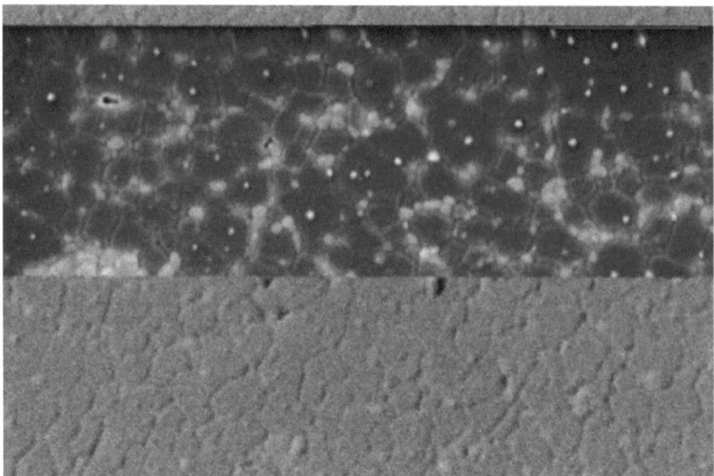

Fig. 7.46 The image illustrates two different scanning modes of a scanning electron microscope (SEM). In the lower part of the image, we can see the relief of the sample. This is obtained using the detection of secondary electrons. In the upper part of the image, we can see light spots surrounded by darker areas. The light spots correspond to the zirconium aggregates in an aluminum matrix, which is obtained using the detection of backscattered electrons (BSEs)

sample lose their energy through collisions with sample atoms before they reach the surface. As a result, AES is a very sensitive technique to probe the chemical composition of only the top 5–10 nm (i.e., 15–30 monolayers). In comparison, the maximum escape depth of secondary electrons has been estimated as 5 nm in metals and 50 nm in insulators.

Not surprisingly, both the beam current (or accelerating voltage) and sample density will greatly affect the *interaction volume* of the bulk sample with primary electrons (Figs. 7.47 and 7.48, respectively). As we saw earlier, the probability for elastic scattering increases with Z^2 (revisit 7.41); hence, as the density of the sample increases, the number of BSEs will increase, reducing the number of electrons that may penetrate to deeper regions of the sample. Similarly, as the energy of the incident beam decreases, fewer inelastic collisions with sample atoms are needed to bring the electrons to rest, thus decreasing the penetration depth into the sample.

The image formed from SEM is primarily the result of secondary electron emission from the sample surface. The Law of Conservation of Energy ensures that any energy lost by the primary beam of electrons must be transferred to the secondary electrons that are ejected from the sample atoms.[64] Those with sufficient energy to traverse the sample surface reach an Everhart–Thornley detector, which consists of a scintillator and photomultiplier tube (PMT).[65] The topographical contrast that arises from an uneven surface is due to a differing number of SEs being released from the sample (Fig. 7.49).

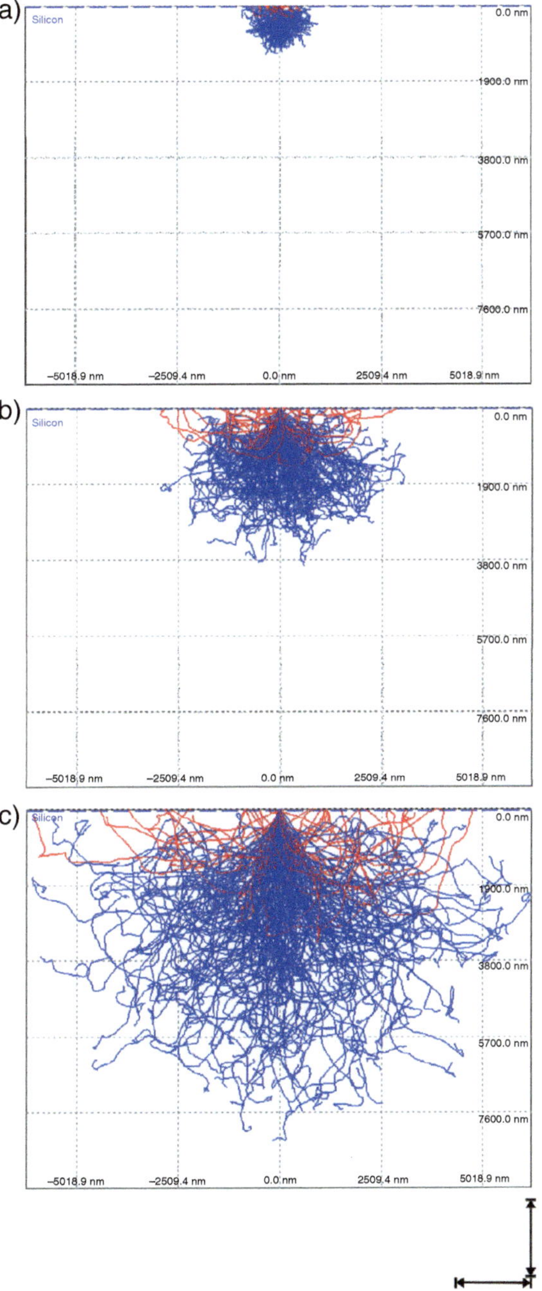

Fig. 7.47 Electron-beam penetration volume resulting from varying the beam current. Shown are X-rays (blue) and backscattered electrons (red) generated from a silicon substrate. The beam currents are **a** 10 keV, **b** 20 keV, and **c** 30 keV[63]

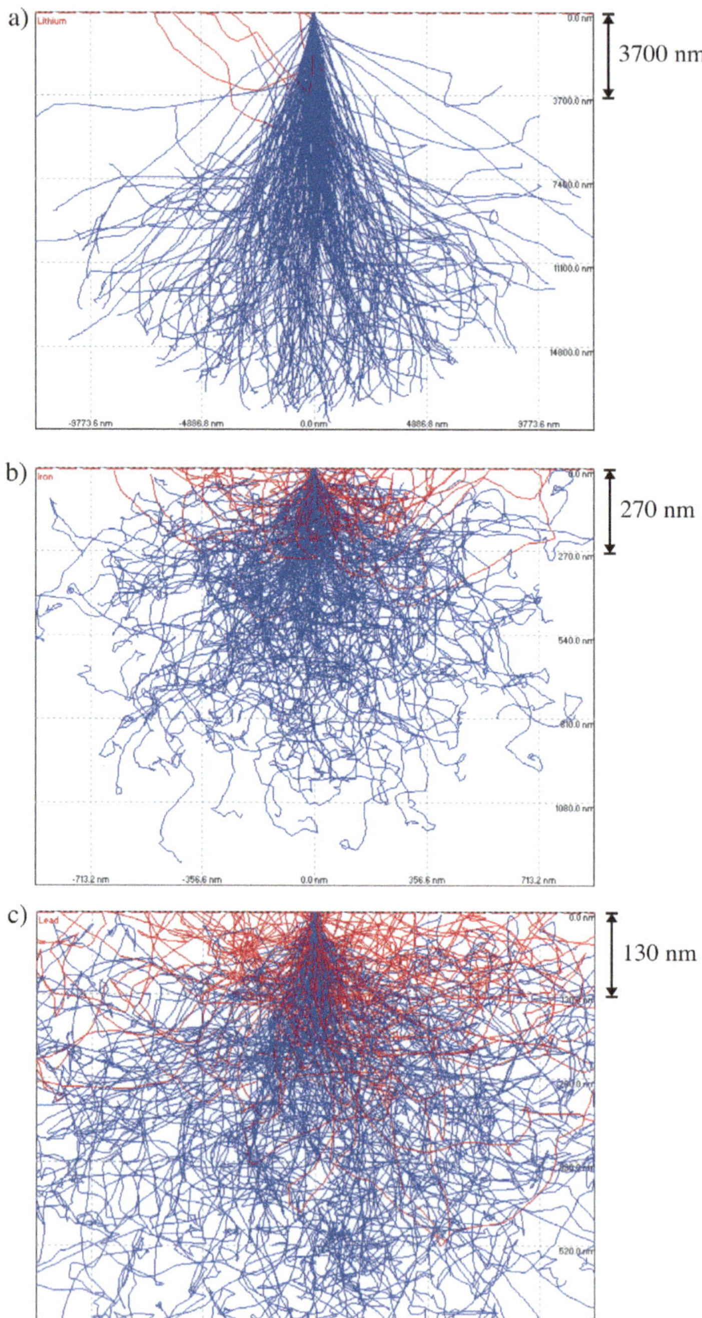

Fig. 7.48 Electron-beam penetration volume resulting from varying the sample density. Shown are X-rays (blue) and backscattered electrons (red) resulting from electron beam impingement on **a** Li, **b** Fe, and **c** Pb. The beam current is 20 keV, and beam diameter 10 nm for all metals[139]

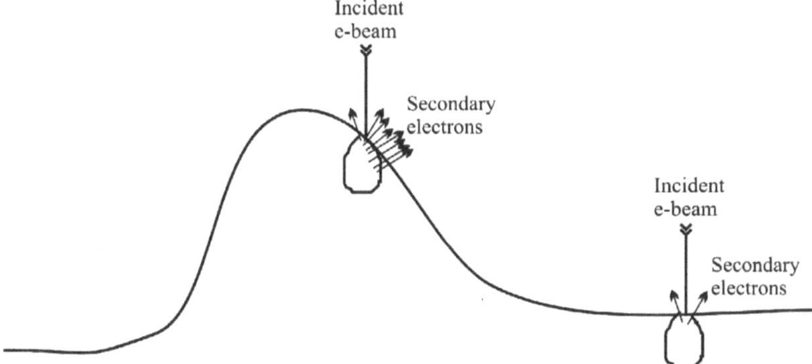

Fig. 7.49 Illustration of topographic imaging of a surface using SEM. When the incident beam strikes the side of a feature, many secondary electrons are released from the side and top of the interaction volume ("edge effect"). In comparison, the interaction of the primary beam with a flat substrate releases significantly fewer secondary electrons that originate near the sample surface. The image contrast results from varying numbers of secondary electrons reaching the detector as the beam is rastered across the sample surface

Structure Determination using SEM

In addition to displaying the familiar bright-field images from secondary electron emission, BSE in a SEM may be used to determine the crystallography of (poly)crystalline samples. This technique is referred to as *electron backscattering diffraction* (EBSD) or *backscattering Kikuchi diffraction* (BKD), used to measure individual crystallite orientations, as well as crystallographic parameters of the sample (e.g., interplane spacings/angles, crystal symmetry elements, etc.). The EBSD patterns are generated from the interaction of the incident electron beam with a highly tilted (*ca.* 70° from horizontal) planar sample.[66]

When electrons impinge on a crystalline sample, they interact with individual lattice planes. When these interactions satisfy the Bragg condition, they exhibit backscattering diffraction and (due to the tilted sample) are directed toward a phosphor screen where the fluorescent pattern is detected by a CCD camera. The resulting pattern consists of a large number of intersecting bands, known as Kikuchi lines, which represent the unique crystallographic properties of the crystal (Fig. 7.50a). Computer software is used to collect/analyze the resulting patterns to determine the crystallography of the material.[67] In association with compositional data from EDS, the exact phase of the material may be identified from a library of known materials. For a more thorough crystallographic description, *orientation imaging microscopy* (OIM) is often carried out in tandem with EBSD. This technique consists of stepping the beam across the sample, with automatic indexing of the resulting EBSD patterns. The resulting OIM map readily reveals the crystallographic orientations and grain boundaries in three-dimensions (Fig. 7.50b)—of use for applications such as sensor/heterogeneous catalyst design and micro/nano defect analyses.

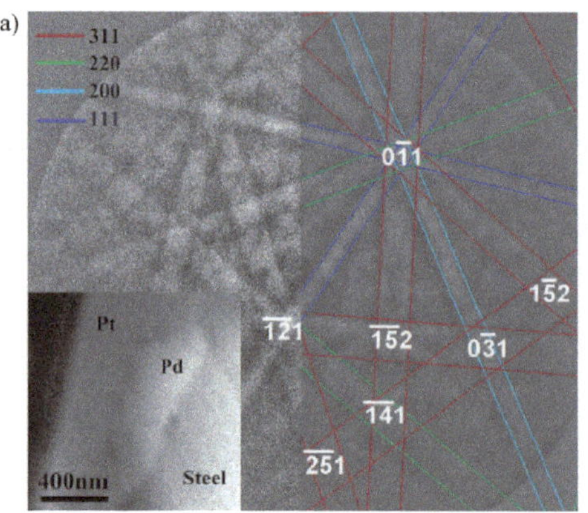

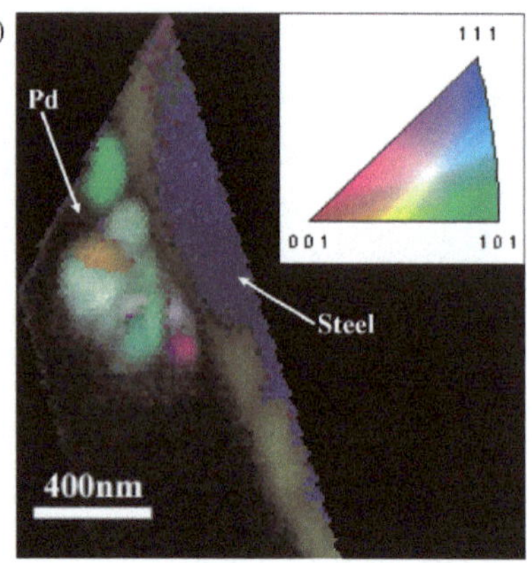

Fig. 7.50 **a** SEM micrograph (left bottom inset) of a Pd particulate selected for crystallographic analysis. Also shown are the crisscrossing Kikuchi lines originating from various crystallographic planes. The indices of each plane are indicated in the color legend at the top-left corner. **b** The orientation imaging microscopic (OIM) profile of the Pd particulate, showing a microcrystalline array (color-coated based on the crystallite orientations). Reproduced with permission from *J. Phys. Chem. B* **2004**, *108*, 556. © 2004 American Chemical Society

Sample Considerations for SEM

Sample preparation for SEM analysis is trivial relative to TEM, with the sample simply deposited onto the top of an adhesive fastened to an aluminum stub holder. Most often, conductive carbon tape is used to sequester the sample; for FESEM, problems with outgasing usually dictates the use of carbon/silver paint, epoxy, or copper/aluminum tape with adequate drying prior to analysis. After allowing the solvent to fully evaporate (if present), a thin conductive film (*ca.* <10 nm) of Au or C is often deposited onto the sample surface.[68] This is especially required if a non-conductive adhesive is used to mount the sample, or if the sample itself is non-conductive. The preparative steps involving conductive mounting materials is important to prevent *charging*—the buildup of electrons on the sample surface, which dramatically affects the imaging ability of the SEM (Fig. 7.51). Sometimes, it is even necessary to use an additional amount of carbon paste or tape to create a conductive path between a tall sample and the aluminum holder. It should be noted that metallic coatings also serve another important use—to increase the SE emission of a sample with a low yield of secondary electrons (e.g., comprising low-*Z* elements).

Electron Microscopy: Elemental Analysis

In addition to imaging applications, both TEM and SEM are also widely used for elemental analysis and chemical mapping of surfaces, using EDS/WDS. However, unlike TEM, facile cross section imaging/chemical mapping may also be performed using specialized sample holders (Fig. 7.52). Besides monitoring characteristic X-rays from sample elements, Auger electron emission may be analyzed—a technique known as *auger electron spectroscopy*, AES (Fig. 7.53a). Analogous to the STEM

Fig. 7.51 SEM images of amorphous carbon nanofibers (nonconductive) grown at room temperature from a dendritic catalyst. Shown is **a** as-formed nanofibers, without a gold coating and **b** after sputtering a thin conductive gold coating on the surface. The uncoated sample exhibited extreme charging (**a**), which thermally degraded the sample and caused movement of the sample during imaging. © Bradley D. Fahlman

extension of TEM, *scanning auger microscopy* (SAM) is also possible, wherein the incident electron beam is scanned across the sample. Perhaps the most intriguing aspect of AES/SAM instruments is the capability of depth-profiling. An argon-ion beam is used to etch away monolayers of the surface, allowing for compositional studies as a function of sample depth (Fig. 7.53b).

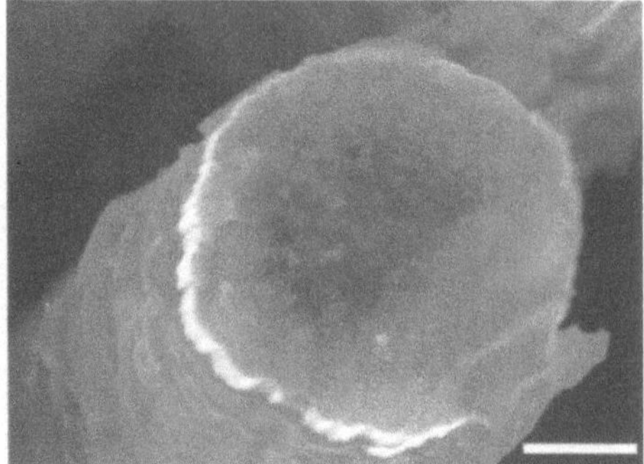

Fig. 7.52 Photograph of a variable-angle SEM sample holder, with an example of a cross-section SEM image (scale bar is 2 μm). © Bradley D. Fahlman

A simple way to distinguish between SEM and SAM is the former collects secondary electrons to form an image; the latter collects Auger electrons for chemical analysis. Recent developments have now afforded dual-detector electron microscopes that are equipped with standard FESEM/EDS capabilities in addition to SAM. This powerful combination offers the direct superimposing of both EDS and SAM chemical maps onto the corresponding high-resolution image.[69] In addition to being more sensitive for light elements (±0.5 at% for Li–U), it should be noted that AES is more suited for surface analysis than EDS. Due to the low kinetic energies of Auger electrons (*ca.* 50 eV–3 keV), chemical information is only obtained from sample depths of *ca.* <50 Å (compared to 1–2 μm for EDS). As such, it is not always possible to coat the sample to prevent charging, which explains why SAM is most often applied for compositional studies of conductive samples.[70]

Environmental Electron Microscopy

Although the analysis of samples in the presence of solvent is normally a *faux pas* for the UHV environment of any electron microscope, there are now instruments known as *environmental electron microscopes* (also known as *in situ* electron microscopes[71]) that are capable of such studies. These instruments have been in development since the early 1970s,[72] for *in situ* studies of virtually any material (wet/dry, insulating/conducting). Before this major technological development, samples such as paints, inks, and biological specimens had to be dried completely to maintain the integrity of the vacuum system. Typically, an environmental electron microscope is not simply a modified EM, but rather a specially designed instrument that is capable of regular imaging, as well as high-resolution[73] operation under atypical conditions such as within liquid media, at high temperature, etc.

The operation of an *environmental SEM* (ESEM) is made possible through use of a differential pumping system (Fig. 7.54) that maintains a UHV environment (10^{-7} Torr) required for the electron gun, while allowing the presence of gases in the sample chamber (10–20 Torr). The pressure and temperature of the sample chamber may be strictly controlled, inducing evaporation or condensation events.

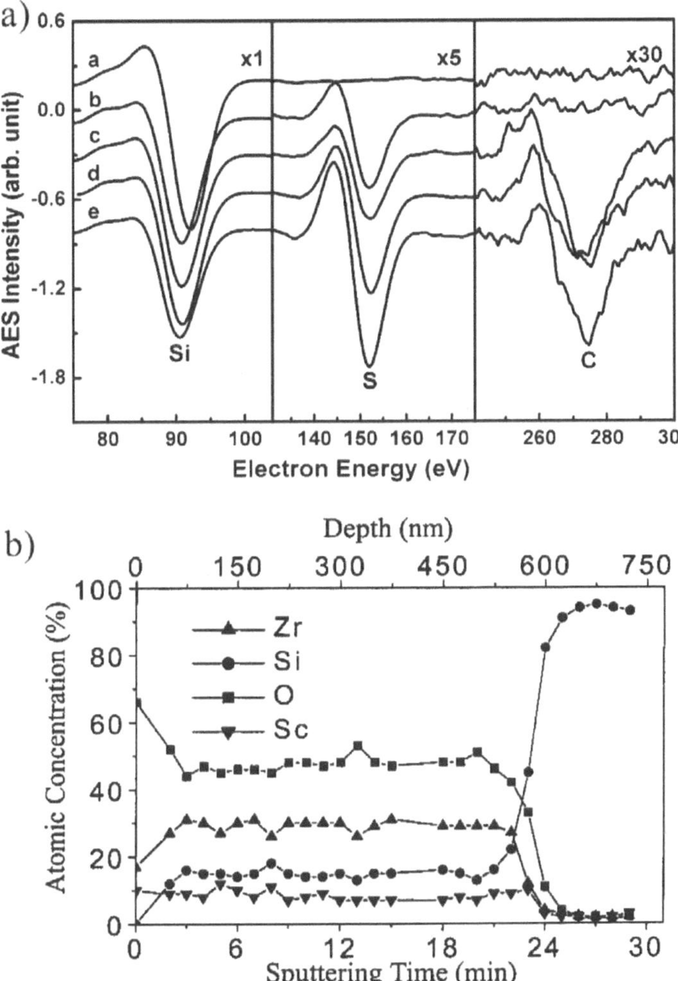

Fig. 7.53 **a** AES analysis of organosulfur compounds adsorbed onto Si(100). Shown are the Si(LVV), S(LVV), and C(KLL) transitions for **a** clean Si(100), and after saturation exposure by **b** H_2S, **c** CH_3SCH_3, **d** CH_3SH, and **e** CH_3SSCH_3. **b** AES depth profile for a $(Sc_2O_3)_{0.08}(ZrO_2)_{0.92}$ thin film deposited onto a Si(100) wafer. This profile shows a homogeneous Sc:Zr:O:Si atomic ratio between 75 and 500 nm, with an interdiffusion layer thickness of <100 nm. By examining the rise of the Si peak, the film thickness can be readily determined. Reproduced with permission from *Chem. Mater.* **2001**, *13*, 372. © 2001 American Chemical Society

The non-vacuum conditions within the sample chamber require a different type of detection system relative to conventional SEMs, referred to as a *gaseous secondary electron detector* (GSED). Due to the energetic nature of the primary beam, there is little scatter from its interaction with the gaseous medium about the sample. *En route* toward the positive electrode, the secondary electrons generated from the sample repeatedly collide with gas molecules, generating a cascade of additional electrons and positive ions. In addition to amplifying the SE signal, the positive ions migrate back to the sample surface where they dissipate the charge buildup—hence, precluding the need for conductive samples or carbon/gold coating. The aperture diameter through which the primary electron beam is passed determines the overall maximum pressure of the sample chamber. For instance, a 0.5 mm aperture dictates a maximum pressure of 10 Torr about the sample; a 1 mm aperture would lower the maximum pressure to 5 Torr.

For in situ TEM, a sample compartment with positive pressure is generated by placing a series of restricting apertures in the column near the pole piece, yielding regions of lower-pressure above/below the sample (Fig. 7.55). It should be noted that *in situ* TEM studies may also be carried out within a conventional instrument. Rather than permanently modifying the TEM column with a differential pumping

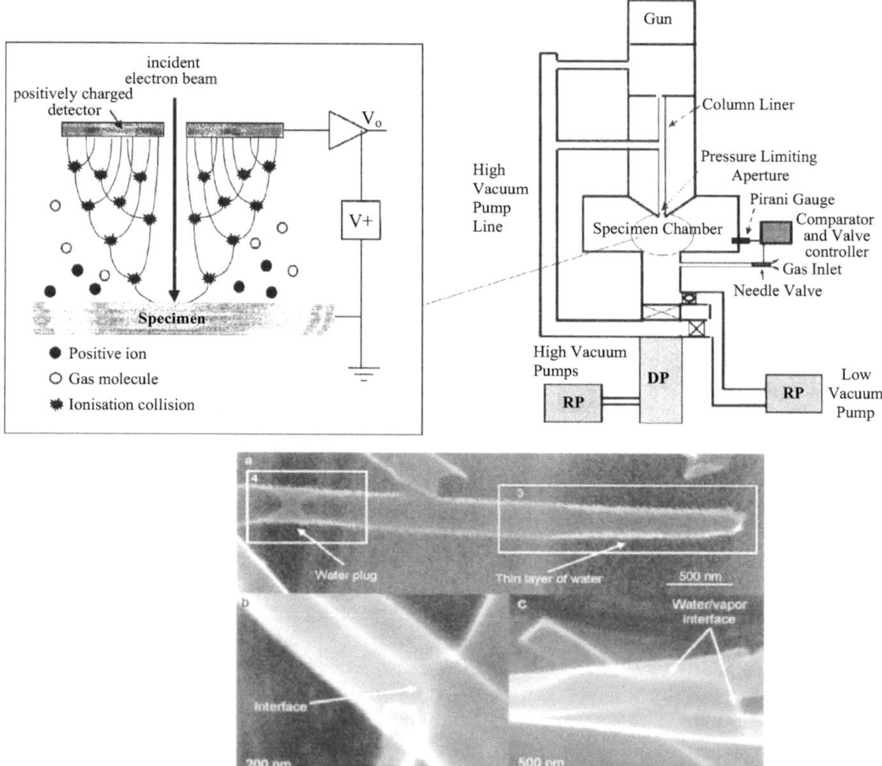

Fig. 7.54 Conventional TEM, illustrating the location where the specimen holder is inserted for *in situ* experiments. Image **c** shows a front view of the pole piece and the narrow pole-piece gap within which the specimen holder shown in **b** needs to be inserted. In the case of aberration-free microscopes, the pole-piece gap can be significantly increased while keeping the resolution high. Reproduced with permission from *MRS Bull.* **2008**, *33*, 83. © 2008 Materials Research Society

system, a sealed chamber with electron-transparent windows is used to maintain a pressurized atmosphere around the sample. TEMs may now be fitted with a number of specialized objective lens pole-pieces that will allow one to image and record real-time events at the atomic level. Gas inlets (partial pressures $\leq$ 15 Torr) may also be affixed, to allow for imaging of specimens under non-vacuum conditions. Examples of possible *in situ* functionalities include:

(i) Heating (up to 1300 K)—e.g., nanoparticle coalescence,[75] real-time growth of 1-D nanostructures.[76]
(ii) Nanoindentation—e.g., indentation response of thin films and nanostructures.[77]
(iii) Scanning tunneling microscopy (STM)[78]—e.g., field emission and electrical properties of carbon nanotubes.[79]
(iv) Tomography[80]—e.g., 3-D visualization of magnetic tunnel junctions.[81]
(v) Liquid-phase[82]—e.g., nanocrystal growth in solution.[83]

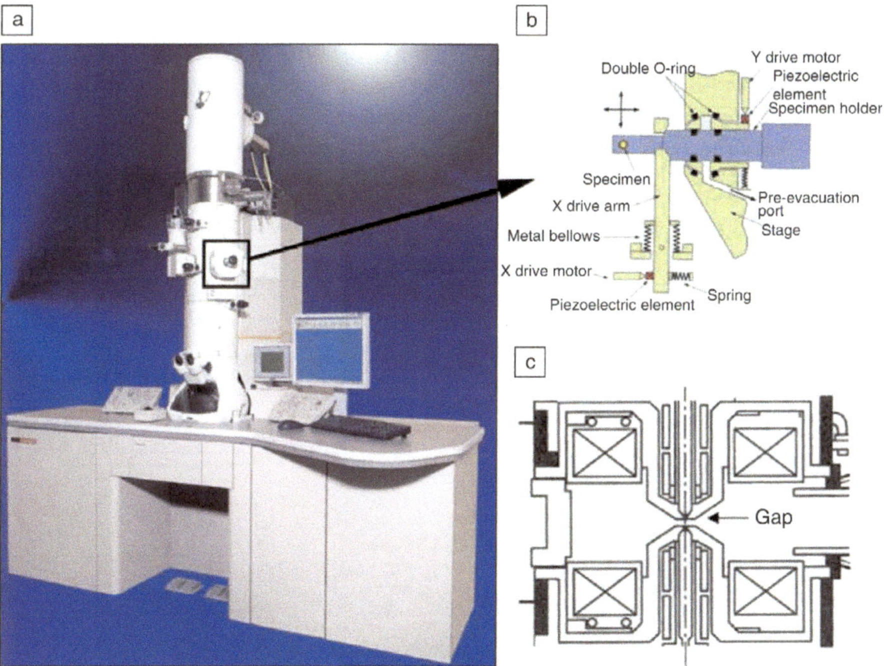

Fig. 7.55 Schematic of the differential pumping arrangement of an ESEM (top right)[74] and the cascade amplification process occurring in the gaseous secondary electron detector (GSED) (top left). Schematic reproduced with permission from Miller, A. F.; Cooper, S. J. *Langmuir* **2002**, *18*, 1310 © 2002 American Chemical Society. The abbreviations RP and DP refer to the rough pump and diffusion pump, respectively. Also shown (bottom) is an intriguing application for ESEM, the in situ imaging of water condensation inside a carbon nanotube (**a**), and underneath/around the CNT (**b**, **c**). Reproduced with permission from *Nano Lett.* **2004**, 4, 989. © 2004 American Chemical Society

7.5 Surface Characterization Techniques Based on Particle Bombardment

7.5.1 Photoelectron Spectroscopy (PES)

The photoelectric effect, first outlined by Einstein in the early 1900s, refers to the ejection of electrons from a surface due to photon impingement. However, it was not until the 1960s that this phenomenon was exploited for surface analysis—a technique referred to as *X-ray photoelectron spectroscopy* (XPS), or *electron spectroscopy for chemical analysis* (ESCA). This technique consists of the irradiation of a sample with monochromatic X-rays (e.g., Al K_α (1.487 keV), Mg K_α (1.254 keV), Ti K_α (2.040 keV)), which releases photoelectrons from the sample surface (Fig. 7.56).[84] Due to the short *free mean path* of the photoelectrons in the solid, this technique provides compositional information from only the top 1–5 nm of a sample.

Each atom in the sample has characteristic binding energies of their inner-shell electrons, referred to as *absorption edges*. In order to excite the electrons, the energy of the incident photons must be at least as large as the binding energy of the electrons. When this energy threshold is exceeded, a large absorption of energy takes place, followed by the release of photoelectrons with excess kinetic energy (7.31) in order to relax the atom back to its stable ground state. Since the binding energy of an

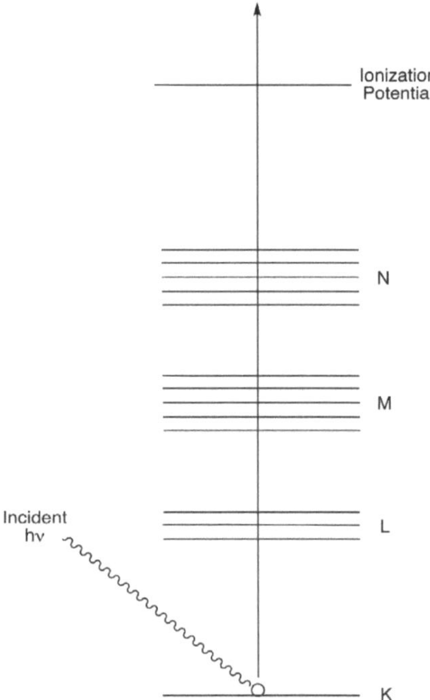

Fig. 7.56 The energy-level diagram for X-ray photoelectron spectroscopy (XPS)

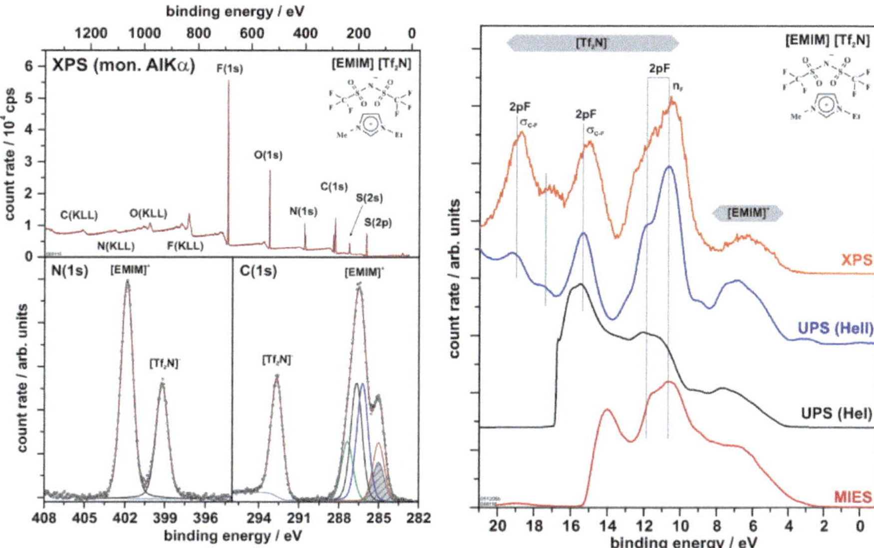

Fig. 7.57 XPS spectrum of an ionic liquid, [EMIM][Tf_2N], detailing the C (1s) and N (1s) regions. Since there are no peaks from the Au substrate, the film thickness is likely > 10 nm. Also shown (right) is the comparison between XPS, ultraviolet photoelectron spectroscopy (UPS, HeI = 21.2 eV, HeII = 40.8 eV radiation), and metastable impact electron spectroscopy (MIES). Whereas XPS and UPS provide information from the first few monolayers of a sample, MIES is used for zero-depth (surface only) analysis, since the probe atoms are excited He atoms that interact with only the topmost layer of sample. Full interpretations for these spectra may be found in the original work: *Langmuir* **2006**, *22*, 7120. © 2006 American Chemical Society

atom is altered by minute changes in its chemical environment (e.g., oxidation state, hybridization/geometry, etc.), XPS provides both elemental quantification and details regarding the chemical environment of the surface atoms (Fig. 7.57)[85]:

$$E_k = h\nu - E_b, \tag{7.31}$$

where E_k is the kinetic energy of the emitted photoelectrons, hv is the energy of the incident photons, and E_b is the binding energy of the inner-shell electrons.

In addition to using X-rays to irradiate a surface, ultraviolet light may be used as the source for photoelectron spectroscopy. This technique, known as *ultraviolet photoelectron spectroscopy* (UPS, Fig. 7.57), is usually carried out using two He lines (HeI at 21.2 eV and HeII at 40.8 eV), or a synchrotron source. This technique is often referred to as "soft PES," since the low photon energy is not sufficient to excite the inner-shell electrons, but rather results in photoelectron emission from valence band electrons—useful to characterize surface species based on their bonding motifs. It should be noted that both UPS and XPS are often performed in tandem with an

Ar^+ source, allowing for chemical analysis of the sample at depths of $\leq 1\ \mu m$ below the surface.

Although a typical XPS detector collects all emitted photons, regardless of their ejection angles, it should be noted that *angle-resolved* XPS (ARPES) and UPS (ARUPS) may also be carried out. By detecting photoelectrons emitted from a surface at different emission angles, one obtains the energy of the electrons as a function of the momentum vector, **k**. This is referred to as "band mapping," since this analysis probes the electronic structure of crystalline materials.[86]

7.5.2 X-ray Absorption Fine Structure (XAFS)

If a tunable X-ray source such as synchrotron radiation is used to generate the incident photons rather than a monochromatic beam, one can select specific absorption edges to gain more detailed information regarding the chemical environment of specific atoms. This technique, known as XAFS, is divided into two spectral regions—XANES and XAFS—which are analogous to ELNES and ELEFS in EELS studies. Whereas XANES is sensitive to the oxidation state and bonding geometry of the probed element, XAFS is useful to determine the distances and co-ordination numbers of neighboring species.[87] The small oscillations that are present in an XAFS spectrum are indicative of the interference effects that occur as the photoelectrons leave the sample surface. As you might expect, this backscattering effect becomes most pronounced with increasing photon energies, and with smaller interatomic distances between the probed atom and its nearest neighbors. As a further extension of this technique, the X-rays may be reflected from a surface at an angle in order to yield even more structural information. This technique is referred to as *total external reflection XAFS*, or REFLEXAFS,[88] particularly useful to determine chemical information from greater sample depths without the need for destructive Ar^+ etching.

In order to illustrate the complementary information yielded from XPS and XAFS, we will consider the work of Gervasini and coworkers, regarding the characterization of CuO catalysts dispersed on silica and silica/alumina supports.[89] The XANES, XAFS, and XPS spectra for these catalyst species are illustrated in Fig. 7.58. The XAFS spectrum (Cu K-edge) for a $Cu/SiO_2/Al_2O_3$ catalyst, calcined in air, has one primary peak at 1.939 Å corresponding to an (octahedral) environment of six O atoms (Fig. 7.58a, thick line).[90] The lack of other O shells suggests that the surface contains isolated Cu ions, likely as a Cu-aluminate phase. In contrast, the same catalyst on a SiO_2 support (Fig. 7.58a, thin line) shows the same octahedral environment of O atoms (1.947 Å), and another large peak corresponding to *ca.* 5 Cu atoms at an average Cu-Cu distance of 2.987 Å. Since this is shorter than the Cu-Cu distances in crystalline CuO, the XAFS spectrum suggests the formation of amorphous CuO aggregates on the SiO_2 support.

After reduction of the catalyst in a H_2 stream at the same temperature as calcination (Fig. 7.58b), the Cu/SiO_2 spectrum shows a large peak at 2.547 Å, corresponding to

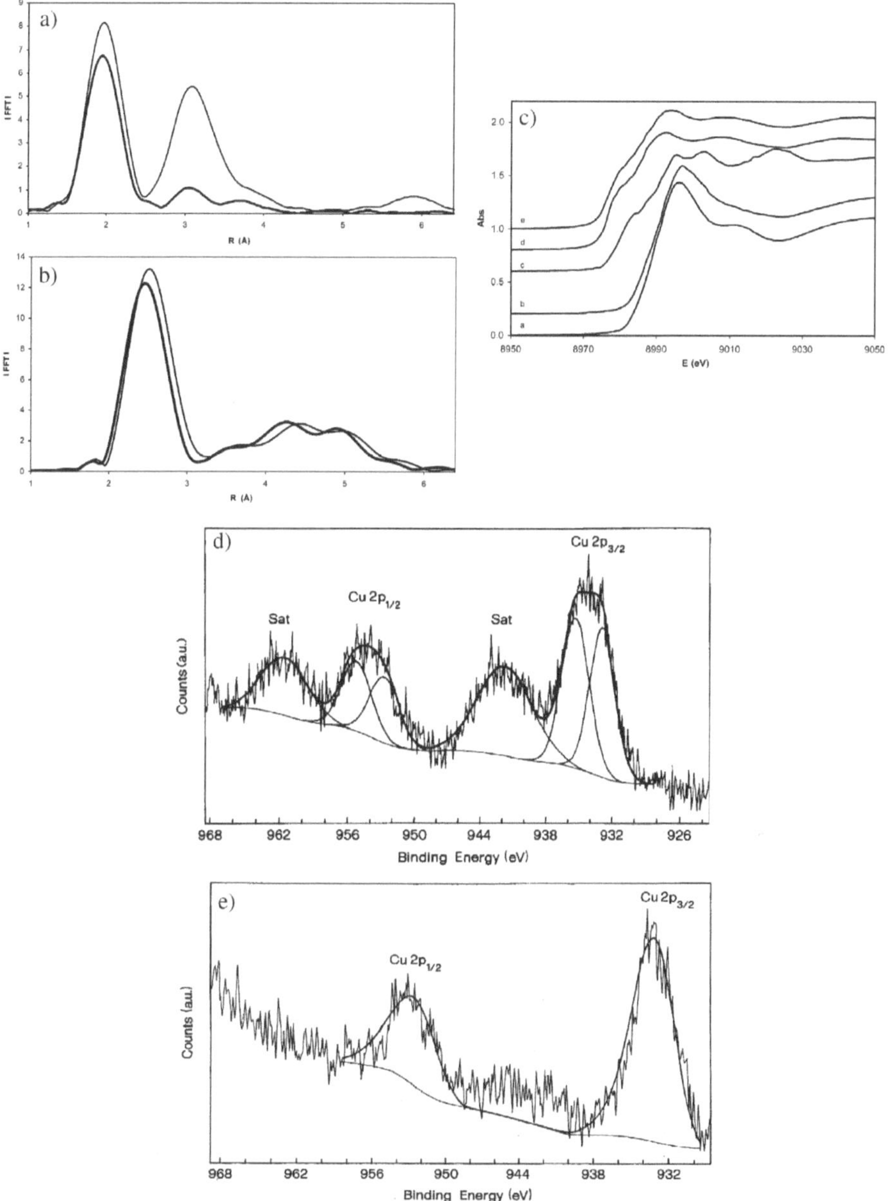

Fig. 7.58 XAFS CASE STUDY: surface characterization of dispersed CuO catalysts on silica and alumina/silica supports. Shown are: **a** EXAFS of the Cu K-edge of the catalyst calcined in air at 543 K. The thick and thin lines indicate Si/Al and Si supports, respectively; **b** EXAFS of the Cu K-edge of the catalyst reduced under a H_2 flow at 543 K; **c** Cu K-edge XANES spectra of the calcined catalyst on **a** SiO_2, **b** SiO_2/Al_2O_3 supports, along with references of **c** Cu foil, **d** Cu_2O, and **e** CuO; **d** XPS spectrum of the Cu 2p core level of the calcined catalyst on a Si support; **e** XPS spectrum of the catalyst on a Si/Al support. Reproduced with permission from *J. Phys. Chem. B* **2006**, *110*, 7851. © 2006 American Chemical Society

ca. 8 Cu atoms. The spectrum closely matches that of the fcc metallic Cu reference, suggesting the presence of small metallic Cu atoms on the surface. On the Si/Al support, the same spectral features are observed; however, line-fitting indicates that the Cu atoms are somewhat electropositive until further reduction takes place at higher temperatures. It was postulated that this resistance toward reduction is likely due to the strong association of surface Cu ions with Lewis acidic aluminum centers of the support.

Not surprisingly, the XANES spectra of Cu/Si and Cu/SiAl (Fig. 7.58c) correspond to the absorption edge of Cu^{2+}. However, the spectra are significantly different from the CuO reference. This suggests the presence of a distorted octahedral geometry about the Cu ions, where the Cu may be bound to three different types of O atoms: (i) from the support, (ii) from surface –OH groups, and (iii) from water molecules.

The XPS spectra (Fig. 7.58d, e) show a broad peak at 934 eV, corresponding to the Cu L-edge of Cu(2p $\rightarrow$ 3d transition). Since the binding energy for Cu^{2+} and Cu^{+} are similar (i.e., 933.6 and 932.5 eV, respectively), the shape of the Cu 2p peaks was examined in detail to determine the valence states of the surface Cu species. Whereas Cu on Si/Al exhibits one contribution, Cu on a SiO_2 support shows the presence of two different Cu species. In particular, this suggests that the Cu^{2+} ions interact with both =O and –OH groups from the silica support. Furthermore, satellite peaks from the spin–orbit components[91] (i.e., Cu($p_{1/2}$) and Cu($p_{3/2}$)) of the Cu(2p) peak are observed only for Cu on a SiO_2 support; this suggests that the Si/Al support contains Cu atoms in a lower valence state, such as Cu^{+} speciation.[92]

Lastly, XPS was also used to glean information regarding the dispersion of Cu on the two surfaces. The Cu($2p_{3/2}$), Si(2s), and Al(2p) peaks were integrated to yield the respective elemental concentrations of each supported catalyst. The molar ratios between the surface species (i.e., Cu(2p)/Si(2s) or Cu(2p)/Si(2s)+Al(2p)) and the total metal species (i.e., Cu/Si or Cu/(Si+Al), as obtained from inductively coupled plasma (ICP) analysis) resulted in values near 1.0 and 0.35 for the SiO_2 and SiO_2/Al_2O_3 supports, respectively. The surface deficiency of Cu on the Si/Al support is proposed to be an artifact of Cu encasement within the pores of aluminum-rich regions, as also suggested from XAFS.

7.5.3 Ion-Bombardment Techniques

Thus far, the majority of surface techniques have employed the "collide and collect" principle, where either electrons or X-rays impinge on the sample, and various emissions are collected/analyzed. However, a number of other ions may also be used for such bombardment-based analyses. For instance, we already saw an example of *metastable impact electron spectroscopy* (MIES; revisit Fig. 7.57), in which metastable He atoms[93] are directed toward a sample. As the atom collides with the sample, the 2s electron tunnels into an empty valence orbital in a sample atom, releasing an Auger electron. A similar surface-relaxation mode may be carried out by incident He ions, which is termed *ion neutralization spectroscopy* (INS). Since

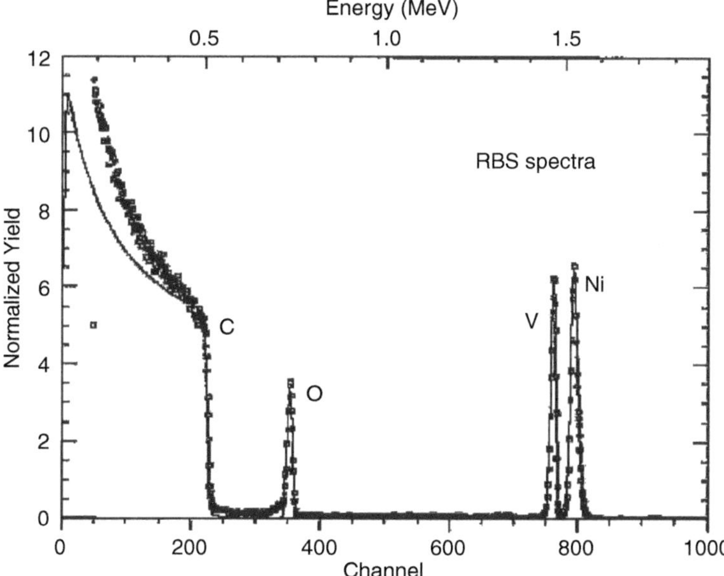

Fig. 7.59 Rutherford backscattering (RBS) spectrum of a $LiNiVO_4$ film (thickness of 240 ± 5 Å) on a carbon substrate. The open squares represent the experimental RBS data, and the continuous line is the simulated data. Reproduced with permission from *J. Phys. Chem. B* **2006**, *110*, 4301. © 2006 American Chemical Society

the incident atoms/ions are not sufficiently energetic, the beam does not penetrate through the sample surface. Hence, the spectra resulting from these techniques provide information regarding the DOS for the topmost surface monolayer of the sample.[94]

If a high-energy beam (2–4 meV) of He (α-particles) or H ions[95] collides with a surface, information may be gained regarding the composition and thickness of a surface coating (Fig. 7.59). This technique is known as *Rutherford backscattering* (RBS), very similar to the original scattering experiments conducted by Rutherford in the early 1900s—essential in the early development of a structural description of the atom. As the energetic ions penetrate the material, they lose energy through collisions with electrons, and (less frequently) with nuclei. When the positively charged He^+ ions approach the nucleus of a sample atom, they will be electrostatically repelled. As you would expect, the energy of the backscattered ions will depend on their incident energy, as well as the mass of the surface atom that causes the scattering. For instance, contact with high-Z atoms such as gold, will result in the energy of the backscattered ions being almost equivalent as the incident beam. By measuring the energy spectrum of the backscattered ions, information on the composition of the elements and their origination depth from the sample may be obtained. Hence, RBS provides facile compositional information regarding a sample, without the need for a reference sample.[96] This technique is most often applied to determine the chemical composition versus depth for heavy elements in a low-Z matrix, amenable for thicknesses of up to 0.5 μm (for He ions) and 10 μm (for H ions).

Based on our previous discussions, it is logical to assume that high-energy ions may dislodge inner-shell electrons, releasing characteristic X-rays (*cf*. EDS). This is the theoretical principle underlying *particle-induced X-ray emission* (PIXE), which uses high-energy H^+ or He^+ ions (at > 2 meV; produced from a van de Graaff accelerator[97]) as the incident beam. This technique offers a nondestructive quantitative analysis of a sample (for elements with $Z \geq 13$), often used by archaeologists and art curators to assist with dating and authenticity assessments. With low detection limits (*ca.* 1–10 ppm), and ability to analyze a number of elements simultaneously, PIXE has also been widely used for the rapid analysis of multilayer thin films for microelectronics and optoelectronics applications. A recent extension, known as micro-PIXE, allows one to focus the ion beam to a diameter of *ca.* 1 μm, to allow for a more localized analysis (Fig. 7.60). The primary benefit of this technique is the wide variety of samples that may be analyzed, from liquids to powders. Although no sample preparation is necessary for PIXE, it should be noted that the analysis is limited to the top 10–50 μm of the sample, depending on its density.

Among the techniques discussed thus far, the best Z-cutoff for elemental composition techniques has been Li (for XPS and AES). One ion-bombardment method, *elastic recoil detection analysis* (ERDA),[98] is able to nondestructively quantify light elements such as hydrogen or deuterium.[99] This technique uses a high-energy ion beam (e.g., He, Au, Cl, C, or O ions), which causes light elements to be elastically ejected from the surface (i.e., "recoiled") due to interactions between the incident beam and their atomic nuclei (Fig. 7.61). By choice of the correct angle between the sample surface and detector, the incident beam is blocked by "stopper foil" to increase the signal/noise ratio for the measurement.

Secondary-Ion Mass Spectrometry (SIMS)

The most common mode of chemical analysis presented herein has been the monitoring of elastically/inelastically scattered or recoiled incident beam species, or the analysis of a secondary emission pattern. In addition to the release of characteristic X-rays, Auger electrons, and photoelectrons, an incident beam may cause ionization of the sample. This technique is known as *secondary-ion mass spectrometry* (SIMS),[100] which represents the most sensitive surface characterization technique developed to date, with detection limits of 10^{10}–10^{15} atoms cm^{-3} (i.e., 0.1 ppb–0.1 ppm) and ability to detect/quantify any element in the periodic table. In contrast to common bulk MS methods such as matrix-assisted laser desorption/ionization (MALDI)[101] and electrospray ionization (ESI),[102] SIMS analyzes samples in their native state without the need for a suitable matrix solution. As a result, SIMS is the best choice to characterize organic-based thin films and polymer surfaces.

The operating principle of SIMS is not unlike other techniques in this section; that is, a high-energy (1–30 keV) ion source is directed onto a sample surface. However, the absorption of this energy by the top *ca.* 50 Å of the sample results in the sputtering of neutral and charged (+/−) species from the surface.[103] These ejected species

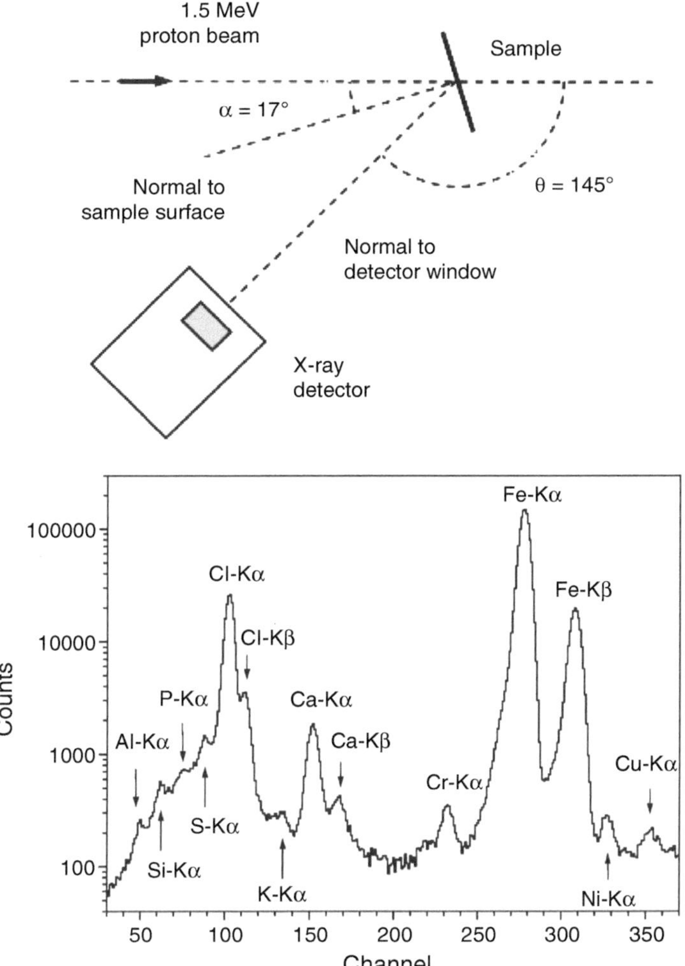

Fig. 7.60 Experimental setup and X-ray spectrum resulting from PIXE analysis of single-walled carbon nanotubes. Reproduced with permission from *J. Phys. Chem. B* **2005**, *109*, 1415. © 2005 American Chemical Society

primarily include neutral atoms and atomic clusters; however, charged molecular fragments are also released from the surface. Analogous to a traditional mass spectrometer, SIMS analyzes these secondary ions based on their relative mass/charge (*m*/*z*) ratios (Fig. 7.62).

Depending on the desired species to be analyzed, a variety of primary ion beams may be used such as Cs^+, O_2^+, O, Ar^+, and Ga^+. Whereas a cationic primary beam such as Cs^+ is used to ionize electronegative elements (e.g., O, C, N, chalcogens, halogens, etc.), oxygen atoms are used to ionize electropositive elements (e.g., Li,

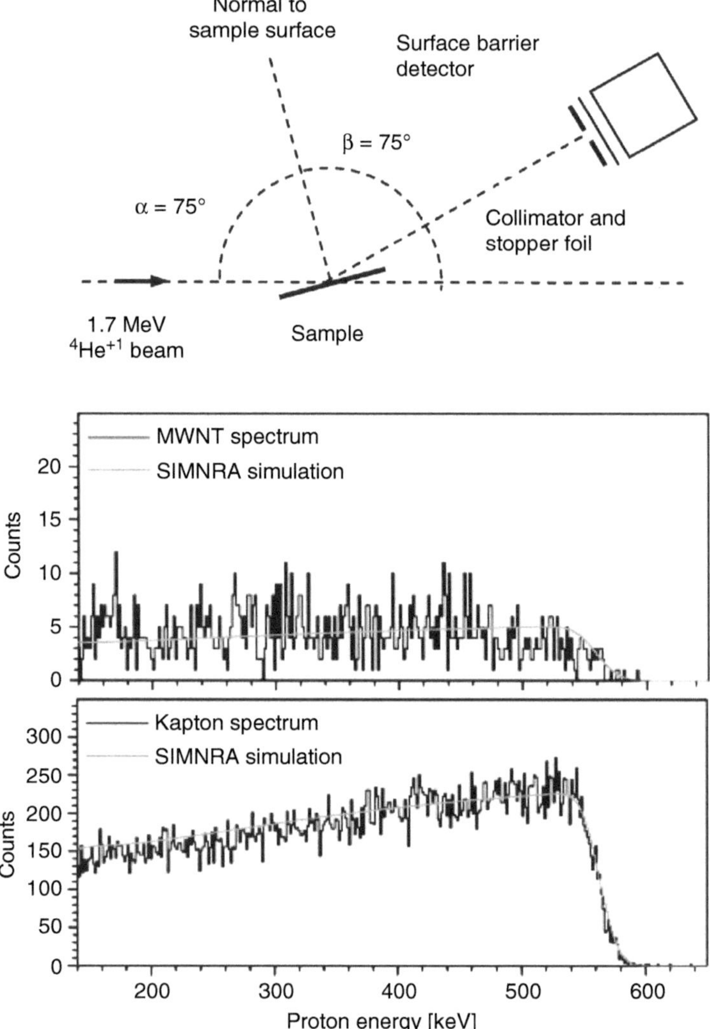

Fig. 7.61 Experimental setup and elastic recoil detection (ERD) spectra for multiwalled nanotubes, and a Kapton polyimide film. A computer program, SIMRA, is useful to simulate ERD as well as other non-RBS data. Reproduced with permission from *J. Phys. Chem. B* **2005**, *109*, 1415. © 2005 American Chemical Society

Mg, Na, B, Al, etc.). Liquid metal ion beams (e.g., Ga, Au/Ge) are used for high-resolution studies, since the beam may be focused to a smaller diameter (<50 nm) for micron and nanoscale analyses.

There are two varieties of SIMS—static and dynamic. *Static SIMS* (often referred to as time-of-flight SIMS, TOF–SIMS) is often the method-of-choice, used for

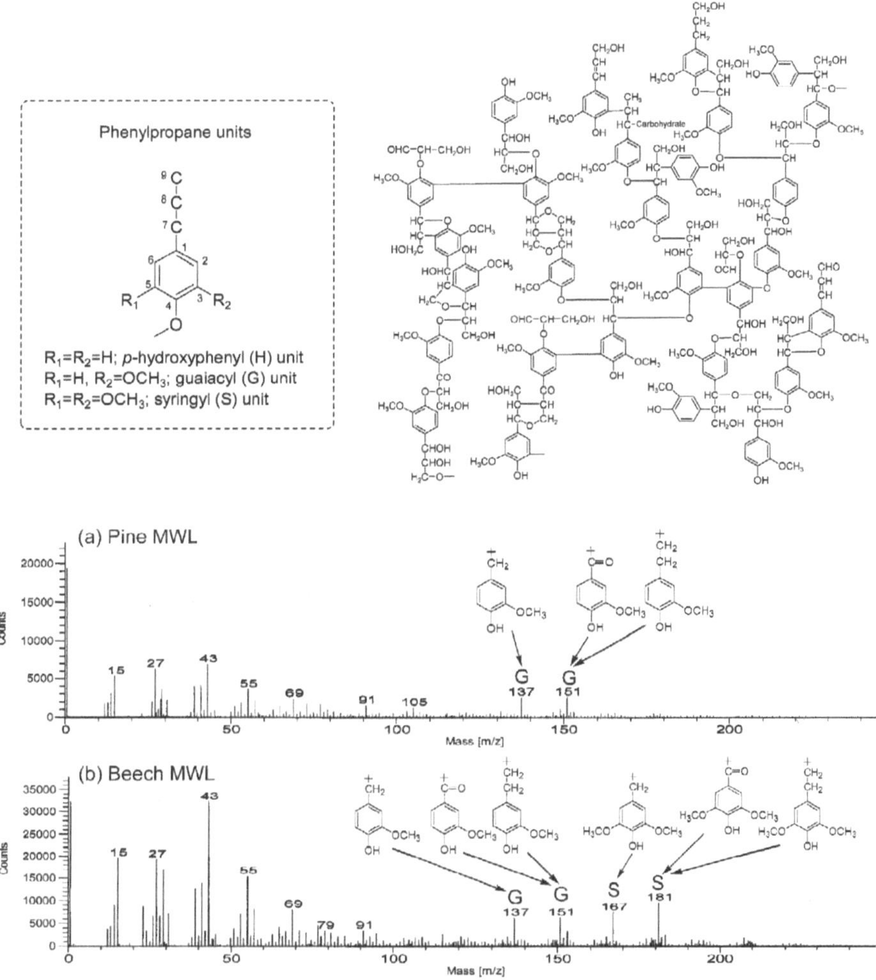

Fig. 7.62 SIMS analysis of lignin, the second most abundant biopolymer in nature, following cellulose. Shown is the phenylpropane subunits and a structural model of softwood lignin (top). The secondary-ion mass spectra of pine (softwood) and beech (hardwood) milled wood lignin (MWL, a solvent-extracted form of lignin from beech wood) are also shown (bottom). Reproduced with permission from *Biomacromolecules* **2005**, *6*, 678. © 2005 American Chemical Society

elemental analysis and imaging of the top 2–3 monolayers of a sample; in comparison, *dynamic SIMS* is used to determine elemental concentrations of the sample, as a function of depth. As such, dynamic SIMS is a destructive technique primarily used for depth profiling, whereas TOF–SIMS does not appreciably deteriorate the surface being analyzed. For instance, due to a slow, controllable sputtering rate, the entire analysis may be performed without removing less than 1/10 of an atomic monolayer.

It should be noted that SIMS will result in surface roughness/cratering and elemental mixing within the outermost monolayers. When the primary beam interacts with surface atoms, the incident energy is transferred to target atoms through a series of binary collisions. Some atoms/fragments receive sufficient energy to be recoiled through the sample surface and are removed from the material. However, other less noticeable effects will result such as the implanting of primary ions and mixing with sample atoms at depths of $\leq$ 10 nm below the surface. The magnitude of these effects is governed by the incident beam,[104] as well as the crystallinity and surface defects of the sample.

Dynamic SIMS typically uses a quadrupole mass analyzer; however, a TOF mass analyzer offers much higher sensitivities (with limitations, vide infra) and mass ranges. In a TOF–SIMS, an ion of known electrical charge, but unknown mass, is accelerated by an electrical field. As a result, all ions of the same charge will have identical kinetic energies. However, the velocity of the ion will depend on the mass/charge ratio, which affects the time it takes for the charged species to reach the detector. That is, light ions will reach the detector before heavier species, allowing for effective mass separation and analysis. It should be noted that each pulse of the primary ion beam produces a full spectrum of secondary ions. Hence, the highest mass resolution is afforded by using the shortest pulse of the primary ion beam, since the time-of-flight for ions may be determined more precisely.

The sputtering event will yield secondary ions with a range of kinetic energies. Whereas atoms will contain their kinetic energy only in translational modes, charged molecular fragments will have the kinetic energy spread over translational, vibrational and rotational modes. By varying the accelerating voltage of the electrostatic ion energy analyzer, one is able to select secondary ions (e.g., monoatomic vs. multiatomic (molecular fragments)) based solely on their translational energies. For instance, ions with a lower energy are deflected at larger angles, which prevents them from passing through to the mass analyzer.

Due to more energetic collisions, an incident beams consisting of either heavier ions (e.g., Au^+ vs. Ga^+), or polyatomic species (e.g., C_{60}^+,[105] Au_3^+, SF_5^+, Bi_3^+), will result in much greater secondary ion yields for high molecular components—especially in the range of 500–5000 Da. For instance, the use of an atomic cluster such as C_{60}^+ results in a significant number of surface atoms being sputtered from the surface, relative to a smaller beam such as Ga^+ (Fig. 7.63). For a 15 keV incident beam, each carbon atom would contain 15,000/60 eV (i.e., 24,000 kJ mol^{-1}, or 50 times greater than the C–C bond strength in C_{60})! Hence, the atomic cluster likely behaves as 60 individual 250 eV atoms with a much broader impact on the surface. However, since the kinetic energy of each C_{60} cluster is much less than a Ga^+ ion (15 keV), the sputtering event for C_{60}^+ is limited to the first few monolayers of the sample.[106]

Another strategy that is used to increase the TOF–SIMS signal for polymeric samples is the application of a metallic coating[107] or nanoparticles[108] onto the sample surface (Fig. 7.64). The signal enhancement is likely a result of facilitated electron loss (7.32), as well as the formation of conjugate ions that contain the neutral polymer

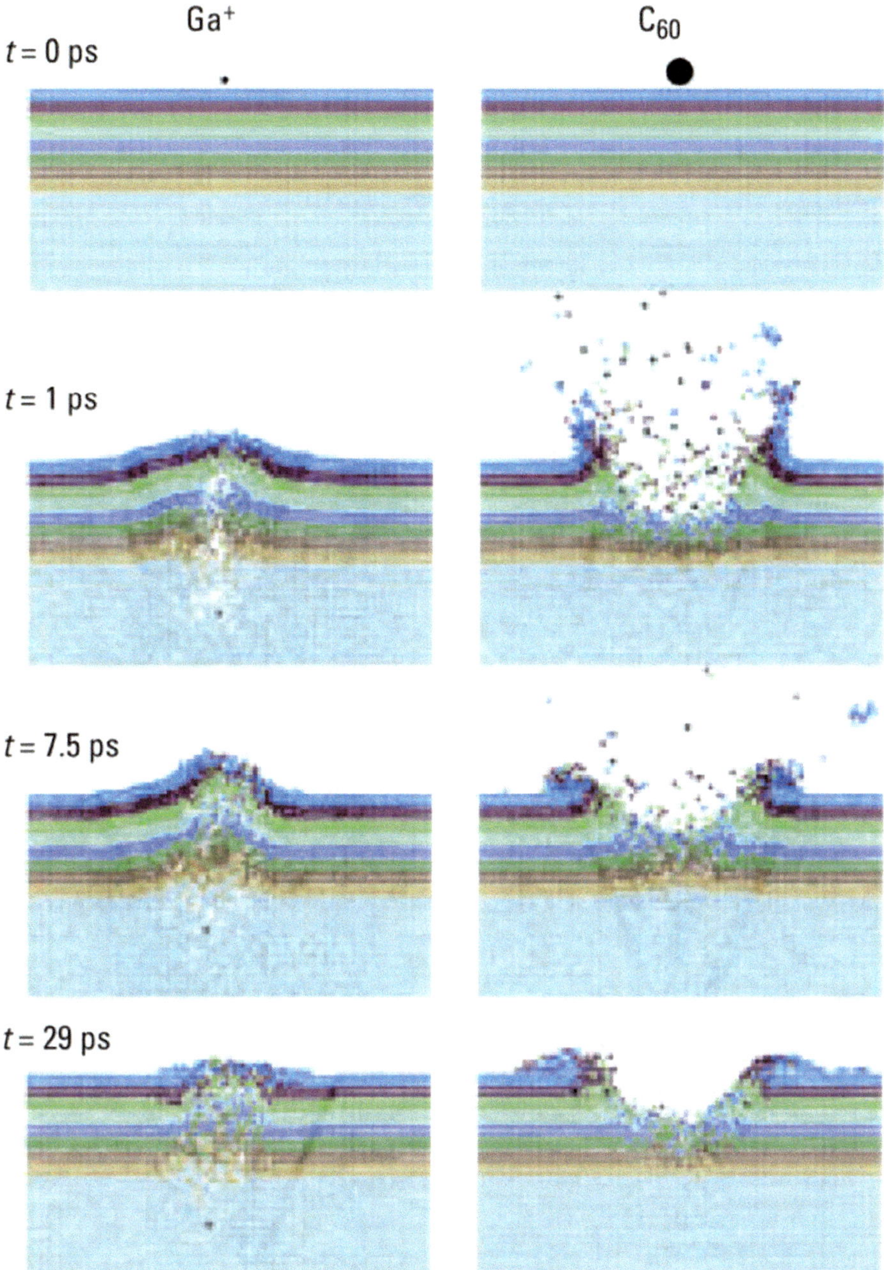

Fig. 7.63 Molecular dynamics simulation of the ion bombardment of a Ag crystal surface with a 15 keV Ga^+ beam, and a 15 keV C_{60} atomic cluster beam. The C_{60} beam results in a larger crater and more material removed from the surface. In contrast, the Ga^+ beam results in a destructive effect at greater sample depths, without successful sputtering of the sample. Reproduced with permission from *Anal. Chem.* **2005**, *77*, 142A. © 2005 American Chemical Society

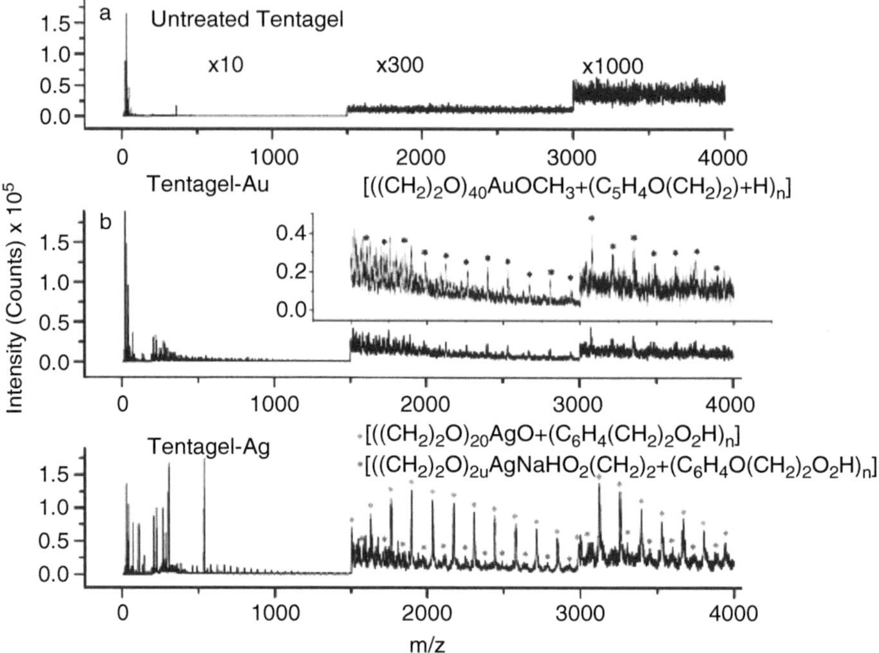

Fig. 7.64 SIMS signal enhancement from metal nanoparticle deposition. Shown are comparative spectra in the high molecular weight range for an untreated polymer (**a**), along with nanoparticle-treated samples (**b**, **c**). Reproduced with permission from *Anal. Chem.* **2006**, *78*, 141. © 2006 American Chemical Society

fragment and metal atoms (7.33) and (7.34). The reaction between the polymer fragment and a proton, either in the gas-phase or at the metallic surface, may also result in signal enhancement (7.35):

$$M \rightarrow [M]^+ + e^- \tag{7.32}$$

$$[M{+}Me]^+ \rightarrow [M]^+ + Me^0 \tag{7.33}$$

$$M + [Me]^+ \rightarrow [M{+}Me]^+ \tag{7.34}$$

$$M + H^+ \rightarrow [M{+}H]^+ \tag{7.35}$$

where M is the neutral polymer fragment; Me, neutral metal atom (present as a film or nanoparticles).

In addition to secondary-ion generation, SIMS also results in the release of secondary electrons near the sample surface. As previously discussed for SEM analysis, this may result in the buildup of a net electric current for nonconductive surfaces. Analogous to microscopy applications, such surface charging will diffuse the primary beam, making it difficult to perform microanalysis using SIMS. Furthermore, charging will deleteriously affect the detection of secondary ions by altering the energy distribution of the recoiled ions. These effects are magnified when a sample contains a dielectric surface and a conductive substrate. Due to the strong induced electric field, mobile metal ions in the substrate may migrate toward the surface. Consequently, the data obtained from depth profiling will no longer reflect the original composition of the sample layers. Since information is gleaned from the uppermost portion of the sample, the application of a simple conductive coating is not feasible for SIMS. Rather, the effects of charging may be minimized by the following techniques:

(i) *Electron bombardment*—typically used in quadrupole (dynamic) SIMS instruments. A low electron beam is used to compensate for a positive charge buildup.
(ii) *Surface conducting grid*—a metallic grid is placed over the sample. When the primary beam impinges this surface, the grid releases secondary electrons that dissipate the buildup of positive charge. Alternatively, a gold or carbon coating may be sputtered onto the sample surface, and then removed from the area to be analyzed—leaving a conductive film that encapsulates the area of interest.
(iii) *Use of negative primary ion beams*—for example, oxygen.

Another widely used application for SIMS is *ion imaging*, which shows secondary ion intensities as a function of spatial location on the sample surface (Fig. 7.65, top).[109] Further, if imaging is performed in tandem with depth profiling, a three-dimensional compositional map of a sample may also be generated (Fig. 7.65, bottom). Two modes of imaging via SIMS are possible:

(i) *Ion microscopy*—the sample is uniformly irradiated and the image is formed directly on a viewing screen via optics similar to SEM (max. lateral resolution of *ca.* 0.5 μm).[110]
(ii) *Ion microprobe*—the ion beam is rastered across the surface (max. lateral resolution of *ca.* 20 nm, using liquid metal ion guns).[111]

For materials that vary composition on extremely sharp depth scales such as ultra-shallow implants and oxynitride barriers in integrated circuits, another technique known as *low-energy electron induced X-ray emission spectrometry* (LEXES) is more appropriate.[112] During depth profiling of lightly-doped Si substrates, SIMS suffers from interference issues due to the background interference from molecular ions of Si. Although energy filtering may be used to improve the signal-to-noise, this limitation significantly affects its measurement precision.[113] In LEXES, the sample is bombarded with low-energy electrons, which causes the release of soft

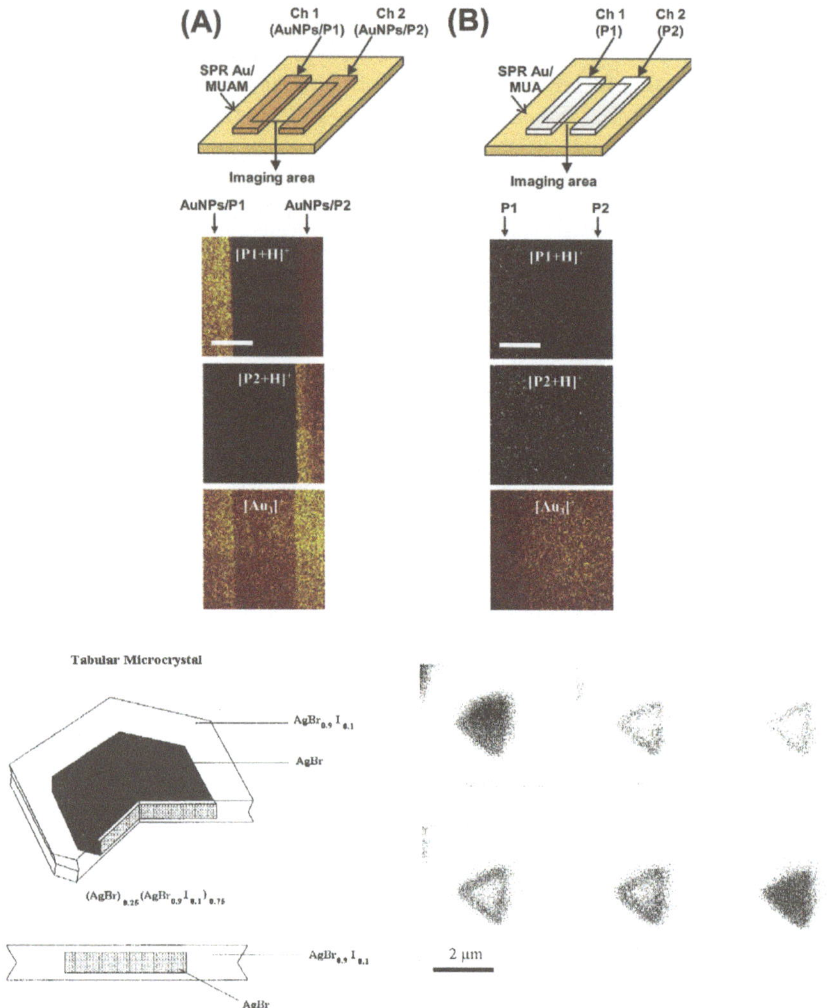

Fig. 7.65 Ion imaging using SIMS. (Top) Positive TOF–SIMS images of fragments from peptides (P1, Ac-PRNYVTP-NH_2; P2, Ac-PRNYpVTP-NH_2) patterned on SAMs, pretreated (**a**) with gold nanoparticles and (**b**) without gold nanoparticles. The lateral region at both sides (both arrows) represents a microfluidic pattern of different peptides (P1 and P2). TOF–SIMS utilized a Au^+ primary ion gun, with an ion dose $<10^{13}$ ions cm^{-2} at a sample area of 300 × 300 μm^2. The scale bar is 100 μm. Images reproduced with permission from *Anal. Chem.* **2006**, *78*, 1913. © 2006 American Chemical Society. (Bottom) Schematic of a silver halide microcrystal with a core–shell structure. The diameter is *ca.* 0.5–2 μm; the thickness is *ca.* 150–400 nm. To the right of the illustration are successive I^- ion images (using a Cs^+ primary beam) for the crystal, shown from the top left corner to the bottom right corner, as a function of depth profiling. Whereas the first and last images show a homogeneous iodine-rich array, the middle images show the I^- distribution localized in the shell of the microcrystal. Images reproduced with permission from *Anal. Chem.* **1997**, *69*, 3772. © 1997 American Chemical Society

X-rays from sample atoms that are detected by wavelength-dispersive spectrometer(s). The analyzed depth may be varied between 1–700 nm, depending on the electron energy and sample being analyzed. Since characteristic X-rays are generated from sample elements, selective elemental analysis is achieved with less background interference.[114]

7.5.4 Atom-Probe Tomography (APT)

Atom-probe microscopy/tomography is based on *field-ion microscopy* (FIM), pioneered by Muller in the late 1950's to image the arrangement of atoms at the surface of a sample. In this technique, a sample is sharpened to a tip of radius <50 nm via electropolishing,[115] FIB,[116] etc. The sample tip is then placed into a UHV chamber, cooled to a temperature of *ca.* 20–100 K, and a given a positive voltage (15–65 V nm^{-1}). The chamber is then backfilled with an imaging gas such as He or Ne. Field-ionization of the imaging gas atoms occurs via quantum mechanical tunneling of the outermost valence electrons of He/Ne into atoms within the sharply pointed tip. The generated He^+/Ne^+ ions are then repelled from the positively charged ions of the nanotip sample and are accelerated toward a negatively charged multichannel plate (MCP) detector. Unlike conventional (electron, light) microscopy, FIM is a lens-less technique that operates via point projection. Accordingly, the resolution is not wavelength-limited and is able to easily reach >2,000,000 X, which is related to the ratio of the tip-to-MCP distance and curvature of the sample microtip.

If the positive charge applied to the sample nanotip is increased, it is possible to field-evaporate sample atoms from the tip as positively charged ions. Field-evaporation may be induced by high-voltage pulsing or pico-/femtosecond laser pulsing. Hence, FIM allows one to this technique allows one to etch away sample atoms one at a time, with positional information. However, if one also includes a time-of-flight (TOF) mass spectrometer, the m/z ratios of individual ions may also be determined, which combines the features of high-resolution TEM (atomic resolution) and secondary ion mass spectrometry (elemental analysis). The combination of these capabilities is referred to as *atom-probe field ion microscopy* (APFIM), or simply *atom-probe microscopy* (APM). Since the positional information may be reconstructed into a 3-D map of the sample, this technique is also referred to as *atom-probe tomography* (APT).[117] Figure 7.66 illustrates the most common type of APT, known as *local-electrode atom-probe* (LEAP) tomography.[118] The power of this technique is amazing; used to illustrate one of the many exciting precedents, Fig. 7.67 shows the spatial positions of individual Ge atoms and P dopants within a single Ge nanowire![119] More recently, Perea et al. have used this technique to determine the location and nearest neighbors of Al within ZSM-5 zeolite crystals.[120]

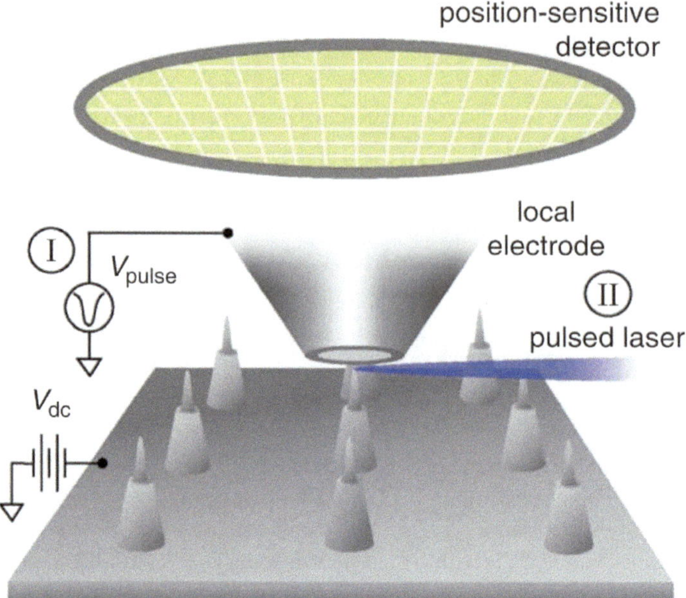

Fig. 7.66 Illustration of a local-electrode atom-probe (LEAP) tomograph employing either voltage, I, or laser pulses, II. The substrate, which contains an array of microtip specimens, is maintained at a positive potential, V_{dc}; the substrate can be translated in the x–y plane. The local electrode is pulsed with a negative voltage, V_{pulse}, to increase the **E**-field at a microtip specimen to the requisite value for field evaporating atoms as ions. Alternatively, a specimen is pulsed with femto- or picosecond laser pulses at a high repetition rate. A field-ion microscope (FIM) looks physically identical to a LEAP tomograph, but it does not have a delay-line detector. The multichannel plate is used to obtain a FIM image of the atoms on the surface of a microtip employing helium or neon as an imaging gas. Thus, a LEAP tomograph also can be used as a FIM. Reproduced with permission from *MRS Bull.* **2009**, *34*, 717. © 2009 Materials Research Society

7.6 Scanning Probe Microscopy (SPM)

The characterization techniques discussed thus far have involved the interaction of an incident beam of radiation (electrons, X-rays, ions, or light) with a sample. Subsequent scattering, transmission, and/or secondary emission is then used to provide an image of the sample, or delineate its chemical composition. This section will discuss a technique that does not involve irradiation, but rather features the scanning of a probe across a surface (hence, termed SPM). A tangible analogy of this technique is the precursor to modern CD players—phonograph/record players (check your local antique shop if you have not seen one before)![121] The heart of a phonograph is a cantilever-supported stylus that followed the grooves on the record. The mechanical movement of the stylus within the grooves results in an electrical response from a piezoelectric crystal, which is converted into sound via the speaker system. Scanning probe microscopy represents a next-generation application for a stylus with a much

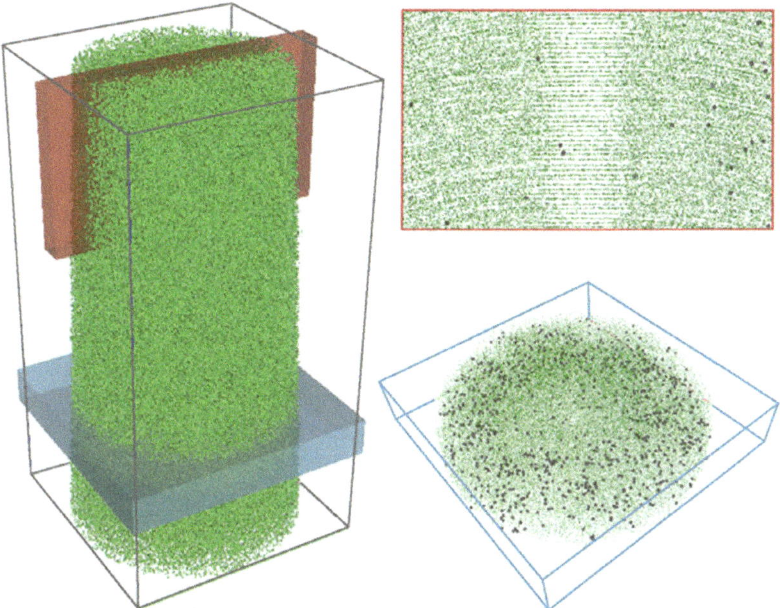

Fig. 7.67 Atom-by-atom mapping of a germanium nanowire by atom probe tomography. Left: 3-D reconstruction of an individual Ge nanowire with each green sphere representing an individual Ge atom. The dimensions are $50 \times 50 \times 100$ nm^3. The region enclosed by the red box is displayed at upper right, with single atomic planes visible in the center of the image. The grey spheres are phosphorous dopant atoms used to control the conductivity. (The dimensions are $5 \times 25 \times 15$ nm^3). The region enclosed by the blue box is displayed in the lower right, revealing an inhomogeneous distribution of phosphorous atoms. (The dimensions are $50 \times 50 \times 10$ nm^3). The 'shell' of enhanced doping results from surface reactions during growth of the nanowire. Reproduced with permission from *Nat. Nanotechnol.* **2009**, *4*, 315. © 2009 Nature Publishing Group

sharper tip (radius of curvature of <50 nm), to follow the surface topography of microscale and nanoscale materials. The resolution of SPM is among the best of all methods to date, with the possibility of atomic scale manipulation/imaging (revisit Fig. 6.18—the manipulation and imaging of individual Xe atoms—performed almost 20 years ago!).[122]

Analogous to record players, a SPM tip is supported by a flexible cantilever. During analysis, the tip is slowly rastered across the surface of a material—either a few Angstroms away from the surface (noncontact mode), or in contact with the sample (contact mode). There are two primary forms of SPM[123]:

(i) *Scanning tunneling microscopy* (STM)—set the precedent for SPM in the early 1980s. The tip is held a few Angstroms from the sample surface during scans.
(ii) *Atomic force microscopy* (AFM)—the tip may interact with the sample surface through either contact or noncontact modes.

Both types of SPM are most often used to provide an image of the surface topography. Although both techniques are frequently carried out under ambient temperatures/pressures, or even in liquids,[124] they may also be performed in an UHV chamber (*ca.* 10^{-11} Torr)[125] in order to prevent sample contamination, prevent oxidation of the surface, or monitor the deposition of sputtered thin films in real time. As a STM tip is scanned across the sample, a quantum mechanical tunneling current is established between the sample surface and a sharpened W or Pt/Ir tip (Fig. 7.68, top). The magnitude of the current is dependent on the distance between the tip and the surface (Δz), as well as the local density of states of the surface. Accordingly, STM is typically performed on conductive and semiconductive surfaces. During topographic imaging, a feedback loop is established to maintain a constant current through varying the distance between the tip and surface ("constant current mode"). In this respect, STM is able to provide real-time, three-dimensional images of the surface with atomic resolution. The use of STM for nanofabrication also represents an important application that is of increasing interest.[126]

In addition to imaging applications, *scanning tunneling spectroscopy* (STS) may be performed, which delineates the local electronic structure of a surface. In this mode, the feedback loop is interrupted, which fixes the distance between tip and sample. A current vs. voltage (*I–V*) curve may then be acquired at a specific (*x*, *y*) position on the surface by ramping the bias voltage and recording the tunneling current. If the *I–V* curves are collected at every point, a three-dimensional map of the electronic structure may be generated.[127]

Rather than monitoring electronic tunneling phenomena, AFM measures the forces between the tip and surface, which depends on the nature of the sample, the probe tip, and the distance between them (Fig. 7.68, bottom).[128] The deflection of the tip in response to surface–tip interfacial forces is recorded by using a laser focused on top of the Si or SiN cantilever and reflected onto a photodetector. The signal emanating from the photodetector is used to generate a surface topographic map, or the direct measurement of surface intermolecular forces. As with STM, a feedback loop is present in the system, which controls the distance between the tip and sample via an electrical current sent to piezoelectric transducers. Such "constant force" scanning is used most frequently, since "constant-height" scanning could result in collisional damage between the tip and surface.

The most common operating modes of AFM are contact, noncontact, and tapping, which are self-explanatory in their manner of interrogation of the surface. In contact-mode AFM, there is a repulsive force between the sample and tip (*ca.* 10^{-9} N); the piezoelectric response of the cantilever either raises or lowers the tip to maintain a constant force. Similarly as STM, the best resolution will be obtained under UHV conditions. That is, in an ambient environment, adsorbed vapors (e.g., N_2, H_2O) form a layer on the surface with a thickness of *ca.* 10–30 monolayers. Consequently, a meniscus will form between the tip and surface, which results in the attraction of the tip toward the surface due to surface tension forces. This force may be neutralized by operating the AFM in a liquid cell, in which the tip is completely immersed in a solvent (Fig. 7.69). It should be noted that frictional forces are not always detrimental;

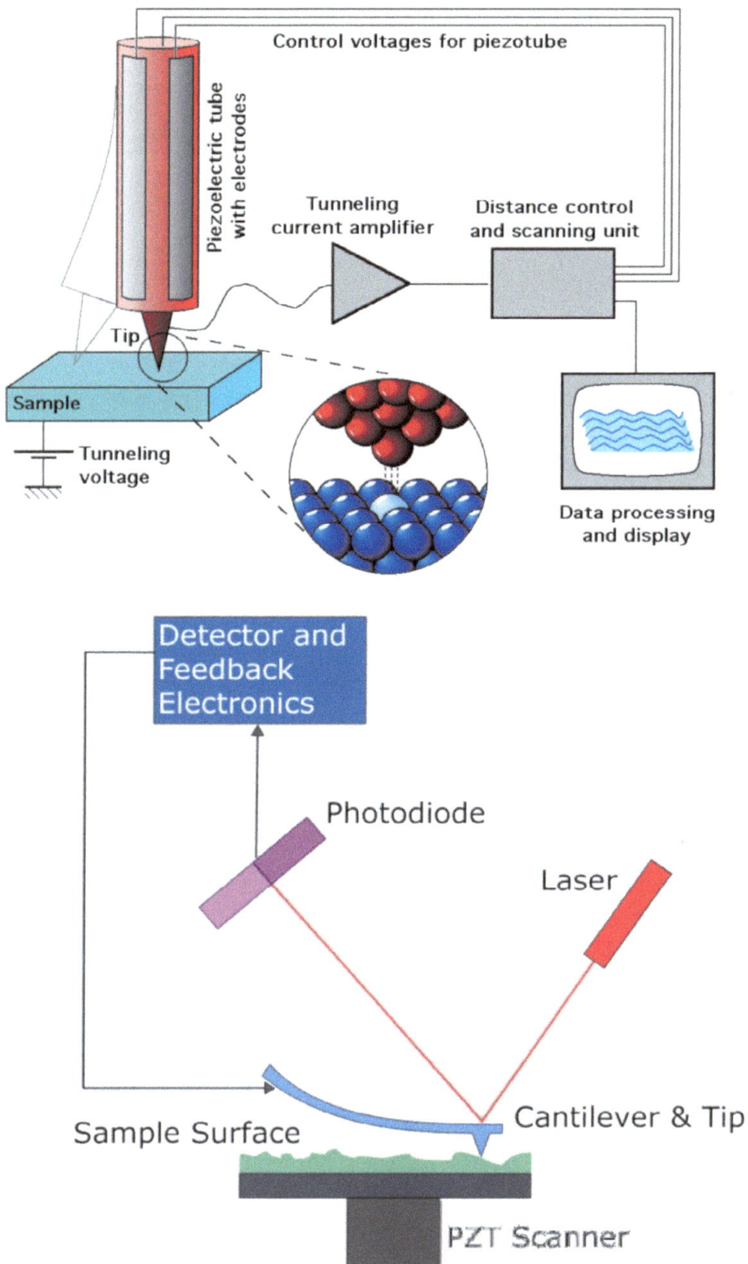

Fig. 7.68 Illustrations of STM (top), and AFM (bottom). Reproduced with permission through the Wikipedia Commons agreement. The illustration of STM is courtesy of Michael Schmid, TU Wien, and is available online at: http://www.iap.tuwien.ac.at/www/surface/STM_Gallery/stm_schematic.html

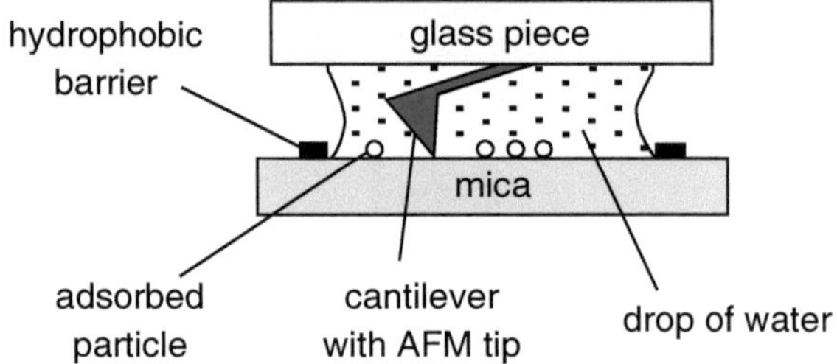

Fig. 7.69 Illustration of an AFM system used for in situ studies within a liquid. Reproduced with permission from *Langmuir* **2006**, *22*, 7254. © 2006 American Chemical Society

lateral force AFM, or the purposeful dragging of the tip along the surface, is useful to determine spatial variations in the composition or phase of a surface.

Noncontact AFM overcomes the frictional and adhesive forces between the tip and sample by hovering the tip a few Angstroms above the surface. In this mode, the attractive van der Waal forces between the tip and surface are monitored. As you might expect, these attractive forces are much weaker than those generated in contact mode, often resulting in lower resolution. In order to improve the sensitivity of the tip, the cantilever is oscillated to better detect surface features through small variations in the oscillation wave characteristics (i.e., phase, amplitude, etc.). It should be noted that noncontact AFM is often preferred over STM (also noncontact) to study "true" molecular assemblies on surfaces. That is, it has been shown that STM imaging may induce irreversible changes in molecular arrangements on surfaces, especially for high-resolution, constant-height imaging.[129]

During tapping AFM, the cantilever oscillation amplitude remains constant when not in contact with the surface. The tip is then carefully moved downward until it gently taps the surface. As the tip passes over an elevated surface feature, the cantilever has less room to oscillate, and the amplitude of oscillation decreases (vice versa for a surface depression). The oscillation frequency is typically 50–500 kHz, with an amplitude of *ca.* 30 nm, which is sufficient to overcome the adhesive forces that are evident in contact (and noncontact) modes (Fig. 7.70). Consequently, the tapping mode is most appropriate for soft samples such as organics, biomaterials, etc.

Without question, AFM exhibits a much greater versatility for surface analysis than STM. In particular, the following variations are possible, through altering the nature of the tip:

(i) *Chemical force microscopy* (CFM)—uses a chemically modified tip to examine interfacial behavior between the sample and functional groups on the tip surface.[130]
(ii) *Magnetic force microscopy* (MFM)—uses a noncontact magnetic-susceptible tip to map the magnetic properties of a surface, with spatial resolutions of <20 nm.[131]
(iii) *Scanning thermal microscopy* (SThM)—uses a resistive Wollaston wire instead of a conventional AFM probe, which acts as a localized heating source and microthermocouple, used to map the thermal conductivity of a surface.[132]

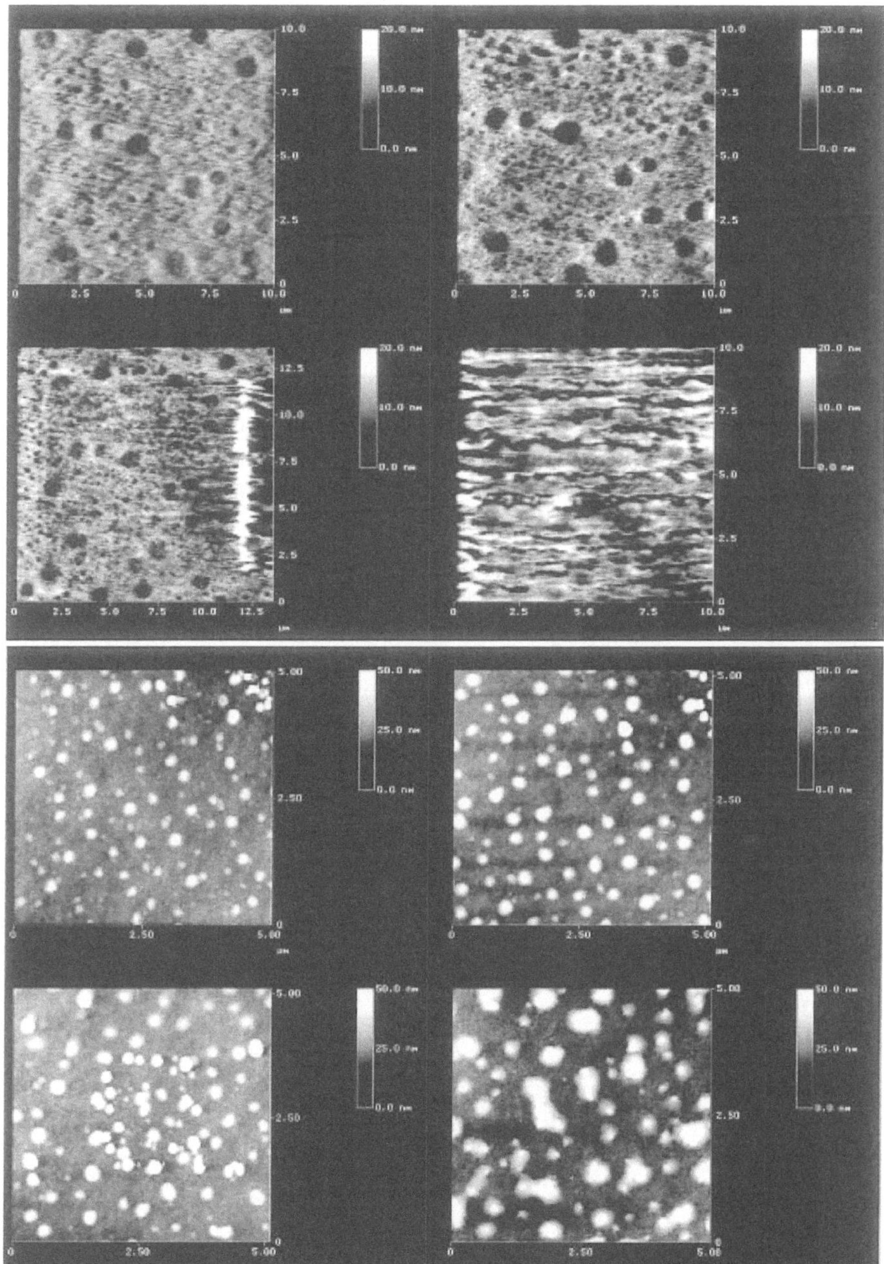

0
2.5
5.0
7.5
10.0
12.5
µm
20.0 nm
10.0 nm
0.0 nm
2.50
5.00
50.0 nm
25.0 nm

◂**Fig. 7.70** Comparative images of contact-mode (top) and tapping-mode (bottom) AFM under an ambient atmosphere. The images represent the surface of freshly cleaved potash–lime–silica glass after the elapsement of time intervals, from upper left to bottom right (top: 150, 264, 264 min (larger scan area), 8 h; bottom: 2, 5, 5 h (larger scan area), 43 h). Humidity-induced pit formation is only observable using contact mode (CM)—note the movement of surface corrosion products by the AFM tip (top, images **c** and **d**). In contrast, tapping-mode (TM) revealed round features that became visible after 2 h of exposure time. These surface features (observable only by TM) are likely swelled glass material due to the breakdown of the silicate network through water adsorption and subsequent surface ion-exchange reactions. Contact-mode AFM is not able to image these surface features, since the tip scrapes away the silicate residue during analysis. Reproduced with permission from *Anal. Chem.* **1997**, *69*, 1012. © 1997 American Chemical Society

(iv) *Scanning electrochemical microscopy* (SECM)—based on the electrochemical interaction between a redox-active species produced at the tip, and the substrate being studied (Fig. 7.71). Hence, this method offers high chemical selectivity, and is able to correlate localized surface features with their chemical reactivities.[133]

In order to carry out the above applications (as well as standard imaging/force measurements), AFM tips may be selected from among a wide variety of sizes and shapes.[134] A discovery that will greatly assist in the characterization of nanomaterials is the fabrication of tips that are terminated with individual gold nanoparticles or single-walled carbon nanotubes (Fig. 7.72).[135] These advanced designs will offer significant improvements in resolution (and artifact generation) over conventional tips.[136] Furthermore, analogous tip designs will allow one to probe the physical, thermal, magnetic, and optical properties of individual nanoarchitectures via SPM.

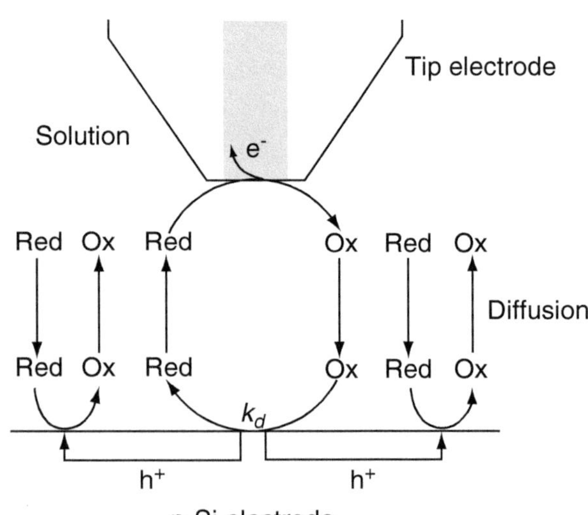

Fig. 7.71 Illustration of scanning electrochemical microscopy (SECM). Reproduced with permission from *Anal. Chem.* **2006**, *78*, 6019. © 2006 American Chemical Society

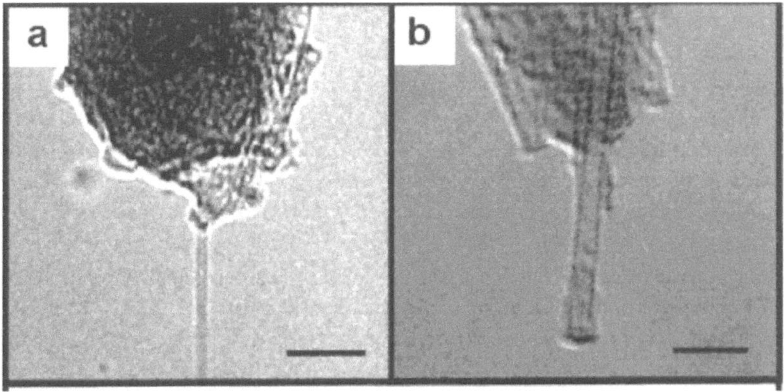

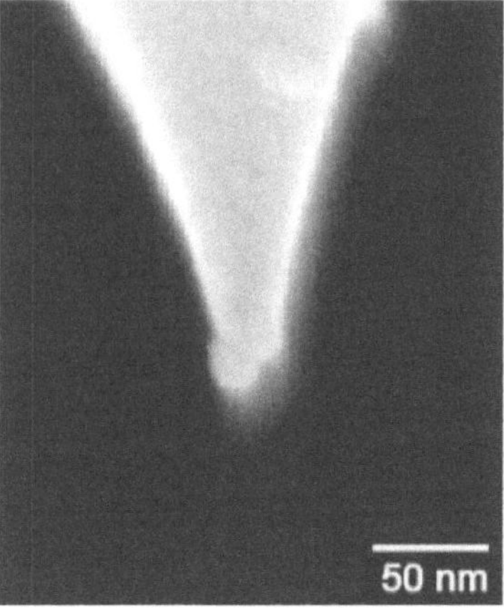

Fig. 7.72 Advanced AFM tip designs. Shown (top) are SWNT-terminated AFM tips (scale bars are 10 nm) and (bottom) an AFM tip terminated with an individual gold nanoparticle (diameter of 14 nm). Nanotube-terminated tips reproduced with permission from *J. Phys. Chem. B* **2001**, *105*, 743. © 2001 American Chemical Society. Nanoparticle-terminated tips reproduced with permission from *Langmuir* **2006**, *22*, 2931. © 2006 American Chemical Society

7.7 Bulk Characterization Techniques

The majority of characterization techniques discussed thus far have been surface-related, with some capable of analyzing sub-surface depths through in situ ion etching. This final section will focus briefly on a selection of common bulk techniques that may be used to characterize as-synthesized materials such as polymers, ceramics, etc. More details on these and other techniques not discussed herein may be found in the "Further Reading" section at the end of this chapter. In particular, these additional resources, as well as countless others online, will highlight solid-state characterization techniques such as:

(i) *Solid-state NMR*—chemical environment of NMR-active nuclei; used to obtain physical, chemical, electronic, and structural information about constituent molecules.
(ii) *Raman spectroscopy*—vibration, stretching, and bending of sample molecules (for the bulk sample, as well as adsorbed surface species); assessing defects in carbon nanotubes.
(iii) *IR spectroscopy* (including surface-characterization modes such as attenuated total reflectance (ATR), diffuse reflectance infrared Fourier transform spectroscopy (DRIFT), and reflection absorption infrared spectroscopy (RAIRS))—complementary to Raman spectroscopy.
(iv) *UV–Vis spectroscopy*—functional group information; sizes of nanoparticles.
(v) *Mass spectrometry* (MS)—information regarding isotopes, mass, and structure.
(vi) *BET*[137] *surface area analysis*—pore size and surface area of powders.
(vii) *Dynamic light scattering* (DLS) and Coulter counting—particle size.
(viii) *Mössbauer spectroscopy*—chemical environment of ^{57}Fe, ^{129}I, ^{119}Sn, or ^{121}Sb atoms in a sample.
(ix) Physical testing techniques (tensile strength, flame retardancy, etc.).

A primary method that is used to characterize the thermal properties of a bulk material is *thermogravimetric analysis* (TGA). This method provides detailed information regarding the thermal stability and decomposition pathway of a material (e.g., stepwise loss of ligands for an organometallic compound), as well as structural information for complex composites (Fig. 7.73). The operating principle of TGA is very simple—the solid is placed in a tiny microbalance pan and heated according to preset ramping conditions. The controlled thermolysis of the compound may be carried out *in vacuo*, or in the presence of a carrier gas such as O_2, N_2, or Ar.

A technique that is often used in tandem with TGA[138] is *differential scanning calorimetry* (DSC).[139] This technique monitors the amount of heat that is required to increase the temperature of a sample, relative to a reference. For example, when a sample undergoes an endothermic phase transition (e.g., melting), heat will be absorbed; conversely, an exothermic event (e.g., crystallization) will require less heat to raise the temperature. Accordingly, DSC is used to determine distinct thermodynamic events, as well as subtle changes such as glass transitions that occur during polymer curing (Fig. 7.74).

The viscoelastic properties of a polymer (high modulus, "glass-like," or low modulus, "rubber-like") may be determined through *dynamic mechanical (thermal) analysis* (DMA or DMTA). This technique involves monitoring the resultant displacement of a polymer following its interaction with an oscillating external force, as the temperature is altered. In addition to readily observing the glass transition, other properties such as stiffness and damping properties are generated from this technique.

An additional method that is used to characterize polymers is *gel-permeation chromatography* (GPC). This technique is a form of *size-exclusion chromatography* (SEC), where components are separated from one another based solely on their sizes

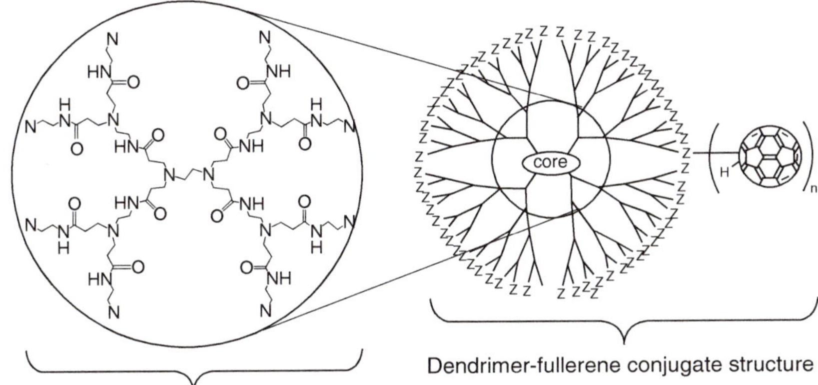

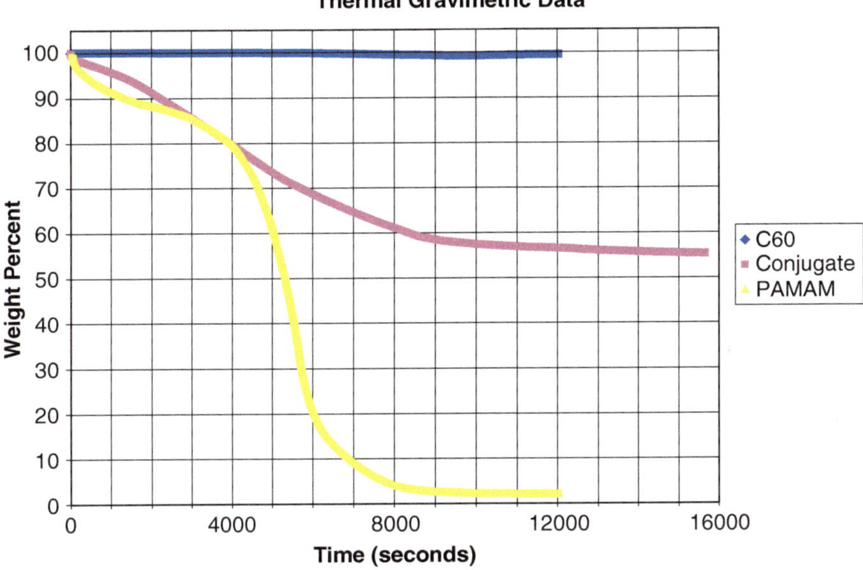

Fig. 7.73 Schematic of a PAMAM dendrimer–fullerene conjugate structure (top), with TGA analysis (bottom). Based on the mass loss of the conjugate, it is suggested that each PAMAM dendrimer is surrounded by 30 fullerene units. Reproduced with permission from *Nano Lett.* **2005**, 5, 1171. © 2005 American Chemical Society

(or hydrodynamic volume). GPC is carried out in the same configuration as other HPLC (high-performance liquid chromatography) methods, with a gel stationary phase and a pressurized liquid mobile phase that elutes the dissolved components from the column. Separation occurs as the solution is passed through the gel, which is typically composed of polystyrene crosslinked with divinylbenzene. While the larger molecules pass through the column without significant retention, the smaller

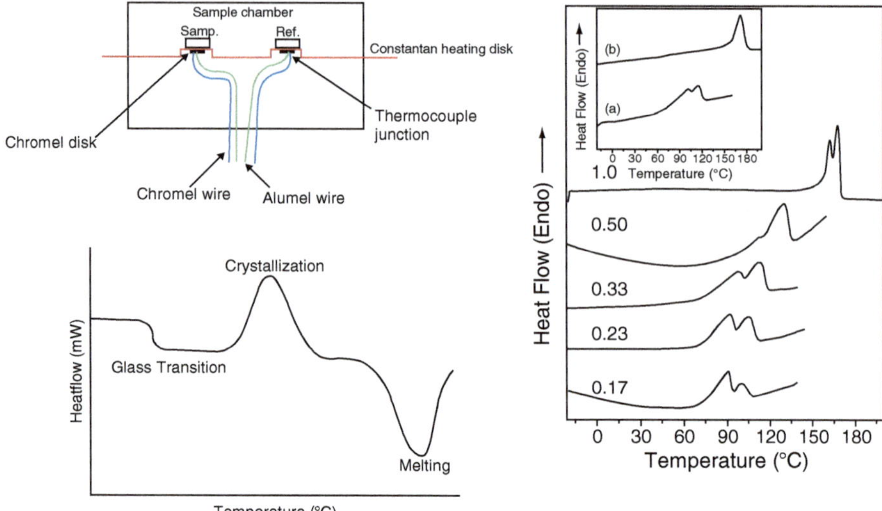

Fig. 7.74 Differential scanning calorimetry (DSC). Shown are **a** schematic of the heat-flux sample chamber; **b** an example of a DSC thermogram, showing endothermic events; **c** DSC thermogram of a poly(vinylidene fluoride)–ethyl acetoacetate polymer–solvent system, showing two melting events for the polymer due to its intermolecular interactions with solvent molecules. The inset shows a comparison between the pure polymer (**b**) and the polymer–solvent (**a**). Reproduced with permission from *Macromolecules* **2006**, *39*, 6110. © 2006 American Chemical Society

molecules are retained by encapsulation within the pores of the gel (*ca.* pore diameters of 100–10^6 nm). The information obtained from gel-permeation chromatographs is mostly used to determine the molecular weight distribution of a polymer, known as the polydispersity index (PDI, Fig. 7.75).

Which Technique Shall I Choose?!

In this final section, we present a flowchart to assist you in choosing the most appropriate technique(s) for materials characterization. Figure 7.76 compares typical analysis depths of various techniques. The choice of characterization method ultimately depends on the sample type (e.g., thin film, bulk crystal, amorphous powder, nanostructure, etc.), the information desired (e.g., elemental compositions, structural information, chemical speciation), and depth of characterization (top few monolayers vs. bulk).

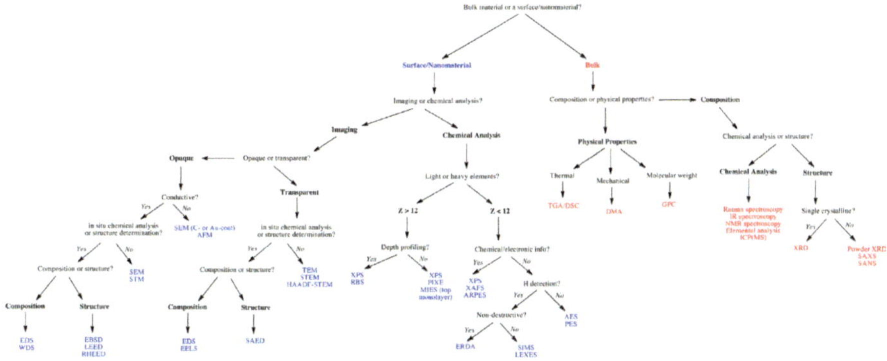

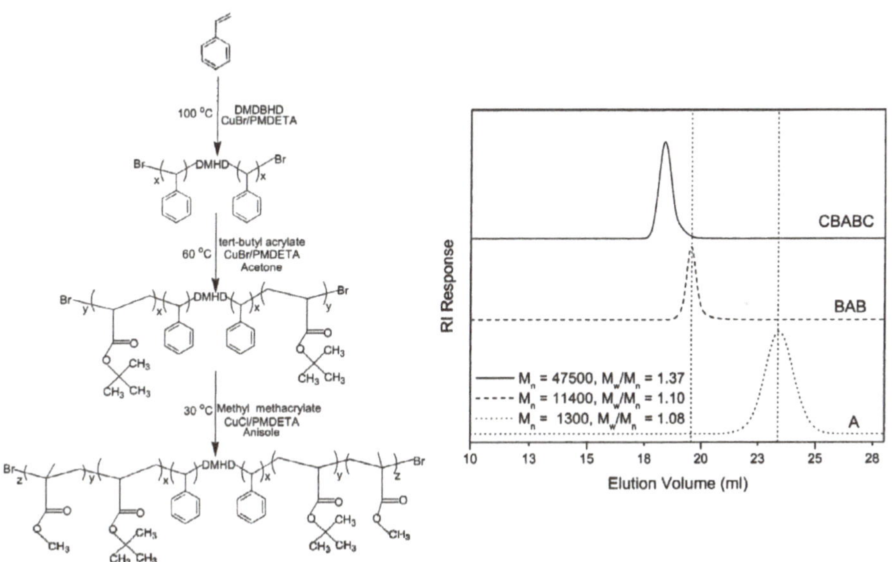

Fig. 7.75 Reaction scheme for the room-temperature synthesis of a CBABC pentablock copolymer, and GPC traces of *a,z*-dibromo-terminated polystyrene (A, dotted line), poly(*t*BA-*b*-styrene-*b*-*t*BA) (BAB triblock copolymer macroinitiator, dashed line), and poly(MMA-*b*-*t*BA-*b*-styrene-*b*-*t*BA-*b*-MMA) (CBABC, solid line). The GPC data shows evidence of a successful polymerization route—low polydispersity index (PDI), and good control over the number-average molecular weight (M_n). Reproduced with permission from *Macromolecules* **2003**, 36, 1039. © 2003 American Chemical Society

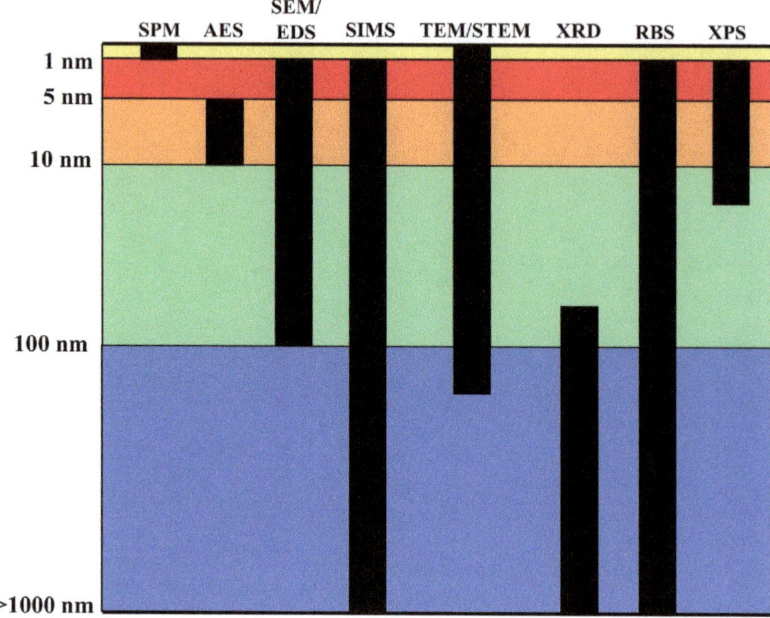

Fig. 7.76 Typical analysis depths for common materials characterization techniques. © Bradley D. Fahlman

Topics for Further Discussion

1. In our discussion of sample staining for TEM analysis, we mentioned that lead citrate was sensitive to CO_2. What is the balanced reaction for generation of the side-product?
2. What is the difference between XPS and AES?
3. What characterization techniques would be best suited for the following:

   (a) Analysis of a thin film (20 nm thickness) for Na content (very accurate and precise value is desired).
   (b) Determination of the Li content of a thin film; comparison of the surface concentration with the Li content at a depth of 100 nm below the surface (diffusion study).
   (c) Image and study the desorption mechanism of SAMs from the surface of Au(111) upon exposure to organic solvents.
   (d) Characterize the film resulting from an attempt to covalently graft poly(ethylene glycol) (PEG) chains on poly(ethylene-*co*-acrylic acid) (EAA) surfaces.
   (e) Characterize the intermediate structures present on a thin film during CVD, at various deposition times.
   (f) Assess whether a new synthetic procedure to grow TiO_2 nanoparticles was successful or not.

(g) Determine the lattice parameters of metallic ReO_3 nanoparticles.
(h) Determine the percentage yield and purity for a batch of single-walled carbon nanotubes.

4. What are the differences between STM and AFM, providing examples of their applications?
5. How would you eliminate the presence of "nanobubbles" on Au surface during AFM imaging? Why is the presence of such adsorbates problematic for SPM studies? (hint: see *Langmuir* **2003**, *19*, 10,510).
6. What are the benefits of using a liquid cell for AFM studies? Are there any limitations of this technique?
7. What is SECM, and what applications would this technique be used for?
8. Describe what is meant by "charging" during SEM analyses, along with methods that are used to circumvent these effects.
9. Why is the yield of Auger electrons most prevalent for elements with low-Z, relative to heavier elements (where X-ray emission is preferred)?
10. Find literature references for tandem UPS/XPS studies of surfaces. What information is gained from each technique?
11. What techniques are used to prepare thin sample sections for TEM analysis?
12. Find literature precedents for tandem RBS/PIXE surface characterization. What information does each technique yield?
13. What are the differences between EDS and WDS?
14. Why does one need a synchrotron source in order to perform XAFS studies?
15. The Getty museum in Los Angeles has contracted your services to determine if a Monet they recently received is authentic. What techniques, including sample preparation, etc. would you use to determine the authenticity of this artwork (handle it carefully!)?
16. Describe the products obtained from the interaction of an electron beam and sample atoms within: (a) a scanning electron microscope, and (b) transmission electron microscope.
17. You have just discovered a new method to grow carbon nanotubes at incredibly low temperatures. The FESEM and TEM images suggest that the as-grown nanostructures are graphitic, and exhibit very desirable aspect ratios. What other characterization techniques should you perform to fully characterize your product? How would you study the growth mechanism (in situ and/or by stopping the growth at varying stages of growth), assuming that you used a CVD-type reaction with an iron-based nanoparticulate catalyst, at a temperature of 200 °C?
18. You have been contracted to design a new polymer that will change its structure in response to temperature changes. What methods would you use to characterize your polymer following its synthesis?
19. You have grown an oxidizable metallic thin film, but must handle it briefly in air before sealing it in a storage container. How would you evaluate the thickness of the native surface oxide, and how deep the oxide diffuses into the thin film, as the sample is annealed?

20. For each of the following, which characterization technique(s) would you use? Make sure you also describe *sample preparation* in your answers:
    (a) To describe the surface morphology (roughness, particle sizes/shapes) of an aluminum oxide thin film grown on a Si substrate
    (b) To determine the elemental composition of a)
    (c) To determine the level of carbon incorporation of a metal film, from the extreme surface (topmost 5 nm), to film depths down to the film/substrate interface. The film thickness is 100 nm
    (d) To determine the morphology of a carbonaceous nanofoam (mostly voidspace, comprised entirely of carbon)
    (e) To prove that you have grown a core/shell nanocluster, with a CdS core and Al_2O_3 shell. The nanoclusters have average diameters of 2 nm, and are suspended in ethanol
    (f) To measure the thickness of the native oxide layer that coats the surface of freshly-grown silicon nanowires. The nanowires are 50 microns in length and have diameters of 10 nm.
    (g) To determine if you have successfully synthesized Si-doped carbon nanotubes. The nanotube diameters are 10–20 nm and have lengths of 50–100 microns.
21. There are reports for the design of miniaturized electron microscopes, that would have the resolution of larger instruments, with a footprint analogous to a light microscope (or even smaller, with rumors of Hitachi working on a pen-sized EM!). Describe some of these designs that involve MEMS fabrication techniques. What limitations will be posed by an instrument of this size?
22. How might you assess the morphological changes of a nanostructured electrode during charging/discharging of a Li-ion battery?
23. What are some techniques you could use to characterize the composition and monodispersity of a single-atom catalyst (SAC), used for a wide variety of energy storage applications?

References and Notes

1. Hackley, V. A.; Clogston, J. D. Measuring the Size of Nanoparticles in Aqueous Media Using Batch-Mode Dynamic Light Scattering, NIST-NCL Joint Assay Protocol, PCC-1, Nov., 2007.
2. The nomenclature for X-ray emission consists of the name of the shell in which the vacancy was created (K, L, M, N), and on the electronic shell that filled the vacancy. For instance, ejection of a K shell electron, filled with a L shell electron is denoted as K_α; if filled with an M shell electron, then K_β is used, and so on.
3. Both flux and brightness are based on a measure of the number of photons per second in a narrow energy bandwidth and in a unit of solid angle in the horizontal and vertical directions. *Flux* is the number of photons per second

passing through a defined area, and is the appropriate measure for experiments that use the entire, unfocused X-ray beam. *Brightness* is a measure of the intensity and directionality of an X-ray beam. It determines the smallest spot onto which an X-ray beam can be focused.

4. A listing of the energies and experimental techniques available from various beamlines at the APS in Argonne National Laboratory: https://www.aps.anl.gov/Beamlines/Directory.
5. Scattering from the nucleus does not contribute to coherent scattering due to its relatively large mass, precluding its oscillation from the impinging of incident X-rays.
6. The cosine of angle ϕ is simply the x component of a unit vector after it is rotated by ϕ around a unit circle. If the vector is rotated at some constant speed, then its x-value will trace out a cosine wave as a function of time, with amplitude of vector length. Waves of different wavelengths or periods would result in the vectors rotating at different speeds; however, X-ray crystallography uses monochromatic photons of a single wavelength.
7. https://scs.illinois.edu/system/files/inline-files/xshellguide.pdf.
8. An excellent summary of crystallography and systematic absences: http://xrayweb.chem.ou.edu/notes/symmetry.html#absence.
9. Another way to state this is that for a single crystal, only a few lattice planes will be oriented at their Bragg angle at any one time.
10. http://www.fhi-berlin.mpg.de/acnew/department/pages/teaching/pages/teaching__wintersemester__2015_2016/frank_girgsdies__peak_profile_fitting_in_xrd__151106.pdf.
11. For a discussion of the Scherrer constant, see: Langford, J. I.; Wilson, A. J. C. *J. Appl. Cryst.* **1978**, *11*, 102.
12. For more information on small-angle scattering, see: Jeffries, C. M. et al. *Nature Rev. Methods Primers* **2021**, *70*, 1.
13. For a nice example that utilizes SAXS, WAXS, and SANS to determine the structural changes of polyethylene chains following annealing, see: Men, Y.; Rieger, J.; Lindner, P.; Enderle, H. -F.; Lilge, D.; Kristen, M. O.; Mihan, S.; Jiang, S. *J. Phys. Chem. B* **2005**, *109*, 16650.
14. For an example of a quantitative SAXS study of a block copolymer-solvent system see: Soni, S. S.; Brotons, G.; Bellour, M.; Narayanan, T.; Gibaud, A. *J. Phys. Chem. B* **2006**, *110*, 15157. An example of the use of SAXS to determine the particle size distribution of nanoparticles, see: Rieker, T.; Hanprasopwattana, A.; Datye, A.; Hubbard, P. *Langmuir* **1999**, *15*, 638.
15. More information on the components and operation of reflection microscopes may be obtained from http://www.microscopyu.com/articles/dic/reflecteddic.html.
16. Synge, E. H. *Philos. Mag.* **1928**, 6, 356.
17. For a thorough review of NSOM, see: Dunn, R. C. *Chem. Rev.* **1999**, *99*, 2891.
18. Mass of a Golf Ball: http://hypertextbook.com/facts/1999/ImranArif.shtml.

19. It is important to note the general trend of decreasing wavelength (and greater resolution) as the velocity of electrons is increased (*i.e.*, higher accelerating voltages).
20. (a) https://www.nist.gov/programs-projects/imaging-toward-neutron-microscope. (b) Cremer, J. T.; Piestrup, M. A.; Gary, C. K.; Pantell, R. H.; Glinka, C. J. *Appl. Phys. Lett.* **2004**, *85*, 494. (c) https://www.nist.gov/programs-projects/towards-neutron-microscope. (d) Liu, D.; Khaykovich, B.; Gubarev, M. V.; Robertson, J. L.; Crow, L.; Ramsey, B. D.; Moncton, D. E. *Nat. Commun.* **2013**, *4*, 2556.
21. https://www.princetoninstruments.com/learn/x-ray-imaging/x-ray-microscopy.
22. Michelson, A. et al. *Science* **2022**, *376*, 203.
23. For example, see: http://accelconf.web.cern.ch/AccelConf/p05/PAPERS/WOAC001.PDF.
24. For an image of the lattice structure and properties of LaB_6, see: https://materialsproject.org/materials/mp-2680/.
25. Perkins, C. L.; Trenary, M.; Tanaka, T.; Otani, S. *Surface Sci.* **1999**, *423*, L222.
26. (a) Shimoi, N.; Sato, Y.; Tohji K. *ACS Appl. Electron. Mater.* **2019**, *1*, 163. (b) Nojeh, A.; Wong, W. K.; Baum, A. W.; Pease, R. F.; Dai, H. *Appl. Phys. Lett.* **2004**, *85*, 112. (c) Irita, M.; Homma, Y. *Surf. Interf. Anal.* **2014**, *46*, 1282.
27. Krauss, A. R.; Auciello, O.; Ding, M. Q.; Gruen, D. M.; Huang, Y.; Zhimov, V. V.; Givargizov, E. I.; Breskin, A.; Chechen, R.; Shefer, E.; Konov, V.; Pimenov, S.; Karabutov, A.; Rakhimov, A.; Suetin, N. *J. Appl. Phys.* **2001**, *89*, 2958, and references therein.
28. Chanana, R. K.; McDonald, K.; Di Ventra, M.; Pantelides, S. T.; Feldman, L. C.; Chung, G. Y.; Tin, C. C.; Williams, J. R. *Appl. Phys. Lett.* **2000**, *77*, 2560.
29. For a more detailed description of resolution and magnification of field-emission electron microscopes, see: (a) Rose, D. J. *J. Appl. Phys.* **1956**, *27*, 215. (b) O'Keefe, M. A. *Ultramicroscopy* **1992**, *47*, 282.
30. Unless otherwise stated, TEM micrographs are collected as bright-field images.
31. Formvar is a copolymer of polyvinyl alcohol, formaldehyde, and polyvinyl acetate.
32. The very same material obtained by reacting normal cotton with nitric acid and sulfuric acid to yield "gun cotton"—;a highly explosive solid that burns with an extremely bright flash leaving behind no residue. Since the thickness is extremely small for the support material and the degree of nitration/sulfation is supposedly less than gun powder compositions, there are no explosion hazards in handling grids. However, the native resin used to make support films is a flammable solid and should be treated with caution.
33. Butvar B98 is a polyvinyl butyral resin containing 20% polyvinyl alcohol. These support films are hydrophilic, hence, they are useful for negative staining methods.

34. The FIB/lift-out technique yields cross sections of both the film and substrate, which maintains the integrity of the interface; in comparison, ultramicrotomy requires the removal of the polymer film and subsequent embedding/sectioning. For a detailed example, see: White, H.; Pu, Y.; Rafailovich, M.; Sokolov, J.; King, A. H.; Giannuzzi, L. A.; Urbanik-Shannon, C.; Kempshall, B. W.; Eisenberg, A.; Schwarz, S. A.; Strzhemechny, Y. M. *Polymer* **2001**, *42*, 1613.
35. Virgilio, N.; Favis, B. D.; Pepin, M. -F.; Desjardins, P. *Macromolecules* **2005**, *38*, 2368.
36. https://www.leica-microsystems.com/products/sample-preparation-for-electron-microscopy/p/leica-em-uc7/.
37. Morniroli, J. -P. "Introduction to Electron Diffraction" in *NATO Series II: Mathematics, Physics and Chemistry* **2006**, *211*, 61.
38. (a) Steeds, J. W.; Morniroli, J. P. *Rev. Mineralogy Geochem.* **1992**, *27*, 37. (b) Tsuda, K.; Mitsuishi, H.; Terauchi, M.; Kawamura, K. *J. Electron Microsc.* **2007**, *56*, 57.
39. An excellent book that contains many high-resolution TEM images and the electron diffraction patterns for a variety of archetypical unit cells see: Sindo, D.; Hiraga, K. *High-Resolution Electron Microscopy for Materials Science*, Springer: New York, 1998.
40. https://www1.udel.edu/pchem/C874/LEED_lecture.pdf.
41. https://phas.ubc.ca/~berciu/TEACHING/PHYS502/PROJECTS/RHEED.pdf.
42. It should be noted that EDS and WDS are often referred to as EDX and WDX, or XEDS and XWDS, respectively.
43. The detection limits for EDS are typically 0.1 at% (for elements with $Z > 10$), whereas WDS is able to detect elements present in concentrations of a few ppm.
44. Electromagnetic lenses suffer from three types of defects: astigmatism, spherical aberrations, and chromic aberrations. Astigmatism may be easily minimized by placing electromagnets, known as stigmator coils, around the column. The current in these coils may be varied, which will reform the distorted electron beam back into a round shape. Chromic aberrations may be reduced by using a field emission source that produces an electron beam with a sharper distribution of electron energies. However, the third type of defect, spherical aberrations (C_s), are not easily circumvented. In fact, this is the primary factor that limits the resolution of electron microscopes to values far below their theoretical values. Recently, techniques have been developed to correct for these defects, often involving the use of multipole lenses (*e.g.*, quadrupole, octapole, etc.) with careful control of their fabrication and operation. For a description of techniques used to design a Cs-corrected HRTEM, see: https://www.salve-project.de/publications/pdf/krivanek_1997.pdf.
45. A collaborative research project between the Lawrence Berkeley National Laboratory, Argonne National Laboratory, Oak Ridge National Laboratory,

and the Univ. of Illinois at Urbana-Champagne developed a transmission electron aberration-corrected microscope that achieved the 0.05 nm resolution target in 2009.

46. Lopatin, S. et al. *Sci. Adv.* **2020**, *6*, 1.
47. Some important precedents that show the utility of *Z*-contrast imaging: (a) Arslan, I.; Yates, T. J. V.; Browning, N. D.; Midgley, P. A. *Science* **2005**, *309*, 2195. (b) Chisholm, M. F.; Kumar, S.; Hazzledine, P. *Science* **2005**, *307*, 701. (c) Nellist, P. D.; Pennycook, S. J. *Science* **1996**, *274*, 413.
48. The photon energies for the light elements are: Be (0.109 keV), B (0.185 keV), C (0.282 keV), N (0.392 keV), O (0.523 keV), and F (0.677 keV). Due to these low energies, the emitted X-rays are easily absorbed by the sample or the components of the detector.
49. Bremsstrahlung is from the German word *bremsen* ("to brake"), and *Strahlung* ("radiation") – thus, "braking radiation." This is characterized by a broad peak that results from deceleration of beam electrons due to scattering from atomic nuclei.
50. Pronounced: *OH-zh* $\overline{a}$.
51. The energy-dispersing device at the heart of EDS is a semiconducting diode. As an incoming X-ray photon impinges the diode, electron–hole pairs are generated, which yields a measurable electrical current. In order to reduce background noise from photons not originating from the sample, the detector is operated at temperatures of *ca.* 140 K. A protective window comprising Be, or more recently, ultrathin windows of BN, diamond, or a supported polymer (paralene, norvar), is used to prevent the condensation of vapors (*e.g.*, water, organics) onto the cooled diode. A "windowless" detector may also be used for EDS; with careful operation, this modification allows the detection of elements down as far as Be—with a maximum efficiency of only 2%. It should be noted that although UHV conditions are used in a TEM/SEM instrument, there will always be a low concentration of vapors—often originating from the rotary pump fluid, or the sample itself due to beam-induced volatilization/decomposition, residual solvent evaporation, etc. The buildup of such a coating on the detector window will reduce the energy of the incoming X-rays, which will drastically reduce the detection sensitivity—especially for low-*Z* elements.
52. It should be noted that reflection techniques are also possible through modifying the angle of approach of the incident electron beam of the TEM. Such analysis is referred to as reflection electron microscopy (REM), and accompanying characterizations such as *reflection high-energy electron diffraction* (RHEED) and *reflection electron energy-loss spectroscopy* (REELS) are also possible, grouped under the umbrella of reflection high-resolution analytical electron microscopy (RHRAEM). (a) For an application of this multifaceted approach to study GaAs(110) surfaces, see: Wang, Z. L. *J. Electron. Microsc. Tech.* **1988**, *10*, 35 (citation found online at: http://www.osti.gov/energycitations/product.biblio.jsp?osti_id=6614173). (b) Additional information may be found in Wang, Z. L. *Reflection Electron Microscopy and Spectroscopy*

for Surface Analysis, Cambridge University Press: Cambridge, UK, 1996. (A book synopsis is found online at: http://www.nanoscience.gatech.edu/zlwang/book/book2_intro.pdf).
53. For a thorough summary/examples of the details generated from EELS spectra, see: Thomas, J. M.; Williams, B. G.; Sparrow, T. G. *Acc. Chem. Res.* **1985**, *18*, 324.
54. (a) Williams, B. G.; Sparrow, T. G.; Egerton, R. F. *Proc. R. Soc. Lond. Ser. A* **1984**, *393*, 409. (b) Sparrow, T. G.; Williams, B. G.; Thomas, J. M.; Jones, W.; Herley, P. J.; Jefferson, D. A. *J. Chem. Soc., Chem. Commun.* **1983**, 1432. (c) Ferrell, R. A. *Phys. Rev.* **1956**, *101*, 554.
55. For example, see: Goris, B.; Guzzinati, G.; Fernandez-Lopez, C.; Perez-Juste, J.; Liz-Marzan, L.; Trugler, A.; Hohenester, U.; Verbeeck, J.; Bals, S.; Van Tendeloo, G. *J. Phys. Chem. C Nanomater. Interfaces* **2014**, *118*, 15356.
56. For a comparison between ELNES, XANES, and the basics of X-ray absorption spectroscopy, see: (a) http://susi.theochem.tuwien.ac.at/reg_user/textbooks/WIEN2k_lecture-notes_2011/Blaha_xas_eels.pdf. (b) http://www.iucr.org/__data/assets/pdf_file/0004/60637/IUCr2011-XAFS-Tutorial_-Ascone.pdf.
57. For an application of EELS in determining the nature of C bonding in amorphous carbon nanotubes see: Hu, Z. D.; Hu, Y. F.; Chen, Q.; Duan, X. F.; Peng, L.-M. *J. Phys. Chem. B.* **2006**, *110*, 8263.
58. (a) For a comprehensive list of tutorials related to XAFS, see: http://gbxafs.iit.edu/training/tutorials.html. (b) For a thorough summary/examples of the details generated from EELS spectra, see: Thomas, J. M.; Williams, B. G.; Sparrow, T. G. *Acc. Chem. Res.* **1985**, *18*, 324.
59. For information on the analysis of surfaces by IR radiation instead of electrons, there exists a complimentary technique known as reflection absorption infrared spectroscopy (RAIRS). For instance, see: (a) Bolina, A. S.; Wolff, A. J.; Brown, W. A. *J. Phys. Chem. B* **2005**, *109*, 16836. (b) Wang, X. D.; Tysoe, W. T.; Greenler, R. G.; Truszkowska, K. *Surf. Sci.* **1991**, *258*, 335. (c) http://www.annualreviews.org/doi/10.1146/annurev.physchem.51.1.381.
60. For example, the development of fibers/fabrics that will actively adsorb and surface deactivate chemical and biological warfare agents, of increasing importance as new modes of terrorist activity continue to emerge. For more information, see: (a) http://web.mit.edu/isn/ (Institute of Soldier Nanotechnologies at M.I.T.). (b) Richards, V. N.; Vohs, J. K.; Williams, G. L.; Fahlman, B. D. *J. Am. Ceram. Soc.* **2005**, *88*, 1973.
61. Wang, Z. L.; Bentley, J.; Evans, N. D. *J. Phys. Chem. B* **1999**, *103*, 751.
62. Some of the (potential) energy of the incident electrons is required to release an outer electron from its valence or conduction band, with the remaining being transferred into the kinetic energy of the ejected secondary electron.
63. For more details, refer to: Goldstein, J.; Newbury, D.; Joy, D.; Lyman, C.; Echlin, P.; Lifshin, E.; Sawyer, L.; Michael, J. *Scanning Electron Microscopy and X-Ray Microanalysis*, 3rd ed., Kluwer: New York, 2003.
64. https://serc.carleton.edu/research_education/geochemsheets/ebsd.html.

65. Simon, C.; Walmsley, J.; Redford, K. *Transmission Electron Microscopy Analysis of Hybrid Coatings*, Proceedings of the 6th International Congress on Advanced Coating Technology, Nuremberg, Germany, April 3–4, 2001.
66. Monte Carlo simulations performed by the author using the program CASINO ("monte CArlo SImulation of electroN trajectory in sOlids"), available free-of-charge on the Internet: http://www.gel.usherbrooke.ca/casino/What.html.
67. For example, see: https://www.bruker.com/products/x-ray-diffraction-and-elemental-analysis/eds-wds-ebsd-sem-micro-xrf-and-sem-micro-ct/esprit-dynamics/overview.html.
68. Both carbon and gold coating are performed using a sputter-coating PVD method. The film of choice is most often C since it is less costly and is transparent to X-rays (for EDS). Gold is used to coat very uneven surfaces and is only useful when EDS is not being performed (strong Au signal would mask other elements present in the sample).
69. Rades, S.; Hodoroaba, V. D.; Salge, T.; Wirth, T.; Lobera, M. P.; Labrador, R. H.; Natte, K.; Behnke, T.; Gross, T.; Unger, W. E. S. *RSC Adv.* **2014**, *4*, 49577.
70. The effect of charging can be controlled by either altering the position of the sample with regard to the incident electron beam, or using an argon ion (Ar^+) gun to neutralize the charge.
71. A recent issue of the MRS Bulletin was devoted to in situ TEM: *MRS Bull.* **2008**, *33*.
72. For a historical background of ESEM development, see: http://www.danilatos.com/.
73. https://nanoscience.ch/wp-content/uploads/sites/8/2017/08/an_introduction_to_esem.pdf.
74. For instance, see: van Huis, M. A.; Kunneman, L. T.; Overgaag, K.; Xu, Q.; Pandraud, G.; Zandbergen, H. W.; Vanmaekelbergh, D. *Nano Lett.* **2008**, *8*, 3959.
75. For example, in situ growth of carbon nanotubes: Lin, M.; Tan, J. P. Y.; Boothroyd, C.; Loh, K. P.; Tok, E. S.; Foo, Y. -L. *Nano Lett.* **2007**, *7*, 2234.
76. For an example of a TEM PicoIndenter®, see: https://www.bruker.com/en/products-and-solutions/test-and-measurement/nanomechanical-instruments-for-sem-tem/hysitron-pi-95-tem-picoindenter.html.
77. https://www.nanoscience.com/techniques/scanning-tunneling-microscopy/.
78. For instance, see: Wang, Z. L.; Poncharal, P.; de Heer, W. A. *Pure Appl. Chem.* **2000**, *72*, 209. May be accessed online at: http://www.iupac.org/publications/pac/2000/pdf/7201x0209.pdf.
79. For more details, refer to: Goldstein, J.; Newbury, D.; Joy, D.; Lyman, C.; Echlin, P.; Lifshin, E.; Sawyer, L.; Michael, J. *Scanning Electron Microscopy and X-Ray Microanalysis*, 3rd ed., Kluwer: New York, 2003.
80. http://hummingbirdscientific.com/products/tomography/.
81. Chiaramonti, A. N.; Schreiber, D. K.; Egelhoff, W. F.; Seidman, D. N.; Petford-Long, A. K. **2008**, *93*, 103113.
82. Klein, K. L.; Anderson, I. M.; De Jonge, N. *J. Microsc.* **2011**, *242*, 117.

83. Zheng, H.; Smith, R. K.; Jun, Y. -W.; Kisielowski, C.; Dahmen, U.; Alivisatos, A. P. *Science* **2009**, *324*, 1309.
84. In contrast to SEM, EDS, and AES that use incident X-rays of *ca.* 10 keV, XPS is less damaging to beam-sensitive samples due to the use of "soft" X-rays, of much less energy (1–1.5 keV).
85. A very detailed example of XPS to distinguish among Li salts for Li-ion battery applications is: Dedryvere, R.; Leroy, S.; Martinez, H.; Blanchard, R.; Lemordant, D.; Gonbeau, D. *J. Phys. Chem. B* **2006**, *110*, 12986.
86. For instance, see: (a) Seah, M. P.; White, R. Surf. Interf. Anal. 2002, 33, 960. (b) Libra, J.; Matolin, V. Surf. Sci. 2006, 600, 2317. (c) https://core.ac.uk/download/pdf/34898173.pdf. (d) In addition to angle-resolved PES, *photoluminescence spectroscopy* is typically used as a nondestructive means to delineate the electronic properties of materials. For more information, see: (i) Glinka, Y. D.; Lin, S.-H.; Hwang, L.-P.; Chen, Y.-T. *J. Phys. Chem. B* **2000**, 104, 8652. (ii) Wu, J.; Han, W.-Q.; Walukiewicz, W.; Ager, J. W.; Shan, W.; Haller, E. E.; Zettl, A. *Nano Lett.* **2004**, 4, 647.
87. For an application example of EXAFS, see: Borgna, A.; Stagg, S. M.; Resasco, D. E. *J. Phys. Chem. B* **1998**, *102*, 5077. An example of XPS and XAFS (XANES and EXAFS), see: Chakroune, N.; Viau, G.; Ammar, S.; Poul, L.; Veautier, D.; Chehimi, M. M.; Mangeney, C.; Villain, F.; Fievet, F. *Langmuir* **2005**, *21*, 6788.
88. For an application example of REFLEXAFS, see: d'Acapito, F.; Emelianov, I.; Relini, A.; Cavotorta, P.; Gliozzi, A.; Minicozzi, V.; Morante, S.; Solari, P. L.; Rolandi, R. *Langmuir* **2002**, *18*, 5277.
89. Gervasini, A.; Manzoli, M.; Martra, G.; Ponti, A.; Ravasio, N.; Sordelli, L.; Zaccheria, F. *J. Phys. Chem. B* **2006**, *110*, 7851.
90. Data analysis represents the most essential and time-consuming aspects of these techniques (as well as EELS). Typically, sample spectra are compared to reference samples that contain the probed element with similar valences and bonding motifs. In this example, Cu metal foil was used for the Cu-Cu interactions, and CuO/Cu_2O were used for the Cu-O contributions. A variety of software programs are used for detailed curve-fitting, in order to obtain information regarding the chemical environment of the sample. For example, see: (a) http://www.dragon.lv/eda/. (b) http://www.xsi.nl/software.html. (c) http://www.xpsdata.com/.
91. For information regarding the quantum mechanical description of spin–orbit splitting, see: (a) http://hyperphysics.phy-astr.gsu.edu/hbase/quantum/sodzee.html#cl. (b) http://www.technology.niagarac.on.ca/people/mcsele/lasers/Quantum.htm.
92. The presence of these satellites are indicative of Cu^{2+}. For instance, see: (a) Espinos, J. P.; Morales, J.; Barranco A.; Caballero, A.; Holgado, J. P.; Gonzalez-Elipe, A. R. *J. Phys. Chem. B* **2002**, *106*, 6921. (b) Morales, J.; Caballero, A.; Holgado, J. P.; Espinos, J. P.; Gonzalez-Elipe, A. R. *J. Phys. Chem. B* **2002**, *106*, 10185. (c) Fuggle, J. C.; Alvarado, S. F. *Phys. Rev. A* **1980**,

22, 1615 (describes the cause of peak broadening in XPS spectra, related to core-level lifetimes).

93. In an excited $2s^1$ state, with an energy of 19.8 eV and a lifetime of *ca.* 4000 s.
94. Many references exist for MIES studies of surfaces, most often carried out in tandem with UPS (to gain information for both the surface and immediate subsurface of the sample). For example, see: (a) Johnson, M. A.; Stefanovich, E. V.; Truong, T. N.; Gunster, J.; Goodman, D. W. *J. Phys. Chem. B* **1999**, *103*, 3391. (b) Kim, Y. D.; Wei, T.; Stulz, J.; Goodman, D. W. *Langmuir* **2003**, *19*, 1140 (very nice work that describes the shortfall of UPS alone, and the utility of a tandem UPS/MIES approach).
95. Note: though conventional RBS is carried out with He^+ ions (which will backscatter from any atom with a greater Z), heavier ions such as C, O, Si, or Cl may be used in order to prevent background backscattering interactions with the matrix. For example, use of incident O ions to eliminate backscattering from lattice O atoms for the RBS analysis of ceramic oxides.
96. Simulations for ion scattering techniques such as RBS are typically compared with actual spectra in order to characterize the surface features. There are many such algorithms; for example: (a) Kido, Y.; Koshikawa, T. *J. Appl. Phys.* **1990**, *67*, 187. (b) Doolittle, L. R. *Nucl. Instrum. Methods* **1986**, *B*15, 227 (RUMP program).
97. Many such systems exist; some examples include: (a) Brookhaven National Laboratory (https://www.bnl.gov/tandem/). (b) Western Michigan University (https://wmich.edu/physics/accelerator). (c) University of Kentucky (http://www.pa.uky.edu/accelerator/).
98. Also known as forward recoil scattering (FRS) or hydrogen forward scattering (HFS).
99. Naab, F. U.; Holland, O. W.; Duggan, J. L.; McDaniel, F. D. *J. Phys. Chem. B* **2005**, *109*, 1415, and references therein.
100. For a very thorough tutorial regarding SIMS, see: http://www.geology.wisc.edu/~wiscsims/HiRes2015/pdfs/talks/HiRes2015_06010920_Kita.pdf.
101. For background information regarding polymer characterization using MALDI, see: (a) Montaudo, G.; Samperi, F.; Montaudo, M. S. Prog. Polym. Sci. 2006, 31, 277. (b) http://ws680.nist.gov/publication/get_pdf.cfm?pub_id=852404.
102. For a thesis that has a nice background on ESI, see: http://uu.diva-portal.org/smash/record.jsf?pid=diva2%3A161937&dswid=7454.
103. Typically, the majority of secondary ions are ejected from the top two or three monolayers (10–20 Å) of the sample.
104. For instance, the ion concentration of the impinging ion beam must be <1% of the number of surface molecules. If this "static limit" is breached, a residue from molecular fragmentation will build up on the surface, which depletes the signal.
105. The use of fullerene ion sources represents an area of increasing interest. For example, see: Cheng, J.; Winograd, N. *Anal. Chem.* **2005**, *77*, 3651, and references therein.

106. Winograd, N. *Anal. Chem.* **2005**, *77*, 142A.
107. For example, see: (a) Delcorte, A.; Medard, N.; Bertrand, P. *Anal. Chem.* **2002**, *74*, 4955. (b) Delcorte, A.; Bour, J.; Aubriet, F.; Muller, J. -F.; Bertrand, P. *Anal. Chem.* **2003**, *75*, 6875.
108. Marcus, A.; Winograd, N. *Anal. Chem.* **2006**, *78*, 141.
109. For example, see: https://www.nrel.gov/docs/fy03osti/33592.pdf.
110. For instance, see: (a) Chandra, S. *Appl. Surf. Sci.* **2003**, *203–204*, 679. (b) Allen, G. C.; Brown, I. T. *Eur. Mass Spectrom.* **1995**, *1*, 493. (c) Chandra, S.; Lorey, D. R. *Int. J. Mass Spectrom.* **2007**, *260*, 90.
111. For example, see: (a) Groner, E.; Hoppe, P. *Appl. Surf. Sci.* **2006**, *252*, 7148. (b) https://shrimprg.stanford.edu/intro-sims.
112. For example, see: https://www.sciencedirect.com/science/article/abs/pii/S0168583X05007317.
113. http://www.geology.wisc.edu/~johnf/g777/MM/Staub-LEXES.pdf.
114. Hombourger, C.; Staub, P. F.; Schuhmacher, M.; Desse, F.; de Chambost, E.; Hitzman, C. *Appl. Surf. Sci.* **2003**, *203–204*, 383.
115. Seidman, D. N. *Annu. Rev. Mater. Res.* **2007**, *37*, 127, and references therein.
116. For example, see: (a) Larson, D. J.; Petford-Long, A. K.; Ma, Y. Q.; Cerezo, A. *Acta. Mater.* **2004**, *52*, 2847. (b) Gault, B.; Menand, A.; de Geuser, F.; Deconihout, B.; Danoix, F. *Appl. Phys. Lett.* **2006**, *88*, 114101.
117. A recent issue of MRS Bulletin is focused on APT: *MRS Bull.* **2009**, *34* (Oct. 2009).
118. Seidman, D. N.; Stiller, K. *MRS Bull.* **2009**, *34*, 717.
119. Perea, D. E.; Hemesath, E. R.; Schwalbach, E. J.; Lensch-Falk, J. L.; Voorhees, P. W.; Lauhon, L. J. *Nature Nanotechnol.* **2009**, *4*, 315.
120. Perea, D. E.; Arslan, I.; Liu, J.; Ristanovic, Z.; Kovarik, L.; Arey, B. W.; Lercher, J. A.; Bare, S. R.; Weckhuysen, B. M. *Nat. Commun.* **2015**, *6*, 7589.
121. Record players are once again rising in popularity, as audiophiles believe the sound quality far surpasses that of digital media.
122. For a recent review of atomic manipulation using AFM, see: Custance, O.; Perez, R.; Morita, S. *Nature Nanotechnol.* **2009**, *4*, 803.
123. Near-field scanning optical microscopy (NSOM) (discussed at the beginning of this chapter) is often grouped alongside other SPM techniques. However, for our discussion, we will focus on AFM and STM since these use *physical* probes to interrogate a surface, rather than focused light.
124. For example, see: Zhang, J.; Chi, Q.; Ulstrup, J. *Langmuir* **2006**, *22*, 6203.
125. For example, see: (a) France, C. B.; Frame, F. A.; Parkinson, B. A. *Langmuir* **2006**, *22*, 7507. (b) Li, W.-S.; Kim, K. S.; Jiang, D. -L.; Tanaka, H.; Kawai, T.; Kwon, J. H.; Kim, D.; Aida, T. *J. Am. Chem. Soc.* **2006**, *128*, 10527. (c) Namai, Y.; Matsuoka, O. *J. Phys. Chem. B* **2006**, *110*, 6451.
126. For a recent precedent, see: Park, J. B.; Jaeckel, B.; Parkinson, B. A. *Langmuir* **2006**, *22*, 5334, and references therein. For a thorough recent review, see: Wan, L. -J. *Acc. Chem. Res.* **2006**, *39*, 334.
127. For example, see: Alam, M. S.; Dremov, V.; Muller, P.; Postnikov, A. V.; Mal, S. S.; Hussain, F.; Kortz, U. *Inorg. Chem.* **2006**, *45*, 2866.

128. Examples of some common forces that may exist between a surface and an AFM tip are Van der Waal, electrostatic, covalent bonding, capillary, and magnetic. In addition to providing information regarding the topography of the surface (constant force mode), forces may be applied to understand the morphology of a surface – for example, to determine the frictional force between the tip and surface, or the elasticity/hardness of a surface feature. For instance, see: Tranchida, D.; Piccarolo, S.; Soliman, M. *Macromolecules* **2006**, *39*, 4547, and references therein.
129. For example, see: O'Dwyer, C.; Gay, G.; Viaris de Lesegno, B.; Weiner, J. *Langmuir* **2004**, *20*, 8172, and references therein.
130. For example, see: (a) Cho, Y.; Ivanisevic, A. *Langmuir* **2006**, *22*, 1768. (b) Poggi, M. A.; Lillehei, P. T.; Bottomley, L. A. *Chem. Mater.* **2005**, *17*, 4289. (c) Gourianova, S.; Willenbacher, N.; Kutschera, M. *Langmuir* **2005**, *21*, 5429.
131. For example, see: (a) Polking, M.; Huan, M. G.; Yourdkhani, A.; Petkov, V.; Kisielowski, C. F.; Volkov, V.; Zhu, Y.; Caruntu, G.; Alivisatos, P. A.; Ramesh, R. *Nat. Mater.* **2012**, *11*, 700. (b) Takamura, Y.; Chopdekar, R. V.; Scholl, A.; Doran, A.; Liddle, J. A.; Harteneck, B.; Suzuji, Y. *Nano Lett.* **2006**, 6, 1287. (c) Li, Y.; Tevaarwerk, E.; Chang, R. P. H. *Chem. Mater.* **2006**, *18*, 2552.
132. For example, see: Zhang, J.; Roberts, C. J.; Shakesheff, K. M.; Davies, M. C.; Tendler, S. J. B. *Macromolecules* **2003**, *36*, 1215, and references therein.
133. For a thorough description of SECM, see: Gardner, C. E.; Macpherson, J. V. *Anal. Chem.* **2002**, *74*, 576A.
134. For instance, see: https://www.brukerafmprobes.com.
135. (a) Nanoparticle-terminated tips: Vakarelski, I. U.; Higashitani, K. *Langmuir* **2006**, *22*, 2931. (b) Nanotube-terminated tips: Hafner, J. H.; Cheung, C. - L.; Oosterkamp, T. H.; Lieber, C. M. *J. Phys. Chem. B* **2001**, *105*, 743. (c) Nanotube-terminated tips: Wilson, N. R.; Macpherson, J. V. *Nano Lett.* **2003**, *3*, 1365.
136. An AFM probe responds to the *average* force between the sample surface and a group of tip atoms that are in close proximity to the surface. In order to image individual atoms by SPM, the surface–tip interactions must be limited to the nearest atom(s) on the tip periphery. Hence, an AFM image will not show individual atoms, but rather an average surface, with its ultimate resolution dependent on the sharpness of the tip structure. In contrast, STM is capable of atomic resolution since the tunneling current passes only through the tip atom that is nearest the sample surface.
137. Named after Brunauer, Emmett, and Teller.
138. Tandem TGA/DSC instruments are commercially available, for example: http://www.tainstruments.com/simultaneous-tgadsc/.
139. An analogous (older) technique is known as *differential thermal analysis* (DTA), which yields the same information as DSC.

Further Reading

1. S.L. Flegler, J.W. Heckman, K.L. Klomparens, *Scanning and Transmission Electron Microscopy: An Introduction* (W. H. Freeman, New York, 1993)
2. Y. Leng, *Materials Characterization: Introduction to Microscopic and Spectroscopic Methods*, 2nd edn. (Wiley, New York, 2013)
3. K. Krishnan, *Principles of Materials Characterization and Metrology* (Oxford University Press, 2021)
4. D.B. Williams, C.B. Carter, *Transmission Electron Microscopy: A Textbook for Materials Science*, 2nd edn. (Springer, New York, 2009)
5. P. Echlin, *Handbook of Sample Preparation for Scanning Electron Microscopy and X-Ray Microanalysis* (Springer, New York, 2009)
6. L.C. Sawyer, D.T. Grubb, G.F. Meyers, *Polymer Microscopy: Characterization and Evaluation of Materials*, 3rd edn. (Springer, New York, 2008)
7. W. Zhou, Z.L. Wang, *Scanning Microscopy for Nanotechnology: Techniques and Applications* (Springer, New York, 2006)
8. J. Goldstein, D. Newbury, J.R. Michael, N.W.M. Ritchie, J.H.J. Scott, D.C. Joy, *Scanning Electron Microscopy and X-Ray Microanalysis*, 4th edn. (Springer, New York, 2018)
9. C.R. Brundle, C.A. Evans, S. Wilson, *Encyclopedia of Materials Characterization – Surfaces, Interfaces, Thin Films* (Elsevier, New York, 1992)
10. D. Campbell, P.A. Pethrick, J.R. White, *Polymer Characterization*, 2nd edn. (Stanley Thornes, Cheltenham, UK, 2000)
11. D. Campbell, *Polymer Characterization: Physical Techniques*, 2nd edn. (CRC Press, New York, 2017)
12. W.J. Criddle, G.P. Ellis, *Spectral and Chemical Characterization of Organic Compounds: A Laboratory Handbook*, 3rd edn. (Wiley, New York, 1990)
13. N.J. Dinardo, *Nanoscale Characterization of Surfaces and Interfaces*, 2nd edn. (Wiley, New York, 2004)
14. D. Brune, R. Hellborg, O. Hunderi, *Surface Characterization: A User's Sourcebook* (Wiley, New York, 1997)
15. A.W. Czanderna, T.E. Madey, C.J. Powell, *Beam Effects, Surface Topography, and Depth Profiling in Surface Analysis (Methods of Surface Characterization)* (Plenum Press, New York, 1998)
16. A.W. Czanderna, D.M. Hercules, *Ion Spectroscopies for Surface Analysis (Methods of Surface Characterization)* (Springer, New York, 1991)
17. D.D. Brandon, W.D. Kaplan, *Microstructural Characterization of Materials* (Wiley, New York, 1999)
18. V. Pecharsky, P. Zavalij, *Fundamentals of Powder Diffraction and Structural Characterization of Materials*, 2nd edn. (Springer, New York, 2008)
19. Y. Waseda, E. Matsubara, K. Shinoda, *X-ray Diffraction Crystallography: Introduction, Examples and Solved Problems* (Springer, New York, 2011)
20. O.H. Seek, B. Murphy, *X-Ray Diffraction: Modern Experimental Techniques* (Pan Stanford, US, 2015)
21. M. Lee, *X-Ray Diffraction for Materials Research: From Fundamentals to Applications* (Apple Academic Press, New York, 2016)
22. R. Cahn, *Concise Encyclopedia of Materials Characterization*, 2nd edn. (Elsevier, San Diego, CA, 2005)
23. N.J. Chou, *Characterization of Polymers (Materials Characterization)* (Momentum Press, New York, 2010)

Appendix A
Timeline of Materials and Technological Discoveries

<50,000 BC Brushes are developed to apply pigment to cave walls
30,000 BC Clothing materials are fabricated from animal skins
24,000 BC Ceramic materials are made from animal fat and bone, mixed with bone ash and clay
20,000 BC Ivory and bone are used to make sewing needles
20,000 BC A nonwoven fabric, later termed felt, is made from compressed wool/hair
10,000 BC Gourds, seapods, bones, and clay are used to make ocarinas or vessel flutes
4000 BC Stones are first used to construct roads in Ur (modern-day Iraq)
3500 BC Copper metallurgy is invented and used to fabricate a variety of materials
3500 BC The first reported use of glass in Egypt and Mesopotamia
3400 BC Linen cloth synthesized from flax is used to wrap mummies in Egypt
3200 BC Bronze is used for weapons and armor
3000 BC Egyptians wear clothing comprised of cotton fibers
3000 BC The Egyptians construct the first stringed musical instrument
3000 BC Soap is first synthesized in Egypt using wood ash and animal fat
2600 BC Silk fibers are used for clothing in China
2000 BC Pewter beginning to be used in China and Egypt
2000 BC The first suspension bridges are used in China and India
1600 BC The Hittites develop iron metallurgy
1600 BC Conceptual designs are invented for bathing suits, fabricated/named bikini in 1946
1300 BC Invention of steel when iron and charcoal are combined properly
1000 BC The abacus is created by the Babylonians
1000 BC Glass production begins in Greece and Syria
900s BC Assyrians develop pontoon rafts for armies to cross rivers
800s BC Spoked wheels are fabricated and used throughout Europe

B. D. Fahlman, *Materials Chemistry*,
https://doi.org/10.1007/978-3-031-18784-1

700 BC Italians invent false teeth
105 BC Paper is first fabricated from bamboo fiber in ancient China
50 BC Glassblowing techniques are developed in Syria

Birth of Christ

590 Chinese scientists discover explosive mixtures consisting of sulfur, charcoal, and saltpeter (potassium nitrate)
618 Paper money is first put into use during the Tang Dynasty of China (618–906)
700s Porcelain is invented in China
747 The first reported air conditioning system comprised of water-powered fan wheels, by Emperor Xuanzong of the Tang Dynasty
1156 First reported synthesis of perfume by Henchum Seiken
1182 The magnetic compass is developed and widely used in China[1]
1249 Gunpowder is designed/synthesized by Rodger Bacon
1280 The cannon is invented in China
1286 Eyeglasses are first used in Venice
1300 Chinese judges wear smoke-colored quartz lenses to conceal their eyes in court
1400 The first use of grenades in France, designed by an unknown inventor
1430 Vision-correcting darkened eyeglasses introduced into China
1450 Crystallo, a clear soda-based glass, is invented by Angelo Barovier
1570 The pinhole camera is invented
1590 Glass lenses are developed in Netherlands and used for the first time in microscopes and telescopes
1593 Galileo invents a water thermometer
1608 The Dutch scientist Hans Lippershey invents the telescope
1612 The Flintlock firearm is developed in France
1621 John Napier invents the slide rule
1643 Torricelli makes the first barometer using mercury in a sealed glass tube
1651 The Dutch scientist Anton van Leeuwenhoek develops a microscope
1668 Isaac Newton invents a reflecting telescope
1709 Gabriel Fahrenheit invents an alcohol thermometer (mercury thermometer developed in 1714)
1710 Bathroom bidet is invented in France
1712 The steam engine is first invented in England
1714 The first patent for a typewriter is awarded in England to Henry Mill
1717 Swim fins (flippers) are invented by Benjamin Franklin
1718 The machine gun is developed in England
1738 William Champion patents a process for the production of metallic zinc by distillation from calamine and charcoal
1749 The lightning rod is invented by Benjamin Franklin
1752 Benjamin Franklin invents the flexible catheter

[1] The magnetic compass may have been first used during the Qin dynasty in China (*ca.* 200 BC).

1760s Benjamin Franklin invents bifocals
1770 First reported use of porcelain false teeth in France
1774 The electric telegraph is developed by Georges Louis Lesage
1776 The swivel chair is invented by Thomas Jefferson
1779 Bry Higgins issued a patent for hydraulic cement (stucco) for use as an exterior plaster
1782 Jacob Yoder builds the first flatboat for freight/passenger transport
1787 The automatic flour mill is invented by Oliver Evans
1789 Chlorine bleach is developed by Claude Louis Berthollet in France
1793 The cotton gin is invented by Eli Whitney
1800 Alessandro Volta makes a Copper/Zinc acid battery
1801 Frederick Graff Sr. invents the common post-type fire hydrant
1805 A self-propelled amphibious vehicle is invented by Oliver Evans
1806 The coffee percolator is invented by Benjamin Thompson Rumford
1808 The first lobster trap is invented by Ebenezer Thorndike
1813 The circular saw, invented by Tabitha Babbitt, is first used in a saw mill
1815 Humphry Davy invents a safety lamp that is used in coal mines without triggering an explosion
1815 Dental floss is invented by Levi Spear Parmly, a New Orleans dentist
1820 Thomas Hancock develops the first elastic fabrics
1821 Thomas Johann Seebeck invents the thermocouple
1823 Charles Macintosh patents a method for making waterproof garments
1824 Patent issued to Joseph Aspdin for the invention of cement
1825 Hans Christian Orsted produces metallic aluminum
1825 William Sturgeon invents the electromagnet
1831 The electric doorbell is invented by Joseph Henry
1834 The threshing machine is invented by Avery and Pitts
1835 Solymon Merrick invents the common household wrench
1836 Samuel Colt invents the revolving firearm (revolver)
1837 Wheatstone and Cooke invent the telegraph
1838 Regnault polymerizes vinylidene chloride via sunlight
1839 Goodyear (US), MacIntosh, and Hancock (England) vulcanize natural rubber
1839 Sir William Robert Grove experimented with the first fuel cell, using hydrogen and oxygen gases in the presence of an electrolyte
1842 The facsimile machine is invented by Alexander Bain
1843 The rotary printing press is invented by Richard Hoe
1849 Ferroconcrete, concrete reinforced with steel, is invented by Monier
1849 The modern gas mask is invented by Lewis Phectic Haslett
1850 The inverted microscope is invented by J. Lawrence Smith
1853 David M. Smith invents the modern wooden clothespin
1855 Bessemer process for mass production of steel patented
1856 Invention of the first synthetic dye, mauveine, by William Henry Perkin
1857 Toilet paper is designed and marketed for the first time
1859 The escalator was invented by Nathan Ames, referred to as "revolving stairs"

1860 Fredrick Walton invents linoleum, comprised of linseed oil, pigments, pine rosin, and pine flour
1861 James Clerk Maxwell demonstrates color photography
1864 Development of flash photography by Henry Roscoe in England
1867 The metallic paper clip was invented by Samuel B. Fay
1867 Barbed wire is invented by Lucien Smith
1872 Asphalt is first developed by Edward de Smedt at Columbia University
1872 Polyvinyl chloride (PVC) is first created by Eugen Baumann
1873 Levi Strauss & Co. begin producing blue-jeans out of durable canvas
1876 Nicolaus Otto invents a gas motor engine
1877 Thomas Edison completes the first phonograph
1881 Alexander Graham Bell builds the first metal detector
1883 Charles Fritts makes the first solar cells using selenium wafers
1883 Warren Johnson invents the first temperature regulating device known as a thermostat
1885 Karl Benz designs and builds the first gasoline-fueled automobile
1885 The first gasoline pump is manufactured by Sylvanus Bowser
1885 George Eastman invents the first flexible photographic film
1887 Contact lenses are invented by Eugen Frick in Switzerland
1888 George Eastman introduces a Kodak camera
1888 The common drinking straw is invented by Marvin Stone
1888 The revolving door is invented by Van Kannel
1890 The zipper is invented by Whitcomb Judson in Chicago, IL
1891 The first commercially produced artificial fiber, Rayon, is invented
1892 Calcium carbide is synthesized, as well as acetylene gas that is generated from the carbide
1896 Henry Ford constructs the first horseless carriage
1898 Nikola Tesla files a patent related to radio-control technology, and demonstrates his concept at the Electrical Exhibition in Madison. Square Garden (New York)
1901 The vacuum cleaner is invented by H. C. Booth
1901 The first mercury arc lamp is developed by Peter Hewitt
1902 August Verneuil develops a process for making synthetic rubies
1902 The neon light is invented in France
1902 The first automatic tea-making machine is invented by A. E. Richardson
1903 Ductile tungsten wire is synthesized by Coolidge
1907 Leo Hendrik Baekeland invents Bakelite (phenol-formaldehyde resins), used in electronic insulation
1908 Cellophane is invented by Brandenberger, a Swiss textile engineer
1909 Leo Baekeland presents the Bakelite hard thermosetting plastic
1909 Synthetic rubber is invented by Fritz Hofmann in Germany
1916 Jan Czochralski invents a method for growing single crystals of metals
1916 Kotaro Honda discovers a strongly magnetic Co/W alloy
1920 Herman Staudinger (Germany) advances the macromolecular hypothesis—the birth of polymer science

1923 Mercedes introduces the first supercharged automobile, the 6/25/40 hp
1924 Corning scientists invent Pyrex, a glass with a very low thermal expansion coefficient
1924 Celanese Corporation commercially produces acetate fibers
1924 The first mobile, two-way voice-based telephone is invented at Bell Labs
1926 Waldo Semon at B.F. Goodrich invents plasticized PVC known as *vinyl*
1929 Polysulfide (Thiokol) rubber is synthesized
1929 Carothers (du Pont) synthesizes the first aliphatic polyesters, establishes the principles of step-growth polymerization, and develops nylon 6,6
1931 Julius Nieuwland develops the synthetic rubber called *neoprene*
1931 Poly(methylmethacrylate) (PMMA) is synthesized
1932 Hans von Ohain and Sir Frank Whittle file patents for the jet engine
1932 Cathode ray tubes (CRTs) are invented by Allen B. Du Mont
1933 Ernest Ruska discovers the electron microscope; magnification of 12,000×
1933 Fawcett and Gibson develop polyethylene (LDPE)
1936 The first programmable computer, the Z1, is developed by Konrad Zuse
1936 Sunglasses become polarized by Ray Ban using a Polaroid filter developed by Edwin H. Land
1937 Polystyrene is developed
1937 Chester Carlson invents a dry printing process commonly called *Xerox*
1938 Roy Plunkett discovers the process for making poly-tetrafluoroethylene, better known as Teflon™
1938 Fiberglass is invented by Russell Slayter
1940 Thomas and Sparka synthesize isobutylene–isoprene rubber
1940 Butyl rubber is synthesized in the US
1941 Canadian John Hopps invents the first cardiac pacemaker
1942 The synthetic fabric, polyester, is invented
1943 The first kidney dialysis machine is developed
1943 Polyurethanes are synthesized by Otto Baeyer
1944 The first plastic artificial eye is developed in the US
1945 Percy Spencer creates the first microwave oven
1946 Mauchly and Eckert develop the first electronic computer ENIAC (Electronic Numerical Integrator and Computer)
1947 The first transistor is invented by Bardeen, Brattain, and Shockley at Bell Labs
1947 The first commercial application of a piezoelectric ceramic (barium titanate) used as a phonograph needle
1947 Invention of magnetic tape for recording applications
1947 Schlack develops epoxy polymeric systems
1951 Individual atoms seen for the first time using the field ion microscope
1950 The first commercial production of acrylic fibers by du Pont
1951 The computer UNIVAC 1 is developed
1951 Polypropylene is developed by Paul Hogan & Robert Banks of Phillips
1952 The first application of antiperspirant deodorant with a roll-on applicator

1953 Karl Ziegler discovers metallic catalysts which greatly improve the strength of polyethylene polymers
1954 Six percent efficiency silicon solar cells made at Bell Labs
1954 Charles Townes and Arthur Schawlow invent the MASER (microwave amplification by stimulated emission or radiation)
1955 Optical fibers are produced
1956 Liquid Paper™ is formulated by Bette Nesmith Graham
1957 Keller first characterizes a single crystal of polyethylene
1958 Bifocal contact lenses are produced
1959 Pilkington Brothers patent the float glass process
1959 The first commercial production of Spandex fibers by du Pont
1960s Polymers are first characterized by GPC, NMR, and DSC
1960 The first working laser (pulsed ruby) is developed by Maimam of Hughes Aircraft Corporation. Javan, Bennet, and Herriot make the first He:Ne gas laser
1960 Spandex fibers are synthesized
1962 The first SQUID superconducting quantum interference device is invented
1962 Polyimide resins are synthesized
1963 The first balloon embolectomy catheter is invented by Thomas Fogarty
1963 Ziegler and Natta are awarded the Nobel Prize for 1950's polymerization studies
1964 Bill Lear (of "Lear Jet" fame!) designs the first eight-track player
1965 A bulletproof nylon fabric, Kevlar, is invented at DuPont
1965 James Russell invents the compact disk
1965 Styrene–butadiene block copolymers are synthesized
1966 Fuel-injection systems for automobiles are developed in the UK
1966 Faria and Wright of Monsanto synthesize and test Astroturf
1967 Keyboards are first used for data entry, replacing punch cards
1968 Liquid crystal display is developed by RCA
1968 Allen Breed invents the first automotive air bag system
1969 The scanning electron microscope (SEM) is first used in laboratories to view cells in 3D
1969 George Smith and Willard Boyle invent charge-coupled devices (CCD) at Bell Labs
1970 The floppy disk (8 in.) is invented by Alan Shugart at IBM
1970 The first microfiber (polyester) is invented by Toray Industries in Japan
The first fabric comprised of microfibers, Ultrasuede, is also introduced
1971 The liquid crystal display (LCD) is invented by James Fergason
1971 The first single chip microprocessor, Intel 4004, is introduced
1971 The video cassette recorder (VCR) is invented by Charles Ginsburg
1971 Hydrogels are synthesized
1972 Motorola demonstrates the use of the first portable cellular phone[2]

[2] The first call that was made with a portable cellular phone was made by Dr. Martin Cooper, who called his rival, Joel Engel, at Bell Labs!

1973 The disposable lighter is invented by Bic
1973 Magnetic resonance imaging (MRI) is invented by Lauterbur and Damadian[3]
1974 Post-it® notes featuring a low-residue adhesive is invented by 3 M
1975 The laser printer is invented
1975 Robert S. Ledley is issued the patent for "diagnostic X-ray systems" (CAT scans)
1976 The inkjet printer is developed by IBM
1977 The Cray-1® supercomputer is introduced by Seymour Cray
1977 Electrically conducting organic polymers are synthesized by Heeger, MacDiarmid, and Shirakawa (Nobel Prize awarded in 2000)
1978 An artificial heart, Jarvik-7, is invented by Robert Jarvik
1978 The first analog video optical disk player is introduced by MCA Discovision
1979 The first cassette Walkman TPS-L2 is invented by Masaru Ibuka of Sony
1980 Compact disk players are introduced by Philips
1981 The world's largest solar-power generating station goes into operation (10 MW capacity)
1981 The scanning tunneling microscope (STM) is invented
1982 The first "personal computer" (PC) is introduced by IBM[4]
1982 Robert Denkwalter et al. from Allied Corporation are granted the first patent for dendrimers
1983 US phone companies begin to offer cellular phone service
1983 Steve Jobs of Apple introduces a new computer featuring the first graphical user interface (GUI), named *The Lisa*
1984 The CD-ROM is invented for computers
1984 The first truly autonomous car was developed at Carnegie Mellon University.
1984 The first clumping kitty litter is invented by biochemist Thomas Nelson
1985 Donald Tomalia and coworkers at Dow Chemical report the discovery of hyperbranched polymers, named *dendrimers*
1986 Synthetic skin is invented by Gregory Gallico, III

[3] The Nobel Prize for Medicine was awarded to Paul C. Lauterbur and Sir Peter Mansfield "for their discoveries concerning magnetic resonance imaging." This announcement failed to acknowledge Raymond V. Damadian who was the first person to propose MRI for medical diagnostics. In an effort to become properly acknowledged, Damadian placed full-page ads in major newspapers such as the *Washington Post*, the *New York Times*, and the *Los Angeles Times*. His fight for proper recognition is not unwarranted; the US National Medal of Technology was awarded to Damadian and Lauterbur in 1988, which recognized both scientists as co-inventors. Unfortunately, in an attempt to "make them [the Stockholm Nobel Prize committee] accountable to world opinion", Damadian may only be remembered for his unprecedented fight for personal recognition, rather than his co-discovery.

[4] The first PC (IBM 5150) retailed for $2,880 and was powered by the Intel 8088 microprocessor comprised of a 3-μm circuit, containing 29,000 transistors. This system was capable of performing 4.8 million cycles per second (4.8 MHz). The first microprocessor, the Intel 4004 developed in 1971, consisted of 2,300 transistors in a 3.5-mm circuit. Amazingly, modern computers costing $2,000 are now capable of performing 2+ billion instructions per second (2 GHz), and feature 100 s of millions of transistors contained within a circuit size of 130 nm! Computational devices have decreased steadily at about five linear dimensions per decade, but are rapidly approaching a barrier. Chapter 6 on nanotechnology will discuss the future of electronic devices.

1987 Bednorz and Muller develop a material that is superconducting at −183 °C
1987 Conducting polymers are developed by BASF
1988 A patent is issued for the Indiglo™ nightlight, consisting of electroluminescent phosphor particles
1989 High-definition television is invented
1989 NEC releases the first "notebook" computer, the NEC Ultralite
1989 A breathable, water- or wind-proof fabric, GORE-TEX®, is introduced
1989 The Intel 486 microprocessor is developed, featuring 1,000,000 transistors
1990 Biotextiles are invented in the US
1991 Iijima of NEC Corporation discovers carbon nanotubes
1991 Sony announces the first carbon anode based commercial Li-ion cell
1992 MiniDiscs (MDs) are introduced by Sony Electronics, Inc
1992 Prof. Jerome Schentag invents a computer-controlled "smart pill," for drug-delivery applications
1993 The Pentium processor is invented by Intel
1994 The first search engine for the World Wide Web is created by Filo and Yang[5]
1994 Lyocell is introduced by Courtaulds Fibers, consisting of a material derived from wood pulp
1994 The first virtual reality headset, the Forte VFX1 was introduced at the Consumers Electronics Show
1995 Nanoimprint lithography is invented by Stephen Chou at Stanford
1995 Digital Versatile Disk or Digital Video Disk (DVD) is invented
1996 The Nobel Prize in Chemistry is awarded to Richard Smalley, Robert Curl, and Harry Kroto for their 1985 discovery of the third form of carbon, known as *buckminsterfullerene* ("bucky balls")[6]
1996 WebTV is invented by Phillips
1996 The Palm Pilot is debuted by 3Com
1997 The gas-powered fuel cell is invented
1997 A fire-resistant building material, Geobond, is patented
1997 The digital video recorder (DVR) is invented by Jim Barton and Mike Ramsay, co-founders of Tivo, Inc
1997 Nokia introduces the Nokia 9000i Communicator. This combines a digital cell phone, hand-held PC, and fax

[5] This was simply a list of other sites called "Jerry's Guide to the World Wide Web". Within 8 months, more than 100,000 people were using this sit as an index to the Web, and it was eventually renamed Yahoo!

[6] This award is considered to have generated a new field known as "nanotechnology," as world-wide exposure was instantly aware of these nanoscale molecules, and other developments in this size regime were found shortly thereafter. It should be noted that the discovery of dendrimers by Denkwalter and coworkers from Allied Corporation was disclosed in 1981, 4 years before buckyballs were discovered. It may be expected that these nanopolymeric materials will be extremely influential toward the nanotechnology revolution (see Chap. 5 for more information on dendritic materials).

1998 Adam Cohen (19 years old!) develops an "electrochemical paint brush" circuit that uses an STM probe to manipulate copper atoms on a silicon surface
1998 Apple computer introduces the iMac
1998 Toyota Motor Corporation releases the Prius—the first mass-produced hybrid low-emission vehicle (LEV)
1998 Television stations in the US began to transition from analog to digital signals
1999 Danish physicist Hau is able to control the speed of light, useful for potential applications in communications systems and optical computers
1999 The chemical ingredient used by mussels to anchor themselves to rocks is discovered, and used to synthesize a waterproof adhesive
1999 Molecular-based logic gates are demonstrated to work better than silicon-based gates, an important precedent in the development of a molecular computer
2000 Intel releases the Pentium IV microprocessor, consisting of 42 million transistors
2000 Motorola releases the i1000 Plus—the first cell phone capable of connecting to the internet
2000 Robotic pets (*e.g.*, Poo-Chi, Tekno) are first introduced
2000 The first generation of "digital jukeboxes", the AudioReQuest ARQ1, retails for $800 and is the first device capable of storing thousands of MP3 songs
2001 The AbioCor self-contained artificial heart is implanted into Robert Tools
2001 SmartShirt sensors, to record and report body diagnostics, are designed by SensaTex, Inc. and Georgia Tech Research Corporation
2001 The bioartificial liver is invented by Kenneth Matsumura
2001 A fuel-cell bicycle is developed by Aprilia
2001 Digital satellite radio is developed by XM and Sirius
2001 SunClean self-cleaning glass is introduced by PPG Industries
2001 A wrinkle-free shirt is developed by Corpo Nove (Italy), consisting of Ti-alloy fibers interwoven with nylon
2001 The self-balancing personal transporter known as the Segway is invented by Dean Karmen
2001 The first armed drone used for combat strike was used in Afghanistan
2002 Clothing comprised of nanowhiskers is invented by Nano-tex, LLC to aid in stain resistance
2002 The lightest substance on Earth, known as *Aerogels*, is developed by NASA
2002 Scientists at SUNY, Buffalo, develop a new type of semiconducting material, GaSb/Mn, that will be used for future spintronics-based devices[7]
2003 Scientists discover a method used to commercially produce spider-web silk[8]

[7] Rather than using electronics to turn switches on/off, spintronic devices use electronic spins to represent information. This will allow these devices to process billions of pieces of information simultaneously, greatly increasing the speed and power of electronic devices. For more information on the future of spintronics, see: http://www.spintronics-info.com/.

[8] Spider-web silk is *ca.* five times stronger than steel by weight, and almost as elastic as nylon. Fibers comprised of the synthetic silk were demonstrated to be stronger than Kevlar, and may be

2003 Nanoparticles are used for the first time for clearcoat paint finishes (PPG – Ceramiclear™)
2003 Nanofilters are used to purity groundwater in Manitoba, Canada
2003 Digital videodisk recorders (DVRs) are introduced
2003 IBM develops the smallest light-emitting transmitter, comprised of carbon nanotubes (CNTs)
2003 Bandages are made from fibrinogen, a soluble protein found in blood
2004 The Blue Gene/L produced by IBM is able to perform 70.7 trillion calculations per second, making it the fastest computer in the world, to date
2004 The eyeware company Oakley develops sunglasses with a built-in audio player
2004 Apple releases the iPod mini—the size of a business card, but able to hold 1,000 songs
2004 A compound in the shape of a Borromean knot is discovered, based on earlier theoretical simulations
2004 Nintendo releases the hand-held gaming system, Nintendo DS
2004 Graphene is discovered by Andre Geim
2005 Carbon nanotubes are synthesized in bulk, and spun into a yarn
2005 iPod Nano and a video-capable iPod are introduced by Apple
2006 High-definition DVD players become commercially available
2006 The Wii video gaming system is released, which detects movement in 3-D
2006 Apple computer introduces MacBook Pro, MacBook, and iMac product lines that contain Intel dual-core chips, the first to contain over one billion transistors
2006 Flat-panel display technologies employing carbon nanotubes are demonstrated
2006 LG designs cellular phone that has a built-in breathalyzer for sobriety testing; this application is also tested as standard equipment for future automobiles
2007 Apple releases the iPhone, which combines cellular phone, internet, and iPod functionalities
2007 LG releases the first dual HD-DVD/Blu-ray high-definition player
2007 D-wave Systems, Inc. unveils the world's first commercially-viable quantum computer
2007 Researchers demonstrate wireless electricity to light a bulb from 7 feet away
2007 A nanowire battery is demonstrated by Dr. Cui at Stanford University
2008 A low-cost solar concentrator is developed at MIT
2008 A bionic contact lens is invented by Babak Parviz
2008 "Buckypaper" is discovered at Florida State University
2008 Nocera and coworkers at MIT develop a new catalyst to efficiently split water into H_2 and O_2 under ambient conditions, which may lead to a new paradigm for the large-scale deployment of solar energy
2008 Chemical vapor deposition is used for the first large-scale growth of graphene

---

useful for biomedical applications such as artificial tendons and ligaments and surgery sutures, as well as lightweight body armor for military applications.

2008 A self-healing rubber is made from vegetable oil
2008 Researchers at the Univ. of Pennsylvania report a robot (ckBot) that re-assembles itself after being dismantled
2009 Tour (Rice) and Dai (Stanford) independently report the first precedents to unzip carbon nanotubes to form graphene nanoribbons
2009 The University of Maryland's Joint Quantum Institute successfully transport data from one atom to another in a container one meter away (the first instance of pseudo-transportation!)
2009 Dow Chemical Co. develops roofing shingles integrated with thin-film solar cells comprised of copper indium gallium diselenide, CIGS
2009 The Boeing 787 Dreamliner, the first jet airliner to use composite materials for most of its fuselage, is developed by Boeing
2009 Self-assembling peptides are used for self-cleaning window applications
2009 The $20 knee is designed by Stanford engineering students
2009 Berkeley researchers create an "invisibility cloak"
2009 18-cm long arrays of SWNTs were synthesized—the longest carbon nanotube array to date
2009 The first Android cell phone is released, based on a Google operating system
2009 Simon Peers and Nicholas Godley unveil an 11-ft-long spider-silk cloth made in Madagascar
2009 The first 3-D digital camera is introduced by Fujifilm
2009 The EnergyHub smart thermostat is developed
2010 Apple releases their first tablet-PC, the iPad
2010 3 M/Littmann develops the first electronic stethoscope
2010 A $35 computer is unveiled in India
2010 Powered exoskeletons are developed to provide mobility assistance for aged and infirmed people
2010 HTC releases the first 4G cellphone, the HTC Evo
2010 The British company Xeros develops a washing machine comprised of nylon beads, requiring 90% less water than traditional machines
2011 The "Smart Bullet" is developed by Allant Tech systems, funded by the United States military; this allows soldiers to measure the distance to a target using a laser range finder, dialing in exactly where the bullet should explode (over/past walls, the corner of buildings, etc.) at precise distances
2011 Scientists at Notre Dame develop "Sunbelievable" solar paint, which incorporates semiconducting nanoparticles that can harness energy from the Sun
2011 Researchers at the University of Michigan report the development of an "optical battery" that harnesses energy from the magnetic component of incident radiation
2011 Namon Nassef demonstrates the "Zero Liquid Discharge Sewage Elimination System (ZLD)", which uses engine heat to oxidize and evaporate toilet, shower, and galley waste from watercraft
2011 An armored glove known as "the BodyGuard" is developed that is equipped with a high-voltage taser, video camera, laser pointer, and flashlight

2012 The "Electronic Dog Nose" is released, which can detect bed bugs using sensors that mimic the noses of bed-bug-sniffing dogs
2012 An inflatable abdominal tourniquet is demonstrated[9]
2012 A recirculating shower concept is reported, which incorporates a water treatment system to continuously capture, clean, and recirculate 70% of the water used during a shower[10]
2012 A prototype for augmented reality contact lenses is released
2013 The "3Doodler" handheld 3-D printer is reported
2013 An edible password pill is designed, which consists of a tiny chip that is powered on using the acid in your stomach. Once activated, it emits a specific 18-bit signal that can be detected by your phone, tablet, or computer
2013 The FDA approves the Argus II device that can restore partial vision to those who have severe retinitis pitmentosa, which can lead to blindness
2013 An artificial pancreas is developed that detects dropping sugar levels and shuts off regular insulin delivery for Type 1 diabetics
2014 Neuromorphic computer chips are developed by Qualcomm, which are designed to process sensory data and react accordingly
2014 The first levitating hoverboard, the Hendo Hoverboard is developed
2014 Contact lenses designed by Google are used to monitor blood sugar levels
2014 Julia Greer at Caltech fabricates ceramic nanolattices of alumina (Al_2O_3) that is one of the strongest and lightest substances ever made
2015 Encrypted mobile payments via Apple Pay is introduced, which combines near-field communication with the Touch ID fingerprint sensor on iPhones and iPads
2015 Sharp introduces the first 8 K television, with 7,680 × 4,320 pixels
2015 The Sorek seawater reverse osmosis (SWRO) desalination plant in Israel provides 627,000 m^3 of drinking water each day, with the lowest energy consumption to date
2015 Researchers in Vienna, Austria, grow 3-D clusters of living neurons from a single skin cell taken from an adult
2015 Tesla releases Autopilot software, allowing thousands of Tesla vehicles to drive themselves in a variety of conditions
2016 Researchers at the University of Washington develop passive Wi-Fi technology that will allow small wireless gadgets to power themselves from nearby radio signals
2016 SpaceX demonstrates the concept of "reusable rockets" with the successful delivery of a payload to the International Space Station, and return first-stage landing on a droneship

[9] Conventional tourniquets don't work around the abdomen since it's impossible to tie them tight enough to stop the blood flow from the aorta. Hence, soldiers with so-called "junctional hemorrhages" have only a few minutes before they bleed to death.

[10] This design is reported to save as much as 20,000–30,000 fewer gallons of water each year for a typical household.

2016 Eric Pop and coworkers at Stanford University fabricate MoS_2 chipswith a thickness of only three atoms, for next-generation transistors
2016 The "AirSelfie" pocket-size drone is demonstrated, which can remotely capture aerial photos and videos
2017 SolarCity's Buffalo factory begins production of 10,000 high-efficiency solar panels per day, representing the largest solar manufacturing plant in North America
2017 Philips IntelliSite Pathology uses a high-resolution scanner to convert pathology specimens into digital images, allowing doctors to send images to colleagues much quicker than traditional methods
2017 The Embrace Neonatal MRI instrument is used for infant scanning
2017 The Nintendo Switch is released, bridging the gap between home and on-the-go gaming
2017 LG releases the Signature TV W, known as the "wallpaper TV" since it is just 2.5 mm thick. It is mounted on any wall using magnets
2018 NASA's Parker Solar Probe makes its 7-yr mission exploring Sun's outer corona. The probe is able to reach top speed of 430,000 mph and withstand temperatures in excess of 2500 °F
2018 The Zephyr S HAPS (Solar High Altitude Pseudo-Satellite) by Airbus stayed aloft for more than 25 days, setting an all-time record
2018 Gorilla Glass 6 is released, which can survive an average of 15 drops on a hard surface from more than 3 ft, twice the durability of Gorilla Glass 5
2018 Audi releases E-Tron virtual mirrors, which replace conventional exterior mirrors with small cameras mounted on angled columns
2018 Carnegie Mellon invents shape-shifting humvee wheels that can convert from circular to triangular tank tread for sand, gravel, and other uneven terrains
2019 Abbott labs release SoToxa, a handheld diagnostic tool that is able to detect the presence of illicit drugs such as opioids, cocaine, and methamphetamines in the field
2019 5G networks are first initiated in metro areas such as New York and Los Angeles, capable of data speeds up to an order of magnitude faster than 4G LTE
2019 Sony unveils the first 16 K television
2019 DJI releases the Osmo Pocket, a mechanically stabilized 4 K camera easily able to be held in the palm of your hand
2019 Chevrolet releases the 8th generation Corvette—the first to feature a mid-engine design
2020 Apple releases the magic keyboard, which converts an iPad into a laptop experience
2020 E Ink's Kaleido technology offers 4,096 colors and better battery life for e-readers
2020 CopenHill in Copenhagen is the cleanest waste-to-energy facility in the world, converting 440,000 tons of trash to power 150,000 homes annually. It also features urban neighborhood activities such as a climbing wall, hiking trails, and a ski slope

2020 The e-OPRA prosthetic anchors directly to the bone via osseointegration, which avoids problems with sores and infections that have plagued traditional socket prosthetics

2021 Sony releases the $\alpha1$ camera, which offers high-resolution (50 MP) and super-fast shooting (30 frames per second)

2021 Apple releases AirTags, which use a low-power Bluetooth connection to find lost devices via the FindMy network

2021 Samsung releases the first folding phone, the Galaxy Z Flip3

2021 MIT develops the Droplette for cosmetic treatments, as well as delivering drugs without needles to treat a variety of skin conditions and even battlefield wounds

2022 DeepMind utilizes AI for protein folding, which opens new paths for drug discovery

2022 Pfizer and other labs introduce a pill for protection against COVID-19 and the latest variants

2022 Lincoln releases the L100 autonomous concept car with a cinema floor

2022 Apple releases all-new Macbook Air and 13-inch MacBook Pro, both featuring the M2 chip built on a 5-nm process by Taiwan Semiconductor Manufacturing Company (TSMC); the 3-nm M2 Pro chip begins production

Appendix B
"There's Plenty of Room at the Bottom"

This speech was given by Richard Feynman on 29 December 1959 at the Annual American Physical Society meeting. This text is provided with permission from the *Engineering and Science* magazine, published quarterly by the California Institute of Technology (http://www.pr.caltech.edu/periodicals/EandS/).

"Imagine experimental physicists must often look with envy at men like Kamerlingh Onnes, who discovered a field like low temperature, which seems to be bottomless and in which one can go down and down. Such a man is then a leader and has some temporary monopoly in a scientific adventure. Percy Bridgman, in designing a way to obtain higher pressures, opened up another new field and was able to move into it and to lead us all along. The development of ever higher vacuum was a continuing development of the same kind.

I would like to describe a field, in which little has been done, but in which an enormous amount can be done in principle. This field is not quite the same as the others in that it will not tell us much of fundamental physics (in the sense of, 'What are the strange particles?') but it is more like solid-state physics in the sense that it might tell us much of great interest about the strange phenomena that occur in complex situations. Furthermore, a point that is most important is that it would have an enormous number of technical applications. What I want to talk about is the problem of manipulating and controlling things on a small scale.

As soon as I mention this, people tell me about miniaturization, and how far it has progressed today. They tell me about electric motors that are the size of the nail on your small finger. And there is a device on the market, they tell me, by which you can write the Lord's Prayer on the head of a pin. But that's nothing; that's the most primitive, halting step in the direction I intend to discuss. It is a staggeringly small world that is below. In the year 2000, when they look back at this age, they will wonder why it was not until the year 1960 that anybody began seriously to move in this direction.

B. D. Fahlman, *Materials Chemistry*,
https://doi.org/10.1007/978-3-031-18784-1

Why cannot we write the entire 24 volumes of the Encyclopedia Brittanica on the head of a pin?

Let's see what would be involved. The head of a pin is a sixteenth of an inch across. If you magnify it by 25,000 diameters, the area of the head of the pin is then equal to the area of all the pages of the Encyclopaedia Brittanica. Therefore, all it is necessary to do is to reduce in size all the writing in the Encyclopaedia by 25,000 times. Is that possible? The resolving power of the eye is about 1/120 of an inch—that is roughly the diameter of one of the little dots on the fine half-tone reproductions in the Encyclopaedia. This, when you demagnify it by 25,000 times, is still 80 angstroms in diameter—32 atoms across, in an ordinary metal. In other words, one of those dots still would contain in its area 1,000 atoms. So, each dot can easily be adjusted in size as required by the photoengraving, and there is no question that there is enough room on the head of a pin to put all of the Encyclopaedia Brittanica.

Furthermore, it can be read if it is so written. Let's imagine that it is written in raised letters of metal; that is, where the black is in the Encyclopedia, we have raised letters of metal that are actually 1/25,000 of their ordinary size. How would we read it? If we had something written in such a way, we could read it using techniques in common use today. (They will undoubtedly find a better way when we do actually have it written, but to make my point conservatively I shall just take techniques we know today.) We would press the metal into a plastic material and make a mold of it, then peel the plastic off very carefully, evaporate silica into the plastic to get a very thin film, then shadow it by evaporating gold at an angle against the silica so that all the little letters will appear clearly, dissolve the plastic away from the silica film, and then look through it with an electron microscope!

There is no question that if the thing were reduced by 25,000 times in the form of raised letters on the pin, it would be easy for us to read it today. Furthermore; there is no question that we would find it easy to make copies of the master; we would just need to press the same metal plate again into plastic and we would have another copy.

How do we write small?

The next question is: How do we *write* it? We have no standard technique to do this now. But let me argue that it is not as difficult as it first appears to be. We can reverse the lenses of the electron microscope in order to demagnify as well as magnify. A source of ions, sent through the microscope lenses in reverse, could be focused to a very small spot. We could write with that spot like we write in a TV cathode ray oscilloscope, by going across in lines, and having an adjustment which determines the amount of material which is going to be deposited as we scan in lines.

This method might be very slow because of space charge limitations. There will be more rapid methods. We could first make, perhaps by some photo process, a screen which has holes in it in the form of the letters. Then we would strike an arc behind the holes and draw metallic ions through the holes; then we could again use our system of lenses and make a small image in the form of ions, which would deposit the metal on the pin.

A simpler way might be this (though I am not sure it would work): we take light and, through an optical microscope running backwards, we focus it onto a very small photoelectric screen. Then electrons come away from the screen where the light is shining. These electrons are focused down in size by the electron microscope lenses to impinge directly upon the surface of the metal. Will such a beam etch away the metal if it is run long enough? I don't know. If it doesn't work for a metal surface, it must be possible to find some surface with which to coat the original pin so that, where the electrons bombard, a change is made which we could recognize later.

There is no intensity problem in these devices—not what you are used to in magnification, where you have to take a few electrons and spread them over a bigger and bigger screen; it is just the opposite. The light which we get from a page is concentrated onto a very small area so it is very intense. The few electrons which come from the photoelectric screen are demagnified down to a very tiny area so that, again, they are very intense. I don't know why this hasn't been done yet!

That's the Encyclopaedia Brittanica on the head of a pin, but let's consider all the books in the world. The Library of Congress has approximately 9 million volumes; the British Museum Library has 5 million volumes; there are also 5 million volumes in the National Library in France. Undoubtedly there are duplications, so let us say that there are some 24 million volumes of interest in the world.

What would happen if I print all this down at the scale we have been discussing? How much space would it take? It would take, of course, the area of about a million pinheads because, instead of there being just the 24 volumes of the Encyclopaedia, there are 24 million volumes. The million pinheads can be put in a square of a thousand pins on a side, or an area of about 3 square yards. That is to say, the silica replica with the paper-thin backing of plastic, with which we have made the copies, with all this information, is on an area of approximately the size of 35 pages of the Encyclopaedia. That is about half as many pages as there are in this magazine. All of the information which all of mankind has every recorded in books can be carried around in a pamphlet in your hand—and not written in code, but a simple reproduction of the original pictures, engravings, and everything else on a small scale without loss of resolution.

What would our librarian at Caltech say, as she runs all over from one building to another, if I tell her that, ten years from now, all of the information that she is struggling to keep track of—120,000 volumes, stacked from the floor to the ceiling, drawers full of cards, storage rooms full of the older books—can be kept on just one library card! When the University of Brazil, for example, finds that their library is burned, we can send them a copy of every book in our library by striking off a copy from the master plate in a few hours and mailing it in an envelope no bigger or heavier than any other ordinary air mail letter.

Now, the name of this talk is 'There is *Plenty* of Room at the Bottom'—not just 'There is Room at the Bottom'. What I have demonstrated is that there *is* room—that you can decrease the size of things in a practical way. I now want to show that there is *plenty* of room. I will not now discuss how we are going to do it, but only what is possible in principle—in other words, what is possible according to the laws of physics. I am not inventing anti-gravity, which is possible someday only if the laws are not what we think. I am telling you what could be done if the laws *are* what we think; we are not doing it simply because we haven't yet gotten around to it.

Information on a small scale

Suppose that, instead of trying to reproduce the pictures and all the information directly in its present form, we write only the information content in a code of dots and dashes, or something like that, to represent the various letters. Each letter represents six or seven 'bits' of information; that is, you need only about six or seven dots or dashes for each letter. Now, instead of writing everything, as I did before, on the *surface* of the head of a pin, I am going to use the interior of the material as well.

Let us represent a dot by a small spot of one metal, the next dash by an adjacent spot of another metal, and so on. Suppose, to be conservative, that a bit of information is going to require a little cube of atoms 5 times 5 times 5—that is 125 atoms. Perhaps we need a hundred and some odd atoms to make sure that the information is not lost through diffusion, or through some other process.

I have estimated how many letters there are in the Encyclopaedia, and I have assumed that each of my 24 million books is as big as an Encyclopaedia volume, and have calculated, then, how many bits of information there are (10^{15}). For each bit I allow 100 atoms. And it turns out that all of the information that man has carefully accumulated in all the books in the world can be written in this form in a cube of material one two-hundredth of an inch wide—which is the barest piece of dust that can be made out by the human eye. So there is *plenty* of room at the bottom! Don't tell me about microfilm!

This fact—that enormous amounts of information can be carried in an exceedingly small space—is, of course, well known to the biologists, and resolves the mystery which existed before we understood all this clearly, of how it could be that, in the tiniest cell, all of the information for the organization of a complex creature such as ourselves can be stored. All this information—whether we have brown eyes, or whether we think at all, or that in the embryo the jawbone should first develop with a little hole in the side so that later a nerve can grow through it—all this information is contained in a very tiny fraction of the cell in the form of long-chain DNA molecules in which approximately 50 atoms are used for one bit of information about the cell.

Better electron microscopes

If I have written in a code, with 5 times 5 times 5 atoms to a bit, the question is: How could I read it today? The electron microscope is not quite good enough, with the greatest care and effort, it can only resolve about 10 angstroms. I would like to try and impress upon you while I am talking about all of these things on a small scale, the importance of improving the electron microscope by a hundred times. It is not impossible; it is not against the laws of diffraction of the electron. The wave length of the electron in such a microscope is only 1/20 of an angstrom. So it should be possible to see the individual atoms. What good would it be to see individual atoms distinctly?

We have friends in other fields—in biology, for instance. We physicists often look at them and say, 'You know the reason you fellows are making so little progress?'

(Actually I don't know any field where they are making more rapid progress than they are in biology today.) 'You should use more mathematics, like we do.' They could answer us—but they're polite, so I'll answer for them: 'What *you* should do in order for *us* to make more rapid progress is to make the electron microscope 100 times better'.

What are the most central and fundamental problems of biology today? They are questions like: What is the sequence of bases in the DNA? What happens when you have a mutation? How is the base order in the DNA connected to the order of amino acids in the protein? What is the structure of the RNA; is it single-chain or double-chain, and how is it related in its order of bases to the DNA? What is the organization of the microsomes? How are proteins synthesized? Where does the RNA go? How does it sit? Where do the proteins sit? Where do the amino acids go in? In photosynthesis, where is the chlorophyll; how is it arranged; where are the carotenoids involved in this thing? What is the system of the conversion of light into chemical energy?

It is very easy to answer many of these fundamental biological questions; you just *look at the thing!* You will see the order of bases in the chain; you will see the structure of the microsome. Unfortunately, the present microscope sees at a scale which is just a bit too crude. Make the microscope one hundred times more powerful, and many problems of biology would be made very much easier. I exaggerate, of course, but the biologists would surely be very thankful to you—and they would prefer that to the criticism that they should use more mathematics.

The theory of chemical processes today is based on theoretical physics. In this sense, physics supplies the foundation of chemistry. But chemistry also has analysis. If you have a strange substance and you want to know what it is, you go through a long and complicated process of chemical analysis. You can analyze almost anything today, so I am a little late with my idea. But if the physicists wanted to, they could also dig under the chemists in the problem of chemical analysis. It would be very easy to make an analysis of any complicated chemical substance; all one would have to do would be to look at it and see where the atoms are. The only trouble is that the electron microscope is one hundred times too poor. (Later, I would like to ask the question: Can the physicists do something about the third problem of chemistry—namely, synthesis? Is there a *physical* way to synthesize any chemical substance?

The reason the electron microscope is so poor is that the f-value of the lenses is only 1 part to 1,000; you don't have a big enough numerical aperture. And I know that there are theorems which prove that it is impossible, with axially symmetrical stationary field lenses, to produce an f-value any bigger than so and so; and therefore the resolving power at the present time is at its theoretical maximum. But in every theorem there are assumptions. Why must the field be symmetrical? I put this out as a challenge: Is there no way to make the electron microscope more powerful?

The marvelous biological system

The biological example of writing information on a small scale has inspired me to think of something that should be possible. Biology is not simply writing information;

it is *doing something* about it. A biological system can be exceedingly small. Many of the cells are very tiny, but they are very active; they manufacture various substances; they walk around; they wiggle; and they do all kinds of marvelous things—all on a very small scale. Also, they store information. Consider the possibility that we too can make a thing very small which does what we want—that we can manufacture an object that maneuvers at that level!

There may even be an economic point to this business of making things very small. Let me remind you of some of the problems of computing machines. In computers we have to store an enormous amount of information. The kind of writing that I was mentioning before, in which I had everything down as a distribution of metal, is permanent. Much more interesting to a computer is a way of writing, erasing, and writing something else. (This is usually because we don't want to waste the material on which we have just written. Yet if we could write it in a very small space, it wouldn't make any difference; it could just be thrown away after it was read. It doesn't cost very much for the material).

Miniaturizing the computer

I don't know how to do this on a small scale in a practical way, but I do know that computing machines are very large; they fill rooms. Why can't we make them very small, make them of little wires, little elements—and by little, I mean *little.* For instance, the wires should be 10 or 100 atoms in diameter, and the circuits should be a few thousand angstroms across. Everybody who has analyzed the logical theory of computers has come to the conclusion that the possibilities of computers are very interesting—if they could be made to be more complicated by several orders of magnitude. If they had millions of times as many elements, they could make judgments. They would have time to calculate what is the best way to make the calculation that they are about to make. They could select the method of analysis which, from their experience, is better than the one that we would give to them. And in many other ways, they would have new qualitative features.

If I look at your face I immediately recognize that I have seen it before. (Actually, my friends will say I have chosen an unfortunate example here for the subject of this illustration. At least I recognize that it is a *man* and not an *apple.*) Yet there is no machine which, with that speed, can take a picture of a face and say even that it is a man; and much less that it is the same man that you showed it before—unless it is exactly the same picture. If the face is changed; if I am closer to the face; if I am further from the face; if the light changes—I recognize it anyway. Now, this little computer I carry in my head is easily able to do that. The computers that we build are not able to do that. The number of elements in this bone box of mine are enormously greater than the number of elements in our 'wonderful' computers. But our mechanical computers are too big; the elements in this box are microscopic. I want to make some that are *sub*microscopic.

If we wanted to make a computer that had all these marvelous extra qualitative abilities, we would have to make it, perhaps, the size of the Pentagon. This has several disadvantages. First, it requires too much material; there may not be enough germanium in the world for all the transistors which would have to be put into this enormous thing. There is also the problem of heat generation and power consumption; TVA would be needed to run the computer. But an even more practical difficulty is that the computer would be limited to a certain speed. Because of its large size, there is finite time required to get the information from one place to another. The information cannot go any faster than the speed of light—so, ultimately, when our computers get faster and faster and more and more elaborate, we will have to make them smaller and smaller.

But there is plenty of room to make them smaller. There is nothing that I can see in the physical laws that says the computer elements cannot be made enormously smaller than they are now. In fact, there may be certain advantages.

Miniaturization by evaporation

How can we make such a device? What kind of manufacturing processes would we use? One possibility we might consider, since we have talked about writing by putting atoms down in a certain arrangement, would be to evaporate the material, then evaporate the insulator next to it. Then, for the next layer, evaporate another position of a wire, another insulator, and so on. So, you simply evaporate until you have a block of stuff which has the elements—coils and condensers, transistors and so on—of exceedingly fine dimensions.

But I would like to discuss, just for amusement, that there are other possibilities. Why can't we manufacture these small computers somewhat like we manufacture the big ones? Why can't we drill holes, cut things, solder things, stamp things out, mold different shapes all at an infinitesimal level? What are the limitations as to how small a thing has to be before you can no longer mold it? How many times when you are working on something frustratingly tiny like your wife's wrist watch, have you said to yourself, 'If I could only train an ant to do this!' What I would like to suggest is the possibility of training an ant to train a mite to do this. What are the possibilities of small but movable machines? They may or may not be useful, but they surely would be fun to make.

Consider any machine—for example, an automobile—and ask about the problems of making an infinitesimal machine like it. Suppose, in the particular design of the automobile, we need a certain precision of the parts; we need an accuracy, let's suppose, of 4/10,000 of an inch. If things are more inaccurate than that in the shape of the cylinder and so on, it isn't going to work very well. If I make the thing too small, I have to worry about the size of the atoms; I can't make a circle of 'balls' so to speak, if the circle is too small. So, if I make the error, corresponding to 4/10,000 of an inch, correspond to an error of 10 atoms, it turns out that I can reduce the dimensions of an automobile 4,000 times, approximately—so that it is 1 mm. across. Obviously, if you redesign the car so that it would work with a much larger tolerance, which is not at all impossible, then you could make a much smaller device.

It is interesting to consider what the problems are in such small machines. Firstly, with parts stressed to the same degree, the forces go as the area you are reducing, so that things like weight and inertia are of relatively no importance. The strength of material, in other words, is very much greater in proportion. The stresses and expansion of the flywheel from centrifugal force, for example, would be the same proportion only if the rotational speed is increased in the same proportion as we decrease the size. On the other hand, the metals that we use have a grain structure, and this would be very annoying at small scale because the material is not homogeneous. Plastics and glass and things of this amorphous nature are very much more homogeneous, and so we would have to make our machines out of such materials.

There are problems associated with the electrical part of the system—with the copper wires and the magnetic parts. The magnetic properties on a very small scale are not the same as on a large scale; there is the 'domain' problem involved. A big magnet made of millions of domains can only be made on a small scale with one domain. The electrical equipment won't simply be scaled down; it has to be redesigned. But I can see no reason why it can't be redesigned to work again.

Problems of lubrication

Lubrication involves some interesting points. The effective viscosity of oil would be higher and higher in proportion as we went down (and if we increase the speed as much as we can). If we don't increase the speed so much, and change from oil to kerosene or some other fluid, the problem is not so bad. But actually we may not have to lubricate at all! We have a lot of extra force. Let the bearings run dry; they won't run hot because the heat escapes away from such a small device very, very rapidly. This rapid heat loss would prevent the gasoline from exploding, so an internal combustion engine is impossible. Other chemical reactions, liberating energy when cold, can be used. Probably an external supply of electrical power would be most convenient for such small machines.

What would be the utility of such machines? Who knows? Of course, a small automobile would only be useful for the mites to drive around in, and I suppose our Christian interests don't go that far. However, we did note the possibility of the manufacture of small elements for computers in completely automatic factories, containing lathes and other machine tools at the very small level. The small lathe would not have to be exactly like our big lathe. I leave to your imagination the improvement of the design to take full advantage of the properties of things on a small scale, and in such a way that the fully automatic aspect would be easiest to manage.

A friend of mine (Albert R. Hibbs) suggests a very interesting possibility for relatively small machines. He says that, although it is a very wild idea, it would be interesting in surgery if you could swallow the surgeon. You put the mechanical surgeon inside the blood vessel and it goes into the heart and 'looks' around. (Of course the information has to be fed out.) It finds out which valve is the faulty one and takes a little knife and slices it out. Other small machines might be permanently incorporated in the body to assist some inadequately-functioning organ.

Now comes the interesting question: How do we make such a tiny mechanism? I leave that to you. However, let me suggest one weird possibility. You know, in the atomic energy plants they have materials and machines that they can't handle directly because they have become radioactive. To unscrew nuts and put on bolts and so on, they have a set of master and slave hands, so that by operating a set of levers here, you control the 'hands' there, and can turn them this way and that so you can handle things quite nicely.

Most of these devices are actually made rather simply, in that there is a particular cable, like a marionette string, that goes directly from the controls to the 'hands'. But, of course, things also have been made using servo motors, so that the connection between the one thing and the other is electrical rather than mechanical. When you turn the levers, they turn a servo motor, and it changes the electrical currents in the wires, which repositions a motor at the other end.

Now, I want to build much the same device—a master–slave system which operates electrically. But I want the slaves to be made especially carefully by modern large-scale machinists so that they are one-fourth the scale of the 'hands' that you ordinarily maneuver. So you have a scheme by which you can do things at one-quarter scale anyway—the little servo motors with little hands play with little nuts and bolts; they drill little holes; they are four times smaller. Aha! So I manufacture a quarter-size lathe; I manufacture quarter-size tools; and I make, at the one-quarter scale, still another set of hands again relatively one-quarter size! This is one-sixteenth size, from my point of view. And after I finish doing this I wire directly from my large-scale system, through transformers perhaps, to the one-sixteenth-size servo motors. Thus I can now manipulate the one-sixteenth size hands.

Well, you get the principle from there on. It is rather a difficult program, but it is a possibility. You might say that one can go much farther in one step than from one to four. Of course, this has all to be designed very carefully and it is not necessary simply to make it like hands. If you thought of it very carefully, you could probably arrive at a much better system for doing such things.

If you work through a pantograph, even today, you can get much more than a factor of four in even one step. But you can't work directly through a pantograph which makes a smaller pantograph which then makes a smaller pantograph—because of the looseness of the holes and the irregularities of construction. The end of the pantograph wiggles with a relatively greater irregularity than the irregularity with which you move your hands. In going down this scale, I would find the end of the pantograph on the end of the pantograph on the end of the pantograph shaking so badly that it wasn't doing anything sensible at all.

At each stage, it is necessary to improve the precision of the apparatus. If, for instance, having made a small lathe with a pantograph, we find its lead screw irregular—more irregular than the large-scale one—we could lap the lead screw against breakable nuts that you can reverse in the usual way back and forth until this lead screw is, at its scale, as accurate as our original lead screws, at our scale.

We can make flats by rubbing unflat surfaces in triplicates together—in three pairs—and the flats then become flatter than the thing you started with. Thus, it is not impossible to improve precision on a small scale by the correct operations. So, when we build this stuff, it is necessary at each step to improve the accuracy of the equipment by working for a while down there, making accurate lead screws, Johansen blocks, and all the other materials which we use in accurate machine work at the higher level. We have to stop at each level and manufacture all the stuff to go to the next level—a very long and very difficult program. Perhaps you can figure a better way than that to get down to small scale more rapidly.

Yet, after all this, you have just got one little baby lathe four thousand times smaller than usual. But we were thinking of making an enormous computer, which we were going to build by drilling holes on this lathe to make little washers for the computer. How many washers can you manufacture on this one lathe?

A hundred tiny hands

When I make my first set of slave 'hands' at one-fourth scale, I am going to make ten sets. I make ten sets of 'hands,' and I wire them to my original levers so they each do exactly the same thing at the same time in parallel. Now, when I am making my new devices one-quarter again as small, I let each one manufacture ten copies, so that I would have a hundred 'hands' at the 1/16th size.

Where am I going to put the million lathes that I am going to have? Why, there is nothing to it; the volume is much less than that of even one full-scale lathe. For instance, if I made a billion little lathes, each 1/4,000 of the scale of a regular lathe, there are plenty of materials and space available because in the billion little ones there is less than 2% of the materials in one big lathe. It doesn't cost anything for materials, you see. So I want to build a billion tiny factories, models of each other, which are manufacturing simultaneously, drilling holes, stamping parts, and so on.

As we go down in size, there are a number of interesting problems that arise. All things do not simply scale down in proportion. There is the problem that materials stick together by the molecular (Van der Waals) attractions. It would be like this: after you have made a part and you unscrew the nut from a bolt, it isn't going to fall down because the gravity isn't appreciable; it would even be hard to get it off the bolt. It would be like those old movies of a man with his hands full of molasses, trying to get rid of a glass of water. There will be several problems of this nature that we will have to be ready to design for.

Rearranging the atoms

But I am not afraid to consider the final question as to whether, ultimately—in the great future—we can arrange the atoms the way we want; the very *atoms*, all the way down! What would happen if we could arrange the atoms one by one the way we want them (within reason, of course; you can't put them so that they are chemically unstable, for example).

Up to now, we have been content to dig in the ground to find minerals. We heat them and we do things on a large scale with them, and we hope to get a pure substance with just so much impurity, and so on. But we must always accept some atomic arrangement that nature gives us. We haven't got anything, say, with a 'checkerboard' arrangement, with the impurity atoms exactly arranged 1,000 angstroms apart, or in some other particular pattern.

What could we do with layered structures with just the right layers? What would the properties of materials be if we could really arrange the atoms the way we want them? They would be very interesting to investigate theoretically. I can't see exactly what would happen, but I can hardly doubt that when we have some *control* of the arrangement of things on a small scale we will get an enormously greater range of possible properties that substances can have, and of different things that we can do.

Consider, for example, a piece of material in which we make little coils and condensers (or their solid state analogs) 1,000 or 10,000 angstroms in a circuit, one right next to the other, over a large area, with little antennas sticking out at the other end—a whole series of circuits. Is it possible, for example, to emit light from a whole set of antennas, like we emit radio waves from an organized set of antennas to beam the radio programs to Europe? The same thing would be to *beam* the light out in a definite direction with very high intensity. (Perhaps such a beam is not very useful technically or economically.)

I have thought about some of the problems of building electric circuits on a small scale, and the problem of resistance is serious. If you build a corresponding circuit on a small scale, its natural frequency goes up, since the wave length goes down as the scale; but the skin depth only decreases with the square root of the scale ratio, and so resistive problems are of increasing difficulty. Possibly we can beat resistance through the use of superconductivity if the frequency is not too high, or by other tricks.

Atoms in a small world

When we get to the very, very small world—say circuits of seven atoms—we have a lot of new things that would happen that represent completely new opportunities for design. Atoms on a small scale behave like *nothing* on a large scale, for they satisfy the laws of quantum mechanics. So, as we go down and fiddle around with the atoms down there, we are working with different laws, and we can expect to do different things. We can manufacture in different ways. We can use, not just circuits, but some system involving the quantized energy levels, or the interactions of quantized spins, etc.

Another thing we will notice is that, if we go down far enough, all of our devices can be mass produced so that they are absolutely perfect copies of one another. We cannot build two large machines so that the dimensions are exactly the same. But if your machine is only 100 atoms high, you only have to get it correct to one-half of one percent to make sure the other machine is exactly the same size—namely, 100 atoms high!

At the atomic level, we have new kinds of forces and new kinds of possibilities, new kinds of effects. The problems of manufacture and reproduction of materials will be quite different. I am, as I said, inspired by the biological phenomena in which chemical forces are used in repetitious fashion to produce all kinds of weird effects (one of which is the author).

The principles of physics, as far as I can see, do not speak against the possibility of maneuvering things atom by atom. It is not an attempt to violate any laws; it is something, in principle, that can be done; but in practice, it has not been done because we are too big.

Ultimately, we can do chemical synthesis. A chemist comes to us and says, 'Look, I want a molecule that has the atoms arranged thus and so; make me that molecule'. The chemist does a mysterious thing when he wants to make a molecule. He sees that it has got that ring, so he mixes this and that, and he shakes it, and he fiddles around. And, at the end of a difficult process, he usually does succeed in synthesizing what he wants. By the time I get my devices working, so that we can do it by physics, he will have figured out how to synthesize absolutely anything, so that this will really be useless.

But it is interesting that it would be, in principle, possible (I think) for a physicist to synthesize any chemical substance that the chemist writes down. Give the orders and the physicist synthesizes it. How? Put the atoms down where the chemist says, and so you make the substance. The problems of chemistry and biology can be greatly helped if our ability to see what we are doing, and to do things on an atomic level, is ultimately developed—a development which I think cannot be avoided.

Now, you might say, 'Who should do this and why should they do it?' Well, I pointed out a few of the economic applications, but I know that the reason that you would do it might be just for fun. But have some fun! Let's have a competition between laboratories. Let one laboratory make a tiny motor which it sends to another lab which sends it back with a thing that fits inside the shaft of the first motor.

High school competition

Just for the fun of it, and in order to get kids interested in this field, I would propose that someone who has some contact with the high schools think of making some kind of high school competition. After all, we haven't even started in this field, and even the kids can write smaller than has ever been written before. They could have competition in high schools. The Los Angeles high school could send a pin to the Venice high school on which it says, 'How's this?' They get the pin back, and in the dot of the 'i' it says, 'Not so hot.'

Perhaps this doesn't excite you to do it, and only economics will do so. Then I want to do something; but I can't do it at the present moment, because I haven't prepared the ground. It is my intention to offer a prize of $1,000 to the first guy who can take the information on the page of a book and put it on an area 1/25,000 smaller in linear scale in such manner that it can be read by an electron microscope.

And I want to offer another prize—if I can figure out how to phrase it so that I don't get into a mess of arguments about definitions—of another $1,000 to the first guy who makes an operating electric motor—a rotating electric motor which can be controlled from the outside and, not counting the lead-in wires, is only 1/64 inch cube.

I do not expect that such prizes will have to wait very long for claimants."

Appendix C
Materials-Related Laboratory Experiments

This section describes the experimental details for six modules that are appropriate for undergraduate curricula. These experiments were reproduced with permission from the Journal of Chemical Education, published by the Division of Chemical Education of the American Chemical Society (http://jchemed.chem.wisc.edu/index.html). Herein, we provide a few representative modules; the breadth of experiments will be expanded in future editions of this textbook, to include other materials classes such as ceramics, metals, polymers, semiconductors, superconductors, and magnetic materials.

C.1 Chemical Vapor Deposition of Carbon Nanotubes

(Fahlman, B. D. *J. Chem. Ed.*, 2002, *79*, 203)

C.1.1 Background Information

There are two primary methods used to produce carbon nanotubes on the laboratory scale: the carbon-arc process and transition metal-catalyzed decomposition of organic precursors. The latter method offers a relatively facile setup and scale-up for large-scale nanotube synthesis. Hence, this method has now become the method of choice for nanotube growth. In 1998, only three reported methods for nanotube growth involved CVD; however, since this time, the number of reported instances using this technology has increased by at least an order of magnitude. This experiment is designed to introduce students to both CVD technology and nanotube characterization techniques by focusing on the variation of nanotubes that may be deposited as a function of the catalyst composition.

B. D. Fahlman, *Materials Chemistry*,
https://doi.org/10.1007/978-3-031-18784-1

Similar catalysts prepared from alumina nanoparticles and ferric nitrate were reported to show the speciation of iron as Fe, FeO, and α-Fe_2O_3 prior to CVD and as α-Fe_2O_3 and FeO after nanotube growth. However, in this reported synthesis, hydrogen gas was used; our method did not use a reducing atmosphere and the iron in these catalysts has been described elsewhere as being present solely as Fe_2O_3 after heating to 1,000 °C for 10 min in an argon atmosphere.

The CVD method used in this experiment to grow nanotubes was accomplished using the pyrolysis of methane, catalyzed by iron-doped alumina catalyst (Eq. 1).

$$CH_4(g) \xrightarrow{Fe_2O_3/Al_2O_3, 1000^\circ C} C(s) + H_2(g) \tag{1}$$

The hydrogen gas produced in the decomposition is thought to eliminate the surface-adsorbed carbon-containing species, $CH_{x(ads)} (x = 0\text{–}3)$ via Eq. 2. This would explain the relative lack of amorphous carbon and graphitic deposits found in our samples (Fig. C.1), a problem that has plagued alternate CVD experiments.

$$\frac{(4-x)}{2} H_2(g) + CH_x(ads) \rightleftharpoons CH_4(g) \tag{2}$$

The deposited nanotubes are found within bundles on the surface of the catalyst particles. Methane is the most difficult hydrocarbon to thermally decompose; hence, precursor decomposition likely takes place through a traditional CVD mechanism of adsorption and surface decomposition/migration rather than preliminary gas-phase pyrolysis. A wealth of information has been reported on the growth of graphitic carbon over metal substrates. Active metals such as iron, nickel, and cobalt are recognized to best catalyze the decomposition of volatile carbon compounds by forming metastable carbide intermediate species. Most importantly, for nanotube growth, these metals allow facile and rapid diffusion of carbon through and over their surfaces. This suggests that if decomposition of the gas-phase organic molecules would have occurred away from the metal surface, only amorphous carbon deposits would have been formed.

Fig. C.1 FESEM image of the raw nanotube felt obtained from catalyst particles containing 7.5 at% Fe

In the first institution of this laboratory module, an assessment of the overall deposition environment of the nanotubes in association with catalyst particles was to be made. Therefore, no purification methods were used to separate nanotubes from catalyst particles. Although methods such as sonication in mineral acids for 15 min have been used to completely remove metal and support particles for MgO-supported nanotubes, this method is not so easily performed for alumina or silica substrates. For these latter supports, more vigorous purification techniques such as prolonged digestion in strong acids or gas-phase methods would need to be utilized to completely separate the nanotubes from the raw felt.

The variance of nanotube diameters with catalyst concentrations is in accord with models for carbon nanotube growth using CVD. It is now well established that the diameter of individual catalyst particles will dictate the diameter of the grown nanotubes. Particles with diameters in the range 1–2 nm catalyze SWNT growth whereas larger particles on the order of 10–50 nm produce multiwall nanotubes (MWNTs). The increase in nanotube diameters with increasing metal concentrations in the catalyst has been attributed to a greater opportunity for the reduction of M^{n+} ions, occurring within a reducing atmosphere, to form relatively large metal particles. However, in our system, such a reducing atmosphere was not used and the increase of nanotube diameters was most likely due to larger supported Fe_2O_3 particles being formed through agglomeration from more concentrated solutions used in the catalyst synthesis. Nevertheless, the physical characteristics of the catalyst particles do not solely govern the nanotube diameter; the partial pressure of the hydrocarbon precursor(s) also plays a role. This is verified from reports by Kong et al. who were able to obtain predominantly SWNTs using Fe_2O_3/alumina catalysts prepared similarly as our 2.4 at% Fe catalysts but using a much greater methane flow (6,150 L min^{-1}).

C.1.2 *Procedure*

CAS Registry Numbers for Chemicals:

Iron(III) nitrate nonahydrate: 7782-61-8
Aluminum oxide nanoparticles: 1344-28-1
Methanol: 67-56-1
Methane: 74-82-8.

Synthesis of the supported catalyst

Materials

Safety glasses
Electronic balance and weighing paper
Fumed alumina nanoparticles (Degussa)
$Fe(NO_3)_3 \cdot 9H_2O$

100-mL round-bottom flasks
Cork flask stands
Methanol.

Rotary evaporator/water bath with appropriate connections.

1. Transfer 1.0 g of fumed alumina nanoparticles to a 100-mL round-bottom flask.
2. Calculate the amount of $Fe(NO_3)_3 \cdot 9H_2O$ necessary to prepare 0.020, 0.050, 0.075, and 0.10 M solutions in methanol (use 30.0 mL as delivered from a volumetric pipette). Accurately weigh out one of these amounts into a clean, dry beaker, add the methanol, and ensure the solid is completely dissolved. Add this solution to the alumina weighed out in step 1. *Note*: Each person in the group will prepare a catalyst from solutions containing a different iron concentration (*i.e.*, unless there is a group of more than four students).
3. Place a magnetic stir bar in the bottom of the round-bottom flask and allow the solution to stir for 90 min at room temperature.
4. Remove the methanol from the solution using a rotary evaporator and a water temperature of *ca.* 50 °C. This should only take on the order of 2–3 min.
5. Place the product in an oven at 130 °C overnight. The solid should now be a brownish-orange color with little/no white particles present in the mixture. Wearing safety goggles and gloves, grind the product into a fine powder using a mortar and pestle within a fume hood.
6. Place the finely ground product in a 20-mL scintillation vial for the CVD growth experiment you will carry out during your next lab section.

CVD growth

Materials

Ceramic boats
Tongs
Heat-proof gloves
Tube furnace
Quartz tube
Methane and argon cylinders with regulators and tubing
Rotometer flow controller.

Oil bubbler with appropriate connections (assembled in a fume hood away from ignition sources).

1. Transfer the synthesized catalyst powder to a ceramic boat and place inside the quartz tube within the tube furnace. Since the final products will be black powders, regardless of the iron content of the catalyst, ensure you record the position of your catalyst in order to differentiate it from your colleagues' samples. Securely fasten the valve joint to the quartz tube with rubber bands to prevent gas leakage during the reaction.
2. Turn on the argon valve and maintain a constant flow (at 50 mL min^{-1}) by monitoring the rotometer beads. Set the furnace to 1,000 °C in all three zones. After

1,000 °C is reached, allow argon to flow through the chamber for 10 additional minutes. Turn on the nitrogen valve and ensure that a stream of nitrogen gas is flowing onto the point where the oil bubbler tubing is fastened to the reaction tube. Failure to do this will cause the tubing to melt due to the high temperature of the evolved gases.
3. Open the methane valve to the same flow rate as argon, close the argon valve, and allow methane to flow for 10 min. During this time, the tubing leading to the bubbler, as well as the mineral oil inside the bubbler, will become black from carbon deposits. After 10 min, reopen the Ar valve, close the methane valve, and allow argon to flow through the chamber. Immediately program the oven to return to room temperature (no cooling ramp is necessary).

Characterization by field-emission SEM

Many techniques are used to characterize carbon nanotubes. Of these, scanning electron microscopy (SEM) is used most frequently to quickly assess the quantity and quality of the nanotubes present in the sample. The extremely fine electron source of the field-emission system to be used in this section will enable the attainment of much higher resolution images than a conventional SEM. Useful magnifications in excess of 200,000 times are often obtainable, which translates to a resolution of 3–5 nm at an accelerating voltage of 30 kV. The high brightness of this source also allows high-resolution imaging and characterization of beam-sensitive materials (*e.g.*, fragile plastics and integrated circuits) at high magnifications in excess of 100,000 times at very low accelerating voltages (*i.e.*, 0.2–5 kV). Insulating samples may also be examined without a conventional conductive coating of gold or carbon.

You will prepare samples for SEM analysis using two methods:

Method 1. Place a small amount (*ca.* 2 mg) of the synthesized material in a glass vial and add 10 mL of methanol. Place the vial in a sonicator for half an hour. Place a drop of the suspension on carbon tape stuck to a mounting pin. Allow the methanol to evaporate completely and insert into the sample chamber. If FESEM images indicate an agglomeration of nanotubes, try preparing your sample from a chloroform suspension.

Method 2. Simply place a small amount (spatula tip) of the black powder onto carbon tape fastened to an aluminum-mounting pin. Tap down the powder with the flat end of the spatula and shake off the excess prior to mounting in the SEM vacuum chamber.

C.2 Supercritical Fluid Facilitated Growth of Copper and Aluminum Oxide Nanoparticles

(Williams, G. L.; Vohs, J. K.; Brege, J. J.; Fahlman, B. D. *J. Chem. Ed.*, 2005, *82*, 771)

The burgeoning field of *nanotechnology* research is focused on objects with architectural dimensions of less than 100 nm. Nanoparticles of metals or metal oxides have a variety of applications for next-generation electronic devices, advanced sensors, inks, antitumor targeting agents, and chemical/biological warfare sensing devices. This experiment is designed to introduce students to many exciting areas of current industrial/research interest including SCF technology, nanoparticle growth, and nanoscale characterization techniques.

C.2.1 Procedure

CAS Registry Numbers for Chemicals

Sodium aluminate: 1302-42-7
Copper (II) chloride dihydrate: 10,125-13-0
1,4-Phenylenediamine: 106-50-3
Ethanol, 200-proof: 64-17-5
Carbon dioxide (bone-dry grade): 124-38-9

Synthesis of aluminum oxide nanoparticles

1. Clean a silicon wafer by rinsing with distilled water, and dry with KimWipes. Place the clean silicon substrate in a tall beaker, and place in the supercritical fluid reactor using long tweezers. With assistance from the instructor, fasten the reactor to the supercritical fluid system.
2. Using solid sodium aluminate powder, prepare a 4% (w/v) aqueous aluminate solution in an 8-mL vial, and shake vigorously until dissolved. Filter this solution into a clean 8-mL vial through a disposable pipette fitted with a cotton plug.
3. In a separate 8-mL vial, place 0.2 g of the surfactant (*ask instructor which one to use*) in 6 mL of heptane (if AOT: sodium bis(2-ethylhexyl)sulfosuccinate) or water (if fluorinated surfactant: ammonium carboxylate perfluoropolyether—$[CF_3O(CF_2CF(CF_3)O)_3CF_2COO]^-[NH_4]^+$). Shake vigorously until the surfactant is completely dissolved. Add two drops of the filtered aluminate solution to the surfactant solution, and shake vigorously until one phase is obtained (*i.e.*, the final solution should not be cloudy).
4. With consultation with the instructor, prime the cosolvent pump and set the appropriate reactor temperature and pressure (see Fig. C.2).
5. Once the reactor pressure and temperature has stabilized, inject the microemulsion solution using the cosolvent pump.
6. After the microemulsion solution has been added to the reactor, allow the pressure to remain at its initial setting. After this time, flush the reactor for 5 min with a dynamic flow of carbon dioxide.
7. Vent the system and disconnect the reactor. Remove the silicon wafer using tweezers and place in a vial until characterization studies are performed. Dip half of the coated wafer in heptane (if AOT was used) or water (if fluorinated surfactant was used) and allow to air dry. Store the coated wafer until analysis by placing it in a plastic weighing tray (polished side up), and place another weighing

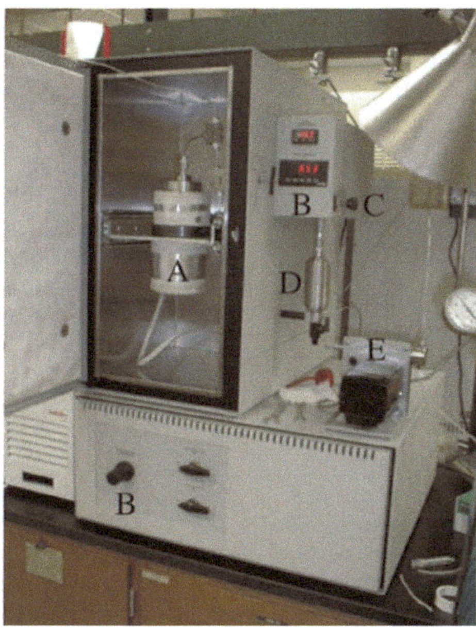

Fig. C.2 Photograph of the supercritical fluid system used for nanoparticle synthesis. Shown is the 300-mL high-pressure reactor (a), with pressure/temperature controllers (b). The system is rated for safe operation at temperatures and pressures below 200 °C and 10,000 psi, respectively. The vessel may be slowly vented, or exposed to a dynamic CO_2 flow, using a multiturn restrictor valve (c), which provides a sensitive control over system depressurization, allowing for the collection of CO_2-solvated species in the stainless steel collector (d). For deposition using the rapid expansion of the supercritical solution (RESS), nanoparticles were blown onto a TEM grid that was placed under the stopcock below D. Also shown is the cosolvent addition pump (e) used for the synthesis of aluminum oxide nanoparticles, capable of delivering liquids into the chamber against a back-pressure of $\leq$5,000 psi

tray on top. Completely tape the two-weighing trays together to prevent dust from forming on the coated wafer.

Synthesis of copper nanoparticles

1. Prepare a 0.05 M copper solution using $CuCl_2 \cdot 2H_2O$ in absolute ethanol (200 proof), and transfer this to a clean 8-mL vial. Also prepare a 0.04 M aqueous solution of 1,4-phenylenediamine, and transfer to a second clean 8-mL vial. Position the vials within the supercritical fluid reactor, avoiding any contact between the solutions. With consultation with your instructor, allow the system to reach the desired pressure and temperature. Maintain these conditions for 45 min.
2. Vent the system and remove the vials from the reactor. Note color changes, turbidity, and phases (*i.e.*, is there one liquid present, or does there appear to be a layer of two liquids?) of both solutions. Remove both vials and keep tightly sealed

until characterization studies are performed. Comment on what you observe in the Results and Discussion section of your formal report.

3. Repeat step 1 using ethanol for both copper and phenylenediamine solutions. This time, when you vent the system after 45 min, place a carbon TEM grid at the base of the collector vessel (the instructor will point this out). This technique, known as *RESS* (rapid expansion of the supercritical solution), is a unique benefit of performing reactions in supercritical fluids. When the system is vented, the gas/liquid carbon dioxide expands, being rapidly converted from its original high-pressure environment to a low-pressure gas. Place the grids in a vial for characterization studies.

C.3 Synthesis and Characterization of Liquid Crystals

(Van Hecke, G. R.; Karukstis, K. K.; Li, H.; Hendargo, H. C.; Cosand, A. J.; Fox, M. M. *J. Chem. Ed*., 2005, *82*, 1349)

This experiment will introduce you to some techniques of chemical synthesis based on acid–base reactions, purification by recrystallization, and brief characterization by melting point. Also on the characterization side, you will investigate the absorption of polarized light by matter. A liquid crystal is a state of matter neither liquid nor crystal but a state in-between. Liquid crystals are often called *mesophases* after the Greek mesos for middle. Normally when a crystal melts it forms an ordinary liquid phase. However, a substance which exhibits liquid crystalline behavior melts at least twice, first into the liquid crystalline or mesophase, and second, into the ordinary liquid.

Molecules in a mesophase exhibit orientation with respect to each other, and, in certain types of liquid crystals, exhibit some extent of ordered position with respect to each other. Thus, mesophases are described by degrees of orientational and positional order. Using ordering as a basis, liquid crystals fall into two types: *nematic* and *smectic*. Nematic mesophases possess only orientation order while smectic mesophases possess orientational and some positional order (Fig. C.3). Also shown in Fig. C.3 is a special type of nematic phase, the *chiral nematic* or *cholesteric* phase (cholesteric is the older and still more common name but chiral nematic is now the correct name). In this phase, not only do the molecules comprising the phase point more or less in the same direction on a local scale, but also on a larger scale that direction changes following a helix. The quantity called *pitch* is defined as the physical distance required for one complete revolution about the optical axis. Cholesteric phases were the first liquid crystals ever observed, and for many years were thought to be a separate type of liquid crystal. Now it is known that what made cholesteric liquid crystalline materials different is the fact that they are composed of optically active or chiral molecules. The cholesteric phase derives its name from the fact that the first substances observed to exhibit what we today call liquid crystallinity were derivatives of the naturally occurring substance cholesterol. While many commercial liquid crystal devices employ the chlolesteric phase to function, cholesterol and its derivatives are important objects of study in medicine since cholesterol derivatives

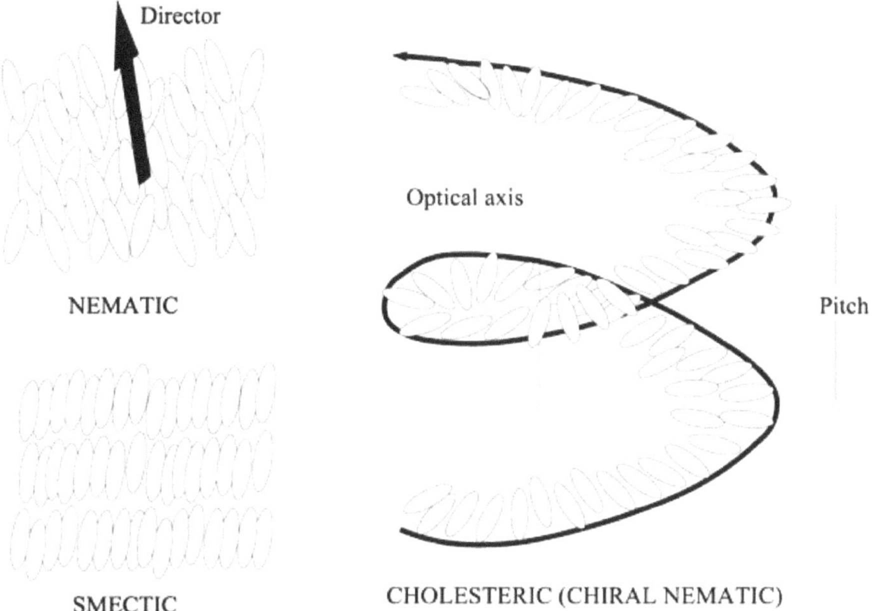

Fig. C.3 Schematic of three mesophases: the nematic and cholesteric (which have only orientation order) and the smectic phase (which has orientational and positional order)

are components of the deposits that form on the wall of arteries and lead to hardening of the arteries.

A nematic mesophase forms when rod-like molecules orient themselves on average in a given direction. The director of the phase is a vector that points in a direction determined by looking at the average of the directions in which all the molecules point or are oriented. The director's orientation is a function of temperature, pressure, and applied electric and magnetic fields. The many practical applications of liquid crystals depend critically on modifying the director by external influences. This experiment will be partly concerned with what factors affect the director of the cholesteric phase, which is the direction in which any local collection of molecules points. What happens to the cholesteric director as a function of temperature will be studied by determining the pitch of the phase as a function of temperature.

Substances in their liquid crystalline phase behave like liquids in that they flow and fill any shape container in which they are placed. However, even though they are fluids, they retain certain optical properties characteristic of a solid, especially those optical properties which are influenced by polarized light. For this experiment how the plane of polarization of plane polarized light changes on passage of such light through a mesomorphic medium is particularly crucial. In the cholesteric mesophase, the director, which remember reflects the local organization of a collection of molecules, follows a helical path. The sense of this helix, left or right, determines the angle of rotation of plane polarized light incident on a cholesteric sample. The pitch of the

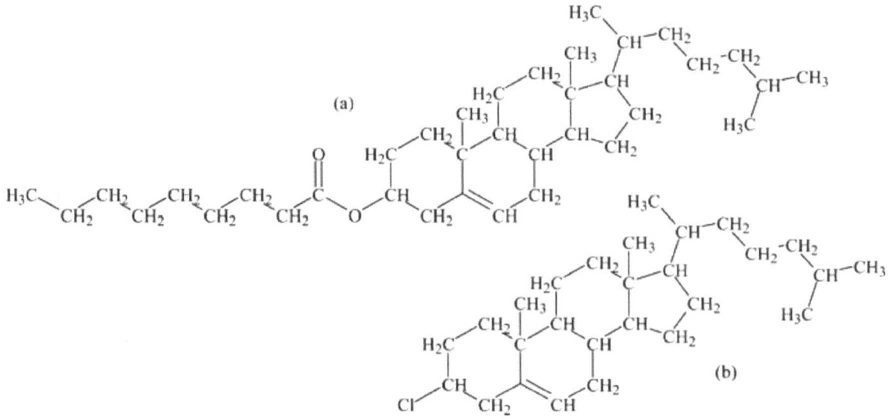

Fig. C.4 Molecular structures of **a** cholesterol nonanoate (ChNon) and **b** cholesteryl chloride (ChCl). Pure cholesterol has the structure of cholesteryl chloride with the −Cl replaced by a −OH group

helix also depends on external influences. In fact, the exact pitch is dependent not only on temperature and pressure, but also in the case of mixtures of liquid crystals, also on composition. You may choose to explore the effect of composition during the second week of this experiment.

During the first week of this experiment, you will synthesize cholesteryl nonanoate, which is an ester derivative of cholesterol, and perform some simple characterizations to test your successful synthesis. Cholesteryl nonanoate (ChNon) is known to exhibit a chiral nematic, *i.e.*, a cholesteric, liquid crystalline phase. While you will synthesize and characterize cholesterol nonanoate, you will actually study mixtures of two liquid crystalline materials each of which is a derivative of cholesterol. The other derivative is cholesteryl chloride (ChCl) whose structure along with that of cholesterol nonanoate is shown in Fig. C.4. The reason to study mixtures here is for convenience of handling. Though both pure materials exhibit liquid crystalline phases (ChNon: at 74 °C, solid:smectic; 80 °C smectic:cholesteric; 93 °C cholesteric:isotropic, and ChCl: about 70 °C, solid:cholesteric), appropriate mixtures are liquid crystalline at much lower and more manageable temperatures, for example, room temperature.

The driving force for the chemical reactions used in the synthetic scheme outlined below is most readily understood in terms of Lewis acid–base chemistry. To begin, cholesterol is dissolved in pyridine. According to the Lewis definition of acids and bases, in which acids and bases are described in terms of their propensity to accept or donate electron pairs, cholesterol is a weak acid (an electron pair acceptor) and pyridine is a strong base (an electron pair donor). These two chemicals, then, undergo an acid–base reaction, transferring a proton (an electron acceptor) to yield the pyridinium ion and a negatively charged cholesterol species (which results as the proton accepts the electrons from the pyridine leaving behind a negative charge on the cholesterol).

The next step involves the addition of nonanoyl chloride and is also understood using the Lewis definition. The carbonyl (C=O) group on nonanoyl chloride is very polar, yielding an electron-deficient carbon, which is then a Lewis acid. The negative cholesterol ion is an electron donor, or Lewis base, and reacts with the positive carbon center to form a new bond and release a chloride ion. This chloride ion is then neutralized by the pyridinium ion in another Lewis acid–base reaction to form pyridine hydrochloride. Pyridine hydrochloride is soluble in water, while the newly synthesized cholesteryl nonanoate is not. Addition of the reaction mixture to aqueous sulfuric acid causes only the desired product, cholesteryl nonanoate, to precipitate. The resulting solid, collected by filtration, contains impurities which you will remove by recrystallization. While the reaction is straightforward, there are certain hazards which must be avoided. Both pyridine and nonanoyl chloride react readily with water in the air, forming the pyridinium ion and nonanoic acid. Even a small amount of decomposition of these essential reactants will cause the intended reaction to fail. For this reason, these reactants should be distilled prior to use, and stored in such a way to minimize their contact with air.

C.3.1 Procedure

CAS Registry Numbers for Chemicals

Cholesterol: 57-88-5
Dry nonanoyl chloride: 764-85-2
Cholesteryl chloride: 910-31-6
Reagent acetone: 67-64-1
1 M sulfuric acid: 7664-93-9
Dry pyridine: 110-86-1
Filter flask (125 mL)
Buchner funnel
Two dram vial with screw cap
Glass stir rod
Graduated cylinder (50 mL)
Stir hot plate with magnetic stirrer
Rubber septum
Disposable syringes and needles
Microscope slides
Mylar film and tape
Digital voltmeter
Polarizer sheet and $\lambda/4$ wave plate
Melting point apparatus

Beckman DU650 spectrophotometers with temperature controllers and cell holders
Abbe refractometers with temperature controllers
Thermocouple.

1. Place about 1.5 g of cholesterol in a 125-mL Erlenmeyer flask. The cholesterol should be recrystallized from acetone prior to acquiring a pure sample. Add 5–8 mL of dry pyridine (obtained by distillation over potassium hydroxide) and a stir bar to flask. Work quickly and carefully when transferring the pyridine to the flask, to minimize water absorption from the ambient atmosphere. Once pyridine is transferred, immediately cap flask with a rubber septum and drying needle.
2. Stir solution in an ice/water bath for about 10 min. The cholesterol should be dissolved. Use a plastic dish for the ice bath container.
3. Ensure that the nonanoyl chloride to be used is dried via distillation *in vacuo*, and stored in a sealed stock bottle for maximum purity. Acid chlorides, such as nonanoyl chloride, readily react with water to form the corresponding carboxylic acid. Withdraw 1.2 mL of nonanoyl chloride from the stock bottle using a fresh, clean/dry plastic syringe fitted with a syringe needle. Pass the syringe needle through the rubber septum on the reaction flask and add the nonanoyl chloride dropwise and slowly. Since the amount of nonanoyl chloride is not the limiting reagent, you do not need to be extremely concerned with the exact amount added provided the amount is adequate. The amount added may be estimated by noting the volume delivered by the syringe and the volume should be about 1.2 mL. A precipitate should form on the addition of the acid chloride.
4. Stir the reaction mixture in an ice bath for 30 min using a magnetic stirrer. Remove solution from ice bath and continue stirring at room temperature for an additional 30 min on the magnetic stirrer.
5. Pour 50 mL of cold 1 M H_2SO_4 into the solution. Stir vigorously by hand using a glass stir rod to break up oil globules. Collect the solid which is your impure product using a typical suction filtration apparatus. Draw air through the sample for several minutes. Visual inspection should suggest when the sample is relatively dry and ready for the next step.
6. To purify your product, recrystallize your collected solid from acetone. To accomplish this, transfer the solid to a 125-mL flask. Add a small amount (20 mL to start) of acetone and a stir bar. Heat to fully dissolve the solid. Add small amounts of acetone, order 1 mL at a time, as needed, to aid in the dissolution. When solid is dissolved, remove from heat. Cool with the aid of an ice bath. Crystals should form. If they do not, try scratching the inside of the flask with a glass rod. Or, boil off some solvent and recool. Remove these crystals by suction filtration just as you did to collect your impure product. Wash these crystals with a cold 5 mL portion of acetone and allow them to dry on the filter paper. Transfer to a preweighed vial. Calculate yield of cholesteryl nonanoate from cholesterol and calculate the conversion of cholesterol to cholesteryl nonanoate.

7. Using appropriate melting point apparatus, characterize the melting behavior of the purified solid. Cholesteryl nonanoate exhibits a chiral nematic, more commonly known as a *cholesteric*, liquid crystalline phase around 85 °C and melts to the isotropic liquid around 93 °C. The cholesteryl nonanoate first melts to form a smectic phase around 75 °C and with further heating transforms to the cholesteric phase around 85 °C.
8. In order to calculate the pitch of the liquid crystal, you will need the average refractive index at several temperatures. Use an Abbe refractometer to measure the refractive indices of your liquid crystal at a number of different temperatures, about every 10° from room temperature to 60–70 °C.
9. As mentioned above you will be studying mixtures of cholesteryl nonanoate and cholesteryl chloride rather than the pure materials. Mixtures are used for the convenience of working at temperatures closer to room temperature. To prepare a mixture of a given composition, use the mole fraction composition scale. A mole fraction of a component in a mixture is defined as the number of moles of that component in the mixture divided by the total number of moles of all components in the mixture. To actually prepare a mixture, add the desired amounts of each component to a screw cap vial. Mix the components physically with a spatula to make a reasonably uniform mixture of the solids. Place the vial in a sample oven whose temperature has been set to about 100 °C and allow the solids to melt and mix. Mix the sample while hot, again with a spatula. At the oven temperature the sample should appear clear since it should be in the isotropic phase. As the sample cools note what happens.
10. Full details regarding the determination of selective reflection and refractive indices may be found online (http://www.jce.divched.org/Journal) in the Supplementary Documentation for the above article.

C.4 Template Synthesis and Magnetic Manipulation of Nickel Nanowires

(Bentley, A. K.; Farhoud, M.; Ellis, A. B.; Lisensky, G. C.; Nickel, A. –M. L.; Crone, W. C. *J. Chem. Ed.*, 2005, *82*, 765)

C.4.1 Procedure

Whatman Anodisc alumina membranes with polypropylene support rings (25-mm membrane diameter) 0.02-μm pore diameter.

Tweezers
Electrical tape
Cu sheet (12″ × 12″ is enough to make approximately 15 substrates)
Nickel wire, 1 mm diameter × 10 m long, 99.5%

AA batteries
Battery holders
Wires with alligator clips
Nickel electroplating solution
Compound optical microscope
Microscope slides and coverslips
Bar magnets
SEM specimen mounts.

Note. The alumina membranes have 20-nm diameter pores on the upper face that quickly widen within 5 μm of the surface to 200 nm throughout the membrane. Thus one face has 20-nm pores and the other has 200-nm pores. The membranes are packed with the 20-nm pores facing upward, but the two faces cannot be easily distinguished visually. However, the clear polypropylene support ring is wider on the 20-nm pore side of the membrane, which is the side that will be coated with the conductive layer.

1. The Cu plates and battery holders should be prepared before the lab period and can be reused indefinitely. Cut rectangles of copper sheet approximately 1.25″ × 2″ and then cut out corners of the rectangle to form the stem that is used for electrical contact. We used a variety of copper sheets that ranged from 0.03″ to 0.06″ in thickness. To assemble the battery holder, first cut a double-ended test lead in half using wire cutters. Then strip the cut ends and solder each lead to the clip on one end of a battery holder.
2. Remove an alumina membrane from the package by holding the polypropylene support ring with tweezers (discard the paper circles separating the membranes from each other). Always use tweezers to hold the membranes by the support ring; the alumina will crack if handled directly. Cracked membranes cannot be used for deposition, as the solution will leak through the membrane and deposit on the copper plate. Dispose of cracked membranes in a glass disposal box. There are two methods of applying a conductive backing to the membranes. If a sputter coater is available (such as those used to coat samples with gold for SEM), then a 250-nm thick layer of silver can be sputtered onto each membrane before the lab period. An alternative method that can be used during the lab period is to have students paint the upward-facing side of the membrane (the side with 20-nm pores; see above) with GaIn eutectic using a cotton-tipped stick. Students will not need to dip the cotton end into the GaIn more than once or twice. The GaIn is applied by gently rubbing the GaIn-coated cotton tip over the membrane surface. Students can check for gaps in the GaIn coating by looking at the noncoated face of the membrane: any areas without GaIn will appear light blue in color. Areas with GaIn will appear white or opaque. We have observed similar success rates with either silver or GaIn as the conductive coating.
3. After the conductive layer is applied, mount the membrane on a copper plate with the conductive, metal-coated side facing the copper by taping the membrane's support ring to the copper using insulating electrical tape, as shown in Fig. C.5

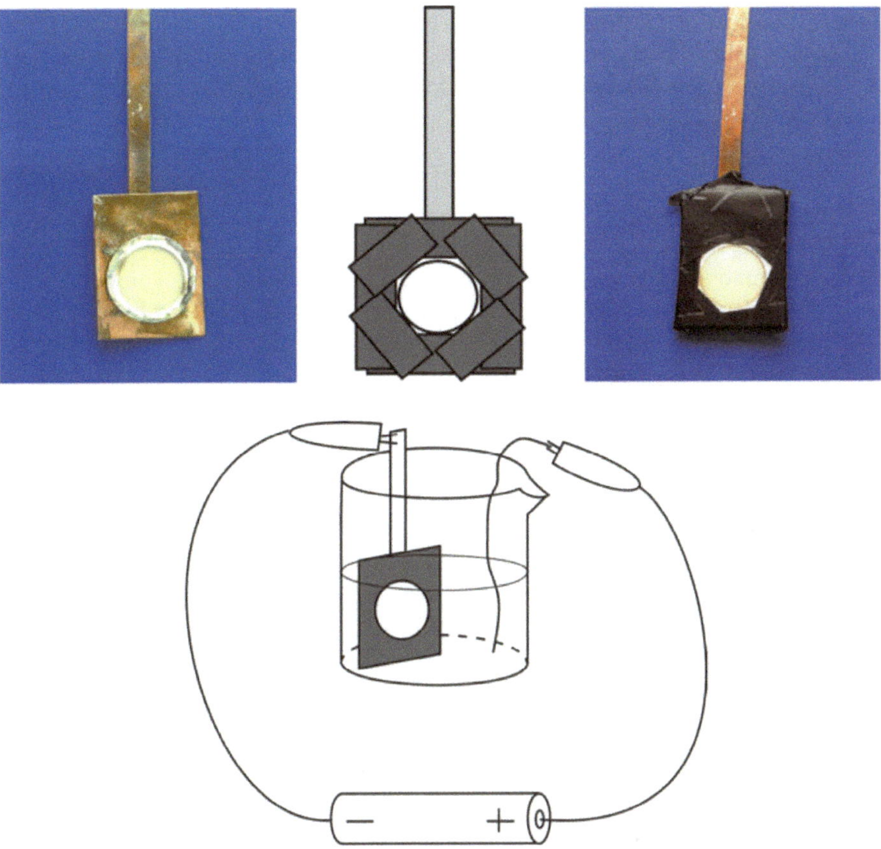

Fig. C.5 (Top) Placement of the membrane on the copper electrode, fastened with electrical tape. (Bottom) The electrochemical cell used to synthesize Ni nanowires

(top). Little if any electric tape should come in contact with the alumina membrane, as it is hard to remove. It is essential to completely cover both sides of the copper plate with the electrical tape to prevent any unwanted electrodeposition on the copper. Leave the stem uncovered so electrical contact can be made with the alligator clip, but this section of the copper should not make contact with the electrolyte.

4. Set a AA battery inside the battery holder, making sure to align the positive and negative ends of the battery with the labels on the holder. Clip the negative lead from the battery to the exposed part of the copper and clip the positive lead to the nickel wire. Place the two electrodes in a 50-mL beaker with the membrane surface facing the nickel wire (Fig. C.5 (bottom)).
5. After the electrodes are arranged in the beaker and connected to the battery, pour in enough Ni-plating solution to just cover the top of the membrane (approximately 40 mL). The electrodeposition of nickel will begin as soon as the solution completes the circuit.

6. As the electrodeposition takes place, the membrane surface changes color from a pale yellow/orange (for silver backing) or yellow/white (for GaIn backing) to black. Allow the cell to sit for 50 min to make nanowires that are ~50 μm long. To make shorter nanowires, stop the reaction sooner. The minimum time required to produce nanowires that are visible using an optical microscope is 10 min.
7. To halt deposition, remove the battery from the cell and disconnect the wires from the electrodes. Remove the Cu plate from the solution and rinse the membrane surface with deionized water. The nickel-plating solution can be returned to its original container and reused indefinitely. We have used the same 500 mL of plating solution for approximately 50 depositions without any noticeable change in quality.
8. Soak the copper plate and nickel-filled membrane in a beaker of acetone for 5 min to dissolve the electrical tape's adhesive. Carefully detach the membrane from the copper plate and the electrical tape. If the copper plate is discolored, it can be cleaned with nitric acid before use in another deposition.
9. To free the nanowires from the alumina template, the GaIn or silver cathode must be removed and the alumina membrane dissolved. To remove the cathode layer, mount the membrane with the GaIn or silver side facing outward on a glass microscope slide using one piece of electrical tape on the support ring. Dip a cotton-tipped stick in concentrated nitric acid and rub it over the surface once or twice. The GaIn or silver will dissolve as it is oxidized and the membrane will appear black as the nickel underneath is exposed.
10. Immediately soak the cotton-tipped stick in water to prevent formation of nitrocellulose. Rinse the membrane surface many times with deionized water to remove excess acid. Peel off the electrical tape that is holding the membrane in place and put the membrane in a 50-mL beaker. Cracks in the membrane are not a concern at this point.
11. Add 5 mL of 6 M NaOH and let the membrane soak for 10 min. Swirl the contents of the beaker occasionally. As the base dissolves the alumina, the support ring will float freely in the solution.
12. Remove the support ring from the solution using tweezers, rinse in water, and discard. Black material will collect at the bottom of the beaker. Use tweezers or a Pasteur pipette to break up any clumps of black material until the solution is a cloudy gray color.
13. Set a magnet next to the side of the beaker and wait 1 min while the nickel nanowires collect against the magnet. Use a Pasteur pipette to remove as much of the alkaline solution as possible (without removing any of the black nanowire material) and then add 15-mL deionized water to the beaker. Resuspend the nanowires in the fresh water using a Pasteur pipette and collect the nanowires against a side again using a magnet. Remove the water again and rinse once more.

14. After removing water from the last washing, add 2–3 mL of ethylene glycol. Other solvents can also be used, such as water or ethanol, but we have found that the wires stay suspended longer in ethylene glycol, allowing the students to better see the response of their wires to magnetic fields using the optical microscope. Transfer the suspension to a small vial. Use a Pasteur pipette or shake the vial to disperse the nanowires; the solution will appear cloudy and gray. The nanowire suspension can be stored indefinitely.
15. Transfer a drop of the nanowire suspension to a microscope slide and cover with a coverslip. If clumps remain in the solution, try to avoid transferring them. Even if the solution does not appear to be very cloudy, there are probably still a large number of nanowires suspended, since each membrane has on the order of one billion pores.
16. To examine the nanowires using SEM, place a drop of nanowires suspended in water directly on an aluminum SEM specimen mount and allow the drop to dry. (To obtain a smoother background in the images, attach a piece of silicon to the specimen mount using carbon tape and place the drop of nanowire suspension on the silicon.) No conductive coating needs to be added. We imaged the nanowires using a LEO FE-SEM with a 3 kV accelerating voltage and 3 mm working distance.
17. To examine the nanowires by X-ray diffraction (XRD) while they are still embedded in the membrane, tape the membrane to a glass microscope slide with the Ni deposition side facing upward. To examine a sample of liberated nanowires, suspend the nanowires in ethanol and place a few drops on a microscope slide. After the solvent evaporates, add a few more drops and repeat this process until there is a visible pile of nanowires on the slide. Our patterns were obtained using a Scintag PADV powder X-ray diffractometer with Cu K-α radiation, a tube voltage of 40 kV and a tube current of 35 mA.

C.5 Introduction to PhotolithographyIntroduction to Photolithography

(Berkowski, K. L.; Plunkett, K. N.; Yu, Q.; Moore, J. S. *J. Chem. Ed.*, 2005)

C.5.1 Procedure

CAS Registry Numbers for Chemicals

Isobornyl acrylate (IBA): 5888-33-5
Bisphenol-A-glycidyldimethyacrylate (Bis-GMA): 1565-94-2
2,2-Dimethoxy-2-phenylacetophenone (DMPA): 24,650-42-8
Oil Red O (Solvent Red 27): 1320-06-5

Oil Blue N (Solvent Blue 14): 2646-15-3
Fluorescent Yellow 3G (Solvent Yellow 98): 12,671-74-8
Ethanol: 64-17-5.

To prevent premature polymerization of the photoresist, store it in an amber vial or a clear vial wrapped with aluminum foil in a chemical refrigerator, if available, and minimize its exposure to light.

Photoresist preparation

1. Use a spatula to mass approximately 2.0 g of Bis-GMA into an amber vial. Bis-GMA is extremely viscous and sticky and you may need an additional spatula to push it to the bottom of the vial.
2. If some of the Bis-GMA sticks to the sides or top of the vial, tightly screw on the vial lid and completely submerge the vial in a beaker full of hot tap water until the Bis-GMA flows to the bottom.
3. Add approximately 3.5 g of IBA to the vial containing Bis-GMA.
4. Add approximately 0.18 g (3% wt) of DMPA to the vial containing IBA and Bis-GMA.
5. Sonicate the mixture for 30 min.
6. Swirl the solution to make sure it is homogeneous. If all components are not dissolved, sonicate the solution for an additional 10 min.
7. Repeat step 6 until the solution is homogeneous.
8. Add the appropriate color dye (1–1.5 mg or a small spatula tip) to the resulting photoresist.
9. Sonicate this solution for 5–10 min.
10. Store the photoresist in an amber vial with a screw-top dropper in a chemical refrigerator, if available. A clear vial wrapped with aluminum foil may also be used.

Photomask preparation

1. Create a 3 × 3 in. black box in Microsoft Word to use as a template.
2. Copy and paste six of these black boxes on a single page to create multiple photomasks.
3. Text-containing photomasks: create a text box, format it for white font and no fill, and center the text box inside the black box to allow at least a 0.5 in. black border surrounding the text.
4. Picture-containing photomasks: manipulate shapes in Microsoft Paint by pasting a white shape on a black background and copy this figure to the center of the black box on your template.
5. Print the photomasks on transparency paper using a black and white printer at a resolution of 600 dpi or higher.

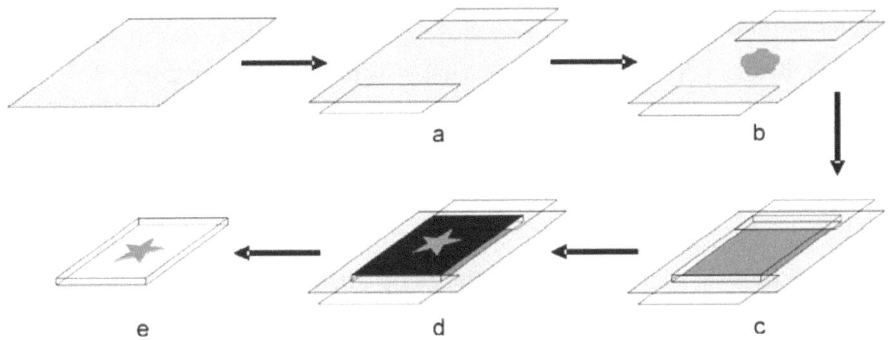

Fig. C.6 Overview of the fabrication process. Two coverslips are placed at the top and bottom of a transparency **(a)** and prepolymer solution is dropped in the center **(b)**. A thick glass slide is positioned on top of the prepolymer solution, allowing it to rest on the coverslips, and causing the prepolymer solution to flow and fill the gap **(c)**. A photomask is aligned on top of the glass slide **(d)**, a second glass slide is placed on top of the photomask to keep it in place, and the exposed photoresist is polymerized using UV light. After rinsing, an insoluble, patterned polymer results **(e)**

Polymer patterning (Fig. C.6).

1. Cut a transparency paper into four equal pieces of 5.5 × 4.25 in. and place one of these flat on the table.
2. Align two thin glass coverslips (22 × 50 mm) on the top and bottom of the transparency.
3. Apply 15–30 drops of the photoresist to the center of the transparency leaving about 2.5 in. between each coverslip and the liquid photoresist.
4. Cover the photoresist with a thick glass slide (75 × 50 × 1 mm) by placing the slide in the center of the transparency and allowing its top and bottom to rest on top of the thin coverslips.
5. Wait until the entire space between the thick glass slide and the transparency is filled with photoresist.
6. Place the photomask on top of the thick glass slide and make sure that photoresist is visible through all parts of the photomask.
7. Place a second thick glass slide on top of the photomask to keep it in place.
8. Polymerize the exposed photoresist by shining UV light (365 nm) from the hand-held lamp 0.5 in. above the sample for 20 s (time may be dependent on the UV lamp).
9. Remove the photomask and carefully peel away the bottom transparency from the thick glass slide.
10. Wash away the unpolymerized photoresist into a beaker with a wash bottle containing ethanol.
11. Dry the resulting polymer by gently dabbing it with a paper towel.

Troubleshooting

1. Photoresist solution is not homogeneous—increase sonication time.
2. Photoresist solution polymerizes in container—store in an amber container or a container wrapped with aluminum foil in a chemical refrigerator, if available.
3. Photoresist does not flow to edges of glass slide or polymer pattern is too thin—increase the initial amount of prepolymer solution on the transparency film.
4. Bubbles form in the polymer product:
   (a) Poke away all bubbles with a pipette tip when photoresist is dropped on the transparency.
   (b) Carefully place the glass slide on top of the photoresist.
5. Incomplete or no polymerization upon exposure to UV light:
   (a) Increase UV exposure time.
   (b) Increase crosslinker concentration (Bis-GMA).
   (c) Increase photoinitiator concentration (DMPA).
   (d) Decrease amount of dye.
6. Brittle or delaminating polymer structures:
   (a) Decrease UV exposure time.
   (b) Decrease crosslinker concentration (Bis-GMA).
   (c) Decrease photoinitiator concentration (DMPA).
7. Polymer forms in areas that were not exposed to UV light:
   (a) Decrease UV exposure time.
   (b) Increase space between structures and letters on photomask.
   (c) Blacken dark areas of photomask with a marker to suppress light transmission.

C.6 Synthesis of Gold Nanoclusters

(Adapted from Weare, W. W.; Reed, S. M.; Warner, M. G.; Hutchison, J. E. *J. Am. Chem. Soc.*, **2000**, *122*, 12890)

Gold nanoclusters are primarily used to catalyze the growth of nanowires (via VLS or SLS routes), and are also utilized for biomedical applications such as targeted drug delivery agents. As discussed in Chap. 6, a number of techniques may be used to synthesize metallic nanoclusters. In this experiment, a gold precursor is transformed into nanoclusters with diameters <2 nm via a two-step reaction. The first step consists of phase-transfer into an organic layer where it reacts with triphenylphosphine. In the second step, the mixture is reduced by sodium borohydride.

C.6.1 Procedure

CAS Registry Numbers for Chemicals

$HAuCl_4 \cdot 3H_2O$: 16,961-25-4
Tetraoctylammonium bromide: 14,866-33-2
Triphenylphosphine: 603-35-0
$NaBH_4$: 16,940-66-2
Toluene: 108-88-3
Hexanes: 110-54-3
Methanol: 67-56-1
Chloroform: 67-66-3
Pentane: 109-66-0
$NaNO_2$, saturated solution: 7632-00-0.

1. Dissolve hydrogen tetrachloroaurate trihydrate (1.00 g, 2.54 mmol) and tetraoctylammonium bromide (1.60 g, 2.93 mmol) in a degassed water/toluene mixture (50 mL/65 mL).
2. When the golden color is transferred into the organic phase, add triphenylphosphine (2.32 g, 8.85 mmol) with vigorous stirring for *ca.* 10 min until the organic phase is white and cloudy.
3. Add aqueous sodium borohydride (1.41 g, 37.3 mmol, dissolved in 10 mL of water immediately prior to use) to the organic phase. The organic phase will immediately turn dark purple. Continue to stir the solution for 3 h under a nitrogen flow.
4. Separate the toluene layer, washing it with water (2×100 mL). Remove the solvent *in vacuo*, or with a stream of nitrogen to yield a black solid.
5. Wash the resulting solid with a series of solvents (hexanes, saturated aqueous sodium nitrite, and a 2:3 methanol:water mixture) to remove the phase-transfer catalyst, byproducts, and unreacted starting materials.
6. Dissolve the solid in chloroform, and slowly add pentane to remove salts such as $AuCl(PPh_3)$. Perform this precipitation procedure three times; after purification, the yield is *ca.* 170 mg of purified nanoparticle from 1 g of $HAuCl_4$.
7. To prepare the sample for TEM analysis, put a drop of the chloroform suspension on a TEM grid and allow the solvent to evaporate. Alternatively, you may spray the solvent suspension onto a grid using an aerosol delivery device.

Nice Papers (One Thorough Review Article) Related to Nanocluster Growth

1. Sun, S.; Zeng, H.; Robinson, D. B.; Raoux, S.; Rice, P. M.; Wang, S. X.; Li, G. *J. Am. Chem. Soc.*, **2004**, 126, 273.
2. Cushing, B. L.; Kolesnichenko, V. L.; O'Connor, C. J. *Chem. Rev.*, **2004**, 104, 3893.

Table C.1 Effect of varying experimental parameters on the formation of porous silicon

Experimental parameter[a]	Resultant porosity	Resultant etching rate	Resulting critical current
HF concentration	Decrease	Decrease	Increase
Current density	Increase	Increase	No effect
Anodization time	Increase	Minimal effect	No effect
Temperature	No effect	No effect	Increase
Wafer doping (p-type)	Decrease	Increase	Increase
Wafer doping (n-type)	Increase	Increase	No effect

[a] Corresponds to an increase in each experimental parameter

C.7 Synthesis of Porous Silicon

V. Lopez, J. P. Badilla, A. Ramirez-Porras and B. D. Fahlman

As you may recall from Chap. 4, bulk Si is an indirect bandgap semiconductor, which limits applications that require radiative emission. However, when bulk Si is electrochemically etched under controlled conditions, a nanoporous material known as *porous silicon* (PS) is generated, which displays a pronounced photoluminescence due to the quantum confinement of photoexcited carriers.[1] Consequently, PS is of great interest for a plethora of applications related to optoelectronics (advanced displays), sensors, biosensors, and photovoltaics. However, the widespread use of PS for applications has been limited due to surface oxidation that readily occurs within minutes of its formation. A number of strategies may be used to passivate the H-terminated Si surface by the covalent attachment of chemical species in order to prevent oxidation.[2] Herein, we describe the formation of PS via electrochemical etching in a HF solution. Though not described here, a great deal of experimental variables may be varied during electrochemical etching (Table C.1), which may be used to control the resultant morphology of the PS for the desired applications.

C.7.1 Procedure

CAS Registry Numbers for Chemicals

Hydrofluoric acid, concentrated (48%): 7664-39-3.
Ethanol: 64-17-5.

V. Lopez · J. P. Badilla · A. Ramirez-Porras (✉)
Centro de Investigación en Ciencia e Ingeniería de Materiales (CICIMA), Universidad de Costa Rica, San Jose, CR, USA

B. D. Fahlman (✉)
Department of Chemistry, Central Michigan University, Mount Pleasant, MI 48,859, USA

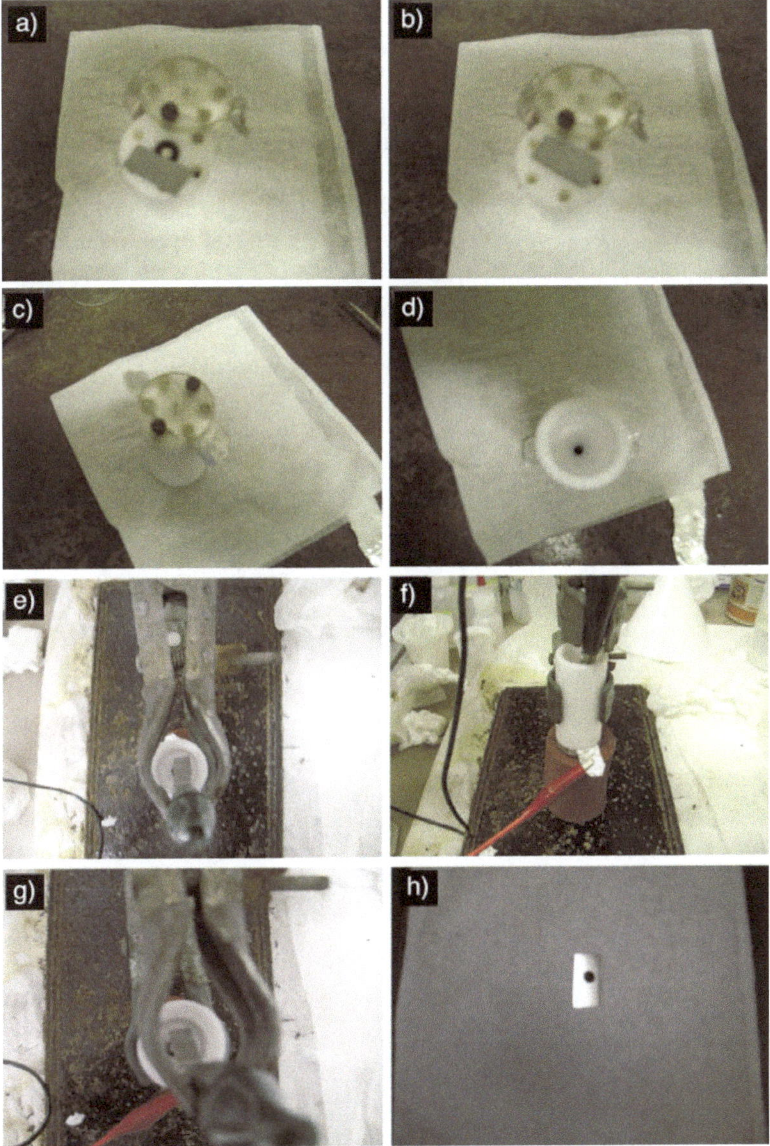

Fig. C.7 Photographs of the experimental setup used to electrochemically etch a porous silicon region on the surface of a single-crystalline Si wafer (steps described in the procedure)

Figure C.7 illustrates the steps required to electrochemically etch Si. A Teflon cell fitted with a Viton o-ring is shown in (a). A silicon wafer is placed polished side down onto the top of the o-ring, (b). A piece of aluminum foil is placed on top of the wafer backside and the plastic backplate is screwed into place, (c). The polished side of the wafer is shown from the top of the Teflon cell in photograph (d). The wafer surface is treated with 10% HF(aq) for 10 min. to remove any native oxide layer, followed by rinsing with water and ethanol. The cell is then filled with an electrolyte consisting of 12.5% HF ($HF:H_2O:EtOH$ of 1:4:3), and a platinum electrode is immersed into the solution, (e). Electrical connections to the platinum and aluminum electrode surfaces are made (note: current flows from the bottom to top) and appropriate current started. Photograph g) shows the presence of tiny bubbles that indicate the electrochemical anodization of the silicon substrate. The final etched wafer is shown in h), which is rinsed with water and ethanol and dried under a flow of nitrogen. Using a current of 54 mA cm^{-2} for 20 min. for a p-type Si(100) substrate with a resistivity of 20–50 Ω cm results in a macroporosity (70%) with pores 2–3 μm in diameter and 40–50 μm depth (*e.g.*, Fig. C.8).

Figure C.8 top–down (top) and cross-section (bottom) SEM images of porous silicon

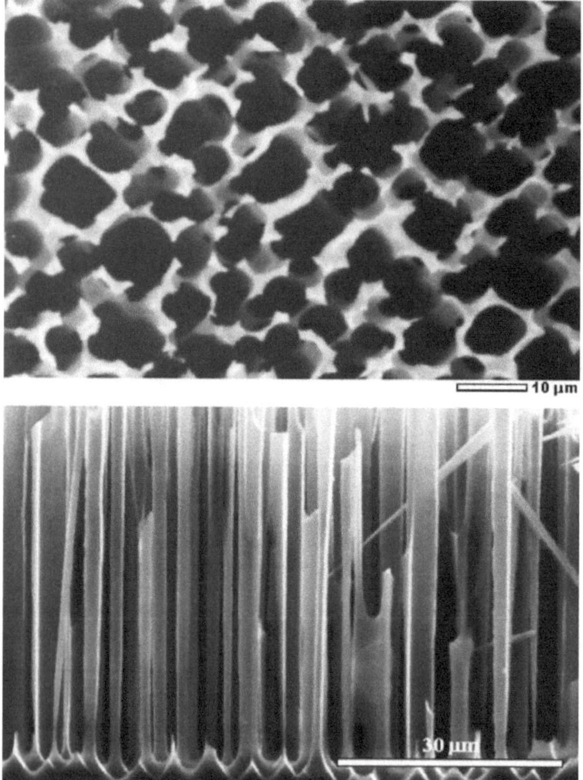

Fig. C.8 Scanning electron micrograph of a representative porous silicon film etched onto p-type Si(100) using the conditions described herein

C.8 Solid–Liquid–Solid (SLS) Growth of Silicon Nanowires

A. Antic, P. Oshel and B. D. Fahlman

Nanowires are crystalline wires with characteristic diameters of less than 100 nm (nm) and comparatively long lengths. Silicon nanowire arrays are of particular interest to progress current microelectronic technology into nanoelectronic systems. The optical and electrical properties of nanowires are based on their morphologies (*i.e.*, composition, crystal structure, and growth orientation), which are primarily based on the production method. Other semiconductors, such as gallium arsenide (GaAs), are also of interest for use in light emitting diodes (LEDs) and lasers.

The most common method used to grow silicon nanowires is by vapor–liquid–solid (VLS), which uses chemical vapor deposition onto a noble metal catalyst such as gold nanoparticles from a gaseous precursor such as SiH_4.[3] Herein, we describe the growth of silicon nanowires (SiNWs) using solid–liquid-solid (SoLS[4]) growth from gold nanoparticulate catalysts. To our knowledge, this is the first report of SoLS being used with nanoparticulate Au catalysts; the other sparse precedents[5] using this technique have used metallic thin films as the catalytic surface. It should be noted that one should also be able to grow SiNWs from other face-centered cubic (FCC) metals; different heating/cooling regimes will be required to determine the most effective size-corrected eutectic for those compositions.

C.8.1 Procedure

CAS Registry Numbers for Chemicals

$HAuCl_4 \cdot 3H_2O$: 16,961-25-4
Sodium citrate: 6132-04-3
Hydrofluoric acid, 48 wt%: 7664-39-3
Ethanol: 64-17-5.

1. Prepare a nanoparticulate gold solution, using the Turkevich method as follows:
    (a) Prepare 500 mL of a 1.0 mM stock solution of $HAuCl_4$ using ultrapure (18 MΩ) water. Note: the $HAuCl_4.xH_2O$ salt is very hygroscopic and must be stored in a dessicator before/after its use.

A. Antic · B. D. Fahlman
Department of Chemistry, Central Michigan University, Mount Pleasant, MI 48,859, USA

P. Oshel
Department of Biology, Central Michigan University, Mount Pleasant, MI 48,859, USA

(b) Prepare a 1% (w/v) stock solution of sodium citrate using ultrapure (18 MΩ) water.
(c) Transfer 20 mL of stock solution (a) into an Erlenmeyer flask. With gentle stirring, heat the solution to boiling.
(d) Add 2 mL of stock solution (b) to the boiling solution. Gold nanoparticles will form as the citrate reduces the Au(III). Remove the solution from heat when the solution has turned red.

2. Remove the native SiO_2 layer from electronic-grade crystalline Si wafer (111) or (100) by immersing in a 1:10 HF:EtOH solution for 30 s., followed by rinsing with DI water. Dry with a KimWipe™.
3. Place 2–3 drops of the Au solution prepared in 1. above onto the wafer via pipette or aerosol spraying, followed by drying by "flash" heating with a hot air gun, toggling the heat on and off and passing the airflow back and forth across the sample.
4. Place the wafer in a ceramic boat inside a horizontal quartz tube within a tube furnace, pre-flushed with Argon for at least 20 min. Re-seal the connections and continue to flush with Argon for at least 30 min. Ramp the temperature of the tube furnace to 1000 °C, and maintain this temperature for at least 15 min.
5. Cool the samples to room temperature within the tube furnace under a flow of Argon. Store samples in an inert-atmosphere glovebox (or dessicator) until characterization measurements are performed. A high yield of silicon nanowires should be evident from scanning electron microscopy (SEM); *e.g.*, Fig. C.9.

Fig. C.9 Scanning electron micrograph of silicon nanowires (SiNWs) grown via SLS at 1000 °C. Image courtesy of Phil Oshel, Department of Biology, CMU

C.9 Synthesis of Ferrofluids

–Some online laboratory modules:

a. https://education.mrsec.wisc.edu/synthesis-of-aqueous-ferrofluid-nanoparticles-buret/.
b. http://chemistry.about.com/od/demonstrationsexperiments/ss/liquidmagnet.htm.

C.10 Metallurgy/Phase Transformations

–Some online laboratory modules:

(a) http://www.iitk.ac.in/mme/test/MME330.pdf.
(b) http://www.csun.edu/~bavarian/mse_528.htm.

C.11 Synthesis of Quantum Dots

–Published online: http://www.chemistrylabmanual.com/uploads/2/5/6/1/25614215/qd_exp_mac.pdf

Notes

1. Halimaoui, A. in *Properties of Porous Silicon,* Canham, L. T., ed., IEE INSPEC—The Institution of Electrical Engineers: London, 1997.
2. For example, see: Wayner, D. D. M.; Wolkow, R. A. *J. Chem. Soc., Perkin Trans.* **2002**, *2*, 23.
3. For example, see: Cui, Y.; Lauhon, L. J.; Gudiksen, M. S.; Wang, J.; Lieber, C. M. *Appl. Phys. Lett.* **2001**, *78*, 2214.
4. Note: the 'SLS' acronym has been reserved for "solution-liquid–solid" growth (for example, see: Trentler, T. J.; Hickman, K. M.; Goel, S. C.; Viano, A. M.; Gibbons, P. C.; Buhro, W. E. *Science* **1995**, *270*, 1791), so we hereby define 'SoLS' as "solid–liquid-solid".
5. For example, see: (a) Yan, H. F.; Xing, Y. J.; Hang, Q. L.; Yu, D. P.; Wang, Y. P.; Xu, J.; Xi, Z. H.; Feng, S. Q. *Chem. Phys. Lett.* **2000**, *323*, 224. (b) Lee, E. K.; Choi, B. L.; Park, Y. D.; Kuk, Y.; Kwon, S. Y.; Kim, H. J. *Nanotechnology* **2008**, *19*, 1.

Index

B. D. Fahlman, *Materials Chemistry*,
https://doi.org/10.1007/978-3-031-18784-1

B

D

E

F

I

Q

R

T